Poplars and Willows

Trees for Society and the Environment

This volume is respectfully dedicated to the memory of Victor Steenackers. Vic, as he was known to his friends, was born in Weelde, Belgium, in 1928. His life was devoted to his family – his wife, Joanna, his 9 children and his 23 grandchildren. His career was devoted to the study and improvement of poplars, particularly through poplar breeding. As Director of the Poplar Research Institute at Geraardsbergen, Belgium, he pursued a lifelong scientific interest in poplars and encouraged others to share his passion. As a member of the Executive Committee of the International Poplar Commission for many years, and as its Chair from 1988 to 2000, he was a much-loved mentor and powerful advocate, spreading scientific knowledge of poplars and willows worldwide throughout the many member countries of the IPC. This book is in many ways part of the legacy of Vic Steenackers, many of its contributing authors having learned from his guidance and dedication. Vic Steenackers passed away at Aalst, Belgium, in August 2010, but his work is carried on by others, including members of his family.

Poplars and Willows

Trees for Society and the Environment

Edited by

J.G. Isebrands

Environmental Forestry Consultants LLC,
New London, Wisconsin, USA

and

J. Richardson

Poplar Council of Canada, Ottawa, Ontario, Canada

Published by
The Food and Agriculture Organization of the United Nations
and
CABI

www.cabi.org

fao.org

CABI is a trading name of CAB International

CABI
Nosworthy Way
Wallingford
Oxfordshire OX10 8DE
UK

Tel: +44 (0)1491 832111
Fax: +44 (0)1491 833508
E-mail: info@cabi.org
Website: www.cabi.org

CABI
38 Chauncey Street
Suite 1002
Boston, MA 02111
USA

Tel: +1 800 552 3083 (toll free)
Tel: +1 (0)617 395 4051
E-mail: cabi-nao@cabi.org

A catalogue record for this book is available from the British Library, London, UK.

Library of Congress Cataloging-in-Publication Data

Poplars and willows: trees for society and the environment / edited by J.G. Isebrands and J. Richardson.
p. cm.
Includes bibliographical references and index.
ISBN 978-1-78064-108-9 (hbk) -- ISBN 978-9251071854 (co publisher fao)
1. Poplar. 2. Willows. 3. Forests and forestry--Environmental aspects. I. Isebrands, J. G. (Judson Gary), 1943- II. Richardson, J. (James), 1941-

SD397.P85P66 2013
635.9'77365--dc23

2013004253

Published jointly by CAB International and FAO
Food and Agriculture Organization of the United Nations (FAO)
Viale delle Terme di Caracalla, 00153 Rome, Italy
website: www.fao.org

ISBN: 978 1 78064 108 9 (CABI)
ISBN: 978 92 5 107185 4 (FAO)

Typeset by SPi, Pondicherry, India.
Printed and bound in the UK by CPI Group (UK) Ltd, Croydon, CR0 4YY.

Commissioning Editor: Rachel Cutts
Editorial Assistant: Emma McCann
Production Editor: Tracy Head

Contents

Contributors

Allegro, G., CRA – Unità di Ricerca per le Produzioni legnose fuori foresta, Strada Frassineto 35, I-15033 Casale Monferrato, Italy (gianni.allegro@entecra.it)

Aronsson, P., Department of Crop Production Ecology, Swedish University of Agricultural Sciences (SLU), PO Box 7043, SE-750 07 Uppsala, Sweden (par.aronsson@slu.se)

Augustin, S., INRA – Centre d'Orléans, UR633 Zoologie Forestière, 2163 Avenue de la Pomme de Pin, CS 40001 Ardon, 45075 Orléans Cedex 2, France (sylvie.augustin@orleans.inra.fr)

Balatinecz, John,* 41 Bloomington Crescent, North York, Ontario, Canada M3N 1R2 (jbalatinecz@rogers.com)

Ball, J.B., Corsham, Wiltshire, SN13 9BB UK (jbball1@yahoo.co.uk)

Brignolas, F., Université d'Orléans, UFR-Faculté des Sciences, UPRES EA 1207 Laboratoire de Biologie des Ligneux et des Grandes Cultures (LBLGC), BP 6759, F-45067, France (franck.brignolas@univ-orleans.fr) and INRA, USC1328 Arbres et Réponses aux Contraintes Hydriques et Environnementales (ARCHE), BP 6759, F-45067, France

Carle, Jim, Forestry Department, FAO, Viale delle Terme di Caracalla, 00153 Rome, Italy (retired). Present address: Mount Maunganui, Bay of Plenty, New Zealand (carle.jim@gmail.com)

Carlson, M., British Columbia Ministry of Forests, Kalamalka Research Station,Vernon, British Columbia, Canada V1B 2C7 (michael.carlson@gov.bc.ca)

Ceulemans, R., Research Group of Plant and Vegetation Ecology, University of Antwerp, Department of Biology, Campus Drie Eiken, Universiteitsplein 1, B-2610 Wilrijk, Belgium (reinhart.ceulemans@ua.ac.be)

Charles, J.G., The New Zealand Institute for Plant and Food Research Limited, Private Bag 92169, Auckland Mail Centre, 1142, Auckland, New Zealand (john.charles@plantandfood.co.nz)

Coleman, M., Department of Forest, Rangeland and Fire Sciences, University of Idaho, Moscow, Idaho, USA (mcoleman@uidaho.edu)

Collins, C.M., Imperial College at Silwood Park, Ascot SL5 7PY, UK (t.collins@ic.ac.uk)

De Boever, Lieven, Belgian Institute of Wood Technology (CTIB/TCHN), Allee Hof ter Vleest 3, B-1070 Bruxelles, Belgium (lieven.deboever@ctib-tchn.be)

Delplanque A., La Jumelière, F-45240 Ligny le Ribault, France (andre.delplanque@dbmail.com)

Dickinson, N., Lincoln University, Department of Ecology, Lincoln 7647, Canterbury, New Zealand (nicholas.dickinson@lincoln.ac.nz)

Dickmann, Donald I., Department of Forestry, Michigan State University, East Lansing, Michigan 48824-1222, USA (dickman1@msu.edu)

*John Balatinecz is the lead author for Chapter 10, but the primary contact for this chapter is J. Richardson (jrichardson@on.aibn.com).

Dimitriou, J., Department of Crop Production Ecology, Swedish University of Agricultural Sciences (SLU), PO Box 7043, SE-750 07 Uppsala, Sweden (ioannis.dimitriou@slu.se)

Doty, S.L., School of Environmental and Forest Sciences, University of Washington, Seattle, Washington 98195, USA (sldoty@u.washington.edu)

Gao, Jian, Büsgen-Institut, Abtlg. Forstbotanik und Baumphysiologie, Georg-August Universität Göttingen, Büsgenweg 2, 37077 Göttingen, Germany. Present addrcss: International Bamboo and Rattan Research Centre, No 8 Futong East Street, Beijing, PR China (gaojian@icbr.ac.cn)

Gardiner, E., Center for Bottomland Hardwood Research, USDA Forest Service, Southern Research Station, Stoneville, Mississippi 38776, USA (egardiner@fs.fed.us)

Gielen, B., Research Group of Plant and Vegetation Ecology, University of Antwerp, Department of Biology, Campus Drie Eiken, Universiteitsplein 1, B-2610 Wilrijk, Belgium (birgit.gielen@ua.ac.be)

Gimenez, R., Plant Protection Area, Faculty of Agronomy, University of Buenos Aires, Av. San Martín 4453, CP1417, Buenos Aires, Argentina (rgimenez@agro.uba.ar)

Heinsoo, K., Institute of Zoology and Botany, Estonian Agricultural University, 51014 Tartu, Estonia (katrin.heinsoo@emu.ee)

Höglund, S., Department of Ecology, Swedish University of Agricultural Sciences, Box 7044, SE-750 07 Uppsala, Sweden (Solveig.Hoglund@ekol.slu.se)

Isebrands, J.G., Environmental Forestry Consultants LLC, New London, Wisconsin, 54961 USA (efcllc@athenet.net)

Jiafu, H., Laboratory for Silviculture and Conservation of Ministry of Education, Forest University Beijing, Beijing 100083, People's Republic of China (hujiafu@hotmail.com)

Jin, Juwan, College of Wood Science and Technology, Nanjing Forestry University, Nanjing 210037, Jiangsu, PR of China (juwanjin@hotmail.com)

Johnson, J.D., Deceased. Formerly at Washington State University, Department of Natural Resource Sciences, Puyallup, Washington, USA

Karnosky, D.F., Deceased. Formerly at Michigan Technological University, Houghton, Michigan, USA.

Kollert, Walter, Secretariat of the International Poplar Commission (IPC), Forestry Department, FAO, Viale delle Terme di Caracalla, 00153 Rome, Italy (walter.kollert@fao.org)

Koo, Y.B., Department of Forest Genetic Resources, Korea Forest Research Institute, 44-3, Omokchun-dong, Suwon, Kyeonggi-do, 441-350, Republic of Korea (ybkoo@foa.go.kr)

Kort, J., Agroforestry Development Centre, Agriculture and Agri-Food Canada, 2 Government Road, Indian Head, Saskatchewan, Canada S0G 2K0 (kortj@agr.gc.ca)

Kuzovkina, Julia, Department of Plant Science, The University of Connecticut, Storrs, Connecticut 06269-4067, USA (jkuzovkina@uconn.edu)

Larsson, S., Department of Ecology, Swedish University of Agricultural Sciences, Box 7044, SE-750 07 Uppsala, Sweden (stig.larsson@ekol.slu.se)

Lebedys, Arvydas, Forestry Department, FAO, Viale delle Terme di Caracalla, 00153 Rome, Italy (arvydas.lebedys@fao.org)

Licht, L., Ecolotree, Inc., 3017 Valley View Lane NE, North Liberty, IA 52317, USA (lou-licht@ecolotree.com)

Luo, Y., Laboratory for Silviculture and Conservation of Ministry of Education, Forest University Beijing, Beijing 100083, People's Republic of China (yqluo@bjfu.edu.cn)

Ma, Qiang, Forestry Department, FAO, Viale delle Terme di Caracalla, 00153 Rome, Italy (qiang.ma@fao.org)

Marron, N., Research Group of Plant and Vegetation Ecology, University of Antwerp, Department of Biology, Campus Drie Eiken, Universiteitsplein 1, B-2610 Wilrijk, Belgium

Present address: UMR 1137 INRA Nancy – Lorraine University, Ecologie et Ecophysiologie Forestières, 54280 Champenoux, France (marron@nancy.inra.fr)

McCracken, A.R., Agri-Food & Biosciences Institute, 18A Newforge Lane, Belfast BT9 5PX, N. Ireland, UK (alistair.mccracken@afbini.gov.uk)

McIvor, I., Plant and Food Research, Private Bag 11 600, Palmerston North, New Zealand (ian.mcivor@plantandfood.ac.nz)

Mertens, P., Département de l'étude du milieu naturel et agricole – Direction du milieu forestier, Avenue Maréchal Juin, 23, B-5030 Gembloux, Belgium (patrick.mertens@spw.wallonie.be)

Nef, L., Forestry Department, UCL, Louvain-la-Neuve, Belgium (lnef@brutele.be)

Newcombe, G., College of Natural Resources, University of Idaho, Moscow, Idaho, USA (georgen@uidaho.edu)

Ostry, M., Northern Research Station, USDA Forest Service, St Paul, Minnesota, USA (mostry@fs.fed.us)

Parra, P., Istituto de Investigación Forestal de Chile – INFOR, Sucre 2397, Ñuñoa, Santiago, Chile (pparra@infor.cl)

Perttu, K., Department of Crop Production Ecology, Swedish University of Agricultural Sciences (SLU), PO Box 7043, SE-750 07 Uppsala, Sweden (kurt.perttu@vpe.slu.se)

Polle, A., Büsgen-Institut, Abtlg. Forstbotanik und Baumphysiologie, Georg-August Universität Göttingen, Büsgenweg 2, 37077 Göttingen, Germany (apolle@gwdg.de)

Ramstedt, M., Department of Forest Mycology and Pathology, Swedish University of Agricultural Sciences, Uppsala, Sweden (mauritz.ramstedt@slu.se)

Richardson, J., 1876 Saunderson Drive, Ottawa, Ontario, Canada K1G 2C5 (jrichardson@on.aibn.com)

Riddell-Black, D., 41 Bradley Road, Nuffield, Henley on Thames, UK RG9 5SG (drusilla@lupus-science.co.uk)

Robinson, B., Soil and Physical Sciences, Lincoln University, PO Box 84, Lincoln 7647, New Zealand (brett.robinson@lincoln.ac.nz)

Rosengren, Linda, Forestry Department, FAO, Viale delle Terme di Caracalla, 00153 Rome, Italy. Present address: WWF Finland, Lintulahdenkatu 10, Helsinki, Finland (linda.rosengren@wwf.fi)

Scarascia-Mugnozza, G., Department of Agronomy, Forestry and Land Use, Agricultural Research Council of Italy, Via del Caravita 7/a I-00186 Roma, Italy (giuseppe.scarascia@entecra.it)

Schroeder, W.R., Agroforestry Development Centre, Agriculture and Agri-Food Canada, 2 Government Road, Indian Head, Saskatchewan, Canada S0G 2K0 (schroederb@agr.gc.ca)

Serapiglia, Michelle J., Cornell University, Department of Horticulture, New York State Agricultural Experiment Station, Geneva, NY 14456, USA (mjs534@cornell.edu)

Singh, A.P., Ecology and Biodiversity Division, Rain Forest Research Institute, PO Box No 136, Jorhat, Assam, India-785001 (singhap@icfre.org; ranoteaps@gmail.com)

Smart, Lawrence B., Cornell University, Department of Horticulture, New York State Agricultural Experiment Station, Geneva, NY 14456, USA (lbs33@cornell.edu)

Stanton, Brian J., GreenWood Resources, Portland, OR 97201, USA (brian.stanton@gwrglobal.com)

Stanturf, John A., Center for Forest Disturbance Science, USDA Forest Service, Athens, GA 30602, USA (drdirt48@gmail.com)

Steenackers, M., Research Institute for Nature and Forest (INBO), Geraardsbergen, Belgium (marijke.steenackers@inbo.be)

van Acker, Joris, Ghent University, Laboratory of Wood Technology, Coupure Links 653, B-9000 Gent, Belgium (joris.vanacker@ugent.be)

van Oosten, Cees, SilviConsult Woody Crops Technology Inc., Nanaimo, BC V9T 4N3, Canada (silviconsult@telus.net)

Volk, T.A., Center for Sustainable and Renewable Energy, State University of New York College of Environmental Sciences and Forestry, Illick Hall, 1 Forestry Drive, Syracuse, NY 13210, USA (tavolk@esf.edu)

Volney, W.J.A., Northern Forestry Centre, 5320 122 Street Northwest Edmonton, Alberta, Canada T6H 3S5 (Jan.Volney@NRCan-RNCan.gc.ca)

Weih, M., Department of Crop Production, PO Box 7043, SE 750 07 Uppsala, Sweden (martin.weih@slu.se)

Yukun, Hua, Wood-based Composites Institute, Nanjing Forestry University, Nanjing 210037, Jiangsu, PR of China (huayukun@njfu.com.cn)

Acknowledgements

The editors wish to express their deep and sincere appreciation to all the many authors who have contributed to the chapters of this book. Nearly 60 authors from 14 countries on five continents have contributed. The work of the lead chapter authors, who in many cases recruited other authors, shaped the individual chapters and collated, edited and revised the contributed texts, is particularly appreciated. Without these dedicated volunteers, all of them experienced poplar and willow scientists and experts, it would not have been possible to create such a massive collection of knowledge and information. To these writers we must add the many willing contributors of photos and figures that enhance this volume so greatly. All of the substantive chapters have been peer reviewed by colleagues to provide quality assurance of the content; most of these reviewers are acknowledged individually by chapter authors. In addition, Don Dickmann and Julia Kuzovkina have reviewed most chapters in relation to issues of the taxonomy of *Populus* and *Salix*, respectively, and their help is appreciated.

The original proposal for creating *Poplars and Willows: Trees for Society and the Environment* was presented in 2002 to the Executive Committee of the International Poplar Commission (IPC), a subsidiary body of the Food and Agriculture Organization (FAO) of the United Nations, as a replacement for the by-then outdated and out-of-print book on poplars and willows published by the FAO in 1980. The committee endorsed the proposal enthusiastically then and, under the continuing leadership of Stefano Bisoffi as Chair, has supported the project strongly throughout its extended gestation. That support is gratefully acknowledged.

Behind the formal endorsement of the IPC Executive Committee, the parent FAO has provided the continuous practical and logistical help, in addition to the financial and administrative backing to the project as co-publisher, that was needed to make it happen. We are deeply appreciative of the vision, patience, wise management and encouragement of Jim Carle, who guided us during the formative years of the project. His successor, Walter Kollert, with kind and gentle prodding, generous support and unfailing understanding and trust, has been largely responsible for enabling us to bring the project to a conclusion that in many ways exceeds our wildest original dreams. In negotiations and interactions between the FAO and the co-publisher, the professional expertise and experience of Rachel Tucker has been greatly appreciated, as has her advice on publication issues. Alberto delLungo has provided invaluable technical and graphical help with ever-gracious expedition, assisted by Roberto Cenciarelli. Graciela Andrade has always been an efficient, generous and smiling guide through the details of the FAO organization and administration.

The editorial and production staff of CABI has been extremely helpful throughout the publication process, responding promptly and with understanding to every enquiry. We are particularly indebted to Commissioning Editor Rachel Cutts, Editorial Assistant Emma McCann, Senior Production Editor Tracy Head and Production Editor Simon Hill, as well as contract copy-editor Chris McEnnerney, for their patience, attention to detail and professional approach.

Finally, we extend the deepest gratitude to our spouses, Laura and Sharon, for their patient encouragement and support throughout this extended project.

J. Richardson and J.G. Isebrands
16 July 2012

1 Introduction

J.G. Isebrands[1]* and J. Richardson[2]
[1]*Environmental Forestry Consultants LLC, New London, Wisconsin, USA;* [2]*Ottawa, Ontario, Canada*

Poplars and willows, the members of the *Salicaceae* family, are trees and shrubs with many valuable characteristics which have led to multiple beneficial uses for society and the environment since the dawn of history. The accumulated global knowledge and information on poplars and willows could fill many volumes. The characteristics which have made them so attractive and useful include fast growth, ease of propagation, propensity to hybridize, pleasing appearance and many uses. As well as providing wood, fibre, fuelwood and other forest products, poplars and willows benefit society in the rehabilitation of degraded land, restoration of forest landscapes and mitigation of climate change. All of these benefits support rural livelihoods and contribute to sustainable development, particularly in developing countries.

Recent evidence suggests that people living along the Euphrates River in the Middle East more than 10,000 years ago used poplars and willows for cooking and heating, and for the construction of their dwellings (Stettler, 2009). Greek gods are said to have woven wreaths of aspen leaves (*Populus tremula*), and Hercules lit a sacrificial fire of aspen wood when he returned from Hades. In 8000 BC, Native American Ojibwe also used poplars and willows for cooking, heating, shelter and medicines (a precursor of the modern-day drug, aspirin) (Hageneder, 2005). Moreover, Gordon (2001) reported that the Third Dynasty of Ur in Mesopotamia between the Tigris and the Euphrates (modern-day Iraq) used poplars and willows for baskets, boats, construction, hoes, ploughs, tool handles, cooking, wattles and animal fodder in 2100–2000 BC (Fig. 1.1). Archaeological studies have shown that poplar was used for cooking, heating and construction during the period between 700 and 200 BC in Youmulakekum, China, just prior to the Han Dynasty (J. Zhang, 2008, personal communication). In 600 AD, the Chinese used poplars for amenity plantings along Xian roadsides and streets, as well as for fuel. In addition, the Hohokam natives used cottonwood and willows for soil stabilization and along irrigation canals in 800 AD in Mexico (now Arizona, USA) (Logan, 2002).

Likewise, willows provided ancient civilizations with many necessities such as furniture, snowshoes, arrow shafts, fish traps, nets and rope, as well as shelter, fences and medicinal remedies (Fig. 1.2). It has been suggested that willow baskets and containers were probably among the first articles manufactured by humans. These articles were used for gathering and carrying food, and for storage (Kuzovkina *et al.*, 2008) (see Fig. 10.13c; Chapter 10, this volume).

*Corresponding author: efcllc@athenet.net

European explorers who travelled to the New World in the 17th and 18th centuries often returned with cuttings of poplars to plant in their home gardens. For example, eastern cottonwood (*Populus deltoides*) was introduced from North America throughout France in the early 1700s under various cultivar names – leading to nomenclature confusion later (Thielges and Land, 1976; Chapter 2, this volume). In the early 1800s, the North American explorers Lewis and Clark relied on cottonwood in their quest for a land/river route from the Atlantic Ocean to the Pacific Ocean. Cottonwood was the source of wood for their canoes, and it was used for cooking, heating and shelter during their successful 2-year transcontinental voyage (DeVoto, 1977).

By the mid-1800s, cultivated poplars were so widespread in Europe that they became the motifs for several French Impressionist painters. Arguably the most famous painter of poplars was the French artist Claude Monet, who started painting poplars in 1858 and began his famous 'poplar' series in 1891, when he painted hundreds of paintings near Giverny, France. A notable story about Monet is that he ended up purchasing the subjects of his poplar paintings from the community when it threatened to harvest them (Tucker, 1989). At about the same time in North America, the Homestead Act of 1862 in the USA and the Dominion Land Act of 1872 in Canada required homesteaders to plant trees and cultivate the soil in exchange for land in the west. These laws encouraged settlers to plant huge areas of native poplars (and some willows) on the prairies of North America, and prompted the formation of shelterbelt programmes in Canada and the USA (Richardson *et al.*, 2007; Chapter 6, this volume).

Cultivation of willow began around 1800, because until then the natural stands of willow had been sufficient to meet the growing demand for baskets throughout Western Europe. The Industrial Revolution and the advent of World War I increased the demand for willows. The light weight and strength of willow baskets made them suitable for carrying food and medical supplies (Kuzovkina *et al.*, 2008).

Up until the early 20th century, most of the poplars and willows that were planted were locally available native plants or putative, spontaneous hybrids of European and North American

Fig. 1.1. Native stand of *Populus euphratica* in central Asia. Photo courtesy of the FAO.

Fig. 1.2. Natural stand of *Salix alba*, Hungary. Photo courtesy of Robert Vidéki, Doronicum Kft., Bugwood.org.

species that had sprung up in European gardens. These hybrids were first described as intercontinental hybrids (now known as Euramerican poplars) in 1755 in France (Dickmann, 2006; Chapter 2, this volume). The popularity and productivity of these hybrids led to a milestone in modern-day poplar culture – the first controlled hybridization of poplar trees by A. Henry in 1912 (Henry, 1914). Subsequently, Stout and Schreiner (1933) began hybridizing poplars at the New York Botanical Garden, USA. These productive hybrids were tested worldwide. Many of them displayed 'hybrid vigour', and some are still in use today. The pioneering works of Henry (1914) and Stout and Schreiner (1933) spawned poplar hybridization programmes throughout the world, especially in Europe (FAO, 1958; Schreiner, 1959). Subsequently, Pauley (1949) is credited with making range-wide collections of eastern cottonwood (*P. deltoides*) seed in the USA for international exchanges with poplar breeders. Pollen was also collected for foreign breeders to use in their hybridization programmes. Eastern cottonwood remains perhaps the most important poplar parent in most worldwide poplar breeding programmes.

The first formal research institute of poplar breeding and culture was founded in 1937 in Italy (FAO, 1958; Dickmann, 2006), the Istituto di Sperimentazione per la Pioppicoltura (ISP) in Casale Monferrato in the Po river valley, under the direction of G. Piccarolo (Fig. 1.3). It pioneered the development of poplar nursery and plantation culture that was necessary to advance poplar culture worldwide (FAO, 1980). In 2004, the institute became the Unita di Ricerca per le Produzioni Legnose Fuori Foresta under CRA, the Italian research council on agriculture.

The increasing worldwide demand for wood products after World War II and the rising world population growth accelerated the spread of fast-growing poplar hybrids to all corners of the world, including China and India (FAO, 1958, 1980; Schreiner, 1959). The successes of the Italian ISP and the increasing demand for wood as well as for pulp and paper led to the establishment of new eastern cottonwood improvement programmes in southern USA (Thielges and Land, 1976). In the southern hemisphere, the increasing demand for matchwood prompted interest in poplar growing. Poplar culture thereby spread to southern latitudes, including Australia, Brazil, Chile, New Zealand and South Africa (Fig. 1.4), where new challenges were encountered (Pryor and Willing, 1965).

Once the ISP and cooperators had demonstrated the rapid growth potential of hybrid poplars, poplar breeding and cultivation spread throughout Europe and North America (Schreiner, 1959; FAO, 1980). Breeding programmes were established in Belgium, France, Holland, Germany, Scandinavia and the UK (Stettler, 2009), in Canada (Richardson *et al.*, 2007) and the USA (Randall, 1976).

Planting of poplars and willows is now even more important globally. For example, although Russia has a huge area of natural poplars and willows, since about 1935 the Russians have planted them in the southern part of the country, where there is a wood deficit and environmental concerns. Systematic genetics and breeding programmes have been established in Russia to improve poplars and willows for

Fig. 1.3. Istituto di Sperimentazione per la Pioppicoltura (ISP), Casale Monferrato, Italy, established in 1939. Photo courtesy of Silvio May.

growth, wood products, biomass and environmental enhancement through shelterbelts and soil erosion control (Tsarev, 2005).

China is recognized as the country with one of the richest resources of natural poplar and willow forests in the world (FAO, 2009), but by 2010 China had also become the world leader in poplar cultivation for timber, fibre, pulp and paper, agroforestry and environmental uses. In 1978, the Chinese government, with the help of international cooperators, established the Three North Shelterbelt Programme across northern China, with the objective of protecting soil and water on agricultural and pastoral lands. By 1991, 20 million ha of tree plantings, comprised mostly of poplar and some willows, were established (Carle and Ma, 2005). This programme soon became the largest poplar planting effort in history, known as the 'Great Green Wall'. In addition, 100,000 ha had been planted in Siyang county in eastern China by 2010, to protect fields from flooding, wind and soil erosion to boost agricultural production (China.org.cn, 2010). When China hosted the 2008 Olympics in Beijing, millions of Chinese citizens planted trees to create a better environment for the Games. Most of these trees were poplars and willows (Beijing Organizing Committee, 2008).

Chile has a long history of planting poplars, starting in the mid-19th century; since about 1985, private industry has contributed significantly to sustainable rural development by planting poplars. El Alamo Agricultural and Forest Company has planted about 3000 ha of poplars in Chile's central valley between the coastal range and the Andes (Ulloa and Villacura, 2005).

In the western USA, private industry has planted large blocks of hybrid poplars since the early 1980s, now totalling nearly 20,000 ha (Stanton *et al.*, 2002). Several companies originated this effort, benefitting from the pioneering poplar genetic and silvicultural work of the University of Washington and Washington State University on *Populus trichocarpa* × *P. deltoides*

Fig. 1.4. Large *Populus deltoides* specimen planted in Hastings, New Zealand. Photo courtesy of J. Richardson.

hybrids. The impressive blocks of poplar plantings east of Portland, Oregon, USA (Plate 17A) are now owned and managed by GreenWood Resources, Inc, for multiple wood products and bioenergy (Stanton *et al.*, 2002; Stettler, 2009).

In parallel with the development of poplar breeding and cultivation programmes in several regions of the world, increasingly urgent concerns appeared in Europe in the 1930s about shortages of timber. Recognizing the potential of poplars, particularly the vigorous growth of hybrids (Dickmann, 2006), to help address these concerns, the Ministry of Agriculture in France established in 1942 a national poplar commission. Other European countries became involved through meetings of specialists and study tours. In 1947, the French Poplar Commission organized an International Poplar Week, with participation from eight countries of Europe. Representatives who met there agreed, with the support of the Food and Agriculture Organization of the United Nations (FAO), to establish an international poplar commission. Thus, the International Poplar Commission (IPC) came into being that year as a subsidiary body of the FAO (FAO, 1958).

Great energy and enthusiasm got the IPC off to a fast and active start. By 1953, six international congresses had been organized in different European countries. The commission expanded quickly in membership and geographical scope, with a poplar conference for the Near and Middle East in 1954 and the First Regional Poplar Conference for Latin America in 1956 in Argentina (FAO, 1958). At the 23rd IPC Session, hosted by China in 2008, 23 member countries and 6 non-member countries were represented by 185 delegates and advisors. The total membership in 2010 was 37 countries. The IPC meets in a full session every 4 years, while its Executive Committee, initially known as the Standing Committee, meets every 2 years.

One of the early results of the teamwork and friendly collaboration among members of the Standing Committee was the production of the book, *Poplars in Forestry and Land Use*, published by the FAO in 1958. The purpose of that volume, as noted by the Chairman of the IPC, Professor P. Guinier, in his introduction to the book, was 'to make available to agriculturists and foresters the results of the research accomplished and to co-ordinate the facts and opinions on which agreement has been reached'. Its contents covered poplar identification and nomenclature; propagation and cultivation; diseases and other damaging agents; and the wood properties and uses of poplars. The book was translated into several languages and was widely used throughout the world as a guide to the development of poplar cultivation (FAO, 1958).

The 1958 edition was so popular that, in less than 15 years, supplies were becoming exhausted and the need for a new edition was seen. Rather than simply reprinting the existing volume, the IPC Executive Committee, under the leadership of Jean Pourtet, decided to revise the book completely to take into account the advances in techniques and the research in genetics, pathology and entomology. With a considerable number of contributions from authors worldwide and with the coordination of Marcel Viart, Vice-Chairman of the IPC Executive Committee, as editor-in-chief, *Poplars and Willows in Wood Production and Land Use* was published by the FAO in 1980. As its title suggests, willows were included, though realizing that information on their cultivation was still limited. The earlier focus on poplar wood as a raw material for industry was tempered by the recognition of other functions of forest trees in plantations and natural ecosystems, such as recreation, enhancing landscapes and 'regenerating the atmosphere' (FAO, 1980).

The 1980 edition also enjoyed great success, but after 20 years, when supplies were again becoming exhausted, it was recognized that its content was out of date, with many recent scientific and technological advances, as well as new developments in the uses of poplars and willows, having taken place. In 2002, the IPC Executive Committee under Dr Stefano Bisoffi proposed the production of a new edition, and this was strongly supported by the FAO. This was to be a new global publication resource for a new age, rather than a reprinting of the previous edition. The objective was to produce a major update of the 1980 edition of the FAO publication on poplars and willows, providing a practical worldwide overview and guide to their basic characteristics, cultivation and use, as well as issues, problems and trends relating to poplars and willows. The scope was intended to be fully worldwide wherever poplars and willows grew and were cultivated. Emphasis on willows was to be increased. A new focus on environmental uses and sustainable rural development was to be incorporated. The goal is a sourcebook and information guide to poplars and willows, which the authors and editors hope the present volume provides.

Overall supervision of the publication project has been provided by the Secretary of the IPC, who is a Senior Forestry Officer with the FAO. Individual chapters were prepared by lead authors, enlisted by the editors on the basis of their scientific and technical knowledge and writing skills, giving primary consideration to knowledge and experience but also taking into account the geographic balance of the writing team. The content of each chapter was prepared by its lead author, with input and contributions from others in an effort to obtain the best information on each topic and good global coverage. All chapters have been peer reviewed, by reviewers selected by the lead authors and the editors. The editors themselves have also reviewed all the chapters, giving particular attention to resolving discrepancies between chapters and striving for a degree of uniformity in style and approach throughout.

The scope of the publication has been expanded beyond that of the 1958 and 1980 volumes. Taxonomy and classification of the *Salicaceae* are addressed in an authoritative manner, with descriptions of all important individual species, but recognizing the differences in philosophy that still exist in these topics. Natural ecosystems are described. Genetic resources of poplars and willows are covered comprehensively, including selection, tree improvement and conservation, as well as the latest developments in genomics. Cultivation of poplars in plantations for industrial uses was the core of previous volumes and continues to be given prominent treatment, with information also on willow cultivation. A major 21st century topic included is the environmental applications of poplars and willows in phytoremediation, rehabilitation of landscapes and for carbon sequestration, biodiversity and urban amenity, among other uses (McCutcheon and Schnoor, 2003). The most important diseases of poplars and willows are described, as well as strategies for their control. Insects and other animal pests of poplars and willows are treated similarly. Wood properties and utilization of poplars and willows for traditional, as well as emerging, new products are presented. Markets, trends and outlook for wood products worldwide are analysed, considering poplars in relation to tree species in general. The benefits of poplars and willows for sustainable land use and rural livelihoods, including agroforestry, food security, poverty alleviation and shelter, particularly in developing countries, are highlighted. Past trends in the understanding and use of the *Salicaceae* are reviewed, current issues – including bioenergy, climate change, genomics, model systems and phytoremediation – are discussed and likely future trends presented in the overall context of the significance and role of poplars and willows in the world in which we live.

References

Beijing Organizing Committee (2008) Beijing citizens join tree-planting campaign. The Beijing Organizing Committee for the Games of the XXIX Olympiad (http://en.beijing2008.cn/97/98/treeplanting.shtml, accessed 20 December 2011).

Carle, J. and Ma, Q. (2005) Challenges of translating science into practice: poplars and other species in the Three North Region of China. *Unasylva* 56, 31–37.

China.org.cn (2010) China's recovered poplar forests boost agriculture (www.china.org.cn/environment/2010-08/05/content_20643229.htm, accessed 20 December 2011).

DeVoto, B. (1997) *The Journals of Lewis and Clark*. Houghton Mifflin Company, New York.

Dickmann, D.I. (2006) Silviculture and biology of short-rotation woody crops in temperate regions: then and now. *Biomass and Bioenergy* 30, 696–705.

FAO (1958) *Poplars in Forestry and Land Use*. Food and Agriculture Organization of the United Nations, Rome.

FAO (1980) *Poplars and Willows in Wood Production and Land Use*. Food and Agriculture Organization of the United Nations, Rome.

FAO (2009) Report, 23rd Session of the International Poplar Commission and 44th Session of its Executive Committee, Beijing, China, 26–30 October 2008. Food and Agriculture Organization of the United Nations, Rome. Available from IPC Secretariat.

Gordon, J.C. (2001) Poplars: trees of the people, trees of the future. *The Forestry Chronicle* 77, 217–219.

Hageneder, F. (2005) *The Meaning of Trees*. Chronicle Books, San Francisco, California.

Henry, A. (1914) A new hybrid poplar. *Gardeners' Chronicle, Series III* 56, 257–258.

Kuzovkina, Y.A., Weih, M., Romero, M.A., Charles, J., Hurst, S., McIvor, I., *et al.* (2008) Salix: botany and global horticulture. *Horticultural Reviews* 34, 447–489.

Logan, M.F. (2002) *The Lessening Stream*. University of Arizona Press, Tucson, Arizona.

McCutcheon, S.C. and Schnoor, J.L. (2003) *Phytoremediation: Transformation and Control of Contaminants*. John Wiley and Sons, Hoboken, New Jersey.

Pauley, S.S. (1949) Forest genetics research: *Populus* L. *Economic Botany* 3, 299–300.

Pryor, L.D. and Willing, R.R. (1965) The development of poplar clones suited to low latitudes. *Silvae Genetica* 14, 123–127.

Randall, W.K. (1976) Progress in breeding the *Aigeiros* poplars. In: Thielges, B.A. and Land, S.B. Jr (eds) *Proceedings: Symposium on Eastern Cottonwood and Related Species*. Louisiana State University, Baton Rouge, Louisiana, pp. 140–150.

Richardson, J., Cooke, J.E.K., Isebrands, J.G., Thomas, B.R. and van Rees, K.C.J. (2007) Poplar research in Canada – a historical perspective with a view to the future. *Canadian Journal of Botany* 85, 1136–1146.

Schreiner, E.J. (1959) *Production of Poplar Timber in Europe and Its Significance and Application in the United States*. Agriculture Handbook 150. USDA Forest Service, Washington, DC.

Stanton, B., Eaton, J., Johnson, J., Rice, D., Schuette, B. and Moser, B. (2002) Hybrid poplar in the Pacific Northwest. *Journal of Forestry* 100, 28–33.

Stettler, R.F. (2009) *Cottonwood and the River of Time*. University of Washington Press, Seattle, Washington.

Stout, A.B. and Schreiner, E.J. (1933) Results of a project in hybridizing poplars. *Journal of Heredity* 24, 216–229.

Thielges, B.A. and Land, S.B. Jr (1976) *Proceedings: Symposium of Eastern Cottonwood and Related Species*. Louisiana State University, Baton Rouge, Louisiana.

Tsarev, A.P. (2005) Natural poplar and willow ecosystems on a grand scale: the Russian Federation. *Unasylva* 56, 10–11.

Tucker, P.H. (1989) *Monet in the 90s*. Museum of Fine Arts, Boston, Massachusetts.

Ulloa, J. and Villacura, L. (2005) Contribution of the private poplar industry in Chile to sustainable rural development. *Unasylva* 56, 12–17.

2 Poplars and Willows of the World, With Emphasis on Silviculturally Important Species

Donald I. Dickmann[1]* and Julia Kuzovkina[2]**
[1]Department of Forestry, Michigan State University, East Lansing, Michigan, USA; [2]Department of Plant Science, The University of Connecticut, Storrs, Connecticut, USA

2.1 Introduction

> while this planet has gone cycling on according to the fixed laws of gravity, from so simple a beginning endless forms most beautiful and most wonderful have been, and are being, evolved.
>
> Charles Darwin, *The Origin of Species*, 1859

If any family of woody plants affirms Darwin's musing, it is the *Salicaceae*. This family – division *Magnoliophyta*, class *Magnoliopsida* (dicots), subclass *Dilleniidae*, order *Salicales* – includes the familiar genera *Populus* (poplars, cottonwoods and aspens) and *Salix* (willows, sallows and osiers).[1] Together, *Populus* and *Salix* comprise 400–500 species (Table 2.1), although there is no agreement among taxonomists as to the exact number. Added to those numbers are countless subspecies, varieties, hybrids and cultivars that together encompass a diversity of morphological forms which, although bordering on the incomprehensible, is beautiful and wonderful none the less.

Although traditional classification placed only *Populus* and *Salix* in the family *Salicaceae*, the similarity of certain floral structures, co-occurrence of salicoid teeth on leaf margins and the glycoside salicin, similarity of leaf and wood anatomy, the hosting of similar rust fungi, and recent plastid DNA evidence suggest a common ancestry between the tropical family *Flacourtiaceae* and the *Salicaceae* (Chase *et al.*, 2002). The Angiosperm Phylogeny Group (2003), therefore, recommended that some genera in the *Flacourtiaceae* be included in the *Salicaceae*, and this proposal has gained acceptance. The reordering of *Salicaceae* places *Populus* and *Salix* in tribe *Saliceae* and a number of genera heretofore in the *Flacourtiaceae* within eight other tribes in the family (Chase *et al.*, 2002). This revised taxonomy greatly expanded the circumscription of the family *Salicaceae sensu lato* to contain 55 genera, while the traditional circumscription of the family included only three to five genera (*Salicaceae sensu stricto*).

The life form of individual plants in tribe *Saliceae* can be tree, shrub or decumbent. Familial characteristics include simple, mostly stipulate, deciduous leaves with alternate or subopposite phyllotaxy and salicoid teeth; dioeciousness; separate flower and vegetative buds; imperfect (unisexual), bracteate flowers borne on catkins (aments); two- to four-valved, dehiscent

*Corresponding author: dickman1@msu.edu (*Populus*)
**Corresponding author: jkuzovkina@uconn.edu (*Salix*)

Table 2.1. Botanical characteristics of *Populus* and *Salix* (modified from FAO, 1980).

Character	Genus *Populus* (poplars, cottonwoods and aspens)	Genus *Salix* (willows, sallows and osiers)
Genome	$2n = 38$ chromosomes; rarely triploid; 485 ± 10 million DNA base pairs, 45,555 nuclear genes, 153 chloroplast and mitochondrial genes	$2n = 38$ chromosomes; diploid to dodecaploid (12×); genome has not been sequenced
Flowers	Appear before leaves, catkins pendulous, wind pollinated. Flowers with oblique perianth, cup-shaped disk without nectaries; bracts irregularly denticulate, shed rapidly; stamens numerous, 5–50, usually with reddish anthers; pollen thin-walled and non-aperturate; stigmas with 4 or more lobes; ovaries with 2, 3 or (rarely) 4 carpels	Appear before, with, or after leaves; catkins mostly erect; insect or wind pollinated. Perianth and disk usually absent but with 1 or 2 nectaries; bracts entire, pubescent, usually persistent; stamens few, 2–12, usually with yellow anthers; pollen thick-walled and tricolpate; stigmas 2-lobed; ovaries with 2 carpels
Fruit	2-, 3- or 4-valved capsule	2-valved capsule
Leaves	Variable in shape – deltoid to cordate to ovate to lancelotate, occasionally palmately lobed; venation palmatopinnate; margins serrate or dentate and glandular. Indeterminate shoots heterophyllous; heteroblasty occurs in some taxa	Never lobed or deltoid, almost always elongate in shape – obovate, oval, ovate-lanceolate, lanceolate or lanceolate-linear; venation pinnate; margins finely serrate or entire, occasionally glandular. Indeterminate shoots homophyllous
Stipules	Not persistent	Sometimes persistent and prominent
Petioles	Long, sometimes flattened transversely; glands may occur at junction of petiole and leaf blade	Short, round in cross section
Buds	Elongated, often pointed, covered by several overlapping scales, sometimes resinous and fragrant; usually divergent from twig; mostly monopodial with prominent terminal bud	Enveloped by a single scale; closely appressed to twig; mostly sympodial and lacking a true terminal bud
Shoots	Moderately stout; brown, purple or red in colour; circular or angular in cross section; lenticels prominent; pith pentagonal in cross section; heterophyllous (dwarf shoots may be present). Many taxa produce root suckers	Slender; green, brown, yellow, orange, purple or red in colour; circular in cross section; pith circular in cross section; homophyllous (does not form brachyblasts). Rarely develop root suckers
Wood	Light (specific gravity 0.31–0.40), straight grained, soft, pale, not durable, often with a disagreeable odour when wet; rays homocellular	Light (specific gravity 0.30–0.42), uniform, straight grained, soft, pale, not durable, tough and shock resistant, odourless; rays heterocellular
Habit	Medium to large trees, rarely shrubs	Extremely variable; can be procumbent plants, multi-stemmed shrubs and medium to large trees
Habitat	Mostly warm and cold temperate regions; common in wetlands, riparian corridors or uplands; few taxa found in tundra and alpine zones	Mostly cold temperate regions; common in wetlands, peatlands, riparian corridors, but uncommon in uplands; abundant in tundra and alpine zones
Number of taxa	22–45	330–500

capsule fruits; tiny seeds attached to a cottony coma; shoots that can show both determinate (fixed) and indeterminate (free) growth patterns; and low-density, diffuse-porous wood. Yet the two genera can be easily separated, based on flower, leaf and bud morphology (Table 2.1; Plates 18A and 18B).

The tribe phylogeny is not altogether clear. Skvortsov (1968, 1999) and Dorn (1976) suggest that *Populus* is evolutionarily more primitive than *Salix*, and certain parallel morphological features point toward willows having arisen from the most advanced group of poplars in section *Populus*. The far greater number of willow taxa also suggests a later phylogeny for this genus. The recent DNA sequencing of the *Populus* genome produced evidence supporting the conclusion that the lineages of *Populus* and *Salix* shared the same large-scale genomic history (Tuskan *et al.*, 2006), although the fossil record showed that these lineages diverged 60–65 million years ago. None the less, some molecular evidence undermined the view that willows emerged from within the genus *Populus* (Eckenwalder, 1996).

Ecologically, members of the *Saliceae* are mostly shade-intolerant colonizers of disturbed wetlands, riparian areas, lakeshores and uplands (Fig. 2.1). Their prodigious production of wind-blown seed, as well as their rapid growth rate, greatly facilitates this colonizing ability. In the absence of disturbances, for example flooding, fire, windstorms, felling by humans or beavers (*Castor* spp.), poplars and willows usually are displaced in a successional sequence by more shade-tolerant broadleaf and coniferous trees. In some wetland and upland habitats, however, certain members of *Saliceae* may form semi-stable communities due to the high frequency of disturbances. Individual poplar and willow stems often are short-lived (40–60 years), although individual specimens of several species can live for 200–300 years in the absence of disturbances (DeBell, 1990; Weisgerber, 1999). The members of *Saliceae* are renowned for their ability to reproduce vegetatively by root collar sprouts, stool shoots, root suckers, layering, cladoptosis and fragmentation, or – with help from humans – stem and root cuttings. Another defining characteristic of the tribe is interspecific hybridization. This recombination commonly occurs in natural

Fig. 2.1. Willow and poplar seedlings invading exposed beach sand on the shore of a freshwater lake. Photo by D. Dickmann.

populations of poplars and, to a more limited extent, in willows, but it also has been exploited for centuries by tree breeders for forestry and horticultural purposes.

Members of the *Saliceae* are indigenous principally in the temperate, boreal and tundra regions of the northern hemisphere (Plates 1–16). A few species are native to the subtropical and tropical regions of North America, Africa and Asia, and a single species of willow occurs naturally in South America. In addition, many species have become naturalized in areas outside their natural range.

The singular combination of the characters of poplars and willows – their rapid growth rate, ease of vegetative propagation, predisposition to hybridize, pleasing appearance and useful wood – have long endeared them to humans. The familiar weeping willows and Lombardy poplar, for example, are the most widely planted ornamental trees in the world (Li, 1996). Willows were mentioned in the Old Testament, and they were highly regarded, for example, 'You shall take on the first day the fruit of goodly trees, branches of palm trees, and boughs of thick trees, and willows of the brook' (Leviticus 23:40).[2] Today, poplars and willows are widely recognized as the trees of choice for intensive, short-rotation forestry in temperate regions (Dickmann, 2006), but short-rotation coppice culture of willow osiers has been practised from time immemorial (Fig. 2.2). The Roman, Pliny the Elder, extolled the usefulness of osiers for vine trellises, rope, agricultural tools, baskets and furniture (Perlin, 1991). The Romans apparently operated their planted willow coppices in a systematic way, calculating yields and labour requirements (Rackham, 1990).

Culture of fast-growing poplars likewise has a long history. In fact, the Latin *Populus* was derived from the Roman appellation *arbor populi* or 'tree of the people', because poplars were so widely planted (Clute, 1943). Another simile with people is the fluttering habit of poplar leaves that are borne on flattened petioles, which in the slightest breeze make a noise likened to talking. Historically, poplars often were part of an agroforestry system and were managed to supply timber, fuelwood, forage and windbreaks. According to Pliny the Elder (Newsome, 1964),

Fig. 2.2. Coppice culture of willow osiers for baskets, wicker furniture and wattle has been ongoing for millennia; *Salix viminalis* (pictured here) has long been one of the favoured 'basket' willows. Photo by J. Kuzovkina.

they were also used as living trellises to support grapes in vineyards. Poplar culture in China also goes back several millennia. In the 2400-year-old book, *Hui Zi*, methods of cutting and layering for planting poplars are described. Another book, *Jin Shu*, published 1500 years ago, mentioned that poplars and pagoda trees (*Styphnolobium japonicum*) were planted along roads in cities for shade. Thus, the contemporary scene that is so familiar across the world – poplar plantings lining a rural track or canal or a willow coppice rising above an adjacent pasture – has roots that go far back into antiquity.

Members of the *Saliceae* also have a dark side. Some people dislike them because of their short lifespan, susceptibility to breakage, production of messy cotton during seed release, aggressive seedling establishment and root sprouting, and vulnerability to insect and disease infestations. Poplars, especially, are not highly regarded as ornamental trees in many places. One prominent horticulturist (Dirr, 1997) summed up his feelings this way: 'In all my traveling and consulting work, I have never recommended, at least when conscious, a poplar.' Some municipalities have passed ordinances banning the planting of poplar trees because of their offensive habits. Poplars and willows are aggressive invaders of disturbed sites, and in certain agricultural and forestry situations they are regarded as weeds. A few species have also become noxious invasives in areas outside their natural range, for example certain willows in Australia and New Zealand.

Our treatment of the *Saliceae* continues with short discussions of systematics and nomenclature, subjects that remain unsettled in this family. Then we introduce in general terms each genus, followed by descriptions of the species in each section of the genus – *Populus* first, then *Salix*. The species descriptions are not intended to be rigorously botanical; that is, they focus on vegetative characteristics and cannot be used for reliable species identification in the field or in herbaria. For that purpose, we recommend a local taxonomic monograph or manual, with keys. Rather, our treatment sets each taxon in its geographic and ecological context, describes certain morphological features, depicts intraspecific variation and discusses uses and commercial importance. All 32 *Populus* species are described. Due to the large number of *Salix* taxa, however, we describe only a 32-species subset of tree and shrub willows that currently are important in forestry, horticulture and environmental applications.

2.1.1 Systematics

The concept of species and the mechanisms by which species evolve continue to be topics of considerable discussion within the biological community (e.g. Baum and Donoghue, 1995; Hey, 2006a). Theoretically, organisms within a species are considered to share a common evolutionary history, a concept that, although somewhat problematic in practice, most biologists agree on. The consensus breaks down on the criteria that can be used to define a species, especially now that molecular-level genomic data are readily available along with traditional morphological and ecological information. None the less, that a species comprises an interbreeding population that is isolated reproductively, and that the offspring of this interbreeding are equally as fit as the parents, seems well established (Coyne and Orr, 2004). Isolation can be based on geographic distance, physiographic barriers, ecological niche separation, flowering phenology, specificity of pollinators, physiological or morphological barriers or differences in ploidy. On a pragmatic level, the population that constitutes a species must possess at least one persistent and readily observable or measurable diagnostic character (Cronquist, 1988; Baum, 1992); the Latin root of species, after all, is the verb *specere* – to look, behold, regard. That measurable or observable diagnostic characters now include gene maps, and the base sequence of genetic and organelle DNA, has further complicated an already complicated business, leading to the rethinking of classical taxonomic ordering in many plant and animal groups, including the *Salicaceae*.

There is one further complication. Darwin's definition of species in *The Origin of Species* (1859), Chapter 2, reads in part: 'a set of individuals *closely resembling* each other ...' (our emphasis). Thus, members of a species are variable, inexact copies of one another, and the characteristics of divergent populations of one species may overlap those of another species; in fact, certain genes may be exchanging between them (Hey, 2006b). This is as true in the *Salicaceae* as in any other family. In our view, some *Populus* and *Salix* systematicists are too absorbed with the resultant variation in morphological minutiae – which in some cases may be programmed by a single gene – leading to a plethora of named species. An example is leaf or twig pubescence – is the presence or absence of pubescence enough to differentiate one species from another, when other morphological traits, ecological habitat, geographic distribution and evolutionary history are similar or even indistinguishable? Linnaeus had it right in his *Philosophia Botanica* (1751): '*Scias characterem non constituere genus, sed genus characterem.*' (It is not the character that makes the genus, but the genus that makes the character.) His maxim applies equally well at the species level.

Designation of species is especially difficult in the *Saliceae*, for several reasons. First, many poplar and willow taxa have a large natural range, which in the extreme may encompass several continents. These taxa naturally segregate into morphologically distinct subspecies and varieties, which, although geographically isolated, may share an evolutionary history and be capable of interbreeding. Should these polymorphic variants really be considered true species? Second, dioecism obligates poplars and willows to be outcrossing. Furthermore, they show few inherent barriers to interspecific and, in some case, intersectional cross-breeding, so they hybridize readily, both naturally and through controlled pollinations. Areas where introgression of species populations occurs create special difficulties (Ronald *et al.*, 1973; Rood *et al.*, 1986; Whitham, 1989). The result of this cross-species gene flow is a multitude of intermediate forms, which in some cases may be either sterile or lacking in fitness. Over the years, species status has been assigned incorrectly to numerous hybrids (known as nothospecies). The common occurrence of polyploidy in *Salix* creates an additional taxonomic problem. Dioeciousness, different phenology for flowers and leaves and the occurrence of large monoclonal stands in some species may make observation of all important traits at one time or on one plant impossible. Leaf heteromorphism (heterophylly and heteroblasty) can be misleading if not recognized by the observer (Eckenwalder, 1980). Thus, a herbarium sheet that contains a single shoot or plant simply is not a reliable basis for species determination in the *Saliceae*.

Based on their manner of dealing with this genetic complexity, taxonomists segregate into two camps (Eckenwalder, 1996). Splitters actively name new taxa. Their tally of *Populus* species can reach as high as 85, with over 60 species in China alone (Fang *et al.*, 1999). The count of *Salix* taxa by the splitter camp approaches 500, and classification is by no means complete. Lumpers tend to be more evolutionarily and ecologically oriented and set rigid standards for the acceptance of new species. Lumpers view many of the taxa proposed by the splitter camp to be nothospecies, either hybrids or polymorphic variants of a single diverse taxon. Whatever their camp, no two authorities seem to agree on the exact number of species in either genus, and systematic arguments can be esoteric and hinge on exceedingly fine points. We have adopted a conservative view of speciation in *Populus* and *Salix* that is based largely on classical taxonomy (Tables 2.2 and 2.3). We recognize, however, that the matter is far from resolved. The discovery of new *Salix* species constantly continues and the ranking of the previously described taxa is evolving, reflecting the complexity of the genus (Table 2.4). In the future, new techniques and data, especially from nuclear and cytological genetics, will resolve some disputed relationships among taxa, leading to new orderings.

2.1.2 Nomenclature

Although some readers may consider this section superfluous, we include it to clarify the exceedingly confused subject of naming, which long has been the bane of serious students and practitioners of *Saliceae* culture. This intractable problem continues to this day. Misnomers abound in the literature of the *Saliceae*, and many of them have been perpetuated for a long time. Some of these misnomers are due to taxonomic ambiguities, others to ignorance of correct nomenclature, and still others to sloppiness on the part of workers in the field. Anyone working in *Saliceae* culture, research or the commercial trade should make a serious attempt to understand the correct nomenclature of his or

Table 2.2. Proposed taxonomic classification of the genus *Populus*.[a]

Section	Taxon	English common name[b]	Notes and synonyms
Abaso	*P. mexicana* Wesmael	Yaqui cottonwood	Monotypic section
Turanga (Afro-Asian poplars)	*P. euphratica* Olivier	Euphrates poplar	Includes *P. diversifolia*
	P. ilicifolia (Engler) Rouleau	Kenyan poplar	Formerly synonymous with *P. euphratica*
	P. pruinosa Schrenk	Desert poplar	Formerly synonymous with *P. euphratica*
Leucoides (Swamp poplars)	*P. glauca* Haines	Asian swamp cottonwood	Formerly *P. wilsonii*
	P. heterophylla Linnaeus	Swamp cottonwood	
	P. lasiocarpa Oliver	Heart-leaf poplar	
Aigeiros (Cottonwoods, black poplar)	*P. deltoides* Marshall	Eastern cottonwood	Includes *P. sargentii*, *P. palmeri* and *P. wislizenii*
	P. fremontii S. Watson	Fremont cottonwood	Includes *P. arizonica*
	P. nigra Linnaeus	Black poplar	
Tacamahaca (Balsam poplars)	*P. angustifolia* James	Narrowleaf cottonwood	
	P. balsamifera Linnaeus	Balsam poplar	Formerly *P. tacamahaca*
	P. cathayana Rehder	Cathay poplar	May be synonymous with *P. suaveolens*; includes *P. purdomii*
	P. ciliata Royle	Himalayan poplar	Heretofore in section *Leucoides*; the former *P. tristis* may be a hybrid with this species
	P. koreana Rehder	Korean poplar	Likely synonymous with *P. suaveolens* or *P. maximowiczii*
	P. laurifolia Ledebour	Laurel poplar	
	P. maximowiczii Henry	Japanese poplar	May be synonymous with *P. suaveolens*; includes *P. ussuriensis*
	P. simonii Carrière	Simon poplar	Includes *P. przewalskii* and *P. kangdingensis*
	P. suaveolens Fischer	Siberian poplar	
	P. szechuanica Schneider	Szechuan poplar	
	P. trichocarpa Torrey & Gray	Black cottonwood	May be synonymous with *P. balsamifera*
	P. yunnanensis Dode	Yunnan poplar	
Populus[c] (White poplars and aspens)	*P. alba* Linnaeus	White poplar	
	P. guzmanantlensis Vazquez & Cuevas	Manantlán poplar	May be synonymous with *P. simaroa*
	P. monticola Brandegee	Baja poplar	Aka *P. brandegeei*; may be naturalized *P. alba* var. *subintegerrima*
	P. simaroa Rzedowski	Balsas poplar	
	P. adenopoda Maximowicz	Chinese aspen	
	P. gamblei Haines	Himalayan aspen	
	P. grandidentata Michaux	Bigtooth aspen	
	P. sieboldii Miquel	Japanese aspen	Includes *P. jesoensis*
	P. tremula Linnaeus	Common aspen	Includes *P. davidiana* and *P. rotundifolia*
	P. tremuloides Michaux	Quaking aspen	

[a]Each of these taxa is described in the text.
[b]Common names vary considerably depending on language and locality.
[c]Formerly *Leuce*.

Table 2.3. A classification of the genus *Salix* as included in major regional floras (authorities for each world region may not agree on taxonomic ordering above the species level).

A. Eurasian Region (includes Russia, Europe, Asia Minor, west and north-east China, Mongolia and North Korea) (Skvortsov, 1968, 1999)

I. Subgenus *Salix*		
Section *Humboldtianae* Pax	Section *Pentandrae* C.K. Schneider	10. *S. fragilis* Linnaeus (lately syn. of *S.* ×*fragilis* or *S. euxina*)
1. *S. acmophylla* Boissier	5. *S. pentandra* Linnaeus	Section *Subalbae* Koidzumi
Section *Amygdalinae* Koch	6. *S. pseudopentandra* Floderus	11. *S. pierotii* Miquel
2. *S. triandra* Linnaeus	7. *S. pentandrifolia* A.K. Skvortsov	**S. babylonica* Linnaeus
3. *S. songarica* Andersson	Section *Salix*	
Section *Urbanianae* C.K. Schneider	8. *S. alba* Linnaeus	
4. *S. cardiophylla* Trautvetter	9. *S. excelsa* S.G. Gmelin (syn. of *S. alba*)	
II. Subgenus *Chamaetia* Nasarov		
Section *Chamaetia* Dumortier	Section *Myrtilloides* Andersson	32. *S. nakamurana* Koidzumi
12. *S. reticulata* Linnaeus	22. *S. myrtilloides* Linnaeus	Section *Myrtosalix* Kerner
13. *S. vestita* Pursh	23. *S. fuscescens* Andersson	33. *S. myrsinites* Linnaeus
14. *S. erythrocarpa* Komarov	Section *Glaucae* Pax	34. *S. berberifolia* Pallas
Section *Retusae* Kerner	24. *S. alatavica* Karelin ex Stschegl.	35. *S. tschuktschorum* A.K. Skvortsov
15. *S. herbacea* Linnaeus	25. *S. glauca* Linnaeus	36. *S. breviserrata* Floderus
16. *S. turczaninowii* Lakschewitz	26. *S. reptans* Ruprecht	37. *S. rectijulis* Ledbour
17. *S. polaris* Wahlenberg	27. *S. pyrenaica* Gouan	38. *S. alpina* Scopoli
18. *S. nasarovii* A.K. Skvortsov	28. *S. arctica* Pallas	39. *S. chamissonis* Andersson
19. *S. nummularia* Andersson	29. *S. ovalifolia* Trautvetter	40. *S. saxatilis* Turczaninow
20. *S. retusa* Linnaeus	30. *S. sphenophylla* A.K. Skvortsov	41. *S. phlebophylla* Andersson
21. *S. serpyllifolia* Scopoli	31. *S. kurilensis* Koidzumi	42. *S. rotundifolia* Trautvetter
III. Subgenus *Vetrix* Dumortier		
Section *Hastatae* Kerner	Section *Glabrella* A.K. Skvortsov	52. *S. myrsinifolia* Salisbury
43. *S. hastata* Linnaeus	48. *S. crataegifolia* Bertoloni	53. *S. apennina* A.K. Skvortsov
44. *S. karelinii* Turczaninow	49. *S. glabra* Scopoli	54. *S. mielichhoferii* Sauter
45. *S. apoda* Trautvetter	50. *S. reinii* Franchet et Savatier ex Seemen	Section *Vetrix* Dumontier
46. *S. fedtschenkoi* Goerz	51. *S. jenisseensis* Floderus	Subsection *Kuznetzowianae* A.K. Skvortsov
47. *S. pyrolifolia* Ledbour	Section *Nigricantes* Kerner	55. *S. kuznetzowii* Lakschewitz ex Goerz

Continued

Table 2.3. Continued.

56. *S. lagerii* Wimmer
SUBSECTION *VULPINAE* KIMURA
57. *S. silesiaca* Willdenow
58. *S. caucasica* Andersson
59. *S. appendiculata* Villars
60. *S. pedicellata* Desfontaines
61. *S. vulpina* Andersson
SUBSECTION *LAEVES* CAMUS
62. *S. caprea* Linnaeus
63. *S. aegyptiaca* Linnaeus
64. *S. cinerea* Linnaeus
65. *S. atrocinerea* Brotero
66. *S. pseudomedemii* Wolf
67. *S. aurita* Linnaeus
68. *S. salviifolia* Brotero
SUBSECTION *SUBSTRIATAE* GOERZ
69. *S. tarraconensis* Pau ex Font
70. *S. starkeana* Willdenow
71. *S. bebbiana* Sargent
72. *S. taraikensis* Kimura
73. *S. abscondita* Larkschewitz
74. *S. iliensis* Regel
75. *S. pseudodepressa* A.K. Skvortsov
Section *Arbuscella* Seringe ex Duby
SUBSECTION *BICOLORES* A.K. SKVORTSOV
76. *S. phylicifolia* Linnaeus
77. *S. basaltica* Coste (syn. of *S. bicolor* subsp. *basaltica*)
78. *S. pulchra* Cham
79. *S. kikodseae* Goerz
80. *S. tianschanica* Regel
81. *S. divaricata* Pallas
SUBSECTION *ARBUSCULAE* HAYEK
82. *S. arbuscula* Linnaeus
83. *S. foetida* Schleicher
84. *S. kazbekensis* A.K. Skvortsov
85. *S. waldsteiniana* Willdenow
86. *S. saposhnikovii* A.K. Skvortsov
87. *S. dshugdshurica* A.K. Skvortsov
88. *S. boganidensis* Trautvetter
89. *S. rhamnifolia* Pallas
Section *Vimen* Dumortier
90. *S. viminalis* Linnaeus
91. *S. turanica* Nasarov
92. *S. armeno-rossica* A.K. Skvortsov
93. *S. schwerinii* E. Wolf
94. *S. dasyclados* Wimmer (lately syn. of *S. gmelinii*)
95. *S. sajanensis* Nasarov
96. *S. argyracea* E. Wolf
97. *S. pantosericea* Goerz
98. *S. udensis* Trautvetter
Section *Subviminales* Schneider
99. *S. gracilistyla* Miquel
Section *Canae* Kerner
100. *S. elaeagnos* Scopoli
Section *Villosae* Rouy
101. *S. lapponum* Linnaeus
102. *S. alaxensis* Coville
103. *S. krylovii* Wolf
104. *S. helvetica* Villars
Section *Lanatae* Koehne
105. *S. lanata* Linnaeus
106. *S. recurvigemmis* A.K. Skvortsov
Section *Daphnella* Seringe ex Duby
107. *S. daphnoides* Villars
108. *S. acutifolia* Willdenow
109. *S. rorida* Lakschewitzv
110. *S. kangensis* Nakai
Section *Incubaceae* Kerner
111. *S. repens* Linnaeus
112. *S. rosmarinifolia* Linnaeus
113. *S. brachypoda* Komarov
Section *Flavidae* Chang
114. *S. gordejevii* Chang
Section *Helix* Dumontier
SUBSECTION *CAESIAE* A.K. SKVORTSOV
115. *S. coesia* Villars
116. *S. kochiana* Trautvetter
SUBSECTION *PURPUREA* A.K. SKVORTSOV
117. *S. purpurea* Linnaeus
**S. koriyanagi* Kimura ex Goerz
118. *S. vinogradovii* A.K. Skvortsov
119. *S. elbursensis* Boissier
120. *S. miyabeana* Seemen
121. *S. amplexicaulis* Bory et Chaubard
122. *S. integra* Thunberg
123. *S. gilgiana* Seemen
SUBSECTION *TENUIJULES* A.K. SKVORTSOV
124. *S. tenuijulis* Ledbour
125. *S. pycnostachya* Andersson
126. *S. olgae* Regel
127. *S. linearifolia* E. Wolf
SUBSECTION *KIRILOWIANAE* A.K. SKVCRTSOV
128. *S. kirilowiana* Stscheglejew
129. *S. niedzwieckii* Goerz
130. *S. capusii* Franchet
131. *S. michelsonii* Goerz ex Nasarov
132. *S. caspica* Pall
133. *S. ledebourana* Trautvetter
Section *Cheilophilae* Hao
134. *S. wilhelmsiana* Marschall a Bieberstein
135. *S. microstachya* Turczaninow ex Trautvetter

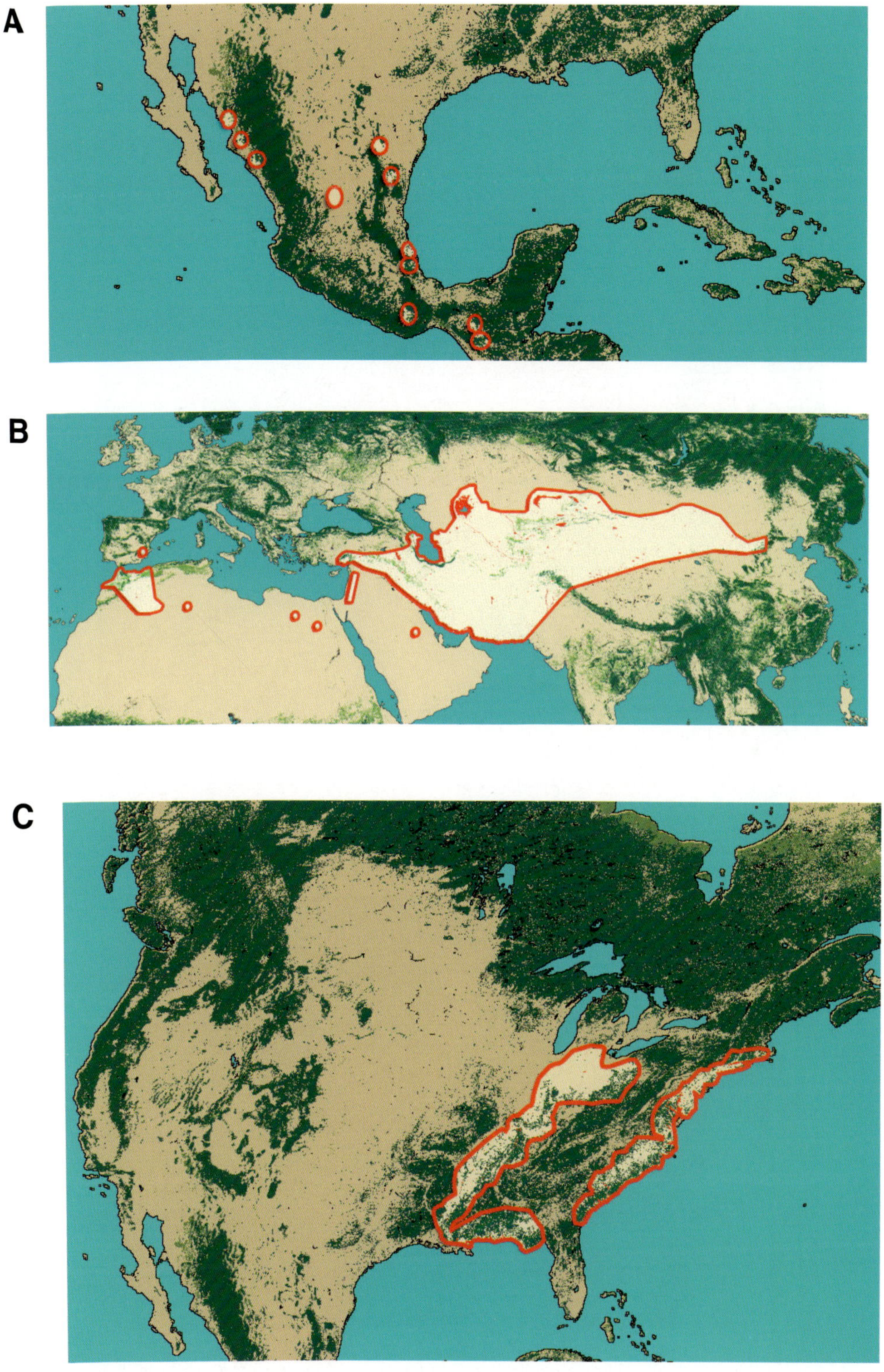

Plate 1. (A) Natural range of *Populus mexicana* in Mexico. Redrawn from Discover Life (http://www.discoverlife.org/). **(B)** Natural range of *Populus euphratica* in Eurasia and Africa. Redrawn from Browicz (1977). **(C)** Natural range of *Populus heterophylla* in eastern North America. Redrawn from US Geological Survey Earth Surface Processes (http://esp.cr.usgs.gov/data/atlas/little/).

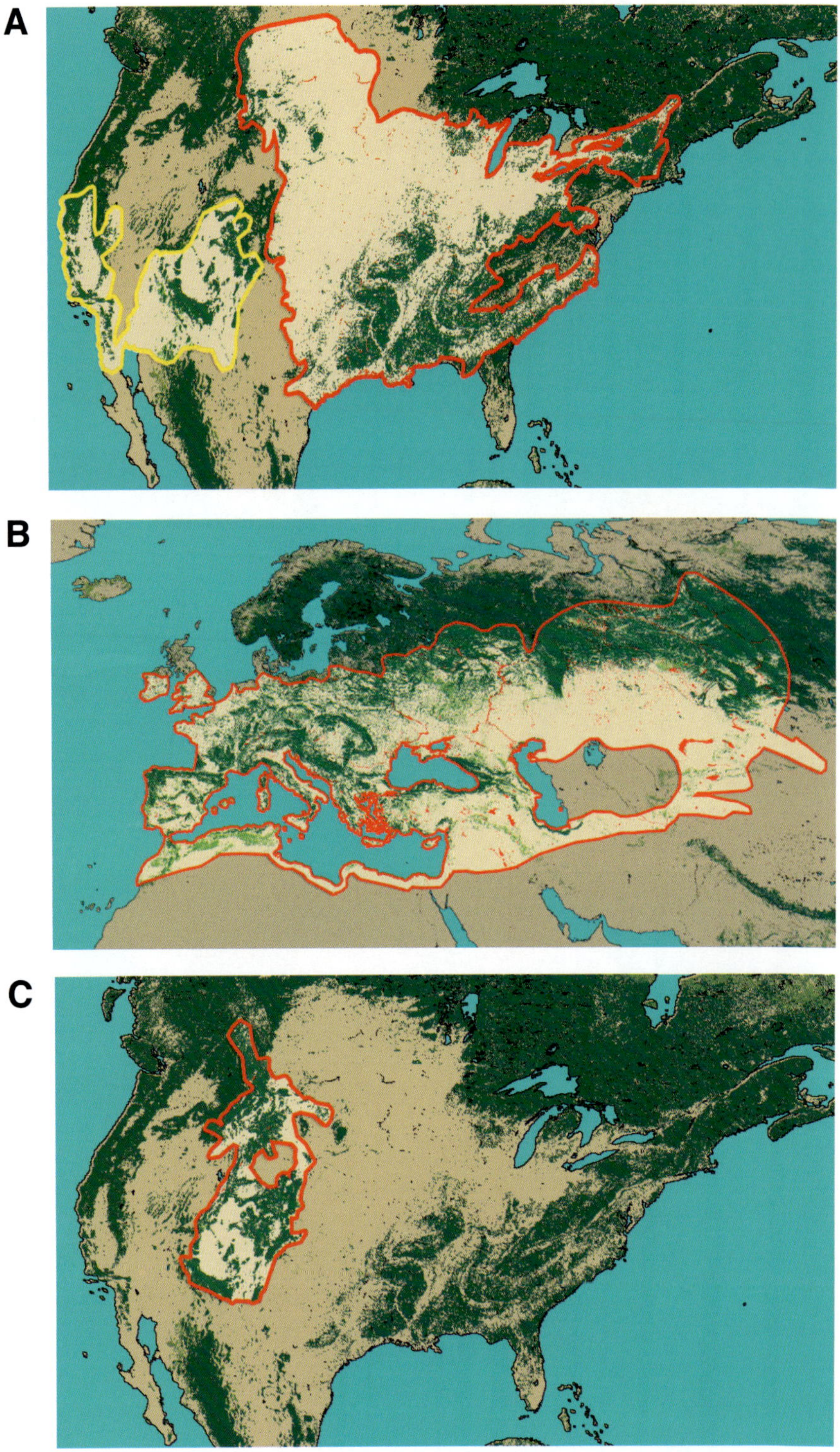

Plate 2. **(A)** Natural ranges of *Populus deltoides* (red) and *Populus fremontii* (yellow) in North America. Redrawn from US Geological Survey Earth Surface Processes (http://esp.cr.usgs.gov/data/atlas/little/). **(B)** Natural range of *Populus nigra* in Eurasia and Africa. Redrawn from Vanden Broeck (2003). **(C)** Natural range of *Populus angustifolia* in western North America. Redrawn from US Geological Survey Earth Surface Processes (http://esp.cr.usgs.gov/data/atlas/little/).

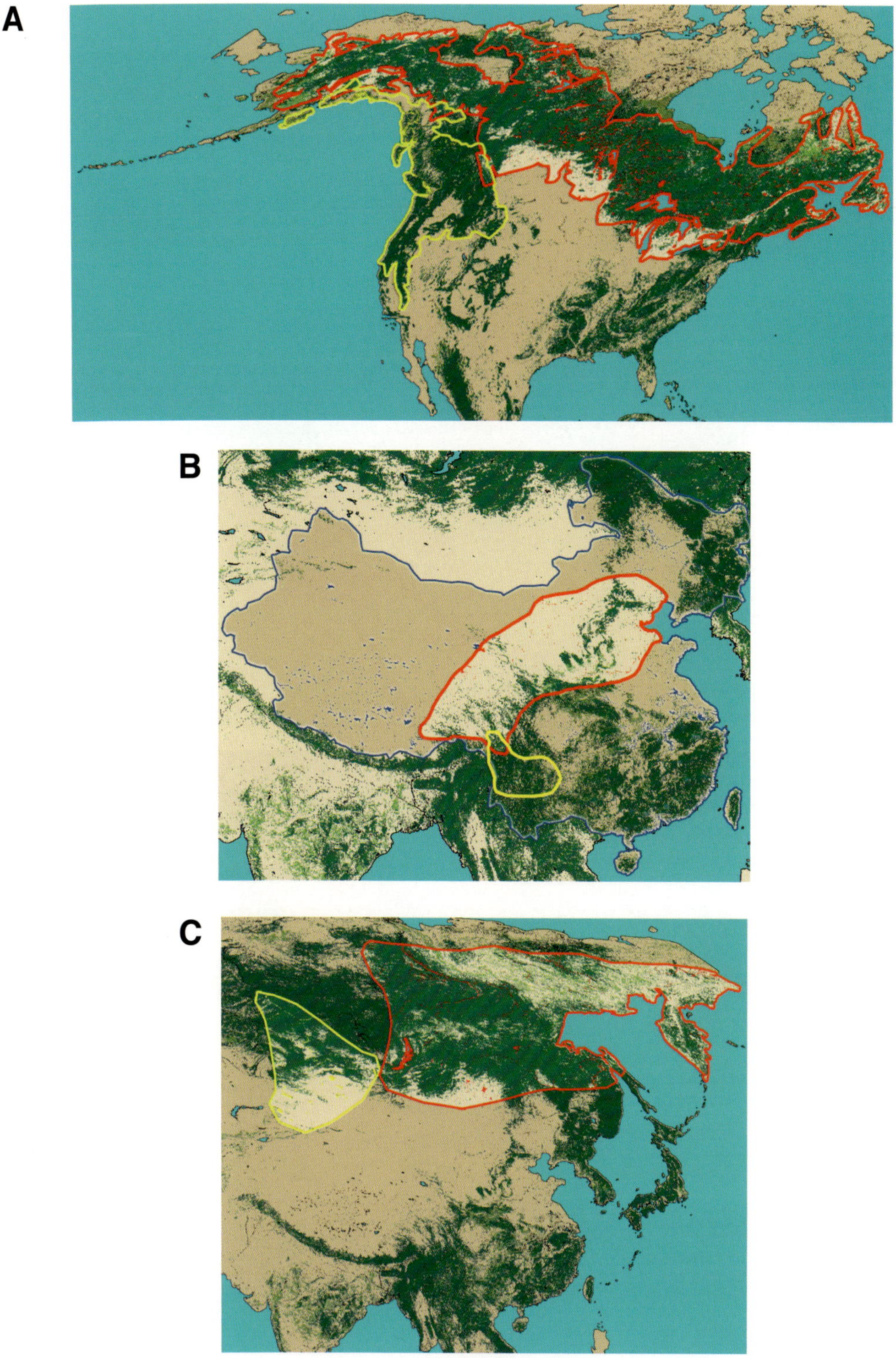

Plate 3. (A) Natural range of *Populus balsamifera* (red) and *Populus trichocarpa* (yellow) in North America. Redrawn from US Geological Survey Earth Surface Processes (http://esp/cr.usgs.gov/data/atlas/little/). **(B)** Natural ranges of *Populus cathayana* (red) and *Populus yunnanensis* (yellow) in China. Redrawn from Weisgerber and Zhang (2005a,b). **(C)** Natural ranges of *Populus laurifolia* (yellow) and *Populus suaveolens* (red) in Asia. Redrawn from Čermák *et al.* (1955).

A

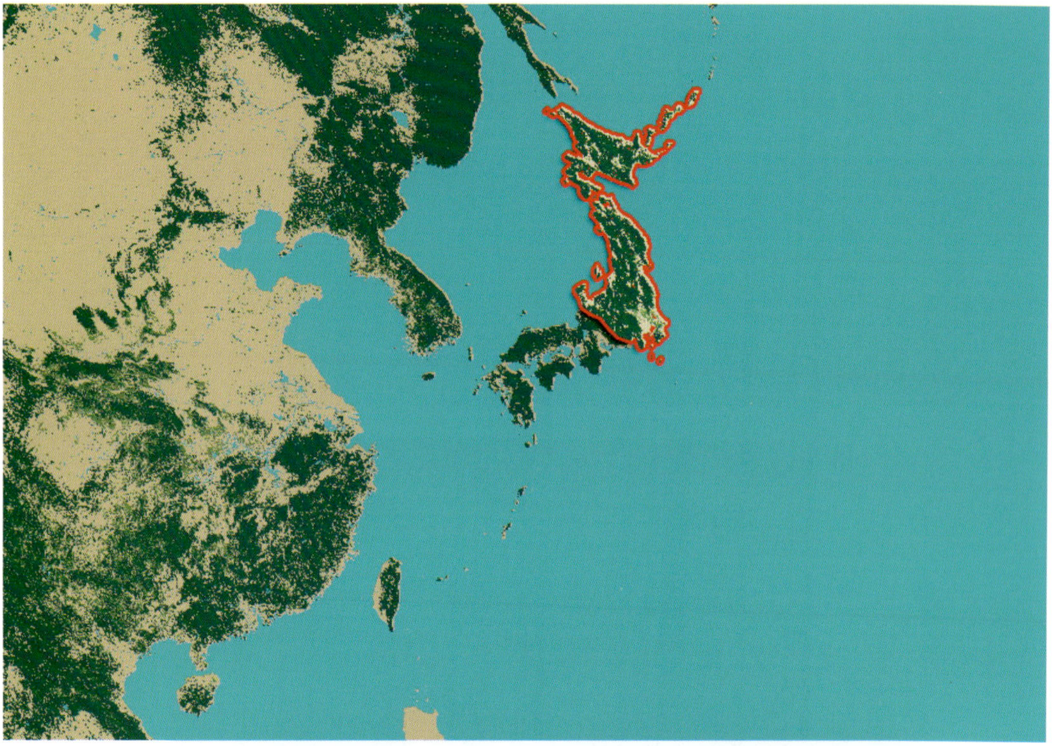

B

Plate 4. (A) Natural range of *Populus maximowiczii* on the islands of Japan. Natural range also includes north-eastern China, eastern Russia and Korea. Redrawn from Hamaya and Inokuma (1957). **(B)** Natural range of *Populus simonii* in Asia. Redrawn from Weisgerber and Han (2001).

A

B

Plate 5. (A) Natural range of *Populus szechuanica* in China. Redrawn from Weisgerber and Han (2001). **(B)** Natural range of *Populus alba* in Eurasia and Africa. Redrawn from Fenaroli and Gambi (1976).

A

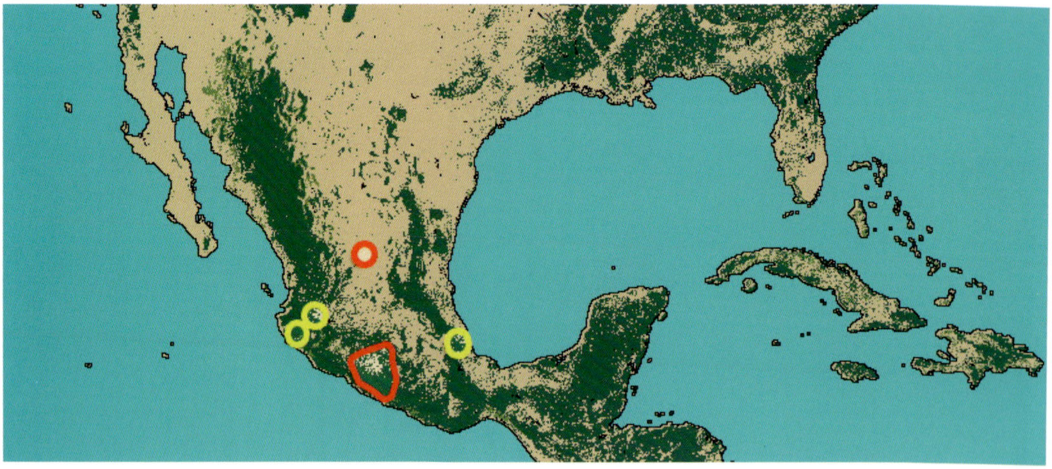

B

Plate 6. **(A)** Natural ranges of *Populus guzmanantlensis* (yellow) and *Populus simaroa* (red) in Mexico. Redrawn from Discover Life (http://www.discoverlife.org/). **(B)** Natural range of *Populus adenopoda* in China, based on descriptive data in Zhenfu *et al.* (1999).

A

B

Plate 7. (A) Natural range of *Populus grandidentata* in eastern North America. Redrawn from US Geological Survey Earth Surface Processes (http://esp.cr.usgs.gov/data/atlas/little/). **(B)** Natural range of *Populus sieboldii* in Japan. Redrawn from Hamaya and Inokuma (1957).

A

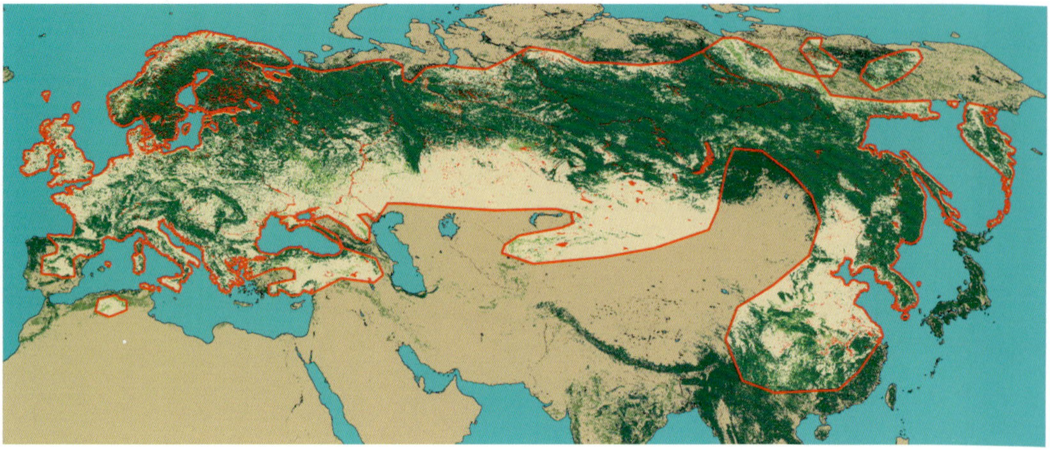

B

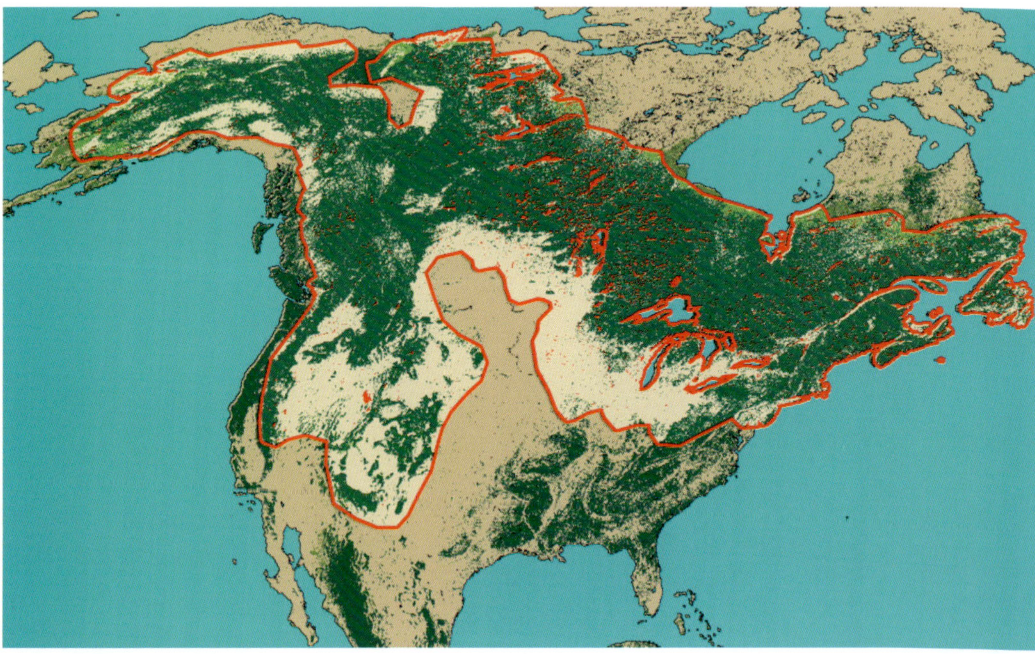

Plate 8. (A) Natural range of *Populus tremula* in Eurasia and Africa. Redrawn from Fenaroli and Gambi (1976). **(B)** Natural range of *Populus tremuloides* in North America. Redrawn from US Geological Survey Earth Surface Processes (http://esp.cr.usgs.gov/data/atlas/little/).

A

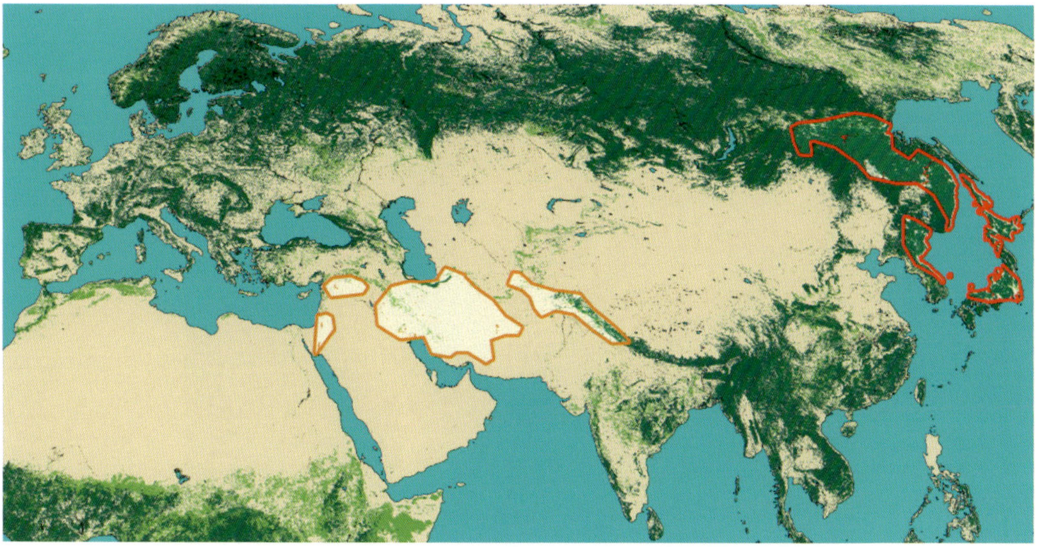

B

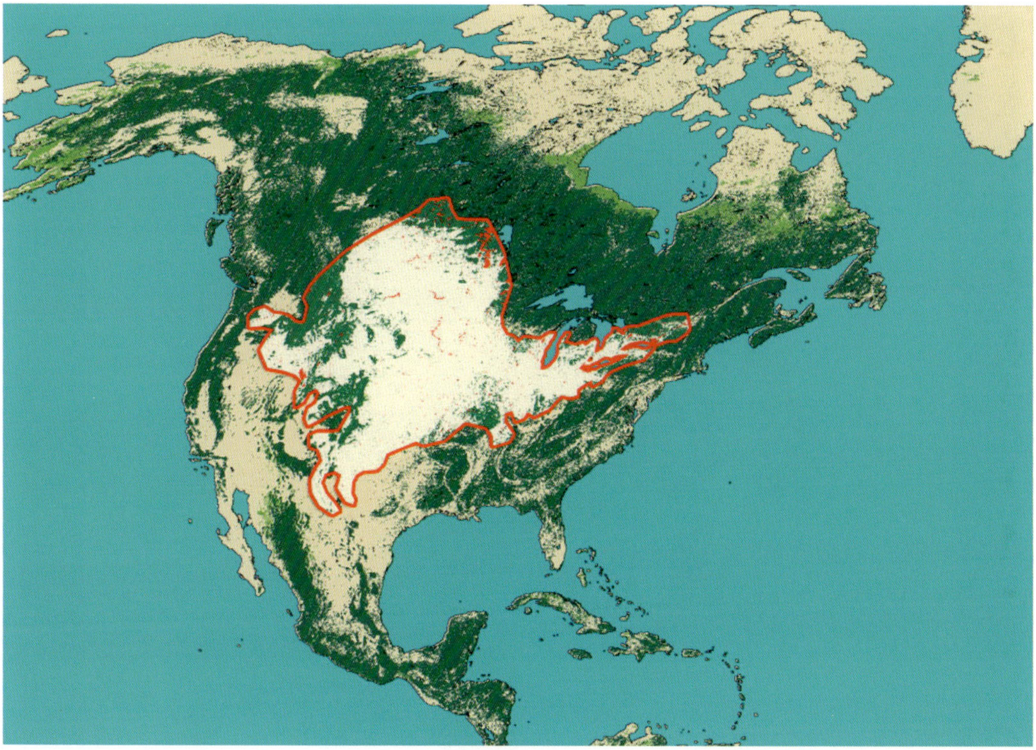

Plate 9. (A) Natural ranges of *Salix acmophylla* (brown) and *Salix cardiophylla* (red) in Eurasia. Redrawn from Skvortsov (1999). **(B)** Natural range of *Salix amygdaloides* in North America. Redrawn from US Geological Survey Earth Surface Processes (http://esp.cr.usgs.gov/data/atlas/little/).

A

B

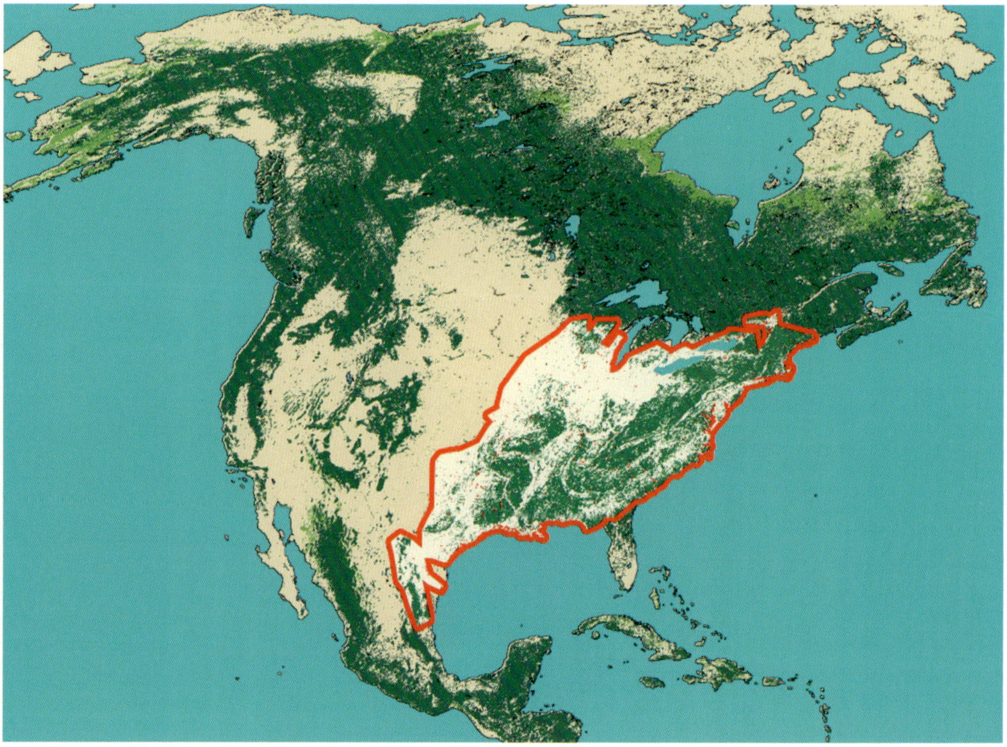

Plate 10. (A) Natural range of *Salix humboldtiana* in Central and South America. Redrawn from Discover Life (http://www.discoverlife.org/). **(B)** Natural range of *Salix nigra* in eastern North America. Redrawn from US Geological Survey Earth Surface Processes (http://esp.cr.usgs.gov/data/atlas/little/).

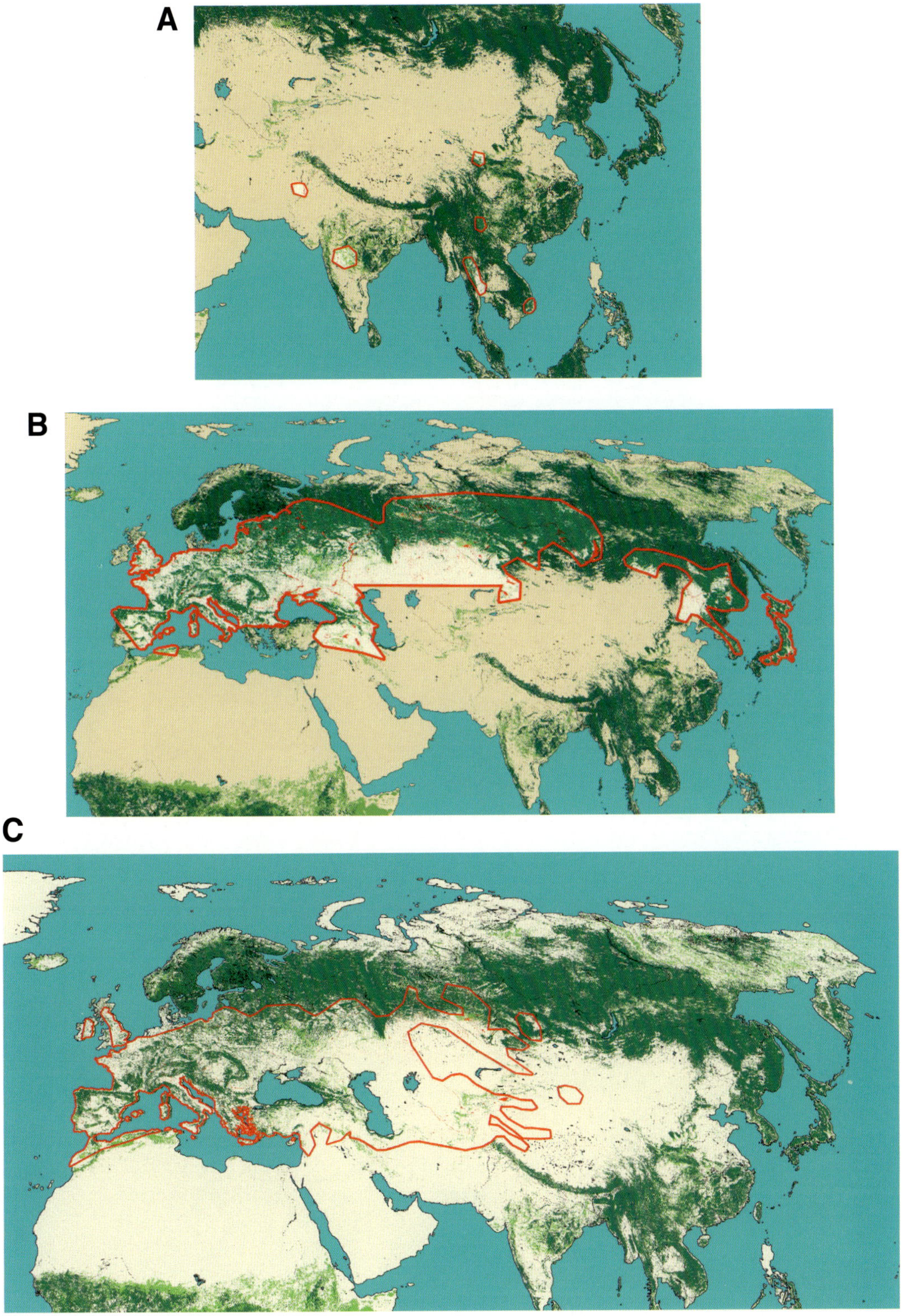

Plate 11. (A) Natural range of *Salix tetrasperma* in Asia. Redrawn from Discover Life (http://www.discoverlife.org/). **(B)** Natural range of *Salix triandra* in Eurasia and Africa. Redrawn from Skvortsov (1999). **(C)** Natural range of *Salix alba* in Eurasia. Redrawn from Skvortsov (1999).

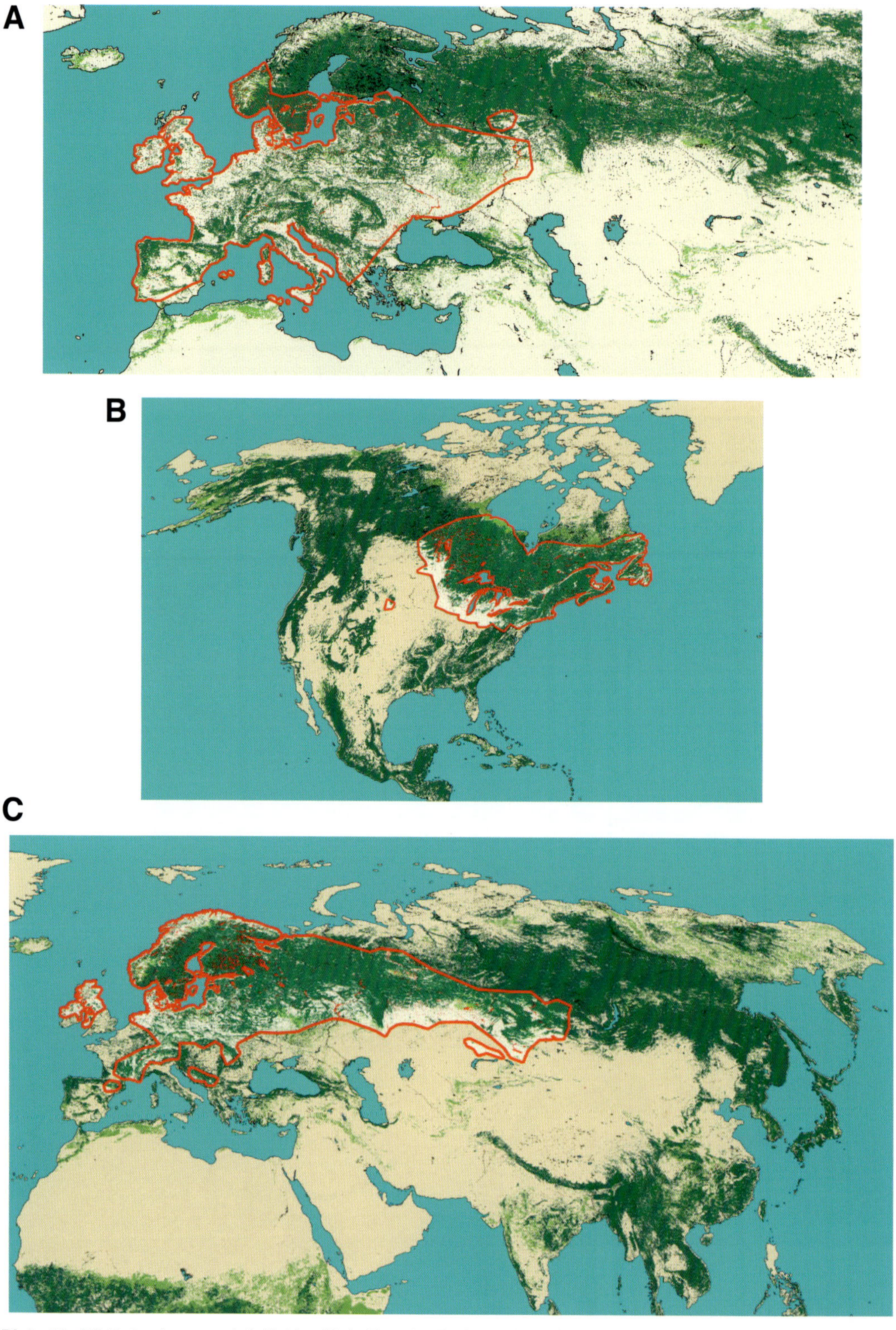

Plate 12. (A) Natural range of *Salix ´ fragilis* in Eurasia. Redrawn from Skvortsov (1999). **(B)** Natural range of *Salix lucida* in North America. Redrawn from US Geological Survey Earth Surface Processes (http://esp.cr.usgs.gov/data/atlas/little/). **(C)** Natural range of *Salix pentandra* in Eurasia. Redrawn from Skvortsov (1999).

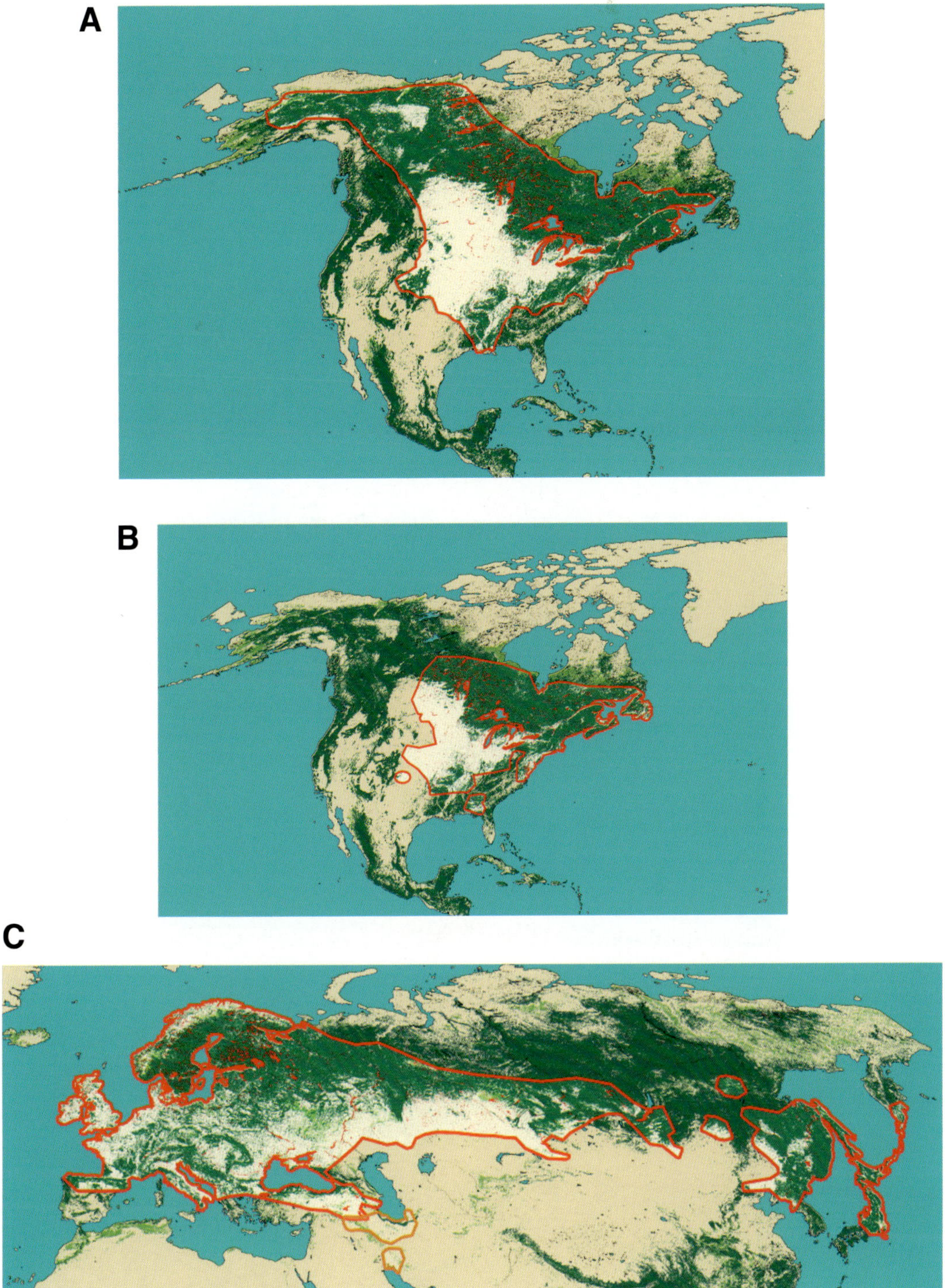

Plate 13. (A) Natural range of *Salix interior* in North America. Redrawn from US Geological Survey Earth Surface Processes (http://esp.cr.usgs.gov/data/atlas/little/). **(B)** Natural range of *Salix eriocephala* in North America. Redrawn from US Geological Survey Earth Surface Processes (http://esp.cr.usgs.gov/data/atlas/little/). **(C)** Natural ranges of *Salix aegyptiaca* (brown) in the Middle East and *Salix caprea* (red) in Eurasia. Redrawn from Skvortsov (1999).

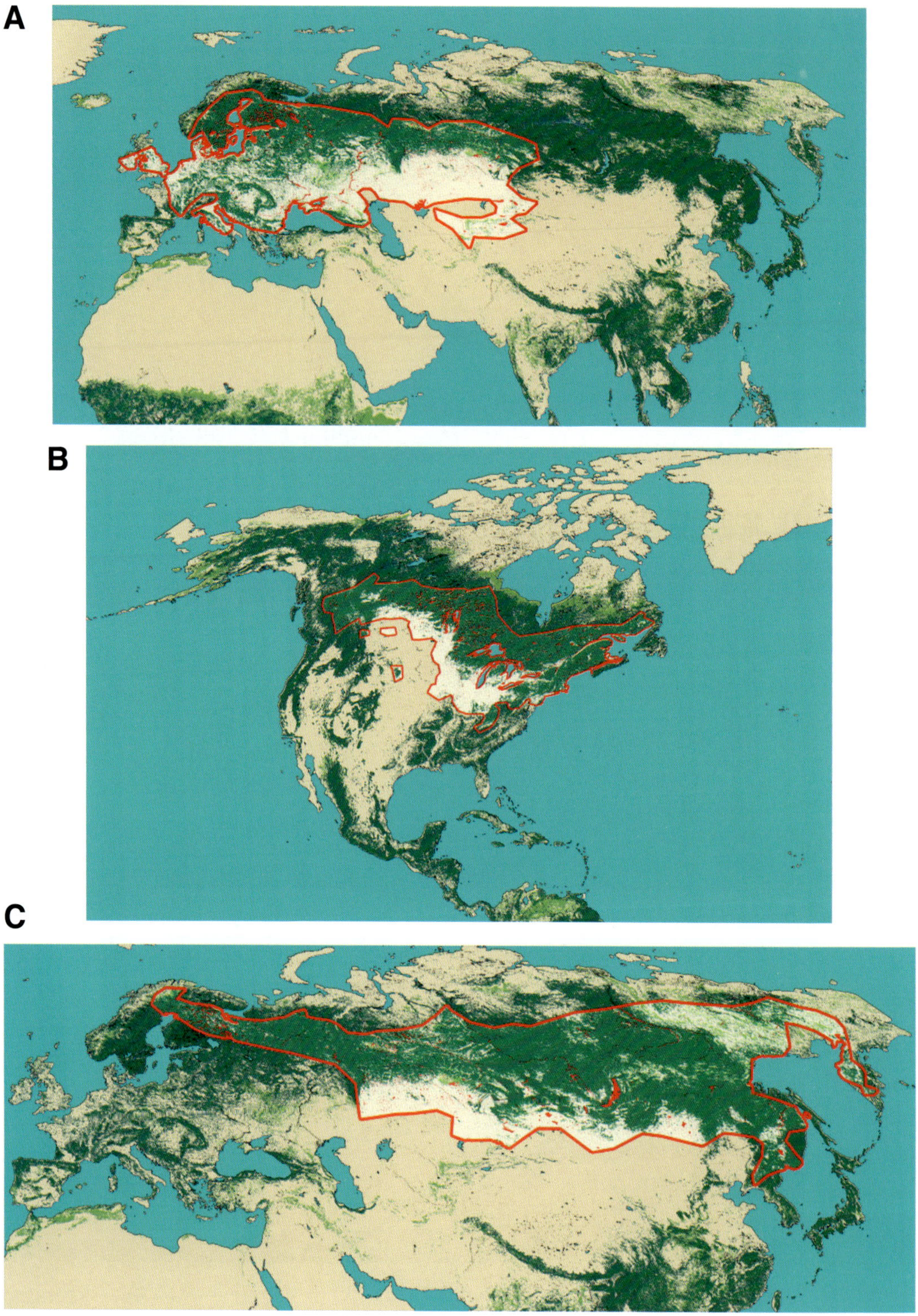

Plate 14. (A) Natural range of *Salix cinerea* in Eurasia. Redrawn from Skvortsov (1999). **(B)** Natural range of *Salix discolor* in North America. Redrawn from US Geological Survey Earth Surface Processes (http://esp.cr.usgs.gov-data/atlas/little/). **(C)** Natural range of *Salix bebbiana* in Eurasia. Redrawn from Skvortsov (1999).

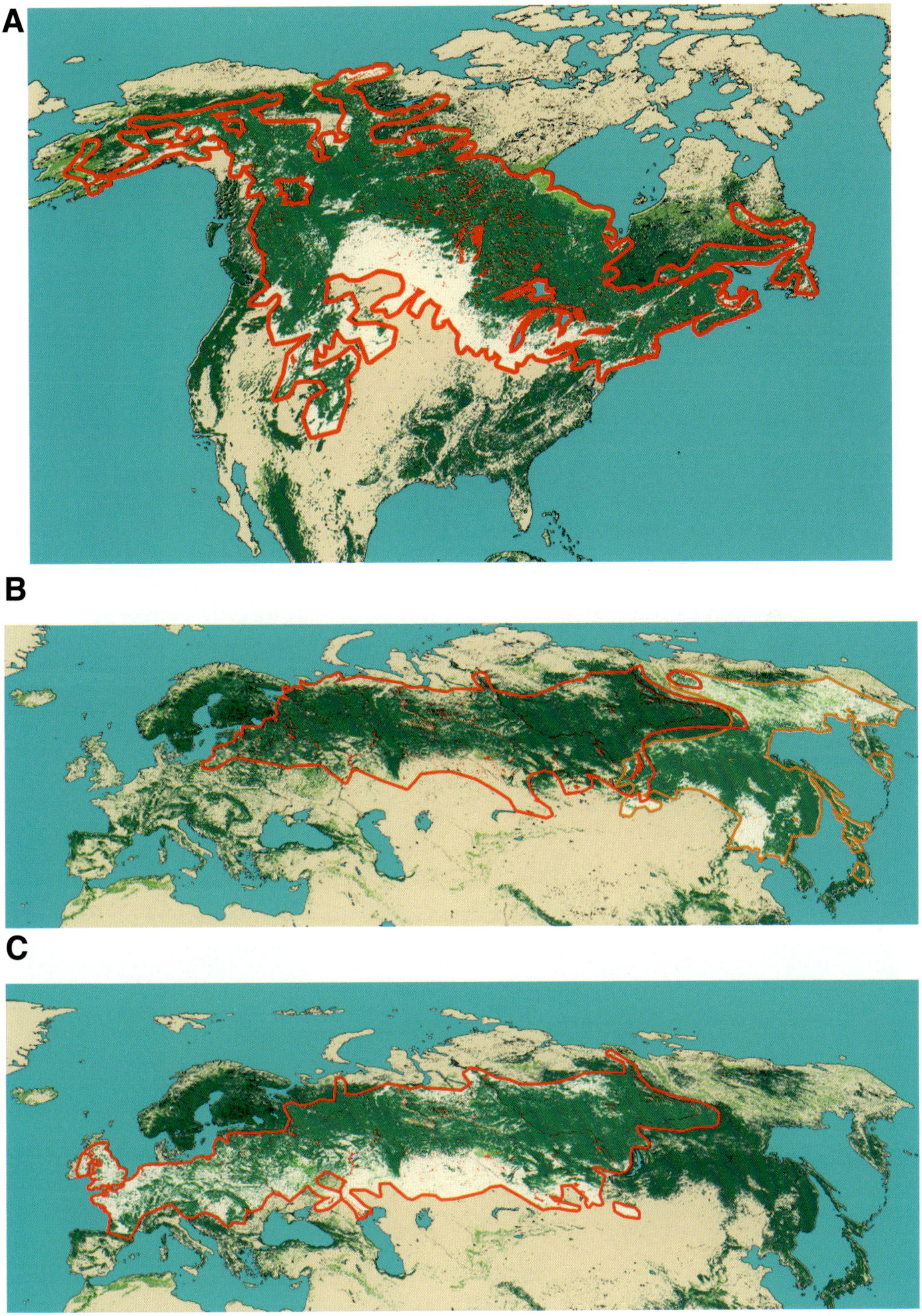

Plate 15. (A) Natural range of *Salix bebbiana* in North America. Redrawn from US Geological Survey Earth Surface Processes (http://www.esp.cr.usgs.gov/data/atlas/little/). **(B)** Natural range of *Salix gmelinii* (red) and *Salix schwerinii* (brown) in Eurasia. Redrawn from Skvortsov (1999). **(C)** Natural range of *Salix viminalis* in Eurasia. Redrawn from Skvortsov (1999).

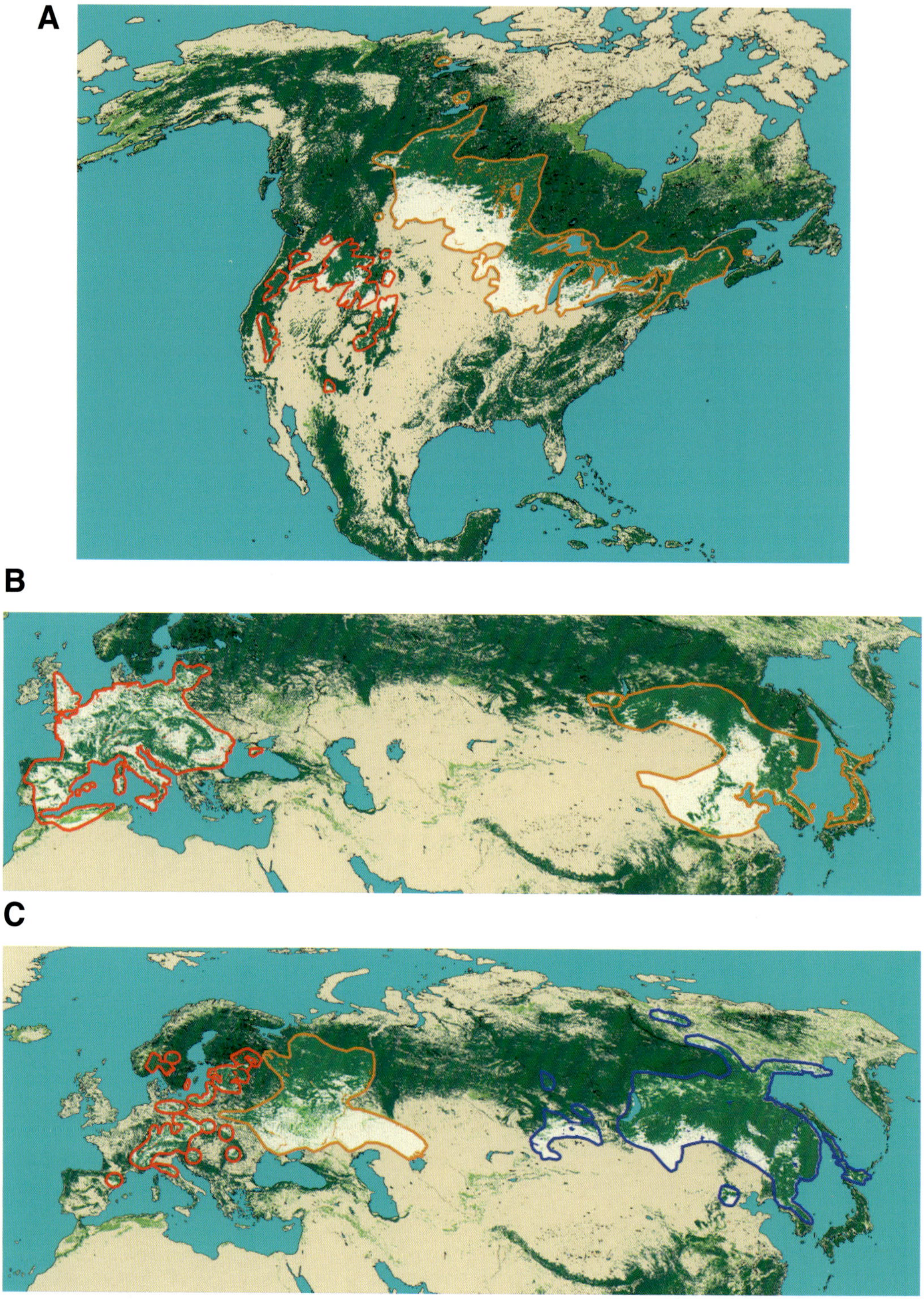

Plate 16. (A) Natural ranges of *Salix geyeriana* (red) and *Salix petiolaris* (brown) in North America. Redrawn from US Geological Survey Earth Surface Processes (http://www.esp.cr.usgs.gov/data/atlas/little/). **(B)** Natural ranges of *Salix miyabeana* (brown) in East Asia and *Salix purpurea* (red) in Eurasia. Redrawn from Skvortsov (1999). **(C)** Natural ranges of *Salix daphnoides* (red) and *Salix acutifolia* (brown) in Eurasia and *Salix rorida* (blue) in Asia. Redrawn from Skvortsov (1999).

Salix species described since 1968 (Skvortsov, 1999)

1. *S. kalarica* (A.K. Skvortsov) Worosch
2. *S. trabzonica* A.K. Skvortsov
3. *S. gracilistyliformis* Korkina
4. *S. vorobievii* Korkina
5. *S. kamtschatica* (A.K. Skvortsov) Worosch
6. *S. jurtzevii* A.K. Skvortsov
7. *S. khokhriakovii* A.K. Skvortsov
8. *S. dailingensis* Y.L. Chou et C.Y. King
9. *S. humaënsis* Y.L. Chou et R.C. Chou
10. *S. liangshuiensis* Y.L. Chou et C.Y. King
11. *S. darpirensis* Jurtzev et Khokhryakov
12. *S. neolapponum* Ch.Y. Yang
13. *S. burqinensis* Ch.Y. Yang
14. *S. paraphylicifolia* Ch.Y. Yang
15. *S. metaglauca* Ch.Y. Yang
16. *S. yanbianica* C.F. Fang et Ch.Y. Yang
17. *S. fimbriata* (A.K. Skvortsov) Czerepanov
18. *S. pseudotorulosa* (A.K. Skvortsov) Czerepanov
19. *S. alexii-skvortsovii* Khokhryakov
20. *S. flabellinervis* Khokhryakov
21. *S. sichotensis* Charkevicz et Vyshin
22. *S. hastatella* Rech. fil.– K. H. Rechinger
23. *S. gussonei* Brullo et Spampinato
24. *S. magadanensis* Nedoluzhko
25. *S. integerrima* (Worosch.) Nedoluzhko
26. *S. xanthicola* Christensen
27. *S. rizeënsis* A. Güner et J. Zelinski
28. *S. brutia* Brullo et Spampinato

B. Region of China (Fang *et al.*, 1999) (the authors do not propose taxa above the section level)

Section *Tetraspermae* C.K. Schneider
1. *S. tetrasperma* Roxburgh
2. *S. araeostachya* C.K. Schneider

Section *Urbanianae* C.K. Schneider
3. *S. maximowiczii* Komarov.

Section *Wilsonia* K.S. Hao ex C.F. Fang et A.K. Skvortsov
4. *S. mesnyi* Hance
5. *S. boseensis* N. Chao
6. *S. kusanoi* C.K. Schneider
7. *S. warburgii* Seemen
8. *S. balansaei* Seemen
9. *S. cavaleriei* H. Leveille
10. *S. chaenomeloides* Kimura
11. *S. wilsonii* Seemen ex Diels
12. *S. neowilsonii* W.P. Fang
13. *S. rosthornii* Seemen
14. *S. chekiangensis* W.C. Cheng
15. *S. dunnii* C.K. Schneider
16. *S. chingiana* K.S. Hao ex C.F. Fang et A.K. Skvortsov
17. *S. tengchongensis* C.F. Fang
18. *S. nankingensis* C. Wang et C.L. Tung

Section *Pentandrae* C.K. Schneider
19. *S. burqinensis* Chang Y. Yang
20. *S. pentandra* Linnaeus
21. *S. paraplesia* C.K. Schneider
22. *S. humaensis* Y.L. Chou et R.C. Chou

Section *Amygdalinae* W. Koch
23. *S. nipponica* Franchet et Savatier
24. *S. songarica* Andersson
25. *S. triandroides* W.P. Fang

Section *Salix*
26. *S. heteromera* Handel-Mazzetti
27. *S. alba* Linnaeus
28. *S. fragilis* Linnaeus (lately syn. of S.×fragilis)
29. *S. chikungensis* C.K. Schneider
30. *S. jishiensis* C.F. Fang et J.Q. Wang
31. *S. weixiensis* Y.L. Chou
32. *S. matsudana* Koidzumi
33. *S. pingliensis* Y.L. Chou
34. *S. sphaeronymphoides* Y.L. Chou
35. *S. chienii* W.C. Cheng
36. *S. babylonica* Linnaeus
37. *S. pseudolasiogyne* H. Leveille
38. *S. capitata* Y.L. Chou et Skvortsov
39. *S. bikouensis* Y.L. Chou
40. *S. eriocarpa* Franchet et Savatier
41. *S. koreensis* Andersson
42. *S. qinghaiensis* Y.L. Chou
43. *S. bangongensis* C. Wang et C.F. Fang
44. *S. sphaeronymphe* Goerz
45. *S. sericocarpa* Andersson
46. *S. pierotii* Miquel
47. *S. longistamina* C. Wang et P.Y. Fu

Section *Fulvopubescentes* C.F. Fang
48. *S. fulvopubescens* Hayata
49. *S. tagawana* Koidzumi
50. *S. doii* Hayata
51. *S. morrisonicola* Kimura
52. *S. takasagoalpina* Koidzumi
53. *S. taiwanalpina* Kimura
54. *S. okamotoana* Koidzumi

Section *Magnificae* C.K. Schneider
55. *S. magnifica* Hemsley
56. *S. pella* C.K. Schneider

Continued

Table 2.3. Continued.

57. *S. moupinensis* Franchet
58. *S. omeiensis* C.K. Scheider
59. *S. xiaoguangshanica* Y.L. Chou et N. Chao
60. *S. radinostachya* C.K. Schneider
61. *S. medogensis* Y.L. Chou
62. *S. contortiapiculata* P.Y. Mao et W.Z. Li

Section *Denticulatae* C.K. Schneider

63. *S. guebriantiana* C.K. Schneider
64. *S. tenella* C.K. Schneider
65. *S. pseudowolohoensis* K.S. Hao ex C.F. Fang et A.K. Skvortsov
66. *S. daguanensis* P.Y. Mao et P.X. He
67. *S. praticola* Handel-Mazzetti ex Enander
68. *S. dissa* C.K. Schneider
69. *S. heishuiensis* N. Chao
70. *S. zhegushanica* N. Chao
71. *S. zangica* N. Chao
72. *S. longiflora* Wallich ex Andersson
73. *S. paratetradenia* C. Wang et P.Y. Fu
74. *S. oreinoma* C.K. Schneider
75. *S. hypoleuca* Seemen ex Diels
76. *S. pseudotangii* C. Wang et C.Y. Yu
77. *S. macroblasta* C.K. Scnheider
78. *S. wangiana* K.S. Hao ex C.F. Fang et A.K. Skvortsov
79. *S. delavayana* Handel-Mazzetti
80. *S. luctuosa* H. Leveille
81. *S. mictotricha* C.K. Schneider
82. *S. rhoophila* C.K. Schneider
83. *S. polyclona* C.K. Schneider
84. *S. tangii* K.S. Hao ex C.F. Fang et A.K. Skvortsov
85. *S. xizangensis* Y.L. Chou
86. *S. cathayana* Diels
87. *S. maerkangensis* N. Chao
88. *S. denticulata* Andersson
89. *S. parvidenticulata* C.F. Fang
90. *S. ochetophylla* Goerz
91. *S. zhouquensis* X.G. Sun
92. *S. etosia* C.K. Schneider
93. *S. minjiangensis* N. Chao
94. *S. austrotibetica* N. Chao

Section *Floccosae* K.S. Hao ex C.F. Fang et A.K. Skvortsov

95. *S. jingdongensis* C.F. Fang
96. *S. faxonianoides* C. Wang et P.Y. Fu
97. *S. microphyta* Franchet
98. *S. kungmuensis* P.Y. Mao et W.Z. Li
99. *S. vaccinoides* Handel-Mazzetti (correct name S. mazzettiana N. Chao)
100. *S. resectoides* Handel-Mazzetti
101. *S. opsimantha* C.K. Schneider
102. *S. floccosa* Burkill
103. *S. zayulica* C. Wang et C.F. Fang
104. *S. coggygria* Handel-Mazzetti
105. *S. annulifera* C. Marquand et Airy-Shaw

Section *Lindleyanae* C.K. Schneider

106. *S. calyculata* J.D. Hooker ex Andersson
107. *S. brachista* C.K. Schneider
108. *S. anticecrenata* Kimura
109. *S. crenata* K.S. Hao ex C.F. Fang et A.K. Skvortsov
110. *S. clathrata* Handel-Mazzetti
111. *S. hirticaulis* Handel-Mazzetti
112. *S. kamanica* C. Wang et P.Y. Fu
113. *S. flabellaris* Andersson
114. *S. souliei* Seemen
115. *S. pilosomicrophylla* C. Wang et P.Y. Fu
116. *S. serpyllum* Andersson
117. *S. ovatomicrophylla* K.S. Hao ex C.F. Fang et A.K. Skvortsov
118. *S. paraflabellaris* S.D. Zhao
119. *S. oreophila* J.D. Hooker ex Andersson
120. *S. kangdingensis* S.D. Zhao et C.F. Fang
121. *S. lindleyana* Wallich ex Andersson
122. *S. glareorum* P.Y. Mao et W.Z. Li
123. *S. scopulicola* P.Y. Mao et W.Z. Li
124. *S. gyirongensis* S.D. Zhao et C.F. Fang
125. *S. ludingensis* T.Y. Ding et C.F. Fang
126. *S. piptotricha* Handel-Mazzetti

Section *Sclerophyllae* C.K. Schneider

127. *S. cupularis* Rehder
128. *S. taipaiensis* Chang Y. Yu
129. *S. maizhokunggarensis* N. Chao
130. *S. occidentalisinensis* N. Chao
131. *S. neoamnematchinensis* T.Y. Ding et C.F. Fang
132. *S. oritrepha* C.K. Schneider
133. *S. gyamdaensis* C.F. Fang
134. *S. atopantha* C.K. Schneider
135. *S. biondiana* Seemen ex Diels
136. *S. qinlingica* C. Wang et N. Chao
137. *S. muliensis* Goerz ex Rehder et Kobuski
138. *S. kongbanica* C. Wang et P.Y. Fu
139. *S. sclerophylloides* Y.L. Chou
140. *S. wuxuhaiensis* N. Chao
141. *S. sclerophylla* Andersson
142. *S. rockii* Goerz ex Rehder et Kobuski
143. *S. juparica* Goerz ex Rehder et Kobuski
144. *S. pseudospissa* Goerz ex Rehder et Kobuski
145. *S. gilashanica* C. Wang et P.Y. Fu
146. *S. dalungensis* C. Wang et P.Y. Fu
147. *S. jinchuanica* N. Chao

Section *Psilostigmatae* C.K. Schneider

148. *S. fargesii* Burkill
149. *S. ernestii* C.K. Schneider
150. *S. eriostachya* Wallich ex Andersson
151. *S. plocotricha* C.K. Schneider
152. *S. phanera* C.K. Schneider
153. *S. phaidima* C.K. Schneider
154. *S. spathulifolia* Seemen ex Diels

155. *S. spodiophylla* Handel-Mazzetti
156. *S. balfouriana* C.K. Schneider
157. *S. argyrophegga* C.K. Schneider
158. *S. salwinensis* Handel-Mazzetti ex Enander
159. *S. psilostigma* Andersson
160. *S. daliensis* C.F. Fang et S.D. Zhao
161. *S. argyrotrichocarpa* C.F. Fang
162. *S. divergentistyla* C.F. Fang
163. *S. daltoniana* Andersson
164. *S. sikkimensis* Andersson
165. *S. nujiangensis* N. Chao
166. *S. bistyla* Handel-Mazzetti
167. *S. fengiana* C.F. Fang et Chang Y. Yang

Section *Eriocladae* K.S. Hao ex C.F. Fang et A.K. Skvortsov

168. *S. inamoena* Handel-Mazzetti
169. *S. amphibola* C.K. Schneider
170. *S. driophila* C.K. Schneider
171. *S. erioclada* H. Leveille et Vaniot
172. *S. dibapha* C.K. Schneider
173. *S. hupehensis* K.S. Hao ex C.F. Fang et A.K. Skvortsov
174. *S. wolohoensis* C.K. Schneider
175. *S. resecta* Diels

Section *Heterochromae* C.K. Schneider

176. *S. heterochroma* Seemen
177. *S. paraheterochroma* C. Wang et P.Y. Fu
178. *S. taishanensis* C. Wang et C.F. Fang
179. *S. trichocarpa* C.F. Fang
180. *S. yadongensis* N. Chao
181. *S. alfredii* Goerz ex Rehder et Kobuski

Section *Retusae* A. Kerner

182. *S. nummularia* Andersson
183. *S. turczaninowii* Lakschewitz

Section *Myrtilloides* Koehne

184. *S. myrtilloides* Linnaeus

Section *Chamaetia* Dumortier

185. *S. vestita* Pursh

Section *Glaucae* Pax

186. *S. glauca* Linnaeus
187. *S. alatavica* Karelin et Kirilow ex Stscheglejew
188. *S. metaglauca* Chang Y. Yang

Section *Diplodictyae* C.K. Schneider

189. *S. arctica* Pallas

Section *Myrtosalix* A. Kerner

190. *S. berberifolia* Pallas
191. *S. rectijulis* Ledebour ex Turczaninow

Section *Hastatae* A. Kerner

192. *S. hastata* Linnaeus
193. *S. karelinii* Turczaninow
194. *S. pyrolifolia* Ledebour
195. *S. fedtschenkoi* Goerz

Section *Arbuscella* Seringe ex Duby

196. *S. saposhnikovii* A.K. Skvortsov
197. *S. paraphylicifolia* Chang Y. Yang
198. *S. tianschanica* Regel
199. *S. divaricata* Pallas

Section *Vetrix* Dumortier

200. *S. taraikensis* Kimura
201. *S. floderusii* Nakai
202. *S. hsinganica* Y.L. Chang et Skvortsov
203. *S. insignis* Andersson
204. *S. caprea* Linnaeus
205. *S. iliensis* Regel
206. *S. cinerea* Linnaeus
207. *S. raddeana* Lakschewitz ex Nasarow
208. *S. sinica* (K.S. Hao ex C.F. Fang et A.K. Skvortsov) G. Zhu
209. *S. wallichiana* Andersson
210. *S. pseudowallichiana* Goerz ex Rehder et Kobuski
211. *S. shandanensis* C.F. Fang

Section *Lanatae* Koehne

212. *S. bhutanensis* Floderus
213. *S. lasiopes* C. Wang et P.Y. Fu

Section *Daphnella* Seringe ex Duby

214. *S. baileyi* C.K. Schneider
215. *S. rorida* Lakschewitz
216. *S. skvortzovii* Y.L. Chang et Y.L. Chou
217. *S. kangensis* Nakai

Section *Argyraceae* Chang Y. Yang

218. *S. argyracea* E.L. Wolf
219. *S. neolapponum* Chang Y. Yang

Section *Vimen* Dumortier

220. *S. characta* C.K. Schneider
221. *S. siuzevii* Seemen
222. *S. sajanensis* Nasarow
223. *S. rehderiana* C.K. Schneider
224. *S. dasyclados* Wimmer (lately syn. of *S. gmelinii*)
225. *S. sachalinensis* F. Schmidt
226. *S. turanica* Nasarow
227. *S. schwerinii* E.L. Wolf

Section *Incubaceae* A. Kerner

228. *S. rosmarinifolia* Linnaeus
229. *S. schugnanica* Goerz

Section *Sieboldianae* (Seemen) C.K. Schneider

230. *S. hylonoma* C.K. Schneider
231. *S. shihtsuanensis* C. Wang et C.Y. Yu
232. *S. yuhuangshanensis* C. Wang et C.Y. Yu

Section *Subviminales* (Seemen) C.K. Schneider

233. *S. rhododendrifolia* C. Wang et P.Y. Fu
234. *S. myrtillacea* Andersson
235. *S. obscura* Andersson
236. *S. lanifera* C.F. Fang et S.D. Zhao
237. *S. taoensis* Goerz ex Rehder et Kobuski
238. *S. gracilistyla* Miquel

Section *Caesiae* A. Kerner

239. *S. caesia* Villars

Continued

Table 2.3. Continued.

240. *S. kochiana* Trautvetter 241. *S. tarbagataica* Chang Y. Yang 242. *S. integra* Thunberg Section *Variegatae* K.S. Hao ex C.F. Fang et A.K. Skvortsov 243. *S. kouytchensis* (H. Leveille) C.K. Schneider 244. *S. variegata* Franchet Section *Cheilophilae* K.S. Hao ex C.F. Fang et A.K. Skvortsov 245. *S. cheilophila* C.K. Schneider 246. *S. cyanolimnea* Hance 247. *S. microstachya* Turczaninow ex Trautvetter 248. *S. wilhelmsiana* Bieberstein	Section *Haoanae* C. Wang et Chang Y. Yang 249. *S. haoana* Fang 250. *S. liouana* C. Wang et Chang Y. Yang 251. *S. permollis* C. Wang et C.Y. Yu 252. *S. pseudopermollis* C.Y. Yu et Chang Y. Yang 253. *S. sinopurpurea* C. Wang et Chang Y. Yang Section *Helix* Dumortier 254. *S. michelsonii* Goerz ex Nasarow 255. *S. capusii* Franchet 256. *S. leveilleana* C.K. Schneider 257. *S. yumenensis* H.L. Yang 258. *S. carmanica* Bornmueller 259. *S. kirilowiana* Stscheglejew 260. *S. pycnostachya* Andersson	261. *S. gracilior* (Siuzev) Nakai 262. *S. yanbianica* C.F. Fang et Chang Y. Yang 263. *S. koriyanagi* Kimura ex Goerz 264. *S. blakii* Goerz 265. *S. tenuijulis* Ledebour 266. *S. lamashanensis* K.S. Hao ex Fang et A.K. Skvortsov 267. *S. caspica* Pallas 268. *S. donggouxianica* C.F. Fang 269. *S. linearistipularis* K.S. Hao 270. *S. sungkianica* Y.L. Chou et Skvortsov 271. *S. suchowensis* W.C. Cheng ex G. Zhu 272. *S. psammophila* C. Wang et Chang Y. Yang Section *Flavidae* Y.L. Chang et Skvortsov 273. *S. gordejevii* Y.L. Chang et Skvortsov
Uncertain species		
274. *S. limprichtii* Pax & Hoffmann 275. *S. longissimipedicellaris* N. Chao ex P.Y. Mao		
C. Region of Japan (Ohashi, 2001)		
I. Subgenus *Pleuradenia* Kimura		
Section *Urbanianae* C.K. Schneider 1. *S. cardiophylla* Trautvetter		
II. Subgenus *Chosenia* (Nakai) H. Ohashi		
2. *S. arbutifolia* Pall		
III. Subgenus *Protitea* Kimura		
3. *S. chaenomeloides* Kimura		
IV. Subgenus *Chamaetia* (Dumortier) Nasarov		
Section *Herbella* Seringe 4. *S. nummularia* Andersson	Section *Myrtilloides* (Borrer) Andersson 5. *S. fuscescens* Andersson	Section *Glaucae* Andersson 6. *S. nakamurana* Koidzumi

V. Subgenus *Salix*		
Section *Triandrae* Dumortier 7. *S. triandra* Linnaeus Section *Subalbae* Koidzumi 8. *S. babylonica* Linnaeus	9. *S. matsudana* Koidzumi 10. *S. eriocarpa* Franchet et Savatier 11. *S. pierotii* Miquel	12. *S. jessoensis* Seemen
VI. Subgenus *Vetrix* Dumortier		
Section *Hastatae* A. Kerner 13. *S. japonica* Thunb 14. *S. shiraii* Seemen 15. *S. rupifraga* Koidzumi 16. *S. sieboldiana* Blume 17. *S. reinii* Seemen Section *Helix* Dumortier 18. *S. miyabeana* Seemen 19. *S. integra* Thunberg	20. *S. koriyanagi* Kimura ex Gorz Section *Incubaceae* Kerner 21. *S. subopposita* Miquel Section *Subviminales* C.K. Schneider 22. *S. gracilistyla* Miquel Section *Hukaoana* Kimura 23. *S. hukaoana* Kimura Section *Daphnella* Seringe 24. *S. rorida* Lakschewitz	Section *Viminella* Seringe 25. *S. schwerinii* E. Wolf 26. *S. udensis* Trautvetter et Mey Section *Cinerella* Seringe 27. *S. taraikensis* Kimura 28. *S. caprea* Linnaeus 29. *S. futura* Seemen 30. *S. vulpina* Andersson
D. African Region (compiled by I.V. Belyaeva; based on Geerinck and Léonard, 1968; Jordaan, 2005; Klopper *et al.*, 2006)		
I. Subgenus *Protitea* Kimura		
Section *Humboldtianae* Andersson 1. *S. acmophylla* Boissier	2. *S. mucronata* Thunberg	
II. Subgenus *Salix*		
Section *Triandrae* (*Amygdalinae* Koch) 3. *S. triandra* Linnaeus	Section *Salix* 4. *S. alba* Linnaeus	
III. Subgenus *Vetrix* Dumortier		
Section *Vulpinae* Kimura 5. *S. pedicellata* Desfontaines Section *Laeves* Camus 6. *S. atrocinerea* Brotero Section *Canae* Kerner	7. *S. elaeagnos* Scopoli Section *Helix* Dumont 8. *S. purpurea* Linnaeus	

Continued

Table 2.3. Continued.

E. North American Region North of Mexico (Argus, 2010)		
I. Subgenus *Protitea* Kimura		
Section *Floridanae* Dorn	2. *S. bonplandiana* Kunth	5. *S. nigra* Marshall
1. *S. floridana* Chapman	3. *S. laevigata* Bebb	6. *S. gooddingii* C.R. Ball
Section *Humboldtianae* Andersson	4. *S. caroliniana* Michaux	7. *S. amygdaloides* Andersson
II. Subgenus *Salix*		
Section *Subalbae* Koidzumi	Section *Salicaster* Dumortier	Section *Maccallianae* Argus
8. *S. babylonica* Linnaeus	11. *S. pentandra* Linnaeus	15. *S. maccalliana* Rowlee
Section *Salix*	12. *S. serissima* Fernald	Section *Triandrae* Dumortier
9. *S. alba* Linnaeus	13. *S. lucida* Muhlenberg	16. *S. triandra* Linnaeus
10. *S. euxina* I.V. Belyaeva	14. *S. lasiandra* Bentham	
III. Subgenus *Longifoliae* Argus		
17. *S. taxifolia* Kunth	20. *S. thurberi* Rowlee	22. *S. columbiana* Argus
18. *S. exigua* Nuttall	21. *S. melanopsis* Nuttall	23. *S. sessilifolia* Nuttall
19. *S. interior* Rowlee		
IV. Subgenus *Chamaetia* Nasarow		
Section *Chamaetia* Dumortier	32. *S. chamissonis* Andersson	41. *S. petrophyla* Rydberg
24. *S. vestita* Pursh	33. *S. fuscescens* Andersson	42. *S. sphenophylla* A.K. Skvortsov
25. *S. reticulata* Linnaeus	34. *S. arctophila* Cockerell	43. *S. cascadensis* Cockerell
26. *S. nivalis* Hooker	ex A. Heller	Section *Myrtilloides* Andersson
Section *Setchellinae* Argus	35. *S. uva-ursi* Pursh	44. *S. pedicellaris* Pursh
27. *S. setchelliana* C.R. Ball	36. *S. phlebophylla* Andersson	45. *S. athabascensis* Raup
Section *Herbella* Seringe	Section *Ovalifoliae* C.K. Schneider	46. *S. chlorolepis* Fernald
28. *S. herbacea* Linnaeus	37. *S. stolonifera* Coville	47. *S. raupii* Argus
29. *S. nummularia* Andersson	38. *S. ovalifolia* Trautvetter	Section *Glaucae* Andersson
30. *S. polaris* Wahlenberg	39. *S. jejuna* Fernald	48. *S. brachycarpa* Nuttall
31. *S. rotundifolia* Trautvetter	Section *Diplodictyae* C.K. Schneider	49. *S. niphoclada* Rydberg
Section *Myrtosalix* A. Kerner	40. *S. arctica* Pallas	50. *S. glauca* Linnaeus

V. Subgenus *Vetrix* Dumortier

Section *Hastatae* A. Kerner
51. *S. commutata* Bebb
52. *S. cordata* Michaux
53. *S. eastwoodiae* Cockerell ex A. Heller
54. *S. wolfii* Bebb
55. *S. orestera* C.K. Schneider
56. *S. myrtillifolia* Andersson
57. *S. pseudomyrsinites* Andersson
58. *S. ballii* Dorn
59. *S. arizonica* Dorn
60. *S. boothii* Dorn
61. *S. barclayi* Andersson
62. *S. pyrifolia* Andersson
63. *S. hastata* Linnaeus
64. *S. farriae* C.R. Ball
65. *S. pseudomonticola* C.R. Ball
66. *S. monticola* Bebb
67. *S. myricoides* Muhlenberg

Section *Cordatae* Barratt ex Hooker
68. *S. eriocephala* Michaux
69. *S. famelica* Argus
70. *S. turnorii* Raup
71. *S. ligulifolia* C.R. Ball C.K. Schneider
72. *S. lutea* Nuttall
73. *S. monochroma* C.R. Ball
74. *S. prolixa* Andersson

Section *Nigricantes* A. Kerner
75. *S. myrsinifolia* Salisbury

Section *Cinerella* Seringe
76. *S. discolor* Muhlenberg
77. *S. hookeriana* Barratt ex Hooker
78. *S. humilis* Marshall
79. *S. scouleriana* Barratt ex Hooker
80. *S. caprea* Linnaeus
81. *S. cinerea* Linnaeus
82. *S. atrocinerea* Brotero
83. *S. aurita* Linnaeus

Section *Fulvae* Barratt
84. *S. bebbiana* Sargent

Section *Phylicifoliae* Andersson
85. *S. drummondiana* Barratt ex Hooker
86. *S. pellita* Bebb
87. *S. pulchra* Chamisso
88. *S. planifolia* Pursh
89. *S. tyrrellii* Raup

Section *Arbuscella* Seringe
90. *S. arbusculoides* Andersson

Section *Candidae* C.K. Schneider
91. *S. candida* Flugge ex Willdenow

Section *Lanatae* Koehne
92. *S. calcicola* Fernald & Wiegand
93. *S. richardsonii* Hooker
94. *S. tweedyi* C.R. Ball

Section *Villosae* Rouy
95. *S. alaxensis* Coville
96. *S. silicicola* Raup
97. *S. barrattiana* Hooker

Section *Viminella* Seringe
98. *S. viminalis* Linnaeus

Section *Canae* A. Kerner
99. *S. elaeagnos* Scopoli

Section *Argyrocarpae* Fernald
100. *S. argyrocarpa* Andersson

Section *Geyerianae* Argus
101. *S. petiolaris* Smith
102. *S. geyeriana* Andersson
103. *S. lemmonii* Bebb

Section *Mexicanae* C.K. Schneider
104. *S. irrorata* Andersson
105. *S. lasiolepis* Bentham
106. *S. tracyi* C.R. Ball

Section *Griseae* Barratt ex Hooker
107. *S. sericea* Marshall

Section *Sitchenses* C.K. Schneider
108. *S. sitchensis* Sanson ex Bongard
109. *S. jepsonii* C.K. Schneider
110. *S. breweri* Bebb
111. *S. delnortensis* C.K. Schneider

Section *Daphnella* Seringe
112. *S. daphnoides* Villars

Section *Helix* Dumortier
113. *S. purpurea* Linnaeus

F. New World Region South of the USA (Argus, 1999; supplemented by I.V. Belyaeva; occurrence data according to G. Argus)

I. Subgenus *Protitea* Kimura

Section *Humboldtianae* Andersson
1. *S. aeruginosa* E. Carranza (Mexico)
2. *S. jaliscana* M.E. Jones (Mexico)
3. *S. bonplandiana* Kunth (Mexico, Guatemala)
4. *S. humboldtiana* Willdenow (from Mexico to central Chile)
5. *S. caroliniana* Michaux (Cuba and Guatemala)

Continued

Table 2.3. Continued.

II. Subgenus *Salix*		
Section *Salix*		
6. *S. ×argentinensis* Ragonese et Alberti	(intersubgeneric cross: *S. babylonica* (Subg. *Salix*, sect. *Subalbae*) × *S. humboldtiana* (Subg. *Protitea*, sect. *Humboldtianae*)	
III. Subgenus *Longifoliae* Argus		
7. *S. taxifolia* Kunth (Arizona, New Mexico, Texas and Mexico)		
8. *S. microphylla* Schlechtendal et Chamisso (Mexico and Guatemala)		
IV. Subgenus *Vetrix* Dumortier		
Section *Cinerella* Seringe		10. *S. mexicana* Seemen (Mexico)
9. *S. paradoxa* Kunth (Mexico)		11. *S. riskindii* Johnston (Mexico)
Section *Mexicanae* C.K. Schneider		12. *S. lasiolepis* Bentham (Mexico)

*Listed without a number in Skvortsov's book as this is an introduced species (Skvortsov, 1968, 1999).

Table 2.4. The list of *Salix* names at the species level (including nothospecies, synonyms, invalidly published and illegitimate names) published with the International Plant Names Index (IPNI) between 1968 and 2012 not included in Table 2.3.[a] The list also shows region of distribution and effective date of publication for each species.

S. anatolica Zielinski and D. Tomasz (Turkey, 2008)
S. angustior (Lautenschlager) Landolt (Switzerland, France, Austria, 2010)
S. ×*ardana* Zielinski and Petrova (Bulgaria, 2006)
S. arrigonii Brullo (Italy, 1993)
S. ×*atroelaeagnos* Serra and M.B. Crespo (Spain, 1995)
S. baladehensis Maassoumi, Moeeni and Rahimin (Iran, 2008)
S. brutia Brullo and Spamp (Italy, 1993)
S. ceretana (P. Montserrat) J. Chmelař (Pyrenees, 1982)
S. chengfui N.Chao (China, 1994) unresolved
S. chingshuishanensis S.S. Ying (Taiwan, 1988)
S. ×*coenocarpetana* Rivas Mart., T.E. Díaz, Fern. Prieto, Loidi and Penas (Spain, 1984)
S. dabeshanensis B.C. Ding and T.B. Chao (China, 1980)
S. elymaitica Maassoumi (Iran, 2009)
S. euxina I.V. Belyaeva (northern Black Sea and Transcaucasia, 2009)
S. fimbriata (A.K. Skvortsov) Baikov (Siberia, 2005) nom. inval.
S. firouzkuhensis Maassoumi (Iran, 2009)
S. ionica Brullo, Scelsi and Spamp (Italy, 2001)
S. issatissensis Maassoumi, Moeeni and Rahimin (Iran, 2008)
S. jichoushanica N. Chao and G.T. Gong (China, 1985)
S. lacus-tari Maassoumi and Kazempour (Iran, 2009)
S. laishuiensis N. Chao and G.T. Gong (China, 1994)
S. ×*legionensis* Llamas and Penas (Spain, 1990)
S. ×*leindanensis* P.I. Mao and H.Li (China, 1993) nom. inval.
S. ×*lochsiensis* D.J. Tennant (Great Britain, 2007)
S. ×*longissima* T.E. Díaz and J. Andrés (Spain, 1987)
S. luzhongensis X.W. Li and Y.Q. Zhu (China, 1993)
S. ×*lyonensis* D.J. Tennant (Great Britain, 2007)
S. mandshurica (Nakai) Petruk (Russian Far East, 2010)
S. ×*meikleana* D.J. Tennant (Scotland, 2007)
S. *mazzettiana* N. Chao (China, 1994) nom. nov. (replacement name for illegitimately published later homonym, *S. vaccinioides*)
S. ×*misaoana* Kimura (Japan, 1988) nom.nov. (replacement name for ×*Toisochosenia tatewakii* Kimura)
S. ×*multidentata* T.E. Díaz and Llamas (Spain, 1987)
S. nepalensis Yonek (Nepal, 2008)
S. oropotamica Brullo, Scelsi and Spamp (Italy, 1996)
S. pattersonii M.C. Johnst (Mexico, 1981)
S. pentandrifolia Sennikov (Caucasus, 2011)
S. ×*permixta* Jeanne Webb (Great Britain, 2008)
S. ×*pormensis* T.E. Díaz and Llamas (Spain, 1987)
S. ×*praegravis* Kimura (Japan, 1985)
S. ×*pseudoelaeagnos* T.E. Díaz and Llamas (Spain, 1987)
S. ×*pseudosalvifolia* T.E. Díaz and E. Puente (Spain, 1987)
S. qamdoensis N. Chao and J. Liu (Tibet, 2001)
S. ×*rijosa* Rivas Martinez, T.E. Díaz, Fern. Prieto, Loidi and Penas (Spain, 1987)
S. rivulicola P.I Mao and W.Z. Li (China, 1987)
S. rizeensis Güner and Ziel (Turkey, 1993)
S. ×*sakamakiensis* Yoshiyama (Japan, 2011)
S. ×*schaburovii* I.V. Belyaeva (Ural, 1998)
S. sinorivularis P.I. Mao and H.Li (China, 1993) nom. inval.
S. tschujensis (Bolschakov) Baikov (Altay, 2005)
S. turuchanensis (A.K. Skvortsov and Kuvaev) Baikov (Siberia, 2005)

Continued

Table 2.4. Continued.

S. tyrrhenica Brullo, Scelsi and Spamp (Italy, 2001)
S. uralicola I.V. Belyaeva (Ural, 2002)
S. ustnerensis (Bolschakov) Baikov (Siberia, 2008)
S. ustnerensis (Bolschakov) Baikov ex A.V. Grebenjuk and Czepinoga (Siberia, 2008)
S. ×*velchevii* Zielinski and Pancheva (Bulgaria, 2006)
S. ×*viridifolia* T.E. Díaz and Penas (Spain, 1987)
S. vyshinii (Nedolushko) Petruk (Russian Far East, 2010)
S. wendtii M.C. Johnston (Mexico, 1981)
S. xanthicola K.I. Christensen (Greece, 1991)
S. ×*zhataica* Efimova, Shurduk and Ahti (Yakutia, 2009)

[a]The authors thank I.V. Belyaeva for extracting this list from the IPNI database. The list was edited by J. Kuzovkina to exclude the species listed in Table 2.3.

her subjects. Often, this means not taking at face value the cultivar name or binomial assigned to a particular poplar or willow that comes into their possession; this name may well be incorrect. The correct name, however, can usually be obtained through a bit of investigation or by contacting a poplar or willow expert.

First and foremost, only recognized taxa warrant a distinct Latin binomial. The full Latin name, which always is italicized, also includes the name of the authority (often abbreviated but not italicized) who first described the taxon. The Latin name of a taxon should be used, with or without authority, at first reference in a scientific publication or an oral presentation. According to the International Code of Plant Nomenclature, the first published description of a taxon (after 1 May 1753) always takes priority; this fundamental rule of systematics has recently led to the rejection of certain long-used binomials in favour of older ones having priority. Examples of a full Latin appellation are *Populus tremula* Linnaeus (usually abbreviated L. for Karl von Linnaeus, or Linné, who first proposed the system of binomial nomenclature) or *Salix eriocephala* Michaux. Often, the authority is omitted, although in a publication, the authority for a particular Latin binomial should be given in a table or at first mention in the text. The authors' names should be standardized according to Brummitt and Powell (1992) or the International Plant Names Index (IPNI) database (http://www.ipni.org/ipni/authorsearchpage.do).

A subspecies is 'an aggregate of local breeding populations of a given species occupying a geographic subdivision of a species range (= geographic race)' (Rieger *et al.*, 1991). Recognized subspecies names are Latinized and italicized; for example, *Populus mexicana* ssp. *dimorpha* (Brand.) Ecken. or simply *P. mexicana dimorpha*. A variety – 'an intraspecific taxon with persistent populational significance' (Cronquist, 1988) – is a more loosely defined and problematic designation of phenotypic variants than a subspecies. They also are designated with a Latinized name; for example, *Salix humilis* var. *microphylla*. Many named varieties, however, are simply commercially propagated clones selected for a particular trait, for example the fastigiate Lombardy poplar, *Populus nigra* 'var. *italica*', or the weeping *P. tremula* 'var. *pendula*', and should be designated as cultivars (see below). Sometimes, the specific name of a subspecies or variety is omitted from a text purposefully or inadvertently, wrongly conveying to these taxa the status of a full species, e.g. '*Salix microphylla*'. This erroneous shortcut is not acceptable because it only leads to more confusion in a subject that is already very confused.

Cultivar (= cultivated variety) names or epithets typically are assigned to clones or propagules cultivated in the commercial trade. After 1958, they are cast in a modern language without numerals, and therefore not italicized. They are placed after the Latin designation and bracketed by single quotes. In the past, the abbreviation cv. sometimes preceded the epithet, but this practice has been abandoned. For example, the epithet for the cultivar of black cottonwood named after the wife of the late North American poplar geneticist, Scott Pauley, is *Populus trichocarpa* Torr. & Gray 'Fritzi Pauley'. The epithet for the Euramerican hybrid *Populus* ×*canadensis* 'Spijk' designates a village in Holland. The willow cultivar *Salix daphnoides* 'Continental Purple' takes its name from the decorative colour of the stems. The epithet can be used alone or simply attached to the genus name, e.g. *Salix* 'Continental Purple'. This abbreviation practice, although lacking critical information about the binomial, is acceptable provided that the rules cited above are followed. Sometimes, the epithet will be modified or changed completely in the commercial trade to make the cultivar more marketable. Numbers are often assigned to cultivars so that they conform to the accession system of an organization, company or agency. These designations are often used in reports, publications and promotional materials without any cross reference to the correct Latin binomial and cultivar name. We highly discourage these confusing practices.

Interspecific hybrids are designated by using the times (×) symbol, not the lower case letter x, e.g. *Populus alba* × *Populus grandidentata* or *Salix alba* × *Salix babylonica*. Many hybrids have been assigned their own Latin binomial; *Populus* ×*rouleauiana* in the case of the previous poplar-aspen hybrid taxon (note that there is no space between the × and the first letter of the Latin specific name). The hybrid binomial can be used in place of the formula designating the two parent species. In the case where a hybrid is bred purposefully, the Latinized hybrid binomial technically applies only to a cross made in a particular direction, although this convention is not strictly followed. Use of the × symbol is absolutely essential when writing a Latin hybrid binomial; the all too common omission of the '×' allows yet another bogus 'species' to enter the literature. For example, writing *P. rouleauiana* by omitting the '×' not only is technically incorrect but also is misleading to uninformed readers.

Hybrid cultivars are designated in the same way as non-hybrid taxa, e.g. *P.* ×*rouleauiana* 'Hansen'. If known varieties, clones, cultivars or hybrids were parents in a hybrid combination, they should be identified in the full Latin formula, e.g. *P. nigra* var. *betulifolia* × *P. trichocarpa* 'Fritzi Pauley'.

In hybrid nomenclature, the maternal parent is listed first in the formula, followed by the paternal parent, e.g. *S. babylonica* (♀) × *S. alba* (♂). If the sexual identity of the parents is unknown, e.g. a spontaneous hybrid, they usually are listed in alphabetical order. Hybrids may be facile in both directions – i.e. either species may function as the maternal (seed-producing) or paternal (pollen-producing) parent – or they may be facile in one direction only, with one species the obligate maternal parent and the other the obligate paternal parent. For example, the well-known hybrid *P. trichocarpa* × *P. deltoides* (*P.* ×*generosa*) almost always requires embryo rescue because of premature dehiscence of the capsule, whereas the reciprocal – *P. deltoides* × *P. trichocarpa* – usually produces viable seed (Stettler *et al.*, 1996; Stanton, 2005). Because many hybrids are fertile, they can generate backcrosses to one or both of the parent species, e.g. (*P. trichocarpa* × *P. deltoides*) × *P. deltoides*. Hybrids also can combine with one another to form advanced-generation multiple hybrids or can combine with another species altogether. In the *Saliceae*, the possibilities – and their associated nomenclatural complexities – are almost endless.

2.2 The Genus *Populus*

2.2.1 Characteristics of poplars

The poplars are taxonomically divided into six sections (Tables 2.2 and 2.5). They are known far and wide for their distinctive qualities – both good and bad. They are deciduous or, rarely, semi-evergreen trees with a wide natural distribution in the northern hemisphere, from the equatorial tropics to the latitudinal and altitudinal limits of tree growth (Plates 1–8). They have also been widely planted throughout the world, including the southern hemisphere.

The earliest recognizable poplar leaves in the fossil record are *Populus wilmattae* Cockerell (aka *Populus cinnamomoides* (Lesquereux) MacGinitie; Eckenwalder, 1980), which has been placed in section *Abaso* (Fig. 2.3). They date from the Late Palaeocene – about 58 million years ago (Collinson, 1992) – but are also relatively abundant in Eocene formations in Utah, Colorado and Wyoming, USA (Manchester *et al.*, 1986), and in Europe (Collinson, 1992). Poplars from section *Leucoides* appeared in the late Eocene, but thereafter the record became more complicated and difficult to interpret (Collinson, 1992; Eckenwalder, 1996). It appears that sections *Tacamahaca* arose in the Late Oligocene,

Table 2.5. Characteristics of the six sections of the genus *Populus* (modified from Eckenwalder, 1996, and FAO, 1980).

Characteristic	Section					
	Abaso Mexican poplar	*Aigeiros* Cottonwoods and black poplar	*Leucoides* Swamp poplars	*Populus* White poplars and aspens	*Tacamahaca* Balsam poplars	*Turanga* Arid and tropical poplars
Leaves (general)	Unifacial; margins finely crenate; heteroblastic	Unifacial or bifacial; margins finely or coarsely crenate; heterophyllous	Bifacial; margins finely crenate	Bifacial; margins finely crenate, sinuate or coarsely dentate; heterophyllous and (rarely) heteroblastic	Bifacial; margins finely crenate; heterophyllous	Unifacial; margins entire or coarsely dentate; heteroblastic
Preformed (early) or juvenile	Lanceolate-linear to lanceolate triangular	Small, deltoid or rhombic, coarsely toothed	Very large, ovate to cordate	Small, ovoid to round; often very coarsely toothed or dentate	Oval to ovate oblong; base rounded to acute	Lanceolate-linear to lanceolate triangular
Neoformed (late) or mature	Long acuminate, broadly ovate or deltoid	Large, cordate, deltoid or ovoid; finely toothed		Large, oblong ovate cordate, finely toothed or lobed	Base cordate	Reniform to rounded-deltoid; dentate at apex
Petiole cross section	Round	Flattened	Round	Round (white poplars) or flattened (aspens)	Round	Round, flattened near lamina
Buds	Dry, yellow, blunt	Large, tan or red, resinous	Brown, slightly resinous	Brown, slightly resinous or non-resinous	Red-brown, resinous, aromatic	Small, dry, downy
Flowers (general)	Disks deciduous, laciniate, irregular; bracts narrow, not ciliate	Disks persistent, entire, regular; bracts broad, not ciliate	Disks persistent, lobed, regular; bracts broad not ciliate; sometimes hermaphroditic	Disks persistent, entire or dentate, oblique; bracts broad, ciliate	Disks persistent, entire, regular; bracts broad, not ciliate	Disks deciduous, laciniate, regular; bracts narrow, not ciliate; sometimes hermaphroditic
Male (♂)	15–40 apiculate stamens	Catkins reddish, 12–60 emarginate stamens	15–35 apiculate stamens	5–25 truncate stamens	10–50 emarginate stamens	Catkins red; 25–30 apiculate stamens
Female (♀)	2 carpels (stigmas)	2–4 carpels (stigmas)	2–4 carpels (stigmas)	2 carpels (stigmas)	2–4 carpels (stigmas)	Catkins large, green, sometimes with subtending leaves; 3 carpels; stigmas large, bright red, sometimes divaricate

Capsule	2-valved; 5–15 seeds per valve	2- to 4-valved; 4–25 seeds per valve	2- to 4-valved; 4–12 seeds per valve	2-valved; 2–15 seeds per valve	2–4 valved; 2–30 seeds per valve	3-valved; 20–40 seeds per valve
Number of taxa[a]	1 (1)	2–8 (3)	3–5 (3)	7–13 (10)	6–15 (12)	2–3 (3)
Vegetative reproduction	Suckering rare; hardwood cutting rooting unknown	Suckering uncommon or absent; hardwood cuttings generally root well	Suckering common; hardwood cuttings do not root well	Suckering profuse; aspen hardwood cuttings do not root; white poplar cuttings root variably	Suckering occasional; hardwood cuttings root very well	Suckering common; hardwood cuttings root variably
Habitat	Riparian areas	Principally riparian areas, swamps and wetlands but occasionally uplands	Swamps and very wet bottomlands	Swamps, wetland borders, riparian areas and uplands	Riparian areas, swamps, wetland borders	Riparian areas in the tropics and arid climates
Geographical distribution	Mexico	Europe, North Africa, Middle East, Central Asia, North America	Eastern North America and Central Asia	Europe, North Africa, Middle East, Asia and North America	Asia, North America	North Africa, East and Central Asia
Silvicultural use	None	Very important	Little or no importance	Very important	Very important	Some importance

[a]Encompasses the range proposed by most poplar taxonomists. The number in parentheses is our proposal (see Table 2.2).

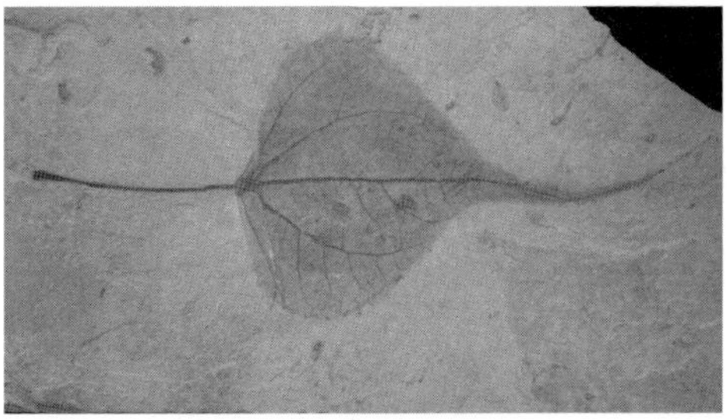

Fig. 2.3. Leaves of *Populus wilmattae* (aka *Populus cinnamomoides*) first appear in the fossil record 58 million years BP. They are especially abundant in Eocene shales from the western USA and Europe. Photo by D. Dickmann.

Aigeiros in the Mid-Miocene and *Populus* in the Mid-Pliocene. More work is needed to clarify these evolutionary relationships.

Poplar trees or clones may be tall and straight (excurrent), multiply branched (deliquescent), twisted and contorted or, rarely, shrub-like in extreme environments. Poplars usually are short-lived compared to other trees, in large measure because they are host to many diseases and insect pests. None the less, certain individual stems can attain ages of 200–300 years, and the root systems of clonal aspens may persist for thousands of years and many successive generations of stems. The rapid growth rate of poplars enables them to reach a large size (Plate 18C). The common cottonwoods of North America (*P. deltoides* and *P. trichocarpa*), the black poplar of Europe (*P. nigra*) and the Asian balsam poplars (*P. maximowiczii*, *P. suaveolens*, *P. szechuanica*, *P. yunnanensis*), for example, can become enormous trees, exceeding 3 m in diameter and 45 m in height.

The cell nucleus of poplars normally contains two sets of 19 ($2n = 38$) chromosomes. Triploids with three sets of chromosomes ($2n = 57$) and tetraploids ($2n = 76$) have also been identified in section *Populus* (Johnsson, 1942; Einspahr *et al.*, 1963; Every and Wiens, 1971). The physical size of the *Populus* genome is remarkably small – six times smaller than *Zea mays* and 40 times smaller than *Pinus taeda*. In addition, the favourable ratio between genetic length and physical length in *Populus* chromosomes makes the genus an attractive choice for genetic mapping and cloning of genes of special importance to forest trees (Bradshaw *et al.*, 2000). In 2004, researchers from around the world completed the sequencing of the genome of *P. trichocarpa*, the first tree and only the third plant to have been sequenced (Tuskan *et al.*, 2006). The size of the *Populus* genome was estimated to be about 485 megabases of DNA. More than 45,000 putative protein-coding genes were identified in the nuclear genome. In addition, the chloroplast and mitochondrial genomes contained 101 and 52 genes, respectively. Genes and gene families linked to lignocellulose wall formation, secondary metabolism, disease resistance, membrane transport and phytohormone biosynthesis and regulation were identified.

All of the taxa in the genus are dioecious, individual trees bearing either staminate (male or pollen-bearing) or pistillate (female or seed-producing) flowers lacking petals and sepals that elongate from axial reproductive buds (see Table 2.5). The small, inconspicuous flowers of both sexes – numbering 30–200 – are borne on pendant catkins (see Table 2.1, Plate 18A and Fig. 2.4). Occasionally, catkins will be hermaphroditic, producing both male and female flowers, or individual trees or clones will bear both male and female catkins (Wyckoff and Zasada, 2007). Production of abnormal flowers can vary from year to year. Catkins typically appear before the leaves in early spring, but unseasonably warm winter temperatures can force them earlier. Flowers of both sexes are borne on cup-shaped disks lacking nectaries; bracts are rapidly deciduous. The number of stamens varies widely among species and ovaries can contain two to four carpels.

After wind pollination, the fruit – an elongated cluster of capsules sometimes likened to a necklace – matures in several weeks to a month or more. Capsules ripen and dehisce into two, three or, rarely, four parts (valves) before leaves are fully developed (Plate 18D) or as late as midsummer, and the cottony seeds take to the air. The spent capsules abscise soon thereafter. Poplar seeds are tiny, averaging about 4000 g^{-1}, but substantial variation (300–16,000 g^{-1}) occurs among species and genotypes within a species; old trees can produce 30–50 million seeds in a single season (Wyckoff and Zasada, 2007). Stands of such trees can create a veritable 'snowstorm' of cotton, which is regarded as a nuisance in urban areas. Therefore, male trees are preferred for horticultural and amenity

Fig. 2.4. The pendant catkins of most poplar species – here male (staminate) catkins of *Populus grandidentata* – appear in early spring before leaves unfold. Photo by D. Dickmann.

plantings; when spent, the pollen-bearing catkins quickly decompose when they fall to the ground. Because seeds are so light and cottony (Fig. 2.5), they can travel for long distances on the wind – 10 km or more (Wyckoff and Zasada, 2007). Secondary transport by moving water can extend this range. As in most tree taxa, seed production of poplars shows a marked periodicity, with several years passing between large or 'bumper' seed crops. Complete reproductive failures, however, are rare.

In most populations of *Populus*, male and female trees tend to be approximately equal in number, but there are exceptions. In the Rocky Mountains of Colorado, USA, for example, female aspen clones predominate at low elevation, but at high elevations nearly all clones are male (Mitton and Grant, 1996). Aspen clones of either sex also may occupy large areas (up to 50 ha), giving the impression that a population is single sexed. Generally, male trees tend to be more precocious that female trees, producing more catkins per tree. Although suggested by Pauley (1949), there is little evidence to substantiate the notion that the greater investment in flowering and fruit production by female trees or clones slows their growth rate compared to males (Saki and Burris, 1985); in fact, sometimes the growth of female clones surpasses that of males (Mitton and Grant, 1996). Finally, some individual trees or clones may be very fruitful and bear many catkins, whereas others are much less fecund.

Barriers to gene flow among *Populus* species in natural populations frequently are ineffective, so spontaneous hybrids are common where species are sympatric, i.e. their natural or planted distribution overlaps (Table 2.6). Interspecific hybridization is common within most *Populus* sections (Fig. 2.6), but intersectional hybrids between taxa in *Aigeiros* and *Tacamahaca* also occur readily in nature. Successful matings across sectional lines in *Abaso*, *Turanga*, *Leucoides* and *Populus*, however, occur with difficulty or not at all (Zsuffa, 1973; Stettler *et al.*, 1980). Natural hybrid populations (hybrid swarms) can be genetically complex and usually are clinal. Along the Danube River in Austria, for example, spontaneous grey poplar hybrids (*P.* ×*canescens* = *P. alba* × *P. tremula*) represent a continuum of genotypes, from intermediate (F_1-like) to highly backcrossed. *P. alba* is the preferred maternal

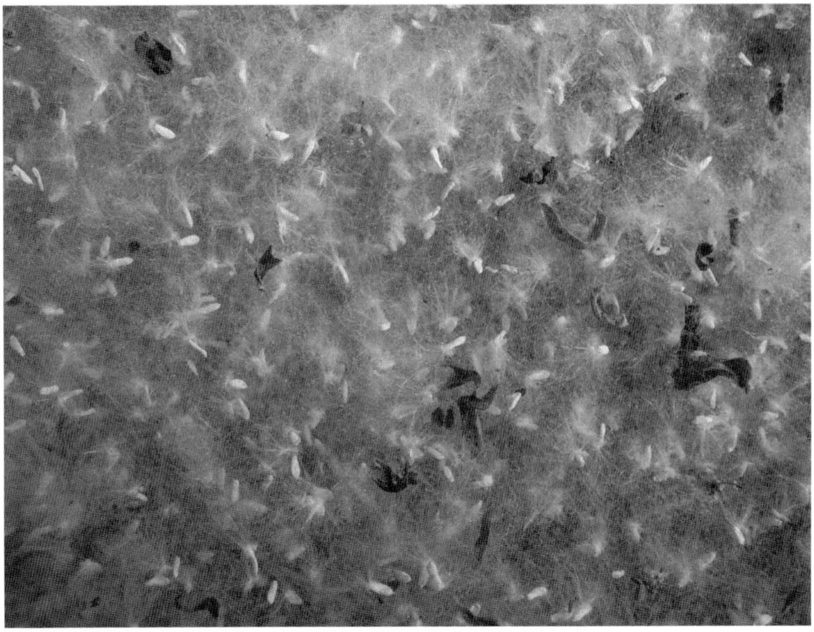

Fig. 2.5. Although poplar and willow seeds – each one attached to a cottony coma – are tiny and retain their viability for only a few weeks, they can travel long distances by wind and water. Photo by D. Dickmann.

parent, and backcrosses are much more frequent to this species than to *P. tremula* (Heinze and Lexer, 2006). Whitham (1989) documented a similar situation in a hybrid swarm of *P. fremontii* × *P. angustifolia* (*P.* ×*hinkeleyana*) along a riparian corridor in northern Utah, USA; in this case, backcrossing occurred only to *P. angustifolia*. The long-term implications of these hybrid swarms for introgression of genes from one taxon into another or evolution of new species are obvious.

Spontaneous hybrids have been cloned and brought into commercial culture for hundreds of years. Additionally, for many decades geneticists have purposely bred hybrids that have been deployed in poplar culture throughout the world. By far the most common of these commercial hybrids is *P. deltoides* × *P. nigra* (*P.* ×*canadensis*, Plate 19A). Tree breeders continue to be active in producing new hybrid cultivars, sometimes using sophisticated microculture and biotechnology techniques to thwart mating barriers.

Morphologically, poplar leaves are distinctive but variable among taxa. They originate from vegetative axial or terminal buds that can be resinous and noticeably fragrant (e.g. section *Tacamahaca*). Leaves usually bear glandular teeth along the margin – which may be fine or very coarse – and often glands at the junction of the petiole and lamina. Stipules are never persistent. Leaves are simple, usually with an elongated or pointed apex, but beyond that there is no common poplar phyllotype. Leaves may be linear, lanceolate, oblong, obovate, deltoid, cordate, rhombic, round, reniform or palmately lobed; they may be longer than wide, wider than long, or equal in both dimensions (Fig. 2.7). Even on the same tree, leaves may differ considerably in size and shape; preformed (early-season) leaves that are present as primordia in the dormant-season bud usually are smaller and distinctly different in shape than neoformed (late) leaves that are initiated by the apical meristem during the growing season (Fig. 2.8). Short shoots and other determinate shoots produce only early leaves. Late leaves are produced on vigorous, indeterminate shoots that elongate throughout the growing season and are typical

Table 2.6. Some naturally occurring hybrids among taxa in the genus *Populus*.[a]

Hybrid parents	Hybrid binomial	English common name and notes
P. adenopoda × *P.* ×*tomentosa*	*P.* ×*pseudotomentosa* Wang & Tung	Backcross hybrid
P. alba × *P. adenopoda*	*P.* ×*tomentosa* Carrière	Peking poplar or Chinese white poplar; aspen parent may be *P. tremula* var. *davidiana*
P. alba × *P. grandidentata*	*P.* ×*rouleauiana* Boivin	
P. alba × *P. tremula*	*P.* ×*canescens* Smith	Grey poplar
P. alba × *P. tremuloides*	*P.* ×*heimburgeri* Boivin	
P. angustifolia × *P. balsamifera*	*P.* ×*brayshawii* Boivin	Brayshaw's poplar
P. angustifolia × *P. deltoides*	*P.* ×*acuminata* Rydberg	Lanceleaf cottonwood; aka *P.* ×*andrewsii*
P. angustifolia × *P. fremontii*	*P.* ×*hinkeleyana* Correll	
P. angustifolia × *P. tremuloides*	*P.* ×*sennii* Boivin	
P. angustifolia × *P. balsamifera* × *P. deltoides*	None	Trihybrid
P. ×*berolinensis* × *P. simonii*	*P.* ×*charbinensis* Wang & Skvortzov	Unverified trihybrid
P. ×*canadensis* × *P. balsamifera*	*P.* ×*rollandii*	
P. deltoides × *P. trichocarpa*	*P.* ×*generosa* Henry	Interamerican poplar; aka *P.* ×*interamericana*
P. deltoides × *P. balsamifera*	*P.* ×*jackii* Sargent	Jack's hybrid poplar or heart-leaf balsam poplar; aka *P. balsamifera* var. *subcordata*, *P. candicans* or *P.* ×*gileadensis*
P. balsamifera × *P. tremuloides*	*P.* ×*dutillyi* Lepage	
P. balsamifera × *P. deltoides* × *P. tremuloides*	*P.* ×*polygonifolia* Bernard	Trihybrid
P. deltoides × *P. nigra*	*P.* ×*canadensis* Moench	Euramerican poplar; aka *P.* ×*euramericana*
P. deltoides × *P. tremuloides*	*P.* ×*bernardii* Boivin	Bernard poplar; may actually be *P.* ×*jackii*
P. fremontii × *P. deltoides*	?	
P. fremontii × *P. nigra*	*P.* ×*inopina* Eckenwalder	
P. grandidentata × *P. tremuloides*	*P.* ×*smithii* Boivin	Aka *P.* ×*barnesii*
P. laurifolia × *P. nigra*	*P.* ×*berolinensis* Dippel	Berlin or Russian poplar; aka *P.* ×*rasumowskyana* or *P.* ×*petrowskyana*
(*P. laurifolia* × *P. nigra*) × *P. balsamifera*	None	Trihybrid
(*P. laurifolia* × *P. nigra*) × *P. deltoides*	None	Trihybrid
P. nigra × *P. simonii*	*P.* ×*xiaohei* Hwang & Liang	Lesser black poplar; aka *P.* ×*gansuensis* or *P.* ×*xiaozhuanica*
P. tremula × *P.* ×*tomentosa*	*P.* ×*hopeiensis* Hu & Chow	Hebei poplar, a trihybrid
P. trichocarpa × *P. fremontii*	*P.* ×*parryi* Sargent	Parry cottonwood

[a]These hybrids have formed spontaneously where the natural range of species is sympatric or where exotic taxa have been planted near natural poplar stands. Wherever two compatible poplar taxa grow together, hybrids usually will form (Eckenwalder, 1996), especially among taxa within a section or between taxa in sections *Aigeiros* and *Tacamahaca*. Documentation of these hybrids varies, and some have not been verified using molecular or cytogenetic techniques.

Fig. 2.6. Wherever *Populus alba* grows it will interbreed spontaneously with local aspens. This is a natural stand of *P. alba* x *Populus grandidentata* (*Populus* ×*rouleauiana*) hybrids established in southern Michigan, USA. Photo by D. Dickmann.

of young trees, coppice sprouts, epicormic shoots and the upper axes of the crown in older trees. Early leaves are more diagnostic morphologically than late leaves, i.e. they tend to be true to the unique phyllotype of a species. In contrast, the shape of late leaves tends to converge among taxa, the notable exception being the maple-like late leaves of *P. alba* and its hybrids (Eckenwalder, 1996). Several taxa, for example *P. mexicana*, *P. euphratica*, *P. ilicifolia* and *P. monticola*, show heteroblastic leaf development, where foliage of young trees up to 10 years of age is distinctly different in venation and shape (usually much narrower) than foliage of mature trees (Eckenwalder, 1980).

Leaf size among taxa also is extremely variable. Species adapted to arid environments produce small pubescent or waxy leaves 5–10 cm^2 in area; species growing in the humid tropics or subtropics, on the other hand, may produce leaves with an area of nearly 500 cm^2. Petioles vary in length from less than 1 cm to nearly 10 cm. Taxa in sections *Aigeiros* and *Populus* have laterally flattened petioles (Plate 19B), giving them an airfoil-like quality that causes leaves to flutter. This trembling or quaking motion has given several poplars their name and has inspired many poetic phrases. Poplar leaves are amphistomatous, although stomatal density is lower on the upper leaf surface. Pubescence (indumentum), if present, may be persistent (*P. alba*) or present primarily on expanding and young leaves (*P. sieboldii*); often, it occurs only along the midribs (*P. maximowiczii*). The presence or absence of pubescence, if used alone, is not a reliable trait to distinguish one species from another. Poplar leaves usually show yellow coloration in autumn, although pale gold, orange or crimson are not uncommon in certain taxa (Plate 19C).

Twig and branch morphology of poplars is quite variable within or among species. Twigs may be glabrous or (when young) pubescent. Trees may produce relatively few coarse branches or many fine branches and twigs. The angle of branches with the stem may be large (sometimes nearly 90°), producing a large, spreading crown; small, producing a narrow, upright or fastigiate crown; or anything in between. In addition to normal proleptic

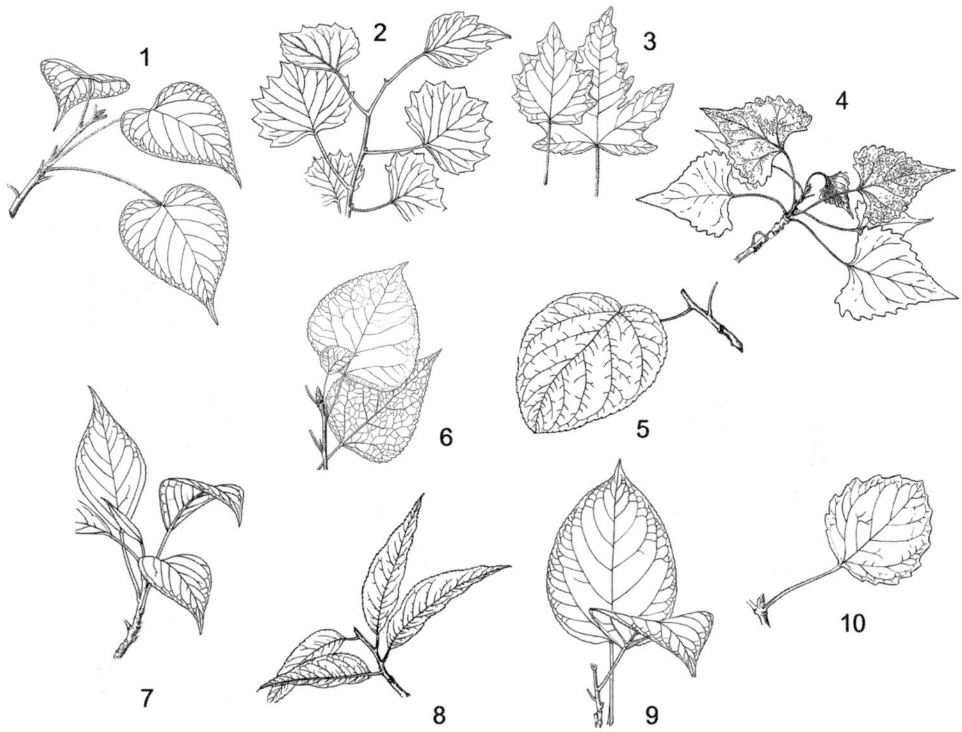

Fig. 2.7. Variation in the shape of mature leaves in *Populus*. 1, *P. ciliata*; 2, *P. euphratica*; 3, *P. alba*; 4, *P. fremontii*; 5, *P. heterophylla*; 6, *P.* ×*jackii*; 7, *P. simonii*; 8, *P. angustifolia*; 9, *P. cathayana*; 10, *P. tremula*. Redrawn from *Flora of China* (www.efloras.org), Sudworth (1908) and Sargent (1949).

branches that grow from buds that have overwintered in a dormant condition, some species and hybrids produce sylleptic branches that elongate from new buds produced during the current growing season. Large branches tend to be brittle and are broken off easily by strong winds, wet snow or ice. In certain species, *P. deltoides* for example, small branches and twigs, often with living leaves attached, break off and fall due to the formation of an abscission layer, a trait unique to these poplars.

The bark of young poplars can be creamy or dirty white, various shades of grey, grey-green, olive-green, orange-brown, red-brown or bronze in colour, and it often remains smooth for many years. Lenticels are prominent on smooth young bark, and they often are diamond shaped (Plate 19F). Corky outgrowths also are common on some poplars. On older trees, the lower bark usually breaks up into coarse, corky or fibrous ridges (Plate 18C). In cottonwoods and balsam poplars, these ridges are very distinctive and prominent, and they may extend well up into the crown.

Poplars are common invaders of disturbed sites (see Fig. 2.1), and they can occur in pure, monotypic stands or in mixed forests with other hardwood and conifer trees. Poplars occupy a variety of ecological habitats. They typically grow in or on the border of alluvial, riparian and wetland habitats, ranging from the far northern latitudes to the tropics. Most poplars (except the aspens) are well adapted to seasonal flooding. They can tolerate temporary anoxic conditions, while the recession or evaporation of surface water and the moist silt or sand left behind provide an ideal – albeit short-lived – environment for germination of seeds, which remain viable for only a few days or weeks. Regeneration is successful when soil moisture remains high enough for roots to grow down in the soil at the same rate that the saturated water front recedes.

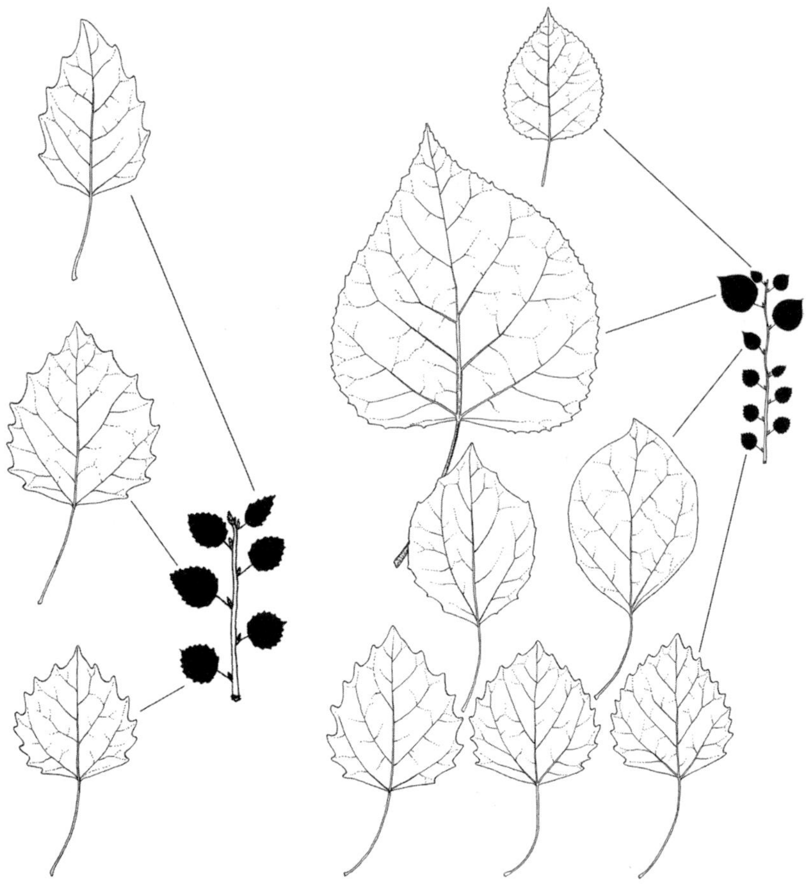

Fig. 2.8. Heterophylly – shown here in *Populus grandidentata* – is a distinctive but taxonomically confusing trait in poplars. Left – determinate shoot with no late-season elongation bearing only coarsely toothed preformed (early) leaves; right – indeterminate shoot with late-season elongation showing basal early leaves and larger, finely toothed late (neoformed) leaves towards the tip. Used with permission from Barnes and Wagner (2004).

Ecologists now recognize that the survival and reproduction of riverine poplars is dependent on periodic floods. In many parts of the world, the damming of rivers and water diversion projects have curtailed flooding, putting these poplars at risk; huge areas once occupied by riverbank or bar poplar groves have become barren or depleted of reproduction (Braatne *et al.*, 1996; Vanden Broeck, 2003; Qi *et al.*, 2005).

In contrast to their wetland relatives, certain poplars are adapted to dry, upland environments. *P. euphratica* and *P. pruinosa* in section *Turanga* can grow in extreme desert environmental conditions, provided their roots have access to a water table (Plate 19D). The aspens (section *Populus*) occupy north-temperate uplands (Fig. 2.9), ranging from wet mesic to xeric. The natural ecological role of this group is the colonization of upland areas disturbed by intense, stand-replacing fires, strong winds, volcanism, avalanches and other disturbances. Their windborne seeds germinate readily on exposed mineral or ash-covered soil, provided adequate moisture is present. Root suckering is another mechanism that allows upland poplars to regain dominance of a disturbed site quickly. Aspens also grow in wet depressions, swampy margins or riparian terraces,

Fig. 2.9. The aspens of section *Populus* are unusual among poplars because they naturally occupy upland habitats, which can be quite dry. Photo by D. Dickmann.

provided some kind of disturbance has allowed them to become established.

Another distinctive feature of certain members of the genus – principally section *Populus* – is their predisposition to form clones by vegetative propagation, principally root suckering (Barnes, 1966). The combination of strong cloning ability and prolific seed production in *Populus* represents a balanced strategy to re-establish in current habitats and invade new ones following major disturbances. The word 'clone' is derived from the Greek, meaning 'a twig or slip'. Clonal plants (ramets) are exact genetic copies of their parent (ortet), unless mutations occur to either parent or progeny. In certain naturally growing aspens, the ortet may have originated millennia ago and the genet has gone through many self-replicating generations (Cottam, 1954; Barnes, 1975). Clones of planted poplar cultivars may go back hundreds of years and many generations to the ortet. Although exact genetic copies of the ortet, clonal ramet phenotypes may not necessarily look exactly like their parent or like one another. Local environmental conditions, disease and insect infestations, browsing, the physiological age of the donor plant (cyclophysis) or the part of the donor plant where the clonal offspring originated (topophysis) may influence the way in which a ramet grows and develops.

Vegetative propagation enables many plant species to survive, compete and reproduce successfully in the ecological habitats they occupy. Cloning can also be a very effective strategy for re-establishment on disturbed habitats or colonization of new habitats. Certain types of vegetative propagation or cloning occur after a poplar tree has been cut down or killed suddenly, by a fire for example. Most poplars will produce copious sprouts from the root collar of a tree that has been killed or felled, although this trait declines as the tree becomes older. Eventually, a few large, surviving root collar sprouts may take on the appearance of mature trees. Young poplars will also produce stool shoots from callus that forms on the surface of cut stumps, a trait that is widely exploited in nurseries for the production of hardwood stem cuttings. Suckers are another way that poplars reproduce vegetatively under natural conditions. These shoots are produced abundantly from adventitious buds on shallow, horizontal roots, especially in the aspens and white poplars, and in this way a single parent tree may produce a dense clonal forest (Fig. 2.10). Suckering occurs principally after a tree is killed, but living trees also will send up suckers from roots that have invaded adjacent open areas, sometimes as far as 30 m from the parent tree.

Certain riparian poplars reproduce vegetatively in unique ways. Cladoptosis occurs when whole lateral twigs, often with leaves attached, are abscised from large trees and then root if they lodge in moist soil (Galloway and Worrall, 1979). More commonly, fragmentation occurs when branches break off trees during dormant season floods or ice scouring and then, if they become covered downstream with silt or sand, root and form new trees (Rood *et al.*, 2003). This last mode of vegetative reproduction represents an ecologically significant means by which riparian poplars become established along the banks of streams or on sandbars after high water recedes.

Of enormous importance for silvicultural and horticultural applications is the establishment of clonal poplar plants with hardwood stem cuttings. This strictly anthropogenic form of vegetative reproduction employs 20- to 30-cm long (sometimes shorter or longer) sections of dormant, 1-year-old woody shoots as planting stock. If these 'sticks' are planted in the spring, they will produce roots quickly from existing primordia in the inner bark and new shoots from the cutting's lateral buds (Fig. 2.11). The resultant clonal plants grow from 1 to 4 m tall in the first growing season, depending on the genotype and local environmental and site conditions. The reliability of adventitious rooting also allows entire young shoots several metres long (sets) to be used, with their basal portion planted as deep as 3 m in the soil. The predisposition of hardwood cuttings to root is a remarkable trait that has allowed the widespread and successful planting of selected pure species and hybrid clones, principally from sections *Aigeiros* and *Tacamahaca*. Hardwood cuttings from other sections of the genus root poorly or not at all because of lack of preformed root primordia. Aspens, especially, cannot be reproduced from hardwood stem cuttings, although they will propagate from root cuttings, leafy softwood cuttings or microculture. Thus, plantations of poplars outside sections *Aigeiros* and *Tacamahaca* are less common worldwide.

Fig. 2.10. The formation of clonal stands by root suckering following timber harvesting or natural disturbances is another distinguishing characteristic of aspens. Photo by D. Dickmann.

Fig. 2.11. Poplars in sections *Aigeiros* and *Tacamahaca* and most willows are routinely propagated using 1-year-old hardwood stem cuttings (a), which can grow several metres tall during the first year (b). Photos by D. Dickmann.

Because they grow so rapidly, natural or planted stands of poplars can produce large volumes of wood in a short period of time. This wood is very versatile and widely used by farmers, crafts people and forest products industries (Chapter 10, this volume). It is light in weight (specific gravity 0.3–0.4), soft, creamy white in colour (except for a dark, pale brown heartwood or wetwood core), straight grained and, because of its diffuse porous anatomy, uniform in texture. Some species readily produce tension wood, which is characterized by lack of cell wall lignification and gelatinous fibres, producing a woolly character to the wood. Most species have low ratings for strength, stiffness, shock resistance and decay resistance, and steam bending. The wood works easily with hand or machine tools, although sharp edges are recommended, and it glues, screws and nails well. Staining can be patchy, but paints and varnish are applied easily. Poplar wood is manufactured into pulp and paper, peeled and sliced veneer, composition boards (including oriented-strandboard, also known as OSB), sawn lumber, crates, boxes, matches, chopsticks, poles, furniture, cabin logs and excelsior (Balatinecz and Kretschman, 2001; see Chapter 10, this volume). In the future, plantation-grown poplars may be an important source of bioenergy. Overall, poplars are not as important as willows in the landscape ornamental trade, although fastigiate cutivars such as *P. nigra* 'Italica' (Fig. 2.12) and *P. alba* 'Bolleana' are very widely planted.

2.2.2 Descriptions of *Populus* species

The horticulturist, Liberty Hyde Bailey (1930), said this about poplars: 'Botanically the poplars are a most bewildering group, and there is wide divergence of opinion and practice as to the limits and names of species.' His words still ring true. The following discussion of 32 *Populus* taxa reflects our conservative approach to species classification. The species list (Table 2.2) represented a rethinking by the senior author of his previous treatments of *Populus* classification (Dickmann and Stuart, 1983; Dickmann, 2001) and was finalized after much study of the world literature and discussions with experts in *Populus* taxonomy. None the less, we include some species of debatable taxonomic validity simply because their binomial is so well known and widely used. This subject is a work in progress, and may always be; the next edition of this book undoubtedly will present a different classification based on new work done in the interim, particularly at the molecular level.

Fig. 2.12. Line plantings of fastigiate poplar cultivars, such as Lombardy (*Populus nigra* 'Italica'), create visually striking patterns on a landscape. Photo by Jean-Pol Grandmont.

Section *Abaso* Eckenwalder

This monotypic section was proposed by Eckenwalder (1977) and comprises *P. mexicana* as the single member (Tables 2.2 and 2.5). It may be the most ancient section in the genus, with leaf specimens clearly related to *P. mexicana* appearing in a 58 million-year-old fossil record from North America (Fig. 2.3).

Populus mexicana **Wesmael, Yaqui Cottonwood.** Previously, this taxon was synonymous with *P. fremontii* because it superficially resembled the *Aigeiros* cottonwoods of the southwestern USA. But Eckenwalder argued that its unique characteristics merited recognition as a distinct taxon in a section of its own. *P. mexicana*, which occurs only in Mexico (Plate 1A), is a medium to large-sized tree (up to 30 m tall) with a spreading, often drooping crown, found in riparian habitats, arroyos and ditches. Although it once formed extensive forests of immense trees, its numbers have been reduced to scattered trees and small groves due to damming and diversion of rivers (Felger *et al.*, 2001).

Yaqui cottonwood exemplifies extreme heteroblastic leaf development. It produces linear, willow-like juvenile leaves for up to 10 years and deltoid to round leaves thereafter (Eckenwalder, 1980). New leaves appear at the time of flowering, while old leaves are still present or abscising. Buds are dry, bright yellow and blunt.

P. mexicana can be divided into two subspecies, one native to the east coast of Mexico – subspecies *mexicana* – and one native to the west coast – subspecies *dimorpha* (Eckenwalder, 1977). Both grow along river flood plains. Subspecies *mexicana* also occurs in the mountains of the states of Oaxaca and Chiapas, making it the most southerly distributed North American poplar. Aside from local consumption, the wood of Yaqui cottonwood has little commercial value. This tree is revered by local people, however, and is often planted around homes and along canals and ditches.

Section *Turanga* Bunge

Turanga comprises three riparian species (Table 2.2), among them the only poplar native solely to the continent of Africa. The intractable nature of *Populus* systematics is exemplified by this section. In spite of as many as ten separate taxa once proposed over the large natural range of *Turanga*, the FAO (1980) considered this section to be monotypic, with *P. euphratica* the sole taxon. There also were proposals to place the poplars in section *Turanga* in a separate genus or subgenus, either *Balsamiflua*, *Euphratodendron* or *Turanga*, because of certain distinctive characteristics of their wood anatomy and flower structure. But these proposals did not gain acceptance in the taxonomic community, the consensus being to retain these singular trees in the genus *Populus* (Browicz, 1977). The primitive floral characteristics, pronounced heteroblasty and adaptation to extreme environments exhibited by members of this section are defining (see Table 2.4).

Populus euphratica **Olivier, Euphrates poplar.** The most well-known and widely distributed member of section *Turanga*, Euphrates poplar, is an extraordinary species. It occurs in a discontinuous natural range from Morocco and Egypt in Africa, through the Middle East to central and western Asia, spanning 115° in longitude (Plate 1B).[3] Its latitudinal and altitudinal ranges also are remarkable – from 48 to 49° north in Kazakhstan to 15° north in Yemen, and from 390 m below sea level in the Dead Sea depression to 4500 m in Kashmir (Browicz, 1977).

The slender, sympodial branches, small, downy buds and dense pubescence on lower leaf surfaces of *P. euphratica* are indicative of the hot, dry environments in which it typically grows. The leathery, heteroblastic leaves are among the most distinctive in the genus, varying in shape from linear or lanceolate on juvenile plants to rounded, ovate or obovate on mature plants (see Fig. 2.7). Mature leaf blades often are wider than they are long and are borne on slender petioles that are equally as long as the blade. The coarse teeth of mature leaves typically are located on the upper (distal) two-thirds to one-half of the blade.

Euphrates poplar typically has a short, stocky bole (up to 1.5 m in diameter) and branchy crown, and often grows in shrubby thickets that may arise from root suckering. Its typical habitat is along watercourses, in riverine tugai forests and in wadis, but it also occurs in oases and on sand dunes. In favourable locations in Central Asia, it attains commercial size (up to 25 m tall) and quality and can live to 200 years of age (Weisgerber, 2000). The fibrous, platy, exfoliating bark on older trees is distinctive. Euphrates poplar grows on a variety of soil types and can tolerate hot, arid, saline conditions, waterlogging and sandstorms. But it cannot be considered a true xerophytic species (Hukin *et al.*, 2005). Gries *et al.* (2003) showed that the growth rate of *P. euphratica* on Taklamakan desert dunes in China (see Plate 19D) decreased appreciably as the depth to groundwater increased from 7 to 23 m. Best growth occurs when groundwater is within 2–4 m from the surface. The requirement of flooding for the establishment of *P. euphratica* seedlings also is well known. The damming of rivers and water diversion projects in several areas in Central Asia have diminished flooding greatly and lowered groundwater tables, leading to depletion of natural stands of Euphrates poplar (Qi *et al.*, 2005). In some localities, the species is facing extirpation.

P. euphratica has great value for environmental stabilization and amelioration; its role in arid-land tugai and oasis ecosystems certainly needs to be conserved. Although it is planted for windbreaks, construction wood and fodder, cuttings root poorly so mass propagation is problematic. But established trees coppice well. Geneticists have viewed this species as a promising parent in hybridization programmes of which the objective is to create tolerance to high solar radiation, heat and soil salinity. Hybrids with *P. deltoides*, *P. nigra* and *P. simonii* have been accomplished using controlled pollinations (Zsuffa *et al.*, 1996), while *P. alba* × *P. euphratica* hybrids have been created using pollination of ovary and ovule microcultures (Mofidabadi *et al.*, 1998).

Populus ilicifolia **(Engler) Rouleau, Kenyan poplar.** The taxonomic placement of this unusual poplar has been very unsettled. Various authorities have placed it in the genus *Celtis*, *Turanga*, *Balsamiflua* or *Tsavo*. Based primarily on floral morphology, Browicz (1966) argued convincingly for full species status for this poplar as *P. ilicifolia*, but recommended that it be placed in a section of its own – *Tsavo*. We follow Eckenwalder (1996), however, and retain this poplar as a distinct taxon in section *Turanga*.

P. ilicifolia occurs naturally in the Tana, Tsavo, Athi/Galana and Uaso Nyiro river systems in Kenya, from near sea level to 1000 m elevation. This natural equatorial range (to 3° south latitude) gives Kenyan poplar the distinction of being the world's southernmost poplar. Unfortunately, its riparian habitat has recently become greatly reduced by human activities, leading the International Union for Conservation of Nature (IUCN) to place this poplar on its 'Red List' of threatened species.

Leaves of Kenyan poplar are deciduous and distinctly heteroblastic: juvenile leaves are linear or narrowly ovate to elliptical and can be toothed or lobed; mature leaves are broadly ovate or obovate, with coarse teeth or lobes. The floral structure is unique in the genus. Female inflorescences – which contain less than ten flowers – produce one to three leaves at their base that are the same size as those on sterile short shoots and persist throughout the growing season. Therefore, Browicz (1966) argued that they should be termed racemes rather than catkins. Equally distinctive, stigmas are divaricate, with two or three lobes with second-order ramifications.

Kenyan poplar occurs in groves on periodically flooded sandbars and in riparian woodlands, where it often is dominant or locally abundant (Wilmot-Dear, 1985). The inevitable damming of the rivers making up its habitat, however, is expected to lead to a decline in the abundance of this species (Hughes, 1990). This poplar roots well from stem cuttings and grows into a large tree up to 30 m in height and 1.5 m in diameter, making it suitable for tropical plantation culture. Because it grows to a large size and the soft, light wood is easily worked, *P. ilicifolia* is fashioned into dugout canoes by local people.

Populus pruinosa **Schrenk, desert poplar.** Once considered a subspecies of *P. euphratica*, this taxon now seems well established (Browicz, 1977). Although its major distinction is the desert habitat where it grows, *P. pruinosa* none the less is a riparian poplar, growing as a small to medium-sized tree or shrub in alluvial tugai, wadi and oasis communities in north-west China, Kazakhstan, Tajikistan, Turkmenistan and Uzbekistan. Preformed leaves are small, reniform in shape (unusual for a poplar), blue-grey and densely tomentose, with entire margins or a few teeth near the apex. Neoformed leaves are elliptic. Young bark is greyish-yellow, but older trees develop a thick, furrowed bark (Fang *et al.*, 1999).

This poplar can tolerate drought, soil salinity, low humidity and high temperatures. Its leaves are well adapted to retard transpiration and minimize heat loading. In the Taklamakan desert in the Xinjiang region of China, *P. pruinosa* grows in conditions where mean annual rainfall is less than 50 mm, sandstorms rage and the salt content of the soil is 2–3% (Weisgerber and Han, 2001). Depth to the water table, however, is critical. Optimum growth occurs when the water table is 1.5–2.5 m below the soil surface; as water tables descend below this optimum level, growth, flowering, suckering and vigour decline (Li *et al.*, 2003b).

The future of this singular species is uncertain. In the northern Amudarya Delta in Uzbekistan, anthropogenic use and overexploitation during the last century have led to the disappearance of 90% of the tugai forest occupied by this species (Rüger *et al.*, 2005). The situation in China is equally grave. Conservation measures, therefore, are desperately needed.

Section *Leucoides* Spach

Section *Leucoides* – the swamp poplars – comprises three species (Table 2.2). None are of great economic value, although they have important ecological functions. One species is native to North America, while the other two are Asian. Unlike poplars in sections *Turanga*, *Aigieros* and *Tacamahaca*, a common trait of poplars in this section is poor rooting of hardwood cuttings, which limits their commercial use (Table 2.5).

Populus glauca **Haines, Asian swamp cottonwood.** Formerly *P. wilsonii* (Table 2.2), this species is similar to *P. lasiocarpa* (see below) and is distributed from 2500 to 3300 m elevation in north-eastern India, Sikkim and the Chinese provinces of Sichuan, Xizang and Yunnan. Leaves are smaller than *P. lasiocarpa*, broadly ovate and cordate at the base. Flowers sometimes are bisexual (Fang *et al.*, 1999). Hybrids with *P. simonii* apparently have been successful but are not widely deployed. Beyond that, little else is known about this poplar outside its native range.

***Populus heterophylla* Linnaeus, swamp cottonwood.** This medium to large tree (up to 40 m tall and 1.5 m in diameter) is distributed discontinuously along the eastern and Gulf coastal plains, Mississippi and Ohio river valleys and the south-western Great Lakes regions of the USA (Plate 1C). Leaves are large, ovoid to cordate (see Fig. 2.7), sparsely pubescent and borne on round petioles. Twigs are tomentose when young, dull brown or grey, with a distinctive orange pith, and bear stout, sparsely resinous, reddish-brown buds. Never common, swamp cottonwood occurs as single scattered trees in mixed-species forests in swamps, sloughs, low-lying areas near tidewater and along river borders, typically on heavy clay soils with high water tables. Swamp cottonwood is among the most flood-tolerant of poplars and will thrive on sites that are too wet for *P. deltoides* (Johnson, 1990). Experience with propagation of swamp cottonwood is minimal, although cuttings are more difficult to root than *P. deltoides*. It does, however, produce root collar sprouts from trees <30 cm in diameter and forms small clonal groups by root suckering (Wagner *et al.*, 1980). Swamp cottonwood is rarely harvested for commercial trade and has received little or no attention from tree geneticists.

***Populus lasiocarpa* Oliver, heart-leaved poplar or Chinese necklace poplar.** Native up to 3500 m elevation in the humid and semi-humid subtropics of China (Guizhou, Hubei, Shaanxi, Sichuan and Yunnan provinces), this striking tree bears huge (up to 35 cm long and 25 cm wide), deeply cordate, bright green leaves that are pubescent beneath. Midribs and petioles are red; twigs are angular, slightly pubescent, and bear large, slightly sticky buds with pubescent basal scales (Fang *et al.*, 1999). Catkins are unusually long (10–15 cm) and even longer in fruit. Bisexual, polygamous flowers and monoeciousness are not uncommon. This poplar may be one of the most ancient and primitive members of the genus (Hong *et al.*, 1987). Heart-leaved poplar grows rapidly and is a very desirable ornamental. Because it does not propagate easily from cuttings and grafts poorly, however, its commercial use is limited. *P. lasiocarpa* is resistant to the rust *Melampsora larici-populina* Kleb. and hybrids with *P. nigra* have been accomplished to capitalize on this trait (Werner and Siwecki, 1994). Hybridization with strong-rooting species like *P. nigra* also might transmit better rooting.

Section *Aigeiros* Duby

Since the dawn of plantation culture, taxa in section *Aigeiros* – both pure species and hybrids – have been the most important members of the genus *Populus* worldwide. Native to Eurasia and North America, clonal poplars in section *Aigeiros* have been and continue to be planted on every continent where trees will grow. When the word ‘poplar’ is used, trees in this section are what come to mind for many people. Ecologically, this section includes some of the major riparian poplars in the northern hemisphere, which, along with willows, are a defining feature of these habitats. The section includes two cottonwoods and black poplar (Table 2.2).

***Populus deltoides* Marshall, eastern cottonwood.** One of the most familiar and silviculturally important poplars, *P. deltoides* is naturally distributed over the eastern, southern and mid-western USA and southern Canada to the foothills of the Rocky Mountains (Plate 2A). But because it has been so widely planted, its range has been expanded to encompass the entire temperate world.

Eastern cottonwood is a medium to very large tree that can reach heights exceeding 45 m and diameters of 3 m or larger. Under closed-canopy forest conditions, it develops a long, straight bole and small, round crown. When growing in the open, however, the stem becomes highly forked (deliquescent) with a massive crown (Plate 18C). Preformed cottonwood leaves are deltoid in shape and prominently toothed, whereas neoformed leaved are more elongated, cordate and finely toothed. All leaves hang from long, flattened petioles that cause them to flutter in the slightest wind. The twigs are stout, angular to ribbed in cross section, and produce slightly resinous, outcurved buds. Bark on young trees is smooth, greenish-yellow to grey, with prominent lenticels, but becomes ashy-grey and deeply furrowed on older trees (Cooper and Van Haverbeke, 1990).

Cottonwood is a tree usually associated with bottomlands, alluvia and riparian corridors, where optimum growth occurs on silty or sandy loam soils. In the lower Mississippi Valley, USA, the best stands occur on the land between the river and levees that is seasonally flooded. The raw sediment left behind after floodwaters recede can be colonized by thousands of cottonwood seedlings per hectare. In the northern part of its range, cottonwood also grows in upland habitats, some of them droughty, and it is a common invader of disturbed sites. It also is a common tree on the sand dunes along the shore of Lake Michigan.

Like other poplars, cottonwood grows in pure stands or in association with other bottomland or pioneer species. The southern cottonwood (subsp. *deltoides*) has been regarded as the fastest-growing native tree in North America. On the best sites in the Mississippi River Delta, height growth can exceed 4 m $year^{-1}$ for the first 5 years (Knowe *et al.*, 1998), and over 145 m^3 of pulpwood can be produced in a 12-year rotation (McKnight, 1970). At maturity, cottonwood attains a large size, reaching heights of over 50 m and diameters of more than 1.5 m.

Cottonwood usually produces good seed crops yearly when mature (Plate 18D) and it will invade newly exposed, moist soils rapidly. The cottony seed can be carried long distances by the wind or on the surface of moving water. Vegetative reproduction by root collar sprouts is vigorous if trees are young when cut. Cottonwood can also be propagated, with high survival rates, by hardwood stem cuttings, although some genotypes root poorly. Therefore, rooting ability is an important selection factor in cottonwood genetic improvement programmes. The northern cottonwood (subsp. *monilifera*) generally does not root as easily from hardwood stem cuttings, nor does it grow as fast as the southern subspecies.

Cottonwood is an important commercial timber species, and it has been widely planted – both within and outside its natural range – for match stock, excelsior, pulpwood, sawtimber, veneer, fuel and fodder. In addition to forest plantings, cottonwood has been used extensively in agroforestry systems, shelterbelts and for reclamation of mine spoils. Its importance as a plantation tree is emphasized by the fact that 47 cultivars currently are registered with the International Poplar Commission (IPC), more than any non-hybrid taxon.[4]

Five natural varieties of cottonwood have been proposed, but Eckenwalder (1977) accepted just three subspecies – *deltoides*, *monilifera* and *wislizeni*. The plains cottonwood (subsp. *monilifera*) is emblematic of the unsettled nature of *Populus* nomenclature. This tree is a defining feature of riparian habitats in the high steppes of west-central North America. It was awarded full-fledged species status (*P. sargentii*) early in the 20th century, then reduced to varietal status as *P. deltoides* var. *occidentalis* (Little, 1979). More recently, Eckenwalder (1977) argued that plains cottonwood so closely resembled the cottonwood of the Great Lakes Region and other northern areas that they all should be included under subsp. *monilifera*.

Two *Aigeiros* cottonwoods indigenous to the south-western USA have also caused taxonomic confusion, but they now are placed within *P. deltoides*. The Rio Grande cottonwood is native to western Texas and New Mexico. Once regarded as a variety of *P. fremontii* or a separate species (*P. wislizeni*), it is now considered to be subspecies *wislizeni* (Eckenwalder, 1977). The other south-western cottonwood, found in central and south-western Texas and northern Mexico, was given the binomial *P. palmeri*. But this rather obscure, medium-sized tree with ovate leaves is now considered synonymous with *P. deltoides*.

Cottonwood has shown a substantial predisposition to hybridize, both naturally and under controlled conditions (Table 2.6). The most important group of hybrids worldwide is *P. deltoides* × *P. nigra*, known collectively as *P.* ×*canadensis*,[5] the Euramerican or Canadian poplars (Plate 19A). Currently, 145 cultivars are registered with the IPC, far more than any other taxon. Natural hybrids between the native black poplar and cottonwoods introduced into Europe by the early botanical explorers of North America (*hybrids spontanés*) were first described in 1755 in France by Duhamel du Monceau. These natural pairings were augmented with controlled crossings, beginning in the 20th century. Euramerican hybrids usually have *P. deltoides* as the maternal parent, because the reciprocal cross is rarely successful. Many of the hybrid poplar clones now under cultivation worldwide are *P.* ×*canadensis*. Intersectional hybrids with *P. trichocarpa* (*P.* ×*generosa* or Interamerican hybrids) or *P. balsamifera* (*P.* ×*jackii*) also have become prevalent in poplar culture. *P.* ×*generosa* is especially important in northwestern North America and, to some extent, in Europe (13 cultivars are currently registered with the IPC).

Most hybrids with *P. deltoides* show hybrid vigour (heterosis). Because of its rapid growth rate, excellent form, site adaptability, ease of propagation and genetic pliability, *P. deltoides* has been – arguably – the most important poplar taxon worldwide for genetic improvement and plantation forestry.

***Populus fremontii* S. Watson, Fremont cottonwood.** This arid-zone poplar is the common, low-elevation (to 1800 m elevation) cottonwood of northern Mexico and the southwestern USA (Plate 2A). Its habitat typically includes canyon bottoms, stream banks and flood plains, large arroyos and irrigation canals (Felger *et al.*, 2001). It can attain heights of 30 m and diameters of 4 m and is the largest native tree in the Sonoran Desert. The branches of Fremont cottonwood bear resinous, pubescent buds and small deltoid leaves (see Fig. 2.7) borne on laterally flattened petioles. Leaves turn bright yellow in the autumn. Preformed leaves have larger marginal teeth than neoformed leaves. The bark on older trees is deeply fissured. This poplar is widely planted as an ornamental and for fuel around ranches and in towns throughout its native region.

Two subspecies are accepted within *P. fremontii* (Eckenwalder, 1977). Subspecies *fremontii* occurs west of the North American continental divide. Subspecies *mesetae* (formerly *P. arizonica*) is indigenous in southern New Mexico, western Texas and the Valley of Mexico; its leaves are more rhombic in shape than subspecies *fremontii*.

***Populus nigra* Linnaeus, black poplar.** The final species in section *Aigeiros* has long been one of the most important poplars, owing to its extensive distribution and commercial deployment. Native to Europe (but not Scandinavia), North Africa and western Asia (Plate 2B), the exact limits of the original natural range of black poplar have been somewhat obscured because of its widespread and ancient cultivation (Weisgerber, 1999). Although the 'type' species for the section, it actually differs in many traits from the two *Aigeiros* cottonwoods of North America. In fact, *P. nigra* more closely resembles section *Tacamahaca* poplars in some traits, and its chloroplast DNA has ties to *P. alba* of section *Populus* (Smith and Sytsma, 1990). Eventually, the North American cottonwoods may have to be placed in a section of their own, leaving *P. nigra* the sole member of section *Aigeiros* (Eckenwalder, 1996).

Black poplar is a large tree, reaching heights of 40 m and diameters over 2 m at maturity; individual specimens can reach 300 years of age (Weisgerber, 1999). It often produces an irregular, branchy crown. The often crooked or swept, buttressed bole can be massive, frequently producing large burls or epicormic branches. Some stands, however, produce straight, well-formed trees, many of which have been propagated vegetatively for commercial plantings (40 *P. nigra* clones are registered currently with the IPC). Ornamental trees frequently are pollarded, which some believe extends their lifespan (Cooper, 2006). The juvenile bark of black poplar is yellow-white, becoming dark and deeply fissured with age.

The leaves of black poplar are typically dimorphic (Plate 18A); preformed leaves are small (15–95 cm^2), distinctly rhombic in shape, dark green in colour, with a slightly lighter undersurface, with finely toothed margins and long, flattened petioles. Neoformed leaves are larger, broader than long, and oval or deltoid in shape. Black poplar twigs are round in cross section, reddish, and produce gummy, usually glabrous, outcurved buds.

Black poplar is an aggressive species that colonizes river flood plains, wastelands and other exposed sites, with the moist, sandy soil exposed after seasonal flooding providing the optimum seedbed. It is a signature species of riparian ecosystems, which are among the most biologically diverse in the area of its natural range. Seed crops are regular and begin to be produced at about 10 years of age. It sprouts vigorously from stumps and, to some extent, suckers from exposed roots or those close to the soil surface. Propagation from stem cuttings is routine, with rooting of 80% or more the norm.

The high human population density in most of its natural range, combined with a long history of exotic poplar introductions that displaced natural *P. nigra* stands, introgression of exotic poplar genes into natural populations and disruption of riparian ecosystems by channelization, canal and dyke construction and dam building threatens the genetic integrity of this species (Vanden Broeck, 2003). In fact, many forest biologists regard *P. nigra* to be one of the most threatened tree species in Europe. Therefore, conservation and restoration of natural riparian ecosystems dominated by black poplar are being given high priority in many European countries (Lefèvre *et al.*, 2001; Cooper, 2006). Inventory and protection of extant natural *P. nigra* populations, establishment of stool beds and seed orchards, planting in riverine habitats and protection from animal damage are being employed. Coordination of these efforts is through EUFORGEN, the European Forest Genetic Resources Programme (Weisgerber, 1999; Vanden Broeck, 2003).

The large geographic distribution of *P. nigra* and its widespread cultivation have given rise to a plethora of named varieties and cultivars. Prominent among them are varieties *betulifolia* and *caudina*, characterized by pubescent young leaves and shoots, and variety *neapolitana*, which lacks pubescence (Weisgerber, 1999). Cultivated clones of black poplar exhibiting a columnar (fastigiate) growth habit – cultivars 'Italica' or Lombardy poplar (see Fig. 2.12), 'Thevestina' and 'Plantierensis' – have been immensely popular for line plantings, roadside trees, windbreaks and ornamental uses. Although afforded variety status by some accounts, these cultivars are likely clonal propagules from a single mutant mother tree (genet).

Lombardy poplar is a male tree, although some female trees have been reported, indicating it may not be a single clone (Li, 1996). Lombardy probably is the oldest and best-known columnar cultivar, having been introduced into cultivation in the Po Valley of Italy early in the 18th century (Zsuffa, 1974; Li, 1996). By 1784, it had spread throughout Europe and even to North America. The dramatic silhouette produced by this distinctively columnar tree has become synonymous with the fastigiate growth habit (see Fig. 2.12). Lombardy poplar has been used extensively throughout the world as a landscape ornamental, and it may be the most widely planted of all poplars. In some places, east-central North America for example, Lombardy and other columnar cultivars succumb to canker diseases and a wetwood bacterium at a very early age, and their planting is discouraged. Breeding programmes with the objective of producing columnar types with a broader genetic diversity, therefore, are needed.

The wood of black poplar is highly regarded, so its culture in plantations is long standing. Because this species is as genetically diverse and pliant as any member of the genus *Populus*, it has been a parent in many hybrid combinations produced spontaneously or by poplar breeders to increase wood production (Table 2.6). The Euramerican clones (*P.* ×*canadensis*) previously discussed under *P. deltoides* have been immensely significant in poplar culture because of their rapid growth rate, straight stem form and resistance to fungal, bacterial and viral diseases. More recently, intersectional hybrids between *P. nigra* and taxa in section *Tacamahaca*, including the Asian *P. maximowiczii*, *P. laurifolia*, *P. simonii*, *P. cathayana* and the North American *P. trichocarpa*, have been successful. Clones of these hybrids show promise for plantation wood production. For example, certain *P. nigra* × *P. maximowizcii* cultivars (six currently registered with the IPC) have been widely deployed in eastern North America, although their susceptibility to wind breakage and canker disease has led to some re-evaluation. In China, *P. simonii* × *P. nigra* has been a favoured combination for a long time (Weisgerber, 1999).

Section *Tacamahaca* Spach

Section *Tacamahaca* – collectively known as the balsam poplars – is taxonomically confusing. Depending on the authority, this section contains as few as 9 (Eckenwalder, 1996) or as many as 40 (Fang *et al.*, 1999) taxa, most of them native to China. Clearly, a thorough taxonomic re-evaluation of this section, employing ecological, morphological, anatomical, phytochemical and cytogenetic criteria, is badly needed. Without such study – which should be multinational and interdisciplinary – the current taxonomic muddle cannot be fully resolved. We will discuss 12 taxa (Table 2.2), reflecting our conservative approach to species designation.

Largely associated with riparian or wetland habitats, balsam poplars are distributed mostly in the northern latitudes, in some cases to the latitudinal or altitudinal limits of tree growth. Yet one species is subtropical. Their ovate-elliptical leaves and resinous, fragrant buds are distinctive. Several species are commercially important. Three species are native to North America, with the remaining being Asiatic. *Tacamahaca* poplars are not indigenous to Europe or Africa.

***Populus angustifolia* James, narrowleaf cottonwood.** This North American *Tacamahaca* poplar currently is of little commercial importance, although it plays a significant ecological role. Widely distributed from southern Canada to northern Mexico (Plate 2C), narrowleaf cottonwood is a medium-sized tree (up to 20 m tall) characteristic of stream flood plains, canyon bottoms and moist upland flats in the foothill mesa and montane life zones of the Rocky Mountain region of North America. It is distinguished from other poplars by its narrow, willow-like leaves (see Fig. 2.7) and the prominently ridged and furrowed bark of older trees. Narrowleaf cottonwood is found in pure stands or growing in association with other intolerant, riparian species. In some situations, it will produce abundant root suckers. Cuttings of narrowleaf cottonwood root readily, and it has been used sparingly as a plantation tree or as an ornamental in the Rocky Mountain region. Natural hybrids of this species with *P. balsamifera*, *P. deltoides* subsp. *monilifera* and *P. fremontii* are common (Table 2.6), in some localities outnumbering the pure species (Brayshaw, 1965a; Rood *et al.*, 1986).

***Populus balsamifera* Linnaeus, balsam poplar.** Occurring transcontinentally across the northern USA and Canada, from the Atlantic Coast to the Rocky Mountains and Alaska (Plate 3A), balsam poplar is the most widely distributed *Tacamahaca* poplar in North America. If one accepts the proposal that *P. trichocarpa* is a subspecies of this taxon (Brayshaw, 1965b), then the range becomes even larger. Balsam poplar grows farther north than any North American poplar, occurring in clonal groves at 68–69°N latitude on the North Slope of Alaska (Bockheim *et al.*, 2003) and as a procumbent plant at timberline in the Brooks mountain range. The Latin binominal *P. balsamifera* (literally 'bearing resin') was used synonymously for eastern cottonwood (now *P. deltoides*) for many years because the description by Linnaeus apparently was too vague to distinguish between the two (Little, 1979).

Balsam poplar has ovate leaves, which are dark lustrous green above and pale metallic green below, often with rusty brown blotches. The leaf margins are finely toothed and the petioles round. The twigs also are round in cross section, with out-curved buds that are saturated with a fragrant, amber-coloured resin. Floral structure is the only way to distinguish this species from *P. trichocarpa* where the ranges of the two species overlap: staminate flowers of *P. balsamifera* show 12–30 stamens and capsules are two-valved, lanceolate and glabrous. The bark of balsam poplar is smooth, greenish to reddish brown, turning grey, with scaly ridges as it ages. The wood is soft, light brown in colour, with grey sapwood. It is considered inferior to other poplars for bleached pulp by the forest industry because of the high resin content and dark colour of the heartwood (Zasada and Phipps, 1990).

Balsam poplar is a medium-sized tree that reaches its largest proportions in north-western North America and Alaska. Its habitat usually is moist or wet soils, such as the borders of streams, lakes, swamps and depressions, although it will also grow on drier sites. Balsam poplar is an intolerant pioneer species that will invade disturbed sites rapidly by suckering or seeding. Although relatively short-lived, it will outlive quaking aspen (*P. tremuloides*) and may eventually dominate on sites where the two coexist. Like many poplars, balsam poplar is succeeded eventually by more tolerant hardwoods and conifers.

Because the species brings adaptability to dry conditions, resistance to cold and ease of propagation by stem or root cuttings, certain hybrids of balsam poplar have been recommended for shelterbelts in the northern North American prairies (Roller *et al.*, 1972), although *P. deltoides* or *P.* ×*canadensis* clones are now favoured. Plantations of pure balsam poplar for wood production are virtually unknown. Hybrid clones of *P. deltoides* × *P. balsamifera* (*P.* ×*jackii*, see Fig. 2.7), however, which arise spontaneously in areas of species sympatry (Table 2.6), have been planted for wood production in eastern Canada. The widely planted female cultivar, balm-of-Gilead (cultivar 'Gileadensis'), is probably a clone of this hybrid (Eckenwalder, 1996), although it has been named variously *P. balsamifera* var. *subcordata*, *P. candicans* (*P.* × *candicans*) or *P.* ×*gileadensis*. It is distinguished by its heart-shaped leaves. Clones of *P. balsamifera* × *P. maximowiczii* (and the reciprocal) have shown promise for forest planting in Quebec, Canada (Eckenwalder, 2001).

Populus cathayana **Rehder, Cathay poplar.** This poplar, although important in its native country, is virtually unknown outside of China, and its taxonomic status is uncertain. Eckenwalder (1996) placed it in the synonymy of *P. suaveolens*. Weisgerber and Zhang (2005a), however, argue that, based on the majority of current taxonomic opinion and available evidence, it should be treated as a species in its own right, although they acknowledge that the matter requires further study.

Occurring from 800 to 3000 m elevation in a wide band from the humid subtropics of south-central China to the north-eastern part of that country (Plate 3B), this balsam poplar has wide ecological amplitude. The typical habitat of Cathay poplar is ravines and alluvial deposits in river valleys, although in mountainous areas it will grow on upland sites, where it spreads mainly by suckering (Fang *et al.*, 1999). Cathay poplar grows up to 30 m tall and 80 cm in diameter but seldom exceeds 100 years in age (Weisgerber and Zhang, 2005a). Young shoots are typically round, olive-green, glabrous and bear resinous buds. Preformed leaves are broadly or narrowly ovate to round (see Fig. 2.7), dark green above and whitish-green below. Neoformed leaves are larger and somewhat heart-shaped. Bark on older trees is greyish white to olive and breaks into blocky ridges at the base of the stem.

In the largely deforested region of central China, Cathay poplar is highly regarded as a source of wood in rural areas because of its rapid growth rate and good form. It establishes well from hardwood cuttings and has been deployed in forest plantations, windbreaks and along streets. Promising hybrids with *P. szechuanica*, *P. deltoides* and *P. nigra* have been selected by Chinese tree improvement workers and deployed in plantations. Somatic hybridization of *P. cathayana* with *P. euphratica* through protoplast fusion in microculture has also been accomplished (Zhuge *et al.*, 2000).

Populus ciliata **Royle, Himalayan poplar.** Reaching a medium to large size (20 m tall), *P. ciliata* is distributed extensively in the mixed forests of the lower slopes (1300–3400 m) of the Himalayan Mountains, from Pakistan east to north Myanmar and the Yunnan province of China (Fang *et al.*, 1999). In certain parts of this range, the Uttar Pradesh of India for example, *P. ciliata* is the only indigenous poplar. It resembles *P. balsamifera* of North America, with ovate–cordate leaves (see Fig. 2.7) that are densely ciliate along the margins and faintly pubescent below. Buds are very resinous. This pioneer tree actively colonizes disturbed sites by windblown seed and root suckers, and reaches its largest size in ravines and alluvial deposits.

Himalayan poplar is economically significant locally. Because it is adapted to upland sites and hardwood cuttings root well, it is being used for afforestation of denuded and abandoned lands in the high hills of India (Kapoor *et al.*, 2004). The hybrid with *P. maximowiczii*, which outperforms both parents in growth and yield, also shows great promise for afforestation (Chauhan *et al.*, 2004). Certain clones and provenances of *P. ciliata*, however, have shown susceptibility to leaf rust caused by *Melampsora* spp., necessitating selection of rust-resistant phenotypes (Gupta *et al.*, 2002). The cultivar 'Tristis No 1' (formerly *P. tristis*) may be a hybrid of this species with *P. nigra* or *P. balsamifera*, but its origins are obscure and warrant further investigation.

Populus koreana **Rehder, Korean poplar.** A species of questionable taxonomic validity, *P. koreana* is very similar to *P. maximowiczii*, differing from the latter mainly by bearing glabrous rather than pubescent leaves and shoots. Eckenwalder (1996), in fact, placed both of these taxa in the synonymy of *P. suaveolens*.

Korean poplar is a large tree that attains heights up to 30 m and diameters to 1.5 m. It occurs naturally in north-east China, the Korean Peninsula and eastern Russia, a range virtually identical to *P. maximowiczii*, save its absence from Japan. The wood of Korean poplar is used locally for construction, pulp and matches. Hybrids with *P. nigra* have been created and are under test.

***Populus laurifolia* Ledebour, laurel poplar.** Another indigenous Asian, laurel poplar ranges from eastern Kazakhstan and north-west China to Mongolia and Siberia (Plate 3C). It occurs in hilly steppe or mountainous regions as gallery forests along stream banks and in flood plains, where it often dominates. Laurel poplar is a modest-sized tree (15–20 m tall), with small, narrowly ovate or broadly lanceolate leaves and slender, sharply angled or winged twigs (Krasnoborov and Malyshev, 2003). Somewhat familiar to North American poplar growers because of its inclusion in the pioneering hybridization work of Stout and Schreiner (1933), laurel poplar has also been cultivated occasionally in Europe.

***Populus maximowiczii* Henry, Japanese poplar.** Long known as *P. suaveolens* until A. Henry pronounced it a distinct taxon in 1913, Eckenwalder (1996, 2001) argued that *P. maximowiczii* should return to the synonymy of *P. suaveolens* as a maritime subspecies. Hamaya and Inokuma (1957) also cast doubts on the legitimacy of this taxon. At this time of writing, the matter remains unresolved, but we include this poplar as a distinct taxon because the binomial is so widely known. The native range of *P. maximowiczii* encompasses low to mid-elevations (to 2000 m) in north-eastern China, eastern Russia and Korea, but it is most commonly associated with northern Japan (Plate 4A). Its typical habitat is river gravel bars, fluvial plains and low terraces, but in Japan it also colonizes volcanic ash on the low slopes of recently active volcanoes (Haruki and Tsuyuzaki, 2001). Japanese poplar is a fast-growing tree that attains heights of up to 30 m and diameters of 2 m, making it one of the largest poplars in eastern Asia. In coastal regions, in fact, seagoing dugout canoes once were fashioned from the largest trees. Leaves are leathery, shiny dark green above and whitish below, with glandular-toothed margins and downy pubescence along the veins. It is one of the first poplars to leaf out in the spring. Shoots of Japanese poplar are round in cross section, densely pubescent and reddish when young, with resinous, fragrant buds.

This taxon is among the best known of the Asian poplars because it has been a parent in many intra- and intersectional hybrids accomplished by poplar geneticists. Common matings with *P. maximowiczii* have been *P. balsamifera*, *P. ciliata*, *P. deltoides*, *P. nigra* and *P. trichocarpa*. Some of the fastest growing hybrid crosses made by the original North American poplar breeding project at the Oxford Paper Company (Stout and Schreiner, 1933) used a Japanese poplar as the female parent. In eastern North America, however, Japanese poplar is highly susceptible to stem cankers caused by *Septoria musiva* Peck, which severely deform and eventually kill trees. However, certain hybrids with *P. nigra* introduced to North America from Europe have shown better canker resistance, although they are not totally immune. The wood of *P. maximowiczii* hybrids is brittle, and damage by wind, snow and ice has occurred in some localities.

***Populus simonii* Carriere, Simon poplar.** This well-known poplar is distributed from sea level to 3000 m in a wide north–south belt in central and eastern China (Plate 4B). It also occurs on the Korean Peninsula and in Mongolia. Simon poplar is a moderate-sized tree with a narrow crown, usually reaching a maximum of 20 m in height and 50 cm in diameter. On moist, fertile sites, however, trees can reach 30 m in height and 1.5 m in diameter. It is drought tolerant and has been described as a xerophyllous species of the dry mesic forest steppe (Yang *et al.*, 1999b). Whereas poplars typically are short-lived, trees of this species over 200 years old have been reported (Weisgerber and Han, 2001). Leaves are small, bright green and narrowly rhomboid to obovate in shape (see Fig. 2.7); branches are reddish brown, conspicuously angled, with sticky, resinous buds (Fang *et al.*, 1999).

Natural stands of *P. simonii* in China have been severely depleted, and some provinces have no natural representatives of the species left (Yang *et al.*, 1999b). Plantation culture of Simon poplar in China goes back over 2000 years, however, and today it is the most widely planted poplar in that country. Northern genotypes are very hardy; protection plantings of them have withstood the severe continental climates of China's northern plains and desertified land in Inner Mongolia (Li *et al.*, 2003a). Simon poplar has also been planted in Europe and North America as an ornamental, especially the drooping cultivar 'Pendula'. This poplar once was deployed in the Canadian shelterbelt programme, although it suffered from winter injury and dieback (Roller *et al.*, 1972), possibly indicating an ill-adapted seed source. Spontaneous hybrids of Simon poplar are common (Table 2.6) and certain clones have been widely planted. Prominent among them is the 'lesser black poplar' (*P. nigra* × *P. simonii*), with Lombardy poplar (*P. nigra* 'Italica') being a common parent. Some advance-generation hybrids are also beginning to be deployed in China, including *P.* ×*canadensis* × *P. simonii*, *P. deltoides* × (*P. simonii* × *P. nigra* 'Italica') and *P. deltoides* × (*P. nigra* × *P. simonii*).

Populus suaveolens **Fischer, Siberian poplar.** Another balsam poplar little known outside its natural range, *P. suaveolens* is a large tree (up to 30 m tall) that carries elliptic leaves – dull green above, grey-white below – and the characteristic cylindrical stems and resinous, aromatic buds of section *Tacamahaca*. Its natural range includes north-central China, Mongolia and eastern Siberia (Plate 3C). Its habitat is sandy and pebbly flood plains and the banks of river valleys, occurring as far north as the montane-taiga and subalpine zone in Siberia, where it is the only indigenous, non-aspen poplar (Krasnoborov and Malyshev, 2003). In its native range, it is considered a valuable tree for planting in hot, dry continental climates.

First described by Fischer in 1841, *P. suaveolens* may be a much more extensively distributed taxon than currently thought. If one recognizes the inevitable polymorphism that occurs within widely distributed *Populus* taxa, then – as suggested by Eckenwalder (1996) – *P. cathayana*, *P. koreana* and *P. maximowiczii* could be included in the synonymy of *P. suaveolens*. The former three taxa, in fact, were first described early in the 20th century, well after Fischer's published description of *P. suaveolens* in 1841. But the matter is far from resolved (Weisgerber and Zhang, 2005a) and begs further study.

Populus szechuanica **Schneider, Szechuan poplar.** Native to mountainous terrain from 1100 to 4600 m elevation in central and southwestern China (Plate 5A), this poplar is little known elsewhere. It reaches the largest dimensions of any Asian poplar – 40 m tall and up to 1.5 m diameter. The large leaves are ovate in shape; branches are angled. The timber of Szechuan poplar is used for construction and furniture in China, and it is planted along roadsides (Fang *et al.*, 1999).

Populus trichocarpa **Torrey & Gray, black cottonwood.** This impressive poplar occurs in Pacific coastal areas from Baja California to Alaska and in adjacent interior regions of northwestern North America (Plate 3A). It epitomizes the taxonomic quandary of whether to lump or split. The vegetative morphology of *P. trichocarpa* is virtually indistinguishable from that of *P. balsamifera*; leaves of both are ovate or ovate-lanceolate, sometimes with a cordate base. There are, however, differences in reproductive structures. The staminate flowers of black cottonwood have two to three times as many stamens as those of balsam poplar, while the capsules of black cottonwood are three-valved (split into three parts) or sometimes four-valved, globose and densely pubescent compared to the two-valved, lanceolate and glabrous capsules of balsam poplar. Where the ranges of the two species overlap in the northern Rocky Mountains of the USA and Canada, and in coastal Alaska (Plate 3A), distinguishing one species from the other may be impossible unless capsules can be examined, and even then many intermediate, introgressive forms exist (Brayshaw, 1965b; Viereck and Foote, 1970). Thus, the proposal of Brayshaw (1965b) to place black cottonwood in the synonymy of *P. balsamifera* as subspecies *trichocarpa* has merit and has gained some acceptance in the scientific community. None the less, because *P. trichocarpa* is such a well-known taxon worldwide, and because it has been used extensively in hybridization programmes and plantation culture, we retain it as a distinct species for this writing.

Black cottonwood is a fast-growing tree of moist alluvial, morainic and outwash habitats, where it grows in pure stands or in association with other low-elevation hardwoods and conifers. This species reaches its best development in climatic regions dominated by moist Pacific Ocean air and occurs up to 1500 m elevation in coastal areas. In interior habitats, it can be found to 2000 m elevation in valleys and canyons. Black cottonwood may grow to the largest size of any poplar; in the Puget Sound area, heights of 50 m and diameters over 1.5 m are not uncommon, with exceptional trees reaching over 60 m in height and nearly 4 m in diameter (Viereck and Foote, 1970; DeBell, 1990). Mature forest-grown trees in coastal habitats develop long, clear boles; from the standpoint of commercial stem form, this species has no equal among North American poplars. The crowns of such trees are typically narrow, cylindrical and round-topped. Trees growing in more arid locations east of the coastal mountain ranges are smaller in stature, with broader, deliquescent crowns (DeBell, 1990).

Like balsam poplar, black cottonwood can be propagated easily from stem cuttings; it sprouts readily from cut stumps and sometimes produces abundant root suckers. Plantation culture of black cottonwood, however, has not been widely practised in its native range, although growth rates can be quite impressive. Rather, during the last decades of the 20th century, clones of Interamerican hybrids (*P.* ×*generosa* = *P. trichocarpa* × *P. deltoides*)[6] developed at the University of Washington began to be planted on a commercial scale in the Pacific Northwest USA and coastal Canada. These Interamerican hybrids are more productive than pure black cottonwood (Heilman and Stettler, 1985; Stettler *et al.*, 1988; see Chapter 4, this volume), and they have become the basis for a new hardwood pulpwood, veneer and sawtimber industry in a region where hardwoods have been little utilized. Plantations of Interamerican hybrids have shown spectacular growth rates on alluvial soils along the lower Columbia River and in irrigated plantations in the high desert east of the Cascade Mountains. After 4 years growth, biomass yields of 50–140 t ha^{-1} can be achieved in intensive culture systems; 15-year sawtimber volumes can be 146 m^3 ha^{-1} or more (Heilman *et al.*, 1990; Scarascia-Mugnozza *et al.*, 1997). Hybrids of black cottonwood with *P. nigra* and *P. maximowiczii* are now entering commercial production to complement the Interamericans. Poplar growers in Europe have also recognized the virtues of black cottonwood, and clones of the pure species and its hybrids, which can be higher-yielding alternatives to traditional Euramerican clones, began to be released for commercial production in the 1970s (Koster, 1972). Currently, 16 clones of *P. trichocarpa* and 13 clones of *P.* ×*generosa* are registered with the IPC.

***Populus yunnanensis* Dode, Yunnan poplar.** The southernmost balsam poplar, this tree is native to the mountains of southern Sichuan, western Guizhou and northern Yunnan provinces in south-western China (Plate 3B), occurring from 1300 to 3700 m elevation. It is characterized by large, dark green, ovate leaves that are whitish to silvery below, often with bright red midribs. At low latitudes, leaves are semi-persistent throughout the year. Petioles are short and often red; twigs are angular and carry buds oozing reddish resin. Bark on older trees is brown, with shallow scaly ridges. Yunnan poplar is adaptable to many soil conditions and thrives in hot climates with long growing seasons. It is rarely found growing naturally in extensive pure stands but rather tends to grow singly or in mixture with sympatric hardwood and coniferous species. Growth rates are rapid – trees can attain heights of 30 m and diameters exceeding 60 cm in 25 years (Weisgerber and Zhang, 2005b).

Although hardwood cuttings of Yunnan poplar root well, it has been planted only to a limited extent in China for wood, agroforestry and ornamental purposes. But it has been introduced successfully in France, India and New Zealand. Hybrids with *P. deltoides* and *P.* ×*canadensis* have been successful, and one clone of each is currently registered with the IPC.

Section *Populus*

This section of the genus (formerly section *Leuce* Duby) is a complex grouping comprising the white poplars (formerly subsection *Albidae* Dode) and aspens (formerly subsection *Trepidae* Dode) (Table 2.2). Like section *Tacamahaca*, the taxonomy of section *Populus* is far from settled, particularly with respect to the Asian aspens and Mexican white poplars. We follow Barnes and Han (1993) and Hamaya and Inokuma (1957), who make a strong case for lumping several putative aspen taxa. Members of section *Populus* are distributed over temperate, boreal and montane reaches of the northern hemisphere, and several are of great economic importance. As a group, they are ecologically distinctive because of their clonal physiognomy and – in the case of aspens – by their common occurrence in upland habitats (see Table 2.5).

***Populus alba* Linnaeus, white poplar.** One of the most distinctive poplars, *P. alba* is widely distributed over northern Africa, southern Europe and west and central Asia (Plate 5B). It varies in form from tall and straight to broad-crowned, crooked and multi-stemmed. White poplar has become naturalized in many areas where it has been introduced, primarily from root suckers, which it produces copiously. Much of the naturalized reproduction of seed origin may actually be spontaneous hybrids with native aspens (see below). White poplar is a striking tree that can grow to a large size – more than 40 m tall and 1 m in diameter (Tsarev, 2005), although it is often smaller. The bark is metallic grey to chalky white on young trees, with distinct diamond-shaped, dark lenticels (see Plate 19F), becoming black and deeply furrowed at the base in older trees. Its leaves are the most distinctive in the genus, varying in shape from the ovoid to deltoid, coarsely toothed preformed leaves to the three- to five-lobed, maple-like neoformed leaves (see Fig. 2.7). Leaves are a rich dark green above, with a covering of thick white felt (indumentum) below, giving them a very attractive appearance. Buds and twigs also are covered with white pubescence.

In bottomland habitats where seasonal variation in water tables is not extreme, white poplar attains magnificent timber proportions. It also grows well on a wide range of site and soil conditions and is regarded as somewhat tolerant of drought, wind, salinity and high temperatures. White poplar has suffered, however, from winter dieback, frost injury and cankers when planted in north temperate areas, presumably owing to Mediterranean seed sources. It produces abundant seed and, like the aspens, suckers vigorously from shallow roots. Hardwood cuttings of white poplar can be used for propagation, although rooting success is very dependent on the genotype of the parent tree. Because of its inherent vigour, site adaptability and ability to root, white poplar has long been considered a superior tree for timber, windbreaks and ornamental purposes. Thirteen *P. alba* cultivars are currently registered with the IPC.

Nine natural varieties of white poplar have been proposed – *globosa, hickeliana, microphylla, nivea, pendula, pyramidalis, richardii, subintegerrima* and *tomentosa* – but several are simply clonal cultivars. Because of its distinctive columnar form, the well-known cultivar 'Bolleana' has been a popular tree for ornamental and line plantings throughout the world and is second only to Lombardy poplar for these purposes. White poplar hybridizes readily with aspens, either spontaneously or via controlled crossing (see Fig. 2.6). The spontaneous hybrid between *P. alba* and *P. tremula* – the grey poplar (*P.* ×*canescens*) – has been known for over two centuries, having arisen spontaneously in regions where the ranges of the parent species are sympatric (see Table 2.6). Grey poplars are intermediate between the two parent species in morphological characteristics and show hybrid vigour (17 cultivars are currently registered with the IPC). They thrive on dry or saline soils better than white poplar.

Peking or Chinese white poplar (*P.* ×*tomentosa*) has been treated by some as a distinct *Populus* species, but it is now considered a hybrid (Yang *et al.*, 1999a), probably between *P. alba* and *P. adenopoda*. Bialobok (1964), however, claimed that the aspen parent was *P. tremula* var. *davidiana*. Peking poplar long has been cultivated in several Chinese provinces (six cultivars are currently registered with the IPC). Fast-growing triploid Peking poplar cultivars, with lower lignin content and longer fibres, have also been produced. Finally, because white poplar was an early emigrant from Europe to North America, hybrids with native aspens are common. *P.* ×*rouleauiana* (*P. alba* × *P. grandidentata*; see Fig. 2.6) and, to a lesser extent, *P.* ×*heimburgeri* (*P. alba* × *P. tremuloides*) have arisen spontaneously in many places in North America (McComb and Hansen, 1954; Spies and Barnes, 1982) and several clones of these taxa have shown promise as plantation trees.

***Populus guzmanantlensis* Vázquez & Cuevas, Manantlán poplar.** This little-known tropical poplar occurs on the lower to mid slopes (800–900 m) of the Sierra de Manantlán in the south-western Mexican state of Jalisco (Plate 6A). It is an occasional tree in mixed, moist semi-deciduous forests, where it can attain heights of 35 m and diameters of 60 cm (Vázquez and Cuevas, 1989). The IUCN has placed this poplar on its 'Red List' of threatened species.

***Populus monticola* Brandegee, Baja poplar.** Endemic to canyon bottoms and oak woodlands from 825 to 1190 m in montane regions of the Mexican states of Baja California Sur, Sonora and Chihuahua, this poplar is a medium- sized tree that can exceed 22 m in height and 1 m in diameter (Felger *et al.*, 2001). Boles can be single and straight or multiple, with bark that is rough and grey down low and whitish with black scarring above. Characterized by strongly heteroblastic leaf development (Eckenwalder, 1980), the juvenile leaves of Baja poplar are narrowly oblong to rhombic, whereas leaves on mature plants are ovate or round. Seasonal heterophylly also occurs on indeterminate shoots. Petioles are laterally compressed. Bailey (1930) suggested that this putative species really was a naturalized population of *P. alba* var. *subintegerrima* introduced by early Spanish settlers. Because *P. alba* has become naturalized in many places in North America, this hypothesis warrants testing using isoenzyme analysis or DNA fingerprinting. The wood of Baja poplar has been used locally for furniture and construction, and it makes a spectacular ornamental if amply supplied with water.

***Populus simaroa* Rzedowski, Balsas poplar.** More widely distributed than Manantlán poplar, this species occurs in scattered locations in the mountains surrounding the Rio Balsas drainage in the Mexican states of Mexico and Guerrero (Plate 6A). *P. simaroa* is unusual because it is deciduous during the summer wet season and leafs out during the winter dry season. *P. simaroa* and *P. guzmanantlensis* are closely related and could be considered varieties or subspecies of a single taxon. Rzedowski (1975) also suggested a relationship to *P. grandidentata*. Because they were first described in 1989 and 1975, respectively, neither *P. simaroa* nor *P. guzmanantlensis* have been studied well enough to make an unequivocal pronouncement about their taxonomic status.

***Populus adenopoda* Maximowicz, Chinese aspen.** Native to mountain slopes from 300 to 2500 m in central and south-eastern China (Plate 6B), *P. adenopoda* is a large tree that can reach 30 m in height (Fang *et al.*, 1999). Its leaves are typically aspen – relatively small, glandular, round to ovate with a pointed apex and borne on a long, laterally flattened petiole. Leaf margins of Chinese aspen are coarsely dentate or rounded, upper leaf surfaces are shiny dark green and the lower leaf surfaces – as well as branches – are densely pubescent when young. The wood is used for timber and pulp.

Two natural varieties are accepted: var. *adenopoda* and var. *platyphylla* (Fang *et al.*, 1999). In addition, a hybrid with *P. alba* (*P.* ×*tomentosa*, Chinese white poplar) is well known and often incorrectly given full species status. This hybrid occurs as a planted tree throughout the natural range of Chinese aspen and is widely used for pulp, timber and as an ornamental and street tree. A backcross hybrid of *P.* ×*tomentosa* with *P. adenopoda* is also cultivated in China.

***Populus gamblei* Haines, Himalayan aspen.** Native to the lower hills of the eastern Himalayas from 400 to 2000 m, this species was first described in 1906 from the vicinity of the city of Darjeeling. It is unrelated to any other Eurasian species, with the possible exception of *P. adenopoda*. Unlike other aspens, which are distinctly north temperate, this species occurs in mixed forests in a subtropical life zone. The fast growth rate of Himalayan aspen has led to interest in it as a plantation species in its native range.

***Populus grandidentata* Michaux, bigtooth aspen.** One of two North American aspens, *P. grandidentata* occurs in the Northeast, Great Lake States and northern Midwest USA and adjacent Canada (Plate 7A). Bigtooth aspen reaches a fairly large size, and on dry sites, foresters consider it superior to the sympatric quaking aspen because of its rapid growth rate and excellent form. Preformed leaves of bigtooth aspen are ovate, silvery on the underside, with margins of coarse, pointed teeth. Neoformed leaves on indeterminate shoots or suckers are larger and more heart shaped, with fine teeth (see Fig. 2.8). Autumn leaves colour to a bright yellow, or occasionally orange or red. Buds diverge from the twigs. Bark on young bigtooth aspen trees is olive-green, brownish-orange, or grey, which, along with its distinctive leaves, readily distinguishes it from its frequent associate, quaking aspen. Bark on the lower bole of mature stems is dark grey and furrowed.

Bigtooth aspen is adapted to well-drained, medium- to coarse-textured upland soils. It also is one of the most intolerant tree species and will only reproduce successfully in openings or areas with a dispersed overstorey. Bigtooth aspen grows rapidly on rich sites and may reach heights of 20 m and diameters of 60 cm in 50 years. Reproduction by root suckers is common, and they can grow to over 2 m during the first growing season (Laidly, 1990). Browsing whitetail deer (*Odocoileus virginianus*) seem to prefer bigtooth aspen suckers to those of quaking aspen, which can partially negate these impressive first-year spurts of growth.

The hybrid of bigtooth aspen with introduced *P. alba* (*P.* ×*rouleauiana*), which often arises spontaneously where the two species grow in proximity (McComb and Hansen, 1954; Spies and Barnes, 1982), is an especially impressive tree that has interested many poplar growers. Where this hybrid has been propagated vegetatively in an efficient, cost-effective manner, it has become widely deployed in plantations.

Populus sieboldii **Miquel, Japanese aspen.** The leaves of a native aspen flutter in every north temperate land mass, and the islands of Japan are no exception. The taxonomic placement of this aspen, however, remains unclear (Hamaya and Inokuma, 1957). In the last analysis, *P. sieboldii* may be recognized as a geographic variant of *P. tremula*. Japanese aspen, which occurs on hillsides and in mountainous regions on all the major islands except southernmost Kyushu (Plate 7B), is a medium-sized tree (up to 25 m tall and 50 cm in diameter), with twigs and buds covered by a semi-persistent white down. Leaves are typically aspen – relatively small and ovate in shape – with a pubescent undersurface when young (Plate 19B). Japanese aspen is a colonizer of highly disturbed sites on the islands, including recently volcanized areas, and can live to be 100 years old or more. Extensive pure stands, however, are rare.

P. sieboldii is an important commercial tree in its native range. Several hybrids are recognized, one of the most promising being *P. sieboldii* × *P. grandidentata*. This hybrid – which has been transformed genetically to reduce lignin content (Tamura *et al.*, 2001) – is being targeted especially for cultivation on dry sites. Hybrids with *P. alba* have also been created to increase rooting of adventitious shoots excised from root segments. An intersectional hybrid with *P. maximowiczii* (Kamabuchi poplar) also exists (Takeda, 1975), but it is unclear whether this is a spontaneous hybrid or one created by Japanese breeders.

Populus tremula **Linnaeus, common or Eurasian aspen.** The archetypal Old World aspen, *P. tremula* occupies the largest natural range of any taxon in the genus and is one of the most widely distributed trees in the world (Plate 8A). In the Russian Federation alone, it occupies 20.6 million ha (Tsarev, 2005). It has been recognized from antiquity for its unique morphological, ecological and economic qualities. With its light grey or greenish-grey bark – often pitted with diamond-shaped lenticels – and fluttering, rattling leaves, *P. tremula* is a familiar part of upland and montane landscapes from the British Isles, through Scandinavia and northern Europe, to the easternmost reaches of China and Russia and the northern islands of Japan. A disjunct population also occurs in Algeria on the African continent.

Not as large as some of its relatives in other sections, Eurasian aspen can reach heights of 25–30 m and trunk diameters of 60 cm, although on stressful sites it can be much smaller. Among the shortest lived of poplars, few trees survive past 100 years. Yet individual clones may persist for millennia through root sprouting. Often a straight, well-formed tree, Eurasian aspen also can be small and contorted on stressful sites.

Preformed leaves – borne on characteristically flattened petioles – are typically small (about 7–30 cm^2), round to broadly ovate, with coarse, sinuous or pointed teeth (see Fig. 2.7). Neoformed leaves are larger, rounded deltoid in shape, with fine teeth. Shoots and leaves may be glabrous or – especially when unfolding – densely pubescent. This variant pubescence, as well as the preformed–neoformed leaf morphology, has been a stumbling block in aspen taxonomy, with several so-called species being simply morphogenic variants of the *P. tremula* archetype.

Eurasian aspen grows on a variety of habitats – ranging from wet mesic to dry mesic or xeric – but it reaches its 'best' development on well-drained, loamy soils high in lime, with a water table within 1.5 m of the surface. This aspen can form extensive pure, even-aged stands following disturbances, although mixed stands of aspen and other overstorey or understorey species are also common. The shallow and widespreading root system produces sucker regrowth of high density (tens of thousands per hectare) if the stand is logged, killed by fire or windthrown. Sucker growth is very rapid (1–2 m) during the first year. Clones resulting from suckering can vary in size from several trees to many thousands.

Sexual reproduction by Eurasian aspen is abundant, and the light cottony seed, which matures in early spring, is carried long distances by the wind. If seeds find a favourable germination environment in a moist, recently disturbed habitat, seedlings will establish abundantly. But if these conditions do not pertain, seed viability is lost quickly and few of the germinants will survive past the seedling stage. Because suckering develops on an established root system, it is the major reproductive mode in many areas.

In certain areas of its natural range, Eurasian aspen has been depleted because of agriculture and human development, but where it is abundant, it is a commercially important tree. Aspens cannot be propagated effectively by dormant hardwood stem cuttings because they lack preformed root primordia, which limits their planting on a commercial scale. Successful trials with hardwood cuttings taken from plants transformed with the *rol*B gene construct, which is known to promote rooting, give hope that this barrier can be overcome (Dai *et al.*, 2004). Young, leafy suckers from root cuttings or leafy shoot tips (softwood cuttings) can be excised and rooted under mist in a controlled environment, and aspens can be mass propagated under microculture (Ahuja, 1984; Barocka *et al.*, 1985), offering other avenues for producing clonal planting stock, provided economic barriers can be overcome.

Several varieties or geographic races of *P. tremula* have been recognized, with some incorrectly afforded species status, e.g. var. *davidiana* as *P. davidiana* (Fang *et al.*, 1999; Weisgerber and Han, 2001). But we follow Barnes and Han (1993) and consider these 'species' to be geomorphic variants of *P. tremula*. Several of the named species or varieties are merely horticultural or forestry cultivars, e.g. 'Glandulosa', 'Pendula' and 'Pyramidalis'. Among the Eurasian aspen hybrids, *P. alba* × *P. tremula* (*P.* ×*canescens*) – the well-known grey poplar – has long been important in plantation forestry (see Table 2.5). The *P. tremula* × *P. tremuloides* (*P.* ×*wettsteinii*) hybrid also shows much promise. Natural *P. tremula* triploids have been recognized for a long time (Johnsson, 1942) and have been used in breeding projects throughout the world. Several advance generation hybrids, for example *P.* ×*canescens* × *P. grandidentata*, are also under trial.

Populus tremuloides **Michaux, quaking or trembling aspen.** The North American counterpart of *P. tremula*, quaking aspen, is the most widely distributed tree species indigenous to North America (Plate 8B). It occurs transcontinentally across the northern USA and Canada to Alaska, and south in the Rocky Mountains, Cascade Mountains and Sierra Nevada Mountains to Mexico. It occupies upland and montane habitats that vary from wet mesic to xeric (see Fig. 2.9). It is the most commercially important poplar in North America, with large volumes harvested each year for wood products. Quaking aspen is a slender tree with straight to crooked form and a small rounded crown. It is small to medium-sized in much of its range, although it can attain heights of 35 m and diameters of nearly 1.3 m in the central Rocky Mountains (Perala, 1990).

The distinctive bark of young quaking aspen is smooth, dirty grey, greenish-white or creamy white, and frequently darkened by warty bands, especially around branch nodes. On old trees, the lower part of the stem will break into dark grey or black shallow ridges. Preformed leaves of quaking aspen are round to oval in shape and similar in size to those of *P. tremula*, but they have finely toothed margins. Neoformed leaves, especially on young suckers, are much larger and more elongated. In the autumn, leaves range in colour from a dull yellow green to bright yellow, although in the Rocky Mountains clones bearing gold and orange leaves are not uncommon. The petiole of quaking aspen, like all aspens, is long and characteristically flattened, causing the leaves to flutter in the wind; hence the common name. The wood is light in colour, soft and straight grained, and it is widely used in North America for pulp, paper, matches, oriented-strandboard, lumber and other wood products.

Quaking aspen is a distinctively clonal species (see Fig. 2.10), although seed reproduction does commonly occur on recently disturbed sites with plentiful moisture. Clones derived from suckering vary from a few trees to stands many hectares in size (Barnes, 1975). In fact, the largest known organism in the world may be a clonal stand of quaking aspen in Utah, USA, which covers 43 ha, contains approximately 47,000 individual stems and weighs an estimated 6 million kg (Mitton and Grant, 1996). Aspen grows rapidly during the first 20 years, generally reaching maturity after 30–40 years. In the Rocky Mountains, quaking aspen reaches its maximum lifespan, with individual trees attaining over 200 years of age (Jones and Schier, 1985). Quaking aspen provides food and prime habitat for a host of animals and birds, and wildlife managers actively promote its reproduction and expansion. None the less, the area occupied by quaking aspen in many parts of the Rocky Mountains recently has declined markedly due to browsing by wild and domestic ungulates and the exclusion of fire (Bartos and Campbell, 1998).

Quaking aspen is one the most genetically variable plants ever studied (Mitton and Grant, 1996). Dissimilarity in growth rate, reproductive and vegetative bud break, autumn leaf coloration, leaf abscission, bark characteristics, stem form and other traits are typical and very visible among clonal stands. As a consequence, quaking aspen at one time or another has been split into as many as four distinct species and 13 varieties or forms. Cooler heads have prevailed, however, and today quaking aspen is considered simply a single, highly polymorphic taxon (Little, 1979; Barnes and Han, 1993).

Natural hybrids between *P. tremuloides* and *P. grandidentata* (*P.* ×*smithii*) have been identified where the natural ranges of the two species overlap (see Table 2.6). These hybrids are not common, however, because *P. tremuloides* generally flowers 1–2 weeks before *P. grandidentata*. Spontaneous or purposefully bred hybrids with other taxa in section *Populus* also occur, *P. tremula* × *P. tremuloides* (*P.* ×*wettsteinii*) for example, a hybrid that has shown promise for forest plantations. Interspecific hybrids of *P. tremuloides* with taxa in other sections, however, are rare. Natural quaking aspen triploids are also well known (Einspahr *et al.*, 1963; Every and Wiens, 1971). Triploid hybrids showed better growth and wood properties than diploids in the northern Lake States, USA (Einspahr *et al.*, 1968), but they succumbed to disease before age 25 (Enebak *et al.*, 1996). Even though many quaking aspen hybrids – as well as clonal selections of the pure species – are promising, the difficulty and expense of producing planting stock vegetatively, as well as severe browsing of young trees by ungulates, has limited forestry plantings in North America. Ornamental planting of quaking aspen in the USA and Canadian West, however, is not uncommon.

2.3 The Genus *Salix*

2.3.1 Characteristics of willows

The genus *Salix* comprises 330–500 species worldwide of deciduous or, rarely, semi-evergreen trees and shrubs (Argus, 1999). Willows are tolerant of a wider range of climates than poplars. Predominantly occurring in temperate and arctic zones, willows are also found in the subtropics and tropics. In many northern floras, the number of willow species outnumbers any other woody genera. They play an important role in the vegetative cover of tundra and areas lying above latitudinal tree lines. Willows are the only woody species in some alpine zones, which are called 'zones of dwarf willows'. Most willows occur in the northern hemisphere, with only a few species indigenous to the southern hemisphere.

Evolutionarily, *Salix* probably arose in the warm temperate or subtropical regions of eastern Asia, where the most apparent links between *Populus* and *Salix* exist (Skortsov, 1968, 1999; Fang, 1987). Advancement then occurred into the tropics and – much more significantly – into temperate and arctic regions. The earliest willow-like leaf fossils resemble those of subgenus *Salix* and occur in Early Eocene formations in North America (Wing, 1981), although diagnostic reproductive structures have not been recovered. Discovery in the Eocene Green River Formation in the western USA of extinct, willow-like fossils assigned the binomial *Pseudosalix handleyi* – twigs with attached leaves, pistillate and staminate flowers, and fruits – casts doubt on whether these Early Eocene leaves represent *Salix* (Boucher *et al.*, 2003). *Pseudosalix* appears to be a linkage taxon between *Salix* and certain tribes now within *Salicaceae* formerly in *Flacourtiaceae*. Fossil leaves of *Salix* are also represented in formations from the Lower Oligocene in Europe, Late Oligocene in Alaska and Late Miocene in Japan (Collinson, 1992).

A close look at the current distribution of willows reflects their evolutionary history, as well as the richness and diversity of the genus (Plates 9–16). The centre of abundance of *Salix* is in China, with 189 endemic species (Fang *et al.*, 1999), followed by the former Soviet Union with about 120 species (Skvortsov, 1968, 1999). There are 113 species inhabiting North America north of Mexico (Argus, 2010), 65 species in Europe and about 280 species in Asia (Argus, 1999). Forty species occur in Mongolia, 30 in Japan, 26 in India, 64 in the region of central Asia including Afghanistan, Pakistan, Azerbaijan, Turkmenistan, Tajikistan, Kyrgyzstan and Uzbekistan. There are 11 native species in the Neotropics (Mexico, Central and South America) (Alford and Belyaeva, 2011) and 8 species in Africa. There are no willows in eastern Brazil or eastern Indonesia. Willows are missing from the Antarctic and occur in Australia and Oceania only as introductions.

The cell nucleus of most *Salix* species contains two sets of 19 chromosomes ($2n = 38$). Species with a base number of 22 have also been reported (Skvortsov, 1968, 1999). Ploidy levels range from diploid ($2n$) to dodecaploid ($12n$). The DNA content of diploid willow species native to Europe was estimated to be 0.76–0.98 pg per diploid nucleus based on flow cytometry (Thibault, 1998).

Willows occur in several life forms: upright trees, shrubs, prostrate plants (Fig. 2.13) or groundcovers, with the majority of taxa occurring as shrubs (Fig. 2.14). Height among tree taxa can be 20–25 m and taller – *S. alba* can reach up to 30 m (Skvortsov, 1968, 1999). Decumbent shrubs, conversely, can be only a few centimetres tall. Diameters of very old trees can sometimes be large, exceeding 3 m, with thick corky, ridged bark (Fig. 2.15). Their habits vary from upright to pendulous to spreading. In nature, clonal propagation is unusual in *Salix* and only species from section *Longifoliae* and *S. setchelliana* form root suckers (Fig. 2.16). A few species (*S. herbacea*) form colonies from rhizomes, and several species, for example *S. humilis*, *S. lucida* and some dwarf willows, create small colonies by layering (Fig. 2.17). Vegetative propagation by broken branches that are brittle at the base and dispersed by water is common for some alluvial species (*S. euxina*, *S.* ×*fragilis* and *S. nigra*). Propagation of most willows by hardwood cuttings is facile because of preformed root primordia on stem nodes, although a few species are not good rooters, e.g. *S. caprea* and *S. scouleriana* (Densmore and Zasada, 1978; Liesebach and Naujoks, 2004).

Sympodial growth is typical for all species (Plate 18B), and only 1- or 2-year-old seedlings retain their terminal bud. The terminal bud dies, followed by the abortion of shoot tips; the clean-cut abscission scar pushed to one side becomes unnoticeable. New growth begins with the development of axillary buds from the preceding year.

A few patterns of shoot development are known for willows. Determinate or fixed growth is typical for most dwarf arctic and alpine species. Under favourable environmental conditions, however, production of neoformed leaves can also take place. Indeterminate or free growth occurs in most lowland species. In addition, sylleptic growth (syllepsis) occurs in a large number of *Salix* species, including all members of section *Longifoliae*. The evergreen *S. humboldtiana* produces leaves and catkins every month of the year (Parolin *et al.*, 2002).

Generally, the length of the growing season is longer for lowland willows than for poplars, athough climatic conditions and inherent differences among species affect the length of the growing season. Some species of southern origin, for example *S. babylonica* and its hybrid *S.* ×*sepulcralis*, retain foliage longer than any other

Fig. 2.13. The dwarf willow *Salix nakamurana* from subgenus *Chamaetia* is cultivated as an ornamental in alpine and rock gardens. Photo by J. Kuzovkina.

Fig. 2.14. The life form of willows ranges from dwarf ground covers to large trees, but most species are shrubs. *Salix petiolaris* (foreground) and *Salix bebbiana* (background) dominate this shrub carr. Photo by D. Dickmann.

Fig. 2.15. The bark of old tree-form willows – here *Salix alba* – is typically broken into corky ridges. Photo by D. Dickmann.

Fig. 2.16. Sandbar willow (*Salix interior*) is one of the few species in the genus that spreads by root suckering (or rhizoblasts), forming dense, shrubby thickets. Photo by D. Dickmann.

Fig. 2.17. Although branch layering is not common in willows, this clone of *Salix lucida* growing on beach sand is spreading slowly by this means. Photo by D. Dickmann.

native willow when cultivated in northern parts of the temperate region.

Most species of *Salix* have a single, cap-like bud scale formed by coalescence of two prophylls, a defining characteristic of the genus (Plate 18B). In subgenus *Protitea*, the margins of the bud scale are free and overlapping on the adaxial side – a characteristic that has been used to distinguish this subgenus. Two kinds of buds are found on willow stems: generative or flower buds and vegetative or leaf buds. A few bud types based on size, shape and position are known: in type 1 (*alba* type), vegetative and reproductive buds are uniform in size and shape and not distinguishable; in type 2 (*arctica* type), a few (usually 2–3 but up to 5 or 6) large apical buds, both vegetative and reproductive, open in spring, followed abruptly by smaller sets of buds that will stay dormant unless exposed to a special treatment; and in type 3 (*caprea* type), large generative buds occur mostly on the upper portion of the branch, with smaller vegetative buds occurring below the inflorescences. Type 3 is typical of many forest species that flower early in the spring (Skvortsov, 1968, 1999).

Generative buds contain embryonic flowers with several rudimentary leaves that form during the preceding year and continue development throughout the winter, resulting in gradual expansion of the inflorescence until it pushes off the bud scale. For most species, flowering takes place from early spring to early summer, but some species flower in autumn, e.g. *Salix variegata*. *Salix* flowers are predominantly insect pollinated (Plate 20A), but wind pollination takes place as well (Argus, 1974; Vroege and Stelleman, 1990; Peeters and Totland 1999; Tamura and Kudo, 2000; Karrenberg *et al.*, 2002).

The sequence of flowering and the development of shoots vary among species, and a few patterns of annual development are recognized. Precocious species, for example *S. caprea* and *S. discolor*, flower before leaves emerge; species with this pattern have typically sessile catkins. In subprecocious species, *S. purpurea* for example, generative budburst and anthesis occur just before the emergence of leaves; catkins of species with this pattern are sessile or borne on short flowering branchlets (these short flowering branches are identical to the proximal end of normal vegetative shoots, but they terminate in a catkin). In coetaneous species, flowering and leaf emergence take place simultaneously, for example *S. alba*, *S. lucida*, *S. nigra*, and the inflorescences are borne on distinct flowering branchlets (Plate 20B). Flowering can also occur throughout the season by syllepsis.

As in poplars, all of the taxa in the genus are dioecious. Single sex flowers are arranged into mostly upright aments or catkins (see Table 2.1; Plates 18B, 20A and B). The perianth in willows is replaced by one to a few nectariferous glands (homologous to the cup-shaped disk in poplars) that rarely connate into a lobed glandular disk (*S. pentandra*). The shape and the number of nectaries vary. A majority of species from subgenus *Vetrix* have one nectary in a flower, while many species from subgenus *Salix* have two nectaries in a male flower. Nectaries are mostly green or yellow, although in some representatives of section *Helix* they are purple or brown.

Each flower in the inflorescence is subtended by a floral bract. These bracts are persistent in staminate flowers, but are deciduous in the pistillate flowers of sections *Humboldtianae*, *Longifoliae*, *Salicaster* and *Salix*. The bracts may be pale or blackish, pubescent or sometimes glabrate. Staminate flowers of most species have two stamens (Plate 18B), with the exception of sections *Humboldtianae* and *Salicaster* with the occurrence of 3–12 stamens. In *Helix*, two stamens are fused into one and partial fusion of stamens takes place in sections *Daphnella*, *Subviminales* and *Vimen*. Pistillate flowers consist of a single pistil (ovary) that may be sessile (*S. purpurea*) or stipitate (born on a stipe or pedicel, as in *S. amygdaloides*), pubescent or glabrous. Styles are connate, partially distinct or distinct; stigmas are mostly two-lobed. Ovule number ranges from 2 to 42 per ovary (Argus, 2010). Male aments usually fall off soon after flowering, while female catkins drop after seed ripening and dispersal.

Willow seeds, containing chlorophyll but no endosperm, are very small (0.8–3 mm) and have limited longevity. Each seed is seated in a ring of fine hairs (Steyn *et al.*, 2004) that facilitate dispersal by wind (Plate 18B). Capsule dehiscence and seed shedding typically occur 3–8 weeks after pollination (Plate 20C), and seeds germinate within a few days on exposed surfaces if moisture is available.

Willow leaves are always simple and never lobed, ranging from narrowly elliptic, or linear, to oblong, ovate, obovate or round (Fig. 2.18). Leaf arrangement is typically alternate, but few species are known with subopposite or nearly opposite arrangement (*S. integra*, *S. purpurea*, *S. subopposita*). Although leaf size can vary considerably, an important quantitative characteristic describing the shape of leaves is the ratio of length to breadth, which varies from 0.7 to 30. The location of the broadest part of the leaf blade (above, about or below the middle of the blade) is also very useful and consistent for shape description. Leaf surfaces may be dull (*S. humilis*, *S. cinerea* L.) or lustrous (*S. lucida*, *S. pentandra*), and the underside of leaf blades may be covered with a whitish, waxy bloom (*S.* ×*fragilis*, *S. discolor*). Leaf pubescence (indumentum) can be used in species identification, although in some species young leaves frequently exhibit indumentum that is lost at maturity. Leaf margins are flat or revolute, due to the development of marginal collenchyma on the upper leaf surface (S. *humilis*, *S. caprea*), entire, serrate or dentate. Occasionally, small glands occur on each denticle, on the leaf margin (sections *Helix* and *Salix*) or close to it (sections *Vetrix* and *Vimen*). Typically, stomata are dense on the lower leaf surface (hypostomatous), although some species are amphistomatous, with stomata also on the upper surfaces (*S. alba*, *S. babylonica*, *S. interior*, *S. nigra*). Venation patterns are characteristic for species or even groups of species: veins may be submerged into the mesophyll and unobtrusive (sections *Hastatae* and *Helix*) or impressed above and raised beneath (*S. bebbiana* and most of the species of subgenus *Vetrix*).

Although not so obvious as in poplars, willow leaf shape may vary along the shoot; a number of reduced leaves occasionally occur on the lower part of a vegetative branchlet and also on flowering branchlets (*S. brachycarpa*, *S. pentandra*). These leaves are called 'proximal' (Argus, 2009) or 'lower' (Skvortsov, 1968, 1999), and they can differ from the later 'normal' leaves in size and shape, toothing, stipules and the presence of stomata in the adaxial epidermis. In addition, small leaf-like bracts, called cataphylls, occur next to the bud scale in some species and develop at the lower end of the branchlet (Sugaya, 1960). They often do not turn green and can abort early. In some species (*S. euxina*, *S. pentandra*) cataphylls exhibit parallel venation.

Vigorous leaves with prominent stipules can occur on coppice or epicormic shoots.

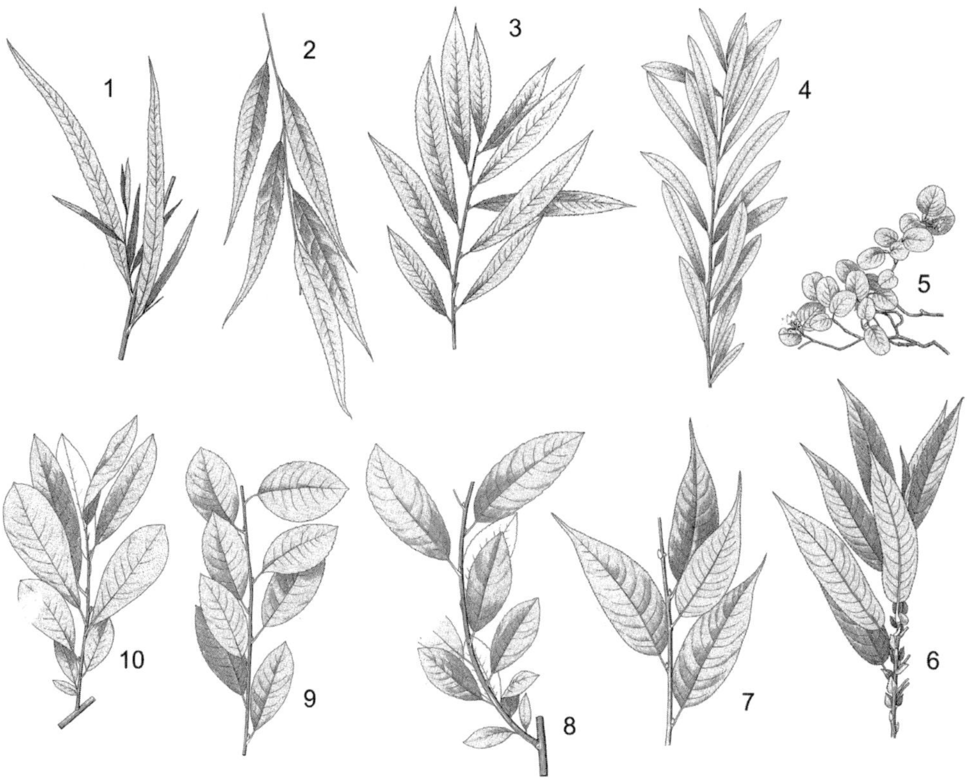

Fig. 2.18. Variation in leaf shape in *Salix*. 1, *S. interior*; 2, *S. babylonica*; 3, *S. alba*; 4, *S. purpurea*; 5, *S. herbacea*; 6, *S. eriocephala*; 7, *S. lucida*; 8, *S. discolor*; 9, *S. bebbiana*; 10, *S. humilis*. Redrawn from Holmgren (1998).

Stipules are prominent in many species (Plate 18B and Fig. 2.19), but in some species they are reduced to tiny rudimentary outgrowths (*S. herbacea*, *S. petiolaris*). The shape of stipules varies from narrowly linear-subulate to rounded, and is an important identification character for many species. In some species, stipules are shed earlier than leaves (caducous), and in others, such as species from the section *Daphnella*, stipules adnate to petioles and shed simultaneously.

Petioles in willows are shorter than in poplars and vary from 1 to 20 mm in length. In the subgenera *Chamaetia* and *Salix*, petioles are channelled; in the subgenus *Vetrix*, they are rounded but never flattened on both sides, as in many poplars. In species from sections *Humboldtianae*, *Salicaster* and *Salix*, petioles have a few pairs of glandular dots at the base of a leaf blade.

Differences in the chemical composition of leaves and stems among genotypes within a species and among species result in variations in palatability and susceptibility to damage by insects, livestock, beaver and browsing wild animals. Some species have relatively high nutritive value and are used as supplementary forage for livestock (*S. alba*, *S. bebbiana*, *S. commutata*, *S. exigua*, *S. geyeriana* and *S. scouleriana*).

A distinctive feature of many willows that has been exploited in cultivation is their ability to resprout from stumps or stools after repeated harvesting (Fig. 2.20). Wide variation in coppicing ability due to genotypic differences in structure and physiology has been observed. Species with high coppicing ability typically belong to the subgenus *Vetrix*, e.g. *S. eriocephala* and *S. viminalis*. Certain bud differentiation patterns and frequent sylleptic sprouting may result in lower coppice response of some tree-formed willows, e.g. *S. amygdaloides* (Sennerby-Forsse and Zsuffa, 1995).

Fig. 2.19. Stipules of some willows – here *Salix eriocephala* – are persistent and prominent, making them a good character for species identification. Photo by D. Dickmann.

Fig. 2.20. A distinctive feature of many willows and poplars that has been exploited in cultivation is their ability to resprout from stumps (stools) after repeated harvesting. Photo by J. Kuzovkina.

Molecular studies indicate that hybridization is an essential source of morphological variability in *Salix*, and extensive gene exchange between species takes place (Argus, 2009). Hybridization and introgression have played major roles in the evolution of some sections (*Longifoliae*), as has polyploidy. About 40% of New World and European species are polyploid (Suda and Argus, 1968, 1969; Argus, 1999; Dobeŝ and Vitek, 2000). A large number of hybrids have been produced by willow breeders through artificial cross-pollination, and many of these hybrids are used in horticultural and silvicultural plantings (Fig. 2.21). Families of species hybrids yield both male and female individuals (Mosseler and Zsuffa, 1989). Many hybrids are fertile and can produce advance generation crosses.

Species of *Salix* differ in their ecological requirements, although the majority of them are adapted to mesic–hydric habitats (Skvortsov, 1968, 1999; Argus, 1986). Riparian or alluvial species growing along river valleys, stream banks and lakeshores require well-aerated substrate and flowing moisture (Plate 20D). Non-riparian species are less exacting in soil aeration requirements and include wetland species that are well adapted to saturated soils. In humid climates, willows also form temporary early successional communities colonizing newly opened upland habitats (roadside ditches, abandoned agricultural fields, railroads, old mine tailings, gravel pits or recently burned, logged or glaciated areas). Moisture availability is an important factor controlling native distribution and abundance of species (*Salix* is possibly derived from the classical word *sallies*: *sal* meaning near and *lis* meaning water; Newsholme, 1992). Plentiful moisture is important during seed germination and early seedling establishment. However, after that stage, constant soil moisture is not as important to the survival of many willow species (Skvortsov, 1968, 1999; Argus, 1986). Some willow species even show xeric traits (*S. humilis*, *S. myricoides*, *S. repens*) and are adapted to drought and heat stress (Plate 20E).

Willows are among the main colonists of water margins, where open and wet habitats create a favourable environment for seed germination. None the less, most willow seedlings do not survive because of water fluctuations, erosion and ice scouring. When established, some species can survive extreme water fluctuations and prolonged periods of partial inundation. Their ability to withstand root exposure and heavy sediment deposition, or burial, as a result of water and wind erosion relies on adventitious rooting of buried stems and the flushing of dormant buds. These adaptations allow willows to survive in the constantly changing environments of flood plains, water margins and sandbars. Long-term survival of tree willows in these environments can produce impressive stands of timber (Fig. 2.22).

Willows have long been useful in human economies. The wood typically is straight grained, sometimes interlocked, with a fine, uniform texture. The heartwood varies from pale reddish brown to greyish brown, with white sapwood. It is light (specific gravity 0.3–0.42), soft and weak, with low shock and decay resistance and poor steam-bending properties. It works and carves easily with hand and machine tools, but sharp edges are required to avoid woolliness. Limited checking occurs during drying. Willow wood glues, screws, nails and finishes satisfactorily. It is used for millwork, packing cases and boxes, artificial limbs, caskets, polo balls, cricket bats, Venetian blinds, poles, turnery, slack cooperage, veneer and inexpensive furniture. Willow wands (osiers) have long been grown for basket making, wicker furniture and wattle construction (see Fig. 2.2). Recently, fast-growing willow plantations (Fig. 2.21) also have become an important source of fibre and bioenergy. The

Fig. 2.21. Dense plantations of willows – many of them hybrids – are grown commercially on short rotations for fibre and bioenergy, with Sweden leading the way. Photo by D. Dickmann.

Fig. 2.22. Old stands of tree willows growing in a flood plain habitat – here *Salix nigra* along the Mississippi River, USA – can obtain impressive timber quality. Photo by B. Lockhart, USDA Forest Service, Bugwood.org.

bark and leaves of willows have been known for millennia for their medicinal properties. In 1828, salicylic acid was isolated from willow bark and it became the precursor of the well-known pain-relieving, fever-reducing and anti-inflammatory drug, aspirin.

Willows are much more important in horticultural applications than poplars. Most people know about 'pussy willows', which are widely collected in the spring for decorative arrangements, but other willow selections are used in the same way. Many cultivars of tree and shrub willows – both pure species selections and hybrids – are also widely planted for ornamentals (Plate 20F), windbreaks, screens and stream bank stabilization. Unfortunately, some willows planted outside their natural range have become naturalized (Fig. 2.23) or, in the worst case, invasive weeds.

2.3.2 Descriptions of *Salix* species

Salix is one of the most difficult and confusing plant genera to understand. Griggs (1905) commented: 'We cannot study willows profitably by tables of dimensions and geometrical descriptions of the leaves any more than we can study the faces of our friends by measuring the length of their noses. As there is a certain almost indefinable individuality in a human face, there is in a species of plants an individuality no less indefinable but no less important.'

Thus, characters considered reliable and easily recognizable in other plant genera are not consistent for willows, and a different approach involving a combination of characters for species diagnoses is required. Skvortsov (1968, 1999) described the phenomenon: 'While infraspecific variability in willows is obvious and striking, differences between species are difficult to understand and articulate.' In addition to remarkable individual variability and polymorphism masking the differences between species and species limits, a number of other factors add to the problems of *Salix* identification. Dioeciousness and asynchronous phenology of flowers and leaves make observation of important traits at one time and on one plant impossible. Vegetative plants often do not exhibit any indication of gender, yet diagnostic keys are rarely written to account for that fact. Phenotypic variability due to developmental stage (the degree of hairiness, presence

Fig. 2.23. The crack willow (*Salix ×fragilis*) and its hybrids have escaped cultivation and have become naturalized in many parts of the world, as have several other willow species. Photo by D. Dickmann.

of stipules and floral bracts, the length of catkins, stipes and ovaries can change with age) and site conditions (moisture, nutrients, shade and exposure) combine to complicate identification. Occurrence of natural interspecific hybrids adds to the uncertainty of the taxonomy of willows. The result is that precise identification is not always possible.

Argus (2009) developed an interactive key to New World and some Eurasian *Salix* which could be used in species identification. The *Flora of China* also has been converted into ActKey for interactive identification (http://flora.huh.harvard.edu/china/delta/www/ident.htm).

A few known classification systems and taxonomic treatments exist that arrange species into subgenera and sections. Reticulate evolutionary pattern, polyploidy, hybridization and introgression often cause difficulties in placing some species into sections as well as in creation of natural groups. The discovery of new *Salix* species and new revisions of the previously described taxa are in progress, proving that *Salix* taxonomy is still far from being a closed book (see Table 2.4). Because of the wide distribution and taxonomic complexity of the genus, a worldwide classification system has yet to be proposed that satisfies all interested parties.

A shortage of taxonomic expertise results in nomenclatural instability of the genus in trade and creates an obstacle for meaningful utilization of species. None the less, several authoritative references and treatments on willows exist, with the most complete, updated and revised classifications included in the following monographs:

- *Willows of Russia and Adjacent Countries* by Alexei K. Skvortsov (1999) – originally published in Russian in 1968 as *Willows of the USSR* – reviews the classification of the genus in the former Soviet Union, and also includes all of Europe, northern Africa, Asia Minor, west and north-east China, Mongolia and North Korea. Skvortsov recognizes three subgenera and 26 sections for that region. Subgenus *Salix* is represented by 6 sections, subgenus *Chamaetia* by 5 sections and subgenus *Vetrix* by 15 sections. This monograph also includes a complete synonymy of 135 species, providing considerable help to resolve nomenclatural and taxonomic problems.

- The most complete classification of *Salix* in North America is 'Infrageneric classification of *Salix* (Salicaceae) in the New World' (1997), 'Classification of *Salix* in the New World' (1999) and the chapter *Salix* in *Flora of North America North of Mexico, Vol 7* (2010) published by George W. Argus. He adapted some of Skvortsov's subsections but moved others into different sections and described new sections to include North American species.
- The treatment of Chinese *Salix* by Wang and Fang (1984) can be consulted for many Asian species. The treatment accepted 37 sections in China, but did not use higher taxa.
- The chapter on the *Salicaceae* by Cheng-fu Fang, Shi-dong Zhao and Alexei K. Skvortsov in *Flora of China, Vol 4* (1999) gives the splitter's view of *Salix* taxonomy. The authors describe 275 species, 189 of which are endemic, but admit at the outset of their discussion that some combining may be warranted.
- 'Saliceae of Japan' by Hiroyoshi Ohashi (2001) provides a systematic treatment of the family for the region. The treatment recognized 30 species and 40 putative natural hybrids in Japan and divides *Salix* into six subgenera (including *Chosenia*, which was newly recognized at the rank of subgenus).

Today's most prominent salicologists were consulted for this work. Listed taxonomic nomenclature above the species level is provided according to Argus (1997, 1999, 2010), Fang *et al.* (1999) and Skvortsov (1968, 1999). Table 2.3 lists species described in these regional treatments for a global overview, although we recognize that interpretation of these listings may be difficult. Compilation of the 'World Checklist of *Salicaceae sensu stricto*' for the 'World Checklist of Dicotyledons' is under way to provide an updated world classification and phylogeny of *Salix* and *Populus* (http://www.kew.org/science-research-data/directory/projects/WClassPhylogSalicaSS.htm) and it is expected to be completed within a few years (I. Belyaeva, personal communications, February 2012). As with poplars, a comprehensive worldwide classification of *Salix* is a work in progress, and may always be.

The earlier classifications recognized four subgenera – *Salix*, *Longifoliae*, *Chamaetia* and *Vetrix* (Skvortsov, 1968, 1999; Argus, 1997, 1999). Ohashi (2000) proposed including two sections which were previously included in the subgenus *Salix* – *Humboldtianae* and *Floridanae* – in the subgenus *Protitea*. Some recent molecular data support this proposition, and the recognition of five subgenera is accepted in the recently published *Flora of North America North of Mexico* (Argus, 2010). Below, we also follow this classification, which includes five subgenera.

Due to the large number of *Salix* taxa worldwide (see Table 2.3), the scope of our descriptions is necessarily limited to 33 species important in plantation research, hybridization and commercial culture (Table 2.7). The subgenus *Chamaetia* is omitted completely here due to its unsuitability for forestry cultivation. This section includes predominantly dwarf shrubs from alpine and arctic habitats; however, a few species are cultivated as ornamentals. Although a few of the species of *Salix* we describe are commercially deployed in the ornamental trade and in production plantations, many species have been utilized on a smaller scale, including traditional uses (basketry, bee-keeping, indoor decorations, etc.), environmental restoration, silvicultural testing and controlled hybridization. Most of the described species are well established taxonomically, and their placement in the genus is settled according to the authorities cited above. None the less, a few nomenclatural uncertainties remain, e.g. the synonymy of *S. babylonica* versus *S. matsudana*, *S. udensis* versus *S. sachalinensis* and *S. cardiophylla* versus *S. maximowiczii*. These uncertainties are a matter of insufficient data and the need to rely on a particular authority. Although a problematic approach, an author's choice of binomial in his or her publications currently may be the deciding factor until additional molecular data become available and consensus among authorities in the taxonomic community is reached.

Subgenus *Protitea* Kimura

The subgenus *Protitea* includes trees or shrubs up to 30 m in height. Representatives of these sections are characterized by bud scales with distinct overlapping adaxial margins, flowers with

Table 2.7. Shrub and tree taxa in the genus *Salix* important in plantation research, hybridization and commercial culture.[a]

Section	Taxon	English common name[b]	Notes and synonyms
Subgenus *Salix*			
Humboldtianae	*S. acmophylla* Boissier		Asia Minor and West Asia
	S. amygdaloides Andersson	Western black or peach-leaved willow	North America, often confused with *S. triandra*
	S. humboldtiana Willdenow	Humboldt willow	Central and South America, sole member of *Saliceae* native to South America
	S. nigra Marshall	Black willow	North America
	S. tetrasperma Roxburgh	Indian or four-seeded willow	Central and East Asia
Amygdalinae	*S. triandra* Linnaeus	Almond-leaved willow	Eurasia, syn. *S. amygdalina*
Urbanianae	*S. cardiophylla* Trautvetter		East Asia, monotypic section that reproduces only by seed, syn. *S. maximowiczii*
Subalbae	*S. babylonica* Linnaeus	Weeping willow	Eurasia, may be synonymous with *S. matsudana*
Salix	*S. alba* Linnaeus	White willow	Europe, Africa and West Asia
	S. euxina I.V. Belyaeva	Euxine willow	Near the Black Sea
	S. ×*fragilis* Linnaeus	Crack or brittle willow	Naturalized in many areas worldwide
Salicaster	*S. lucida* Muhlenberg	Shining or American bay willow	North America
	S. pentandra Linnaeus	Bay or laurel-leaved willow	Eurasia
Subgenus *Longifoliae*			
Longifoliae	*S. interior* Rowlee	Sandbar or narrow-leaved willow	North America, *S. exigua* is a closely related species from western North America
Subgenus *Vetrix*			
Cordatae	*S. eriocephala* Michaux	Heart-leaved willow	North America
Cinerella	*S. aegyptiaca* Linnaeus	Armenian or musk willow	Asia Minor, syn. *S. medemii*
	S. caprea Linnaeus	Goat or pussy willow	Eurasia
	S. cinerea Linnaeus	Grey willow or common sallow	Eurasia
	S. discolor Muhlenberg	American pussy willow	North America
Fulvae	*S. bebbiana* Sargent	Bebb, long-beaked or ochre- flowered willow	Eurasia and North America
Viminella	*S. gmelinii* Pallas	Gmelin's willow	Eurasia, syn. *S. dasyclados*, *S. burjarica*
	S. schwerinii Wolf	Schwerin willow	East Asia, syn. *S. kinuyanagi*

Continued

Table 2.7. Continued.

Section	Taxon	English common name[b]	Notes and synonyms
	S. udensis Trautvetter		East Asia, syn. *S. sachalinensis* or *S. siuzevii*
	S. viminalis Linnaeus	Common osier or basket willow	Eurasia
Subviminales	*S. gracilistyla* Miquel	Japanese pussy or rosegold willow	East Asia
Geyerianae	*S. geyeriana* Andersson	Geyer willow	Western North America
	S. petiolaris Smith	Meadow or slender willow	North America
Helix	*S. koriyanagi* Kimura		Korea, syn. *S. purpurea* var. *japonica*
	S. miyabeana Seemen	Miyabe willow	East Asia
	S. purpurea Linnaeus	Purple willow or purple osier	Northern Africa and Europe
Daphnella	*S. acutifolia* Willdenow	Pointed-leaf willow	Eastern Europe
	S. daphnoides Villars	Violet willow	Europe
	S. rorida Lakschewitz		Central and East Asia

[a]Each of these taxa is described briefly in the text. [b]Common names vary considerably depending on language and locality.

more than two stamens and generally a tropical distribution (Argus, 2010).

Section *Humboldtianae* Andersson

This section comprises 9–14 medium-sized tree species (see Table 2.3) with distinctive free overlapping (not connate) bud scale margins. Both generative and vegetative buds are sharply pointed and small. Male flowers have more than two stamens. Old bark of mature specimens has coarse longitudinal fissures. Members of this section are mostly spread throughout the tropics and subtropics.

***Salix acmophylla* Boissier.** The distribution of this willow is restricted to arid tropical and subtropical continental regions in Asia Minor and Central Asia (Plate 9A), where it grows along riverbanks and irrigation ditches in the piedmont or mountains, reaching 2000–2200 m elevation in Afghanistan, Pakistan and northern India. *S. acmophylla* is a medium-sized tree 8–10 m tall, with a rounded crown and rough bark with irregular vertical cracks. Its pendulous branches are often slender, red and glabrous, bearing acute small buds. The glabrous linear-lanceolate leaves, up to 15 cm long, are very long tapering, with entire or slightly serrate margins. There are four to six stamens in male flowers. It is a commonly cultivated willow in the region of its distribution, although it is not cold hardy in temperate climates. In the western Himalayas, where it is planted in dry inner valleys along irrigation channels, *S. acmophylla* is used for basketry, fuel and building materials. In Pakistan, it is used by carpet weavers for dyeing, as fodder for camels and cattle, a decoction of bark is used as a febrifuge, but the wood is of little use (Khan, 1967). It is known to hybridize with *S. alba* (Skvortsov, 1968, 1999).

***Salix amygdaloides* Andersson, western black or peach-leaved willow.** A North American species occupying a more northern range than *S. nigra*, *S. amygdaloides* is distributed transcontinentally from New York and Quebec to British Columbia and south to Texas, Arizona and Kentucky (Plate 9B). It typically occurs in wetlands and along rivers, streams and lakes; it grows on sandy, silty or gravelly substrates. *S. amygdaloides* is tolerant of wet and occasionally flooded soils and holds banks from washing away.

A tree up to 20 m tall, the common name – western black willow – suggests its affinity with *S. nigra*, but its habit and morphology differ considerably from that species. It is generally single stemmed, with a narrow, rounded crown and straight trunk, with clean branches and smooth, brown bark. The winter buds of *S. amygdaloides* are nearly twice as large as those of *S. nigra*, with a much lighter base where they are protected by the petiole. Mature leaves on slender and terete petioles are thickly glaucous underneath. There are three to seven stamens in male flowers.

S. amygdaloides is confused occasionally with *S. triandra* (syn. *S. amygdalina* L.), but it is considerably different in bud, flower and bark structure. Hybrids with *S. nigra* and *S. caroliniana* form in areas where the species are sympatric (Argus, 1986). This species has been deployed in biomass production trials in the USA (Kopp *et al.*, 2001), although it has a lower re-sprouting ability than the shrubby willows (Sennerby-Forsse and Zsuffa, 1995).

Salix humboldtiana **Willdenow, Humboldt willow.** This large tree willow is unique in the *Saliceae*, being the only species in the family indigenous to South America, where it grows along watercourses in Chile, Uruguay, Argentina (Plate 20D) and southern Brazil in warm temperate and subtropical regions (Plate 10A). It also occurs in Central America, one of only three willow taxa found there (Dorn, 1976). It is limited in its range of distribution due to susceptibility to frost. Humboldt willow is distinguished by a fastigiate crown, thick scaly bark, long, narrow glabrous leaves and male flowers containing four to nine stamens. Although it is of little economic importance, it is occasionally used in land reclamation projects in Argentina and in hybridization work in Chile, Argentina and India. *Salix* ×*argentinensis* Ragonese is a group of hybrids between *S. babylonica* and *S. humboldtiana*, and a number of its clones have been selected and cultivated extensively in Argentina. The most widely cultivated clones are 'Hibrido', 'Mestizo' and 'Mestizo Amos' (FAO, 1980).

Salix nigra **Marshall, black willow.** This species from eastern North America (*S. gooddingii* Ball is its western counterpart) occupies a wide area from southern Canada to northern Florida, west to Minnesota and Texas (Plate 10B). As a pioneer species, black willow often forms extensive stands along river margins and flood plains on wet, fine-textured alluvial soils. It may occur along the edges of ponds, swamps and roadside ditches and is well adapted to stagnant water.

S. nigra commonly can grow to over 20 m tall (maximum 43 m), with an irregular crown and a trunk diameter of 1 m or more. This normally deliquescent willow can produce an upright stem in closed stands (see Fig. 2.22) – especially in the Mississippi River Delta – but commonly grows in clumps with crooked and leaning trunks (Pitcher and McKnight, 1990). Within the first 5 years of growth, *S. nigra* develops thick, rough, furrowed bark with wide ridges on the lower trunk, rendering it low in palatability to browsing animals. Its reddish brown branches are very brittle at the base. In autumn, most of the season's twigs drop off, hindering stem elongation and making trees thick and stubby. The long, narrow foliage of black willow is uniformly green on both sides.

The wood of black willow is very light (specific gravity 0.34–0.41), soft and straight grained. During the 18th and early 19th centuries, it was used to make fine charcoal for black gunpowder, and it was used extensively for artificial limbs (Pitcher and McKnight, 1990). Black willow is relatively resistant to *Cytospora* canker and has received some attention from tree geneticists. It also has been deployed in biomass production trials in the USA (Kopp *et al.*, 2001).

Salix tetrasperma **Roxburgh, Indian or four-seed willow.** This tree is a paleotropical species inhabiting low elevations in India and eastern Asia (Plate 11A). It is closely related to *S. bonplandiana* Kunth, a neotropical species distributed in Mexico and Guatemala (Skvortsov, 1999). It is found in subtropical areas of India along the banks of rivers and streams, in wet swampy places, ascending up to 2100 m elevation. The species is polymorphous and a few varieties are distinguished. North-western plants are distinguished from southern and eastern representatives by their longer and narrower leaves.

Salix tetrasperma can be up to 10 m tall, with silky pubescent young shoots. Linear-lanceolate leaves have an acuminate apex and serrated margin. Its catkins, both male and female, are very long, up to 10 cm. There are typically eight stamens in male flowers. Glands in male flowers are usually connected, forming a lobed disk. Indian willow occasionally flowers in November or December before the expansion of new leaves, exhibiting precocious development, which is unusual for the representatives of this subgenus.

Indian willow is frost hardy. It coppices and pollards well. The heartwood is not long-lasting and large trees are usually hollow. *S. tetrasperma* is cultivated sporadically in the foothills and inner valleys in northern India and is used as fodder for sheep and goats (Singh, 1982). To a lesser extent, it is also used for timber, fuel and baskets. Its red, even-grained, soft wood is used for posts, planking and matches. Bark is used for tanning and as a febrifuge (Khan, 1967). This species is currently hybridized with *S. alba*, *S. babylonica* and *S. humboldtiana* to create clones suitable for cultivation from the tropical to the temperate regions of India.

Subgenus *Salix*

The subgenera *Salix* along with *Protitea* are primitive subgenera in the genus, being sharply isolated morphologically from the more advanced willows and sharing many common characteristics with *Populus* (Skvortsov, 1968, 1999; Dorn, 1976) (Table 2.8). It comprises predominantly trees or robust shrubs, with typically narrow, serrated leaves, rather loose catkins that appear with the leaves on leafy stalks and pale bracts without pigmentation, which in female flowers abscise by the time capsules ripen. The vegetative and generative buds are of similar appearance. All members of this subgenus have glands on the channelled petiole near the leaf base that are weakly developed in section *Subalbae*. There are two nectaries in the male flower and usually one in the female flower.

The subgenus *Salix* is a heterogeneous group divided into sections containing flowers with several stamens and sections with only two stamens. This subgenus is well represented on all continents (see Tables 2.3 and 2.7) and is as commonly cultivated in southern regions as poplar. Arborescent willows can attain a large size – up to 20–30 m in height – with growth rates comparable to poplars. Frequently seen along watercourses and growing in fertile soils of river valleys, representatives of this subgenus are vital to the economy of many southern and arid regions, where they are used for timber, fuel, basket making, prevention of soil erosion, fodder, shelter for domestic animals and ornamentals.

Section *Amygdalinae* Koch

This section appears to contain only Old World species (Table 2.3). The bark of old branches in this section characteristically exfoliates in patches of irregular shape, resembling *Chosenia*.

***Salix triandra* Linnaeus, almond-leaved willow (syn. *S. amygdalina* L.).** A willow with a broad distribution in Eurasia (Plate 11B), this species grows on river and stream banks and occasionally on secondary habitats such as ditches and gullies. Its vertical distribution ranges from lowland to moderate mountain elevations; in the Pyrenees, it ascends to 1700 m. In China, Japan and on the Korean Peninsula, it only grows on the foothills. Almond-leaved willow is a tall shrub or occasionally small tree up to 3 m tall, with short stems and wide crowns. Narrowly cylindrical and somewhat pendulous serotinous catkins are borne on long, leafy stalks. Male flowers contain three stamens.

Almond-leaved willow is an extremely variable species. Two distinct forms are known: form *discolor*, with conspicuous glaucous bloom on the underside of leaves, and form *concolor*, with a green underside. Populations from the Caucasus and Iran that have been described as *S. medwedewii* Dode – characterized by their delicate and slender shoots and catkins, with smaller buds and leaves – should probably be treated as a subspecies of *S. triandra* (Skvortsov, 1968, 1999).

S. triandra, along with *S. viminalis* and *S. purpurea*, is one of the most important species used in basketry. It produces high-quality, hard and pliable rods containing little pith. Almond-leaved willow is also used as a parent in current breeding programmes in Europe. It is planted in China to protect embankments and as a nectariferous plant (Fang *et al.*, 1999).

Section *Urbanianae* Schneider

This monotypic section (see Table 2.3) comprises the only representative in the genus that reproduces exclusively by seeds (Skvortsov, 1968, 1999). Probably because of this surprising feature, it has only limited economic importance.

***Salix cardiophylla* Trautvetter (syn. *S. maximowiczii* Komarov).** A massive tree up to 30 m tall, with a trunk up to 1 m in diameter,

Table 2.8. Some evolutionarily primitive and advanced characteristics in *Salix* (from Skvortsov, 1968, 1999).

Primitive characters	Advanced characters
Alluvial	Non-alluvial habitats
Habit: erect trees	Habit: shrubs or dwarf shrubs
Bud scale margins distinct	Bud scale margins connate, scale cap-like
Lower cataphylls broad, their veins parallel	Lower cataphylls narrow, their veins pinnate, as in regular leaves
Bud size gradation of type 1 (*alba*)	Bud size gradation of type 2 (*arctica*) or (*caprea*)
Petioles channelled above, glandular at leaf base	Petioles convex above, eglandular
Young leaves produce odorous pitch	Leaves not pitchy
Leaves acuminate	Leaves obtuse or short-pointed
Veins prominent neither beneath nor above, leaves flat	Veins impressed above, prominent beneath; leaf margins revolute
Leaf denticles small and uniform	Leaf denticles coarse and irregular or lacking
Glands marginal	Glands submarginal or extramarginal
Distinct hypodermal layer in mesophyll	Hypodermis not distinct
Catkins narrowly cylindrical, long, sparsely flowered, more or less drooping	Catkins more stout and short, erect, compactly flowered
Bract connate at base to ovary stipe, abaxial nectary, and stamens	Bracts quite distinct
Bracts colourless, abscising in female catkins after flowering	Bracts coloured (brown or black), persistent
Bracts puberulent on the inside, particularly at base	Bracts clothed with long trichomes, mostly at apex
Nectaries two, or three, or glandular disk replacing individual nectaries	Solitary adaxial nectary
Stamens multiple (three or more), their number fluctuating	Stamens three or two, their number constant; further evolution leading toward coalescence of stamen filaments
Stamen filaments comparatively short, pubescent	Stamen filaments comparatively long, glabrous
Anthers small, not pigmented	Anthers large, pigmented
Ovaries stipitate	Ovaries sessile
Styles partially or entirely distinct, separated	Styles entirely connate

this willow is distributed throughout the Russian Far East, North Korea, north-east China and Japan (Plate 9A). Its leaves are elliptic to oblong-ovate and branchlets are bright crimson with a pruinose bloom (Newsholme, 1992). Its bark becomes deeply fissured with age. It is used for timber and matchwood, as a nectariferous plant and as an ornamental. Skvortsov (1968, 1999) and Ohashi (2001) believe that *S. maximowiczii* is synonymous with *S. cardiophylla*, although *S. maximowiczii* is accepted in *Flora of China* (Fang *et al.*, 1999).

Section *Subalbae* Koidzumi

This East Asian section includes trees of moderate to large size with generative buds larger than vegetative, two stamens in the male flower, petioles without glands, and persistent bracts. It includes only two species (see Table 2.3).

***Salix babylonica* Linnaeus, weeping willow (syn. *S. matsudana* Koidzumi).** One of the best known trees in the world, weeping willow's broad spreading crown, lanceolate leaves (Fig. 2.18) and long pendulous branches are distinct. It is a medium to large tree that commonly reaches 15–18 m in height, with a trunk to 80 cm in diameter. When grown in favourable conditions of moist, sandy depressions, however, it attains large size, reaching up to 20 m in height and more than 2 m in trunk diameter. Its original range comprised river valleys in arid and semi-arid regions of central and north China, but the exact limits of this range are now obscured. Although it is relatively sensitive to winter cold, it is still one of the most widely cultivated trees worldwide. In Eurasia, the northern limit of its successful cultivation concurs with the northern margin of commercial peach production and includes southern England, Belgium, southern Germany, the Czech Republic, Hungary, southern Romania, the Crimea, Caucasus, Uzbekistan, Piedmont Kirghizia, north-east China and the Korean Peninsula (Skvortsov, 1968, 1999). In North America, the sympatry with commercial peach production would also hold.

A few clones are present in cultivation. A female clone with weeping branches was probably introduced to Europe from the Near East in the 17th century and is still the only one that appears to grow everywhere in Europe and North America. Santamour and McArdle (1988) proposed 'Babylon' as a cultivar name for the clone that represented a highly atypical selection from the species. It had been introduced at various points along the ancient trade route through southern Asia to the Near East, and to Europe circa 1730. There are a few additional clones cultivated in Middle Asia, including some that are male.

Non-weeping forms are common in China and regarded as *S. matsudana* (Hankow willow). Some botanists, including Skvortsov (1968, 1999) and Santamour and McArdle (1988), believe that *S. matsudana* is synonymous with *S. babylonica*, but Fang *et al.* (1999) consider them to be different and distinct species. I.V. Belyaeva treats *S. babylonica* as the correct name and *S. matsudana* as its synonym in *World Checklist for Salicaceae sensu stricto* (personal communication).

In addition to *S. babylonica*, there are other weeping willows that are less sensitive to cold and more suitable for cultivation in north temperate zones. Among them, *S.* ×*sepulcralis* nothovar. *chrysocoma* Meikle, a hybrid of *S. alba* var. *vitellina* and *S. babylonica*, is very popular in cultivation for its bright yellow branches and weeping habit. *S.* ×*sepulcralis* Simonkai, a hybrid of *S. alba* var. *alba* × *S. babylonica*, is also an attractive tree, though with a less pendulous habit than *S.* ×*sepulcralis* nothovar. *chrysocoma* and with olive-brown twigs (Meikle, 1984). In addition, *S.* ×*pendulina* Wenderoth, a hybrid of *S. babylonica* with *S. euxina*, as well as its varieties *S.* ×*pendulina* var. *elegantissima* and *S.* ×*pendulina* var. *blanda* that probably originated in Germany, are common in cultivation. *S.* ×*pendulina* var. *elegantissima* is a tree with the extreme weeping habit of *S. babylonica*, while *S.* ×*pendulina* var. *blanda* (Wisconsin weeping willow or Thurlow weeping willow) is a tree with less pendulous branches.

S. babylonica has three often cultivated, non-weeping ornamental forms that are products of ancient Chinese selection: 'Tortuosa', an upright female clone with irregularly twisted branches; 'Crispa' (or *S. babylonica* var. *annularis*), a clone with leaves curved into rings; and 'Umbraculifera', a cultivar with a fan-shaped, subglobose, dense crown that is called bread willow in China because of its resemblance to a loaf of bread. Planted on every continent, *S. babylonica* is used for timber, weaving wicker baskets, afforestation and as a nectariferous plant, in addition to its ornamental uses.

Section *Salix*

This section includes only three large or moderate-sized tree species (see Table 2.3) with coarse, fissured bark and two stamens in male flowers.

***Salix alba* Linnaeus, white willow (syn. *S. excelsa* Gmelin).** One of the best known and largest willows, attaining heights of 30 m and trunks up to 1 m or more in diameter, white willow is characterized by graceful ascending or frequently weeping crowns. It is widely distributed throughout Europe, Asia Minor and western Siberia, and is also present in northern Africa (Plate 11C), exhibiting a uniform set of characteristics across its entire range. Its branchlets, buds and narrow leaves (Plate 20F) are generally densely pubescent with silky hairs, giving it a silvery appearance that makes it easily visible from a distance; hence its common name. *S. alba* is a keystone species of riverbank communities, where it occasionally forms groves extending for many kilometres.

White willow is widely cultivated and has been introduced far beyond its natural range. In many areas, it is one of the most common plants cultivated near human habitations in parks, estates, residential lots and on riverbanks. It does not, however, seem to escape and naturalize as readily as *S.* ×*fragilis*, the hybrid crack willow (*S. alba* × *S. euxina*).

S. alba has been in cultivation for many centuries and, as a result, numerous cultivars have been selected and bred for ornamental characteristics. These variants of white willow were often treated as subspecies, varieties or cultivars. The most popular selections include *S. alba* 'Vitellina' (*S. alba* var. *vitellina* (Linnaeus) Stokes), with bright yellow stem colour that occurs as a group of mainly female clones and *S. alba* 'Chermesina' (*S. alba* var. *chermesina* Hartig) (Plate 20F), with bright orange-red colour on young stems. These ornamental selections probably originated in central and southern Europe because of the southern patterns of their phenology; in temperate climates, they continue growing throughout the entire season, producing three or four generations of sylleptic shoots (Skvortsov, 1968, 1999). *S. alba* 'Sericea' (*S. alba* var. *sericea* Gaudin) is a very ornamental cultivar producing silver-appearing leaves due to intense pubescence on both sides of mature foliage (Plate 20F).

S. alba is an important commercial timber species. The wood is used for archery bows, hoops, canoes, wooden kitchen utensils, sandals, charcoal and log cabins (Nasarow, 1936, 1970). *S. alba* 'Caerulea' (*S. alba* var. *caerulea* (Smith) Smith) – the cricket-bat willow – is an extensively cultivated, vigorous variety, with dark brown branches and coarsely toothed leaves which are sparsely pubescent underneath. It provides high-quality wood for this specialized use. According to Warren-Wren (1972), only female trees of this cultivar produce timber of the required standard for cricket bats. White willow is also used for wicker baskets and as a nectariferous plant. Young branches and leaves are readily eaten by goats. *S. alba* can succeed in drier soils and its hybrids with *S. babylonica* are suggested as potential browse for ewes in dry summer conditions (Pitta *et al.*, 2005). The bark of white willow is also used for medicinal purposes; for tannin to dye silk, wool and leather; for production of a buff varnish for painting; and for rope making.

White willow frequently hybridizes with *S. euxina*, and the taxon with that parentage is now treated as *S. ×fragilis* (Belyaeva, 2009) (previously *S. ×rubens* Schrank). Hybrids with *S. euxina* are often observed in secondary and disturbed habitats and in areas with no remaining natural habitat. *S. alba* also hybridizes with *S. acmophylla*, *S. pentandra* and *S. babylonica*, and its numerous hybrids are frequently used in breeding programmes.

***Salix euxina* I.V. Belyaeva, Euxine willow (formerly *S. fragilis*).** The name of crack willow was often attached to a complex group of willows and for some time the binomial *S. fragilis* had been applied to a hybrid group and to a species from Asia Minor. A recent publication by Belyaeva (2009) clarifies what should be understood by the name *S. fragilis* and how this name should be applied more appropriately to the widely distributed hybrid crack willow. This hybrid should be named *S. ×fragilis*, while the pure species is now described as *S. euxina*.

Skvortsov (1968, 1973, 1999) proposed that the original distribution of this species was rather narrow, only near the Black Sea, including northern Asia Minor and the Armenian High Plateau, where it grew on the banks of mountain streams. The name – S. *euxina* – derived from Euxeinos Pontos, the ancient Greek name for the Black Sea, reflects the distribution of the species (Belyaeva, 2009).

A modest-sized tree attaining 15–20 m in height and 1 m in diameter, Euxine willow has a wide crown and deeply fissured bark on old trunks. Its leaves are glabrous and its smooth, olive-green branchlets are very brittle at the base; the tree propagates itself easily by broken branches (fragmentation). This species, which has naturalized in Europe, hybridizes easily with *S. alba*. However, it occurs much more rarely than its hybrid *S. ×fragilis* (approximately only one-tenth of specimens represent pure crack willow (Belyaeva, 2009)).

An ornamental clone of Euxine willow, *S. euxina* 'Bullata' (formerly *S. fragilis* 'Bullata' or *S. fragilis* var. *sphaerica* Hryn.), with a short trunk and very dense spherical crown, originated in the Baltic countries in the late 18th century and is currently popular in Europe.

***Salix ×fragilis* Linnaeus, crack willow, the hybrid of *S. alba* and *S. euxina*.** This is one of the most commonly cultivated and naturalized tree willows. The name *S. 'fragilis'* was frequently used for both – *S. euxina* and its hybrids with *S. alba* – prior to the publication by I.V. Belyaeva and her description of *S. euxina* (Belyaeva, 2009). As *S. euxina*, *S. ×fragilis* has branches which are highly brittle at the base, and frequently reproduces by stem fragmentation. However, its branches and branchlets are hairy or glabrescent in age (glabrous in *S. euxina*), leaf blades not glaucous beneath, with uniformly dense stomata on both surfaces (glaucous and hypostomatous for *S. euxina*) and pistillate catkins slender and loosely flowered (*S. euxina* has stout and moderately densely flowered catkins). Importantly, most of the living and herbarium material labelled as *S. fragilis* is likely hybrids of *S. alba* and *S. euxina* and for them the binomial '*S.* ×*fragilis*' should be applied. Thus, all herbarium and living specimens labelled as *S. fragilis* need to be revised.

There are a few clones of *S.* ×*fragilis* in cultivation. Some of these clones with S. *alba* have been previously assigned specific binomials (*S.* ×*rubens* Schrank and *S.* ×*viridis* Franchet) which should now be considered as synonyms of *S.* ×*fragilis*. *S.* ×*fragilis* is common in wet places, near water and on residential lots. As with white willow, *S.* ×*fragilis* and its hybrids occasionally form extensive stands extending for many kilometres along rivers (Plate 12A). It is planted occasionally for reforestation, used extensively along streams to hold earth and as a windbreak. *S.* ×*fragilis* is widely naturalized in various parts of the world. It became widespread in Europe, Asia, Africa and North and South America (see Fig. 2.23). In Australia and New Zealand, it is considered to be a noxious invasive plant (Cremer, 2003).

This nothospecies is also used as a nectariferous plant, for fuel and construction timber, and the young branches and leaves provide food for goats. It also has been deployed in biomass production trials in the USA (Kopp *et al.*, 2001).

Section *Salicaster* Dumortier

This section consists of six to eight species distributed in the temperate regions of Eurasia and North America (see Table 2.3), with two distinct groups of species that can be considered subsections. One group includes the Eurasian species *S. pentandra*, *S. pseudopentandra*, *S. pentandroides* and *S. paraplesia* from China and the North American *S. serissima*. The second group includes the North American species *S. lasiandra* and *S. lucida*. The representatives of this section are trees – or under unfavourable conditions, shrubs – that are characterized by lustrous leaves and always with conspicuous glands at the apex of the petioles. The numerous leaf glands on young expanding leaves produce an aromatic, viscous gum resembling that of certain poplars. Species belonging to this section have buds that are ovoid, blunt or rounded, and lustrous brown. Coetaneous or serotinous male catkins are dense and stout, with numerous (three to eight) bright yellow stamens. There are two nectaries in male flowers, and they occasionally form a cup-like structure around the base of the stamens.

***Salix lucida* Muhlenberg, shining or American bay willow.** This shrub or small tree grows up to 6–8 m tall, with smooth or slightly scaly bark and yellowish or green, smooth polished branchlets (Plate 20B). It resembles *S. pentandra* in most characters, but with ferruginous hairs on unfolding leaves, long attenuate tail-like leaf tips and prominent stipules (Plate 20B). It is an eastern North American species distributed from Newfoundland to Saskatchewan, south to West Virginia and Nebraska (Plate 12B), where it occurs on margins of streams and lakeshores (see Fig. 2.17), wet meadows and swamps.

Shining willow is an attractive plant – characterized by highly glossy branches and leaves – with similar ornamental characteristics to *S. pentandra*. It has received some attention in hybridization work. *S. lasiandra* is a western North American taxon, closely related to and weakly delimited from *S. lucida*, that New Zealand geneticists view as a promising parent in hybridization programmes. Shining willow has been deployed in biomass production trials in the USA (Kopp *et al.*, 2001).

***Salix pentandra* Linnaeus, bay or laurel-leaved willow.** *S. pentandra* is a small tree, or occasionally a large shrub, attaining 15–18 m in height in favourable conditions, with spreading branches forming a rounded, broad crown. The bark of old trunks of this species is dark brown or grey, with coarse longitudinal fissures, while the bark of branchlets is glabrous, shining brown or reddish. Leathery, lustrous ovate-elliptic leaves are very ornamental. These leaves become black on drying and have a distinct bitter taste. Leaf blades are 2.5–3 times as long as broad and have a pale and obscurely veined undersurface. Staminate cylindrical aments are very fragrant. Large mature capsules ripen at the end of summer – later than other willows – and after fruiting they are retained on the tree well into winter. Its seeds remain viable and germinate the following spring, which is unusual for lowland willows. The morphological characters of this species are rather consistent across its entire range, although the populations from west European mountains have more stout buds resembling those of *S. pentandroides* and could be treated as a subspecies (Skvortsov, 1968, 1999).

S. pentandra is common throughout the northern and central parts of Europe and western Asia (Plate 12C). It is usually associated with grassy and *Sphagnum* bogs, wet meadows and swampy valleys, inundated areas and damp wood clearings with high water tables. In the steppe belt of the southern limits of its range, it occurs in lower flood plains and in shallow depressions amidst inland sandy territories. In the mountains, *S. pentandra* occurs near the altitudinal limits of tree growth.

Because of its use for basket making and excellent ornamental characteristics, *S. pentandra* has been planted widely outside of its natural range, including North America, where it rarely escapes from cultivation. It is used in parks, cemeteries and private estates because of its dark green, lustrous, laurel-like leaves and attractive catkins. It is a good late honey plant. Its bark is used for tanning and the leaves are used for the production of a yellow dye (Nasarow, 1936, 1970). The twigs from clones 'Lumley' and 'Patent Lumley' are suitable for coarse wickerwork. The wood is denser than other willows and is suitable for fuel. As compared to other species, bay willow is more difficult to propagate by seed and cuttings and it also grows more slowly (Nasarow, 1936, 1970). This species has low palatability and is used for erosion control in New Zealand.

The most common hybrid of *S. pentandra* is *S.* ×*ehrhartiana* G. Meyer (*S. alba* × *S. pentandra*), Ehrhart's willow, a graceful male plant with slightly glossy twigs and pubescent young leaves that later become glabrous, an inherited trait from *S. alba*. It has been planted on stream banks and as an ornamental tree. Another common hybrid is *S.* ×*meyeriana* Rostkovius ex Willdenow (*S. pentandra* × *S. euxina*), a shiny-leaved or pointed-leaved willow with highly polished twigs and glabrous leaves. It is likely that the name '*S. pentandra*' is mistakenly applied to this hybrid in living and herbarium specimens in North America (Zinovjev, 2011). It can be distinguished from the species by its relatively narrow cylindrical catkins. A hybrid with *S. triandra* is also known (Stace, 1997).

Subgenus *Longifoliae* (Andersson) Argus

This subgenus, which contains but one section, comprises eight New World species with distinctive morphological features. Similar to the poplars in section *Populus*, *Longifoliae* willows propagate by root shoots (also called rhizoblasts or root suckers), forming dense thickets. Their stigmas are deciduous after flowering, and their linear, amphistomatous leaves are similar to those of *Chosenia*. The species of the subgenus *Longifoliae* frequently produce sylleptic vegetative and reproductive shoots, similar to *Populus*. Another distinctive feature of this subgenus is branched catkins, not known elsewhere in the *Salicaceae sensu stricto*.

The species identification in this subgenus is rather difficult, as they 'seem to form a syngameon of poorly resolved semispecies' (Argus, 2010). In addition to examination of live and herbarium specimens, molecular, cytological, hybridization and common garden studies may be required to confirm species identities.

Section *Longifoliae* Andersson

***Salix interior* Rowlee, narrow-leaved or sandbar willow.** This familiar North American colonial shrub occurs over most of northern North America, from Alaska to New Brunswick, and south to Colorado, Texas and Mississippi (Plate 13A). It grows 4–6 m tall, with slender twigs, a flaky epidermis, and narrowly elliptic or linear leaves, 16–18 times as long as broad (Plate 20C), that create a filtered shade. The characteristic habit of this plant is clumps about 2 m tall (occasionally 4–9 m tall), consisting of a thicket of slim stems that are all clonal root suckers (see Fig. 2.16). These thickets can be recognized from a considerable distance, giving a pleasing softness to the landscape. The winter buds are small and the twigs resemble those of black willow because the bark is about the same colour. Many of the buds are defective and drop off early in autumn, and their places are taken by small lateral buds that develop on each side of the old scar – an interesting and distinctive characteristic. Flowers emerge with leaves, and capsules ripen in early to midsummer (Plate 20C).

Sandbar willow aggressively colonizes gravel and sandbars of river flood plains, exposed beaches, mud banks of streams, ponds, drainage ditches and along roadsides, binding the soil by its dense root system. The unstable habitat created by erosion and mechanical damage often prevents the species from reaching its full potential. Although this species is tolerant of flooding, it does not compete well with species less tolerant of flooding, and rarely reaches the higher levels of the canopy (Argus, 1986). *S. interior* also grows well in sandy soil and occurs on sand dunes, where it is used for erosion control. It has also been used in biomass production trials (Kopp *et al.*, 2001) and artificial hybridization work. Hybrids with *S. eriocephala* have been deployed in biomass trials in eastern North America. Coyote willow (*S. exigua*) is a closely related species distributed in western North America (Argus, 1986). Some taxonomists, Voss (1985) for example, combine the two species together as *S. exigua*, although Argus recommends treating them as separate species (Argus, 2010).

Subgenus *Vetrix* Dumortier

More than two-thirds of the species in the genus *Salix* belong in this large and diverse taxon. It comprises mostly shrubs and small trees, with petioles lacking glands at the base of the blade, generative buds considerably larger than vegetative buds and mostly precocious catkins. Male and female flowers usually have pigmented and persistent floral bracts and a single nectary. There are two stamens in male flowers that occasionally are connate.

Section *Cordatae* Barratt

This New World section includes six species, only one of which has any commercial importance.

***Salix eriocephala* Michaux, heart-leaved willow.** This willow is distributed over a wide range of central and eastern North America (Plate 13B). It is commonly found along streams and on gravelly or rocky riverbanks, but also on moist but well-drained sites such as marshy fields or mixed mesophytic woods.

A sprawling shrub up to 6 m tall, with branches looping over on to the ground, the young twigs and bud scales of *S. eriocephala* are puberulent to pubescent. The inner membranaceous layer of the bud scale is free and separate from the outer layer of the bud scale. Developing leaves are reddish. Mature leaves are often cordate at the base, narrowly oblong and serrated (Fig. 2.18), with tomentose petioles and prominent persistent, rounded stipules (see Fig. 2.19). Flowers appear with leaves exhibiting coetaneous development. The pistils of heart-leaved willow are glabrous, with distinct styles and subtended by bracts with long hairs.

The flexible and tough rods of *S. eriocephala* make it an important species for basket production. It was imported to Europe during the last century and propagated as basket willow under the name of *S. americana* (Mosseler, 1987). Nowadays, it is widely planted within and outside its natural range. The basket willow clone 'Rouge d'Orleans' is characterized by a slender, arched branching habit and narrow leaves (Newsholme, 1992). Another widely used basket clone is cultivar 'Americana', a hybrid of *S. eriocephala* and *S. petiolaris*. Heart-leaved willow has been the subject of intense genetic testing in the USA, and clones and hybrids of it are currently under trial for biomass production (Kopp *et al.*, 2001).

Section *Cinerella* Dumortier

This large section comprises at least 30 species that are widely distributed across forested areas of the temperate and boreal Old and New World. The representatives are small or medium-sized trees or large shrubs that often have longitudinally striated wood under the bark of shoots and occasionally recurved generative buds that are usually much larger than vegetative buds. Broad leaves with conspicuous underside reticulation have convex petioles and frequently inequilateral stipules. Precocious or subprecocious, densely pubescent catkins consist of flowers with a solitary short nectary.

***Salix aegyptiaca* Linnaeus, Armenian or musk willow (syn. *S. medemii* Boissier).** This small tree, 8–10 m tall, with thick branches, resembles *S. caprea* in its morphology and ecology. But unlike goat willow (see below), it is propagated easily from cuttings. It is distributed over a small region in the Caucasus and parts of Turkey, Iran and Turkmenistan (Plate 13C). The habitats of this species include lighted forests on the slopes and banks of streams, and as secondary vegetation in clearings, forest edges and residential lots. This willow is cultivated in some locations in Azerbaijan and Middle Asia, Afghanistan and Pakistan. The Latin name derives from a cultivated specimen from Egypt. Armenian willow is favoured for cultivation as a nectariferous and ornamental plant. Male catkins were commonly used in the East for making a drink.

***Salix caprea* Linnaeus, goat or pussy willow.** This small or medium-sized tree – or, if damaged, occasionally a shrub – grows to 15 m tall with 2–3 cm thick branches and irregularly fissured bark. It is widely distributed in Europe and Asia (Plate 13C). Its branchlets are greenish on one side and light brown on the other. The decorticated wood of young stems is typically smooth. Its leaves are large and broadly elliptic, resembling the leaves of apple trees, with a velvety underside and prominent veins (Plate 18B). This species does not root as easily as most willows.

S. caprea is a species that grows on well-drained soils, often on forest edges, dry meadows, roadsides and secondary habitats, avoiding the saturated soils of wetlands. It is an aggressive invader of disturbed sites; for example, it commonly colonized bombed sites during World War II (Meikle, 1984). It grows up to the timberline in mountains. It thrives in dry soil and has been cultivated through the ages. It is easily recognized in the leafless early spring landscape by its conspicuous and abundant catkins on bare stems, and because of this splendid display, it is often planted on residential lots. Its large, chestnut-brown catkins supply the pussy willows for decorating European churches on Palm Sunday. It also is used as a nectariferous plant.

The ornamental cultivar ‘Pendula’ is common in the trade. Hybrids with *S. cinerea* are frequent and are intermediate between the two parent species in morphological characteristics (Meikle, 1984).

***Salix cinerea* Linnaeus, grey willow or common sallow.** A shrub, 3–5 m tall, with an orbicular crown, this willow is naturally distributed over a large area in Europe and western Asia (Plate 14A). Its branches are often conspicuously ribbed with longitudinal ridges that can be seen even without peeling the bark. Branchlets are dull, greenish grey and densely pubescent. Its thick, wrinkled, dull green leaves, often with slightly twisted apex, are densely pubescent and reticulated underneath with persistent stipules. Grey willow grows in many low-lying damp situations in waterlogged and reduced soils such as wetlands, moist depressions, ditches, embankments, banks of stagnant or slow water bodies, and forest edges. Similar to goat willow, *S. cinerea* is a pioneer species that rapidly invades disturbed sites. Unlike goat willow, however, the distribution of *S. cinerea* never reaches the latitudinal or altitudinal limits of tree growth.

S. cinerea ‘Tricolor’, an ornamental cultivar with speckled white-red and green leaves, originated in Germany. Some hybrids of grey willow that are in cultivation include *S.* ×*smithiana* Willdenow, a hybrid of *S. cinerea* and *S. viminalis* that is commonly used as a rootstock for ornamental willows, and *S.* ×*hirtei* Strahler (*S. viminalis* × *S. cinerea* × *S. aurita*). They are hardy, vigorous, upright shrubs that make fast-growing windbreaks and thrive in poor soil. They have also been planted for biomass production in Europe (Heaton *et al.*, 2001).

***Salix discolor* Muhlenberg, American pussy willow.** This common North American pussy willow typically grows in wet places, swampy thickets and on shores in northern regions across most of the continent (Plate 14B). It is rarely found growing naturally in extensive stands, but rather occurs as single plants. Pussy willow grows 2–4 m tall, with straight, upright stems. Bark of old trees is divided by shallow fissures into thin, plate-like oblong scales. The decorticated wood is smooth or has a few short, indistinct ridges. Stout branchlets are dark brown to yellowish, with deciduous hairs early in the season, later becoming glabrous. Elliptic, lanceolate or obovate leaves (Fig. 2.18) are glaucous, with prominent irregular venation underneath and covered with ferruginous hairs when young. As its name implies, the very large, well filled, purple-brown generative buds produce attractive woolly catkins in early spring that are popular in decorative arrangements. Capsules ripen before leaves emerge (Plate 20C).

S. discolor has been investigated for its potential for biomass production (Kopp *et al.*, 2001), because it seems better adapted to heavy, poorly drained soils than *S. viminalis*. It is also widely used for wetland restoration projects.

Section *Fulvae* Barratt

Salix bebbiana **Sargent, Bebb, long-beaked or ochre-flowered willow.** One of the most widely distributed willows, the natural range of *S. bebbiana* encompasses both Old and New Worlds – northern and western North America, northern Europe and Russia, northern China and northern Korea (Plates 14C and 15A). This shrub, which exhibits uniform morphological characteristics everywhere, grows to 6 m tall, with a short stem and stout ascending branches that form a broad round head (see Fig. 2.14). Its bark is reddish, olive-green or grey tinged, with shallow fissures, and the decorticated wood has short ridges. Mature generative buds are pubescent, with depressed margins along the ribs and flat beaks. The ovate leaves of *S. bebbiana* are distinctive, with entire, coarsely serrated or crenate margins and very pronounced, intricately meshed veins that are impressed above and raised below. The upper surfaces of the leaf blades are dull and lightly pubescent, the lower leaf surfaces whitish and densly pubescent. Leaf apices are slightly twisted or folded. The flowers are coetaneous and appear just before or with leaves (Plate 20A). Its long-beaked pistils are subtended by pale bracts.

Bebb willow typically occurs along the borders of deciduous woods in North America, often in open, xeric habitats. It is also common in wet thickets, carrs (see Fig. 2.14), prairies and along stream banks and lakeshores. In Eurasia, it occurs in light forests, such as pine (*Pinus*) or larch (*Larix*), often on stony slopes and on various secondary habitats, where it grows on a variety of soils, even those that are infertile and dry. *S. bebbiana* is remarkable for its drought tolerance and, because of this trait, it has been included in biomass trials (Kopp *et al.*, 2001), although no commercial deployment is known at this time.

This small section probably also includes *S. starkeana*.

Section *Viminella* Seringe

This section comprises 12–13 Old World species distributed in temperate forested regions, typically on alluvial habitats. Representatives of this section are trees or tall shrubs with long flexible shoots. Leaves are typically narrow, pubescent and silvery underneath, with many parallel, prominent veins and linear or falcate stipules. Floriferous buds are conspicuously different from vegetative buds: oval to nearly cylindrical (not flattened), with apices either straight or bent toward shoots. Precocious or subprecocious flowers have a solitary rectangular or linear nectary, 0.6–1.5 mm long, and elongate tapering stigmas.

Salix gmelinii **Pallas, Gmelin's willow (syn. *S. dasyclados* Wimmer).** This largely Russian willow has a distribution close to that of *S. viminalis* (Plate 15B). A tree or shrub up to 10 m tall, its stems can reach 80–90 cm in diameter, with thick, straight, grey-brown branches. Grown in favourable locations in northern Russia, it can attain heights up to 20 m along river valleys, roadside ditches and on well-drained, moist, aerated substrates. Specimens in the Polar Urals grow north of the latitudinal limits of *Larix* and *Betula* (Skvortsov, 1968, 1999). The decorticated wood of woolly-stemmed willow has few ridges and its 2–3 cm thick branchlets are densely clothed with pubescence. Leaves are broadly lanceolate, up to 12 cm long and 2.5 cm wide, widest toward the apex, with the primary vein apparent on both sides. Leaf margins are entire or with rare glands. Stipules are large and persistent. Petioles embrace generative buds. Catkins are long and densely sericeous.

There are many common hybrids in Western Europe resembling *S. gmelinii*, and for that reason many authors treat *S. gmelinii* as a hybrid (Newsholme, 1992). However, normal seed production, absence of hybrid segregation, a specific ecological niche and vast distribution make this assumption inaccurate. Skvortsov (1968, 1999) suggests that the name *S. dasyclados* be replaced by *S. burjatica* because the originally described specimen might be of hybrid origin. In fact, the *S. burjatica* binomial is widely used by researchers in European publications. Belyaeva and Sennikov (2008) have resurrected the '*Salix gmelinii* Pallas' name and use it as the correct name for *S. dasyclados*, while *S. dasyclados* and *S. burjatica* are considered as synonyms.

S. gmelinii is used for basket making and embankment protection, and for biomass production (Kopp *et al.*, 2001).

Salix schwerinii **Wolf, Schwerin willow (syn. *S. kinuyanagi* Kimura).** This very vigorous shrub or small tree is an East Asian species distributed in the Russian Far East, Mongolia, north-east China, North Korea and Japan (Plate 15B). It grows along rivers and streams on alluvial deposits and can reach 12 m in height, with grey fissured bark and 2- to 3-year-old shoots that are tawny-brown. Its long, narrow leaves are similar to *S. viminalis*, with the underside densely clothed with silvery hairs that are parallel to the midrib.

Extensively cultivated for coarse basketry and furniture in Japan under its synonymous name *S. kinuyanagi*, it is also an ornamental willow often planted in parks because of its graceful, wide-spreading branching habit and silvery leaves. *S. schwerinii* has been a parent in some hybridization work accomplished by geneticists in Europe for biomass production (Verwijst, 2001).

Salix udensis **Trautvetter (syn. *S. sachalinensis* Schmidt, *Salix siuzevii* Seemen).** Usually a tall shrub or small tree 8–10 m in height, with long, red-brown spreading branches, specimens as tall as 30 m are encountered on the Kamchatka Peninsula. Its native distribution also includes the Russian Far East, north-east China, North Korea and Japan. The habitat of *S. udensis* includes the banks of rivers, streams and ditches, as well as other damp habitats. It differs from other species in this section by its sparse leaf pubescence, short nectaries and elongated stipules.

Some botanists, including Skvortsov (1968, 1999) and Ohashi (2001), believe that *S. sachalinensis* is synonymous with *S. udensis*. *S. sachalinensis* is accepted by Fang *et al.* (1999) and is the name that is widely used in US publications. The author's choice of binomial in any specific publication will be the deciding factor until a precedent is established. Skvortsov (1999) believes that *S. siuzevii* also is synonymous with *S. udensis*, but Fang *et al.* (1999) consider *S. udensis* and *S. siuzevii* to be different species. I.V. Belyaeva uses *S. udensis* as the correct name and *S. sachalinensis* as its synonym in *World Checklist for Salicaceae sensu stricto* (personal communication).

The distinctive male cultivar 'Sekka' (Japanese fantail willow), with contorted and remarkably flattened and fasciated stems, is very popular in flower arrangements. *S. udensis* is also used to protect embankments, for weaving wicker articles and as a nectariferous plant, as well as a parent in current breeding programmes in North America.

Salix viminalis **Linnaeus, common osier or basket willow.** This well-known tall shrub or multi-stemmed tree grows up to 6–8 m tall, with long, flexible, olive or grey-brown branches. Basket willow is distributed over wide regions in Eurasia, except the Far East (Plate 15C). It grows on fresh alluvial substrates along riverbanks and ditches, and on sandy soils with sufficient moisture. Its narrow leaves, up to 20 cm long, with conspicuously revolute or undulate margins are dark green above, with silky pubescence underside. Marginal leaf glands are distinctly removed from the leaf margin to the upper leaf surface. Its densely sericeous catkins contain flowers with brown bracts.

Long a favourite basket species (see Fig. 2.2), during colonial times it was introduced to North and South America for coarse basketry. The colour of its branches is reminiscent of *S. triandra*, another species that is popular for basket production. However, *S. viminalis* is more vigorous and hardy, producing straight and flexible shoots up to 4 m in length, and its rods contain a high proportion of pith. It is considered one of the best willows for the consolidation of banks and dykes, and is also used for timber and reforestation. In the northern parts of Siberia, the bark of *S. viminalis* is used for making fishing nets and its foliage is used to feed livestock (Nasarow, 1936, 1970).

S. viminalis hybridizes freely with many species, e.g. *S.* ×*smithiana* (*S. cinerea* × *S. viminalis*), *S.* ×*hirtei* (*S. viminalis* × *S. cinerea* × *S. aurita*), *S.* ×*sericans* Tausch (*S. caprea* × *S. viminalis*), *S.* ×*fruticosa* Doell (*S. aurita* L. × *S. viminalis*), *S.* ×*friesiana* Andersson (*S. repens* × *S. viminalis*), *S.* ×*mollissima* Hoffman (*S. triandra* × *S. viminalis*), *S.* ×*rubra* Hudson (*S. purpurea* × *S. viminalis*) and *S.* ×*forbyana* Smith (*S. cinerea* × *S. purpurea* × *S. viminalis*). Combinations with *S. gmelinii* (formerly *S. dasyclados*) dominate in European hybridization programmes. Many clones of this species, and some of its hybrids, are planted as short-rotation coppice for biomass production (Verwijst, 2001). It is the most important species in Europe for commercial deployment and shows great future potential there and in North America (Kopp *et al.*, 2001).

Section *Subviminales* Schneider

This is a very small Eurasian section consisting of just two species – *S. gracilistyla* described below and *S. blinii* from South Korea – that are closely related to the species in the previous section (Skvortsov, 1968, 1999).

Salix gracilistyla **Miquel, Japanese pussy or rosegold willow.** A tall, spreading shrub or small tree, this willow's large, beaked, generative buds produce large, densely pubescent catkins in early spring. The underside of its leaves have conspicuous parallel veins due to dense pubescence along the veins. Stipules are prominent and petioles embrace the generative buds. Male flowers contain nearly connate filaments. Styles are very long, up to 4 mm. Its distribution encompasses eastern Asia, the Russian Far East, Japan, the Korean Peninsula and far north-eastern China, where it grows along streams and rivers.

S. gracilistyla is grown to protect embankments, for weaving wicker baskets and as an ornamental plant, due to its arching habit and outstanding display of large catkins.

Section *Geyerianae* Argus

This New World section comprises three species with narrowly elliptic or linear leaves that are glaucous, with thick waxy bloom underneath, and subprecocious or coetaneous catkins that appear just below the tip of the previous year's shoot.

Salix geyeriana **Andersson, Geyer willow.** A shrub up to 4 m tall that has pruinose branches and linear-lanceolate leaves, this willow is indigenous to western North America (Plate 16A) but virtually unknown outside of its native range. It has been planted for restoration of fluvial mine tailings in the western USA because of its documented metal tolerance. In general, Geyer willow is recognized as an early colonizer of disturbed areas (Fisher *et al.*, 2000).

Salix petiolaris **Smith, meadow or slender willow.** This mid-sized or tall shrub grows in wet meadows, carrs (see Fig. 2.14), wetlands, shores and stream banks in north-eastern North America (Plate 16A). It can reach heights of 6 m, with erect stems and branches that are puberulent at first, then becoming glabrous and are flexible or somewhat brittle at the base. Its narrowly lanceolate leaves have slender petioles that are up to 15 mm long. The hybrid cultivar 'Americana' (*S. eriocephala* × *S. petiolaris*) is widely used for basket production. Meadow willow has also been deployed in biomass production trials in the USA (Kopp *et al.*, 2001).

Section *Helix* Dumont

This large and very diverse Old World section comprises about 30 species. There are only five species native to Europe, with the remainder restricted to Central and eastern Asia. This section is one of the most complex and difficult to treat because its taxa often differ from each other only in somewhat subtle characteristics. Representatives of this section have flexible slender branches, narrow, flat leaves without prominent veins and non-revolute margins. Its distinctive characteristics include stomata often located on the upper leaf surface; precocious to serotinous catkins which are mostly narrowly cylindrical; and flowers with short, subsquare nectaries, connate stamen filaments and short styles and stigmas. Members of this section are widely distributed in warm regions of the temperate zone, including arid steppe areas, and several are of economic importance.

Salix koriyanagi **Kimura (syn. *S. purpurea* var. *japonica* Nakai).** Although closely related to and very much like *S. purpurea*, the geographical distribution of the two species is very different – *S. koriyanagi* is a native of Korea rather than Europe or Africa. It is grown extensively in Japan to produce fine flexible rods for fine basketry and as an ornamental plant.

Salix miyabeana **Seemen, Miyabe willow.** Distributed in the Russian Far East, Japan, Korea, north-east and north China and Mongolia (Plate 16B), the habitat of this willow is banks of rivers and streams, flood plains and damp meadows. A tall, erect shrub or small tree up to 6 m tall, Miyabe willow has grey branches and pale brown branchlets. Lanceolate leaves up to 15 cm long, with numerous stomata on the upper surface, have fully developed linear-subulate stipules. Leaf margins are distinctly serrate to the very base of the blade. Mature leaves are conspicuously thick and callous. Filaments in male catkins are completely connate. This fast-growing species has been deployed in biomass plantation culture in eastern North America (Kopp *et al.*, 2001), and promising hybrids of this species have been selected by willow breeders.

***Salix purpurea* Linnaeus, purple willow or purple osier.** This graceful, medium-sized to tall shrub or small tree can grow up to 4 m but usually is 1–2 m tall. Its native distribution includes northern Africa and Europe (Plate 16B), where it is confined mostly to the banks of rivers and streams, damp meadows and moist sand. The branches of purple osier are long, slender, brown to olive-green and 0.6–1.5 cm thick; branchlets are reddish brown to olive and completely glabrous, smooth and shiny; the wood is bright yellow beneath the bark. Leaves are occasionally distinctly opposite and subopposite and variable in length, with very short petioles (see Fig. 2.18). Sessile catkins appear before the leaves. Nectaries are mostly bright purple. Filaments in male flowers are connate with purple anthers, hence the name of the species.

S. purpurea is commonly cultivated across all of Europe. It was introduced to North America during colonial times for basketry, where it became naturalized. It is one of the most adaptable willows, tolerating extremes of heat, wetness and drought. Many clones and hybrids such as 'Dicky Meadows', 'Red Buds' and *S.* ×*rubra* provide excellent material for fine basketry because the rods are tough, hard and slender.

This species is very ornamental and many cultivars are known. 'Pendula' is an elegant cultivar with long pendulous branches; 'Nana' is a low-growing, very fine-textured clone; and 'Eugene' is a cultivar with branches ascending steeply into a conical shape. Purple willow is used for hedges and for consolidation of coastal sands. It is particularly well adapted to drought and soil salinity and establishes easily on sandy soils. Planted in combination with tall species such as *S. viminalis*, this willow produces efficient and attractive windbreaks (Newsholme, 2002). *S. purpurea* is extremely bitter tasting and unpalatable, supposedly limiting the damage from insects and animals. Opossum (*Trichosurus vulpecula*) -resistant cultivars are being used for hillside stabilization projects in New Zealand. American geneticists view this species as a promising parent for biomass breeding (Kopp *et al.*, 2001).

Section *Daphnella* Seringe

This small Eurasian section contains only four species, characterized by frequently pruinose shoots, large generative buds and lanceolate or linear-lanceolate leaves with persistent stipules. The bark is usually lemon-yellow inside.

***Salix acutifolia* Willdenow, pointed-leaf willow.** A tall shrub or a tree of 6–10 m, with long and slender drooping branches, this willow is distributed throughout Europe and western Asia (Plate 16C), where it grows on sandy deposits along river valleys. In the steppe belt, it also occurs in sandy areas away from river valleys. *S. acutifolia* closely resembles *S. daphnoides* in many traits. Its leaves are glabrous and narrow – up to 15 times as long as broad. In north-western Russia and the Baltic Republics, *S. acutifolia* is cultivated even more frequently than *S. daphnoides* (see below). Because of its very precocious and densely hairy catkins, it is used to decorate churches and cemeteries for Palm Sunday, and it is often planted alongside ponds because of its elegantly drooping habit. Pointed-leaf willow is also used for sand fixing in the forest-steppe and steppe zones of southern European Russia, Ukraine and Kazakhstan. It is frost hardy and also tolerates the hot summer of the sandy deserts of Central Asia.

***Salix daphnoides* Villars, violet willow.** This straight-stemmed and sometimes pyramidal tree or tall shrub grows up to 15 m tall. It has a scattered distribution in Europe (Plate 16C), where it grows on the alluvial banks of rivers in mountains and occasionally along the largest lowland rivers. Violet willow gained its common name because of its deep purple branchlets covered with white bloom. Young shoots and young leaves are pubescent. Leaves are 3–4 times as long as wide, with lanceolate, acute stipules. Historically, *S. daphnoides* has been widely planted as an ornamental, for wickerwork and as a nectariferous plant.

***Salix rorida* Lakschewitz.** This large tree can grow 15–20 m tall, with trunks to 1–2 m in diameter and deeply cracking bark (Nasarow, 1936, 1970). It is distributed throughout eastern Russia, Korea and Japan along the banks of rivers, where it grows solitarily or in small groves (Plate 16C). Unlike the two previous species, *S. rorida* has broad, round stipules. It is grown to protect embankments, for timber, for weaving wicker articles and as a nectariferous plant. People in the Far East used large trunks of *S. rorida* to make canoes.

Acknowledgements

A number of colleagues from around the world provided helpful discussions during the preparation of this monograph, especially concerning *Populus* and *Salix* speciation. We are especially indebted to George Argus, Burton Barnes, Irina Belyaeva, Brian Stanton, Reinhard Stettler, Horst Weisgerber and Allan Wilkinson for their reviews of the manuscript. We also thank Alberto Del Lungo and Lei Chen at the FAO, Rome, for preparing the range maps. Jim Carle at the FAO, Rome, has also been very supportive. Special thanks to our technical editors – Jim Richardson and Jud Isebrands – for their steadfast vision and gentle but firm encouragement during the long gestation of this work.

Notes

[1] A third monotypic, willow-like genus in *Salicaceae* from Asia – *Chosenia* – was proposed by the Japanese botanist, Nakai (cf. Rehder, 1990; Fang *et al.*, 1999). It differs from other members of *Salix* because of its pendant male catkins, filament bases fused to bracts, two nearly free styles and the absence of nectariferous flower glands. As such, it may represent a linkage organism between *Populus* and *Salix*. Based on the work of Hjelmqvist (1948) and others, however, we treat this taxon as *Salix arbutifolia*.

[2] Some scholars think that these 'willows' may in fact have been *Populus euphratica*, but that speculation is beyond resolving.

[3] Using genetic fingerprinting, Fay *et al.* (1999) identified a putative natural population in Spain as a clone that probably was introduced.

[4] The IPC maintains a searchable database of registered *Populus* cutivars on its website (www.fao.org/forestry/ipc/en).

[5] According to the rules of plant nomenclature, *Populus* ×*euramericana* Guinier, a commonly used binomial for this hybrid and the one once adopted by the IPC, does *not* take precedence over that proposed in 1785 by Moench (or Mönch) – *Populus* ×*canadensis*. Boom (1957) argued this point quite effectively. Moench first described this now ubiquitous hybrid, and his binomial should be used.

[6] In 1972, van Broekhuizen proposed the Latin binomial *Populus* ×*interamericana* for hybrids of *Populus trichocarpa* × *Populus deltoides*, and this appellation has been widely used, including by the IPC. However, Henry (1914) assigned *Populus* ×*generosa* to this hybrid and, although less descriptive, by the rules of botanical nomenclature this name takes precedence (Eckenwalder, 1984; Rehder, 1990).

References

Ahuja, M.R. (1984) A commercially feasible micropropagation method for aspen. *Silvae Genetica* 33, 174–176.

Alford, M.H. and Belyaeva, I.V. (2011) Neotropikey family data: Family *Salicaceae sensu lato* (including *Samydaceae* and the majority of former *Flacourtiaceae*) (http://www.kew.org/science/tropamerica/neotropikey/families/Salicaceae.htm, accessed 24 February 2013).

Angiosperm Phylogeny Group, The (2003) An update of the Angiosperm Phylogeny Group classification for the orders and families of flowering plants: APG II. *Botanical Journal of the Linnean Society* 141, 399–436.

Argus, G.W. (1974) An experimental study of hybridization and pollination in *Salix* (willows). *Canadian Journal of Botany* 52, 1613–1619.
Argus, G.W. (1986) The genus *Salix* (*Salicaceae*) in the Southeastern United States. *Systematic Botany Monographs 9.* American Society of Plant Taxonomists, Laramie, Wyoming.
Argus, G.W. (1997) Infrageneric classification of *Salix* (*Salicaceae*) in the New World. *Systematic Botany Monographs 52.* American Society of Plant Taxonomists, Laramie, Wyoming.
Argus, G.W. (1999) Classification of *Salix* in the New World. Version 5: 5 July 1999. *Botanical Electronic News (BEN)* 227 (http://www.ou.edu/cas/botany-micro/ben/ben227.html, accessed 12 February 2012).
Argus, G.W. (2009) Guide to the interactive identification of native and naturalized New World *Salix* using INTKEY (DELTA) (http://137.229.141.57/wp-content/uploads/2011/02/GuideToSalixInteractiveIdentification-21Apr09.pdf, accessed 12 February 2012).
Argus, G.W. (2010) *Salix.* In: Flora of North America Editorial Committee (eds) *Flora of North America North of Mexico, Vol 7.* Oxford University Press, New York.
Bailey, L.H. (1930) *The Standard Cyclopedia of Horticulture, Vol III.* The Macmillan Company, New York.
Balatinecz, J.J. and Kretschman, D.E. (2001) Properties and utilization of poplar wood. In: Dickmann, D.I., Isebrands, J.G., Eckenwalder, J.E. and Richardson, J. (eds) *Poplar Culture in North America.* National Research Council of Canada Research Press, Ottawa, pp. 277–291.
Barnes, B.V. (1966) The clonal growth habit of American aspens. *Ecology* 47, 439–447.
Barnes, B.V. (1975) Phenotypic variation of trembling aspen in western North America. *Forest Science* 21, 319–328.
Barnes, B.V. and Han, F. (1993) Phenotypic variation of Chinese aspens and their relationships to similar taxa in Europe and North America. *Canadian Journal of Botany* 71, 799–815.
Barnes, B.V. and Wagner, W.H. Jr (2004) *Michigan Trees, Revised and Updated.* University of Michigan Press, Ann Arbor, Michigan.
Barocka, K.H., Baus, M., Lontke, E. and Sievert, F. (1985) Tissue culture as a tool for *in vitro* mass propagation of aspen. *Zeitschrift für Pflanzenzuchtung* 94, 340–343.
Bartos, D.L. and Campbell, R.B. Jr (1998) Decline of quaking aspen in the interior West – examples from Utah. *Rangelands* 20, 17–24.
Baum, D.A. (1992) Phylogenetic species concepts. *Trends in Ecology and Evolution* 7, 1–2.
Baum, D.A. and Donoghue, M.J. (1995) Choosing among alternative 'phylogenetic' species concepts. *Systematic Botany* 20, 560–573.
Belyaeva, I. (2009) Nomenclature of *Salix fragilis* L. and a new species, *S. euxina* (Salicaceae). *Taxon* 58, 1344–1348.
Belyaeva, I. and Sennikov, A. (2008) Typification of Pallas' names in *Salix. Kew Bulletin* 63, 277–287.
Bialobok, S. (1964) Studies on *Populus tomentosa* Carr. *Arboretum Kornickie* 9, 5–35.
Bockheim, J.G., O'Brien, J.D., Monroe, J.S. and Hinkel, K.M. (2003) Factors affecting the distribution of *Populus balsamifera* on the North Slope of Alaska, USA. *Arctic, Antarctic, and Alpine Research* 35, 331–340.
Boom, B.K. (1957) *Populus canadensis* Moench versus *P. euramericana* Guinier. *Acta Botanica Neerlandica* 6, 54–59.
Boucher, L.D., Manchester, S.R. and Judd, W.S. (2003) An extinct genus of *Saliceae* based on twigs with attached flowers, fruits, and foliage from the Eocene Green River Formation of Utah and Colorado, USA. *American Journal of Botany* 90, 1389–1399.
Braatne, J.H., Rood, S.B. and Heilman, P.E. (1996) Life history, ecology, and conservation of riparian cottonwoods in North America. In: Stettler, R.F., Bradshaw, H.D. Jr, Heilman, P.E. and Hinckley, T.M. (eds) *Biology of Populus and Its Implications for Management and Conservation.* National Research Council of Canada Research Press, Ottawa, pp. 57–85.
Bradshaw, H.D. Jr, Ceulemans, R., Davis, J. and Stettler, R. (2000) Emerging model systems in plant biology: poplar (*Populus*) as a model forest tree. *Journal of Plant Growth Regulators* 19, 306–313.
Brayshaw, T.C. (1965a) *Native Poplars of Southern Alberta and Their Hybrids.* Publication 1109. Canada Department of Forestry, Ottawa.
Brayshaw, T.C. (1965b) The status of the black cottonwood (*Populus trichocarpa* Torrey and Gray). *Canadian Field Naturalist* 79, 91–95.
Browicz, K. (1966) *Populus ilicifolia* (Engler) Rouleau and its taxonomic position. *Acta Societas Botanicorum Poloniae* 35, 325–335.
Browicz, K. (1977) Chorology of *Populus euphratica* Oliver. *Arboretum Kornickie* 22, 5–26.
Brummitt, R.K. and Powell, C.E. (1992) *Authors of Plant Names.* Royal Botanical Gardens, Kew, UK.

Čermák, K., Hofman, V., Krečmer, J., Čabart, J. and Syrovy, I. (1955) *Lesnicky a Mysivecky Atlas.* Ustředni Správa Geodesie a Kratografie, Praha.

Chase, M.W., Zmarzty, S., Lledo, M.D., Wurdack, K.J., Swensen, S.M. and Fay, M.F. (2002) When in doubt, put it in the *Flacourtiaceae*: a molecular phylogenetic analysis based on plastid *rbcL* DNA sequences. *Kew Bulletin* 57, 141–181.

Chauhan, N., Negi, M.S., Sabharwai,V., Khurana, D.K. and Lakshmikumaran, M. (2004) Screening interspecific hybrids of *Populus* (*P. ciliata* × *P. maximowiczii*) using AFLP markers. *Theoretical and Applied Genetics* 108, 951–957.

Clute, W.N. (1943) Meaning of plant names XCII: willows and poplars. *American Botanist* 49, 78–84.

Collinson, M.E. (1992) The early fossil record of the Saliceae. *Proceedings of the Royal Society of Edinburgh* 98B, 155–167.

Cooper, D.T. and Van Haverbeke, D.F. (1990) *Populus deltoides* Bartr. ex Marsh. Eastern Cottonwood. In: Burns, R.M. and Honkala, B.H. (eds) *Silvics of North America. Vol 2: Hardwoods.* Agriculture Handbook 271. USDA Forest Service, Washington, DC, pp. 530–543.

Cooper, F. (2006) *The Black Poplar: History, Ecology and Conservation.* Windgather Press, Macclesfield, UK.

Cottam, W.P. (1954) Prevernal leafing of aspen in Utah mountains. *Journal of the Arnold Arboretum* 35, 239–250.

Coyne, J.A. and Orr, H.A. (2004) *Speciation.* Sinauer Associates, Sunderland, Massachusetts.

Cremer, K.W. (2003) Introduced willows can become invasive pests in Australia. *Biodiversity* 4, 17–24.

Cronquist, A. (1988) *The Evolution and Classification of Flowering Plants,* 2nd edn. The New York Botanical Garden, Bronx, New York.

Dai, W., Cheng, Z.-M. and Sargent, W.A. (2004) Expression of the *rol*B gene enhances adventitious root formation in hardwood cuttings of aspen. *In Vitro Cellular and Developmental Biology – Plant* 40, 366–370.

Darwin, C. (1859) *On the Origin of Species by Means of Natural Selection*, 1st edn. Watts and Company, London.

DeBell, D.S. (1990) *Populus trichocarpa* Torr. & Gray. Black Cottonwood. In: Burns, R.M. and Honkala, B.H. (eds) *Silvics of North America. Vol 2: Hardwoods.* Agriculture Handbook 271. USDA Forest Service, Washington, DC, pp. 570–576.

Densmore, R. and Zasada, J. (1978) Rooting potential of Alaskan willow cuttings. *Canadian Journal of Forest Research* 8, 477–479.

Dickmann, D.I. (2001) An overview of the genus *Populus.* In: Dickmann, D.I., Isebrands, J.G., Eckenwalder, J.E. and Richardson, J. (eds) *Poplar Culture in North America.* National Research Council of Canada Research Press, Ottawa, pp. 1–42.

Dickmann, D.I. (2006) Silviculture and biology of short-rotation woody crops in temperate regions: then and now. *Biomass and Bioenergy* 30, 696–705.

Dickmann, D.I. and Stuart, K.W. (1983) *The Culture of Poplars in Eastern North America.* Michigan State University, East Lansing, Michigan.

Dirr, M.A. (1997) *Dirr's Hardy Trees and Shrubs: an Illustrated Encyclopedia.* Timber Press, Portland, Oregon.

Dobeŝ, C. and Vitek, E. (2000) *Documented Chromosome Number Checklist of Austrian Vascular Plants.* Museum of Natural History, Vienna.

Dorn, R.D. (1976) A synopsis of American *Salix. Canadian Journal of Botany* 54, 2769–2789.

Eckenwalder, J.E. (1977) North American cottonwoods (*Populus*, Salicaceae) of sections *Abaso* and *Aigeiros. Journal of the Arnold Arboretum* 58, 194–208.

Eckenwalder, J.E. (1980) Foliar heteromorphism in *Populus* (*Salicaceae*), a source of confusion in the taxonomy of Tertiary leaf remains. *Systematic Botany* 5, 366–383.

Eckenwalder, J.E. (1984) Natural intersectional hybridization between North American species of *Populus* (Salicaceae) in sections *Aigeiros* and *Tacamahaca.* II. Taxonomy. *Canadian Journal of Botany* 62, 325–335.

Eckenwalder, J.E. (1996) Systematics and evolution of *Populus.* In: Stettler, R.F., Bradshaw, H.D. Jr, Heilman, P.E. and Hinckley, T.M. (eds) *Biology of Populus and Its Implications for Management and Conservation.* National Research Council of Canada Research Press, Ottawa, pp. 7–32.

Eckenwalder, J.E. (2001) Descriptions of clonal characteristics. In: Dickmann, D.I., Isebrands, J.G., Eckenwalder, J.E. and Richardson, J. (eds) *Poplar Culture in North America.* National Research Council of Canada Research Press, Ottawa, pp. 331–382.

Einspahr, D.W., van Buijtenen, J.P. and Peckham, J.R. (1963) Natural variation and heritability in triploid aspen. *Silvae Genetica* 12, 51–58.

Einspahr, D.W., Benson, M.K. and Peckham, J.R. (1968) Wood and pulp properties of 5-year-old diploid, triploid, and triploid hybrid aspen. *Tappi* 51, 72–75.

Enebak, S.A., Ostry, M.E., Wycoff, G.W. and Li, B. (1996) Mortality of hybrid triploid aspen in Wisconsin and Upper Michigan. *Canadian Journal of Forest Research* 26, 1304–1307.

Every, A.D. and Wiens, D. (1971) Triploidy in Utah aspens. *Madroño* 21, 138–147.

Fang, C.-F. (1987) On the distribution and origin of *Salix* in the World. *Chinese Acta Phytotaxonomy Sinica* 25, 307–312. (In Chinese, English abstract)

Fang, Z.-F., Zhao, S.-D. and Skvortsov, A.K. (1999) *Salicaceae.* In: Wu, Z.-Y. and Raven, P.H. (eds) *Flora of China, Vol. 4.* Science Press, Beijing and Missouri Botanical Garden Press, St Louis, Missouri, pp. 162–274.

Fay, M.F., Lledo, M.D., Kornblum, M.M. and Crespo, M.B. (1999) From the waters of Babylon? *Populus euphratica* in Spain is clonal and probably introduced. *Biodiversity and Conservation* 8, 769–778.

Felger, R.S., Johnson, M.B. and Wilson, M.F. (2001) *The Trees of Sonora, Mexico.* Oxford University Press, New York.

Fenaroli, L. and Gambi, G. (1976) *Alberi: Dendroflora Italica.* Museo Tridentino di Scienze Naturali, Trento, Italy.

Fisher, K.T., Brummer, J.E., Leininger, W.C. and Heil, D.M. (2000) Interactive effects of soil amendments and depth of incorporation on Geyer willow. *Journal of Environmental Quality* 29, 1786–1793.

Food and Agriculture Organization (FAO) of the United Nations (1980) *Poplars and Willows in Wood Production and Land Use.* Forestry Series No. 10. FAO, Rome.

Galloway, G. and Worrall, J. (1979) Cladoptosis: a reproductive strategy in black cottonwood. *Canadian Journal of Forest Research* 9, 122–125.

Geerinck, D. and Léonard, J. (1968) Une nouvelle famille pour la flore Phanérogamique de la République Démocratique du Congo: Salicacées. *Bulletin de la Société Royale de Botanique de Belgique* 101, 17–22.

Gries, D., Zeng, F., Foetzki, A., Arndt, S.K., Bruelheide, H., Thomas, F.M., Zhang, X. and Runge, M. (2003) Growth and water relations of *Tamarix ramosissima* and *Populus euphratica* on Taklamakan desert dunes in relation to depth to a permanent water table. *Plant Cell and Environment* 26, 725–736.

Griggs, R.F. (1905) *The Willows of Ohio.* Proceedings of the Ohio State Academy of Science V. IV(6), Special papers No. 11. Spahr and Glennn Printers, Columbus, Ohio.

Gupta, A., Seghal, R.N., Panwar, P. and Thakur, I.K. (2002) Selection of rust (*Melampsora* spp.) resistant trees of *Populus ciliata* in Himachal Pradesh. *Indian Journal of Forestry* 25, 96–98.

Hamaya, T. and Inokuma, T. (1957) *Native Species of* Populus *in Japan.* Forestry Agency, Ministry of Agriculture and Forestry, Tokyo, Japan.

Haruki, M. and Tsuyuzaki, S. (2001) Woody plant establishment during the early stages of volcanic succession on Mount Usu, northern Japan. *Ecological Research* 16, 451–457.

Heaton, R.J., Slater, F.M., Randerson, P.F. and Samuel, R.W. (2001) The influence of fertilization on the yield of short rotation willow coppice in the uplands of mid-Wales. *Aspects of Applied Biology* 65, 77–82.

Heilman, P.E. and Stettler, R.F. (1985) Genetic variation and productivity of *Populus trichocarpa* T. and G. and its hybrids. II. Biomass productivity in a four-year plantation. *Canadian Journal of Forest Research* 15, 384–388.

Heilman, P.E., Stettler, R.F., Hanley, D.P. and Carkner, R.W. (1990) *High Yield Hybrid Poplar Plantations in the Pacific Northwest.* Pacific Northwest Extension Publication PNW-356. USDA Forest Service, Portland, Oregon.

Heinze, B. and Lexer, C. (2006) Natural hybrid zones as a laboratory of evolution: *Populus* ×*canescens* (= *P. alba* × *P. tremula*) along the Austrian Danube. Abstract of presentation at International Poplar Symposium IV. Nanjing Forestry University, Nanjing, China, pp. 42.

Henry, A. (1914) A new hybrid poplar. *Gardeners' Chronicle, Series III* 56, 257–258.

Hey, J. (2006a) On the failure of modern species concepts. *Trends in Ecology and Evolution* 21, 447–450.

Hey, J. (2006b) Recent advances in assessing gene flow between diverging populations and species. *Current Opinion in Genetics and Development* 16, 592–596.

Hjelmqvist, H. (1948) Studies on the floral morphology and phylogeny of the *Amentiferae. Botanica Notiser Supplement* 2, 1–117.

Holmgren, N.H. (1998) *Illustrated Companion to Gleason and Cronquist's Manual.* The New York Botanical Garden, New York.

Hong, T., Ma, Z.-L. and Chen, J.-S. (1987) Floral morphology of *Populus lasiocarpa* and its phylogenetic position in *Populus. Acta Botanica Sinica* 29, 236–241.

Hughes, F.M.R. (1990) The influence of flooding regimes on forest distribution and composition in the Tana River floodplain, Kenya. *Journal of Applied Ecology* 27, 475–491.

Hukin, D., Cochard, H., Dreyer, E., LeThiec, D. and Bogeat-Triboulot, M.B. (2005) Cavitation vulnerability in roots and shoots: does *Populus euphratica* Olivia, a poplar from arid areas of Central Asia, differ from other poplar species? *Journal of Experimental Botany* 56, 2003–2010.

Johnson, R.L. (1990) *Populus heterophylla* L. Swamp cottonwood. In: Burns, R.M. and Honkala, B.H. (eds) *Silvics of North America. Vol 2: Hardwoods.* Agriculture Handbook 271. USDA Forest Service, Washington, DC, pp. 551–554.

Johnsson, H. (1942) Cytological studies of triploid progenies of *Populus tremula. Hereditas* 28, 306–312.

Jones, J.R. and Schier, G. A. (1985) Growth. In: DeByle, N.V. and Winokur, R.P. (eds) *Aspen: Ecology and Management in the Western United States.* General Technical Report RM-119. USDA Forest Service, Rocky Mountain Research Station, Fort Collins, Colorado, pp. 19–24.

Jordaan, M. (2005) FSA Contributions 18. *Salicaceae. Bothalia* 35, 7–20.

Kapoor, K.S., Verma, R.K. and Rawat, R.S. (2004) Performance of different provenances of *Populus ciliata* Wall. ex Royale in Kullu valley, Himachal Pradesh. *Indian Journal of Forestry* 27, 57–61.

Karrenberg, S., Kollmann, J. and Edwards, P.J. (2002) Pollen vectors and inflorescence morphology in four species of *Salix. Plant Systematics and Evolution* 235, 181–188.

Khan, A.H. (1967) *Salix in Pakistan.* West Pakistan Agricultural University Press, Lyalpur, Pakistan.

Klopper, R.R., Chatelain, C., Banninger, V., Habashi, C., Steyn, H.M., de Wet, C., *et al.* (2006) *Checklist of the Flowering Plants of Sub-Saharan Africa. Report No 42.* Southern African Botanical Diversity Network, Pretoria, South Africa.

Knowe, S.A., Foster, G.S., Rousseau, R.J. and Nance, W.L. (1998) Height-age and height-diameter relationships for monocultures and mixtures of eastern cottonwood clones. *Forest Ecology and Management* 106, 115–123.

Kopp, R.F., Smart, L.B., Maynard, C.A., Isebrands, J.G., Tuskan, G.A. and Abrahamson, L.P. (2001) The development of improved willow clones for eastern North America. *Forestry Chronicle* 77, 287–292.

Koster, R. (1972) Eleven new poplar clones: an introduction. Communication No. 16. Department of Silviculture, Agricultural University, Wageningen, Netherlands.

Krasnoborov, I.M. and Malyshev, L.I. (eds) (2003) *Flora of Siberia. Vol 5: Saliceae – Amaranthaceae.* Science Publishers, Inc, Enfield, Massachusetts.

Laidly, P.R. (1990) *Populus grandidentata* Michx. Bigtooth aspen. In: Burns, R.M. and Honkala, B.H. (eds) *Silvics of North America. Vol 2: Hardwoods.* Agriculture Handbook 271. USDA Forest Service, Washington, DC, pp. 544–550.

Lefèvre, F., Kajba, D., Heinze, B., Rotach, P., de Vries, S.M.G. and Turok, J. (2001) Black poplar: a model for gene resource conservation in forest ecosystems. *Forestry Chronicle* 77, 239–244.

Li, H.-L. (1996) *Shade and Ornamemtal Trees: Their Origin and History.* University of Pennsylvania Press, Philadelphia, Pennslyvania.

Li, F.R., Zhang, H., Zhao, L.Y., Shirato, Y. and Wang, X.Z. (2003a) Pedoecological effects of a sand-fixing poplar (*Populus simonii* Carr.) forest in a desertified sandy land of Inner Mongolia, China. *Plant and Soil* 256, 431–442.

Li, Z.J., Liu, J.-P., Yu, J. and Zhou, Z.-L. (2003b) Investigation on the characteristics of biology and ecology of *Populus euphratica* and *Populus pruinosa. Xibei Zhiwu Xuebao* 23, 1292–1296.

Liesebach, M. and Naujoks, G. (2004) Approaches on vegetative propagation of difficult-to-root *Salix caprea. Plant Cell, Tissue and Organ Culture* 79, 239–247.

Linnaeus, C. (1751) *Philosophia Botanica.* Stockholm and Amsterdam.

Little, E.L. Jr (1979) *Checklist of United States Trees.* Agriculture Handbook 541. USDA Forest Service, Washington, DC.

McComb, A.L. and Hansen, N.J. (1954) A naturally occurring aspen-poplar hybrid. *Journal of Forestry* 52, 528–529.

McKnight, J.S. (1970) Planting cottonwood cuttings for timber production in the South. Research Paper SO-60. USDA Forest Service, New Orleans, Louisiana.

Manchester, S.R., Dilcher, D.L. and Tidwell, W.D. (1986) Interconnected reproductive and vegetative remains of *Populus* (*Salicaceae*) from the middle Eocene Green River Formation, northeastern Utah. *American Journal of Botany* 73, 156–160.

Meikle, R.D. (1984) *Willows and Poplars of Great Britain and Ireland.* Botanical Society of the British Isles, London.

Mitton, J.B. and Grant, M.C. (1996) Genetic variation and natural history of quaking aspen. *BioScience* 46, 25–31.

Mofidabadi, A.J., Modri-rahmati, A.R. and Tavesoil, A. (1998) Application of ovary and ovule culture in *Populus alba* L. × *P. euphratica* Oliv. hybridization. *Silvae Genetica* 47, 332–334.

Mosseler, A. (1987) Interspecific hybridization and reproductive barriers between some North American willow species. PhD thesis, University of Toronto, Toronto, Ontario, Canada.

Mosseler, A. and Zsuffa, L. (1989) Sex expression and sex ratios in intra- and interspecific hybrid families of *Salix* L. *Silvae Genetica* 38, 12–17.

Nazarow, M.I. (1936) *Salix.* In: Komarov, V.L. (ed.) *Flora of the USSR. Volume V.* Ixdatel'stvo Akademii Hauk SSSR, Moskva-Leningrad, Russia. (In Russian)

Nazarow, M.I. (1970) *Salix.* In: Komarov, V.L. (ed.) *Flora of the USSR. Volume V.* Israel Program for Scientific Translations, Jerusalem.

Newsholme, C. (1992) *Willows: the Genus Salix.* B.T. Batsford Ltd, London.

Newsholme, C. (2002) *Willows, the Genus Salix.* Timber Press, Portland, Oregon.

Newsome, J. (ed.) (1964) *Pliny's Natural History: a Selection from Philemon Holland's Translation.* Clarendon Press, Oxford, UK.

Ohashi, H. (2000) A systematic enumeration of Japanese *Salix* (*Saliceae*). *Journal of Japanese Botany* 75, 1–41.

Ohashi, H. (2001) *Saliceae* of Japan. *Science Reports of the Tohoku University, 4th series, Biology* 40, 269–396.

Parolin, P., Oliveira, A.C., Piedade, M.T.F., Wittmann, F. and Junk, W.J. (2002) Pioneer trees in Amazonia floodplains: three key species form monospecific stands in different habitats. *Folia Geobotanica* 37, 225–238.

Pauley, S.S. (1949) Forest-tree genetics research: *Populus* L. *Economic Botany* 3, 299–330.

Peeters, L. and Totland, O. (1999) Wind to insect pollination ratios and floral traits in five alpine *Salix* species. *Canadian Journal of Botany* 77, 556–563.

Perala, D.A. (1990) *Populus tremuloides* Michx. Quaking aspen. In: Burns, R.M. and Honkala, B.H. (eds) *Silvics of North America. Vol 2: Hardwoods.* Agriculture Handbook 271. USDA Forest Service, Washington, DC, pp. 555–569.

Perlin, J. (1991) *A Forest Journey: the Role of Wood in the Development of Civilization.* Harvard University Press, Cambridge, Massachusetts.

Pitcher, J.A. and McKnight, J.S. (1990) *Salix nigra* Marsh. Black willow. In: Burns, R.M. and Honkala, B.H. (eds) *Silvics of North America. Vol 2: Hardwoods.* Agriculture Handbook 271. USDA Forest Service, Washington, DC, pp. 768–772.

Pitta, D.W., Barry, T.N., Lopez-Villalobos, N. and Kemp, P.D. (2005) Effect on ewe reproduction of grazing willow fodder blocks during drought. *Animal Feed Science and Technology* 120, 217–234.

Qi, F., Wei, L., Jianhua, A., Yonghong, S., Yewu, Z., Zongqiang, C., *et al.* (2005) Environmental effects of water resource development and use in the Tarim River basin of northwestern China. *Environmental Geology* 48, 202–210.

Rackham, O. (1990) *Trees and Woodland in the British Landscape.* J.M. Dent and Sons, London.

Rehder, A. (1990) *Manual of Cultivated Trees and Shrubs Hardy in North America*, 2nd edn (reprinted). Dioscorides Press, Portland, Oregon.

Rieger, R., Michaelis, A. and Green, M.M. (1991) *Glossary of Genetics.* Springer-Verlag, Berlin.

Roller, K.J., Thibault, D.H. and Hidahl, V. (1972) Guide to the identification of poplar cultivars on the prairies. Publication 1311. Canadian Forest Service, Ottawa.

Ronald, W.G., Lenz, L.M. and Cumming, W.A. (1973) Biosystematics of the genus *Populus* L. I. Distribution and morphology of native Manitoba species and variants. *Canadian Journal of Botany* 51, 2431–2442.

Rood, S.B., Campbell, J.S. and Despins, T. (1986) Natural poplar hybrids from southern Alberta. I. Continuous variation for foliar characteristics. *Canadian Journal of Botany* 64, 1382–1388.

Rood, S.B., Kalischuk, A.R., Polzin, M.L. and Braatne, J.H. (2003) Branch propagation, not cladoptosis, permits dispersive, clonal reproduction of riparian cottonwoods. *Forest Ecology and Management* 186, 227–242.

Rüger, N., Schlüter, M. and Matthies, M. (2005) A fuzzy habitat suitability index for *Populus euphratica* in the northern Amudarya delta (Uzbekistan). *Ecological Modelling* 184, 313–328.

Rzedowski, J. (1975) Three new Mexican dicotyledons of possible ornamental interest. *Boletin de la Sociedad Botanica de Mexico* 35, 37–49.

Saki, A.K. and Burris, T.A. (1985) Growth in male and female aspen clones: a twenty-five-year longitudinal study. *Ecology* 66, 1921–1927.

Santamour, F.S. and McArdle, A.J. (1988) Cultivars of *Salix babylonica* and other weeping willows. *Journal of Arboriculture* 14, 180–184.

Sargent, C.S. (1949) *Manual of the Trees of North America.* Dover Publications, New York.

Scarascia-Mugnozza, G.E., Ceulemans, R., Heilman, P.E., Isebrands, J.G., Stettler, R.F. and Hinckley, T.M. (1997) Production physiology and morphology of *Populus* species and their hybrids grown under short rotation. II. Biomass components and harvest index of hybrid and parental species clones. *Canadian Journal of Forest Research* 27, 285–294.

Sennerby-Forsse, L. and Zsuffa, L. (1995) Buds structure and resprouting in coppiced stools of *Salix viminalis* L., *S. eriocephala* Michx., and *S. amygdaloides* Anders. *Trees* 9, 224–234.

Singh, R.V. (1982) *Fodder Trees of India.* Oxford and IBH Publishing Co., New Delhi.

Skvortsov, A.K. (1968) *Willows of the USSR.* Nauka, Moscow. (In Russian)

Skvortsov, A.K. (1973) Present distribution and probable original range of the crack willow (*Salix fragilis* L.). Problems in biogeocoenology, geobotany and botanical geography. Nauka, Leningrad, Russia, pp. 263–280. (In Russian)

Skvortsov, A.K. (1999) *Willows of Russia and Adjacent Countries. Taxonomical and Geographical Revision* (English translation of 1968 Russian edition). University of Joensuu, Joensuu, Finland.

Smith, R.L. and Sytsma, K.J. (1990) Evolution of *Populus nigra* (Sect. *Aigeiros*): introgression hybridization and the chloroplast contribution of *Populus alba* (Sect. *Populus*). *American Journal of Botany* 77, 1176–1187.

Spies, T.A. and Barnes, B.V. (1982) Natural hybridization between *Populus alba* and the native aspens in southeastern Michigan. *Canadian Journal of Forest Research* 12, 653–660.

Stace, C. (1997) *New Flora of the British Isles.* Cambridge University Press, New York.

Stanton, B.J. (2005) The effect of reciprocal hybridization on reproduction of the intersectional cross, *Populus ×generosa. Forest Genetics* 12, 131–140.

Stettler, R.F., Koster, R. and Steenackers, V. (1980) Interspecific crossability studies in poplars. *Theoretical and Applied Genetics* 58, 273–282.

Stettler, R.F., Fenn, R.C., Heilman, P.E. and Stanton, B.J. (1988) *Populus trichocarpa × P. deltoides* hybrids for short-rotation culture: variation patterns and 4-year field performance. *Canadian Journal of Forest Research* 18, 745–753.

Stettler, R.F., Zsuffa, L. and Wu, R. (1996) The role of hybridization in the genetic manipulation of *Populus.* In: Stettler, R.F., Bradshaw, H.D. Jr, Heilman, P.E. and Hinckley, T.M. (eds) *Biology of Populus and Its Implications for Management and Conservation.* National Research Council of Canada Research Press, Ottawa, pp. 87–112.

Steyn, E.M.A., Smith, G.F. and Van Wyk, A.E. (2004) Functional and taxonomic significance of seed structure in *Salix mucronata* (*Salicaceae*). *Bothalia* 34, 53–59.

Stout, A.B. and Schreiner, E.J. (1933) Results of a project in hybridizing poplars. *Journal of Heredity* 24, 216–229.

Suda, Y. and Argus, G.W. (1968) Chromosome numbers of some North American *Salix. Brittonia* 20, 191–197.

Suda, Y. and Argus, G.W. (1969) Chromosome numbers of some North American Arctic and Boreal *Salix. Canadian Journal of Botany* 47, 859–862.

Sudworth, G.B. (1908) *Forest Trees of the Pacific Slope.* USDA Forest Service, Washington, DC.

Sugaya, S. (1960) Bearing of the cataphyllotaxy on the interpretation of the nectary structures in the flowers of the *Saliceae. Scientific Reports of Tohoku University, 4th series (Biology)* 26, 9–24.

Takeda, A. (1975) Studies on the cold resistance of forest trees IV. Relation between bud dormancy and frost hardiness in Kamabuchi poplar. *Bulletin of the Faculty of Agriculture Mie University* 48, 261–274.

Tamura, S. and Kudo, G. (2000) Wind pollination and insect pollination of two temperate willow species, *Salix miyabeana* and *Salix sachalinensis. Plant Ecology* 147, 185–192.

Tamura, T., Morohoshi, N. and Yasuda, S. (2001) Suitability of peroxidase-suppressed transgenic hybrid aspen (*Populus sieboldii × P. grandidentata*) for pulping. *Holzforschung* 55, 335–339.

Thibault, J. (1998) Nuclear DNA amount in pure species and hybrid willows (*Salix*): a flow cytometric investigation. *Canadian Journal of Botany* 76, 157–165.

Thomé, O.W. (1885) *Flora von Deutschland Österreich und der Schweiz.* Verlag von Fr. Eugen Köhler, Gera-Untermhaus, Germany.
Tsarev, A.P. (2005) Natural poplar and willow ecosystems on a grand scale: the Russian Federation. *Unasylva* 221, 10–11.
Tuskan, G.A., DiFazio, S., Jansson, S., Bohlmann, J., Grigoriev, I., Hellsten, U., *et al.* (2006) The genome of black cottonwood, *Populus trichocarpa* (Torr. and Gray). *Science* 313, 1596–1604.
Van Broekhuizen, J.T.M. (1972) Morphological description and identification of a number of new commercial poplar clones. Communication 16. Department of Silviculture, Agricultural University Wageningen, Netherlands.
Vanden Broeck, A. (2003) EUFORGEN Technical guidelines for genetic conservation and use for European black poplar (*Populus nigra*). International Plant Genetic Resources Institute, Rome.
Vázquez, G.J.A. and Cuevas, R. (1989) Una nueva especie tropical de *Populus* (*Salicaceae*) de la Sierra de Manantlán, Jalisco, México. *Acta Botanica Mexico* 8, 39–45.
Verwijst, T. (2001) Willows: an underestimated resource for environment and society. *Forestry Chronicle* 77, 281–285.
Viereck, L.A. and Foote, J.M. (1970) The status of *Populus balsamifera* and *P. trichocarpa* in Alaska. *Canadian Field Naturalist* 84, 169–173.
Voss, E.G. (1985) Michigan Flora, Part II: Dicots (*Saururaceae-Cornaceae*). Bulletin 59. Cranbrook Institute of Science, Bloomfield Hills, Michigan.
Vroege, P.W. and Stelleman, P. (1990) Insect and wind pollination in *Salix repens* L. and *Salix caprea* L. *Israel Journal of Botany* 39, 125–132.
Wagner, W.H., Daniel, T.F., Beitel, J.M. and Crispin, S.R. (1980) Studies on *Populus heterophylla* in southern Michigan. *Michigan Botanist* 19, 269–275.
Wang, C. and Fang, C.F. (1984) *Saliceae. Flora Republicae Popularis Sinicae* 20. Science Press, Beijing, pp. 1–403.
Warren-Wren, S.C. (1972) *The Complete Book of Willows.* A.S. Barnes and Company, New York.
Weisgerber, H. (1999) *Populus nigra Linné, 1753. Enzyklopädie der Holzgewächse 16.* Wiley-VCH, Weinheim, Germany.
Weisgerber, H. (2000) *Populus euphratica Olivier, 1801. Enzyklopädie der Holzgewächse 22.* Wiley-VCH, Weinheim, Germany.
Weisgerber, H. and Han, Y. (2001) Diversity and breeding potential of poplar species in China. *Forestry Chronicle* 77, 227–237.
Weisgerber, H. and Zhang, Z. (2005a) *Populus cathayana Rehder, 1931. Enzyklopädie der Holzgewächse 41.* Nicol, Hamburg, Germany.
Weisgerber, H. and Zhang, Z. (2005b) *Populus yunnanensis Dode, 1905. Enzyklopädie der Holzgewächse 39.* Nicol, Hamburg, Germany.
Werner, A. and Siwecki, R. (1994) Embryogenesis and characteristics of hybrids between resistant to poplar rust *Populus lasiocarpa* and the susceptible *Populus nigra. Arboretum Kornickie* 39, 145–154.
Whitham, T.G. (1989) Plant hybrid zones as sinks for pests. *Science* 244, 1490–1493.
Wilmot-Dear, C.M. (1985) *Flora of Tropical East Africa.* A.A. Balkema, Boston, Massuchusetts.
Wing, S.L. (1981) A study of paleoecology and paleobotany in the Willowwood Formation (Early Eocene, Wyoming). PhD dissertation, Yale University, New Haven, Connecticut.
Wyckoff, G.W. and Zasada, J.C. (2007) *Populus* L. In: Bonner, F.T. and Nisley, R.T. (eds) *Woody Plant Seed Manual.* USDA Forest Service, Washington, DC.
Yang, K.-X., Ti, Z.-Z. and Yi, Z.-Z. (1999a) Cytogenetic studies on the origin of Chinese white poplar. *Journal of Beijing Forestry University* 21, 6–10.
Yang, Z.X., Li, G.T. and Gao, Z.H. (1999b) A survey of *Populus simonii* resources in four provinces of northwest China (in Chinese). *World Forestry Research* 12, 49–53.
Zasada, J.C. and Phipps, H.M. (1990) *Populus balsamifera* L. Balsam poplar. In: Burns, R.M. and Honkala, B.H. (eds) *Silvics of North America. Vol 2: Hardwoods.* Agriculture Handbook 271. USDA Forest Service, Washington, DC, pp. 518–529.
Zhuge, Q., Huang, M.R. and Wang, M.X. (2000) A study on poplar somatic hybridization. *Journal of Nanjing Forestry University* 24, 6–10.
Zinovjev, A. (2011) *Salix* ×*meyeriana* (= *S. pentandra* × *S. euxina*) – a forgotten willow in Eastern North America. *Phytotaxa* 22, 57–60.
Zsuffa, L. (1973) A summary review of inter-specific breeding in the genus *Populus* L. In: Fowler, D.P. and Yeatman, C.W. (eds) *Proceedings of the 14th Meeting of the Committee on Forest Tree Breeding in Canada.* Canadian Forest Service, Fredericton, New Brunswick, Canada, pp. 107–123.

Zsuffa, L. (1974) The genetics of *Populus nigra* L. *Annales Forestales* 6, 29–53.
Zsuffa, L., Giordano, E., Pryor, L.D. and Stettler, R.F. (1996) Trends in poplar culture: some global and regional perspectives. In: Stettler, R.F., Bradshaw, H.D. Jr, Heilman, P.E. and Hinckley, T.M. (eds) *Biology of Populus and Its Implications for Management and Conservation.* National Research Council of Canada Research Press, Ottawa, pp. 515–539.

3 Ecology and Physiology of Poplars and Willows

J. Richardson,[1]* J.G. Isebrands[2] and J.B. Ball[3]
[1]*Ottawa, Ontario, Canada;* [2]*Environmental Forestry Consultants LLC, New London, Wisconsin, USA;* [3]*Corsham, Wiltshire, UK*

3.1 Introduction

Much has been learned about the ecology and physiology of poplars and willows since the previous editions of the FAO series on poplar and willow production were published (FAO, 1958, 1980). Forest ecology is the study of relationships between all the organisms, i.e. plants, insects, mammals and birds, in the forest and the relationship between them and the physical environment (Spurr and Barnes, 1980). Tree physiology is the study of a plant's life processes and their functions in relation to the environment (Landsberg, 1986). Thus, the total environment of poplars and willows growing in nature is a complex interaction of their genetic constitution with the physical and biological elements in the ecosystem (Fowells and Means, 1990; Farmer, 1996). It is impossible to know all aspects of every poplar and willow growing in nature, but amazingly, poplars and willows have managed to be ecologically resilient and able to adapt to their ever-changing environment for millions of years (Farmer, 1996; Kuzovkina, 2010). In this chapter, we review the natural occurrence, life history and present status of selected ecologically important species of poplars and willows around the world. It is our belief that more knowledge and understanding of how these trees and shrubs grow under different environmental conditions will allow future generations to make more rational decisions on how to respond to changing environmental conditions. Future forest managers will be faced with making decisions on how to improve wildlife habitat, soil erosion control, water production, biomass production and/or respond to and mitigate the effects of climate change as well as human activity impacts. They should make these decisions based on knowledge and understanding of ecological and physiological processes rather than by chance (Spurr and Barnes, 1980; Landsberg, 1986; Kimmins, 1987; Kozlowski *et al.*, 1991; Hinckley *et al.*, 1992; Farmer, 1996).

Readers interested in more details on the ecology and silvics of poplars and willows beyond the scope of this chapter should refer to Maini (1968), Boyce (1976), Burns and Honkala (1990), Mitchell *et al.* (1992), Stettler *et al.* (1996), Dickmann *et al.* (2001b) and Kuzovkina *et al.* (2008). Those interested in more general references on forest ecology and physiology should consult Spurr and Barnes (1980), Landsberg (1986), Kimmins (1987) and Kozlowski *et al.* (1991).

*Corresponding author: jrichardson@on.aibn.com

3.2 *Populus euphratica* Olivier, Euphrates Poplar

3.2.1 Natural occurrence

Populus euphratica Olivier has a very extensive but discontinuous natural distribution, from Morocco and Egypt in Africa (and possibly Spain) through the Middle East to Central Asia, including the Xinjiang Autonomous Region of China and the Himalayan valleys of India and Pakistan (Plate 1B; see Chapter 2, this volume). It occurs from 48 to 49°N in Kazakhstan to 15°N in Yemen and from below sea level in the Dead Sea area to 4000 m in the Himalayas. Estimates of the natural area of *P. euphratica* are incomplete, but it is clear that the area is shrinking. Wang *et al.* (1996) estimated about 400,000 ha in China, but this is much less than the estimate of 568,000 ha by Viart (1988) just a few years earlier.

Although originally growing in dense forests (ICRAF, 2011), *P. euphratica* is now typically found in linear stands along riverbanks and gallery forests, and on islands in rivers, but it can tolerate extremely arid conditions in desert areas provided its roots can reach water in the subsoil layers. It also tolerates seasonally waterlogged soils as well as highly saline conditions. The high temperatures in much of its natural range (more than 50°C in India) are also tolerated well. This wide ecological amplitude is one of the most defining characteristics of the species (Viart, 1988; Wang *et al.*, 1996).

Other tree and shrub species commonly associated with *P. euphratica* include water-demanding species such as *Tamarix* spp. and *Salix* spp., as well as others (Wang *et al.*, 1996). These associations were more common when dense forests were prevalent (ICRAF, 2011).

3.2.2 Life history

Natural reproduction of *P. euphratica* is through root suckers or seed. Seedlings germinate and grow rapidly on fresh alluvial soil after flooding recedes on riparian sites. They require much light for normal development (ICRAF, 2011). On desert conditions in China, growth is best when groundwater is within 2–4 m of the surface, but decreases considerably when the groundwater is deeper (Gries *et al.*, 2003).

The form of *P. euphratica* is normally short boled, twisted and heavily branched. Along the Euphrates River in Turkey, *P. euphratica* reaches an average diameter of 30 cm and a height of 30 m. In the Goksu River area, diameters and heights are less, 18–20 cm and 8–10 m, respectively (F. Toplu, 2008, unpublished results). Similar ranges in productivity on different site conditions are reported from Iran – diameters ranging from 10 to 27 cm and heights from 3 to 12 m. The average density of stands in Khuzistan province was 320–350 stems ha^{-1} and mean annual diameter growth 9.5 mm (M. Calagari, 2008, unpublished results). Under favourable conditions, diameters as great as 80 cm have been reported (CABI, 2012a).

3.2.3 Status

Despite its vast natural range, the status of *P. euphratica* is severely threatened as a result of many centuries of overexploitation for fuelwood, fodder, shelter and clearing for agriculture (Viart, 1988). In Kazakhstan, the area of older stands decreased between 1969 and 1979, and most remaining stands were composed of young trees originating from root suckers (Wang *et al.*, 1996).

The importance of *P. euphratica* as a colonizing woody species on exposed soil and as a windbreak have been recognized, as well as the value of this extremely diverse and tolerant genetic resource, resulting in some efforts at conservation (see Chapter 4, this volume). In China, a major *in situ* conservation project is under way in the Tarim River nature reserve in Xinjiang Autonomous Region (Yimit *et al.*, 2006). In Central Asia, tugai forests, of which *P. euphratica* is a major constituent, are being studied and protected (Schluter *et al.*, 2006).

3.3 *Populus deltoides* Marshall, Eastern Cottonwood

3.3.1 Natural occurrence

Eastern cottonwood occurs widely throughout the USA and Canada, from the south-eastern USA in Georgia and Florida, north to Quebec and Ontario, west along the Gulf of Mexico to

Texas, north to Colorado, Wyoming and Montana in the USA and farther north to Manitoba, Saskatchewan and Alberta in Canada. In the western part of the range, the tree was formerly known as the plains cottonwood, *P. deltoides* var. *occidentalis* (Plate 2A; see Chapter 2, this volume).

Eastern cottonwood typically occurs along streams, rivers and flood plains, but is also found on sandy and rocky sites near the Great Lakes and in semi-arid areas in the western parts of its range. Because of its wide range, eastern cottonwood is subject to temperatures ranging from as high as 46°C in the south to as low as –45°C in the north and west. It occurs in areas with growing seasons less than 100 frost-free days to more than 200 frost-free days, but does not do well in regions with little frost. Rainfall in its range varies from 380 to 1500 mm, but because it occurs along streams, it receives its moisture from deep aquifers, allowing it to survive semi-arid conditions.

Its best growth is on moist, well-drained fine sands or silt loams near streams, but it survives on infertile sands and clays, making it very adaptable to varying conditions. When it grows on slopes it is usually confined to lower slopes with access to moisture.

Eastern cottonwood occurs in pure stands along streams and in mixed stands on upland sites. It is associated with a number of forest cover types including black ash – American elm – red maple, bur oak, river birch – sycamore, silver maple – maple, sweet gum – willow oak, sycamore – sweetgum and black willow (Cooper and van Haverbeke, 1990).

3.3.2 Life history

Eastern cottonwood is dioecious, having both male and female clones. Flowering occurs from February to April before leaves appear, depending on location in the range. Time of flowering varies dramatically within a stand, and northern trees flower at lower temperatures than southern trees. Trees are 4–5 years old when they flower and flowering occurs almost every year.

Seed production therefore begins at age 4–5 and increases with age. Large trees produce millions of seeds in a year, which are disseminated by wind about 2 months after flowering. Female clones are considered a nuisance in some urban areas because of their huge 'cotton' crops, which can be allergenic. Seeds often fall in water and are carried long distances before being deposited. Seeds must reach a favourable seedbed before germinating and 90% germination is possible on good conditions such as on moist, exposed silt loam. When conditions are right for germination, seedlings germinate by the millions and occupy very large areas.

Eastern cottonwood readily sprouts vegetatively from low-cut stumps up to age 25. It is typically a fast-growing tree that can become very large at maturity. It often lives to be over 200 years old and can attain heights of up to 53–58 m and diameters from 120 to 300 cm. At age 5, it can attain 30 cm in diameter and at age 9 more than 30 m in height, with over 4 m height growth per year (Dickmann and Stuart, 1983). Root systems are generally shallow and wide spreading but can be deep on sandy loam soils (Farrar, 1995). Eastern cottonwood is very intolerant to shade. Thus, pure stands are more common than mixed stands. Its rapid growth does allow it to outgrow competitors, except where there is prolonged flooding.

Although it grows rapidly, eastern cottonwood is subject to numerous damaging agents. Insect pests include the clearwing borer (*Paranthrene* spp.), cottonwood leaf beetle (*Chrysomela* spp.), cottonwood borer (*Plectrodera* spp.), cottonwood twig borer (*Gypsonoma* spp.) and poplar borer (*Saperda* spp.) (Cooper and van Haverbeke, 1990). Common diseases are *Fusarium* canker and *Cytospora* on poor sites. *Melampsora* leaf rust is very damaging on more difficult sites, especially in the autumn and is under strong genetic control. *Melampsora* is seriously damaging to cottonwood culture in the southern hemisphere (Pryor and Willing, 1965). *Marssonina* leaf spot is damaging to susceptible clones in the south-eastern part of the range. *Septoria* canker is not so much a problem with pure eastern cottonwood as it is with its hybrids. Cottonwood across its range is subject to animal damage from deer and rodents.

3.3.3 Status

Natural populations of eastern cottonwood remain abundant throughout its range, except where man-made dams have interfered with natural reproduction (Braatne *et al.*, 1996; Stettler, 2009). Riverine systems more than ever before provide opportunities for spontaneous hybridization with naturally co-occurring populations of *Populus* species that have been useful in the study of ecological and evolutionary concepts (Whitham *et al.*, 2001). Eastern cottonwood remains the most important parental species for worldwide *Populus* hybridization programmes (Dickmann, 2006; Chapter 4, this volume).

3.4 *Populus nigra* Linnaeus, Black Poplar

3.4.1 Natural occurrence

Black poplar, *P. nigra* L., has a very extensive range throughout Europe – except for the Nordic countries – and from North Africa to West and Central Asia, including the Caucasus and large parts of the Middle East (see Chapter 2, this volume; Plate 2B). It occurs from 64°N in Siberia to 30°N in Pakistan, and from sea level to 4000 m in elevation. However, it is considered to be on the verge of extinction over much of its natural range, particularly in Western Europe, due to the impact of high human population densities (Vanden Broeck, 2003).

Black poplar grows best in temperate climates with rainfall in spring and autumn. It can tolerate dry summers and does not grow well in areas with high rainfall (>1000 mm annually). It can survive on a wide range of different soil conditions, from stony, poor or dry soils to heavy clay soils, even tolerating waterlogging, but does not grow well on such conditions. It grows best on deep, medium-texture soils with pH between 5.5 and 7.5 and high fertility, such as on flat or gently sloping areas near the base of south-facing slopes (CABI, 2012b).

Until loss of habitat started in the 17th century, black poplar was an important constituent of the natural flood plain forests on the banks of rivers in Europe (see Plate 17C). Its present distribution extends from isolated trees in areas most heavily impacted by human intervention to large blocks of pure or mixed stands in less heavily impacted areas (see also Chapter 2, this volume).

3.4.2 Life history

P. nigra produces regular seed crops starting at about 10 years of age. Seeds are disseminated by wind and water, and germinate readily in moist sandy soil exposed after seasonal inundation on river flood plains, which it is therefore able to colonize aggressively. However, anoxic conditions will prevent seedling root growth and establishment (Vanden Broeck, 2003). It also sprouts vigorously from stumps, roots easily from fallen trees, broken roots and branches, particularly at the juvenile stage, and can reproduce to some extent by means of suckers from shallow or exposed roots (Cagelli *et al.*, 1998).

Mature trees have broad, dense crowns, with more or less straight stems bearing thick knots and heavy branches. Older individuals can reach 20–40 m in height and 60–100 (or even 200) cm in diameter. Black poplar trees may live up to 400 years, though 200–300 years is more normal (Allegri, 1971). It may be tall and straight, or broad crowned or crooked, or even with multiple stems.

3.4.3 Status

P. nigra is a pioneer species growing mainly along rivers. Its natural populations are, to a large extent, threatened by river engineering, habitat destruction and other human activities (EUFORGEN, 2009) (Plate 19E). It is considered to be on the verge of extinction in a large part of its natural range. In the UK, it has been reduced from a useful provider of products for the agricultural community before 1800 to an estimated remnant of only 7000 trees (Cottrell, 2004).

The importance of European flood-plain forests including *P. nigra*, as among the most diverse ecosystems in Europe and potential centres for biodiversity, is increasingly recognized (Vanden Broeck, 2003; Toplu, 2005). Interest is growing in conserving and restoring such riparian ecosystems for the natural control of flooding, to help control the diffuse pollution of water and because of their potentially high amenity value. The FLOBAR 1 and 2 projects funded by the European Union (EU) are investigating how to restore the biological function of flood-plain forests by restoring the physical processes that drive them (Cottrell, 2004).

The *Populus nigra* Network was established in 1994 as part of the European Forest Genetic Resources Programme (EUFORGEN). The aim of the EUFORGEN Networks is to conserve, at a European level, the still existing genetic resources available and, where possible, to maintain or restore the genetic diversity through a number of measures (EUFORGEN, 2009, 2011). A number of European countries, including Austria, Croatia, France, Germany, Hungary, Italy, the Netherlands, Spain and Turkey, have either *in situ* or *ex situ* conservation programmes for *P. nigra* (Chapter 4, this volume). Croatia, in particular, is actively conserving its remaining *P. nigra* populations, which, though extensively fragmented, are considered well preserved by European standards (Kajba *et al.*, 2006). Conservation work in Turkey has identified five new distribution areas of *P. nigra* in eastern Anatolia (Toplu, 2005).

3.5 *Populus balsamifera* Linnaeus, Balsam Poplar

3.5.1 Natural occurrence

Balsam poplar occurs across North America and is one of the most widely distributed North American tree species. It grows in 31 US states and in 13 Canadian provinces and territories (USDA Forest Service, 2012a). It does not occur west of the Cascade Mountains, where *P. trichocarpa* predominates (Viereck and Foote, 1970); however, in North America it does grow farther north than any other tree species, even occurring in disjunct stands along rivers on the Alaskan North Slope above the treeline (Edwards and Dunwoodie, 1985) (Plate 3A; Chapter 2, this volume).

Balsam poplar grows commonly in alluvial ecosystems that are subject to flooding. Its maximum stand development occurs along flood plains in Alaska, the Yukon, Northwest Territories, British Columbia and Alberta (Zasada and Phipps, 1990). It grows best in soil consisting of river sediment and organic matter. Its survival during periodic flooding is due to preformed root primordia along balsam poplar stems (Farmer *et al.*, 1989). Balsam poplar also occurs on upland sites with glacial till soil, on outwash sand and on loess, although it does better on upland sites with wet spring weather. However, it can also grow on dry, sandy, north-facing slopes in Canada, as well as in areas near warm springs in the northern permafrost zone.

Over its wide natural range, balsam poplar is subject to huge temperature differences. The lowest temperatures are from −18° to −62°C and the highest temperatures range from 30° to 44°C (CABI, 2012c). Balsam poplar has a very high frost resistance that accounts for its northernmost distribution (Peterson and Peterson, 1992; Farmer, 1996). The annual precipitation in its range varies from 150 to 300 mm in Alaska and Yukon to 1400 mm in the Maritime Provinces in the east. Annual snowfall is lowest in interior Alaska and highest in Newfoundland. Frost-free periods vary from less than 75 days to 160 days, with the longest growing seasons in the south and the shortest in the north (Zasada and Phipps, 1990).

Balsam poplar is found mainly in mixed forest stands where other species eventually dominate. It occurs with white spruce, trembling aspen, jack pine, red spruce-balsam fir, northern white cedar, black ash, American elm and red maple on wetter sites. On upland sites, it also co-occurs with alpine fir, white birch and black spruce. It does grow in pure stands along rivers, where it is associated with alder, willows and white spruce. It is associated with a variety of shrub species on low sites, including red-osier dogwood, hazel, cranberry bush and raspberry.

3.5.2 Life history

Balsam poplar is a dioecious species with a ratio of male to female clones of about 1:1. Flowering occurs before leaf-out, typically in April and May, but not until June and July in the northern part of its range. Balsam poplar reaches flowering age by 10 years and normally produces large seed crops, depending on the year. Seed dissemination occurs typically in May and June and lasts about 2 weeks. Seeds remain viable for 4–6 weeks and are dispersed long distances by wind and water. Cooler and drier conditions favour seed viability. Its seed does not require a dormancy period and can germinate in a wide range of temperatures. Germination occurs over a wide range of environmental conditions.

The height of first-year seedling growth depends on stand density. It ranges from 2 to 32 cm, depending on environmental conditions and location. Balsam poplar is a prolific sprouter, with new stems originating from broken stems or roots, or from buried stems and branches. It usually occurs in stands with multiple clonal groups of mixed gender. Balsam poplar roots readily from stem cuttings and can produce roots from stems as old as 10–15 years (Zasada and Phipps, 1990). It readily produces suckers after the parent tree is disturbed. Density of suckers is greatest on sites with organic soils. Therefore, suckering is an important factor in establishment of balsam poplar on disturbed sites.

Balsam poplar can become a very large tree under good environmental conditions. Its size ranges from 20 to 30 m in height and from 40 to 180 cm in diameter. In the northern part of its range, it is often the largest tree in mixed 80- to 100-year-old forest stands. But there is clonal variation in growth rate (Peterson and Peterson, 1992). Most clonal stands have an age span of only 5 years, but they can span 50–60 years. Stand density of balsam poplar varies markedly with stand history. It often comprises only 5–10% of the mixed stand (Zasada and Phipps, 1990).

As an early successional clonal species, balsam poplar has low shade tolerance and rapid juvenile growth. It typically has a short lifespan and is replaced easily by more shade-tolerant associates. Balsam poplar is an early dominant species in riparian areas with willows and alders for about 20 years. In mixed upland stands, it is dominant for about 50 years and then is replaced by white spruce. There are reports of 200-year-old balsam poplar stands in the far northern areas of the Yukon. It can persist for over 230 years on the Alaskan North Slope (Edwards and Dunwoodie, 1985).

Balsam poplar is considered one of the most fire-resistant species in the boreal forest. Older trees have thick bark that affords them fire resistance; its ability to produce sprouts and suckers enables it to recover from fire. However, severe fires will kill balsam poplar stands (USDA Forest Service, 2012a). Although flooding of rivers can help establish balsam poplar stands, changes in river channels can destroy them (Braatne *et al.*, 1996).

The species is attacked by some insects, including poplar and willow borer (*Cryptorhynchus*), bronze poplar borer (*Agrilus*) and poplar borer (*Saperda*). These pests are probably the most damaging. It is also occasionally attacked by forest tent caterpillar (*Malacosoma*) and aspen leaf beetle (*Chrysomela*), but they are not primary damaging agents. The most common decay-causing diseases of mature balsam poplars are *Phellinis*, *Pholiata*, *Corticium* and *Bjerkandra*, but damage varies by site and environmental conditions. *Septoria* leaf spot is damaging to seedlings (Zasada and Phipps, 1990). Moose, deer and elk browse on balsam poplars to some extent. Hares and rodents browse on saplings and small trees, often girdling the stem, but the trees usually re-sprout to form new stems (USDA Forest Service, 2012a). Balsam poplars have a high resin content, so they are less palatable to browsing animals than other poplars.

Boreal forests that contain a significant component of balsam poplars support a very large variety of wild animals and birds. But these wildlife species are probably more dependent on trembling aspen than on balsam poplar, again because of the high resin content of the leaves and stems (Peterson and Peterson, 1992).

3.5.3 Status

Balsam poplar is an important component of the North American boreal forest. It is an important riparian species which helps stabilize river and stream banks. It re-colonizes sites that are disturbed by fire and/or logging. And, it provides important habitat for a wide variety of wildlife. Balsam poplar continues to be abundant across its range, largely due to its re-sprouting ability. It is not threatened or endangered, but there are concerns over the effects of climate change, pollution, oil exploration and forest stand conversion in the future (USDA Forest Service, 2012a). Balsam poplar's presence has been decreased by the effects of man-made dams on rivers and streams (Braatne *et al.*, 1996). Balsam poplar hybridizes naturally with *P. angustifolia*, *P. deltoides* and sometimes *P. tremuloides* in the western USA and Canada (Dickmann and Stuart, 1983; Rood *et al.*, 1985; Chapter 2, this volume).

3.6 *Populus maximowiczii* Henry, Japanese Poplar, and *Populus suaveolens* Fischer, Siberian Poplar

3.6.1 Taxonomy

The taxonomy of the east Asian balsam poplars, *P. maximowiczii* Henry, sometimes known as Japanese poplar, and *P. suaveolens* Fischer, sometimes known as Siberian poplar (or Mongolian poplar), has not been finally resolved. However, there is some suggestion that *P. maximowiczii* may eventually be recognized as a maritime subspecies of *P. suaveolens* (Eckenwalder, 2001) (see Chapter 2, this volume, for discussion of the taxonomy of these species). For the purposes of describing natural ecosystems, the two are considered here together under the name *P. suaveolens*, while recognizing that in breeding programmes the *P. maximowiczii* name is encountered most often (Chapter 4, this volume).

3.6.2 Natural occurrence

P. suaveolens has an extensive natural distribution, from north-central China, through Mongolia, east Siberia and Korea, to northern Japan (Plates 3C and 4A; Chapter 2, this volume). It occurs from 38°N in central China and Japan to almost the Arctic Circle in eastern Siberia, and from 100°E near Lake Baikal east to the Bering Strait. In elevation, it is found from sea level in Russia to 1400 m in Korea and 2000 m farther south in China (Vorob'ev, 1968; Flora of China Editorial Committee, 2003).

Over its wide range, *P. suaveolens* grows in a broad spectrum of habitats differing in respect to climate, soil moisture and soil fertility. Generally, the climate in its native range is temperate and continental or monsoon (northern type) in nature, with cold, dry winters and warm, moist summers, rain falling particularly in late summer. Mean maximum temperatures of the hottest month range from 18° to 25°C and mean minimum temperatures of the coldest month from −3° to −30°C. Annual rainfall ranges from 600 to 1400 mm (CABI, 2012d).

P. suaveolens grows as a solitary tree or in groups along mountain rivers and streams (CABI, 2012d). In the north, it is found on alluvial flood plains in even-aged stands of pure poplar or mixed with willow or *Chosenia* (another genus in the *Salicaceae* family) (Vasil'ev, 1967). Farther south, it grows in stands with coniferous species such as *Picea jezoensis*, *Pinus koraiensis* and *Abies sachalinensis* (Ageenko, 1969; Kojima, 1979), or with broadleaved species such as ash and elm.

A wide variety of soil conditions is tolerated, but best growth occurs on fresh to moist, well-drained loamy sands, which are often stratified alluvial deposits with pH 5–6, low humus content and sharp decrease in available phosphorus and potassium with increasing depth. However, minerals in river water and sediments compensate for any deficiency (Ageenko, 1969). On mountain slopes, *P. suaveolens* is sometimes found on brown forest soils with a thick humus layer (CABI, 2012d). In northern Japan and the Kuril Islands, the soils originate from volcanic ash and have substantial available phosphorus (Shoji, 1993). Regrowth after volcanic eruptions is by vegetative means. In the north, *P. suaveolens* may grow on soils with a layer of permafrost (CABI, 2012d).

3.6.3 Life history

Natural reproduction of *P. suaveolens* is through wind-dispersed seeds or root and shoot cuttings. However, natural regeneration by seeds is limited by the fact that the species is intolerant of shade. It regenerates best in open areas after floods or wildfires. Average stem densities of 14,000 ha^{-1} were recorded in vegetatively regenerated stands on mountain slopes in Japan and of 28,000 ha^{-1} with seedling regeneration (Tsuyuzaki and Haruki, 1996). Initial growth rates can be more than 1 m $year^{-1}$ height increment in 15- to 20-year-old stands and average volume increment of 15 m^3 ha^{-1} $year^{-1}$ (Ageenko, 1969).

Within much of its natural range, *P. suaveolens* may be one of the few woody plants reaching tree size. It reaches heights of 30 m and diameters of 2 m, making it one of the largest – as well as the fastest growing – poplars in East Asia (Dickmann, 2001) (see Fig. 4.10, Chapter 4, this volume, which shows the very closely related *P. maximowiczii*). The oldest trees may be as much as 200–300 years old, or even 400 years in the north (CABI, 2012d).

3.6.4 Status

The rapid growth and large size at maturity of *P. suaveolens* have attracted the interest of poplar breeders in other parts of the world to the species, mainly under the name of *P. maximowiczii*. Its genetic resources are not seen as being under any kind of threat and no conservation efforts have been undertaken. Exploration and documentation of its range of variation should be pursued and taxonomic identity issues resolved.

3.7 *Populus trichocarpa* Torrey and Gray, Black Cottonwood

3.7.1 Natural occurrence

Black cottonwood is the largest hardwood tree in western North America. It occurs from Kodiak Island at 62°N latitude south through Alaska, British Columbia, Washington, Oregon and California to Baja California in Mexico (latitude 31°N). It is also found inland in Alberta, Saskatchewan, Idaho and Montana. There are scattered populations in Wyoming, Utah, Nevada and North Dakota, USA (Plate 3A; Chapter 2, this volume).

It grows in climates varying from semi-arid to humid, but is best known for its growth in the humid coastal zones of the Pacific Northwest of Canada and the USA. Black cottonwood grows in temperature extremes from 16°C in the north to 47°C in the south. It tolerates minimum temperatures of 0° to –47°C. It grows in areas with precipitation from 250 to 3050 mm, of which one-third is snow in the inland areas. It tolerates a frost-free period of 70 days in the interior and can grow in the south, where there are more than 260 frost-free days a year.

Black cottonwood grows on a wide range of soils and topography. It prefers moist, loose, porous sandy or gravelly soils along riverbanks, but can grow on clay soils on uplands. High soil acidity restricts its growth. It grows from sea level to 1500 m elevation in the mountains. It grows in mixtures with tree willows in black cottonwood – willow forest cover types. It can grow with red alder, Douglas fir, western hemlock, western red cedar, Sitka spruce, grand fir, bigleaf maple and Oregon ash (DeBell, 1990).

3.7.2 Life history

Black cottonwood is dioecious, with both male and female clones. This feature makes it ideal for hybridization with other poplar species (Stettler, 2009). Flowering occurs at about 10 years and appears from late March to late May in coastal areas to mid-June in northern and inland stands.

Seed is disseminated in late May to late June in coastal areas and up to late July in inland mountains. It is transported easily by wind and water. The seed germinates in high percentages, but is viable for only 2 weeks to 1 month. Germination rate is high on moist bottomlands along streams.

Black cottonwood sprouts readily from cut stumps, and occasionally from roots. It is known for its ability to propagate vegetatively from small shoots and branches, thereby colonizing sandbars along streams and rivers. Mature black cottonwood trees are known to reach 40 m in height and diameters of more than 120 cm in natural stands. Trees are long-lived, surviving for more than 150 years (DeBell, 1990; Farrar, 1995). However, the period of active growth is considerably less, with trees often maturing in 60 years.

The species is very shade intolerant and grows best in full light. When seedlings are established in large numbers, they thin out in only 5 years due to competition. In nature, it grows rapidly and often outgrows competitors in mixed stands. In natural stands of black cottonwood, damaging agents are minimal. However, the increase in black cottonwood plantings has increased the incidence of insects and diseases. For example, *Septoria* canker was largely absent in the Pacific Northwest until the 21st century. *Septoria* canker does occur in black cottonwood east of the Cascade Mountains (Newcombe, 1996; Newcombe *et al.*, 2001; Chapter 8, this volume). Moreover, *Melampsora* rust is increasing in natural stands. A native poplar insect, the clearwing moth (*Paranthrene* spp.), has also increased in the Pacific Northwest riparian habitats with increasing monoculture (Brown *et al.*, 2006). Black cottonwood stands are subject to the usual problems of browsing from deer and rodents. Frost and ice damage and wind damage are also common in natural stands of black cottonwood because it is taller than surrounding trees (DeBell, 1990).

3.7.3 Status

Black cottonwood remains a dominant species along riverine systems throughout its range. Natural populations are abundant, except where humans have interfered with rivers by damming (Braatne *et al.*, 1996; Stettler, 2009). Black cottonwood hybridizes freely with other poplar species where they occur together. Natural hybrids of *P. angustifolia*, *P. balsamifera*, *P. deltoides* and *P. fremontii* occur regularly (Dickmann and Stuart, 1983; DeBell, 1990; Whitham *et al.*, 2001; Chapter 2, this volume).

3.8 *Populus alba* Linnaeus, White Poplar

3.8.1 Natural occurrence

Populus alba L. is found in Mediterranean and temperate forest ecosystems from central and southern Europe to West and Central Asia and northern Africa (Plate 5B; Chapter 2, this volume). It is typically riparian and occurs in Europe in linear formation along rivers or as isolated trees, due mainly to human influence.

In the Italian peninsula, *P. alba* is present in all regions in a variety of edaphic and climatic conditions from sea level to low mountain sites. Its wide ecological amplitude is considered to have influenced the formation of different adaptive traits within this species (M. Sabatti, 2005, unpublished information). A large variation in measured traits has been observed among and within provenances in Italy, showing the potential of the species for relevant progress in selection and breeding. However, the area of natural populations of *P. alba* in Italy has been estimated as only a few thousand hectares (M. Sabatti, 2005, unpublished information).

More specifically, *P. alba* grows best in a climate which is not too severe, on sites with full-light conditions and well-drained, deep, silt or sandy-silt soils (Gathy, 1970). In bottomland habitats where seasonal variation in water tables is not extreme, white poplar attains magnificent timber proportions. In such sites, the soil structure can appear very variable, due to the effect of periodic river flooding, presenting alternating silty and sandy horizons and leading to hydromorphic soils in formations called pseudogley (M. Sabatti, 2005, unpublished information).

From a phytosociological point of view, *P. alba* stands in Italy are considered to form an alliance occurring within the moister parts of the Querco-Fagetea class. The guide species of this alliance are *P. alba* and *P. nigra*, as well as sometimes the natural hybrid *P.* ×*canescens* (*P. alba* × *P. tremula*). The associated ground flora is determined by the impact of flooding and human activities, so that plant biodiversity is generally reduced and consists mainly of *Salix* spp. (M. Sabatti, 2005, unpublished information).

P. alba is regarded as somewhat tolerant of high temperatures, salinity, wind and drought. However, in Italy at least, it suffers from low temperatures and frost (M. Sabatti, 2005, unpublished information).

3.8.2 Life history

P. alba produces abundant seed and also reproduces by means of suckers, which develop copiously and vigorously from its shallow roots. As a pioneer species, it can colonize bare soil and, in Italy, it is generally more abundant along secondary river streams than on the main rivers (M. Sabatti, 2005, unpublished information). It hybridizes spontaneously with native aspen (*P. tremula*), where they occur together, resulting in the grey poplar (*P.* ×*canescens*) or, where the aspen parent is *P. adenopoda* or *P. tremula* var. *davidiana*, producing the Peking or Chinese white poplar (*P.* ×*tomentosa*) (Stettler *et al.*, 1996; Chapter 2, this volume). Grey poplar is more tolerant of drought and salinity than *P. alba*.

As a mature tree, it may be tall and straight, or broad crowned or crooked, or even with multiple stems. Individual trees have been found in the Voronezh region of the Russian Federation of more than 40 m in height and 1 m in diameter (Tsarev, 2005) (see Fig. 4.25, Chapter 4, this volume). Average figures for growth rates in natural stands are not readily available, but in plantations in the Province of Lucca (Tuscany, Italy), mean annual increments of 20 $m^3\ ha^{-1}$ (maximum 38–39 $m^3\ ha^{-1}$ on the best sites) with a rotation of about 20 years have been reported (M. Sabatti, 2005, unpublished information).

3.8.3 Status

P. alba is a unique pioneer species of riparian ecosystems, contributing to the natural control of flooding and water quality. Flood-plain forests are among the most diverse ecosystems in Europe and are increasingly recognized as centres for biodiversity. Today, there is a real interest in the restoration of riparian ecosystems for the natural control of flooding and also because the river borders can serve as corridors through which larger nature areas are connected.

White poplar is also among the most threatened tree species in Europe, due to alteration of riparian ecosystems by human activities, including urbanization and flood control. Regulation of floods has altered the regeneration capacities of the species and favoured the succession of poplar stands by hardwood forests. Although it may still regenerate locally with great success, there have been significant reductions in populations in some regions. However, in Italy, the area occupied by *P. alba*, after shrinking in the 20th century due to strong competition from agriculture and other land uses, is now considered to be stable (M. Sabatti, 2005, unpublished information).

In 1999, the activities of the *Populus nigra* Network, established as part of EUFORGEN, were extended to include *P. alba*. The aim of the EUFORGEN Networks is to conserve, at a European level, the still existing genetic resources available and, where possible, to maintain or restore the genetic diversity through a number of measures (EUFORGEN, 2009). In Spain, an *in situ P. alba* conservation programme is managed through EUFORGEN (Alba, 2000). An *ex situ* conservation clone bank has been established in Hungary (see Chapter 4, this volume). At the University of Tuscia in Viterbo, Italy, an *ex situ P. alba* conservation programme has been under way since 1988, with 350 genotypes assembled from provenances throughout the Italian peninsula and established in common gardens for the evaluation of genetic diversity in morphological and physiological traits (Sabatti, 1994; Sabatti *et al.*, 2001).

Basic research on *P. alba*, *P. tremula* and hybrid swarms of *P.* ×*canescens* which are endemic to Austria is expected to lead to the identification of unique stands of mixed taxonomic status that warrant conservation. This work explores the re-colonization of these species from disconnected refugia in central Europe after glaciation (Fussi *et al.*, 2010).

3.9 *Populus tremula* Linnaeus, Common Aspen

3.9.1 Natural occurrence

Common aspen or Eurasian aspen has the largest native range of any species in the genus: from 40° to 70°N latitude. It grows from the Atlantic Ocean in the UK, the Channel Islands and Ireland eastward to central Siberia, China and the central islands of Japan, as well as south to Algeria in North Africa. There is some controversy about its classification in China, where it is sometimes referred to as *P. davidiana* (von Wuehlisch, 2009) (Plate 8A; Chapter 2, this volume).

P. tremula grows on a wide variety of soils ranging from shallow and rocky to loamy sand and heavy clay. It grows on well-drained or seasonally waterlogged, nutrient-poor soils, with pH ranging from acid to alkaline. It grows best on moist soils that are well aerated, where annual precipitation exceeds evapotranspiration, and it will tolerate up to 2 months of rainfall less than 40 mm (von Wuehlisch, 2009; CABI, 2012e). It grows at a wide range of elevations from sea level to 1600 m in the Pyrenees, 1900 m in the Caucasus and 2000 m in the Alps. Common aspen can tolerate extreme cold (–30°C) in Siberia and maximum temperatures of 30°C or more in the southern part of its range.

Aspen occurs with a great number of overstorey and understorey species throughout its range. Its co-occurring species are too numerous to list here in total, but include *Abies*, *Alnus*, *Betula*, *Picea*, *Pinus*, *Quercus* and *Tilia* spp. in European Russia (Tsarev, 2005). It occurs with *Abies*, *Betula*, *Fraxinus* and *Tilia* in the overstorey, and with *Corylus* and *Sorbus* in the understorey (Kull and Niinemets, 1998).

3.9.2 Life history

P. tremula is a dioecious species with abundant seed production. Seeds are dispersed by wind, but do not remain viable for long and do not store well. Germination requires moist soil conditions and seedling mortality is high, so seed production is not the primary means of reproduction (CABI, 2012e).

Trees of common aspen do not grow as large as those of other members of the genus and are short-lived. Individual trees can reach 40 m in height and 60 cm diameter on good sites, but typically are much smaller when growing on stressful sites.

Common aspen is a prolific sprouter; it is disturbance adapted and is a primary successional species after fire, logging or other disturbance. Rapid growth from seedlings or suckers continues for 20 years before levelling off. Root suckers are produced, such that when the parent stem is damaged, the suckers form clones. These may reach an age of 200 years, with mature stands reproducing vegetatively (von Wuehlisch, 2009).

Because it sprouts vegetatively, it often forms pure stands. In some areas, its presence has decreased due to human activity. But it can reproduce from stem or root cuttings, which are used for artificial regeneration.

An early successional species, *P. tremula* has low shade tolerance. It grows rapidly in open areas and has its best growth in areas with warm summers, such as the Mediterranean region (CABI, 2012e). Common aspen traditionally regenerates after fire, but fire suppression has changed the natural dynamics. Regeneration after fire requires moist soil conditions.

Common aspen is attacked by a vast number of pests and diseases. Common insect pests are *Phleomyzus* aphids, *Chrysomela* leaf beetle, clearwing moth (*Paranthrene*), European tussock moth (*Orgyia*), poplar borer (*Saperda*), *Tremex* wasps and *Melasoma* mites. Common diseases include *Armillaria* root rot, *Hypoxylon* stem canker, *Melampsora* leaf rust, *Phellinus* heart rot, *Pollaccia* shoot blight and *Xanthomonas* bacterial canker (von Wuehlisch, 2009; CABI, 2012e; Chapters 8 and 9, this volume).

Rodents and ungulates, such as roe deer and moose, as well as hares, all browse common aspen heavily. Studies have shown, however, that repeated browsing by animals has little effect on the survival and reproduction of stands of common aspen, due to their suckering ability. In fact, common aspen has a high ecological value. Many birds and mammals benefit from aspen stands and forests, which provide them with habitat and food.

3.9.3 Status

In general, common aspen is not threatened or endangered over its extensive range. There is much genetic diversity in most populations. However, in isolated regions, agricultural practices and land use may have decreased the genetic diversity of the species, and in some regions common aspen may be considered threatened. Gene conservation activities are under way to ensure the long-term diversity of the species (von Wuehlisch, 2009).

3.10 *Populus tremuloides* Michaux, Quaking or Trembling Aspen

3.10.1 Natural occurrence

Quaking aspen is one of the most widely distributed tree species in the world. It occurs in 37 US states, 12 Canadian provinces and territories and in Mexico (Perala, 1990; USDA Forest Service, 2012b) (Plate 8B; Chapter 2, this volume).

It grows over its wide range on a great variety of soils, from shallow and rocky to deep, loamy sands and heavy clays (Perala, 1990). It can colonize very poor soils, including volcanic cones, rock outcrops, glacial outwash and landslides. But, the growth of quaking aspen is affected strongly by the soils and environmental conditions. Quaking aspen requires good soil drainage; high water tables or deep water tables limit its growth. It grows from sea level to more than 3000 m elevations in the mountains of Colorado, USA. It is common at higher elevations in the arid regions of the USA and Mexico, and it grows best on warm, southern exposures in Alaska and western Canada. Over much of western North America it occurs in small patches, but in Minnesota, Wisconsin and Michigan, where it is an important commercial species, it grows in dense stands (Zasada *et al.*, 2001; USDA Forest Service, 2012b). It also grows in pure stands in Alaska, Utah, Colorado,

Maine and central Canada (Peterson and Peterson, 1992; USDA Forest Service, 2012b).

P. tremuloides grows in a wide variety of climate conditions because of its range. It grows in temperature extremes ranging from −57° to 41°C and grows within the warmest permafrost zones of Alaska and Canada. Altitude is an important determinant of its distribution in the Rocky Mountains, where it occurs in a narrow elevation range from 2000 to 3350 m (Perala, 1990). Quaking aspen grows where annual precipitation exceeds evapotranspiration. It occurs in 180 mm annual precipitation in Alaska, to 3000 mm in the Maritime Provinces of eastern Canada. In summary, quaking aspen is limited by water surplus and then by minimum and maximum temperature (Shields and Bockheim, 1981).

Quaking aspen grows in association with a large number of trees and shrubs over its range (Perala, 1990; CABI, 2012f; USDA Forest Service, 2012b). However, it often grows in pure stands because of its clonal growth habit (Barnes, 1966; Steneker, 1973). It is a major component of the aspen (eastern), aspen (western) and white spruce-aspen forest cover type. It occurs with many forest species, including mixed northern hardwoods such as paper birch, sugar maple, red maple and especially the closely associated poplar species bigtooth aspen (*P. grandidentata*) in eastern North America (Laidly, 1990; Zasada *et al.*, 2001). It co-occurs with a huge number of shrub species including willows across its range and depending on region (Zasada *et al.*, 2001; USDA Forest Service, 2012b).

3.10.2 Life history

Quaking aspen is a dioecious species with long flower catkins that appear in April and May in the eastern part of its range to May and June in the higher elevations of the west. Flowers appear prior to leaf-out. There is clonal variation in flowering, and temperatures of 12°C for 6 days are necessary for aspen to flower (Perala, 1990). Male to female sex ratios vary in clones of quaking aspen from 3:1 to 1:1 (Peterson and Peterson, 1992); male clones are more common at high elevations and female clones are more common at lower elevations, with female clones growing faster (Grant and Mitton, 1979).

Seed crops of quaking aspen vary from year to year, with good crops every 4–5 years. Trees flower by age 10 and continue to flower to age 50–70 years. Seed dispersal is by wind or water and may last for 3–5 weeks. Viability of seeds is high, but of short duration. Under good conditions, they remain viable for 2–4 weeks.

Seedling growth is slow after germination, attaining 15–30 cm in height in the first year; but seedlings have long (20–25 cm) taproots. Height growth is rapid for the first 20 years, but then slows. Most aspen trees are part of a clone formed from the original parent seedling (Barnes, 1966; Sheppard, 1993). Quaking aspen often reproduces from root sprouts or suckers, with mature stands produced by the suckering.

Quaking aspen is a small to medium-sized tree that is short-lived, but some clones are known to live for thousands of years (CABI, 2012f). However, quaking aspen can attain a height of over 36 m and a diameter of 130 cm under favourable conditions. Mature stands are typically 20–25 m in height, with diameters of 18–30 cm. Mature aspen trees are comprised of both short and long shoots (Peterson and Peterson, 1992). Quaking aspen is very intolerant of shade throughout its range and usually occurs in pure stands as a result of disturbance such as logging, fire or browsing. However, quaking aspen is highly susceptible to fire, which can kill mature trees and kill or injure their roots, depending on intensity. On the other hand, it is an aggressive pioneer species, especially after fire (USDA Forest Service, 2012b).

There are a large number of damaging agents. It is severely damaged by *Venturia* shoot blight periodically and *Marssonina* leaf spot, especially in the western range. It is subject to *Septoria* leaf spot, as well as to infection by *Melampsora* leaf rust in the autumn, although often not seriously. It is damaged by numerous stem and root canker diseases, including *Hypoxylon* canker (*Entoleuca*), *Ceratocystis* canker, *Cytospora* canker, *Dothichiza* canker and trunk rot (*Phellinus*) (Chapter 8, this volume).

Quaking aspen is attacked by a wide variety of defoliators, borers and sucking insects. In the west, the forest tent caterpillar (*Malacosoma*) defoliates large areas of aspen. Outbreaks may last 2–3 years. The large aspen tortrix (*Choristoneura*) attacks aspen throughout its range, but typically in more northern regions. In eastern North America, the gypsy moth (*Lymantria*) preferentially attacks aspen. Aspen is also attacked by leaf miners such as *Phyllocnistis* and *Phyllonorycter*, which are very common in the Midwestern USA and are damaging if occurrence continues for several years. The cottonwood leaf beetle (*Chrysomela*) is another defoliating insect attacking aspen. The most damaging boring insect is the poplar borer (*Saperda*). The total number of damaging insects is too numerous to list here (Perala, 1990; Peterson and Peterson, 1992; Chapter 9, this volume).

Aspen is a favourite browse for a wide variety of animals. Young trees are damaged or killed by rodents such as mice, rabbits and hares. Larger browsing animals including mule deer, white-tailed deer, elk and moose cause frequent damage to aspen. Beavers often cut down large clones of aspen near waterways, but trees normally re-sprout. Porcupines damage aspen trees by feeding on tree crowns. At the same time, quaking aspen forests provide essential habitat for a large variety of birds and mammals. It is important for breeding and nesting sites, foraging and rest for many animals. Young stands are most important for browsing, as aspen grows out of reach for ungulates in 6–10 years. It is an important source of protein in winter months for many animals. Beavers use aspen for food, but also to construct their lodges and dams from trees along waterways.

Quaking aspen is also very important for many birds. For example, the ruffed grouse uses aspen for foraging, courting, breeding and nesting throughout is range. Quaking aspen is managed for ruffed grouse habitat in the Midwestern North America, as the grouse depends on multi-age stands for its livelihood (Peterson and Peterson, 1992). A large number of other birds depend on quaking aspen throughout its range (USDA Forest Service, 2012b).

3.10.3 Status

The status of quaking aspen is by no means threatened or endangered. However, there is recent evidence that it is being affected by climate change in a couple of ways (Peterson and Peterson, 1992; see Chapter 7 this volume). Shields and Bockheim (1981), working in the Great Lakes region of the USA, found that longevity of aspen clones decreased with increasing mean annual temperature. Similarly, aspen decline in the mountain region of the western USA appears to be climate related (Knight, 2001). If the earth continues to warm, aspen health and productivity may be compromised. Moreover, increasing carbon dioxide and pollutants such as ground-level ozone caused by man-made activities has been shown to decrease quaking aspen growth and productivity by increasing the incidence of damaging insects and diseases of aspen (Percy *et al.*, 2002). Also, increased large-scale fires and fire incidence associated with climate change may affect the extent and growth of western quaking aspen stands. On the other hand, studies have shown that quaking aspen is also a major carbon sink in the boreal forest that may help mitigate climate change in the future (Peterson and Peterson, 1992). Despite these concerns, there is evidence that quaking aspen has been subjected and adapted to climate change for thousands of years and will likely be an important component of North American forests for years to come.

3.11 Other Chinese Poplars

In China, there are many more native poplar species than the ones featured above in this chapter. The variation in adaptability to extreme environmental conditions in some of the Chinese poplars is impressive. They grow from 80 annual frost-free days in Inner Mongolia at 54°N latitude to 25°N latitude in the semi-humid tropics in the south. Some are able to tolerate dry conditions with less than 50 mm of rain a year, where there are frequent sandstorms and alkaline soils with a high salt content. In altitude, they range from 1300 to 4300 m (Weisgerber *et al.*, 1995; Weisgerber and Han, 2001).

Several of the species have promising growth potential and are important ecologically in the region of occurrence. These include *P. adenopoda* (Chinese aspen), *P. simonii* (Simon poplar), *P. szechuanica* (Szechuan poplar) and *P. yunnanensis* (Yunnan poplar) (Chapter 2, this volume). *P. adenopoda* becomes a large tree in the high elevation mountains of central and south-eastern China. *P. simonii* has a widespread distribution throughout China and occurs in 18 provinces from north-east to south-east; moreover, it occurs from sea level to 3000 m elevation and can tolerate temperatures from −35°C to 28°C (CABI, 2012g). Its growth rate and adaptability have led to widespread plantings throughout China, but natural stands have been severely depleted by drought and anthropogenic factors (Weisgerber *et al.*, 1995). *P. szechuanica* is a little known poplar native to the mountains of central and south-western China and is the largest Chinese poplar. *P. yunnanensis* is native to south-western China, growing in low to high latitudes, and has the ability to grow well in conditions of combined drought and salinity. It is unique in that female trees of the species are declining more rapidly than male trees in nature. The problem is so severe that female trees of the species are rare. Thus, *P. yunnanensis* is now on the threatened and endangered species list. This sex-specific difference raises concerns about the sustainability of the species in nature in the region (Chen *et al.*, 2010).

Climate change and anthropogenic factors have led concerned forest scientists throughout the world to pursue conservation programmes to preserve genetic resources of these native Chinese poplar species (Sigaud, 2003). International efforts are under way to collect, propagate and conserve these ecologically valuable species before they are lost (Weisgerber *et al.*, 1995; Weisgerber and Han, 2001).

3.12 *Salix humboldtiana* Willd., Humboldt's Willow

3.12.1 Natural occurrence

Salix humboldtiana Willd. is the only species of the *Salicaceae* found in South America. Its natural range extends from southern Mexico throughout Central America to the west side of South America in Colombia, Ecuador, Bolivia and Peru and south to the central region of Chile. It also occurs east of the Andes in Argentina, Uruguay and southern Brazil (US Department of Agriculture, 2012) (Plate 10A; Chapter 2, this volume).

S. humboldtiana is a species of warm temperate and subtropical climates, tolerating only very light frosts (−3°C) (Belov, 2009). It is found in river valleys and along watercourses for the most part, though with more widespread occurrence in Argentina (FAO, 1980) (Plate 20D; Chapter 2, this volume). In Chile, where it is found at low altitudes in the interior valleys of the Andes and between 500 and 2000 m elevation in the coastal mountains, it grows in water or has its roots in permanent water along watercourses and on lakeshores and riverbanks (Belov, 2009).

There are no reported estimates of the extent of the natural areas of *S. humboldtiana*. It does not appear to form continuous stands of any extent.

3.12.2 Life history

S. humboldtiana is not as strongly light-demanding as other riparian willow species, requiring some shade (20–40%) to grow well. At maturity, it reaches tree size up to 15 m (Belov, 2009).

3.12.3 Status

Unlike other *Salix* species described in this chapter, *S. humboldtiana* has attracted little or no economic interest. A few natural varieties have been described (FAO, 1980) and it has been used in hybridization work (Chapter 2, this volume). It has been introduced to other parts of the world, including Europe, India and Australasia. In Australia, *S. humboldtiana* has been declared a weed of national significance (Australian Weeds Committee, 2012).

3.13 *Salix nigra* Marshall, Black Willow

3.13.1 Natural occurrence

Black willow occurs throughout the eastern USA and adjacent parts of Canada. It is found in 37 US states, 4 Canadian provinces and Mexico (Pitcher and McKnight, 1990; USDA Forest Service, 2012c) (Plate 10B; Chapter 2, this volume).

It is found along river and stream margins in alluvial soils and battures. It grows almost anywhere there are favourable light and moisture conditions, even if there is periodic flooding.

However, it can grow in almost any soil once established (McLeod and McPherson, 1973; CABI, 2012h).

Over its wide range, black willow is subjected to wide temperature extremes from 46°C in the south to −50°C in the north; its distribution is thought to be independent of temperature. It occurs in areas where growing seasons are less than 100 frost-free days to those with over 200 frost-free days. It grows best in regions with average rainfall of 1300 mm, where 500 mm of that occurs during the growing season (Pitcher and McKnight, 1990).

Black willow is a short-lived pioneer species and is a co-dominant associated with other riparian species in cover types such as river birch-sycamore, cottonwood, sweetgum, American elm, baldcypress and tupelo. It grows best along the lower Mississippi River and the Gulf of Mexico, and is often associated with smaller species such as red maple, boxelder, red mulberry, *Salix interior* and *Salix exigua*.

3.13.2 Life history

Black willow is dioecious, with both male and female trees; however, males and females are indistinguishable, except during flowering and fruiting. Flowering typically begins in February in the southern part of the range and extends through June in the north. Flowers appear prior to leafing out. Pollination is by insects seeking nectar, but also occurs by wind. Trees flower and produce seed by age 10 or younger; they continue to produce for 20–70 years, with large quantities of seeds almost every year. The seed is disseminated by both wind and water. Seedling development occurs on moist and wet soil, and seed is viable for only a short period of 12–24 h. Seedlings are established only if mineral soil is in the open and wet. Black willow seedling establishment and distribution is thereby limited by soil moisture at the time of seed dispersal (McLeod and McPherson, 1973; Dorn, 1976). Once established, seedlings grow rapidly, often reaching 1.2 m in the first year (Pitcher and McKnight, 1990). Black willow sprouts prolifically from low-cut stems.

Growing in natural stands along the Mississippi River, black willow can become large trees, usually up to 20–40 m tall, with diameters of 20–120 mm. Trees growing in the northern part of the range and on poor sites in the southern part of the range are smaller. Trees growing in pure natural stands are self-pruning and become handsome, but open-grown trees are often very limby. Black willow is relatively short-lived, averaging only 55 years; some trees have been known to reach 70 years. It tends to be shallow rooted and to produce adventitious roots and suckers. Black willow usually grows in dense stands and is very intolerant of shade; this leads to high mortality at a young age.

Few insects cause serious damage to black willow. The trees are attacked by forest tent caterpillar (*Malacosoma*), gypsy moth (*Lymantria*), cottonwood leaf beetle (*Chrysomela*), willow sawfly (*Nematus*) and imported willow leaf beetle (*Plagiodera*). They are also subject to stem borers such as cottonwood borer (*Plectrodera*). Black willow also has canker diseases such as *Phytophthora* and *Cytospora*. Leaf rust caused by *Melampsora* is common in young stands throughout its range. Black willow is not very fire resistant, so hot fires kill entire stands (Pitcher and McKnight, 1990). It provides important wildlife habitat for both game and non-game birds and animals. However, it is not a preferred browse species for rodents or deer (USDA Forest Service, 2012c).

3.13.3 Status

Black willow stands and individuals are present throughout its range, but in some areas man-made dams have influenced seedling establishment, just as reported in eastern cottonwood (Braatne *et al.*, 1996). It is not threatened or endangered but has not been widely protected, and has not undergone significant tree improvement. However, it does not readily hybridize naturally and hybrids are difficult to identify.

3.14 *Salix alba* Linnaeus, White Willow

3.14.1 Natural occurrence

White willow, *S. alba* L., has a wide natural distribution throughout Europe, except for the Nordic countries, its range extending from the

Mediterranean – including the North African coast in Morocco and Algeria – through central Russia to the Chinese border, Iran and Asia Minor (Plate 11C; Chapter 2, this volume). It occurs from sea level to 2400 m in elevation, reaching the highest elevations in the most southerly parts of its range. White willow is widely cultivated within and beyond its natural range, so it is difficult to distinguish its natural range of distribution from its naturalized range (CABI, 2012i).

S. alba is a species of temperate climates with mild winters, warm summers and short summer drought. Occasional snowfalls can damage branches and crowns. However, it has a wide tolerance for semi-arid climates and relatively cold winters. It is a riparian species, of river valleys and wetlands, preferring to grow in sandy, silty or calcareous soils, or even relatively poor soils, with a high water table. It can tolerate periodic flooding, coastal sand dunes or saline soils (CABI, 2012i). Its wide tolerance of soil conditions is, however, dependent on the roots having access to sufficient moisture (FAO, 1980).

It is difficult to obtain estimates of the extent of the natural areas of *S. alba*. In the 1970s, the most important stands occurred in the valleys of the Danube and its tributaries, particularly the Drava, extending to an estimated 20,000 ha in the former Yugoslavia and 80,000 ha in Romania, where pure stands were found on the lower reaches of the Danube. Elsewhere, *S. alba* grew in mixtures with poplars and other broadleaved species (FAO, 1980).

3.14.2 Life history

S. alba produces seed regularly, but the seeds are short-lived and very difficult to store. They germinate and seedlings become established readily on fresh flood-deposited soils along rivers and streams, if light and moisture are plentiful. It is a strongly light-demanding, shade-intolerant colonizer of open areas (CABI, 2012i). It coppices readily and suckers abundantly, characteristics which are the basis for its propagation in cultivation. Mature trees can reach 30 m in height and 1 m in diameter (FAO, 1980).

3.14.3 Status

Like a number of other willows, *S. alba* is probably better known in cultivation than as a component of natural ecosystems. As a result of centuries of cultivation, many cultivars have been developed and are in use primarily for ornamental purposes (Chapter 2, this volume). It also hybridizes readily with several other *Salix* species, including *S. fragilis*. However, unlike other riparian species such as *P. nigra*, threatened with extinction in nature because of destruction of their natural habitat, there are no serious concerns for the conservation of natural populations of *S. alba* in the early 21st century.

3.15 *Salix interior* Rowlee, Sandbar Willow or Narrow-leaved Willow, and *Salix exigua* Nutt., Coyote Willow

3.15.1 Natural occurrence

Sandbar willow, also known as narrow-leaved willow, is a shrub willow that occurs transcontinentally across North America. It grows in 34 US states and 9 Canadian provinces (Plate 13A; Chapter 2, this volume). The closely related species *S. exigua*, coyote willow, is often grouped (sometimes mistakenly) with sandbar willow and occurs in 12 western states of the USA and three provinces of western Canada (Argus, 1986; see Chapter 2, this volume). Notably, it co-occurs with sandbar willow in Saskatchewan, Alberta and British Columbia (USDA Natural Resources Conservation Service, 2012a). In this section, we treat sandbar and coyote willow together, as they have such similar plant characteristics and life history, although sandbar willow is considered an eastern species and coyote willow a western species.

Sandbar willow and coyote willow grow on sandy or gravelly soils along streams, rivers and shorelines (see Fig. 2.16, Chapter 2, this volume). They are always found near water and are not well adapted to upland drier sites (USDA Natural Resources Conservation Service, 2012a). They occur on high to low elevations north to south, and east to west gradients, as well as on steep to shallow riparian terrains. Coyote willow only grows at elevations below

2700 m. Sandbar and coyote willows grow across a huge range of temperatures and precipitations because of their wide range. Temperatures range from −18° to −62°C in the northern part of their distribution and 30–44°C in the south. Annual precipitation ranges from 150 to 300 mm in Alaska and the Yukon territories to 1400 mm in the Maritime Provinces of the east. Frost-free growing periods for these species range from less than 75 days in the north to more than 200 days in the south.

Sandbar willow is classified as an obligate or facultative wetland species (Collet, 2004). It is often used by ecologists and conservationists as an indicator species of wetlands (Johnston, 1993). It co-occurs with many wetland trees, shrubs and plants in deep, wet lowlands, overflow areas, wet meadows and other areas with a high water table. Some of the tree species it co-occurs with are paper birch (*Betula papyrifera*), green ash (*Fraxinus pennsylvanica*), white cedar (*Thuja occidentalis*) and American elm (*Ulmus americana*). It co-occurs with many other *Salix* species including *S. amygdaloides*, *S. bebbiana*, *S. discolor*, *S. eriocephala*, *S. lucida*, *S. nigra* and *S. petiolaris*. It also co-occurs with the shrub species such as red-osier dogwood (*Cornus sericea*) and other dogwoods (Johnston, 1993). It occurs in pure stands and thickets where conditions are favourable (USDA Natural Resources Conservation Service, 2012a). In dry riparian areas, the coyote willow co-occurs with trees and shrubs, including Fremont cottonwood (*P. fremontii*), Goodding willow (*S. gooddingii*), salt cedar (*Tamarix gallica*) and mesquite (*Prosopis*), as well as herbaceous plants such as *Carex*, *Festuca* and *Pluchera*.

3.15.2 Life history

Sandbar willow and coyote willow are dioecious shrubs with both male and female plants. Flowering begins at a young age; each year, the female plants produce thousands of seeds. At maturity, the parachute-like seeds are carried up to hundreds of metres in the air. Seeds that land in water are carried downstream, floating for up to several days. But willow seeds have no food reserves, so they die in several days without suitable habitat (Argus, 2006). Germination occurs in 12–14 h if seeds are constantly moist (USDA Natural Resources Conservation Service, 2012a). Sandbar and coyote willows require moist soil from flooding or capillary wetting to establish on flood plains. Timing of declining streamflow is critical to survival. Both sandbar and coyote willows root readily from disturbed stems or roots. Beaver-cut stems and branches root easily and sucker to form thickets along streams (see Fig. 2.16, Chapter 2, this volume). They also spread aggressively from stream banks to other favourable adjacent sites, where they are invasive in some agricultural areas.

These willows are short-lived shrubs, reaching from 1 to 6 m in height. They grow very rapidly when their roots are near water, but they also decline rapidly without water. They are very shade intolerant, and decline rapidly if shaded by co-occurring species. However, they can survive for years without seedling establishment, because of their vegetative reproduction capacity.

Sandbar and coyote willows are attacked by a large number of insects and diseases. Insects include willow leaf blotch miner (*Micrurapteryx*) and numerous aphids, willow gall insects and scale insects (USDA Natural Resources Conservation Service, 2012a; see Chapter 9, this volume). They are also attacked by many diseases, including the canker disease (*Cytospora*), *Melampsora* leaf rust and tar spot disease (*Rhytisma*).

Rabbits and many ungulates, such as mule deer, white-tailed deer, elk and moose browse on sandbar and coyote willow twigs, foliage and bark. Beavers consume and/or browse willow branches, and several species of birds eat the buds and twigs. In the western USA and Canada, coyote willow is browsed heavily by livestock (USDA Natural Resources Conservation Service, 2012a).

Sandbar and coyote willows provide important riparian ecosystem services including stream bank stabilization, water quality improvement, flood abatement and essential fish and wildlife habitat. They provide especially important habitat for breeding birds (USDA Natural Resources Conservation Service, 2012a).

3.15.3 Status

Sandbar and coyote willows are, in general, neither threatened nor endangered, because of their huge range in North America. But, there are some localized areas where they are considered threatened. For example, sandbar willow is threatened in Connecticut and Massachusetts due to alteration of riparian zones by human activity. Sandbar willow is an endangered species in Maryland, USA. Riparian forests have been altered and fragmented throughout North America by human intervention. These activities have affected sandbar and coyote willow populations seriously. Livestock grazing is a serious factor in western riparian regions. Overgrazing by livestock and wildlife has affected numerous coyote willow populations.

3.16 *Salix eriocephala* Michaux, Heart-leaved Willow

3.16.1 Natural occurrence

Heart-leaved willow, also sometimes known as Missouri River willow, occurs in the eastern and north central region of North America. It has been reclassified as *S. eriocephala* after taxonomic controversy; previously, it was also known as *S. cordata* and/or *S. rigida* (Argus, 1980) (Plate 13B; Chapter 2, this volume). Heart-leaved willow grows in 35 US states and 8 Canadian provinces. It is noteworthy that it has not been identified in the US states of Mississippi, North Carolina or South Carolina (USDA Natural Resources Conservation Service, 2012b).

Heart-leaved willow grows in a variety of habitats, including open woodlands, riparian meadows, streamside edges, sloughs, ditches and wetlands (Johnston, 1993; Illinois Wildflowers, 2012; USDA Natural Resources Conservation Service, 2012b). It grows across a range of temperatures from −38°C in the north to 44°C in the south. It needs a minimum of 75 frost-free days to survive. It grows best in a rather narrow annual precipitation range from 200 to 500 mm, but will grow in precipitation up to 1400 mm in the Canadian Maritime Provinces (USDA Natural Resources Conservation Service, 2012b). Heart-leaved willow grows best in acidic soils from pH 4.0 to 7.0 and will not tolerate highly alkaline soils. Its preference is for wet to moist soils containing loam, calcareous sand and/or gravel (Illinois Wildflowers, 2012).

It is found mainly in mixed wetland stands, where it co-occurs with many woody and herbaceous species. In wetlands, it co-occurs with several other shrub *Salix* species as well as dogwood (*Cornus*) species. It also co-occurs with tree species such as birch (*Betula*), ash (*Fraxinus*), white cedar (*Thuja occidentalis*) and American elm (*Ulmus americana*) (Johnston, 1993).

3.16.2 Life history

Heart-leaved willow is a dioecious shrub with male and female catkins on separate shrubs. Flowering occurs in April or May, depending on location; in northern regions, flowering may be in June (Native Plant Data Base, 2012). The flowering period lasts for 1–2 weeks. Green to brown flowers are replaced by seed capsules that split open to release tiny, cottony seeds. Seeds are distributed by wind and water. The shrub spreads by re-seeding itself (Illinois Wildflowers, 2012).

This willow has high seedling vigour, and mature plants typically reach heights of 1–4 m. However, at maturity, it can reach heights of over 12 m in 20 or more years. Its growth form is branchy, especially near the stem base, with long, unbranched stems above. It is a rhizomatous plant with a shallow (0.5 m) branched root system. It also sprouts readily from stem cuttings (Sagebud, 2012).

Though it prefers full sunlight conditions, heart-leaved willow is more tolerant of shade than most co-occurring shrub willows (Illinois Wildflowers, 2012). It has low drought tolerance and medium fire tolerance; it is not fire resistant.

Like other shrub willows, it is attacked by many insects and diseases. Feeding insects include the willow leaf miner (*Micrurapteryx*) and gall-forming sawflies (*Phyllocolpa*) (Fritz and Kaufman, 1993). Wood-boring insects attacking heart-leaved willows include willow borer (*Agrilus*) and poplar and willow borer (*Cryptorhynchus*); beetles attacking it include willow leaf beetle (*Chrysomela*) and striped willow beetle (*Disonycha*) and various aphids also attack (Illinois Wildflowers, 2012; Chapter 9, this volume). Heart-leaved willow is also damaged by willow diseases, including willow blight (*Glomerella*), *Melampsora* leaf rust (Bennett *et al.*, 2011), willow scab (*Venturia*) and *Botryosphaeria* canker (Willowpedia, 2012).

Heart-leaved willow is a favourite food source for many animals and birds. White-tailed deer and elk browse the stems and leaves; beavers eat the bark and wood, and its leaves are eaten by some turtles. Many birds, including ruffed grouse, eat its buds and flowers. Other birds use this willow for nesting sites. Heart-leaved willow is an important nectar and pollen source for beneficial insects, including bees, flies and butterflies (Illinois Wildflowers, 2012).

3.16.3 Status

Heart-leaved willow occurs widely throughout its range and in general is not threatened or endangered. However, it is listed as threatened in the US state of Indiana and endangered in Florida, presumably because of alteration of riparian zone by human activity, like the other shrub willow species. It hybridizes readily with other willow species in nature (Argus, 2006).

3.17 *Salix viminalis* Linnaeus, Common Osier or Basket Willow

3.17.1 Natural occurrence

Common osier or basket willow, *S. viminalis* L., has a very wide natural distribution throughout Eurasia, except for the Nordic countries and the Alps, extending over about 140 degrees of longitude, from France to eastern Siberia, north-west China and the Altai. To the north, it reaches the forest tundra boundary in Russia, extending beyond that in the valleys of major rivers (Skvortsov, 1999) (Plate 15C; Chapter 2, this volume). In Western Europe, it has been cultivated for so long that naturalization has taken place, making it difficult now to know where it may have occurred originally (Newsholme, 1992). It is not clear, for example, whether it is native to the British Isles (CABI, 2012j). It occurs from sea level to 600 m in elevation in the Urals, to 900 m in the Carpathians, and even to 1800 m in the Kaba river valley in the Altai. Its range has been extended in cultivation to many other parts of the world, including North America, South America (Brazil and Chile), Pakistan, India and Australasia (CABI, 2012j).

S. viminalis is a species of cool temperate or boreal climates, tolerating cold winters with temperatures dropping as low as –54°C and warm, but not very hot, summers (mean maximum temperature of the hottest month 22°C) (CABI, 2012j). Although it grows on plains and mountains, it is primarily a valley or riparian species, most commonly found close to water on riverbanks, next to lakes or in marshes (Skvortsov, 1999). Precipitation amounts are not critical as it can tolerate long periods without rain, provided there is sufficient soil moisture. Osier is found on fresh alluvial substrates which may be sandy or pebbly, with plentiful groundwater, but it grows best on sandy loams and loams (Sidorov, 1978; Skvortsov, 1999). It can tolerate limestone and calcareous soils as well as acid soils, but on the latter may develop fragile twigs (CABI, 2012j).

There are no reported estimates of the extent of natural areas of *S. viminalis*. As noted above, in the western part of its range, it is in any case difficult or impossible to distinguish naturalized occurrences from its original native area.

3.17.2 Life history

S. viminalis regenerates readily and rapidly. It is a light-demanding species, growing in nature only on unshaded sites (CABI, 2012j). It grows rapidly to form a tall shrub or multi-stemmed small tree, reaching at most 6–8 m in height (Chapter 2, this volume).

3.17.3 Status

Variation among natural populations of *S. viminalis* has been studied in Russia, including in the Urals (Skvortsov, 1968, 1999; Shaburov and Belyaeva, 1991). It is widely cultivated and used for basketry, stream bank protection and for biomass production for energy. It hybridizes readily with many other *Salix* species, a characteristic which has been widely exploited in cultivation. There are no serious concerns for conservation of natural populations.

3.18 Physiology of Poplars and Willows

There have been major advancements in the knowledge of the physiological processes of poplars and willows since the publication of previous FAO volumes on poplars and willows (FAO, 1958, 1980). There have been thousands of publications on the subject since 1980, and entire books have been written on the physiological processes associated with poplar and willow growth and yield. These publications are both fundamental and applied in scope and are so numerous that we make no attempt to review comprehensively the vast literature on the subject here.

In this chapter, our goal is to provide an overview of the recent history and developments in the knowledge of the physiological processes of poplars and willows without great detail. We also provide the reader who seeks details on individual processes with a summary table with information sources on those processes (Table 3.1). That table is structured to provide sources of information on the various physiological processes while scaling up from the molecular level to the stand level.

Early research on the physiology of poplar and willows focused on experimental approaches to the basic ecological requirements of plant growth (FAO, 1980). These requirements included light, temperature, water, nutrients and oxygen. They were usually studied singularly on site-specific local materials as they varied in relation to the natural environment (Fowells and Means, 1990).

Poplars and willows are diverse pioneer species that have the inherent ability to grow rapidly over a range of environmental conditions. This capability makes them ideal for environmental applications (Chapter 6, this volume); it is a result of their ability to capture light efficiently through the process of photosynthesis and take up large quantities of water and nutrients from the soil.

The quantity and display of leaf area in poplars and willows are among the most important determining factors in their rapid growth and productivity (Ceulemans, 1990). Moreover, poplars and willows have inherently high photosynthetic rates per unit leaf area (Nelson, 1984; Ceulemans and Isebrands, 1996). In addition, their annual whole tree photosynthesis is high because they have a combination of high leaf area index and high autumnal photosynthesis due to late-season leaf retention (Nelson and Isebrands, 1983). There is also genetic variation in poplars and willows in leaf area, leaf display and photosynthesis rate that allows them to be amenable to genetic selection and breeding for specific applications (Ceulemans and Isebrands, 1996). Thus, there is potential for genetic manipulation of carbon allocation patterns among clones. For example, clones with high numbers of first-year sylleptic branches are more productive than others because they have higher quantities of carbon exported to the lower stem and roots during the growing season (Scarascia-Mugnozza *et al.*, 1999).

Poplars and willows are native to riparian zones and are known for their inherent ability to take up large quantities of water and nutrients (Stettler, 2009). Poplars have been shown to take up between 20 and 50 kg of water day^{-1}, depending on their age and size (Hinckley *et al.*, 1994). This uptake translates into annual water use of between 2010 and 2260 mm ha^{-1} year^{-1} by age 8, which is comparable to perennial agricultural crops (Braatne, 1999; Dickmann *et al.*, 2001b).

Growth of poplars and willows depends on a favourable nutrient availability in the soil. However, they do grow across a wide range of soil qualities, including soils with varying pH levels and salinities (FAO, 1980). Poplars and willows have a high nutrient requirement to maintain maximum productivity (Ericsson *et al.*, 1992; Isebrands, 2007). Nitrogen is often the most limiting nutrient, and poplars grow best in soils where pH is between 6.0 and 7.5. Willows can tolerate higher soil pH and salinity, which makes them useful for soil remediation applications (Mirck and Volk, 2010).

The quantity of nitrogen required to sustain poplar growth varies by region and species and ranges from 100 to 275 kg ha^{-1} year^{-1} (Stanturf *et al.*, 2001). Coleman *et al.* (2006) found that the most efficient post-establishment fertilization schedule for poplar plantings was multiple small dose amendments of 50 kg ha^{-1} of nitrogen to maintain optimum leaf nitrogen over the course of the season.

Table 3.1. List of selected representative references with information on the physiology of poplars and willows from molecular to stand level.

Subject	Reference
1. Molecular level	
Populus	
Herbicide tolerance	Donahue *et al.*, 1994
Physiological traits	Bradshaw *et al.*, 2000
Physiological traits	Brunner *et al.*, 2004
Nitrogen	Cooke and Weih, 2005
Nitrogen	Cooke *et al.*, 2005
Physiological traits	*New Phytologist*, 2005
Water use	Berta *et al.*, 2010
Salinity	Chen and Polle, 2010
Physiological traits	Dillen *et al.*, 2010
Nutrition, hormones	Polle and Douglas, 2010
2. Organ level	
Populus	
Photosynthesis	Larson *et al.*, 1972
Photosynthesis	Nelson, 1984
Carbon, nitrogen	Dickson, 1989
Water use	Blake and Tschaplinski, 1992
Photosynthesis	Dunlap *et al.*, 1993
Cambium physiology	Larson, 1994
Water, climate change	van Volkenburgh and Taylor, 1996
Photosynthesis	Dillen *et al.*, 2011
3. Tree level	
Populus	
Water use	Bialobok, 1976
Photosynthesis, carbon allocation	Isebrands and Nelson, 1983
Photosynthesis, carbon allocation	Nelson and Isebrands, 1983
Photosynthesis	Isebrands *et al.*, 1988
Water use	Tschaplinski and Blake, 1989
Physiological traits	Ceulemans, 1990
Photosynthesis	Michael *et al.*, 1990
Photosynthesis	Ceulemans and Saugier, 1991
Carbon allocation, roots	Friend *et al.*, 1991
Physiological traits, genetics	Dickmann, 1991
Root physiology	Dickmann and Pregitzer, 1992
Water use	Blake *et al.*, 1996
Photosynthesis, carbon allocation	Ceulemans and Isebrands 1996
Physiological traits, stress	Hinckley, 1996
Wood formation	Telewski *et al.*, 1996
Carbon allocation	Scarascia-Mugnozza *et al.*, 1999
Physiological traits	Dickmann *et al.*, 2001b
Water use	DesRochers *et al.*, 2007
Physiological traits	Gornall and Guy, 2007
Water use	Kort and Blake, 2007
Photosynthesis, drought	Silim *et al.*, 2009
Salix	
Root physiology	Dickmann and Pregitzer, 1992
Physiological traits	Taylor *et al.*, 2003
Physiological traits	Orlovic *et al.*, 2006

Continued

Table 3.1. Continued.

Subject	Reference
4. Stand level	
Populus	
Yield physiology	Isebrands *et al.*, 1983
Light interception, productivity	Cannell *et al.*, 1988
Physiological traits	Hinckley *et al.*, 1989
Ecophysiology	Isebrands *et al.*, 1990
Ecophysiology	Mitchell *et al.*, 1992
Water use	Hinckley *et al.*, 1994
Physiological traits	Ceulemans *et al.*, 1996
Production physiology	Heilman *et al.*, 1996
Riparian, water use	Kranjcec *et al.*, 1998
Coppice physiology	Allen *et al.*, 1999
Production physiology	Ceulemans and Deraedt, 1999
Riparian physiology	Braatne, 1999
Water use	Gochis and Cuenea, 2000
Water use	Vose *et al.*, 2000
Physiological traits, climate change	Gielen and Ceulemans, 2001
Ecophysiology, riparian	Rood *et al.*, 2003
Coppice physiology	Rae *et al.*, 2004
Physiological traits	Dickmann, 2006
Physiological models	Buck-Sorlin *et al.*, 2008
Water use	Zhang *et al.*, 2008
Riparian, water use	Hultine *et al.*, 2010
Nitrogen, climate change	Rennenberg *et al.*, 2010
Salix	
Light interception	Cannell *et al.*, 1987
Light interception, productivity	Cannell *et al.*, 1988
Nutrition	Ericsson *et al.*, 1992
Ecophysiology	Mitchell *et al.*, 1992
Coppice physiology	Sennerby-Forsse *et al.*, 1992
Physiological traits	Ceulemans *et al.*, 1996
Carbon allocation	Ericsson *et al.*, 1996
Yield physiology	Tharakan, 1999
Ecophysiology	Kuzovkina *et al.*, 2008
Salinity	Mirck and Volk, 2010
Salinity	Hangs *et al.*, 2011

The recent proliferation of information on the physiology of poplars and willows was prompted by the premise that we must know and understand how trees grow in order to improve silvicultural practices and tree genetics (Kramer, 1986; Dickmann, 1991; Stettler *et al.*, 1996; Dickmann *et al.*, 2001b). The new approach for studying poplar and willow physiology was patterned after the highly successful physiological-based agronomic crop improvement programmes where crop physiologists, agronomists and geneticists worked together in multi-disciplinary teams toward a common goal. That approach was based on knowledge of the growth, development and functioning of leaves, as leaf production and performance constituted the biological basis of all agricultural production (Dale and Milthorpe, 1983). The leaf is the plant organ where nearly all molecular, biochemical and physiological processes occur in poplars and willows. The aggregation of leaves in these species, including crown architecture, canopy structure and leaf area development, are related intimately to their growth and productivity.

A forerunner of much of the physiological work on poplar was the pioneering work by Larson and co-workers in the 1960s and 1970s, who studied the role of leaves in the complex structural–functional relationships of wood formation (Larson, 1994; reviewed by Telewski *et al.*, 1996). That work demonstrated the importance of the leaf in relation to the meristematic activity of the cambium and to secondary wood formation in the stem and roots of poplars (Larson and Isebrands, 1974; Larson, 1983). Once this relationship was understood, it helped multidisciplinary research team efforts at numerous institutions to study the genetic variation in the structural and functional determinants of the productivity of poplars and willows (Isebrands *et al.*, 1983; Ceulemans, 1990; Hinckley, 1996; Stettler *et al.* 1996; Dickmann, 2006).

The agronomic approach to crop physiology with poplar and willow culture followed when they began to be grown as a short-rotation woody crop for wood and energy (Stettler *et al.*, 1996; Zsuffa *et al.*, 1996; Dickmann, 2006). The worldwide interest in short-rotation woody crops thereby led to much more physiological research on poplar and willows (Mitchell *et al.*, 1992).

3.18.1 Overview of poplar and willow physiology

Isebrands *et al.* (1983) first described baseline integrated physiological information on crown morphology, photosynthesis and photosynthetic distribution in relation to biomass yields in poplars with contrasting growth patterns. Then, Hinckley *et al.* (1989) outlined the role of leaf and root morphology, anatomy, stomatal behaviour, turgor maintenance, canopy architecture, photosynthesis and carbon allocation in hybrid poplars displaying hybrid vigour in the field. Ceulemans (1990), having worked with Hinckley, summarized how leaf physiological traits as well as whole tree and canopy structural traits related to biomass productivity in highly productive poplar clones. Soon after, Dickson and Isebrands (1991) also described how leaf development was important in regulating shoot and plant development in poplars with respect to their response to environmental stresses.

Mitchell *et al.* (1992) published a comprehensive review of poplar and willow ecophysiology following an International Energy Agency (IEA) Bioenergy Conference in Uppsala, Sweden. At that time, their book provided an up-to-date summary of what was known on the ecophysiology of short-rotation crops. It included overviews of poplar and willow growth dynamics, including nutrition, water relations, carbon allocation and coppice physiology, as well as modelling physiological growth processes.

A book, *Biology of Populus and Its Implications for Management and Conservation* by Stettler *et al.* (1996), followed after an International Poplar Symposium in Seattle, Washington, USA, that was sponsored by the International Union of Forestry Research Organizations (IUFRO). The book had an entire section on the physiology of growth, productivity and stress response in poplars (Hinckley, 1996). That section included chapters on leaf growth physiology, wood formation, root physiology, carbon allocation, water relations, stress physiology and production physiology, and the link between physiology and molecular genetics. In the book, Heilman *et al.*'s chapter on production physiology describes how production physiology is concerned with factors that influence and control the productivity of poplars grown in stands, including solar energy capture and conversion in relation to environmental and genetic factors (Heilman *et al.*, 1996).

Another book, *Poplar Culture in North America* (Dickmann *et al.*, 2001a), was published in conjunction with a meeting of the International Poplar Commission in Portland, Oregon, USA. In that book, Dickmann *et al.* (2001b) reviewed the latest knowledge of the physiological ecology of poplars, including photosynthesis, respiration, hormones, water relations and nutrition in relation to growth (Fig. 3.1).

Recently, in a chapter of the book *Genetics and Genomics of Populus* by Jansson *et al.* (2010), Dillen *et al.* (2010) explained the vigorous growth performance of *Populus* as a result of its high photosynthetic carbon uptake, leaf area development, sylleptic branch production and seasonal regulation of hormones.

Because of the long history of physiological studies, knowledge of poplar/willow growth processes has become important for use in computer modelling of the growth process of poplars and willows to simulate stand production in the field (Mitchell *et al.*, 1992; Host and Isebrands, 1994; Ceulemans *et al.*, 1996). This approach has further enhanced our knowledge of poplar and willow tree growth and yield prediction.

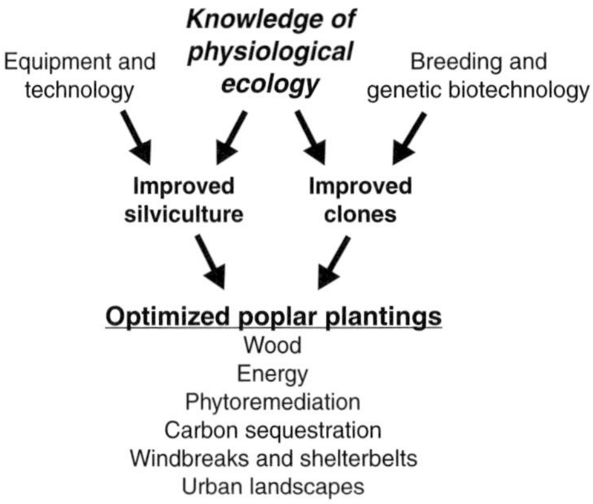

Fig. 3.1. Knowledge of physiological ecology is the foundation for optimized poplar and willow culture. Adapted with permission from Dickmann *et al.* (2001b).

Knowledge of poplar and willow physiological processes was prominent in helping advance a new developing applied scientific discipline for soil and water remediation called phytoremediation (Chapter 6, this volume). This development takes advantage of the ability of poplars and willows to grow rapidly, take up large volumes of water and break down toxic chemical pollutants through their inherent biochemical processes and microbial associates (Newman *et al.*, 1997; Doty, 2008).

Other advancements that have increased the physiological knowledge of poplars and willows are in the scientific disciplines of global climate change and pollution (Isebrands *et al.*, 2000; *New Phytologist*, 2005) (Chapter 7, this volume). Research in both scientific disciplines has led to numerous publications on the physiology of poplars and willows because they are such important components of so many natural ecosystems worldwide.

In the 1990s, there were major technological advances in molecular tools for genetic engineering and genomics (Bradshaw and Strauss, 2001; Joshi *et al.*, 2011) that have propelled poplars and willows to an elevated status as 'model biological materials'. These advances, coupled with the increased physiological and biochemical knowledge on poplars and willows, have helped increase research funding for multi-disciplinary teams to study the poplar and willow genome (Dickmann, 2006; Tuskan *et al.* 2006).

References

Ageenko, A.S. (1969) *Forests of the Far East.* Lesnaya Promyshlennost, Moscow.

Alba, N. (2000) Conservacion de Recursos Geneticos del Genero Populus en Espana. *Forest Systems* 9, 45–57.

Allegri, E. (1971) Identification of species and varieties of poplar indigenous in Italy. *Annali dell' Istituto Sperimentale per la Selvicoltura, Arezzo* 2, 1–62.

Allen, S., Hall, R. and Rosier, P. (1999) Transpiration by two poplar varieties grown as coppice for biomass production. *Tree Physiology* 19, 493–501.

Argus, G.W. (1980) The typification and identity of *Salix eriocephala* Michx (*Salicaceae*). *Brittonia* 32, 170–177.

Argus, G.W. (1986) The genus *Salix* (*Salicaceae*) in the southeastern United States. *Systematic Botany Monographs 9.* American Society of Plant Taxonomists, Ann Arbor, Michigan.

Argus, G.W. (2006) *Guide to Salix (Willow) in the Canadian Maritime Provinces (New Brunswick, Nova Scotia, and Prince Edward Island).* Canadian Museum of Nature, Ottawa, Ontario, Canada.

Australian Weeds Committee (2012) Chilean pencil willow. *Salix humboldtiana 'Pyramidalis'* (http://www.weeds.org.au/cgi-bin/weedident.cgi?tpl=plant.tpl&ibra=all&card=T30, accessed 26 January 2012).

Barnes, B.V. (1966) The clonal habit of American aspens. *Ecology* 47, 439–447.

Belov, M. (2009) Chileflora. *Salix humboldtiana* (http://www.chileflora.com/florachilena/FloraEnglish/HighResPages/EH0571.htm, accessed 26 January 2012).

Bennett, C., Aime, M.C. and Newcombe, G. (2011) Molecular and pathogenic variation within *Melampsora* on *Salix* in western North America reveals numerous cryptic species. *Mycologia* 103, 1004–1018.

Berta, M., Giovannelli, A., Sebastiani, F., Camussi, A. and Racchi, M.L. (2010) Transcriptome changes in the cambial region of poplar (*Populus alba*) in response to water stress. *Plant Biology* 12, 341–354.

Bialobok, S. (1976) The Poplars – *Populus* L. US Department of Agriculture Translation No TT 75-54073. Warsaw, Poland, pp. 156–159.

Blake, T. and Tschaplinski, T. (1992) Water relations. In: Mitchell, C.P., Ford-Robertson, J.B., Hinckley, T.M. and Sennerby-Forsse, L. (eds) *Ecophysiology of Short Rotation Forest Crops.* Elsevier Applied Science, London, pp. 66–94.

Blake, T.J., Sperry, J.S., Tschapinski, T.J. and Wang, S.S. (1996) Water relations. In: Stettler, R.F., Bradshaw, H.D. Jr, Heilman, P.E. and Hinckley, T.M. (eds) *Biology of Populus and Its Implications for Management and Conservation.* National Research Council of Canada Research Press, Ottawa, pp. 401–422.

Boyce, S.G. (1976) Ecology of natural stands of cottonwood. In: Thielges, B.A. and Land, S.B. Jr (eds) *Proceedings: Symposium on Eastern Cottonwood and Related Species, September 28–October 2, 1976, Greenville, Mississippi.* Louisiana State University, Baton Rouge, Louisiana, pp. 11–19.

Braatne, J. (1999) Biological aspects of hybrid poplar cultivation on floodplains in Western North America: a review. EPA Document No 910-R-99-02. US Environmental Protection Agency, Washington, DC.

Braatne, J.H., Rood, S.B. and Heilman, P.E. (1996) Life history, ecology and conservation of riparian cottonwoods in North America. In: Stettler, R.F., Bradshaw, H.D. Jr, Heilman, P.E. and Hinckley, T.M. (eds) *Biology of Populus and Its Implications for Management and Conservation.* National Research Council of Canada Research Press, Ottawa, pp. 57–85.

Bradshaw, H.D. and Strauss, S.H. (2001) Breeding strategies for the 21st century: domestication of poplar. In: Dickmann, D.I., Isebrands, J.G., Eckenwalder, J.E. and Richardson, J. (eds) *Poplar Culture in North America.* National Research Council of Canada Research Press, Ottawa, pp. 383–394.

Bradshaw, H.D., Ceulemans, R., Davis, J. and Stettler, R.F. (2000) Emerging model systems in plant biology. *Journal of Plant Growth Regulation* 19, 306–313.

Brown, J.J., Kittelson, N.T., Hannon, E.R. and Walsh, D.B. (2006) An endemic population of western poplar clearwing moths (Lepidoptera: Sesiidae) invades a monoculture of hybrid poplar. *Journal of Economic Entomology* 99, 771–779.

Brunner, A.M., Busov, V.B. and Strauss, S. (2004) Poplar genome sequence: functional genomics in an ecologically dominant plant species. *Trends in Plant Science* 9, 49–56.

Buck-Sorlin, G., Kniemeyer, O. and Kurth, W. (2008) A model of poplar (*Populus* sp.) physiology and morphology based on relational growth grammars. In: Deutsch, A., Bravo De La Parra, R., De Boer, R., Diekmann, O., Jagers, P., Kisdi, E., *et al.* (eds) *Mathematical Modeling of Biological Systems. Volume II.* Birkhäuser, Boston, Massachusetts, pp. 325–334.

Burns, R.M. and Honkala, B.H. (eds) (1990) *Silvics of North America. Volume 2, Hardwoods.* Agriculture Handbook 654. USDA Forest Service, Washington, DC.

CABI (2012a) *Populus euphratica.* Forestry Compendium (http://www.cabi.org/fc/, accessed 21 December 2011).

CABI (2012b) *Populus nigra.* Forestry Compendium (http://www.cabi.org/fc/, accessed 21 December 2011).

CABI (2012c) *Populus balsamifera* (balm of Gilead). Forestry Compendium (http://www.cabi.org/fc/, accessed 20 January 2012).

CABI (2012d) *Populus maximowiczii.* Forestry Compendium (http://www.cabi.org/fc/, accessed 3 February 2012).

CABI (2012e) *Populus tremula.* Forestry Compendium (http://www.cabi.org/fc/, accessed 12 February 2012).

CABI (2012f) *Populus tremuloides* (trembling aspen). Forestry Compendium (http://www.cabi.org/fc/, accessed 28 January 2012).

CABI (2012g) *Populus simonii.* Forestry Compendium (http://www.cabi.org/fc/, accessed 12 February 2012).

CABI (2012h) *Salix nigra* (black willow). Forestry Compendium (http://www.cabi.org/fc/, accessed 13 January 2012).

CABI (2012i) *Salix alba.* Forestry Compendium (http://www.cabi.org/fc/, accessed 17 January 2012).

CABI (2012j) *Salix viminalis.* Forestry Compendium (http://www.cabi.org/fc/, accessed 26 January 2012).

Cagelli, L., Lefèvre, F. and Bisoffi, S. (1998) Black poplar (*Populus nigra*) (*Salicaceae* family). *Sherwood – Foreste ed Alberi Oggi* 4, 43–47.

Cannell, M.G.R., Milne, R., Sheppard, L.J. and Unsworth, M.H. (1987) Radiation interception and productivity of willow. *Journal of Applied Ecology* 24, 261–278.

Cannell, M.G.R., Sheppard, L.J. and Milne, R. (1988) Light use efficiency and woody biomass production of poplar and willow. *Forestry* 61, 125–136.

Ceulemans, R. (1990) Genetic variation in functional and structural productivity determinants in poplar. University of Antwerpen, Thesis Publishers, Amsterdam, Netherlands.

Ceulemans, R. and Deraedt, W. (1999) Production physiology and growth potential of poplars under short rotation forestry. *Forest Ecology and Management* 121, 9–23.

Ceulemans, R. and Isebrands, J.G. (1996) Carbon acquisition and allocation. In: Stettler, R.F., Bradshaw, H.D. Jr, Heilman, P.E. and Hinckley, T.M. (eds) *Biology of Populus and Its Implications for Management and Conservation.* National Research Council of Canada Research Press, Ottawa, pp. 355–399.
Ceulemans, R. and Saugier, B. (1991) Photosynthesis. Chapter 2. In: Raghavendra, A. (ed.) *Physiology of Trees.* John Wiley and Sons, New York, pp. 21–50.
Ceulemans, R., McDonald, A.J.S. and Pereira, J.S. (1996) A comparison among eucalyptus, poplar and willow characteristics with reference to coppice growth process modeling approach. *Biomass and Bioenergy* 112, 215–231.
Chen, L., Zhang, S., Zhao, H., Korpilainen, H. and Li, C. (2010) Sex-related adaptive responses to interaction of drought and salinity in *Populus yunnanensis. Plant, Cell and Environment* 33, 1767–1778.
Chen, S. and Polle, A. (2010) Salinity tolerance in *Populus. Plant Biology* 12, 317–333.
Coleman, M.D., Tolsted, D., Nichols, T., Johnson, W.D., Wene, E.G. and Houghtaling, T. (2006) Post-establishment fertilization of Minnesota hybrid poplar plantations. *Biomass and Bioenergy* 30, 740–749.
Collet, D.M. (2004) *Willows of Interior Alaska.* US Fish and Wildlife Service, Alaska Region, Anchorage, Alaska.
Cooke, J.E.K. and Weih, M. (2005) Nitrogen storage and seasonal nitrogen cycling in *Populus*: bridging molecular physiology and ecophysiology. *New Phytologist* 167, 19–30.
Cooke, J.E.K., Martin, T.A. and Davis, J.M. (2005) Short term physiological and developmental responses to nitrogen availability in hybrid poplar. *New Phytologist* 167, 41–52.
Cooper, D.T. and van Haverbeke, D.F. (1990) *Populus deltoides* Bartr. ex Marsh. Eastern cottonwood. In: Burns, R.M. and Honkala, B.H. (eds) *Silvics of North America. Volume 2, Hardwoods.* Agriculture Handbook 654. USDA Forest Service, Washington, DC, pp. 530–543.
Cottrell, J. (2004) Conservation of black poplar (*Populus nigra* L.). Information Note FCIN57. Forestry Commission, Edinburgh, UK.
Dale, J.E. and Milthorpe, F.L. (1983) *The Growth and Functioning of Leaves.* Cambridge University Press, Cambridge, UK.
DeBell, D.S. (1990) *Populus trichocarpa* Torr. and Gray. Black cottonwood. In: Burns, R.M. and Honkala, B.H. (eds) *Silvics of North America. Volume 2, Hardwoods.* Agriculture Handbook 654. USDA Forest Service, Washington, DC, pp. 570–576.
DesRochers, A., van den Driessche, R. and Thomas, B.R. (2007) The interaction between nitrogen source, soil pH, and drought in the growth and physiology of three poplar clones. *Canadian Journal of Botany* 85, 1046–1057.
Dickmann, D.I. (1991) Role of physiology in forest tree improvement. *Silva Fennica* 25, 248–256.
Dickmann, D.I. (2001) An overview of the genus *Populus.* In: Dickmann, D.I., Isebrands, J.G., Eckenwalder, J.E. and Richardson, J. (eds) *Poplar Culture in North America.* National Research Council of Canada Research Press, Ottawa, pp. 1–42.
Dickmann, D.I. (2006) Silviculture and biology of short-rotation woody crops in temperate regions: then and now. *Biomass and Bioenergy* 30, 696–705.
Dickmann, D.I. and Pregitzer, K.S. (1992) The structure and dynamics of woody plant root systems. In: Mitchell, C.P., Ford-Robertson, J.B., Hinckley, T.M. and Sennerby-Forsse, L. (eds) *Ecophysiology of Short Rotation Forest Crops.* Elsevier Applied Science, London, pp. 95–123.
Dickmann, D.I. and Stuart, K.W. (1983) *The Culture of Poplars in Eastern North America.* Michigan State University, East Lansing, Michigan.
Dickmann, D.I., Isebrands, J.G., Eckenwalder, J.E. and Richardson, J. (eds) (2001a) *Poplar Culture in North America.* National Research Council of Canada Research Press, Ottawa.
Dickmann, D.I., Isebrands, J.G., Blake, T.J., Kosala, K. and Kort, J. (2001b) Physiological ecology of poplars. In: Dickmann, D.I., Isebrands, J.G., Eckenwalder, J.E. and Richardson, J. (eds) *Poplar Culture in North America.* National Research Council of Canada Research Press, Ottawa, pp. 77–118.
Dickson, R.E. (1989) Carbon and nitrogen allocation in trees. *Annals of Forest Science* 46, 631s–647s.
Dickson, R.E. and Isebrands, J.G. (1991) Leaves as regulators of stress response. In: Mooney, H.A., Winner, W.E. and Pell, E.J. (eds) *Response of Plants to Multiple Stresses.* Academic Press, San Diego, California, pp. 3–34.
Dillen, S.Y., Rood, S.B. and Ceulemans, R. (2010) Growth and physiology. In: Jansson, S., Bhalerao, R.P. and Groover, A.T. (eds) *Genetics and Genomics of Populus.* Springer Publishers, New York, pp. 39–66.

Dillen, S.Y., Monclus, R., Barbareaux, C., Bastien, C., Ceulemans, R., Dreyer, E., *et al.* (2011) Is the ranking of poplar genotypes for leaf carbon isotope discrimination stable across sites and years in two different full sub families? *Annals of Forest Science* 68, 1265–1275.

Donahue, R.A., Davis, T.D., Michler, C.H., Riemenschneider, D.E., Carter, D.R., Marquardt, P.E., *et al.* (1994) Growth photosynthesis and herbicide tolerance of genetic modified hybrid poplar. *Canadian Journal of Forest Research* 24, 2377–2383.

Dorn, R.D. (1976) A synopsis of American *Salix. Canadian Journal of Botany* 54, 2769–2789.

Doty, S.L. (2008) Tansley review: enhancing phytoremediation through the use of transgenics and endophytes. *New Phytologist* 179, 318–333.

Dunlap, J.M., Braatne, J.H., Hinckley, T.M. and Stettler, R.F. (1993) Interspecific variation in photosynthetic traits in *Populus trichocarpa. Canadian Journal of Botany* 71, 1304–1311.

Eckenwalder, J.E. (2001) Descriptions of clonal characteristics. In: Dickmann, D.I., Isebrands, J.G., Eckenwalder, J.E. and Richardson, J. (eds) *Poplar Culture in North America.* National Research Council of Canada Research Press, Ottawa, pp. 331–335.

Edwards, M.E. and Dunwoodie, P.W. (1985) Dendrochronological and palynological observations of *Populus balsamifera* in northern Alaska, USA. *Arctic and Alpine Research* 17, 271–278.

Ericsson, T., Rytter, L. and Linder, S. (1992) Nutritional dynamics and requirements of short rotation forests. In: Mitchell, C.P., Ford-Robertson, J.B., Hinckley, T.M. and Sennerby-Forsse, L. (eds) *Ecophysiology of Short Rotation Forest Crops.* Elsevier Applied Science, London, pp. 35–65.

Ericsson, T., Rytter, L. and Vapaavuori, E. (1996) Physiology of carbon allocation in trees. *Biomass and Bioenergy* 111, 115–117.

EUFORGEN (2009) *Conservation of Genetic Resources of Black and White Poplar in Europe.* European Forest Genetic Resources Programme (EUFORGEN), Rome, Italy.

EUFORGEN (2011) European Forest Genetic Resources Programme, past activities (1990–2009) (http://www.euforgen.org/about_euforgen/past_activities_1990_2009.html, accessed 5 January 2012).

FAO (1958) *Poplars in Forestry and Land Use.* Food and Agriculture Organization of the United Nations, Rome.

FAO (1980) *Poplars and Willows in Wood Production and Land Use.* FAO Forestry Series No 10. Food and Agriculture Organization of the United Nations, Rome.

Farmer, R.E. Jr (1996) The genecology of *Populus.* In: Stettler, R.F., Bradshaw, H.D. Jr, Heilman, P.E. and Hinckley, T.M. (eds) *Biology of Populus and its Implications for Management and Conservation.* National Research Council of Canada Research Press, Ottawa, pp. 33–55.

Farmer, R.E. Jr, Freitag, M. and Garlick, K. (1989) Genetic variation and 'C' effects in balsam poplar rooting. *Silvae Genetica* 39, 62–65.

Farrar, J.F. (1995) *Trees in Canada.* Fitzhenry and Whiteside Ltd, Markham, Ontario, Canada.

Flora of China Editorial Committee (2003) Flora of China Web. Harvard University Herbaria (http://flora.huh.harvard.edu/china/, accessed 3 February 2012).

Fowells, H.A. and Means, J.E. (1990) The tree and its environment. In: Burns, R.M. and Honkala, B.H. (eds) *Silvics of North America. Volume 2, Hardwoods.* Agriculture Handbook 654. USDA Forest Service, Washington, DC, pp. 1–11.

Friend, A.L., Scarascia-Mugnozza, G., Isebrands, J.G. and Heilman, P.E. (1991) Quantification of two-year-old poplar root systems: morphology, biomass, and 14C distribution. *Tree Physiology* 8, 109–119.

Fritz, R.S. and Kaufman, S.R. (1993) Variable enemy impact at three scales: herbivore species, plant species, and plant genotype. *Oikos* 68, 463–472.

Fussi, B., Lexer, C. and Heinze, B. (2010) Phylogeography of *Populus alba* (L.) and *Populus tremula* (L.) in Central Europe: secondary contact and hybridization during recolonisation from disconnected refugia. *Tree Genetics and Genomes* 6, 439–450.

Gathy, P. (1970) Plaidoyer pour une essence oubliée: le peuplier blanc (*Populus alba* L.). *Bulletin de la Société Royale Forestière de Belgique* 6, 265–271.

Gielen, B. and Ceulemans, R. (2001) The likely impact of rising atmospheric CO_2 on natural and managed *Populus.* A literature review. *Environmental Pollution* 115, 335–358.

Gochis, D.J. and Cuenea, R.H. (2000) Plant water use and crop curves in hybrid poplars. *Journal of Irrigation and Drainage Engineering* 126, 206–214.

Gornall, J.L. and Guy, R.D. (2007) Geographic variation in ecophysiological traits of black cottonwood (*Populus trichocarpa*). *Canadian Journal of Forest Research* 85, 1202–1213.

Grant, M.C. and Mitton, J.B. (1979) Elevational gradients in adult sex ratios and sexual differentation in vegetative growth rates of *Populus tremuloides.* Michx. *Evolution* 33, 914–918.
Gries, D., Zeng, F., Foetzki, A., Arndt, S.K., Bruelheide, H., Thomas, F.M., *et al.* (2003) Growth and water relations of *Tamarix ramosissima* and *Populus euphratica* on Taklamakan desert dunes in relation to depth to a permanent water table. *Plant Cell and Environment* 26, 725–736.
Hangs, R.D., Schoenau, J.J., Van Rees, K.C.J. and Steppuhn, H. (2011) Examining the salt tolerance of willow (*Salix* spp.) bioenergy species for use on salt-affected agricultural lands. *Canadian Journal of Plant Science* 91, 509–517.
Heilman, P.E., Hinckley, T.M., Roberts, D.A. and Ceulemans, R. (1996) Production physiology. In: Stettler, R.F., Bradshaw, H.D. Jr, Heilman, P.E. and Hinckley, T.M. (eds) *Biology of Populus and Its Implications for Management and Conservation.* National Research Council of Canada Research Press, Ottawa, pp. 459–489.
Hinckley, T.M. (1996) Physiology of growth productivity and stress response. In: Stettler, R.F., Bradshaw, H.D. Jr, Heilman, P.E. and Hinckley, T.M. (eds) *Biology of Populus and Its Implications for Management and Conservation.* National Research Council of Canada Research Press, Ottawa, pp. 277–281.
Hinckley, T.M., Ceulemans, R., Dunlap, J.M., Figliola, A., Heilman, P.E., Isebrands, J.G., *et al.* (1989) Physiological, morphological, and anatomical components of hybrid vigor in *Populus.* In: Kreeb, K.H., Richter, H. and Hinckley, T.M. (eds) *Structural and Functional Responses to Environmental Stresses.* SPB Academic Publishing, The Hague, Netherlands, pp. 199–217.
Hinckley, T.M., Braatne, J., Ceulemans, R., Clum, P., Dunlap, J., Newman, D., *et al.* (1992) Growth dynamics and canopy structure. In: Mitchell, C.P., Ford-Robertson, J.B., Hinckley, T.M. and Sennerby-Forsse, L. (eds) *Ecophysiology of Short Rotation Crops.* Elsevier Applied Science, London, pp. 1–34.
Hinckley, T., Brooks, J.R., Cermak, J., Ceulemans, R., Kucera, J., Meinzer, F.C., *et al.* (1994) Water flux in hybrid poplar stand. *Tree Physiology* 14, 1005–1018.
Host, G.E. and Isebrands, J.G. (1994) An interregional validation of ECOPHYS, a growth process model of juvenile poplar clones. *Tree Physiology* 14, 933–945.
Hultine, K.R., Bush, S.E. and Ehleringer, J.R. (2010) Ecophysiology of riparian cottonwood and willow before, during, and after two years of soil water removal. *Ecological Applications* 20, 347–361.
ICRAF (2011) *Populus euphratica.* AgroForestryTree Database. World Agroforestry Centre (http://www.worldagroforestrycentre.org/sea/products/afdbases/af/asp/SpeciesInfo.asp?SpID=1335, accessed 21 December 2011).
Illinois Wildflowers (2012) Heart-leaved willow. *Salix rigida* (www.illinoiswildflowers.info/trees/plants/hl_willow.html, accessed 7 February 2012).
Isebrands, J.G. (2007) Best Management Practices Poplar Manual for Agroforestry Applications in Minnesota (Final). Environmental Forestry Consultants LLC, New London, Wisconsin (http://www.extension.umn.edu/distribution/naturalresources/00095.html, accessed 26 July 2013).
Isebrands, J.G. and Nelson, N.D. (1983) Distribution of 14C-labelled photosynthates within intensively cultured *Populus* clones during the establishment year. *Physiologia Plantarum* 59, 9–18.
Isebrands, J.G., Nelson, N.D., Dickmann, D.I. and Michael, D.A. (1983) Yield physiology of short rotation intensively cultured poplars. In: Hansen, E.A. (ed.) *Intensive Plantation Culture: 12 Years Research.* General Technical Report NC-91. USDA Forest Service, North Central Research Station, St Paul, Minnesota, pp. 77–93.
Isebrands, J.G., Ceulemans, R. and Wiard, B. (1988) Genetic variation in photosynthesis traits among *Populus* clones in relation to yield. *Plant Physiology and Biochemistry* 26, 427–437.
Isebrands, J.G., Dickson, R.E. and Ceulemans, R. (1990) Dynamics of ecophysiological processes in tree crowns and forest canopies. *Tree Physiology* 7, 1–367.
Isebrands, J.G., Dickson, R.E., Rebbeck, J. and Karnosky, D.F. (2000) Interacting effects of multiple stresses on growth and physiological processes in northern forest trees. In: Mickler, R.A., Birdsey, R.A. and Hom, J. (eds) *Responses of Northern US Forests to Environmental Change.* Springer-Verlag, New York, pp. 149–180.
Jansson, S., Bhalerao, R.P. and Groover, A.T. (2010) *Genetics and Genomics of Populus.* Springer, New York.
Johnston, C.A. (1993) Shrub species as indicators of wetland sedimentation. *Wetlands* 23, 911–920.
Joshi, C.P., DiFazio, S.P. and Kole, C. (2011) *Genetics, Genomes and Breeding of Poplar.* Taylor and Francis, Abingdon, UK.

Kajba, D., Gracan, J., Ivankovic, M., Bogdan, S., Gradecki-Postenjak, M., Littvay, T., *et al.* (2006) [Conservation of forest genetic resources in Croatia] *Glasnik za Sum. Pokuse* 5, 235–249.
Kimmins, J.P. (1987) *Forest Ecology.* MacMillan Publishing Company, New York.
Knight, D.H. (2001) Summary: Aspen decline in the West? In: Sheppard, W.D., Binckley, D., Bartos, D.L., Stohlgren, T.J. and Eskew, L.G. (eds) *Sustaining Aspen in Western Landscapes.* Symposium Proceedings RMRS-P-18. USDA Forest Service, Rocky Mountain Research Station, Fort Collins, Colorado, pp. 441–446.
Kojima, S. (1979) Biogeoclimatic zones of Hokkaido Island, Japan. *Journal of the College of Liberal Arts, Toyama University* 12, 97–141.
Kort, J. and Blake, T.J. (2007) Sap velocity as an indicator of diurnal and long-term hydraulic resistance in mature 'Walker' hybrid poplar trees. *Canadian Journal of Botany* 85, 1033–1040.
Kozlowski, T.T., Kramer, P.J. and Pallardy, S.G. (1991) *The Physiological Ecology of Woody Plants.* Academic Press, San Diego, California.
Kramer, P.J. (1986) The role of physiology in forestry. *Tree Physiology* 2, 1–16.
Kranjcec, J., Mahoney, J.M. and Rood, S.B. (1998) The responses of three riparian cottonwood species to water table decline. *Forest Ecology and Management* 110, 77–87.
Kull, O. and Niinemets, U. (1998) Distribution of leaf photosynthetic properties in tree canopies: comparison of species with different shade tolerance. *Functional Ecology* 12, 472–479.
Kuzovkina, J. (2010) Willows beyond bioenergy: *Salix* for ecological engineering and as a specialty crop in low input systems. In: Scarascia-Mugnozza, G. and Stanton, B. (eds) *Book of Abstracts. Fifth International Poplar Symposium. Poplars and Willows: From Research Models to Multipurpose Trees for a Bio-based Society.* September 20–25, 2010, Orvieto, Italy. Research Council for Agriculture, Rome, Italy, pp. 9 (http://www.iufro.org/publications/proceedings/proceedings-meetings-2010/, accessed 22 February 2013).
Kuzovkina, Y.A., Weih, M., Romero, M.A., Charles, J., Hurst, S., McIvor, I., *et al.* (2008) *Salix*: botany and global horticulture. *Horticultural Reviews* 34, 447–489.
Laidly, P.R. (1990) *Populus grandidentata.* Bigtooth aspen. In: Burns, R.M. and Honkala, B.H. (eds) *Silvics of North America. Volume 2. Hardwoods.* Agriculture Handbook 654. USDA Forest Service, Washington, DC, pp. 544–550.
Landsberg, J.J. (1986) *Physiological Ecology of Forest Production.* Academic Press, London.
Larson, P.R. (1983) Primary vascularization and the siting of primordia. In: Dale, J.E. and Milthorpe, F.L. (eds) *The Growth and Functioning of Leaves.* Cambridge University Press, Cambridge, UK, pp. 25–51.
Larson, P.R. (1994) *The Vascular Cambium.* Springer-Verlag, New York.
Larson, P.R. and Isebrands, J.G. (1974) Anatomy of the primary–secondary transition zone in stems of *Populus deltoides. Wood Science Technology* 8, 11–26.
Larson, P.R., Isebrands, J.G. and Dickson, R.E. (1972) Fixation patterns of 14C within developing leaves of eastern cottonwood. *Planta* 107, 301–341.
McLeod, K.W. and McPherson, J.K. (1973) Factors limiting the distribution of *Salix nigra. Bulletin of the Torrey Botanical Club* 100, 102–110.
Maini, J.S. (1968) Silvics and ecology of *Populus* in Canada. In: Maini, J.S. and Cayford, J. (eds) *Growth and Utilization of Poplars in Canada.* Canada Department of Forestry and Rural Development, Ottawa, pp. 20–69.
Michael, D.A., Dickmann, D.I., Isebrands, J.G. and Nelson, N.D. (1990) Photosynthesis patterns during the establishment year within two *Populus* clones with contrasting morphology and phenology. *Tree Physiology* 6, 22–27.
Mirck, J. and Volk, T.A. (2010) Response of three shrub willow varieties (*Salix* spp.) to stormwater treatments with different concentrations of salts. *Bioresource Technology* 101, 3484–3492.
Mitchell, C.P., Ford-Robertson, J.B., Hinckley, T.M. and Sennerby-Forsse, L. (1992) *Ecophysiology of Short Rotation Forest Crops.* Elsevier Applied Science, New York.
Native Plant Data Base (2012) *Salix eriocephala* (http://nativeplants.evergreen.ca/search/view-plant.php?ID=00610, accessed 7 February 2012).
Nelson, N.D. (1984) Poplars are not inherently low in photosynthetic capacity. *Photosynthetica* 18, 600–605.
Nelson, N.D. and Isebrands, J.G. (1983) Late-season photosynthesis and photosynthate distribution in an intensively-cultured *Populus nigra* × *laurifolia* clone. *Photosynthetica* 17, 537–549.
Newcombe, G. (1996) The specificity of fungal pathogens of *Populus.* In: Stettler, R.F., Bradshaw, H.D. Jr, Heilman, P.E. and Hinckley, T.M. (eds) *Biology of Populus and Its Implications for Management and Conservation.* National Research Council of Canada Research Press, Ottawa, pp. 223–246.

Newcombe, G., Ostry, M., Hubbes, M., Perinet, P. and Mottet, M.J. (2001) Poplar diseases. In: Dickmann, D.I., Isebrands, J.G., Eckenwalder, J.E. and Richardson, J. (eds) *Poplar Culture in North America*. National Research Council of Canada Research Press, Ottawa, pp. 249–276.
Newman, L., Strand, S., Choe, N., Duffy, J., Ekuan, G., Ruszaj, M., *et al.* (1997) Uptake and biotransformation of trichloroethylene by hybrid poplars. *Environmental Science Technology* 31, 1062–1067.
New Phytologist (2005) Poplar functional genomics. *New Phytologist* 167, 9–170.
Newsholme, C. (1992) *Willows: the Genus Salix*. B.T. Batsford, London.
Orlovic, S., Pajevic, S., Klasnja, B., Galic, Z. and Markovik, M. (2006) Variability of physiological and growth of white willow (*Salix alba*) clones. *Genetika* 38, 145–152.
Perala, D.A. (1990) *Populus tremuloides*, quaking aspen. In: Burns, R.M. and Honkala, B.H. (eds) *Silvics of North America. Volume 2. Hardwoods.* Agriculture Handbook 654. USDA Forest Service, Washington, DC, pp. 555–569.
Percy, K.E., Awmack, C.S., Lindroth, R.L., Kubiske, M.E., Kopper, B.J., Isebrands, J.G., *et al.* (2002) Altered performance of forest pests under atmospheres enriched by CO_2 and O_3. *Nature* 402, 403–407.
Peterson, E.B. and Peterson, N.M. (1992) Ecology, management, and use of aspen and balsam poplar in the prairie provinces, Canada. Special Report 1. Forestry Canada, Edmonton, Alberta, Canada.
Pitcher, J.A. and McKnight, J.S. (1990) *Salix nigra* Marsh, black willow. In: Burns, R.M. and Honkala, B.H. (eds) *Silvics of North America. Volume 2, Hardwoods.* Agriculture Handbook 654. USDA Forest Service, Washington, DC, pp. 768–772.
Polle, A. and Douglas, C. (2010) The molecular physiology of poplars; paving the way for knowledge-based biomass production. *Plant Biology* 12, 239–241.
Pryor, L.D. and Willing, R.R. (1965) The development of poplar clones suited to low latitudes. *Silvae Genetica* 14, 123–127.
Rae, A.M., Robinson, K.M., Street, N.R. and Taylor, G. (2004) Morphological and physiological traits influencing biomass productivity in short-rotation coppice poplar. *Canadian Journal of Forest Research* 34, 1488–1498.
Rennenberg, H., Wildhagen, H. and Ehiting, B. (2010) Nitrogen nutrition in poplar trees. *Plant Biology* 12, 275–291.
Rood, S.B., Campbell, J.S. and Despins, T. (1985) Natural poplar hybrids from southern Alberta. I. Continuous variation for foliar characteristics. *Canadian Journal of Botany* 64, 1382–1388.
Rood, S.B., Braatne, J.H. and Hughes, F.M.R. (2003) Ecophysiology of riparian cottonwoods streamflow dependency, water relations and restoration. *Tree Physiology* 23, 1113–1124.
Sabatti, M. (1994) Ecologia genetica in popolazioni naturali di *Populus alba* L.: produttività, fenologia ed efficienza di uso idrico. PhD dissertation, University of Tuscia, Viterbo, Italy.
Sabatti, M., D'Ovidio, R., Tanzarella, O.A. and Scarascia Mugnozza, G.E. (2001) Assessment of geographic variation by RAPD markers among Italian open-pollinated progenies of *Populus alba* L. *Genetic Resources and Crop Evolution* 48, 423–428.
Sagebud (2012) Missouri River willow (*Salix eriocephala*) (http://www.sagebud.com/missouri-river-willow-salix-eriocephala, accessed 7 February 2012).
Scarascia-Mugnozza, G., Hinckley, T.M., Stettler, R.F., Heilman, P.E. and Isebrands, J.G. (1999) Production physiology and morphology of *Populus* species and their hybrids grown under short rotation. III. Seasonal carbon allocation patterns from branches. *Canadian Journal of Forest Research* 29, 1419–1432.
Schluter, M., Ruger, N., Savitsky, A.G., Novikova, N.M., Matthies, M. and Lieth, H. (2006) TUGAI: an integrated simulation tool for ecological assessment of alternative water management strategies in a degraded river delta. *Environmental Management* 38, 638–653.
Sennerby-Forsse, L., Ferm, A. and Kauppi, A. (1992) Coppice ability and sustainability. In: Mitchell, C.P., Ford-Robertson, J.B., Hinckley, T.M. and Sennerby-Forsse, L. (eds) *Ecophysiology of Short Rotation Forest Crops*. Elsevier Applied Science, London, pp. 146–184.
Shaburov, V.I. and Belyaeva, I.V. (1991) Comparative analysis of the technical qualities of the rods in *Salix viminalis* stands in the Central Urals. *Rastitel'nye Resursy* 27, 99–102.
Sheppard, W.D. (1993) Initial growth, development and clonal dynamics of regenerated aspen in the Rocky Mountains. Research Paper RM 324.USDA Forest Service, Rocky Mountain Research Station, Fort Collins, Colorado.
Shields, W.J. and Bockheim, J.G. (1981) Deterioration of trembling aspen clones in the Great Lakes region. *Canadian Journal of Forest Research* 11, 530–537.
Shoji, S. (1993) Studies on early weathering of volcanic ash and rehabilitation of soil environment. Report No. 04453126 for the Grant-in-Aid of Scientific Research from the Ministry of Education, Japan.

Sidorov, A.I. (1978) *Tannidnie iwi.* Lesnaya promishlennost, Moscow.

Sigaud, P. (2003) Genetic conservation and use of poplars in China with special attention to northern China. In: Ducci, P. (ed.) *Proceedings of First International Conference on the Future of Poplar Culture.* Commissione Nationale del Pioppo, Rome, pp. 4–13.

Silim, S., Nash, R., Reynard, D. and Schroeder, W. (2009) Leaf gas exchange and water potential responses to drought in nine poplar (*Populus* spp.) with contrasting drought tolerance. *Trees* 23, 959–969.

Skvortsov, A.K. (1968) *Willows of the USSR.* Nauka, Moscow.

Skvortsov, A.K. (1999) *Willows of Russia and Adjacent Countries. Taxonomical and Geographical Revision.* (English translation of 1968 Russian edition). University of Joensuu, Joensuu, Finland.

Spurr, S.H. and Barnes, B.V. (1980) *Forest Ecology*, 3rd edn. John Wiley and Sons, New York.

Stanturf, J.A., van Oosten, C., Netzer, D.A., Coleman, M.D. and Portwood, C.J. (2001) Ecology and silviculture of poplar plantations. In: Dickmann, D.I., Isebrands, J.G., Eckenwalder, J.E. and Richardson, J. (eds) *Poplar Culture in North America.* National Research Council of Canada Research Press, Ottawa, pp. 153–206.

Steneker, G.A. (1973) The size of trembling aspen (*Populus tremuloides* Michx.) clones in Manitoba. *Canadian Journal of Forest Research* 3, 472–478.

Stettler, R.F. (2009) *Cottonwood and the River of Time.* University of Washington Press, Seattle, Washington.

Stettler, R.F., Bradshaw, H.D. Jr, Heilman, P.E. and Hinckley, T.M. (eds) (1996) *Biology of Populus and Its Implications for Management and Conservation.* National Research Council of Canada Research Press, Ottawa.

Taylor, G., Robinson, K. and Karp, A. (2003) *Identifying Physiological Traits for Yield of Willow Biomass.* University of Southampton, Southampton, UK.

Telewski, F.W., Aloni, R. and Sauter, J.J. (1996) Physiology of secondary tissues of *Populus.* In: Stettler, R.F., Bradshaw, H.D. Jr, Heilman, P.E. and Hinckley, T.M. (eds) *Biology of Populus and Its Implications for Management and Conservation.* National Research Council of Canada Research Press, Ottawa, pp. 301–329.

Tharakan, P.J. (1999) *Clonal Performance Evaluation and Yield Physiology of Willow and Poplar Bioenergy Crops.* M.S. thesis. State University of New York, College Environmental Science and Forestry, Syracuse, New York.

Toplu, F. (2005) Breeding and conservation of black poplar (*Populus nigra*) gene resources in Turkey. *Unasylva* 56, 26–30.

Tsarev, A.P. (2005) Natural poplar and willow ecosystems on a grand scale: the Russian Federation. *Unasylva* 56, 10–11.

Tschlaplinski, T.J. and Blake, T.J. (1989) Water relations and photosynthetic capacity as determinants of productivity in hybrid poplar cultivars. *Canadian Journal of Botany* 67, 1689–1697.

Tsuyuzaki, S. and Haruki, M. (1996) Tree regeneration patterns on Mount Usu, northern Japan, since the 1977–78 eruptions. *Vegetatio* 126, 191–198.

Tuskan, G.A., DiFazio, S., Jansson, S., Bohlmann, J., Grigoriev, I., Hellsten, U., *et al.* (2006) The genome of black cottonwood, *Populus trichocarpa* (Torr. & Gray). *Science* 313, 1596–1604.

US Department of Agriculture (2012) USDA Germplasm Resources Information Network (http://www.ars-grin.gov, accessed 26 January 2012).

USDA Forest Service (2012a) *Populus balsamifera* subspecies *balsamifera* (http://www.fs.fed.us/database/feis/plants/tree/popbalb/all.html, accessed 20 January 2012).

USDA Forest Service (2012b) *Populus tremuloides,* trembling aspen (http://www.fs.fed.us/database/feis/plants/tree/poptre/all.html, accessed 28 January 2012).

USDA Forest Service (2012c) *Salix nigra* (http://www.fs.fed.us/database/feis/plants/tree/salnig/all.html, accessed 13 January 2012).

USDA Natural Resources Conservation Service (2012a) *Salix exigua*, Coyote willow (http://plants.usda.gov/plantguide/pdf/cs_saex.pdf, accessed 26 July 2013).

USDA Natural Resources Conservation Service (2012b) *Salix eriocephala* Michx. Missouri River willow (http://plants.usda.gov/core/profile?symbol=SAER, accessed 26 July 2013).

Vanden Broeck, A. (2003) EUFORGEN technical guidelines for genetic conservation and use for European black poplar (*Populus nigra*). International Plant Genetic Resources Institute, Rome.

van Volkenburgh, E. and Taylor, G. (1996) Leaf growth physiology. In: Stettler, R.F., Bradshaw, H.D. Jr, Heilman, P.E. and Hinckley, T.M. (eds) *Biology of Populus and Its Implications for Management and Conservation.* National Research Council of Canada Research Press, Ottawa, pp. 283–299.

Vasil'ev, N. (1967) Flood plain forests of the River Khor basin (Khabarovskiy Krai). In: *Results of Forest Study of the Far East.* Vladivostok, USSR, pp. 95–98.

Viart, M. (1988) Mini-monograph on *Populus euphratica.* Document submitted to the 18th Session of the International Poplar Commission, Beijing, China, September 1988. Available from IPC Secretariat, FAO, Rome.

Viereck, L.A. and Foote, J.M. (1970) The status of *Populus balsamifera* and *P. trichocarpa* in Alaska. *Canadian Field Naturalist* 84, 169–173.

von Wuehlisch, G. (2009) EUFORGEN technical guidelines for genetic conservation and use of Eurasian aspen (*Populus tremula*). Diversity International, Rome.

Vorob'ev, D. (1968) *Wild-growing Trees and Bushes of the Far East.* Nauka, Leningrad, USSR.

Vose, J., Swank, W., Harvey, G., Clinton, B. and Sobek, C. (2000) Leaf water relations and sapflow in eastern cottonwood (*Populus deltoides*) trees planted for phytoremediation of a groundwater pollutant. *International Journal of Phytoremediation* 2, 53–73.

Wang, S., Chen, B. and Li, H. (1996) *Euphrates Poplar Forest.* China Environmental Science Press, Beijing.

Weber, J.C., Stettler, R.F. and Heilman, P.E. (1985) Genetic variation and productivity of *Populus trichocarpa* and its hybrids. I. Morphology and phenology of 50 native clones. *Canadian Journal of Forest Research* 15, 376–383.

Weisgerber, H. and Han, Y. (2001) Diversity and breeding potential of poplar species in China. *Forestry Chronicle* 77, 227–237.

Weisgerber, H., Kownatzki, D. and Mussong, M. (1995) Natural poplar resources in China and their significance for breeding and afforestation. *Silvae Genetica* 44, 298–303.

Whitham, T.G., Floate, K.D., Martinsen, G.D., Driebe, E.M. and Keim, P. (2001) Ecological and evolutionary implications of hybridization: *Populus*–herbivore interactions. In: Stettler, R.F., Bradshaw, H.D. Jr, Heilman, P.E. and Hinckley, T.M. (eds) *Biology of Populus.* National Research Council of Canada Research Press, Ottawa, pp. 247–275.

Willowpedia (2012) Diseases (http://willow.cals.cornell.edu/Research/Diseases.html, accessed 8 February 2012).

Yimit, H., Ayup, M., Wang, G.Z., Luo, H. and Ebeidulla, D. (2006) *P. euphratica* ecosystem fragility and protecting strategy on Tarim *P. euphratica* nature reserve in Xinjiang. In: Gao, W. and Ustin, S.L. (eds) *Remote Sensing and Modeling of Ecosystems for Sustainability III.* Proceedings, SPIE, Vol 6298. SPIE Digital Library, Bellingham, Washington, pp. 62981N.

Zasada, J.C. and Phipps, H.M. (1990) *Populus balsamifera.* Linnaeus. Balsam poplar. In: Burns, R.M. and Honkala, B.H. (eds) *Silvics of North America. Volume 2. Hardwoods.* Agriculture Handbook 654. USDA Forest Service, Washington, DC, pp. 518–529.

Zasada, J.C., David, A.J., Gilmore, D.W. and Landhäusser, S.M. (2001) Ecology and silviculture of natural stands of *Populus* species. In: Dickmann, D.I., Isebrands, J.G., Eckenwalder, J.E. and Richardson, J. (eds) *Poplar Culture in North America.* National Research Council of Canada Research Press, Ottawa, pp. 119–151.

Zhang, X., Meng, T. and Kang, E. (2008) Conversion of water consumption of a single tree and a forest stand of *Populus euphratica. Forestry Studies in China* 10, 81–87.

Zsuffa, L., Giordano, E., Pryor, L.D. and Stettler, R.F. (1996) Trends in poplar culture: some global and regional perspectives. In: Stettler, R.F., Bradshaw, H.D. Jr, Heilman, P.R. and Hinckley, T.M. (eds) *Biology of Populus and Its Implications for Management and Conservation.* National Research Council of Canada Research Press, Ottawa, pp. 515–539.

4 The Domestication and Conservation of *Populus* and *Salix* Genetic Resources

Brian J. Stanton,[1]* Michelle J. Serapiglia[2] and Lawrence B. Smart[2]
[1]GreenWood Resources, Portland, Oregon, USA; [2]Cornell University, Department of Horticulture, New York State Agricultural Experiment Station, Geneva, New York, USA

4.1 Introduction

There are many similarities in the ecology and genetics of *Populus* and *Salix* as closely related genera in the family *Salicaceae*. In this chapter, we will highlight those commonalities as well as the unique characteristics of the diversity of poplars and willows throughout the world. The genetic resources of both genera are organized into sections and species (see Chapter 2, this volume) and structured by an additional hierarchy of intraspecific differentiation including subspecies or geographic varieties, populations, families and, finally, the individual genotype (Schreiner, 1970, 1974; Steenackers, 1996; Trybush *et al.*, 2008). There are 32 species in the genus *Populus* (Eckenwalder, 1996; Allaby, 1998). In sharp contrast are the more than 350 species that have been described for the genus *Salix* (Argus, 1997; Skvortsov, 1999). With the advent of molecular techniques, there is now a better understanding of the distribution of allelic variation associated with ecologically and commercially important traits distributed within the wild and domesticated stocks of the genetic resources of both genera (Rönnberg-Wästljung and Gullberg, 1996; Rönnberg-Wästljung, 2001; Hanley, 2003; Cervera *et al.*, 2004; Wegrzyn *et al.*, 2010).

Taken as a whole, poplars and willows are a most remarkable group of trees and shrubs that have been associated with humankind since antiquity, a consequence of their riparian ecology and humanity's early riverine civilizations. Settlements in Mesopotamia, the Indus Valley, the Nile Valley and the Yellow River basin all relied variously on *P. euphratica*, *P. tremula*, *P. nigra* and *P. alba* for timber, fuel, animal forage and protection from wind and sun. *S. bebbiana*, *S. discolor*, *S. exigua*, *S. interior*, *S. lucida*, *S. nigra* and *S. petiolaris* were valued for their medicinal and structural properties by native American tribes (Erichsen-Brown, 1979), while *S. babylonica*, the well-known weeping willow tree, has long been an iconic feature of Asian art and culture.

The natural range of *Populus* and *Salix* spans an impressive ecological amplitude, primarily across the North American, European and Asian land masses – from the subtropics to the boreal forests and arctic tundra, riparian to montane ecosystems and the man-made environment of modern agriculture. As a consequence, poplar and willow geneticists – those responsible for conserving and domesticating germplasm of *Populus* and *Salix* – have an especially broad mandate: to study the genetic diversity of natural

*Corresponding author: brian.stanton@gwrglobal.com

populations and be familiar with all the modern tools for genetic improvement in order to serve specific societal needs (Kuzovkina and Quigley, 2005; Kuzovkina *et al.*, 2008; Kuzovkina and Volk, 2009, Stanton *et al.*, 2010).

Genetic resources are managed for the dual purpose of domestication and conservation (Steenackers, 1996) (Plate 17C). Domestication is the process of exploiting and manipulating genetic variation at multiple levels to breed commercial cultivars for wood-based commodities, energy feedstock and environmental services (Libby, 1973; Dickmann *et al.*, 1994; Bradshaw and Strauss, 2001). The goal of conservation is the maintenance or expansion of genetic variation among and within species. It is an indispensable foundation for future domestication programmes (Ledig, 1986; Tanksley and McCouch, 1997; Storme *et al.*, 2004). Conservation practices also maintain a reservoir of diversity to provide continuity in ecosystem function in the face of an ever-changing environment (Berrang *et al.*, 1986; Hughes *et al.*, 2008).

Throughout the Americas, Asia, Australasia and Europe, *Populus* and *Salix* are being managed by the 37 member nations of the International Poplar Commission (IPC), among others, using widely different approaches and technologies that range from traditional hybridization for cultivar development to molecular investigations into the structure of the genome. This chapter reviews the principal domestication and conservation programmes of 24 of the most active IPC nations plus two other non-member countries with dynamic programmes. Most of this work involves 12 species in the genus *Populus* that are noteworthy for their commercial and ecological values. They are the North American species *P. balsamifera*, *P. deltoides*, *P. trichocarpa* and *P. tremuloides* and the Eurasian species *P. alba*, *P. cathayana*, *P. ciliata*, *P. euphratica*, *P. maximowiczii*, *P. nigra*, *P. simonii* and *P. tremula*. Within the genus *Salix*, 10 species – *S. caprea*, *S. dasyclados*,[1] *S. eriocephala*, *S. koriyanagi*, *S. miyabeana*, *S. purpurea*, *S. udensis*, *S. schwerinii*, *S. triandra* and *S. viminalis* – are being utilized in developing the world's renewable energy industry, while three others – *S. alba*, *S. babylonica* (synonym *S. matsudana*) and *S. nigra* – are favoured for timber products. The chapter first reviews the genetic systems of *Populus* and *Salix* and how they form the basis for the various domestication and conservation options available today. The second part of the chapter provides a description of the rich diversity of programmes focused on *Populus* and *Salix* genetics around the globe. The chapter includes a listing of the world's leading *Populus* and *Salix* geneticists and breeders who have contributed material.

4.2 Overview of Domestication and Conservation Approaches

Populus domestication has a history of nearly 100 years, beginning with Henry's (1914) work at the Royal Botanic Gardens, Kew, in the UK, and the work of Stout and Schreiner (1933) and Stout *et al.* (1927) at the New York Botanical Garden in the USA. Other early domestication efforts include those of Wettstein-Westersheim (1933) in Germany, Al'benskii and Delitsina (1934) in Russia, Heimburger (1936) in Canada, Jacometti (1934, 1937) in Italy and Houtzagers (1952) in the Netherlands. *Salix* domestication traces to the hybridization studies of Nis Heribert-Nilsson in Sweden (Heribert-Nilsson, 1918), along with Nilsson and Hakansson's cytological work in the 1930s (Nilsson, 1931; Hakansson, 1933, 1938). In the UK, H.P. Hutchinson began work in willow conservation and breeding in the 1920s at the Long Ashton Research Station that was continued by K.G. Stott for the following 30 years (Newsholme, 1992; Stott, 1992).

4.2.1 Genetic systems

Genetic systems are the manner in which a species' heritable material is organized and transmitted from generation to generation (Grant, 1975). The process of domestication is partially determined by the relatively unique genetic systems of *Populus* and *Salix*. These systems enable the creation of new genotypes to allow for adaptation to changing environments while preserving the adaptability of the current generation (Zsuffa *et al.*, 1984; Mitton and Grant, 1996; Arens *et al.*, 1998; Martinsen *et al.*,

2001; Karp *et al.*, 2011). Open recombination is promoted by:

1. Strict outcrossing enforced by dioecism (male and female flowers produced on separate plants) (Plate 18A and B).
2. Fecund seed production owing to inflorescences containing 30–40 pistillate flowers or 60–80 staminate flowers (Fechner, 1972; Boes and Strauss, 1994).
3. Effective gene flow by virtue of abundant production of airborne pollen in *Populus*. *Salix* relies primarily on insect pollination, but some species can also be wind pollinated (Argus, 1974; Tamura and Kudo, 2000).
4. Small seeds with attached fibres that facilitate long-distance transport by air and water (Weber and Stettler, 1981). Under the right conditions, seeds of both genera germinate promptly and sizeable populations establish with regularity.
5. Relatively large basic chromosome number (n = 19) that enhances gametic recombination during reduction division (Wu and Stettler, 1994). Diploidy is the rule in *Populus*, while ploidy levels in *Salix* range from diploid to dodecaploid (Zsuffa *et al.*, 1984). Polyploidy has been suggested as a route to speciation in *Salix* (Dorn, 1976). Evidence of this is the limited occurrence of polyploidy in the more primitive subgenera, whereas about 40% of the species in the more advanced subgenus *Vetrix* are polyploids (Dorn, 1976).
6. A substantial portion of genetic variation allocated within populations that may be associated with gametic disequilibrium among physically unlinked alleles (Weber and Stettler, 1981; Keller *et al.*, 2010). The importance of geographic differentiation among *Populus* and *Salix* populations notwithstanding, the significant amount of genetic variation in adaptive traits that resides within breeding populations becomes available during reproduction for recombination in forming novel genotypes (Weber *et al.*, 1985; Foster, 1986; Lascoux *et al.*, 1996; Lin *et al.*, 2009; Sulima *et al.*, 2009).
7. Factors of adaptation inherited as composite traits, the individual components of which may be matched to different aspects of the environment. For example, in autumnal phenology in *Populus*, the cessation of annual shoot growth is tied to changes in day length, while temperature modifies the sensitivity to day-length signals when influencing the duration of bud formation (Rohde *et al.*, 2011a, b).

Vegetative propagation helps to perpetuate a genotype beyond the normal lifespan of a typical early-successional tree, thereby slowing the evolutionary rate and perpetuating hybrids and polyploids of limited fertility for prolonged opportunities to reproduce sexually (Salick and Pfeffer, 1999; Hardig *et al.*, 2000). Vegetative regeneration or cloning is highly developed in both *Populus* and *Salix* and counterbalances the open recombination system preserving the parent generation's characteristics. Examples of *Populus* clonal propagation from roots, stumps and branches include: (i) *P. tremuloides* forms extensive clonal stands by suckering from root sprouts on upland sites (Frey *et al.*, 2003); (ii) *P. balsamifera* colonizes grasslands by suckering from roots of trees growing in surrounding forests (Little and Dale, 1999); and (iii) *P. trichocarpa* initiates stands along river courses via detached branches (Rood *et al.*, 2003a) (Fig. 4.1). Many populations of *Salix* are also clonal in nature, particularly those in highly disturbed areas where establishment may occur through the rooting of fallen branches and, for a small number of species such as *S. exigua*, via root suckering.

One key feature of *Populus* and *Salix* genetic systems that has affected domestication programmes is the capacity for interspecific hybridization. Natural cross-species hybridization is almost the rule wherever poplar species of the same or closely related sections are in contact and may involve two to four species in a given area (Floate, 2004). The process often entails backcrossing to one or more of the contributing species, introgressing genes into their genomes. Such hybrid swarms play an important ecological role for local invertebrates and avifauna (Whitham *et al.*, 1999) and may also determine the composition of understorey plant communities (Lamit *et al.*, 2011). Persistent hybrid swarms, such as those that occur in *P.* ×*canescens* (Lexer *et al.*, 2005), are a force that further opens the recombination system of the participating species to new alleles and coevolved sets of genes. This phenomenon occurs in nature with the following notable examples: (i) *P.* ×*canescens*, the hybrid of *P. alba* and *P. tremula* (van Loo *et al.*, 2008); (ii) *P.* ×*smithii*, the hybrid of

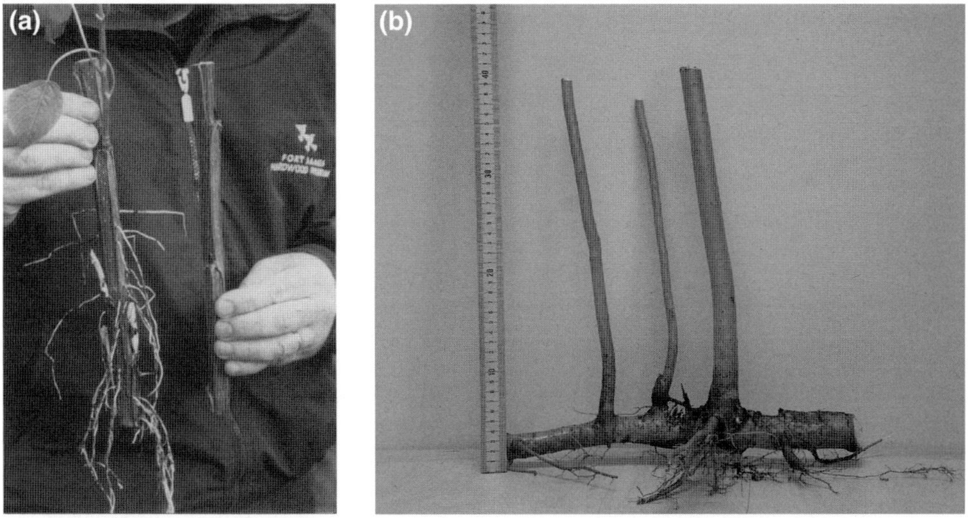

Fig. 4.1. Vegetative regeneration is well developed in *Populus*. *Populus ×generosa* (*Populus deltoides* × *Populus trichocarpa* (a) propagates from hardwood stem cuttings. Photo courtesy of Rich Shuren, GreenWood Resources. *Populus tremuloides* (b) propagates by suckering from roots. Photo courtesy of Andrew David, University of Minnesota.

P. grandidentata and *P. tremuloides* (Pregitzer and Barnes, 1980); (iii) *P. ×parryi*, the hybrid of *P. fremontii* and *P. trichocarpa* (Eckenwalder, 1984); and (iv) *P. ×jackii*, the hybrid of *P. deltoides* and *P. balsamifera* (Gom and Rood, 1999). The identical phenomenon is observed in *Salix* where members of the same section commonly hybridize when growing sympatrically, to the extent that discrimination between populations can be difficult (Brunsfeld *et al.*, 1992; Purdy and Bayer, 1995). An example is *S. ×fragilis* (formerly *S. ×rubens*), a common, naturally occurring interspecific hybrid resulting from the cross combination of *S. alba* and *S. fragilis*. The two species and their hybrid progeny can be found throughout Europe in overlapping habitats, creating interesting population structures and introgression patterns that confound the identification of pure species and hybrids (Triest *et al.*, 1999; Meneghetti *et al.*, 2007). Another example of a hybrid swarm is *S. starkeana* × *S. bebbiana* in Russia. Natural interspecific hybridization, though observed in many *Salix* species, is limited to specific populations consisting of taxonomically similar species with overlapping floral phenology growing within close proximity (Mosseler and Papadopol, 1989).

4.2.2 Breeding strategies

Populus and *Salix* domestication strategies are based on the genetic systems described above by frequently incorporating both interspecific hybridization and vegetative propagation. Genotypic variance is maximized through the combination of genomes of distinct species and exploited through clonal selection operating across the entire range of genetic variation (Riemenschneider *et al.*, 2001a; Stanton *et al.*, 2010). *Populus* interspecific hybridization is the main breeding approach in programmes operating at latitudes higher than about 25–32° of the equator. This may be tied to the fact that the process of seasonal growth cessation exhibits clinal variation associated with latitude: taxa from higher latitudes undergo dormancy induction under the influence of a relatively longer photoperiod (Farmer, 1993; Hall *et al.*, 2007), and hybrids formed from them have not always performed well at lower latitudes, perhaps because their annual growth cycle concludes too early to take full advantage of the growing season. Interspecific hybridization in poplar is often replaced by intraspecific recurrent breeding of *P. deltoides* using provenances from lower latitudes to maintain phenological adaptation to a short autumnal photoperiod (Farmer, 1996; Howe and Brunner, 2005). This is less of a concern with *Salix*, species of which do not set a terminal bud near the end of the season and thus can continue growing in a day-length independent manner.

First-generation hybridization (F_1) is used most often as the controlled breeding strategy as it frequently results in heterosis for growth rate (Stettler *et al.*, 1988; Larsson, 1998, 2001; Li *et al.*, 1998; Kopp *et al.*, 2001; Lindegaard *et al.*, 2001; Chauhan *et al.*, 2004; Smart and Cameron, 2008; Stanton *et al.*, 2010; Karp *et al.*, 2011). Advanced-generation breeding into the second (F_2) interspecific generation has not been as widely reported (Stettler *et al.*, 1996). Although transgressive segregants are occasionally found, the mean performance of the F_2 generation is less than the F_1 and is not commonly pursued; a reduction in heterozygotic loci along with the disruption of species-specific linkage groups or gene complexes during F_1 hybrid gametogenesis may explain the diminished vigour.

Backcrossing is a more common approach to advanced-generation breeding and is generally used to introduce a single, highly heritable trait from a donor species to improve a recurrent species that is otherwise suitable, save a missing characteristic. In the north central region of the USA, *P. deltoides* × *P. maximowiczii* F_1 hybrids are backcrossed to *P. deltoides* selections to introduce the strong adventitious rooting ability of *P. maximowiczii* into the recurrent *P. deltoides* parent, while maintaining the latter species' resistance to *Septoria* stem canker (Ostry and Berguson, 1993). In China's Jiangsu province, *P.* ×*canadensis* (*P. deltoides* × *P. nigra*, synonym *P.* ×*euramericana*) hybrids are backcrossed to southern provenances of *P. deltoides* to exploit interspecific heterosis, while maintaining adaptation to the local photoperiod. In Sweden and the UK, *S. viminalis* × *S. schwerinii* hybrids have been backcrossed to *S. viminalis* to add improvements in stem form and yield to superior rust resistance achieved in the first-generation cross. Hybridization involving three or more species has also been pursued as an advanced generation strategy in both *Populus* (e.g. complexes of *P.* ×*petrowskyana* × *P. maximowiczii* or *P.* ×*jackii* × *P. maximowiczii*), but especially with *Salix*. Examples of complex *Salix* interspecific taxa are: (i) (*S. koriyanagi* × *S. purpurea*) × *S. miyabeana*; (ii) (*S. udensis* (formerly *S. sachalinensis*) × *S. miyabeana*) × (*S. viminalis* × (*S. schwerinii* × *S. viminalis*)); and (iii) ((*S. dasyclados* × *S. viminalis*) × *S. dasyclados*) × (*S. viminalis* × (*S. schwerinii* × *S. viminalis*)). The development of genomic tools – molecular maps combined with high-resolution chromosome haplotyping for marker-assisted selection and genomic selection protocols – will enable such advanced-generation strategies to become more widely used in both poplar and willow (Peleman and van der Voort, 2003; Zhang *et al.*, 2006; Hanley *et al.*, 2007; Berlin *et al.*, 2010). Willow breeding may also involve hybridization of species differing in ploidy level to increase performance; diploid *S. viminalis* or *S. purpurea* selections have been crossed with tetraploid selections of *S. miyabeana* and *S. udensis* (formerly *S. sachalinensis*) to produce triploid progeny of reduced fertility. This latter feature is important as it lessens their invasiveness potential (Smart and Cameron, 2008).

Populus and *Salix* breeding programmes have often concentrated on the selection of highly ranked individuals from segregating populations created from the repeated hybridization of currently available breeding stock, without a parallel effort to develop a new generation of parent breeders. This is a short-term domestication approach that can only sustain genetic gains insofar as the current parental generation has been fully exploited (Thielges, 1985). Over time, gains accrue solely from an increase in selection intensity, which accumulates throughout the period that such populations are developed and evaluated. Phenotypic assortative mating schemes have been recommended as a technique to enhance the segregation of progeny populations, and thereby the results of short-term breeding programmes (Foster, 1993).

Alternatively, a long-term improvement approach incorporates the recurrent breeding of the parental species to improve their hybridizing quality necessary for sustained genetic advancement (Dieters *et al.*, 1997). Reciprocal recurrent selection is the ideal long-term procedure that improves both parental additive and non-additive gene effects that are captured when the F_1 generation is created (Riemenschneider *et al.*, 2001a). Implementation of the three components of a reciprocal recurrent effort in an interspecific hybridization programme – breeding and selection of parent species A and species B, and their F_1 hybrid generation – may be prohibitively expensive. Polymix breeding combined with paternity analysis may increase the efficiency of interspecific progeny testing relative to traditional full-sib breeding. This can occur when parents used in the polymix do not vary substantially in differential reproductive success, as has been demonstrated in *P.* ×*canadensis* hybridization (Wheeler *et al.*, 2006). However, a lack of full reproductive crossability between species can complicate greatly the estimation of parental interspecific hybrid breeding values (Stanton, 2005). Therefore, *Populus* programmes may substitute intraspecific breeding values as a guide for recombining parental populations (Bisoffi and Gullberg, 1996). Selection for general combining ability may enhance the economics of a long-term improvement programme dependent on the correlations between parental breeding values in pure species and interspecific combinations (Dungey, 2001; Kerr *et al.*, 2004). In *Salix*, long-term, recurrent selection programmes will be developed for parental species or closely related species hybrids such as *S. purpurea* × *S. koriyanagi*. Through the development of genomic tools, marker-assisted selection and genomic selection approaches for parental species should become available within the next few years.

4.2.3 Controlled crossing technique and crossability

Although some poplar breeders in Italy and China conduct hybrid crosses on orchard trees in arboreta, the controlled breeding of poplar and willow is usually conducted in glasshouses using procedures as previously described (Stanton and Villar, 1996; Kopp *et al.*, 2002a) (Fig. 4.2). Dormant 0.9–1.2 m floral cuttings are collected and stored at –2.2°C to meet chilling requirements for poplar, with length dependent on the provenance of the breeding stock. (Populations from southern latitudes may have more stringent requirements (Farmer, 1964).) No chilling requirements seem necessary for *Salix*. Staminate inflorescences are commonly forced in water culture in both genera at temperatures varying between diurnal and nocturnal cycles.

Populus pollen is extracted from ripened and nearly ripened stamens, screened through 80–100 mesh sieves, dried over desiccant for 24 h at room temperature and refrigerated at 1.1°C until used within 3–5 weeks. Storage at –80°C is required to maintain pollen viability over the long term. *Salix* pollen is extracted from mature catkins using toluene, collected on filter paper and dried briefly by vacuum filtration, then stored at –20°C (Mosseler, 1989; Kopp *et al.*, 2002a). Atomizers, imitation bee sticks or artists' brushes are used in pollination. Pistillate parents of poplar are typically maintained in greenhouse isolation chambers, while for willows, simply maintaining male and female shoots in separate glasshouses or forcing them sequentially has been sufficient to avoid unintended cross-pollination.

Seed production in *Populus* section *Populus* and in most *Salix* species can be accomplished using cuttings propagated in water culture, because of a relatively brief period of seed maturation (i.e. 4 weeks in section *Populus* and 12–40 days in *Salix*) (Einspahr and Winton, 1976). By contrast, the 8–20 week maturation period typical of *Populus* sections *Aigeiros* and *Tacamahaca* necessitates a functional root system achieved by rooting the floral cuttings. Difficult-to-root species such as *P. deltoides* are grafted on to potted rootstock (Farmer and Nance, 1968) or rooted using a rooting hormone and soil warming.

Dependent on the compatibility of the cross, a single catkin can produce only a few seeds or well over 100 seeds. *Salix and Populus* seed matures in one season, does not exhibit physiological dormancy, remains viable for only a short period and germinates readily on mineral soil without stratification. True leaves and sufficiently developed root systems appear within 3–4 weeks, allowing seedlings to be transplanted into larger growing containers. Seedlings are large enough to be moved to a pot yard or planted in the field within another 4–6 weeks.

To shorten the breeding cycle, flower induction methods have proven successful with plants grown from cuttings of mature *P. deltoides* specimens (Yuceer *et al.*, 2003). These techniques could lead to a shortening of generation intervals if similarly successful with juvenile specimens. Similar research has not been pursued for *Salix* because it is not unusual for fast-growing willow seedlings to produce floral buds in their first or second growing season.

Crossability relationships in *Populus* as described by Zsuffa (1975) still hold for the most part. Species of section *Populus* are effectively isolated from sections *Aigeiros* and *Tacamahaca* by the failure to establish a chemical 'dialogue' between elongating pollen tubes and surrounding

Fig. 4.2. Various controlled crossing techniques are used in breeding *Populus deltoides*. Controlled-pollinated seed can be produced on rooted stock (a) (photo courtesy of Brian Stanton, GreenWood Resources), or using grafted floral scions (b) (photo courtesy of Lorenzo Vietto, CRA-PLF Agriculture Research Council).

stylar tissue (Stettler *et al.*, 1980; Gaget *et al.*, 1984; Villar *et al.*, 1987). However, the close crossing relationship between sections *Aigeiros* and *Tacamahaca* is now thought to be limited, with success dependent on the direction of reciprocal crosses (Zsuffa *et al.*, 1999; Stanton, 2005). As a consequence, various *in vitro* embryo rescue techniques have been developed for breeding recalcitrant *Tacamahaca–Aigeiros* intersectional combinations (Kouider *et al.*, 1984; Savka *et al.*, 1987; Raquin *et al.*, 1993; Mofidabadi *et al.*, 1998).

Extensive *Salix* hybridization spanning the seminal work in Sweden to later investigations in Canada revealed that species belonging to the same section hybridized readily, and a few species from different sections and different ploidy levels could also cross under artificial conditions (Heribert-Nilsson, 1918; Argus, 1974). Crosses between the subgenera, *Salix* and *Vetrix*, were also described in the original hybridization reports (Heribert-Nilsson, 1918). Zsuffa *et al.* (1984) provided an in-depth table for most of the artificial hybrids for *Salix* reported in past literature. Mosseler (1989) tested the crossability of eight species, *S. amygdaloides*, *S. bebbiana*, *S. discolori*, *S. eriocephala*, *S. exigua*, *S. lucida*, *S. pellita* and *S. petiolaris*. Crosses that produced progeny with high viability were combinations of *S. pellita* with *S. discolor* and *S. petiolaris* with *S. bebbiana*. Most other crosses produced no viable seed or seed with limited viability. Current willow breeding programmes continue to expand the breadth of the crossability of *Salix* as novel species combinations are attempted.

4.2.4 Testing, selection and deployment

The traits targeted for *Populus* and *Salix* improvement are generally classified as either: (i) agronomic, for example yield, stem form, pest resistance, tolerance of cold, flooding, salinity and drought, wind firmness, adventitious rooting; or (ii) wood quality, for example specific gravity, fibre length, cell wall thickness, cellulose and lignin content. All have exhibited varying magnitudes of genetic variation and have responded well to clonal selection (Dunlap and Stettler, 1996; Petit-Conil *et al.*, 1997; Rae *et al.*, 2004; Rowland *et al.*, 2004; Marron *et al.*, 2007; Kovacevic *et al.*, 2008; Serapiglia *et al.*, 2008, 2009; Rood *et al.*, 2010; Dillen *et al.*, 2011; Zalesny and Zalesny, 2011). Today, there is increasing emphasis on selection for ecophysiological characteristics such as water-use efficiency (Karp and Shield, 2008; Fichot *et al.*, 2010) and tolerance of winter injury (Cocozza *et al.*, 2009). Index selection has been recommended to select effectively for a multiplicity of component traits that collectively define *Populus* and *Salix* ideotypes (Riemenschneider *et al.*, 1992; Dickmann and Keathley, 1996).

Disease resistance plays a universal role of singular importance in defining the long-term sustainability of *Populus* and *Salix* cultivars (Plates 23C, 25A). The pathogens of poplar with most significant impact include *Discosporium* canker, *Entoleuca* canker (formerly *Hypoxylon*), *Marssonina* leaf spot, *Melampsora* leaf rust, *Septoria* canker, *Venturia* shoot blight and *Xanthomonas* bacterial canker (Newcombe, 1996; Newcombe *et al.*, 2001; Feau *et al.*, 2010). The components of disease resistance are best understood in the *Melampsora* leaf rust patho-system that involves both major and minor genes (Pinon, 1992; Newcombe *et al.*, 1996; Newcombe, 1998). Major gene resistance is expressed in the isolation of the infection by the host's hypersensitive response; if such a response is lacking, the rate of infection spreads and sporulation is controlled by the host's quantitative resistance mechanism conditioned by polygenic systems (Dowkiw *et al.*, 2003). Rust caused by *Melampsora* spp. (Plate 23A) is also among the most serious diseases impacting *Salix* bioenergy crops and can reduce biomass yields by as much as 40% (Pei *et al.*, 2005). A significant amount of research in the UK has studied the pathotypes involved in the willow rust complex and host reaction (McCracken and Dawson, 1992; Pei *et al.*, 2008, 2010; Bayon *et al.*, 2009; Karp *et al.*, 2011). *S. viminalis* is highly susceptible to the specific pathotypes associated with that region; however, non-native Asian species *S. schwerinii* and *S. udensis* (formerly *S. sachalinensis*) are highly resistant to *Melampsora* rusts in Europe (Gullberg and Ryttman, 1993; Pei *et al.*, 1996, 2010). Less information is available for the *Melampsora* pathotypes affecting willow species in North America. Previous studies have shown that the North American native, *S. eriocephala*, is susceptible to *Melampsora* infection (Fritz *et al.*, 1994; Roche and Fritz, 1998; Cameron *et al.*, 2008). *Melampsora* resistance in *Salix* is highly heritable, making breeding for resistance a relatively easy process (Pei *et al.*, 1996, 2008, 2010).

Insect resistance is also vitally important in developing *Populus* and *Salix* cultivars. A good example of the impact that selection can have on domestication success is the value of *P. maximowiczii* in breeding resistance against poplar-willow borer (*Cryptorhynchus lapathi*) (Hannon *et al.*, 2008; Broberg *et al.*, 2010). The relative resistance among species and hybrids to herbivores is demonstrably variable (Fritz, 1999), and chemistry is a key determinant of susceptibility. *Populus* species produce various phenylpropanoid derivatives, including phenolic glycosides and condensed tannins (Lindroth *et al.*, 1988; Lindroth and Hwang, 1996), leaf surface chemicals (Lin *et al.*, 1998) and enzyme-based defences that deter herbivores (Constabel *et al.*, 2000). Interspecific *Salix* hybridization incorporating species with contrasting defensive chemistries (e.g. *S. caprea* = high tannin, low phenolic glucosides, and *S. repens* = low tannin, high phenol glucosides) resulted in F_1 and F_2 progeny that produced an array of defensive chemistry at concentrations intermediate to those of either parent (Hallgren *et al.*, 2003). While more research is needed to define resistance mechanisms better, species that produce phenolic glycosides and higher concentrations of tannins are generally more resistant to insects than species with lower levels (Kelly and Curry, 1991; Orians *et al.*, 1997; Fritz *et al.*, 2001; Albrectsen *et al.*, 2004, 2007).

Because the experimental unit in both *Populus* and *Salix* genetic testing is often the individual genotype as a clonal propagule, the size of potential research plots is often exceedingly large compared to seedling-based programmes in which the experimental unit is a full-sib family (Shaw and Hood, 1985). Consequently, multi-stage test protocols are frequently used to manage large experimental clonal populations (Libby, 1987; Lindegaard and Barker, 1997; Smart and Cameron, 2008). Populations are truncated sequentially as selection proceeds across stages, along with an increase in replication and plot size to minimize intergenotypic competition effects, as well as a broadening of test sites to assess the importance of genotype–environment interactions. Testing culminates in yield trials. A critical aspect is how best to allocate selection intensity among traits across test stages (Bisoffi, 1989; Riemenschneider *et al.*, 2001a). One important difference between poplar and willow selection practices is that willows have been targeted more commonly as a bioenergy crop in which selection for apical dominance gives way to the selection for coppice ability.

One objective for clonal selection programmes is to have a diverse pool of operational genotypes to minimize the risk of plantation failures due to atypical weather conditions and evolving pathogen virulence (Libby, 1982; Roberds and Bishir, 1997). Selection and deployment of durable resistance combined with the maintenance of reasonable levels of genetic diversity are essential to balance profitability and disease management objectives (Feau *et al.*, 2010). The general recommendation is that 15–30 clones are an appropriate number of selections for most clonal programmes (Libby, 1982; Park, 2002). These are best developed as dynamic groups into which new selections are regularly infused (Stelzer, 1997). An alternative strategy is to identify genotypes of diverse parentage or taxa that have similar growth requirements and complementary growth patterns that can be deployed successfully as multi-clonal cultivar mixtures (Huhn, 1985; Huehn, 1988; Foster *et al.*, 1998; McCracken *et al.*, 2011). But today, nearly all *Populus* and *Salix* production plantations are established as a mosaic of monoclonal blocks (see Plate 21A), despite the concern that pests may spread more rapidly than they would in plantations of cultivar mixtures (Ledig, 1986). One exception is the UK, where polyclonal mixtures of willow are common and thought necessary to achieve acceptable yields (McCracken and Dawson, 1997; McCracken *et al.*, 2005, 2011).

4.2.5 Certification, regulation and international trade

The commercialization of newly developed cultivars is regulated to varying extents throughout the world. The European Union (EU) has the most sophisticated system – the Organization for Economic Cooperation and Development (OECD) – which stipulates the conditions under which both individual cultivars and mixtures can be marketed and traded among its member states (Nanson, 2001; Muhs, 2008). OECD regulations define the way in which cultivars are named and identified, the manner in which mixtures are composed to maintain genetic diversity, the extent of performance testing for qualification under separate trading categories (e.g. Qualified and Tested), and the maximum number of years and land area individual cultivars and mixtures can be commercialized before renewing their marketing licence. Approved materials are registered on a national list maintained by a designated authority within each country. In the People's Republic of China, the commercial release of new cultivars is approved by the State Forestry Administration and granted only after performance tests have been conducted for a minimum of 5 years in field trials established at multiple locations following a defined experimental design. Other countries such as the USA and India have no similar regulations, while some Canadian provinces disallow the use of exotic *Populus* or *Salix* cultivars on public lands.

Protection of intellectual property rights is increasingly an important issue with breeders who develop clonally propagated willow and poplar cultivars. The International Union for the Protection of New Varieties of Plants, headquartered in Geneva, Switzerland, numbers 70 member countries that follow recognized principles for providing intellectual property rights to the developers of new cultivars. In North America, the US Patent and Trademark Office provides for patent protection of asexually propagated genotypes. Additionally, the US Department of Agriculture offers a certificate of ownership for plant cultivars under the Plant Variety Protection Act. Certificates are limited to seed-propagated varieties and, therefore, mostly do not apply to clonal cultivars.

There is great value in extending interspecific hybridization programmes of selected provenances of several of the key species to many countries beyond their native ranges. This has created a need for the international exchange of a variety of reproductive materials – pollen, seed and vegetative propagules. While the first two have a comparatively low risk of carrying unwanted pests into importing countries, the probability of inadvertently introducing injurious organisms when importing vegetative propagation stock – stem or root cuttings – is appreciably higher. The International Plant Protection Convention of 1997 therefore provided phytosanitary protocols to limit this risk (Schrader and Unger, 2003). These include requiring exporting countries to: (i) inspect and certify that shipments are free of organisms listed on import permits issued by the receiving countries; (ii) declare that the shipment is practically free of all other pests; and (iii) apply any required phytosanitary treatments prior to shipment. Post-entry responsibilities of the receiving country include inspection of the shipment, and in some cases, quarantine and follow-up close monitoring (Fig. 4.3). Perhaps the safest method of shipping plant material is in sterile culture.

4.2.6 Molecular approaches

Populus

Novel approaches to *Populus* domestication have considered polyploidy (Mashkina and Isakov, 2002), cell and protoplast fusion (Wakita *et al.*, 2005), somaclonal variation (Antonetti and Pinon, 1993) and mass propagation via organogenesis (Thakur *et al.*, 2005), micropropagation (Louis and Eils, 1997) and somatic embryogenesis (Cheema, 1989). Genomics and genetic transformation are now the primary areas in which new domestication tools are being developed (Boerjan, 2005; Merkle and Nairn, 2005; Wegrzyn *et al.*, 2010). The commercial use of genetic transformation is challenged by governmental regulations, the certification standards of the Forest Stewardship Council (Strauss *et al.*, 2001) and societal perceptions of the technology (Strauss *et al.*, 1997; Mathews and Campbell, 2000; Neumann *et al.*, 2007). The challenge for genomics is mainly the difficulty inherent in the process of fine-mapping genes and the economic efficiency of marker-aided breeding. The latter was advanced greatly in 2006 with the sequencing of the *Populus* genome (Wullschleger *et al.*, 2002; Tuskan *et al.*, 2006).

Genetic transformation is well developed in *Populus* and can improve upon conventional

Fig. 4.3. A Chilean plant quarantine station. Many countries permit the importation of *Populus* reproductive material with appropriate phytosanitary controls of treatment, inspection and monitoring to avoid accidental pest introductions. Photo courtesy of Brian Stanton, GreenWood Resources.

hybridization through the introduction of specific genes and new genetic function into highly productive genotypes with minimal disruption of genetic quality (Kim *et al.*, 1997). *Populus* has been modified for herbicide resistance (Donahue *et al.*, 1994; Meilan *et al.*, 2002), lignin content (Boerjan *et al.*, 1997; Dinus, 2000), leaf beetle resistance (Ellis and Raffa, 1997), growth rate (Shani *et al.*, 2000; Jing *et al.*, 2004), sterility (Brunner *et al.*, 2007) and phytoremediation (Doty *et al.*, 2000, 2007; Merkle *et al.*, 2007, Doty, 2008). Transformation protocols typically rely on the soil bacterium, *Agrobacterium tumefaciens*, to introduce foreign DNA into the *Populus* genome (De Block, 1990; Confalonieri *et al.*, 1994; Ma *et al.*, 2004). Field trials of genetically modified varieties have been conducted in North America with *P. deltoides* and *P.* ×*generosa* (*P. deltoides* × *P. trichocarpa*) (TGERC, 2000), and in Europe with *P. tremula*, *P. tremuloides*, *P. deltoides* and *P.* ×*canescens* (Pilate *et al.*, 2002; McCord and Gartland, 2003). In Asia, varieties of *P. simonii* × *P. nigra*, *P. nigra* and advanced-generation *P.* ×*tomentosa* (*P. alba* × *P. tremula* var. *davidiana*) hybrids transformed with *Bt*-atratoxin fusion protein genes that confer resistance to leaf defoliators have been approved for production use (Cao *et al.*, 2010; Hu *et al.*, 2010). Commercial application of this technology will require clones of proven plantation performance, efficient transformation and shoot regeneration protocols, reliable storage methods for transgenic lines and faithful transgene expression throughout ontogeny (Pilate *et al.*, 1997; Ma *et al.*, 2004; Tsai and Hubscher, 2004). An ongoing research priority is to increase *Populus* transformation efficiency (Song *et al.*, 2006). The use of genetically modified cultivars has raised concerns over potential ecological effects if transgenic plantations reproduce with wild relatives (Slavov *et al.*, 2004). For that reason, a major goal has been the study of floral regulation (Yuceer *et al.*, 2002) and ultimately transformation for reproductive sterility or transgene excision (Strauss *et al.*, 1995; Meilan and Strauss, 1997; Brunner *et al.*, 2007).

Genomic science is the study of an organism's genetic structure. The science is most advanced in *Populus*, where important gene loci have been mapped across the 19 chromosomes that comprise its genome. Genomic maps of *Populus* have relied on a variety of biochemical markers, for example amplified fragment length polymorphism (AFLP) and randomly amplified polymorphic DNA (RAPD), to locate loci affiliated with traits of commercial importance, including growth rate (Bradshaw and Stettler, 1995), disease resistance (Villar *et al.*, 1996; Tabor *et al.*, 2000), vegetative phenology (Frewen *et al.*, 2000; Ingvarsson *et al.*, 2006) and wood formation (Sterky *et al.*, 1998). Such maps have been constructed for a range of species including *P. deltoides*, *P. trichocarpa*, *P. adenopoda* and *P. nigra* (Cervera *et al.*, 2001, 2004; Gaudet *et al.*, 2008).

The genomics field is especially important in *Populus* hybrid domestication, because it can provide a better understanding of first generation interspecific heterosis and epigenetics (Varshney *et al.*, 2005). A more established application of this technology is the identification or certification of commercial varieties using AFLP, RAPD and microsatellite markers (Castiglione *et al.*, 1993; Sigurdsson *et al.*, 1995; Rahman and Rajora, 2002; Bekkaoui *et al.*, 2003; Rajora and Rahman, 2003; Fossati *et al.*, 2005). Although these markers have proven their worth in determining individual clonal identity, their utility in multi-species breeding programmes has been evaluated more recently. Liesebach *et al.* (2010) were able to use ten nuclear microsatellite loci to identify clones belonging to *Aigeiros*, *Tacamahaca*, *Populus* and intersectional *Aigeiros–Tacamahaca* hybrids. Single nucleotide polymorphisms (SNPs) – variations in individual nucleotides at set positions in the genome – are now in vogue as molecular markers. Talbot *et al.* (2011) used a suite of 26 SNPs to resolve the taxonomy of complex hybrid combinations resulting from interspecific hybridization of *P. balsamifera*, *P. deltoides*, *P. laurifolia* and *P. nigra*. Furthermore, combining chloroplast SNP markers with polymerase chain reaction-restriction fragment length polymorphism (PCR-RFLP) markers is a novel approach that appears to have good applications to meeting the needs of multi-species breeding programmes (Schroeder *et al.*, 2011).

The importance of clonal and pedigree identification notwithstanding, the more substantive contribution of molecular markers will be the development of genomic-assisted breeding programmes that allow for more precise and efficient evaluations of breeding and segregating populations than otherwise if based solely on standard phenotype evaluation. Previously, work in this area concentrated on selection within interspecific pedigrees, for example backcross and F_2, with high rates of linkage disequilibrium and stable marker–trait relationships (Strauss *et al.*, 1992; Bradshaw and Stettler, 1995; Zhang *et al.*, 2006). But now, association genetics studies are the preferred route to the identification of gene function, and thus possibly for genomic-assisted breeding for *Populus* improvement (Varshney *et al.*, 2005; Neale, 2007). Newly developed automated technologies for large-scale gene sequencing are now capable of mapping allelic variation across approximately 85% of the *Populus* genome's estimated 45,000 genes. This approach seeks to deduce gene function through the joint associations of phenotypes and functional markers generated from SNPs (Neale and Savolainen, 2004; Cronk, 2005; Varshney *et al.*, 2005; Gonzalez-Martinez *et al.*, 2007; Wegrzyn *et al.*, 2010). Functional markers are those considered tightly linked to, or a part of, segregating alleles that can be tracked throughout generations in a recurrent breeding programme, as the marker–trait associations are substantially unaltered by gametic recombination (Neale, 2007). The technique is based on the idea that phenotypes and SNP markers are linked tightly and associated statistically with one of the two chromosomes of a homologous pair, i.e. a haplotype. The success of detecting such marker–phenotype associations and their utility in a genomic-assisted hybridization effort ultimately depends on the extent of linkage disequilibrium in the discovery population, the extent of SNP variation, the degree to which SNPs control the expression of important traits and the stability of the marker–phenotype associations across pedigrees and environments.

Association mapping has been applied to the study of candidate genes. Now it is also applied in the form of genome-wide association studies that are based on scanning an entire genome for SNPs associated with particular phenotypes (Tuskan *et al.*, 2011). Conceivably, SNPs may also be used to guide the composition of breeding populations based on genetic distance among individual members (Rahman and

Rajora, 2002). This application may figure in the identification of unique haplotypes or recombinants that conceivably could increase the yield of heterotic offspring in interspecific crosses (Kelleher *et al.*, 2007). The use of molecular information to maximize hybridization response would improve the design of breeding orchards for increased genetic divergence based on phenotypic data alone, as has been done for *P. deltoides* using hierarchal cluster analysis for wood quality and growth traits (Pande and Dhiman, 2011).

The technologies of genetic transformation and genomics intersect when genetic maps are aligned with physical maps to locate candidate genes that can be cloned for transformation projects (Meilan and Strauss, 1997; Busov *et al.*, 2005). The close ratio of 200 kB of physical map distance per centimorgan of recombination frequency in *Populus* allows for efficient positional cloning (Stirling *et al.*, 2001). Beyond the importance to domestication and cultivar development, knowledge of the *Populus* genome and the ease with which it can be transformed are, together, a powerful research tool into the study of epigenetics, ontogenetic development and the physiology of adaptation (Brunner *et al.*, 2004).

Molecular technology has also been applied to the study of the genus' taxonomy and evolutionary history (Yin *et al.*, 2004). Separate sections are now thought to be appropriate for the classification of *P. nigra* and *P. deltoides* based on AFLP markers (Cervera *et al.*, 2005), confirming earlier conclusions based on analyses of allozyme variation (Rajora and Zsuffa, 1990). The distinctiveness of *P. nigra* among other species of section *Aigeiros* also shows in the nucleotide sequences of their chloroplast DNA (Hamzeh and Dayanandan, 2004). The molecular study of chloroplast DNA variation has shown that, following glaciation, the *P. nigra* route of recolonization of Europe occurred from two main refugia in Spain and Italy and the Balkans (Cottrell *et al.*, 2005). Similar research into the evolution of the *P. alba*, *P. tremula* and *P.* ×*canescens* hybrid complex determined that the Danube River was an important corridor for the introgression of *P. alba* into *P. tremula* during post-glacial European colonization (Fussi *et al.*, 2010). The genetic diversity within Italian *P. alba* populations is associated with the extent of introgression from *P. tremula* (Castiglione *et al.*, 2010). In North America, *P. balsamifera* came to establish its present-day distribution by migration from the central to the northern and eastern portions of its range following glaciation (Keller *et al.*, 2010). Conversely, *P. trichocarpa*, the sibling species of *P. balsamifera*, recolonized the Pacific Northwest region of the North American continent from multiple glacial refugia (G.T. Slavov *et al.*, 2011, unpublished data).

Salix

Molecular approaches to *Salix* domestication have targeted marker development for linkage and quantitative trait loci (QTL) mapping. The double pseudo-test cross method has been used to generate genetic linkage maps from several different mapping populations, and a number of QTLs for important agronomic traits have been mapped (Hanley *et al.*, 2002; Tsarouhas *et al.*, 2002, 2003, 2004; Rönnberg-Wästljung *et al.*, 2003, 2005; Weih *et al.*, 2006). Recently, QTLs for enzymatic saccharification yield were identified, providing evidence for the genetic control of sugar release and implications for breeding genotypes for biofuel production (Brereton *et al.*, 2010). AFLP markers have been used to predict within-family variability for height growth (Kopp *et al.*, 2002b). Molecular markers have been developed as a fingerprinting method to discriminate among related individuals, since morphological characteristics are unreliable (Trybush *et al.*, 2008; King *et al.*, 2010).

Salix linkage maps based on two mapping populations of *S. viminalis* have been aligned to the *P. trichocarpa* genome, confirming the high level of genomic conservation and macrosynteny between the two species (Hanley *et al.*, 2007; Berlin *et al.*, 2010). However, sequence divergence between the two genera is too great to allow direct transfer of molecular markers, requiring the development of a reference genome sequence for willow. Efforts to sequence the *S. purpurea* genome are currently in progress at the US Department of Energy Joint Genome Institute as a collaborative effort between Cornell University, Oak Ridge National Laboratory and the J. Craig Venter Institute. To complement the next generation sequencing of the genome, bacterial artificial chromosome (BAC) libraries have been developed using both random-shear technology and enzyme digests. In addition, transcriptome sequencing from eight different *S. purpurea* tissue samples has been performed to elucidate regulatory gene networks involved with key traits and developmental processes in willow.

While genetic transformation of *Populus* has been used extensively for functional genomics and trait improvement, such is not the case for *Salix*. Although there are reports of the successful transformation of *Salix* (Vahala *et al.*, 1989; Rocha, 1991; Xing, 1995), an efficient transformation and regeneration system for shrub willows has yet to be optimized. *Agrobacterium*-mediated transformation has been successful in both *S. viminalis* and *S. lucida* callus, but plants have not been regenerated. Moreover, particle bombardment transformation of *S. alba* has not been successful (Devantier *et al.*, 1993). *In vitro* regeneration of willow tissue is rather difficult, with only a few reports of success (Stoehr *et al.*, 1989; Xing, 1995; Lyyra *et al.*, 2006).

4.2.7 Conservation

Populus

The conservation of *Populus* genetic resources enhances the ability of a species to adapt to future changes in climate, pathogens and hydrologic conditions, and to recover from past constraints imposed by human activities (Bajaj, 1979). Construction of levees, dams and revetments along many of the world's major rivers has reduced the natural cycle of flooding essential to the seedling regeneration of *Populus* stands, and consequently has resulted in depleted genetic resources (Rood and Mahoney, 1990; de Vries, 2001; Rood *et al.*, 2003b). Conservation programmes maintain populations of suitable size and structure to sustain diversity in the face of such challenges. The future of *Populus* domestication programmes is therefore reliant on the success of today's conservation activities (Legionnet and Lefevre, 1996). The case for conservation is represented best by *P. tremuloides*. For some time, it has been known that changes in air quality have altered this species' population diversity and genetic structure (Berrang *et al.*, 1986; Karnosky *et al.*, 1989; Lindroth *et al.*, 2001), and such changes will likely alter the genetic make-up of future populations in view of genotypic variation in ozone response (McGrath *et al.*, 2010). Similarly, canopy decline and mortality in *P. tremuloides* populations in western North America that are under strong genetic control could potentially alter the genetic architecture of future stands (St Clair *et al.*, 2010). Climate change and increased temperature may further impact population structure by increasing the incidence of *Entoleuca mammata* stem canker, the species' most serious pathogen (Ostry and Anderson, 2009).

Conservation involves both *ex situ* and *in situ* efforts (Ledig, 1986). The *ex situ* approach includes long-term storage of seed and pollen (Tauer, 1979; Walters *et al.*, 2011), cryopreservation of dormant buds (Ahuja, 1988) and cell cultures (Towill, 2002; Jokipii *et al.*, 2004; Tsai and Hubscher, 2004) and the preservation of wild collections in cultivated arboreta. Alternatively, *in situ* conservation efforts involve the identification, cataloguing, assessment and protection of native populations in natural reserves (Benea, 2003). Population sizes should be sufficiently large to support overall habitat function of the conserved resources (Whitham *et al.*, 2003).

The European Forest Genetic Resources Programme (EUFORGEN) to conserve *P. nigra* and *P. alba* is perhaps the most advanced *Populus* conservation programme comprising both *ex situ* and *in situ* components. Technical guidelines for *in situ P. nigra* conservation have been published (Lefevre *et al.*, 2001). Additionally, core clonal collections of both species have been secured (Vietto and Bianco, 2004) (Fig. 4.4). The People's Republic of China has initiated a major *in situ* conservation effort for *P. euphratica*, with the establishment of the Tarim River nature reserve in Xinjiang Autonomous Region (Zhang *et al.*, 2005b; Yimit *et al.*, 2006). Populations of *P. nigra* of considerable ecological value are also found in Xinjiang (Fig. 4.5). In Central Asia, the study and protection of the tugai forests serves as an *in situ* conservation of *P. euphratica*, one of the tugai's main components (Ruger *et al.*, 2005; Schluter *et al.*, 2006).

Gene flow from neighbouring plantations of exotic species or interspecific hybrids can alter the genetic composition of *in situ* reserves. The most extensively studied example is the dynamics of gene flow between natural stands of *P. nigra* and *P.* ×*canadensis* plantations in Europe (Cagelli and Lefevre, 1995; Vanden Broeck *et al.*, 2005;

Fig. 4.4. *Populus nigra* clonal archive at Casale Monferrato, Italy. *Ex situ* propagation efforts to preserve the genetic resources of this important Eurasian species, as well as *Populus alba*, is a priority conservation imperative in Europe. Photo courtesy of Lorenzo Vietto, CRA-PLF Agriculture Research Council.

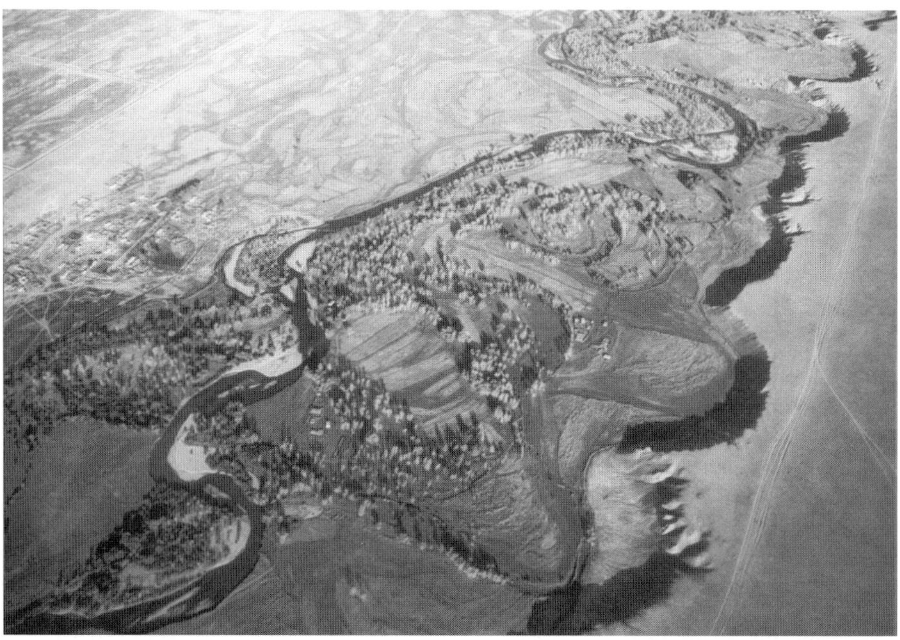

Fig. 4.5. *Populus nigra* growing along the Ulungur River in China's Xinjiang Autonomous Region. Located at the eastern periphery of its natural geographic range, such *P. nigra* populations are especially valuable for their adaptation to unique environmental conditions. Photo courtesy of Marc Villar, National Agricultural Research Institute, France.

Rathmacher *et al.*, 2010). Introgression takes place predominantly through backcrossing of *P.* ×*canadensis* males to female *P. nigra* genotypes (Heinze, 2008; Smulders *et al.*, 2008a; Ziegenhagen *et al.*, 2008). A molecular study of 44 *P. nigra* genotypes along the Rhine River and its tributaries in the Netherlands revealed that nearly one-half of the sample was affected by natural hybridization with *P. deltoides* and *P.* ×*canadensis* (Smulders *et al.*, 2008b). Chenault *et al.* (2011) determined that hybridization of native stocks of *P. nigra* with the widely distributed Lombardy poplar cultivar (*P. nigra* var. *italica*) was also commonplace. In eastern Canada, the introgression risk to *P. deltoides* and *P. balsamifera* posed by plantations established with complex *Populus* hybrids of *P. trichocarpa*, *P. maximowiczii* and *P. nigra* is greatest for small peripheral populations (Meirmans *et al.*, 2010). And in the Canadian prairie, the phenology of native and exotic poplars is synchronized sufficiently to enable interspecific hybridization, assuming good seedling establishment and survival (Huybregts *et al.*, 2007).

The ramifications of gene flow affect not only the population genetics of individual species, but also the integrity of their communities. For example, in the south-west of the USA, the genetic composition of foundation *Populus* taxa is now known to affect the diversity of its community's species: the richness of associated plants making up the understorey decreases as the overstorey component changes from *P. angustifolia* to interspecific *P. angustifolia* × *P. fremontii* hybrids to *P. fremontii* (Lamit *et al.*, 2011). Moreover, the capacity of riparian systems to respond to shifts in environmental conditions is a reflection of the heritability of net primary productivity of *Populus* species and their interspecific hybrids (Lojewski *et al.*, 2009).

Molecular methods have been applied to the development of both *in situ* and *ex situ* conservation efforts. Winfield *et al.* (1998) used molecular tools to identify specific *P. nigra* genotypes for *ex situ* collections in which maximum genetic diversity was desired. Molecular tools have also proven useful in confirming genetic identity in *P. nigra in situ* reserves subject to potential introgression with *P. deltoides* (Tabbener and Cottrell, 2003; Holderegger *et al.*, 2005) and in assessing levels of genetic diversity in conserved populations of *P. ciliata* (Rajagopal *et al.*, 2000), *P. nigra* (van der Schoot *et al.*, 2000; Pospiskova and Salkova, 2006; Smulders *et al.*, 2008a), *P. tremuloides* (Dayanandan *et al.*, 1998) and *P. alba* (Sabatti *et al.*, 2001). Finally, molecular tools have been employed successfully in ecological studies of *P.* ×*canescens* hybrid swarms. These are maintained in central Europe, principally by introgression of *P. tremula* into *P. alba* via *P. tremula* pollen (Lexer *et al.*, 2005), while in northern Italy, the *P.* ×*canescens* hybrid swarm is maintained mainly by backcrossing to *P. alba* (Fossati *et al.*, 2004).

Salix

A worldwide effort to conserve *Salix* collections as a repository of genetic variation and as a source of breeding stock is critical for the future domestication efforts of shrub willows for renewable energy and environmental engineering applications. Among the largest *Salix* germplasm repositories is the Botanic Garden of the Urals Branch of the Russian Academy of Sciences, which holds 2000 accessions representing 147 species and 136 hybrids from around the world. The National Willow Collection at Rothamsted Research in the UK numbers over 1300 genotypes, representing approximately 100 species (Karp *et al.*, 2011). The willow clone archive at the Swedish University of Agricultural Sciences holds approximately 600 genotypes from collections throughout Europe, Russia and Siberia, and Asia. In North America, the Chadwick Arboretum at the Ohio State University holds over 600 *Salix* accessions. Smaller willow living collections are also found throughout the USA and Canada at various arboreta and botanical gardens.

Like *Populus*, *Salix* genetic resources are threatened by the loss of wetlands to industrial and urban development (Mitsch and Grosselink, 2000) (Fig. 4.6). Conservation efforts include the reintroduction of native species to wetland habitats through seeding, although this practice has not proven to be reliable (Reinartz and Warne, 1993; Grabau *et al.*, 2011). Schroeder *et al.* (2009) proposed harvesting willow biomass as a wetland management strategy in Canada, providing a renewable source of bioenergy and offering protection from nutrient contamination from surrounding agricultural fields. Utilizing willows for stream bank stabilization has become common practice in the UK and

Fig. 4.6. Endangered *Salix cordata* (dune willow) in a protected area on the shore of Lake Ontario in New York, USA. Photo courtesy of Lawrence Smart, Cornell University.

has offered a very successful solution to wetland restoration (Anstead and Boar, 2010). Willow species introduced into New Zealand during European settlement has resulted in their widespread establishment along riverbanks and swamps. Now, they have become an integral part of the landscape, providing erosion control, riverbank stabilization and filtration of agricultural runoff (Kuzovkina *et al.*, 2008).

4.3 Worldwide Domestication and Conservation Programmes

4.3.1 Asia

The People's Republic of China, Japan, Korea and India are four Asian nations that maintain membership in the IPC and have active domestication and conservation programmes. (While Russia and Turkey carry out *Populus* and *Salix* programmes, their activities take place mainly in their European sectors and are discussed elsewhere.) In China, in particular, significant plantation expansion is predicted due to the growing demand for paper and paperboard products throughout Asia, along with the increased demand for veneer and medium-density fibreboard. Overwhelmingly, China has the largest area of poplar and willow plantations in Asia, nearly 5 million ha combined (FAO, 2008).

People's Republic of China

POPULUS. China is home to a large number of native species, including *P. alba*, *P. adenopoda*, *P. cathayana*, *P. ciliata*, *P. tremula* var. *davidiana*, *P. euphratica*, *P. koreana*, *P. maximowiczii*, *P. nigra*, *P. simonii*, *P. suaveolens*, *P. szechuanica*, *P.* ×*tomentosa*, *P. tremula* and *P. yunnanensis*. The period from 1950 to 2010 has witnessed a tremendous increase in Chinese poplar cultivation for a variety of uses, including timber production, agroforestry and ecological protection (Gwyther, 2006; Zhang and Song, 2006). Throughout this period, China has often relied on the extensive use of a limited number of clonal varieties deployed across large areas, at times leading to widespread plantation failures (Weisgerber and Han, 2001). During the 1970s, as plantation development accelerated, introductions of Italian *P.* ×*canadensis* cultivars, 'I-214', 'I-9805', 'I-9807', 'Neva', 'Bellota' and 'Guariento', and *P. deltoides* cultivars, 'I-63/51' (synonym 'Lux'), 'I-72/58' (synonym 'San Martino') and 'I-69/55' (synonym 'Harvard') gained widespread use.[2] Although the propagation and breeding of many of these cultivars continues, a priority has been placed on the development of new cultivars to support major plantation programmes in the Yellow and Yangtze River basins and elsewhere (Li *et al.*, 2005a).

The important taxa used in Chinese domestication practices can be classified according to the following five regions.

1. North-east (Liaoning, Jilin, Heilongjiang provinces) – *P. deltoides* × *P. cathayana*, *P. nigra* × *P. cathayana*, *P. deltoides* × *P. simonii*, *P.* ×*canadensis*, *P. deltoides* × *P. suaveolens*, *P. deltoides* × *P. maximowiczii*, *P. deltoides* × *P. koreana*.
2. North central (Shandong, Hebei, Henan provinces) – *P.* ×*canadensis*, *P.* ×*tomentosa*.
3. North-west (Xinjing Autonomous Region, Ningxia Autonomous Region, Shanxi, Gansu provinces) – *P. deltoides* × *P. simonii*, *P. deltoides* × *P. cathayana*, *P.* ×*canadensis*, *P.* ×*tomentosa*, *P. deltoides*.
4. South-east (Jiangsu, Anhui, Hunan provinces) – *P. deltoides*, *P.* ×*canadensis*.
5. South-west (Yunnan, Sichuan provinces) – *P. deltoides* × *P. yunnanensis*, *P. deltoides* × *P. szechuanica*, *P.* ×*canadensis*.

The Chinese Academy of Forestry, Beijing Forestry University and Nanjing Forestry University have all conducted breeding programmes over the past 50 years. The Chinese Academy of Forestry has been instrumental in developing the second generation of Italian *P. deltoides* cultivars, 'I-63/51' and 'I-69/55'. Work at the academy also includes the *P.* ×*canadensis*, the F_1 intraspecific *P. deltoides* (Zhonglin and J-series) and the *P. deltoides* × *P. cathayana* (Xifeng series) hybrid taxa. Beijing Forestry University has focused mainly on breeding *P.* ×*tomentosa* triploid hybrids for the Yellow River basin. In the south-east, Nanjing Forestry University has led the domestication of *P. deltoides* for the Yangtze River basin (Fig. 4.7). Early work concentrated on the hybridization of Italian *P. deltoides* cultivars 'I-63/51' and 'I-69/55' (NL series). Over 200 *P. deltoides* open-pollinated seedlots imported from the south-eastern USA are undergoing replicated field evaluation. Moreover, various provincial forestry institutes are also actively involved in *Populus* domestication, including: (i) Liaoning Poplar Research Institute (hybridization of *P. deltoides* 'Shanhaiguanensis', *P.* ×*canadensis* and native selections of *P. cathayana* and *P. simonii*); (ii) Shanxi Provincial Forestry Institute (hybridization of *P. deltoides* and *P. cathayana* 'Changanesis'); (iii) Forestry Institute of Heilongjiang province (hybridization of *P. simonii* and *P. nigra*); and (iv) Forestry Institute of Baicheng Prefecture (*P. nigra* × *P. simonii* hybridization).

Poplar domestication in China values stress adaptability and insect resistance as much as productivity (Qin *et al.*, 2003). Tolerance of saline soils in the Yellow River delta is a priority, as are drought and alkaline-tolerant cultivars for the Inner Mongolian frontier (Chen *et al.*, 2002; X. Zhang *et al.*, 2005). Cultivar evaluation for drought tolerance involves clonal testing for belowground biomass allocation (Gong *et al.*, 2011). Naturally occurring hybrids between *P. simonii* and *P. nigra* and artificial hybrids formed by crossing *P. deltoides* and *P. nigra* with *P. cathayana* and *P. simonii* have been identified that are tolerant to cold, drought and saline conditions. Likewise, insect-resistant germplasm is important for the western frontier, where many stem-boring species are a significant challenge, e.g. *Anoplophora glabripennis*, *Cryptorhynchus lapathi*, *Paranthrene tabaniformis*, *Clostera anachoreta*, *Apocheima* sp. and *Saperda* sp. (Yin and Lu, 2005). Genetic transformation for resistance to boring and defoliating insects is also being studied (Hu *et al.*, 2001; Chen *et al.*, 2005; Lin *et al.*, 2006). The Chinese government approved genetically transformed poplars for commercial use in 2005 (Stone, 2008).

The future of China's domestication effort will focus on intra- and interspecific breeding of native species in sections *Populus* and *Tacamahaca*, as well as exotic sources of *P. deltoides* and *P. nigra* in view of the importance of the *P.* ×*canadensis* taxon (Li *et al.*, 2005b). Although good provenance research is generally lacking, many of China's native species have extensive geographic ranges and substantial levels of within-species variation that will support national and provincial breeding

Fig. 4.7. *Populus deltoides* undergoing evaluation at Nanjing Forestry University in Jiangsu province in the Yangtze River basin. Photo courtesy of Mick O'Neill, New Mexico State University.

programmes (Weisgerber and Han, 2001). Thus, less reliance on North American and European plant material is expected over time (Guo and Zhang, 2010; Gong *et al.*, 2011). There has been a moderate amount of plus tree selection and breeding in *P. simonii*, *P. cathayana*, *P. koreana* and *P. yunnanensis*, but most controlled breeding to date has been limited to non-recurrent F_1 hybridization with *P. deltoides* and *P. nigra*.

China's national germplasm conservation programme highlights its *Populus* genetic resources (Gao *et al.*, 2001). An *ex situ* effort is being used to conserve native sources of *P. tremula* var. *davidiana*, *P. simonii*, *P. cathayana*, *P. nigra* and *P.* ×*tomentosa*, as well as exotic collections of *P. deltoides* and *P. nigra* at national gene bank conservation centres in Shandong, Henan, Jiangsu and Hubei provinces. China's major *in situ* conservation effort involves the inventory and study of *P. euphratica* stands in the northwest Tarim River basin (Y.M. Zhang *et al.*, 2005) (Fig. 4.8).

SALIX. There are over 250 species of willow found in China, mainly in the Yangtze River valley and in parts of northern China. Approximately 50 species are used in the country's breeding programmes in support of a variety of plantation operations. The largest area of willow cultivation is found in the south-eastern provinces of Jiangsu, Anhui, Jiangxi and Hunan, where plantations are managed for multiple markets. Plantations are also established in the provinces of Shandong Henan and Liaoning, mostly for the wicker industry. Equally important are large windbreaks of shrub willow planted in the arid and semi-arid sandy frontier in the west and north of China to reduce wind erosion and the occurrence of sandstorms.

Salix domestication in China is conducted primarily by the Jiangsu Academy of Forestry. A collection of 2000 accessions representing over 40 species from the temperate and the subtropical zones and the Tibetan Plateau has been established in Jiangsu as the major genetic

Fig. 4.8. *Populus euphratica* is especially important to China's *Populus* gene conservation efforts in view of its adaptability to droughty and saline sites. Pictured here is a stand growing in the Gobi desert near Ejin Qi, Inner Mongolia. Photo courtesy of Bill Schroeder, Agriculture and Agri-Food Canada.

resource for domestication and conservation projects. Other institutions – the Chinese Academy of Forestry, Nanjing Forestry University, Northwest Agriculture and Forestry University, Jiangxi Academy of Forestry, Hunan Academy of Forestry – are also involved in breeding and cultivar selection. The State Forestry Administration maintains a central database for all the new cultivars.

S. matsudana, *S. babylonica*, *S. integra*, *S. suchowensis*, *S. leucopithecia* and *S. alberti* are native to China and feature prominently in the country's breeding programmes, along with non-natives *S. nigra*, *S. koreensis*, *S. purpurea*, *S. viminalis* and *S. alba*. Commercial varieties developed by the Jiangsu Academy of Forestry include those for timber production in the Yangtze basin (e.g. *Salix* ×*jiangsuensis* 'J172' and *S.* ×*jiangsuensis* 'J795'), as well as cultivars for biomass production (e.g. *S.* ×*jiangsuensis* 'J52-2'). Mean annual increments have reached 18 m^3 ha^{-1} for timber production and 40 green t ha^{-1} for biomass production. Adaptability to drought, salinity and flooding, and resistance to diseases and insects are priority selection criteria. Cultivars are also selected for use in phytoremediation projects, the basket and wicker industry and for the ornamental landscape industry. Good examples of selections for the wicker industry are *S. suchowensis* × *S. integra* 'J8-26' and *S. integra* × *S. suchowensis* 'J9-6'. Both are characterized by long, supple shoots with few lateral branches. Examples of ornamental selections are *S.* × *aureo-pendula* 'J1011', a weeping form with golden branches, and *S. turanica* × *S.* ×*leucopithecia* '887', a shrub form with large colourful buds.

India

POPULUS. Six *Populus* species – *P. alba*, *P. ciliata*, *P. euphratica*, *P. gamblei*, *P. glauca* and *P. suaveolens* – are native to India's Himalayan region. Poplar is cultivated in the northernmost mountain states of Jammu and Kashmir, Himachal Pradesh, Uttrakhand and in the plain states of Punjab, Haryana and Uttar Pradesh. The plains region at the Himalayan foothills is India's major commercial poplar-growing region, compromising approximately 90% of the country's total poplar plantation area. Here, *P. deltoides* is overwhelming the preferred taxon. In the subtropical zone up to 650 m elevation and in the adjoining plains, *P. deltoides* and *P.* ×*canadensis* are the preferred taxa. *P. deltoides* and *P. ciliata* × *P. maximowiczii* interspecific hybrids are favoured in the subhumid zone of 650–1800 m elevation. *P. alba* is grown mainly in the cold desert areas of Jammu, Kashmir and Himachal Pradesh. *P. ciliata* is grown above 1800 m in the temperate wet zone, but cultivation here also incorporates selections of *P. yunnanensis*, *P. alba*, *P. nigra* and *P. ×canadensis*. *P. ciliata*, *P. nigra* and *P. alba* are grown in the temperate zone above 2200 m.

P. deltoides from the southern US provenances is the most important among all taxa bred in India (Fig. 4.9). It is widely cultivated above 28°N latitude for veneer logs, paper pulp, matchsticks and packing cases, etc. (Puri *et al.*, 2002). Production cultivars are 'G-48', imported from Australia, 'Udai', 'WSL-22', 'WSL39' and 'WSL-32', developed by the Western India Match Company Limited (WIMCO), and 'S7C15' and 'S7C8', imported from the Texas Forest Service, USA.

Populus genetic research and operational breeding is led by the Y.S. Parmar University of Horticulture and Forestry at Solan and the Western India Match Company (Khurana and Narkhede, 1995). Domestication is structured along three lines: (i) provenance trials, plus tree selection, open-pollinated progeny trials and clonal selection of *P. ciliata*; (ii) progeny trials of open-pollinated *P. deltoides* seed collections from the southern USA; and (iii) interspecific hybridization and clone testing of the *P. ciliata* × *P. maximowiczii* taxon. Taxa of lesser interest include the first interspecific hybrid generations of *P. deltoides* × *P. ciliata* and *P. deltoides* × *P. yunnanensis* and *P. ×canadensis* backcrosses to *P. deltoides*. A goal is to develop climatic adaptability within the four discrete elevation zones through selection for appropriate timing of spring phenology. Additional improvement priorities are adventitious rooting; resistance to stem-boring insects, for example *Apriona cinerea*, shoot-boring insects, for example *Eucosma glaceta*, and leaf-defoliating insects, for example *Clostera cupreata*; *Melampsora larici-populina* leaf rust resistance; and drought tolerance (Singh, 2000). Work continues to develop molecular markers for: (i) effective hybridizations based on estimated genetic distance of parental breeders; and (ii) cultivar identity control and protection.

Fig. 4.9. Western India Match Company's second-stage *Populus deltoides* × *P. deltoides* clonal trial. Multiple-stage test protocols are commonly used in poplar improvement programmes. Photo courtesy of Ramesh Dhiman, Western India Match Company.

Korea

POPULUS. Five taxa are native to the Korean Peninsula: *P. maximowiczii*, *P. koreana*, *P. simonii*, *P. tremula* var. *davidiana* and *P. tremula* var. *glandulosa*. Korean industries dependent on poplar include matchstick, plywood, pulp and paper, and medium-density fibreboard, and cogeneration. Domestication began in the 1950s with the introduction of more than 300 *P.* ×*canadensis* varieties from Italy. Eventually, cultivars 'I-214' and 'I-476' achieved nationwide use during the 1970s, until *Marssonina brunnea* and *Melampsora larici-populina* epidemics resulted in their replacement by locally developed *P.* ×*canadensis* cultivar 'Eco28'.

Today, the Korea Forest Research Institute at Suwon leads the country's domestication effort. Commercial hybrids are developed and deployed mostly as polyclonal varieties of three interspecific hybrid taxa: (i) *P. alba* × *P. tremula* var. *glandulosa* (Hyun-sashi series); (ii) *P. nigra* × *P. maximowiczii* (Yang-hwang-chol series); and (iii) *P. koreana* × *P. nigra* (Suwon series). The Hyun-sashi series actually encompasses four polyclonal varieties each composed of 4–20 individual clones bred from native *P. tremula* var. *glandulosa* and Italian *P. alba* selections. The Yang-hwang-chol series is a single polyclonal variety of eight clones selected from the first interspecific generation of *P. nigra* and *P. maximowiczii*. The Suwon series is made up of a single clonal variety developed from interspecific hybridization between *P. koreana* and *P. nigra*. A recurrent breeding and varietal selection programme has been initiated for *P. tremula* var. *davidiana* to identify selected clones for upland sites.

SALIX. There are approximately 35 species of *Salix* native to the Korean Peninsula. Domestication research at the Kangwon National University is evaluating domesticated cultivars and ones suited for short-rotation coppice throughout the region. The two main species are *S. caprea* and *S. gracilistyla*. *S. caprea* is a candidate for re-foresting marginal land, while *S. gracilistyla* is a candidate for cultivation of abandoned farmland. A core collection of genotypes of both species has been characterized for biomass production and phytoremediation applications. The ecological role of *Salix nipponica* and *S. koreensis* in riparian settings is also being studied at the university.

Japan

POPULUS. Japan's native *Populus* species are *P. maximowiczii*, *P. sieboldii* and *P. tremula* var. *davidiana*. The Oji Paper Company has been a leader in *Populus* domestication in Japan for over 40 years, originally in support of papermaking operations. The project continues with a goal of developing commercial planting stock for upland sites in both Japan and China. This has developed along three lines: first was the field identification of superior *P. maximowiczii* phenotypes; followed by controlled crossing and progeny testing for growth, form and pest resistance; and, finally, clonal evaluation. Six selected clones have received official registration. The second effort had female *P. maximowiczii* selections hybridized with *P. deltoides*, *P.* ×*generosa* and *P. ciliata* male selections using pollen imported from the USA, Canada and India: two *P. maximowiczii* × *P. deltoides* clones from this effort have been registered and deployed to the mountainous region of China's Hubei province based on heterosis in growth, rust resistance and cold tolerance (Fig. 4.10).

The third emphasis has involved interspecific hybridization using the endemic species *P. tremula* var. *davidiana* and *P. sieboldii* and *P. alba* selections from an established land race as maternal parents in crosses with Canadian, Italian and German sources of *P. tremula*, *P. grandidentata* and *P. alba*. Eight cultivars belonging to the *P. tremula* var. *davidiana* × *P. tremula*, *P. tremula* var. *davidiana* × *P. grandidentata* and *P. tremula* var. *davidiana* × *P. alba* taxa have been selected for commercial planting. Over the past 30 years, Oji Paper's procurement of raw material has shifted from domestic to foreign countries, in some cases accompanied by *Populus* breeding in support of amelioration plantings. As an example, hybrid breeding of *P.* ×*tomentosa* and *P. alba* var. *pyramidalis* has been initiated to develop drought tolerance for environmental plantings in China's Liaoning province.

4.3.2 North America

Populus domestication in North America was promoted originally as a fibre resource for the

Fig. 4.10. Superior *Populus maximowiczii* phenotype in China's Liaoning province. *P. maximowiczii* is increasing in importance to worldwide poplar breeding programmes when used in interspecific combination with *Populus deltoides*, *Populus trichocarpa* and *Populus nigra*. Photo courtesy of Man Shengjun, Liaoning Poplar Research Institute.

pulp and paper industry, while willows were cultivated for basket making. Poplar and willow are now viewed as the premier woody perennial feedstock crops for the fast-growing cellulosic renewable fuels industry, suitable for either biochemical (Sannigrahi and Ragauskas, 2010) or thermochemical conversion processes (Scott *et al.*, 2000). The environmental amelioration industry is also actively developing poplar and willow for phytoremediation of contaminated sites, the land application of municipal and industrial effluent and for agricultural streamside filter plantings (Felix *et al.*, 2008; Kuzovkina and Volk, 2009). Regionally, poplar is also being developed for various non-structural, specialty solid wood markets. Both Canada and the USA are actively engaged in the domestication and conservation of poplars and willows (Richardson *et al.*, 2007). *Populus* programmes prominently feature *P. deltoides*, probably the continent's most important species in view of its use in domestication projects around the globe. *P. tremuloides* is also quite important for its expansive transcontinental range and its use in both US and Canadian breeding programmes, as well as those in Finland and Germany. Willow breeding programmes utilize a wide range of hybrids, although breeding in Canada has centred on its native species, primarily *S. eriocephala*.

USA

POPULUS. Eight species of *Populus* – *P. angustifolia*, *P. balsamifera*, *P. deltoides*, *P. fremontii*, *P. grandidentata*, *P. heterophylla*, *P. tremuloides* and *P. trichocarpa* – are native to the USA. Work with native and exotic species occurs nearly exclusively in three regions featuring the following taxa:

1. Lower Mississippi River Valley – *P. deltoides* × *P. deltoides*.
2. Central and North Central – *P. deltoides* × *P. deltoides*, *P.* ×*canadensis*, *P. nigra* × *P. maximowiczii*, *P. tremuloides* × *P. tremula*, *P. alba* × *P. grandidentata*, *P.* ×*canescens*.
3. Pacific Northwest – *P.* ×*generosa*, *P.* ×*canadensis*, *P. deltoides* × *P. maximowiczii*.

Domestication was initiated in the lower Mississippi River Valley in the 1960s between 30° and 35°N latitudes when the US Forest Service (USFS) launched an intensive clonal improvement programme for *P. deltoides* to support the pulp and paper and matchstick industries. (Interspecific hybrids historically have not performed well in this region, perhaps due to their poor adaptation to the photoperiod of lower latitudes and susceptibility to *Septoria* stem canker.) The USFS Stoneville, Mississippi, Experimental Station released 14 *P. deltoides* clones for commercial use in 1970 (Land, 1974). Many of these, for example 'ST66', 'ST70' and 'ST72', were related and not all proved to be superior throughout a full rotation across the south-eastern USA. Accordingly, the USFS continued to screen large clonal collections, eventually testing 3700 genotypes for adventitious rooting, growth rate, disease resistance, stem form and wood density (Fig. 4.11). Selections totalling 198 were subsequently moved forward into a second stage of testing in 1980, in clonal trials replicated

Fig. 4.11. Clone trial of *Populus deltoides*, perhaps the most important species for worldwide *Populus* domestication programmes. Much of the world's sampling of *P. deltoides* has relied on provenances from the lower Mississippi River Valley. Photo courtesy of Randy Rousseau, Mississippi State University.

at Wickliffe, Kentucky, Fitler, Mississippi and Profit Island, Louisiana. The objective was to assess the extent of genotype × environmental interaction, as well as to compose a subset for advanced-generation breeding (Cooper, 1980). (Industrial breeding programmes of Westvaco Corporation and GreenWood Resources have since developed the second generation of the collection, with third-generation development expected over the next 5–10 years.) Other significant efforts in the southern USA include: (i) Oklahoma State University's collection of 450 genotypes from the xeric south-western portion of the species range (Nelson and Tauer, 1987); and (ii) a clonal collection of 626 *P. deltoides* genotypes assembled by Mississippi State University from six physiographic regions encompassing the south-eastern portion of the species' range from the Atlantic coast west to east Texas (Land *et al.*, 2001).

Iowa State University has also maintained a long-standing interest in the domestication of *P. deltoides* for the central USA in the vicinity of 42°N latitude. (*Septoria* canker disease has eliminated the use of most interspecific hybrids in this region.) Improvement is sought in *Melampsora* leaf rust resistance, growth rate and suitability for use in phytoremediation projects (Tabor *et al.*, 2000; Riemenschneider *et al.*, 2001b; Zalesny *et al.*, 2005a). Selection for adventitious rooting is relaxed, as *P. deltoides* planting stock is normally produced using rooted cuttings grown in high-density nursery beds (Zalesny *et al.*, 2005b). On drier upland sites, the putative *P. alba* × *P. grandidentata* hybrid cultivar 'Crandon' has recorded the highest biomass production rates in this region (Goerndt, 2005). Consequently, breeding and clonal testing of the *P. alba* × *P. grandidentata* taxon is being conducted, as is development of cost-effective vegetative propagation and coppice management schemes (Hall *et al.*, 1990; Green, 1998).

In the north central region at approximately 46°N latitude, *P. deltoides* × *P. deltoides* and *P.* ×*canadensis* are the taxa favoured by the pulp and paper and renewable energy industries, although the most widespread commercial cultivar ('NM6') belongs to the *P. nigra* × *P. maximowiczii* taxon (Netzer *et al.*, 2002). The predominant commercial *P.* ×*canadensis* cultivars are 'Baden 431' (synonym 'DN2'), 'Gelrica' (synonym 'DN5'), 'Eugenei' (synonym 'DN34'), 'Robusta' (synonym 'DN17') and 'Raverdeau' (synonym 'DN182'). Ongoing domestication projects at the University of Minnesota's Natural Resources Research Institute are pursuing the *P.* ×*canadensis* and *P. nigra* × *P. maximowiczii* taxa in addition to recurrent *P. deltoides* breeding (Riemenschneider *et al.*, 2001a). Families are bred annually in progeny sizes of 45–90 full-sibs. These are observed for 3 years for growth rate, disease resistance and survival, after which 30 seedlings per family are cloned and evaluated for adventitious rooting, growth rate, resistance to *Septoria* canker and *Melampsora* leaf rust and stem form.

The University of Minnesota has also been leading hybridization studies in section *Populus* through its Aspen and Larch Genetics Project and Cooperative. Started in 1952 in support of the pulp and paper industry, the cooperative today is breeding aspen for the oriented strand-board and renewable energy industries. (Aspen is an ideal candidate for biomass plantings by virtue of the ease with which high-density stands can be regenerated from root suckers after harvest.) Early research studied the effects of interspecific hybridization and ploidy level on stand productivity (Einspahr and Benson, 1964). Current research addresses the basis of interspecific heterosis in the diploid *P.* ×*wettsteinii* taxon (*P. tremuloides* × *P. tremula* and reciprocal) and the estimation of breeding values for growth rate, wood quality and *Entoleuca mammata* resistance (Li and Wyckoff, 1991; Li *et al.*, 1993, 1998; Li, 1995; David and Anderson, 2002). Since 2007, the effort has been extended to the development of the second interspecific *P.* ×*wettsteinii* generation, along with the development of an inexpensive method of vegetative propagation (Fig. 4.12).

In the Pacific Northwest, poplar domestication was started by the University of Washington and Washington State University during the1970s. Today, over 20,000 ha are being managed for the pulp and paper, timber and environmental services industries. Early investigations revealed that the first interspecific *P.* ×*generosa* generation consistently outperformed biomass production of native *P. trichocarpa*, as well as the second and backcross interspecific generations of *P. deltoides* and *P. trichocarpa* (Stettler *et al.*, 1988). Selected F_1 *P.* ×*generosa*, for example '11-11' and '49-177', and *P. deltoides* × *P. maximowiczii* clonal varieties are used along the lower Columbia River flood plain of western Oregon and Washington, while *P.* ×*canadensis* F_1 clonal selections, for example 'NE367', are preferred on the leeward side of the Cascade Mountains in the shrub-steppe environment of the arid mid Columbia River basin. Industrial breeding programmes are now the leading source of new F_1 interspecific varietal selections, as well as second-generation parental selections of *P. deltoides*, *P. nigra* and *P. trichocarpa* (Riemenschneider *et al.*, 2001a). Improvements are being sought in yield, resistance to *Venturia* shoot blight, *Melampsora* leaf rust, resistance to *Cryptorhynchus lapathi*, stem form, wood specific gravity and wind firmness.

Fig. 4.12. An F_1 superior *Populus* ×*wettsteinii* (*Populus tremuloides* × *Populus tremula*) variety growing in northern Minnesota. The *P.* ×*wettsteinii* taxon is used in both North America and Europe. Photo courtesy of Andrew David, University of Minnesota.

SALIX. Historically, the cultivation of willows in the USA has focused on shrub species for basket making or ornamental landscaping applications. More recently, domestication has expanded to cultivar selection for use in both power plants and co-generation plants. Nearly all of this work has been conducted in the north-east at the State University of New York (SUNY). Initially, in the 1990s, willow germplasm was imported from Canada and Sweden for field evaluation in New York. Much of this material – *S. eriocephala*, *S. viminalis*, *S. viminalis* × *S. schwerinii* – developed pest problems and a regional breeding programme was initiated at the College of Environmental Science and Forestry at SUNY. A large and diverse collection of willows was assembled from provenances throughout the north-east and mid-west, with *S. purpurea* and *S. eriocephala* constituting the majority of the sample (Kopp, 2000; Kopp *et al.*, 2001; Smart *et al.*, 2005; Smart and Cameron, 2008). Molecular marker analyses later showed high levels of genetic diversity and a relatively high degree of heterozygosity to be typical of the two species (Kopp *et al.*, 2002b; Lin *et al.*, 2009).

By 2007, the willow collection had expanded to over 700 accessions, representing more than 20 species and interspecific hybrids.

The yield of the best current commercial shrub willow cultivars approximates 12 dry t ha^{-1} $year^{-1}$ (Volk *et al.*, 2011). Some of the commercial cultivars in use in the north-east are: *S.* ×*dasyclados* 'SV1', *S. eriocephala* 'S25', *S. miyabeana* 'SX64', *S. udensis* (formerly *S. sachalinensis*) 'SX61' and *S. caprea* hybrid 'S365'. Improved cultivars are in the offing: more than 600 families and 2500 progeny from crosses completed in 1998 and 1999 have been evaluated in family screening trials at SUNY. Moreover, a new generation of genotypes that were bred between 2001 and 2005 is undergoing testing in a 76-clone trial in Geneva, New York. Based on early measurements of growth, the top genotype has 20% greater total basal area than the best commercial cultivar in the second year after coppicing, indicating that further yield enhancements are likely once new selections are released.

Ongoing *Salix* breeding in the north-east is now located at Cornell University's New York State Agricultural Experiment Station. Improvement priorities are adaptability to low pH and poorly drained soils, cold tolerance, rust resistance and insect resistance. Sterile varieties or ones with reduced fertility are also sought after. The strategy to achieve these objectives is to maintain long-term recurrent selection and improvement of tetraploid genotypes (*S. miyabeana* and hybrids with *S. udensis* (formerly *S. sachalinensis*)) in parallel with the breeding and selection of diploid genotypes (*S. purpurea*, *S. viminalis* hybrids, *S. integra*, *S. koriyanagi* and related species). High-yielding, disease- and pest-resistant cultivars will be selected from among the progeny of the tetraploid–diploid crosses.

Canada

POPULUS. *P. angustifolia*, *P. balsamifera*, *P. grandidentata*, *P. tremuloides*, *P. trichocarpa* and *P. deltoides* are native to Canada. *P. balsamifera* and *P. tremuloides* are renowned for their extensive transcontinental ranges. Canadian domestication programmes are most active in the provinces of Quebec, Saskatchewan and Alberta. In the Canadian prairie, a growing need for *Populus* domestication is short-rotation, intensive culture systems for fibre and biomass production on abandoned agricultural lands to revitalize rural communities (Amichev *et al.*, 2010).

In the east, Quebec's Ministry of Natural Resources has worked on the improvement of *P. deltoides*, *P. balsamifera*, *P. maximowiczii*, *P. trichocarpa* and *P. nigra* since 1969, resulting in the release of 40 cultivars for use throughout the province's five bioclimatic zones (Périnet, 2007). The principal domestication approach encompasses: (i) non-recurrent F_1 hybrid breeding; (ii) three-way hybridization; and (iii) recurrent intraspecific breeding of *P. deltoides*, *P. maximowiczii*, *P. trichocarpa* and *P. nigra*. F_1 hybrid taxa of *P. maximowiczii* in combination with *P. balsamifera* and *P. trichocarpa* and complex *P. maximowiczii* crosses with *P.* ×*generosa*, *P.* ×*jackii* and *P.* ×*berolinensis* (*P. laurifolia* × *P. nigra* var. *italica*) all show the best growth and adaptability to the acidic and less fertile forest soils and colder areas in the northern bioclimatic region. In the St Lawrence Valley of southern Quebec, emphasis has been placed on selecting clones resistant to *Septoria* canker from a variety of taxa, including *P.* ×*canadensis*, *P. deltoides* × *P. maximowiczii* and the three-way cross combination of *P.* ×*canadensis* and *P. maximowiczii* (Fig. 4.13). The *P. nigra* × *P. maximowiczii* and the *P. balsamifera* × *P. maximowiczii* taxa are reserved for the less fertile, higher elevation forested sites in the province. Growth rate, cold hardiness, stem and crown form, disease and insect resistance, site adaptability and wood quality are all priority improvement traits. Evaluation of seedling populations begins in nurseries from which superior phenotypes are cloned for multiple-stage field testing. After 5 years, the best subset is then re-tested with increased replication for 15 years and then final selections are made.

Agriculture and Agri-Food Canada's (AAFC) Agroforestry Development Centre located at Indian Head, Saskatchewan, has a history of over 60 years of poplar hybridization and selection (Cram, 1960). Cultivar development historically emphasized environmental applications, e.g. shelterbelts, intercropping, control of odours and particulates from farm feedlots and poultry houses. Commercial varieties in use are 'Walker', 'Assiniboine', 'Manitou', 'Katepwa', 'Hill' and 'Okanese'. With the exception of 'Okanese', all have arisen from a *P. deltoides* seedlot putatively pollinated by a *P.* ×*petrowskyana*

Fig. 4.13. A 10-year-old commercial plantation of *Populus deltoides* × *Populus maximowiczii* '3442' established by Norampac on marginal farmland in Quebec's Temiscouata–Bas-Saint-Laurent region. Selection within this taxon features resistance to winter sunscald and *Septoria* canker. Photo courtesy of Pierre Périnet, Quebec Ministry of Natural Resources and Wildlife.

hybrid (*P. nigra* × *P. laurifolia*). 'Okanese' was the result of a backcross between 'Walker' and *P.* ×*petrowskyana*. Today, the Agroforestry Development Centre has expanded its breeding to a new class of cultivars of diverse genetic backgrounds suitable for short-rotation commodity and biomass production and dry land farming. The improvement strategy entails: (i) species and provenance selection; (ii) interspecific crossing of native and Asian species; (iii) initial laboratory tests for cold tolerance (thermography) and water-use efficiency (carbon isotope discrimination); and (iv) multiple-stage field evaluations of families selected from the laboratory tests. Field evaluations begin with 1-year nursery tests from which within-family selections are made based on height, diameter and leaf area index. These are cloned the following year for multiple-site field trials and evaluated for cold and drought tolerance, crown and stem form, growth rate, the tendency to produce multiple stems and sylleptic branches, and resistance to *Melampsora* leaf rust and *Septoria* canker. A base population of 106 families represented by over 16,000 genotypes is currently being evaluated. The population is comprised mainly of F_1 taxa bred from *P. balsamifera*, *P. deltoides*, *P. laurifolia*, *P. maximowiczii*, *P. nigra* and *P. simonii*, as well as a number of advanced-generation taxa developed from several complex hybrid types (e.g. *P. deltoides* × *P.* ×*petrowskyana* and *P.* ×*petrowskyana* × *P. maximowiczii*) (Fig. 4.14). The AgroForestry Development Centre is participating in an extensive *ex situ P. balsamifera* conservation effort involving 39 populations selected from six longitudinal transects across Canada. A parallel programme is being developed for *P. deltoides* var. *monilifera* with a sampling of germplasm from the South Saskatchewan and Missouri rivers.

In the province of Alberta, Alberta-Pacific Forest Industries Inc. (Al-Pac) – a manufacturer of high-quality pulp for the photographic papers market – and the Western Boreal Aspen Cooperative lead industrial poplar improvement programmes for the region between 52° and 54°N latitude. The Al-Pac project involves a variety of first-generation cross combinations among *P. balsamifera*, *P. deltoides*, *P. laurifolia*,

Fig. 4.14. A superior selection of the *Populus ×petrowskyana × Populus maximowiczii* three-way cross in the Canadian prairie region of Saskatchewan. Photo courtesy of Bill Schroeder, Agriculture and Agri-Food Canada, Agroforestry Development Centre.

P. maximowiczii, *P. nigra* and *P. simonii*. Testing is conducted in stages, with selections made at age nine, approximately one-half the length of a commercial rotation. The initial selection criteria are resistance to *Septoria* canker and cold and drought hardiness; later-stage selections are based on volume, adventitious rooting, pulp quality and stem form. Phenotypic stability across a range of sites and management practices, for example spacing and fertilization, is evaluated as a final check prior to commercial deployment. Clonal propagation uses stem cuttings produced in stoolbeds, as well as branch cuttings collected in young commercial plantations. Current cultivars in use are *P. deltoides* × *P.* ×*petrowskyana* selections 'Walker', 'Brooks 6' and 'Katepwa', *P. balsamifera* × *P. simonii* cultivar 'P38' and *P.* ×*jackii* cultivar 'Northwest'. *P. tremuloides*' genetic improvement is also being addressed, both through recurrent intraspecific breeding (Thomas *et al.*, 1997) and interspecific hybridization with Finnish sources of *P. tremula* and Chinese sources of *P. tremula* var. *davidiana*.

Alberta's Western Boreal Aspen Cooperative is a collaboration of paper and timber industries working on the domestication of section *Populus* using the endemic *P. tremuloides* and non-native sources of *P. tremula*, *P. tremula* var. *davidiana*, *P. grandidentata*, *P. alba* and *P.* ×*canescens*. Initially, clonal evaluation of superior *P. tremuloides* phenotypes from the wild was featured as a short-term improvement approach. Now, candidates for clonal testing are taken from pedigreed populations at age three using a combined family selection technique. Clonal evaluations are conducted at age seven in field trials using incomplete block designs replicated across multiple sites. Survival and height comprise metric selection criteria. Ease of propagation is considered a threshold trait. Pest resistance is not a strong selection emphasis. The cooperative's *P. tremuloides* provenance trials, breeding populations and family trials are being maintained as an *ex situ* conservation effort. Additionally, the long-term *P. tremuloides* domestication strategy employs a multiple-population breeding plan to direct and enhance within-species genetic variation (Namkoong and Koshy, 1997). Finally, the University of Alberta conducts research on the genetic structure of *in situ P. tremuloides* reserves as affected by climate change.

SALIX. Canada's University of Toronto initiated willow breeding in North America in the early 1980s under the direction of Louis Zsuffa (Zsuffa, 1988). *S. amygdaloides*, *S. bebbiana*, *S. discolor*, *S. eriocephala*, *S. exigua*, *S. lucida*, *S. pellita* and *S. petiolaris* were the seminal species in this project. Early work verified species differences in biomass productivity, moisture content and specific gravity (Mosseler *et al.*, 1988). Furthermore, research was completed in species crossability relationships, interspecific pollination barriers and hybrid performance (Mosseler and Papadopol, 1989; Mosseler and Zsuffa, 1989; Mosseler, 1990). More and more, there is interest in joint environmental–biomass production projects that take advantage of willow's capacity for high levels of biomass yield coupled with its significant rates of water and nutrient uptake. AAFC's Agri-Environment Services Branch at

Indian Head, Saskatchewan, is now screening willow genotypes for such multi-purpose applications. Testing is done through traditional field experimentation, but also extends to laboratory screening for cold tolerance, nutrient-use efficiency and water-use efficiency.

The Indian Head station started a willow germplasm collection in 2005 encompassing *S. amygdaloides*, *S. bebbiana*, *S. discolor*, *S. eriocephala* and *S. petiolaris* from natural stands in the provinces of New Brunswick, Quebec, Ontario and Saskatchewan. A total of 55 populations and 244 individual genotypes has been sampled. Controlled breeding was initiated in 2008 based on intra- and interspecific crosses between geographic diverse populations of *S. dasyclados*, *S. discolor*, *S. eriocephala*, *S. lucida* and *S. scouleriana*.

4.3.3 Europe

The founding meeting of the IPC was organized and hosted by the French Poplar Commission in 1947, followed the next year by a meeting in Italy. This legacy is evident today in the many European IPC member nations that are active in the management of *Populus* and *Salix* genetic resources, the most notable of which are Austria, Belgium, Croatia, Finland, France, Germany, Hungary, Italy, the Netherlands, Russia, Serbia, Spain, Sweden, Turkey and the UK. Non-member nations Latvia and Poland are very active in *Salix* domestication. Some of these countries are developing new commercial varieties, while others are working primarily on advancing poplar and willow molecular science. A growing interest in both genera is due to strong support within the EU for short-rotation forestry as a source of renewable energy feedstock (Mitchell *et al.*, 1999; Kauter *et al.*, 2003; Bunn *et al.*, 2004; Pellis *et al.*, 2004; Aylott *et al.*, 2008) and environmental services (Riddell-Black, 1998). Life cycle analyses for poplar-derived ethanol in the EU have forecasted reductions in global warming and the rate of depletion of the ozone layer, although acidification and eutrophication may materialize (Gonzalez-Garcia *et al.*, 2009). The emergence of the renewable energy industry has reinvigorated many European breeding and clonal development programmes. The European effort is noteworthy for its level of cooperation among nations and the close integration of basic investigations and applied breeding programmes. These are most evident in the study and conservation of *P. nigra* under the EUFORGEN network and the former EUROPOP project (Van Dam and Bordacs, 2002).

Austria

POPULUS. *P. nigra* domestication in Austria has a history of nearly 50 years, though many of the earliest selections are no longer in commercial use. The Federal Research and Training Centre for Forests, Natural Hazards and Landscape in Vienna leads in the development of commercial varieties in support of Austria's renewable biomass energy industry. A clonal collection assembled from stands throughout the country during the 1990s is being studied for geographic patterns of climatic adaptability and the mechanism of quantitative *Melampsora* rust resistance. General trends indicate that *P.* ×*canadensis* clones bred for warmer regions, for example Hungary and northern Italy, are adapted to eastern Austria, while *Tacamahaca* hybrids developed for cooler regions in north-western Europe do well in the Austrian west and at higher elevations. (The pattern of climatic adaptability of native genetic resources is likely much finer, perhaps, on a scale of hundreds of kilometres.) Controlled hybridization is an ongoing effort, with superior *P. nigra* phenotypes from natural stands used in developing the *P.* ×*canadensis* and the *P. nigra* × *P. maximowiczii* taxa. The *P.* ×*generosa* taxon is also being bred in Austria. Since the *P. nigra* programme is at an early domestication stage, the maintenance of a broad genetic basis is emphasized. Accordingly, *in situ* conservation of native *P. nigra* stands in riparian set-asides is being arranged with local landowners and regional governments. Mass selection is used to truncate seedling populations for replicated clonal testing for coppicing ability and disease resistance in high-density stands. Recurrent breeding programmes are being initiated for both *P. nigra* and *P. deltoides*.

P. alba, *P. tremula* and hybrid swarms of *P.* ×*canescens* are also endemic to Austria and the subject of intense molecular study of introgression, adaptation and speciation (Lexer *et al.*, 2005, 2007; van Loo *et al.*, 2008; Fussi *et al.*, 2010) (Fig. 4.15). This basic research is also

Fig. 4.15. A natural clone of *Populus tremula* near Harbach in Waldviertel, Austria. This clone has established through root suckering and is identified by its distinctive stem form and bark colour. Molecular investigations into the natural hybridization of *P. tremula*, *Populus alba* and hybrid swarms of *Populus* ×*canescens* is an interest of Austrian poplar geneticists studying the process of adaptation and speciation in the genus. Photo courtesy of Berthold Heinze, The Federal Research Centre for Forestry.

expected to lead to the identification of unique stands of mixed taxonomic status that warrant conservation as well as the identification of commercial *P.* ×*canescens* cultivars for the nation's future bioenergy programme.

Belgium

POPULUS. Poplar domestication in Belgium has a productive record dating to 1948. The project focused primarily on the *P.* ×*generosa* taxon, one achievement of which was the release of cultivars 'Beaupre' and 'Unal' in the late 1960s (Steenackers *et al.*, 1996). The Research Institute for Nature and Forest (INBO) and the Research Centre for Nature, Forests, and Wood (CRNFB) operate Belgium's poplar domestication and conservation efforts as poplar cultivation now supplies about one-half of the nation's annual hardwood harvest. The main selection criteria are adventitious rooting, growth rate, stem form, photoperiodic adaptation, climatic and soil adaptation, and disease resistance (*Melampsora larici-populina*, *Marssonina brunnea* and *Xanthomonas populi*). Resistance breeding features selection for quantitative resistance; this is now a higher priority than selection for complete resistance in view of the breakdown of clonal resistance to *Melampsora larici-populina* during the 1980s and 1990s that eliminated the *P.* ×*generosa* selections used throughout north-western Europe. The failure of selected *P.* ×*generosa* cultivars further shifted reliance to the *P.* ×*canadensis* taxon and its superior tolerance to *Melampsora* infection. Wood quality is also a feature of the breeding programme due to poplar's importance in plywood and dimensional wood markets.

The Belgian domestication programme encompasses four species: *P. nigra*, *P. deltoides*, *P. trichocarpa* and *P. maximowiczii* (Fig. 4.16). Up to 100 intra- and interspecific controlled crosses are made annually in which *P.* ×*canadensis*, *P.* ×*generosa* backcrosses to *P. deltoides*, and the F_1 hybridization of *P. maximowiczii* with *P. deltoides*

Fig. 4.16. Full-sib progeny test of *Populus deltoides* × *P. deltoides* in Belgium. Recurrent breeding of parental species is a key component of a complete interspecific hybridization programme necessary for sustained genetic advancement. Photo courtesy of Marijke Steenackers and Boudewijn Michiels, Research Institute for Nature and Forest.

and *P. trichocarpa* are featured. Between 5000 and 10,000 genotypes are under evaluation in multiple-stage field tests during any given year. The INBO programme releases fully tested cultivars on a continuous basis following tests of 15–20 year's duration in plots replicated across soil types and climate zones. Six recently commercialized selections include three of the *P.* ×*canadensis* taxon ('Muur', 'Oudenberg', 'Vesten'), one of the *P.* ×*generosa* backcross taxon ('Grimminge') and two representing the *P. trichocarpa* × *P. maximowiczii* taxon ('Bakan', 'Skado'). Selections from the *P.* ×*generosa* taxon have not yet materialized because of the difficulty in developing durable *Melampsora* resistance based on polygenic systems. INBO is expanding its clonal archives of *P. deltoides* and *P. maximowiczii* genotypes under international exchange programmes and recently expanded the scope of the breeding programme to *P.* ×*canescens*. The Flanders Institute for Biotechnology is assisting the INBO breeding programme by developing a marker-assisted selection protocol to accelerate the evaluation of wood quality and disease resistance.

Croatia

POPULUS. There are 16 registered cultivars in Croatian *Populus* culture. These include selections of *P.* ×*canadensis*, for example 'I-214', 'Pannonia', 'Tiepolo', 'M-1', 'L-12', 'BL Costanzo', 'Triplo'; *P. deltoides*, for example 'San Martino', '457', '710', 'S1-8' and 'S6-36'; and *P. alba*. Ongoing cultivar evaluation efforts are determining the suitability of clones selected for long rotations when used in short-rotation biomass production systems. Evaluation criteria are growth rate, coppicing ability and disease resistance (*Marssonina*, *Melampsora*, *Dothichiza*). Most conservation work in Croatia is concentrated on *P. nigra* and its two subspecies. Although much of the original *P. nigra* distribution has been somewhat fragmented, most remaining populations are considered well preserved by European standards. Subspecies *nigra* grows along the Mura, Sava, Drava and Danube rivers, while subspecies *caudina* – considered a xeromorphic form – is found in the Neretva river drainage (Kajba *et al.*, 2004, 2006). The University of Zagreb leads both *in situ* and *ex situ* conservation programmes (Fig. 4.17).

Fig. 4.17. Croatian *Populus* conservation efforts maintain the composition of mixed riparian associations of *Populus nigra*, *Populus alba* and hybrid swarms of *Salix alba* and *Salix fragilis*, as seen here in this mixed-aged stand along the Danube River. Photo courtesy of Davorin Kajba, Faculty of Forestry, University of Zagreb.

Croatian conservation has featured selection of locally adapted superior *P. nigra* phenotypes for the breeding, testing and propagation of improved planting stock for the nation's riparian restoration programme (Kajba *et al.*, 2006). Stands targeted for restoration are those considered to be of a sufficient size to avoid a loss of genetic diversity through genetic drift over time. A secondary conservation goal is ultimately to replace *P.* ×*canadensis* plantations with *P. nigra* restoration plantings in higher-quality riparian habitats involving 50-year conservation easements. A plan to apply molecular tools to monitor the rate of introgression in restored stands is being developed. Artificial *P.* ×*canadensis* plantations will still be managed for wood production across one-quarter of the riparian zone, but will use *P. nigra* buffers to isolate nearby native *P. nigra* populations from hybrid reproduction. An *ex situ* conservation programme was launched in 1995 with the establishment of a clonal archive at Cakovec containing 83 *P. nigra* genotypes propagated clonally from mature trees of the Sava, Drava and Mura river populations. In 1998, a second archive of 101 genotypes was established at Darda. A similar *ex situ* conservation programme is being planned for *P. alba*.

Finland

POPULUS. Interspecific hybridization of Finland's native aspen, *P. tremula*, and the North American species, *P. tremuloides*, was started in the 1950s to support the Finnish matchstick industry. The work lost momentum during the 1970s, but was restarted in the mid-1990s when the pulp and paper industry identified aspen as an important component of their future fibre supply strategy. Intraspecific *P. tremula* breeding and interspecific hybridization with Canadian *P. tremuloides* provenances now figure prominently in Finland's national long-term forest tree breeding strategy under the direction of The Finnish Forest Research Institute (METLA) (Fig. 4.18).

Domestication started with open-pollinated *P. tremula* seed collections from 1000 superior phenotypes from both native stands and genetic trials. Combined family and within-family

Fig. 4.18. A *Populus* ×*wettsteinii* (*Populus tremula* × *Populus tremuloides*) clonal test in Finland established using vegetative propagation of succulent axillary shoots. Photo courtesy of Egbert Beuker, The Finnish Forest Research Institute.

selection for growth and form, wood and fibre quality, *Venturia tremulae* resistance and amenability to micropropagation was then used to identify individuals for clonal testing (Yu *et al.*, 2001). Vegetative propagation via root cuttings has been developed for commercial nursery production, and although more economical than micropropagation, the methodology encounters more clonal variation and remains under development (Stenvall, 2006). Finnish requirements for clone testing mandate a 13-year test rotation and four replicates of 25-tree plots. Twenty-five F_1 *P.* ×*wettsteinii* genotypes have been released for cultivation in southern Finland, with additional releases scheduled for 2014. The pace of domestication and plantation development may be intensified as aspen propagation and plantation management techniques become increasingly refined. A significant inducement to commercial plantations would be coppice management for energy production.

France

POPULUS. The National Institute for Agricultural Research (INRA) and the Association of Forests and Cellulose (AFOCEL), a research organization of the French pulp and paper industry, are leaders in French poplar domestication and conservation. Over 60 commercial nurseries now trade in 25 different *Populus* cultivars, the most common ones belonging to the *P.* ×*canadensis* taxon, for example 'I-214', 'Triplo', 'I-45/51', of Italian origin, 'Dorskamp', 'Flevo' and 'Koster', of Dutch origin, and 'Ghoy' and 'Raspalje' of French origin (Paillassa, 2004). Nursery production of *P.* ×*generosa* cultivars is in decline today because of *Melampsora* infection.

The French poplar programme under INRA's direction has been very active in studying genetic variation in the *Populus–Melampsora* pathosystem (Legionnet *et al.*, 1999). The work provided the first evidence for the existence of physiological races within *Melampsora allii-populina* (Pinon, 1992; Frey and Pinon, 1997). Molecular investigations have expanded the pathology work, showing the possibility of developing selection criteria based on race-specific, quantitative resistance components (Goue-Mourier *et al.*, 1996; Lefevre *et al.*, 1998; Dowkiw and Bastien, 2007; Dowkiw *et al.*, 2010). Quantitative rust resistance mechanisms are also being studied through association genetics and gene expression analysis (Dowkiw *et al.*, 2003). Marker-assisted selection techniques for *Melampsora* leaf rust resistance are in development; a major QTL that accounts for more than half of the total variation in uredia size in *P. trichocarpa* has been identified (Jorge *et al.*, 2005). Collaborators in Belgium have identified a similar QTL in *P. nigra*. Work in the genomics of lignin and tension wood formation and the genetics of drought tolerance is also under way (Pilate *et al.*, 2004; Plomion *et al.*, 2006; Dillen *et al.*, 2011).

INRA has established a *P. nigra ex situ* conservation programme encompassing 350 genotypes. Additionally, an inventory of natural populations along the Loire River and other drainages has been completed using molecular tools to quantify genetic diversity among populations (Plate 17B). Research at INRA places importance on monitoring the degree of gene flow between cultivated interspecific hybrids and native *P. nigra* populations (Cagelli and Lefevre, 1995; Imbert and Lefevre, 2003).

Germany

POPULUS. Germany's *Populus* domestication and conservation efforts have been led by the Institute for Forest Genetics and Forest Tree Breeding at Grosshansdorf since 1948 and by the Northwest German Forest Research Station located in Hann. Münden since 1955 (Mohrdiek, 1979; Melchior, 1981). Section *Populus* is the focus of the Grosshansdorf programme that was initiated with various interspecific F_1 hybrid taxa bred from superior phenotypes of *P. tremula*, *P. tremuloides*, *P. alba*, *P.* ×*canescens* and *P. grandidentata* from native stands and cultivated plantings throughout Germany. Hybridization of *P. tremula* and *P. tremuloides* is now the featured improvement strategy, as heterosis is frequently observed (Melchior, 1985) (Fig. 4.19). To promote the commercial use of this hybrid vigour, investigation of methods of mass vegetative propagation is pursued (Muhs, 1998). Research in *Populus* genetic transformation is also conducted; Grosshansdorf was responsible for the first experimental release of genetically modified *P.* ×*wettsteinii* varieties in 1996 in Germany (Fladung *et al.*, 1996; Fladung and Muhs, 1999). An interspecific full-sib *P.* ×*wettsteinii* progeny has been approved for trading under the 'tested' category as 'Holsatia' (Melchior *et al.*,

Fig. 4.19. A 20-year-old clonal field trial of *Populus* ×*wettsteinii* (*Populus tremula* × *Populus tremuloides*) at Vorwerksbusch, Reinbek, Germany. Top-rated genotypes have a stem diameter of 38 cm and heights over 30 m when grown with 4 × 4 m spacing. Photo courtesy of Georg von Wuehlisch, Federal Research Centre for Forestry and Forest Products.

1987). Similarly, two backcross *P.* ×*wettsteinii* families have received approval to be traded under the name 'Vorwerksbusch' (Muhs and Melchior, 1986). Finally, 14 superior *P.* ×*wettsteinii* clonal selections have been assembled into a varietal mixture that has received approval for commercial trade in the category 'tested' under the name 'Grosshansdorf' (Muhs, 1987).

The Northwest German Forest Research Station at Hann. Münden is breeding intersectional *Aigeiros–Tacamahaca* hybrids with a concentration on resistance to *Pollaccia elegans*, *Xanthomonas populi*, *Dothichiza* spp., *Marssonina brunnea* and *Melampsora* spp. Various cross combinations of *P. maximowiczii*, *P. trichocarpa*, *P. nigra* and *P. deltoides* have led to new cultivars suitable for deployment in varietal mixtures of five to ten genotypes characterized by compatible growth curves, high productivity and phenotypic stability (Weisgerber, 1993). The current priority is the selection of cultivars for high-yield, short-rotation biomass production (Fig. 4.20). Six hundred *P. nigra* genotypes are maintained in an *ex situ* conservation programme. An *in situ P. nigra* conservation effort involves an inventory of native stands and incorporates isozymes and microsatellite marker technology to confirm species identity while assessing population diversity. Populations of the Rhine River Valley are not strongly differentiated (Gebhardt *et al.*, 2001).

Hungary

POPULUS. The poplar estate in Hungary is quite large, with approximately 120,000 ha established with mostly *P.* ×*canadensis* cultivars approved by the country's Forest Research Institute. These include 'Agathe F', 'Blanc du Poitou', 'H-328', 'I-45/51', 'I-154', 'I-214', 'I-273', 'Koltay', 'Kopecky', 'Luisa Avanzo', 'Marilandica', 'Pannonia', 'Sudar', 'Parvifol', 'Rabamenti' and 'Robusta' (Fig. 4.21). Most have been in use for 20 years or longer and three – 'I-214', 'Pannonia' and 'Robusta' – occupy nearly one-half of the poplar land base. Cultivars of both *P.* ×*canadensis* ('I-214', 'NE229') and *P.* ×*generosa* ('Unal') have been identified for coppice biomass production (Marosvolgyi *et al.*, 1999).

Fig. 4.20. A test plot of *Populus maximowczii* × *Populus trichocarpa* 'Matrix 11' being evaluated for biomass production by the Northwest German Forest Research Station at Hann. Münden, Germany. Intrasection *Tacamahaca* hybrids grown in high-density stands for bioenergy feedstock regenerate vigorously by coppice management. Photo courtesy of Brian Stanton, GreenWood Resources.

Fig. 4.21. A 5-year-old Hungarian plantation of *Populus ×canadensis* (*Populus deltoides × Populus nigra*) 'Koltay'. Photo courtesy of Jozsef Kiss, Institute of Genetics and Biotechnology, Saint Stephanus University, Hungary.

The relatively narrow genetic base, along with the need to develop new cultivars for short-rotation biomass production and droughty alkaline soils, has stimulated ongoing cultivar development at the Forest Research Institute's Sarvar Experimental Station. Newly selected *P.* × *canadensis* cultivars ('Sv2-24' and 'Sv1-64') have shown an increase in radial stem growth of 40–50% relative to standard cultivar 'I-214'. *P. alba* is also undergoing domestication for both wood production (cultivars 'H758' and 'H425-4') and agroforestry (*P. alba* × *P. alba* 'Bolleana', *P. alba* 'H427-3', *P. alba* × *P. grandidentata* 'H422-9') (Redei, 2000). *Ex situ* conservation clone banks of *P. nigra* and *P. alba* have been established with 5800 accessions from previous collection and breeding activities. *Populus* planting in Hungary's natural reserves has been restricted lately to native species.

The Institute of Genetics and Biotechnology of Saint Stephanus University in Godollo is developing biotechnological approaches to *Populus* domestication for environmental services. Projects include the *in vitro* evaluation of *P. ×canescens* and *P. nigra* transgenic lines for phytoextraction of zinc and copper for the remediation of bauxite mine spoils (Bordacs *et al.*, 2002; Bittsanszky *et al.*, 2005; Foyer and Noctor, 2005; Gyulai *et al.*, 2005) and techniques to screen for adaptability to arbuscular mycorrhizal fungi colonization (Takacs *et al.*, 2005). The institute is also studying somaclonal variation in *P. nigra* and *P. deltoides* using anther culture to develop novel genetic variation for Hungary's conventional *Populus* breeding programme (Kiss *et al.*, 2001).

Italy

POPULUS. Italian poplar breeding began in Villafranca Piemonte in 1922 at the Pignatelli Estate and later with the establishment of the Poplar Improvement Institution in 1931. *P. ×canadensis* 'I214', perhaps the most widely planted *Populus* clone worldwide, was developed here under Giovanni Jacometti in 1929 (Fig. 4.22). The Villafranca Institution was replaced in 1939 when the Poplar Research Institute – Istituto di Sperimentazione per la Pioppicoltura, or ISP – was established at Casale Monferrato in the Po River Valley. Today, the institute is organized within the Ministry of Agriculture's Research Council, and through its Research Unit for Intensive Wood Production (CRA-PLF) leads *Populus* domestication programmes in four programmatic areas: (i) conservation and evaluation

Fig. 4.22. Global *Populus* domestication programmes frequently rely on broadly adapted selections, perhaps the best example of which is the Italian *Populus* ×*canadensis* (*Populus deltoides* × *Populus nigra*) cultivar 'I-214' bred by Giovanni Jacometti over 80 years ago. Photo courtesy of Lorenzo Vietto, CRA-PLF Agriculture Research Council, Italy.

of genetic resources; (ii) controlled hybridization and varietal development for improved growth rate, wood quality and soil adaptability; (iii) breeding for resistance to *Melampsora larici-populina*, *Melampsora allii-populina*, *Marssonina brunnea*, *Venturia populina* and *Discosporium populeum*, and the woolly aphid, *Phloeomyzus passerinii*; and (iv) the development of genetic transformation technology and marker-aided selection for varietal development.

Interspecific hybridization of the *P.* ×*canadensis* taxon and recurrent breeding of parental species *P. deltoides* and *P. nigra* are central to the Italian CRA-PLF programme (Bisoffi and Gullberg, 1996). In the mid-1980s, 95 *P. deltoides* female genotypes and 147 male *P. nigra* genotypes were hybridized using a polycross and common tester mating design to create a segregating F_1 population, from which seven elite selections were made and registered in 2011 with the National Register of Basic Forest Materials. The *P. deltoides* and *P. nigra* parents of highest general combining ability were then identified in the early 1990s and used to breed 22 *P.* ×*canadensis* clones that were undergoing testing for adaptability to a range of environmental conditions in Europe. An F_2 population has also been developed using 19 *P.* ×*canadensis* genotypes and a multiple population breeding approach structured along the lines of pathogen resistance. A second cycle of intraspecific *P. deltoides* (34 genotypes) and *P. nigra* (49 genotypes) breeding is under way. Emphasis has been placed on suitability for both biomass energy plantings as well as veneer plantations. Field evaluation trials are established with 12 replications of approximately 25 genotypes in single-tree plots.

Currently, 43 clones selected from ISP are permanently recognized by the Italian National Register of Basic Forest Materials, among which are seven designated solely for short-rotation use. Most belong to the *P.* ×*canadensis* F_1 and backcross generations, although the *P. deltoides*, *P. alba*, *P. deltoides* × *P. maximowiczii* and several complex hybrid taxa are also represented. Well-known *P.* ×*canadensis* cultivars attributed to the institute include 'I-214', 'I-154', 'I-488', 'I-45/51', 'Neva', 'Taro', 'Timavo', 'Triplo', 'Mella', 'Brenta', 'Soligo' and 'Lambro' (Regione Emilia-Romagna, 1999). *P. deltoides* cultivars include 'San Martino', 'Lux', 'Onda', 'Dvina', 'Oglio' and 'Lena'. CRA-PLF is also responsible for *P. alba* × *P. alba* cultivar 'Villafranca' and *P. deltoides* × *P. maximowiczii* cultivar 'Eridano'. Two recent releases targeted specifically for biomass production are *P.* ×*canadensis* 'Orion' and *P. deltoides* × *P. deltoides* 'Baldo'. Also of note is the work of the Forest Research and Development Centre in Rome during the 1980s with the *P.* ×*canadensis* taxon; notable cultivar releases include 'Luisa Avanzo', 'Cima', 'Bellotto' and 'Guariento'; the latter, along with *P.* ×*canadensis* 'Neva' and 'I-214', are widely planted in China.

Alasia Franco Vivai is a commercial nursery that conducts a large *Populus* controlled hybridization and clonal selection programme to support biomass plantings throughout the continent (Fig. 4.23). Native populations of *P. alba*, *P. deltoides*, *P. nigra* and *P. trichocarpa* have been sampled intensively and brought together in large seedling archives at a field-testing facility in Savigliano. Superior full-sibs identified in intra- and interspecific crossing programmes were then expanded with additional cross-pollinations to increase within-family selection intensities. Interspecific hybridization focuses on the *P.* ×*canadensis* and *P.* ×*generosa* taxa. Field testing involves a multiple-stage approach to the improvement of disease resistance (*Marssonina brunnea*, *Melampsora larici-populina*, *Discosporium populeum* and poplar mosaic virus), growth rate and adaptation to local photoperiods. Newly released cultivars show biomass production levels of 40 dry t ha^{-1} achieved during 2-year coppice rotations on sites of good to moderately good agronomic quality (Paris *et al.*, 2011).

The Department of Forest Environment and Resources of the University of Tuscia at Viterbo has been conducting an *ex situ P. alba* conservation programme since 1988. *P. alba* genotypes totalling 350 have been assembled from provenances throughout the Italian peninsula and established in common gardens for the evaluation of genetic diversity in morphological and physiological traits (Sabatti, 1994; Sabatti *et al.*, 2001). Similarly, a *P. nigra ex situ* conservation programme has been established in collaboration with the Italian Forest Service encompassing about 350 genotypes. The university has also cooperated in the evaluation of *P. alba* genotypes for commercial biomass production; *P. alba* selections 'Marte' and 'Saturno' have received provisional registration with the Italian National Register of Basic Forest Materials (Mareschi *et al.*, 2005). Three-generation pedigrees (F_1, F_2 and backcrosses) of *P. alba* and *P. nigra* have also been created for the molecular study of the inheritance of several

Fig. 4.23. Alasia Franco Vivai *Populus* ×*canadensis* (*Populus deltoides* × *Populus nigra*) clonal field trial in Kwidzyn, Poland, showing a superior 1-year-old bioenergy selection. Alasia Franco Vivai is the largest developer of new plantation poplar cultivars for Europe. Photo courtesy of Brian Stanton, GreenWood Resources.

adaptive traits (Beritognolo *et al.*, 2008; Gaudet *et al.*, 2008). The University of Tuscia is also a leading research centre for the study of sex determination in the genus; the critical locus has been located on linkage group XIX (Gaudet *et al.*, 2008; Paolucci *et al.*, 2010).

SALIX. Twelve *Salix* species are found growing in Italy across a diversity of sites from the alluvial to the mountainous. *S. alba* and *S. triandra* are commonly found in northern Italy on alluvial sites. Not as common, *S. elaeagnos* and *S. purpurea*, are also found on alluvial sites, but at higher elevations or on plains where they tolerate drier summer months. *S. pedicellata* and *S. triandra* are found in the south of Italy and are adapted to warmer temperatures. *S. cinerea* is frequently found growing in areas that exhibit prolonged flooding. *S. apennina*, *S. appendiculata*, *S. caprea*, *S. helvetica* and *S. foetida* are adapted to the colder, windy areas of alpine environments.

Willow has a long history of cultivation for firewood, vineyard poles, wicker products and as ornamentals. The main area of willow production is located in northern Italy's Po Valley. Plantings of *S. alba* var. *vitellina* and *S.* ×*fragilis* are utilized to produce wicker furniture in the Veneto region. The main focus now is to breed and select clones for short-rotation production of bioenergy feedstock. Experimental, high-density, short-rotation biomass plantations established in Italy's Piedmont, Lombardy and Umbria regions exhibit reasonably high production rates. There is also interest in exploiting willow for veneer and sawlog production, as well as its capacity for several environmental applications, including the restoration of quarries and landslides, the stabilization of eroded riverbanks and the phytoremediation of contaminated sites (Zacchini *et al.*, 2009).

CRA-PLF's Mezzi Experimental Station maintains a willow germplasm collection of 590 genotypes representing 30 species, principally of European and Asian provenances, including *S. jessoensis* and *S. babylonica*. This collection builds on Italy's first collection assembled by Sivio May in 1957 (May, 1962, 1981). The collection also includes full-sib *S. alba* families (Vietto and Cagelli, 2006). The Mezzi Experimental Station manages the collection as a conservation bank and also offers seed, pollen and floral branches to worldwide breeding programmes. A second Italian repository of *Salix* germplasm is an alpine collection that has been established by Veneto Agricoltura in the Veneto region.

Over the past 20 years, CRA-PLF has maintained a breeding and selection programme for three main species: *S. alba*, *S. babylonica* and *S. matsudana* (Bisoffi *et al.*, 1991; Bisoffi and Gullberg, 1996). The selection objective is biomass yield under short-rotation coppice systems. Interspecific progeny of *S. alba* bred with *S. matsudana*, *S. jessoensis* and *S. babylonica* have displayed fast growth, drought tolerance, pest resistance and good coppice ability. Clonal progeny are banked at the Mezzi Experimental Station in advance of commercial propagation.

S. alba is the only willow taxon included in the EU directive's list of reproductive forestry materials that are regulated in Italy. *S. alba* cultivars must undergo 10 years of field experimentation using a prescribed series of trials to determine productivity, pest reactions and wood quality. Open-pollinated *S. matsudana* clones 'S76-008', 'S78-003', 'Drago' and 'Levante', *S. babylonica* × *S. alba* clone '131-25' and *S. alba* clone 'S164-017' are in experimental development. 'Drago' and 'Levante' have been patented by the European Community Plant Variety Office for commercialization in Europe.

Latvia

SALIX. Latvia is not a member of the IPC but is actively involved in *Salix* domestication. Latvia's main export commodities include wood chips and pellets. Over the past 20 years, as much as 1.5 million ha of agricultural land have been fallowed or taken out of crop production. This has created an opportunity to grow short-rotation woody crops on a sizeable land base in service of Latvia's domestic and export markets. Lativia joined the EU in 2004 and many farmers considered willow production as a source of biofuel feedstock. However, Latvian farmers have been reluctant to establish woody biomass plantations, because of the delay in revenue owing to a 5-year crop rotation and a heretofore volatile wood market.

Research is being conducted at the Latvia State Forest Research Institute to utilize wastewater sludge to increase the productivity of short-rotation crop plantations on abandoned farmland. The following cultivars developed in Sweden and marketed by Agrobransle are featured in Latvian willow production: *S. schwerinii* × *S. viminalis* 'Tora', (*S. schwerinii* × *S. viminalis*) × *S. viminalis* 'Torhild' and *S. viminalis* × (*S. schwerinii* × *S. viminalis*) 'Sven'. Ash from the conversion of bioenergy is used in increasing plantation productivity and guaranteeing system sustainability. 'Tora' and 'Torhild' are the most promising clones producing an estimated 28–31 dry t ha^{-1} over a 3-year rotation cycle.

The Netherlands

POPULUS. *Populus* cultivation is commonly encountered throughout the Netherlands, although interest among the country's farmers is lessening. Approximately 16,000 ha of plantations and 15,000 ha of roadside and field-side plantings constitute the country's poplar estate. Despite the decrease in plantation development, domestication remains an ongoing national emphasis, continuing an effort that began in 1948. Over this period, controlled hybridization has resulted in about 500,000 genotypes of various taxa and generations bred from provenance collections of *P. nigra*, *P. deltoides* and *P. trichocarpa*. First-generation hybridization of the *P.* ×*canadensis* and *P.* ×*generosa* taxa has predominated. There is also keen interest in interspecific hybridization of *P. maximowiczii*.

About 10,000 genotypes have been screened for resistance to *Melampsora larici-populina* and *Marssonina brunnea* and are undergoing second-stage field evaluation for *Xanthomonas populi* resistance and growth rate. The performance – yield, stem form, branch architecture and wind tolerance – of selections considered for commercial use are verified in trials utilizing large monoclonal plots lasting 15–20 years (Fig. 4.24). The latest release includes four *P.* ×*canadensis* cultivars ('Albelo', 'Degrosso', 'Polargo' and 'Sanosol'), many of which have shown broad adaptability in tests throughout Europe. These join the widely used *P.* ×*canadensis* cultivar, 'Koster', that continues to exhibit good performance since its release in 1989. The Netherlands is active in conserving its native *P. nigra* genetic resource. Several hundred genotypes have been collected, identified with molecular markers and maintained in a gene bank at the Centre for Genetic Resources at Wageningen. The collection supports both habitat restoration projects as well as controlled hybridization programmes.

Fig. 4.24. *Populus* domestication programmes in the Netherlands have traditionally featured selections for veneer log rotations. Shown is selected *Populus* ×*canadensis* (*Populus deltoides* × *Populus nigra*) cultivar 'Polargo' at age 20, with excellent stem form, crown structure and growth rate. Photo courtesy of Sven de Vries, Centre for Genetic Resources, Netherlands.

Poland

SALIX. Willow has been grown for decades in Poland for wattle, vegetation filters and for river-bank stabilization. Now, interest in willow as an energy crop is rising steadily, featuring *S. viminalis* and its hybrids. Approximately 6000 ha are being used to grow willow feedstock plantings for domestic energy markets: large-scale power plants, combined heat and power units, pellets and briquettes. Energy plantings are rather small for the most part, distributed throughout the country and established on a variety of soil types. Despite the progress that has been made, willow cultivation for energy is still regarded as a new agricultural venture attended by several risks such as high establishment costs, lack of harvesting equipment and insufficient production profitability. At the same time, and on a smaller scale, willow is being used to phytoremediate industrial waste sites (Szczukowski *et al.*, 2004).

Willow research and breeding over the past 20 years has been led by the Department of Plant Breeding and Seed Production in the University of Warmia and Mazury in Olsztyn. Approximately 150 genotypes of several *Salix* species are used for controlled breeding (Szczukowski *et al.*, 2004). The improvement process emphasizes pest resistance and soil adaptability. In 2003, the first three Polish varieties of *S. viminalis*, 'Start', 'Sprint' and 'Turbo', were registered in the Research Centre for Cultivar Testing. Other studies at the University of Warmia and Mazury are addressing yield as a function of different coppice rotations. In one study, a 21% increase in yield was observed for those harvested on a 4-year cycle compared to the cumulative yield of four annual harvests (Stolarski *et al.*, 2008).

Russia

POPULUS. Most *Populus* domestication work in Russia is concentrated west of the Ural Mountains. Naturally occurring members of the genus include the Eurasian species, *P. alba*, *P. nigra* and *P. tremula*, and the Asian species, *P. laurifolia*, *P. maximowiczii* and *P. suaveolens*. *P. tremula* predominates with 21.5 million ha of natural stands. Its domestication has a rather lengthy history in Russia, dating to the mid-1930s (Al'benski and Delitsina, 1934). The main centres of poplar research today are: (i) the Voronezh State Forest Technical Academy; (ii) the Central Research Institute of Forest Genetics and Breeding, Petrozavodsk State University; and (iii) the Karelian Research Centre of the Institute of Forestry of the Russian Academy of Science. Planting programmes rely mainly on cultivars selected from the *P. alba* × *P. alba*, for example 'Sowietica Pyramidalis', 'Bolleana Kamyshinsky', 'Bolide', 'Veduga'; *P. nigra*, for example 'Rozier', 'Thevestina', 'Pioner'; and the *P.* ×*canadensis* taxa, for example 'Brabantica', 'Bachelieri', 'Gelrica', 'Marilandica', 'Regenerata', 'Robusta', 'Sacrau-59', 'Serotina'. Superior parental phenotypes of *P. alba*, *P. nigra* and *P. tremula* have been identified for ongoing intraspecific, i.e. *P. alba* × *P. alba*, *P. nigra* × *P. nigra*, *P. tremula* × *P. tremula*, and interspecific, i.e. *P.* ×*canadensis* and *P.* ×*canescens*, hybridization (Fig. 4.25). Superior clonal selections are sought with improvements in growth rate, decay resistance, for example *Fomes igniarius* and *Phellinus tremulae*, wood quality and stem and crown form. Genetic gains are sought in spring and autumn phenology to develop adaptability to the steppe and semi-desert regions of the north Caspian region.

SALIX. The Russian *Salix* genetic resource is considerable in terms of land area and its breadth of species. About 5.3 million ha of natural stands contain 135 *Salix* species (Skvortsov, 1968). The most widespread distribution belongs to species in the subgenus *Salix* and includes *S. alba*, *S. babylonica*, *S. fragilis*, *S. pentandra* and *S. triandra* (Nazarov, 1936; Morozov, 1966; Skvortsov, 1968). Russian species belonging to the subgenus *Vetrix* are *S. acutifolia*, *S. caprea*, *S. cinerea*, *S. dasyclados*, *S. myrsinifolia*, *S. purpurea*, *S. schwerinii* and *S. viminalis*. Finally, *S. glauca*, *S. herbacea*, *S. myrtilloides*, *S. polaris* and *S. reticulata*, *S. retusa* from the subgenus *Chamaetia* are important components of the country's genetic resource. The most valuable cultivars combine frost resistance and high rates of biomass productivity, and are grown for energy feedstock, tannin, withes and ornamental and environmental applications. A research priority is the restoration and conservation of natural stands, and creation of new collections for *ex situ* conservation. More than 200 species and cultivar accessions reside in the combined *Salix* collections of Russia's main botanical gardens and academic institutions.

A variety of organizations in Russia have undertaken *Salix* hybridization and selection programmes to improve biomass production, stem form, rod quality for wicker use, tannin content and quality, ornamental value and frost resistance (Veresin *et al.*, 1974; Starova, 1980; Chumakov, 1989; Maximenko, 2002; Tsarev, 2005). The Voronezh Central Research Institute of Forest Genetics and Breeding is one of the leading research institutions. A large number of species have been brought into the country's breeding programme, including frost-resistant *S. acutifolia*, *S. caprea*, *S. dasyclados*, *S. pentandra*, *S. purpurea*, *S. schwerinii*, *S. triandra* and *S. viminalis*.

Fig. 4.25. *Populus alba* breeding constitutes a major portion of Russia's *Populus* domestication programme. Accordingly, identification of superior *P. alba* phenotypes is a national priority. This includes specimens such as this one found in the Khoper State Reserve in the Voronezh region that measured 43 m in height and 110 cm in breast-height stem diameter at approximately 130 years of age. Photo courtesy of Anatoly Tsarev, Petrozavodsk State University.

Serbia

POPULUS. The Institute of Lowland Forestry and Environment at Novi Sad (formerly the Poplar Research Institute) leads Serbia's poplar domestication programme. *P.* ×*canadensis* is the featured taxon, although the institute has also worked with *P.* ×*canescens* (Guzina and Vujovic, 1986). Evaluation of *P.* ×*canadensis* seedling populations begins with nursery observations for disease and insect resistance and adventitious rooting. Resistance to aphids (*Aphidoidea* spp.), poplar leaf beetle (*Chrysomela populi*), poplar leaf miner (*Leucoptera sinuella*), black stem disease (*Dothichiza populea*), leaf spot (*Marssonina brunnea*) and leaf rust (*Melampsora* spp.) are all considered. A special project is to identify drought-tolerant genotypes through the use of indirect selection criteria, including leaf water potential, transpiration intensity, net photosynthetic rate, leaf area index and root–shoot ratio. Selected seedlings are replicated clonally in field trials across different soil types to assess the magnitude of genotype × environmental interactions. The final evaluation phase is conducted over a 15-year rotation and includes biomass productivity, wood density, calorific value and several chemical and mechanical wood properties. Testing also covers the identification of genotypes for environmental services. Criteria include the ability for phytoextraction of heavy metals, phytoremediation of nitrates and polycyclic aromatic hydrocarbons, and the presence of phytotoxic responses (Pilipovic *et al.*, 2005).

Spain

POPULUS. Spain's involvement with *Populus* domestication dates from the 1950s when *P.* ×*canadensis* cultivars 'Campeador', 'Negrito de Granada', 'Blanquillo', 'Canada Blanco' and 'Santa Fe' gained widespread use in the country's veneer and pulpwood industries. Later, an effort was launched to evaluate *P.* ×*canadensis* cultivars from other European programmes (Padro, 1987, 1992), as well as to develop a national breeding programme for *P. deltoides* and the endemic *P. nigra* and *P. alba* (Padro, 1987, 1992; Alba, 1992; Gonzalez-Antonanzas *et al.*, 2000; Grau *et al.*, 2000). Multiple-stage testing had relied on field trials of both randomized complete block and incomplete block designs and 12 years of evaluation (Padro, 1992; Maestro *et al.*, 2001a). With the growing importance of the bioenergy industry, clones are evaluated for calorific value and ash content, along with the standard array of agronomic characteristics (Sixto *et al.*, 2006, 2007; Alba *et al.*, 2007). Breeding of the native *P. nigra* and *P. alba* for adaptability to droughty and saline soils is in place for ecological restoration projects (Alba, 1992; Alba *et al.*, 2001; Sixto *et al.*, 2005). *P. nigra* cultivars 'Bordils' and 'Poncella' and *P. alba* cultivars 'Siberia Extremena' and 'Platero' are used in restoration projects where *P.* ×*canadensis* hybrids are disallowed (Maestro and Alba, 2008).

Spain's *in situ P. nigra, P. alba* and *P. tremula* conservation programmes are managed through their EUFORGEN association (Alba, 2000; Maestro *et al.*, 2001b) (Fig. 4.26). Molecular markers are used in: (i) cultivar identification (Alvarez *et al.*, 2000; De Lucas *et al.*, 2008); (ii) study of *P. alba* geographic variation patterns across river systems and elevation zones (Alba *et al.*, 2000, 2001); and (iii) assessment of allelic diversity within a core conservation collection of 26 *P. tremula* cultivars (Gomez *et al.*, 2003). Several academic and research institutions (UMA, IIAG-CSIC) have advanced *Populus* transformation protocols, along with techniques for molecular and physiological analysis of transgenic lines (Couselo and Corredoira, 2004; Jing *et al.*, 2004).

Sweden

POPULUS. Hybrid poplars have been cultivated in Sweden since the 1930s. There is renewed interest in *Aigeiros–Tacamahaca* intersectional

Fig. 4.26. A native riparian stand of *Populus alba* growing near Beas de Segura in the province of Jaen, Spain. Conservation of Spain's three endemic poplar species – *Populus alba*, *Populus nigra* and *Populus tremula* – is noteworthy, as each has an exceedingly broad geographic distribution that encompasses the Iberian Peninsula in the extreme south-west of Europe. Photo courtesy of Nuria Alba, Forest Ecology and Genetics, Centre of Forest Research, INIA.

hybridization to produce superior cultivars for short-rotation, renewable energy feedstock production to replace the *P. maximowiczii* × *P. trichocarpa* cultivar 'NE42', a staple of the *Populus* nursery industry (Perttu, 1998; Telenius, 1999). Test results indicate that stands of 'NE42' and the *P.* ×*generosa* cultivar 'Boelare' and others could produce 70–105 t ha^{-1} after 10–15 years when grown for biomass (Johansson and Karacic, 2011). The *P.* ×*generosa* taxon has not performed well, however, owing to extensive frost damage and stem canker, both likely a result of the poor adaptation of the *P. trichocarpa* parental sources, i.e. coastal Oregon and Washington provenances from 42° to 49°N latitude, to the conditions of the Swedish planting locations (56° to 69°N latitude) (Christersson, 1996). Therefore, in 1994, the Department of Short Rotation Forestry of the Swedish University of Agricultural Sciences in Uppsala acquired a new *P. trichocarpa* provenance collection from British Columbia in the vicinity of 53° to 58°N latitude (Christersson, 2006). Fifty of the best performing clones from preliminary trials are being retested at five locations throughout southern and central Sweden for growth rate, phenology and resistance to insect, disease, frost and drought. Initial results indicate good adaptability to the country's environments (Fig. 4.27). Ultimately, *P. trichocarpa* selections may be used directly as propagation stock or indirectly as breeding stock in developing interspecific *P.* ×*generosa* cultivars. The same 50 clones are also undergoing genetic transformation studies at the Umeå Scientific Plant Centre (Sundberg *et al.*, 1997). Target traits are biomass production and wood quality. Umeå is also active in identifying genes involved in xylogenesis that may affect the efficiency of energy conversion (Hertzberg *et al.*, 2001; Schrader *et al.*, 2004). Finally, interest is developing in the *P.* ×*wettsteinii* taxon for bioenergy production due to its strong root-suckering ability (Rytter, 2002; Rytter and Stener, 2003).

SALIX. *Salix* species constitute part of the natural flora in Sweden's temperate and arctic regions, and their products have been used in the country for centuries. During the 1700s, *S. viminalis* was introduced to Sweden from continental Europe and since then it has been cultured for basket making, fencing and small-scale

Fig. 4.27. A 1-year-old *Populus trichocarpa* clone trial located near Halmsted, Sweden, at approximately 57°N latitude. The test is one in a series of second-stage trials replicated across five locations in the southern and central regions of Sweden. The provenance pictured here is from Prince George, British Columbia, Canada (53°51′N latitude). Photo courtesy of Lars Christersson, Swedish University of Agricultural Sciences.

husbandry (Verwijst, 2001). Large-scale implementation of *Salix* culture in Sweden as a renewable energy source was brought about by the petroleum crisis of the early 1970s, and its adoption facilitated by the many district heating systems distributed throughout Sweden (Hoffmann and Weih, 2005). Sweden is one of the few countries in the world where willow cultivation exceeds that of poplar, with a total of 14,000 ha under cultivation.

A *Salix* genetic improvement effort was launched in 1978 with the identification of superior clones from native stands (Sennerby-Forsse *et al.*, 1983). The clone archive at the Swedish University of Agricultural Sciences (SLU) at Uppsala holds 600 *Salix* accessions collected across a broad region of Europe and Asia, including western Russia and Siberia. This collection constitutes a major resource utilized at SLU for its recurrent selection programme for enhanced biomass production (Gullberg, 1993) (Fig. 4.28). A second *Salix* breeding effort was undertaken in 1987 by Svalof Weibull AB, a private plant-breeding company that transitioned to Lantmannen SW Seed in 2010 (Larsson, 1997). Two major species – *S. viminalis* and *S. dasyclados* – are featured in the Swedish breeding programmes, but *S. caprea*, *S. daphnoides*, *S. ericocephala*, *S. schwerinii* and *S. triandra* are also recruited into hybridizations. The Swedish breeding programme focuses mainly on yield, but also on resistance to leaf rust, insects and frost. The breeding strategy in place is designed to release new varieties regularly, thereby minimizing the development of disease outbreaks associated with an over-reliance on a limited range of cultivars. Presently, seven willow cultivars are registered in Sweden for commercial use.

Fig. 4.28. Dr Ioannis Dimitriou stands next to a section of the willow collection at the Swedish University of Agricultural Sciences (SLU) in Uppsala. Photo courtesy of Lawrence Smart, Cornell University.

The commercial breeding programme is closely linked to SLU's molecular and ecophysiological research activities, e.g. Weih *et al.* (2008). Genetic linkage maps are available for the key species (Ronnberg-Wastljung, 2001; Tsarouhas *et al.*, 2002; Ronnberg-Wastljung *et al.*, 2003; Berlin *et al.*, 2010). Also, genetic markers for various agronomic traits including growth, phenology, drought, cold and pest resistance have been identified (Tsarouhas *et al.*, 2003, 2004; Ronnberg-Wastljung *et al.*, 2005, 2006, 2008; Weih *et al.*, 2006). Studies of key physiological traits – phenology, cold hardiness, water-use efficiency, tolerance of drought and heat, site adaptability – are being carried out as an adjunct to the country's selection practices (Nordh and Verwijst, 2004; Wikberg and Ögren, 2004, 2007; Nordh, 2005; Weih and Nordh, 2005; Weih *et al.*, 2006; Wikberg, 2006; Linderson *et al.*, 2007; Weih and Ronnberg-Wästljung, 2007; Weih and Bonosi, 2009; Bonosi *et al.*, 2010).

Turkey

POPULUS. *P. nigra*, *P. tremula*, *P. euphratica* and *P. alba* are native to Turkey, with *P. nigra* the most widely distributed species represented by its three subspecies, *nigra*, *caudina* and *usbekistanica*. The Poplar and Fast Growing Forest Trees Research Institute at Izmit is the leading institution that oversees the introduction and evaluation of exotic germplasm as well as Turkey's own hybridization work (Isik and Toplu, 2004). The national hybridization programme involves intra- and/or interspecific crosses among *P. deltoides*, *P. maximowiczii*, *P. nigra*, *P. trichocarpa* and *P.* ×*canadensis*. Genetic gains are being pursued in yield, adventitious rooting, wood quality, disease resistance

(*Xanthomonas populi*, *Marssonina brunnea*, poplar mosaic virus) and frost resistance (Toplu, 2005). Over the past 35 years, 350 selections of *P. deltoides* and various *Populus* interspecific taxa have been imported from Italy, the USA, the Netherlands, France and Belgium. Commercial clones used in Turkish plantations are *P.* ×*canadensis* cultivars 'I-214' and 'I-45/51', *P. deltoides* cultivars 'Samsun' (synonym 'I-77/51') and 'Izmit' and *P. nigra* cultivars 'Afghanica', 'Kocabey', 'Gazi', 'Anadolu', 'Geyve' and 'Behicbey' (Ayan and Sivacioglu, 2006). Ten *P. deltoides* × *P. deltoides* selections undergoing evaluation at the Izmit institute show potential yield improvements relative to the standard *P. deltoides* variety 'Samsun' (Ozel *et al.*, 2010). Production *P. nigra* plantations have relied to a very great extent on 'Kocabey' (Fig. 4.29). An *ex situ P. nigra* conservation programme is being led by the Izmit institute in which over 1000 genotypes have been archived in clonal nursery plantings at Ankara-Behicbey in Central Anatolia (Toplu, 2005).

UK

SALIX. The UK is a world leader in willow domestication. Much of today's breeding and selection activity began following World War II, when interest in fast-growing willow plantings grew as a source of pulp for the paper industry. Later, willow biomass was identified as a renewable energy source as a series of oil crises and coal shortages struck the country during the 1970s and 1980s. The Long Ashton Research Station in Somerset directed *Salix* domestication from 1922 through 2002, when projects and its germplasm collection were moved to Rothamsted Research. The Rothamsted collection is considered of international importance and is probably one of the world's largest, holding 1500 accessions representing 100 pure species and numerous interspecific taxa from Europe, Asia and North America. It is a dynamic repository making additional collections and sharing in the accessions of other institutions. It also now contains many of Europe's energy cultivars (Fig. 4.30).

The highest yielding cultivar grown in the UK is 'Resolution', a cross between two *S. viminalis* × *S. schwerinii* hybrids. Despite the success of 'Resolution', there is now an initiative to add more diversity to the country's programme, employing taxa beyond the *S. viminalis* × *S. schwerinii* taxon. Two very successful varieties of the *S. viminalis* × *S. dasyclados* taxon – 'Ashton Stott' and 'Ashton Parfitt' – were produced in 1988. Two other commercial taxa of unique construction –

Fig. 4.29. A 9-year-old stand of *Populus nigra* cultivar 'Kocabey' (synonym 'Tr. 77/10'). This selection is the foremost clone grown in commercial plantations in Central Anatolia, Turkey. Photo courtesy of Ferit Toplu, Poplar and Fast Growing Forest Trees Research Institute, Turkey.

Fig. 4.30. Willow breeders and geneticists, Ian Shield, Steve Hanley and William Macalpine, continue to characterize and exploit the UK National Willow Collection housed at Rothamsted Research. Photo courtesy of Lawrence Smart, Cornell University.

S. rehderiana × *S. dasyclados* and *S. eriocephala* × *S. dasyclados* – show substantial yield capabilities, with individual hybrid selections approaching 20 dry t ha^{-1} $year^{-1}$.

Ongoing diversity breeding is largely a reflection of the major threat posed to willow culture by evolving *Melampsora* pathogenicity, coupled with the reality that the majority of cultivars marketed for bioenergy throughout Europe rely on rust resistance derived from one Siberian *S. schwerinii* genotype. *Melampsora* resistance appears to be conveyed by a major gene modified by several minor genes. It is presumed that one resistance mechanism will not provide durable resistance; 'Ashton Stott' has become progressively more susceptible, and has lately been withdrawn from sale in the UK. Thus, breeding now relies increasingly on alternative sources of rust resistance such as *S. hookeriana* from North America and *S. miyabeana* and *S. udensis* (formerly *S. sachalinensis*) from Asia. Generally, accessions from Europe are more susceptible than those from elsewhere (Pei *et al.*, 2004).

Rust-resistance breeding at Rothamsted is supported by a strong research effort in molecular genetics. QTLs for rust resistance have been identified for *S. schwerinii* (Hanley *et al.*, 2011), as well as QTLs associated with yield components and saccharification rates (Brereton *et al.*, 2010). Furthermore, Rothamsted manages 12 different mapping families and an association mapping population consisting of *S. viminalis* accessions from across Europe.

4.3.4 South America

The genus *Populus* is not found in the southern hemisphere, with the exception of populations of *P. ilicifolia* native to Kenya and Tanzania. Similarly, only two *Salix* species – *S. humboldtiana* and *S. mucronata* – are native to the southern hemisphere. As exotic plantation species, however, *Populus* and *Salix* are yielding exceedingly good results in South America, especially in Argentina and Chile. This is significant, because South American nations are known globally as leaders in the production, marketing and science of exotic plantation forestry. The plantation production of Monterey pine (*Pinus radiata*) and various eucalyptus species (*Eucalyptus* spp.) throughout Argentina, Brazil, Chile and Uruguay has made significant contributions to the global wood products industry over the past 30 years. *Populus* and *Salix* plantations have not approached the economic importance of these two genera, but may do so some day. Historically, poplar has been grown in South America for veneer logs for the matchstick industry. This market remains a strong outlet, although poplar, along with willow, is now being grown increasingly for the pulp and paper and medium-density fibreboard industries. Initiatives are also under way to grow selections from both genera as feedstock for the renewable energy industry.

Argentina

POPULUS. *Populus* is grown in Argentina principally in the Paraná River delta along the Atlantic

Coast in the vicinity of 34°S latitude for the production of logs for both veneer and saw mills and for chips for pulp and paper and medium-density fibreboard mills. The delta's preferred taxon is *P. deltoides* from south-eastern USA provenances because of its resistance to *Septoria* stem canker and adaptation to the local photoperiod. Argentina's National Institute for Agricultural Technology (INTA) leads the country's *P. deltoides* domestication effort. Selected cultivars have originated from the USA's Mississippi River Valley and include 'ST66', 'ST67', 'ST91' and 'ST109'. Additional *P. deltoides* commercial selections are Italian cultivar 'I-72/51', Australian cultivars '106/60' and '129/60' and a local *P.* ×*canadensis* cultivar 'Ragonese INTA 22' (Fig. 4.31).

Growth rate, stem form and resistance to *Septoria* canker, *Melampsora medusae* leaf rust and wind throw are of paramount importance (Ares and Gutierrez, 1996). Resistance to *Platypus mutatus* stem borer is a selection criterion of increasing importance (Alfaro *et al.*, 2007). An INTA priority is to improve yield while increasing the diversity of *P. deltoides* germplasm originating in Mississippi, Oklahoma and Texas provenances of the USA. The main breeding strategy is *P. deltoides* recurrent breeding, although backcrossing of *P.* ×*canadensis* to *P. deltoides* is being explored. A 14,000-genotype *P. deltoides* collection from Mississippi is currently undergoing field evaluation, with 59 selections at an advanced stage of yield testing. Poplar cultivation is also prominent in the western province of Mendoza at approximately 33°S latitude, where over 16,000 ha are being managed, mainly with *P.* ×*canadensis* cultivars 'Conti-12' and 'Guardi', but also including *P. deltoides* cultivars 'Catfish 2', 'Catfish 5', 'Harvard' and 'Alton'.

SALIX. *Salix* plantations cover 51,000 ha in Argentina, the large majority of which is found in the Paraná River delta (SAGPyA, 2004). Argentina's only native willow, *S. humboldtiana*, has heretofore not been considered suitable for commercial production, although it is included in ongoing improvement efforts. Traditionally, willow plantation growing stock relied on genotypes introduced from other countries to meet Argentina's burgeoning wood needs (Cerrillo, 2005). Genotypes introduced directly for commercial usage belonged to two main taxa, *S. babylonica* var. *sacramenta* and *S. alba* var. *calva*;

Fig. 4.31. *Populus* cultivation in Argentina employs cultivars of the *Populus deltoides* and *Populus* ×*canadensis* taxa. *P. deltoides* selections are most commonly encountered in the Paraná River delta, where this stand of Italian *P. deltoides* cultivar 'I-72/51' was photographed. Photo courtesy of Brian Stanton, GreenWood Resources.

these were also employed in breeding programmes between 1950 and 1970. Between 1986 and 1989, a base of new genotypes was developed by interspecific hybridization among eight species – *S. alba*, *S. amygdaloides*, *S. babylonica*, *S. bondplandiana*, *S. ×fragilis*, *S. humboldtiana*, *S. matsudana* and *S. nigra* – funded by Argentina's CIEF project.

In 2004, a *Salix* domestication programme was initiated for the Paraná River delta as a collaborative effort of the Secretaría de Agricultura, Ganadería, Pesca y Alimentación (SAGPyA), INTA and the private company, Papel Prensa SA. Two hundred and forty genotypes, the majority belonging to the *S. matsudana* × *S. alba* taxon, are undergoing evaluation for growth rate, stem form, wood specific gravity, wood colour and fibre length. Improved selections are targeted for three main markets: newsprint, particleboard and sawn wood products. They would replace the cultivars currently in use in delta plantations – *S. babylonica* × *S.alba* 'A131-25' and 'A131-27' and *S. matsudana* × *S. alba* 'NZ26992', 'NZ26993' and 'A13-44' (Ragonese, 1989). Research at INTA is also focusing on improving willow micropropagation techniques (Garay *et al.*, 2005).

Chile

POPULUS. *Populus* has been cultivated in Chile's central valley between 32° and 36°S latitude in the Maule Region since 1939 for the manufacture of quality matchsticks and a variety of specialty products for domestic and international markets. The country's largest industrial operation, CAF El Alamo, is composed of 3000 ha managed with furrow irrigation (Ulloa and Villacura, 2005). Plantations have also been established along the flood plains of the coastal rivers in the vicinity of 39°S latitude near Valdivia. Both areas of operation have relied nearly exclusively on Italian *P. ×canadensis* cultivars 'I-154', 'I-214' and 'I-488', and *P. deltoides* cultivar 'I-63/51' (Fig. 4.32). Since 2001, a new initiative in poplar domestication has been directed by the University of Talca's Poplar Technology Centre (CTA) to expand the kinds of taxa available to Chilean *Populus* culture. The *P. ×generosa* taxon and assorted hybrids among *P. maximowiczii*, *P. trichocarpa* and *P. deltoides* are the main ones presently being considered by this initiative (Gonzalez, 2007). CTA imported over 2600 genotypes from North America and

Fig. 4.32. An exceedingly homogeneous stand of *Populus deltoides* cultivar 'I-63/51' in Chile's central valley being managed for veneer logs. The logs harvested from such stands contribute to manufacturing efficiencies because of the uniformity of their wood properties. Photo courtesy of Brian Stanton, GreenWood Resources.

Europe between 1999 and 2001. By 2002, clonal trials were established at several locations between 32° and 37°S latitude in the central valley under a project funded by the Chilean National Science Foundation. The main selection criteria are growth rate, stem form, wood density, resistance to aphids (*Chaitophorus leucomelas*) and leaf rust (*Melampsora medusae*) (Ramirez *et al.*, 2004). Selection for phytoremediation of mine tailings from the Chilean copper industry is a special feature of CTA's domestication project. CTA is also developing restricted maximum likelihood models to analyse clonal variation in populations replicated in time and space. GreenWood Resources Chile, SA is breeding and selecting cultivars of the *P.* ×*generosa* and the *P. deltoides* × *P. maximowiczii* taxa for a 7200 ha biomass project in Chile's Bío-Bío Region.

4.3.5 Australasia

Populus was introduced into Australasia between 1840 and 1850 (McIvor *et al.*, 2011). Despite its lengthy history, the genus has never been regarded throughout the continent as a commercial hardwood due, in part, to the impact of *Melampsora* leaf rust epidemics beginning 35 years ago. Moreover, Australia's well-developed *Pinus* plantation industry has reduced the need for poplar plantations. Currently, New Zealand is the only country with an active *Populus* domestication programme. There, poplar domestication has promoted the genus' use in environmental services. This involves breeding and selection, as well as the introduction of exotic cultivars following stringent phytosanitary import requirements. As with *Populus*, *Salix* was brought into New Zealand during the mid-1800s. Eleven species and assorted hybrids have since become naturalized and are now used in securing erodible soils and riverbanks (Harman, 2004).

New Zealand

POPULUS. There are approximately 500 ha in New Zealand managed for poplars using commercial *P. deltoides* × *P. yunnanensis* cultivars 'Veronese' and 'Kawa', *P.* ×*canadensis* cultivars 'Selwyn', 'Weraiti', 'Dudley' and 'Otahuao', and the *P.* ×*canadensis* × *P. yunnanensis* cultivar 'Toa'. Ongoing domestication seeks additional cultivars from different taxa that are suitable for use in hillside and bank stabilization and soil conservation projects and as windbreaks (Wilkinson, 2000; Dodd *et al.*, 2008). Today, the New Zealand Institute for Plant and Food Research is leading the breeding and selection. Initially, 343 clones were introduced from other countries over a 22-year period, from which 11 cultivars were identified for environmental plantings (Wilkinson, 2000). The Institute for Plant and Food Research maintains a collection of 200 clones for controlled breeding and research, including a long-term *P. trichocarpa* recurrent breeding programme (Wilkinson, 2000; McIvor *et al.*, 2011). Additionally, the *P. maximowiczii* × *P. nigra* taxon is being pursued because of its good resistance to *Melampsora* leaf rust and possum browsing (McIvor *et al.*, 2011). Resistance to *Marssonina* spp. is also an important selection criterion in New Zealand (Spiers, 1998). Recently two *P. maximowiczii* × *P. nigra* cultivars – 'NZ93-005-09' and 'NZ93-005-010' – were recommended for commercial release for farm forestry use, based on their broad climatic adaptability (Fig. 4.33).

4.4 Outlook

Since the publication of the FAO's *Poplars and Willows in Wood Production and Land Use* in 1980, the role that *Populus* and *Salix* genetic resources have assumed in the overall management of commercial operations has grown considerably as a consequence of: (i) the growing demand for a wide array of wood products delivered from intensively managed plantations, including fuelwood, pulping fibres, composites and veneers and sawn wood products; (ii) new technologies to convert biomass into liquid fuels; (iii) the use of poplars and willows by the environmental services industry in the remediation of contaminated soils, the treatment of municipal and industrial effluent, containment of landfill leachates and the sequestration of greenhouse gases (Felix *et al.*, 2008); and (iv) the reliance on agroforestry systems for the dual production of food and fibre crops (Rivest *et al.*, 2009). The long-term growth of these industries will remain highly dependent on how successfully *Populus* and *Salix* genetic resources are managed for the continued production of improved cultivars.

Fig. 4.33. *Populus maximowiczii* × *Populus nigra* varietal planting for hillside stabilization in New Zealand. This taxon is becoming increasingly important to the country's *Populus* domestication efforts, because it adapts to the region's diverse climatic and elevation zones. Photo courtesy of Ian McIvor, New Zealand Institute for Plant and Food Research Limited.

Such growth will also be impacted by the continuing destruction and fragmentation of riparian ecosystems and the *Populus* and *Salix* habitats they support. Paralleling this is a growing global movement to restore more natural river functions and the revitalization of wetlands. Profound influences like these point to the need for establishing a conservation imperative for the *Salicaceae*.

Fulfilment of several key programmatic and scientific domestication opportunities will ensure the future of sustainable *Populus* and *Salix* cultivation. Among these is less reliance on the repeated propagation and deployment of a limited number of cultivars, some of which have been in service for 50 years and more. This is, in many cases, a consequence of limited access to improved breeding populations of both native and exotic species. Thus, a coordinated plan of international cooperation for reciprocal hybridization services fostered by the exchange of pollen and seed, perhaps best facilitated by the poplar and willow genetics working groups of the FAO's IPC and the International Union of Forest Research Organizations (IUFRO), would help in the development of needed germplasm to ensure worldwide sustainable wood production. Any exchange arrangement could also become part of a cooperative population breeding strategy. This would be challenging from both a funding and a logistical perspective. It could, however, be extremely valuable by augmenting commonplace short-term, non-recurrent hybridization programmes with a plan that warrants long-term genetic advancement. Related to this is the need for a standardized and tractable experimental design protocol for multiple-stage field evaluation programmes. Finally, more widespread use of newer linear mixed models in analyses of genetic performances would be equally beneficial.

New genomic tools should help realize gains from operational poplar and willow improvement programmes. However, a clear path toward practical application of genomic selection is lagging. Genomic tools will prove most worthwhile for managing recurrent breeding populations if they facilitate the estimation of parental combining abilities and genetic distances among parents that can be used to identify unique haplotypes, leading to recombination for increased heterosis. Regardless of how genomic tools are used, field testing will continue to be an irreplaceable part of any genetic improvement programme – there is no substitute for it.

Beyond their utility in applied hybridization, molecular tools will become especially noteworthy for the insights they should provide into the genetic determinants of heterosis, the covariance structure of composite traits, and the genetic basis of genotype × environmental interactions. A similar opportunity resides in the application of genetically transformed cultivars. This field has advanced greatly, to the point where long-term research trials, while a necessary step to commercialization, are constrained by the regulatory process and the need to provide adequate containment controls. Devising a means to meet these needs without unduly hindering the science of poplar and willow genetic transformation should now become a priority.

Finally, a universal strategy for the characterization and conservation of critical *Populus* and *Salix* genetic resources, perhaps modelled on the integrated European *P. nigra* programme, should be proposed. Here again, the use of molecular tools to measure population heterogeneity and identify genes and gene frequencies relevant to operational breeding efforts is important. In view of limited funds for conservation initiatives, work in this arena will, by necessity, need to assess and prioritize species and populations carefully for their commercial and ecological values and the integrity of their current habitats. A partial list of critical *Populus* species at present includes *P. alba*, *P. ciliata*, *P. deltoides*, *P. euphratica*, *P. maximowiczii*, *P. nigra*, *P. simonii*, *P. tremula*, *P. tremuloides* and *P. yunnanensis*. *Salix* species of increasing societal value are *S. caprea*, *S. dasyclados*, *S. eriocephala*, *S. integra*, *S. purpurea*, *S. miyabeana*, *S. schwerinii* and *S. viminalis*. Comprehensive ecological studies of reasonably intact ecosystems of these and other species that still exist in places of North America, Europe and Asia should provide good guidance for sound conservation efforts in the more domesticated landscapes. A growing area of investigation is the interplay of the genomics of species that dominate their habitat compared with lesser-ranked species, while analysing mutualism, symbiosis, herbivory or parasitism (Whitham *et al.*, 2008). Such relationships are based in molecular and genomic evolution that ultimately defines ecosystem functionality, as seen with *P. angustifolia* and its associated species. This is a novel concept with far-reaching implications for effectively utilizing and conserving intact genetic resources in the *Salicaceae* (Whitham *et al.*, 2008).

Acknowledgements

We gratefully acknowledge the constructive reviews of this chapter by the following colleagues: Drs Reinhard F. Stettler, University of Washington (Professor Emeritus); Donald I. Dickmann, Michigan State University (Professor Emeritus); Stefano Bisoffi, Italian Ministry of Agriculture; Steven H. Strauss, Oregon State University; Berthold Heinze, The Federal Research and Training Centre for Forests, Austria; Marc Villar, The National Agricultural Research Institute, France; Jud G. Isebrands, Environmental Forestry Consultants, LLC; Randall J. Rousseau, Mississippi State University; Barbara R. Thomas, University of Alberta (Adjunct Professor); Francisco Zamudio-Arancibia, University of Talca, Chile; Li Shanwen, GreenWood Resources, China; and David B. Neale, University of California, Davis.

We also extend appreciation to the following contributors for sharing their experience and knowledge of *Populus* and *Salix* genetic resources in their respective countries: Silvia Cortizo, Teresa Cerillo, National Institute for Agricultural Technology, Argentina; Francisco Zamudio, University of Talca, Chile; Zhang Qiwen, Su Xiaohua, Chinese Academy of Forestry; Zhang Zhiyi, Beijing Forestry University, China; Man Shengjun, Liaoning Poplar Research Institute, China; Li Shanwen, GreenWood Resources, China; Shi Shizheng, Jiangsu Academy of Forestry, China; Devinder Khurana, Dr Y.S. Parmar University, India; Ramesh Dhiman, Western India Match Company Limited; Takayoshi Takeda, Oji Paper Company, Limited, Japan; Yeong Bon Koo, Korea Forest Research Institute; Hyun-Seok Lee, Chan-Hoon Ahn, Jae-Soon Yi, Kangwon National

University, Korea; Eun-Ju Cheong, USDA-ARS National Germplasm Resources Laboratory, USA; Ian McIvor, Plant and Food Research Limited, New Zealand; Berthold Heinze, Federal Research and Training Centre for Forests, Austria; Patrick Mertens, Ministère de la Région Wallonne, Belgium; Marijke Steenackers, Boudewijn Michiels, Institute for Forestry and Game Management, Belgium; Davorin Kajba, Faculty of Forestry, University of Zagreb, Croatia; Egbert Beuker, Finnish Forest Research Institute, Finland; Marc Villar, National Institute for Agricultural Research, France; George von Wuehlisch, Federal Research Centre for Forestry and Forest Products, Germany; Alwin Janssen, Northwest German Forest Research Station; Joszef Kiss, Institute of Genetics and Biotechnology, Saint Stephanus University, Hungary; Lorenzo Vietto, Agricultural Research Council, Casale Monferrato, Italy; Fabrizio Nardin, GreenWood Resources, Italy; Maurizio Sabatti, Tuscia University, Italy; Sara Bergante, Laura Rosso, Gianni Facciotto, Agricultural Research Council, Research Unit for Intensive Wood Production, Italy; Dagnija Lazdina, State Forest Research Institute, Latvia; Sven M.G. de Vries, Centre for Genetic Resources, the Netherlands; Marius Stolarski, University of Warmia and Mazury, Poland; Ewa Krasuska, Institute for Fuels and Renewable Energy, Poland; Anatoly Tsarev, Petrozavodsk State University, Russia; Sasa Orlovic, Institute of Lowland Forestry and Environment, Serbia; Nuria Alba, Forest Ecology and Genetics, Forest Research Centre, INIA, Spain; Lars Christersson, Martin Weih, Swedish University of Agricultural Sciences; Ian Shield, Centre for Bioenergy and Climate Change, Rothamsted Research, UK; Ferit Toplu, Poplar Research Institute, Turkey; Bill Schroeder, Dan Walker, Garth Inouye, Agriculture and Agri-Food Canada, Agroforestry Development Centre, Canada; Pierre Périnet, Quebec Ministry of Natural Resources and Wildlife, Canada; Jean Brouard, Western Boreal Aspen Cooperative, Canada; Barb Thomas, Genstat Consulting, Canada; Randy Rousseau, Mississippi State University, USA; Bernie McMahon, Andrew David, University of Minnesota, USA; Rick Hall, Iowa State University, USA.

Finally, B. Stanton thanks Ms Suzanne Cushing for copy-editing and proofreading.

Notes

[1]Recent taxonomic work has proposed the name *Salix gmelinii* Pallas in place of *Salix dasyclados* Wimmer (see Chapter 2, this volume); however, the *S. dasyclados* name is used throughout this chapter as being more widely recognized by willow breeders.

[2]The poplar cultivar ‘I-72/58’ (synonym ‘San Martino’) was originally thought to be a *Populus* ×*canadensis* hybrid but is now believed to be a *Populus deltoides* selection based on AFLP analysis (Lorenzo Vietto, CRA-PLF Agriculture Research Council, Italy, 2011, personal communication).

References

Ahuja, M.R. (1988) Differential growth response of aspen clones stored at sub-zero temperatures. In: Ahuja, M.R. (ed.) *Somatic Cell Genetics of Woody Plants.* Kluwer Academic Publications, Dordrecht, the Netherlands, pp. 173–180.

Alba, N. (1992) Mejora genetica de *Populus alba* L. In: Padró, A. (ed.) *Proceedings 19th Session of the International Poplar Commission, Zaragoza, Spain, 22–25 September, 2002,* Volume 2. International Poplar Commission, Salamanca, Spain, pp. 157–168.

Alba, N. (2000) Conservacion de Recursos Geneticos del Genero *Populus* en Espana. *Investigacion Agraria – Sistemas y Recursos Forestales* 2, 45–57.

Alba, N., Agundez, D. and Alia, R. (2000) Genetic variation in *Populus alba* L.: a comparison of isozyme markers and quantitative traits. In: Isebrands, J.G. and Richardson, J. (eds) *21st Session of the International Poplar Commission (IPC 2000). Poplar and Willow Culture: Meeting The Needs of Society and The Environment.* General Technical Report NC-215. US Department of Agriculture, Forest Service, North Central Research Station, St Paul, Minnesota, p. 3.

Alba, N., Maestro, C., Agundez, D. and Notivol, E. (2001) Advances in the preservation of *Populus nigra* L. In: van Dam, B. and Bordacs, S. (eds) *Proceedings of an International Symposium: Genetic Diversity in River Populations of European Black Poplar. Implications for Riparian Eco-system Management.* Szekszard, Hungary, 16–20 May 2001. Verlag C. Nyomda, Budapest, pp. 125–136.

Alba, N., Godoy, N. and Sixto, H. (2007) Growth and potential production of different *Populus alba* clones for biomass. In: Maniatis, K., Grimm, H.P., Helm, P. and Grassi, A. (eds) *Proceedings of Fifteenth European Biomass Conference and Exhibition, Berlin, Germany.* WIP-Renewable Energies, Munich, Germany, pp. 727–729.

Al'benskii, A.V. and Delitsina, A.V. (1934) Experiments and investigations. All-Union Institute of Forest Cultivation and Forest Melioration, Moscow, Russia. 2nd Issue, pp. 107–109 (*Plant Breeding Abstracts* 5, 62).

Albrectsen, B.R., Gardfjell, H., Orians, C.M., Murray, B. and Fritz, R.S. (2004) Slugs, willow seedlings and nutrient fertilization: intrinsic vigor inversely affects palatability. *Oikos* 105, 268–278.

Albrectsen, B.R., Guiterrez, L., Fritz, R.S., Fritz, R.D. and Orians, C.M. (2007) Does the differential seedling mortality caused by slugs alter the foliar traits and subsequent susceptibility of hybrid willows to a generalist herbivore? *Ecological Entomology* 32, 211–220.

Alfaro, R.I., Humble, L.M., Gonzalez, P., Villaverde, R. and Allegro, G. (2007) The threat of the ambrosia beetle *Megaplatypus mutatus* (Chapuis) (=*Platypus mutatus* Chapuis) to world poplar resources. *Forestry* 80, 471–479.

Allaby, M. (ed.) (1998) *A Dictionary of Plant Sciences.* Oxford University Press, Reading, UK.

Alvarez, A., Cervera, M.T., Agúndez, D., Alba, N., Martinez Zapater, J.M. and Grau, J.M. (2000) Identification of different commercial clones of *Populus* using AFLP. In: Isebrands, J.G. and Richardson, J. (eds) *21st Session of the International Poplar Commission (IPC 2000). Poplar and Willow Culture: Meeting The Needs of Society and The Environment.* General Technical Report NC-215. US Department of Agriculture, Forest Service, North Central Research Station, St Paul, Minnesota, p. 5.

Amichev, B.Y., Johnston, M. and Van Rees, K.C.J. (2010) Hybrid poplar growth in bioenergy production systems: biomass prediction with a simple process-based model (3PG). *Biomass and Bioenergy* 34, 687–702.

Antonetti, P.L.E. and Pinon, J. (1993) Somaclonal variation in poplar. *Plant Cell, Tissue, and Organ Culture* 35, 99–106.

Anstead, L. and Boar, R.R. (2010) Willow spiling: peview of streambank stabilisation projects in the UK. *Freshwater Reviews* 3, 33–47.

Arens, P., Coops, H., Jansen, J. and Vosman, B. (1998) Molecular genetic analysis of black poplar (*Populus nigra* L.) along Dutch rivers. *Molecular Ecology* 7, 11–18.

Ares, A. and Gutierrez, L. (1996) Selection of poplar clones for the Lower Valley of the Colorado River, Argentina. *Forestry* 69, 75–82.

Argus, G.W. (1974) An experimental study of hybridization and pollination in *Salix* (willow). *Canadian Journal of Botany* 52, 1613–1619.

Argus, G.W. (1997) Infrageneric classification of *Salix* (*Salicaceae*) in the New World. *Systematic Botany Monographs* 52, 1–121.

Ayan, S. and Sivacioglu, A. (2006) Review of the fast growing forest tree species in Turkey. *Boletin del CIDEU* 2, 57–71.

Aylott, M.J., Casella, E., Tubby, I., Street, N.R., Smith, P. and Taylor, G. (2008) Yield and spatial supply of bioenergy poplar and willow short-rotation coppice in the UK. *New Phytologist* 178, 358–370.

Bajaj, Y.P.S. (1979) Technology and prospects of cryopreservation of germplasm. *Euphytica* 28, 267–285.

Bayon, C., Pei, M.H., Ruiz, C., Hunter, T. and Karp, A. (2009) Genetic structure and population dynamics of a heteroecious plant pathogen *Melampsora larici-epitea* in short-rotation coppice willow plantations. *Molecular Ecology* 18, 3006–3019.

Bekkaoui, F., Mann, B. and Schroeder, B. (2003) Application for the identification and management of hybrid poplar accessions. *Agroforestry Systems* 59, 53–59.

Benea, V.I. (2003) Necessity of protection and conservation of native poplar genetic resources – *Populus alba* L., *Populus* ×*canescens* S.M., *Populus nigra* L., *Populus tremula* L., in the world context. *Revista Padurilor* 118, 10–16.

Beritognolo, I., Sabatti, M., Brosché, M. and Scarascia Mugnozza, G. (2008) Functional genomics to discover genes for salt tolerance in annual and perennial plants. In: Abdelly, C., Ozturk, M., Ashraf, M. and Grignon, C. (eds) *Biosaline Agriculture and Salinity Tolerance in Plants.* Birkhäuser Verlag AG, Basel, Switzerland, pp. 273–286.

Berlin, S., Lagercrantz, U., von Arnold, S., Ost, T. and Ronnberg-Wastljung, A. (2010) High-density linkage mapping and evolution of paralogs and orthologs in *Salix* and *Populus*. *BMC Genomics* 11, 129 (doi:10.1186/1471-2164-11-129).

Berrang, P., Karnosky, D.F., Mickler, R.A. and Bennett, J.P. (1986) Natural selection for ozone tolerance in *Populus tremuloides*. *Canadian Journal of Forest Research* 16, 1214–1216.

Bisoffi, S. (1989) Recent developments in poplar selection and propagation techniques. In: Hessian Forest Research Station (ed.) *Proceedings of IUFRO Working Party S2.02.10, October 2–6, 1989*. Institute of Forest Tree Breeding of the Hessian Forest Research Station and Research Institute of Fast Growing Tree Species, Hann. Münden, Germany, pp. 18–45.

Bisoffi, S. and Gullberg, U. (1996) Poplar breeding and selection strategies. In: Stettler, R.F., Bradshaw, H.D. Jr, Heilman, P.E. and Hinckley, T.M. (eds) *Biology of Populus and its Implications for Management and Conservation*. National Research Council of Canada Research Press, Ottawa, pp. 139–158.

Bisoffi, S., Cagelli, L., Baralis, P., Calligari, P., Girino, P.V., Giorcelli, A., *et al.* (1991) Progetto: 'Miglioramento genetico del *Salix alba*' – Protocollo del progetto approvato il 07-1990 – Istituto di Sperimentazione per la Pioppicoltura, Casale Monferrato, Italy, pp. 30

Bittsanszky, A., Komíves, T., Gullner, G., Gyulai, G., Kiss, J., Heszky, L., *et al.* (2005) Ability of transgenic poplars with elevated glutathione content to tolerate zinc (2+) stress. *Environment International* 31, 251–254.

Boerjan, W. (2005) Biotechnology and the domestication of forest trees. *Current Opinion in Biotechnology* 16, 159–166.

Boerjan, W., Baucher, M., Chabbert, B., Petit-Conil, M., Leple, J.-C., Pilate, G., *et al.* (1997) Genetic modification of lignin biosynthesis in quaking aspen and poplar. In: Klopfenstein, N.B., Chun, Y.W., Kim, M.-S. and Ahuja, M.R. (eds) *Micropropagation, Genetic Engineering, and Molecular Biology of Populus*. Section V, Chapter 26. General Technical Report RM-GTR-297. USDA Forest Service, Rocky Mountain Forest and Range Experiment Station, Fort Collins, Colorado, pp. 193–205.

Boes, T.K. and Strauss, S.H. (1994) Floral phenology and morphology of black cottonwood, *Populus trichocarpa* (*Salicaceae*). *American Journal of Botany* 81, 562–567.

Bonosi, L., Ghelardini, L. and Weih, M. (2010) Growth responses of 15 *Salix* genotypes to temporary water stress are different from the responses to permanent water shortage. *Trees* 24, 843–854.

Bordacs, S., Borovics, A. and Bach, I. (2002) Genetic diversity of natural populations and gene bank of black poplar in Hungary. In: van Dam, B.C. and Bordacs, S. (eds) *Genetic Diversity in River Populations of European Black Poplar. Proceedings of the International Symposium, Szesksard, Hungary, 16–20 May, 2001*. Verlag C. Nyomda, Budapest, pp. 93–106.

Bradshaw, H.D. Jr and Stettler, R.F. (1995) Molecular genetics of growth and development in *Populus*. IV. Mapping QTLs with large effects on growth, form, and phenology traits in a forest tree. *Genetics* 139, 963–973.

Bradshaw, H.D. Jr and Strauss, S.H. (2001) Breeding strategies for the 21st century: domestication of poplar. In: Dickmann, D.I., Isebrands, J.G., Eckenwalder, J.E. and Richardson, J. (eds) *Poplar Culture in North America*. National Research Council of Canada Research Press, Ottawa, pp. 383–394.

Brereton, N.J.B., Pitre, F.E., Hanley, S., Ray, M.J., Karp, A. and Murphy, R.J. (2010) QTL mapping of enzymatic saccharification in short rotation coppice willow and its independence from biomass yield. *Bioenergy Research* 3, 251–261.

Broberg, C.L., Inkster, J.A.H. and Borden, J.H. (2010) Phenological and chemical differences among hybrid poplar clones (*Salicaceae*) varying in resistance to *Cryptorhynchus lapathi* (L.) (Coleoptera: Curculionidae). *Biochemical Systematics and Ecology* 38, 29–48.

Brunner, A.M., Busov, V.B. and Strauss, S.H. (2004) Poplar genome sequence: functional genomics in an ecologically dominant plant species. *Trends in Plant Science* 9, 49–56.

Brunner, A.M., Li, J., DiFazio, S.P., Shevchenko, O., Montgomery, B.E., Mohamed, R., *et al.* (2007) Genetic containment of forest plantations. *Tree Genetics and Genomes* 3, 75–100.

Brunsfeld, S.J., Soltis, D.E. and Soltis, P.S. (1992) Evolutionary patterns and processes in *Salix* sect. *Longifoliae*: evidence from chloroplast DNA. *Systematic Botany* 17, 239–256.

Bunn, S.M., Rae, A.M., Herbert, C.S. and Taylor, G. (2004) Leaf-level productivity traits in *Populus* grown in short rotation coppice for biomass energy. *Forestry* 77, 307–323.

Busov, V.B., Brunner, A.M., Meilan, R., Filichkin, S., Ganio, L., Gandhi, S., *et al.* (2005) Genetic transformation: a powerful tool for dissection of adaptive traits in trees. *New Phytologist* 167, 9–18.

Cagelli, L. and Lefevre, F. (1995) The conservation of *Populus nigra* L. and gene flow with cultivated poplars in Europe. *Forest Genetics* 2, 135–144.

Cameron, K.D., Phillips, I.S., Kopp, R.F., Volk, T.A., Maynard, C.A., Abrahamson, L.P., *et al.* (2008) Quantitative genetics of traits indicative of biomass production and heterosis in 34 full-sib F_1 *Salix eriocephala* families. *Bioenergy Research* 1, 80–90.

Cao, C.-W., Liu, G.-F., Wang, Z.-Y., Yan, S.-C., Ma, L. and Yang, C.-P. (2010) Response of the gypsy moth, *Lymantria dispar* to transgenic poplar, *Populus simonii* × *P. nigra*, expressing fusion protein gene of the spider insecticidal peptide and *BT*-toxin C-peptide. *Journal of Insect Science* 10, Article 200, pp.1–13.

Castiglione, S., Wang, G., Damiani, G., Bandi, C., Bisoffi, S. and Sala, F. (1993) RAPD fingerprints for identification and for taxonomic studies of elite poplar (*Populus* spp.) clones. *Theoretical and Applied Genetics* 87, 54–59.

Castiglione, S., Cicatelli, A., Lupi, R., Patrignani, G., Fossati, T., Brundu, G., *et al.* (2010) Genetic structure and introgression in riparian populations of *Populus alba* L. *Plants Biosystems* 144, 656–668.

Cerrillo, T. (2005) Subprograma Sauces en el Delta. Libro: Mejores Árboles para Más Forestadores. PPMPM, Convenio SAGPyA – INTA. Publicado por Proyecto Forestal de Desarrollo. Dic. 2005. Buenos Aires.

Cervera, M.-T., Storme, V., Ivens, B., Gusmao, J., Liu, B.H., Hostyn, V., *et al.* (2001) Dense genetic linkage maps of three *Populus* species (*Populus deltoides*, *P. nigra*, and *P. trichocarpa*) based on AFLP and microsatellite markers. *Genetics* 158, 787–809.

Cervera, M.T., Sewell, M.M., Faivre-Rampant, P., Storme, V. and Boerjan, W. (2004) Genome mapping in *Populus*. In: Kumar, S. and Fladung, M. (eds) *Molecular Genetics and Breeding of Forest Trees*. The Haworth Press, Philadelphia, Pennsylvania, pp. 387–410.

Cervera, M.T., Storme, V., Soto, A., Ivens, B., Van Montagu, M., Rajora, O.P., *et al.* (2005) Intraspecific and interspecific genetic and phylogenetic relationships in the genus *Populus* based on AFLP markers. *Theoretical and Applied Genetics* 111, 1440–1456.

Chauhan, N., Negi, M.S., Sabharwal, V., Khurana, D.K. and Lakshmikumaran, M. (2004) Screening interspecific hybrids of *Populus* (*P. ciliata* × *P. maximowiczii*) using AFLP markers. *Theoretical and Applied Genetics* 108, 951–957.

Cheema, G.S. (1989) Somatic embryogenesis and plant regeneration from cell suspension and tissue cultures of mature himalayan poplar (*Populus ciliata*). *Plant Cell Reports* 8, 124–127.

Chen, J., Dai, L.-Y., Wang, X.-P., Tian, Y.-C. and Lu, M.-Z. (2005) The *cry3Aa* gene of *Bacillus thuringiensis* Bt886 encodes a toxin against long-horned beetles. *Applied Microbiology and Biotechnology* 67, 351–356.

Chen, S., Li, J., Fritz, E., Wang, S. and Huttermann, A. (2002) Sodium and chloride distribution in roots and transport in three poplar genotypes under increasing NaCl stress. *Forest Ecology and Management* 168, 217–230.

Chenault, N., Arnaud-Haond, S., Juteau, M., Valade, R., Almeida, J.-L., Villar, M., *et al.* (2011) SSR-based analysis of clonality, spatial genetic structure and introgression from the Lombardy poplar into a natural population of *Populus nigra* L. along the Loire River. *Tree Genetics and Genomes* 7, 1249–1262.

Christersson, L. (1996) Future research on hybrid aspen and hybrid poplar cultivation in Sweden. *Biomass and Bioenergy* 11, 109–113.

Christersson, L. (2006) Biomass production of intensively grown poplars in the southernmost part of Sweden: observation of characters, traits and growth potential. *Biomass and Bioenergy* 30, 497–508.

Chumakov, V.V. (1989) Recommendation on assortment and technology of creation of plantations of high tannin willows in the European part of RSFSR. Ministry of Forestry of RSFSR, Moscow. (In Russian)

Cocozza, C., Lasserre, B., Giovannelli, A., Castro, G., Fragnelli, G. and Tognetti, R. (2009) Low temperature induces different cold sensitivity in two poplar clones (*Populus* ×*canadensis* Moench 'I-214' and *P. deltoides* Marsh. 'Divina'). *Journal of Experimental Botany* 60, 3655–3664.

Confalonieri, M., Balestrazzi, A. and Bisoffi, S. (1994) Genetic transformation of *Populus nigra* by *Agrobacterium tumefaciens*. *Plant Cell Reports* 13, 256–261.

Constabel, C.P., Yip, L., Patton, J.J. and Christopher, M.E. (2000) Polyphenol oxidase from hybrid poplar. Cloning and expression in response to wounding and herbivory. *Plant Physiology* 124, 285–295.

Cooper, D.T. (1980) Study plan: cooperative cottonwood advanced clonal tests. Report FS-SO-1402-1.19. USDA Forest Service, Southern Forest Experiment Station, Pineville, Louisiana.

Cottrell, J.E., Krystufek, V., Tabbener, H.E., Milner, A.D., Connoly, T., Sing, L., *et al.* (2005) Postglacial migration of *Populus nigra* L.: lessons learnt from chloroplast DNA. *Forest Ecology and Management* 219, 293–312.

Couselo, J.L. and Corredoira, E. (2004) Transformacion genetica de *Populus tremula* × *tremuloides* con la secuencia AtPCS1 para su uso en programas de fitorremediacion. *Revista Academia Galega Ciencias* 23, 79–94.
Cram, W.H. (1960) Performance of seventeen poplar clones in south central Saskatchewan. *The Forestry Chronicle* 36, 204–224.
Cronk, Q.C.B. (2005) Plant eco-devo: the potential of poplar as a model organism. *New Phytologist* 166, 39–48.
David, A. and Anderson, P. (2002) Aspen and Larch Genetics Cooperative 2002 Annual Report. Report 13. Department of Forest Resources, University of Minnesota, St Paul, Minnesota.
Dayanandan, S., Rajora, O.P. and Bawa, K.S. (1998) Isolation and characterization of microsatellites in trembling aspen (*Populus tremuloides*). *Theoretical and Applied Genetics* 96, 950–956.
De Block, M. (1990) Factors influencing the tissue culture and the *Agrobacterium tumefaciens*-mediated transformation of hybrid aspen and poplar clones. *Plant Physiology* 93, 1110–1116.
De Lucas, A.I., Santana, J.C., Recio, P. and Hildalgo, E. (2008) SSR-based tool for identification and certification of commercial *Populus* clones in Spain. *Annals of Forest Science* 65, Article 107, 2–7.
de Vries, S.M.G. (2001) Conservation of natural ecosystems of poplar and willows. *The Forestry Chronicle* 77, 255–257.
Devantier, Y.A., Moffatt, B., Jones, C. and Charest, P.J. (1993) Microprojectile-mediated DNA delivery to the *Salicaceae* family. *Canadian Journal of Botany* 71, 1458–1466.
Dickmann, D.I. and Keathley, D.E. (1996) Linking physiology, molecular genetics, and the *Populus* ideotype. In: Stettler, R.F., Bradshaw, H.D. Jr, Heilman, P.E. and Hinckley, T.M. (eds) *Biology of Populus and its Implications for Management and Conservation.* National Research Council of Canada Research Press, Ottawa, pp. 491–514.
Dickmann, D.I., Gold, M.A. and Flore, J.A. (1994) The ideotype concept and the genetic improvement of tree crops. *Plant Breeding Reviews* 12, 163–193.
Dieters, M.J., Nikles, D.G., Toon, P.G. and Pomroy, P. (1997) Genetic parameters for F_1 hybrids of *Pinus caribaea* var. *hondurensis* with both *Pinus oocarpa* and *Pinus tecunumanii. Canadian Journal of Forest Research* 27, 1024–1031.
Dillen, S.Y., Monclus, R., Barbaroux, C., Bastien, C., Ceulemans, R., Dreyer, E., *et al.* (2011) Is the ranking of poplar genotypes for leaf carbon isotope discrimination stable across sites and years in two different full-sib families? *Annals of Forest Science* 68, 1265–1275.
Dinus, R.J. (2000) Genetic modification of short rotation poplar biomass feedstock for efficient conversion to ethanol. Report ORNL/Sub99-4500007253/1. Oak Ridge National Laboratory, Oak Ridge, Tennessee.
Dodd, M.B., Thorrold, B.S., Quinn, J.M., Parminter, T.G. and Wedderburn, M.E. (2008) Improving the economic and environmental performance of a New Zealand hill country farm catchment: 1. Goal development and assessment of current performance. *New Zealand Journal of Agricultural Research* 51, 127–141.
Donahue, R.A., Davis, T.D., Michler, C.H., Riemenschneider, D.E., Carter, D.R., Marquardt, P.E., *et al.* (1994) Growth, photosynthesis, and herbicide tolerance of genetically modified hybrid poplar. *Canadian Journal of Forest Research* 24, 2377–2383.
Dorn, R.D. (1976) A synopsis of American *Salix. Canadian Journal of Botany* 54, 2769–2789.
Doty, S.L. (2008) Enhancing phytoremediation through the use of transgenics and endophytes. *New Phytologist* 179, 318–333.
Doty, S.L., Shang, T.Q., Wilson, A.M., Tangen, J., Westergreen, A.D., Newman, L.A., *et al.* (2000) Enhanced metabolism of halogenated hydrocarbons in transgenic plants containing mammalian cytochrome P450 2E1. *Proceedings of the National Academy of Sciences* 97, 6287–6291.
Doty, S.L., James, C.A., Moore, A.L., Vajzovic, A., Singleton, G., Ma, C., *et al.* (2007) Enhanced phytoremediation of volatile environmental pollutants with transgenic trees. *Proceedings of the National Academy of Sciences* 104, 16816–16821.
Dowkiw, A. and Bastien, C. (2007) Presence of defeated qualitative resistance genes frequently has major impact on quantitative resistance to *Melampsora larici-populina* leaf rust in *P.* ×*interamericana* hybrid poplars. *Tree Genetics and Genomes* 3, 261–274.
Dowkiw, A., Husson, C., Frey, P., Pinon, J. and Bastien, C. (2003) Partial resistance to *Melampsora larici-populina* leaf rust in hybrid poplars: genetic variability in inoculated excised leaf disk bioassay and relationship to complete resistance. *Phytopathology* 93, 421–427.
Dowkiw, A., Voisin, E. and Bastien, C. (2010) Potential of Eurasian poplar rust to overcome a major quantitative resistance factor. *Plant Pathology* 59, 523–534.

Dungey, H.S. (2001) Pine hybrids – a review of their use performance and genetics. *Forest Ecology and Management* 148, 243–258.
Dunlap, J.M. and Stettler, R.F. (1996) Genetic variation and productivity of *Populus trichocarpa* and its hybrids. IX. Phenology and *Melampsora* rust incidence of native black cottonwood clones from four river valleys in Washington. *Forest Ecology and Management* 87, 233–256.
Eckenwalder, J.E. (1984) Natural intersectional hybridization between North American species of *Populus* (*Salicaceae*) in sections *Aigeiros* and *Tacamahaca*. I: population studies of *P. xparryi*. *Canadian Journal of Botany* 62, 317–324.
Eckenwalder, J.E. (1996) Systematics and evolution of *Populus*. In: Stettler, R.F., Bradshaw, H.D. Jr, Heilman, P.E. and Hinckley, T.M. (eds) *Biology of Populus and its Implications for Management and Conservation*. National Research Council of Canada Research Press, Ottawa, pp. 732.
Einspahr, D.W. and Benson, M.K. (1964) Production and evaluation of aspen hybrids. *Journal of Forestry* 62, 806–809.
Einspahr, D.W. and Winton, L.L. (1976) Genetics of quaking aspen. Research Paper WO-25. USDA Forest Service, Washington, DC.
Ellis, D.D. and Raffa, K.F. (1997) Expression of transgenic *Bacillus thuringiensis* δ-endotoxin in poplar. In: Klopfenstein, N.B., Chun, Y.W., Kim, M.-S. and Ahuja, M.R. (eds) *Micropropagation, Genetic Engineering, and Molecular Biology of Populus*. General Technical Report RM-GTR-297. USDA Forest Service, Rocky Mountain Forest and Range Experiment Station, Fort Collins, Colorado, pp. 178–186.
Erichsen-Brown, C. (1979) *Medicinal and Other Uses of North American Plants: a Historical Survey With Special Reference to the Eastern Indian Tribes*. General Publishing Company, Toronto, Ontario, Canada.
FAO (1980) *Poplars and Willows in Wood Production and Land Use*. FAO Forestry Series No 10. FAO/IPC, Rome, 328 pp.
FAO (2008) Synthesis of Country Progress Reports received, prepared for the 23rd Session of the International Poplar Commission, jointly hosted by FAO and by the Beijing Forestry University, the State Forest Administration of China and the Chinese Academy of Forestry; Beijing, China, 27–30 October 2008. International Poplar Commission, Working, Paper IPC/6. Forest Management Division, FAO, Rome.
Farmer, R.E. Jr (1964) Cottonwood flowering as related to cold requirement of flower buds. *Forest Science* 10, 296–299.
Farmer, R.E. Jr (1993) Latitudinal variation in height and phenology of balsam poplar. *Silvae Genetica* 42, 148–153.
Farmer, R.E. Jr (1996) The genecology of *Populus*. In: Stettler, R.F., Bradshaw, H.D. Jr, Heilman, P.E. and Hinckley, T.M. (eds) *Biology of Populus and its Implications for Management and Conservation*. National Research Council of Canada Research Press, Ottawa, pp. 33–35.
Farmer, R.E. and Nance, W.L. (1968) Crossing eastern cottonwood in the greenhouse. *Proceedings of the International Plant Propagators Society* 17, 333–338.
Feau, N., Mottet, M.-J., Périnet, P., Hamelin, R.C. and Bernier, L. (2010) Recent advances related to poplar leaf spot and canker caused by *Septoria musiva*. *Canadian Journal of Plant Pathology* 32, 122–134.
Fechner, G.H. (1972) Development of the pistillate flower of *Populus tremuloides* following controlled pollination. *Canadian Journal of Botany* 50, 2503–2509.
Felix, E., Tilley, D.R., Felton, G. and Flamino, E. (2008) Biomass production of hybrid poplar (*Populus* sp.) grown on deep-trenched municipal biosolids. *Ecological Engineering* 33, 8–14.
Fichot, R., Chamaillard, S., Depardieu, C., Le Thiec, D., Cochard, H., Barigah, T.S., *et al.* (2010) Hydraulic efficiency and coordination with xylem resistance to cavitation, leaf function, and growth performance among eight unrelated *Populus deltoides* × *Populus nigra* hybrids. *Journal of Experimental Botany* 62, 2093–2106.
Fladung, M. and Muhs, H.-J. (1999) Untersuchungen zur Stabilität und Expressivitätfremder Gene in Aspenklonen (*Populus tremula* und *P. tremula* × *P. tremuloides*) unter Freilandbedingungen. In: Schiemann, J. (ed.) *Freisetzungsbegleitende Sicherheitsforschung mit Gentechnischveränderten Pflanzen und Mikroorganismen*. Proceedings of the BMBF Workshop, Braunschweig, Germany, 25–26 May1998. BEO, Jülich, Germany, pp. 91–100.
Fladung, M., Muhs, H.-J. and Ahuja, M.R. (1996) Morphological changes in transgenic *Populus* carrying the rolC gene from *Agrobacterium rhizogenes*. *Silvae Genetica* 45, 349–354.
Floate, K. (2004) Extent and patterns of hybridization among the three species of *Populus* that constitute the riparian forest of southern Alberta, Canada. *Canadian Journal of Botany* 82, 253–264.

Fossati, T., Patrignani, G., Zapelli, I., Sabatti, M., Sala, F. and Castiglione, S. (2004) Development of molecular markers to assess the level of introgression of *Populus tremula* into *P. alba* natural populations. *Plant Breeding* 123, 383–385.
Fossati, T., Zapelli, I., Bisoffi, S., Micheletti, A., Vietto, L., Sala, F., *et al.* (2005) Genetic relationships and clonal identity in a collection of commercially relevant poplar cultivars assessed by AFLP and SSR. *Tree Genetics and Genomes* 1, 11–20.
Foster, G.S. (1986) Provenance variation of eastern cottonwood in the lower Mississippi valley. *Silvae Genetica* 35, 32–38.
Foster, G.S. (1993) Selection and breeding for extreme genotypes. In: Ahuja, M.R. and Libby, W.J. (eds) *Clonal Forestry I Genetics and Biotechnology.* Springer-Verlag, Germany, pp. 50–67.
Foster, G.S., Rousseau, R.J. and Nance, W.L. (1998) Eastern cottonwood clonal mixing study: intergenotypic competition effects. *Forest Ecology and Management* 112, 9–22.
Foyer, C.H. and Noctor, G. (2005) Redox homeostasis and antioxidant signaling: a metabolic interface between stress perception and physiological responses. *Plant Cell* 17, 1866–1875.
Frewen, B.E., Chen, T.H.H., Howe, G.T., Davis, J., Rohde, A., Boerjan, W., *et al.* (2000) Quantitative trait loci and candidate gene mapping of bud set and bud flush in *Populus. Genetics* 154, 837–845.
Frey, B.R., Lieffers, V.J., Landhausser, S.M., Comeau, P.G. and Greenway, K.J. (2003) An analysis of sucker regeneration of trembling aspen. *Canadian Journal of Forest Research* 33, 1169–1179.
Frey, P. and Pinon, J. (1997) Variability in pathogenicity of *Melampsora allii-populina* expressed on poplar cultivars. *European Journal of Forest Pathology* 27, 397–407.
Fritz, R.S. (1999) Resistance of hybrid plants to herbivores: genes, environment, or both? *Ecology* 80, 382–391.
Fritz, R.S., Nichols-Orians, C.M. and Brunsfeld, S.J. (1994) Interspecific hybridization of plants and resistance to herbivores: hypotheses, genetics, and variable responses in a diverse herbivore community. *Oecologia* 97, 106–117.
Fritz, R.S., Hochwender, C.G., Lewkiewicz, D.A., Bothwell, S. and Orians, C.M. (2001) Seedling herbivory by slugs in a willow hybrid system: developmental changes in damage, chemical defense, and plant performance. *Oecologia* 129, 87–97.
Fussi, B., Lexer, C. and Heinze, B. (2010) Phylogeography of *P. alba* (L.) and *Populus tremula* (L.) in Central Europe: secondary contact and hybridisation during recolonisation from disconnected refugia. *Tree Genetics and Genomes* 6, 439–450.
Gaget, M., Said, C. and Dumas, C. (1984) Pollen–pistal interactions in interspecific crosses of *Populus* (sections *Aigeiros* and *Leuce*): pollen adhesion, hydration and callose responses. *Journal of Cell Science* 72, 173–184.
Gao, W., Rao, V.R. and Zhou, M.-D. (eds)(2001) *Plant Genetic Resource Conservation and Use in China.* Proceedings of National Workshop on Conservation and Utilization of Plant Genetic Resources, 25–27 October 1999, Beijing, China. Institute of Crop Germplasm Resources, CAAS and IPGRI Office for East Asia, Beijing.
Garay, M.R., Noseda, P.A., Cortizo, S., Mujica, G., Franzone, P. and Rios, R. (2005) Cultivo *in vitro* de Salicaceas. Poster presented at III Congreso Forestal Argentino y Latinoamericano. Corrientes, Argentina.
Gaudet, M., Jorge, V., Paolucci, I., Beritognolo, I., Scarascia Mugnozza, G. and Sabatti, M. (2008) Genetic linkage maps of *Populus nigra* L. including AFLPs, SSRs, SNPs, and sex trait. *Tree Genetics and Genomes* 4, 25–36.
Gebhardt, K., Pohl, A. and Vornam, B. (2001) Genetic inventory of black poplar populations in the upper Rhine floodplains: conclusions for conservation of an endangered plant species. In: van Dam, B.C. and Bordacs, S. (eds) *Genetic Diversity in River Populations of European Black Poplar. Proceedings of the International Symposium, Szesksard, Hungary, 16–20 May, 2001.* Verlag C. Nyomda, Budapest, pp. 145–156.
Goerndt, M.E. (2005) Growing short-rotation woody biomass in a plantation setting on marginal lands in Iowa. MSc. thesis, Iowa State University, Ames, Iowa.
Gom, L.A. and Rood, S.B. (1999) The discrimination of cottonwood clones in a mature grove along the Oldman River in southern Alberta. *Canadian Journal of Botany* 77, 1084–1094.
Gomez, A., Manzanera, J.A., Aguiriano, E., Grau, J.A. and Bueno, M.A. (2003) SSR markers for monitoring an *in vitro* core collection of *Populus tremula. Silvae Genetica* 52, 224–229.
Gong, J.R., Zhang, X.S. and Huang, Y.M. (2011) Comparison of the performance of several hybrid poplar clones and their potential suitability for use in northern China. *Biomass and Bioenergy* 35, 2755–2764.

Gonzalez, F. (2007) Forestry option: the poplars take off. *Lignum* 17, 38–43.

Gonzalez-Antonanzas, F., Grau, J.M., Sixto, H. and Montoto, J.L. (2000) Comparison of new *P.* ×*interamericana* clones in medium altitude areas in Spain. In: Isebrands, J.G. and Richardson, J. (eds) *21st Session of the International Poplar Commission (IPC 2000). Poplar and Willow Culture: Meeting The Needs of Society and The Environment.* General Technical Report NC-215. US Department of Agriculture, Forest Service, North Central Research Station, St Paul, Minnesota, p. 63.

Gonzalez-Garcia, S., Gasol, C.M., Gabarrell, X., Rieradevall, J., Moreira, M.T. and Feijoo, G. (2009) Environmental profile of ethanol from poplar biomass as transport fuel in southern Europe. *Renewable Energy* 35, 1014–1023.

Gonzalez-Martinez, S.C., Wheeler, N.C., Ersoz, E., Nelson, C.D. and Neale, D.B. (2007) Association genetics in *Pinus taeda* L. I. Wood property traits. *Genetics* 175, 399–409.

Goue-Mourier, M.C., Faivre-Rampant, P., LeGuerroue, B., Lefevre, F. and Villar, M. (1996) Molecular and genetic approaches to rust resistance (*Melampsora* sp.) in poplars (*Populus* sp.). In: Ahuja, M.R., Boerjan, W. and Neale, D.B. (eds) *Somatic Cell Genetics and Molecular Genetics of Trees.* Kluwer Academic Publishers, Dordrecht, the Netherlands, pp. 249–253.

Grabau, M.R., Milczarek, M.A., Karpiscak, M.M., Raulston, B.E., Garnett, G.N. and Bunting, D.P. (2011) Direct seeding for riparian tree re-vegetation: small-scale field study of seeding methods and irrigation techniques. *Ecological Engineering* 37, 864–872.

Grant, V. (1975) *Genetics of Flowering Plants.* Columbia University Press, New York.

Grau, J.M., Gonzalez-Antonanzas, F., Sixto, H. and Hernandez, E. (2000) Comparison of known poplar clones in medium altitude areas in Spain. In: Isebrands, J.G. and Richardson, J. (eds) *21st Session of the International Poplar Commission (IPC 2000). Poplar and Willow Culture: Meeting The Needs of Society and The Environment.* General Technical Report NC-215. US Department of Agriculture, Forest Service, North Central Research Station, St Paul, Minnesota, p. 65.

Green, B.K. (1998) Variation in rooting potential and shoot growth of full-sib hybrid aspen. MSc. thesis, Iowa State University, Ames, Iowa.

Gullberg, U. (1993) Towards making willows pilot species for coppicing production. *The Forestry Chronicle* 69, 721–726.

Gullberg, U. and Ryttman, H. (1993) Genetics of field resistance to *Melampsora* in *Salix viminalis. European Journal of Forest Pathology* 23, 75–84.

Guo, X.-Y. and Zhang, X.-S. (2010) Performance of 14 hybrid poplar clones grown in Beijing, China. *Biomass and Bioenergy* 34, 906–911.

Guzina, V. and Vujovic, M. (eds) (1986) *Poplars and Willows in Yugoslavia.* Poplar Research Institute, Novi Sad, Serbia.

Gwyther, L.E. (2006) Spreading agroforestry for sustainability: a comparative view of Shandong and Sichuan Provinces. *Journal of Forestry* 104, 324–327.

Gyulai, G., Humphreys, M., Bittsanszky, A., Skot, K., Kiss, J., Sko, L., *et al.* (2005) AFLP analysis and improved phytoextraction capacity of transgenic *gshI*-poplar clones (*Populus* ×*canescens* L.) for Copper *in vitro. Zeitschrift für Naturforschung* 60c, 300–306.

Hakansson, A. (1933) The conjugation of chromosomes in a *Salix* hybrid. *Hereditas* 18, 199–214.

Hakansson, A. (1938) Cytological studies on bastards of *Salix. Hereditas* 24, 1–32.

Hall, R.B., Colletti, J.P., Schultz, R.C., Faltonson, R.R., Kolison, S.H. Jr, Hanna, R.D., *et al.* (1990) Commercial-scale vegetative propagation of aspens. In: USDA Forest Service (ed.) *Proceedings of the Aspen Symposium, Duluth, MN, 25–27 July 1989.* General Technical Report NC-140. USDA Forest Service, North Central Forest Research Station, St Paul, Minnesota, pp. 211–219.

Hall, D., Luquez, V., Garcia, V.M., St Onge, K.R., Jansson, S. and Ingvarsson, P.K. (2007) Adaptive population differentiation in phenology across a latitudinal gradient in European aspen (*Populus tremula* L.): a comparison of neutral markers, candidate genes and phenotypic traits. *Evolution* 61, 2849–2860.

Hallgren, P., Ikonen, A., Hjalten, J. and Roininen, H. (2003) Inheritance patterns of phenolics in F1, F2, and back-cross hybrids of willows: implications for herbivore responses to hybrid plants. *Journal of Chemical Ecology* 29, 1143–1158.

Hamzeh, M. and Dayanandan, S. (2004) Phylogeny of *Populus* (*Salicaceae*) based on nucleotide sequences of chloroplast *TRNT-TRNF* region and nuclear rDNA. *American Journal of Botany* 91, 1398–1408.

Hanley, S.J. (2003) Genetic mapping of important agronomic traits in biomass willow. PhD thesis, University of Bristol, Bristol, UK.

Hanley, S.J., Barker, J.H.A., Van Ooijen, J.W., Aldam, C., Harris, S.L., Åhman, I., *et al.* (2002) A genetic linkage map of willow (*Salix viminalis*) based on AFLP and microsatellite markers. *Theoretical and Applied Genetics* 105, 1087–1096.

Hanley, S.J., Mallott, M.D. and Karp, A. (2007) Alignment of a *Salix* linkage map to the *Populus* genomic sequence reveals macrosynteny between willow and poplar genomes. *Tree Genetics and Genomes* 3, 35–48.

Hanley, S.J., Pei, M.H., Powers, S.J., Ruiz, C., Mallott, M.D., Barker, J.H.A., *et al.* (2011) Genetic mapping of rust resistance loci in biomass willow. *Tree Genetics and Genomes* 7, 597–608.

Hannon, E.R., Kittelson, N.T., Eaton, J.A. and Brown, J.J. (2008) Screening hybrid poplar clones for susceptibility to *Cryptorhynchus lapathi* (Coleoptera: Curculionidae). *Journal of Economic Entomology* 101, 199–205.

Hardig, T.M., Brunsfeld, S.J., Fritz, R.S., Morgan, M. and Orians, C.M. (2000) Morphological and molecular evidence for hybridization and introgression in a willow (*Salix*) hybrid zone. *Molecular Ecology* 2000, 9–24.

Harman, H.M. (2004). Feasibility of biological control of grey willow *Salix cinerea*. DOC Science Internal Series 183. Department of Conservation, Wellington, New Zealand.

Heimburger, C. (1936) Report on poplar hybridization 1936. *The Forestry Chronicle* 12, 285–290.

Heinze, B. (2008) Genetic traces of cultivated hybrid poplars in the offspring of native *Populus nigra* in Austria. *Preslia* 80, 365–374.

Henry, A. (1914) The artificial production of vigorous trees. *Journal of the Department of Agriculture and Technical Institute, Irish Free State* 15, 34–52.

Heribert-Nilsson, N. (1918) Experimentelle studien uber variabilitat, Spaltung, Artibildung, und evoluation in der gattung *Salix*. Lunds Universitats Arsskrift N. F. Avd 2.

Hertzberg, M., Aspeborg, H., Schrader, J., Andersson, A., Erlandsson, R., Blomqvist, K., *et al.* (2001) A transcriptional roadmap to wood formation. *Proceedings of the National Academy of Sciences* 98, 14732–14737.

Hoffmann, D. and Weih, M. (2005) Limitations and improvement of the potential utilisation of woody biomass for energy derived from short rotation woody crops in Sweden and Germany. *Biomass and Bioenergy* 28, 267–279.

Holderegger, R., Angelone, S., Brodbeck, S., Csencsics, D., Gugerli, F., Hoebee, S.E. *et al.* (2005) Application of genetic markers to the discrimination of European black poplar (*Populus nigra*) from American black poplar (*P. deltoides*) and hybrid poplars (*P. ×canadensis*) in Switzerland. *Trees* 19, 743–748.

Houtzagers, G. (1952) Forest genetics and poplar breeding in the Netherlands. II. Poplar breeding in the Netherlands. *Euphytica* 1, 161–174.

Howe, G.T. and Brunner, A.M. (2005) An evolving approach to understanding plant adaptation. *New Phytologist* 167, 1–5.

Hu, J.J., Tian, Y.C., Han, Y.F., Li, L. and Zhang, B.E. (2001) Field evaluation of insect-resistant transgenic *Populus nigra* trees. *Euphytica* 121, 123–127.

Hu, J., Yang, M. and Lu, M. (2010) Advances in biosafety studies on transgenic insect-resistant poplars in China. *Biodiversity Science* 18, 336–345.

Huehn, M. (1988) Multiclonal mixtures and number of clones I. Number of clones and yield stability (deterministic approach without competition). *Silvae Genetica* 37, 67–73.

Hughes, A.R., Inouye, B.D., Johnson, M.T.J., Underwood, N. and Vellend, M. (2008) Ecological consequences of genetic diversity. *Ecology Letters* 11, 609–623.

Huhn, M. (1985) Theoretical studies on the necessary number of components in mixtures. 1. Number of components and yield stability. *Theoretical and Applied Genetics* 70, 383–389.

Huybregts, A.A., Thomas, B.R. and Dancik, B.P. (2007) Flowering phenology and seed viability of native and non-native poplars in north-central Alberta. *The Forestry Chronicle* 83, 239–246.

Imbert, E. and Lefevre, F. (2003) Dispersal and gene flow of *Populus nigra* (*Salicaceae*) along a dynamic river system. *Journal of Ecology* 91, 447–456.

Ingvarsson, P.K., Garcia, M.V., Hall, D., Luquez, V. and Jansson, S. (2006) Clinal variation in *phyB2*, a candidate gene for day-length-induced growth cessation and bud set, across a latitudinal gradient in European aspen (*Populus tremula*). *Genetics* 172, 1845–1853.

Isik, F. and Toplu, F. (2004) Variation in juvenile traits of natural black poplar (*Populus nigra* L.) clones in Turkey. *New Forests* 27, 175–187.

Jacometti, G. (1934) Ricerche e studi sul pioppo. In: *La 3° Mostra Forestale e Montana Arnaldo Mussolini.* Comitato Nazionale Forestale, Roma, pp. 3–15.

Jacometti, G. (1937) I nuovi pioppi italiana. *Atti del Convegno di Pioppicoltori, Comitato Nazionale Forestale, Casale Monferrato* 11.7, 39–50.

Jing, Z.P., Gallardo, F., Pascual, M.B., Sampalo, R., Romero, J., Torres de Navarra, A., *et al.* (2004) Improved growth in a field trial of transgenic hybrid poplar overexpressing glutamine synthetase. *New Phytologist* 164, 137–145.

Johansson, T. and Karacic, A. (2011) Increment and biomass in hybrid poplar and some practical implications. *Biomass and Bioenergy* 35, 1925–1934.

Jokipii, S., Ryynanen, L., Kallio, P.T., Aronen, T. and Haggman, H. (2004) A cryopreservation method maintaining the genetic fidelity of a model forest tree, *Populus tremula* L. × *Populus tremuloides* Michx. *Plant Science* 166, 799–806.

Jorge, V., Dowkiw, A., Faivre-Rampant, P. and Bastien, C. (2005) Genetic architecture of qualitative and quantitative *Melampsora larici-populina* leaf rust resistance in hybrid poplar: genetic mapping and QTL detection. *New Phytologist* 167, 113–127.

Kajba, D., Ballian, D., Idzojtic, M. and Bogdan, S. (2004) The differences among hairy and typical European black poplars and the possible role of the hairy type in relation to climatic changes. *Forest Ecology and Management* 197, 279–284.

Kajba, D., Gracan, J., Ivankovic, M., Bogdan, S., Gradecki-Postenjak, M., Littvay, T., *et al.* (2006) [Conservation of forest genetic resources in Croatia]. *Glasnik za Sum. Pokuse*, pos. izd. 5, 235–249.

Karnosky, D.F., Berrang, P.C., Scholz, F. and Bennett, J.P. (1989) Variation in and natural selection for air pollution tolerances in trees. In: Scholz, F., Gregorius, H.R. and Rudin, D. (eds) *Genetic Effects of Air Pollutants in Forest Tree Populations*. Springer-Verlag, Berlin, pp. 29–37.

Karp, A. and Shield, I. (2008) Bioenergy from plants and the sustainable yield challenge. *New Phytologist* 179, 15–32.

Karp, A., Hanley, S.J., Trybush, S.O., Macalpine, W., Pei, M. and Shield, I. (2011) Genetic improvement of willow for bioenergy and biofuels. *Journal of Integrative Plant Biology* 53, 151–165.

Kauter, D., Lewandowski, I. and Claupein, W. (2003) Quantity and quality of harvestable biomass from *Populus* short rotation coppice for solid fuel use – a review of the physiological basis and management influences. *Biomass and Bioenergy* 24, 411–427.

Kelleher, C.T., Chiu, R., Shin, H., Bosdet, I.E., Krzywinski, M.I., Fjell, C.D., *et al.* (2007) A physical map of the highly heterozygous *Populus* genome: integration with the genome sequence and genetic map and analysis of haplotype variation. *The Plant Journal* 50, 1063–1078.

Keller, S.R., Olson, M.S., Silim, A., Schroeder, W. and Tiffin, P. (2010) Genomic diversity, population structure, and migration following rapid range expansion in the balsam poplar, *P. balsamifera*. *Molecular Ecology* 19, 1212–1226.

Kelly, M.T. and Curry, J.P. (1991) The influence of phenolic compounds on the suitability of 3 *Salix* species as hosts for the willow beetle *Phratora vulgatissima*. *Entomologia Experimentalis et Applicata* 61, 25–32.

Kerr, R.J., Dieters, M.J. and Tier, B. (2004) Simulation of the comparative gains from four different hybrid tree breeding strategies. *Canadian Journal of Forest Research* 34, 209–220.

Khurana, D.K. and Narkhede, S. (1995) Poplar improvement in Himachal Pradesh (India). In: Khurana, D.K. (ed.) *Poplars in India: Recent Research Trends*. International Development Research Centre, New Dehli and Dr Y.S. Parmar University of Horticulture and Forestry, Nauni (Solan), India, pp. 7–40.

Kim, M.-S., Klopfenstein, N.B. and Chun, Y.W. (1997) *Agrobacterium*-mediated transformation of *Populus* species. In: Klopfenstein, N.B., Chun, Y.W., Kim, M.-S. and Ahuja, M.R. (eds) *Micropropagation, Genetic Engineering, and Molecular Biology of Populus*. General Technical Report RM-GTR-297. USDA Forest Service, Rocky Mountain Forest and Range Experiment Station, Fort Collins, Colorado, pp. 51–59.

King, R.A., Harris, S.L., Karp, A. and Barker, J.H.A. (2010) Characterisation and inheritance of nuclear microsatellite loci for use in populations studies of the allotetraploid *Salix alba–Salix fragilis* complex. *Tree Genetics and Genomes* 6, 247–258.

Kiss, J., Kondrak, M., Torjek, O., Kiss, E., Gyulai, G., Mazik-Tokei, K., *et al.* (2001) Morphological and RAPD analysis of poplar trees of anther culture origin. *Euphytica* 118, 213–221.

Kopp, R.F. (2000) Genetic improvement of *Salix* using traditional breeding and AFLP fingerprinting. PhD thesis, State University of New York, College of Environmental Science and Forestry, Syracuse, New York.

Kopp, R.F., Smart, L.B., Maynard, C.A., Isebrands, J.G., Tuskan, G.A. and Abrahamson, L.P. (2001) The development of improved willow clones for eastern North America. *The Forestry Chronicle* 77, 287–292.

Kopp, R.F., Maynard, C.A., Rocha de Niella, P., Smart, L.B. and Abrahamson, L.P. (2002a) Collection and storage of pollen from *Salix* using organic solvents. *American Journal of Botany* 89, 248–252.

Kopp, R.F., Smart, L.B., Maynard, C.A., Tuskan, G.A. and Abrahamson, L.P. (2002b) Predicting within-family variability in juvenile height growth of *Salix* based upon similarity among parental AFLP fingerprints. *Theoretical and Applied Genetics* 105, 106–112.

Kouider, M., Skirvin, R.M., Saladin, K.P., Dawson, J.O. and Jokela, J.J. (1984) A method to culture immature embryos of *Populus deltoides in vitro*. *Canadian Journal of Forest Research* 14, 956–958.

Kovacevic, B., Guzina, V., Kraljevic-Balalic, M., Ivanovic, M. and Nikolic-Doric, E. (2008) Evaluation of early rooting traits of eastern cottonwood that are important for selection tests. *Silvae Genetica* 57, 13–21.

Kuzovkina, Y.A. and Quigley, M.F. (2005) Willows beyond wetlands: uses of *Salix* L. species for environmental projects. *Water, Air, and Soil Pollution* 162, 183–204.

Kuzovkina, Y.A. and Volk, T.A. (2009) The characterization of willow (*Salix* L.) varieties for use in ecological engineering applications: co-ordination of structure, function and autecology. *Ecological Engineering* 35, 1178–1189.

Kuzovkina, Y.A., Weih, M., Romero, M.A., Charles, J., Hurst, S., McIvor, I., *et al.* (2008) *Salix*: botany and global horticulture. In: Janick, J. (ed.) *Horticultural Reviews*. John Wiley and Sons, Hoboken, New Jersey, pp. 447–489.

Lamit, L.J., Wojtowicz, T., Kovacs, Z., Wooley, S.C., Zinkgraf, M., Whitham, T.G., *et al.* (2011) Hybridization among foundation tree species influences the structure of associated understory plant communities. *Botany* 89, 165–174.

Land, S.B. Jr (1974) Forest tree improvement: Mississippi certifies nation's first 'blue tag'. *Journal of Forestry* 72, 353.

Land, S.B. Jr, Stine, M., Ma, X., Rockwood, D.L., Warwell, M.V. and Alker, G.R. (2001) A tree improvement program for eastern cottonwood in the southeastern United States. In: Dean, J.F.D. (ed.) *Proceedings 26th Biennial Southern Forest Tree Improvement Conference*. University of Georgia, Athens, Georgia, pp. 84–93.

Larsson, S. (1997) Commercial breeding of willow for short rotation coppice. *Aspects of Applied Biology* 49, 215–218.

Larsson, S. (1998) Genetic improvement of willow for short-rotation coppice. *Biomass and Bioenergy* 15, 23–26.

Larsson, S. (2001) Commercial varieties from the Swedish willow breeding programme. *Aspects of Applied Biology* 65, 193–198.

Lascoux, M., Thorsen, J. and Gullberg, U. (1996) Population stucture of a riparian willow species, *Salix viminalis* L. *Genetics Research, Cambridge* 68, 45–54.

Ledig, F.T. (1986) Conservation strategies for forest gene resources. *Forest Ecology and Management* 14, 77–90.

Lefevre, F., Goue-Mourier, M.C., Faivre-Rampant, P. and Villar, M. (1998) A single gene cluster controls incompatibility and partial resistance to various *Melampsora laricia-populina* races in hybrid poplars. *Phytopathology* 88, 156–163.

Lefevre, F., Barsoum, N., Heinze, B., Kajba, D., Rotach, P., de Vries, S.M.G., *et al.* (2001) *In situ* conservation of *Populus nigra*. EUFORGEN Technical Bulletin. International Plant Genetic Resources Institute, Rome.

Legionnet, A. and Lefevre, F. (1996) Genetic variation of the riparian pioneer tree species *Populus nigra* L. I. Study of population structure based on isozymes. *Heredity* 77, 629–637.

Legionnet, A., Muranty, H. and Lefevre, F. (1999) Genetic variation of the riparian pioneer tree species *Populus nigra*. II. Variation in susceptibility to the foliar rust *Melampsora larici-populina*. *Heredity* 82, 318–327.

Lexer, C., Fay, M.F., Joseph, J.A., Nica, M.-S. and Heinze, B. (2005) Barrier to gene flow between two ecologically divergent *Populus* species, *P. alba* (white poplar) and *P. tremula* (European aspen): the role of ecology and life history in gene introgression. *Molecular Ecology* 14, 1045–1057.

Lexer, C., Buerkle, C.A., Joseph, J.A., Heinze, B. and Fay, M.F. (2007) Admixture in European *Populus* hybrid zones makes feasible the mapping of loci that contribute to reproductive isolation and trait differences. *Heredity* 98, 74–84.

Li, B. (1995) Aspen improvement strategies for western Canada – Alberta and Saskatchewan. *The Forestry Chronicle* 71, 720–724.

Li, B. and Wyckoff, G.W. (1991) A breeding strategy to improve aspen hybrids for the University of Minnesota aspen/larch genetics cooperative. In: Hall, R.B., Hanna, R.D. and Nyong'o, R.N. (eds)

Joint Proceedings of International Energy Agency Joint Meeting of the Task V Activity Groups on Exchange of Genetic Material, Pest/Disease Management, and Joint Trials of Alnus, Populus, and Salix, 22–27August 1991, Ames, Iowa, USA and a workshop of the Task II Activity Group on Evaluation of Alnus Species and Hybrids, 8–10 August 1988, Vancouver, BC, Canada. Iowa State University, Ames, Iowa, pp. 33–41.

Li, B., Wyckoff, G.W. and Einspahr, D.W. (1993) Hybrid aspen performance and genetic gains. *Northern Journal of Applied Forestry* 10, 117–122.

Li, B., Howe, G.T. and Wu, R. (1998) Developmental factors responsible for heterosis in aspen hybrids (*Populus tremuloides* × *P. tremula*). *Tree Physiology* 18, 29–36.

Li, S., Zhang, Z., Luo, J., He, C., Pu, Y. and An, X. (2005a) Progress and strategies in cross breeding of poplars in China. *Forestry Studies in China* 7, 54–60.

Li, S., Zhang, Z., He, C., An, X., Yu, Z. and Li, B. (2005b) Variation analysis of seed and seedling traits of cross combination progenies in *Populus*. *Forestry Studies in China* 7, 61–69.

Libby, W.J. (1973) Domestication strategies for forest trees. *Canadian Journal of Forest Research* 3, 265–276.

Libby, W.J. (1982) What is a safe number of clones per plantation? In: Heybroek, H.M., Stephan, B.R. and von Weissenberg, K. (eds) *Resistance to Diseases and Pests in Forest Trees. Topic 7. Proceedings of the Third International Workshop on the Genetics of Host–Parasite Interactions in Forestry.* Centre for Agricultural Publishing and Documentation, Wageningen, the Netherlands, pp. 342–360.

Libby, W.J. (1987) Testing for clonal forestry. *Annales Forestales* 13, 69–75.

Liesebach, H., Schneck, V. and Ewald, E. (2010) Clonal fingerprinting in the genus *Populus* L. by nuclear microsatellite loci regarding differences between sections, species, and hybrids. *Tree Genetics and Genomes* 6, 259–269.

Lin, J., Gibbs, J.P. and Smart, L.B. (2009) Population genetic structure of native versus naturalized sympatric shrub willows (*Salix*; *Salicaceae*). *American Journal of Botany* 96, 771–785.

Lin, S., Binder, B.F. and Hart, E.R. (1998) Insect feeding stimulants from the leaf surface of *Populus*. *Journal of Chemical Ecology* 24, 1781–1790.

Lin, S.-Z., Zhang, Z.-Y., Zang, Q. and Lin, Y.-Z.(2006) Progress in the study of molecular genetic improvements of poplar in China. *Journal of Integrative Plant Biology* 48, 1001–1007.

Lindegaard, K.N. and Barker, J.H.A. (1997) Breeding willows for biomass. *Aspects of Applied Biology* 49, 155–162.

Lindegaard, K.N., Parfitt, R.I., Donaldson, G., Hunter, T., Dawson, W.M., Forbes, E.G.A., *et al.* (2001) Comparative trials of elite Swedish and UK biomass willow varieties. *Aspects of Applied Biology* 65, 183–192.

Linderson, M.L., Iritz, Z. and Lindroth, A. (2007) The effect of water availability on stand-level productivity, transpiration, water use efficiency and radiation use efficiency of field-grown willow clones. *Biomass and Bioenergy* 31, 460–468.

Lindroth, R.L. and Hwang, S.Y. (1996) Diversity, redundancy, and multiplicity in chemical defense systems of aspen. In: Romeo, J.T., Saunders, J.A. and Barbosa, P. (eds) *Phytochemical Diversity and Redundancy in Ecological Reactions.* Plenum, New York, pp. 25–56.

Lindroth, R.L., Scriber, J.M. and Hsia, M.T.S. (1988) Chemical ecology of the tiger swallowtail – mediation of host use by phenolic glycosides. *Ecology* 69, 814–822.

Lindroth, R.L., Roth, S. and Nordheim, E.V. (2001) Genotypic variation in response of quaking aspen (*Populus tremuloides*) to atmospheric CO_2 enrichment. *Oecologia* 126, 371–379.

Little, L.R. and Dale, M.R.T. (1999) A method for analysing spatio-temporal pattern in plant establishment, tested on a *Populus balsamifera* clone. *Journal of Ecology* 87, 620–627.

Lojewski, N.R., Fischer, D.G., Bailey, J.K., Schweitzer, J.A., Whitham, T.G. and Hart, S.C. (2009) Genetic basis of aboveground productivity in two native *Populus* species and their hybrids. *Tree Physiology* 29, 1133–1142.

Louis, K.A. and Eils, L.E. (1997) Application of tissue culture systems for commercial plant production. In: Klopfenstein, N.B., Chun, Y.W., Kim, M.-S. and Ahuja, M.R. (eds) *Micropropagation, Genetic Engineering, and Molecular Biology of Populus.* General Technical Report RM-GTR-297. USDA Forest Service, Rocky Mountain Forest and Range Experiment Station, Fort Collins, Colorado, pp. 236–240.

Lyyra, S., Lima, A. and Merkle, S.A. (2006) *In vitro* regeneration of *Salix nigra* from adventitious shoots. *Tree Physiology* 26, 969–975.

Ma, C., Strauss, S.H. and Meilan, R. (2004) *Agrobacterium*-mediated transformation of the genome-sequenced poplar clone, Nisqually-1 (*Populus trichocarpa*). *Plant Molecular Biology Reporter* 22, 311a–311i.

McCord, S. and Gartland, K. (eds) (2003) *Forest Biotechnology in Europe: The Challenge, The Promise, The Future. Appendix 2: GM Tree Field Trials in Europe.* The Institute of Forest Biotechnology, Raleigh, North Carolina.

McCracken, A.R. and Dawson, M. (1992) Clonal response in *Salix* to *Melampsora* rusts in short rotation coppice plantations. *European Journal of Forest Pathology* 22, 19–28.

McCracken, A.R. and Dawson, W.M. (1997) Growing short rotation coppice willow (*Salix*) in clonal mixtures as a method of reducing the impact of rust disease caused by *Melampsora epitea* var. *epitea. European Journal of Forest Pathology* 27, 319–328.

McCracken, A.R., Dawson, W.M. and Carlisle, D. (2005) Short-rotation coppice willow mixtures and rust disease development. In: Pei, M.H. and McCracken, A.R. (eds) *Rust Diseases of Willow and Poplar.* CAB International, Cambridge, Massachusetts, pp. 185–194.

McCracken, A.R., Walsh, L., Moore, P.J., Lynch, M., Cowan, P., Dawson, M., *et al.* (2011) Yield of willow (*Salix* spp.) grown in short rotation coppice mixtures in a long-term trial. *Annals of Applied Biology* 159, 229–243.

McGrath, J.M., Karnosky, D.F. and Ainsworth, E.A. (2010) Spring leaf flush in aspen (*Populus tremuloides*) clones is altered by long-term growth at elevated carbon dioxide and elevated ozone concentrations. *Environmental Pollution* 158, 1023–1028.

McIvor, I.R., Hedderley, D.I., Hurst, S.E. and Fung, L.E. (2011) Survival and growth to age 8 of four *Populus maximowiczii* × *P. nigra* clones in field trials on pastoral hill slopes in six climatic zones of New Zealand. *New Zealand Journal of Forestry Science* 41, 151–163.

Maestro, C. and Alba, N. (2008) Material forestal de reproducción de *Populus* autócotonos: propuestas para la restauración de riberas. *Cuadernos de la Sociedad Española de Ciencias Forestales* 24, 57–62.

Maestro, C., Notivol, E. and Plana, L. (2001a) Resultados preliminares de la introducción experimental de nuevos clones de chopo italianos y españoles en el valle del Cinca (Huesca). Actas del I Simposio del Chopo de Castilla y León, Zamora, Spain, 9–11 May 2001. Editorial Junta de Castilla y León, Spain, pp. 411–415.

Maestro, C., Alba, N. and Agúndez, D. (2001b) Conservación *ex situ* de *Populus nigra* en España: Caracterización morfológica e isoenzimática de las colecciones establecidas. Actas del I Simposio del Chopo de Castillo y León, Zamora, Spain, 9–11 May 2001. Editorial Junta de Castilla y León, Spain, pp. 401–409.

Mareschi, L., Paris, P., Sabatti, M., Nardin, F., Giovanardi, R., Manazzone, S., *et al.* (2005) Le nuove varietà di pioppo da biomassa garantiscono produttività interessanti. *Informatore Agrario* 18, 49–53.

Marosvolgyi, B., Halupa, L. and Wesztergom, I. (1999) Poplars as biological energy sources in Hungary. *Biomass and Bioenergy* 16, 245–247.

Marron, N., Dillen, S.Y. and Ceulemans, R. (2007) Evaluation of leaf traits for indirect selection of high yielding poplar hybrids. *Environmental and Experimental Botany* 61, 103–116.

Martinsen, G.D., Whitham, T.G., Turek, R.J. and Keim, P. (2001) Hybrid populations selectively filter gene introgression between species. *Evolution* 55, 1325–1335.

Mashkina, O.S. and Isakov, Y.N. (2002) Genetic improvement of poplar. *Lesovedenie* 3, 68–73.

Mathews, J.H. and Campbell, M.M. (2000) The advantages and disadvantages of the application of genetic engineering to forest trees: a discussion. *Forestry* 73, 371–380.

Maximenko, A.P. (2002) Plantation cultivation and use of *Salix* biomass. *Kubansky uchebnik.* Krasnodar, Russia. (In Russian)

May, S. (1962) Una nuova pianta di importanza economica nella coltura del legno: il Salice. *Bollettino dell' Agricoltura* 96, 163–174.

May, S. (1981) I salici arborei. In: Giordano, E. (ed.) *Le risorse forestali interne.* Proceedings of 1° Congresso Nazionale. 'Il legno nelle attività economiche del Paese', 1–4 December, 1981, Rome. Ministero dell'agricultura e delle foreste in collaboration with the AFI, Italy, pp. 103–105.

Meilan, R. and Strauss, S.H. (1997) Poplar genetically engineered for reproductive sterility and accelerated flowering. In: Klopfenstein, N.B., Chun, Y.W., Kim, M.-S. and Ahuja, M.R. (eds) *Micropropagation, Genetic Engineering, and Molecular Biology of Populus.* General Technical Report RM-GTR-297. USDA Forest Service, Rocky Mountain Forest and Range Experiment Station, Fort Collins, Colorado, pp. 212–219.

Meilan, R., Han, K.-H., Ma, C., DiFazio, S.P., Eaton, J.A., Hoien, E.A., *et al.* (2002) The *CP4* trangene provides high levels of tolerance to Roundup® herbicide in field-grown hybrid poplars. *Canadian Journal of Forest Research* 32, 967–976.

Meirmans, P.G., Lamothe, M., Gros-Louis, M.-C., Khasa, D., Périnet, P., Bousquet, J., *et al.* (2010) Complex patterns of hybridization between exotic and native North American poplar species. *American Journal of Botany* 97, 1688–1697.
Melchior, G.H. (1981) Aufgaben und Ziele des Instituts für Forstgenetik und orstpflanzen-züchtung unter besonderer Berücksichtigung von Pappeln und anderen Laubbaumarten. *Die Holzzucht* 35, 7–11.
Melchior, G.H. (1985) Züchtung von Aspen und Hybridaspen und ihre Perspektiven für die Praxis. *Allgemeine Forst- und Jagdzeitung* 156, 112–122.
Melchior, G.H., Krusche, D. and Sachsse, H. (1987) Wachstum, holztechnische Eigenschaften und züchterische Aspekte der Hybridaspenfamilie Holsatia. *Die Holzzucht* 41, 25–29.
Meneghetti, S., Barcaccia, G., Paiero, P. and Lucchin, M. (2007) Genetic characterization of *Salix alba* L. and *Salix fragilis* L. by means of different PCR-derived marker systems. *Plant Biosystems* 141, 283–291.
Merkle, S.A. and Nairn, C.J. (2005) Hardwood tree biotechnology. *In Vitro Cellular and Develelopmental Biology* 41, 602–619.
Merkle, S.A., Andrade, G.M., Nairn, C.J., Powell, W.A. and Maynard, C.A. (2007) Restoration of threatened species: a noble cause for transgenic trees. *Tree Genetics and Genomes* 3, 111–118.
Mitchell, C.P., Stevens, E.A. and Watters, M.P. (1999) Short-rotation forestry – operations, productivity and costs based on experience gained in the U.K. *Forest Ecology and Management* 121, 123–136.
Mitsch, W.J. and Grosselink, J.G. (2000) *Wetlands.* John Wiley and Sons, New York.
Mitton, J.B. and Grant, M.C. (1996) Genetic variation and the natural history of quaking aspen: the ways in which aspen reproduces underlie its great geographic range, high levels of genetic variability, and persistence. *Bioscience* 46, 25–31.
Mofidabadi, A.J., Modir-Rahmati, A.R. and Tavesoli, A. (1998) Application of ovary and ovule culture in *Populus alba* L. × *P. euphratica* Oliv. hybridization. *Silvae Genetica* 47, 332–334.
Mohrdiek, O. (1979) *Progeny Tests with Leuce Poplars in Germany: Crossings Within and Between Species and Backcrossings.* Wiedebusch, Hamburg, Germany.
Morozov, I.R. (1966) The determination of willow of USSR and their cultivation. In: *Forest Industry.* Moscow, Russia. (In Russian)
Mosseler, A. (1989) Interspecific pollen–pistil incongruity in *Salix. Canadian Journal of Forest Research* 19, 1161–1168.
Mosseler, A. (1990) Hybrid performance and species crossability relationships in willows (*Salix*). *Canadian Journal of Botany* 68, 2329–2338.
Mosseler, A. and Papadopol, C.S. (1989) Seasonal isolation as a reproductive barrier among sympatric *Salix* species. *Canadian Journal of Botany* 67, 2563–2570.
Mosseler, A. and Zsuffa, L. (1989) Sex expression and sex ratios in intra- and interspecific hybrid families of *Salix* L. *Silvae Genetica* 38, 12–17.
Mosseler, A., Zsuffa, L., Stoehr, M.U. and Kenney, W.A. (1988) Variation in biomass production, moisture content, and specific gravity in some North American willows (*Salix* L.). *Canadian Journal of Forest Research* 18, 1535–1540.
Muhs, H.-J. (1987) Genetik und züchtung von forstpflanzen vor der einführung neuer biotechnologischer methoden. *Swiss Biotech* 6, 14–16.
Muhs, H.-J. (1998) 50 Jahre Institut für Forstgenetik und Forstpflanzenzuchtung im Arboretum 'Tannenhoft'. In: Liesebach, M. and Stephan, B.R. (eds) *Tannenhöft – 90 Jahre Arboretum – 50 Jahre Institut für Forstgenetik und Forstpflanzenzuchtung.* Bundesforschungsanstalt Forst- und Holzwirtschaft, Hamburg, Germany, pp. 117–140.
Muhs, H.-J. (2008) Regulations for clonal forest reproductive material moving in international trade. Special Report to the International Poplar Commission. FAO, Rome.
Muhs, H.-J. and Melchior, G.H. (1986) Some results of F_1-hybrids and backcrossed hybrids of *Populus tremula* and *Populus tremuloides*. In: Mitchell, C.P., Nilsson, P.O. and Zsuffa, L. (eds) *Proceedings of the Joint International Energy Agency/Forestry Energy Programme and FAO/Cooperative Network on Rural Energy, Forest Energy Conference and Workshops on: Research in Forestry for Energy, Rungstedgaard, Denmark, 28–30 October 1985.* Uppsatseroch Resultater Nr. 49. Sveriges Lantbruksuniversitet, Garpenberg, Sweden, pp. 449.
Namkoong, G. and Koshy, M.P. (1997) Managing the genetic variance. In: Matyas, C. (ed.) *Perspectives of Forest Genetics and Tree Breeding in a Changing World.* IUFRO World Series Vol 6. IUFRO Secretariat, Vienna, pp. 9–16.
Nanson, A. (2001) The new OECD scheme for the certification of forest reproductive materials. *Silvae Genetica* 50, 181–2187.

Nazarov, M.I. (1936) *Salix*. In: Komarov, V.L. (ed.) *Flora of the USSR. Volume V.* Ixdatel'stvo Akademii Hauk SSSR, Moskva-Leningrad, Russia, pp. 24–214. (In Russian)
Neale, D.B. (2007) Genomics to tree breeding and forest health. *Current Opinions in Genetics and Development* 17, 539–544.
Neale, D.B. and Savolainen, O. (2004) Association genetics of complex traits in conifers. *Trends in Plant Science* 9, 325–330.
Nelson, C.D. and Tauer, C.G. (1987) Genetic variation in juvenile characters of *Populus deltoides* Bartr. from the southern great plains. *Silvae Genetica* 36, 216–221.
Netzer, D.A., Tolsted, D.N., Ostry, M.E., Isebrands, J.G., Riemenschneider, D.E. and Ward, K.T. (2002) Growth, yield, and disease resistance of 7- to 12-year-old poplar clones in the north central United States. USDA General Technical Report NC-229. USDA Forest Service, North Central Forest Experiment Station, St Paul, Minnesota.
Neumann, P.D., Krogman, N.T. and Thomas, B.R. (2007) Public perceptions of hybrid poplar plantations: trees as an alternative crop. *International Journal of Biotechnology* 9, 468–483.
Newcombe, G. (1996) The specificity of fungal pathogens of *Populus*. In: Stettler, R.F., Bradshaw, H.D. Jr, Heilman, P.E. and Hinckley, T.M. (eds) *Biology of Populus and its Implications for Management and Conservation*. National Research Council of Canada Research Press, Ottawa, pp. 223–246.
Newcombe, G. (1998) Association of *Mmd1*, a major gene for resistance to *Melampsora medusae* f. sp. *deltoidae*, with quantitative traits in poplar rust. *Phytopathology* 88, 114–121.
Newcombe, G., Bradshaw, H.D. Jr, Chastagner, G.A. and Stettler, R.F. (1996) A major gene for resistance to *Melampsora medusae* f. sp. *deltoidae* in a hybrid poplar pedigree. *Phytopathology* 86, 87–94.
Newcombe, G., Ostry, M., Hubbes, M., Périnet, P. and Mottet, M.-J. (2001) Poplar diseases. In: Dickmann, D.I., Isebrands, J.G., Eckenwalder, J.E. and Richardson, J. (eds) *Poplar Culture in North America*. National Research Council of Canada Research Press, Ottawa, pp. 249–276.
Newsholme, C. (1992) *Willows: the Genus Salix*. Timber Press, Portland, Oregon.
Nilsson, N.H. (1931) The genesis of a whole cinera – similar typus from bastard *Salix viminalis* × *caprea*. *Hereditas* 15, 309–319.
Nordh, N.-E. (2005) Long term changes in stand structure and biomass production in short rotation willow coppice. Doctoral thesis, Swedish University of Agricultural Sciences, Uppsala, Sweden.
Nordh, N.-E. and Verwijst, T. (2004) Above-ground biomass assessments and first cutting cycle production in willow (*Salix* sp.) coppice – a comparison between destructive and non-destructive methods. *Biomass and Bioenergy* 27, 1–8.
Orians, C.M., Huang, C.H., Wild, A., Dorfman, K.A., Zee, P., Dao, M.T.T., *et al.* (1997) Willow hybridization differentially affects preference and performance of herbivorous beetles. *Entomologia Experimentalis et Applicata* 83, 285–294.
Ostry, M.E. and Anderson, N.A. (2009) Genetics and ecology of the *Entoleuca mammata–Populus* pathosystem: implications for aspen improvement and management. *Forest Ecology and Management* 257, 390–400.
Ostry, M.E. and Berguson, W.E. (1993) Selecting hybrid poplars to reduce disease risk may also reduce biomass yield. *Tree Planters' Notes* 44, 128–131.
Ozel, H.B., Ertekin, M. and Tunctaner, K. (2010) Genetic variation in growth traits and morphological characteristics of eastern cottonwood (*Populus deltoides* Bartr.) hybrids at nursery stage. *Scientific Research and Essays* 5, 962–969.
Padro, A. (1987) Creación y selección de nuevos híbridos euramericanos de chopo en el marco de la mejora del género *Populus* en España. PhD thesis, E.S.I.M., Madrid.
Padro, A. (1992) *Clones de chopo para el valle medio del Ebro.* Editorial Gobierno de Aragón, Departamento de Agricultura, Ganadería y Montes, Zaragoza, Spain.
Paillassa, E. (2004) Where to find poplar cultivars for 2004–05 plantations. *Forêt-Entreprise* 159, 47–51.
Pande, P.K. and Dhiman, R.C. (2011) Performance and variability patterns in wood properties and growth traits in the parents, F_1 and F_2 generation hybrid clones of *Populus deltoides*. *Journal of Forestry Research* 22, 379–385.
Paolucci, I., Gaudet, M., Jorge, V., Beritognolo, I., Terzoli, S., Kuzminsky, E., *et al.* (2010) Genetic linkage maps of *Populus alba* L. and comparative mapping analysis of sex determination across *Populus* species. *Tree Genetics and Genomes* 6, 863–875.
Paris, P., Mareschi, L., Sabatti, M., Pisanelli, A., Ecosse, A., Nardin, F., *et al.* (2011) Comparing hybrid *Populus* clones for SRF across northern Italy after two biennial rotations: survival, growth, and yield. *Biomass and Bioenergy* 35, 1524–1532.

Park, Y.S. (2002) Implementation of conifer somatic embryogenesis in clonal forestry: technical requirements and deployment considerations. *Annals of Forest Science* 59, 651–656.

Pei, M.H., Royle, D.J. and Hunter, T. (1996) Pathogenic specialisation of *Melampsora epitea* var. *epitea* on *Salix. Plant Pathology* 45, 679–690.

Pei, M.H., Ruiz, C., Bayon, C. and Hunter, T. (2004) Rust resistance in *Salix* to *Melampsora larici-epitea. Plant Pathology* 53, 770–779.

Pei, M.H., Ruiz, C., Bayon, C., Hunter, T. and Lonsdale, D. (2005) Pathogenic variation in poplar rust *Melampsora larici-populina* from England. *European Journal of Plant Pathology* 111, 147–155.

Pei, M.H., Lindegaard, K., Ruiz, C. and Bayon, C. (2008) Rust resistance of some varieties and recently bred genotypes of biomass willows. *Biomass and Bioenergy* 32, 453–459.

Pei, M.H., Shield, I., Macalpine, W., Lindegaard, K.N., Bayon, C. and Karp, A. (2010) Mendelian inheritance of rust resistance to *Melampsora larici-epitea* in crosses between *Salix sachalinensis* and *S. viminalis. Plant Pathology* 59, 862–872.

Peleman, J.D. and van der Voort, J.R. (2003) Breeding by design. *Trends in Plant Science* 8, 330–334.

Pellis, A., Laureysens, I. and Ceulemans, R. (2004) Growth and production of a short rotation coppice culture of poplar. I. Clonal differences in leaf characteristics in relation to biomass production. *Biomass and Bioenergy* 27, 9–19.

Périnet, P. (2007) The poplar breeding program in Quebec. In: Périnet, P., Perron, M. and Belanger, P. (eds) *Poplar Culture: a Collaborative Effort from Clone to Mill. 2007 Annual Meeting of the Poplar Council of Canada.* Ministère des Ressources naturelles et de la Faune du Québec, Direction de la recherche forestière, Québec, Canada, pp. 11–12.

Perttu, K.L. (1998) Environmental justification for short-rotation forestry in Sweden. *Biomass and Bioenergy* 15, 1–6.

Petit-Conil, M., de Choudens, C. and Chantre, G. (1997) Selection of poplar clones for thermochemical pulping. *Pulp and Paper Canada* 98, 58–61.

Pilate, G., Ellis, D.D. and Hawkins, S. (1997) Transgene expression in field-grown poplar. In: Klopfenstein, N.B., Chun, Y.W., Kim, M.-S. and Ahuja, M.R. (eds) *Micropropagation, Genetic Engineering, and Molecular Biology of Populus.* Section V, Chapter 26. General Technical Report RM-GTR-297. USDA Forest Service, Rocky Mountain Forest and Range Experiment Station, Fort Collins, Colorado, pp. 84–89.

Pilate, G., Guiney, E., Holt, K., Petit-Conil, M., Lapierre, C., Leple, J.-C., *et al.* (2002) Field and pulping performances of transgenic trees with altered lignification. *Nature Biotechnology* 20, 607–612.

Pilate, G., Dejardin, A., Laurans, F. and Leple, J.-C. (2004) Tension wood as a model for functional genomics of wood formation. *New Phytologist* 164, 63–72.

Pilipovic, A., Nikolic, N., Orlovic, S., Petrovic, N. and Krstic, B. (2005) Cadmium phytoextraction potential of poplar clones (*Populus* spp.). *Zeitschrift fur Naturforschung* 60, 247–251.

Pinon, J. (1992) Variability in the genus *Populus* in sensitivity to *Melampsora* rusts. *Silvae Genetica* 41, 25–34.

Plomion, C., Lalanne, C., Claverol, S., Meddour, H., Kohler, A., Bogeat-Triboulot, M.-B., *et al.* (2006) Mapping the proteome of poplar and application to the discovery of drought-stress responsive proteins. *Proteomics* 6, 6509–6527.

Pospiskova, M. and Salkova, I. (2006) Population structure and parentage analysis of black poplar along the Morava River. *Canadian Journal of Forest Research* 36, 1067–1076.

Pregitzer, K.S. and Barnes, B.V. (1980) Flowering phenology of *Populus tremuloides* and *P. grandidentata* and the potential for hybridization. *Canadian Journal of Forest Research* 10, 218–223.

Purdy, B.G. and Bayer, R.J. (1995) Allozyme variation in the Athabasca sand dune endemic, *Salix silicola*, and the closely related widespread species, *S. alaxensis. Systematic Botany* 20, 179–190.

Puri, S., Swamy, S.L. and Jaiswal, A.K. (2002) Evaluation of *Populus deltoides* clones under nursery, field and agrisilviculture system in subhumid tropics of central India. *New Forests* 23, 45–61.

Qin, G., Jiang, Y. and Qiao, Y. (2003) Resistance to adversity of new poplar clones. *Forestry Studies in China* 5, 18–21.

Rae, A.M., Robinson, K.M., Street, N.R. and Taylor, G. (2004) Morphological and physiological traits influencing biomass productivity in short-rotation coppice poplar. *Canadian Journal of Forest Research* 34, 1488–1498.

Ragonese, A. (1989) Fitotecnia en sauces. Presentation. In: *Actas Primeras Jornadas sobre Silvicultura y Mejoramiento del Género Salix, CIEF*, Noviembre 1989, Buenos Aires.

Rahman, M.H. and Rajora, O.P. (2002) Microsatellite DNA fingerprinting, differentiation, and genetic relationships of clones, cultivars, and varieties of six poplar species from three sections of the genus *Populus. Genome* 45, 1083–1094.

Rajagopal, J., Bashyam, L., Bhatia, S., Khurana, D.K., Srivastava, P.S. and Lakshmikumaran, M. (2000) Evaluation of genetic diversity in the Himalayan poplar using RAPD markers. *Silvae Genetica* 49, 60–66.
Rajora, O. and Rahman, M. (2003) Microsatellite DNA and RAPD fingerprinting, identification and genetic relationships of hybrid poplar (*Populus* × *canadensis*) cultivars. *Theoretical and Applied Genetics* 106, 470–477.
Rajora, O.M.P. and Zsuffa, L. (1990) Allozyme divergence and evolutionary relationships among *Populus deltoides, P. nigra*, and *P. maximowiczii. Genome* 33, 44–49.
Ramirez, C.C., Zamudio, F., Verdugo, J.V. and Nunez, M.E. (2004) Differential susceptibility of poplar hybrids to the aphid *Chaitophorus leucomelas* (Homoptera: Aphididae). *Journal of Economic Entomology* 97, 1965–1971.
Raquin, C., Troussard, L. and Villar, M. (1993) In-ovary embryo culture as a tool for poplar hybridization. *Canadian Journal of Botany* 71, 1271–1275.
Rathmacher, G., Niggermann, M., Kohnen, M., Ziegenhagen, B. and Bialozyt, R. (2010) Short-distance gene flow in *Populus nigra* L. accounts for small-scale spatial genetic structures: implications for *in situ* conservation measures. *Conservation Genetics* 11, 1327–1338.
Redei, K. (2000) Early performance of promising white poplar (*Populus alba*) clones in sandy ridges between the rivers Danube and Tisza in Hungary. *Forestry* 73, 407–413.
Regione Emilia-Romagna Assessorato Agricoltura (1999) Il Pioppo. Supplemento ad Agricoltura No 4. Regione Emilia-Romagna, Bologna, Italy.
Reinartz, J.A. and Warne, E.L. (1993) Development of vegetation in small created wetlands in southeastern Wisconson. *Wetlands* 13, 153–164.
Richardson, J., Cooke, J.E.K., Isebrands, J.G., Thomas, B.R. and Van Rees, K.C.J. (2007) Poplar research in Canada – a historical perspective with a view to the future. *Canadian Journal of Botany* 85, 1136–1146.
Riddell-Black, D. (1998) Development of a water industry manual for biosolids use in short rotation forestry. *Biomass and Bioenergy* 15, 101–107.
Riemenschneider, D.E., McMahon, B.G. and Ostry, M.E. (1992) Use of selection indices to increase tree height and to control damaging agents in 2-year-old balsam poplar. *Canadian Journal of Forest Research* 22, 561–567.
Riemenschneider, D.E., Stanton, B.J., Vallée, G. and Périnet, P. (2001a) Poplar breeding strategies. In: Dickmann, D.I., Isebrands, J.G., Eckenwalder, J.E. and Richardson, J. (eds) *Poplar Culture in North America.* National Research Council of Canada Research Press, Ottawa, pp. 43–76.
Riemenschneider, D.E., Berguson, W.E., Dickmann, D.I., Hall, R.B., Isebrands, J.G., Mohn, C.A., *et al.* (2001b) Poplar breeding and testing strategies in the north-central U.S.: demonstration of potential yield and consideration of future research needs. *The Forestry Chronicle* 77, 245–253.
Rivest, D., Cogliastro, A., Vanasse, A. and Olivier, A. (2009) Production of soybean associated with different hybrid poplar clones in a tree-based intercropping system in southwestern Quebec, Canada. *Agriculture, Ecosystems and Environment* 131, 51–60.
Roberds, J.H. and Bishir, J.W. (1997) Risk analyses in clonal forestry. *Canadian Journal of Forest Research* 27, 425–432.
Rocha, S.P. (1991) Micropropagation and *Agrobacterium* transformation of willow (*Salix lucida* Muhl.). MSc. thesis, State University of New York, College of Environmental Science and Forestry, Syracuse, New York.
Roche, B.M. and Fritz, R.S. (1998) Effects of host plant hybridization on resistance to willow leaf rust caused by *Melampsora* spp. *European Journal of Forest Pathology* 28, 259–270.
Rohde, A., Bastien, C. and Boerjan, W. (2011a) Temperature signals contribute to the timing of photoperiodic growth cessation and bud set in poplar. *Tree Physiology* 31, 472–482.
Rohde, A., Storme, V., Jorge, V., Gaudet, M., Fabbrini, V.F., Ruttink, T., *et al.* (2011b) Bud set in poplar – genetic dissection of a complex trait in natural and hybrid populations. *New Phytologist* 189, 106–121.
Rönnberg-Wästljung, A.C. (2001) Genetic structure of growth and phenological traits in *Salix viminalis. Canadian Journal of Forest Research* 31, 276–282.
Rönnberg-Wästljung, A.C. and Gullberg, U. (1996) Genetic relationships between growth characters in *Salix viminalis* grown in Sweden. *Theoretical and Applied Genetics* 93, 15–21.
Rönnberg-Wästljung, A.C., Tsarouhas, V., Semerikov, V. and Lagercrantz, U. (2003) A genetic linkage map of a tetraploid *Salix viminalis* × *S. dasyclados* hybrid based on AFLP markers. *Forest Genetics* 10, 185–194.

Rönnberg-Wästljung, A.C., Glynn, C. and Weih, M. (2005) QTL analyses of drought tolerance and growth for a *Salix dasyclados* × *Salix viminalis* hybrid in contrasting water regimes. *Theoretical and Applied Genetics* 110, 537–549.
Rönnberg-Wästljung, A.C., Åhman, I., Glynn, C. and Widenfalk, O. (2006) Quantitative trait loci for resistance to herbivores in willow: field experiments with varying soils and climate. *Entomologia Experimentalis et Applicata* 118, 163–174.
Rönnberg-Wästljung, A.C., Samils, B., Tsarouhas, V. and Gullberg, U. (2008) Resistance to *Melampsora larici-epitea* leaf rust in *Salix*: analyses of quantitative trait loci. *Journal of Applied Genetics* 49, 321–331.
Rood, S.B. and Mahoney, J.M. (1990) Collapse of riparian poplar forests downstream from dams in western prairies: probable causes and prospects for mitigation. *Environmental Management* 14, 451–464.
Rood, S.B., Kalischuk, A.R., Polzin, M.L. and Braatne, J.H. (2003a) Branch propagation, not cladoptosis, permits dispersive, clonal reproduction of riparian cottonwoods. *Forest Ecology and Management* 186, 227–242.
Rood, S.B., Braatne, J.H. and Hughes, F.M.R. (2003b) Ecophysiology of riparian cottonwoods: stream flow dependency, water relations and restoration. *Tree Physiology* 23, 1113–1124.
Rood, S.B., Nielsen, J.L., Shenton, L., Gill, K.M. and Letts, M.G. (2010) Effects of flooding on leaf development, transpiration, and photosynthesis in narrowleaf cottonwood, a willow-like poplar. *Photosynthesis Research* 104, 31–39.
Rowland, D.L., Sher, A.A. and Marshal, D.L. (2004) Inter- and intra-population variation in seedling performance of Rio Grande cottonwood under low and high salinity. *Canadian Journal of Forest Research* 34, 1458–1466.
Ruger, N., Schulter, M. and Matthies, M. (2005) A fuzzy habitat suitability index for *Populus euphratica* in the northern Amudarya delta (Uzbekistan). *Ecological Modeling* 184, 313–328.
Rytter, L. (2002) Nutrient content in stems of hybrid aspen as affected by tree age and tree size, and nutrient removal with harvest. *Biomass and Bioenergy* 23, 13–25.
Rytter, L. and Stener, L.-G. (2003) Clonal variation in nutrient content in woody biomass of hybrid aspen (*Populus tremula* L. × *P. tremuloides* Michx.). *Silva Fennica* 37, 313–324.
Sabatti, M. (1994) Ecologia genetica in popolazioni naturali di *Populus alba* L.: produttività, fenologia ed efficienza di uso idrico. PhD thesis, University of Tuscia, Viterbo, Italy.
Sabatti, M., D'Ovidio, R., Tanzarella, O.A. and Scarascia Mugnozza, G.E. (2001) Assessment of geographic variation by RAPD markers among Italian open-pollinated progenies of *Populus alba* L. *Genetic Resources and Crop Evolution* 48, 423–428.
SAGPyA (2004) Estadisticas forestales 2004. Dirección de Forestación, Secretaría de Agricultura, Ganadería, Pesca y Alimentación, Ministerio de Economía, Buenos Aires.
St Clair, S.B., Mock, K.E., LaMalfa, E.M., Campbell, R.B. and Ryel, R.J. (2010) Genetic contributions to phenotypic variation in physiology, growth, and vigor of western aspen (*Populus tremuloides*) clones. *Forest Science* 56, 222–230.
Salick, J. and Pfeffer, E. (1999) The interplay of hybridization and clonal reproduction on the evolution of willows. *Plant Ecology* 141, 163–178.
Sannigrahi, P. and Ragauskas, A.J. (2010) Poplar as a feedstock for biofuels: a review of compositional characteristics. *Biofuels, Bioproducts, Biorefineries* 4, 209–226.
Savka, M.A., Dawson, J.O., Jokela, J.J. and Skirvin, R.M. (1987) A liquid culture method for rescuing immature embryos of eastern cottonwood. *Plant Cell and Tissue Organ Culture* 10, 221–226.
Schluter, M., Ruger, N., Savitsky, A.G., Novikova, N.M., Matthies, M. and Lieth, H. (2006) TUGAI: an integrated simulation tool for ecological assessment of alternative water management strategies in a degraded river delta. *Environmental Management* 38, 638–653.
Schrader, G. and Unger, J.-G. (2003) Plant quarantine as a measure against invasive alien species: the framework of the international plant protection convention and the plant health regulations in the European Union. *Biological Invasions* 5, 357–364.
Schrader, J., Nilsson, J., Mellerowicz, E., Berglund, A., Nilsson, P., Hertzberg, M., *et al.* (2004) A high-resolution transcript profile across the wood-forming meristem of poplar identifies potential regulators of cambial stem cell identity. *Plant Cell* 16, 2278–2292.
Schreiner, E.J. (1970) Genetics of eastern cottonwood. USDA Forest Service Research Paper WO-11. USDA Forest Service, Washington, DC.
Schreiner, E.J. (1974) Poplars can be bred to order for mini-rotation fiber, timber and veneer production and for amenity plantings. In: *Proceedings of the 21st Northeastern Forest Tree Improvement Conference,*

27–30 August, 1973, Fredericton, NB, Canada. USDA Forest Service, Northeastern Forest Experiment Station, Upper Darby, Pennsylvania, pp. 85–96.

Schroeder, H., Hoeltken, A.M. and Fladung, M. (2011) Differentiation of *Populus* species using chloroplast single nucleotide polymorphism (SNP) markers – essential for comprehensible and reliable poplar breeding. *Plant Biology* 14, 374–381.

Schroeder, W., Kort, J., Savoie, P. and Preto, F. (2009) Biomass harvest from natural willow rings around prairie wetlands. *Bioenergy Research* 2, 99–105.

Scott, D.S., Paterson, L., Piskorz, J. and Radlein, D. (2000) Pretreatment of poplar wood for fast pyrolysis: rate of cation removal. *Journal of Analytical and Applied Pyrolysis* 57, 169–176.

Sennerby-Forsse, L., Sirén, G. and Lestander, T. (1983) Results from the first preliminary test with short rotation willow clones. SEF Project Technical Report 30. Swedish University of Agricultural Sciences, Uppsala, Sweden.

Serapiglia, M.J., Cameron, K.D., Stipanovic, A.J. and Smart, L.B. (2008) High-resolution thermogravimetric analysis for rapid characterization of biomass composition and selection of shrub willow varieties. *Applied Biochemistry and Biotechnology* 145, 3–11.

Serapiglia, M.J., Cameron, K.D., Stipanovic, A.J. and Smart, L.B. (2009) Analysis of biomass composition using high-resolution thermogravimetric analysis and percent bark content for the selection of shrub willow bioenergy crop varieties. *Bioenergy Research* 2, 1–9.

Shani, Z., Dekel, M., Jensen, C.S., Tzfira, T., Goren, R., Altman, A., *et al.* (2000) *Arabidopsis thaliana endo-1,4-β-glucanese (cel1)* promoter mediates *uidA* expression in elongating tissues of aspen (*Populus tremula*). *Journal of Plant Physiology* 156, 118–120.

Shaw, D.V. and Hood, J.V. (1985) Maximizing gain per effort by using clonal replicates in genetic tests. *Theoretical and Applied Genetics* 71, 392–399.

Sigurdsson, V., Anamthawat-Jonsson, K. and Sigurgeirsson, A. (1995) DNA fingerprinting of *Populus trichocarpa* clones using RAPD markers. *New Forests* 10, 197–206.

Singh, A.P. (2000) Relative natural resistance of *Populus deltoides* clones against defoliator *Clostera cupreata* (Lepidoptera: Notodontidae) in northern India. *Agroforestry Systems* 49, 319–326.

Sixto, H., Grau, J.M., Alba, N. and Alia, R. (2005) Response to sodium chloride in different species and clones of genus *Populus* L. *Forestry* 78, 93–104.

Sixto, H., Barrio, M., Alba, N., Hernández, M.J., Montoto, J.I., Roig, S., *et al.* (2006) Poplar trials in Spain for biomass as a renewable energy source. In: Nanjing Forestry University (ed.) *Abstracts, Fourth International Poplar Symposium, Meeting the Needs of a Growing World Through Poplar and Willow Science, Nanjing, China, 5–9 June 2006.* Nanjing Forestry University, Nanjing, China, pp. 113–114.

Sixto, H., Hernandez, M.J., Barrio, M., Carrasco, J. and Canellas, I. (2007) Plantaciones del genero *Populus* para la produccion de biomasa con fines energeticos: revision. *Investigacion Agraria: Sistemas y Recursos Forestales* 16, 277–294.

Skvortsov, A.K. (1968) *Willows of the USSR. Systematical Review.* Nauka, Moscow.

Skvortsov, A. (1999) *Willows of Russia and Adjacent Countries: Taxonomical and Geographical Revision* (English translation of 1968 Russian edition). University of Joensuu, Joensuu, Finland.

Slavov, G.T., DiFazio, S.P. and Strauss, S.H. (2004) Gene flow in forest trees: gene migration patterns and landscape modelling of transgene dispersal in hybrid poplar. In: den Nijs, H.C.M., Bartsch, D. and Sweet, J. (eds) *Introgression from Genetically Modified Plants into Wild Relatives.* CAB International, Wallingford, UK, pp. 89–106.

Smart, L.B. and Cameron, K.D. (2008) Genetic improvement of willow (*Salix* spp.) as a dedicated bioenergy crop. In: Vermerris, W.E. (ed.) *Genetic Improvement of Bioenergy Crops.* Springer Science, New York, pp. 347–376.

Smart, L.B., Volk, T.A., Lin, J., Kopp, R.F., Phillips, I.S., Cameron, K.D., *et al.* (2005) Genetic improvement of shrub willow (*Salix* spp.) crops for bioenergy and environmental applications in the United States. *Unasylva* 56, 51–55.

Smulders, M.J.M., Cottrell, J.E., Lefevre, F., van der Schoot, J., Arens, P., Vosman, B., *et al.* (2008a) Structure of the genetic diversity in black poplar (*Populus nigra* L.) populations across European river systems: consequences for conservation and restoration. *Forest Ecology and Management* 255, 1388–1399.

Smulders, M.J.M., Beringen, R., Volosyanchuk R., Vanden Broeck, A., van der Schoot, J., Arens, P., *et al.* (2008b) Natural hybridisation between *Populus nigra* L. and *P.* ×*canadensis* Moench. Hybrid offspring competes for niches along the Rhine river in the Netherlands. *Tree Genetics and Genomes* 4, 663–675.

Song, J., Lu, S., Chen, Z.-Z., Rodrigo, L. and Chiang, V.L. (2006) Genetic transformation of *Populus trichocarpa* genotype Nisqually-1: a functional genomic tool for woody plants. *Plant and Cell Physiology* 47, 1582–1589.

Spiers, A.G. (1998) *Melampsora* and *Marssonina* pathogens of poplars and willows in New Zealand. *European Journal of Forest Pathology* 28, 233–240.

Stanton, B.J. (2005) The effect of reciprocal hybridization on reproduction of the intersectional cross, *Populus × generosa*. *Forest Genetics* 12, 131–140.

Stanton, B.J. and Villar, M. (1996) Controlled reproduction of *Populus*. In: Stettler, R.F., Bradshaw, H.D. Jr, Heilman, P.E. and Hinckley, T.M. (eds) *Biology of Populus and its Implications for Management and Conservation*. National Research Council of Canada Research Press, Ottawa, pp. 113–138.

Stanton, B.J., Neale, D.B. and Li, S. (2010) *Populus* breeding: from the classical to the genomic approach. In: Jansson, S., Bhalerao, R.P. and Groover, A.T. (eds) *Genetics and Genomics of Populus*. Springer, New York, pp. 309–348.

Starova, N.V. (1980) Selection of *Salicaceae*. In: *Forest Industry*. Moscow. (In Russian)

Steenackers, J., Steenackers, M., Steenackers, V. and Stevens, M. (1996) Poplar diseases, consequences on growth and wood quality. *Biomass and Bioenergy* 10, 267–274.

Steenackers, V. (1996) Towards a global management of poplar genetic resources. Presentation at Twentieth Session of the International Poplar Commission, Budapest, Hungary 1–4 October 1996. International Poplar Commission. Available from IPC Secretariat, FAO, Rome.

Stelzer, H.E. (1997) Evaluating genetic diversity concerns in clonal deployments. *Canadian Journal of Forest Research* 27, 438–441.

Stenvall, N. (2006) Multiplication of hybrid aspen (*Populus tremula* L. × *P. tremuloides* Michx.) from cuttings. Academic Dissertation, University of Helsinki, Faculty of Agriculture and Forestry, Helsinki.

Sterky, F., Regan, S., Karlsson, J., Hertzberg, M., Rohde, A., Holmberg, A., *et al.* (1998) Gene discovery in the wood-forming tissues of poplar: analysis of 5,692 expressed sequence tags. *Proceedings of the National Academy of Sciences* 95, 13330–13335.

Stettler, R.F., Koster, R. and Steenackers, V. (1980) Interspecific crossability studies in poplars. *Theoretical and Applied Genetics* 58, 273–282.

Stettler, R.F., Fenn, R.C., Heilman, P.E. and Stanton, B.J. (1988) *Populus trichocarpa* × *Populus deltoides* hybrids for short rotation culture: variation patterns and 4-year field performance. *Canadian Journal of Forest Research* 18, 745–753.

Stettler, R.F., Zsuffa, L. and Wu, R. (1996) The role of hybridization in the genetic manipulation of *Populus*. In: Stettler, R.F., Bradshaw, H.D. Jr, Heilman, P.E. and Hinckley, T.M. (eds) *Biology of Populus and its Implications for Management and Conservation*. National Research Council of Canada Research Press, Ottawa, pp. 87–112.

Stirling, B., Newcombe, G., Vrebalov, J., Bosdet, I. and Bradshaw, H.D. Jr (2001) Suppressed recombination around the MXC3 locus, a major gene for resistance to poplar leaf rust. *Theoretical and Applied Genetics* 103, 1129–1137.

Stoehr, M.U., Cai, Y. and Zsuffa, L. (1989) *In vitro* plant-regeneration via callus-culture of mature *Salix exigua*. *Canadian Journal of Forest Research* 19, 1634–1638.

Stolarski, M., Szczukowski, S., Tworkowski, J. and Klasa, A. (2008) Productivity of seven clones of willow coppice in annual and quadrennial cutting cycles. *Biomass and Bioenergy* 32, 1227–1234.

Stone, R. (2008) China plans $3.5 billion GM crops initiative. *Science* 321, 1279.

Storme, V., Vanden Broeck, A., Ivens, B., Halfmaerten, D., Van Slycken, J., Castiglione, S., *et al.* (2004) *Ex-situ* conservation of black poplar in Europe: genetic diversity in nine gene bank collections and their value for nature development. *Theoretical and Applied Genetics* 108, 969–981.

Stott, K.G. (1992) Willows in the service of man. *Proceedings of the Royal Society of Edinburgh* 98B, 169–182.

Stout, A.B. and Schreiner, E.J. (1933) Results of a project in hybridizing poplars. *Journal of Heredity* 24, 216–229.

Stout, A.B., McKee, R.H. and Schreiner, E.J. (1927) The breeding of forest trees for pulpwood. *Journal of the New York Botanical Garden* 28, 49–63.

Strauss, S.H., Lande, R. and Namkoong, G. (1992) Limitations of molecular-marker-aided selection in forest tree breeding. *Canadian Journal of Forest Research* 22, 1050–1061.

Strauss, S.H., Rottmann, W.H., Brunner, A.M. and Sheppard, L.A. (1995) Genetic engineering of reproductive sterility in forest trees. *Molecular Breeding* 1, 5–26.

Strauss, S.H., Know, S.A. and Jenkins, J. (1997) Benefits and risks of transgenic Roundup® ready cottonwoods. *Journal of Forestry* 95, 12–19.

Strauss, S.H., Campbell, M.M., Pryor, S.N., Coventry, P. and Burley, J. (2001) Plantation certification and genetic engineering: FSCs ban on research is counterproductive. *Journal of Forestry* 99, 4–7.
Sulima, P., Przyborowski, J.A. and Zaluski, D. (2009) RAPD markers reveal genetic diversity in *Salix purpurea* L. *Crop Science* 49, 857–863.
Sundberg, B., Tuominen, H., Nilsson, O., Moritz, T., Little, C.H.A., Sandberg, G., *et al.* (1997) Growth and development alteration in transgenic *Populus*: status and potential applications. In: Klopfenstein, N.B., Chun, Y.W., Kim, M.-S. and Ahuja, M.R. (eds) *Micropropagation, Genetic Engineering, and Molecular Biology of Populus.* General Technical Report RM-GTR-297. USDA Forest Service, Rocky Mountain Forest and Range Experiment Station, Fort Collins, Colorado, pp. 74–83.
Szczukowski, S., Tworkowski, J. and Stolarski, M. (2004) *Willow for Energy.* Wydawnictwo Plantpress Sp. z o.o., Krakow, Poland.
Tabbener, H.E. and Cottrell, J.E. (2003) The use of PCR based DNA markers to study the paternity of poplar seedlings. *Forest Ecology and Management* 179, 363–376.
Tabor, G.M., Kubisiak, T.L., Klopfenstein, N.B., Hall, R.B. and McNabb, H.S. Jr (2000) Bulked segregant analysis identifies molecular markers linked to *Melampsora medusae* resistance in *Populus deltoides. Phytopathology* 90, 1039–1042.
Takacs, T., Radimszky, L. and Nemeth, T. (2005) The arbuscular mycorrhizal status of poplar clones selected for phytoremediation of soils contaminated with heavy metals. *Zeitschrift fur Naturforschung* 60, 357–361.
Talbot, P., Thompson, S.L., Schroeder, W. and Isabel, N. (2011) An efficient single nucleotide polymorphism assay to diagnose the genomic identity of poplar species and hybrids on the Canadian prairies. *Canadian Journal of Forest Research* 41, 1102–1111.
Tamura, S. and Kudo, G. (2000) Wind pollination and insect pollination of two temperate willow species, *Salix miyabeana* and *Salix sachalinensis. Plant Ecology* 147, 185–192.
Tanksley, S.D. and McCouch, S.R. (1997) Seed banks and molecular maps: unlocking genetic potential from the wild. *Science* 277, 1063–1066.
Tauer, C.G. (1979) Seed tree, vacuum, and temperature effects on eastern cottonwood seed viability during extended storage. *Forest Science* 25, 112–114.
Telenius, B.F. (1999) Stand growth of deciduous pioneer tree species on fertile agricultural land in southern Sweden. *Biomass and Bioenergy* 16, 13–23.
TGERC (2000) Tree Genetic Engineering Research Cooperative Annual Report 1999–2000. Oregon State University, Corvallis, Oregon.
Thakur, A.K., Sharma, S. and Srivastava, D.K. (2005) Plant regeneration and genetic transformation studies in petiole tissue of Himalayan poplar (*Populus ciliata* Wall.). *Current Science* 89, 664–668.
Thielges, B.A. (1985) Breeding poplars for disease resistance. FAO Forestry Paper 56. FAO, Rome.
Thomas, B.R., MacDonald, S.E. and Dancik, B.P. (1997) Variance components, heritabilities, and gain estimates for growth chamber and field performance of *Populus tremuloides*: growth performance. *Silvae Genetica* 46, 317–326.
Toplu, F. (2005) Breeding and conservation of black poplar (*Populus nigra*) gene resources in Turkey. *Unasylva* 56, 26–30.
Towill, L.E. (2002) Crypreservation of plant germplasm: introduction and some observations. In: Towill, L.E. and Bajaj, Y.P.S. (eds) *Cryopreservation of Plant Germplasm II.* Biotechnology in Agricultural and Forestry Series, Volume 50. Springer Verlag, London, pp. 3–21.
Triest, L., De Greef, B., Vermeersch, S., Van Slycken, J. and Coart, E. (1999) Are *Salix alba* and *Salix fragilis* a hybrid species complex? *Plant Systematics and Evolution* 215, 169–187.
Trybush, S., Jahodova, S., Macalpine, W. and Karp, A. (2008) A genetic study of a *Salix* germplasm resource reveals new insights into relationships among subgenera, sections and species. *Bioenergy Research* 1, 67–79.
Tsai, C.J. and Hubscher, S.L. (2004) Cryopreservation in *Populus* functional genomics. *New Phytologist* 164, 73–81.
Tsarev, A.P. (2005) Basic directions and results of willows breeding in Russia. *Transactions of Forest Engineering Faculty of Petrozavodsk State University* 5, 123–131. (In Russian)
Tsarouhas, V., Gullberg, U. and Lagercrantz, U. (2002) An AFLP and RFLP linkage map and quantitative trait locus (QTL) analysis of growth traits in *Salix. Theoretical and Applied Genetics* 105, 277–288.
Tsarouhas, V., Gullberg, U. and Lagercrantz, U. (2003) Mapping of quantitative trait loci controlling timing of bud flush in *Salix. Hereditas* 138, 172–178.

Tsarouhas, V., Gullberg, U. and Lagercrantz, U. (2004) Mapping of quantitative trait loci (QTLs) affecting autumn freezing resistance and phenology in *Salix. Theoretical and Applied Genetics* 108, 1335–1342.

Tuskan, G.A., DiFazio, S., Jansson, S., Bohlmann, J., Grigoriev, I., Hellsten, U., *et al.* (2006) The genome of black cottonwood, *Populus trichocarpa* (Torr. & Gray). *Science* 313, 1596–1604.

Tuskan, G., Slavov, G., DiFazio, S., Muchero, W., Pryia, R., Schackwitz, W., *et al.* (2011) *Populus* re-sequencing: towards genome-wide association studies. *Bio Med Central Proceedings* 5(Suppl. 7), I21.

Ulloa, J. and Villacura, L. (2005) Contribution of a private poplar industry in Chile to sustainable rural development. *Unasylva* 56, 12–17.

Vahala, T., Stabel, P. and Eriksson, T. (1989) Genetic transformation of willows (*Salix* spp.) by *Agrobacterium tumefaciens. Plant Cell Reports* 8, 55–58.

Van Dam, B.C. and Bordacs, S. (eds) (2002) *Genetic Diversity in River Populations of European Black Poplar. Implications for Riparian Eco-system Management.* Csiszar Nyomda Ltd, Budapest.

Vanden Broeck, A., Villar, M., Van Bockstaele, E. and Van Slycken, J. (2005) Natural hybridization between cultivated poplars and their wild relatives: evidence and consequences for native poplar populations. *Annals of Forest Science* 62, 601–613.

van der Schoot, J., Pospiskova, M., Vosman, B. and Smulders, M.J.M. (2000) Development and characterization of microsatellite markers in black poplar (*Populus nigra* L.). *Theoretical and Applied Genetics* 101, 317–322.

van Loo, M., Joseph, J.A., Heinze, B., Fay, M.F. and Lexer, C. (2008) Clonality and spatial genetic structure in *Populus* ×*canescens* and its sympatric backcross parent *P. alba* in a central European hybrid zone. *New Phytologist* 177, 506–516.

Varshney, R.K., Graner, A. and Sorrells, M.E. (2005) Genomics-assisted breeding for crop improvement. *Trends in Plant Science* 10, 621–630.

Veresin, M.M., Tsarev, A.P. and Sivolapov, A.I. (1974) High-quality stands of poplars and treelike willows in Hopersky State Reserve. In: Environment Conservation Society, Council of Voronezh region (eds) *Natural Resources of the Voronezh Area and Their Protection.* Publishing house of Voronezh University, Voronezh, Russia, pp. 65–67. (In Russian)

Verwijst, T. (2001) Willows: an underestimated resource for environment and society. *The Forestry Chronicle* 77, 281–285.

Vietto, L. and Bianco, B. (2004) Progress on national activities on gene conservation of black poplar (*Populus nigra* L.) and white poplar (*Populus alba* L.) in Italy. In: Koskela, J., de Vries, S.M.G., Kajba, D. and von Wuhlisch, G. (eds) *Populus nigra Network.* European Forest Genetic Resources Programme (EUFORGEN). International Plant Genetic Resources Institute, Rome, pp. 45–53.

Vietto, L. and Cagelli, L. (2006) Raccolta, conservazione e gestione del germoplasma delle *Salicaceae.* Manuale per la raccolta, studio, conservazione e gestione del germoplasma. APAT 140–147.

Villar, M., Gaget, M., Said, C., Knox, R.B. and Dumas, C. (1987) Incompatibility in *Populus*: structural and cytochemical characteristics of the receptive stigmas of *Populus alba* and *P. nigra. Journal of Cell Science* 87, 483–490.

Villar, M., Lefevre, F., Bradshaw, H.D. Jr and Teissier du Cros, E. (1996) Molecular genetics of rust resistance in poplars (*Melampsora larici-populina* Kleb/*Populus* sp.) by bulked segregant analysis in a 2×2 factorial mating design. *Genetics* 143, 531–536.

Volk, T.A., Abrahamson, L.P., Cameron, K.D., Castellano, P., Corbin, T., Fabio, E., *et al.* (2011) Yields of willow biomass crops across a range of sites in North America. *Aspects of Applied Biology* 112, 67–74.

Wakita, Y., Yokota, S., Yoshizawa, N., Katsuki, T., Nishiyama, Y., Yokoyama, T., *et al.* (2005) Interfamilial cell fusion among leaf protoplasts of *Populus alba*, *Betula platyphylla* and *Alnus firma*: assessment of electric treatment and *in vitro* culture conditions. *Plant Cell, Tissue and Organ Culture* 83, 319–326.

Walters, C., Hill, L.M., Volk, G.M. and Haiby, K. (2011) 'Intermediate' seed storage physiology: *Populus* as a natural model system of a complex trait. Tenth Conference of the International Society for Seed Science, Costa do Sauipe, Brazil. *Informativo ABRATES* 21, 280.

Weber, J.C. and Stettler, R.F. (1981) Isoenzyme variation among ten populations of *Populus trichocarpa* Torr. et Gray in the Pacific Northwest. *Silvae Genetica* 30, 82–87.

Weber, J.C., Stettler, R.F. and Heilman, P.E. (1985) Genetic variation and productivity of *Populus trichocarpa* and its hybrids I. Morphology and phenology for 50 native clones. *Canadian Journal of Forest Research* 15, 376–383.

Wegrzyn, J.L., Eckert, A.J., Choi, M., Lee, J.M., Stanton, B.J., Sykes, R., *et al.* (2010) Association genetics of traits controlling lignin and cellulose biosynthesis in black cottonwood (*Populus trichocarpa, Salicaceae*) secondary xylem. *New Phytologist* 188, 515–532.

Weih, M. and Bonosi, L. (2009) Assessment of genotype ranking in long-term biomass production of *Salix* based on juvenile plant traits: breeding implications. *Bioenergy Research* 2, 29–36.

Weih, M. and Nordh, N.-E. (2005) Determinants of biomass production in hybrid willows and the prediction of field performance from pot studies. *Tree Physiology* 25, 1197–1206.

Weih, M. and Rönnberg-Wästljung, A.C. (2007) Shoot biomass is related to vertical leaf nitrogen gradient in *Salix* canopies. *Tree Physiology* 27, 1551–1559.

Weih, M., Rönnberg-Wästljung, A.-C. and Glynn, C. (2006) Genetic basis of phenotypic correlations among growth traits in hybrid willow (*Salix dasyclados* × *S. viminalis*) grown under two water regimes. *New Phytologist* 170, 467–477.

Weih, M., Didon, U.M.E., Rönnberg-Wästljung, A.C. and Björkman, C. (2008) Integrated agricultural research and crop breeding: allelopathic weed control in cereals and long-term productivity in perennial biomass crops. *Agricultural Systems* 9, 99–107.

Weisgerber, H. (1993) Poplar breeding for the purpose of biomass production in short rotation periods in Germany: problems and first findings. *The Forestry Chronicle* 69, 727–729.

Weisgerber, H. and Han, Y. (2001) Diversity and breeding potential of poplar species in China. *The Forestry Chronicle* 77, 227–237.

Wettstein-Westersheim, W. (1933) Die kreuzungsmethode und die beschreibung von F_1 – bastarden bei *Populus*. [The method of hybridization and the description of F_1 hybrids of poplars.] *Zeitschrift fur Zuchtung* 18, 597–626. (*Plant Breeding Abstracts* 4, 163–164.)

Wheeler, N., Payne, P., Hipkins, V., Saich, R., Kenny, S. and Tuskan, G. (2006) Polymix breeding with paternity analysis in *Populus*: a test for differential reproductive success (DRS) among pollen donors. *Tree Genetics and Genomes* 2, 53–60.

Whitham, T.G., Martinsen, G.D., Floate, K.D., Dungey, H.S., Potts, B.M. and Keim, P. (1999) Plant hybrid zones affect biodiversity: tools for a genetic-based understanding of community structure. *Ecology* 80, 416–428.

Whitham, T.G., Young, W.P., Martinsen, G.D., Gehring, C.A., Schweitzer, J.A., Shuster, S.M., *et al.* (2003) Community and ecosystem genetics: a consequence of the extended phenotype. *Ecology* 84, 559–573.

Whitham, T.G., DiFazio, S.P., Schweitzer, J.A., Shuster, S.M., Allan, G.J., Bailey, J.K., *et al.* (2008) Extending genomics to natural communities and ecosystems. *Science* 320, 492–495.

Wikberg, J. (2006) Water relations in *Salix* with focus on drought responses. Doctoral thesis, Swedish University of Agricultural Sciences, Uppsala, Sweden.

Wikberg, J. and Ögren, E. (2004) Interrelationships between water use and growth traits in biomass-producing willows. *Trees* 18, 70–76.

Wikberg, J. and Ögren, E. (2007) Variation in drought resistance, drought acclimation and water conservation in four willow cultivars used for biomass production. *Tree Physiology* 27, 1339–1346.

Wilkinson, A.G. (2000) Introduced forest trees in New Zealand; recognition, role and seed source. 17. The Poplars *Populus* spp. FRI Bulletin No 124. New Zealand Forest Research Institute, Rotorua, New Zealand.

Winfield, M.O., Arnold, G.M., Cooper, F., Le Ray, M., White, J., Karp, A., *et al.* (1998) A study of genetic diversity in *Populus nigra* subsp. *betulifolia* in the upper Severn area of the UK using AFLP markers. *Molecular Ecology* 7, 3–10.

Wu, R. and Stettler, R.F. (1994) Quantitative genetics of growth and development in *Populus*. I. A three-generation comparison of tree architecture during the first 2 years of growth. *Theoretical and Applied Genetics* 89, 1046–1054.

Wullschleger, S.D., Jansson, S. and Taylor, G. (2002) Genomics and forest biology: *Populus* emerges as the perennial favorite. *The Plant Cell* 14, 2651–2655.

Xing, Z. (1995) Genetic transformation and regeneration of willows (*Salix* spp.). PhD dissertation, State University of New York, College of Environmental Science and Forestry, Syracuse, New York.

Yimit, H., Ayup, M., Wang, G.Z., Luo, H. and Ebeidulla, D. (2006) *P. euphratica* ecosystem fragility and protecting strategy on Tarim *P. euphratica* nature reserve in Xinjiang. In: Gao, W. and Ustin, S.L. (eds) *Remote Sensing and Modeling of Ecosystems for Sustainability III. Proceedings of SPIE* 6298, (doi:10.1117/12.684944).

Yin, T.M., DiFazio, S.P., Gunter, L.E., Riemenschneider, D. and Tuskan, G.A. (2004) Large-scale heterospecific segregation distortion in *Populus* revealed to be a dense genetic map. *Theoretical and Applied Genetics* 109, 451–463.

Yin, W. and Lu, W. (2005) Review of tree selection and afforestation for control of Asian longhorned beetle in north China. Forest Health and Biosecurity Working Paper FBS/ 7E. FAO, Rome.

Yu, Q., Mantyla, N. and Salonen, M. (2001) Rooting of hybrid clones of *Populus tremula* L. × *P. tremuloides* Michx. by stem cuttings derived from micropropagated plants. *Scandinavian Journal of Forest Research* 16, 238–245.

Yuceer, C., Harkess, R.L., Land, S.B. Jr and Luthe, D.S. (2002) Structure and developmental regulation of *CONSTANS-LIKE* genes isolated from *Populus deltoides. Plant Science* 163, 615–625.

Yuceer, C., Kubiske, M.E., Harkess, R.L. and Land, S.B. Jr (2003) Effects of induction treatments on flowering in *Populus deltoides. Tree Physiology* 23, 489–495.

Zacchini, M., Pietrini, F., Scarascia Mugnozza, G., Iori, V., Pietrosanti, L. and Massacci, A. (2009) Metal tolerance, accumulation and translocation in poplar and willow clones treated with cadmium in hydroponics. *Water and Soil Pollution* 197, 23–24.

Zalesny, R.S. Jr and Zalesny, J.A. (2011) Clonal variation in lateral and basal rooting of *Populus* irrigated with landfill leachate. *Silvae Genetica* 60, 35–44.

Zalesny, R.S. Jr, Bauer, E.O., Hall, R.B., Zalesny, J.A., Kunzman, J., Rog, C.J., *et al.* (2005a) Clonal variation in survival and growth of hybrid poplar and willow in an *in situ* trial on soils heavily contaminated with petroleum hydrocarbons. *International Journal of Phytoremediation* 7, 177–197.

Zalesny, R.S. Jr, Hall, R.B., Bauer, E.O. and Riemenschneider, D.E. (2005b) Soil temperature and precipitation affect the rooting ability of dormant hardwood cuttings of *Populus. Silvae Genetica* 54, 47–58.

Zhang, D., Zhang, Z. and Yang, K. (2006) QTL analysis of growth and wood chemical content traits in an interspecific backcross family of white poplar (*Populus tomentosa* × *P. bolleana*) × *P. tomentosa. Canadian Journal of Forest Research* 36, 2015–2023.

Zhang, X., Wu, N. and Li, C. (2005) Physiological and growth responses of *Populus davidiana* ecotypes to different soil water contents. *Journal of Arid Environments* 60, 567–579.

Zhang, Y. and Song, C. (2006) Impacts of afforestation, deforestation, and reforestation on forest cover in China from 1949 to 2003. *Journal of Forestry* 104, 383–387.

Zhang, Y.M., Chen, Y.N. and Pan, B.R. (2005) Distribution and floristics of desert plant communities in the lower reaches of Tarim River, southern Xinjiang, People's Republic of China. *Journal of Arid Environments* 63, 772–784.

Ziegenhagen, B., Gneuss, S., Rathmacher, G., Leyer, I., Bialozyt, R., Heinze, B., *et al.* (2008) A fast and simple genetic survey reveals the spread of poplar hybrids at a natural Elbe river site. *Conservation Genetics* 9, 373–379.

Zsuffa, L. (1975) A summary review of interspecific breeding in the genus *Populus* L. In: Fowler, D.P. and Yeatman, C.W. (eds) *Proceedings of the Fourteenth Canadian Tree Improvement Association Meeting, Fredericton, NB, Canada, August 1973.* Part 2. Canadian Forest Service, Ottawa, pp. 107–123.

Zsuffa, L. (1988) A review of the progress in selecting and breeding North American *Salix* species for energy plantations at the Faculty of Forestry, University of Toronto, Canada. In: Eriksson, G. (ed.) *Proceedings of Willow Breeding Symposium, Uppsala, Sweden, August 31–September 1, 1987.* Research Notes 41. Department of Forest Genetics, Swedish University of Agricultural Sciences, Uppsala, Sweden, pp. 41–51.

Zsuffa, L., Mosseler, A. and Raj, Y. (1984) Prospects for interspecific hybridization in willow for biomass production. In: Perttu, K. (ed.) *Ecology and Management of Forest Biomass Production Systems.* Swedish University of Agricultural Sciences, Uppsala, Sweden, pp. 261–281.

Zsuffa, L., Lin, D. and Payne, P. (1999) One-way crossing barriers in some interspecific crosses of *Aigeiros* and *Tacamahaca* poplars. *The Forestry Chronicle* 75, 833–836.

5 Operational Poplar and Willow Culture

John A. Stanturf[1]* and Cees van Oosten[2]

[1]*Center for Forest Disturbance Science, USDA Forest Service, Athens, Georgia, USA;*
[2]*SilviConsult Woody Crops Technology Inc., Nanaimo, British Columbia, Canada*

5.1 Introduction

The fast-growing species of poplars and willows have been cultivated for millennia for commodity uses, generally for their wood (FAO, 1980; Dickmann, 2006). Poplars and willows were economically important in rural areas because of their ease of propagation and adaptation to areas generally too wet for farming, e.g. flood plains and along watercourses. They provided wood for many rural uses and were particularly important for basketry. Poplar was widely used for industrial and commercial packaging, especially fresh agricultural products and foodstuffs, before development and widespread adoption of corrugated paperboard. For example, before powdered eggs were available for the bakery industry, cottonwood boxes were preferred for shipping raw eggs because the wood was non-aromatic and did not impart an odour to the eggs. The wood of poplar and willow continues to have many commodity and specialized uses, from matchsticks, paper pulp, plywood and other composite boards, to willow cricket bats (Table 5.1; Chapter 10, this volume).

Exploitation of natural poplar stands to meet heavy demands for wood during wartime in the 20th century, as well as demand from post-war industrialization, spurred the development of industrial plantations in Europe, initially in Italy and France (FAO, 1980; Castro and Zanuttini, 2008). This plantation development in the 1950s was facilitated by the intentional introduction of North American clones into European breeding programmes (FAO, 1980; Zsuffa *et al.*, 1996), although introductions were made much earlier (Pourtet, 1976). Depletion of natural stands of aspen and cottonwood in the USA through exploitation, conversion to agricultural land and alteration of natural river flows (Sternitzke, 1976), combined with the energy crisis of the 1970s, generated interest in poplar plantations for both roundwood and bioenergy. Interest in bioenergy waned as oil prices subsided, but current concerns for climate change caused by the release of fossil carbon into the atmosphere has renewed interest in short-rotation poplars and willows for bioenergy production (Berndes *et al.*, 2001; Zerbe, 2006; Christersson, 2008; Bergante and Facciotto, 2011; Volk *et al.*, 2011a, b; Zalesny *et al.*, 2011; Tullus *et al.*, 2012) and integrated systems for producing energy and commodities (Kelley, 2006).

Although poplars and willows occur as natural stands, this chapter is limited to the production and utilization of these species in plantations

*Corresponding author: drdirt48@gmail.com

(Plate 21A). Forest plantations are forests comprised of at least 40% planted trees that have ongoing management to maintain a composition of almost exclusively planted trees. Plantations are often characterized by few species planted at regular spacing and growing in straight rows, although there are many variations (Zhang and Stanturf, 2008). As discussed later, in several jurisdictions in Canada and the USA, poplar plantations are considered agricultural crops. Industrial uses refer to production plantations of poplars and willows managed primarily to produce commodities such as roundwood or fuelwood, as opposed to primarily environmental benefits (Isebrands and Karnosky, 2001). Environmental or conservation plantations differ from production plantations by virtue of their primary purpose; they may still be characterized as regularly spaced and a single age-class, although more complex plantations (mixed species, multiple age-classes) have been advocated for conservation purposes. Many plantings such as windbreaks serve a double purpose: they protect crops but are also used to manufacture goods. For example, in the Patagonia region of Argentina, when windbreaks are replaced, the harvested wood is used to manufacture boxes for fruit and vegetables, as well as other incidental uses. The techniques used to establish plantations, whether for production or conservation purposes, are based on the same technology.

Poplars and willows grow rapidly, propagate readily and lend themselves to improvement by conventional tree breeding

Table 5.1. Reported industrial uses of poplars and willows.

Use	Products	Poplar	Willow
Bioenergy	Firewood, chips, pellets, charcoal	India, Italy, Serbia, Montenegro, Turkey, China, Canada, USA	Sweden, Russia, China, USA, Canada, New Zealand
Pulpwood	Paper, rayon	India, USA, Canada, Argentina, Sweden, Estonia, Finland, Italy, France, Serbia, Montenegro, Iran, China, Korea, New Zealand	Russia, China, Argentina
Panels	Plywood, oriented strand board (OSB), medium-density fibreboard (MDF), particleboard, basin board, laminated wood	India, USA, Canada, Chile, Argentina, Serbia, Montenegro, Turkey, Korea, Italy	
Sawn wood	Lumber, rough construction, veneer	India, USA, Canada, Argentina, Chile, Sweden, Russia, Italy, France, Serbia, Montenegro, Turkey, Iran, Uzbekistan, China, Italy	
Veneer (light packaging and matches)	Packing cases, crates, fruit and vegetable boxes, wood wool, basketry, matches	India, Argentina, Brazil, Chile, France, Serbia, Montenegro, Iran, Uzbekistan, New Zealand, Italy	Chile, Russia
Specialty products	Ice-cream spoons, chopsticks, toothpicks, sporting goods, pencils, furniture, moulding, tool handles, tannins	India, USA, Chile, Turkey, Iran, Uzbekistan, China, Korea, New Zealand	India
Agriculture	Fodder, vermi-compost	India	New Zealand
Poles, posts and other products	Fence posts, supports for agricultural crops, roof rafters	India, Turkey, China, Chile, Argentina	

as well as biotechnology (Stettler *et al.*, 1996a; Dickmann *et al.*, 2001; see also Chapter 4, this volume); these characteristics make poplars and willows attractive for cultivation in shorter rotations than is feasible for other species. Various terms have been used to differentiate short-rotation forests from other planted forests, including short-rotation forestry, SRF; short-rotation woody crops, SRWC; short-rotation intensive culture, SRIC (considered an agronomic crop in Canada); and fast-wood. Short-rotation forests typically use high-density, single-species plantings that may include coppicing, with rotation lengths of less than 10–30 years (Mitchell, 1992; Zsuffa *et al.*, 1993; Makeschin, 1999; Cossalter and Pye-Smith, 2003; Weih, 2004). In some countries, a distinction is made between forest plantations of poplars and short-rotation plantings because of land ownership, for example, in Canada, SRIC is applied to hybrid poplar and willow plantings on farmland, or because of preferential tax or regulatory treatment for an agricultural, as opposed to a forestry land use, for example, in Oregon in the USA and in some Canadian provinces, rotations of 12 years or less qualify as a farm use; in Washington state, USA, it is 15 years or less. In some European countries, herbicides are permitted on farmland but not on forests. Such terminology distinctions are largely ignored in this chapter.

The objective of this chapter is to provide a global overview of poplar and willow culture, pointing out commonalities as well as differences. After a brief overview, the chapter is divided between *Populus* species and *Salix* species and follows the sequence of establishment, tending and production. The countries with the largest area of poplar plantations are China (7.6 million ha), France (236,000 ha), Turkey (125,000 ha), Hungary (109,000 ha), Spain (105,000 ha), Italy (101,000 ha), Sweden (49,000 ha) and Romania (48,000 ha). Seven other countries have significant areas of poplar plantations (Table 5.2). The countries with the most planted willow are China (438,000 ha), Argentina (56,400 ha), Romania (19,500 ha), New Zealand (20,000 ha) and Sweden (11,000 ha). Russia and Estonia also have sizeable willow plantings (Table 5.2). Other countries known to cultivate poplars and willows are listed, although data are unavailable.

Table 5.2. Countries with significant areas of planted poplar and willow (FAO, 2008, 2012).

Country	Poplar (1000 ha)	Willow (1000 ha)
Europe		
Belgium	33	0.02
Bulgaria	18.9	0.1
Croatia	13	3.6
Estonia	1	<1
France	236	
Germany	10	
Hungary	109.3	
Italy	101.4	20
The Netherlands	31	
Romania	47.9	19.5
Russia	31	1
Serbia[a]	33[a]	6.9[a]
Spain	105	0.7
Sweden	49.3	11.1
Switzerland	5.4	
UK	1.3	2
Asia/Africa		
China	7570	437.6
India[b]	0.7	1
Iran	150	10
Turkey	125	
Uzbekistan	3	
Egypt	7	
Morocco	4.3	
Australia/New Zealand		20
North America		
Canada	44.1	
USA	45	0.3
South America		
Argentina	40.5	56.4
Brazil	5.5	
Chile	5	3

[a]Serbia and Montenegro statistics are combined totals (prior to 2006).
[b]India's agroforestry area of poplars is over 300,000 ha.

5.2 Poplar

5.2.1 Stand establishment

Planting material

The naturally occurring poplars in riverine environments provided wood for rural economies in antiquity, specifically *P. euphratica*, *P. tremula*, *P. nigra* and *P. alba*. With the advent of industrial plantations, poplar species have been established in countries beyond their natural range. Often, plantations were developed using the limited number of genotypes that were available. For example, China has relied on a limited number of clones including six *P.* ×*canadensis* cultivars and two *P. deltoides* cultivars (Q.-W. Zhang and J.-H. Li, 2005, unpublished report); in India, the main poplars grown for industrial purposes cover approximately two dozen clones of *P. deltoides* (R.C. Dhiman, 2005, unpublished report). Nevertheless, many countries today have active breeding programmes (Fig. 5.1) to produce material for industrial plantations (Chapter 4, this volume). These programmes rely on 12 species that include North American (*P. balsamifera*, *P. deltoides*, *P. trichocarpa* and *P. tremuloides*) and Eurasian species (*P. alba*, *P. cathayana*, *P. ciliata*, *P. euphratica*, *P. maximowiczii*, *P. nigra*, *P. simonii* and *P. tremula*) and hybrids of these species (Chapter 4, this volume). Introduction of the North American *P. deltoides* and *P. trichocarpa* into European domestication programmes spurred development of many of the hybrids in current use worldwide (FAO, 1980; Bisoffi and Gullberg, 1996).

The ability of poplars in the sections *Aigeiros* and *Tacamahaca* for vegetative propagation has facilitated the selection and breeding of superior poplar genotypes. Despite some barriers to hybridization among sections (Stettler *et al.*, 1996b), intersectional hybrids between *Aigeiros* and *Tacamahaca* have been economically important (Eckenwalder, 2001; Zalesny *et al.*, 2011). Although hybrids are common within and among species of the section *Populus*, they do not hybridize readily with other sections (Eckenwalder, 1984). Species of the *Aigeiros* and *Tacamahaca* sections are easy to propagate through asexual means, usually by vegetative propagation of unrooted dormant stem cuttings or sets (also called whips), but poor rooting ability may disqualify some genotypes (Zalesny and Zalesny, 2009). Eastern cottonwood (*P. deltoides*, *Aigeiros* section) displays great variability in

Fig. 5.1. A clonal production nursery of elite *Populus* ×*canadensis* genotypes in Shandong Province, China. Photo courtesy of GreenWood Resources.

rooting ability. Interspecific hybrids within and between the *Aigeiros* and *Tacamahaca* sections usually root well (Eckenwalder, 2001). Poplars in section *Populus* (the white poplars and aspens) are difficult to propagate from stem cuttings, as are the interspecific hybrids between *P. tremuloides* and *P. tremula*, and planting stock is generally produced from seed or root cuttings (Stanturf *et al.*, 2001; Stenvall *et al.*, 2006).

Traits targeted for selection are agronomic (or silvicultural, depending on your perspective), wood quality and disease resistance (Chapter 4, this volume). Silvicultural traits of importance include stem form, yield and rooting ability, generally; for particular locations, cold tolerance, wind firmness and salt tolerance have been desirable traits. For industrial purposes, wood quality traits such as specific gravity, fibre length, cell wall thickness and lignin content have been critical traits depending on use requirements (for pulping or veneer, generally). Disease resistance has been of great importance and sometimes insufficiently regarded in the introduction of new genotypes to an area, leading to plantation failures. The pathogens of greatest economic importance are *Melampsora* leaf rust and *Septoria* stem canker (Chapter 8, this volume). Other serious pathogens include *Discosporium* canker, *Hypoxylon* canker, *Marssonina* leaf spot, *Venturia* shoot blight and *Xanthomonas* bacterial canker (Cellerino, 1999; Newcombe *et al.*, 2001; Giorcelli *et al.*, 2008). Details on poplar breeding strategies can be found in Chapter 4, this volume.

Establishing plantations

Plantation establishment requires the production of appropriate planting material that is adapted to the conditions of the available sites. Sites must be prepared and some adverse conditions may require amelioration prior to planting. Management objectives, markets, costs and regulation may all factor into decisions made in the process of establishing poplar plantations. Because these factors vary among poplar growing regions and over time, there is no single method of poplar culture (Stanturf *et al.*, 2001), but one commonality is the criticality of controlling competing vegetation during establishment.

Planting stock types

The aspen and *P. alba* (section *Populus*) stock type is a rooted plant, either bare root or container. It can be produced from seed or root cuttings. Poplars in the other sections are propagated easily from unrooted stem or branch cuttings. Poplar planting stock can be produced as several different types that can be differentiated between rooted and unrooted material. Length of the outplanted material is another distinguishing characteristic; cuttings tend to be 1 m or less, sets (also called whips or stakes) vary in length from 1.5 m to as long as 5 or 6 m. Thicker tall material may be called a pole. Rooted cuttings are also called barbatelles and rooted sets are called stecklings. In India, rooted sets are called entire transplants (ETPs). Choice of planting stock is a function of what material is available, its quality (how well nurseries can produce the material) and ease of planting, as well as management objectives and site conditions (Table 5.3). The ease of planting is a factor not only as a cost consideration but also how well material will be planted operationally; large material may be difficult to handle or require extra effort to prepare an adequate planting hole, thus poorly supervised planting operations may impact survival. Site conditions of importance are soil moisture and temperature, competing vegetation and threat of browsers (both wild ungulates and domestic livestock).

Unrooted dormant cuttings of poplars (other than the aspens and *P. alba* in the section *Populus*, which do not root readily) are produced from 1-year-old stem material, varying in length from mini (2–3 cm) to regular cuttings (15 cm to a maximum of about 1 m long). When planted in soil, adventitious roots grow from stem pieces; usually, it is recommended that viable buds be present for stems to form. Unrooted dormant sets can be cut from 1- or 2-year-old dormant material, but roots develop better from 1-year-old material. Unrooted sets also require buds for new stems to develop. In Italy and France, plantations are established mostly with 2-year-old unrooted poles 6–8 m long, sometimes with 1-year-old poles (A. Berthelot, 2005, unpublished

Table 5.3. Conditions under which certain poplar stock types can be used.

	Unrooted stock				Rooted stock		
					Bare root		
	Cuttings		Sets	Poles	Small	Sets	Container
Density of plantation	>1500 stems ha^{-1}	700–1500 stems ha^{-1}	<400 stems ha^{-1}	<700 stems ha^{-1}	>700 stems ha^{-1}	<400 stems ha^{-1}	>700 stems ha^{-1}
Plantation purpose	Biomass	Fibre and solid wood	Solid wood	Solid wood	Fibre and solid wood	Solid wood	Research trials; new stoolbeds; extreme drought conditions at planting
Soil moisture conditions	Good	Good	Excellent	Excellent	Good	Good	Needs irrigation if planted in full leaf
Weed control	Excellent	Excellent	Reasonable	None to reasonable	Excellent	Reasonable	Excellent
Threat of browsers	Variable	High to low	High	None to high	Low	High	Low

report; G. Picchi, 2005, unpublished report). Poles are unbranched and if roots are present, they are removed. In Argentina and Chile, 1-year-old unrooted poles 3–4 m long are planted to allow earlier introduction of cattle into the plantation (Fig. 5.2), which also controls competing vegetation (S. Cortizo and R. Suarez, 2005, unpublished report).

Rooted stock is produced by planting unrooted dormant cuttings or sets in a nursery bed and allowing them to grow a viable root system. Rooted stock can be either bare root or container (in Canada, they may be called plug or PSB).[1] Container plants are produced from seed, stem or root cuttings (aspen) or small, single-bud hardwood or greenwood cuttings (Fig. 5.3). These plants are usually dormant when planted, but they can be planted after breaking dormancy in the same growing season, if done immediately and there is sufficient time remaining in the season to develop an adequate root system (van Oosten, 2006).

Fig. 5.2. Hand-planting 2-year-old whips in Chile. Whips are planted 80 cm deep in auger-drilled holes at 6 × 6 m spacing. Note the stumps from the previous stand. Photo courtesy of J. Stanturf.

Stock production systems

Most poplar planting stock is produced in nurseries; a stoolbed nursery is common for producing unrooted material (FAO, 1980; Stanturf *et al.*, 2001). A stool is a stump from which new sprouts emerge. Stools may be started from any stock type, but normally dormant cuttings are used. Stoolbed nurseries usually are established on very good sites and may have irrigation. Stoolbed structure and layout vary regionally. For the best production of the healthiest stock, the lifespan of a stoolbed should be limited to 3–7 years. Dormant material harvested from branches of young plantations instead of stoolbeds is known as serial cuttings. Rooted cuttings and sets can be grown from unrooted dormant cuttings in a nursery bed. They are lifted after a year and outplanted as dormant rooted sets; sometimes, they may be transplanted and grown an additional year to develop a larger caliper. Container stock is often begun in a greenhouse and finished outdoors (van Oosten, 2006).

Fig. 5.3. A single-bud cutting about 2 months old, designated PSB 415, produced by the plug-Styroblock® system developed in Canada. This tree was started in the greenhouse and meets the target standard of 5–7 mm caliper and 50 cm height. Photo courtesy of C. van Oosten.

Unrooted dormant cuttings, sets and poles

Unrooted material is most commonly produced in a stoolbed. Cuttings may be planted manually or mechanically. For cuttings, stools are cut back annually to a height of 5–15 cm in the dormant season, thus producing 1-year-old sprouts every year. When very tall planting stock is required, the stools are cut back every other year, to produce a 2-year old pole. Harvested sprouts are sawn into cuttings or sets while dormant and must be refrigerated and remain dormant while waiting outplanting. Storage is in coolers or freezers, depending on the length of storage. The density of the stools in beds varies among regions and determines the caliper of the sprouts and controls the number of viable buds. Each cutting or set must have dormant viable buds. In North America, spacing is typically 0.3 m × 0.3 m, or slightly less than 0.1 m^2 per stool (Stanturf *et al.*, 2001). In Argentina, cuttings are planted 0.4–0.5 m apart within the row and rows are 0.7–0.8 m apart (S. Cortizo and R. Suarez, 2005, unpublished report). When the stools are planted too widely, the sunlight that penetrates the canopy stimulates buds to develop into sylleptic branches, rendering the sprouts useless for cuttings.

Poles as produced in France and Italy are planted as cuttings at different spacing to produce 1- or 2-year-old material (Frison, 1997; A. Berthelot, 2005, unpublished report). For 1-year-old poles, cuttings are planted at 1.8 m × 0.6 m spacing, or wider for low vigour clones. After one growing season, the poles are pruned and harvested by sawing at the base (for unrooted poles) or by extracting them with the root system. For 2-year-old poles, a wider spacing is used, for example 2.6 m × 0.5 m, and pruning is done annually. Poles are harvested during the dormant season and generally planted within a few days. Poles can be stored under refrigeration for a few months but must be handled carefully to avoid damaging the bark. Poles are also planted in Chile (Ulloa and Villacura, 2005) and recently also in the western USA (Fig. 5.4).

Weed control strategies

Competition from weeds is a serious threat during the establishment of new stoolbeds. Herbicides provide the most effective control of weeds (Stanturf *et al.*, 2001). Mulching can be

Fig. 5.4. (a) Poles (2 years old) are freshly cut from donor fields and (b) planted at 540 stems ha^{-1} (approximately 6.1 × 3 m spacing) on a hybrid poplar farm near Boardman, Oregon, USA. Photo courtesy of C. van Oosten.

used to control weeds, but they re-establish over time and the mulch can create habitat for rodents. Sawdust has been used as mulch, but it will tie up available nitrogen and can acidify the soil. A nursery may be included in a crop rotation with cereals, allowing use of broadleaved-specific herbicides and thereby reducing the weed load for the poplar rotation. During site preparation for a stoolbed, grasses and broadleaved weeds can be controlled effectively with herbicides that would damage the poplars. After cuttings are planted, a pre-emergent or pre-bud break herbicide application may be used. Choice of herbicides will vary by location, soil texture and pH, weed species and local regulations. Some specific practices can be found in regional guides, e.g. Stanturf *et al.* (2001), DEFRA (2002) and van Oosten (2006).

Mechanical methods may be practical when spacing is sufficiently wide, for example in the production of poles (Fig. 5.5). Up to four passes with cultivators is common in France, along with pre-emergent and growing-season herbicides (A. Berthelot, 2005, unpublished report). In closely spaced stoolbeds, manual labour is common to hand-weed portions of stoolbeds. Weeding needs decline rapidly when the stock fully occupies the stoolbed and shades out the weeds. Later leaf litter forms a layer of mulch, which suppresses weeds effectively (Scarcella *et al.*, 2011).

Fertilization and irrigation

Fertilization and irrigation schedules are very specific to local conditions and aim to avoid nutrient deficiencies and moisture stress in stoolbeds. Often, a balanced application of nutrients at the start of the growing season is sufficient. Direct foliar applications of nutrients can correct nutrient imbalances that develop during the growing season. Excess nitrogen is to be avoided because an oversupply can cause increased weed competition, promote formation of sylleptic branches and can delay the onset of dormancy. The same principles apply to irrigation, where the aim is to provide just enough water to maintain even growth. Over-irrigation can also promote sylleptic branch formation. Water should be withheld late in the growing season to promote hardening off and to avoid frost damage.

Crop health, protection and hygiene

Even though poplars are susceptible to a myriad of insects and pathogens (Ostry *et al.*, 1989; Mattson *et al.*, 2001; Duplessis *et al.*, 2009; Chapters 8 and 9, this volume), there is great variability among clones in their resistance (Robison and Raffa, 1998; Nordmann *et al.*, 2005). The most serious disease and pest problems facing the nursery grower are leaf rusts, blackstem diseases, cottonwood leaf beetle and

Fig. 5.5. A stoolbed of poles in Chile in their second growing season, grown in tight spacing (approximately 1.5 × 0.6 m), one pole per stool, is of *Populus deltoides* × *Populus nigra* hybrids. Photo courtesy of C. van Oosten.

stem borers (Mattson *et al.*, 2001; Newcombe *et al.*, 2001; Stanturf *et al.*, 2001). High stoolbed densities favour foliage diseases such as *Melampsora* rusts, especially with overhead irrigation. Clones with normally low susceptibility in plantations may develop serious problems in stoolbeds. Protection strategies are a combination of chemical control, cultural practices and use of resistant clones. If *Melampsora* rust causes early defoliation, cuttings in this physiologically weakened state are more vulnerable to blackstem disease and high levels of mortality. Blackstem diseases are caused by a number of organisms (*Cytospora chrysosperma*, *Phomopsis oblonga*, *Cryptodiaporthe populea* (*Dothichiza populea*) and *Colletotrichum gloeosporioides*) that are opportunistic on stressed plants. Blackstem is often considered a storage disease, and although improper storage can cause the disease to spread, it usually starts in a stressed plant well before it is put into storage. Stress can occur in the stoolbed because of drought, insufficient light or nutrients, frost damage, insect damage or leaf diseases such as *Melampsora* rust. The disease spreads and usually leads to poor growth, and often mortality. Diseased cuttings become a source of inoculum, and inadequate culling worsens the condition. Removing suppressed stems at or prior to harvest is the most effective control. The stressed stem usually is lighter in colour and smaller than its neighbours (van Oosten, 2006).

The cottonwood leaf beetle, or CLB (*Chrysomela scripta*), is the most serious insect threat in stoolbeds and is a serious pest in plantations. The CLB defoliates developing leaves and, in extreme cases, feeds on the woody part of the stem. Successful control has been achieved in the USA with several commercial insecticides, including several Bt (*Bacillus thuringiensis*) products (Stanturf *et al.*, 2001), but there may not be registered products in other countries (van Oosten, 2006).

Stem borers native to North America include the poplar borer (*Saperda calcarata*). It is a large beetle that bores into the stems of young trees with a diameter around 10 cm (Newcombe *et al.*, 2001; van Oosten, 2006). The larvae bore into the wood and create galleries that weaken the stem, often leading to breakage. Pesticides registered in Canada and the USA can control this pest (Stanturf *et al.*, 2001; van Oosten, 2006), and it is important to remove dead infected stems as they provide a habitat for the borer.

Unrooted dormant branch cuttings

Dormant material can be harvested from branches of young plantations instead of stoolbeds. These are known also as serial cuttings. First-order branches near the top of the tree produce vigorous cuttings of sufficient diameter. Riparian cottonwoods, i.e. *P. deltoides*, naturally reproduce asexually by branch breakage and crown damage. In plantations of *P. trichocarpa* × *P. deltoides* hybrids, sylleptic branches can be used for small-diameter cuttings to establish stoolbeds. Sylleptic branches from the previous year grow to a reasonable size the second year, but only the 1-year-old portion of these branches is used. Branch cuttings also must be stored in coolers or freezers until planting.

Rooted dormant cuttings

Bareroot dormant cuttings can be used to establish widely spaced plantations for solid wood products. This system of plant production is expensive and labour-intensive. Plants usually begin with unrooted cuttings planted in the nursery; after growing 1 year, bareroot plants with the root systems intact are excavated for outplanting. Root systems may be trimmed to a manageable size at the nursery. Often, the tops are also trimmed for easier handling or to balance top and roots and avoid planting stress (Grossnickle, 2005; DesRochers and Tremblay, 2009). Bareroot stock is lifted while the trees are dormant. Large stock (especially if kept in the nursery bed for 2 years) can be several metres tall with large caliper; it cannot be stored easily and must to be transported and planted immediately.

In India, the most common method for establishing poplar plantations uses ETPs that are bareroot saplings 4–6 m tall with 5–10 cm caliper (R.C. Dhiman, 2005, unpublished report). Material is collected from previously lifted ETPs and cuttings of at least 20–25 cm with five active buds are stored until planted in the nursery bed at 80 × 60 cm spacing. Cuttings are treated with fungicide and insecticide and irrigated weekly until the monsoon rains arrive. During the growing season, the nursery is hoed manually at least twice and shoots are singled at least once. Earth is mounded 15–20 cm around the base of the shoot to provide anchoring during the rainy (monsoon) season. Because of the warm Indian climate, the apical bud on very tall plants may not develop and will retard the growth of the plant; thus, the practice is to remove 30–40 cm of the stem above a healthy bud (Chandra, 2011).

Container stock

Materials that can be produced in a container nursery may be grown from single-bud or small stem cuttings, seed, or root cuttings. Dormant single-bud hardwood cuttings are used for clones that are difficult to propagate or if only a limited amount of material is available, such as from a breeding programme. This method is expensive and labour-intensive but can be used to multiply a single mother plant quickly into thousands of identical plants. Uses include establishing a new stoolbed with an improved genotype or for experimental purposes (Stanturf *et al.*, 2001). In the Prairie region of Canada, unrooted single-bud or small cuttings are placed in PSB containers in late spring. They start off in a greenhouse and are placed outside later in the summer under full sunlight prior to lifting, packaging and storing (van Oosten, 2006).

For the hard-to-propagate aspens (section *Populus*) and their hybrids, only rooted plants, either bareroot or container, can be used to establish plantations. Micropropagated hybrid aspen (*P.* ×*wettsteinii*) has been used in Estonia after developing in the nursery bed for 1 year (K. Jürgens, K. Heinsoo and A. Tullus, 2005, unpublished report). Rooted container stock is widely used in Canada and Finland (E. Beuker, 2005, unpublished report), although bareroot stock is beginning to gain acceptance in parts of Canada. Container stock is produced from dormant root cuttings or seed and grown in containers in a greenhouse in order to produce fully rooted plants with soil for outplanting. The container crop is initiated in the late winter in the greenhouse, and grows during the spring and summer into large plants with well-developed root systems. The containers are placed outdoors during the summer and overwintered in a cooler or freezer (Stanturf *et al.*, 2001; van Oosten, 2006).

Site requirements

Poplars generally grow best under high light intensity and warm temperatures during the growing season. Soil texture and drainage are two of the most important site factors for a successful plantation (Baker and Broadfoot, 1979). The influence of soil texture and drainage condition on site quality for poplar is summarized in Table 5.4. They prefer alluvial soils that are well aerated, have sufficient moisture and nutrients, are sufficiently deep (>l m to the water table), have a medium texture (sand/loam) and have a soil pH in the 5.0–7.5 range (Baker and Broadfoot, 1979). While droughty soils should be avoided, supplemental irrigation has been successful even on deep sands (Gallagher *et al.*, 2006; Robison *et al.*, 2006). In northern climes, sandy soils warm earlier in the spring and favour growth of hybrid aspen, but this advantage may be offset by the risk of drought conditions later in the growing season (Bergante *et al.*, 2010; Scotti *et al.*, 2010; Tullus *et al.*, 2012).

Saturation and waterlogging during the growing season cause anaerobic conditions to develop in soils that starve the root systems of oxygen, leading to drought-like symptoms. Most poplar clones cannot tolerate anaerobic conditions for very long into the spring months and must have well-aerated soils by the beginning of summer to survive and thrive. Younger trees are more vulnerable. Some clones do not tolerate saturated soil conditions in the winter very well either.

Heavy soils (clay, clay loam and silty clay loam textures) are considered less favourable for poplar growth than soils with lower clay content (Stanturf *et al.*, 2001). Because finer textured soils generally have poor aeration and poor drainage, they restrict equipment access during wet periods, making weed control difficult. Survival is reduced and growth during the first few years can be disappointing. The lack of rapid

Table 5.4. The influence of soil texture and drainage condition on site quality (very good to poor) for poplar. Shaded fields indicate potential to improve suitability through ditching, installing drain tile, subsoiling, or some combination. (Sources: Baker and Broadfoot, 1979; Dickmann and Stuart, 1983; Stanturf *et al.*, 2001.)

	Natural drainage class		
Dominant profile textures	Well and moderately well drained	Somewhat poorly drained	Poorly and very poorly drained
Fine clay (>60% clay)	Fair	Fair	Poor
Clay (40–60%)	Fair	Fair	Poor
Clay loam and silty clay loam	Good	Poor	Poor
Loam and silt loam	Good to very good	Fair	Poor
Loam and silt loam 25–50 cm over well-decomposed peat	Good to very good	Poor	Poor
Loam and silt loam marbled with well-decomposed peat	Good to very good	Fair to good	Poor
Sandy loam	Very good	Fair to good	Poor
Loamy sand	Very good	Fair to good	Poor
Sand	Poor	Fair	Poor
Sandy loam 35–100 cm over clay	Very good	Fair	Poor
Sandy loam 50–100 cm over loam to clay loam	Very good	Fair	Poor
Sandy loam 50–100 cm over sand	Good	Very good	Poor
Loamy sand 35–100 cm over clay	Very good	Fair	Poor
Sand to loamy sand 50–100 cm over loam to clay loam	Very good	Very good	Poor
Sand to loamy sand 100–150 cm over loam to clay	Good	Very good	Poor
Organic (muck, peat)	N/A	N/A	Poor to fair

growth and early crown closure leads to an abundance of weed competition, slowing tree growth even more. Recent advances in pre-emergent herbicides and application technology have improved weed control, enabling poplars to be established successfully on sites with as much as 90% clay (Stanturf *et al.*, 2001).

Poplars generally are intolerant of saline conditions; the North American species *P. trichocarpa* is extremely intolerant of salt and so are its hybrids; *P. deltoides* is slightly less intolerant (Stanturf *et al.*, 2001). Salt damage to the trees resembles desiccation damage. Physiologically, the tree suffers from drought stress. Leaves remain small and yellowish-green. Sometimes, the leaf edges become necrotic. The condition worsens as summer drought sets in, resulting in tree mortality. Sensitivity to salinity should be a concern to growers who rely on irrigation or fertigation to manage their poplar crop, and adequate drainage must be provided along with sufficient water to flush salts through the rooting zone. In countries with Mediterranean or semi-arid climates, soils tend to have higher pH and may develop salinity problems under impeded drainage. In Italy, for example, growth limitations develop when soil conductivity is >1.5 dS m^{-1}, and >4 dS m^{-1} is considered a severe limitation (G. Picchi, 2005, unpublished report; Scotti *et al.*, 2010). Similar recommendations are made for the Prairie Provinces of Canada (van Oosten, 2006). Breeding for salt tolerance is important for use of poplars in phytoremediation but not for industrial plantations (Chapter 4, this volume).

Poplars can perform well on shallow soils, although better sites generally are preferred. During extended drought, shallow soils may have insufficient moisture storage, leading to reduced growth or mortality. Shallow soils also may present a windthrow problem. Shallowness of the rooting zone can be caused by a high water table that does not retreat during the summer, an impermeable soil layer, bedrock, soils that are naturally very compact or compaction resulting from heavy machine traffic. Various clones may show differential resistance to windthrow (Chapter 4, this volume), and fertigation practices may affect root development that is maladapted to the prevailing winds (Coleman, 2007).

Peat soils are commonly thought to be poor sites for poplars. Even in Finland, where many conifer plantations are established on peaty mineral soils, truly peat (organic) soils are avoided (Stanturf *et al.*, 2001). Peats are usually waterlogged and very acidic, but there are exceptions. Weed control on peat soils can be challenging. Access may be difficult at critical times due to waterlogging, precluding mechanical control. Soils with high organic matter content will bind and render ineffective many pre-emergent herbicides. Artificial drainage may be the key to successful poplar management on these soils. Several sites with a high peat component in the northwestern USA are reasonably well drained and support good growth of hybrid poplar (Stanturf *et al.*, 2001). Windthrow damage is a real threat, especially if water tables are shallow, but some poplar clones are well suited to these conditions and hardly pose a serious windthrow problem.

Site preparation

Poplars are very intolerant of shading, thus it is essential to get a poplar plantation off to a faster start than competing vegetation (Stanturf *et al.*, 2001). Proper site preparation for planting is essential to the successful establishment of poplar plantations and to provide easy access for essential weed control. Without adequate site preparation, the survival and growth of poplars may be diminished drastically (FAO, 1980; Dickmann and Stuart, 1983; Facciotto, 1998; Stanturf *et al.*, 2001). A thorough evaluation of soil and site conditions will aid in the selection of appropriate treatments that will result in reduced planting costs, more effective herbaceous weed control and reduced damage to young poplars in mechanical cultivation. Some sites will have impervious soil layers; mechanical disruption of these layers will improve internal drainage and aeration. Plantations may be established on open pasture or agricultural land, cutover natural stands or prior plantations. On prior pasture or farmland, site preparation can be very simple. On cutover forest or prior plantations, site preparation becomes complex and very expensive due to stumps, logging debris and heavy vegetation.

Agricultural sites

Agricultural land may have supported a cover of crops before conversion to poplar plantations, and often these sites have a history of good weed control and adequate fertilization. Former pasture or grass hay sites may require herbicide treatments to kill grasses and tillage to break up sod. Open agricultural land is commonly prepared using combinations of conventional and minimum tillage methods, such as disking, chisel ploughing, subsoiling and mowing (Fig. 5.6). Many poplar growers have added herbicide treatments to their arsenal of site preparation tools in order to reduce early weed competition, and in the process have reduced mechanical weed control. In Sweden, herbicides are applied in the autumn before ploughing and again in the spring after harrowing, before or just at planting (M. Ramstedt, 2005, unpublished report). In the Canadian province of Quebec, herbicides are not allowed, requiring tall, rooted whips ('stecklings') to be planted. In the delta region of Argentina, weeds are crushed with a heavy iron or wooden cylinder, or simply burned (S. Cortizo and R. Suarez, 2005, unpublished report). Where mechanical planting or irrigation is practised, as in Italy, land levelling may be used (ISP, 2002; G. Picchi, 2005, unpublished report). Raised beds or bedding is relatively new to poplar culture in North America but has a long history of success in pine plantation culture on poorly drained sites. Bedding has been used extensively in Washington State but may have contributed to instability during windstorms, although this seems to depend on the clone. In China, bedding and surface drainage ditches may be used on poorly drained sites (Q.-W. Zhang and J.-H. Li, 2005, unpublished report). In countries where poplars are grown with agricultural crops, including India and China, there is no separate preparation for the poplar (R.C. Dhiman, 2005, unpublished report; Q.-W. Zhang and J.-H. Li, 2005, unpublished report).

Deep ploughing or subsoiling is recommended under some conditions, especially for soils with higher clay content or impermeable subsoil layers. Ripping or subsoiling along the planting row is common in the USA, often using modified conventional farm equipment (Stanturf *et al.*, 2001). In Italy, both deep ploughing and ripping (70–100 cm) are practised, but ripping is preferred because it does not turn under the fertile topsoil (Colorio *et al.*, 1996; Facciotto, 1998; G. Picchi, 2005, unpublished report). Ripping is usually followed by ploughing to 30 or 40 cm depth. In France, deep

Fig. 5.6. Site preparation by agricultural tractor in India; the operator is cross-disking the site. Older plantations are in the background. Photo courtesy of R.C. Dhiman.

ploughing and herbicides are recommended but seldom used (A. Berthelot, 2005, unpublished report). Where poles are planted, only individual planting spots may be prepared using an auger or hydraulic excavator (mechanical digger).

Forestry sites

Preparation of sites after timber harvest is generally more involved. The longer the previous rotation, the larger, and more troublesome, will be the material still on the site. New growth of herbaceous and woody vegetation, stumps, roots and compaction from logging traffic can further complicate this process. Site preparation after harvest seeks to facilitate planting by removing debris or reducing its size and incorporating it into the soil. Conventional land clearing methods such as shearing, raking, piling and burning have not changed much over the years. These are still the preferred methods used in the southern USA (Stanturf *et al.*, 2001). After shorter-rotation poplar plantations, less intensive, more cost-effective means of clearing sites are possible. In many countries, planting is between the rows of the previous stand. Site preparation between existing stumps has been successful, using an orchard flail to reduce woody debris, followed by a rototiller or a modified pavement grinder to grind and incorporate debris further into the soil. This leaves stumps intact and sprouts are controlled with herbicides or mowing. Alternatively, stumps may be treated with herbicide such as glyphosate immediately after harvest.

Stumps may also be removed along with other logging debris and competing vegetation using an excavator. This was the practice in the Canadian province of British Columbia on lands that previously supported a poplar plantation or a mixed stand of conifers and black cottonwood. Due to its expense, this has been replaced by leaving stumps in the ground and preparing individual planting spots using a small excavator with a modified brush rake (Fig. 5.7). The humus layer and mineral soil are mixed in this process. On poorly drained land, planting mounds are created. In Italy and China, agricultural crops commonly follow poplar plantations so that stumps and other debris must be removed. In Italy, branches and tops are concentrated and burned, comminuted and incorporated, or chipped and sold for bioenergy (ISP, 2002; G. Picchi, 2005, unpublished report). Stumps are removed with a special machine (levaceppi) or destroyed with a grinder. Fine roots are removed by harrowing. In China, stumps and large roots are removed and an agricultural crop or green manure grown for 1 or 2 years before replanting with poplar (Q.-W. Zhang and J.-H. Li, 2005, unpublished report).

Planting

Planting is a crucial phase of plantation establishment, and only quality planting stock should be used. Hand planting of unrooted dormant cuttings is common in many poplar growing regions where adequate weed control is possible. Although either seedlings or cuttings can be used, cuttings are preferred because they survive and grow as well as seedlings and cost less to produce and plant. Additionally, genetically superior clones can be expanded more rapidly through vegetative propagation. The desired length for cuttings for planting varies from 15 to 45 cm. Optimum cutting size is from 1 to 2 cm in top diameter. Cuttings with a top diameter larger than 2 cm are excellent planting stock but are hard to handle. Cuttings may be planted by inserting them directly into moist soil using a narrow planting spade or a dibble (Fig. 5.8) or by pushing them directly into soft, well-prepared soil. Planting depth varies by conditions; as a general rule, more of the cutting is below ground than above, so that roots have adequate moisture to develop. In most countries, at least one bud is above ground; more than one bud will develop into unwanted sprouts, necessitating expensive removal. In Russia, where planting unrooted cuttings follows immediately after frozen soil thaws, it may be necessary to use a Kolesov sword, a type of planting spade (A. Tsarev, 2005, unpublished report).

In coastal areas in the Canadian province of British Columbia, long cuttings (up to 1 m) are inserted at least 30 cm into the soil. The remainder is above ground and prevents immediate shading by small weeds, and the cuttings are visible during subsequent maintenance. This method may require singling to a single leader in the first or second growing season. In Argentina, one corporate plantation operation

Fig. 5.7. (a) Reforestation site preparation using spot mounding with hydraulic excavator, Kingcome Island, British Columbia, Canada. The harvested stand was a mix of cottonwood (*Populus trichocarpa*), red alder (*Alnus rubra*), western hemlock (*Tsuga heterophylla*) and red cedar (*Thuja plicata*). Low-intensity management of hybrid poplar, planted with unrooted whips approximately 1.5–1.8 m long. (b) Second year after planting (note whip next to hardhat, circled). (c) Aerial view of stand when approximately 5 years old. Residual trees are red cedar. Photos courtesy of Kruger Products LP (a), C. van Oosten (b and c).

in the humid pampas region of the province of Buenos Aires uses unrooted dormant cuttings with a length of 1.2–1.3 m. These are deep-planted 1 m below ground, with 0.2–0.3 m above ground. Planting is achieved with a hollow dibble connected to a water supply that squirts water at high pressure to form the planting hole. This method also delivers sufficient water for immediate growth in this rain-uncertain but very fertile region of deep loess soils.

Fig. 5.8. Planting dibble with step used for hand-planting bareroot *Populus* cuttings. The string on the ground marks the row. Photo courtesy of J. Stanturf.

Planting rooted material requires greater effort than unrooted cuttings. If rooted cuttings are small enough, they can be planted with a narrow tree-planting spade (van Oosten, 2006). Planting spades and dibbles are also suitable for planting container stock. Dibbles may have a foot-step to help drive it into the soil and also serve as a guide to planting depth. The shape and size of the dibbles are close to the dimensions of the root plug of the container stock (van Oosten, 2006).

Machine planting requires a well-prepared site that may be harrowed or cultivated prior to planting (DEFRA, 2002). Machine planting methods are not widespread and vary by country. In Italy, for short-rotation coppice plantings for biomass, a specially designed machine cuts and drives cuttings into the soil for their entire length (Balsari *et al.*, 2003a, b; G. Picchi, 2005, unpublished report). In Canada, mechanical tree planters pull a knife or rolling coulter blade through the soil, followed by a trencher. The planting trench produced is deep enough to accommodate the root system (van Oosten, 2006). Larger material such as whips and poles may require an auger that can be hand-held or machine mounted (Fig. 5.9).

Fig. 5.9. Auger planting poplar poles in China. Photo courtesy of J.G. Isebrands.

Poles are planted in Italy into augered holes about 30 cm in diameter, to a depth of 70 cm for 1-year-old poles and 120 cm for 2-year-old poles (ISP, 2002; G. Picchi, 2005, unpublished report). In India, ETP planting holes are 10 cm in diameter and augered to a depth of 1 m (Fig. 5.10). In western Argentina, bareroot sets are planted by shovel or with an auger (S. Cortizo and R. Suarez, 2005, unpublished report).

Season

Poplars are usually planted during the dormant season. In North America, this means when soil temperatures have warmed to 10–15°C, which varies from March to late April and extends to late May or mid-June (Zalesny *et al.*, 2004). At this temperature, roots are initiated in unrooted cuttings. In Sweden, soil temperatures above 5°C are recommended for planting (M. Ramstedt, 2005, unpublished report). In southern hemisphere countries such as Argentina and Chile, the corresponding winter months are June–September (S. Cortizo and R. Suarez, 2005, unpublished report; Ulloa and Villacura, 2005). Although later planting is possible in some years, it is generally not recommended. In Finland, however, planting in summer and even into autumn has been successful in experiments (E. Beuker, 2005, unpublished report). Delaying planting until after bud break of surrounding vegetation in the spring has been successful and affords an opportunity to plant sites that remain wet throughout the winter and flood during the normal planting season. Delayed planting is advantageous on low, wet sites but should not be used on drier ridges unless irrigation is available.

Cuttings should remain in freezer storage until planting (Stanturf *et al.*, 2001). In drier regions, harvested poplar whips and cuttings should be soaked in fresh water for a minimum of 2 days to prevent them from drying out during storage and planting. Cuttings or whips should not be exposed to drying conditions during transport to planting sites. Exposure to light for extended periods before planting is also harmful. When planting will be delayed until after the start of the normal growing season, cuttings must be kept in freezer storage at −2 to −4°C.

Fig. 5.10. Hand-augering a planting hole for an entire transplant (ETP), as used in India; ETP are rooted sets. Photo courtesy of R.C. Dhiman.

Proper spacing is needed to achieve maximum production in industrial plantations and marking planting spots is considered part of the planting process, even though marking may be done in autumn before planting in late winter. Cross-marking is common whether planting is by hand or machine. The planting row may be ripped in the autumn prior to planting or in the spring. In the southern USA, a slow-release nitrogenous fertilizer is placed at the bottom of the rip (Stanturf *et al.*, 2001). Cross-marking can be accomplished by various methods, including painting (van Oosten, 2006) or the planting crew using spacing guides. In fertigated hybrid poplar plantations in the western USA, planting spots are determined by the placement of the emitters; the cutting is planted at a wet spot (J. Eaton, 2005, unpublished report).

Espacement and planting density

Three considerations guide the choice of spacing in industrial poplar plantations: (i) product objective; (ii) weed control; and (iii) cost of planting material (Weih, 2004). Because survival is generally 90% or better, seldom is fill-in planting required. Where pulpwood is the objective, most plantations are established with spacing from 2.1 × 3.0 m to 4 × 4 m; most cottonwood and aspen plantations in the USA, for example, have been planted at something close to 3 × 3 m spacing (Stanturf *et al.*, 2001; J. Eaton, 2005, unpublished report; Zalesny *et al.*, 2011). Early spacing trials with both *P. deltoides* in the south (Gascon and Krinard, 1976; Anderson and Krinard, 1985) and *P. trichocarpa* in the west (DeBell, 1990) established that 3.7 × 3.7 m spacing was a good compromise for pulpwood and sawlog management. Spacing rows at least 3 m apart allows access by commercial farm equipment, which is essential for mechanical or chemical weed control (Stanturf *et al.*, 2001). Regular spacing, usually square, allows space for cross-cultivation, but indications are that rectangular spacing may produce higher yields, at least in very high-density biomass plantings. This could be due to more rapid crown differentiation, and a shorter time growth is checked by competition (DeBell *et al.*, 1996).

Wider spacing is common for sawlog rotations (Fig. 5.11). In Italy, for example, wide spacing was common until recently, with as few as 200 plants ha^{-1}, or about 7 m square (Frison and Facciotto, 1993; G. Picchi, 2005, unpublished report; Castro and Zanuttini, 2008). In Argentina, where livestock grazing in plantations is common, spacing ranges from 4 × 4 m to 6 × 6 m (S. Cortizo and R. Suarez, 2005, unpublished report). In the irrigated hybrid poplar plantations on the Columbia River plateau of eastern Washington and Oregon, pulpwood spacing was 3.05 × 2.29 m or 1450 stems ha^{-1}, but this spacing was found to be too narrow and the distance between trees in the row was widened to 4.58 m for sawlog management (J. Eaton, 2005, unpublished report), while maintaining the between-row spacing at 3.05 m to facilitate the existing irrigation drip-line spacing. Recently, planting has included between-row spacing of 6.10 m and in-row spacing of 3.05 m for approximately 540 stems ha^{-1}; this allows two irrigation driplines per tree row. Economic analysis of stand densities suggests that even wider spacing may be warranted.

Fig. 5.11. Sawlog-size stand of poplar in Chile, ready for manual felling. Trees have been pruned to 8 m, beginning in the second growing season, to increase value. Photo courtesy of J. Stanturf.

The tendency of some cultivars to develop sylleptic branches, especially in the lower bole, dictates caution in using wider spacing, unless pruning is feasible. Experience from experimental and operational plantings of *P. deltoides* in the southern USA illustrates the effect of spacing (Anderson and Krinard, 1985). Generally, diameter increased as spacing increased, from 3.7 × 3.7 m to 7.4 × 7.4 m. All spacing intervals were thinned at least once, except the two widest. Initial spacing has no effect on the rate at which diameter growth peaks, generally by the third or fourth year (Krinard and Johnson, 1984). Sawtimber yields were greatest for stands spaced 7.4 × 7.4 m. Wider spacing, however, required intensive pruning to maintain quality and more weed control to establish plantations successfully (Stanturf *et al.*, 2001).

Interest in bioenergy production has spurred research on spacing to achieve maximum biomass production, generally at narrower spacing than pulpwood plantations. Spacing may be as close as 0.5–1.5 m to increase biomass production per unit area and reduce weed competition (Weih, 2004). For operational bioenergy plantations in Italy, for example, densities depend on cutting cycles (Bisoffi and Facciotto, 2000; G. Picchi, 2005, unpublished report). For a 1-year cycle, a double-row design is used, with 0.75 m within the twin rows, 2.8 m between the sets of twins and 0.4 m between the twin rows, resulting in 14,000 stems ha^{-1}. For longer cutting cycles, spacing varies between 2.8 × 0.6 m (2-year cycle) and 3 × 2 m (5-year cycle). The double-row system of planting cuttings that has been used so successfully with willows has received only limited attention with *Populus* in North America (Zalesny *et al.*, 2011), but is widely used in short-rotation coppice systems for bioenergy (DEFRA, 2002). Other planting designs incorporating row crops (both food and bioenergy) are under investigation. In Russia, in the forest-steppe zone, wide spacing is the norm (6 × 6 m) with agricultural crops raised between the rows until canopy closure (A. Tsarev, 2005, unpublished report).

Coppice

The coppice system of natural regeneration offers an inexpensive alternative to replanting a second rotation stand by utilizing the ability of poplars to sprout readily from stump or root collar. Coppice management, however, is uncommon in industrial poplar plantations. Most poplar growers continually replace old planting stock with genetically improved stock; thus, coppice is unattractive even for pulpwood production. If sawlogs or veneer logs are the product goal, replanting remains the best option because of poor stem form in coppice, and stumps of larger trees sprout less vigorously. Coppice may, however, be economically attractive for bioenergy production and for non-industrial private landowners because of lower establishment costs, although poplars do not coppice as well as willow (Tubby and Armstrong, 2002). In bioenergy plantations, cutting after the first growing season to stimulate profuse sprouting has the advantage of capturing the site even if first year survival is low (Tullus *et al.*, 2012). This is standard practice in willow bioenergy plantations (Danfors *et al.*, 1997; Abrahamson *et al.*, 2002; DEFRA, 2002).

Coppicing works best when cutting occurs during the dormant season, when root reserves of carbohydrate are greatest, but this may also be the season when soil strength is lowest, limiting the weight of harvesting machinery (Verani *et al.*, 2008). Machinery limitations currently favour short cutting cycles, for example 1–2 years in Italy, for bioenergy plantations (Spinelli *et al.*, 2009).

Some non-industrial landowners in the southern USA have used coppicing for pulpwood production (Stanturf *et al.*, 2001). Plantations harvested during the winter months are typically those that may be targeted for coppicing. Harvest should begin no later than age 10 in the rotation to ensure vigorous sprouting. Often, there is a proliferation of shoots that arises from a single stump, and how these shoots are treated can potentially affect growth, yield and average tree size through the second rotation. Because of multiple sprouting, it has been customary to thin stumps back to two sprouts in the winter after the third growing season, removing up to ten sprouts from each stump. Without this cleaning step, yields of the coppice rotation will be half or less than the first rotation because of small stem size. Another coppice approach that has been tested is to fell every other row in the first-rotation harvest. After it is clear that sprouting has been successful, usually after one or two growing seasons, the residual trees are harvested in the summer to discourage sprouting. In this way, even multiple sprouts on a stump will have sufficient growing space to develop to merchantable size.

Clonal deployment and risk management

Disease and pest resistance are critical concerns in poplar breeding programmes, and clonal deployment strategies to minimize the risk of plantation failure should be the norm in industrial plantations (Ramstedt, 1999; Mattson *et al.*, 2001; Zalesny *et al.*, 2011). However, the practice is monoclonal plantings to facilitate cultivation, health monitoring, inventory and harvest scheduling (Zalesny *et al.*, 2011). The standard practice is to plant a mixture of monoclonal blocks, with some growers taking care to plant different genotypes in adjacent blocks and others paying little attention to diversity among adjacent clones (Zalesny *et al.*, 2011). Monoclonal blocks are easier to manage and, as yet, polyclonal mixes have not shown any sustained yield benefit (DeBell and Harrington, 1997; Knowe *et al.*, 1998). Mixtures require clones with similar growth characteristics at a given site to avoid some clones out-competing the other clones (Verwijst, 2001).

In many countries, poplar plantations are stocked with material from a limited number of clones. Often, this was due to limitations on the number of clones available, with even fewer performing successfully (R.C. Dhiman, 2005, unpublished report). Breeding programmes have expanded the genetic diversity of planting stock (Chapter 4, this volume), and multiple clones may be deployed yearly after testing under local conditions (e.g. Rédei, 2000; Coyle *et al.*, 2006). In the USA, for example, 3–16 different clones may be deployed annually; block size ranges from 2 to 45 ha (J. Eaton, 2005, unpublished report). In Canada, forest management regulations vary among provinces and according to land ownership (public or Crown lands versus private (Plate 21B)). In general, there are few restrictions on the deployment of hybrid poplars, except monoclonal blocks cannot exceed 10 ha in size on publicly owned forestland in the province of British Columbia; there are no such restrictions on private farmland. Several other provinces have restrictions on the deployment of hybrid poplars on public forestland. The limited number of clones suitable for the Prairie Provinces underscores the risk of planting only the 'best' performers; guidelines based on common sense suggest monoclonal blocks should be 20 ha or smaller on private land (van Oosten, 2006).

5.2.2 Stand tending

Competition control

Competition in any form will affect poplar plantation survival and growth. Control of competing vegetation is critical to establish poplar plantations successfully (Von Althen, 1981; Hansen and Netzer, 1985; Schuette and Kaiser, 1996) and poplars require full sunlight, adequate water and nutrients to realize their maximum growth potential (Demeritt, 1990). Substandard weed control in the first several years of plantation establishment can lower production at rotation significantly. For these reasons, aggressive weed control must begin by controlling perennial grasses and broadleaved weeds in the site preparation phase. Ignoring this important aspect increases costs and leads to unpredictable results once the plantation is established. Aggressive weed control continues during establishment and may continue longer in sawlog rotations with wider spacing that require more time for crowns to close and shade out weeds. Different strategies for controlling competing vegetation are used in industrial plantations, depending on local conditions and traditions. Herbicides and mechanical methods, in combination and alone, are common. Hand hoeing or tall planting stock (whips or poles) are preferred alternatives where labour is inexpensive, herbicides are unavailable (or too expensive), or both. Livestock grazing after successful establishment is also used to control weeds (Fig. 5.12). Any or all of these methods may be used at various developmental stages of the poplar plantation. Because the methods and timing of application are so varied, only representative examples will be given.

In China and India, poplars are often grown in combination with agricultural crops and benefit from weed control for the companion crop (Fig. 5.13). In China, mowing or scything is common on wetter sites; disk cultivation is common on drier sites (Q.-W. Zhang and J.-H. Li, 2005, unpublished report). Ploughing at least twice a year is common in India (R.C. Dhiman,

Fig. 5.12. Cattle grazing in a hybrid poplar stand in Chile. Grazing controls weeds and the stand provides thermal cover as well as forage. Livestock are excluded from young plantations to avoid damage. Photo courtesy of J. Stanturf.

Fig. 5.13. Poplars planted on a farm in China; poplars are integrated into smallholder farms in many Asian countries. Photo courtesy of J. Stanturf.

2005, unpublished report). In Italy and France, where spacing is relatively wide, two to four passes of a disk cultivator annually for 4–5 years is alternated by shallow ploughing in the winter (Facciotto, 1998; A. Berthelot, 2005, unpublished report; G. Picchi, 2005, unpublished report). Directed spraying of glyphosate during the first 2–3 years is followed by disking between the rows.

The practice in North America is more intensive, although fewer herbicides are available for poplar plantations in Canada than in the USA (Thomas *et al.*, 2000). Competition control strategies vary by region and depend on annual rainfall, soils and herbicide registration (Stanturf *et al.*, 2001). Herbicides requiring soil incorporation are usually applied prior to planting. Other herbicides are sprayed directly over newly planted dormant stock. These herbicides usually do not provide complete control throughout the growing season. Nearly all regions use a pre-emergent herbicide prior to or immediately following planting. In the southern USA, newly planted eastern cottonwoods are sprayed in bands 0.9 m in width as needed directly over the dormant tree rows with oxyfluorfen alone or in combination with herbicides such as imazaquin (Stanturf *et al.*, 2001). The strategy in the Midwest is to broadcast apply herbicides such as imazaquin and pendimethalin over entire plantations of dormant, newly planted hybrid poplar cuttings (Stanturf *et al.*, 2001). One industrial plantation owner in the state of Minnesota successfully uses flumioxazin as a pre-emergent herbicide at the time of planting, followed by broadcasting a tank mix of pendimethalin and imazaquin at the start of the second growing season when the trees are still dormant. In the Pacific Northwest, weed control strategies vary, depending on the local rainfall patterns. Extremely low rainfall areas east of the Cascade Mountains are often irrigated and herbicides such as trifluralin are soil-incorporated prior to tree planting, although this is more difficult in the second rotation where stumps limit working the soil. West of the Cascades, glyphosate and metsulfuron methyl in tank mix are applied for site preparation as a post-emergent the season before planting, a tank mix of glyphosate and sulfometuron methyl as a pre-emergent in a band directly before planting and as a pre-emergent in a directed basal spray prior to bud break in young plantations (J. Eaton, 2005, unpublished report). As companies seek certification, however, use of herbicides may become more limited.

Poplars typically are grown on sites that have recently been in agriculture and the weed complex is herbaceous broadleaves and grasses, although persistent woody vines are a problem in the southern USA (Stanturf *et al.*, 2001). Control of existing weeds can be done by applications of non-residual herbicides such as glyphosate, alone or in combination with 2,4-D. This is usually done the year prior to plantation establishment, before mechanical site preparation begins. After planting, options are more limited because poplars are generally sensitive to herbicides and mechanical damage. Certain herbicides, however, can still be applied effectively while poplars are still dormant. Several other herbicides are registered in the USA and Canada to control grasses (Canada: sethoxydim and fluazifop-P-butyl; USA: quizalofop P-ethyl) and some broadleaved weeds when trees are actively growing (Canada and the USA: clopyralid). Generally, these herbicides can be applied either directed to the base of the trees or right over the actively growing trees. Some equipment used after planting includes various hooded (shielded) and broadcast sprayers, backpack sprayers, cultivators and hand weeding. Cultivation equipment must be kept shallow enough to avoid root damage to the poplars, usually no deeper than 5 cm. Cultivators with guide wheels can control the depth of cultivation accurately. Care must also be taken to avoid damage from tool bars or other equipment to the bark and buds of young trees.

In Canada, where fewer herbicides have been approved for poplar plantations than in the USA, the general strategy in forest plantations is to plant unrooted or rooted whips (stecklings) to gain an advantage over competing vegetation. For short-rotation, intensively managed hybrid poplar plantations, integrated weed control using a few recently labelled pre-emergent herbicides (oxyfluorfen and flumioxazin) and mechanical cultivation can be used, except in the province of Quebec, where herbicides are not allowed. In Quebec, rooted whips (stecklings) are used to improve plantation success in the absence of herbicide use.

Mulches have been tested and may have some use in smaller intensive plantings for bioenergy and fertigated sites near mills or power plants (DEFRA, 2002; Robison *et al.*, 2006). Synthetic and paper mulches have been tested (Thomas *et al.*, 2001; Shogren and Rousseau, 2005; Geyer *et al.*, 2006), primarily to avoid using herbicides. In agroforestry systems, intercropping has a weed control effect (Delate *et al.*, 2005) and is widely used in India (R.C. Dhiman, 2005, unpublished report) and China (Q.-W. Zhang and J.-H. Li, 2005, unpublished report).

Nutrition

Poplars are demanding of high nutrient levels and generally are established on relatively fertile sites. In some places, municipal effluent and biosolids have been used; biosolids have been beneficial especially on marginal soils. Use of treated wastewater is more common in willow bioenergy plantations (see below). Nevertheless, nutrient limitations may occur as related to high inherent requirements due to high productivity of poplars, limited availability of native soil nutrients and imbalance among essential nutrients (Stanturf *et al.*, 2001). The commonest form of fertilization is at planting, in the planting hole or trench. Applications of potassium and phosphorus are common in India (R.C. Dhiman, 2005, unpublished report), limited applications of NPK in Canada (Thomas *et al.*, 2000; DesRochers *et al.*, 2006; van Oosten, 2006), and nitrogen in the southern USA (Stanturf *et al.*, 2001) and east of the Cascade Mountains in the west (J. Eaton, 2005, unpublished report). Phosphorus fertilization enhances early growth of aspen and hybrid aspen in Canada (van den Driessche *et al.*, 2003, 2005; Liang and Chang, 2004). The benefit generally comes from rapid height growth above competing vegetation. The application of nitrogen may continue throughout the rotation, especially on coarser textured soils or those with low organic matter content (Einspahr and Wycoff, 1978). The highest levels of poplar productivity have been obtained when N supply is adequate and other nutrients are kept in balance with N to avoid relative deficiencies (Stanturf *et al.*, 2001). Although growth on some sites has been shown to respond to other nutrients, it is most important to provide nitrogen as the main element limiting poplar growth.

Fertilizer recommendations focused on N generally consider foliar N levels of 2 and 3% as critical; in other words, levels below this suggest that N should be added (Dickmann and Stuart, 1983; Hansen, 1993). Growth rates are known to increase at higher foliar concentrations (Jia and Ingestad, 1984; Coleman *et al.*, 1998), but these levels are difficult to achieve operationally. The amount of N and other nutrients required to support optimum growth is shown in Table 5.5. The critical foliar concentration level may vary with genotype because of differences in N use efficiency (Blackmon *et al.*, 1979; Heilman, 1985). These estimates demonstrate the very high N requirement of rapidly growing poplar, especially hybrid poplars, compared with other forest types (Heilman and Xie, 1993). The high nutrient demand is due to the young age of intensively managed poplar plantations and their high productivity. Peak demand occurs by age 5 or 6 years (Nelson *et al.*, 1987). Many sites with high native soil fertility do not respond to fertilization, indicating the site supply capacity is adequate to meet even the high nutrient requirements of poplar. None the less, nutrients not adequately supplied by the site must be supplemented through fertilization if optimum growth rates are to be maintained. Maintaining balance between N and other essential nutrients is critical for achieving optimum production. For example, poplar stands may not respond to N additions unless accompanied by additions of P, K or other nutrients (Blackmon, 1976).

Nutrients besides N may improve poplar growth (Stanturf *et al.*, 2001), including phosphorus (P), potassium (K), calcium (Ca) and micronutrients such as boron (B), molybdenum (Mo) and zinc (Zn). Other micronutrients may be required to maintain optimum balance on certain sites, and not all sites will respond even to added N (Stanturf *et al.*, 2001). Nutrients can be applied separately or with N in fertilizer blends. Phosphorus may be limiting on sites such as the coarse-textured, well-drained soils used for fertigation systems, highly weathered soils or upland marine and some alluvial soils. Phosphorus applied at planting will encourage root development. It will persist and become slowly available for several years (possibly even through the rotation) because of mineral fixation with iron, aluminum and calcium, as well as immobilization in organic matter. Super-phosphate can be broadcast along with N, but fertilizer use

Table 5.5. Amount of nutrients required for sustained growth of poplar species and their hybrids compared with an average of temperate deciduous and conifer forest types.

Genotype	Age (year)	NPP[a] (Mg ha^{-1} year^{-1})	Requirement (kg ha^{-1} year^{-1}) N	P	K	Ca	Mg	Reference
P. deltoides	4–6	17	102	11.5	88	151	17.9	Cited by Bernier, 1984
P. deltoides	7	17	107	11	91	157	18	Nelson *et al.*, 1987
P. trichocarpa	4	7–18	95–159					Heilman and Stettler, 1985
P. trichocarpa × *deltoides*	4	27–28	271–276					Heilman and Stettler, 1985
P. ×*canadensis*[b]	4	11	168					Heilman and Stettler, 1985
P. ×*canadensis*[b]	1–2	12–24	182–246	20–36	113–171	121–237	38	Cited by Bernier, 1984
Temperate deciduous	30–120	10	98	7.2	48	56	10.4	Cited by Bernier, 1984
Temperate conifers	15–450	8.3	46	5.5	28	20	4.6	Cited by Bernier, 1984

[a]Net primary productivity (NPP) includes belowground and aboveground biomass, including foliar mass.
[b]Synonym *P. ×euramericana*.

efficiency can be low if roots have not fully exploited the site, and soluble P exposed to a large reaction surface on soil particles is easily fixed. Granular super-phosphate, alone or in a mixture with N, may be banded and incorporated along planting rows or placed in a patch directly below the cutting at establishment and improves efficiency of use. Another approach is to inject a mixture of N and P where the base of the cutting will be during the subsoiling/row marking operation (Fig. 5.14). This places the nutrients at an optimal location for tree roots and out of the reach for shallow-rooted competing vegetation.

Poplars grown on farms often benefit from fertilizing with animal manure. For example, in Iran a green manure cover crop is grown and ploughed under with 30–40 t of decayed animal manure before planting (P. Nejad, B. Reza, H. Hasan, H. Sabeti and A. Babaipour, 2005, unpublished report). Additional manure may be added to the planting hole and around the base of the plant. Later, compost with added inorganic N may be incorporated. In Argentina and other countries where livestock are grazed in poplar plantations, it is believed that animal manure contributes nutrients to the trees (S. Cortizo and R. Suarez, 2005, unpublished report).

Irrigation and drainage

Poplar requires adequate moisture throughout the growing season and even on the best sites may experience periodic dry conditions. The capital expenditure necessary for irrigation usually is not justified, however. The exceptions are fibre farms established on coarse soils to provide all-weather wood supply to pulp mills (Stanturf *et al.*, 2001; Stanton *et al.*, 2002; J. Eaton, 2005, unpublished report; Gallagher *et al.*, 2006; Robison *et al.*, 2006) and semi-arid and Mediterranean environments such as east of the Cascade Mountains in the western USA (Stanton *et al.*, 2002; J. Eaton, 2005, unpublished report), the steppes of Russia (A. Tsarev, 2005, unpublished report), Uzbekistan (G. Vildanova and K. Tolipov, 2005, unpublished report), India (R.C. Dhiman, 2005, unpublished report), China (Q.-W. Zhang and J.-H. Li, 2005, unpublished report), Mendoza and Rio Negro provinces in western Argentina (S. Cortizo and R. Suarez, 2005, unpublished report), Chile (Ulloa and Villacura, 2005) and Italy (ISP, 2002; G. Picchi, 2005, unpublished report). Irrigation technology varies from gravity-fed ditches to drip systems. For example, the practice in India is to fill

Fig. 5.14. Mounder and fertilizer dispenser pulled by a farm tractor to prepare for planting on former agricultural sites. The fertilizer dispenser is mounted behind a subsoiler shank, which is followed by mounding. Fertilizer is fed by gravity continuously at the bottom of the rip. The disks pull soil into a mound that serves to mark the planting row. Cuttings (45 cm in length) are planted by hand 30 cm deep, right on to the fertilizer. The fertilizer was mono-ammonium phosphate (11-52-0) applied at 220–340 kg ha^{-1} to supply between 50 and 75 kg P ha^{-1}. Photo courtesy of Alberta-Pacific Forest Industries Inc.

planting holes partially with soil and then flood irrigate the field; water accumulates in the planting holes, which are later filled with soil. Irrigation continues as needed, every 15 days during the first growing season or according to the needs of the companion agricultural crop. Irrigation is less frequent from the second growing season through to the end of the rotation (R.C. Dhiman, 2005, unpublished report). In Italy, both gravity-fed and sprinkler irrigation are used in intensively cultured plantations but not in biomass plantings. Gravity-fed systems (Fig. 5.15) require essentially level fields and require large amounts of water even with careful monitoring of plant needs (ISP, 2002; G. Picchi, 2005, unpublished report). Sprinkler systems can be used on any kind of site and use less water. In Russia, poplar is irrigated every 14–15 days in the summer and monthly in spring and autumn. The amount of water is lower in the first 2 years of stand development (3500 m^3 ha^{-1} $year^{-1}$) and increases in later years (up to 5000 m^3 ha^{-1} $year^{-1}$). Irrigation may be gravity-fed or sprinkler (A. Tsarev, 2005, unpublished report).

The drip irrigation systems used in the desert regions east of the Cascade Mountains in the western USA (Fig. 5.16) require a major capital investment and careful monitoring and control (J. Eaton, 2005, unpublished report). Annual precipitation is only 20 cm and falls mostly during the dormant season, so virtually all crop needs must be supplied by irrigation. Water is drawn from the Columbia River and its tributaries and pumped through buried pipe to the drip distribution system. Water demand is predicted from transpiration models using data from soil moisture sensors. Emitters at each plant deliver metered applications of water and fertilizer, and sometimes insecticide (J. Eaton, 2005, unpublished report). Demand by young plants is low (15–25 cm ha^{-1} for 1-year-old trees), but increases rapidly until levelling off at canopy closure (80–100 cm ha^{-1} annually).

Poplar plantations are often established on alluvial sites and excess water may affect management activities and growth. Drainage or levee systems in some cases may have been established for agricultural purposes. In the southern USA in the flood plain of the Mississippi

Fig. 5.15. Gravity-fed irrigation of hybrid poplar in Chile. Gravity-fed systems require essentially level fields and large amounts of water. Photo courtesy of J. Stanturf.

Fig. 5.16. A new plantation of dripline-irrigated hybrid poplar at the start of the first growing season at Boardman, Oregon, USA, against a backdrop of an older irrigated plantation. Photo courtesy of C. van Oosten.

River, eastern cottonwood (*P. deltoides*) plantations have been established both in the protected area behind levees and within the unprotected area flood plain (Stanturf *et al.*, 2001). Similarly in the Paraná River delta region of Argentina, channels and ditches are constructed to remove excess water from plantings on higher ground and behind levees (S. Cortizo and R. Suarez, 2005, unpublished report). In many countries such as France, wetlands are now protected and drainage consists of cleaning existing ditches prior to planting (A. Berthelot, 2005, unpublished report). In Iran where flash flooding threatens agricultural crops, water diversion channels funnel floodwaters to poplar plantations (Sagheb-Talebi, 2005).

Protection

Poplar plantations are susceptible to a host of pests and diseases and the best strategy is prevention rather than correction. Literally hundreds of diseases and insects affect poplars (Giorcelli *et al.*, 2008; see also Chapters 8 and 9, this volume), and disease and insect resistance are key components of breeding programmes. The main emphasis in breeding programmes is for disease resistance (van Oosten, 2006; Chapter 4, this volume) and screening especially for *Melampsora* rust resistance should be carried out before a newly introduced clone is deployed in a region. Nevertheless, disease and insect pests likely will adapt and overcome resistance or tolerance, hence the importance of a clonal deployment strategy. Other protection needs include animal pests and fire.

Diseases

The first line of defence against diseases is breeding for resistance (Duplessis *et al.*, 2009; Chapters 4 and 8, this volume). Planting resistant clones that are otherwise adapted to the site and maintaining healthy trees is essential to obtain the benefit of breeding programmes. *Melampsora* leaf rust is the most serious foliar disease in North and South America and Europe, but seems to be of only minor importance in India. The rust causes premature defoliation and decreased growth and may weaken young plants, leading to mortality (Cellerino, 1999). There are several species of *Melampsora* rusts, and poplar clones differ greatly in their susceptibility. While the best defence is to plant resistant clones, low-density plantings afford some protection, as is done in European plantations (van Oosten, 2006). *Melampsora* rust can be problematic in stoolbeds and can seriously affect many hybrid clones that are normally considered tolerant in plantation settings. In Canada, one fungicide (tebuconazole) was recently approved to control this rust in stoolbeds and intensively managed plantations. Protection strategies in Russia against *Melampsora* and other leaf diseases is to gather leaf litter at the end of the growing season and burn it, followed by ground application of 3–5% Bordeaux mixture in May (A. Tsarev, 2005, unpublished report).

The most serious disease of hybrid poplars in North America is *Septoria* leaf spot and canker caused by the fungus *Septoria musiva*, which is also present in Argentina and southern Brazil (van Oosten, 2006; Isebrands, 2007). High densities (greater than 800 stems ha^{-1}) are conducive to spread of this disease that begins as circular necrotic spots on the leaves and progresses to cankers at the base of branches (Isebrands, 2007). Several formulations of one registered fungicide are available in Canada (thiophanate-methyl) (van Oosten, 2006) to control *Septoria* leaf spot but not in the USA (Isebrands, 2007). *P. deltoides* is not susceptible to this *Septoria* damage, but many of its hybrids are. In Canada, susceptible native *P. balsamifera* in a plantation setting was reported in the province of Alberta, and in 2006 *Septoria musiva* stem cankers were reported in the province of British Columbia (Callan *et al.*, 2007) on several hybrids of *P. trichocarpa* × *P. maximowiczii* grown in a stoolbed; this was the canker's first occurrence west of the Rocky Mountains in North America. Subsequent surveys found several hybrids of *P. trichocarpa* × *P. deltoides* to be affected in stoolbeds and nearby stands. The canker disease has now also been confirmed as present on the native *P. trichocarpa* in south-western British Columbia, although many phenotypes appear to be resistant. *Marssonina* leaf spot, however, can be a serious problem in nurseries, stoolbeds and dense plantings of the North American poplar species, including trembling aspen (*P. tremuloides*). Selection and breeding of hybrid poplar clones with diverse parentage and planting resistant clones is one defence; another is to disk between tree rows after leaf fall to decrease inoculum (Isebrands, 2007). Early spring (April) application with a systemic pesticide is used in Italy for non-resistant clones (ISP, 2002; G. Picchi, 2005, unpublished report), and in Canada several formulations of one registered fungicide (thiophanate-methyl) are available to control this disease.

The most significant disease of poplar in India is the southern leaf blight, *Bipolaris maydis*, that also attacks some strains of maize (R.C. Dhiman, 2005, unpublished report). The disease is a problem during the monsoon season; leaves of infected trees dry and crumble. The *P. deltoides* 'G3' clone, once widely planted, is especially susceptible and no longer planted (Chandra, 2001). Breeding and screening for resistance is the main form of defence (R.C. Dhiman, 2005, unpublished report).

Insects

The major insect pests of North American poplar plantations (Chapter 9, this volume) include defoliating insects such as cottonwood leaf beetle, or CLB (*Chrysomela scripta*), poplar tent maker (*Clostera inclusa*); borers such as the cottonwood twig borer (*Gypsonoma haimbachiana*), cottonwood clearwing borer (*Paranthrene dollii*), cottonwood borer (*Plectrodera scalator*) and exotics (Haack, 2006), and poplar borer (*Saperda calcarata*) and poplar-willow borer, or PWB (*Cryptorhynchus lapathi*) (van Oosten, 2006); and aphids, mites and leafhoppers (Morris *et al.*, 1975; Solomon, 1985; Coyle *et al.*, 2005). Grasshoppers (various species) and sawflies (*Nematus* spp.) have been recent and recurrent problems in the province of Alberta, Canada, in new and young plantations. A frequent monitoring schedule should be used to control these insects prior to large infestations (Coyle *et al.*, 2002, 2008). Labelled general-purpose insecticides such as carbaryl or *Bacillus thuringiensis* (*Bt*) may be applied to control some of these pests. In addition, the CLB is controlled in irrigated hybrid poplar plantations with a nicotine-based systemic insecticide, imidacloprid (J. Eaton, 2005, unpublished report).

Stem borers have taken on greater importance in western North America in recent years, as poplar management has focused on producing sawlogs. The three most important pests are the western poplar clearwing moth (*Paranthrene tabaniformis*), PWB (*Cryptorynchus lapathi*) and the carpenterworm moth (*Prionoxystus robiniae*). All form galleries that severely degrade the lumber produced. Experience suggests that hybrids with *P. deltoides* × *P. nigra* parentage are more resistant than *P. trichocarpa* × *P. deltoides* or *P. trichocarpa* × *P. nigra* parentage (J. Eaton, 2005, unpublished report). In Italy, the PWB is treated by spraying the tree trunk for the first 3 years with pyrethroid or organophosphorus insecticides (ISP, 2002; G. Picchi, 2005, unpublished report). These insecticides are also used to treat the poplar borer, *Saperda carcharias*.

The large-scale afforestation effort of drylands in northern China has resulted in novel pest outbreaks (Chapter 7, this volume). Large monocultures, limited genetic material and suboptimal growing conditions have been implicated in outbreaks of Asian longhorn beetle (*Anoplophora glabripennis*) since 1998 (Chapter 7, this volume). The beetle is not native to most of the affected areas, hence lacking in natural enemies. It has spread to 13 provinces in northern China (Pan, 2005), with widespread mortality of thousands of hectares of poplar plantations (see Chapter 7, this volume).

In South America, east of the Andes, the ambrosia beetle, *Megaplatypus mutatus* (also identified as *Platypus mutatus*) causes serious damage, resulting in major degrade of the wood and often stem breakage. The economic damage is in the outer shell of the stemwood, which contains the most valuable veneer and lumber grades. It is associated with a blue stain that degrades the wood and is reported to affect many hardwood trees, including *Salix*. This insect was recently found in Italy. In Argentina, this ambrosia beetle causes serious damage to poplar plantations due to breakage and reduced vigour (Alfaro, 2003; Giménez and Etiennot, 2003).

Animals

Poplars are a preferred browse for most cervid species (deer, elk and moose) and may cause establishment failure, especially of smaller plantations subject to high browsing pressure. Deterrents such as electric fences and repellents may reduce browsing to tolerable levels. Trees may grow out of the reach of deer if browsing pressure is low, by the end of the second growing season (Netzer, 1984), but for several years will remain susceptible to bucks rubbing during the rutting season. Large mammal browsing can be so serious that the landowner is left with only two options: fence or forget growing poplar. In cutover forest stands, slash can be bulldozed into brush fences 3 m or higher (McKnight, 1970). Electric fencing is another option, but requires continual maintenance while plants are susceptible. A five-strand fence, with the lowest strand 25 cm off the ground and the other strands 30 cm apart above it, has worked in the north-eastern USA in forest clearcuts (Brenneman, 1982). Other options are available, including a more expensive woven wire fence (Dickmann and Lantagne, 1997). To be effective, at least two tiers of 1.2-m woven wire are required. Stay wires (no wider than 15 cm apart), a third tier of fencing, or a strand or two of barbed wire will be needed to keep deer from penetrating.

Small mammals are also localized problems. Periodically high vole (*Microtus* spp.) populations can be a problem even in older (3- and 4-year-old) plantations with grass or snow cover that provides protection from predators. Voles feed on roots and lower stems, which can lead to heavy tree mortality. In the north-western USA (west of the Cascades) and coastal south-western British Columbia (Canada), the main problem with voles occurs after canopy closure, when shading reduces alternate food sources such as grasses. Grass control can prevent this problem, although mice and voles can still cause trouble under snow cover. Beaver (*Castor canadensis*), porcupine (*Erethizon dorsatum*), squirrels (*Sciuridae* spp.), hare (*Lepus* spp.) and rabbits (*Sylvilagus* spp.) are occasional problems in North American plantations, especially by stripping bark from young trees. Damage can be minimized by effective weed control and by planting large stock (Stanturf *et al.*, 2001). In France, plantations are planted with individual tree protection against roe deer (*Capreolus capreolus*), coypus (*Myocastor coypus*, an introduced rodent species from South America also known as river rat) and rabbits (A. Berthelot, 2005, unpublished report). In the UK, fencing must be buried and turned outward to deter rabbits (DEFRA, 2002). In northern Europe, damage to hybrid aspen stands from deer and moose is controlled by fencing, at least during the first rotation in coppice stands (Tullus *et al.*, 2012).

Livestock grazing in plantations must be controlled to prevent damage to young plants. In Argentina, this is accomplished by planting large material (poles) and excluding livestock for the first year or two (S. Cortizo and R. Suarez, 2005, unpublished report). Blue bull or nilgai (*Boselaphus tragocamelus*) is an indigenous antelope of central and northern India that damages nurseries and newly planted areas by trampling and breaking (Dhiman, 2004). They are controlled by watchmen and scarecrows and by pungent repellents.

Stand improvement

Three stand improvement treatments are practised in poplar plantations: singling, thinning and pruning. By far the most common practice is pruning in sawlog and veneer management systems.

Singling

Removing multiple stems in order to concentrate growth on a single stem is a common practice in many poplar growing regions where pulpwood, veneer or sawlogs are the management objective. Generally, singling is done at the beginning of the second growing season, allowing one stem to express dominance or other desirable characteristics (R.C. Dhiman, 2005, unpublished report). Because the singling treatment is labour-intensive, growers in many regions seek to avoid it by planting cuttings deeply enough that only a single bud develops (Stanturf *et al.*, 2001). On the other hand, in drier regions, development of multiple buds is an insurance that a stem will develop and singling is less expensive than fill-in planting (P. Nejad, B. Reza, H. Hasan, H. Sabeti and A. Babaipour, 2005, unpublished report). In biomass plantations, multiple stem development is often an objective (DEFRA, 2002).

Thinning

Thinning is uncommon in industrial plantations; initial planting density is generally the target harvest density. Limited thinning has been used to convert stands from pulpwood to sawlog management or other situations where rotations have been lengthened (Stanton *et al.*, 2002; J. Eaton, 2005, unpublished report). Early plantations of native eastern cottonwood (*P. deltoides*) in the southern USA envisioned opportunities to produce sawlogs and veneer within

20–30 years of planting, and thinning trials were conducted (Stanturf *et al.*, 2001). Cottonwood is characterized by very rapid diameter and height growth in the early years, and plantations must be managed aggressively to maintain this rapid growth and avoid stagnation. Timing of thinning treatments will be determined largely by initial spacing, which is affected by site quality, establishment practices and survival. Initial spacing has no effect on the rate at which diameter growth peaks, generally by the third or fourth year (Krinard and Johnson, 1984). Because cottonwood cannot tolerate side competition, it responds poorly to release following crowding. Wide spacing with pruning of the lower branches or closer spacing accompanied by early thinning is necessary to maintain rapid growth of individual trees.

Thinning has become popular in Argentina as the demand for solid wood has increased (S. Cortizo and R. Suarez, 2005, unpublished report). Initial stand density of 625 stems ha^{-1} is reduced to between 278 and 430 stems ha^{-1} (Borodowski and Suárez, 1999; Borodowski *et al.*, 2005). While thinning is often precommercial, some Indian growers will harvest the largest stems and allow unmerchantable stems to grow another year or two to reach merchantable diameter (R.C. Dhiman, 2005, unpublished report). Rising pulpwood demand in India has prompted some farmer-growers to increase initial density above 400–500 stems ha^{-1} in hopes of commercial thinning for pulpwood after 3 or 4 years.

Pruning

Pruning to reduce knots and increase wood quality is common in veneer and sawlog management systems, especially where initial planting density is low. Pruning usually begins early and no later than the third growing season. The goal in Italy is 5 m of branch-free stem at harvest, thus pruning to a height of 6–7 m is needed (Facciotto, 1999; G. Picchi, 2005, unpublished report). The lower 3 m is veneer quality and the upper 2 m of the log is taken for solid wood products. Pruning is progressive and every year all branches are removed below the point where bole diameter reaches 10 cm. Initial pruning (education pruning) begins in the dormant season after the second year, to shape the stem and eliminate double apex shoots and large branches. Cleaning pruning begins at the same time and continues for 5 years, gradually removing all branches less than 60 mm in diameter. Hydraulic shears and small chainsaws on long poles are used (Fig. 5.17); workers may be raised to the higher levels on hydraulic lifts mounted on agricultural tractors (Fig. 5.18). Because planting stock may be 1- or 2-year-old poles, different strategies are followed. For 1-year-old poles, more shaping may be required since branches develop on the upper two-thirds of the pole (3–4 m tall). The taller 2-year-old poles (6–8 m) develop branches in the upper half of the stem and require mostly cleaning pruning. Such intensive pruning may reduce growth rates but is more than offset by the higher value of the final assortment. Similar regimes are followed in France (A. Berthelot, 2005, unpublished report), Argentina (S. Cortizo and R. Suarez, 2005, unpublished report) and Chile (Ulloa and Villacura, 2005; Baettig *et al.*, 2010): pruning to 7–8 m, beginning in the second year.

Fig. 5.17. Pruning lower limbs with a small chainsaw mounted on a pole. Pruning to reduce knots and increase wood quality is common in sawlog management systems, especially where initial planting density is low. Pruning usually begins early and no later than the third growing season. Photo courtesy of J. Stanturf.

Fig. 5.18. Two workers in tandem, pruning taller trees from hydraulic lifts mounted on an agricultural tractor. Initial pruning was accomplished from the ground using hydraulic shears or small chainsaws on long poles. Photo courtesy of J. Stanturf.

Pruning in North America is limited to pruning of lower branches in Canadian plantations to allow equipment access and in sawlog management of irrigated hybrid poplar plantations east of the Cascade Mountains in the USA (J. Eaton, 2005, unpublished report). Pruning in the eastside plantations uses a three-lift system that prunes up the stem to approximately 8 m. Pruning begins at the start of the third growing season and is completed within 3 years. Pruning during the growing season reduces epicormic branching. All pruning is by hand; lower levels with lopping shears and upper lifts with pruning poles. Generally, 50–60% of live crown remains after pruning in order to avoid growth loss, although this may differ among clones (J. Eaton, 2005, unpublished report).

5.2.3 Production

Rotation length

Rotation lengths vary according to the time needed for trees to meet product requirements, with sawlog and veneer rotations the longest and pulpwood and bioenergy the shortest. Pulpwood and chip-and-saw rotations are determined by the time needed to reach minimum piece size for economical harvesting and handling. While rotation lengths generally are longer for sawlogs and veneer production, in the western USA a maximum rotation length of 12–15 years is due partly to local (state) regulations on maximum rotation length to qualify as an agricultural tree crop, and therefore exempt from more restrictive forest practice regulations (J. Eaton, 2005, unpublished report). Bioenergy rotations are fixed by limits on the size of material that can be harvested economically by available equipment (Weih, 2004; Verani *et al.*, 2008). The very short-rotation poplar coppice bioenergy systems using modified silage harvesters typically are on 2- or 3-year cutting cycles (Verani *et al.*, 2008; Spinelli *et al.*, 2009). Rotation lengths vary by poplar growing region because of different management regimes, climate and growing conditions and product objectives (Table 5.6). Markets for poplar wood in Serbia and Montenegro are unstable and plantations are carried longer, from 22 years on optimal growing sites along the Danube River to 32 years on suboptimal sites along the Sava River (S. Orlovic, B. Klasnja, Z. Galic, L.P. Pajnik and P. Pap, 2005, unpublished report).

Survival

Industrial poplar plantations typically have high survival rates when properly managed. This requires that clones are adapted to the site and competing vegetation is controlled, especially in the first 2 years (Stanturf *et al.*, 2001; Weih, 2004). Contributing factors are adequate site preparation and quality planting stock. Expected survival in many countries is greater than 90%, unless a major disturbance occurs such as windstorms or growing-season flooding. If poles are planted at wide spacing, there is an opportunity to replant if initial survival falls below 90%. In Argentina, the threshold is greater than 15% mortality in the first year (S. Cortizo and R. Suarez, 2005, unpublished report). In France, dead trees may be replaced in the first 2 or 3 years (A. Berthelot, 2005, unpublished report). In countries where greater planting density is the practice, it is uncommon to replant dead cuttings.

Table 5.6. Poplar yield, stand density and rotation length.

	Stand density (stems ha^{-1})	Rotation length (years)	Mean annual increment (m^3 ha^{-1} $year^{-1}$)	Diameter at harvest (cm)	Height at harvest (m)	Basal area at harvest (m^2 ha^{-1})	Yield (m^3 ha^{-1})	Yield, green (Mg ha^{-1})	Source
Canada									
British	280–450	33	15						van Oosten, 2008
Columbia	550	15	25						
Quebec	1111	15	12–20						van Oosten, 2008
Alberta	816–1111	18	16–18						van Oosten, 2008
Ontario	1111	15	12–15						C. van Oosten, 2005 (unpublished)
USA									
Westside PNW	750–768	6–8	25–30						J. Eaton, 2005 (unpublished)
Eastside	1450	6–8	42	17–22	22–25				J. Eaton, 2005 (unpublished)
PNW	725	10–15	32			40–45			
South	745	8–10		25–30	30	30–34	250–350		J. Eaton, 2005 (unpublished)
Midwest	1100	12		20–25	18–20			270	J. Eaton, 2005 (unpublished)
South America									
Chile	278	12	23						Ulloa and Villacura, 2005
Argentina	278–625	10–16	20–21	35–45	28–30				S. Cortizo and R. Suarez, 2005 (unpublished)
Europe									
Italy	291	10.7	19	28	24	18			Coaloa, 1999; G. Picchi, 2005 (unpublished)
France	200	15–25	15	40	30–35	20–30			A. Berthelot, 2005 (unpublished)
	2000	7–9	26	12–15	15–18	27–31			
Spain	250–505	13	16						Barrio-Anta *et al.*, 2008
Sweden	5000	15–20						136	M. Ramstedt, 2005 (unpublished)
Finland	1000	20–30	12		20				E. Beuker, 2005 (unpublished)
Russia	500	21	20	27	36				A. Tsarev, 2005 (unpublished)
Serbia	278–555	22–32					200–500		S. Orlovic *et al.*, 2005 (unpublished)
Montenegro	278–555	22–32					200–500		S. Orlovic *et al.*, 2005 (unpublished)
Asia									
India	400–500	6–8	20–30						R.C. Dhiman, 2005 (unpublished)
China	277–333	12–13	18–27						Q.W. Zhang and J.H. Li, 2005 (unpublished)
(south)	833–1111	6–7	22.5–27						

Growth and yield

Many factors influence the growth of poplars in plantations, including species or clone (Zabek and Prescott, 2006), site quality (Baker and Broadfoot, 1979), climate (Chandra, 2011) and spacing. Tree growth is not uniform, however, even when individuals are all from the same clone. Poplars are extremely intolerant of shading, such that crowns of eastern cottonwood (*P. deltoides*) do not touch even in densely spaced plantations. Poplar clones vary in their tolerance of shading; some can be planted closer together than others, a concept expressed as 'stockability' (DeBell *et al.*, 1989). After establishment, the amount of growing space available to an individual tree dominates stand yield and significantly influences the average size stem attained by harvest age (Ranney *et al.*, 1987; DeBell *et al.*, 1997a, b). Although site potential is relatively fixed, at least under a given management intensity, the time required to achieve culmination of mean annual biomass increment can be influenced by manipulating growing space available to individual stems (Stanturf *et al.*, 2001). Even more importantly, the time required to reach a minimum or average stem size can be influenced by manipulating growing space, nutrients and water (Stanturf *et al.*, 2001).

Growth rates of poplar in industrial plantations are among the highest in the world (Dickmann and Stuart, 1983; Boysen and Strobl, 1991; Christie, 1994; Stanturf *et al.*, 2001; Stanton *et al.*, 2002; Weih, 2004; Dickmann, 2006; Zalesny *et al.*, 2011; Tullus *et al.*, 2012), and even higher growth rates can be obtained in experimental and bioenergy plantings. Directly extrapolating from small research plots to operational yield expectations, however, is dangerous. For example, an experiment with a hybrid poplar clone in small plots determined mean production values at age 4 to be 28 Mg ha^{-1} year^{-1} (Heilman and Stettler, 1985). Another experiment using larger plots (DeBell *et al.*, 1996) attained growth equal to or better than other studies with the same clone, but estimated yield to be 18 Mg ha^{-1} year^{-1}. Yield predictions have been published for individual trees or for stands (e.g. Christie, 1994), but should be used with caution outside of the region in which they were developed (Zabek and Prescott, 2006). Individual-tree and stand-level regression equations have been published for poplars (Krinard, 1988; Tuskan and Rensema, 1992; Clendenen, 1996; Lodhiyal and Lodhiyal, 1997; Scarascia-Mugnozza *et al.*, 1997; Kort and Tornock, 1999; Netzer and Tolsted, 1999; Stanturf *et al.*, 2001; Aylott *et al.*, 2008). A stand-level equation for *P. deltoides* in the southern USA (Cao and Durand, 1991a) uses site index equations (Cao and Durand, 1991b) to scale up the individual-tree volume equations of Krinard (1988) and assumes that yields reflect planting site-adapted clones.

Improved genetic material and advances in establishment, tending and protection of poplar has produced significant gains over time. Growth of natural stands of eastern cottonwood (*P. deltoides*), black cottonwood (*P. trichocarpa*) and aspen (*P. tremula*, *P. tremuloides*) and representative yields from industrial plantations (Table 5.7) provide a baseline for comparing growth and yield currently achieved in industrial poplar plantations (see Table 5.6). The highest values reported for operational plantation culture are from hybrid poplar in the western USA under irrigation with mean annual increment of 42 m^3 ha^{-1} (Stanton *et al.*, 2002; J. Eaton, 2005, unpublished report). Clonal trials generally provide estimates of aboveground dry matter, including branches but not leaves, expressed as Mg ha^{-1} year^{-1}. This indicates biological potential (Table 5.8) but must be interpreted with regard to species, length of growing season, density and rotation length. Industrial plantations in the USA have achieved sustainable yields of from 10 to 20 Mg ha^{-1} year^{-1} on an oven-dry basis, and potential yields have been estimated as 18, 27 and 40 Mg ha^{-1} year^{-1} for the Midwest, South and Northwest regions, respectively (Volk *et al.*, 2011a, b). Doubling yields will require advanced breeding and appropriate silvicultural techniques.

Harvesting and processing

Harvesting methods vary from fully mechanized in some regions of North America and Europe to fully hand labour in India (R.C. Dhiman, 2005, unpublished report) and combinations such as hand-felling and motorized skidding. In the western USA, mechanized equipment with hydraulic shears or hot saws fell, accumulate

Table 5.7. Growth and yield estimates of natural and managed poplar stands.

Stand type	Stand age	Mean annual volume increment (m^3 ha^{-1})	Reference
Natural *P. deltoides*, southern USA, 1900	10	12.6	Williamson, 1913
	20	17.1	Williamson, 1913
Natural *P. trichocarpa*, north-west USA, best site	112	5.5	Smith, 1980
Planted *P. tremuloides*	15	6.4–7.4	Einspahr, 1984[a]
P. tremula forest stand	80	14.8	Opdahl, 1992
Italian poplar plantations (clone 'I-214'), good sites, medium spacing, 1960s	12	36	Prevosto, 1965
Italian poplar plantations (clone 'I-214', 'Neva', 'Dvina', 'Lena'; 278–320 stems ha^{-1})	7–13	7.9–35	Facciotto *et al.*, 2003
P. trichocarpa plantations, 1970s	24	15–20	Smith, 1980; Murray and Harrington, 1983
Intensively managed poplar plantations, USA, 1980s	10–20	7–25	Dickmann and Stuart, 1983
Irrigated hybrid poplar plantations, USA, 2005	6	42	J. Eaton, 2005 (unpublished)

[a]Representative yields; from Tullus *et al.*, 2012.

and bunch the trees that are then skidded or forwarded to the roadside for loading or processing (J. Eaton, 2005, unpublished report). One location east of the Cascade Mountains has a central processing site; after whole trees are felled and skidded to the roadside, they are loaded and transported to the processing site. Stems are then scanned and merchandized; sawlogs are sent to a sawmill and the rest of the stem is debarked and ground into chips. Residual material is consolidated and then shipped for composting or bioenergy production. In some other plantations, portable debarking and chipping machines process stems at the roadside (Fig. 5.19) to minimize skidding to a central chipper (J. Eaton, 2005, unpublished report). In south-western British Columbia, Canada, similar methods are used, including mechanical harvesters with directional felling heads (Thomas *et al.*, 2000). In some cases, material is moved with hydraulic excavators equipped with a loading grapple that picks up the wood and swings it in the direction of the road. This method is called 'hoe-chucking'. Logs are transported to the mill for debarking and chipping; some merchandizing of veneer grade logs occurs. Although hoe-chucking is sometimes used in short-rotation plantations, it is expensive and causes unwanted stem breakage with smaller diameter material.

Felling and delimbing are manual processes in Argentina (S. Cortizo and R. Suarez, 2005, unpublished report), France (A. Berthelot, 2005, unpublished report) and Italy (G. Picchi, 2005, unpublished report; Castro and Zanuttini, 2008). In France, logs are extracted by skidders or forwarders (A. Berthelot, 2005, unpublished report); in Argentina, they are loaded on to trailers by a crane and transported (Fig. 5.20). Debarking is done at the wood yard or mill. In Italy, logs are concentrated using skidders or loaders with a hydraulic boom (G. Picchi, 2005, unpublished report). If the buyer is a large company, each log is scaled in the woods and delimbed and bucked (sectioned). Branches, tops and unmerchantable logs are chipped and loaded directly into trucks or piled on-site for later transport. Use of feller-bunchers or processors is increasing in France and Italy; the reduction in cost achieved with mechanization offsets the less efficient selection of assortments (Spinelli *et al.*, 2004). Harvesting in bioenergy coppice plantations uses modified silage harvesters (Spinelli *et al.*, 2009).

Harvesting is a fully manual process in India (R.C. Dhiman, 2005, unpublished

Table 5.8. Yields achieved in experimental plantings.

Location	Clone/ species[a]	Age (years)	Density (stems ha^{-1})	Productivity (Mg ha^{-1} $year^{-1}$)	Reference
West USA	T	2	111,111	13.4–20.9	Heilman *et al.*, 1972
West USA/ Canada	T	Multiple	6,944–111,111	9.0–11	Smith and DeBell, 1973
West USA	T	8	6,944–111,111	5.8–9.7	Heilman and Peabody, 1981
UK	T	5	40,000	9–10	Cannell and Smith, 1980
West USA	T, TD	4	6,944	5.2–27.8	Heilman and Stettler, 1985
West USA	T, TD	4	6,944	11.3–12.6	Heilman and Stettler, 1990
West USA	T, TD	7	10,000	11–18	DeBell *et al.*, 1996
West USA	T, TD	4	10,000	14–35	Scarascia-Mugozza *et al.*, 1997
UK	T, TD	4	2,500–10,000	13.6	Armstrong *et al.*, 1999
Sweden	T,TD	13	6,944	8.0	Christersson, 2006
West USA	TD	5	5,917–308,642	6.4 (high density) – 30 (low density)	DeBell *et al.*, 1993
West USA	TD	7	2,500–40,000	10.1–18.2	DeBell *et al.*, 1996
West Canada	TD	4–13	1,111	9.2–13.6	Zabek and Prescott, 2006
France	TD	2–3	Multiple	0.6–3.5	Auclair and Bouvarel, 1992
Belgium	TD	Multiple	15,625	30	Pontailler *et al.*, 1999
South USA	D	5–20	1,111	10–11	Switzer *et al.*, 1976
South USA	D	5	10,000	13.3	Dowell *et al.*, 2009
India	D	5–8	400	9–14	Lodhiyal *et al.*, 1995
China	D, DN	4	Multiple	17.4–23.4	Fang *et al.*, 2007
South USA	DN	5	10,000	7.4	Dowell *et al.*, 2009
North-east USA	DN	4	17,313–694,444	7.7	Bowersox and Ward, 1976
North-central USA	DN	4	26,875–189,036	11.3–13.8	Ek and Dawson, 1976
East Canada	DN	Multiple	Multiple	5–19	Anderson, 1979
North-central USA	DN	7–12	1,736	4.7–10	Netzer *et al.*, 2002
North-central USA	DN, B	Multiple	Multiple	4.9–12.8	Strong and Hansen, 1993
North-central USA	B	5	6,944	4.2	Isebrands *et al.*, 1982
Finland	B	6	5,102–20,408	4.2	Ferm *et al.*, 1989
UK	B	5	4,444	10.2–16.2	Proe *et al.*, 2002
Czech Republic	N, MT	4–7	2,268	7.6–9.4	Benetka *et al.*, 2007
Canada	MN	4	18,146	16.62–18.05	Labrecque and Teodorescu, 2005
Belgium	Multiple	6	4,444–17,778	10.8	Laureysens *et al.*, 2003
Belgium	Multiple	4	10,000	2.8–11.4	Laureysens *et al.*, 2004
Belgium	Multiple	Multiple	10,000	9.7	Laureysens *et al.*, 2005
Belgium	Multiple	3	4,444–17,778	2.8–9.7	Al Afas *et al.*, 2007
Belgium	Multiple	11	4,444–17,778	13.3–14.6	Al Afas *et al.*, 2008
Chile	Multiple	6	7,500	4–8	Baettig *et al.*, 2010
Italy	Multiple	4	5,917	6.5–17.75	Paris *et al.*, 2011
Italy	D, DN	9	10,000	2.7–14.4	Bergante and Facciotto, 2011

Continued

Table 5.8. Continued.

Location	Clone/ species[a]	Age (years)	Density (stems ha^{-1})	Productivity (Mg ha^{-1} year^{-1})	Reference
Italy	Multiple	5	1,143–1,667	7.9–25.0	Facciotto and Bergante, 2011
Czech Republic	Multiple	6	5,495	8.1–13.9	Trnka *et al.*, 2008
Europe	Multiple	Multiple	Multiple	10–15	Dillen *et al.*, 2007
Sweden	Multiple	Multiple	410–2,500	3–10	Christersson, 2010
Germany	Aspens and hybrids	10	4,167–5,556	4.7–12.4	Liesebach *et al.*, 1999
Germany	HA	23	830–6,900	9.0	Johansson, 1976
Sweden	HA	9	5000	7.9	Karačić *et al.*, 2003
Estonia	HA	10	1,100–1,300	1.6–3.8	Tullus *et al.*, 2012

[a]Species and clone designations as follows: T, *Populus trichocarpa*; D, *Populus deltoides*; N, *Populus nigra*; B, *Populus balsamifera*; M, *Populus maximowiczii*; letters together are hybrids, e.g. TD = *P. trichocarpa* × *P. deltoides*; HA = hybrid aspen, *Populus* ×*wettsteinii*.

Fig. 5.19. A portable debarking and chipping machine is processing stems at the roadside to minimize skidding to a central chipper. Residual material is consolidated and then shipped for composting or bioenergy production. Photo courtesy of BoiseCascade.

report). The base of each tree is excavated slightly and side roots are severed by axe. As the tree falls, the supporting roots are cut. The fallen stem is sectioned into standard-length logs (2.5 m) and delimbed by axe. Logs are sorted by girth (oversize are >60 cm in the middle of the log, undersize are <50 cm). Debarked wood with a 20–50 cm girth may be sold as pulpwood. Roots, bark and branches are sold for firewood.

Fig. 5.20. Poplar logs are loaded on to trailers by a crane and transported to the mill. Felling, delimbing and bucking are done manually, by chainsaw crews. Photo courtesy of J. Stanturf.

Transportation

Truck transportation is the main transportation mode in industrial operations. Where plantations are located close to a major river system, barges may be used to transport chips or logs to the mill, for example in the USA (J. Eaton, 2005, unpublished report), Serbia and Montenegro (S. Orlovic, B. Klasnja, Z. Galic, L.P. Pajnik and P. Pap, 2005, unpublished). In the Delta region of eastern Argentina, most transport is by barge (Fig. 5.21). Short-distance hauling of chips by farm tractor occurs in Italy (G. Picchi, 2005, unpublished report); long-distance hauling by rail is used somewhat in France, India and China (A. Berthelot, 2005, unpublished report; R.C. Dhiman, 2005, unpublished report; Q.-W. Zhang and J.-H. Li, 2005, unpublished report). Short logs may be transported by truck. Farmers in India may transport wood by bullock carts or, rarely, by bicycle-rickshaw (R.C. Dhiman, 2005, unpublished report).

Storage

Poplar is generally used within a few days of harvesting. The fibre oxidizes and discolours on exposure to the air; the discoloration lessens the brightness advantage for pulping as compared to other species. Higher-quality lumber can be produced from processing fresh logs because drying tends to cause checking on the log ends. Dry logs also take longer to cut, slowing throughput and raising operating costs. Match factories in India may store poplar logs under sprinklers for 3 months or longer to soften them for easier peeling.

5.3 Willow

Willows (*Salix* spp.) have long been cultivated for wickerwork, stakes, specialty products and fuelwood (FAO, 1980). Industrial cultivation of willows was concentrated in the Danube basin in Europe (Marković, 1986) and the Paraná delta in South America (FAO, 1980). Today, the primary industrial uses are for wicker furniture and baskets, and interest has renewed in willow for biofuels, particularly in northern Europe and North America. Cultivation of willow for the manufacture of cricket bats continues as a specialty product in India. Willows have been planted in many countries for soil conservation, especially stream bank stabilization. Recently in New Zealand, willows have been planted in browsing blocks for livestock fodder and managed as a grazing system, using *S. matsudana* × *S. alba* 'Tangoio' (I. McIvor and

Fig. 5.21. Where plantations are located close to a major river system, barges may be used to transport chips or logs to the mill. Most transport is by barge in the Paraná delta of Argentina; here poplar logs are being loaded on to a barge. Photo courtesy of J. Stanturf.

I. Nicholas, 2005, unpublished report). Many of the techniques used in poplar culture apply to willow, especially coppice methods. Willow, especially the species with shrub growth forms, has very high juvenile growth rates and the fastest growth rates under boreal conditions (Christersson *et al.*, 1993; Labrecque and Teodorescu, 2003) and vigorous sprouting from stumps (Ceulemans *et al.*, 1996). These characteristics have made them attractive for bioenergy and phytoremediation applications (Kuzovkina and Quigley, 2005) and in combination (Mirck *et al.*, 2005; Ruttens *et al.*, 2011).

5.3.1 Stand establishment

Planting material

Industrial plantings of willow are predominantly species of shrub willows. Clones of *S. viminalis*, *S. schwerinii* and *S. gmelinii* and hybrids predominate in northern Europe (Larsson, 1998). A nascent willow bioenergy industry in the northern USA relies on the native *S. eriocephala* and the widespread *S. purpurea* that was introduced by European settlers to New York State for basket making in the 1700s (Smart *et al.*, 2005). European clones are not planted in the USA because they are very susceptible to the potato leafhopper (*Empoasca fabae* Harris), a pest of lucerne. Clones with *S. viminalis* pedigree are especially susceptible; resistance to this pest can be bred by crossing with Asian clones (Keoleian and Volk, 2005). An active breeding programme incorporates many clones of these species as well as Eurasian species (Smart and Cameron, 2008). Efforts to revive the Chilean basket willow sector rely on the naturalized *S. viminalis* (Abalos Romero, 2005).

Commercial willow bioenergy plantations in Europe rely on material produced by specialist growers in nursery beds and supplied as 1-year-old whips (rods) for mechanical planting. Because these improved clones are protected by European plant breeders' rights, producing material for self-use or sale is illegal (DEFRA, 2002), except that filling gaps with material cut from an existing crop is allowed (Caslin *et al.*, 2010). Sweden is the leader in willow bioenergy cultivation in Europe (Wright, 2006) and many of the commercially available clones come from a Swedish breeding programme,[2] including 'Tora', 'Sven', 'Torhild', 'Tordis', 'Olof', 'Gudrun' and 'Inger' (Table 5.9). Older Swedish clones such as 'Jorr' and 'Joruun' are less productive and have poorer disease resistance (Caslin *et al.*, 2010). A European breeding programme based at Rothamsted Research in the UK has released commercial clones 'Nimrod', 'Resolution', 'Discovery', 'Endeavour', 'Beagle' and 'Terra Nova'. The Swedish programme has focused on *S. viminalis* and its hybrids with *S. schwerinii* (Table 5.9), but the European programme encompasses more species (Caslin *et al.*, 2010).

In northern Europe, willows are planted as cuttings or whips (rods), depending on the type of machinery available. Cuttings are 18–20 cm long and cut fresh from whips procured from a licensed producer; planting whips are cut and trimmed willow stems 1.5–3 m long. Whips are harvested from 1-year-old material while dormant and either planted immediately or stored at −2° to −4°C (DEFRA, 2002). In the USA, unrooted dormant stem cuttings 20–25 cm long or whips greater than 1.3–2 m long (Abrahamson *et al.*, 2002) are used, with size limitations dictated by the planting machinery used. In New Zealand where plantings may be in small areas on farms, 25 cm unrooted dormant cuttings or 1 m stakes are used (I. McIvor and I. Nicholas, 2005, unpublished report).

Site requirements

Willow plantations grow mostly on marginal agricultural soils and are often integrated into farms. Site requirements for willows are similar to those for poplars (Table 5.10), although native willows are adapted to wetter sites than native poplars, growth on poorly drained soils may be non-economic (Abrahamson *et al.*, 2002). Nevertheless, when plantings are for both wastewater treatment and other environmental purposes in combination with bioenergy production, site requirements may be less restrictive. Clay soils with good aeration may be suitable, although establishment may be slower and early growth lower, but once established they may be highly productive (Abrahamson *et al.*, 2002; DEFRA, 2002). Trafficability may be a concern on clay soils, especially if the site is prone to flooding in winter. Growth on organic soils in Sweden has been acceptable, but difficulty in controlling weeds and frost-prone landscape positions render them less suited for commercial plantations (Lantmännen Agroenergi, no date).

Site preparation

Effective weed control is just as critical for willow plantings as it is for poplars. If good weed control until canopy closure (1–2 years) is absent, failure is a likely prospect (Abrahamson *et al.*, 2002). Both mechanical and chemical treatments are used, beginning at the end of the growing season prior to planting. Mowing

Table 5.9. Commercially available *Salix* clones in Northern Europe in 2006[a] (http://pohjonen.org/veli/vpapps/2006-agrob-salix-clones.pdf).

Willow clone name	Parentage
'Doris'	*S. dasyclados*
'Gudrun'	*S. dasyclados*
'Inger'	*S. triandra* × *S. viminalis*
'Jorr'	*S. viminalis* × *S. viminalis*
'Karin'	(*S. schwerinii* × *S. viminalis*) × *S. viminalis*) × *S. burjatica*)
'Olof'	(*S. viminalis* × (*S. schwerinii* × *S. viminalis*))
'Tora'	*S. viminalis* × *S. schwerinii*
'Sherwood'	(*S. viminalis* × *S. eriocephala*) × (*S. schwerinii* × *S. viminalis*)
'Sven'	*S. viminalis* × (*S. schwerinii* × *S. viminalis*)
'Tordis'	(*S. schwerinii* × *S. viminalis*) × *S. viminalis*
'Torhild'	(*S. schwerinii* × *S. viminalis*) × *S. viminalis*

[a]SalixEnergi Europa (http://www.salixenergi.se/) is the current owner of rights/royalty for willow cuttings from the Swedish programme in the whole of Europe.

Table 5.10. Soil characteristics suitable for willow bioenergy crops (adapted from Abrahamson *et al.*, 2002).

Soil characteristic	Suitable	Unsuitable
Texture	Loams, sandy loams, loamy sands, clay loams and silt loams; clay soils with adequate aeration[a]	Coarse sand, heavy clay soils without adequate aeration
Structure	Well developed to single grain	Massive or structureless
Drainage	Imperfectly to moderately well drained	Excessively well or very poorly drained
pH	5.5–8.0	More acid than 5.5 or more basic than 8.0
Depth (of rooting)	Greater than or equal to 46 cm	Less than 46 cm

[a]Establishment may be more difficult and early growth slower on these soils.

to remove hay or brush, including baling, and removal of excessive vegetation may be necessary prior to herbicide application (Abrahamson *et al.*, 2002). Depending on the weed complex, one or two applications of a glyphosate-based herbicide are needed beginning in late summer or early autumn (DEFRA, 2002). To control perennial broadleaved weeds, a tank mix with 2,4-D and glyphosate may be necessary (Abrahamson *et al.*, 2002). Aggressive weed control must therefore start with controlling perennial grasses and broadleaved weeds in the site preparation phase; ignoring this important aspect leads to increased costs and unpredictable results once the plantation is established.

Subsoiling to a depth of 40 cm followed by ploughing to a depth of 25 cm is the next step (Abrahamson *et al.*, 2002; DEFRA, 2002), followed by cross-disking (Abrahamson *et al.*, 2002). Soils prone to erosion can be protected by sowing an annual cover crop such as winter rye (*Secale cereale* L.) that must be killed prior to planting (Abrahamson *et al.*, 2002). In the USA, cultivation (cultimulching) just before planting is recommended (Abrahamson *et al.*, 2002). In the UK and Ireland, suitable sites can be ploughed and power-harrowed in mid-March, 6 weeks before planting. A second or third application of glyphosate kills any germinated weeds. This approach is not practical on heavy clay soils, and these sites should be power-harrowed just before planting (DEFRA, 2002; Caslin *et al.*, 2010). Protruding rocks (more than 5 cm) should be removed from the site to avoid interference with mechanical planters and damage to expensive saw blades on harvesting equipment.

Planting

Bioenergy plantings generally follow the method developed in Sweden that uses a double or twin-row system. Planting density in Sweden has decreased from 20,000 cuttings ha^{-1} in the early 1990s to around 12,000 stems ha^{-1} today. Spacing between individual rows in each set of double rows is 0.75 m; the sets of double rows are 1.5 m apart. The in-row plant spacing is 0.75 m. Commercial plantings in Northern Ireland and the USA retain the earlier practice of planting 18,000 stems ha^{-1} to attain a final density of 15,000 stems ha^{-1}; row spacing remains the same, but spacing between plants is reduced to 0.6 m (Abrahamson *et al.*, 2002; Caslin *et al.*, 2010). The lower planting density produces thicker stems and thus larger material when chipped (DEFRA, 2002). Biomass yield is highly correlated with planting density; studies have shown that yield plateaus above 20,000 or 25,000 stems ha^{-1} (depending on the clone) and there is a sharp decline below 10,000 stems ha^{-1} (Bergkvist and Ledin, 1998).

These row spacings accommodate the use of step planters (Fig. 5.22) such as the Swedish machine designed by Salix Maskiner (Caslin *et al.*, 2010). Step planters plant two double rows in a single pass and automatically cut whips 1.5–2.5 m in length into cuttings 18–20 cm long. Whips are fed into the machine manually and guided by two belts. In the planting mechanism, the whip is cut to the desired length and the cutting is inserted vertically into a slit in the soil made by a coulter; the machine firms the soil around the cuttings (Abrahamson *et al.*, 2002; DEFRA, 2002; Caslin *et al.*, 2010). Another Swedish machine,

Fig. 5.22. The double-row planting system accommodates use of step planters that can plant two double rows in a single pass. Whips are fed into the machine manually and cut automatically by the planting mechanism into cuttings 18–20 cm long, which are inserted vertically into a slit in the soil made by a coulter; the machine firms the soil around the cuttings. Photo courtesy of N. Thevathasan.

the Fröebbesta planter, uses 25-cm-long cuttings that are fed into a planting tube manually and driven by hydraulic-powered rubber wheels into an open slit made by the machine. The slit is closed around the cutting by two packing wheels (Abrahamson *et al.*, 2002). The smaller, more manoeuvrable Fröebbesta planter can be used for planting small areas or riparian buffers (Abrahamson *et al.*, 2002) and modified cabbage planters have been used in the UK for small areas (DEFRA, 2002). Application of a pre-emergent herbicide immediately after planting is recommended (Abrahamson *et al.*, 2002; DEFRA, 2002; Caslin *et al.*, 2010). Oxyfluorfen and simazine have been used, but others are being tested (Abrahamson *et al.*, 2002). In Canada, the herbicide, flumioxazin, was recently approved for use as a pre-emergent in willow crops.

Clonal deployment and risk management

Similar to poplars, willows are susceptible to rust pathogens and only resistant clones should be planted. In the UK, where the moist maritime climate favours the pathogen, planting at least six clones from different breeding programmes is recommended. Planting should be in intimate mixtures (DEFRA, 2002; Tubby and Armstrong, 2002; Caslin *et al.*, 2010; Wickham *et al.*, 2010). This is accomplished using the step planter by planting short runs (10–15 cuttings) of a single clone, followed by a short run of another, randomly selected clone out of the six or more (Caslin *et al.*, 2010). In the north-eastern USA where rust has not yet become a major problem, planting clones in small blocks of a few double rows to 1 ha in size has been recommended (Abrahamson *et al.*, 2002).

5.3.2 Stand tending

Competition control

Typically, 90% survival is obtained as long as weed control is effective (Abrahamson *et al.*, 2002). If herbicide treatments are not totally effective, some mechanical cultivation may be needed (Abrahamson *et al.*, 2002; Lantmännen Agroenergi, no date), but may not be effective under moister climatic conditions (Caslin *et al.*, 2010). Grasses may be controlled with suitable post-emergent herbicides without harming the willow (Abrahamson *et al.*, 2002). Some broad-leaved weeds (such as Canada or creeping thistle – *Cirsium arvense*) can be controlled effectively with minimal or no injury to the willow with the post-emergent herbicide, clopyralid. Growth during the first season will vary by clone, rainfall and site conditions, but willow should be at least 1 m tall at the end of the season and may reach 4 m, with 1–3 sprouts per cutting (Abrahamson *et al.*, 2002; DEFRA, 2002).

Cutback

During the first dormant season following planting, the willow should be cut back to within 3–10 cm of ground level to encourage sprouting (Abrahamson *et al.*, 2002; DEFRA, 2002; Caslin *et al.*, 2010). Cutting needs to be accomplished before bud break using a modified, reciprocating-type mower that gives a clean cut, without tearing or pulling the cutting from the soil (Abrahamson *et al.*, 2002; DEFRA, 2002; Caslin *et al.*, 2010). Multiple stems will emerge, up to 5–20, depending on the clone. Canopy closure should be achieved within a few months of active growth. Under some conditions, a contact herbicide may be needed to control weeds that have established during the previous year (DEFRA, 2002).

Nutrition

Willow has higher demand for nutrients than many tree or shrub species, yet lower than most agricultural crops. On most sites where willow plantations will be established, sufficient nutrients are available that no fertilization is required during the first establishment year. Indeed, fertilization may simply stimulate competing vegetation and hinder effective weed control (Abrahamson *et al.*, 2002; DEFRA, 2002; Caslin *et al.*, 2010). Most published work on willow nutrition is based on early clones and may underestimate nutrient requirements of the higher-yielding clones now in commercial use (Caslin *et al.*, 2010). Estimates of crop requirements based on nutrient removals in harvested material are in the range of 150–400 kg N ha^{-1}, 180–250 kg P ha^{-1} and 24–48 kg K ha^{-1} over a 3-year rotation (Caslin *et al.*, 2010).

Nitrogen fertilization has been shown to increase yields significantly under experimental conditions (Adegbidi *et al.*, 2001; Mola-Yudego and Aronsson, 2008). In Sweden, sewage biosolids applications are common in commercial willow plantations, but it is doubtful that nitrogen requirements are being met (Aronsson *et al.*, 2002; Lantmännen Agroenergi, no date). One complicating factor is the difficulty of applying fertilizer beyond the first year without damaging the crop (Abrahamson *et al.*, 2002). Sewage biosolids applied in liquid form through a dribble bar can be applied until the coppice reaches 2.5 m in height, or possibly in the second year of a rotation (DEFRA, 2002). Slow-release fertilizers (mineral as well as composted sewage biosolids) applied after cutback to meet total rotational demand may be a solution (DEFRA, 2002). Fertilizer recommendations for the UK are based on the site nutrients available in the soil (Table 5.11).

Protection

In northern Europe, *Melampsora* rust is the most important disease of willow coppice systems (DEFRA, 2002; Toome *et al.*, 2006; Caslin *et al.*, 2010; Chapter 8, this volume). Climatic conditions favour rapid infection that leads to premature defoliation and entry of secondary pathogens through unprotected leaf scars (Caslin *et al.*, 2010). In addition to lowering yields directly from the defoliation, the secondary infections from dieback organisms (*Fusarium* spp. or *Glomerella* spp.) can cause sufficient damage to shots and stools that mortality ensues. Control with fungicides is possible, but in some countries may be deemed economically unfeasible or environmentally undesirable (Caslin *et al.*, 2010). In Canada, the fungicide, tebuconazole, was recently approved for *Melampsora* rust control in willow crops; use of this fungicide is primarily foreseen for stoolbed production.

Table 5.11. Fertilization recommendations (kg ha^{-1}) for short-rotation willow in the UK; soil index refers to amount of site nutrient availability. Soil index = 1 is low site nutrients and responsive to fertilization (adapted from Wickham *et al.*, 2010.)

Soil index	Nitrogen (N)	Phosphorus (P)	Potassium (K)
1	130	34	155
2	100	24	135
3	75	0	120
4	40	0	0

Although little to no rust has been reported, the expectation is that *Melampsora* rust will eventually gain a foothold in expanding willow crops. Where rust is well entrenched and resistant clones are available, they should be deployed. Since new rust races will appear over time, even clones considered resistant will fall victim and need to be replaced with new clones.

Willow beetles (Chrysomelids) are the major economic pest problem in northern European plantations (Caslin *et al.*, 2010; Chapter 9, this volume). Both adults and larvae feed on the leaves and can reduce yield by as much as 40% (DEFRA, 2002). Even though damage may appear visually to be severe, defoliation experiments have shown that effect on yield is minimal if <30% of the leaf surfaces are damaged (Caslin *et al.*, 2010). There can be two generations per year but there is significant year-to-year variation in populations. The adults overwinter in rotting wood and under the bark of trees in forest areas around the coppice. Although it is not feasible to treat established plantations, some control is possible by spraying insecticides around the border of a plantation when the beetles are re-colonizing the coppice in the spring. Because some clones are more resistant to beetle damage than others, planting mixtures has been effective in limiting economic effects (DEFRA, 2002; Caslin *et al.*, 2010).

5.3.3 Production

Harvest cycle

Coppice growth is rapid after cutback, particularly in the third and fourth growing season (2 and 3 years after coppice). Thus, a 4-year harvest cycle is common in commercial plantations. Harvest may be delayed, however, if initial establishment is poor or growth is slower than average due to the high cost of the harvest operation (Abrahamson *et al.*, 2002; DEFRA, 2002; Caslin *et al.*, 2010). Swedish practice is to stage the initial harvest after growth has accumulated to 25 Mg ha^{-1} (oven dry) or until the largest shoots exceed 6 cm in basal diameter. This usually requires 3–4 years (Lantmännen Agroenergi, no date). In Chilean basket willow plantations, rotations may be 9–11 years with annual harvesting (Abalos Romero, 2005).

One advantage of coppice systems is that several crops can be harvested from the same root system, thereby avoiding several repeats of site preparation and establishment costs. Yields will plateau after the second harvest cycle but can be maintained for up to 7–10 harvest cycles (Abrahamson *et al.*, 2002; Caslin *et al.*, 2010). Similarly, in basket willow plantations in Chile, plantings are expected to produce for 10–12 years (Abalos Romero, 2005). On the other hand, improved planting material with greater productivity and disease resistance may justify more frequent replanting (Wickham *et al.*, 2010).

Growth and yield

Yield is usually given on the basis of oven dry, aboveground matter to standardize comparisons. Harvesting occurs in the dormant season after leaf fall, and the harvested material accounts for about 60% of total annual net productivity (Caslin *et al.*, 2010). Average annual growth of 10–20 Mg ha^{-1} $year^{-1}$ has been reported from experiments with even higher growth rates (30 Mg ha^{-1} $year^{-1}$) recorded from irrigated and fertilized research plots (Labrecque and Teodorescu, 2003, 2005; Larsson *et al.*, 2003; Szczukowski *et al.*, 2005; Arevalo *et al.*, 2007; Stolarski *et al.*, 2007; Aylott *et al.*, 2008; Cerrillo *et al.*, 2008; Mola-Yudego and

Aronsson, 2008; Fillion *et al.*, 2009; Mola-Yudego, 2010; Tullus *et al.*, 2012). Commercial yields of 10–12 Mg ha^{-1} year^{-1} are probably a good benchmark for current levels of production (Mola-Yudego, 2010; Volk *et al.*, 2011a), and advances in breeding and optimization of management systems, including matching clones to sites, should increase commercial yields closer to what is attainable in experimental plantings. For example, in New York State in the USA, second harvest cycle yields of experimental trials increased by 18–62% compared to the first harvest cycle yields. More recent trials with advanced material are yielding 20–40% more than unimproved standard clones (Volk *et al.*, 2011a).

Further analysis of yield data from four consecutive harvest cycles from the US trial, combined with yield data from the first harvest cycle from a network of trials across the USA and Canada, provided a comparison of old versus new willow clones (Volk *et al.*, 2011b). The overall yield increase from the first to the second harvest cycle was 23% for four commercial clones. By the fourth harvest cycle, these same clones showed an overall increase of 30.8% over the first harvest cycle with a yield of 23.4–32.4 Mg ha^{-1} year^{-1} (oven dry). In the network's first harvest cycle trials, the top three new clones had a 13.9% greater yield (11.5 Mg ha^{-1} year^{-1} oven dry) than the older three reference clones. Increases in yield are not all due to improved genotypes. Some of these increases can be attributed to improved crop management practices, especially weed control and site factors. The impact on yield from factors such as disease and insect pressure, winter dieback, predation by various animals, for example deer, rabbits, etc., are still poorly understood and require better quantification.

Harvesting and processing

Harvesting methods vary according to available machinery and end-user requirements. Willow may be harvested as whips (rods), billets, chips or round bales. Whips up to 8 m long are produced loose and must be collected; they are often bundled if transported some distance. In Chile, whips (called switches) are sorted by size and colour and bundled; bundles may weigh as much as 50 kg (Abalos Romero, 2005). Bundler harvesters in bioenergy plantations cut whole stems, bind them and re-cut them into 2.5-m-long bundles (DEFRA, 2002).

Direct-chip harvesting is preferred for bioenergy production because chip quality is better if fresh material is chipped, as opposed to dried rods or bundles (DEFRA, 2002). Silage harvesters with specially designed cutting heads cut, chip and blow material into wagons in one continuous operation (Fig. 5.23). Most direct-chip harvesters have been designed to cut a double row in a single pass (Caslin *et al.*, 2010), but at least one machine cuts a single row and can be used to cut across rows, if necessary (Abrahamson *et al.*, 2002). While direct-chip harvesters are the most efficient harvesting system, drying the fresh chips poses some difficulty and chip quality is degraded unless moisture is removed soon after harvest (DEFRA, 2002; Caslin *et al.*, 2010). Chips are optimally 5 × 5 × 5 cm in size and moisture content must be lowered from 45–60% at harvest to below 30% (DEFRA, 2002). Billets are cut stems 5–10 cm long; they are produced by harvesters that cut the stems whole, re-cut into billets and blow the material into accompanying trailers (DEFRA, 2002; Caslin *et al.*, 2010). They are modified from sugarcane harvesters and the larger the size of the billets as compared to chips, the more air space there is between the pieces, which improves circulation and promotes natural drying. As with whip harvesters, chipping dry material reduces quality.

A relatively new development is to cut and bale willow biomass into round bales (Fig. 5.24). Willow shoots are cut and shredded into smaller pieces and baled in one operation by a modified agricultural hay baler. The main advantages of this system are lower capital costs and greater flexibility. The baler method harvests biomass as a stand-alone operation without the immediate need for transport capability. Storing bales until they are needed for processing increases flexibility in scheduling harvests. Bales can be stored either on site or at the processing plant; standard farm equipment can handle the bales.

Drying and storage are the weak links in the willow bioenergy supply chain (Caslin *et al.*, 2010; Wickham *et al.*, 2010), except when produced as round bales. Chips require immediate use or drying to avoid decomposition and degradation of quality, i.e. caloric value. Some agricultural facilities such as ventilated grain floors

Fig. 5.23. Silage or maize harvesters with specially designed cutting heads cut, chip and blow willow material into wagons in one continuous operation. Direct-chip harvesting is preferred for bioenergy production because chip quality is better if fresh material is chipped, as opposed to dried whips or bundles. Shown here is a Claas Jaguar direct-chip harvester. Photo courtesy of Claas Group.

with heated air can reduce moisture levels in fresh chips to acceptable moisture content, and low-cost methods with forced ventilation drying have been demonstrated (Caslin *et al.*, 2010). Whips and billets can be stored for several months under ambient air conditions and chipped at lower moisture contents than fresh stems (Wickham *et al.*, 2010). Round bales dry down naturally in the field or at the processing plant. When stored on site, the benefits are lower transport weights and thus decreased transportation costs

Processing whips for basket willow production can be done by the grower or an intermediary (Abalos Romero, 2005). Stems are harvested beginning after one growing season, although initial yield is low, and commercial production begins after two growing seasons. Cut shoots are referred to as switches and range from 0.6 cm to 6 m in length; diameters (at the thick end) range from 0.4 to 3 cm. Bark stripping, drying and sorting are the postharvest treatments. Traditional bark stripping consists of standing the switches in water until shoots emerge in the spring and then stripping the bark off by hand, with knives. Large-scale processing consists of boiling the switches and then stripping them with machines (Fig. 5.25). After air-drying, switches are sorted and bundled according to length, diameter and defects (Abelos Romero, 2005).

At some point, a final harvest will be made and it will be necessary to prepare the site for replanting with willow or another crop. The older the plantation at final harvest, the larger the root system and the more difficult it will be to prepare the site. In the UK and Ireland, willow plantations are expected to be followed by a return to grass or row crops (DEFRA, 2002; Caslin *et al.*, 2010). After the final winter harvest, the stools are allowed to re-sprout. After the sprouts have reached 30–50 cm in height, they are sprayed with a translocated herbicide such as glyphosate to kill the stool. After sufficient time for the herbicide to be absorbed and translocated (usually a minimum of 2 weeks), the stool and sprouts are mulched and incorporated into

Fig. 5.24. Willow bale produced by the 'Willow Harvester' (a), a prototype bale harvester developed through Université Laval, Agriculture and AgriFood Canada (AAFC) and Natural Resources Canada. These bales are typically 1.20 m in width. The 'Willow Harvester' is a prototype; other balers have been produced commercially and several have been used to bundle biomass from harvest residue in the southern pine region of the USA. 'Biobaler' harvesting 3-year-old coppice willow at Conestota, New York (b). Photos courtesy of C. van Oosten (a) and J. Richardson (b).

the surface layer of the soil. The majority of the root system is left in place to decompose and soil structure is not disrupted. Grass is sown into the soil and grown for a year or two before the field is placed back into production (DEFRA, 2002; Caslin *et al.*, 2010). In Sweden, the stools remain in place until the spring after the final harvest. The actively growing sprouts are sprayed with

Fig. 5.25. Large-scale processing of basket willow; switches are stripped using specialized machines. Fresh material has been boiled, giving it uniform colour. The next stage is air-drying. Photo courtesy of J. Stanturf.

herbicide such as glyphosate. After the sprouts die, the land is worked with a heavy disk that breaks up the stools and severs the large roots without raking them to the surface. The stand can be replanted to another willow rotation or converted back to agriculture (Lantmännen Agroenergi, no date).

Acknowledgements

Many people contributed to this chapter by providing information and insights into the production of *Populus* and *Salix* in industrial plantations. Although much of the country-specific detail was omitted in order to keep the chapter to a reasonable size, we hope the salient points have been retained. We would like to thank especially the editors, Jud Isebrands and Jim Richardson, for their patience, support and advice. Our heartfelt thanks go to the following: Jake Eaton, GreenWood Resources, USA; Brian Stanton, GreenWood Resources, USA; Gianni Picchi, Scuola Superiore S. Anna, Italy; Alain Berthelot, FCBA Institut Technologique, France; Egbert Beuker, Finnish Forest Research Institute METLA, Finland; Ramesh Diman, Wimco Seedlings Limited, India; Anatoly Tsarev, Petrozavodsk State University, Russia; Silvia Cortizo, University of Buenos Aires, Argentina; Raúl Suarez, Núcleo de Extensión Forestal (NEF) Delta, Argentina; K. Jürgens, Estonian University of Life Sciences, Estonia; Katrin Heinsoo, Estonian University of Life Sciences, Estonia; Arvo Tullus, Estonian University of Life Sciences, Estonia; Mauritz Ramstedt, Swedish Agricultural University, Sweden; Ian McIvor, HortResearch, New Zealand; Ian Nicholas, Ensis, New Zealand; Pajand Nejad, Swedish Agricultural University, Sweden; Baqeri Reza, Tehran University, Iran; Habibi Hasan, Tehran University, Iran; H. Sabeti, Ministry of Agriculture, Iran;

Arwand Babaipour, Ministry of Agriculture, Iran; Sasa Orlovic, Institute of Lowland Forestry and Environment, Serbia; Bojana Klasnja, Institute of Lowland Forestry and Environment, Serbia; Zoran Galic, Institute of Lowland Forestry and Environment, Serbia; Leopold Poljakovic Pajnik, Institute of Lowland Forestry and Environment, Serbia; Predrag Pap, Institute of Lowland Forestry and Environment, Serbia; Gulya Vildanaova, Tashkent State Agrarian University, Uzbekistan; K. Tolipov, Forestry Department, Uzbekistan; Qiwen Zhang, Chinese Academy of Forestry, China; Jin-hua Li, Chinese Academy of Forestry, China; Yeong-Bon Koo, Korea Forest Research Institute, Republic of Korea; Dan Netzer, US Forest Service, USA; Mark Coleman, University of Idaho, USA; and C. Jeffrey Portwood, Forestar Group, USA.

Notes

[1] PSB stands for 'plug Styroblock®'; most container stock in Canada uses the Styroblock® tray system (van Oosten, 2006).

[2] SalixEnergi Europa (http://www.salixenergi.se/) is the current owner of rights/royalty for willow cuttings from the Swedish programme in the whole of Europe.

References

Abalos Romero, M.I. (2005) Towards development of the Chilean basket willow sector. *Unasylva* 231, 40–46.

Abrahamson, L.P., Volk, T.A., Kopp, R.E., White, E.H. and Ballard, J.L. (2002) *Willow Biomass Producer's Handbook*. State University of New York College of Environmental Science and Forestry, Syracuse, New York.

Adegbidi, H.G., Briggs, R.D, Volk, T.A., White, E.H., Abrahamson, L.P. and Bickelhaupt, D.H. (2001) Biomass and nutrient export by willow clones in experimental bioenergy plantations in New York. *Biomass and Bioenergy* 20, 389–398.

Al Afas, N.A., Marron, N. and Ceulemans, R. (2007) Variability in *Populus* leaf anatomy and morphology in relation to canopy position biomass production and varietal taxon. *Annals of Forest Science* 64, 521–532.

Al Afas, N.A., Marron, N., Van Dongen, S., Laureysens, I. and Ceulemans, R. (2008) Dynamics of biomass production in a poplar coppice culture over three rotations (11 years). *Forest Ecology Management* 255, 1883–1891.

Alfaro, R.I. (2003) The 'grand forest borer' *Platypus mutatus* (= *sulcatus*): an important pest of poplar culture in Argentina. A plan of action. *SAGPyA Forestal* 28, 11–18.

Anderson, H.W. (1979) Biomass production of hybrid poplar in mini-rotation. In: Fayle, D.C.F., Zsuffa, L. and Anderson, H.W. (eds) *Poplar Research, Management and Utilization in Canada*. Forest Research Information Paper No 102. Ontario Ministry of Natural Resources, Toronto, Ontario, Canada.

Anderson, W.C. and Krinard, R.M. (1985) The investment potential of cottonwood sawtimber plantations. In: Shoulders, E. (ed.) *Proceedings of the Third Biennial Southern Silvicultural Research Conference*. General Technical Report SO-54. USDA Forest Service Southern Forest and Range Experiment Station, New Orleans, Louisiana, pp. 190–197.

Arevalo, C.B.M., Volks, T.A., Bevilacqua, E. and Abrahamson, L. (2007) Development and validation of aboveground biomass estimations for four *Salix* clones in central New York. *Biomass and Bioenergy* 31, 1–12.

Armstrong, A., Johns, C. and Tubby, I. (1999) Effects of spacing and cutting cycle on the yield of poplar grown as an energy crop. *Biomass and Bioenergy* 17, 305–314.

Aronsson, P., Heinsoo, K., Perttu, K. and Hasselgren, K. (2002) Spatial variation in above-ground growth in unevenly wastewater-irrigated willow *Salix viminalis* plantations. *Ecological Engineering* 19, 281–287.

Auclair, D. and Bouvarel, L. (1992) Intensive or extensive coppice cultivation of short rotation hybrid poplar coppice on forest land. *Bioresource Technology* 42, 53–59.

Aylott, M.J., Casella, E., Tubby, I., Street, N.R., Smith, P. and Taylor, G. (2008) Yield and spatial supply of bioenergy poplar and willow short-rotation coppice in the UK. *New Phytologist* 178, 358–370.

Baettig, R., Yáñez, M. and Albornoz, M. (2010) Short rotation woody crops of hybrid poplars for bioenergy in Chile: state of the art. *Bosque* 31, 89–99.

Baker, J.B. and Broadfoot, W.M. (1979) A practical field method of site evaluation for commercially important hardwoods. General Technical Report SO-36. USDA Forest Service Southern Forest and Range Experiment Station, New Orleans, Louisiana.

Balsari, P., Airoldi, G. and Facciotto, G. (2003a) A machine to cut poplar sets into cuttings. In: Ciccarese, L., Lucci, S. and Mattsson, A. (eds) *Proceedings of the Conference Nursery Production and Stand Establishment of Broad-Leaves to Promote Sustainable Forest Management*. Dalarna University, Sweden, pp. 3–7.

Balsari, P., Airoldi, G. and Facciotto, G. (2003b) Operative and economic evaluation of machines for planting cuttings. In: Ciccarese, L., Lucci, S. and Mattsson, A. (eds) *Proceedings of the Conference Nursery Production and Stand Establishment of Broad-Leaves to Promote Sustainable Forest Management*. Dalarna University, Sweden, pp. 9–16.

Barrio-Anta, M., Sixto-Blanco, H., Cañellas-Rey De Viñas, I. and Castedo-Dorado, F. (2008) Dynamic growth model for I-214 poplar plantations in the northern and central plateau in Spain. *Forest Ecology and Management* 255, 1167–1178.

Benetka, V., Vratny, F. and Salkova, I. (2007) Comparison of productivity of *Populus nigra* with an interspecific hybrid in a short rotation coppice on marginal areas. *Biomass and Bioenergy* 31, 367–374.

Bergante, S. and Facciotto, G. (2011) Nine years of measurements in Italian SRC trial with 14 poplar and willow clones. In: Faulstich, M., Ossenbrink, H., Dallemand, J.F., Baxter, D., Grassi, A. and Helm, P. (eds) *Proceedings of the 19th European Biomass Conference and Exhibition.* ETA-Florence Renewable Energies, Florence, Italy, pp. 178–182.

Bergante, S., Facciotto, G. and Minotta, G. (2010) Identification of the main site factors and management intensity affecting the establishment of short-rotation-coppices (SRC) in Northern Italy through stepwise regression analysis. *Central European Journal of Biology* 5, 522–530.

Bergkvist, P. and Ledin, S. (1998) Stem biomass yields at different planting designs and spacing in a willow coppice system. *Biomass and Bioenergy* 14, 149–156.

Berndes, G., Azar, C., Kåberger, T. and Abrahamson, D. (2001) The feasibility of large-scale lignocellulose-based bioenergy production. *Biomass and Bioenergy* 20, 371–383.

Bernier, B. (1984) Nutrient cycling in *Populus*: a literature review with implications in intensively-managed plantations. IEA/ENFOR Report 6. Canadian Forestry Service, Ottawa, Ontario.

Bisoffi, S. and Facciotto, G. (2000) I cedui a turno breve (SRF). SHERWOOD-*Foreste ed Alberi Oggi* 59, 21–23.

Bisoffi, S. and Gullberg, U. (1996) Poplar breeding and selection strategies. In: Stettler, R.F., Bradshaw, H.D. Jr, Heilman, P.R. and Hinckley, T.M. (eds) *Biology of Populus and Its Implications for Management and Conservation.* National Research Council of Canada Research Press, Ottawa, Ontario, pp. 139–158.

Blackmon, B.G. (1976) Response of *Aigeiros* poplars to soil amelioration. In: Thielges, B.A. and Land, S.B. Jr (eds) *Proceedings: Symposium on Eastern Cottonwood and Related Species.* Louisiana State University, Division of Continuing Education, Baton Rouge, Louisiana, pp. 344–358.

Blackmon, B.G., Baker, J.B. and Cooper, D.T. (1979) Nutrient use efficiency by three geographic sources of eastern cottonwood. *Canadian Journal of Forest Research* 9, 532–534.

Borodowski, E.D. and Suárez, R.O. (1999) Raleo en plantaciones de *Populus* spp. *SAGPyA Forestal* 10, 2–5.

Borodowski, E., Cortizo, S., Mema, V. and Landi, L. (2005) Raleo tradicional en el Delta del Paraná: implicancias en ocho clones de álamo. Actas en CD de las XX Jornadas Forestales de Entre Ríos, Concordia, Entre Ríos, Argentina.

Bowersox, T.W. and Ward, W.W. (1976) Growth and yield of close spaced young hybrid poplars. *Forest Science* 22, 449–454.

Boysen, B. and Strobl, S. (1991) *A Growers Guide to Hybrid Poplar.* Ontario Ministry of Natural Resources, Toronto, Ontario, Canada.

Brenneman, R. (1982) Electric fencing to prevent deer browsing on hardwood clearcuts. *Journal of Forestry* 80, 660–661.

Callan, B.E., Leal, I., Foord, B., Dennis, J.J. and van Oosten, C. (2007) *Septoria musiva* isolated from cankered stems in hybrid poplar stool beds, Fraser Valley, British Columbia. *Pacific Northwest Fungi* 2, 1–9.

Cannell, M. and Smith, R.L. (1980) Yields of mini-rotation in close-spaced hardwoods in temperate regions: review and appraisal. *Forest Science* 26, 415–428.

Cao, Q.V. and Durand, K.M. (1991a) A growth and yield model for improved eastern cottonwood plantations in the lower Mississippi Delta. *Southern Journal of Applied Forestry* 15, 213–216.

Cao, Q.V. and Durand, K.M. (1991b) Site index curves for eastern cottonwood plantations in the lower Mississippi Delta. *Southern Journal of Applied Forestry* 15, 28–30.

Caslin, B., Finnan, J. and McCracken, A. (2010) *Short Rotation Coppice Willow Best Practice Guidelines.* Teagasc Crops Research Centre, Carlow, Ireland and Agri-Food and Bioscience Institute, Belfast, Northern Ireland, UK (http://www.teagasc.ie/publications/2011/314/WillowBestPractice.pdf, accessed 24 May 2012).

Castro, G. and Zanuttini, R. (2008) Poplar cultivation in Italy: history, state of the art, perspectives. In: Van Acker, J. and Fioravanti, M. (eds) *Proceedings of the Cost Action E44 Final Conference in Milan on a European Wood Processing Strategy: Future Resources Matching Products and Innovations.* Ghent University, Ghent, Belgium, pp. 141–154.

Cellerino, G.P. (1999) Review of fungal diseases in poplar (http://www.fao.org/DOCREP/004/AC492E/AC492E00.HTM, accessed 24 May 2012).

Cerrillo, T., Facciotto, G. and Bergante, S. (2008) Biomass production of different willow's combinations – preliminary results. In: *Proceedings 16th European Conference and Exhibition, From Research to Industry and Markets.* ETA-Florence Renewable Energies, Florence, Italy, pp. 567–569.

Ceulemans, R., McDonald, A.J.S. and Pereira, J.S. (1996) A comparison among eucalypt, poplar and willow characteristics with particular reference to a coppice, growth-modeling approach. *Biomass and Bioenergy* 11, 215–231.

Chandra, J.P. (2001) Scope of poplar cultivation. *Indian Forester* 127, 51–60.

Chandra, J.P. (2011) Development of poplar based agroforestry system. *Indian Journal of Ecology* 38, 11–14.

Christersson, L. (2006) Biomass production of intensively grown poplars in the southernmost part of Sweden: observations of characters, traits, and growth potential. *Biomass and Bioenergy* 30, 497–508.

Christersson, L. (2008) Poplar plantations for paper and energy in the south of Sweden. *Biomass and Bioenergy* 32, 997–1000.

Christersson, L. (2010) Wood production potential in poplar plantations in Sweden. *Biomass and Bioenergy* 34, 1289–1299.

Christersson L., Sennerby-Forsse, L. and Zsuffa, L. (1993) The role and significance of woody biomass plantation in Swedish agriculture. *The Forestry Chronicle* 69, 687–693.

Christie, J.M. (1994) Provisional yield tables for poplar in Britain. Technical Paper 6. Forestry Commission, Alice Holt, UK.

Clendenen, G.W. (1996) Use of harmonized equations to estimate above-ground woody biomass for two hybrid poplar clones in the Pacific Northwest. *Biomass and Bioenergy* 11, 475–482.

Coaloa, D. (1999) Condizioni favorevoli all'espansione della pioppicoltura. *L'Informatore Agrario* 36, 79–82.

Coleman, M. (2007) Spatial and temporal patterns of root distribution in developing stands of four woody crop species grown with drip irrigation and fertilization. *Plant and Soil* 299, 195–213.

Coleman, M.D., Dickson, R.E. and Isebrands, J.G. (1998) Growth and physiology of aspen supplied with different fertilizer addition rates. *Physiologia Plantarum* 103, 513–526.

Colorio, G., Beni, C., Facciotto, G., Allegro, G. and Frison, G. (1996) Influenza del tipo di lavorazione preimpianto su accrescimento e stato sanitario del pioppo. *L'Informatore Agrario* 52, 51–57.

Cossalter, C. and Pye-Smith, C. (2003) *Fast-Wood Forestry.* Center for International Forestry Research, Jakarta.

Coyle, D.R., McMillin, J.D., Hall, R.D. and Hart, E.R. (2002) Cottonwood leaf beetle (Coleoptera: Chrysomelidae) defoliation impact on *Populus* growth and above-ground volume in a short-rotation woody crop plantation. *Journal of Economic Entomology* 4, 293–300.

Coyle, D.R., Nebeker, T.E., Hart, E.R. and Mattson, W.J. (2005) Biology and management of insect pests in intensively managed North American hardwood forest system. *Annual Review of Entomology* 50, 1–29.

Coyle, D.R., Coleman, M.D., Durant, J.A. and Newman, L.E. (2006) Survival and growth of 31 *Populus* clones in South Carolina. *Biomass and Bioenergy* 30, 750–758.

Coyle, D.R., Hart, E.R., McMillin, J.D., Rule, L.C. and Hall, R.B. (2008) Effects of repeated cottonwood leaf beetle defoliation on *Populus* growth and economic value over an 8-year harvest rotation. *Forest Ecology and Management* 255, 3365–3373.

Danfors, B., Ledin, S. and Rosenqvist, H. (1997) *Energiskogsodling-Handledning för Odlare* (*Short-Rotation Willow Coppice Growers Manual*). Swedish Institute of Agricultural Engineering, Uppsala, Sweden.

DeBell, D.S. (1990) *Populus trichocarpa* Torr. & Gray. In: Burns, R.M. and Honkala, B.H. (eds) *Silvics of North America. Vol 2 Hardwoods.* Agriculture Handbook 654. USDA Forest Service, Washington, DC, pp. 570–576.

DeBell, D.S. and Harrington, C.A. (1997) Productivity of *Populus* in monoclonal and polyclonal blocks at three spacings. *Canadian Journal of Forest Research* 27, 978–985.

DeBell, D.S., Harms, W.R. and Whitesell, C.D. (1989) Stockability: a major factor in productivity differences between *Pinus taeda* plantations in Hawaii and the southeastern United States. *Forest Science* 35, 708–719.

DeBell, D.S., Clendenen, G.W. and Zasada, J.C. (1993) *Populus* biomass: comparison of wood grass versus wider spaced short rotation systems. *Biomass and Bioenergy* 4, 305–313.

DeBell, D.S., Harrington, C.A., Clendenen, G.W. and Zasada, J.C. (1996) Tree growth and stand development in short rotation *Populus* stands: 7-year results of two clones at three spacings. *Biomass and Bioenergy* 11, 253–269.

DeBell, D.S., Harrington, C.A., Clendenen, G.W. and Zasada, J.C. (1997a) Tree growth and stand development in four *Populus* clones in large monoclonal plots. *New Forests* 14, 1–18.

DeBell, D.S., Harrington, C.A., Clendenen, G.W., Radwan, M.A. and Zasada, J.C. (1997b) *Increasing the Productivity of Short-Rotation Populus Plantations.* Final Report to the Bioenergy Feedstock Development Program, Oak Ridge National Lab, Oak Ridge, Tennessee (https://bioenergy.ornl.gov/reports/debell/contents.html, accessed 24 May 2012).

DEFRA (2002) *Growing Short Rotation Coppice, Best Practice Guidelines.* Department for Environment, Food and Rural Affairs, London.

Delate, K., Holzmueller, E., Frederick, D.D., Mize, C. and Brummer, C. (2005) Tree establishment and growth using forage covers in an alley-cropped system in Midwestern USA. *Agroforestry Systems* 65, 43–52.

Demeritt, M.E. (1990) Poplar hybrids. In: Burns, R.M. and Honkala, B.H. (eds) *Silvics of North America. Vol 2 Hardwoods.* Agriculture Handbook 654. USDA Forest Service, Washington, DC, pp. 577–582.

DesRochers, A. and Tremblay, F. (2009) The effect of root and shoot pruning on early growth of hybrid poplars. *Forest Ecology and Management* 258, 2062–2067.

DesRochers, A., van den Driessche, R. and Thomas, B.R. (2006) NPK fertilization at planting of three hybrid poplar clones in the boreal region of Alberta. *Forest Ecology and Management* 232, 216–225.

Dhiman, R.C. (2004) Damage of poplar nurseries and young field plantations by Blue Bull. *Indian Forestry* 130, 467–468.

Dickmann, D.I. (2006) Silviculture and biology of short-rotation woody crops in temperate regions: then and now. *Biomass and Bioenergy* 30, 696–705.

Dickmann, D.I. and Lantagne, D. (1997) Planting oaks for timber and other uses. North Central Regional Extension Publication No 605. Michigan State University Extension, East Lansing, Michigan.

Dickmann, D.I. and Stuart, K.W. (1983) *The Culture of Poplars in Eastern North America.* Michigan State University Press, East Lansing, Michigan.

Dickmann, D.I., Isebrands, J.G., Eckenwalder, J.E. and Richardson, J. (eds) (2001) *Poplar Culture in North America.* National Research Council of Canada Research Press, Ottawa, Ontario.

Dillen, S.Y., Marron, N., Bastien, C., Ricciotti, L., Salani, F., Sabatti, M., *et al.* (2007) Effects of environment and progeny on biomass estimations of five hybrid poplar families grown at three contrasting sites across Europe. *Forest Ecology and Management* 252, 12–23.

Dowell, R.C., Gibbins, D., Rhoads, J.L. and Pallardy, S.G. (2009) Biomass production physiology and soil carbon dynamics in short-rotation-grown *Populus deltoides* and *P. deltoides* × *P. nigra* hybrids. *Forest Ecology and Management* 257, 134–142.

Duplessis, S., Major, I., Martin, F. and Séguin, A. (2009) Poplar and pathogen interactions: insights from *Populus* genome-wide analyses of resistance and defense gene families and gene expression profiling. *Critical Reviews in Plant Sciences* 28, 309–334.

Eckenwalder, J.E. (1984) Natural intersectional hybridization between North American species of *Populus* (*Salicaceae*) in sections *Aigeros* and *Tacamahaca*. II. Taxonomy. *Canadian Journal of Botany* 62, 325–335.

Eckenwalder, J.E. (2001) Descriptions of clonal characteristics. In: Dickmann, D.I., Isebrands, J.G., Eckenwalder, J.E. and Richardson, J. (eds) *Poplar Culture in North America.* National Research Council of Canada Research Press, Ottawa, Ontario, pp. 331–382.

Einspahr, D.W. (1984) Production and utilization of triploid hybrid aspen. *Iowa State Journal of Research* 58, 401–409.

Einspahr, D.W. and Wycoff, G. (1978) Growth response of hybrid aspen to intensive forest management. *Tappi Journal* 61, 49–52.

Ek, A.R. and Dawson, D.H. (1976) Actual and projected yields of *Populus* Tristis #1 under intensive culture. *Canadian Journal of Forest Research* 6, 132–144.

Facciotto, G. (1998) Le lavorazioni del suolo in pioppicoltura. SHERWOOD – *Foreste ed Alberi Oggi* 31, 39–44.

Facciotto, G. (1999) La potatura del pioppeto. SHERWOOD – *Foreste ed Alberi Oggi* 42, 31–36.

Facciotto, G. and Bergante, S. (2011) Selezione di cloni di pioppo per la destinazione da biomassa. In: *Lo sviluppo delle colture energetiche in Italia.* Il contributo dei progetti di ricerca SUSCACE e FAESI. Research Unit for the Production of Wood Outside Forests, Casale Monferrato, Italy, pp. 31–44.

Facciotto, G., Minotta, G. and Zambruno, G.P. (2003) Analisi della produttività dei cloni di pioppo Dvina, Lena e Neva. In: Caivano, F. and Girardi, T. (eds) *Atti del IV Congresso 'Meridiani e foreste'.* Società Italiana di Selvicoltura ed Ecologia Forestale 4. Università degli Studi della Tuscia v. San Camillo de Lellis, Viterbo, Italy, pp. 151–158.

Fang, S., Xue, J. and Tang, L. (2007) Biomass production and carbon sequestration potential in poplar plantations with different management patterns. *Journal of Environmental Management* 85, 672–679.

FAO (1980) *Poplars and Willows in Wood Production and Land Use.* Food and Agriculture Organization of the United Nations, Rome.

FAO (2008) Synthesis of Country Progress Reports, 23rd Session International Poplar Commission, Beijing, China. FAO Working Paper IPC/6E. Rome.

FAO (2012) Synthesis of Country Progress Reports, 24th Session International Poplar Commission, Dehradun, India. FAO Working Paper IPC/12E. Rome.

Ferm, A., Hytönen, J. and Vuori, J. (1989) Effect of spacing and nitrogen fertilization on the establishment and biomass production of short rotation poplar in Finland. *Biomass* 18, 95–108.

Fillion, M., Brisson, J., Teodorescu, T.I., Sauvé, S. and Labrecque, M. (2009) Performance of *Salix viminalis* and *Populus nigra* × *Populus maximowiczii* in short rotation intensive culture under high irrigation. *Biomass and Bioenergy* 33, 1271–1277.

Frison, G. (1997) Il vivaio per il pioppo. *L'Informatore Agrario* 53, 53–57.

Frison, G. and Facciotto, G. (1993) *La densità d'impianto e i suoi riflessi produttivi in pioppicoltura.* Ed. L'Informatore Agrario, Verona, Italy.

Gallagher, T., Shaffer, B. and Rummer, B. (2006) An economic analysis of hardwood fiber production on dryland irrigated sites in the US Southeast. *Biomass and Bioenergy* 30, 794–802.

Gascon, R.J. Jr and Krinard, R.M. (1976) Biological response of plantation cottonwood to spacing, pruning, thinning. In: Thielges, B.A. and Land, S.B. Jr (eds) *Proceedings: Symposium on Eastern Cottonwood and Related Species.* Louisiana State University, Division of Continuing Education, Baton Rouge, Louisiana, pp. 385–391.

Geyer, W.A., Atchison, R.L. and Carlisle, J. (2006) Evaluations of synthetic mulches on the establishment and growth of cottonwood. *Journal of Sustainable Agriculture* 28, 145–156.

Giménez, R.A. and Etiennot, A.E. (2003) Host range of *Platypus mutatus* (Chapuis, 1865) (Coleoptera: Platypopidae). *Entomotropica* 18, 89–94.

Giorcelli, A., Allegro, G. and Gennaro, M. (2008) Emerging pests and diseases in poplar cultivation in Italy. In: *Abstracts of Submitted Papers, Poplars, Willows and People's Wellbeing.* International Poplar Commission 23rd Session, Beijing (ftp://ftp.fao.org/docrep/fao/011/k3334e/k3334e.pdf, accessed 24 May 2012).

Grossnickle, S.C. (2005) Importance of root growth in overcoming planting stress. *New Forests* 30, 273–294.

Haack, R.A. (2006) Exotic bark- and wood-boring Coleoptera in the United States: recent establishments and interceptions. *Canadian Journal of Forest Research* 36, 269–288.

Hansen, E.A. (1993) A guide for determining when to fertilize hybrid poplar plantations. Research Paper NC-319. USDA Forest Service North Central Forest Experiment Station, St Paul, Minnesota.

Hansen, E.A. and Netzer, D.A. (1985) Weed control in short-rotation intensively cultured poplar plantations. Research Paper NC-260. USDA Forest Service North Central Forest Experiment Station, St Paul, Minnesota.

Heilman, P.E. (1985) Sampling and genetic variation of foliar nitrogen in black cottonwood and its hybrids in short rotation. *Canadian Journal of Forest Research* 15, 1137–1141.

Heilman, P. and Peabody, D.V. Jr (1981) Effect of harvest cycle and spacing on productivity of black cottonwood in intensive culture. *Canadian Journal of Forest Research* 11, 118–123.

Heilman, P.E. and Stettler, R.F. (1985) Genetic variation and productivity of *Populus trichocarpa* and its hybrids. II. Biomass production in a 4-year plantation. *Canadian Journal of Forest Research* 15, 384–388.
Heilman, P.E. and Stettler, R.F. (1990) Genetic variation and productivity of *Populus trichocarpa* and its hybrids. IV Performance of short rotation coppice. *Canadian Journal of Forest Research* 20, 1257–1264.
Heilman, P.E. and Xie, F. (1993) Influence of nitrogen on growth and productivity of short-rotation *Populus trichocarpa* × *Populus deltoides* hybrids. *Canadian Journal of Forest Research* 23, 1863–1869.
Heilman, P.E., Peabody, D.V., DeBell, D.S. and Strand, R.F. (1972) A test of close-spaced, short rotation culture of black cottonwood. *Canadian Journal of Forest Research* 2, 456–459.
Isebrands, J.G. (2007) *Best Management Practices Poplar Manual for Agroforestry Applications in Minnesota (Final).* Environmental Forestry Consultants LLC, New London, Wisconsin (http://www.extension.umn.edu/distribution/naturalresources/00095.html, accessed 24 May 2012).
Isebrands, J.G. and Karnosky, D.F. (2001) Environmental benefits of poplar culture. In: Dickmann, D.I., Isebrands, J.G., Eckenwalder, J.E. and Richardson, J. (eds) *Poplar Culture in North America.* National Research Council of Canada Research Press, Ottawa, Ontario, pp. 207–218.
Isebrands, J.G., Ek, A.R. and Meldahl, R.S. (1982) Comparison of growth model and harvest yields of short rotation intensively cultured *Populus*: a case study. *Canadian Journal of Forest Research* 12, 58–63.
ISP (2002) Pioppicoltura. Produzioni di qualità nel rispetto dell'ambiente. Diffusioni Grafiche. Istituto di Sperimentazione per la Pioppicoltura, Casale Monferrato, Italy.
Jia, H. and Ingestad, T. (1984) Nutrient requirement and stress response of *Populus simonii* and *Paulownia tomentosa. Physiologia Plantarum* 62,117–124.
Johansson, H. (1976) Das Produktionspotential der Hybridaspe (*Populus tremula* × *tremuloides*) in Sudschweden [Growth potential of hybrid aspen (*Populus tremula* × *tremuloides*) in southern Sweden]. *Die Holzzucht* 11, 19–22.
Karačić, A., Verwijst, T. and Weih, M. (2003) Above-ground woody biomass production of short-rotation *Populus* plantations on agricultural land in Sweden. *Scandinavian Journal of Forest Research* 18, 427–437.
Kelley, S.S. (2006) Forest biorefineries: reality, hype, or something in between? *PaperAge* March/April, 46–48.
Keoleian, G.A. and Volk, T.A. (2005) Renewable energy from willow biomass crops: life cycle energy, environmental and economic performance. *Critical Reviews in Plant Sciences* 24, 385–406.
Knowe, S.A., Foster, G.S., Rousseau, R.J. and Nance, W.L. (1998) Height–age and height–diameter distributions for monocultures and mixtures of eastern cottonwood. *Forest Ecology and Management* 106, 115–123.
Kort, J. and Tornock, R. (1999) Carbon reservoir and biomass in Canadian prairie shelterbelts. *Agroforestry Systems* 44, 175–186.
Krinard, R.M. (1988) Volume equations for plantation cottonwood trees (*Populus deltoides*). Research Note SO-347. USDA Forest Service Southern Forest and Range Experiment Station, New Orleans, Louisiana.
Krinard, R.M. and Johnson, R.L. (1984) Cottonwood plantation growth through 20 years. General Technical Report SO-212. USDA Forest Service Southern Forest and Range Experiment Station, New Orleans, Louisiana.
Kuzovkina, Y.A. and Quigley, M.F. (2005) Willows beyond wetlands: uses of *Salix* L. species for environmental projects. *Water, Air and Soil Pollution* 162, 183–204.
Labrecque, M. and Teodorescu, T.I. (2003) High biomass yields achieved by *Salix* clones in SRIC following two 3-year coppice rotations on abandoned farmland in southern Quebec, Canada. *Biomass and Bioenergy* 25, 135–146.
Labrecque, M. and Teodorescu, T.I. (2005) Field performance and biomass production of 12 willow and poplar clones in short-rotation coppice in southern Quebec (Canada). *Biomass and Bioenergy* 29, 1–9.
Lantmännen Agroenergi (no date) *Manual for SRC Willow Growers.* Lantmännen Agroenergi AB/Salix, Orebro, Sweden (http://www.agroenergi.se/PageFiles/135/ENG_Odlarmanual%20Salix%20%28Revidering%29.pdf?epslanguage=sv, accessed 24 May 2012).
Larsson, S. (1998) Production – *Salix.* In: *Bioenergy from Agriculture* – Salix, *Straw Fuels and Biogas.* SLF Report No 31. Swedish Farmers Association Research Foundation, Stockholm, pp. 17–23. (In Swedish)
Larsson, S., Cuingnet, C., Clause, P., Jacobsson, I., Aronsson, P., Perttu, K., *et al.* (2003) Short-rotation willow biomass plantations irrigated and fertilized with wastewaters. Sustainable Urban Renewal and Wastewater Treatment Report No 37. Danish Environmental Protection Agency, Copenhagen.

Laureysens, I., Deraedt, W., Indeherberge, T. and Ceulemans, R. (2003) Population dynamics in a 6-year-old coppice culture of poplar. I. Clonal differences in stool mortality, shoot dynamics, and shoot diameter distribution in relation to biomass production. *Biomass and Bioenergy* 24, 81–95.

Laureysens, I., Bogaert, J., Blust, R. and Ceulemans, R. (2004) Biomass production of 17 poplar clones in a short rotation coppice culture on a waste disposal site and its relation to soil characteristics. *Forest Ecology and Management* 187, 295–309.

Laureysens, I., Pellis, A., Willems, J. and Ceulemans, R. (2005) Growth and production of short rotation coppice culture of poplar. III. Second rotation results. *Biomass and Bioenergy* 29, 10–21.

Liang, H. and Chang, S. (2004) Response of trembling and hybrid aspens to phosphorus and sulfur fertilization in a Gray Luvisol: growth and nutrient uptake. *Canadian Journal of Forest Research* 34, 1391–1399.

Liesebach, M., von Wuehlisch, G. and Muhs, H.-J. (1999) Aspen for short-rotation coppice plantations on agricultural sites in Germany: effects of spacing and rotation time on growth and biomass production of aspen progenies. *Forest Ecology and Management* 121, 25–39.

Lodhiyal, L.S. and Lodhiyal, N. (1997) Variation in biomass and net primary productivity in short rotation high density central Himalayan poplar plantations. *Forest Ecology and Management* 98, 167–179.

Lodhiyal, L.S., Singh, R.P. and Singh, S.P. (1995) Structure and function of an age series of poplar plantations in central Himalaya: I Dry matter dynamics. *Annals of Botany* 76, 191–199.

McKnight, J.S. (1970) Planting cottonwood cuttings for timber production in the South. General Technical Report SO-60. USDA Forest Service Southern Forest and Range Experiment Station, New Orleans, Louisiana.

Makeschin, F. (1999) Short rotation forestry in Central and Northern Europe – introduction and conclusions. *Forest Ecology and Management* 121, 1–7.

Marković, J. (1986) Poplar and willow plantations. In: Guzina, V. (ed.) *Poplars and Willows in Yugoslavia.* Institut za Topolarstvo (Poplar Research Institute), Novi Sad, Yugoslavia, pp. 36–44.

Mattson, W.J., Hart, E.A. and Volney, W.J.A. (2001) Insect pests of *Populus*: coping with the inevitable. In: Dickmann, D.I., Isebrands, J.G., Eckenwalder, J.E. and Richardson, J. (eds) *Poplar Culture in North America.* National Research Council of Canada Research Press, Ottawa, Ontario, pp. 219–248.

Mirck, J., Isebrands, J.G., Verwijst, T. and Ledin, S. (2005) Development of short-rotation willow coppice systems for environmental purposes in Sweden. *Biomass and Bioenergy* 28, 219–228.

Mitchell, C. (1992) Ecophysiology of short rotation forest crops. *Biomass and Bioenergy* 2, 25–37.

Mola-Yudego, B. (2010) Regional potential yields of short rotation willow plantations on agricultural land in Northern Europe. *Silva Fennica* 44, 63–76.

Mola-Yudego, B. and Aronsson, P. (2008) Yield models for commercial willow biomass plantations in Sweden. *Biomass and Bioenergy* 32, 829–837.

Morris, R.C., Filer, T.H., Solomon, J.D., McCracken, F.I., Overgaard, N.A. and Weiss, M.J. (1975) Insects and diseases of cottonwood. General Technical Report SO-8. USDA Forest Service Southern Forest and Range Experiment Station, New Orleans, Louisiana.

Murray, M.D. and Harrington, C.A. (1983) Growth and yield of a 24-year-old black cottonwood plantation in western Washington. *Tree Planters' Notes* 34, 3–5.

Nelson, L.E., Switzer, G.L. and Lockaby, B.G. (1987) Nutrition of *Populus deltoides* plantations during maximum production. *Forest Ecology and Management* 20, 25–41.

Netzer, D.A. (1984) Hybrid poplar plantations outgrow deer browsing effects. Research Note NC-325. USDA Forest Service North Central Forest Experiment Station, St Paul, Minnesota.

Netzer, D.A. and Tolsted, D.N. (1999) Yields of ten and eleven year-old hybrid poplars in the North Central United States. Final report to the US Department of Energy, Biofuels Feedstock Development Program, USDA Forest Service North Central Research Station, Rhinelander, Wisconsin.

Netzer, D.A., Tolsted, D.N., Ostry, M.E., Isebrands, J.G., Riemenschneider, D.E. and Ward, T.T. (2002) Growth, yield, and disease resistance of 7 to 12 year old poplar clones in the North Central United States. General Technical Report NC-229. USDA Forest Service North Central Forest Experiment Station, St Paul, Minnesota.

Newcombe, G., Ostry, M., Hubbes, M., Perinet, P. and Mottet, M.-J. (2001) Poplar diseases. In: Dickmann, D.I., Isebrands, J.G., Eckenwalder, J.E. and Richardson, J. (eds) *Poplar Culture in North America.* National Research Council of Canada Research Press, Ottawa, Ontario, pp. 249–276.

Nordmann, E.E., Robison, D.J., Abrahamson, L.P. and Volk, T.A. (2005) Relative resistance of willow and poplar biomass production clones across a continuum of herbivorous insect specialization: univariate and multivariate approaches. *Forest Ecology and Management* 217, 307–318.

Opdahl, H. (1992) Site-index, growth and yield in aspen (*Populus tremula* L.) stands in south Norway. *Communications of Skogforsk* 44, 1–44. (In Norwegian with English summary)

Ostry, M.E., Wilson, L.F., McNabb, H.S. Jr and Moore, L.M. (1989) *A Guide to Insect, Disease, and Animal Pests of Poplars.* Agriculture Handbook 677. USDA Forest Service, Washington, DC.

Pan, H.Y. (2005) Review of the Asian longhorned beetle – research, biology and management in China. Working paper FBS/6E, FAO, Rome.

Paris, P., Mareschi, L., Sabatti, M., Pisanelli, A., Ecosse, A., Nardin, F., *et al.* (2011) Comparing hybrid *Populus* clones for SRF across northern Italy after two biennial rotations: survival, growth and yield. *Biomass and Bioenergy* 35, 1524–1532.

Pontailler, J.Y., Ceulemans, R. and Guittet, J. (1999) Biomass yield of poplar after five 2-year coppice rotations. *Forestry* 72, 157–163.

Pourtet, J. (1976) The International Poplar Commission – its role in world poplar culture. In: Thielges, B.A. and Land, S.B. Jr (eds) *Proceedings: Symposium on Eastern Cottonwood and Related Species.* Louisiana State University, Division of Continuing Education, Baton Rouge, Louisiana, pp. 31–37.

Proe, M., Griffiths, J.H. and Craig, J. (2002) Effect of spacing, species, and coppicing in leaf area, light interception, and photosynthesis in short rotation forestry. *Biomass and Bioenergy* 23, 315–326.

Prevosto, M. (1965) L'accrescimento del pioppo euroamericano 'I/-214' nei diversi ambient della pianura Lombardo-piemontese in relazione alla spaziatura e al turno. Ente Nationale per la Cellulosa e per la Carta, Rome, Italy. (Cited in FAO (1980) *Poplars and Willows in Wood Production and Land Use.* Food and Agriculture Organization of the United Nations, Rome.)

Ramstedt, M. (1999) Rust disease on willows – virulence variation and resistance breeding strategies. *Forest Ecology and Management* 121, 101–111.

Ranney, J.W., Wright, L.L. and Layton, P.A. (1987) Hardwood energy crops: the technology of intensive culture. *Journal of Forestry* 85, 17–28.

Rédei, K. (2000) Early performance of promising white poplar (*Populus alba*) clones in sandy ridges between the rivers Danube and Tisza in Hungary. *Forestry* 73, 407–413.

Robison, D.J. and Raffa, K.F. (1998) Productivity, drought tolerance, and pest status of hybrid *Populus*: tree improvement and silvicultural implications. *Biomass and Bioenergy* 14, 1–20.

Robison, T.L., Rousseau, R.J. and Zhang, J. (2006) Biomass productivity improvement for eastern cottonwood. *Biomass and Bioenergy* 30, 735–739.

Ruttens, A., Boulet, J., Weyens, N., Smeets, K., Adriaensen, K., Meers, E., *et al.* (2011) Short rotation coppice culture of willows and poplars as energy crops on metal contaminated agricultural soils. *International Journal of Phytoremediation* 13(S1), 194–207.

Sagheb-Talebi, K. (2005) Rehabilitation of temperate forests in Iran. In: Stanturf, J.A. and Madsen, P. (eds) *Restoration of Boreal and Temperate Forests.* CRC Press, Boca Raton, Florida, pp. 397–407.

Scarascia-Mugozza, G.E., Ceulemans, R., Heilman R., Isebrands, J.G., Stettler, R.F. and Hinckley, T.M. (1997) Production physiology and morphology of *Populus* species and their hybrids grown under short rotation II. Biomass components and harvest index of hybrid and parental species clones. *Canadian Journal of Forest Research* 27, 285–294.

Scarcella, M., Palumbo, D., Giorcelli, A. and Gennaro, M. (2011) Guida al diserbo del pioppo in ambiente meridionale. In: *Lo sviluppo delle colture energetiche in Italia.* Il contributo dei progetti di ricerca SUSCACE e FAESI. Research Unit for the Production of Wood Outside Forests, Casale Monferrato, Italy, pp. 89–97.

Schuette, B. and Kaiser, C. (1996) Weed control strategies for SRIC hybrid poplars: farmer's perspective. In: Stokes, B.J. (ed.) *Proceedings Short Rotation Woody Crops Operations Working Group Meeting, Paducah, Kentucky, USA, September 23–25, 1996* (http://www.woodycrops.org/NR/rdonlyres/BF9B2067-FDB0-49B0-9543-8EEA03A415FD/1662/Proceedings2.pdf, *pp.* 112–116, accessed 24 May 2012).

Scotti, C., Facciotto, G. and Canestrale, R. (2010) Soil map of pedological limitations to the growth of poplar clones for biomass production in Emilia-Romagna Region (Italy). In: *Proceedings of the 18th European Biomass Conference and Exhibition.* ETA-Florence Renewable Energies, Florence, Italy, pp. 448–451.

Shogren, R.L. and Rousseau, D.J. (2005) Field testing of paper/polymerized vegetable oil mulches for enhancing growth of eastern cottonwood trees for pulp. *Forest Ecology and Management* 208, 115–122.

Smart, L.B. and Cameron, K.D. (2008) Genetic improvement of willow (*Salix* spp.) as dedicated bioenergy crop. In: Vermerris, W. (ed.) *Genetic Improvement of Bioenergy Crops.* Springer, New York, pp. 377–396.

Smart, L.B., Volk, T.A., Lin, J., Kopp, R.F., Phillips, I.S., Cameron, K.D., *et al.* (2005) Genetic improvement of shrub willow (*Salix* spp.) crops for bioenergy and environmental applications in the United States. *Unasylva* 221, 51–55.

Smith, J.H.G. (1980) Growth and yield of poplar in British Columbia. Paper presented at the 1980 meeting of the Poplar Council of Canada; cited in DeBell, D.S. (1990) *Populus trichocarpa* Torr. & Gray. In: Burns, R.M. and Honkala, B.H. (eds) *Silvics of North America. Vol 2 Hardwoods.* Agriculture Handbook 654. USDA Forest Service Washington, DC, pp. 570–576.

Smith, J.H.G. and DeBell, D.S. (1973) Opportunities for short rotation culture and complete utilization of seven Northwestern tree species. *The Forestry Chronicle* 49, 31–34.

Solomon, J.D. (1985) Impact of insects on growth and development of young cottonwood plantations. Research Paper SO-217. USDA Forest Service Southern Forest and Range Experiment Station, New Orleans, Louisiana.

Spinelli, R., Cosentino, G. and Nati, C. (2004) Pioppicoltura, il processore può danneggiare il legno? *Alberi e territorio* 6, 50–52.

Spinelli, R., Nati, C. and Magagnotti, N. (2009) Using modified foragers to harvest short-rotation poplar plantations. *Biomass and Bioenergy* 33, 817–821.

Stanton, B., Eaton, J., Johnson, J., Rice, D., Schuette, B. and Moser, B. (2002) Hybrid poplar in the Pacific Northwest: the effects of market-driven management. *Journal of Forestry* 100, 28–33.

Stanturf, J.A., van Oosten, C., Netzer, D.A., Coleman, M.D. and Portwood, C.J. (2001) Ecology and silviculture of poplar plantations. In: Dickmann, D.I., Isebrands, J.G., Eckenwalder, J.E. and Richardson, J. (eds) *Poplar Culture in North America.* National Research Council of Canada Research Press, Ottawa, Ontario, pp. 153–206.

Stenvall, N., Haapala, T. and Pulkkinen, P. (2006) The role of a root cutting's diameter and location on the regeneration ability of hybrid aspen. *Forest Ecology and Management* 237, 150–155.

Sternitzke, H.S. (1976) Impact of changing land use on Delta hardwood forests. *Journal of Forestry* 74, 25–27.

Stettler, R.F., Bradshaw, H.D. Jr, Heilman, P.R. and Hinckley, T.M. (eds) (1996a) *Biology of Populus and Its Implications for Management and Conservation.* National Research Council of Canada Research Press, Ottawa, Ontario.

Stettler, R.F., Zsuffa, L. and Wu, R. (1996b) The role of hybridization in the genetic manipulation of *Populus.* In: Stettler, R.F., Bradshaw, H.D. Jr, Heilman, P.R. and Hinckley, T.M. (eds) (1996) *Biology of Populus and Its Implications for Management and Conservation.* National Research Council of Canada Research Press, Ottawa, Ontario, pp. 87–112.

Stolarski, M., Szczukowski, S., Tworkowski, J. and Kopaczel, M. (2007) Profitability of willow production in short cycles in the low Vistula Valley. *Polish Journal of Natural Sciences* 22, 172–182.

Strong, T.F. and Hansen, E.A. (1993) Hybrid spacing/productivity relations in short rotation intensive culture plantations. *Biomass and Bioenergy* 4, 255–261.

Switzer, G.L., Nelson, L.E. and Baker, J.B. (1976) Accumulation and distribution of dry matter and nutrients in *Aegeiros* poplar plantations In: Thielges, B.A. and Land, S.B. Jr (eds) *Proceedings: Symposium on Eastern Cottonwood and Related Species.* Louisiana State University, Division of Continuing Education, Baton Rouge, Louisiana, pp. 359–369.

Szczukowski, S., Stolarski, M., Tworkowski, J., Przyborowski, J. and Klasa, A. (2005) Productivity of willow coppice plants grown in short rotations. *Plant Soil Environment* 51, 423–430.

Thomas, K.D., Comeau, P.G. and Brown, K.R. (2000) The silviculture of hybrid poplar plantations. Extension Note 47. British Columbia Ministry of Forests Research Program, Victoria, British Columbia, Canada.

Thomas, K.D., Reid, W.J. and Comeau, P.G. (2001) Vegetation management using polyethylene mulch mats and glyphosate herbicide in a coastal British Columbia hybrid poplar plantation – 4-year growth response. *Western Journal of Applied Forestry* 16, 26–30.

Toome, M., Heinsoo, K. and Luik, A. (2006) Abundance of willow rust (*Melampsora* sp.) on different willow clones in Estonian energy forest plantations. *Proceedings of the Estonian Academy of Sciences Biology Ecology* 55, 308–317.

Trnka, M., Trnka, M., Fialová, J., Koutecky, V., Fajman, M., Žalud, Z. and Hejduk, S. (2008) Biomass production and survival rates of selected poplar clones grown under a short-rotation system on arable land. *Plant Soil Environment* 54, 78–88.

Tubby, I. and Armstrong, A. (2002) Establishment and management of short rotation coppice. Practice Note 7, Forestry Commission Forest Research, Edinburgh, UK.

Tullus, A., Rytter, L., Tullus, T., Weih, M. and Tullus, H. (2012) Short-rotation forestry with hybrid aspen (*Populus tremula* L. × *P. tremuloides* Michx.) in northern Europe. *Scandinavian Journal of Forest Research* 27, 10–29.

Tuskan, G.A. and Rensema, T.R. (1992) Clonal differences in biomass characteristics, coppice ability, and biomass prediction equations among four *Populus* clones grown in North Dakota. *Canadian Journal of Forest Research* 22, 348–354.

Ulloa, J. and Villacura, L. (2005) Contribution of a private poplar industry in Chile to sustainable rural development. *Unasylva* 221, 12–17.

van den Driessche, R., Rude, W. and Martens, L. (2003) Effect of fertilization and irrigation on growth of aspen (*Populus tremuloides* Michx.) seedlings over three seasons. *Forest Ecology and Management* 186, 381–389.

van den Driessche, R., Niemi, F. and Charleson, L. (2005) Fourth year response of aspen seedlings to lime, nitrogen and phosphorus applied at planting and 1 year after planting. *Forest Ecology and Management* 219, 216–228.

van Oosten, C. (2006) Hybrid Poplar Crop Manual for the Prairie Provinces. Report for the Saskatchewan Forest Centre. SilviConsult Woody Crops Technology, Inc., Nanaimo, British Columbia, Canada (http://www.poplar.ca/upload/documents/cropman.pdf, accessed 19 December 2012).

van Oosten, C (2008) Activities related to poplar and willow cultivation and utilization in Canada. Canadian Report to the 23rd Session, International Poplar Commission, Beijing, China, for the period 2004–2007 (http://www.poplar.ca/upload/documents/ipccan2008.pdf, accessed 19 December 2012).

Verani, S., Sperandia, G., Picchio, R., Spinelli, R. and Picchi, G. (2008) Field handbook, poplar harvesting. Working Paper IPC/8. FAO, Rome.

Verwijst, T. (2001) Willows: an underestimated resource for environment and society. *The Forestry Chronicle* 77, 281–285.

Volk, T.A., Buford, M.A., Berguson, B., Caputo, J., Eaton, J., Perdue, J.H., *et al.* (2011a) Woody feedstocks – management and regional differences. In: Braun, R., Karlen, D. and Johnson, D. (eds) *Sustainable Alternative Fuel Feedstock Opportunities, Challenges and Roadmaps for Six US Regions.* Proceedings of the Sustainable Feedstocks for Advance Biofuels Workshop. Soil and Water Conservation Society, Ankeny, Iowa, pp. 120–141.

Volk, T.A., Abrahamson, L.P., Cameron, K.D., Castellano, P., Corbin, T., Fabio, E., *et al.* (2011b) Yields of willow biomass crops across a range of sites in North America. In: Booth, E., Halford, N., Sheild, I., Taylor, G., Turley, D. and Voigt, T. (eds) *Aspects of Applied Biology 112, Biomass and Energy Crops IV.* Association of Applied Biologists, Warwick, UK, pp. 67–74.

Von Althen, F.W. (1981) Site preparation and post-planting weed control in hardwood afforestation: white ash, black walnut, basswood, silver maple, hybrid poplar. Report 0-X-325. Canada Department of the Environment, Canadian Forest Service, Great Lakes Forest Research Centre, Sault Ste Marie, Ontario, Canada.

Weih, M. (2004) Intensive short rotation forestry in boreal climates: present and future perspectives. *Canadian Journal of Forest Research* 34, 1369–1378.

Wickham, J., Rice, B., Finnan, J. and McConnon, R. (2010) *A Review of Past and Current Research on Short Rotation Coppice in Ireland and Abroad.* Council for Forest Research and Development (COFORD), Dublin.

Williamson, A.W. (1913) Cottonwood in the Mississippi Valley. Bulletin 24. US Department of Agriculture, Washington, DC.

Wright, L. (2006) Worldwide commercial development of bioenergy with a focus on energy crop-based projects. *Biomass and Bioenergy* 30, 706–714.

Zabek, L.M. and Prescott, C.E. (2006) Biomass equations and carbon content of aboveground leafless biomass of hybrid poplar in coastal British Columbia. *Forest Ecology and Management* 223, 291–302.

Zalesny, R.S. Jr and Zalesny, J.A. (2009) Selecting *Populus* with different adventitious root types for environmental benefits, fiber, and energy. In: Niemi, K. and Scagel, C. (eds) *Adventitious Root Formation of Forest Trees and Horticultural Plants – From Genes to Applications.* Research Signpost, Kerala, India, pp. 359–384.

Zalesny, R.S. Jr, Bauer, E.O. and Riemenschneider, D.E. (2004) Use of belowground growing degree days to predict rooting of dormant hardwood cuttings of *Populus. Silvae Genetica* 53, 154–160.

Zalesny, R., Cunningham, M., Hall, R., Mirck, J., Rockwood, D., Stanturf, J., *et al.* (2011) Woody biomass from short rotation energy crops. In: Zhu, J.Y., Zhang, X. and Pan, X. (eds) *Sustainable Production of*

Fuels, Chemicals, and Fibers from Forest Biomass. ACS Symposium Series. American Chemical Society, Washington, DC, pp. 27–63.

Zerbe, J.I. (2006) Thermal energy, electricity, and transportation fuels from wood. *Forest Products Journal* 56, 6–14.

Zhang, D. and Stanturf, J. (2008) Forest plantations. In: Jørgensen, S.E. and Fath, B. (eds) *Encyclopedia of Ecology.* Academic Press, Oxford, UK, pp. 1673–1680.

Zsuffa, L., Sennerby-Forsse, L., Weisgerber, H. and Hall, R.B. (1993) Strategies for clonal forestry with poplars, aspens and willows. In: Ahuja, M.R. and Libby, W.J. (eds) *Clonal Forestry II. Conservation and Application.* Springer Verlag, Berlin, pp. 91–119.

Zsuffa, L., Giordano, E., Pryor, L.D. and Stettler, R.F. (1996) Trends in poplar culture: some global and regional perspectives. In: Stettler, R.F., Bradshaw, H.D. Jr, Heilman, P.R. and Hinckley, T.M. (eds) *Biology of Populus and Its Implications for Management and Conservation.* National Research Council of Canada Research Press, Ottawa, Ontario, pp. 515–539.

6 Environmental Applications of Poplars and Willows

J.G. Isebrands,[1]* P. Aronsson,[2] M. Carlson,[3] R. Ceulemans,[4] M. Coleman,[5] N. Dickinson,[6] J. Dimitriou,[2] S. Doty,[7] E. Gardiner,[8] K. Heinsoo,[9] J.D. Johnson,[10] Y.B. Koo,[11] J. Kort,[12] J. Kuzovkina,[13] L. Licht,[14] A.R. McCracken,[15] I. McIvor,[16] P. Mertens,[17] K. Perttu,[2] D. Riddell-Black,[18] B. Robinson,[6] G. Scarascia-Mugnozza,[19] W.R. Schroeder,[12] J. Stanturf,[20] T.A. Volk[21] and M. Weih[2]

[1]Environmental Forestry Consultants LLC, USA; [2]Swedish University of Agricultural Sciences, Sweden; [3]British Columbia Ministry of Forests, Canada; [4]University of Antwerp, Belgium; [5]University of Idaho, USA; [6]Lincoln University, New Zealand; [7]University of Washington, USA; [8]USDA Forest Service, USA; [9]Estonian Agricultural University, Estonia; [10]Deceased. Formerly at Washington State University, USA; [11]Korindo Group, Indonesia; [12]Agriculture and Agri-Food Canada, Canada; [13]The University of Connecticut, USA; [14]Ecolotree, Inc., USA; [15]Agri-Food & Biosciences Institute, UK; [16]Plant and Food Research, New Zealand; [17]Direction du milieu forestier, Belgium; [18]Lupus-Science, UK; [19]Agricultural Research Council of Italy, Italy; [20]USDA Forest Service, USA; [21]State University of New York, USA

6.1 Introduction

Poplars and willows have been planted for environmental purposes for millennia. There are reports that poplars were planted to improve the human environment 4000 years ago in the third dynasty of Ur, for streamside stabilization 2000 years ago in what is now the south-western USA by native North Americans and for urban amenities by the early Chinese dynasties (see Chapter 1, this volume). Early settlers in Europe and North America planted poplars and willows (and other species) to provide shelter and to protect crops. There were also a significant number of linear plantings of poplars in cities for protection, visual screens and aesthetics (FAO, 1980; Isebrands and Karnosky, 2001).

For most of the 20th century, the primary focus of poplar plantings was for wood and fibre production (FAO, 1958, 1980) (see also Chapter 5, this volume). However, in the late 20th century and the 21st century, the focus of poplar and willow plantings has shifted toward ecosystem and environmental services (Costanza *et al.*, 1997; USDA Forest Service, 2011). Ecosystem services are the goods and services trees provide to society, including watershed services, nutrient cycling, waste management, carbon storage, scenic landscapes, biodiversity and wildlife habitat. In the past, these benefits were valued as public goods and difficult to assess economically. However, as the world population grows, they are now considered vital to human health and livelihoods (USDA Forest Service, 2011; Zalesny, 2011).

*Corresponding author: efcllc@athenet.net

Poplars and willows are making an important contribution to ecosystem services worldwide. In this chapter, we provide a practical worldwide overview of the environmental applications of poplars and willows. Our goal is to synthesize the latest knowledge on these applications with respect to sustainable livelihoods, land use and restoration. The applications covered include land protection, watershed stabilization, waste management and other ecosystem services.

6.2 Windbreaks and Shelterbelts

6.2.1 Introduction

J.G. ISEBRANDS

Windbreaks and shelterbelts have a long history dating back to early agricultural settlement, and poplars and willows have been an important part of these plantings. According to the FAO/IUFRO Forestry Terminology Committee (Ford-Robertson, 1971), windbreaks are a strip of trees and/or shrubs planted to alter windflow and microclimates around farmsteads, homes, orchards and feedlots. Shelterbelts are similar but more extensive, designed to alter windflow and microclimates around agricultural fields (Ford-Robertson, 1971; Helms, 1998). The first systematic planting of shelterbelts was by German immigrants in the Russian Steppes in 1789 to protect fields. The term 'shelterbelt' can be traced back to 1833, and since then there have been thousands of kilometres planted throughout the world (Stoeckler and Williams, 1949). In modern times, poplars and willows have been planted in windbreaks and shelterbelts, especially in the Prairie Provinces of western Canada, the Plains States of the USA and very extensively in northern China (Isebrands, 2007; Richardson *et al.*, 2007).

6.2.2 Worldwide overview

J. KORT AND W.R. SCHROEDER

Poplars and willows are widely used in windbreaks and 'shelterbelts' (i.e. 'field windbreaks' or 'hedgerows') throughout the world's temperate regions (van Eimern *et al.*, 1964). The terms 'windbreak' and 'shelterbelt' are used interchangeably as general terms in some regions and they are hereafter used interchangeably in this chapter for convenience.

Poplars or willows are most suitable for a windbreak if they are adapted to a particular region and provide a good structure on which its effectiveness depends – namely its height, porosity, orientation and the distance between windbreaks. The distance protected by a windbreak is proportional to its height (Tabler, 1980), while the percentage reduction of wind depends on porosity (Heisler and DeWalle, 1988). Dense windbreaks reduce wind speed more but may have greater turbulence and a decreased zone of protection (Heisler and DeWalle, 1988), so a good windbreak design should result in the right porosity for the function it is expected to perform. Porous windbreaks are best for wind erosion control (Hagen, 1976), while dense windbreaks improve the microclimate for crop growth in their sheltered zone (Rosenberg, 1974). Dense windbreaks are also most effective for trapping blowing soil or snow (Scholten, 1988), but may increase diseases like mildew or create frost pockets (Rosenberg, 1974). For livestock protection, dense windbreaks reduce wind chill in winter, improving weight gains and survival of newborn animals, while tall, spreading windbreaks provide more shade in summer.

Willow or poplar species or clones can be used to create a variety of windbreak structures. Shrubby species or coppice practices can be used to create a dense, multi-stemmed windbreak suitable for trapping snow or soil, especially if more than one row is used. On the other hand, tall, widely-spaced poplar or willow trees result in porous windbreaks with a great distance of even wind protection across the field (George *et al.*, 1963). As deciduous trees, poplars have greater porosity in the winter than in the summer, with over 70% wind reduction achieved by a foliated poplar windbreak compared to 25% in the leafless state (Vézina, 1994). Fan *et al.* (2010) modelled the three-dimensional aerodynamic structure of Chinese four-row shelterbelts of hybrids of *Populus* ×*beijingensis* W.Y. Hsu, showing that the greatest component of windbreak drag was due to the leaves, while most of the volume was in the trunk. Although their rapid growth makes poplar and willow species attractive for quick windbreak protection, their short lifespan, high moisture requirements, spreading root systems and their negative effects on nearby crops sometimes cause land managers to favour longer-lived, less competitive species, especially in areas of limited moisture. Their susceptibility to drought may result in serious dieback (Rood *et al.*, 2000). Crop growth reductions near the trees have been attributed to moisture limitation (Sharma *et al.*, 2001), shade (Thevathasan and Gordon, 2004) and phytotoxic phenols leached from leaves (Singh *et al.*, 1998), although Pandey (2008) stated that *P. deltoides* shelterbelts had less negative impact on crops than *Eucalyptus* shelterbelts in Haryana state in India. Such drawbacks may make poplars or willows undesirable in some shelterbelts, but land managers need to balance this with their more rapid growth and shelter and their potential economic value.

Where soils and climate are suitable for these species, they may be preferred to other species or used in combination, especially if they have economic, aesthetic or other benefits in addition to their windbreak function. When the value of harvested poplars or willows is significant, using them in windbreaks often allows the land manager to maximize value from the land (Puri and Nair, 2004). Aesthetic considerations are important for poplar and willow windbreaks on the Dutch landscape (Londo *et al.*, 2004), while willows in shelterbelts have also been recommended because of their early pollen production (Vézina, 1994).

The use of poplars and willows in windbreaks in different regions and countries depends on climatic, economic and social or cultural conditions or traditions. This has resulted in a great variety of practices, including poplars for farmyards in the North American Great Plains (Fig. 6.1), pollarded willows along Dutch highways, poplar forest belts on the Russian Steppes or intensively cultivated poplar field boundaries in the Punjab and Haryana states of India. Some of these regional practices are described below.

In North America, poplar hybrids and willows are recommended in western Canada and the US Great Plains States, as one of the rows of multiple-row windbreaks for farmyard protection (Fig. 6.2), because of their rapid

Fig. 6.1. Windbreak comprised of multiple species, including poplars, in Saskatchewan, Canada. Photo courtesy of Agriculture and AgriFood Canada, Indian Head, Saskatchewan, Canada.

Fig. 6.2. Multiple-row poplar shelterbelt in Saskatchewan, Canada. Photo courtesy of Agriculture and AgriFood Canada, Indian Head, Saskatchewan, Canada.

height growth (Isebrands and Karnosky, 2001; Agriculture and Agri-Food Canada, 2005), and they are similarly used in eastern regions (Vézina, 1994; Volk *et al.*, 2004). Schroeder and Kort (2007) have reviewed the advantages of poplars and willows in North America. They give details of shelterbelt design and function. Slower-growing, longer-lived species complement the rapid growth of the poplar over time by increasing both windbreak density and the windbreak's lifespan. Kenney (1992) suggested that deciduous hardwoods were not suitable for windbreaks because of their excessive porosity in winter, but George *et al.* (1963) suggested that poplar windbreaks on the US Great Plains would result in even snow coverage across the fields. Willows are used for windbreaks in special circumstances and are especially recommended in wet sites where frequently anoxic soil conditions limit the number of species that can be used. They have tree or shrub forms, some of which produce valuable shoots for the floral trade or for willow chairs or other crafts (Labrecque and Teoderescu, 2004). As shrubs, they can be used to develop roadside barriers that are effective in trapping snow which would otherwise cause traffic hazards. Because selected species or clones are amenable to coppicing, their current use as a bioenergy crop presents possibilities for the long-term management of willow windbreaks which can be coppiced periodically for bioenergy.

In Russia, many thousands of hectares of poplar shelterbelts exist on the steppes, dating from early efforts in 1946–1953 or more recent efforts since 1966 (Schroeder and Kort, 1989). Many of these are in the form of wide, multiple-row shelterbelts that are as much as 500 m apart.

In the UK, Stott and Belcher (1978) considered poplars to be better than conifers for windbreaks because the conifers resulted in turbulence, frost pockets and an increase in the incidence of disease. In France, poplars are considered to be variable in form, so that windbreaks of various porosities can be designed with heights of 20–25 m and rapid growth, while shelterbelt willows are considered to have value as habitat for pollen production and for the floral trade (IDF, 1981). In the Netherlands, Londo *et al.* (2004) concluded that the best configurations of coppice willows for bioenergy production were ecological upland corridors that provided multiple environmental benefits such as the protection of biodiversity. Danish shelterbelts normally include 10–14 species and are multi-row plantings, designed in such a way as to allow for species succession (Als, 1990). They typically include both willows and poplars, which tend to be dominant in the early years of the shelterbelt but, because of their shorter lifespan, eventually yield to longer-lived species.

In South America, poplars and willows are increasingly being used in windbreaks in Chile and in Argentina, where 1500 km of poplar and willow windbreaks provide shelter for important irrigated crops in South Patagonia (Peri and Bloomberg, 2002).

The level of adoption of poplar trees in windbreaks is likely greatest in northern India and in northern China. In India, poplar shelterbelts and plantations have 'spread across the region like a storm' (Singh, 2004) since the 1970s, especially in the northern states of Punjab and Haryana (Dhanda *et al.*, 2004). According to Pandey (2008), poplar species are not among the top ten tree species planted outside forests in India, but are locally important in the Punjab and Hiryana states, comprising 21% of the non-forest trees growing in Punjab. Poplar trees growing in single-row shelterbelts were found to have less negative impact on adjacent crops than eucalyptus trees, being complementary rather than competitive with crops like winter wheat, and have been promoted for this purpose. The high level of adoption has been, in part, due to government promotion and programming, but also because the value of the wood in short rotations, combined with the production of winter crops when the poplar trees are dormant, appears to be a viable practice that reduces the competitive interactions between the crops and the trees (Puri and Nair, 2004). But Puri (2004) cautioned that the advertised benefits of agroforestry systems might be overstated as a development vehicle and for the alleviation of poverty if the right tree and crop species combinations were not used or if the intensiveness of management was insufficient. According to Pandey (2008), large-scale tree planting on private land resulted in lower prices from pulp and paper mills, but poplars were important for fuelwood or for income. Pandey (2008) said that the development of market mechanisms and price stability would increase the use of poplar shelterbelts as standard practice for farmers in northern India.

In the Three North Region of China, many poplar shelterbelts have been planted since 1978 under the Three North Shelterbelt Programme to develop the 'Great Green Wall' (Moore and Russell, 1990; Carle and Ma, 2005). These plantings have included 'forest-net' shelterbelts, in which poplar shelterbelts have been planted in a grid to protect crops or animals from all directions (Fig. 6.3), as well as wood production shelterbelts which are wider, consisting of six or more rows of poplar trees (Moore and Russell, 1990). The entire area addressed by the programme is over 4 million km^2, with over 20 million ha of mostly hybrid poplars as of 1991, most of which are in afforestation plantations (Carle and Ma, 2005). According to Fan *et al.* (2010), there are about 2.2 million ha of hybrid poplar shelterbelts in China (Fig. 6.4). This large-scale use of poplars has been accompanied by research activities to develop superior genetic materials and agronomic practices (Carle and Ma, 2005).

In summary, poplars and willows make good windbreaks if properly designed and managed and if they meet the social and economic needs of the region in which they are grown. The needs may be economic – the production of biomass for bioenergy, pulp or lumber integrated with crop production benefits. They may also be social – the protection of soils, biodiversity or people. Social uses of windbreaks also include the protection of roads, homes and gardens, and they may also be planted for their aesthetic value. Poplar and willow windbreaks are used more in regions where the need is supported by programmes or policies, resulting in several areas in the world where they are adopted in large numbers, most notably in China and India.

6.2.3 Shrub willow living snow fences

T.A. VOLK

In areas where snowfall is prevalent, snow blowing across open fields can create dangerous road conditions for the public, increase the number of accidents and injuries and create expensive, time-consuming and challenging situations for road crews to ameliorate. Snow and ice removal costs in the USA exceed US$2 billion each year,

Fig. 6.3. Shelterbelt of poplars in Three North Region in Inner Mongolia, China. Photo courtesy of J.G. Isebrands.

Fig. 6.4. Shelterbelt in Jiangxi Province, China. Photo courtesy of J.G. Isebrands.

while indirect costs related to corrosion and environmental impacts from snow removal and control activities have been estimated to add another US$5 billion each year. Factoring in costs associated with accidents and injuries would further increase this figure (Tabler, 2003).

The threshold wind speed at which snow will begin to move is around 15 km h^{-1}, and the work ability of wind speed is proportional to the cube of the wind speed (Tabler, 2003), so slight reductions in wind speed can have significant impacts on snow movement and distribution. Structural snow fences have been used for a long time to reduce wind speeds and control blowing and drifting snow along roadways and other key locations. Various types of structural snow fences have been used to reduce blowing snow, including solid wood Wyoming snow fences, slatted wood, porous plastic and, most recently, three-dimensional structures like 'snow snakes' (Tabler, 2006).

Structural snow fences can reduce blowing and drifting snow immediately after they are installed and are an effective choice in some situations, but they have a number of limitations.

One is the high establishment and maintenance costs. One study in Wyoming, USA, showed that 1.6 km of snow fence reduced snow and ice removal costs by US\$14,497 year^{-1} and saved US\$8,256 year^{-1} in vehicle accident costs (Daigneault and Betters, 2000). However, the calculated cost of a double-row, slatted wooden snow fence was US\$10,153 km^{-1} to establish and US\$5,390 km^{-1} to maintain each year. Many structural snow fences like this have to be installed, taken down and stored each year and have a projected effective lifespan of just a few years. Wyoming snow fences are more permanent structures and are also effective, but cost US\$54,112 km^{-1} to establish and US\$1,641 km^{-1} year^{-1} for maintenance (Daigneault and Betters, 2000). Even with these high establishment and maintenance costs, the cost–benefit ratio of these structural snow fences was still greater than two. Another limitation of structural snow fences is that shorter plastic and wooden slatted snow fences can quickly become buried in drifting snow, making them ineffective for the remainder of the winter, and they are they are aesthetically unappealing.

An alternative approach to controlling blowing and drifting snow, as well as providing additional benefits to landowners and the environment, is to design and install living snow fences. These are plantings of trees, shrubs and/or native grasses a short distance upwind of roads, homes, farmsteads, communities or other important facilities (Gullickson *et al.*, 1999). The use of living snow fences to reduce the blowing and drifting snow into roadways and other transportation corridors is not a new concept. In the early 1900s, a number of railway companies planted living snow fences throughout the western USA. By 1915, one company had planted over 500,000 seedlings, though many died during the droughts of 1929–1933 (Perko, 1995). Interest in living snow fences has grown again recently because of the increased costs associated with setting up and maintaining snow fences. Living snow fences can be cheaper to install and maintain than structural snow fences, have a greater height, and therefore can capture more snow. For example, a living snow fence consisting of two rows of conifers and a row of deciduous shrubs was estimated to cost US\$12,700 km^{-1} to install but only US\$207 km^{-1} year^{-1} to maintain (Daigneault and Betters, 2000). A 1.2-m-high snow fence can capture snow up to 1.5 t m^{-1} along its length, while a living snow fence with an effective height of 2.4 m can capture up to 6.8 t m^{-1} (Tabler, 2003). In addition, living snow fences are more aesthetically pleasing, and they have the potential to provide benefits such as wildlife habitat, CO_2 capture and woody biomass for renewable energy and other products.

While living snow fences have many positive attributes, they have some limitations and are often misunderstood. Much of the previous work on living snow fences has been done using slow-growing species that require two or more widely spaced rows for effective control and can take 6–20 years to become effective (Gullickson *et al.*, 1999; Daigneault and Betters, 2000; Tabler, 2003). In addition, living snow fences comprised primarily of trees can outgrow their effectiveness over time, as large openings form near the ground, allowing increased wind speed and reducing snow capture. These traditional designs for living snow fences require large areas, which is a significant limitation in areas where roadside rights-of-way are usually narrow and landowners are less willing to set aside wide strips of land.

There are several options available to overcome these limitations, including the use of a single or closely spaced double row of fast-growing willow or other shrubs. A living snow fence composed of a single row of uncoppiced Streamco willow (*Salix purpurea*) established in central New York State in 1993 began capturing snow after the second growing season and had met its design expectations within 5 years, despite several drier than average growing seasons (Dickerson and Barber, 1999). The landowner, Steve Butts, is also very pleased with the results and has seen a reduction in accidents on his stretch of highway from about seven to ten per year before the willow snow fence was established to zero to one per year after its establishment (S. Butts, 2008, personal communication). Over the past few years at least 12 shrub willow living snow fence demonstrations have been installed across New York State as collaborative projects between state and county Departments of Transportation, Soil and Water Conservation Districts (SWCD), the Thruway Authority, the USDA Natural

Resources Conservation Service and local landowners. Initial results indicate that this approach is effective at controlling blowing and drifting snow 2–3 years after establishment and has relatively low installation and maintenance costs (Fig. 6.5). An important factor in the lower establishment costs is that willow can be established using unrooted dormant cuttings, which are cheaper than bareroot seedlings or plants with root balls.

The most important characteristics for effective living snow fences are the high density of stems and branches during the winter, good height growth, relatively uniform density along the length of the plant and an upright form. Many willows and other shrubs inherently possess several of these characteristics. One-year-old willow on a 2-year-old root system had dormant season densities ranging from 25 to 40%, depending on the variety. Two years after coppicing, the dormant season density was 42–56% (T.A. Volk, 2008, unpublished data) (Fig. 6.6). Many structural snow fences are designed with 50% density. The density of willow snow fences can be varied by selecting varieties that produce different numbers of stems and have different growth habits (Tharakan *et al.*, 2005), changing the spacing between plants, by coppicing to alter the number of stems and degree of side branching and by varying the number of rows planted. Rates of establishment can be modified by changing the size of planting stock, correctly matching plant species to site conditions, which can often be quite harsh near roadways, and altering soil conditions.

An appealing attribute of living snow fences is that they are made up of living plants, and if properly installed, they will be in place and function for decades. However, because they are living plants, they require more planning and care during installation to be successful compared to structural snow fences. Planning should include an assessment of blowing snow conditions at the problem area, determination of the best location and orientation, evaluation of growing conditions for plants, selection of the right plant material and proper site preparation for planting (Gullickson *et al.*, 1999). Some varieties of willow that are vigorous in high-density plantings for biomass production have not been effective in living

Fig. 6.5. A double-row shrub willow living snow fence in central New York State. The shrub willows are 2 years old aboveground on a 3-year-old root system. Photo courtesy of Mark Appleby, SUNY-ESF.

Fig. 6.6. Dormant season density of a shrub willow living snow fence that is 2-year-old aboveground growth on a 3-year-old root system. Photo courtesy of T.A. Volk.

snow fences because the stems tend to spread outward rather than grow upright. In extreme cases, some varieties have grown horizontally rather than vertically when planted in single-row living snow fences. The experience gained from the living willow snow fence demonstration and research projects so far has resulted in several conclusions. Good weed control is paramount in the successful establishment of willow snow fences. The cost of landscape fabric (weed mat) should be considered necessary rather than optional. The use of larger cuttings appears to lead to appreciably earlier snow fence establishment while remaining practical in terms of equipment management. Some varieties of willow should not be used in future snow fence designs because of their tendency to develop a spreading canopy, susceptibility to heavy deer browsing and damage by some generalist insects. Other varieties and species of shrub willow show great promise as material for living willow snow fences due to their relatively unpalatable nature to herbivores, rapid growth and multiple stem growth pattern. Further testing and developing of this application for shrub willows in ongoing.

6.3 Soil Erosion Control and Riparian Buffers

6.3.1 Use of poplar and willow to create forested riparian buffers in the Lower Mississippi Alluvial Valley

E. GARDINER AND J. STANTURF

Background

Flanking the third largest river in the world, the watershed, flood plain and delta of the Mississippi River encompass nearly 41% of North America. The lower flood plain and delta, which comprise the 11 million ha Lower Mississippi Alluvial Valley (LMAV), hold rich alluvial soils that support vast stands of native *Salicaceae* (Fig. 6.7). Several species of poplar (*Populus* spp.) and willow (*Salix* spp.) are native to the LMAV, but eastern cottonwood (*P. deltoides*) and black willow (*S. nigra*) are unquestionably the most dominant *Salicaceae* in the region. Though their range extends beyond the region, these species are most productive on the alluvial soils of the LMAV, where

Fig. 6.7. Black willow (*Salix nigra*) colonizing a sandbar in the Mississippi River, Washington County, Mississippi, USA. Photo courtesy of J. Stanturf.

eastern cottonwood has attained heights over 50 m and black willow has attained heights over 40 m in natural stands (Cooper, 1990; Pitcher and McKnight, 1990).

Historical use of poplar and willow

As commerce developed along the Mississippi River during the 1800s and early 1900s, settlement patterns concentrated deforestation and development along the main river channel and its tributaries and distributaries. These settlement patterns created the need to stabilize banks and construct levees to protect developed property along watercourses. Black willow saplings, because of their pliable stems and their abundance along watercourses, were woven into mattresses to stabilize eroding banks and provided foundations for the construction of jetties and levees (Barry, 1997) (Fig. 6.8). The engineering utility of black willow for bank stabilization is still recognized, but more recent techniques employ the use of willow posts and long cuttings to establish living revetments for stabilization of eroding stream banks (Schaff *et al.*, 2003; Martin *et al.*, 2005).

Poplar and willow riparian buffers

Row crop agriculture is the primary land use in the LMAV, and because of this the principal thrusts behind the establishment of forested riparian buffers in this region is to lessen the impacts of soil erosion and agricultural pollution on water quality. Established riparian buffers function to accomplish these objectives on agricultural landscapes through various mechanisms that include increasing water infiltration into soil, reducing sediment loading to streams and filtering runoff of agricultural nutrients and chemicals such as nitrogen and phosphorus. Eastern cottonwood plantings, in particular, were effective in improving surface water runoff and groundwater quality on a silty loam soil in the LMAV (Thornton *et al.*, 1998). The US government recognizes the benefits of riparian buffers to water quality and maintains a voluntary programme providing private landowners an incentive to establish and manage forested riparian buffers. Through Conservation Practice 22 of the Conservation Reserve Program, the Farm Service Agency of the US Department of Agriculture will cost-share with landowners the establishment of riparian buffers on qualified agricultural areas (www.fsa.usda.gov).

Fig. 6.8. Construction of black willow mats in 1915 for riverbank stabilization in the LMAV. Photo courtesy of J. Stanturf.

Eastern cottonwood has been used more than black willow for creating riparian buffers in the LMAV, primarily because of historical management and markets for the species in the region. An advantage of the past development of the eastern cottonwood industry in the region is that superior clones are available for deployment on specific soils. However, several ecological and silvical characteristics of eastern cottonwood and black willow make both species desirable as dominant tree species on riparian buffers. These species are ideally suited for use in the LMAV because they are native to the region and planting stock is readily available. Both species establish readily from cuttings, are fast growing and will develop forest structure and canopy on alluvial sites relatively quickly. Furthermore, both species can be established on degraded agricultural sites and will tolerate soil flooding and sediment deposition (Broadfoot and Williston, 1973; McKnight *et al.*, 1981; Hook, 1984). As a footnote to the above, in India poplars also are planted along the banks of rivers originating in the Himalayas, to capture soil and debris during periods of flooding (Dhiman, 2012) (Fig. 6.9).

Co-benefits of establishing poplar and willow riparian buffers

The co-benefits of established riparian buffers often provide the additional incentive necessary to encourage landowners to remove land from agricultural production to install forested buffers along drainages (Fig. 6.10). Several co-benefits related to ecology and forest management can be realized through the establishment of well-designed eastern cottonwood or black willow riparian buffers in the LMAV. For example, Gardiner *et al.* (2004) noted that eastern cottonwood plantations provided a favourable understorey microenvironment for the regeneration of other native woody vegetation, thereby providing a viable afforestation practice for facilitating forest restoration (Fig. 6.11). Twedt and Portwood (1997) and Hamel (2003) demonstrated the value of eastern cottonwood plantations as habitat for neotropical and other migratory birds. Additionally, forested riparian

Fig. 6.9. Poplar trees are planted along rivers in India that flood frequently. The trees are often flooded up to 2 m in depth and after flooding there are deep silt deposits. Photo courtesy of R. Dhiman.

Fig. 6.10. Eastern cottonwood (*Populus deltoides*) established as a riparian buffer along the margin of a wetland in the Lower Mississippi Alluvial Valley, Sharkey County, Mississippi, USA. Photo courtesy of J. Stanturf.

buffers can be designed to provide landscape corridors for terrestrial animals and improve invertebrate and fish habitat in adjacent streams.

Linked with ecological benefits, rapidly growing eastern cottonwood and black willow plantations can be used to sequester atmospheric

Fig. 6.11. Eastern cottonwood plantation established as a riparian buffer and interplanted with other native bottomland hardwood species to enhance co-benefits of increased woody plant diversity, etc., Sharkey County, Mississippi, USA. Photo courtesy of J. Stanturf.

carbon quickly (Stanturf *et al.*, 2003). Black willow, a species very tolerant of hydric soil conditions, is especially suited for this role on riparian sites too wet to sustain high productivity of other species. As incentive to landowners, recent changes to the Conservation Reserve Program allow for marketing carbon sequestered in eastern cottonwood plantations used as nurse crops for other bottomland hardwood species. Additionally, well-managed plantations can produce fibre or timber and could potentially produce feedstock for biofuel production (Stanturf and Portwood, 1999; Stanturf *et al.*, 2003).

6.3.2 Streamside restoration and stabilization with riparian buffers in the Pacific Northwest, USA

J.D. JOHNSON

Poplars and willows are well suited for use in riparian buffers and for streamside restoration and stabilization projects in the Pacific Northwest USA. They are native riparian species, having evolved to take advantage of conditions of high water table and periodic flooding. They are often used in combination with the more water-tolerant willows planted along stream edges where the water table is at or near the surface and the poplars are planted upslope from the willows where the water table is lower. The rapid growth of poplars and willows allows them to capture the planting site from competing vegetation, and if whips are used, they will be above the competing vegetation and can shade out the competition as the tree canopy develops (Fig. 6.12). In many instances where slower-growing native species are used, they often are out-competed by aggressive invasive plant species. In the Pacific Northwest, two troublesome invasives are reed canarygrass (*Phalaris arundinacea* L.) and Himalayan blackberry (*Rubus discolor*) that are capable of overgrowing slower-growing native riparian species.

Functions of riparian buffers include sediment reduction, bank stabilization, nutrient removal, pesticide barrier and breakdown, shade and large woody debris recruitment. Of these buffer functions, only the first, sediment reduction, is not an attribute of poplars and

Fig. 6.12. A hybrid poplar buffer established from whips planted in April. The photo was taken in September and the trees are over 5 m tall. Photo courtesy of J. Johnson.

willows, because surface roughness is required to slow surface water flow that facilitates sedimentation. The other five functions, however, are ones that poplars and willows exhibit. Their extensive root systems consist of strong, woody roots needed for bank stabilization and a large mat of fine roots used for nutrient removal and pesticide breakdown. Nitrate uptake from groundwater by a hybrid poplar buffer exhibited a several hundredfold decrease from the adjacent crop into the trees, reducing the level to less than 1 ppm. Poplars' tall stature and dense canopy, as well as rapid stem growth, contribute to their ability to shade streams, which reduces their warming, and as the trees mature and topple, they contribute large woody debris to streams, creating a series of riffs and pools needed for fish habitat.

6.3.3 Erosion control in New Zealand

I. McIVOR

Poplars and willows are used extensively in New Zealand to reduce soil erosion on hill-country farms (Wilkinson, 1999). Out of 11.2 million ha of the North Island in pasture, some 3.7 million ha (32%) require significant soil conservation measures to attain biophysical sustainability. The stabilizing effects of trees in mass soil movement and fluvial processes are well known. The most common tree species used for the stabilization and rehabilitation of the North Island's erodible land are poplars and willows, together with localized timber blocks, usually *Pinus radiata* (Thompson and Luckman, 1993). Poplars and willows are favoured for hillside stabilization because of their rapid development and ease of planting (Fig. 6.13). At planting densities required for hillside stabilization, pasture receives sufficient light to develop underneath the trees. Poplars reduce erosion by drying out the soil and binding soil particles with their extensive root systems. Guevara-Escobar *et al.* (2002) demonstrated that evaporation from widely spaced poplar–pasture systems was significantly greater than from pasture alone. Initially, poplars and willows should be densely planted to ensure rapid ramification of the slope (McIvor *et al.*, 2008). Trees can then be thinned selectively as the root systems develop (McIvor *et al.*, 2009). Individually protected poplar and

Fig. 6.13. Earthflow stabilization using poplar and willow on a hillside near Gisborne, North Island, New Zealand. Photo courtesy of I. McIvor.

willow poles allow land users to plant trees directly into areas susceptible to erosion without having to exclude grazing stock such as sheep and young cattle (Guevara-Escobar *et al.*, 2007). The tree spacing required for slope stabilization ranges from 25 to 156 stems ha^{-1}, depending on the severity and proximity to the actively eroding area. At this spacing, animal production remains the dominant enterprise. Poplars are planted most effectively on the sides of gullies, rather than the gully floor. Here, their deep roots access the subsurface flows that can trigger erosion events. Not all regions and hill slopes are suitable for the establishment of poplar and willow species. Steep gradients, water stress, continuing soil disturbance and desiccating winds inhibit tree establishment and/or subsequent growth on upper hill slopes in many of New Zealand's eastern regions.

Land-use change has been shown to change the soil C stocks, with broadleaf plantations resulting in higher C stocks than pine plantations (Guo and Gifford, 2002). The new Emissions Trading Scheme, introduced in New Zealand in 2008, is providing a large financial incentive for forestation (Maclaren *et al.*, 2008) and has resulted in a marked increase in poplar and willow plantings on erodible pastoral land.

6.3.4 Riverbank stabilization in New Zealand

I. McIVOR AND B. ROBINSON

In New Zealand, planting willows is a valuable tool for use in river engineering. The willows' high root density stabilizes riverbanks and prevents the river from changing course. Periodic 'layering' can promote dense stands of willows that are less likely to succumb to floods. Layering involves cutting larger trees so that the trunk lies on the moist riverbank and subsequently forms shoots and roots along its entire length. Willows improve water quality by intercepting contaminants in runoff or subsurface flow. They also provide a physical barrier to prevent stock from entering the waterways (Fig. 6.14), where they would otherwise cause edge damage by trampling and contamination by urination and defecation. Glova and Sagar (1994) reported a greater abundance and diversity of benthic invertebrates, together with greater numbers of large brown

Fig. 6.14. Riparian planting of *Salix schwerinii*, *Salix matsudana* × *S.alba* and *Populus deltoides* along the Selwyn River, Coe's Ford, New Zealand. Photo courtesy of I. McIvor.

Fig. 6.15. Poplars and willows make palatable stock fodder on New Zealand farms. Carterton, New Zealand. Photo courtesy of I. McIvor.

trout, in willow-protected streams. However, in small streams, willows can reduce biodiversity due to shading out of aquatic flora (Lester *et al.*, 1994). Judicious clone selection is critical for river engineering. It is imperative that the willows do not invade downstream or adjacent areas. Male clones that do not fragment are ideal. *S. fragilis* and other species are noxious weeds in many Australasian waterways, due to vegetative reproduction from fragile shoots (Wilkinson, 1999) that are broken off by the water current.

Poplar and willow from riverbanks as supplementary stock fodder in New Zealand

Poplars and willows used for hillside and riverbank stabilization also provide supplementary stock fodder during times of drought (Wilkinson, 1999). Some poplar and willow clones can maintain production in water stress conditions that result in pasture dieback (McIvor *et al.*, 2005), particularly when managed via pollarding. The foliage and small twigs provide an emergency food source for both cattle (Fig. 6.15) and sheep (Hathaway, 1987; Douglas *et al.*, 1996). Feeding poplars and willows to stock has proven health benefits. Nelson *et al.* (1984) and Barry and Kemp (2001) attributed an improvement in growth and fecundity to high protein, tannin or trace element concentrations in poplar leaves. Moore *et al.* (2003) demonstrated that willow feed reversed weight loss under severe drought conditions and stock progressively ate thicker branches as pasture became scarce. High levels of tannins in willow leaves may effectively de-worm stock. High levels of tannins may also reduce the nitrogen concentration in the urine of ruminants (Carulla *et al.*, 2005). This has environmental benefits due to reduced nitrate leaching from urine patches.

Feeding poplars and willows to stock may alleviate trace element deficiencies. Many New Zealand pasturelands are deficient in cobalt, zinc and copper (Lee *et al.*, 1999). Poplars and willows can have leaf cobalt and zinc concentrations that are six times higher than pasture growing in the same environment (Robinson *et al.*, 2005). However, poplars and willows may also introduce toxic trace elements into the animal's diet. Most New Zealand pasturelands have elevated cadmium concentrations due to repeated applications of cadmium-rich superphosphate fertilizer (Bramley, 1990). Robinson *et al.* (2000) showed that the commonly used New Zealand varieties of poplar, 'Kawa' (*P. deltoides* × *P. yunnanensis*), and willow, 'Tangoio' (*S. matsudana* × *S. alba*), accumulated cadmium at levels of up to 14 $\mu g\ g^{-1}$ in the dry leaves when grown in a soil containing just 0.6 $\mu g\ g^{-1}$ of this element. This concentration is above levels (1–5 $\mu g\ g^{-1}$) shown to affect livestock adversely (Underwood and Suttle, 1999).

6.4 Land Restoration

J.G. ISEBRANDS

Poplars and willows are pioneering species and some of the first species to revegetate surface mine spoils in the northern hemisphere (Brenner *et al.*, 1984; Chapter 3, this volume). Hybrid poplars, cottonwoods and willows historically have been used to restore surface mine spoils (Hart and Byrnes, 1960; Lumme and Tormala, 1988) and other marginal soils artificially (Misra and Tewari, 1999) for much of the 20th century and beyond (Funk, 1960; Limstrom, 1960). There are many thousands of hectares of former mine spoils worldwide that are in need of restoration (Knabe, 1964; Rockwood *et al.*, 2006).

The early efforts on revegetating mine spoil lands focused on species selection. Many species were tested and successes were dependent on the soil and environmental conditions of the site. Hybrid poplars and willows almost always exhibited the most rapid early growth (Hart and Byrnes, 1960), but performance depended on the clone (Davidson and Davis, 1972; Bungart *et al.*, 2001). Clonal performance was often tied to growth and disease and insect resistance (Davis, 1964; Lumme and Tormala, 1988), as well as regeneration methods. Davidson and Davis (1972) recommended coppice regeneration for the restoration and stand conversion of mine spoils.

The success of regenerating strip-mined land has been attributed to the increased number of microorganisms in the rhizosphere associated with poplar and willow roots (Cundell, 1977). Moreover, some hybrid poplars grown on strip-mined lands have been found to be useful for pulpwood and lumber products (Davidson, 1979).

But despite all these years, more genetic selection and breeding is needed to improve poplar and willow performance and success on these lands (Davidson, 1979; Lumme and Tormala, 1988; Rockwood *et al.*, 2006).

One of the most recent applications of using poplars and willows for reclamation and restoration of strip-mined land is in the coal and oilsands region of western North America (Figs 6.16 and 6.17). Government agencies in Canada and the USA require mining companies to revegetate the mine spoils (and areas disturbed by mining activity) with native species after mining is completed. Therefore, there is a major effort to grow native poplars and willows in greenhouses and nurseries for planting on such land (Fig. 6.18). Focus has been on replanting native species such as aspen (*P. tremuloides*) and balsam poplar (*P. balsamifera*), but these efforts have only just begun (Richardson, 2012) (Fig. 6.19).

6.5 Phytoremediation

6.5.1 Introduction

J.G. ISEBRANDS

Phytoremediation is a general term coined in the early 1990s for an emerging green technology using plants to clean up – or 'remediate' – contaminated soil, sediments, groundwater, surface water and air by removing, degrading and containing toxic chemicals (US EPA, 1998, 2000; Licht and Isebrands, 2005). Phytoremediation technologies primarily use six mechanisms to accomplish clean-up goals:

1. Phytoextraction: the uptake and translocation of contaminants from groundwater into plant tissue.
2. Phytovolatilization: the transfer of contaminants to air via plant transpiration.

Fig. 6.16. Coal strip mine restoration site near Genesee generating station in Alberta, Canada. Photo courtesy of J. Richardson.

3. Rhizosphere degradation: breakdown of contaminants within the rhizosphere, i.e. soil surrounding roots, by microbes.
4. Phytodegradation: the breakdown of contaminants within plant tissue.
5. Phytostabilization: the stabilization of contaminants in the soil and groundwater through absorption and accumulation on to plant roots.
6. Hydraulic control: intercepting and transpiring large quantities of water to contain and control migration of contaminants.

Poplars and willows are some of the most preferred tree species for phytoremediation because they grow rapidly, have many and deep roots and take up large quantities of water and nutrients (Isebrands and Karnosky, 2001; Licht and Isebrands, 2005). They not only take up substantial quantities of water and nutrients but also they provide root surface area for beneficial microbes and mycorrhizae that perform phytoremediation functions.

An International Phytotechnology Society has emerged since 2000 to promote phytotechnologies for cleaning up environmental contamination problems. They have also published a journal since 2002, called the *International Journal of Phytoremediation*, in which the latest applications of phytoremediation are published. In this short time frame, hundreds of articles on the use of poplars and willows for environmental applications have been published. In addition, there has recently been a comprehensive overview published on phytoremediation that features many case studies involving poplars and willows (McCutcheon and Schnoor, 2003).

In this chapter, we give an overview of some examples of the environmental applications of poplars and willows in some of the International Poplar Commission (IPC) member countries. The following contributions are presented in alphabetical order of country.

Fig. 6.17. Oilsand strip mine site near Fort McMurray, Alberta, Canada. Photo courtesy of D. Riddell-Black.

Fig. 6.18. Native balsam poplar grown for strip mine and oilsand land reclamation at Smoky Lake Forest Nursery in Alberta, Canada. Photo courtesy of J.G. Isebrands.

Fig. 6.19. Revegetation of oilfield site with native poplar near Fort McMurray, Alberta, Canada. Photo courtesy of D. Riddell-Black.

6.5.2 Belgium

R. CEULEMANS

Phytoremediation – heavy metals

The phytoremediation potential of a poplar or willow plantation depends on both the biomass productivity rate and the concentration of the element (heavy metals) in the biomass. In the study of Laureysens *et al.* (2004a, 2005), the phytoremediation potential of 13 different poplar clones was examined, together with the analysis of the canopy profiles of heavy metals as well as differences in concentrations among leaves, stems and bark. In terms of productivity, clones 'Wolterson' (*P. nigra* (N)), 'Fritzi Pauley' (*P. trichocarpa* (T)) and 'Balsam Spire' (*P. trichocarpa* × *P. balsamifera* (T × B)) showed the highest aboveground woody biomass production (wood + bark), averaging 18, 15 and 14 Mg ha^{-1}, respectively, after 2 years. In combination with its relatively high Al and Zn concentration in wood, this clone showed potential for the phytoextraction of both metals (Al and Zn) (Table 6.1). Clones 'Fritzi Pauley' (T), 'Columbia River' (T), 'Trichobel' (T × T) and 'Balsam Spire' (T × B) also had a relatively high biomass production, i.e. 15, 12, 13 and 15 Mg ha^{-1} (2-year period), respectively. In combination with a relatively high wood and bark metal concentration, 'Trichobel' (T × T) showed potential for Al phytoextraction, while 'Balsam Spire' (T × B) showed potential for Cd and Zn uptake (Tables 6.1 and 6.2).

Leaf, wood and bark concentrations

Variations in leaf concentrations between clones were high for all metals, ranging between 112 and 174 μg g^{-1} for Al, 3.07 and 8.26 μg g^{-1} for Cd and 411 and 695 μg g^{-1} for Zn in mature leaves (Fig. 6.20). Analyses showed that there was a significant clonal variation for mature and senescing leaves for Fe and Pb; for Cu, clonal variation was significant for all three leaf ages. One single clone containing the highest concentration of all metals at the same time was not found. Generally, clonal rankings in leaf concentration were significantly different among metals and among leaf ages per metal (Fig. 6.20).

Table 6.1. Metal concentrations in wood of 2-year old poplar stems of different clones harvested in August (Aug) and November (Nov).

		Al (μg g⁻¹)		Cd (μg g⁻¹)		Fe (μg g⁻¹)		Mn (μg g⁻¹)		Zn (μg g⁻¹)	
Clone	Parentage[a]	Aug	Nov	Aug	Nov	Aug	Nov	Aug	Nov	Aug	Nov
'Balsam Spire'	T × B	8.1	11.2	1.11	2.86	6.7	43.6	2.0	1.6	29	39
'Beaupré'	T × D	10.0	16.3	2.36	2.64	10.8	7.1	5.1	1.7	24	32
'Hazendans'		16.5	15.3	1.63	1.01	31.3	77.3	9.6	7.6	31	33
'Hoogvorst'		13.6	15.6	1.94	1.11	8.1	15.8	2.5	3.5	34	37
'Raspalje'		9.5	10.5	1.74	3.18	8.0	13.9	3.8	5.0	38	26
'Unal'		8.3	15.5	2.77	2.63	22.0	6.5	2.1	2.3	26	29
'Columbia River'	T	18.3	12.7	1.43	2.33	18.4	8.7	2.0	1.8	39	37
'Fritzi Pauley'		13.3	11.1	1.11	0.80	58.7	9.2	2.3	2.2	26	28
'Trichobel'		7.9	19.9	1.22	0.70	5.3	20.6	1.4	1.9	26	34
'Gaver'	D × N	8.1	20.4	2.77	2.26	4.9	25.8	2.4	7.5	35	31
'Gibecq'		8.9	14.2	1.63	0.29	4.5	9.6	1.0	2.4	31	51
'Primo'		12.6	12.7	2.88	3.29	5.2	11.7	2.3	2.6	34	54
'Wolterson'	N	13.4	36.8	–	0.91	47.2	59.3	4.6	19.7	42	50

[a]T, *Populus trichocarpa*; B, *Populus balsamifera*; D, *Populus deltoides*; N, *Populus nigra*.

Table 6.2. Mean metal content (SE) per stool and per hectare for six poplar clones in a short-rotation coppice culture. The 2-year old shoots were harvested in November.

Clone	Al (μg stool⁻¹)	Cd (μg stool⁻¹)	Zn (μg stool⁻¹)	Al (g ha⁻¹)	Cd (g ha⁻¹)	Zn (g ha⁻¹)
'Balsam Spire'	976 (116)	64 (7)	2213 (264)	8.7 (1.3)	0.57 (0.09)	19.8 (3.0)
'Fritzi Pauley'	1578 (190)	14 (2)	1570 (191)	13.9 (2.0)	0.13 (0.02)	13.9 (2.0)
'Gaver'	188 (12)	55 (8)	740 (58)	1.5 (0.1)	0.43 (0.10)	6.0 (0.3)
'Hazendans'	370 (64)	17 (9)	1065 (190)	3.2 (0.6)	0.16 (0.09)	9.2 (1.8)
'Trichobel'	596 (144)	13 (3)	1072 (261)	5.5 (1.6)	0.12 (0.03)	9.8 (2.9)
'Wolterson'	648 (35)	51 (2)	2641 (81)	5.9 (0.2)	0.47 (0.02)	24.2 (0.5)

Little or no Co, Cr, Cu, Ni or Pb was accumulated in the wood, and the concentration of Cr in bark was also below the detection limit (0.1 μg l^{-1}). The metal concentrations in bark were significantly higher than the concentrations in wood (Table 6.1), both when samples were collected in August and in November ($P < 0.001$). On average, the Al concentration in bark was ten times the concentration in wood; likewise, the bark concentration of Mn and Zn was six times the wood concentration. The bark concentration of Fe and Cd was, respectively, four and three times higher than the wood concentration, on average.

When sampled in November, clone 'Fritzi Pauley' (T) showed a mean Al concentration in wood of 90 μg g^{-1}, while for clones 'Wolterson' (N) and 'Balsam Spire' (T × B), wood Al concentration averaged 34 and 64 μg g^{-1}, respectively (Table 6.1). For Cd, the concentration averaged 2.2 and 3.3 μg g^{-1} in clones 'Wolterson' (N) and 'Balsam Spire' (T × B), respectively; for Zn, the concentration averaged 147 and 144 μg g^{-1} in clones 'Wolterson' (N) and 'Balsam Spire' (T × B), respectively (Table 6.1). Clone 'Fritzi Pauley' (T) had a mean Cd and Zn concentration of, respectively, 0.7 and 92 μg g^{-1}. Furthermore, metal content per plot was correlated significantly with wood dry mass and total biomass production, but not with bark dry mass. For Cd and Zn, a significant correlation between metal content per plot and number of shoots was found, because clones 'Wolterson' (N) and 'Balsam Spire' (T × B) had the highest Cd and Zn concentration and accumulation (Table 6.1). These results suggest that selection and improvement of poplar clones for phytoextraction should focus on biomass production, stool survival and metal concentration; population dynamics should not be taken into account.

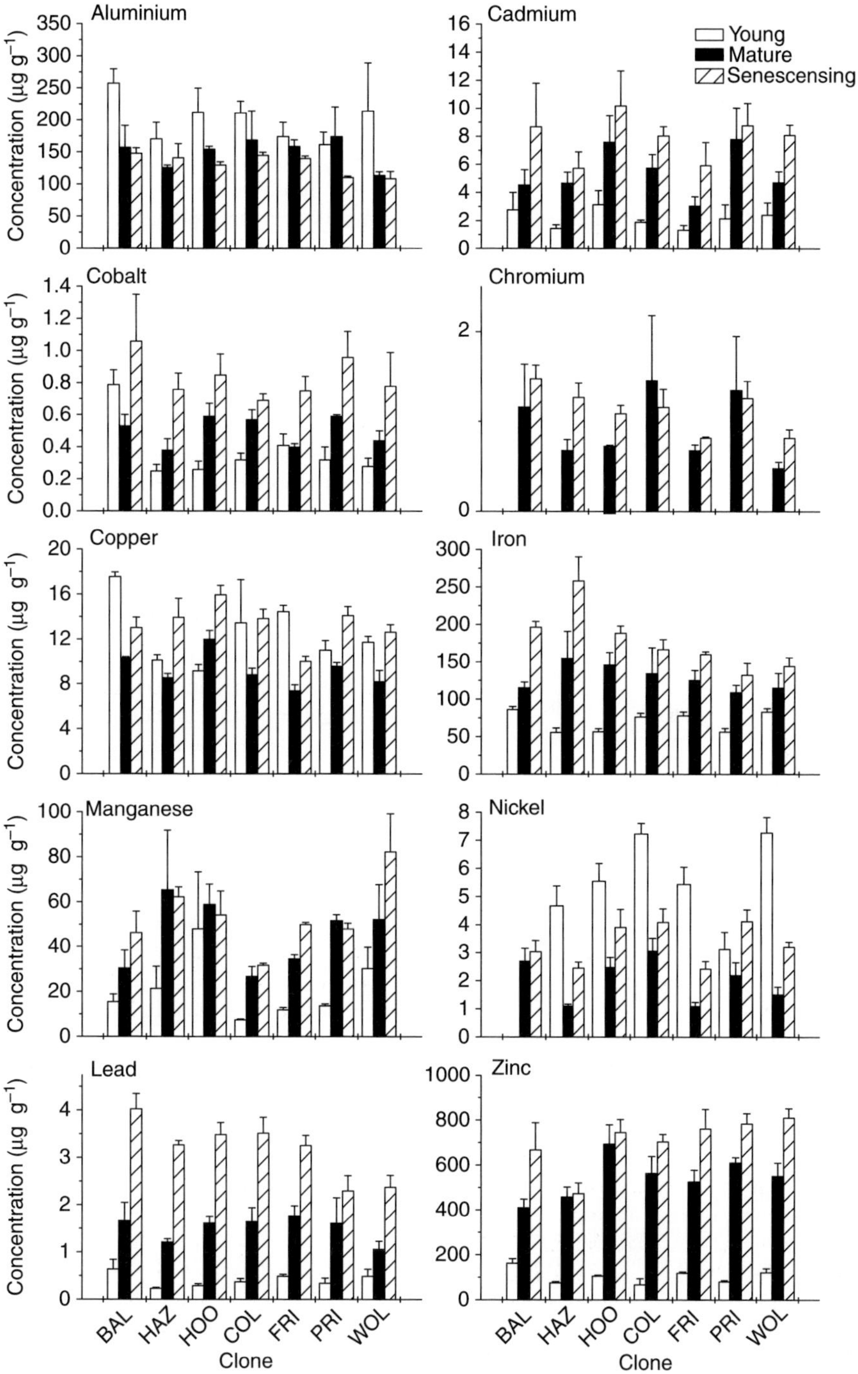

Fig. 6.20. Mean metal concentration in young, mature and senescing poplar leaves for clones 'Balsam Spire' (BAL), 'Hazendans' (HAZ), 'Hoogvorst' (HOO), 'Fritzi Pauley' (FRI), 'Primo' (PRI) and 'Wolterson' (WOL). Mean values of replicates and their standard error bars are presented.

When the transfer coefficients (= plant tissue concentration/total soil concentration) were calculated, Cd, Zn and Cu showed to be taken up most easily by poplar: Cd > Zn > Cu > Mn > Co, Ni > Pb > Cr > Fe > Al. Transfer coefficients differed among leaves, wood and bark, due to the different tissue metal concentrations, but the sequence in the transfer coefficients was similar for these three organs.

Phytoextraction potential

In contrast to the metal content per plant (or stool), the metal content per plot (or per unit of ground area) represents the real phytoextraction potential (Table 6.2), as it includes stool mortality. Clone 'Fritzi Pauley' (T) showed the highest Al accumulation over 2 years, 1.4 kg ha^{-1} (Table 6.2). For Al, metal content per plot was correlated significantly with woody dry mass ($r = 0.513$) and woody biomass production ($r = 0.587$); no significant correlation was found with bark dry mass, number of shoots or mean shoot diameter. Clones 'Wolterson' (N) and 'Balsam Spire' (T × B) showed the highest accumulation of Cd, 47 and 57 g ha^{-1}, respectively, and of Zn, 2.4 and 2.0 kg ha^{-1}, respectively (Table 6.2). For Cd, metal content per plot was correlated significantly with wood dry mass ($r = 0.472$), woody biomass production ($r = 0.503$) and number of shoots per stool ($r = 0.719$) and per plot ($r = 0.706$). For Zn, metal content was correlated significantly with wood dry mass ($r = 0.746$), woody biomass production ($r = 0.806$), stool mortality ($r = -0.543$) and number of shoots per stool ($r = 0.492$) and per plot ($r = 0.526$).

Significant clonal variation in uptake and accumulation was observed for most metals (Tables 6.1 and 6.2) that were analysed, as also shown for willow (e.g. Rachwal *et al.*, 1992; Watson *et al.*, 1999; Aronsson and Perttu, 2001; Pulford *et al.*, 2001). However, clones with the highest concentration of all metals were not found, confirming earlier observations for willow (Riddell-Black, 1994; Pulford *et al.*, 2002).

After 2 years, clone 'Fritzi Pauley' showed the highest accumulation of Al, averaging 1.4 kg ha^{-1}, while clones 'Wolterson' and 'Balsam Spire' showed phytoextraction potential for Cd and Zn, averaging, respectively, 47 and 57 kg ha^{-1} for Cd and 2.4 and 2.0 kg ha^{-1} for Zn (Table 6.2). Several studies have shown that clones with a high uptake of a combination of several metals have not yet been identified (Riddell-Black, 1994; Pulford *et al.*, 2001). This is probably due to the antagonistic properties of several metals. Likewise, hyperaccumulators accumulate only one or a limited number of metals. However, the uptake by these plants is much higher in comparison with poplar or willow. Therefore, Ernst (1996) suggested using these short-rotation coppice (SRC) cultures on slightly contaminated soils. The trees will take up part of the heavy metals and additionally will stabilize the soil, reducing metal leaching and dust blow. The combination of wood for energy production with phytoremediation will make both economically more feasible. Metals in the biomass remain in the ashes or are filtered to avoid translocation of the heavy metal pollutants to the atmosphere (Punshon and Dickinson, 1997).

We did not study root accumulation, although many studies have shown that most metals accumulate in the roots (Kabata-Pendias and Pendias, 1984; Landberg and Greger, 1996). This would imply that for phytoremediation purposes roots need to be ploughed up after the last rotation cycle rather than rototilled and left in the soil. Root metal concentrations and possible clonal differences could be the objects of further research. We have, however, shown that poplar SRC offers possibilities for phytoremediation of slightly contaminated soils (Fig. 6.21).

6.5.3 Canada

M. CARLSON

Canada has had active phytoremediation projects with poplars and willows since before 1990. Emphasis has been on wastewater treatment, hydrocarbon remediation and remediation of solvents and heavy metals on brownfields. Probably the most notable and publicized early phytoremediation project in Canada was the municipal effluent irrigation project in Vernon, British Columbia. In 1988, the joint project between the city of Vernon, the British Columbia Forest Service and the British Columbia Ministry of the Environment began irrigating hybrid cottonwood trees and other tree species on a 10 ha site near Vernon (Carlson, 1992). The wastewater-treated poplars grew very rapidly and

Fig. 6.21. View of the experimental short-rotation coppice culture with 17 poplar clones in Boom (province of Antwerpen, Belgium). The poplars were planted in 1996 and managed on 3-year coppice rotations for more than 10 years. As the plantation was established on slightly polluted wasteland soils (enhanced heavy metals in the soil), the significant phytoextraction potential of the different clones has been quantified. Photo courtesy of R. Ceulemans.

after 5 years the plantation average was 10.6 cm in diameter and 13.6 m in height (Fig. 6.22). Irrigation rates increased from 30 cm in the first year to over 75 cm at year five. Annual concentrations of effluent were 158 kg ha^{-1} nitrogen, 60 kg ha^{-1} phosphorus and 120 kg ha^{-1} potassium. The project has been very successful and has received broad support from the public. The residents enjoy the green landscape, enhanced wildlife viewing, hiking opportunities and cost savings for effluent disposal. Twenty years after the project was initiated, the poplars were harvested and the planting recycled (Fig. 6.23). The site was cleared and a new irrigation system installed. The intention was to plant new hybrid poplar whips at 5 × 5 m spacing to initiate the next generation of Vernon's municipal wastewater programme with poplars.

The Vernon project prompted other municipalities to consider irrigating poplars and willows with treated municipal sewage wastewater. A nationwide Canadian Biomass Innovation Network project, led by Natural Resources Canada, was initiated in Alberta (Krygier, 2011). The project, managed by the Canadian Forest Service, consisted in 2011 of six research and demonstration sites near Edmonton, Alberta. The first site, located at Whitecourt, Alberta, focused on using willow and poplar clones for wastewater disposal and developing an alternative energy source for the region. Clones are monitored for growth, biomass yield, pest incidence, heavy metal accumulation and soil and groundwater quality. After the first rotation, yield was increased with irrigation and no adverse soil chemistry was detected. Enhanced soil nitrogen availability was found after biosolids application. Cost savings were demonstrated as well, thereby prompting expansion at the other sites. Several other private wastewater projects and landfill vegetative caps were initiated using poplars and willows in British Columbia (Passive Remediation Systems, 2012).

Since the late 1990s, there have been phytoremediation projects in Saskatchewan, Canada, using poplars and willows to remediate petroleum-hydrocarbon-contaminated sites and landfill covers (University of Saskatchewan, 2012). University scientists working with Environment Canada, with support from petroleum providers, investigated the potential of using native poplars and willows for petroleum-contaminated site remediation. Emphasis has been on selecting those that exhibit early growth and degrade the

Fig. 6.22. Wastewater-irrigated hybrid poplar plantation, Vernon, British Columbia, Canada. Photo courtesy of M. Carlson.

Fig. 6.23. Regrowth of hybrid poplar plantation irrigated with wastewater, Vernon, British Columbia, Canada. Photo courtesy of M. Carlson.

contaminants of concern. Initially, there was not enough knowledge on which poplar or willow clone would be best in the various regions. Another early problem, not unique to Canada, was that commercial quantities of the better poplar and willow clones were not available for the public. This problem has improved with the availability of more plant materials from commercial nurseries.

Another major centre for phytoremediation research and development in Canada has been at the Plant Biology Research Institute (IRBV) in the Montreal Botanical Garden in Montreal, Quebec. The institute has become a major leader nationally and internationally for research and applications on reclaiming and restoring abandoned contaminated urban sites, known as 'brownfields'. The institute staff uses poplars and willows (and some other plants) to remediate such sites for public use in the Montreal region. Most of these sites have compacted soils, poor drainage and a suite of contaminants in the soil and water. These contaminants often include a mixture of both organic and inorganic compounds. The focus has been on establishing willows and poplars on these difficult sites. The sites that are being restored successfully include those on which wastewater sludge is applied and heavy metal contamination is problematic (Labrecque *et al.*, 1995). The group has employed both willow and poplar coppice as vegetative filters to contain soil and water contamination

(Guidi *et al.*, 2008). They have shown that both heavy metal and nutrient uptake in willows and poplars are enhanced by mycorrhizal fungi (Bissonnette *et al.*, 2010; Fillion *et al.*, 2011). They also work on new ways to establish willows and poplars for different environmental applications on difficult sites (Teodorescu *et al.*, 2011). These establishment techniques have proven successful on petroleum-contaminated sites in Quebec (Guidi *et al.*, 2012). These efforts have used green technology to reclaim and restore brownfields in Quebec.

6.5.4 Estonia

K. HEINSOO

The first seven energy forest plantations were established in Estonia in 1993–1995, within the framework of scientific cooperation of multiple Estonian research institutions and the Swedish University of Agricultural Sciences, in order to promote research and application of fast-growing, short-rotation forestry in Estonia. The total area of these plantations was 3.1 ha comprised of *S. viminalis* and *S. dasyclados* clones originating from the Swedish Energy Forest Programme (Koppel *et al.*, 1996). Since that time, the number of plantations for research purposes has increased twice, based on the planting material originating from Sweden, the UK and local sources. Moreover, in 2008, the area of the plantations was more than ten times larger than in 1998. After 2005, the first *Salix* plantations of private stakeholders were established in Estonia. There was a very high interest among local farmers to start such alternative crop cultivation, and the Estonian Ministry of Agriculture planned to start subsidizing the establishment of short-rotation plantations (SRPs) of 300 ha year^{-1} from 2009. Therefore, a very rapid increase in the SRP area in Estonia is predicted up to 2020.

Because of the local climate, *Salix* has been considered as the most promising tree species for SRP in Estonia, with the average annual biomass growth in fertile soils of 10–12 dry t ha^{-1}. But production can be less due to the lack of soil nutrients, water and weed control, along with frequent late spring frosts and browsing by different mammals (mainly beavers).

At first, the SRPs in Estonia were planted to supply local heating boilers with wood chips, but there were also stakeholders who were interested in producing pellets from this raw material. In both cases, high yield can be achieved only by additional fertilization of the plantations. In order to decrease the costs of fertilization, scientists from the Estonian University of Life Science implemented different applied studies using municipal wastewater or sludge instead of mineral fertilizers. The first results have been promising and are discussed hereafter as three case studies. Larger-scale usage of these methods will depend on sceptical local legislation and the problems of the environmental monitoring procedures required by the Estonian Ministry of Environment.

Aarike case study

The Aarike plantation was first established in spring 1995 with cuttings of *S. viminalis* clone '78183' (clone numbers correspond to the Swedish clone numbering system), but for several reasons was mostly replanted 3 years later with *S. dasyclados* clone '79097'. This plantation was designed specially to purify municipal wastewater originating from a retirement home (approximately 25 persons) in a local village, and the area of the plantation was 180 m^2. The wastewater from the home's different buildings, with a total volume of 20 m^3 day^{-1}, flowed into three septic tanks. After flowing through the septic tanks, the water was fed by gravitational flow on to the SRP. Before planting, the SRP area was isolated from the deeper groundwater by layers of heavy clay and gravel. A serpentine ditch of wastewater divided the system into three blocks, each consisting of four beds constructed with a 20 cm layer of filter material with a 10 cm humus layer on top.

During subsequent years, different components of wastewater (BOD_7, NH_4-N, NO_3-N, Kjeldahl-N, PO_4-P, total-P) were analysed periodically from both the inflow and outflow points of this vegetation filter. The results revealed that the purification process was sufficient for biological oxygen demand (BOD), with the value in the outflow exceeding the regulatory limits only rarely during the winter months. The uptake of N compounds was limited due to the poor nitrification process and the purification efficiency of P that was low in all steps (Table 6.3; for details, see Kuusemets *et al.*, 2001). The last result was not unexpected, because the Aarike vegetation filter was, due to limited available land, designed much smaller (at 7 m^2 per population equivalent (p.e.)) than the recommended design criterion (20 m^2 per p.e.) for surface-flow constructed wetlands in cold climate conditions (Brix, 1994).

The willow biomass estimates and chemical analyses of different tissues revealed that only a limited proportion of the above-mentioned elements were restored in the *Salix* plants (Table 6.4). There is strong evidence that most of the pollutants were stored in the filter material or utilized by the soil microbes (Heinsoo and Koppel, 2003).

Nõo case study

The Nõo plantation is situated on a 0.91 ha area of pseudo-podzolic soil that was originally rich in potassium and phosphorus but slightly acidic, with low nitrogen content. This SRP was planted in spring 1995 with several clones of *S. viminalis* ('78195', '78183', '78021', '78012', '82007' and '78101') and one clone of *S. dasyclados* ('81090') growing in different plots. In May 2001, after the winter harvesting of the SRP, half of each plot was supplied with the composted sewage sludge originating from the Tartu wastewater treatment plant. The sludge amount (6.3 t per 0.44 ha) was calculated based on the limiting factor P that should not exceed 200 kg ha^{-1} if the next rotation period was assumed to last for 4 years. During this rotation period, the shoot production was estimated annually from randomly selected and marked plants in both sludge-treated and control sections of each clone plot. In subsequent years, sewage sludge almost doubled the shoot biomass for most of the clones, even though one of them ('78183') was re-harvested in the following winter (Fig. 6.24). Most probably, the exceptional decrease in shoot production in clone '78012' can be explained by the extremely low survival rate of the plants in the sludge-treated section of the plot. Throughout this experiment, we also analysed gravitational water quality from both SRP sections at depths of 10 and 40 cm. The BOD_7, N and P in the water from 40 cm depth did not exceed the limits for wastewater purification systems in Estonia. An increase in BOD_7 at the 10 cm depth was only found during the first year after sludge treatment. By contrast, the amount of P at this depth was higher during the second year of experiment than could be explained by the additional dissolution of some P compounds to the gravitational water over this time.

Table 6.3. Average wastewater purification efficiency (%) in Aarike SRP.

	Period					Period average
	1995	1996	1997	1998	1999	
BOD_7	60	72	60	60	88	75
Total N	23	29	35	41	28	32
Total P	14	19	20	18	9	14

Table 6.4. Estimated N and P removal with the biomass during Aarike SRP harvesting.

	N	P
Purification efficiency (%)	32	14
Annual removal (kg)	35	2.1
Concentration in shoots (%)	0.74	0.07
Stored in shoots (kg $year^{-1}$)	1.14	0.11
Removal in biomass (%)	3	5

Kambja case study

The Kambja SRP was one of the three prototypes that were established in Estonia in 2003 within an EC LIFE Environment project 'Estwaste' (Aasamaa *et al.*, 2010). The main goal of this activity was to find an inexpensive and efficient wastewater purification method for those communities in rural areas whose

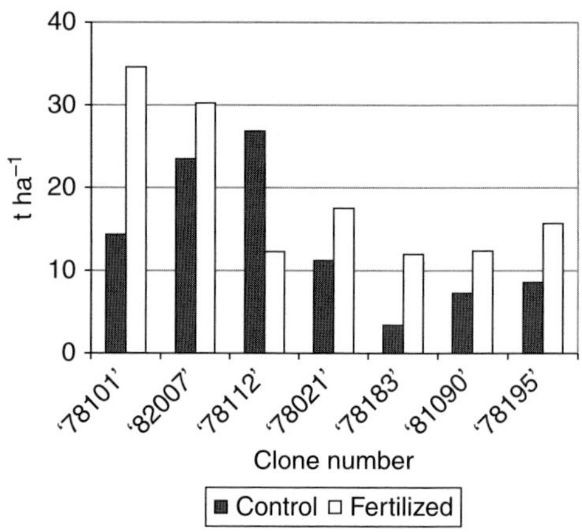

Fig. 6.24. Estimated biomass yields of sewage-sludge-treated and control plots in Nõo SRP in autumn 2003.

previous wastewater purification plants had been amortized since the large economic changes in Estonia in 1995. The first step of wastewater purification at the Kambja site was designed to be carried out in the mechanical filter and wastewater storage ponds only. During the vegetation period, this pre-purified wastewater was to be pumped to the SRP through a specially designed irrigation network. During the winter months, this water was planned to run through the previously existing additional bioponds. The wastewater load of the prototype was approximately 1000 p.e. and the SRP was comprised of 16.1 ha of *Salix*, *Populus tremula* f. *gigas* and *Alnus incana* plants, of which 11 ha was supplied with the irrigation pipes network (Truu *et al.*, 2009). The water quality was monitored monthly at different parts of the prototype and the shoot yield estimated annually.

The results of this project affirmed that a vegetation filter could be an efficient wastewater purification option in Estonian weather conditions (Fig. 6.25). The pollutant contents (both organic matter and N) had already started to decrease very rapidly in the storage ponds. However, the outflow from the second biopond had too much N and P to meet the limits for water that may be discharged to natural waterbodies in Estonia (Table 6.5). This problem was solved by discharging the water to the SRP (Heinsoo and Holm, 2008). The gravitational water collected from a depth of 40 cm never exceeded the above-mentioned limits. The additional SRP area of 5.1 ha could be used later for other purposes, e.g. sewage sludge disposal if bioponds need purification. *P. tremula* and *A. incana* had only limited growth during the first years of the SRP as a result of weed problems due to the added water and nutrients. Therefore, we are not able to recommend these species in an SRP with wastewater application in the Estonian climate.

The established prototypes also offered a good overview of problems with wastewater purification systems. One of the main issues is to keep the water distribution as homogeneous as possible without significant increase in pipe network costs (Aronsson *et al.*, 2002). For economic reasons, the maximum wastewater load to the SRP should be estimated. The most critical point is to create a practical method of water collection and monitoring from different parts and depths of the SRP. This is also needed in order to diminish the environmental risks and to permit control of the activities performed by the local environment protection authorities.

Fig. 6.25. As a result of wastewater irrigation, the height of the most promising *Salix* clones at Kambja exceeded 6 m as early as the third summer. Photo courtesy of K. Heinsoo.

Table 6.5. Average concentrations of pollutants in different parts of the Kambja prototype.

	Monitoring point			
Indicator	No. 1	No. 2	No. 3	No. 4
BOD_7 (mg O l^{-1})	63.9	28.0	21.6	–
Total N (mg l^{-1})	25.1	18.9	16.5	7.5
Total P (mg l^{-1})	3.4	3.2	2.7	0.7

6.5.5 Italy

G. SCARASCIA-MUGNOZZA

The enhanced level of pollutants in soil and water due to industrialization is one of the major environmental problems at global scale.

In particular, cadmium is considered one of the most dangerous heavy metals, having toxic effects on plants and animals. Cadmium (Cd) enters the environment from industrial processes, heating systems, urban traffic and phosphate fertilizers; another source of Cd in the soil is the mineralization of rocks (Rauser and Muwly, 1995). Typical symptoms of Cd phytotoxicity are chlorosis, growth inhibition and respiratory and nitrogen metabolism changes, as well as low biomass accumulation. Besides, exposure to Cd causes reductions in water and nutrient uptake and photosynthesis (Sanità di Toppi and Gabbrielli, 1999; Pietrini *et al.*, 2003). To remove Cd and other pollutants from the contaminated areas, unconventional techniques that use biological processes have been applied successfully. In particular, plants can be used for removing heavy metals from soil and accumulating them in the harvestable parts. This technology, called phytoextraction (Kumar *et al.*, 1995; Raskin *et al.*, 1997), is less expensive and environmentally disruptive than conventional remediation systems that consist mainly of the excavation and incineration of soil (Cunningham and Ow, 1996). Other advantages of utilizing plants to clean up contaminated areas are the production of biomass and landscape restoration. The efficiency of phytoextraction depends largely on several plant characteristics such as the capability to hyperaccumulate metals, and also on the non-essential ones, fast growth, a deep and extensive root system and the ability to translocate metals in the aerial parts. Since about 2000, forest trees have been studied to assess their potential to remediate heavy metal contaminated sites (Pulford and Watson, 2003). Some aspects of forest tree biology and cultivation appear promising for a phytoremediation strategy. Among them are the large biomass yield that can be used for energy production, an extended and deep root apparatus, a low impact on trophic chains, the capability of some tree species to grow in marginal soils and other ecological benefits. With respect to hyperaccumulating plants, metal uptake by trees is reported to be less remarkable but, on an area basis, the removal of heavy metals from soil could be more effective due to the biomass production.

Several studies have focused on the potential of willows and poplars to be used for phytoextraction (Riddell-Black, 1994; Pulford *et al.*, 2002; Laureysens *et al.*, 2004b). In fact, these *Salicaceae* are reported to be adapted to grow in severe soil conditions (pioneer species) that characterize contaminated areas, besides their ability to accumulate heavy metals (Pulford and Watson, 2003). Moreover, cultural management of willows and poplars by means of SRC cultures is another promising aspect to be considered for phytoremediation strategies (Ceulemans *et al.*, 1992; Scarascia-Mugnozza *et al.*, 1997; Perttu, 1999). The availability of clones selected for high biomass production and disease resistance is another remarkable aspect in utilizing these *Salicaceae* for phytoextraction. Most of the studies conducted on trees show that tolerance to heavy metals depends on their compartmentalization in the roots and low translocation to the leaves. This is probably the major constraint to overcome for a more efficient utilization of these species to clean up soils from metal contamination. Moreover, a significant clonal variation in heavy metal accumulation was found in poplar and willow (Watson *et al.*, 1999; Laureysens *et al.*, 2004a). In this context, screening of clones characterized by different biomass production for heavy metal accumulation and distribution among the organs could be very effective in selecting plant material with important traits for phytoextraction.

The emphasis of the institute scientific group and co-workers at Consiglio Nationale delle Ricerche (CNR) has been to find plants that extract heavy metals efficiently from contaminated soils and water and to study the association of plants and soil bacteria for degradation of chlorinated hydrocarbons. The results of a hydroponic screening study helped identify promising Italian clones for further testing (Zacchini *et al.*, 2009). Pietrini *et al.* (2010) conducted a screening study of those poplar clones that could be used for Cd phytoremediation.

The results are shown in Table 6.6 and in Plate 21C. These show that clones with high values of net photosynthesis have higher values of total dry mass and Cd content. The efficiency of Photosystem II (PSII) was not particularly affected by Cd treatment and only 'A4A' showed a significant decrease in this parameter. The chlorophyll content observed seemed unrelated to the maintenance of a high photosynthetic capacity. In conclusion, 'SS5'

Table 6.6. Net photosynthesis (A) measured at light intensity of 300 μmol photons m^{-2} s^{-1} (growth conditions), chlorophyll fluorescence ratio of dark adapted leaves (Fv/Fm), total chlorophyll content, total dry mass and cadmium content of poplar and willow clones treated with 50 μM cadmium solution for 3 weeks.

Clones	A (μmol m^{-2} s^{-1})	Fv/Fm (rel. units)	Chlorophylls (a+b) (μg cm^{-2})	Total dry mass (g per plant)	Cadmium content (mg per plant)
'11-5'	6.15 b	0.74 a	25.99 c	11.15 bcd	7.01 bcd
'A4A'	1.03 c	0.47 b	7.96 f	8.24 cd	5.79 cde
'I-214'	4.33 b	0.67 a	14.90 e	16.92 a	8.54 abc
'Luisa Avanzo'	5.29 b	0.73 a	39.94 a	14.50 ab	7.25 bcd
'Lux'	4.97 b	0.68 a	9.32 f	11.99 bc	9.81 ab
'Nigra Poli'	0.55 c	0.72 a	20.36 d	6.69 d	5.27 de
'58-861'	2.28 c	0.74 a	30.03 bc	9.99 cd	2.57 e
'SS5'	8.47 a	0.72 a	31.65 b	18.24 a	11.18 a

Differences among clones were assessed using analysis of variance (ANOVA) and the means were compared using the LSD test. Means (n = 5) in the same column followed by the same letters are not significantly different at the $P < 0.05$ level.

showed the best performance for Cd uptake and biomass production. Among poplar clones, 'Luisa Avanzo' and 'I-214' showed favourable data for dry mass and Cd content, while 'A4A' was the clone with the lowest values of all parameters considered.

In a related study, Pietrini *et al.* (2010) showed the spatial distribution of Cd in poplar and willow leaves and how that impacted photosynthesis under different strategies.

An earlier Italian study by Sebastiani *et al.* (2004) investigated the heavy metal accumulation and growth in poplar clones exposed to industrial wastes. The CNR group and co-workers have investigated the linkage between bioenergy production in poplar and willow plantings with soil and wastewater phytoremediation throughout Italy (Paris *et al.*, 2009), an effort that includes selecting Italian clones for use in nitrogen removal, heavy metal removal and for animal waste disposal.

The CNR group has worked closely with scientists at the University of Parma, Italy, on improving the growth, physiological and molecular traits for the phytoremediation of heavy metals and chlorinated organic pollutants using the latest molecular and physiological tools (Marmiroli *et al.*, 2011). The group at Parma has been active in the International Phytotechnology Society and in 2010 hosted the first international conference of the society in Europe.

6.5.6 Korea

Y.B. KOO

Research on phytoremediation began in the early 1990s in Korea. Even though phytoremediation studies were conducted before 1990, it was limited only to removing heavy metals from herbaceous plants as an indoor study and investigating the degree of heavy metal accumulation of plants in the contaminated site of an abandoned mine. From the middle of the 1990s, poplar, a fast-growing species, has been used to remove contaminants and cover hazardous waste sites. Research has been conducted on heavy metal absorption ability, adaptation at a waste landfill site, leachate absorption ability and absorption and adaptation to livestock wastewater of poplars.

Leachate removal and reclamation of a landfill site

An experimental planting of *Populus* ×*euramericana* was established on the Nanjido Waste Landfill site in 1994 and 1995 to evaluate the suitability of the poplar for landfill reclamation. Tree growth was measured and the heavy metal concentrations of soil, landfill leachate and wood were evaluated after harvesting the poplars in 1996. The average height and breast-height diameter were 7.8 m and 9.2 cm, respectively.

This impressive growth performance was much superior to that of trees in the general plantation. The survival rate was also very impressive. The heavy metal concentrations of both the soil and the poplar wood were higher than those reported for other native sites. It is not clear whether the soil brought in to cover the site was from the source of high heavy metal concentration. The high growth rate, high survival percentages and the ability of the poplar to take up and immobilize heavy metals suggest that *P.* ×*euramericana* may be a good species for landfill reclamation (Koo *et al.*, 1997).

To identify poplar species suitable for landfill reclamation, different levels of diluted leachate were used to irrigate poplar trees that were planted into pots with a clay and sand 1:1 ratio. Height and diameter in three species were not significantly different among the different concentrations of leachate treatments. The average evapotranspiration amount of three other species and *P. alba* × *P. tremula* cv. 'glandulosa' were 361 ml day^{-1} and 409 ml day^{-1}, respectively. *P. alba* × *P. tremula* cv. 'glandulosa' was much superior to other species.

Four fast-growing tree species, *P. alba* × *P. tremula* cv. 'glandulosa', *P. nigra* × *P. maximowiczii*, *P.* ×*euramericana* and *Paulownia tomentosa* were used to compare the phytoremediation efficiency on leachate and contaminated soil. The growth performance and contaminant absorption capacity of these species were measured after treating leachate. On the 50% diluted leachate treatment, the hybrid poplar (*P. alba* × *P. tremula* cv. 'glandulosa') showed the best height of 105 cm. *Paulownia tomentosa* showed the highest biomass production, followed by *P. alba* × *P. tremula* cv. 'glandulosa', *P.* ×*euramericana* and *P. nigra* × *P. maximowiczii*. The transpiration rate and stomatal conductance of the hybrid poplar (*P. alba* × *P. tremula* cv. 'glandulosa') were higher than those of other species. The total nitrogen content of *Paulownia tomentosa* was 72,640 mg kg^{-1} in dry weight, and it was higher than that of other species. The heavy metal and salinity content (Na and Cl) of *Paulownia tomentosa* were lower than those of *Populus* spp. (Koo *et al.*, 1998, 1999).

Poplar and willow clones were planted to identify species and/or clones suitable for landfill reclamation at the Kimpo Metropolitan Landfill site in 1997. Growth performance, vitality and visible foliar injury by pollutants, fungi and insects were investigated for 5 years for ten clones from four poplar species and two clones from a willow species. The average survival rate of poplar and willow clones was decreased drastically from 90% in 1997 to 57% in 2001. Among the poplar species, *P. alba* × *P. tremula* cv. 'glandulosa' showed the highest survival rate of 69% and 'Clivus', one of the clones from the hybrid poplar, showed the highest survival rate of 84% (Koo *et al.*, 2002).

Livestock wastewater uptake

Two-month-old rooted cuttings of *P. alba* × *P. tremula* cv. 'glandulosa', *P.* ×*euramericana* and *P. nigra* × *P. maximowiczii* clones were exposed to livestock wastewater, one of the major water pollutants, and to groundwater in order to determine the effects of livestock wastewater on growth response and absorption capacity of the species. For this purpose, five clones of each species were used. In all the species, the height growth of rooted cuttings was better in the livestock wastewater treatment than in groundwater. Of all the poplar species we compared, the height growth was best in *P. alba* × *P. tremula* cv. 'glandulosa'. Aboveground biomass such as leaf and shoot dry weight of all the species increased in the livestock wastewater treatment, while root dry weight decreased. In all the poplar species, the amount of livestock wastewater absorbed was less than that of groundwater. *P. alba* × *P. tremula* cv. 'glandulosa' had the best absorption capacity for livestock wastewater among the three poplar species. The '72-16' clone, one of the *P. alba* × *P. tremula* cv. 'glandulosa' clones, showed the best absorption capacity.

Five 1-year-old clones of each of *P. alba* × *P. tremula* cv. 'glandulosa', *P. nigra* × *P. maximowiczii* and *P.* ×*euramericana* were irrigated with livestock wastewater. Total nitrogen (N), total phosphorus (P) and P_2O_5 content in soil increased with livestock wastewater treatment, but nitrate-N and ammonium-N content in soil did not increase with livestock wastewater treatment as compared with those for groundwater treatment. Total P content in plant tissues decreased with livestock wastewater treatment, while total N content largely increased in comparison with those for groundwater treatment. *P. alba* × *P. tremula* cv. 'glandulosa' was best in dry weight, total amount of nitrogen absorbed per tree and water-use efficiency among the three poplar species (Yeo *et al.*, 2002, 2003).

One-year-old rooted cuttings of *P. alba* × *P. tremula* cv. 'glandulosa', *P. nigra* × *P. maximowiczii* and *P.* ×*euramericana* were also planted in the field close to a milk cow ranch and irrigated with livestock wastewater. The concentration of N and P in the soil treated with livestock wastewater was higher than that in the soil without such treatment. In the livestock wastewater treatment group, *P.* ×*euramericana* showed the best performance in height and breast-height diameter. On the other hand, *P. alba* × *P. tremula* cv. 'glandulosa' showed the best result in shoot dry weight and breast-height diameter, and dry weight of shoots was increased with livestock wastewater treatment. In conclusion, *P. alba* × *P. glandulosa* cv. 'glandulosa' was more tolerant to livestock wastewater than *P.* ×*euramericana* or *P. nigra* × *P. maximowiczii*. Total N and total P concentrations in plant shoots with livestock wastewater were higher than those of shoots without livestock wastewater. Three-year-old *P. alba* × *P. tremula* cv. 'glandulosa' contained 247 g of nitrogen in leaf and stem. The absorption of livestock wastewater increased sharply with age. At age 3, one individual tree absorbed 604 l of livestock wastewater during one growing season.

Other contaminants

Two-month-old rooted cuttings of *P. alba* × *P. tremula* cv. 'glandulosa', *P. nigra* × *P. maximowiczii*, *P.* ×*euramericana*, *P. deltoides* and *P. koreana* × *P. nigra* var. 'Italica' were cultivated at different salt concentrations (0.0, 0.1, 0.5 and 1.0% NaCl) for 60 days. Growth performance, dry biomass and the number of leaves were inhibited drastically with the increase of salinity over all species. The survival rate of the five different poplars reached 70% at 0.1% NaCl treatment, but most of the poplars died with serious visible injury at 0.5 and 1.0% NaCl treatments during the period of the treatment. Consequently, poplars seemed to survive at the lower concentration of 0.1% of NaCl. Height growth and dry biomass productivity of *P. deltoides* at 0.1% NaCl were higher than those of untreated trees. Na^+ contents in leaf, stem and root increased with increase in salinity. Na^+ contents in leaf and stem at 0.1% NaCl treatment were highest in *P. deltoides*. K^+ contents in leaf and root and Ca^{2+} content in root showed a tendency to decrease with increasing NaCl concentration (Yeo *et al.*, 1999).

Rooted cuttings of *P. alba* × *P. tremula* cv. 'glandulosa' and germinated seedlings of *Betula schmidtii* were planted in pots and irrigated with lead (Pb)-containing water for 60 days. In both tree varieties, growth inhibition was observed in 800 and 1500 ppm of $Pb(NO_3)_2$. Most of the Pb was accumulated in plant roots and only a small portion was transported to the shoots. The translocation rates of Pb for *B. schmidtii* and *P. alba* × *P. tremula* cv. 'glandulosa' were 1.6–2.6% and 1.2–1.6%, respectively. The maximum Pb content accumulated in shoots was 468.0 mg kg^{-1} dry weight in *P. alba* × *P. tremula* cv. 'glandulosa' and 602.0 mg kg^{-1} dry weight in *B. schmidtii*. Although tolerance to lead was generally higher in *B. schmidtii* than in *P. alba* × *P. tremula* cv. 'glandulosa', the highest tolerance to lead was observed in the *P. alba* × *P. tremula* cv. 'glandulosa' clone, '72-16'. We think that *P. alba* × *P. tremula* cv. 'glandulosa' and *B. schmidtii* look promising for phytoextraction based on their Pb uptake ability, high biomass production and suitability for large-scale cultivation (Yeo *et al.*, 2001).

6.5.7 New Zealand and Australia

B. ROBINSON

Introduction

Poplars and willows are ubiquitous in the New Zealand rural landscape. In Australia, the use of willows is limited due to the perception that they are weeds. The negative environmental effects attributed to willows are the obstruction and diversion of streams, displacement of native vegetation and an exacerbation of Australia's chronic water shortage due to their high rates of growth and transpiration (Cremer, 2003). In New Zealand, some willow species, especially *S. fragilis* and *S. cinerea*, have become weeds in wetland areas (Lester *et al.*, 1994).

The period 2000–2010 in Australasia has seen an increase in phytoremediation using poplars and willows. The trees are planted on contaminated sites, with the aim of reducing environmental risk. Commercial phytoremediation in Australasia employs poplars and willows as biopumps to reduce contaminant mobility and enhance the *in situ* degradation of some organic contaminants. Phytoremediation using poplars and willows offers a low-cost means of maintaining Australasia's 'clean-green' image abroad. Due to their rapid establishment and high evapotranspiration, poplars and willows are effective in reducing the water flux through contaminated material. This results in less contaminant moving off site and creates an aerobic environment in the root zone that favours the degradation of some pollutants. Poplars are particularly suited to remediation of organochlorines (Ferro *et al.*, 2000), polycyclic aromatic hydrocarbons (PAHs), excess ammonia and nitrates, and the immobilization of heavy-metal contaminants (Robinson *et al.*, 2003).

New Zealand's environmental legislation, the Resource Management Act (RMA) 1992, favours risk-reducing technologies such as phytoremediation. Regulation of contaminated sites is based on their effects on human health and the surrounding ecosystems, rather than on the levels of contaminants contained therein. Thus, phytoremediation using poplars and willows aids compliance with the RMA by reducing contaminant mobility, even if the total levels of contaminant in the soil remain unchanged. Unlike New Zealand, Australia has no overarching environmental legislation. Rather, disparate bills have been passed that address specific environmental issues. These may vary between states. Willows are being used for some wastewater and biosolids applications in Australia, however (Laidlaw *et al.*, 2012).

Most contaminated sites in New Zealand are associated with agricultural and silvicultural production. An estimated 50,000 disused sheep-dipping sites contain elevated levels of persistent pesticides such as dieldrin and sodium arsenate. Numerous sites associated with timber processing contain high levels of wood preservatives. In addition to agricultural and silvicultural contaminated sites, Australia has over 2 million ha of opencast mining and many contaminated sites associated with smelting and processing. Both countries face environmental issues associated with urban development, especially the disposal and treatment of sewage sludge and burgeoning landfills.

New Zealand has a temperate oceanic climate with a high rainfall. Meteorological conditions are ideal for poplar and willow growth, thus making phytoremediation viable for many contaminated sites. However, the high rainfall–evapotranspiration ratio limits the effectiveness of poplars and willows to eliminate contaminant leaching. On the other hand, Australia often suffers from drought and associated soil salinity, which affects plant growth negatively. Such conditions are ideal for poplars and willows to mitigate leaching. Phytoremediation is well suited for the extensive, low-value contaminated sites in Australasia. The low population densities of both New Zealand (14.8 people km^{-1}) and Australia (2.4 people km^{-1}) reduce the pressure for the rapid remediation of such sites. Here, we detail two phytoremediation case studies using both poplars and willows.

Case study 1. Phytoremediation of a contaminated wood-waste pile using poplar

Many New Zealand timber products contain pentachlorophenol (PCP), boron or copper-chromium-arsenic (CCA) to protect the wood against decay. High concentrations of these preservatives occur in wood treatment and wood-waste disposal sites, and pose a risk to receiving waters through leaching. One such site is located at the base of the Coromandel Peninsula, New Zealand (37.2°S, 175.6°E). Sawdust and yard scrapings dumped over 30 years from 1966 produced a 3.6 ha pile, with an average depth of 15 m. Geotechnical engineering ensures no surface or groundwater enters the pile. A holding pond collects leachate that results from high rainfall events. Vegetation had failed to establish naturally. Consequently, evaporation from the pile was negligible, indicated by the presence of saturated material at depths greater than 20 mm.

The annual rainfall of 1135 mm caused regular leaching from the pile into a local stream. Boron-rich leachate raised the stream concentration above the New Zealand Drinking Water Standard (1.4 mg l^{-1}), especially in the summer months when stream flow was low. The site thus violated New Zealand's effects-based

RMA and the local authority demanded that the site be remediated.

The landowners, in collaboration with the local authority, chose phytoremediation to manage the site. They based their decision on model predictions of the cost and effect of the remediation (Robinson *et al.*, 2003). In July 2000, Plant and Food Research Limited implemented a 1 ha trial using ten poplar and willow clones and two species of *Eucalyptus*. Two *P. deltoides* hybrid clones were the best candidates for phytoremediation based on survival, biomass production and boron uptake. There was a large variation in the tolerance of the trees to the wood-waste environment (Fig. 6.26).

In July 2001, the remainder of the pile was planted at a density of 7000 trees ha^{-1}. Fertilizers were added periodically. A pump recirculated leachate that occurred during the winter months and following high rainfall events. This leachate served as irrigation during the summer months. After 3 years of growth, the poplars had formed a closed canopy over 50% of the pile (Plate 21D). This figure increased to 80% the following year, with the tallest trees exceeding 8 m in height.

Before planting, the bare sawdust pile discharged boron-rich leachate during all months of the year. Following phytoremediation, the trees reduced the drainage to the 3 winter months. Summer is the greatest concern for waterway contamination because low stream flows result in less contaminant dilution. During winter, drainage may be released into a nearby stream at times of high flow, when the risk of exceeding the New Zealand Drinking Water Standard is minimal (Robinson *et al.*, 2007).

The poplar leaves at this site had average copper and chromium concentrations of 6.6 and 4.9 mg kg^{-1} dry mass. Arsenic concentrations were below detection limits (1 mg kg^{-1}). These low leaf concentrations will not facilitate the entry of CCA into the food chain. Before abscission, the average leaf boron concentration

Fig. 6.26. A field trial of poplar and willow clones as well as *Eucalyptus* and *Acacia* on a contaminated wood-waste pile, North Island, New Zealand. Note the range in biomass production and chlorosis. The trees in the photograph are 2 years old. Photo courtesy of B. Robinson.

was nearly 700 mg kg^{-1} on a dry matter basis, over 28 times higher than the boron concentration in the sawdust (40 mg kg^{-1} dry matter). While boron is a contaminant at high concentrations, it is also an essential plant and animal micronutrient (Banuelos and Ajwa, 1999) that is deficient for plant growth in many soils. Given the low concentrations of other contaminants, harvested poplar material could be applied to nearby horticultural land that is deficient in this element. Periodic coppicing of the poplars could therefore remove boron from the site. Since the leaves contain most of the boron, coppicing should occur before abscission.

The cost of phytoremediation was NZ$200,000, including a 5-year site maintenance plan. Scientific costs, including the trial planting material and chemical analyses, were higher than the planting and maintenance of the site. The costs of poplar and willow phytoremediation should decrease as an effective clone bank and better site management techniques are developed. Nevertheless, planting trials are essential to optimize the poplar clone and soil amendment regime, especially on non-soil media such as mine tailings. By comparison, the estimated cost of capping the site was NZ$1.2 million.

Case study 2. Phytoremediation of a disused sheep-dip site using willows

New Zealand has an estimated 50,000 disused sheep-dip sites. The surrounding soils contain elevated levels of arsenic, organochlorines and organophosphates. Land-use changes can result in the incorporation of these sites into residential or intensive agricultural zones. Robinson (1995) found a dieldrin plume in groundwater near one such site. Often, sheep dips were installed near wells or streams to facilitate preparation and disposal of the pesticide solution. The measurement of elevated dieldrin concentrations in a well led to the discovery of a disused sheep-dipping site in a nearby asparagus field. Soil analyses revealed dieldrin concentrations from 10 to 70 mg kg^{-1} over 100 m^2. The Dutch Intervention Value for dieldrin in soil is 4 mg kg^{-1} (VROM, 2000). Soil arsenic concentrations were not significantly higher than on adjacent areas, indicating that no arsenic pesticides were used at this site.

In September 2001, this 100 m^2 site was planted in willow clones. By January 2005, the average height of the trees was over 5 m.

Parallel experiments using soil collected from this sheep-dip site showed that willows effected a 20% degradation over a period of 5 months. However, this degradation rate cannot be extrapolated to estimate a clean-up time because the bioavailability of dieldrin bound to different soil fractions is unknown. The biological activity, as measured by dehydrogenase activity, was six times higher in the root zone of willows compared to a pasture species. Previous studies (Eriksson *et al.*, 2000) have shown that biological activity leads to a greater rate of decomposition of some contaminants. Phytoremediation on this sheep-dip site prevents the cultivation and harvesting of asparagus, promotes the degradation of dieldrin and reduces dieldrin leaching by enhanced evapotranspiration. The management of this site will include periodic coppicing of the willows to maintain a high stem density. Periodic soil analyses will reveal the long-term effect of the trees on the soil dieldrin concentration.

Little is known about the role of roots on the *in situ* degradation of organochlorine pesticides or PAHs. Remediation systems therefore have the potential to be enhanced greatly by developing clones that promote degradation or investigating the use of soil amendments that enhance the performance of poplars and willows in this role.

In both New Zealand and Australia, experimental and commercial phytoremediation using poplars has occurred on disused industrial sites with high soil loadings of lead, organochlorine pesticides and PAHs. In all cases, the primary role of the trees is to immobilize and, in some cases, degrade soil contaminants.

The growing populations and economies of Australia and New Zealand create new environmental pressures, especially due to intensive farming and urban development. With additional research combined with successful marketing, poplar and willow phytoremediation will be a valuable tool to combat future environmental degradation.

New developments and challenges

POPLARS AND WILLOWS AS NITROGEN SPONGES. Intensive dairy farming in New Zealand requires the disposal of large volumes of N-rich effluent that is often pumped on to pasture to improve growth. This can contribute to the contamination of receiving waters with nitrates. The ability of poplars and willows to coppice repeatedly makes them promising candidates for use in effluent irrigated systems, either for direct irrigation (Fig. 6.27) or as riparian buffer zone plantings to capture nitrogen from seepage or re-irrigated tile drainage. As with poplars and willows used for hillside stabilization, harvested tree material from effluent-irrigated blocks may provide valuable stock fodder.

A field trial was conducted on a dairy farm in southern Wairarapa in September 2001. Three blocks (225 m^2), each of Argyle (*P. deltoides* × *P. nigra*) poplars and Tangoio (*S. matsudana* × *S. alba*) willows, were planted as 1.2 m stakes. The trees were coppiced annually and the experiment concluded in the autumn of 2004. One block of each species was irrigated with fresh dairy effluent at rates of 2.5 mm and 5 mm a week. One block received no effluent. The biomass production of both species increased two- and threefold on the respective effluent application of 2.5 mm and 5 mm a week. This rapid growth resulted in soft lignified tissue, with a low density. While such material is unsuitable as timber, the soft tissues may improve palatability for stock fodder. At the highest treatment, the trees removed nitrogen at a rate of over 400 kg ha^{-1} $year^{-1}$, over twice the rate of pasture growing under similar conditions (150 kg ha^{-1} $year^{-1}$). Small blocks of poplars and willows may be more effective than pasture at reducing N-leaching because of their deeper root systems and re-evaporation of some incident rainfall (Roygard *et al.*, 1999).

NEED FOR NEW VARIETIES OF POPLARS AND WILLOWS. Poplar and willow breeding has focused on selecting for drought tolerance, disease resistance, lowered wind damage and possum (a marsupial herbivore) resistance (McIvor, 2008). However, future breeding could also focus on developing varieties for biomass production in degraded environments. This not only includes sites earmarked for phytoremediation but also marginal agricultural lands where the trees may be used as stock fodder. Varieties may be developed that take up limited amounts of cadmium while accumulating high concentrations of essential trace elements such as zinc and cobalt. Analyses of various poplar and willow clones have shown that high accumulation rates of essential trace elements are not necessarily correlated with high cadmium uptake (Robinson *et al.*, 2005).

Fig. 6.27. Dairy shed effluent irrigation on to a stand of *Salix matsudana* × *S. alba* near Carterton, North Island, New Zealand. Photo courtesy of B. Robinson.

Conclusions

Poplars and willows have proven environmental and economic benefits in Australasia. There are a growing number of potential roles for these trees. Erosion prevention and riverbank stabilization are still the most important uses of poplars and willows (Section 6.3, this chapter). Emerging roles, with successful case studies, include phytoremediation, wastewater management, supplementary stock fodder, production of industrial chemicals and carbon sequestration. A limiting factor in the use of willows in Australasia is the perception that these trees are weeds and that native species should be replacing them in the landscape. Nevertheless, many of the aforementioned roles cannot be filled by native species within a short time frame, and within an agricultural system, which are often requirements. Research in 2010 focused on developing systems where poplars and willows were combined with native species to enhance the economic and ecological value of farming systems.

6.5.8 Sweden

P. ARONSSON, J. DIMITRIOU AND K. PERTTU

Background

Cultivation of short-rotation willow coppice was introduced in Sweden after the oil crises in the 1970s, with the intention of replacing fossil fuels by renewable energy sources. Comprehensive research, to identify fast-growing species to be grown intensively for energy purposes, suggested that willows in coppice systems were the most suitable for this (Sirén *et al.*, 1987). During the first decade, the land available for short-rotation willow crops was abandoned farmland and appropriate wetlands. In the middle of the 1980s, however, it became obvious that the surplus production of agricultural crops left an opening for the use of more productive farmland. The research then had also shown that willow cultivation on wetlands was extremely difficult for several reasons, e.g. low soil pH and spring frosts. During the first half of the 1990s, investigations had shown that willow crops were suitable also for phytoremediation applications (Aronsson and Perttu, 1994), and that a combination with biomass production for energy purposes was a cost-effective method for wastewater treatment (Rosenqvist *et al.*, 1997; Dimitriou *et al.*, 2009a).

Commercial short-rotation willow coppice crops in Sweden

About 12,000 ha of short-rotation willow coppice crops are grown on Swedish arable land using mainly different clones and hybrids of *S. viminalis*, *S. dasyclados* and *S. schwerinii* (Dimitriou *et al.*, 2011). In 2012, there was one ongoing commercial willow breeding programme in Europe. The breeding was initiated at SLU, Uppsala, Sweden, in the late 1970s and was then commercialized by Svalöf Weibull AB in 1987 (Lindegaard *et al.*, 2001). The breeding programme was later run by the Swedish company, Lantmännen Agroenergi AB, aiming at introducing new varieties (clones) on the European market. During 2005, some 30 clones were registered with breeder's rights by Lantmännen Agroenergi AB, of which seven varieties were being marketed in Sweden in 2012 (see further at http://www.agrobransle.se). In the UK, the IACR-Long Ashton at Bristol managed a willow breeding programme funded by the European Willow Breeding Partnership owned by SW Seed, IACR-Long Ashton and Murray Carter Ltd. This programme lasted from 1996 to 2003 and resulted in six new varieties released on the market with breeder's rights (Stig Larsson, Lantmännen Agroenergi AB, 2010, personal communication).

Willow cultivation is fully mechanized, from soil preparation, planting and management to harvesting and combustion. In the initial phase, approximately 12,000 cuttings ha^{-1} are planted in double rows, to facilitate management (weeding, fertilization, etc.) and harvesting (Dimitriou *et al.*, 2009b). Despite a considerable fertilization effect, Swedish commercial willow plantations are rarely fertilized (Mola-Yudego and Aronsson, 2008). This is most likely due to the high costs for fertilizers (around 1 kg^{-1} N) and the relatively low price for the willow chips (around 33–40 t^{-1} dry matter). However, municipal sewage sludge is applied to most willow plantations, usually 1 year after planting and then repeatedly after the harvests. This reduces the need for additional phosphorus fertilization, but also generates income for the farmers receiving the sludge (Dimitriou and Rosenqvist, 2011). The willows are harvested every 3–5 years, during the non-growing season, preferably when the soil is frozen, using specially designed machines, i.e. converted Claas Jaguar corn harvesters. The aboveground biomass is chipped on site, then stored or transported directly to and burned in combined heat and power plants.

Examples of large-scale phytoremediation systems in Sweden

MUNICIPAL WASTEWATER. Municipal wastewater, i.e. wastewater from flushing toilets, contains nitrogen and phosphorus and is in most cases a well-balanced nutrient solution suitable for fertilizing of plants. For sanitary reasons, however, non-food and non-fodder crops, such as willow cultivations, are preferred (WHO, 2006). During the 1990s, large willow plantations equipped with drip irrigation or sprinkler systems positioned low to the ground were established adjacent to wastewater treatment plants to improve the efficiency of nitrogen treatment while producing biomass irrigated with wastewater. It was assumed that with a growth rate of 10 dry t ha^{-1} $year^{-1}$ and a shoot nitrogen concentration of 0.5% (Aronsson, 2000), 50 kg ha^{-1} $year^{-1}$ of nitrogen would be removed from the field at harvest. Research has shown, however, that nitrogen retention in short-rotation willow coppice can be more than 200 kg ha^{-1} $year^{-1}$ because of denitrification and long-term binding of nitrogen in the soil (Mortensen *et al.*, 1998; Aronsson and Bergström, 2001; Dimitriou and Aronsson, 2004, 2011).

Wastewater irrigation of willows is practised in an elegantly designed system at Enköping (about 20,000 inhabitants) in central Sweden (80 km west of Stockholm) (Plate 21E). The nitrogen-rich wastewater from dewatering of sludge, which formerly was treated in the wastewater plant, is now distributed to an adjacent 75 ha willow plantation during the growing season (Dimitriou and Aronsson, 2005). The applied wastewater corresponds to a load of approximately 150 kg ha^{-1} $year^{-1}$ of nitrogen applied to the plantation by use of a drip irrigation system. The growth has been recorded annually and has been found to be about 10 dry t ha^{-1} $year^{-1}$. Nitrogen leaching to groundwater has also been recorded, and once the willows had established properly, leaching was negligible (unpublished data). Substantial efforts have also been made to quantify the N_2O emissions from the system.

LANDFILL LEACHATE. Landfill leachate (water that has percolated through a landfill) is usually treated together with municipal wastewater in conventional treatment plants. This is generally costly and involves high energy consumption. Therefore, landfill operators are becoming increasingly interested in alternative solutions for on-site treatment. One method is to aerate the leachate and then use it to irrigate short-rotation willow coppice, either on restored parts of the landfills or on adjacent arable fields. The main advantage of this method is the low establishment costs compared with conventional engineered systems (Rosenqvist and Ness, 2004).

A willow plantation established on a restored cover of the landfill decreases leachate formation by means of high evapotranspiration. A near-zero net discharge of landfill leachate *can* be achieved by recycling the leachate into a short-rotation willow coppice plantation, even in the humid climatic conditions of northern Europe. Simultaneously, hazardous compounds in the leachate, for example ammonium and a range of potentially toxic organic substances, are taken up by the willows or transformed and retained in the soil–plant system (Dimitriou and Aronsson, 2007). A high concentration of ammonium in water is an environmental hazard, but nitrification in the soil is usually highly efficient, and thus leaching and discharge of ammonium should be avoidable in such treatment systems. The high concentration of various salts (usually dominated by the seemingly harmless NaCl) often are more problematic. Most salts are not possible to treat in a soil–plant system. Instead, such salts can only be diluted in time and space, but eventually they will reach the recipient water body. Still, the salt concentration needs to be addressed when designing a treatment system, since plants may suffer from either too high ionic strength in the root zone or from direct toxic effect caused by plant uptake of, for example, sodium. In addition, spray irrigation of landfill leachate may cause leaf necrosis and should be avoided. There are clear clonal differences as regards the plants' ability to cope with high salt concentrations (Dimitriou *et al.*, 2006a), and this needs to be considered when selecting varieties for establishing a system for phytoremediation of landfill leachate.

There were about 20 sites in Sweden in 2012 where landfill leachate was used to irrigate short-rotation willow coppice in sprinkler or drip irrigation systems. For example, at Upplands-Bro in central Sweden, a system operated by the company, Ragnsells Avfallsbehandling AB, stores and aerates the landfill leachate in ponds and then pumps it into a 5 ha short-rotation willow coppice field, which is irrigated daily during the growing season with approximately 2–3 mm of wastewater (Dimitriou *et al.*, 2003). Results from irrigating willows with the leachate on both field and controlled conditions indicated that, with careful planning, successful treatment of leachate combined with enhanced biomass could be achieved (Aronsson *et al.*, 2010; Dimitriou and Aronsson, 2010). The landfill operator intended to expand the extent of on-site treatment and planned to treat the entire volume of landfill leachate by irrigation of mainly willows, but also conventional forest.

INDUSTRIAL WASTEWATER. Large quantities of industrial wastewater are produced in Sweden after wet storage (sprinkling) of wood in sawmills and pulp mills (Jonsson, 2004). Sprinkling is carried out in summer to protect stored wood from damage by insects and fungi and from drying cracks. A medium-sized sawmill in Sweden consumes approximately 100,000 m^3 of water annually for watering the stored wood, and large amounts of runoff water from log yards need treatment. This is due to the fact that it contains a range of organic compounds, originating from the tree bark, as well as substantial amounts of phosphorus originating both from the bark and from soil particles attached to the logs or to the tyres of the trucks transporting the logs. Wastewater produced after rainfall or snowmelt can pollute neighbouring catchments or groundwater if it is left untreated. Until recently, in most cases such water has been disposed of in rivers or lakes. Treatment by irrigation of trees or perennial grasses has been tested as an alternative to constructed wetlands. Sandy soils are especially capable of retaining dissolved organic compounds, whereas the retention of phosphorus was higher in a clayey soil (Jonsson *et al.*, 2004). In the long run, the accumulated load of phosphorus needs to be considered since a soil might be phosphorus-saturated and start leaching considerable amounts to ground and drainage water.

SEWAGE SLUDGE AND WOOD ASH. Most Swedish willow plantations are fertilized regularly with municipal sewage sludge. This takes place after harvest, when spreading can be undertaken using ordinary agricultural machinery. Application of sludge to willow plantations is far less problematic compared to application to food crops, due to sanitary and public esteem reasons. In addition, the heavy metal concentration of Swedish sewage sludge used to be high, resulting in a build-up of the soil pool of metals when applied to arable land, and this has affected public opinion about sludge use in agriculture.

Usually, sewage sludge is highly imbalanced in terms of plant nutrients, with much higher phosphorus (P) content in relation to nitrogen (N) and potassium (K). The N is also mainly organically bound and contributes little to the N supply of the plants. Wood ash from the combustion of various wood fuels, on the other hand, contains both K and P but hardly any N. Thus, mixing sewage sludge with wood ash will result in a more balanced PK-fertilizer, which can replace conventional fertilizers (Dimitriou *et al.*, 2006b; Adler *et al.*, 2008). Ultimately, any biofuel-based energy system must include recycling of plant nutrients in the ash in order to be sustainable.

Despite a dramatic improvement of the heavy-metal concentration in Swedish sewage sludge, it still poses a long-term problem. From the point of view of human health, cadmium (Cd) is the most problematic metal. Willows have been shown to take up and accumulate substantial amounts of Cd, and this fact has attracted much attention in the perspective of phytoremediation of contaminated soils (Perttu *et al.*, 2002). However, operators of heating plants using willow chips were concerned because of this, since the result could be wood ash of lower quality. Recent research has shown that growing willows results in a net removal of Cd from the soil which is in the order of 5–10 g ha^{-1} $year^{-1}$ (Hasselgren, 1999; Klang-Westin and Perttu, 2002). The current regulation on the application of sewage sludge to farmland allows an annual Cd load of 0.75 g ha^{-1} $year^{-1}$. Thus, Cd does not pose a long-term problem in sludge-fertilized willow plantations. In fact, theoretically, during the estimated 25-year lifespan of a willow plantation, the net Cd removal can bring soil Cd levels back to pre-industrial levels (Perttu *et al.*, 2002). The situation is not as positive as regards the other heavy metals. These are not nearly as problematic for human health as Cd, but could pose long-term sustainability problems. These metals are not taken up by the willow plants as efficiently as Cd. The uptake of metals has proved to be clone specific (Landberg and Greger, 1996), and thus there might be opportunities for breeding efforts towards maximized metal uptake.

Cadmium and other heavy metals will remain in the different ash fractions (mainly in the fly ash) in the heating plant and will need further attention to be recycled back to arable land. It is technically relatively easy to clean ash from heavy metals, but this environmental service is not being paid for today and therefore heavy-metal-contaminated fly ash (normally about 15% of the total ash produced) is usually disposed of in landfills.

Conclusions

When used for phytoremediation, short-rotation willow coppice crops offer advantages, such as high biomass yields and removal of hazardous compounds through frequent harvests. The high evapotranspiration rate and root tolerance of willows to flooding conditions allow high irrigation rates. In addition, short-rotation willow stands are capable of restoring polluted sites by taking up substantial amounts of heavy metals such as cadmium, as well as retaining large amounts of nutrients in the soil–plant system. Besides removing hazardous compounds successfully, willow coppice phytoremediation systems utilize the nutrients and water applied to increase biomass production. Large-scale systems provide ecologically sound and economically competitive alternative treatment solutions.

6.5.9 UK

Wastewater and Biosolids

A.R. McCRACKEN

INTRODUCTION

Fast-growing energy crops such as SRC willow (*Salix* spp.) and short-rotation forestry (SRF) poplar (*Populus* spp.) are particularly well suited for phytoremediation and offer opportunities for the management of high nutrient wastewater streams and biosolids in Europe (Aronsson and Perttu, 2001). Their rapid growth rate and uptake of large volumes of water potentially enables SRC willow and poplar to absorb nitrogen (N), and to a lesser extent phosphorus (P), from the soil. Furthermore, willow can also effectively take up heavy metals such as zinc and cadmium (Riddell-Black, 1994). This characteristic has been utilized for the extraction of cadmium and zinc from contaminated brownfield sites and has also been achieved using *Salix*, *Populus* and *Alnus* in England, UK (French *et al.*, 2006). In Sweden, SRC willow systems have been developed at several sites for the commercial management of pollutants (Mirck *et al.*, 2005)

EFFLUENTS (WASTEWATER)

Nutrient removal. Energy crops, and especially SRC willow, have high water use, due to their long growing season and relatively deep rooting systems (Jørgensen and Schelde, 2001). This, along with good nutrient-use efficiency (NUE), makes willow an ideal candidate to be irrigated with high nutrient effluents. In an early trial in Poland, Kowalik and Randerson (1994) irrigated four willow species with four levels of municipal wastewater at Osobowice near Wroclaw. For all four species, there was an increase in yield following irrigation, although at the highest level of effluent application (equivalent to >1120 kg N ha^{-1}) the differences were not significant, indicating that the plants may have become nutrient saturated. They suggested that at this particular site, an irrigation load of 1000 mm $year^{-1}$ (525 kg N ha^{-1}), i.e. 50 mm $week^{-1}$, as a maximum, would provide an adequate supply of nutrients for plant growth and also increase the efficiency of N and P removal and improve the quality of effluent leaving the plot. Often, a limiting factor on how much wastewater can be irrigated to a site will be hydraulic loading and plant evapotranspiration rate. In Northern Ireland, due to a relatively high rainfall and high soil water retention, the maximum amount of effluent which can be applied at most sites is around 650 mm $year^{-1}$, although this can vary significantly, depending on specific site characteristics.

In Sweden, when willow stands were irrigated with landfill leachate at an equivalent rate of 1600 kg N ha^{-1} $year^{-1}$, there was a 93% reduction in the N content of the leachate over a 10-year period (Duggan, 2005). Willow stands were responsible for a 96.8–99.9% ammonium removal efficiency and a 43.4–93.3% reduction in total nitrogen in leachates (Hasselgren, 1998), with nitrogen removal being estimated to be 100 kg N ha^{-1} $year^{-1}$ (Borjesson, 1999). Similarly, Brierley *et al.* (2001) reported that SRC willow removed 90% of the mass of N from landfill leachates. In a study in Wales, leachate was added to SRC willow at a rate equivalent to an N concentration of 225 kg N ha^{-1} $year^{-1}$. Based on an N content of 0.5–0.7% N of harvested willow biomass (Scholz and Hellebrand, 2003) and an annual harvest of 10 dry t ha^{-1} $year^{-1}$, this equated to an annual offtake of between 50 and 70 kg N ha^{-1} $year^{-1}$ (Jones *et al.*, 2006). This result would indicate a significant imbalance in N inputs and outputs. However, when a full mass balance of nutrient flows is undertaken for SRC willow, many other routes of N metabolism are identified (Jones *et al.*, 2006).

Polishing. In 1998, an SRC willow plantation was established adjacent to the wastewater treatment works (WWTW) at Culmore, County Londonderry, Northern Ireland, as part of the EU-Fair5 project. Six treatments were imposed: (1–3) three rates of wastewater irrigation from the neighbouring WWTW; (4) clean water; (5) sewage sludge; and (6) zero application control. A full description of the project and the results from 4 years' data are given in Larsson *et al.* (2003). The growth of the willow was increased significantly by the nutrients available in the wastewater. Analysis of both the groundwater and soil water suggested that the impact of treatments, particularly at high levels of nitrogen and phosphorus, was limited even when the application of water and nutrients exceeded the requirements of the plants (Table 6.7). When the wastewater treatment effects were calculated using a mass balance technique on the willow–soil system, it was estimated that 67–74% BOD, 52–75% total nitrogen and 90–98% total phosphorus were removed. The report (Larsson *et al.*, 2003) concluded that the management of a wastewater irrigation system according to water and nutrient requirements of the SRC willow was possible without any negative environmental impacts with regard to oxygen-demanding substances and eutrophying components. However, elevated nitrogen leaching may occur from willow coppice after irrigation with very nitrogen-rich wastewater applied over a short period, e.g. worst-case conditions of 320 kg N ha^{-1} during 8 days (Dimitriou and Aronsson, 2004).

When using untreated primary municipal effluent there are genuine risks in relation to

Table 6.7. Concentration of constituents (means July 1999–April 2000) in superficial groundwater at Culmore, Northern Ireland (mg l^{-1}). Mean values of wastewater (WW) and pure water (PW) are included for comparison (Larsson *et al.*, 2003).

	WW	1 PE[a] WW	2 PE WW	3 PE WW	PW	1 PE PW	Sludge	Control
pH	6.9	6.4	6.3	6.4	7.1	6.3	6.4	6.4
BOD	106	32	35	30	3.6	31	31	31
COD	245	171	149	196	13	126	119	177
N-total	19	6.5	4.5	3.6	2.7	4	4.8	3.3
NH_4-N	18	1.6	1.6	1.6	1.8	1.6	1.7	1.5
NO_3-N	0.53	4.9	2.9	2	0.92	2.4	3.1	1.7
P-total	12	1.3	1.3	0.89	0.02	1	1.3	1.25
PO_4-P	2.0	0.57	0.5	0.65	–	0.48	0.49	0.57
K	11	3.8	4.4	4.3	1.9	3.3	5.2	2.3
Cr	215	91	99	149	24	56	122	58
Cd (μg^{-1})	0.018	0.018	0.017	0.016	0	0.016	0.017	0.014
Pb (μg^{-1})	0.15	0.16	0.28	0.18	0.22	0.18	0.19	0.16
Zn (μg^{-1})	120	70	69	110	25	67	58	70
Cu (μg^{-1})	15	40	39	51	7	40	52	30

[a]PE, Potential evapotranspiration.

transmission of animal or human pathogen via aerosols or contamination of the plantation water systems, including surface water, due to high hydraulic loads. At the Culmore site, there was some contamination of the superficial groundwater by indicator faecal microorganisms, which escalated with increasing application rates of the wastewater. Some of this was related to soil texture that had a high proportion of clay. In soils with high clay, cracks can occur, which allows for the rapid and unhindered transportation of water through the soil profile.

A further trial was established in 2002 close to the WWTW in Culmore as part of a EU Life project called 'Water Renew' (http://www.afbini.gov.uk/ANSWER). Primary effluent was applied to SRC willow, poplar and, for comparison, grass. Nitrogen uptake and removal from the system were monitored carefully. There was no significant movement of nutrients through the soil in the poplar or willow plantations at the rate of application, that is approximately 250 kg N ha^{-1} (Werner and McCracken, 2008). Grass plots are less secure, especially at the end of the growing season. The ability of the woody crops to deal with the high volumes and high nutrient content of the effluent during the period when the soils are virtually saturated and the plants are not actively transpiring is an important part of the investigation.

Economics. An economic assessment of the use of wastewater irrigation of willow in Northern Ireland (Rosenqvist and Dawson, 2005) concluded that the added value of the bioremediation of wastewater using SRC willow had the potential to radically alter the economic sustainability of the crop. Similar results had been reported from Sweden some years previously (Rosenqvist *et al.*, 1997), particularly allowing for the large cost of the conventional removal of nitrogen and phosphorus from effluent.

Despite all of the benefits of using willow for the cost-effective treatment of nutrient-rich wastewaters, uptake of the technology has been limited even in countries such as Sweden. The reasons for this have been due to various barriers such as institutional, structural and technical/geographical (Börjesson and Berndes, 2006). As issues including the implementation of the EU Nitrates Directive (91/676/EC), the EU Water Frameworks Directive (2000/60/EC), increasing concern about climate change and the difficulties of handling waste become more pressing, the use of SRC willow as a means of bioremediation should become more widespread. In many situations, SRC willow is best suited to the tertiary treatment of effluents from small inefficient treatment works or septic tanks.

BIOSOLIDS

Composition. Biosolids (sludge) are the solid residue generated during the treatment of domestic sewage in a water treatment works. Municipal biosolids typically have a dry matter content of just over 30%, a neutral pH and relatively high levels of nitrogen and phosphorus (Table 6.8). Levels of other elements including

Table 6.8. A typical analysis (based on dry weight) of biosolids from Culmore Wastewater Treatment Works taken in September 2006.

Dry matter	31.34%	Calcium	8158 mg kg^{-1}
pH	7.0	Magnesium	1059 mg kg^{-1}
Conductivity	1337 mS cm^{-1}	Potassium	8334 mg kg^{-1}
Ash	29.59%	Phosphorus	9250 mg kg^{-1}
Nitrogen (TON)	0.815%	Sodium	539 mg kg^{-1}
Ammonia (DM)	0.858%	Aluminium	7592 mg kg^{-1}
Nitrogen in DM	2.231%	Boron	9.9 mg kg^{-1}
BD	695 q l^{-1}	Cadmium	0.4 mg kg^{-1}
		Cobalt	4.9 mg kg^{-1}
		Copper	69.8 mg kg^{-1}
		Iron	4330.3 mg kg^{-1}
		Lead	32.2 mg kg^{-1}
		Manganese	121.2 mg kg^{-1}
		Molybdenum	0
		Zinc	35.2 mg kg^{-1}

heavy metals may vary considerably, depending on the intake of the water treatment works, and will change with seasonal factors.

Disposal. Currently, routes for the disposal of biosolids include incineration, landfill, anaerobic digestion, and in some countries, application to land. There is, however, enormous potential in applying biosolids to fast-growing energy crops such as SRC willow, and there are many environmental benefits of applying biosolids to SRC willow, which include:

- diversion of organic waste from landfill, incineration and transportation over large distances;
- diversion of the food safety fears associated with the recycling of biosolids to food and feed crops;
- enhanced energy security through displacement of heating oil and fossil fuels used for the manufacture of artificial fertilizer;
- provision of zero carbon renewable fuel and the associated reduction in CO_2 emissions;
- enhanced soil carbon sequestration through willow root biomass and soil injection of organic waste;
- new agricultural diversification, provision of rural employment and reconnection of urban–rural divide;
- increased biodiversity;
- compliant and sustainable waste management; and
- improved water quality.

Injection into SRC willow. In a commercial trial in Northern Ireland, untreated municipal biosolids have been injected into the soil using a specifically adapted piece of machinery (Fig. 6.28). A track is opened up to a depth of approximately 20 cm and the biosolids extruded into it. A second attachment immediately covers the biosolids so that at no time is it on the surface, reducing odours to zero and significantly reducing health risks from pathogens. A loading of 74 t ha^{-1} can be applied in a single pass, which is best done in the first year, soon after coppicing. Work is currently being undertaken to increase the clearance height of the machine so that biosolids can be applied at any time during the first year of regrowth without significant damage to the plants. In the Northern Ireland trial, biosolids have been applied to SRC willow at the following (equivalent) rates: 37 and 74 t ha^{-1} in a single pass in July 2005, 118 t ha^{-1} in two passes in July and December 2005 and 128 t ha^{-1} in three passes in July and December 2005 and March 2006. The water in boreholes positioned in each plot is sampled at monthly intervals and

Fig. 6.28. Biosolids injection into freshly coppiced SRC willow. Photo courtesy of Rural Generation Ltd.

structured soil samples taken every 6 months. By the end of 2006, no significant increases in water or soil nitrogen or phosphorus were observed. Similarly, there has been no evidence of a build-up of coliform bacteria in the soil or soil water.

In a recent study in Sweden, a mixture of biosolids and wood-ash mixtures were applied to short-rotation willow coppice (Dimitriou *et al.*, 2006b). The application of biosolids/ash mixtures resulted in changes to the soil pH and subsequently to the metal solubility in the soil. A decrease in solubility, particularly of Cd, was not, however, correlated with reduced uptake of the metal by the willow. The effectiveness of SRC willow in accumulating metal contaminants is a combination of the ability to absorb them from the soil and a function of the amount of biomass they produce. Differences in uptake of heavy metals could be related directly to the genetic differences between willow genotypes, and it has been suggested that some of the newer, faster growing and more productive genotypes may be even more effective in the bioremediation of heavy metals (Dimitriou *et al.*, 2006b). Furthermore, as mentioned previously, there could be added benefits in changing to a 2-year harvest cycle, thus increasing the off-take of biomass and contaminants from the site. However, it should also be noted that Cd concentrations will be greatest in the foliage and so the crop may have to be harvested while still green in order to maximize off-take of the contaminants from the site.

The application of biosolids offers a very significant economic benefit for the growing of SRC willow, through a gate fee paid by the water treatment works for disposal of the material. In addition, the willow has shown significant yield increases in response to readily available nitrogen that may offset the yield penalties of moving to a 2- rather than 3-year harvest cycle. Changes in local, national or European legislation, however, may mean that, in the future, biosolids have to be pretreated before application to the soil. Pretreatment could involve heat to kill pathogens or perhaps storage for 3 months, again to effect a pathogen kill. Treatment of biosolids with lime to raise the pH to 12 will result in the destruction of pathogenic bacteria after 24 h (Grabow *et al.*, 1978). There may be serious implications to plant growth from applying biosolids with such a high pH to the soil.

Nitrogen and phosphorus. The information on N requirements of a productive SRC willow plantation is quite varied. However, Labrecque and Teodorescu (2003) calculated that with a yield of 20 dry t ha^{-1} year^{-1} in the third rotation of *S. viminalis*, it would require around 180 kg N annually. If excessive nutrients are applied to a crop, whether in an inorganic form, in semi-solid materials or in liquids, the key risks are:

- leakage of nitrogen (N), which could contaminate the groundwater systems; and
- unacceptable build-up of phosphorus (P) in the soil and/or phosphorus runoff contaminating waterways.

In many of the studies carried out on the fate of nitrogen and/or phosphorus in SRC willow and other SRF crops, the nutrients have been in inorganic form. In these situations, leaching of N has been reported, especially in the establishment year, e.g. Labrecque *et al.* (1997), Hasselgren (1998), Mortensen *et al.* (1998), Alker (1999), Aronsson *et al.* (2000), Aronsson and Bergström (2001), Dimitrou and Aronsson (2004), Godley *et al.* (2004), Goodlass *et al.* (2007), Werner and McCracken (2008) and McCracken *et al.* (2009). In established crops, there are consistently low levels of leaching of N from SRC willow, even when quite high N additions have been made (Mortensen *et al.*, 1998; Aronsson *et al.*, 2000; Heller *et al.*, 2003). The dangers of leakage of N from biosolids are considerably less than from inorganic fertilizers or even effluents.

Nitrogen leaching. It is a legitimate concern that if N application rates are significantly surplus to crop requirements, there will be an accumulation of N in the soil, leading to the excess N leaching from the system and causing contamination of the groundwater. However, numerous studies have demonstrated that the levels of N leakage from established SRC willow are low to zero, in a range of soils, even at excessively high application rates far above normal crop requirements (Dimitriou and Aronsson, 2004). High fertilization rates (up to 153 kg N ha^{-1} year^{-1}) to willow growing on light textured soils resulted in remarkably low concentrations in groundwater at a depth of 2 m (Aronsson *et al.*, 2000). N application rates equivalent to 110–244 kg N ha^{-1} year^{-1} to SRC willow growing in lysimeters in a range of soils from clay to sandy had low to negligible leaching during the second and third year after establishment (Aronsson and Bergström, 2001). In a field study in Northern Ireland using municipal sewage effluent, Werner and McCracken (2008) detected no elevation in N levels in soil or groundwater over a 3-year irrigation period. Nitrogen retention, defined as the difference between nitrogen input through fertilization and nitrogen leaching in vegetation filter plants, is high (Aronsson, 2001).

Nitrogen loading. Not all the nitrogen applied to a crop will be taken up into that crop. In agricultural systems where nutrients are applied according to crop need, typical fertilizer use efficiency for nitrogen is 30%, i.e. 30% of the applied nitrogen is present in aboveground biomass. The other 70% is either present in roots, bound to soil organic matter, used by soil microbial biomass processes or lost through organic nitrogen mineralization and ammonia volatilization, nitrate leaching and denitrification.

The UK *Fertiliser Manual* (Anon., 2010) suggests a yearly off-take of 3 kg N t^{-1} biomass. At 30% availability for a 12 t ha^{-1} $year^{-1}$ yield, this is a nitrogen off-take of 120 kg N ha^{-1} $year^{-1}$. Other work suggests that the typical willow stem N content is 3–4 kg dry t^{-1} wood where no fertilizer has been applied. Stem N concentrations where nutrients are not limited have been measured from 6 to 25 kg N dry t^{-1}. Under these circumstances, annually exported N in harvested stems would be 60–250 kg, assuming a productivity of 10 dry t $year^{-1}$ and ha^{-1}. The lower range of 60 kg ha^{-1} $year^{-1}$ would suggest an application rate of 200 kg N ha^{-1} $year^{-1}$.

CONCLUSIONS

SRC willow and poplar offer realistic and practical approaches for the bioremediation of wastewater, effluents and sewage biosolids. A wide spectrum of *Salix* spp. and *Populus* spp. are efficient at water uptake and the utilization of high levels of nitrogen, which in turn results in increased biomass yields. While less effective in phosphorus uptake, willow and poplar may also have a role in the management of high phosphorus effluents. There is extensive evidence to show that willow is especially good at the extraction of cadmium and zinc from contaminated waste. However, there are issues as to where the heavy metals are accumulated and it may be necessary to harvest plants while leaves are still attached in order to achieve maximum off-take from the site. The application of biosolids to growing willow crops can be carried out during the first year of regrowth. By reducing the harvest cycle from 3 to 2 years, it will be possible to get maximum levels of biosolids to the crop. Short-term studies have indicated that there are minimal problems associated with the build-up of nutrients in the soil or with microbial contamination of groundwater. However, in the application of both biosolids and effluent there are concerns about the long-term impact on the soil, and it is essential that trials are monitored over prolonged periods. This is particularly important where biosolids have been pretreated, resulting in high pH and/or changes in structure or microbiological content.

Phytoremediation and carbon sequestration of degraded lands

N. DICKINSON AND D. RIDDELL-BLACK

Soils of brownfield, urban and industrial areas in the UK provide a large-scale opportunity to use phytoremediation, but the focus here should be on the more realistic possibilities of risk-managed phytostabilization and monitored natural attenuation. The wider practical applications of phytoremediation have a huge scope for cross-cutting other environmental agendas with synergies that involve the recovery and provision of services from degraded landscapes and contaminated soils. Additional focus on biomass energy, improved biodiversity, watershed management, soil protection, carbon sequestration and improved soil health is required for the justification and advancement of phytotechnologies.

Some highly productive non-hyperaccumulator species of plants such as species of *Salix* and *Populus* accumulate concentrations of metals in aboveground parts at significantly higher than the normal ranges found in most plants (Pulford and Watson, 1996, 2003). These plants may provide highly productive crops combined with significantly elevated metal concentration in aboveground tissues. We may be able to identify some genotypes that tend to accumulate high concentrations of more mobile trace elements. Particular examples have been identified already.

Trace elements of concern pertain both to the more phytotoxic elements, for example Zn, Cu, Ni, that may restrict plant productivity and to the more zootoxic elements, for example Pb, Cd, As, that may present a human health risk. Many previously developed brownfield or urban soils affected by waste disposal or aerial deposition from industrial fallout, transport or less obvious diffuse sources of pollution may also meet the just-above-the-contamination-threshold criterion for phytoextraction, but in this case it is generally only the zootoxins that are of concern. These have a very localized and heterogeneous spatial dispersion of pollutants, sometimes with an exaggerated perception of contamination risk (French *et al.*, 2006).

Of the non-hyperaccumulator species most studied in the context of phytoextraction, a combination of high productivity and elevated uptake of Cd and Zn has been identified in various taxa of *Salix* and of B and Se in *Populus* (McLeod and Ciravolo, 1998; Robinson *et al.*, 2000). Both genera are usually planted from stem cuttings, producing clonal plantations that would be expected to show reduced variability between plants compared to plants raised from seed, although this is hardly the case. In one study by French *et al.* (2006), a three-parent hybrid (*S.* ×*calodendron*) of *S. viminalis*, *S. caprea* and *S. cinerea* was identified that generally performed the best in terms of yield and metal uptake, but other studies have found elevated uptake in a range of different species and varieties – in fact, productivity seems to be the most important trait, as several willows in some situations appear to transfer relatively high concentrations of Cd and Zn to aboveground tissues (Dos Santos *et al.*, 2007).

ENHANCED UPTAKE OF METALS USING CHELATES. Many studies have now shown that metal uptake by plants can be accelerated by applying a range of chelating agents including EDTA and EDDS to soil contaminated with Pb and Cu and low molecular weight organic acids such as oxalic acid and citric acid to improve the removal of Cd, Cr and Ni. In one study, the application of EDTA and NH_4Cl to soils as mobilizing agents increased the uptake of metals into *Populus* but induced toxicity symptoms in the crop (Komarek *et al.*, 2007). In another, EDDS enhanced Cu, Cd and Zn uptake in *Salix* (Meers *et al.*, 2007). The same chemicals are well known to increase leaching of metals from soils, which may be unavoidable, thus potentially mobilizing metals towards groundwater. Recent reviews have argued that phytoextraction remain separate from chelate-assisted phytoextraction.

RISK MANAGEMENT AND PHYTOEXTRACTION. Environmental concerns are associated with enhanced mobilization of metals into plants. Food chain risk is the most obvious, and there is some evidence, for example, that cadmium transfer occurs to leaf-feeding invertebrates and to willow-feeding birds. However, higher trophic level accumulation of metals has otherwise seldom been found, except in animals feeding from the soil and the decomposer food web. Nevertheless, leaf fall always occurs to some extent during the growth cycle, prior to harvest, and this may transfer metals from deeper soil layers to the surface in the longer term. Harvest and subsequent disposal of phytoextractive plants potentially also poses onward problems: for example, combustion of biomass may disperse into the atmosphere, unless combustion temperatures are low and particulate emissions are controlled carefully. Incineration, ashing and other methods of contaminated crop disposal have been discussed elsewhere (Sas-Nowosielska *et al.*, 2004).

PHYTOSTABILIZATION OF BROWNFIELD SOILS. There is a much more extensive and well-established knowledge of how to mitigate toxicity in contaminated soil and how to establish a vegetation cover on contaminated land, with many case studies being well documented since at least 1970. A vegetation cover simply may reduce the wind blow of metal-contaminated soil as dust particulates, which can be the largest source of human health risk. Plant evapotranspiration may influence soil hydraulics sufficiently to prevent metals moving towards groundwater (Robinson *et al.*, 2007). In selecting plant species for phytostabilization, it has been shown that some species of *Populus* are unsuitable because of their uptake of high foliar concentrations of Cd (Mertens *et al.*, 2004). On the other hand, *S. caprea* is a stress tolerator and one of the first and only woody plants to colonize metal-contaminated land naturally in northern

Europe. It has been found to translocate high concentrations of Cd (116 mg kg $^{-1}$) and Zn (4680 mg kg $^{-1}$) to its foliage (Unterbrunner *et al.*, 2007).

CARBON SEQUESTRATION ON DEGRADED LANDS. Soil provides the main terrestrial storage of carbon, containing more carbon that the atmosphere and vegetation combined. The decline of stocks of soil carbon in agricultural land is well known and most obviously related to modern agricultural practices and less use of recycled organics. However, in older soils, especially where forests have been converted to agriculture, there appears to be a real risk that a fresh carbon supply stimulates the decomposition of carbon buried in deeper layers (Fontaine *et al.*, 2007). Various forms of organic matter amendments, including composts and sludges, form an important component of many site remediations of contaminated land which often contain younger soils without deep buried layers of organic carbon (Lal, 2007). Clearly, organic amendments have the potential to replenish depleted carbon stocks in degraded landscapes, thereby sequestering significant amounts of atmospheric carbon. Understanding the effects of organic carbon additions and different soil organic fractions on metal mobility has already received considerable attention, but modelling the retention of carbon is only just beginning to receive attention. It is important to understand how it is possible to encourage the long-term accumulation of the recalcitrant humified carbon compounds that become slow-cycling storage carbon.

6.5.10 USA

Overview

S. DOTY

Phytoremediation is the use of plants for the treatment of environmental pollutants. Plants act as solar-powered pump-and-treat systems as they pull up water-soluble contaminants through the roots and translocate them through the plant tissue, where they can be metabolized, sequestered or volatilized (reviewed in McCutcheon and Schnoor, 2003; Pilon-Smits and Freeman, 2006; Vangronsveld *et al.*, 2009; Dhankher *et al.*, 2011). Poplar and willow have been used successfully for the remediation of a variety of environmental pollutants (Rockwood *et al.*, 2004). The rapid growth, high biomass, extensive roots, low maintenance costs and adaptability of these tree species make them ideal plants for phytoremediation projects (Stettler *et al.*, 1996). For example, these trees have far more massive root systems than most herbaceous plants, reaching several metres. A 5-year-old poplar tree can take up 100 l of water day $^{-1}$; at this rate, 1 ha of poplar trees could remove 1.12 kg of a low-level contaminant (1 ppm) in just 1 year (Stomp *et al.*, 1994). Both poplar and willow can re-sprout after the aboveground biomass is removed (coppiced), with little disturbance to the site. This is advantageous because some inorganic contaminants such as metals could be harvested regularly, or flowering could be prevented. Not only does it have an inherent capability of taking up and metabolizing pollutants but also poplar is amenable to genetic transformation methods to enhance that ability drastically. In addition, both tree species harbour a wide variety of microorganisms that improve plant growth and may also assist in the metabolism of pollutants.

INHERENT ABILITY OF POPLAR AND WILLOW TO DEGRADE ORGANIC POLLUTANTS. Low molecular weight (MW) organic compounds such as trichloroethylene (TCE), carbon tetrachloride (CT), chloroform (CF) and benzene are serious environmental pollutants. Most are known or suspected carcinogens, neurotoxins and hepatotoxins. TCE is the most common pollutant at US Environmental Protection Agency Superfund sites, so it has received significant research attention. In field studies, it was demonstrated that hybrid poplar trees (*P. trichocarpa* × *deltoides*) can take up and metabolize TCE (Gordon *et al.*, 1997; Newman *et al.*, 1999) and CT (Wang *et al.*, 2004). When the poplar trees were exposed to TCE at levels typical of those in polluted groundwater, the trees were able to take up over 99% of the TCE. Less than 9% of the TCE taken up was transpired, as detected by leaf bag experiments. In order to determine if poplar cells had an inherent ability to degrade TCE and CT or if microorganisms were responsible for the degradation, studies were conducted with pure poplar cells in cell suspension cultures. When these poplar cell cultures were exposed to TCE, the same metabolites were seen that had been seen in the whole-plant studies (Newman *et al.*, 1997; Shang *et al.*, 2001; Shang and Gordon, 2002). Similarly, when

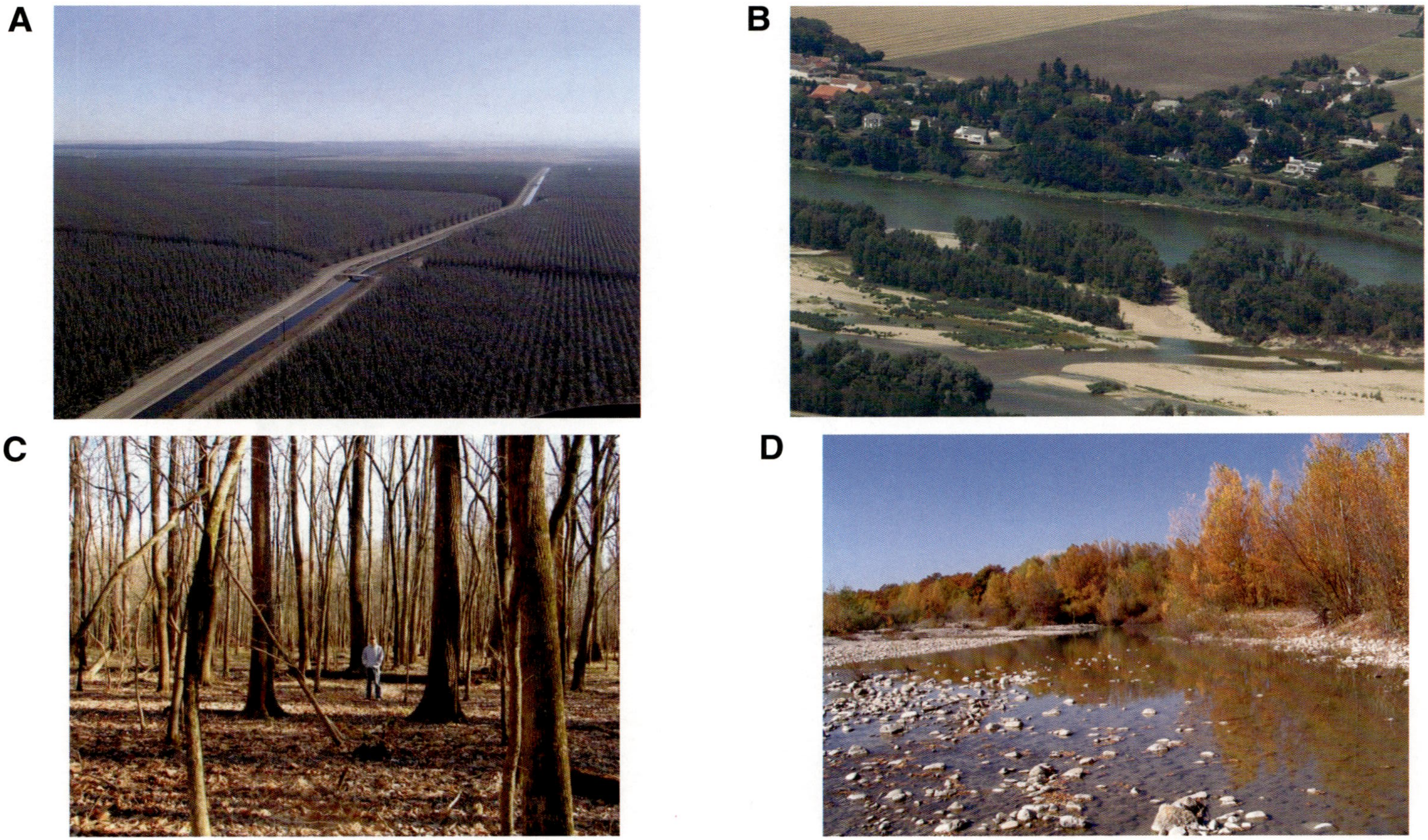

Plate 17. (A) Extensive poplar plantings near Boardman, Oregon, USA. Photo courtesy of Jake Eaton, GreenwoodResources. **(B)** *Populus nigra in situ* conservation unit (Island of Mareau aux Prés, Natural Reserve of St-Mesmin) along the Loire River near Orleans, France. *Populus* genetic resources are often fragmented by agriculture and other human activities. Conservation of this important species has worldwide implications. Photo courtesy of Catherine Bastien, The National Agricultural Research Institute. **(C)** Undomesticated populations are indispensable to the initiation of breeding and varietal development as well as gene conservation programmes. Good examples include *Populus deltoides* from the Mississippi River valley (left) and *Populus nigra* (right) from the Paglia River in central Italy. Photos courtesy of Randy Rousseau, Mississippi State University (left) and Maurizio Sabatti, University of Tuscia (right).

Plate 18. (A) Botanical characteristics of a representative poplar, *Populus nigra* (black poplar). 1, Vegetative twig showing rhombic preformed leaf and deltoid neoformed leaf; 2, twig with true terminal bud and male (staminate) catkins in anthesis; 3, male flower and anthers; 4, twig with true terminal bud and female (pistillate) catkins in anthesis; 5, female flower; 6, maturing two-valved capsules: dehiscing capsule (right) shows seeds with cottony coma; 7, seed. Modified from Thomé (1885). **(B)** Botanical characteristics of a representative willow, *Salix caprea* (goat willow). 1, Vegetative twig showing mature leaves and stipules; 2, expanding floral buds (note single bud scale); 3, twig with male (staminate) catkins in anthesis; 4, male flower; 5, twig with female (pistillate) catkin in anthesis; 6, female flower; 7, immature capsule; 8, dehiscing two-valved capsule; 9, seed with cottony coma. Modified from Thomé (1885). **(C)** Specimens of several poplar species, such as this *Populus deltoides,* can grow to large size and may live for several hundred years. Photo courtesy of R. Miller. **(D)** Dehiscing capsules of *Populus deltoides*; the cottony seed soon will take to the air. Photo courtesy of D. Dickmann.

Plate 19. (A) The most widely planted hybrid poplar taxon worldwide is *Populus xcanadensis* (*Populus deltoides* x *Populus nigra*), the Euramerican or Canadian poplar, shown here in Belgium. **(B)** The flattened, airfoil-like petioles of poplars in sections *Aigeiros* and *Populus* – shown here by *Populus sieboldii* – cause leaves to characteristically flutter in the wind. **(C)** The usual autumn coloration of poplar leaves is yellow or gold. **(D)** Poplars in section *Turanga* survive in very arid environments, provided their roots can reach the water table. This grove of *Populus euphratica* grows in the Taklamakan Desert in Xinjiang, China. **(E)** Riparian *Populus nigra* impacted by human activity, Turkey. **(F)** Like *Populus alba* shown here, diamond-shaped lenticels and warty outgrowths mark the smooth bark of many poplars. Photos courtesy of D. Dickmann (A, C, F), Maki Laboratory, Tohoku University, Japan (B), Yong-Ling Ow (D) and F. Toplu (E).

Plate 20. **(A)** A pistillate catkin of *Salix bebbiana* in anthesis. Unlike the wind-pollinated poplars, the flowers of most willows are pollinated by both insects and wind. **(B)** Staminate catkins of *Salix lucida* in anthesis. Note their upright posture and – in this species – the long, foliaceous floral branchlets. **(C)** Reproductive phenology varies among willows, e.g. capsule ripening and dehiscence of *Salix discolor* (left) occur in spring prior to leaf out, whereas in *Salix interior* (right) they occur during early summer after leaves are fully developed. **(D)** *Salix humboldtiana* growing in a seasonally flooded riparian habitat along the Paraná River, Argentina. **(E)** *Salix myricoides* is adapted to drought and heat stress; here it grows on a sand dune on the shore of Lake Michigan, USA. **(F)** Willows are popular ornamentals. Left: Another popular and colourful selection of white willow is *Salix alba* 'Chermesina'. Right: Intense pubescence on both sides of mature foliage produces the striking, silver appearance of *Salix alba* 'Argentea'. Photos courtesy of D. Dickmann (A, B, C) and J. Kuzovkina (D, E, F).

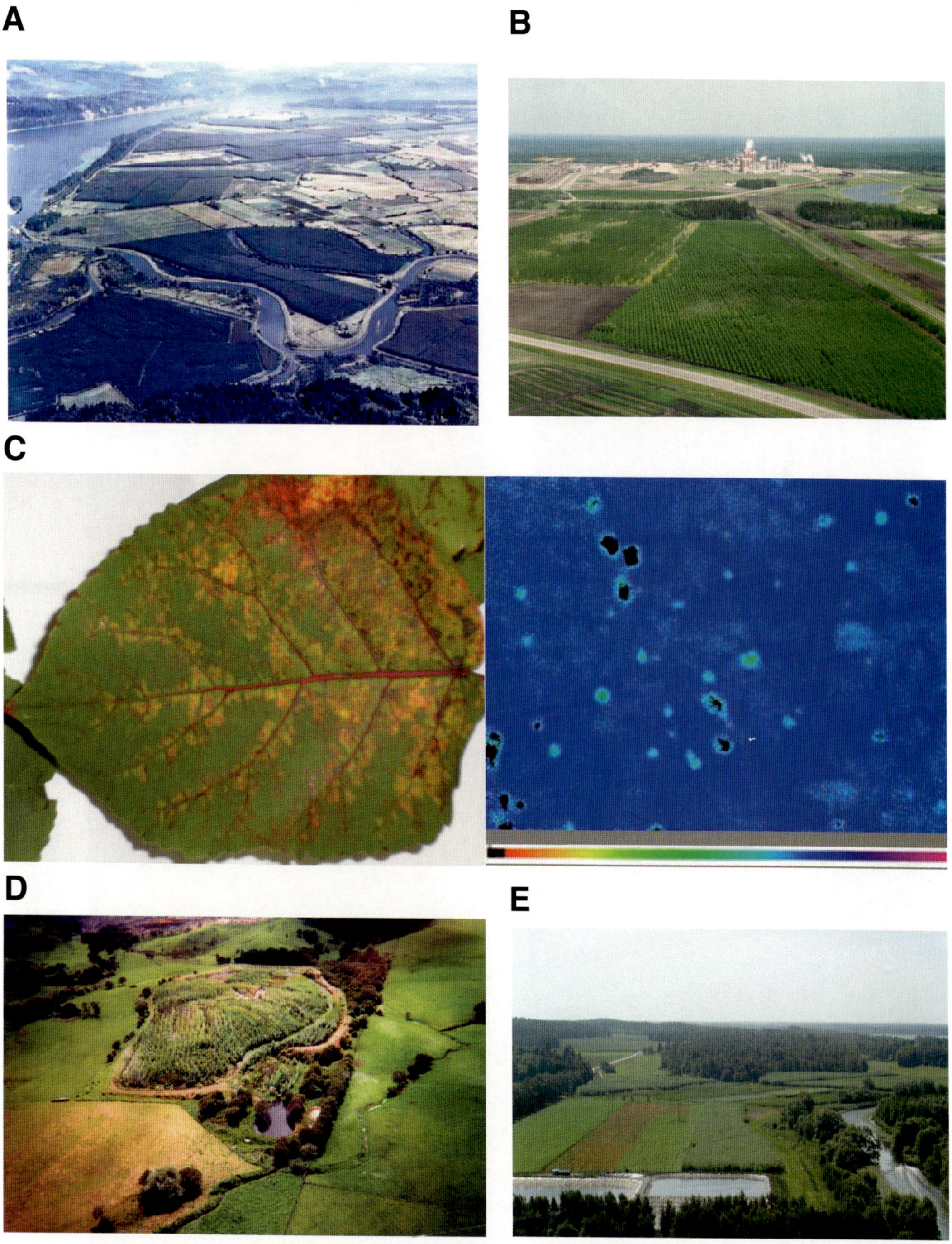

Plate 21. **(A)** Aerial view of intensively managed *Populus* x*generosa* plantations along the lower Columbia River flood plain west of the Cascade Mountains (Oregon and Washington, USA). **(B)** Large block planting of poplar on private land at Alberta-Pacific Forest Products mill, Canada. **(C)** Effects of cadmium treatment on poplar leaf (left) and chlorophyll fluorescence image (Fv/Fm) after dark adaptation of the same leaf (right). The false colour code depicted at the bottom of the image ranges from Fv/Fm = 0.00 (black) to Fv/Fm = 1.00 (pink). **(D)** Phytoremediation of wood-waste pile, New Zealand. The poplars, mostly *Populus xeuramericana* and *Populus alba* hybrids, are 3 years old. **(E)** Panoramic view of collection ponds for wastewater from wastewater treatment facilities in Enköping, Sweden and short-rotation willow coppice field irrigated with the water. Photos courtesy of GreenWood Resources (A), B. Thomas (B), G. Scarascia-Mugnozza (C), B. Robinson (D) and P. Aronsson (E).

Plate 22. (A) Succession of black cottonwood (*Populus trichocarpa*) and willow (*Salix* spp.) on a gravel bar in the Carbon River, Pierce County, Washington, USA. **(B)** The twelve treatment rings of the aspen FACE experiment are shown here. **(C)** Black cottonwood (*Populus trichocarpa*: left) and willow (*Salix* spp.: right) leaves showing smaller individual leaves on the left in each photograph resulting from poor fertility. Photos courtesy of Jon D. Johnson (A, C) and R. Anderson (B).

Plate 23. (A) *Melampsora* leaf rust on willow. **(B)** Defoliation of highly susceptible poplar by *Melampsora* leaf rust. **(C)** Leaf disease of poplar caused by *Melampsora medusae*. **(D)** *Melampsora medusae* uredinia on poplar. **(E)** Defoliation of poplar clone (left) affected by *Marssonina brunnea*. **(F)** Leaf spots caused by *Marssonina brunnea*. **(G)** Stem lesions caused by *Marssonina brunnea*. Photos courtesy of M. Ramstedt (A) and M. Ostry (B, C, D, E, F, G).

A

B

C

D

E

Plate 24. **(A)** *Venturia* shoot blight of poplar. **(B)** *Venturia* leaf blight. **(C)** *Fusicladium* willow scab. **(D)** Bronze leaf disease of hybrid aspen. **(E)** Leaf symptoms of bronze leaf disease caused by *Apioplagisostoma populi*. Photos courtesy of M. Ostry (A, B, D, E) and M. Ramstedt (C).

Plate 25. **(A)** *Septoria* canker on poplar. **(B)** *Septoria* leaf spot on poplar. **(C)** Stem breakage of poplar at *Septoria* canker. **(D)** Stem breakage at *Hypoxylon* canker. **(E)** *Hypoxylon* canker on aspen. **(F)** Perithecia of *Entoleuca mammata*. Photos courtesy of M. Ostry.

Plate 26. **(A)** Stem necrosis on willow caused by *Pseudomonas syringae*. **(B)** Young bacterial stem canker. **(C)** Old bacterial canker and rough bark of affected stem. **(D)** Bacterial droplet of a young stem canker. Photos courtesy of M. Ramstedt (A) and M. Ostry (B, C, D).

A

B

C

D

Plate 27. **(A)** Damage caused by leaf feeders – *Phratora vitellinae* larvae skeletonizing leaf. **(B)** Damage caused by leaf feeders – *Byctiscus populi* adult and feeding damage. **(C)** Damage caused by leaf miners – poplar leaves mined by individual *Phyllonorycter* sp. larvae. **(D)** Damage caused by leaf feeders – *Byctiscus populi* adult on 'rolled' leaf containing eggs. Photos courtesy of A. Delplanque (A, B, D) and L. Nef (C).

A

B

C

D

E

F

Plate 28. (A) Damage caused by gall formers. Galls caused by *Aceria parapopuli*. **(B)** Damage caused by gall formers. Galls caused by *Mordwilkoja vagabundus*. **(C)** Damage caused by gall formers. Galls caused by *Pemphigus spirothecae* on *Populus nigra*. **(D)** Damage caused by bud and young shoot feeders. *Rabdophaga rosaria* on *Salix myrsinifolia*. **(E)** Damage caused by bud and young shoot feeders. *Byctiscus betulae* on *Salix* sp. **(F)** Damage caused by bud and young shoot feeders. *Gypsonoma aceriana* damage to young poplar shoot. Photos courtesy of L. Nef (A, B), S. Augustin (C, D, E) and A. Delplanque (F).

Plate 29. (A) Disease vector, *Phytobia cambii.* **(B)** Disease vector, *Xyleborus dispar*. **(C)** Disease vector, *Rhytidodus decimus*. **(D)** Damage caused by mammals and birds, pileated woodpecker. Photos courtesy of M. Martinez (A), Daniel Adam, Office National des Forêts, Bugwood.org (B), A. Delplanque (C) and J. Charles (D).

Plate 30. (A) Biological control of insect pests. *Megarhyssa praecellens* parasitizing larva of *Tremex fuscicornis*. **(B)** Biological control of insect pests. Tachinid pupa from *Phratora* larvae. **(C)** Biological control of insect pests. *Arma custos* larva feeding on *Chrysomela tremulae*. **(D)** Biological control of insect pests. *Episyrphus balteatus* larva feeding on *Chrysomela populi* eggs. **(E)** Biological control of insect pests. *Beauveria bassiana* on *Chrysomela tremulae*. Photos courtesy of P. Parra Sanhueza (A) and A. Delplanque (B, C, D, E).

A

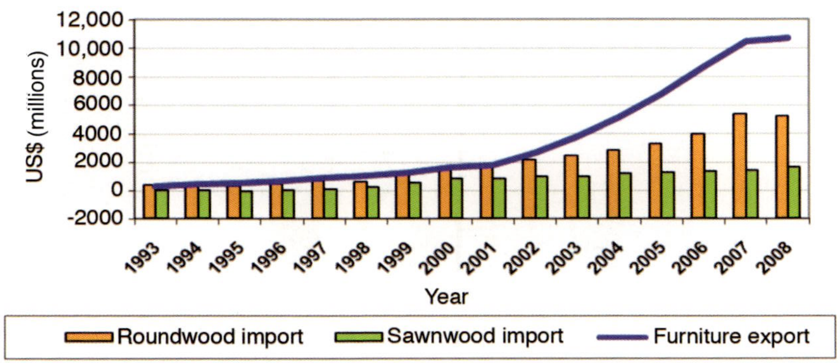

B

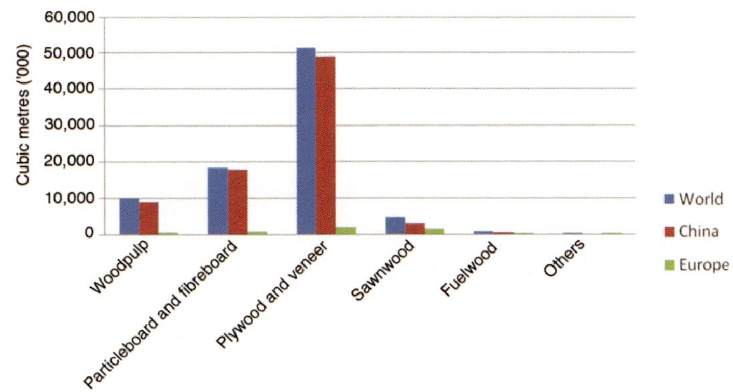

C

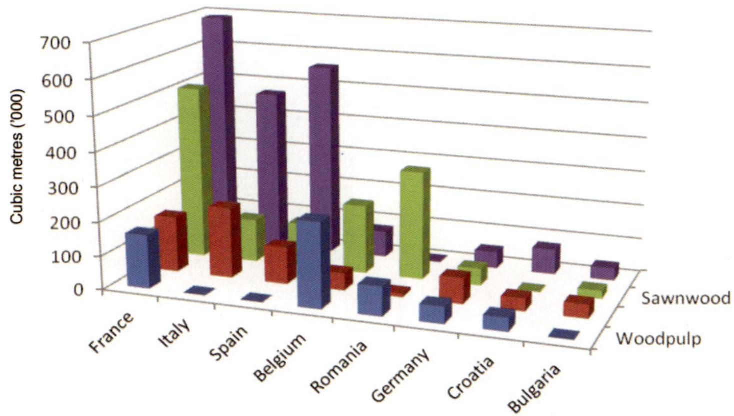

D

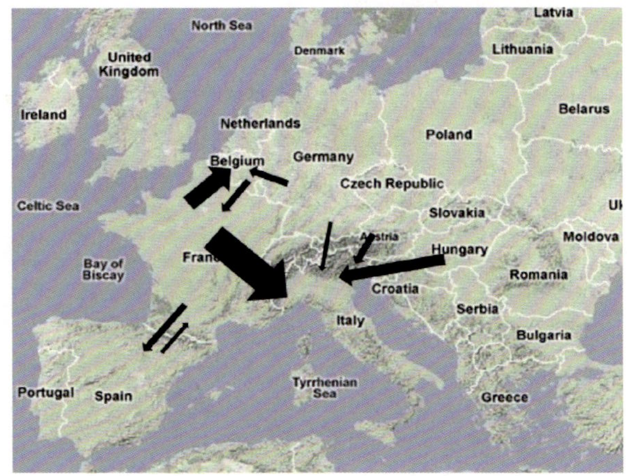

Plate 31. **(A)** Trade trend of roundwood, sawn wood and furniture in China. Data source: SFA (2006, 2009a). **(B)** Importance of China in the production of poplar products in the world. Data source: FAO (2008). **(C)** Production of processed poplar products, 2007. Data source: FAO (2008). **(D)** Major flows of poplar roundwood and sawn wood in Europe, 2010.

A

B

C

D

E

Plate 32. (A) The production of willow for various types of furniture and baskets. **(B)** Poplar shelterbelts protect against hot, dry winds and increase humidity, soil moisture and crop yields. **(C)** *Populus alba* and *Salix babylonica* shelter a remote government outpost at around 3000 m in the Andes of Argentina. **(D)** A living, green wall of poplar trees fends off the encroaching desert in Inner Mongolia, northern China. **(E)** Poplars (and willows), trees for society and the environment. Photos courtesy of Fairchild Farms, Canada (A), FAO/Three North Shelterbelt Bureau (B), FAO/J. Carle (C), J.E. Jacquot, http://www.treehugger.com/files/2007/06/living_green_wall.php (D) and FAO (E).

poplar culture cells were dosed with CT, metabolism was clearly evident. The metabolism of TCE and CT by poplar cells is similar to that in mammalian cells. In mammals, the first step in the pathway is initiated by the cytochrome P450 2E1 enzyme, resulting in the TCE metabolites, chloral, trichloroethanol and trichloroacetic acid, and the CT metabolites, chloroform and carbon dioxide (Wang *et al.*, 2002). When the poplar culture cells were exposed to radiolabelled TCE (Newman *et al.*, 1997) or CT (Wang *et al.*, 2002), low levels of radiolabelled carbon dioxide were produced, indicating that poplar cells had the capacity to mineralize these pollutants. Not only do the poplar and mammalian pathways result in the same metabolites but also the CT (Wang *et al.*, 2002) reaction and TCE (S. Doty, 2012, unpublished results) reactions are blocked by the same inhibitors as in mammals. Therefore, the reactions are carried out by similar enzymes.

Most phytoremediation studies that investigated removal of TCE used one or two genotypes of poplar. In a recent study funded by the US National Science Foundation, 9 poplar and 12 willow varieties were chosen for their previous success in phytoremediation efforts or local native significance, and experiments were conducted to compare toxicity, uptake and degradation of TCE (Miller *et al.*, 2011). Although many of the genotypes removed TCE from solution, there was a wide range in the ability of plants to degrade TCE. A wild willow clone showed the highest level of TCE metabolism. There was a six-fold range in the ability of five different clones of *P. deltoides* to degrade TCE. It was speculated that differences in the expression of key enzymes involved in TCE metabolism might explain the different abilities of varieties of the same species.

Another important class of environmental pollutants for which poplar and willow can be used for remediation are the explosives, 2,4,6-trinitrotoluene (TNT), hexahydro-1,3,5-trinitro-1,3,5-triazine (RDX) and octahydro-1,3,5,7-tetranitro-1,3,5,7-tetraazocine (HMX). More than 100 military bases and explosives manufacturing facilities in the USA are contaminated with these chemicals. The groundwater at these sites is contaminated; therefore, the health risk is spread beyond the military bases themselves (Rivera *et al.*, 1998). Research with aquatic plants demonstrated that plants could transform (metabolize) TNT in the absence of microorganisms (Hughes *et al.*, 1997). Both poplar and willow have been used in munitions remediation research. Hybrid poplar (*P. deltoides* × *P. nigra*) was able to take up TNT from hydroponic solution but did not seem to translocate it (Thompson *et al.*, 1998). Using radiolabelled TNT, the authors demonstrated that about 75% of the radiolabel remained in the root tissue even after 42 days, while 10% was translocated to the foliage (Thompson *et al.*, 1998). In soil, the hybrid poplar was less able to remove the TNT, leaving behind 75% of the TNT after a 20-day period. The hybrid poplar metabolized the TNT to 2-ADNT and 4-ADNT and to a number of unidentified compounds. In a study comparing phytoremediation of TNT by hybrid willow (*Salix* clone 'EW-20') and Norway spruce (*Picea abies*), it was shown that nearly half of the above-ground radiolabelled TNT was in bark-free wood of willow compared with about 60% of it in the older needles of spruce (Schoenmuth and Pestemer, 2004). Both tree species readily metabolized TNT.

Polycyclic aromatic hydrocarbons (PAHs) are also prevalent environmental pollutants. PAHs are characterized by low solubility, high soil sorption, hydrophobicity and long half-life for the higher MW PAHs. This class of aromatic hydrocarbons consists of three or more fused benzene rings in linear, angular or cluster arrangements. Some examples of the EPA priority PAHs and their ring structure include naphthalene, anthracene and phenanthrene (3 rings each), fluoranthene, pyrene and benz[a]anthracene (4 rings each), benzo[a]pyrene (5 rings) and benzo[ghi]perylene (6 rings). Since PAHs are lipophilic, adsorption to root surfaces may be another important first step in phytoremediation (Schwab *et al.*, 1998). Ballach *et al.* (2003), Wittig *et al.* (2003) and Kuhn *et al.* (2004) conducted a three-part investigation into the use of poplar cuttings for PAH removal. *P. nigra* cuttings caused a reduction in the amounts of the PAHs, anthracene, phenanthrene, pyrene, fluoranthene, chrysene and benzo[a]pyrene. An extensive field study was conducted using poplar trees to reduce PAH concentration in groundwater (Widdowson *et al.*, 2005). The researchers determined that concentrations began to fall at the time the poplar roots reached the saturated zone, approximately 1 year after planting, and that a variety of factors including rhizospheric microorganisms, plant uptake, phytovolatilization and biodegradation contributed to the decrease in PAH concentration.

Willow has also been used in phytoremediation of this class of pollutants, but with variable success. A stand of willow (*S. viminalis* L. Orm) was used in a field trial for remediation of a site contaminated with PAHs, mineral oil and heavy metals (Vervaeke *et al.*, 2003). After 1.5 years, the plot that was planted with willow removed 23% of the PAHs compared with the unplanted plot that removed 32%. Perhaps the failure of the willow to remove the PAHs at this site was due to the compounding problems of heavy metals and mineral oil. In a study using PAH-contaminated soil from a gas plant, Spriggs *et al.* (2005) determined that black willow (*S. nigra* Marshall) outperformed poplar, ash and the unplanted controls in the degradation of PAHs. In a hydroponic study, willow plants were exposed to a variety of PAHs. Naphthalene killed the plants, while benzo(a)pyrene and phenanthrene had no effect on willow growth. However, in a related study, several willow clones took up naphthalene readily but stalled after 3 days with phenanthrene and pyrene due to severe phytotoxicity (Z. Khan and S. Doty, 2012, unpublished results). Given the wide genetic diversity in willow species, it is not surprising that different varieties would have greatly different abilities in the tolerance and degradation of toxic pollutants.

Poplar and willow have been tested for the ability to take up and degrade a variety of other organic pollutants, including methyl-tert-butyl ether (MTBE), petroleum, ethylene dibromide, dibromochloropropane, pentachlorophenol, trichloroethane, formaldehyde and chlorinated benzenes and toluenes (Newman *et al.*, 1998). In a laboratory-scale phytoremediation study with willow (*S. babylonica*), Corseuil and Moreno (2001) found that willow cuttings could remove more than 99% of the ethanol and benzene in less than 1 week from hydroponics. In a screening study to select clones for petroleum remediation, poplar and willow had high survival rates in soil that was heavily contaminated with petroleum hydrocarbons. For example, the commercial poplar clones, 'NM6' and 'DN34', had 88% and 89% survivability, respectively, as 20 cm cuttings in the contaminated soil. The willow clones, 'Sx61' and 'SV1', had survivability of 63% and 49%, respectively. Overall, poplar and willow have excellent potential to be used successfully for the phytoremediation of a wide range of organic pollutants.

PHYTOREMEDIATION OF INORGANIC ENVIRONMENTAL POLLUTANTS WITH POPLAR AND WILLOW. Remediation of metals presents a different challenge, since the pollutants cannot be metabolized but must instead be translocated to the foliage, where it is harvested more easily or volatilized, such as in the case of mercury. Although most research in this area has focused on natural hyperaccumulating plants, both poplar and willow have been used with some success, as their higher biomass may compensate for the lack of hyperaccumulation ability. In a review by Pulford and Watson (2003), willow is specifically suggested for the phytoremediation of heavy-metal-contaminated lands. Since successful phytoremediation of inorganics relies on the ability of the plant to regrow readily after the upper foliage is harvested to remove the extracted metals, willow is especially well suited for this type of remediation. In a paper by French *et al.* (2006), *Salix*, *Populus* and *Alnus* were compared in a study on the remediation of brownfield land. The five willow clones and *Larix* were able to concentrate copper. Four of the *Salix* clones also concentrated cadmium and zinc up to 13 times higher than the soil concentration levels. Since willow is a fast-growing, high biomass tree, these data are encouraging that cadmium, one of the most serious metal pollutants, could be remediated successfully.

ENHANCING PHYTOREMEDIATION CAPABILITY USING TRANSGENICS. Although trees are capable of reducing the levels of pollutants at contaminated sites, the rates of pollutant removal are not high enough to be of practical value in many circumstances. Other reasons for the genetic engineering of plants for phytoremediation are that some pollutants are too toxic or at too high concentrations for the plants to survive; the plant species that can metabolize the pollutant are not suited for the climate or environment of the contaminated site; or there is no known plant that can remediate a particular chemical. In a modelling study of phytoremediation effectiveness, it was shown that the plantation size that would be required to remediate a plume of

TCE effectively could be quite large and was dependent on a variety of hydrologic factors as well as the length of the dormancy period of the trees (Matthews *et al.*, 2003). A variety of genetic strategies for enhancing phytoremediation have been proposed (Stomp *et al.*, 1994). A simple strategy is to infect the plants with *Agrobacterium rhizogenes*, a soil bacterium that transfers DNA encoding auxin synthesis genes into plant cells. Once the genes are incorporated into the plant genome, the plant expresses the auxins and root development is initiated (Costantino *et al.*, 1994). Poplar is susceptible to *A. rhizogenes*, resulting in a larger root mass (Pythoud *et al.*, 1987). However, these plants are often dwarfed with altered leaf morphology, limiting their practical use.

A more direct method for enhancing the effectiveness of phytoremediation is to overexpress in transgenic trees the genes involved in metabolism or transport of specific pollutants (reviewed in Doty, 2008; Dowling and Doty, 2009; Dhankher *et al.*, 2011). This can be readily achieved by using *Agrobacterium tumefaciens*-mediated plant transformation. Depending on the hybrid and particular clone, reasonable transformation frequencies can be achieved (Han *et al.*, 2000). Increasing the metabolism of the pollutant can cause a strong enough concentration gradient that allows the plant to remove far more of the pollutant than a non-transgenic plant removes. For example, when the cytochrome P450 2E1 gene (*CYP2E1*) was overexpressed in tobacco plants, the transgenics removed 98% of the ethylene dibromide, a substrate of P450 2E1, compared with 63% removal by the null vector control plants (Doty *et al.*, 2000). When the gene was overexpressed in hybrid poplar (*P. tremula* × *alba*), TCE metabolism was strongly enhanced (Doty *et al.*, 2007). Other substrates of P450 2E1 include carbon tetrachloride, benzene and chloroform. The transgenic poplar also removed these chemicals at greater rates than did the control plants (Doty *et al.*, 2007).

Phytoremediation of nitroaromatics is improved significantly with transgenic plants due to the phytotoxicity of these pollutants (Rylott and Bruce, 2008). When bacterial nitroreductase (nfsI) was overexpressed in tobacco plants, the transgenic plants were more tolerant to higher levels of TNT and could metabolize it at far greater rates than the control plants (Hannink *et al.*, 2001). Another bacterial nitroreductase gene (*pnrA*) was introduced into poplar, resulting in increased tolerance to TNT (van Dillewijn *et al.*, 2008). Another common explosive is RDX. A gene (*xplA*) from an RDX-degrading bacterium was introduced into *Arabidopsis* plants and the resulting transgenic plants tolerated and removed high levels of RDX (Rylott *et al.*, 2006). Even stronger improvements in RDX removal were achieved when *xplA* and *xplB*, two genes involved in bacterial degradation of RDX, were introduced into transgenic plants (Jackson *et al.*, 2007). Since military training ranges are often co-contaminated with both TNT and RDX, phytoremediation plants will need tolerance to both pollutants. When both bacterial genes (*nfsI* and *xplA*) involved in TNT and RDX metabolism were co-introduced into hybrid poplar, the plants removed TNT and RDX from solution more rapidly than the control plants (S. Doty, 2012, unpublished data).

Phytoremediation of toxic metals is also improved with transgenics (reviewed in Eapen and D'Souza, 2005; Meagher and Heaton, 2005; Dhankher *et al.*, 2011). Many of the genes involved in metal uptake, translocation and sequestration are being identified using the model plant, *Arabidopsis*, and using hyperaccumulating plants. The phytoremediation potential of these natural hyperaccumulators is limited by their small size, slow growth rates and limited growth habitat (Meagher and Heaton, 2005). Therefore, if the genes were transferred to plant species such as poplar and willow, with their high biomass and extensive root systems, significant removal of the heavy metals should be achieved. Yellow poplar (Rugh *et al.*, 1998) and cottonwood (Che *et al.*, 2003) have been transformed with genes to reduce the toxicity of mercury, resulting in tolerance to higher levels of mercury than the control plants. As with RDX, adding both bacterial genes involved in the metabolism of the pollutant (*merA* and *merB*) in transgenic plants had better results. Transgenic eastern cottonwood trees expressing both genes were highly tolerant to organic mercury, demonstrating the potential for their use in the phytoremediation of this important pollutant (Lyyra *et al.*, 2007).

ENHANCING PHYTOREMEDIATION CAPABILITY USING BACTERIAL ENDOPHYTES. Recently, attention has been focused on the role of endophytic bacteria on phytoremediation (reviewed in Newman and Reynolds, 2005; Doty, 2008; Weyens *et al.*, 2009a; Khan and Doty, 2011)). The term 'endophytic' refers to microbes living within plant tissues rather than rhizospheric bacteria living on or around the plant roots. Some endophytes are diazotrophic and can provide fixed nitrogen to the host plant (Reinhold-Hurek and Hurek, 1998; Doty, 2011). Some of these nitrogen-fixing bacteria have been isolated from wild poplar and willow in their native riparian habitat (Doty *et al.*, 2009). Endophytes can enhance plant growth and increase plant resistance to pathogens, drought and even herbivores (Selosse *et al.*, 2004). Plants can harbour dozens of different symbiotic or neutral bacterial species within the stems and roots, and this microbial community can be altered according to the environmental conditions. For example, at a petroleum-contaminated site, the genes encoding enzymes involved in petroleum degradation were more prevalent in the bacteria from the root interior than from the surrounding soil (Siciliano *et al.*, 2001). Surprisingly, this selection was plant species specific. In other words, some plant species seemed to have the ability to recruit, or selectively expand, the necessary bacteria to remove pollutants. How some plant species are able to recruit the necessary bacteria at a given site is currently an unexplored field of research. In a field test of the phytoremediation of a BTEX plume by poplar, the endophytic and rhizospheric bacteria associated with the trees were credited with the success of the remediation (Barac *et al.*, 2009). The population of BTEX-degrading microorganisms rose and fell with the concentration of the pollutant.

Recently, a novel endophyte of hybrid poplar was isolated that could degrade TCE rapidly (Kang *et al.*, 2012). The molar ratio of TCE removal to chloride generation suggested that this endophyte completely degraded TCE. This was the first report demonstrating that a naturally occurring poplar endophyte could degrade TCE rapidly and aerobically without the addition of toxic-inducing substrates. The strain also produced high levels of plant hormones that promoted root growth (J.W. Kang, 2012, unpublished data).

Not every bacterium with the necessary pollutant-degrading capacity has the ability to grow within the plants where the contamination is present. For this reason, a great deal of work has been done to provide the microbes that can live in a given site with the ability to degrade the pollutant (reviewed in Romantschuk *et al.*, 2000). In a ground-breaking study, the concept of engineering endophytes for phytoremediation was proven to be successful (Barac *et al.*, 2004). The catabolic plasmid from a relative of a yellow lupine endophyte was transferred conjugatively to the natural endophyte, providing the genes for toluene degradation. When yellow lupine plants were inoculated with this altered endophyte, the plants had an enhanced tolerance of toluene. This clear protective effect was only obtained when the natural endophyte was provided with the catabolic plasmid. The original host of the plasmid did not confer this effect since it was apparently unable to establish the necessary relationship with the plant. Conjugation of the required plasmid to native endophytes in plants was demonstrated with poplar trees (Taghavi *et al.*, 2005). Although the trees were inoculated with a toluene-degrading endophyte, the original inoculum was not found in the trees, but rather the genes responsible for the pollutant degradation had transferred to the endophytes already in the poplar. This natural conjugative transfer resulted in increased tolerance to toluene and reduced phytotranspiration of the pollutant.

A field test of poplar inoculated with an engineered TCE-degrading endophyte resulted in reduced evapotranspiration of TCE (Weyens *et al.*, 2009b). Trees in a TCE-contaminated field site were inoculated with cultures of the bacterium. At the end of the growing season, the levels of unaltered TCE transpired from inoculated poplar trees was less than that from uninoculated trees, suggesting that the TCE was metabolized more fully when the trees were colonized by the bacterium.

Endophytes can be engineered to harbour genes for both organic and metal detoxification. A *Burkholderia cepacia* strain containing the genes for TCE metabolism as well as for nickel resistance, and sequestration was used on yellow lupine as a model plant (Weyens *et al.*, 2010). The colonized plants had increased root mass compared to controls when both groups were exposed to TCE and nickel. There was a trend towards decreased phytovolatilization of TCE, although it was not statistically significant. Since many polluted sites are contaminated with both organics and metals, this research is an important step forward in improving phytoremediation.

A poplar-associated bacterium, *Methylobacterium* sp. strain BJ001, degraded TNT, RDX and HMX (van Aken *et al.*, 2004). This pink-pigmented symbiotic bacterium mineralized approximately 60% of the RDX and HMX to carbon dioxide in about 2 months. It is possible that this endophyte of hybrid poplar (*P. deltoides* × *P. nigra* 'DN34') assists in the phytoremediation of nitroaromatic pollutants within the tree.

Endophytes may assist in the phytoremediation of recalcitrant PAHs. A strain of *Pseudomonas putida* containing genes for the degradation of naphthalene protected pea plants from the phytoxic effects of this PAH (Germaine *et al.*, 2009). The inoculated seeds had higher germination rates in soil contaminated with naphthalene, and the colonized plants removed more of the pollutant from the soil and were healthier than the uninoculated controls. Natural endophytes of poplar and willow were isolated that could grow on PAHs (Z. Khan, 2012, unpublished data). One of these endophytes provided strong protection for willow exposed to toxic levels of phenanthrene (Z. Khan *et al.*, 2012, unpublished results). Therefore, there is great potential for endophyte-assisted phytoremediation of this class of pollutants.

SUMMARY. Poplar and willow are being used successfully in phytoremediation applications for some important classes of pollutants. With their high transpiration rates, deep roots, inherent biochemical abilities and amenability to coppicing, the *Salicaceae* family is especially well suited for remediation. As our understanding of the genes involved in the degradation of specific pollutants grows, the ability to increase greatly the success and speed of phytoremediation will continue. Especially in cases where the pollutant is extremely phytotoxic, such engineering strategies may be necessary. An alternative method for improving phytoremediation, the use of microbial plant partners, may also help us reach the same goals. As microbes with the necessary genes for engineering endophytes are identified, and as natural endophytes with pollutant-degrading abilities are isolated, further advances in the phytoremediation of both organics and heavy metals may be achievable.

Metal resistance and accumulation in North American willow species

J. KUZOVKINA

While the resistance to some metals has been documented for a few European *Salix* species, there is very limited knowledge about the potential of North American species of *Salix* for phytoremediation. The use of native species for environmental projects is a high-profile issue in North America, as it decreases the ecological risks associated with the introduction and possible invasion of alien species into new environments (Kuzovkina *et al.*, 2008). With the total number of willow species growing throughout North America at about 103 (Argus, 1999; Chapter 2, this volume), there is a possibility for a broad screening of candidates for environmental applications that are indigenous to North America (Kuzovkina and Quigley, 2005).

The research at the Ohio State University extended the study of willows' response to heavy metals to New World species (Kuzovkina and Quigley, 2004a; Kuzovkina *et al.*, 2004a). The efficacy for the phytoremediation of five willow species was tested by studying copper and cadmium uptake in a greenhouse hydroponic system (Fig. 6.29). The willow species used in the study were *S. discolor* Muhl., *S. eriocephala* Michx., *S. exigua* Nutt., *S. nigra* Marsh. and *S. lucida* Muhl. Hardwood cuttings of uniform 20 cm length were hydroponically rooted in half-strength Hoagland's nutrient solution for 5 weeks prior to the beginning of the experiment. Each cutting was mounted into a plastic pot cover to prevent algal growth and set into a pot containing 900 ml of constantly aerated solution. After 5 weeks, when the root systems were well developed, the hydroponic solution was replaced with half-strength Hoagland's nutrient solution containing either 5 or 25 μM additional Cu or Cd (added as $CuSO_4$ or $CdSO_4$). The experiment continued for 28 days after the addition of metals.

Different species of willow, as well as some clones, varied considerably in their metal translocation patterns and their ultimate resistance

Fig. 6.29. The greenhouse hydroponic system used for screening five North American willow species for their resistance to copper and cadmium. Photo courtesy of J. Kuzovkina.

to heavy metals. The differences between species in sensitivity to high metal content ranged from the stimulation of root and shoot growth to severe inhibition of growth. *Salix* species were less sensitive to Cd than to Cu, and plant growth for most species was not inhibited, even at high concentrations. Growth and transpiration for most species were not decreased by either 5 μM copper or 25 μM of cadmium in the solutions. In *S. exigua* and *S. eriocephala*, 25 μM copper caused foliar injury and reduced dry weight for all species after 21 days. Inhibition of growth in Cd treatments was evident only for *S. lucida*. In contrast, growth of *S. nigra* and *S. exigua* was stimulated even at high Cd concentrations. The copper content of aerial tissues was relatively lower than that of cadmium, while cadmium appeared to be more mobile within the plant. For most species, the highest Cd content was found in wood, while intermediate in roots and lowest in shoots. For Cu treatments, the trend was different and the highest amount of metal was found in roots, while intermediate in wood and least in shoots. The amount of copper found in new growth in 25 μM treatments was lower than that in 5 μM solution.

The results indicate that *S. nigra* is a promising North American species for phytoremediation research because of its high total metal content in plant tissues and its capacity to maintain high biomass during the experiment, especially in Cd treatments (see Fig. 2.22, this volume). *S. exigua* (see Fig. 2.16, this volume) exhibited resistance to Cd but not to Cu. Future field study needs to be conducted to confirm the findings and feasibility of phytoremediation technology using these species.

Stress tolerance in North American willow species

J. KUZOVKINA

Various environmental applications of willows are currently under way in an array of ecotechnological projects aiming to alleviate environmental degradation, to control the cycling of nutrients and contaminants and to provide value-added products (Volk *et al.*, 2006). The sustainability of any constructed ecosystem is dependent on optimum plant performance that is influenced by species autoecology. Plant genotypes should closely match the local climatic and microclimatic conditions (Isebrands and Karnosky, 2001), and plants should be tolerant of the adverse conditions that frequently occur in many degraded landscapes.

Another direction of the research is the selection of *Salix* species that are tolerant of various environmental stresses, with priority given to indigenous North American species. A few studies have identified significant differences among North American *Salix* species in reaction to soil compaction, flooding and ozone (Kuzovkina and Quigley, 2004b; Kuzovkina *et al.*, 2004b). The current research extends the study to the response to water stress of native willow species. Drought poses a serious challenge to plant development, limiting successful establishment and growth, especially in unfavourable soils. Many degraded landscapes that include brownfields, mines, industrial spoils, overburdens, quarries and waste sites are often characterized by shallow and compacted soils with very limited water availability (Chapter 7, this volume).

Though the majority of willows belong to the mesic-hydric type of vegetation, some North American species, such as *S. humilis* and *S. myricoides*, exhibit xeric traits and are better adapted to drought and heat stress (Cowles, 1991). Field observations suggest that other willow species are found on a variety of sites ranging from wetlands and flood plains to mesic or even xeric upland areas. This variation in suitable habitats for different species is under investigation to document a range of drought tolerances in native *Salix* species and to identify the drought resistance mechanism that is present in some species. Greenhouse experiments are being conducted to carry out a screening test to identify the length of drought periods that separate different genotypes, to monitor important plant physiological parameters and to identify drought-sensitive and drought-resistant genotypes. The information on species drought tolerance is very important in the context of various applications including phytoextraction, biofiltration, revegetation of degraded land and bioenergy, to ensure the compatibility between candidate species to site conditions and project objectives, as well as in the context of hybridization work.

Shrub willows for phytoremediation

T.A. VOLK

Shrub willows have numerous inherent characteristics that make them a good choice for phytoremediation, including rapid juvenile growth rates, vigorous coppicing ability that is maintained even after multiple harvests, ease of establishment from unrooted cuttings, tolerance of high planting densities, high degree of genetic diversity and potential for rapid genetic improvement. In addition, willows' perennial nature, extensive and diffuse root systems, high transpiration rates and tolerance of waterlogged conditions make them potentially beneficial for a wide range of other applications. Years of research and development of shrub willow-based biomass production systems in North America (Kenney *et al.*, 1990; Volk *et al.*, 2006) and Europe (Armstrong, 1999; Verwijst, 2001) have expanded the knowledge base about the biology, ecology and management of shrub willows. This information has been used to develop new applications for shrub willows in the north-eastern USA, including phytoremediation (Licht and Isebrands, 2005).

Shrub willows are being used in the USA to remediate and contain sites contaminated with various industrial wastes (Licht and Isebrands, 2005; Mirck *et al.*, 2005). Willows have been shown to uptake heavy metals and organics from soils (i.e. phytoextraction) (Riddell-Black *et al.*, 1997), facilitate the breakdown of organics to non-toxic compounds (i.e. rhizodegradation) (Ebbs *et al.*, 2003) and control water dynamics, including contaminated groundwater flow and water penetration into soils via evapotranspiration (i.e. phytovolatilization and hydraulic control) (Corseuil and Moreno, 2001). Many of the characteristics that make shrub willows effective in biomass production systems are also beneficial for phytoremediation systems. Since willows have developed as pioneer species, they have the ability to survive in relatively hostile, disturbed and wet sites. In addition, they have a high capacity to transpire water (Chapter 3, this volume), which is a beneficial attribute in phytoremediation systems. A broad gene pool (there are over 330 species of willow across the world (Argus, 1999), with many more natural and human-developed species hybrids) provides opportunities to screen and develop willow to grow on a wide range of sites and produce specific phytoremediation effects.

Several phytoremediation projects using willow and hybrid poplar (*Populus* spp.) are currently under way in the north-eastern USA (Table 6.9). Most trials in New York, USA, are related to using willow to control site water problems, either through the management of

Table 6.9. Willow phytoremediation trials in the north-eastern USA (after Volk *et al.*, 2006).

Trial location	Year established	Phytoremediation method	Site contaminants	Number of clones	Planting density (plants ha^{-1})
Utica, New York (Jackson, 2000)	1999	Rhizodegradation	PAHs[a]	8	108,000
Rochester, New Hampshire	2000	Hydraulic control	PAHs	8	36,000
Fort Drum, New York (Kornacki, 2005; Salladin, 2005)	2001	Hydraulic control, rhizodegradation, phytovolatilization	PAHs, herbicides	20	161,000
Solvay, New York (Johnson, 2005; Farber, 2006)	2003	Hydraulic control	Chloride and other salts	40	15,400
Yorktown, Virginia	2004	Hydraulic control	PAHs	8	36,000
Syracuse, New York (Purdy, 2006)	2006	Phytoextraction	Arsenic	4	Greenhouse trial

[a]PAHs, Polycyclic aromatic hydrocarbons.

water entering into the contaminated site (Solvay trial) or by controlling contaminated groundwater chemistry and flow (Rochester, Fort Drum and Yorktown trials). Recently, greenhouse studies examining the potential of willow to remediate arsenic-contaminated soil have been initiated (Purdy, 2006). Arsenic contamination is a widespread problem because more than 40 t of arsenical pesticides were applied annually to farmland, especially apple and other orchards, in the USA in the 1930s and 1940s, and there has been limited redistribution of this material since that time (Renshaw *et al.*, 2006).

Three of the ongoing trials are operational-scale case studies where the willow plantings are expected to contribute to site clean-up through various phytoremediation processes. Since a phased approach was not possible with these trials, a fail-safe design was used to establish the trials. Site preparation was intensive, and unique problems were solved in specific ways at each site. Planting densities were high and a set of known, plastic (wide ecological amplitude) varieties were used, so that if one or more variety failed (up to 50%), the system would likely still function in relation to phytoremediation processes. After 1–3 years of experience, there have not been any large-scale failures in terms of plant mortality. All of the operational trials have monitoring schemes to collect data on willow survival, growth and their impact on a site's contamination levels and/or hydrology. Monitoring results have been turned into action, including the replacement of poor varieties with new ones or the expansion of proven varieties and the adoption of new cultural techniques associated with planting, site preparation and tending.

The two other trials – Fort Drum and the Solvay Wastebeds – have used a phased approach to test, refine and develop a system that can be applied at an operational scale at each site. In the trial at Fort Drum, the goal is to reduce the flow of contaminated water moving through seeps from a landfill using a shrub willow-based phytoremediation system. The system was developed over several years, testing a number of willow varieties and different planting designs. The willows were established in or near seeps in soils that were poorly or very poorly drained, so the challenge was to get the material to thrive in these wet, contaminated conditions. The existing willow biomass establishment system was transformed over several years to create a system tailored specifically to the site's conditions and specific phytoremediation clean-up goals on the site. Different planting designs that were developed and tested included the use of cardboard rings filled with soil, planting boxes constructed from lumber or earthen berms. The first growing season after installing these three designs, survival was high (<93%) for all of them (Salladin, 2005). The aboveground biomass production in the planting boxes and tubes was similar and greater than that of the biomass of the willows planted in earthen berms, (Salladin, 2005).

During 2004 and 2005, piezometer measurements showed that the water table in the area where the willows had been established had been lowered slightly during the growing season, probably due to the increased evapotranspiration from the willows (Thompson, 2006). Water table depths and plant growth were to be monitored as the shrub willows grew over the successive few years. The success of the planting systems over the first few years and the indication that the willows had already had an impact on the water table resulted in plans to deploy this system over a greater area at the site.

The project on the Solvay wastebeds focused on using shrub willows as an alternative cover to a standard geomembrane cap. The project's goal was to minimize the amount of water that percolated into the wastebeds and ultimately decrease the amount of leachate, which had high concentrations of chlorine (Cl) generated from the wastebeds, to reduce the impact on groundwater and surface water in the region. Secondary goals for this project were to produce woody biomass for the renewable energy market developing in the region and to transform the wastebeds into a productive community asset.

The Solvay wastebeds are a by-product of over 100 years of production of chlorine and alkali, which is fundamental to the chemical industry. The process was developed by Ernest Solvay in the 1860s and produced chlorine (Cl_2), soda ash (Na_2CO_3) and caustic soda (NaOH) (Michalenko, 1991). For the production of soda ash (Na_2CO_3), abundant and inexpensive supplies of limestone, salt, water, a reliable and robust process and space to deposit the waste were needed. Many of these features were present on the western shore of Onondaga Lake near Syracuse, New York, so the Solvay Process Company established a soda ash plant there in 1884 and ran it until 1986.

The volume of waste generated by the production of Na_2CO_3 using the Solvay process is enormous: for the production of 0.91 Mt of soda ash, about 10 m^3 of liquid waste is created, containing approximately 0.91 Mt of $CaCl_2$ and 0.45 Mt of NaCl (Michalenko, 1991) and other by-products. This material was deposited into sedimentation basins that were surrounded by berms. By the time the production process ended, the wastebeds covered approximately 600 ha of land 16–21 m deep. Some of this area was converted to alternative uses such as parking areas for the New York State Fairgrounds, construction and debris landfill and the development of malls and other facilities, but six wastebeds remained, covering about 222 ha.

The material in the wastebeds is a harsh environment for plants to become established and to thrive in. Greater than 70% of the Solvay waste consists of silt-size particles made up of calcium and magnesium salts, with Ca making up greater than 86% of the cation exchange capacity (CEC) (Michalenko, 1991). The pH of the material ranges from mid-8 in the top 20 cm to greater than 11 at depths of 40 cm or more.

The first step in developing this system was to screen 38 shrub willow and two hybrid poplar varieties from the SUNY-ESF collection to determine which ones would be the most successful on the wastebeds. Previous studies showed that *Salix* (*S. alba*, *S. bebbiana*, *S. discolor*, *S. purpurea* and *S. rigida*) and *Populus* (*P. deltoides*, *P. tremuloides* and *P. ×canescens*) species successfully colonized the wastebeds as the age since deposition increased (Hewlett, 1956). Solvay waste was collected from an area on the wastebeds that had been amended with biosolids in the early 1990s and from another area that was unamended Solvay waste. A 1:1 ratio of ProMix and fine sand served as a control. The willows and hybrid poplar were planted as 12-cm-long cuttings in tubes and grown in a greenhouse for 11 weeks. For 27 of these varieties, biomass was greatest in the amended Solvay waste treatment, indicating that this was a good growth medium. Biomass was greatest in the Promix and sand treatment for 11 other varieties, while two had the highest production on the unamended Solvay waste. In the amended Solvay waste, ten varieties had better biomass production than clone 'SV1', which is a high biomass-producing standard used in various screening trials and the SUNY-ESF breeding programme (Smart *et al.*, 2005). There was a fourfold difference in the root:shoot ratios among these higher producing varieties, ranging from 0.05 to 0.22. While aboveground biomass production was good, the varieties with low root:shoot ratios may be susceptible to the dry conditions found on the wastebeds during certain parts of the growing season.

The positive growth results from the greenhouse screening trial then prompted the design and installation of two subsequent trials, a greenhouse trial to examine the effect of different organic amendments and a field trial on the wastebeds with a limited number of varieties. For the organic amendment greenhouse trial, 'SV1' was used as a standard and two varieties that had higher aboveground biomass than 'SV1' in the screening trial but very different root:shoot ratios were selected. Clone '9882-34' had a low root:shoot ratio (0.06), '9871-31' had a high root:shoot ratio (0.22), while 'SV1' was intermediate (0.17).

This trial used three different organic amendments, Anhueser Busch biosolids (ABB), Bristol Meyer Squibb biosolids (BMS) and lime-stabilized Syracuse metrosludge biosolids (MBS). These were mixed at a 1:1 dry weight ratio to two different depths of mixing, either half the depth of the 34-cm-deep pot or for the total depth of the pot. After 15 weeks in the greenhouse (Fig. 6.30), the total aboveground biomass of the three varieties grown in the ABB and BMS amendments was significantly greater than willow grown in MBS or unamended Solvay waste for both mixing depths (Farber, 2006). The pH of MBS was 12.3 compared to 5.6 for ABB and 6.1 for BMS, which was an important factor influencing growth, since the unamended Solvay waste had a pH of 8.3. Depth of incorporation of ABB or BMS did not affect above- or belowground biomass for '9871-31' or '9882-34'. Based on the results from this trial, a field experiment was established with two willow varieties and three organic amendments in the spring of 2006 (ABB, unstabilized MBS and composted yard waste from an adjacent village). Due to changes in manufacturing processes, the BMS amendment was no longer available. First-year growth data were very encouraging, with height growth exceeding 2 m on many of the treatments.

The first field trial on the wastebeds involved planting a subset of successful willow varieties from the screening trial in areas where the amended and unamended Solvay waste was collected. For the area with amended Solvay waste, ten different willow varieties and two cutting lengths (25 cm and 50 cm) were used in a replicated trial. After the first growing season, survival was greater than 80% for all varieties and treatments, except for two varieties planted with 25-cm cuttings, and aboveground growth

Fig. 6.30. Shrub willows after 15 weeks of growth in a greenhouse trial to test the effect of three different organic amendments and two mixing rates. Photo courtesy of T. Volk.

was good. During the second growing season, survival declined significantly for two of the varieties. This pattern emphasizes the need for long-term and consistent monitoring of phytoremediation field trials to avoid system failures after initial successes. For the varieties that survived, aboveground biomass production was good on the amended Solvay waste. After two growing seasons, aboveground biomass exceeded 20 t ha^{-1} oven-dry for four of the varieties planted with 50-cm-long cuttings. These growth rates are comparable to trials in central New York on agricultural soils (Volk *et al.*, 2006). Production with the shorter cuttings exceeded 15 t ha^{-1} oven-dry for three of the varieties.

The focus of this project was to develop an alternative cap using shrub willows. In order to assess the potential for such a system, it is important to monitor components of the water budget and then model the long-term effect of the willows on the site's water budget, as described in Mirck and Volk (2010). Monitoring of weather data, soil moisture content, throughfall and sapflow (Fig. 6.31) occurred at the site. Data collected were used as input to the SHAW (simultaneous heat and water) model. Initial modelling efforts indicated that an alternative cap could reduce percolation significantly over a 28-year period (Johnson, 2005). These model runs illustrate that it will take several years for a willow cap to become established and fully functional but that, once they are established, they are robust and functional. Work was ongoing to calibrate this model using data collected from the site in order to reflect the water budget dynamics on the site as accurately as possible.

Wastewater treatment

J.D. JOHNSON

As environmental regulations in the USA become more stringent, municipal wastewater treatment facilities are turning to poplar plantations to deal with wastewater reuse in lieu of disposing into adjacent rivers and streams (Fig. 6.32). Typical facilities must dispose of millions of litres of wastewater annually that contain high

Fig. 6.31. Three-year-old willow trial on the Solvay wastebeds where sapflow, throughfall, soil moisture and various plant characteristics are being measured as part of the effort to develop an alternative vegetative cover for the wastebeds. Photo courtesy of T. Volk.

Fig. 6.32. Aerial view of the City of Woodburn, Oregon, USA, hybrid poplar stands used for treating municipal wastewater. Photo courtesy of J. Johnson.

quantities of both nitrogen and phosphorus. One of the early examples of using poplars for wastewater treatment by a municipality was at Woodburn, Oregon, USA (Zodrow, 1999; Isebrands and Karnosky, 2001; City of Woodburn, 2012).

Hybrid poplars are especially well suited for treating this wastewater because their extensive fine root systems readily take up the nitrogen and phosphorus, which is used in growth, and their canopy's large leaf area transpires large volumes of water into the atmosphere (Smesrud *et al.*, 2000). Treatment facilities in the Pacific Northwest of the USA estimate annual transpiration of a closed canopy hybrid poplar stand to be a little over 10 million l ha^{-1} $year^{-1}$. This amount is probably even higher when the wastewater is applied through sprinklers that increase evaporation before the water hits the ground. In drier and warmer climates, application rates can be higher. One issue that often develops over several years of applying wastewater to poplar stands is nutrient deficiencies. Under typical waste treatment processes, many of the nutrients are removed and so the application of high volumes of low nutrient-containing water can lead to nutrient deficiencies through leaching and growth dilution. An inexpensive remedy adopted by many wastewater treatment facilities is to apply nutrient-rich biosolids, a by-product of the treatment process, annually to the poplar stand. In many states, application of biosolids to food crops is not allowed, making its disposal problematic and potentially costly. Wastewater treatment with poplars is now popular is other regions of the USA.

Evapotranspiration covers

L. LICHT

Evapotranspiration (ET) covers, or vegetative caps, are being used increasingly at municipal solid waste landfills, hazardous waste sites and mine sites (Rock, 2012). The primary objectives of these vegetative caps are to minimize water percolation into the buried landfill waste via a 'sponge and pump' mechanism and to prevent surface soil erosion on the cap (Licht *et al.*, 2001). The plants, which are traditionally grasses, shrubs or trees, take up precipitation for growth and release it back to the atmosphere by transpiration (Fig. 6.33). The design of the ET cover provides water storage capacity and evapotranspiration to control moisture and percolation into the underlying waste. The primary tree species used for ET covers are poplars and willows (US EPA, 2003; Rock, 2012). The first ET cover using trees was in 1990 in Beaverton, Oregon, USA, using 1.5-m-long hybrid poplar whips (Licht *et al.*, 2001). There have been a number of successful ET covers since 1990 (Rock, 2012). Nixon *et al.* (2001) have shown that success depends on the proper choice of plant material for the site, as well as good site management. Licht and Isebrands (2005) showed that an ET cap in the state of Washington using hybrid poplar trees was highly successful when properly installed, and additional wildlife benefits were seen as an advantage with the system. Likewise, Abichou *et al.* (2012) found that an ET cover vegetated with native cottonwood was a feasible alternative for use in Florida, USA. Biosolids are often used as soil amendments to enhance ET cover establishment (Felix *et al.*, 2008). Of the 217 ET projects reviewed by Rock (2012), many are using poplars and willows. It is too soon to determine what the long-term performance of these alternative systems will be, but use of native poplars and willows is likely to cause less disturbance to the surrounding ecosystem (US EPA, 2003).

Other research and information

J.G. ISEBRANDS

There are many other research institutions and government agencies in the USA doing fundamental and applied research on phytotechnologies (Rockwood *et al.*, 2004; Strycharz and Newman, 2009). Most of them are working with poplars and willows, and the number of institutions and scientists are too numerous to include here. However, there are several institutions that should be mentioned. Firstly, the US Environmental Protection Agency Service Center in Cincinnati, Ohio, serves as a database manager and clearinghouse for all phytotechnology research and applications in the USA. Their website has hundreds of references to the use of poplar and willow for all aspects of phytoremediation

Fig. 6.33. Buffer planting on Chanute US Air Force Base, Illinois – 3-year hybrid poplar planted to intercept water. Photo courtesy of L. Licht.

(for further information see http://clu-in.org). Secondly, one of the leaders in the phytoremediation of explosives and organic solvents is at the University of Iowa, Iowa City, Iowa. The programme leader for that group is J.L. Schnoor, who is co-author of a state-of-the-art book on phytoremediation (McCutcheon and Schnoor, 2003) and a world expert on the use of poplars and willow for remediation of these compounds in soil and water. Their website refers to hundreds of fundamental and applied articles from their group through the history of phytoremediation in the USA (for further information see http://www.instantref.com/ECSEI/Schnoor-CV.pdf), many of which are on poplars (and willows), e.g. Burken and Schnoor (1998). Lastly, one of the other leaders in the phytoremediation of poplars and willows is in the USDA Forest Service, Northern Research Station in Rhinelander, Wisconsin. The group has a long history of fundamental and applied research projects and is currently led by R.S. Zalesny and co-workers. Their group has carried out extensive research on the interaction of genetics and the environment on phytotechnology applications. This includes work on the irrigation of poplars and willows with landfill leachate (Zalesny and Bauer, 2007a; Zalesny *et al.*, 2007a), choosing genotypes for landfill covers (Zalesny and Bauer, 2007b; Zalesny *et al.*, 2007b), opportunities for utilizing treated wastewater (Zalesny *et al.*, 2011) and clonal variation in rootability with irrigated landfill leachate (Zalesny and Zalesny, 2011). For further information see http://nrs.fs.fed.us/people/Zalesny.

6.6 Ecosystem Services

6.6.1 Biodiversity, environment and landscape

M. WEIH

Tree plantations can have positive or negative effects on biodiversity, depending on location, management and previous land use (Cossalter and Pye-Smith, 2003), and studies on biodiversity in plantations of fast-growing trees often arrive at contradictory conclusions, especially when different kinds of organisms are considered (Hartley, 2002). Plantations of *Salix* and other fast-growing trees grown on agricultural land can improve biodiversity at the landscape level, in particular if the plantations are established instead of cultures of cereals and spruce or fallow ground in a homogeneous agricultural landscape. For example, compared to managed coniferous forests and farmland in boreal Sweden, young poplar and willow plantations, especially if not too large in size, have been concluded to increase vascular-plant diversity (Gustafsson, 1987; Weih *et al.*, 2003). Similar to the observations on floras, fauna diversity (birds and mammals) is frequently found to be higher in willow and hybrid poplar stands compared to agricultural croplands (Weih and Nordh, 2007; Weih, 2008, 2009). Thus, the more extensive management of tree plantations compared to intensively managed cereal crops can improve habitat quality for many organisms, including plants and birds (Christian *et al.*, 1998; Berg, 2002; Weih *et al.*, 2003; Dhondt *et al.*, 2004). In addition, plantations of fast-growing trees appear to have a potential as important habitats for gamebirds (Sage and Robertson, 1994). Plantations of poplars or willows can also affect soil properties positively compared to conventional agriculture. For example, carbon sequestration and water-holding capacity were found to increase in formerly arable soils that were planted with fast-growing willow and poplar for 6–10 years (Kahle *et al.*, 2005). Many concerns are raised regarding the impact of plantations of fast-growing trees on the landscape (Skärbäck and Becht, 2005). However, if used creatively as part of active landscape analysis and design, plantations of fast-growing trees can improve greatly the visual and recreational values of a landscape and, particularly, plantations of relatively small size can improve the aesthetic perception of homogeneous agricultural landscapes by adding variation and structure (Rode, 2005).

6.6.2 Carbon sequestration

M. COLEMAN AND J.D. JOHNSON

A relatively new use for hybrid poplars is to reduce atmospheric carbon dioxide by sequestering it into tree biomass. Due to their rapid growth, hybrid poplars are ideally suited for this use and a number of poplar companies have been exploring the possibility of selling their carbon stores to various industries. Where greatest gains can be made are on marginal agricultural or pasture lands that do not require supplemental irrigation (irrigation requires expenditure of energy from fossil fuels). In 2000, nearly 5.3 million ha were identified in the Pacific Northwest of the USA (Idaho, Oregon and Washington) that potentially could support hybrid poplar plantations for carbon sequestration (J. Johnson, 2000, unpublished data). Estimation of total carbon sequestration rates range from 11 to 20 Mg ha^{-1} $year^{-1}$, depending on clone and climate, resulting in total carbon sequestration after 8 years of between 90 and 160 Mg ha^{-1}. In addition to tree carbon, soil carbon under hybrid poplar stands was found to increase from between 9% in heavier soils to 62% in sandy soils, compared to adjacent soils with annual cropping. Values for soil carbon ranged from 4 to 15 Mg ha^{-1}, with the lower value being found in sandy soil. Hybrid poplar plantations grown for long-term carbon storage could play a very important role in slowing the increase in atmospheric carbon dioxide (Tuskan and Walsh, 2001). For example, the potential of planting poplars on farmland in India for carbon sequestration is being investigated (Gera, 2012).

Coleman *et al.* (2004) compared soil carbon of short-rotation poplar plantings with adjacent agricultural crops and woodlots in Minnesota, USA. They found greater soil carbon in poplars than in paired agricultural crops and concluded that short-rotation poplars offered opportunities for carbon sequestration, as well as erosion control and wildlife habitat improvement in the central USA. Moreover, Sanchez *et al.* (2007) working in the southeastern USA found that native cottonwood when irrigated and fertilized had higher soil carbon than other tree species. These results from a different region also suggest that poplars grown under short rotation offer carbon sequestration opportunities. But, the key advantage with these plantings would still come from the displacement of fossil fuels over long periods (Tuskan and Walsh, 2001).

6.6.3 Poplar growing in the environment of the Walloon region, Belgium

P. MERTENS

Poplar growing in the Walloon region of Belgium is ubiquitous and is often as single trees or small groups (Mertens, 2002). They are usually present in non-forested townships and villages. Poplar culture in these areas is in semi-open environments characterized by ecosystems that provide shelter for herbaceous and shrub undergrowth (Mertens, 1999). These stands provide for a multitude of environmental benefits and ecosystem services as well as traditional social-economic value for landowners, such as wood products. They provide soil erosion control, heterogeneous corridors and edges and landscape aesthetics (Fig. 6.34), biodiversity, fixation of nutrients such as nitrogen and phosphorus and habitat for animals and avifauna

Fig. 6.34. Poplars grown in semi-open environments improve landscape aesthetics. Photo courtesy of P. Mertens.

(Mertens, 1999). These unique semi-open systems provide long-term benefits where human intervention is only needed for occasional commercial operations, leaving the open space to spontaneous native vegetation growth in the rural landscape.

6.6.4 Other emerging ecosystem services opportunities

J.G. ISEBRANDS

There are other intangible benefits from poplar and willow culture that often go unrecognized. They include wildlife benefits such as a greater diversity of non-game animals, invertebrates and fish species when compared to agricultural crops. Poplars and willows also provide winter habitat for upland birds and game animals. Numerous studies have shown the positive effects of maintaining heterogeneous planting edges and multiple age-class vegetation on small mammal populations, breeding bird diversity and breeding bird habitat (Isebrands, 2007).

Another important environmental benefit of poplar and willow plantings in agricultural regions can be with livestock operations. The number of large livestock operations (i.e. cattle, pigs and poultry) is increasing worldwide with human population growth. Some of the most challenging problems with livestock operations are in odour control and animal waste management. Multi-species shelterbelts that include poplars and willows can help mitigate livestock operation problems (Tyndall and Colletti, 2001, 2006; Malone, 2002). The tree plantings dilute manure-generated odour compounds in the atmosphere, deposit odorous dust by decreasing wind speeds, physically intercept dust and absorb volatile odour compounds. At the same time, they provide both visual and sound barriers from the livestock operations, which are appreciated by neighbours (Fig. 6.35).

Fig. 6.35. View from motorway of multi-species shelterwood including poplar surrounding a turkey-rearing facility in western Minnesota, USA. Photo courtesy of J. Isebrands.

Acknowledgements

We would to thank the following scientists for their valuable contributions of information and photographs for this chapter (in alphabetical order): R. Dhiman, C. Hendrickson, V. Iori, D. Karnosky, M. Labrecque, A. Massacci, J. Mirck, F. Pietrini, L. Poppy, T. Punshon, D. Rockwood, S. Rood, L. Sebastiani, M. Utmazian, V. Zacchini and R. Zalesny, Jr. Special thanks go to D.I. Dickmann, K. Perttu, D. Riddell-Black and R.F. Stettler for their helpful discussions and encouragement during the course of this chapter preparation. And finally, we are greatly indebted to S.K. O'Leary for her clerical and moral support during this entire arduous process.

References

Aasamaa, K., Heinsoo, K. and Holm, B. (2010) Biomass production, water use and photosynthesis of *Salix* clones grown in a wastewater purification system. *Biomass and Bioenergy* 34, 897–905.

Abichou, T., Musagasa, J., Yuan, L., Chanton, J., Tawfiq, K., Rockwood, D., *et al.* (2012) Field performance of alternative landfill covers vegetated with cottonwood and *Eucalyptus* trees. *International Journal of Phytoremediation* 14, 47–60.

Adler, A., Dimitriou, I., Aronsson, P., Verwijst, T. and Weih, M. (2008) Wood fuel quality of two *Salix viminalis* stands fertilised with sludge, ash and sludge–ash mixtures. *Biomass and Bioenergy* 32, 914–925.

Agriculture and Agri-Food Canada (2005) *Planning Farm Shelterbelts for the Prairies.* Agriculture and Agri-Food Canada, PFRA Shelterbelt Centre, Indian Head, Saskatchewan, Canada.

Alker, G.R. (1999) Phytoremediation of nutrient rich wastewaters and leachates using *Salix.* PhD thesis, Imperial College, University of London, London.

Als, C. (1990) The shelterbelt planting in Denmark. In: Xiang, K.-F., Shi, J.-C., Baer, N.W. and Sturrock, J.W. (eds) *Protective Plantation Technology.* Publishing House of Northeast Forestry University, Harbin, China, pp. 20–34.

Anon. (2010) *Fertiliser Manual* (RB209), 8th edn. The Stationery Office, Norwich, UK.

Argus, G.W. (1999) Classification of *Salix* in the New World. Version 5: 5 July 1999. *Botanical Electronic News (BEN)* No 227 (http://www.ou.edu/cas/botany-micro/ben/ben227.html, accessed 12 June 2012).

Armstrong, A. (1999) Establishment of short rotation coppice. Practice Note. Forestry Commission, Edinburgh, UK.

Aronsson, P. (2000) Nitrogen retention in vegetation filters in short-rotation willow coppice. Doctoral thesis, *Acta Universitatis Agriculturae Sueciae, Silvestria* 161 ISBN 91-576-5895-1.

Aronsson, P.G. (2001) Dynamics of nitrogen leaching and ^{15}N turnover in intensively fertilized and irrigated basket willow grown in lysimeters. *Biomass and Bioenergy* 21, 143–154.

Aronsson, P.G. and Bergström, L.F. (2001) Nitrate leaching from lysimeter-grown short rotation willow coppice in relation to N-application, irrigation and soil type. *Biomass and Bioenergy* 21, 155–164.

Aronsson, P. and Perttu, K. (eds) (1994) *Willow Vegetation Filters for Municipal Wastewaters and Sludges. A Biological Purification System.* Proceedings of a study tour, conference and workshop in Sweden, 5–10 June 1994. Report 50. Swedish University of Agricultural Sciences, Department of Ecology and Environmental Research, Section of Short Rotation Forestry.

Aronsson, P. and Perttu, K. (2001) Willow vegetation for wastewater treatment and soil remediation combined with biomass production. *The Forestry Chronicle* 77, 293–299.

Aronsson, P.G., Bergström, L.F. and Elowson, S.N.E. (2000) Long-term influence of intensively cultured short-rotation willow coppice on nitrogen concentrations in groundwater. *Journal of Environmental Management* 58, 135–145.

Aronsson, P., Heinsoo, K., Perttu, K. and Hasselgren, K. (2002) Spatial distribution of biomass productivity in unevenly wastewater-irrigated willow plantations. *Ecological Engineering* 19, 281–287.

Aronsson, P., Dahlin, T. and Dimitriou, I. (2010) Treatment of landfill leachate by irrigation of willow coppice – plant response and treatment efficiency. *Environmental Pollution* 158, 795–804.

Ballach, H.-J., Kuhn, A. and Wittig, R. (2003) Biodegradation of anthracene in the roots and growth substrate of poplar cuttings. *Environmental Science and Pollution Research* 10, 308–316.

Banuelos, G.S. and Ajwa, H.A. (1999) Trace elements in soils and plants: an overview. *Journal of Environmental Science and Health Part A – Toxic/Hazardous Substances & Environmental Engineering* 34, 951–974.

Barac, T., Taghavi, S., Borremans, B., Provoost, A., Oeyen, L., Colpaert, J.V., *et al.* (2004) Engineered endophytic bacteria improve phytoremediation of water-soluble, volatile, organic pollutants. *Nature Biotechnology* 22, 583–588.

Barac, T., Weyens, N., Oeyen, L., Taghavi, S., van der Lelie, D., Dubin, D., *et al.* (2009) Field Note: Hydraulic containment of a BTEX plume using poplar trees. *International Journal of Phytoremediation* 11, 416–424.

Barry, J.M. (1997) *Rising Tide: the Great Mississippi Flood of 1927 and How it Changed America.* Touchstone, New York.

Barry, T. and Kemp, P. (2001) Ewes respond to poplar feed. *Tree Feed* 1, 2–3.

Berg, Å. (2002) Breeding birds in short-rotation coppices on farmland in central Sweden. The importance of *Salix* height and adjacent habitats. *Agriculture, Ecosystems and Environment* 90, 265–276.

Bissonette, L., St-Arnaud, M. and Labrecque, M. (2010) Phytoextraction of heavy metals by two *Salicacae* clones in symbiosis with arbuscular mycorrhizal fungi during the second year of a field trial. *Plant and Soil* 332, 55–67.

Borjesson, P. (1999) Environmental effects of energy crop cultivation in Sweden – i: identification and quantification. *Biomass and Bioenergy* 16, 137–154.

Börjesson, P. and Berndes, G. (2006) The prospects for willow plantations for wastewater treatment in Sweden. *Biomass and Bioenergy* 30, 428–438.

Bramley, R.G.V. (1990) Cadmium in New Zealand agriculture. *New Zealand Journal of Agriculture Research* 33, 505–519.

Brenner, F.J., Werner, M. and Pike, J. (1984) Ecosystem development and natural succession in surface coal mine restoration. *Minerals and the Environment* 6, 10–22.

Brierley, E.D.R., McDevitt, J.E., Thorn, P., Tyrrel, S.F. and Stephens, W. (2001) Application of landfill leachate to willow short rotation coppice. In: Bullard, M.J., Christian, D.G., Knight, J.D., Lainsbury, M.A. and Parker, S.R. (eds) *Biomass and Energy Crops II. Aspects of Biology* 65. University of York, York, UK, pp. 321–328.

Brix, H. (1994) Constructed wetlands for municipal wastewater treatment in Europe. In: Mitsch,W. (ed.) *Global Wetlands: Old World and New.* Elsevier, Amsterdam, pp. 325–334.

Broadfoot, W.M. and Williston, H.L. (1973) Flooding effects on southern forests. *Journal of Forestry* 71, 584–587.

Bungart, R., Grunewald, H. and Huttl, R.F. (2001) Productivity and water budget of two poplar clones on a mine spoil in the Lusatian lignite mining region. *Forstwissenschaftliches Centralblatt* 120, 125–138.

Burken, J.G. and Schnoor, J.L. (1998) Predictive relationships for uptake of organic contaminants by hybrid poplar trees. *Environmental Science and Technology* 32, 3379–3385.

Carle, J. and Ma, Q. (2005) Challenges of transplanting science into practice: poplars and other species in the Three North Region of China. *Unasylva* 56, 31–37.

Carlson, M. (1992) Municipal effluent irrigation of fast-growing hybrid poplar plantations near Vernon, British Columbia. *The Forestry Chronicle* 68, 206–208.

Carulla, J.E., Kreuzer, M., Machmuller, A. and Hess, H.D. (2005) Supplementation of *Acacia mearnsii* tannins decreases methanogenesis and urinary nitrogen in forage-fed sheep. *Australian Journal of Agriculture Research* 56, 961–970.

Ceulemans, R., Scarascia-Mugnozza, G., Wiard, B.M., Braatne, J.H., Hinckley, T.M., Stettler, R.F., *et al.* (1992) Production physiology and morphology of *Populus* species and their hybrids grown under short rotation. I. Clonal comparisons of 4-year growth and phenology. *Canadian Journal of Forest Research* 22, 1937–1948.

Che, D., Meagher, R.B., Heaton, A.C., Lima, A., Rugh, C.L. and Merkle, S.A. (2003) Expression of mercuric ion reductase in Eastern cottonwood (*Populus deltoides*) confers mercuric ion reduction and resistance. *Plant Biotechnology Journal* 1, 311–319.

Christian, D.P., Hoffman, W., Hanowski, J.M., Niemi, G.J. and Beyea, J. (1998) Bird and mammal diversity on woody biomass plantations in North America. *Biomass and Bioenergy* 14, 395–402.

City of Woodburn (2012) Woodburn wastewater treatment facilities (www.ci.woodburn.or.us/?q=tags/woodburn-wastewater-treatment-plant, accessed 31 July 2013).

Coleman, M.D., Isebrands, J.G., Tolsted, D.N. and Tolbert, V.R. (2004) Comparing soil carbon of short rotation poplar plantations with agricultural crops and woodlots in the North Central United States. *Ecological Management* 33 (Supplement 1), S299–S308.

Cooper, D.T. (1990) *Populus deltoides* Bartr. ex Marsh. var. deltoides, Eastern Cottonwood. In: Burns, R.M. and Honkala, B.H. (eds) *Silvics of North America – Volume 2, Hardwoods.* Agricultural Handbook 654. USDA Forest Service, Washington, DC, pp. 530–537.

Corseuil, H.X. and Moreno, F.N. (2001) Phytoremediation potential of willow trees for aquifers contaminated with ethanol-blended gasoline. *Water Research* 35, 3013–3017.
Cossalter, C. and Pye-Smith, C. (2003) *Fast-wood Forestry: Myths and Realities.* Center for International Forestry Research (CIFOR), Jakarta.
Costantino, P., Capone, I., Cardarelli, M., De, P.A., Mauro, M.L. and Trovato, M. (1994) Bacterial plant oncogenes: the rol genes' saga. *Genetica* 94, 203–211.
Costanza, R., d'Arge, R., deGroot, R., Farber, S., Grasso, M., Hannon, B., *et al.* (1997) The value of the world's ecosystem services and natural capital. *Nature* 387, 252–260.
Cowles, H.C. (1991) The ecological relations of the vegetation on the sand dunes of Lake Michigan. In: Real, L.A. and Brown, J.H. (eds) *Foundations of Ecology.* The University of Chicago Press, Chicago, Illinois, pp. 28–55.
Cremer, K.W. (2003) Introduced willows can become invasive pests in Australia. *Biodiversity* 4, 17–24.
Cundell, A.M. (1977) The role of microorganisms in the revegetation of strip-mined land in the Western United States. *Journal of Range Management* 30, 299–305.
Cunningham, S.D. and Ow, D.W. (1996) Promises and prospects of phytoremediation. *Plant Physiology* 110, 715–719.
Daigneault, W. and Betters, D.R. (2000) A comparison of the economic efficiency of living and artificial snow fence designs for road protection. *Western Journal of Applied Forestry* 15, 70–74.
Davidson, W.H. (1979) Hybrid poplar pulpwood and lumber from a reclaimed strip mine. Research Note NE-232. USDA Forest Service, Northeastern Forest Experiment Station, Broomall, Pennsylvania.
Davidson, W.H. and Davis, G. (1972) Sprouting of thinned hybrid poplars on bituminous strip-mine spoils in Pennsylvania. Research Note NE-147. USDA Forest Service, Northeastern Forest Experiment Station, Upper Darby, Pennsylvania.
Davis, G. (1964) Second-year results of hybrid poplar test plantings on bituminous strip mine spoils in Pennsylvania. Research Note NE-19. USDA Forest Service, Northeastern Forest Experiment Station, Upper Darby, Pennsylvania.
Dhanda, R.S., Gill, R.I.S. and Dogra, A.S. (2004) Growth and productivity of poplar-based agri-silviculture system for diversification and food security in northwest India. In: Allen, S.C. (ed.) *Working Together for Sustainable Land-use Systems. Book of Abstracts for the 1st World Agroforestry Congress, Orlando, Florida, USA, June 27–July 2, 2004.* University of Florida, Gainesville, Florida, pp. 50.
Dhankher, O.P., Pilon-Smits, E.A., Meagher, R.B. and Doty, S.L. (2011) Biotechnological approaches for phytoremediation. In: Altman, A. and Hasegawa, P.M. (eds) *Plant Biotechnology and Agriculture: Prospects for the 21st Century.* Elsevier, Amsterdam, pp. 309–328.
Dhiman, R. (2012) Status of poplar culture in India. *ENVIS Forestry Bulletin* 12, 15–32.
Dhondt, A.A., Wrege, P.H., Sydenstricker, K.V. and Cerretani, J. (2004) Clone preference by nesting birds in short-rotation coppice in plantations in central and western New York. *Biomass and Bioenergy* 27, 429–435.
Dickerson, J.A. and Barber, A. (1999) Developing living snow fence practices in the Northeast. In: Buck, L.E. and Lassoie, J.P. (eds) *Proceedings 5th Conference Agroforestry in North America.* Cornell University, Ithaca, New York, pp. 122–123.
Dimitriou, I. and Aronsson, P. (2004) Nitrogen leaching from short-rotation willow coppice after intensive irrigation with wastewater. *Biomass and Bioenergy* 26, 433–441.
Dimitriou, I. and Aronsson, P. (2005) Willows for energy and phytoremediation in Sweden. *Unasylva* 56, 47–50.
Dimitriou, I. and Aronsson, P. (2007) Landfill leachate treatment on short-rotation willow coppice: resource instead of waste? In: Lehmann, E.C. (ed.) *Landfill Research Focus.* Nova Publishers, New York, pp. 263–291.
Dimitriou, I. and Aronsson, P. (2010) Landfill leachate treatment with willows and poplars – efficiency and plant response. *Waste Management* 30, 2137–2145.
Dimitriou, I. and Aronsson, P. (2011) Wastewater and sewage sludge application to willows and poplars grown in lysimeters – plant response and treatment efficiency. *Biomass and Bioenergy* 35, 161–170.
Dimitriou, I. and Rosenqvist, H. (2011) Sewage sludge and wastewater fertilisation of short rotation coppice (SRC) for increased bioenergy production – biological and economic potential. *Biomass and Bioenergy* 35, 835–842.
Dimitriou, I., Aronsson, P. and Weih, M. (2003) Evaluation of leaf length and fluctuating asymmetry as stress indicators after irrigation of willow with landfill leachate. In: Christensen, T.H., Cossu, R. and

Stegmann, R. (eds) *Proceedings of the Sardinia 2003, Ninth International Waste Management and Landfill Symposium, Sardinia, Italy.* CISA Publisher, Padova, Italy, pp. 594–595.

Dimitriou, I., Aronsson, P. and Weih, M. (2006a) Stress tolerance of five willow clones after irrigation with different amounts of landfill leachate. *Bioresource Technology* 97, 150–157.

Dimitriou, I., Eriksson, J., Adler, A., Aronsson, P. and Verwijst, T. (2006b) Fate of heavy metals after application of sewage sludge and wood-ash mixtures to short-rotation willow coppice. *Environmental Pollution* 142, 160–169.

Dimitriou, I., Baum, C., Baum, S., Busch, G., Schulz, U., Köhn, J., *et al.* (2009a) The impact of short rotation coppice (SRC) cultivation on the environment. *Landbauforschung – vTI Agriculture and Forestry Research* 3(59), 159–162.

Dimitriou, I., Busch, G., Jacobs, S., Schmidt-Walter, P. and Lamersdorf, N. (2009b) A review of the impacts of short rotation coppice cultivation on water issues. *Landbauforschung – vTI Agriculture and Forestry Research* 3(59), 197–206.

Dimitriou, I., Rosenqvist, H. and Berndes, G. (2011) Slow expansion and low yields of willow short rotation coppice in Sweden; implications for future strategies. *Biomass and Bioenergy* 35, 4613–4618.

Dos Santos Utmazian, M.N., Wieshammer, G., Vega Montoya, R. and Wenzel, W.W. (2007) Hydroponic screening for metal resistance and accumulation of cadmium and zinc in twenty clones of willows and poplars. *Environmental Pollution* 148, 155–165.

Doty, S.L. (2008) Tansley Review: Enhancing phytoremediation through the use of transgenics and endophytes. *New Phytologist* 179, 318–333.

Doty, S.L. (2011) Nitrogen-fixing endophytic bacteria for improved plant growth. In: Maheshwari, D.K. (ed.) *Bacteria in Agrobiology.* Springer, Germany, pp. 183–199.

Doty, S.L., Shang, Q.T., Wilson, A.M., Tangen, J., Westergreen, A., Newman, L.A., *et al.* (2000) Enhanced metabolism of halogenated hydrocarbons in transgenic plants containing mammalian P450 2E1. *Proceedings of the National Academy of Sciences of the USA* 97, 6287–6291.

Doty, S.L., James, C.A., Moore, A.L., Vajzovic, A., Singleton, G.L., Ma, C., *et al.* (2007) Enhanced phytoremediation of volatile environmental pollutants with transgenic trees. *Proceedings of the National Academy of Sciences of the USA* 104, 16816–16821.

Doty, S.L., Oakely, B., Xin, G., Kang, J.W., Singleton, G., Khan, Z., *et al.* (2009) Diazotrophic endophytes of native black cottonwood and willow. *Symbiosis* 47, 23–33.

Douglas, G.B., Bulloch, B.T. and Foote, A.G. (1996) Cutting management of willows (*Salix* spp.) and leguminous shrubs for forage during summer. *New Zealand Journal of Agriculture Research* 39, 175–184.

Dowling, D.N. and Doty, S.L. (2009) Improving phytoremediation through biotechnology. *Current Opinion in Biotechnology* 20, 204–206.

Duggan, J. (2005) The potential for landfill leachate treatment using willows in the UK – a critical review. *Resources, Conservation and Recycling* 45, 97–113.

Eapen, S. and D'Souza, S.F. (2005) Prospects of genetic engineering of plants for phytoremediation of toxic metals. *Biotechnology Advances* 23, 97–114.

Ebbs, S., Bushey, J., Poston, S., Kosma, D., Samiotakis, M. and Dzombak, D. (2003) Transport and metabolism of free cyanide and iron cyanide complexes by willow. *Plant, Cell and Environment* 26, 1467–1478.

Eriksson, M., Dalhammar, G. and Borg-Karlson, A.K. (2000) Biological degradation of selected hydrocarbons in an old PAH/creosote contaminated soil from a gas work site. *Applied Microbiology and Biotechnology* 53, 619–626.

Ernst, W.H.O. (1996) Bioavailability of heavy metals and decontamination of soils by plants. *Applied Geochemistry* 11, 163–167.

Fan, Z.P., Gao, J.G., Zeng, D.H., Zhou, X.H. and Sun, X.K. (2010) Three-dimensional (3D) structure model and its parameters for poplar shelterbelts. *Science China Earth Sciences* 53, 1513–1526.

FAO (1958) *Poplars in Forestry and Land Use.* Food and Agriculture Organization of the United Nations, Rome.

FAO (1980) *Poplars and Willows in Wood Production and Land Use.* FAO Forestry Series No 10. Food and Agriculture Organization of the United Nations, Rome.

Farber, M.E. (2006) The effect of organic amendments on shrub willow growth in Solvay waste material. MSc. thesis, State University of New York College of Environmental Science and Forestry, Syracuse, New York.

Felix, E., Tilley, D.R., Felton, G. and Flamino, E. (2008) Biomass production of hybrid poplar (*Populus* spp.) grown on deep trenched municipal biosolids. *Ecological Engineering* 33, 8–14.

Ferro, A., Chard, B., Gefell, M., Thompson, B. and Kjelgren, R. (2000) Phytoremediation of organic solvents in groundwater: pilot study at a superfund site. In: Wickramanayake, G.B., Gavaskar, A.R., Alleman, B.C. and Maga, V.S. (eds) *Bioremediation and Phytoremediation of Chlorinated and Recalcitrant Compounds. Volume 2.* Battelle Press, Columbus, Ohio, pp. 461–466.
Fillion, J., Brisson, J., Guidi, W. and Labrecque, M. (2011) Increasing phosphorus removal in willow and poplar vegetative filters using arbuscular mycorrhizal fungi. *Ecological Engineering* 37, 199–205.
Fontaine, S., Barot, B., Barre, P., Bdioui, N., Mary, B. and Rumpel, C. (2007) Stability of organic carbon in deep soil layers controlled by fresh carbon supply. *Nature* 450, 277–280.
Ford-Robertson, F.C. (1971) *Terminology of Forest Science, Technology Practice, and Products.* Multilingual Forestry Terminology Series No 1. Society of American Foresters, Washington, DC, 348 pp.
French, C.L., Dickinson, N.M. and Putwain, P.D. (2006) Woody biomass phytoremediation of contaminated brownfield land. *Environmental Pollution* 141, 387–395.
Funk, D. (1960) A revised bibliography of strip mine reclamation. Paper No 35. USDA Forest Service, Central States Forest Experiment Station, St Paul, Minnesota.
Gardiner, E.S., Stanturf, J.A. and Schweitzer, C.J. (2004) An afforestation system for restoring bottomland hardwood forests: biomass accumulation of Nuttall oak seedlings interplanted beneath eastern cottonwood. *Restoration Ecology* 12, 525–532.
George, E.J., Broberg, D. and Worthington, D.L. (1963) Influence of various types of field windbreaks on reducing wind velocity and depositing snow. *Journal of Forestry* 61, 345–349.
Gera, M. (2012) Poplar culture for speedy carbon sequestration in India: a case study from Terai Region of Uttarakhand. *ENVIS Forestry Bulletin* 12, 75–83.
Germaine, K.J., Keogh, E., Ryan, D. and Dowling, D.N. (2009) Bacterial endophyte-mediated naphthalene phytoprotection and phytoremediation. *FEMS Microbiology Letters* 296, 226–234.
Glova, G.J. and Sagar, P.M. (1994) Comparison of fish and macroinvertebrate standing stocks in relation to riparian willows (*Salix* spp.) in 3 New Zealand streams. *New Zealand Journal of Marine Freshwater Research* 28, 255–266.
Godley, A., Alker, A., Hallett, J., Marshall, R. and Riddell-Black, D. (2004) Landfill leachate nutrient recovery by willow short rotation coppice. *Arboriculture Journal* 28, 253–279.
Goodlass, G., Green, M., Hilton, B. and McDonough, S. (2007) Nitrate leaching from short rotation coppice. *Soil Use and Management* 23, 178–184.
Gordon, M.P., Choe, N., Duffy, J., Ekuan, G., Heilman, P., Muiznieks, I., *et al.* (1997) Phytoremediation of trichloroethylene with hybrid poplars. In: Kruger, E.L., Anderson, T.A. and Coats, J.R. (eds) *Phytoremediation of Soil and Water Contaminants.* American Chemical Society, Washington, DC, pp. 177–185.
Grabow, W.O.K., Middendorff, I.G. and Basson, N.C. (1978) Role of lime treatment in the removal of bacteria, enteric viruses and coliphages in waste water reclamation plant. *Applied Environmental Microbiology* 35, 663–669.
Guevara-Escobar, A., Kemp, P.D., Mackay, A.D. and Hodgson, J. (2002) Soil properties of a widely spaced, planted poplar (*Populus deltoides*)-pasture system in a hill environment. *Australian Journal of Soil Research* 40, 873–886.
Guevara-Escobar, A., Kemp, P.D., Mackay, A.D. and Hodgson, J. (2007) Pasture production and composition under poplar in a hill environment in New Zealand. *Agroforestry Systems* 69, 199–213.
Guidi, W., Piccioni, E. and Bonari, E. (2008) Evapotranspiration and crop coefficient of poplar and willow short rotation coppice used as a vegetative filter. *Bioresource Technology* 94, 4832–4840.
Guidi, W., Kadri, H. and Labrecque, M. (2012) Establishment techniques for using willow for phytoremediation on a former oil refinery in southern Quebec: achievements and constraints. *Chemistry and Ecology* 28, 49–64.
Gullickson, D., Josiah, S.J. and Flynn, P. (1999) Catching the snow with living snow fences. University of Minnesota Extension Service Publication MI7311-S, St Paul, Minnesota, 127 pp.
Guo, L.B. and Gifford, R.M. (2002) Soil carbon stocks and land use change: a meta analysis. *Global Change Biology* 8, 345–360.
Gustafsson, L. (1987) Plant conservation aspects of energy forestry – a new type of land use in Sweden. *Forest Ecology and Management* 21, 141–161.
Hagen, L.J. (1976) Windbreak design for optimum wind erosion control. In: Tinus, R.W. (ed.) *Shelterbelts on the Great Plains. Proceedings of Symposium, April, 1976, Denver, Colorado, USA.* GPAC Publication 78. Great Plains Agricultural Council, Lincoln, Nebraska, pp. 31–36.
Hamel, P.B. (2003) Winter bird community differences among methods of bottomland hardwood forest restoration: results after seven growing seasons. *Forestry* 76, 189–197.

Han, K.-H., Meilan, R., Ma, C. and Strauss, S.H. (2000) An *Agrobacterium tumefaciens* transformation protocol effective on a variety of cottonwood hybrids (genus *Populus*). *Plant Cell Reports* 19, 315–320.

Hannink, N., Rosser, S.J., French, C.E., Basran, A., Murray, J.A.H., Nicklin, S., *et al.* (2001) Phytodetoxification of TNT by transgenic plants expressing a bacterial nitroreductase. *Nature Biotechnology* 19, 1168–1172.

Hart, G. and Byrnes, W.R. (1960) Trees for strip-mined lands. Paper No 136. USDA Forest Service, Northeastern Forest Experiment Station, Upper Darby, Pennsylvania.

Hartley, M.J. (2002) Rationale and methods for conserving biodiversity in plantation forests. *Forest Ecology and Management* 155, 81–95.

Hasselgren, K. (1998) Use of municipal waste products in energy forestry: highlights from 15 years of experience. *Biomass and Bioenergy* 15, 71–74.

Hasselgren, K. (1999) Utilization of sewage sludge in short-rotation energy forestry: a pilot study. *Waste Management and Research* 4, 251–262.

Hathaway, R.L. (1987) Willows for the future. Publication 53. Soil Conservation Centre, Ministry of Works and Development, Palmerston North, New Zealand.

Heinsoo, K. and Holm, B. (2008) Use of municipal wastewater and composted wastewater sludge in willow short rotation coppice in Estonia. In: Sarsby, R.W. and Meggyes, T. (eds) *Proceedings of the International Conference of Construction for a Sustainable Environment: Construction for a Sustainable Environment, Vilnius, Lithuania.* CRC Press, Boca Raton, Florida, pp. 463–470.

Heinsoo, K. and Koppel, A. (2003) *Nutrient Removal Capacity of Willow Short Rotation Forest – Estonian Case Study.* Publicationes Instituti Geographici Universitatis Tartuensis 94. Tartu University Press, Tartu, Estonia, pp. 102–105.

Heisler, G.M. and DeWalle, D.R. (1988) Effects of windbreak structure on wind flow. *Agriculture, Ecosystems and Environments* 22/23, 41–69.

Heller, M.C., Keoleian, A. and Volk, T.A. (2003) Life cycle assessment of willow bioenergy cropping system. *Biomass and Bioenergy* 25, 147–165.

Helms, J.A. (1998) *The Dictionary of Forestry.* Society of American Foresters, Bethesda, Maryland, 210 pp.

Hewlett, J.D. (1956) The development of vegetation on the Solvay Wastebeds. MSc. thesis, State University of New York College of Environmental Science and Forestry, Syracuse, New York.

Hook, D.D. (1984) Waterlogging tolerance of lowland tree species of the South. *Southern Journal of Applied Forestry* 8, 136–149.

Hughes, J.B., Shanks, J., Vanderford, M., Lauritzen, J. and Bhadra, R. (1997) Transformation of TNT by aquatic plants and plant tissue cultures. *Environmental Science and Technology* 31, 266–271.

IDF (1981) *La réalisation pratique des haies brise-vent et bandes boisées.* Institut pour le Développement Forestier, Paris.

Isebrands, J.G. (2007) *Best Management Practices Poplar Manual for Agroforestry Applications in Minnesota (Final).* Environmental Forestry Consultants LLC, New London, Wisconsin (http://www.extension.umn.edu/distribution/naturalresources/00095.html, accessed 24 May 2012).

Isebrands, J.G. and Karnosky, D.F. (2001) Environmental benefits of poplar culture. In: Dickmann, D.I., Isebrands, J.G., Eckenwalder, J.E. and Richardson, J. (eds) *Poplar Culture in North America.* National Research Council of Canada Research Press, Ottawa, Ontario, pp. 207–218.

Jackson, S. (2000) Establishing willow, poplar and other vegetation on a brownfield site in Utica, NewYork, USA. MSc. thesis, State University of NewYork College of Environmental Science and Forestry, Syracuse, New York.

Jackson, R.G., Rylott, E.L., Fournier, D., Hawari, J. and Bruce, N.C. (2007) Exploring the biochemical properties and remediation applications of the unusual explosive-degrading P450 system XplA/B. *Proceedings of the National Academy of Sciences of the USA* 104, 16822–16827.

Johnson, A.R. (2005) Assessing performance of a *Salix* based evapotranspiration (ET) cover using the SHAW model. MSc. thesis, State University of New York College of Environmental Science and Forestry, Syracuse, New York.

Jones, D.L., Williamson, K.L. and Owen, A.G. (2006) Phytoremediation of landfill leachate. *Waste Management* 26, 825–837.

Jonsson, M. (2004) Wet storage of roundwood – effects on wood properties and treatment of run-off water. PhD thesis, Swedish University of Agricultural Sciences, Uppsala, Sweden.

Jonsson, M., Dimitriou, I., Aronsson, P. and Elowson, T. (2004) Effects of soil type, irrigation volume and plant species on treatment of log yard run-off in lysimeters. *Water Research* 38, 3473–3644.

Jørgensen, U. and Schelde, K. (2001) Energy crop water and nutrient use efficiency. International Energy Agency, IEA Bioenergy Task 17, Short Rotation Crops (http://infohouse.p2ric.org/ref/17/16275.pdf, accessed 1 June 2012).

Kabata-Pendias, A. and Pendias, K. (1984) *Trace Elements in Soils and Plants.* CRC Press Inc., Boca Raton, Florida.

Kahle, P., Baum, C. and Boelcke, B. (2005) Effect of afforestation on soil properties and mycorrhizal formation. *Pedosphere* 15, 754–760.

Kang, J.W., Khan, Z. and Doty, S.L. (2012) Biodegradation of trichloroethylene (TCE) by an endophyte of hybrid poplar. *Applied and Environmental Microbiology* 78, 3504–3507.

Kenney, W.A. (1992) The role of *Salicaceae* species in windbreaks. *The Forestry Chronicle* 68, 209–213.

Kenney, W.A., Sennerby-Forsse, L. and Layton, P. (1990) A review of biomass quality research relevant to the use of poplar and willow for energy conversion. *Biomass* 21, 163–188.

Khan, Z. and Doty, S.L. (2011) Endophyte assisted phytoremediation. *Current Topics in Plant Biology* 12, 97–105.

Klang-Westin, E. and and Perttu, K. (2002) Effect of nutrient supply and soil cadmium removal by willow. *Biomass Bioenergy* 23, 415–426.

Knabe, W. (1964) A visiting scientist's observations and recommendations concerning strip-mined reclamation in Ohio. *Ohio Journal of Science* 64, 132–157.

Komarek, M., Tiustos, P., Szakova, J. and Chrastny, V. (2007) The use of poplar during a two-year induced phytoextraction of metals from contaminated agricultural soils. *Environmental Pollution* 158, 2428–2438.

Koo, Y.B., Lee, S.K., Kim, P.G., Byun, K.O. and Woo, S.Y. (1997) Growth and absorption capacity of heavy metals of *Populus euramericana* at Nanjido landfill. *Poplars* 14, 25–32.

Koo, Y.B., Noh, E.R., Woo, S.Y. and Lee, S.K. (1998) Using poplar trees as a landfill cover and leachate treatment. *Poplars* 15, 19–29.

Koo, Y.B., Kim, I.S., Yeo, J.K., Lee, S.K. and Joo, C.H. (1999) Phytoremediation of landfill leachate and contaminated soil by fast growing tree species. *Poplars* 16, 26–36.

Koo, Y.B., Yeo, J.K., Kim, I.S., Kim, T.S., Kim, Y.J. and Yeo, I.S. (2002) Adaptation test of poplar and willow clones at Kimpo Metropolitan landfill. *Journal of the Korean Forestry Society* 91, 405–411.

Koppel, A, Perttu, K. and Ross, J. (1996) Estonian energy forest plantation – general information. In: Perttu, K. and Koppel, A. (eds) *Short Rotation Willow Coppice for Renewable Energy and Improved Environment.* Report 57. Swedish University of Agricultural Sciences, Department of Short Rotation Forestry, Uppsala, Sweden, pp. 159–165.

Kornacki, A.H. (2005) First-year survival and growth of select willow clones in saturated, PAH contaminated seeps. MSc. thesis, State University of New York College of Environmental Science and Forestry, Syracuse, New York.

Kowalik, P.J. and Randerson, P.F. (1994) Nitrogen and phosphorus removal by willow stands irrigated with municipal waste water – a review of the Polish experience. *Biomass and Bioenergy* 6, 133–139.

Krygier, R.F. (2011) Production of short rotation woody biomass for energy feedstocks as a cost-effective solution for disposal of industrial and municipal biosolids and municipal sewage wastewater. In: Mirck, J. and Schroeder, W. (eds) *Poplars and Willows on the Prairies. Collection of Abstracts.* Joint Conference Poplar Council of Canada, IPC Environmental Applications Working Party, and Poplar Council of the United States, Edmonton, Alberta, Canada, 18–24 September 2011. Poplar Council of Canada, Edmonton, Alberta, Canada, pp. 44.

Kuhn, A., Ballach, H.-J. and Wittig, R. (2004) Studies in the biodegradation of 5 PAHs (phenanthrene, pyrene, fluoranthene, chrysene and benzo(a)pyrene) in the presence of rooted poplar cuttings. *Environmental Science and Pollution Research* 11, 22–32.

Kumar, P.B, Dushenkov, V., Motto, H. and Rasakin, I. (1995) Phytoextraction: the use of plants to remove heavy metals from soils. *Environmental Science and Technology* 29, 1232–1238.

Kuusemets, V., Heinsoo, K., Sild, E. and Koppel, A. (2001) Short rotation willow plantation for wastewater purification: case study at Aarike, Estonia. In: Villacampa, Y., Breddia, C.A. and Uso, J.L. (eds) *Ecosystems and Sustainable Development II.* WIT Press, Southampton, UK, pp. 61–68.

Kuzovkina, Y.A. and Quigley, M.F. (2004a) *Metal Resistance and Accumulation in North American Willow (Salix L.) Species.* Proceedings of the 22nd Session of the International Poplar Commission, Santiago, Chile. FAO, Rome, pp. 151.

Kuzovkina, Y.A. and Quigley, M.F. (2004b) *Differential Response of Willow (Salix L.) Species to Acute Ozone Exposure.* Proceedings of the 22nd Session of the International Poplar Commission, Santiago, Chile. FAO, Rome, pp. 31.

Kuzovkina, Y.A. and Quigley, M.F. (2005) Willows beyond wetlands: uses of *Salix* L. species for environmental projects. *Water, Air and Soil Pollution* 162, 183–204.

Kuzovkina, Y.A., Knee, M. and Quigley, M.F. (2004a) Cadmium and copper uptake and translocation of five *Salix* L. species. *International Journal of Phytoremediation* 6, 269–287.

Kuzovkina, Y.A., Knee, M. and Quigley, M.F. (2004b) Soil compaction and flooding effects on the growth of twelve *Salix* L. species. *Journal of Environmental Horticulture* 22, 155–160.

Kuzovkina, Y.A., Weih, M., Abalos Romero, M., Charles, J., Hurst, S., McIvor, I., *et al.* (2008) *Salix*: botany and global horticulture. *HortReviews* 34, 447–489.

Labrecque, M. and Teodorescu, T.I. (2003) High biomass yield achieved by *Salix* clones in SRIC following two 3-year coppice rotations on abandoned farmland in southern Quebec, Canada. *Biomass and Bioenergy* 25, 135–146.

Labrecque, M. and Teodorescu, T.I. (2004) From living walls to wood panels: multiple uses of willow stems produced from short rotation culture in Southern Québec, Canada. Working Report to International Poplar Commission 22nd Session, Santiago, Chile, November 29–December 2, 2004. Available from IPC Secretariat, FAO, Rome.

Labrecque, M., Teodorescu, T.I. and Daigle, S. (1995) Effect of wastewater sludge on growth and heavy metal bioaccumulation on two *Salix* species. *Plant and Soil* 171, 303–316.

Labrecque, M., Teodorescu, T.I. and Diagle, S. (1997) Biomass productivity and wood energy of *Salix* species after 2 years growth in SRIC fertilized with wastewater sludge. *Biomass and Bioenergy* 12, 409–417.

Laidlaw, W.S., Arndt, S.K., Huynh, T.T., Gregory, D. and Baker, A.J. (2012) Phytoextraction of heavy metals by willows growing in biosolids under field conditions. *Journal of Environmental Quality* 41, 134–143.

Lal, R. (2007) Soil science and carbon civilization. *Soil Science Society of America Journal* 71, 1425–1437.

Landberg, T. and Greger, M. (1996) Differences in uptake and tolerance to heavy metals in *Salix* from unpolluted and polluted areas. *Applied Geochemistry* 11, 175–180.

Larsson, L., Cuingnet, C., Clause, P., Jacobsson, P., Dawson, M., Wilson, F., *et al.* (2003) Short rotation willow biomass plantations irrigated and fertilised with wastewater. *Sustainable Urban Renewal and Wastewater Treatment* No 37. Danish Environmental Protection Agency, Copenhagen.

Laureysens, I., Blust, R., De Temmerman, L., Lemmens, C. and Ceulemans, R. (2004a) Clonal variation in heavy metal accumulation and biomass production in a poplar coppice culture. I. Seasonal variation in leaf, wood and bark concentrations. *Environmental Pollution* 131, 485–494.

Laureysens, I., Bogaert, J., Blust, R. and Ceulemans, R. (2004b) Biomass production of 17 poplar clones in a short-rotation coppice culture on a waste disposal site and its relation to soil characteristics. *Forest Ecology and Management* 187, 295–309.

Laureysens, I., De Temmerman, L., Hastir, T., Van Gysel, M. and Ceulemans, R. (2005) Clonal variation in heavy metal accumulation and biomass production in a poplar coppice culture. II. Vertical distribution and phytoextraction potential. *Environmental Pollution* 133, 541–551.

Lee, J., Masters, D.G., White, C.L., Grace, N.D. and Judson, G.J. (1999) Current issues in trace element nutrition of grazing livestock in Australia and New Zealand. *Australian Journal of Agriculture Research* 50, 1341–1364.

Lester, P.J., Mitchell, S.F. and Scott, D. (1994) Effects of riparian willow trees (*Salix fragilis*) on macroinvertebrate densities in 2 small central Otago, New Zealand streams. *New Zealand Journal of Marine Freshwater Research* 28, 267–276.

Licht, L.A. and Isebrands, J.G. (2005) Linking phytoremediated pollutant removal to biomass economic opportunities. *Biomass and Bioenergy* 28, 203–218.

Licht, L., Aitchison, E., Schnabel, W., English, M. and Kaempf, M. (2001) Landfill capping within woodland ecosystems. *Practice Periodical of Hazardous, Toxic, and Radioactive Waste Management* 5, 175–184.

Limstrom, G. (1960) *Forestation of Strip Mined Land in the Central States*. Agricultural Handbook. USDA Forest Service, Washington, DC.

Lindegaard, K., Parfitt, R.I., Donaldson, G., Hunter, T., Dawson, W.M., Forbes, E.G.A., *et al.* (2001) Comparative trials of Swedish and UK biomass willow varieties. In: Bullard, M.J., Christian, D.G., Knight, J.D., Lainsbury, H.A. and Parker, S.R. (eds) *Aspects of Applied Biology 65, Biomass and Energy Crops II*. Association of Applied Biologists, Warwick, UK, pp. 183–198.

Londo, M., Roose, M., Dekker, J. and de Graaf, H. (2004) Willow short-rotation coppice in multiple land-use systems: evaluation of four combination options in the Dutch context. *Biomass and Bioenergy* 27, 205–211.

Lumme, I. and Tormala, T. (1988) Selection of fast growing willow (*Salix* spp.) clones for short rotation forestry on mined peatlands in northern Finland. *Silva Fennica* 22, 67–88.
Lyyra, S., Meagher, R.B., Kim, T., Heaton, A., Montello, P., Balish, R.S., *et al.* (2007) Coupling two mercury resistance genes in eastern cottonwood enhances the processing of organomercury. *Plant Biotechnology Journal* 5, 254–262.
McCracken, A.R., Werner, A., Johnston, C., Moore, P., Doherty, L., Lynch, M., *et al.* (2009) Potential of using energy crops as biological filters for the bioremediation of sewage and industrial effluents. In: Horan, N. (eds) *Proceedings of the 3rd European Water and Wastewater Management Conference, Birmingham, UK, September 2009.* Aqua Enviro Ltd, Wakefield, UK Paper 45, 1–10.
McCutcheon, S.C. and Schnoor, J.L. (2003) *Phytoremediation, Transformation and Control of Contaminants.* John Wiley and Sons, Inc., Hoboken, New Jersey.
McIvor, I. (2008) Breeding poplars and willows. *New Zealand Tree Grower* 29, 27–32.
McIvor, I., Cumming, H. and Hurst, S. (2005) Response of four *Salix* species to soil water deficit. *Agronomy New Zealand* 35, 74–80.
McIvor, I.R., Douglas, G.B., Hurst, S.E., Hussain, Z. and Foote, A.G. (2008) Structural root growth of young Veronese poplars on erodible slopes in the southern North Island, New Zealand. *Agroforestry Systems* 72, 75–86.
McIvor, I.R., Douglas, G.B. and Benavides, R. (2009) Coarse root growth of Veronese poplar trees varies with position on an erodible slope in New Zealand. *Agroforestry Systems* 76, 251–264.
McKnight, J.S., Hook, D.D., Langdon, O.G. and Johnson, R.L. (1981) Flood tolerance and related characteristics of trees of the bottomland forests of the southern United States. In: Clark, J.R. and Benforado, J. (eds) *Wetlands of Bottomland Hardwood Forests.* Elsevier Scientific Publishing Company, Amsterdam, pp. 28–69.
Maclaren, P., Manley, B., Andrews, C., Branch, C., Coles, R., Cooper, B., *et al.* (2008) Impact of the New Zealand Emissions Trading Scheme on forest management. *New Zealand Journal of Forestry* 53, 33–39.
McLeod, K.W. and Ciravolo, T.G. (1998) Boron tolerance and potential boron removal by bottom land tree seedlings. *Wetlands* 18, 431–436.
Malone, G.W. (2002) An on farm demonstration of planting trees around poultry houses. *Poultry Science* 81(Supplement 1), 126.
Marmiroli, M., Pietrini, F., Maestri, E., Zacchini, M., Marmiroli, N. and Massacci, A. (2011) Growth, physiological and molecular traits in *Salicaceae* trees investigated for phytoremediation of heavy metals and organics. *Tree Physiology* 31,1319–1334.
Martin, L.T., Pezeshki, S.R. and Shields, F.D. Jr (2005) Soaking treatment increases survival of black willow posts in a large-scale field study. *Ecological Restoration* 23, 748–763.
Matthews, D.W., Massmann, J. and Strand, S.E. (2003) Influence of aquifer properties on phytoremediation effectiveness. *Ground Water* 41, 41–47.
Meagher, R.B. and Heaton, A.C.P. (2005) Strategies for the engineered phytoremediation of toxic element pollution: mercury and arsenic. *Journal of Industrial Microbiology and Biotechnology* 32, 502–513.
Meers, E., Vandecasteele, B., Ruttens, A., Vangrouveld, J. and Tack, F.M.G. (2007) Potential of five willow species (*Salix* spp.) for phytoextraction of heavy metals. *Environmental and Experimental Botany* 60, 57–68.
Mertens, P. (1999) Environmental aspects of the Sambro-Meuse poplar culture. *Biomass and Bioenergy* 16, 275–280.
Mertens, P. (2002) La populiculture et son environnement Wallon. *Silva Belgica* 1, 35–47; 2, 34–48.
Mertens, J., Vervaeke, P., De Schrijver, A. and Layssaert, S. (2004) Metal uptake by young trees from dredged brackish sediment: limitations and possibilities for phytoextraction and phytostabilisation. *Science of The Total Environment* 326, 209–215.
Michalenko, E.M. (1991) Pedogenesis and invertebrate microcommunity succession in immature soils originating from chlor-alkali wastes. PhD thesis, State University of New York College of Environmental Science and Forestry, Syracuse, New York.
Miller, R.S., Khan, Z. and Doty, S.L. (2011) Comparison of trichloroethylene toxicity, removal, and degradation by varieties of *Populus* and *Salix* for improved phytoremediation applications. *Journal of Bioremediation and Biodegradation* S7-001 (doi: 10.4172/2155-6199.S7-001).
Mirck, J. and Volk, T.A. (2010) Seasonal sap flow of four *Salix* varieties growing on the Solvay wastebeds in Syracuse, NY, USA. *International Journal of Phytoremediation* 12, 1–23.
Mirck, J., Isebrands, J.G., Verwijst, T. and Ledin, S. (2005) Development of short-rotation willow coppice systems for environmental purposes in Sweden. *Biomass and Bioenergy* 28, 219–228.

Misra, P.N. and Tewari, S.K. (1999) On the performance of poplars (*Populus deltoides*) on marginal soils in northern India. *Biomass and Bioenergy* 16, 257–262.

Mola-Yudego, B. and Aronsson, P. (2008) Yield models for commercial willow biomass plantations in Sweden. *Biomass and Bioenergy* 32, 829–837.

Moore, R. and Russell, R. (1990) The 'Three Norths' forest protection system – China. *Agroforestry Systems* 10, 71–88.

Moore, K.M., Barry, T.N., Cameron, P.N., Lopez-Villalobos, N. and Cameron, D.J. (2003) Willow (*Salix* sp.) as a supplement for grazing cattle under drought conditions. *Animal Feed Science Technology* 104, 1–11.

Mortensen, J., Nielsen, K.H. and Jørgensen, U. (1998) Nitrate leaching during establishment of willow (*Salix viminalis*) on two soil types and two fertilization levels. *Biomass and Bioenergy* 15, 457–466.

Nelson, N.D., Sturos, J.A., Fritschel, P.R. and Satter, L.D. (1984) Ruminant feed stuff from commercial foliage of hybrid poplars grown under intensive culture. *Forest Products Journal* 34, 37–44.

Newman, L.A. and Reynolds, C.M. (2005) Bacteria and phytoremediation: new uses for endophytic bacteria in plants. *Trends in Biotechnology* 23, 6–8.

Newman, L., Strand, S., Choe, N., Duffy, J., Ekuan, G., Ruszaj, M., *et al.* (1997) Uptake and biotransformation of trichloroethylene by hybrid poplars. *Environmental Science and Technology* 31, 1062–1067.

Newman, L., Doty, S.L., Gery, K.L., Heilman, P., Muiznieks, I., Shang, Q.T., *et al.* (1998) Phytoremediation of organic contaminants: a review of phytoremediation research at the University of Washington. *Journal of Soil Contamination* 7, 531–542.

Newman, L., Wang, X., Muiznieks, I., Ekuan, G., Ruszaj, M., Cortellucci, R., *et al.* (1999) Remediation of trichloroethylene in an artificial aquifer with trees: a controlled field study. *Environmental Science and Technology* 33, 2257–2265.

Nixon, D.J., Stephens, W., Tyrrel, S.F. and Bierley, E.D.R. (2001) The potential for short rotation energy forestry on restored landfill caps. *Bioresource Technology* 77, 237–245.

Pandey, D. (2008) Trees outside the forest (TOF) resources in India. *International Forestry Review* 10, 125–133.

Paris, P., Massacci, A., Aromolo, R., Ecosse, A., Bianconi, D. and Scarascia-Mugnozza, G. (2009) Linking wood bioenergy production and willow plantations with soil and wastewater phytoremediation in Italy. *Abstracts 1st Conference on Advances in Wastewater Treatment and Reuse.* University of Tehran, Tehran, p. 186 (Abstract).

Passive Remediation Systems (2012) Projects (http://prsi.ca/projects.html, accessed 7 June 2012).

Peri, P.L. and Bloomberg, M. (2002) Windbreaks in southern Patagonia, Argentina: a review of research on growth models, windspeed reduction, and effects on crops. *Agroforestry Systems* 56, 129–144.

Perko, D.J. (1995) Living snow fence: friend or foe? In: Rietveld, W.J. (ed.) *Agroforestry and Sustainable Systems Symposium Proceedings.* USDA Forest Service General Technical Report RM-GTR-261, Ft Collins, Colorado, pp. 43–46.

Perttu, K.L. (1999) Environmental and hygienic aspects of willow coppice in Sweden. *Biomass and Bioenergy* 16, 291–297.

Perttu, K., Eriksson, J., Greger, M., Göransson, A., Blombäck, K., Klang-Westin, E., *et al.* (2002) Content and fluxes of cadmium in the soil–willow-system. Final report. Statens Energimyndighet, Eskilstuna, Sweden.

Pietrini, F., Iannelli, M.A., Pasqualini, S. and Massacci, A. (2003) Interaction of cadmium with glutathione and photosynthesis in developing leaves and chloroplasts of *Phragmites australis* (Cav.) Trin. ex strudel. *Plant Physiology* 133, 829–837.

Pietrini, F., Zacchini, M., Iori, V., Pietrosani, L., Ferretti, M. and Massacci, A. (2010) Spatial distribution of cadmium in leaves and its impact on photosynthesis: examples of different strategies in willow and poplar clones. *Plant Biology* 12, 355–363.

Pilon-Smits, E.A.H. and Freeman, J.L. (2006) Environmental cleanup using plants: biotechnological advances and ecological considerations. *Frontiers in Ecology and the Environment* 4, 203–210.

Pitcher, J.A. and McKnight, J.S. (1990) *Salix nigra* Marsh., black willow. In: Burns, R.M. and Honkala, B.H. (eds) *Silvics of North America – Volume 2, Hardwoods.* Agricultural Handbook 654. USDA Forest Service, Washington, DC, pp. 768–772.

Pulford, I.D. and Watson, C. (1996) Phytoremediation of heavy metals. *Environmental Pollution* 92, 241–246.

Pulford, I.D. and Watson, C. (2003) Phytoremediation of heavy metal-contaminated land by trees – a review. *Environment International* 29, 529–540.

Pulford, I.D., Watson, C. and McGregor, S.D. (2001) Uptake of chromium by trees: prospects for phytoremediation. *Environmental Geochemistry and Health* 23, 307–311.

Pulford, I.D., Riddell-Black, D. and Stewart, C. (2002) Heavy metal uptake by willow clones from sewage sludge-treated soil: the potential for phytoremediation. *International Journal of Phytoremediation* 4, 59–72.

Punshon, T. and Dickinson, N.M. (1997) Mobilisation of heavy metals using short-rotation coppice. *Aspects of Applied Biology* 49, 285–292.

Purdy, J. (2006) Assessing the potential of shrub willow (*Salix* spp.) and the role of fungal symbiosis in phytoremediation applications. MSc. thesis, State University of New York College of Environmental Science and Forestry, Syracuse, New York.

Puri, S. (2004) How can we feign sustainability and development of agricultural lands through agroforestry practices in India? In: Allen, S.C. (ed.) *Working Together for Sustainable Land-use Systems. Book of Abstracts for the 1st World Agroforestry Congress, Orlando, Florida, June 27–July 2, 2004.* University of Florida, Gainesville, Florida, p. 273 (Abstract).

Puri, S. and Nair, P.K.R. (2004) Agroforestry research for development in India: 25 years of experiences of a national program. *Agroforestry Systems* 61, 437–452.

Pythoud, F., Sinkar, V.P., Nester, E.W. and Gordon, M.P. (1987) Increased virulence of *Agrobacterium rhizogenes* conferred by the vir region of pTiBo542: application to genetic engineering of poplar. *Bio/Technology* 5, 1323–1327.

Rachwal, L., De Temmerman, L.O. and Istas, J.R. (1992) Differences in the accumulation of heavy metals in poplar clones of various susceptibility to air pollution. *Arboretum Kornickie* 37, 101–111.

Raskin, I., Smith, R.D. and Salt, D.E. (1997) Phytoremediation of metals: using plants to remove pollutants from the environment. *Current Opinion in Biotechnology* 8, 221–226.

Rauser, W.E. and Muwly, P. (1995) Retention of cadmium in roots of maize seedlings. Role of complexation by phytochelatins and related thiol peptides. *Plant Physiology* 109, 195–202.

Reinhold-Hurek, B. and Hurek, T. (1998) Life in grasses: diazotrophic endophytes. *Trends in Microbiology* 6, 139–144.

Renshaw, C.E., Bostick, B.C., Feng, X., Wong, C.K., Winston, E.S., Karimi, R., *et al.* (2006) Impact of land disturbance on the fate of arsenidal pesticides. *Journal of Environmental Quality* 35, 61–67.

Richardson, J. (2012) Poplars and willows on the Prairies. In: Poplar Council of Canada (ed.) *Poplar Council of Canada Newsletter.* Winter 2011. Poplar Council of Canada, Edmonton, Alberta, Canada, pp. 1–7 (www.poplar.ca/upload/documents/n201112e.pdf, accessed 31 July 2013).

Richardson, J., Cooke, J.E.K., Isebrands, J.G., Thomas, B.R. and Van Rees, K.C.J. (2007) Poplar research in Canada – a historical perspective with a view to the future. *Canadian Journal Botany* 85, 1136–1146.

Riddell-Black, D. (1994) Heavy metal uptake by fast growing willow species. In: Aronsson, P. and Perttu, K. (eds) *Willow Vegetation Filters for Municipal Wastewaters and Sludges.* Report Volume 50. Swedish University Agricultural Sciences, Uppsala, Sweden, pp. 145–151.

Riddell-Black, D., Pulford, I.D. and Stewart, C. (1997) Clonal variation in heavy metal uptake by willow. *Aspects of Applied Biology* 49, 327–334.

Rivera, R., Medina, V.F., Larson, S.L. and McCutcheon, S.C. (1998) Phytotreatment of TNT-contaminated groundwater. *Journal of Soil Contamination* 7, 511–529.

Robinson, B.H., Mills, T.M., Petit, D., Fung, L.E., Green, S.R. and Clothier, B.E. (2000) Natural and induced cadmium-accumulation in poplar and willow: implications for phytoremediation. *Plant and Soil* 227, 301–306.

Robinson, B., Green, S., Mills, T., Clothier, B., van der Velde, M., Laplane, R., *et al.* (2003) Phytoremediation: using plants as biopumps to improve degraded environments. *Australian Journal of Soil Research* 41, 599–611.

Robinson, B., Mills, T., Green, S., Chancerel, B., Clothier, B., Fung, L., *et al.* (2005) Trace element accumulation by poplars and willows used for stock fodder. *New Zealand Journal of Agriculture Research* 48, 489–497.

Robinson, B.H., Green, S.R., Chancerel, B., Mills, T.M. and Clothier, B.E. (2007) Poplar for the phytomanagement of boron contaminated sites. *Environmental Pollution* 150, 225–233.

Robinson, P. (1995) The fate of Vetrazin (cyromazine) during wool scouring and its effects on the aquatic environment. PhD thesis, Lincoln University, Christchurch, New Zealand, 180 pp.

Rock, S. (2012) Evapotranspiration (ET) covers. *International Journal of Phytoremediation* Supplement 1, 1–93.

Rockwood, D.L., Naidu, C.V., Carter, D.R., Rahmani, M., Spriggs, T.A., Lin, C., *et al.* (2004) Short-rotation woody crops and phytoremediation: opportunities for agroforestry? *Agroforestry Systems* 61, 51–63.
Rockwood, D.L., Carter, D.R., Langholtz, M.H. and Stricker, J.A. (2006) *Eucalyptus* and *Populus* short rotation woody crops for phosphate mine lands in Florida, USA. *Biomass and Bioenergy* 30, 728–734.
Rode, M. (2005) Energetische nutzung von biomasse und der naturschutz. *Natur und Landschaft* 80, 403–412. (In German, with English summary)
Romantschuk, M., Sarand, I., Petanen, T., Peltola, R., Jonsson-Vihanne, M., Koivula, T., *et al.* (2000) Means to improve the effect of *in situ* bioremediation of contaminated soil: an overview of novel approaches. *Environmental Pollution* 107, 179–185.
Rood, S.B., Patino, S., Coombs, K. and Tyree, M.T. (2000) Branch sacrifice: cavitation-associated drought adaptation of riparian cottonwoods. *Trees* 14, 248–257.
Rosenberg, N.J. (1974) *Microclimate: the Biological Environment.* John Wiley and Sons, New York.
Rosenqvist, H. and Dawson, M. (2005) The economics of using wastewater irrigation of willow in Northern Ireland. *Biomass and Bioenergy* 29, 83–89.
Rosenqvist, H. and Ness, B. (2004) An economic analysis of leachate purification through willow-coppice vegetation filters. *Bioresource Technology* 94, 321–329.
Rosenqvist, H., Aronsson, P., Hasselgren, K. and Perttu, K. (1997) Economics of using municipal wastewater irrigation of willow coppice crops. *Biomass and Bioenergy* 12, 1–8.
Roygard, J.K.F., Green, S.R., Clothier, B.E., Sims, R.E.H. and Bolan, N.S. (1999) Short rotation forestry for land treatment of effluent: a lysimeter study. *Australian Journal of Soil Research* 37, 983–991.
Rugh, C.L., Senecoff, J.F., Meagher, R.B. and Merkle, S.A. (1998) Development of transgenic yellow poplar for mercury phytoremediation. *Nature Biotechnology* 16, 925–928.
Rylott, E.L. and Bruce, N.C. (2008) Plants disarm soil: engineering plants for the phytoremediation of explosives. *Trends in Biotechnology* 27, 73–81.
Rylott, E.L., Jackson, R.G., Edwards, J., Womack, G.L., Seth-Smith, H.M., Rathbone, D.A., *et al.* (2006) An explosive-degrading cytochrome P450 activity and its targeted application for the phytoremediation of RDX. *Nature Biotechnology* 24, 216–219.
Sage, R.B. and Robertson, P.A. (1994) Wildlife and game potential of short rotation coppice in the UK. *Biomass and Bioenergy* 6, 41–48.
Salladin, B.W. (2005) Establishing select willow clones on a water-saturated, PAH contaminated site using various innovative site improvement techniques. MSc. thesis, State University of New York College of Environmental Science and Forestry, Syracuse, New York.
Sanchez, F.G., Coleman, M., Garten, C.T. Jr, Luxmoore, R.J., Stanturf, J., Trettin, C., *et al.* (2007) Soil carbon after 3 years, under short rotation woody crops grown under varying nutrient and water availability. *Biomass and Bioenergy* 31, 793–801.
Sanità di Toppi, L. and Gabbrielli, R. (1999) Responses to cadmium in higher plants. *Environmental and Experimental Botany* 41,105–130.
Sas-Nowosielska, A., Kucharski, R., Malkowski, E., Pogrezeba, M., Kuperberg, J.M. and Krynski, K. (2004) Phytoextraction crop disposal – an unsolved problem. *Environmental Pollution* 128, 373–379.
Scarascia-Mugnozza, G., Ceulemans, R., Heilman, P.E., Isebrands, J.G., Stettler, R.F. and Hinckley, T.M. (1997) Production physiology and morphology of *Populus* species and their hybrids grown under short rotation. II. Biomass components and harvest index of hybrid and parental species clones. *Canadian Journal of Forest Research* 27, 285–294.
Schaff, S.D., Pezeshki, S.R. and Shields, F.D. Jr (2003) Effects of soil conditions on survival and growth of black willow cuttings. *Environmental Management* 31, 748–763.
Schoenmuth, B.W. and Pestemer, W. (2004) Dendroremediation of trinitrotoluene (TNT). Part 2: fate of radio-labelled TNT in trees. *Environmental Science and Pollution Research* 11, 331–339.
Scholten, H. (1988) Snow distribution on crop fields. *Agriculture, Ecosystems and Environments* 22/23, 363–380.
Scholz, V. and Hellebrand, H.J. (2003). Clean energy from farmland – long-term results from practically orientated field trials. Institute of Agricultural Engineering, Potsdam, Germany.
Schroeder, W.R. and Kort, J. (1989) Shelterbelts in the Soviet Union. *Journal of Soil and Water Conservation* 44, 130–134.
Schroeder, W. and Kort, J. (2007) An overview of shelterbelts in modern agriculture. In: Olivier, A. and Campeau, S. (eds) *When Trees and Crops Get Together.* Proceedings 10th North American Agroforestry Conference, Quebec City, Quebec, Canada. University of Laval, Quebec City, Quebec, Canada, pp. 3–14.

Schwab, A.P., Al-Assi, A.A. and Banks, M.K. (1998) Adsorption of naphthalene onto plant roots. *Journal of Environmental Quality* 27, 220–224.
Sebastiani, L., Scebba, F. and Toynetti, R. (2004) Heavy metal accumulation and growth in poplar clones Eridano (*Populus deltoides* × *P. maximowiczii)* and I-214 (*P.* ×*euramericana*) exposed to industrial waste. *Environmental and Experimental Botany* 52, 79–88.
Selosse, M.-A., Baudoin, E. and Vandenkoornhuyse, P. (2004) Symbiotic microorganisms, a key for ecological success and protection of plants. *Comptes Rendus Biologies* 327, 639–648.
Shang, Q.T. and Gordon, M.P. (2002) Transformation of [14C] trichloroethylene by poplar suspension cells. *Chemosphere* 47, 957–962.
Shang, Q.T., Doty, S.L., Wilson, A.M., Howald, W.N. and Gordon, M.P. (2001) Trichloroethylene oxidative metabolism in plants: the trichloroethanol pathway. *Phytochemistry* 58, 1055–1065.
Sharma, N.K., Samra, J.S. and Singh, H.P. (2001) Influence of boundary plantation of poplar (*Populus deltoides*) on soil water use and water use efficiency of wheat. *Agricultural Water Management* 51, 173–185.
Siciliano, S.D., Fortin, N., Mihoc, A., Wisse, G., Labelle, S., Beaumier, D., *et al.* (2001) Selection of specific endophytic bacterial genotypes by plants in response to soil contamination. *Applied and Environmental Microbiology* 6, 2469–2475.
Singh, H.P., Kohli, R.K. and Batish, D.R. (1998) Effect of poplar (*Populus deltoides*) shelterbelt on the growth and yield of wheat in Punjab, India. *Agroforestry Systems* 40, 207–213.
Singh, K.J. (2004) Agroforestry and cropping pattern as basis for land reclamation in northern India. In: Allen, S.C. (ed.) *Working Together for Sustainable Land-use Systems. Book of Abstracts for the 1st World Agroforestry Congress, Orlando, Florida, June 27–July 2, 2004.* University of Florida, Gainesville, Florida, p. 148 (Abstract).
Sirén, G., Sennerby-Forsse, L. and Ledin, S. (1987) Energy plantations – short-rotation forestry in Sweden. In: Hall, D.O. and Overend, R.P. (eds) *Biomass: Regenerable Energy.* John Wiley and Sons Ltd, Chichester, UK, pp. 119–143.
Skärbäck, E. and Becht, P. (2005) Landscape perspective on energy forests. *Biomass and Bioenergy* 28, 151–159.
Smart, L.B., Volk, T.A., Lin, J., Kopp, R.F., Phillips, I.S., Cameron, K.D., *et al.* (2005) Genetic improvement of shrub willow (*Salix* spp.) crops for bioenergy, bioproducts, and environmental applications. *Unasylva* 56, 51–55.
Smesrud, J., Dickey, J., Asare, S., Cox, A., Lanier, A., Jordahl, J., *et al.* (2000) Beneficial reuse of landfill leachate with hybrid poplar. General Technical Report NC-215. USDA Forest Service, North Central Research Station, St Paul, Minnesota.
Spriggs, T.A., Banks, K. and Schwab, P. (2005) Phytoremediation of polycyclic aromatic hydrocarbons in manufactured gas plant-impacted soil. *Journal of Environmental Quality* 34, 1755–1762.
Stanturf, J.A. and Portwood, C.J. (1999) Economics of afforestation with eastern cottonwood (*Populus deltoides*) on agricultural land in the Lower Mississippi Alluvial Valley. In: Haywood, J.D. (ed.) *Proceedings of the Tenth Biennial Southern Silvicultural Research Conference.* General Technical Report SRS-30. USDA Forest Service, Washington, DC, pp. 66–72.
Stanturf, J.A., Gardiner, E.S. and Schoenholtz, S. (2003) Interplanting for bioenergy and riparian restoration in the USA. In: Nicholas, I.D. (ed.) *Proceedings of the Conference on Short Rotation Crops for Bioenergy.* IEA Bioenergy Task 30, pp. 241–249 (http://www.shortrotationcrops.org/PDFs/2003Mtg/29-StanturfetalNZ.pdf, accessed 2 June 2012).
Stettler, R.F., Bradshaw, H.D., Heilman, P.E. and Hinckley, T.M. (1996) *Biology of Populus and its Implications for Management and Conservation.* National Research Council of Canada Research Press, Ottawa, Ontario.
Stoeckler, J.G. and Williams, R.A. (1949) Windbreaks and shelterbelts. In: Stefferud, A. (ed.) *Trees – Yearbook of Agriculture.* US Department Agriculture, Washington, DC, pp. 191–199.
Stomp, A.M., Han, K.H., Wilbert, S., Gordon, M.P. and Cunningham, S.D. (1994) Genetic strategies for enhancing phytoremediation. *Annals of the New York Academy of Sciences* 721, 481–491.
Stott, K.G. and Belcher, A.R. (1978) Living windbreaks: a review of work at Long Ashton. *ARC Research Review* 4, 68–75.
Strycharz, S. and Newman, L. (2009) Use of native plants for remediation of trichloroethylene. I. Deciduous trees. *International Journal Phytoremediation* 11, 150–170.
Tabler, R.D. (1980) Self-similarity of wind profiles in blowing snow allows outdoor modeling. *Journal of Glaciology* 26, 421–434.

Tabler, R.D. (2003) Controlling blowing and drifting snow with snow fences and road design. Final Report, National Cooperative Highway Research Program. NCHRP Project 20-7(147). Tabler and Associates, Niwot, Colorado, 307 pp.

Tabler, R.D. (2006) Three dimensional roughness elements for snow retention. Final Report FHWA-WY-06/04F. State of Wyoming Department of Transportation, Cheyenne, Wyoming, 170 pp.

Taghavi, S., Barac, T., Greenberg, B., Borremans, B., Vangronsveld, J. and van der Lelie, D. (2005) Horizontal gene transfer to endogenous endophytic bacteria from poplar improves phytoremediation of toluene. *Applied and Environmental Microbiology* 71, 8500–8505.

Teodorescu, T.I., Guidi, W. and Labrecque, M. (2011) The use of non-dormant rods as planting material: a new approach to establishing willow for environmental applications. *Ecological Engineering* 37, 1430–1433.

Tharakan, P.J., Volk, T.A., Nowak, C.A. and Abrahamson, L.P. (2005) Morphological traits of 30 willow clones and their relationship to biomass production. *Canadian Journal Forest Research* 35, 421–431.

Thevathasan, N.V. and Gordon, A.M. (2004) Ecology of tree intercropping in the north temperate region: experiences from southern Ontario, Canada. *Agroforestry Systems* 61, 257–268.

Thompson, P.L., Ramer, L.A. and Schnoor, J.L. (1998) Uptake and transformation of TNT by hybrid poplar trees. *Environmental Science and Technology* 32, 975–980.

Thompson, R.C. and Luckman, P.G. (1993) Performance of biological erosion control in New Zealand soft rock hill terrain. *Agroforestry Systems* 21, 191–211.

Thompson, S.E. (2006) Phytoremediation pilot study at the old sanitary landfill, Fort Drum, NY. Semiannual progress report for July–December 2005. Malcolm Pirnie Inc, White Plains, New York.

Thornton, F.C., Joslin, J.D., Bock, B.R., Houston, A., Green, T.H., Schoenholtz, S., *et al.* (1998) Environmental effects of growing woody crops on agricultural land: first year effects on erosion, and water quality. *Biomass and Bioenergy* 15, 57–69.

Truu, M., Truu, J. and Heinsoo, K. (2009) Changes in soil microbial community under willow coppice: the effect of irrigation with secondary-treated municipal wastewater. *Ecological Engineering* 35, 1011–1020.

Tuskan, G.A. and Walsh, M.E. (2001) Short rotation woody crop systems, atmospheric carbon dioxide, and carbon management: a US case study. *The Forestry Chronicle* 77, 259–264.

Twedt, D.J. and Portwood, C.J. (1997) Bottomland hardwood reforestation for neotropical migratory birds: are we missing the forest for the trees? *Wildlife Society Bulletin* 25, 647–652.

Tyndall, J. and Colletti, J. (2001) Shelterbelts, livestock odor mitigation and sustainable agriculture: a research framework. In: Schroeder, W. and Kort, J. (eds) *Temperate Agroforestry: Adoptive and Mitigative Roles in a Changing Physical and Socio-Economic Climate.* Prairie Farm Rehabilitation Administration, Indian Head, Saskatchewan, Canada, pp. 207–220.

Tyndall, J. and Colletti, J. (2006) Mitigating swine odor with strategically designed shelterbelt systems: a review. *Agroforestry Systems* 69, 45–65.

Underwood, E.J. and Suttle, N.F. (1999) *The Mineral Nutrition of Livestock*, 3rd edn. CAB International, Wallingford, UK.

University of Saskatchewan (2012) Phytoremediation of petroleum hydrocarbons (http://phytopet.usask.ca/, accessed 8 June 2012).

Unterbrunner, R., Puschenreiter, M., Sommer, P., Wieshammer, G., Tiustos, P., Zupan, M., *et al.* (2007) Heavy metal accumulation in trees growing on contaminated sites in Central Europe. *Environmental Pollution* 148, 107–114.

USDA Forest Service (2011) Ecosystem services (http://www.fs.fed.us/ecosystemservices/, accessed 29 May 2012).

US EPA (1998) A citizen's guide to phytoremediation. EPA–542-F-98-011, Office of Solid Waste and Emergency Response, Washington, DC.

US EPA (2000) Introduction to phytoremediation. EPA/600/R-99/107. US Environmental Protection Agency, Cincinnati, Ohio.

US EPA (2003) Evapotranspiration landfill cover systems fact sheet. EPA 542-F-03-015. US Environmental Protection Agency, Cincinnati, Ohio, 12 pp.

van Aken, B., Yoon, J.M. and Schnoor, J.L. (2004) Biodegradation of nitro-substituted explosives 2,4,6-trinitrotoluene, hexahydro-1,3,5-trinitro-1,3,5-triazine, and octahydro-1,3,5,7-tetranitro-1,3,5-tetrazocine by a phytosymbiotic *Methylobacterium* sp. associated with poplar tissues (*Populus deltoides* × *nigra* DN34). *Applied and Environmental Microbiology* 70, 508–517.

van Dillewijn, P., Couselo, J., Corredoira, E., Delgado, A., Wittich, R.-M., Ballester, A., *et al.* (2008) Bioremediation of 2,4,6-trinitrotoluene by bacterial nitroreductase expressing transgenic aspen. *Environmental Science and Technology* 42, 7405–7410.

van Eimern, J., Karschon, R., Razumova, L.A. and Robertson, G.W. (1964) Windbreaks and shelterbelts. Technical Note 59. WMO – No 147.TP.70. World Meteorological Organization, Geneva.

Vangronsveld, J., Herzig, R., Weyens, N., Boulet, J., Adriaensen, K., Ruttens, A., *et al.* (2009) Phytoremediation of contaminated soils and groundwater: lessons from the field. *Environmental Science and Pollution Research International* 16, 765–794.

Vervaeke, P., Luyssaert, S., Mertens, J., Meers, E., Tack, F.M. and Lust, N. (2003) Phytoremediation prospects of willow stands on contaminated sediment: a field trial. *Environmental Pollution* 126, 275–282.

Verwijst, T. (2001) Willows: an underestimated resource for environment and society. *The Forestry Chronicle* 77, 281–285.

Vézina, A. (1994) Les haies brise-vent pour la protection des aires de travail, des bâtiments et des pâturages. In: Desmarais, C. (ed.) *Les actes de la Colloque sur L'Arbre en ville et à la champagne: les pratiques de végétalisation Montreal, Quebec, Canada, Nov 2–3, 1994.* Fondation Louis-de-Gonzague-Fortin, La Pocatière, Québec, Canada, pp. 67–93.

Volk, T.A., Abrahamson, L.P., Merrit, J. and Baker, T. (2004) Capturing the snow with fast-growing willow living snowfences. Presented September 16, 2004 at ROW 8, Saratoga Springs, New York (http://www.esf.edu/willow/PDF/Presentations/Snowfences.pdf, accessed 31 May 2012).

Volk, T.A., Abrahamson, L.P., Nowak, C.A., Smart, L.B., Tharakan, P.J. and White, E.H. (2006) The development of short-rotation willow in the northeastern United States for bioenergy and bioproducts, agroforestry and phytoremediation. *Biomass and Bioenergy* 30, 715–727.

VROM (2000) Circular on target values and intervention values for soil remediation. Ministry of Housing, Spatial Planning and Environment, Department of Soil Protection (VROM), The Hague, Netherlands.

Wang, X., Gordon, M.P. and Strand, S.E. (2002) Mechanism of aerobic transformation of carbon tetrachloride by poplar cells. *Biodegradation* 13, 297–305.

Wang, X., Dossett, M.P., Gordon, M.P. and Strand, S.E. (2004) Fate of carbon tetrachloride during phytoremediation with poplar under controlled field conditions. *Environmental Science and Technology* 38, 5744–5749.

Watson, C., Pulford, I.D. and Riddell-Black, D. (1999) Heavy metal toxicity responses of two willow (*Salix*) varieties grown hydroponically: development of a tolerance screening test. *Environmental Geochemistry and Health* 21, 359–364.

Weih, M. (2008) Short rotation forestry (SRF) on agricultural land and its possibilities for sustainable energy supply. *TemaNord* 2008: 543 (http://www.norden.org/pub/sk/showpub.asp?pubnr=2008:543, accessed 10 June 2012).

Weih, M. (2009) Willow short rotation coppice commercially grown on agricultural land in Sweden – possibilities for improvement of biodiversity and landscape design. Technical Review No 4. IEA Bioenergy Task 30 (www.shortrotationcrops.org/pdfs/tech review no. 04 for webfinal.pdf, accessed 10 June 2012).

Weih, M. and Nordh, N.-E. (2007) Biomass production with fast-growing trees on agricultural land in cool-temperate regions: possibilities, limitations, challenges. In: Clarkson, P.A. (ed.) *Environmental Research Advances.* Nova Science Publishers, New York, pp. 155–170.

Weih, M., Karacic, A., Munkert, H., Verwijst, T. and Diekmann, M. (2003) Influence of young poplar stands on floristic diversity in agricultural landscapes (Sweden). *Basic and Applied Ecology* 4, 149–156.

Werner, A. and McCracken, A. (2008) The use of short rotation coppice poplar and willow for the bioremediation of sewage effluent. In: Booth, E., Green, M., Karp, A., Shield, I., Stock, D. and Turley, D. (eds) *Biomass and Energy Crops III, Aspects of Biology* 90. Association of Applied Biologists, Warwick Enterprise Park, Wellesbourne, Warwick, UK, pp. 317–324.

Weyens, N., van der Lelie, D., Taghavi, S. and Vangronsveld, J. (2009a) Phytoremediation: plant–endophyte partnerships take the challenge. *Current Opinion in Biotechnology* 20, 248–254.

Weyens, N., van der Lelie, D., Artois, T., Smeets, K., Taghavi, S., Newman, L., *et al.* (2009b) Bioaugmentation with engineered endophytic bacteria improves contaminant fate in phytoremediation. *Environmental Science and Technology* 43, 9413–9418.

Weyens, N., Croes, S., Dupae, J., Newman, L., van der Lelie, D., Carleer, R., *et al.* (2010) Endophytic bacteria improve phytoremediation of Ni and TCE co-contamination. *Environmental Pollution* 158, 2422–2427.

WHO (2006) *Guidelines for the Safe Use of Wastewater, Excreta, and Greywater. Volume 2: Wastewater Use in Agriculture.* World Health Organization, Geneva, Switzerland.

Widdowson, M.A., Shearer, S., Andersen, R.G.T. and Novak, J.T. (2005) Remediation of polycyclic aromatic hydrocarbon compounds in groundwater using poplar trees. *Environmental Science and Technology* 39, 1598–1605.

Wilkinson, A.G. (1999) Poplars and willows for soil erosion control in New Zealand. *Biomass and Bioenergy* 16, 263–274.

Wittig, R., Ballach, H.-J. and Kuhn, A. (2003) Exposure of the roots of *Populus nigra* L. cv. Loenen to PAHs and its effect on growth and water balance. *Environmental Science and Pollution Research* 10, 235–244.

Yeo, J.K., Koo, Y.B. and Kim, I.S. (1999) Effect of NaCl salinity on growth, and cation accumulation of 5 poplar species. *Forest Research Institute Journal of Forest Science* 61, 9–17.

Yeo, J.K., Kim, I.S., Koo, Y.B. and Lee, J.C. (2001) Uptake and tolerance to lead in *Populus alba* × *glandulosa* and *Betula schmidtii. Journal of the Korean Forestry Society* 90, 600–607.

Yeo, J.K., Kim, I.S., Koo, Y.B., Kim, T.S. and Son, D.S. (2002) Absorption ability and tolerance to livestock waste water of poplar species and clones. *Journal of the Korean Society of Waste Management* 19, 912–920.

Yeo, J.K., Kim, I.S., Koo, Y.B., Kim, Y.J. and Joo, S.H. (2003) Growth and absorption of livestock waste water of poplar species at test plantation. *Journal of the Korean Society of Waste Management* 20, 742–749.

Zacchini, M., Pietrini, F., Scarascia-Mugnozza, G., Iori, G., Pietrsanti, L. and Massacci, A. (2009) Metal tolerance, accumulation, and translocation in poplar and willow clones treated with cadmium in hydroponics. *Water, Air and Soil Pollution* 197, 23–34.

Zalesny, J.A., Zalesny, R.S. Jr, Coyle, D.R. and Hall, R.B. (2007a) Growth and biomass of *Populus* irrigated with landfill leachate. *Forest Ecology and Management* 248, 143–152.

Zalesny, J.A., Zalesny, R.S. Jr, Wiese, A.H. and Hall, R.B. (2007b) Choosing tree genotypes for phytoremediation of landfill leachate using phyto-recurrent selection. *International Journal of Phytoremediation* 9, 513–530.

Zalesny, R.S. Jr (2011) Ecosystem services associated with phytotechnologies. Presentation at 8th Conference of the International Phytotechnology Society: Putting Plants to Work Where We Live, Labor, Breathe, and Play. Portland, Oregon.

Zalesny, R.S. Jr and Bauer, E.O. (2007a) Selecting and utilizing *Populus* and *Salix* for landfill covers: implications for leachate irrigation. *International Journal of Phytoremediation* 9, 497–511.

Zalesny, R.S. Jr and Bauer, E.O. (2007b) Evaluation of *Populus* and *Salix* continuously irrigated with landfill leachate I. Genotype-specific elemental phytoremediation. *International Journal of Phytoremediation* 9, 281–306.

Zalesny, R.S. Jr and Zalesny, J.A. (2011) Clonal variation in lateral and basal rooting of *Populus* irrigated with landfill leachate. *Silvae Genetica* 60, 35–44.

Zalesny, R.S. Jr, Stanturf, J.A., Evett, S.R., Kandil, N.F. and Soriano, C. (2011) Opportunities for woody crop production using treated wastewater in Egypt. I. Afforestation strategies. *International Journal of Phytoremediation* 13(S1), 102–121.

Zodrow, J.J. (1999) Recent applications of phytoremediation technologies. *Remediation Journal* 9, 26–36.

7 Abiotic Stresses

N. Marron,[1,2*] B. Gielen,[1] F. Brignolas,[3] J. Gao,[4,9] J.D. Johnson,[5] D.F. Karnosky,[6] A. Polle,[4] G. Scarascia-Mugnozza,[7] W.R. Schroeder[8] and R. Ceulemans[1]

[1]Research Group of Plant and Vegetation Ecology, University of Antwerp, Belgium; [2]Present address: UMR 1137 INRA Nancy – Lorraine University, France; [3]Université d'Orléans, France and INRA, USC1328 Arbres et Réponses aux Contraintes Hydriques et Environnementales (ARCHE), France; [4]Büsgen-Institut, Abtlg. Forstbotanik und Baumphysiologie, Georg-August Universität Göttingen, Germany; [5]Deceased. Formerly at Washington State University, USA; [6]Deceased. Formerly at Michigan Technological University, USA; [7]Department of Agronomy, Forestry and Land Use, Agricultural Research Council of Italy, Italy; [8]Agroforestry Development Centre, Agriculture and Agri-Food Canada, Canada; [9]Present address: International Bamboo and Rattan Research Centre, Beijing, PR China

7.1 Introduction

Abiotic stresses are usually defined as non-living environmental factors that have harmful effects on organisms. They are generally classified in two categories according to their origin: (i) edaphic stresses, such as water deficit, salinity, lack of nutrients or soil pollution; and (ii) atmospheric stresses, such as ozone, elevated carbon dioxide (CO_2), increased temperature, frost or high irradiance. Abiotic stress conditions cause extensive losses to agricultural production worldwide. Individually, stress conditions such as drought, salinity or heat have been the subject of intense research. However, in the field, trees as well as other plants are routinely subjected to a combination of different abiotic stresses. In drought-stricken areas, for example, many trees encounter a combination of drought and other stresses, such as heat or salinity. Recent studies have revealed that the molecular and metabolic response of trees to drought and heat is unique and cannot be directly extrapolated from the response to each of these different stresses applied individually (Mittler, 2006).

Members of the *Salicaceae* family are known to be particularly sensitive to environmentally induced stresses, the cumulative effects of which determine their distribution, reproductive success and productivity (Neuman *et al.*, 1996). However, within this overall sensitivity, the range of variation in terms of tolerance, i.e. the severity of the impact on biomass production, is generally very wide in the family. At this time, the selection efforts for the creation of new varieties are essentially focused on traits such as resistance to diseases, in particular foliar rust, productivity and wood quality. Tolerance to environmental constraints is not yet taken into account as a selection criterion for new genotypes in breeding programmes. However, such criteria have to be considered with much more

*Corresponding author: marron@nancy.inra.fr

attention because of two current tendencies: (i) climate changes; and (ii) the present extension of willow and poplar cultivation for bioenergy production purposes towards non-alluvial zones more frequently subjected to drought and less fertile than riverside zones, in such a way as to avoid competition for land use with food agriculture (Dreyer *et al.*, 2004).

This chapter focuses on the responses and resistance mechanisms used by poplars and willows to survive a variety of environmental stresses. The chapter is notably composed of two tabular bibliographical compilations of scientific studies carried out on drought and elevated carbon dioxide responses of poplars and willows between the beginning of the 1990s and 2006. These tables present the objectives, the growth conditions, the measured traits and the main conclusions of hundreds of studies carried out worldwide during the last decades and dealing with drought and elevated CO_2 conditions.

7.2 Edaphic Stresses

7.2.1 Water deficit

N. Marron and F. Brignolas

Context

Fresh water is one of the most unequally distributed resources around the world and this disparity is likely to be aggravated by the current climate changes. At continental and regional scales, more intense and longer droughts have indeed been observed over wide areas since the 1970s. Increased drying linked with higher temperatures and decreased precipitation has contributed to changes in drought, and it is very likely that extreme events will become frequent in the future (IPCC, 2007). Even if poplars are considered as the most sensitive trees to water deficit, the wide diversity of the 29 species of the *Populus* genus in terms of resistance strategies has made possible the colonization of many types of ecosystems, including very dry areas and deserts. Furthermore, from a farming point of view, poplars are among the fastest growing trees in temperate latitudes, but their high productivity is associated with high water requirements. As a consequence, their productivity may be strongly limited by water availability (Tschaplinski *et al.*, 1994; Zsuffa *et al.*, 1996). To sustain the extension of poplar cultivation from flood plains and bottomlands to uplands where soil water availability is subject to seasonal changes, more water-use-efficient hybrids are required. At whole plant level, water-use efficiency (WUE) is defined as the ratio between biomass production and water consumption. The identification of poplars combining satisfactorily high productivity and high WUE would be a considerable advantage in moderately drought-constrained areas (Braatne *et al.*, 1992; Marron *et al.*, 2005; Monclus *et al.*, 2005, 2006).

The study of drought resistance is an ambitious objective because it depends on numerous complex and often interdependent traits. The identification of these traits is not yet complete and the approach used by scientists is often empirical. It consists of subjecting homogeneous plant material to water withholding and to record the subsequent molecular (transcriptomics and proteomics analysis), cellular (growth reduction, changes in metabolic activities) and whole-plant modifications (water flows, photosynthesis, organ growth, etc.). However, these modifications are more an adjustment of the plant functioning in response to constraining conditions than a real adaptive response associated to a given level of resistance. Poplar is one of the model plants for which catalogues of complete genes are available (with *Arabidopsis* and rice, for instance) and for which it is possible to lead integrative biology studies, combining ecophysiology and genomics. For instance, for clone 'Beaupré' (*Populus* ×*canadensis*), at an early stage after the beginning of the constraining episode, and before any effect on growth, photosynthesis and water potential, a very neat response in terms of DNA transcription was observed (Dreyer *et al.*, 2004).

Strategies of drought resistance

A crucial step in the colonization of terrestrial environments by plants has been the evolution of mechanisms that enable plants to control their water loss while continuing to fix carbon dioxide in photosynthesis. This step has been so important because the availability of water is probably the key factor determining plant distribution and survival in natural ecosystems, and it is also the most important limiting factor in agricultural production. All mechanisms that tend to maintain plant survival or productivity under conditions of limited soil water supply can be described as drought resistance mechanisms (Passioura, 2002). Ecophysiologists have shown that different strategies can contribute to explaining drought resistance (Jones, 1993).

The first strategy consists of avoiding water deficit. Drought avoidance involves completion of the life/reproductive cycle during favourable conditions, limitation of transpiration and maximization of root uptake, and would include: (i) perennial/deciduous plants that remain dormant during drought; and (ii) species of arid environments with permanent access to the water table (phreatophyte species such as *Populus euphratica*). These mechanisms allow plant survival at the expense of biomass production.

The second strategy consists of tolerating water deficit. Drought tolerance is prevalent under temperate climates, in plants for which drought occurs at random, and encompasses mechanisms allowing the maintenance of water flow, gas exchange and cell turgor under drought conditions. Consequently, these mechanisms not only allow plant survival during drought but also the preservation of biomass production. The concept of tolerance to water deficit, when applied to cultivated tree species such as poplars, has been defined as the ability to limit the decrease in biomass production in response to a moderate water deficit (Passioura, 2002).

Additionally, some authors consider that diversity in WUE could be associated with diversity in drought resistance, and they classify the ability to show high WUE in an independent third category of drought resistance strategy, i.e. efficiency mechanisms. The aim of this third strategy is to optimize the utilization of resources under drought conditions, and especially water (Jones, 1992, 2004). At the whole-plant level, WUE is defined as the ratio between biomass production and water consumption. At the leaf level, intrinsic WUE is defined as the ratio between assimilation and stomatal conductance. Although the relationship between WUE and drought resistance is known to be variable among species, identification of poplars combining a satisfactorily high productivity and a high WUE would be a considerable advantage in moderately dry areas (Braatne *et al.*, 1992; Marron *et al.*, 2005; Monclus *et al.*, 2005, 2006). In spite of an overall sensitivity to drought in all the species belonging to the *Populus* genus, a very wide diversity in their drought tolerance levels, in their response patterns to water deficit as well as in their WUE, has been reported (Pallardy and Kozlowski, 1981; Gebre and Kuhns, 1991; Liu and Dickmann, 1992b; Brignolas *et al.*, 2000; Marron *et al.*, 2003, 2005; Zhang *et al.*, 2004; Monclus *et al.*, 2005, 2006).

Chronology of the response to drought

Globally, poplar response to drought involves different mechanisms that can be ranked into short-, medium- and long-term processes. The earliest plant response to drought is the reduction of cell growth and the decrease in stomatal conductance (Blake *et al.*, 1996) (Fig. 7.1, Table 7.1a). Cell water uptake, responsible for growth, is linked to the difference in water potential between the xylem and the tissues in elongation, and to the water conductance of the tissues. Cell expansion occurs when water penetrates into the cell and allows expansion while maintaining cell turgor above a threshold (Lockhart, 1965). So, during drought, cell growth reduction can be due to: (i) decreased turgor; (ii) a reduction in cell wall extensibility; or (iii) an increase in the cell wall deformation threshold (Passioura and Fry, 1992). The relative importance of these three parameters is very variable, but it seems that cell wall extensibility is the most important in many situations (Cosgrove, 1993). Stomatal closure is a very quick and flexible process limiting the risk of xylem cavitation, i.e. loss of xylem conductivity, under low water potential due to reduced soil water availability (Harvey and Van den Driessche, 1997; Sperry *et al.*, 2002) (Table 7.1b). When the soil water content decreases, the tension responsible for water rising in the plant increases and can cause a disruption of the water columns, which can lead to the phenomenon of cavitation and to irreversible embolism (Table 7.1c). Among woody species of the temperate zone, poplars display the highest vulnerability to cavitation, which tends to confirm their strong susceptibility to drought (Fig. 7.2). Stomatal closure can be

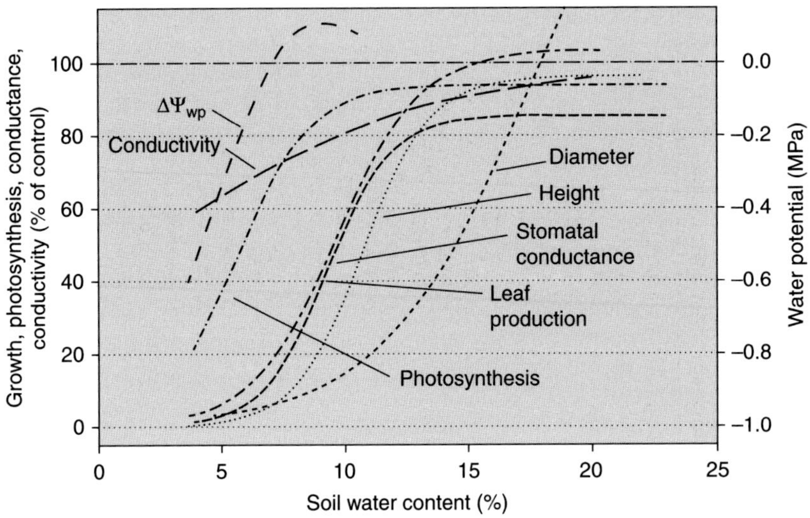

Fig. 7.1. Response of different processes to the decrease of soil water availability in young plants of *Populus euphratica*: growth in diameter and height, stomatal conductance, leaf production, photosynthesis, stem hydraulic conductivity and leaf water predawn potential (Ψ_{wp}) (Dreyer *et al.*, 2004).

induced by an increase in the content of abscisic acid (ABA) in the vicinity of the guard cells of the stomata. This phytohormone is released by the roots and transported by the rising sap flow (Davies and Zhang, 1991). However, an important variability of stomatal sensitivity has been observed among poplars. For example, some *P. trichocarpa* ecotypes and *P. koreana* × *P. trichocarpa* 'Peace' are known to present stomata that are insensitive to exogenous ABA applications (Hinckley and Braatne, 1994; Ridolfi *et al.*, 1996).

At medium term, an osmotic adjustment can occur, i.e. an increase in the osmotic potential due to an accumulation of solutes or to a reduction in cell volume (Morgan, 1984) (Table 7.1d). Cell growth is mainly dependent on cell turgor, the two major components of which are water and osmotic potentials. Under drought conditions, transpiration results in a (passive) decrease of water and osmotic potentials caused by water loss. To counterbalance this, an accumulation of solutes can occur, resulting in an (active) reduction of the osmotic potential, and so in an increase in cell turgor due to water penetration into the cell. The accumulated solutes are various and belong to diverse biochemical families. However, proline and sugars are known to be the main contributors to osmotic adjustment (Costa *et al.*, 1998). For various poplar species, it has been shown that sucrose is the main contributor to osmotic adjustment during drought, followed by glucose, fructose and *myo*-inositol (Gebre *et al.*, 1994; Pelah *et al.*, 1997; Sibout and Guerrier, 1998). However, it has also been shown that the contribution of the sugar accumulation under drought conditions is often small and that other drought-resistance mechanisms contributed to the genotypic differences in the field (Gebre *et al.*, 1998).

At longer term, drought is responsible for morphological and anatomical changes resulting from the reduction in cell expansion and from modifications of carbon allocation to the different organs of the plant (Table 7.1e). Roots are favoured at the expense of stems and leaves. Consequently, the ratios of roots:stem and roots:foliage increase and the leaf area is reduced, due to a decrease in both the individual size and number of the leaves (Braatne *et al.*, 1992; Liu and Dickmann, 1993; Van Splunder *et al.*, 1996; Ibrahim *et al.*, 1997; Tschaplinski *et al.*, 1998; Marron *et al.*, 2003). These modifications have the dual advantage of improving root water uptake and reducing transpiration and subsequent risks of cavitation (Braatne *et al.*, 1992; Tschaplinski *et al.*, 1998). Three phenomena

Table 7.1a. Overview of recent available literature on drought response of *Populus* and *Salix* experiments in ISI (Thompson Institute) scientific journals (period 1998–2006) and of former keystone studies (1987–1997). (a) Leaf structure, physiology and growth. Studies are ordered according to the age of the plant material and then chronologically (year of publication). Abbreviations used are listed at the end of the table.

Species	Plant material	Experiment location	Growth conditions	Substrate	Water deficit intensity	Objectives of the study	Traits examined	Main conclusions	References
P. trichocarpa *P. deltoides* *P. trichocarpa* × *P. deltoides*	Young cuttings	Seattle, Washington, USA	Gh	Pumice/peat/ fine bark	0 < fraction of transpirable soil water < 1	Assessment of the relationships between soil water, leaf growth, transpiration and plant water balance during a complete drought cycle	LA/g_s/T/SD/ LAnat/g_h	Maintenance of growth differ between parents and F_1 in a pattern characteristic of an overdominant mode of inheritance – F_1 hybrids were more drought resistant than either parental species	Braatne *et al.*,1992
P. tremula	4- to 5-week-old plantlets	Rehovot, Israel	IvC		20 and 30% plant water loss	Description of the cloning, sequence analysis, isolation and characterization of the SP1 protein and its stress responsiveness, boiling solubility and oligomeric structure	EG/[SP1]	SP1 proteins are hydrophilic and remain soluble on boiling – they represent a new class of protein involved in the plant's response to abiotic stress	Wang *et al.*, 2002

Continued

Table 7.1a. Continued.

Species	Plant material	Experiment location	Growth conditions	Substrate	Water deficit intensity	Objectives of the study	Traits examined	Main conclusions	References
P. koreana × *P. trichocarpa*	2-month-old plants	Tsukuba, Japan	Pt	Gravel/ vermiculite	$-0.74 < \psi_{leaf} < -1.27$ MPa	Examination of the induction responses of leaves of well-watered and dehydrated plants grown at high-light or low-light regime	GE/g_s	Mild water stress may have reduced steady-state assimilation and g_s, but it had little impact on the photosynthetic induction response in high-light leaves	Tang and Liang, 2000
P. ×*canadensis*	2-month-old cuttings	Orléans, France	Gh	Sand/peat moss/clay	$-3.5 < \psi_{wp} < -0.4$ MPa	Test of the drought behaviour of two contrasting clones in terms of leaf growth and water relations	LA/LN/LG	The growth advantage of clone 'Luisa Avanzo' under control conditions appeared to be counterbalanced by its higher susceptibility to water stress	Brignolas *et al.*, 2000
P. ×*canadensis*	2-month-old cuttings	Orléans, France	Gh	Blond peat/ brown peat/horse manure/ compost	$-2.20 < \psi_{wp} < -0.65$ MPa	Assessment of the impact of successive drought and re-watering cycles on plant water relations, growth and SLA parameters	g_s/LG/SLA/ DLE	Differences in drought impact between clones is attributable mainly to differences in plasticity after re-watering rather than to clonal differences in drought responses	Marron *et al.*, 2003

P. trichocarpa × *P. koreana* *P.* ×*canadensis*	4- to 5-month-old plants	Nancy, France	GC	Sand/blond peat	$-1.4 < \psi_{wp} < -0.2$ MPa	Detection of possible stomata-independent limitations of photosynthesis in response to an imposed short-term drought	GE/Y/J/$\delta^{13}C$	Reduced CO_2 influx was a major cause of the limitation of net CO_2 assimilation during drought in 'Robusta' but not in 'Peace', where drought probably reversibly reduced the apparent carboxylation efficiency of Rubisco	Ridolfi and Dreyer, 1997
P. deltoides	6-month-old coppiced shoots	Oracle, Arizona, USA	SRC	Bare soil/organic matter	$0.10 <$ soil volumetric water content $< 0.27\ m^3\ m^{-3}$	Study of the effects of elevated CO_2 and VPD on isoprene emission rates during drought stress	GE/[Isop]/CL	Drought and high VPD dramatically increased the proportion of assimilated carbon lost as isoprene	Pegoraro *et al.*, 2004
P. deltoides	9-month-old trees	Oracle, Arizona, USA	IFB		$-1.50 < \psi_{leaf} < -0.70$ MPa	Understanding of leaf-level responses to drought stress and evaporative demand under elevated CO_2	LMA/LAI/SA/Ht/Cc/T	The transpiration responses at high VPD in the presence of high SWC and throughout the low SWC treatment suggest some hydraulic limitations to water use occurred	Engel *et al.*, 2004

Continued

Table 7.1a. Continued.

Species	Plant material	Experiment location	Growth conditions	Substrate	Water deficit intensity	Objectives of the study	Traits examined	Main conclusions	References
P. tremula × *P. alba*	9- to 12-month-old cuttings		GC	Peat-based commercial growth medium	$-3 < \psi_{soil} < 0$ MPa	Comparison of the ecophysiological performance of transgenic poplars overexpressing the pine GS1 gene with non-transgenic plants before, during and after the transient imposition of water stress	GE/F/[Chl]/[Gly]/[GS]/[GOGAT]/[Rub]	The overexpression of pine cytosolic GS1 enhanced sustained photosynthetic electron transport capacity during severe stomatal limitation	El-Khatib *et al.*, 2004
S. psammophila	1-year-old seedlings	Ordos Sandland Ecological Station, China	Gh	Sand	19.7% of SWC decrease	Investigation of how a 50% decrease in precipitation affects growth and physiology and how physiology and structure is adjusted to the low precipitation conditions	Y/Ht/LN/LA/SLA/RSR/BA/WUE/g_s/GE/T	Instantaneous WUE was not affected by the water deficit – SLA and RSR were decreased in response to drought	Xiao *et al.*, 2005
P. trichocarpa	1-year-old trees	Puyallup / Wenatchee, Washington, USA	P		200 < average annual precipitation < 2030 mm	Epidermal and stomatal cell traits examination on late leaves of 40 black cottonwood clones originating from a mesic and a xeric river valley	LAnat/SD/LA	Acclimation to the hotter, drier summer climate was evident, as the clones generally had smaller epidermal cell diameter and higher cell density and abaxial stomatal density	Dunlap and Stettler, 2001

P. deltoides	1-year-old coppiced plants	Oracle, Arizona, USA	IFM	Silt loam	0.31 < soil volumetric water content < 0.41 $m^3 m^{-3}$	Understanding of the relationship between CO_2, water availability and isoprene emission	[Isop]/GE	Water limitation can override the inhibitory effect of elevated CO_2	Pegoraro *et al.*, 2005a
P. deltoides	1-year-old coppiced plants	Oracle, Arizona, USA	IFM	Silt loam	0.13 < soil volumetric water content < 0.27 $m^3 m^{-3}$	Exploration of the relationship between isoprene uptake and atmospheric CO_2 concentration and drought	[Isop]/K	Drought suppressed the sink capacity, but the full sink capacity of dry soil was recovered within a few hours on re-wetting	Pegoraro *et al.*, 2005b
P. deltoides	2-year-old plants	Stony Brook, New York, USA	Gh		$-1.3 < \psi_{leaf} < -0.5$ MPa	Examination of the role of alternative carbon sources in isoprene production during conditions of water stress and high leaf temperature	GE/$\delta^{13}C/g_s$/ [Isop]	Across water- and heat-stress experiments' allocation of photosynthate was correlated negatively to the ratio of isoprene emission to photosynthesis	Funk *et al.*, 2004
P. grandidentata	Adult trees	Pennsylvania, USA	F		$-0.8 < \psi_{wp} < -0.2$ MPa	Measurements of gas exchange, water potential, leaf structure, leaf nitrogen and the microenvironment during a seasonal drought	GE/N/LA/ SLA/ LAnat	Significant amount of plasticity for most gas exchange and leaf structural parameters	Abrams and Mostoller, 1995

Continued

Table 7.1a. Continued.

Species	Plant material	Experiment location	Growth conditions	Substrate	Water deficit intensity	Objectives of the study	Traits examined	Main conclusions	References
P. tremula	Adult trees	Järvselja, Estonia	F	Loamy soil	$-4 < \psi_{soil} < 0$ MPa	Study of the limitations caused by variations in leaf temperature and water availability on photosynthetic electron transport rates	J/N/[Pig]	Chlorophyll fluorescence restricts potential carbon gain under conditions of water limitation less than does stomatal conductance	Niinemets *et al.*, 1999a
P. tremula	Adult trees	Järvselja, Estonia	F	Loamy soil	$-4 < \psi_{soil} < 0$ MPa	Investigation of morphology, chemical composition and photosynthetic capacity of leaf laminas along a canopy light gradient	LG/$\delta^{13}C$/SLA/J/N/C/[Lignin]	Leaf water stress is a major factor altering foliage structure and assimilative capacity	Niinemets *et al.*, 1999b
P. alba	Adult trees	Nikola Tesla-A, Serbia	PTS	Fluvisol	Leaf water saturation deficit = 16.4	Ecophysiological research on the ash deposits covering ten plant species	Y/[B, Cu, Mn, Zn, Pb, Cd]	High boron concentrations in *P. alba* tissues (superior to 100 $\mu g\ g^{-1}$) – low photosynthetic efficiency during the period of summer drought (0.43–0.62)	Pavlović, *et al.*, 2004

P. tremula	Adult trees	Järvselja, Estonia	F	Loamy soil	$-4 < \psi_{soil} < 0$ MPa	Test of the hypothesis that water stress effects significantly modify canopy gradients in intercellular CO_2 mole fractions	[Sugar]/g_s/ GE	Stress effects influence C_i more strongly than within-canopy light gradients – leaves acclimated to different water stress conditions may regulate water use largely independent of foliar photosynthetic potentials	Niinemets *et al.*, 2004
P. fremontii *S. gooddingii*	Adult trees	Wickenburg, Arizona, USA	F		$-1.3 < \psi_{wp} < -0.3$ MPa	Investigation of the physiological and growth responses of riparian trees to groundwater availability	GE/SG/BG/ $\delta^{13}C$	*S. gooddingii* appears to be more sensitive than *P. fremontii* to declines in water availability and is more responsive to increased water availability	Horton *et al.*, 2001a
P. fremontii *S. gooddingii*	Adult trees	Maricopa County, Arizona, USA	D		131 < annual precipitation < 386 mm	Investigation of leaf gas exchange responses to leaf temperature, VPD and predawn and midday shoot water potential of two native Sonoran Desert riparian tree species	GE/SLA	High VPD had a smaller effect on leaf gas exchange in willow than in cottonwood – willow had a less negative ψ_{wp} threshold for stomatal closure than cottonwood	Horton *et al.*, 2001b

Continued

Table 7.1a. Continued.

Species	Plant material	Experiment location	Growth conditions	Substrate	Water deficit intensity	Objectives of the study	Traits examined	Main conclusions	References
P. fremontii *S. gooddingii*	Adult trees	Bill Williams River and Hassayampa River, Arizona, USA	AV		$-4.0 < \psi_{wp} < -0.5$ MPa	Investigation of the physiological response to groundwater availability along gradients of depth to groundwater at two rivers	$GE/g_s/S$	The study provides estimates of the range of depth to groundwater that can maintain healthy, mature trees	Horton *et al.*, 2001c
P. tremula	Adult trees	Järvselja, Estonia	F	Loamy soil		Investigation of the responses of photosystem II quantum yield to light availability in the short and in the long term	$Y/J/q_p/q_{np}/$ [Chl]	Foliar water stress scales positively with long-term quantum flux density – inverse patterns of variation in water and light availabilities in the canopy result in a greater decline in assimilation than is predicted by decreases in stomatal conductance alone	Niinemets and Kull, 2001

S. gordejevii *S. microstachya*	Adult trees	Hunshandak Sandand, Inner Mongolia, China	D	Sand	Wetland/ lowland/ dunes	Investigation of the maximum quantum efficiency of 99 native plant species distributed in fixed sand dunes, lowland and wetland	Y	Based on maximum quantum yield measurements, *S. gordejevii* is among the species recommended as ideal species for ecological restoration in degraded sandland ecosystems	Li *et al.*, 2004c
S. gordejevii *S. babylonica*	Adult trees	Horqin Sandy Land, Inner Mongolia, China	N	Sand	$-1.25 < \pi_O < 0$ MPa	Examination of the heat and drought tolerance of *S. gordejevii* compared with *S. babylonica*	Y/EL/F	*S. gordejevii* has a higher capacity for drought tolerance and a lower capacity for heat tolerance than *S. babylonica*, indicating that *S. gordejevii* inhabits desert environments mainly due to its high drought tolerance	Yang *et al.*, 2004

Continued

Table 7.1a. Continued.

Species	Plant material	Experiment location	Growth conditions	Substrate	Water deficit intensity	Objectives of the study	Traits examined	Main conclusions	References
P. tremuloides	80-year-old trees	Saskatchewan, Canada	F		$-1.75 < \psi_{leaf} < -0.20$ MPa	Monitoring of photosynthesis and transpiration over a 42-day midsummer period	GE/g_s	Strong reductions in carbon uptake may be expected at the more extreme values of vapour pressure deficit that occur during periods of regional drought, even if soil water is not locally limiting	Hogg *et al.*, 2000

Growth conditions: AV, alluvial valley; D, desert; F, forest; GC, growth chamber; Gh, greenhouse; IFB, intensive forest biome; IFM, intensive forest mesocosm; IvC, *in vitro* culture; N, nursery; P, plantation; Pt, phytotron; PTS, power thermal station; SRC, short-rotation culture. Traits examined: BA, biomass allocation; BG, branch growth; C, carbon; Cc, stem circumference/diameter; Chl, chlorophylls; $\delta^{13}C$, carbon isotope composition; CL, carbon loss; EG, expression of genes; DLE, duration of leaf expansion; EL, electrolyte leakage; F, fluorescence; g, conductance; GE, gas exchange; g_h, hydraulic conductance; Gly, glycine; g_s, stomatal conductance; GOGAT, glutamine oxoglutarate aminotransferase; GS, glutamine synthase; Ht, stem height; Isop, isoprene; J, rate of photosynthetic electron transport; K, soil activity factor; LA, leaf area; LAI, leaf area index; LAnat, leaf anatomy; LG, leaf growth; LMA, leaf mass area; LN, number of leaves; N, nitrogen content; Pig, pigments; π_o, osmotic potential; q_p, photochemical quenching; q_{np}, non-photochemical quenching; RSR, root to shoot ratio; Rub, rubisco; S, survival; SA, sapwood area; SD, stomatal density/dimension; SG, stem growth; SLA, specific leaf area; SWC, soil water content; T, transpiration; VPD, vapour pressure deficit; WUE, water-use efficiency; Y, quantum yield; Ψ, water potential; [ ], content/concentration.

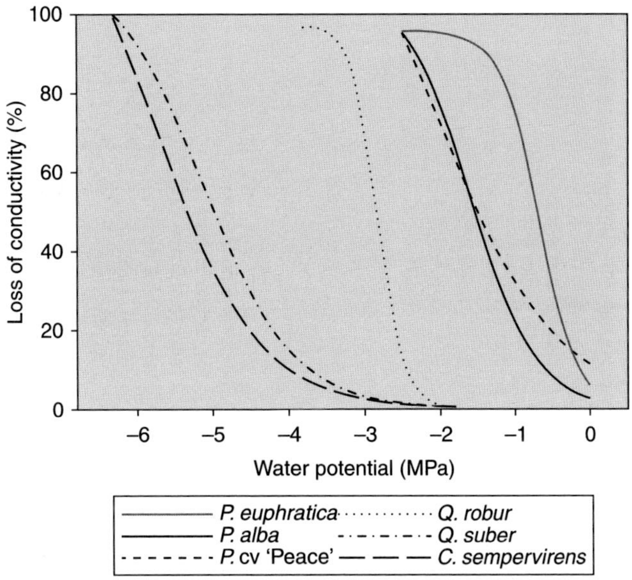

Fig. 7.2. Vulnerability to cavitation according to water potential for *Populus euphratica*, *Populus alba*, *Populus koreana* × *Populus trichocarpa* 'Peace', *Quercus robur*, *Quercus suber* and *Cupressus sempervirens* (Hukin *et al.*, 2005).

aiming to reduce leaf area coexist in poplar: (i) loss of leaves; (ii) reduction in the individual size of leaves; and (iii) decrease in the number of simultaneously expanding leaves. Leaf loss is commonly observed under natural conditions, notably for *P. trichocarpa* and *P. deltoides* (Cooper and Van Haverbeke, 1990; DeBell, 1990). However, leaf loss is a drastic strategy and the first response to water deficit is often a reduction in leaf growth. A reduction in the number of leaves dominates the adaptive response of the leaf area to drought, but a reduction in the individual leaf size and a coexistence of both strategies have also been observed in poplar (Liu and Dickmann, 1993; Ibrahim *et al.*, 1997; Marron *et al.*, 2003). A combined reduction in cell growth and cell division is responsible for the decrease in leaf size (Dale, 1988). Leaf structural modifications are also commonly observed under drought conditions. Indeed, drought frequently causes a reduction in the specific leaf area (SLA, defined as the ratio between leaf area and leaf dry weight) due to an increase of leaf density (Niinemets, 2001; Marron *et al.*, 2003). This density increase is linked to thicker cell walls and to smaller and more packed cells in the leaves expanded under drought conditions (Niinemets, 1999). An improved water status has been shown to occur for species with low SLA (i.e. high leaf density), but the physiological basis of the relationships is still unknown (Nautiyal *et al.*, 2002).

All the previously listed mechanisms aim to limit water losses and to improve water uptake, but some of them provoke damaging consequences, which the plant has to confront (Tables 7.1f and g). Due to stomatal closure, a slight water deficit decreases CO_2 absorption necessary for photosynthesis. If drought conditions persist, a direct inhibition of photosynthesis occurs. In both cases, the result is an exposure of the chloroplast to an excess of excitation energy. Several protection processes exist, such as leaf movements, aiming to reduce light absorption and increase overall heat dissipation in the light-harvesting antenna (Chaves and Oliveira, 2004). In addition, limitation of CO_2 fixation provides an insufficient sink for electrons generated in the electron-transport chains (ETC). In this case, alternative outlets for electrons gain in importance and lead to overproduction of reactive oxygen species (ROS) and to potential oxidative damage (Edreva, 2005). Under such conditions, oxygen acts as an alternative electron acceptor,

Table 7.1b. Overview of recent available literature on drought response of *Populus* and *Salix* experiments in ISI (Thompson Institute) scientific journals (period 1998–2006) and of former keystone studies (1987–1997). (b) Water relations – stomatal behaviour. Studies are ordered according to the age of the plant material and then chronologically (year of publication). Abbreviations used are listed at the end of the table.

Species	Plant material	Experiment location	Growth conditions	Substrate	Water deficit intensity	Objectives of the study	Traits examined	Main conclusions	References
P. × *canadensis* *P.* 'Tristis' × *P. balsamifera*	Young cuttings	East Lansing, Michigan, USA	Gh	Sandy-loam soil	$-0.5 < \psi_{soil\ matric} < 0$ MPa	1. Comparison of soil drying and flooding in the ability to induce ABA accumulation 2. Exploration of the role of leaf-N in modifying stomatal response 3. Examination of the relationship between stomatal conductance and ABA accumulation	[ABA]/GE	Contrasting physiological strategies between clones for survival under prolonged drying conditions	Liu and Dickmann, 1992a
P. angustifolia *P. balsamifera* *Salix drummondiana* *Salix exigua* *Salix lutea*	Young saplings and seedlings	Lethbridge, Alberta, Canada	Rz/Gh	Coarse gravel/ medium sand/ fine gravel	0, 1, 2, 3, 4 or 8 cm day^{-1} water table decline	Investigation of the relative tolerances of saplings and seedlings of different willow and cottonwood species to different rates of water table decline	Ht/RL/BA/S	Willow and cottonwood saplings similarly affected by abrupt water table decline, but willow seedlings slightly more vulnerable than cottonwood seedlings	Amlin and Rood, 2002
P. trichocarpa	7-week-old cuttings	Toronto, Canada	Gh		$-2.8 < \psi_{epidermis} < -0.4$ MPa	Examination of the role of ABA in the mediation of stomatal responses to low water potential	[ABA]/g_s/g_{leaf}	The unresponsiveness of *P. trichocarpa* stomata to water potentials is not due to the inability of this species to produce an increased concentration of ABA in response to water stress	Schulte and Hinckley, 1987

S. nigra	2- to 6-month-old cuttings	Columbia, Missouri, USA	Gh	Sand/peat moss/ silt-loam	$-3 < \psi_{wp} < 0$ MPa	Comparison of the patterns of water relations, xylem sap ABA concentration and stomatal aperture in drought-sensitive black walnut and black willow, less drought-sensitive sugar maple and drought-tolerant white oak	g_s/[ABA]	In the early stages of drought, increased ABA concentration in the xylem sap of black willow was probably of root origin and provided a signal to the shoot of the water status of the roots	Loewenstein and Pallardy, 1998
P. trichocarpa × *P. koreana* *P.* ×*canadensis*	4- to 5-month-old plants	Nancy, France	GC	Sand/peat	$-0.8 < \psi_{wp} < -0.2$ MPa	Report of the leaf-age dependency of the lack of sensitivity of the stomata of clone 'Peace' to exogenous abscisic acid	g_s/T/$[Ca]_{xylem}$	Drought control of stomatal conductance in clone 'Peace' is ABA-independent and could involve calcium ions	Ridolfi *et al.*, 1996
P. ×*canadensis*	6-month-old plants	Montpellier, France	Gh		$-1.5 < \psi_{wp} < -0.1$ MPa	Joint analysis of the stomatal control of different species under naturally fluctuating evaporative demand and soil water status	g_s/$[ABA]_{xylem}$	ABA content in xylem was related to soil water status with common relationships for different experimental conditions, but with markedly different responses among species	Tardieu and Simonneau, 1998
P. trichocarpa × *P. deltoides* *P. deltoides* × *P. nigra*	1-year-old coppiced plants	Swanbourne, UK	SRC	Clay loam	$0.12 <$ SWC $< 0.40\%$	Report of field measurements of transpiration during the summer of 1994 for two contrasting clones, 'Beaupré' and 'Dorskamp'	Cc/T/g_s	Leaf conductance declines slightly with increasing atmospheric VPD in both clones, but only in 'Beaupré' did leaf conductance decrease as soil water deficit increased	Allen *et al.*, 1999

Continued

Table 7.1b. Continued.

Species	Plant material	Experiment location	Growth conditions	Substrate	Water deficit intensity	Objectives of the study	Traits examined	Main conclusions	References
P. ×*canadensis* *P.* 'Popularis'	1-year-old cuttings	Beijing, China	Gh	Loam soil	$-0.6 < \psi_{wp} < -0.1$ MPa	Investigation of the effects of water stress and external ABA supply on shoot growth, stomatal conductance and water status of a drought-sensitive and a drought-tolerant genotype	SG/GE/g_s	Sensitivity of poplar roots to variation in soil water content varies by clone – a rapid short-term accumulation of ABA in shoots in response to water stress may contribute to drought tolerance	Chen *et al.*, 1997
P. trichocarpa × *P. deltoides* *S. dasyclados*	3-year-old coppiced plants	Hunstrete, UK	P	Clay loam	450 < SWC < 600 mm	Measurement of the transpiration rates of SRC grown at a site with a more freely draining soil	LA/LAI/T/g_s/SCA	The high stomatal conductances were maintained even when atmospheric humidity deficits and soil water deficits were large	Hall *et al.*, 1998
P. trichocarpa × *P. balsamifera*	6-year-old trees	Reading, UK	P	Sandy loam	$-1.3 < \psi_{leaf} < -0.2$ MPa	Description of the diurnal and seasonal patterns of sap flow of isolated poplar trees growing close to the water table in a mild, temperate climate	SF/g_s	Poplar trees took 15–60% of water transpired from groundwater, with the proportion increasing as the soil in the unsaturated zone dried out	Zhang *et al.*, 1999
P. fremontii *P. angustifolia* *P. fremontii* × *P. angustifolia*	10-year-old cuttings	Ogden, Utah, USA	P		$-0.9 < \psi_{wp} < -0.4$ MPa	How do some riparian poplars respond to pulse increases in water availability in previously dry zones?	T/g_h/LA/SA	Trees were insensitive to water addition to the surface soil that were twice the magnitude of whole-tree transpiration rates	Cox *et al.*, 2005

P. tremula	Adult trees	Järvselja, Estonia	F	Loamy soil	$-4 < \psi_{soil} < 0$ MPa	Study of the control of stomatal conductances imposed by soil water availability and foliage acclimation to long-term integrated irradiance	π_o/g_s/SLA/LWC/[Osmo]/$[ABA]_{xylem}$	No evidence that ABA increased with advancing water stress – stomatal sensitivity to ABA is not constant along the canopy light gradient	Niinemets *et al.*, 1999c
P. tremula	Adult trees	Tartu, Estonia	F		$-0.7 < \psi_{xp} < 0$ MPa	Examination of the correlations between leaf ABA content, stomatal conductance, hydraulic conductance and photosynthetic characteristics	[ABA]/GE/g_s/g_h	ABA correlated with stomatal conductance, stomatal sensitivity to an increase in leaf water potential, shoot hydraulic conductance and photosynthesis	Aasamaa *et al.*, 2002
P. euphratica	Adult trees	Qira Oasis, China	O		$-3.0 < \psi_{leaf} < -0.2$ MPa	Test of the hypothesis that growth of *P. euphratica* in a Chinese desert depends on vertical distance to a permanent water table	SG/g_s/g_h	Stomatal conductance of *P. euphratica* was more strongly reduced in response to decreasing leaf water potential than that of *Tamarix ramosissima*	Gries *et al.*, 2003
P. tremula	Adult trees	Järvselja, Estonia	F	Podzolic soil	$-2.5 < \psi_{shoot} < -0.3$ MPa	Determination of the characteristics of drought acclimation – clarification of the role of ABA in these strategies	[ABA]/g_s/g_h/$[Osmo]_{leaf}$	Leaf ABA content not significantly changed – xylem ABA content increased during the drought period	Aasamaa *et al.*, 2004

Continued

Table 7.1b. Continued.

Species	Plant material	Experiment location	Growth conditions	Substrate	Water deficit intensity	Objectives of the study	Traits examined	Main conclusions	References
P. grandidentata	Adult trees	Pellston, Michigan, USA	F		5 < SWC < 25%	Detailed understanding of how PAR, VPD and soil water interact to control transpiration	T/SF	Moderate increases in stomatal sensitivity to VPD during dry periods	Bovard *et al.*, 2005
P. deltoides	Adult trees	Green River, Utah/ Yampa River, Colorado, USA	AV		$-0.5 < \psi_{wp} < -0.15$ MPa	Understanding of the impacts of reduced peak flow and soil water recharge on the physiological functioning, morphology and future of Fremont cottonwood forest	D/g_s/RG	Water relations at the leaf and stem level are currently similar for the two rivers due to structural adjustments	Williams and Cooper, 2005

Growth conditions: AV, alluvial valley; F, forest; GC, growth chamber; Gh, greenhouse; O, oasis; P, plantation; Rz, rhizopods; SRC, short-rotation culture. Traits examined: ABA, abscisic acid; BA, biomass allocation; Cc, stem circumference/diameter; D, population density; g, conductance; GE, gas exchange; g_h, hydraulic conductance; g_s, stomatal conductance; Ht, stem height; LA, leaf area; LAI, leaf area index; LWC, leaf water content; Osmo, osmoticum; PAR, photosynthetically active radiation; π_o, osmotic potential; RG, root growth; RL, root length; S, survival; SA, sapwood area; SCA, stem cross-sectional area; SF, sap flux; SG, stem growth; SLA, specific leaf area; SWC, soil water content; T, transpiration; VPD, vapour pressure deficit; Ψ_{xp}, xylem water potential; [ ], content/concentration.

Table 7.1c. Overview of recent available literature on drought response of *Populus* and *Salix* experiments in ISI (Thompson Institute) scientific journals (period 1998–2006) and of former keystone studies (1987–1997). (c) Water relations – hydraulic components. Studies are ordered according to the age of the plant material and then chronologically (year of publication). Abbreviations used are listed at the end of the table.

Species	Plant material	Experiment location	Growth conditions	Substrate	Water deficit intensity	Objectives of the study	Traits examined	Main conclusions	References
S. viminalis *S. viminalis* × *S. schwerinii* *S. purpurea*	Young cuttings	Umeå, Sweden	GC	Fertilized peat/ granular soil conditioner	20 < SWC < 100%	Assessment of the range of variation in hydraulic properties and stomatal regulation – assessment of the resulting variation in drought responses – investigation of the degree to which drought resistance can be enhanced	LA/LW/g_h/ SG/PLC/g_s	Drought resistance was related negatively to maximum growth yields – because the level of drought resistance was related negatively to maximum stomatal conductance, growth may have been affected adversely as a result of reduced photosynthesis	Wikberg and Ögren, 2004
P. deltoides	2-month-old scions	Burlington, Vermont, USA	Gh	Peat/ vermiculite mix	−2.5 < ψ < −0.5 MPa	Does cutting stems in air introduce unnatural embolism into the xylem at the cut surface?	PLC	Sharp increase of conductivity loss below water potentials of −1 MPa, with 100% loss by −2 MPa	Tyree *et al.*, 1992

Continued

Table 7.1c. Continued.

Species	Plant material	Experiment location	Growth conditions	Substrate	Water deficit intensity	Objectives of the study	Traits examined	Main conclusions	References
P. euphratica *P. alba* *P. trichocarpa* × *P. koreana*	8- to 16-week-old cuttings	Champenoux, France	Gh	Sand/peat	4, 5, 7.5 and 10% of soil volumetric water content	Test whether the hydraulic architecture of *P. euphratica* differs from that of other poplars	PLC/g_s	From the point of view of its hydraulic architecture, *P. euphratica* displays a very poor level of drought tolerance, confirming its phreatophytic habit	Hukin *et al.*, 2005
P. trichocarpa	3-month-old cuttings	Pullman, Washington, USA	Gh	Peat/moss/ vermiculite mix	$-2.9 < \psi < -0.6$ MPa	Determination of variations in drought-induced xylem cavitation, xylem air-entry points, stomatal behaviour and hydraulic conductivity in four populations	PLC/g_s/g_h	Interpopulation differences in resistance to drought-induced xylem cavitation, stomatal behaviour and hydraulic conductivity within *P. trichocarpa*	Sparks and Black, 1999

P. trichocarpa *P. deltoides* *P. trichocarpa* × *P. deltoides*	3-month-old cuttings	Victoria, British Columbia, Canada	Gh	Sand	$-1.91 < \psi_{xp} < -0.76$ MPa	Examination of N and K nutrition on drought and cavitation resistance	GE/WUE/PLC/LA/SD	Drought-resistant clones exhibited similar WUE to drought-susceptible clones, but had smaller, more numerous stomata and greater leaf retention under drought conditions	Harvey and van den Driessche, 1999
P. tremuloides	3.5-month-old seedlings	Edmonton, Alberta, Canada	GC	Nutrient solutions	Mild drought: ψ_{shoot} = –1.1 MPa – severe drought: ψ_{shoot} = –2.3 MPa	Test of the hypothesis that mild drought, contrary to a severe one, increases root hydraulic conductivity	g_h/g_s/RVFD	Aquaporin-mediated transport is important in the regulation of root water flow under drought – root water flow properties are affected strongly by the stress level	Siemens and Zwiazek, 2004
P. tremuloides	4-month-old seedlings	Edmonton, Alberta, Canada	GC	Washed coarse sand	Mild drought: ψ_{shoot} = –1.3 MPa – severe drought: ψ_{shoot} = –2.5 MPa – stress recovery: ψ_{shoot} = –1.8 MPa	Test of the hypothesis that mild and severe levels of water deficit would have opposite effects on root water flow properties	g_h/g_s/RVFD/R_{root}/AE	Root water flow properties were modified differently depending on the water deficit stress level	Siemens and Zwiazek, 2003

Continued

Table 7.1c. Continued.

Species	Plant material	Experiment location	Growth conditions	Substrate	Water deficit intensity	Objectives of the study	Traits examined	Main conclusions	References
P. trichocarpa × *P. koreana* *P.* ×*canadensis*	4- to 5-month-old plants	Nancy, France	GC	Sand/peat	10% < SWC	Comparison of the xylem hydraulic properties of two clones known to differ in their ability to close their stomata during a period of drought	PLC	Efficient stomatal regulation is essential for maintaining the integrity of xylem sap flow under drought conditions – 'Peace' shows an exception to the general rule of stomatal control of embolism	Cochard *et al.*, 1996
P. nigra *P.* ×*canadensis* *S. alba*	5- to 10-year-old trees	Garonne river, France	AV		0 < daily rainfall < 18 mm	Test of the active sapwood depth of poplar – comparison of the sap flow of the three species – comparison of the sap flows at two distinct ages	SF/D/Ht/Cc/SG	Under the worst conditions, e.g. the especially long drought in summer 1998, the values of sap flux density dropped to 40 dm^3 day^{-1}, and even to 12 dm^3 day^{-1} for a few days	Lambs and Muller, 2002

P. tremula *S. caprea*	8- to 10-year-old saplings	Tartu, Estonia	P	Clay loam	Mild water deficit	Study of the relationship between shoot hydraulic conductance and stomatal sensitivity to changes in leaf water status	GE/g_h/g_s	Hydraulic conductance in the tree shoots changed significantly during only a few days of mild water stress or starvation in darkness	Aasamaa and Sõber, 2001
P. fremontii	23- to 30-year-old trees	Bernardo, New Mexico, USA	F	Sand/gravel	$-2.25 < \psi_{xp} < 0$ MPa	1. Construction of vulnerability curves 2. Calculation of native embolism 3. Calculation and report of ψ_{cav}	PLC/PNE	This population is very vulnerable to cavitation – native state embolism is between 19 and 42%	Leffler *et al.*, 2000
P. deltoides *P. balsamifera* *P. angustifolia*	Adult trees 1-year-old stems	Lethbridge, Alberta, Canada	AV		Range of xylem water potentials	Examination of vulnerability to cavitation in three riparian cottonwood species	PLC	The three species are the most vulnerable tree species reported so far in North America	Tyree *et al.*, 1994
P. tremula	Adult trees	Rapolano Terme, Italy	AV		$-2 < \psi_{xp} < 0$ MPa	Investigation of how proximity to natural CO_2 springs affected the seasonal patterns of xylem embolism	PLC/Cc/k	The interaction with seasonal stress events might influence strongly the competitive ability in a global change scenario	Tognetti *et al.*, 1999a

Continued

Table 7.1c. Continued.

Species	Plant material	Experiment location	Growth conditions	Substrate	Water deficit intensity	Objectives of the study	Traits examined	Main conclusions	References
P. fremontii *S. gooddingii*	Adult trees	Cienega Creek Natural Preserve, Arizona, USA	D		$-10 < \psi_{xp} < 0$ MPa	Evaluation of how the limitation of xylem pressure by cavitation corresponded with plant distribution along a moisture gradient	PLC/T/k	Cavitation vulnerability limits plant distribution by defining maximum drought tolerance across habitats and influencing competitive ability of drought tolerant species in mesic habitats	Pockman and Sperry, 2000
P. tremuloides	Adult trees	Medicine Bow Mountains, Wyoming, USA	F	Inceptisol	$-1.7 < \psi_{wp} < -0.2$ MPa	Evaluation of seasonal patterns of water use at the canopy level and mechanisms underlying contrasting responses to seasonal drought	SF/g_h/T/LAI/D/SCA	*P. tremuloides* showed less sensitivity to soil moisture than the other species, with relatively high sap flux continuing late into the season and intermediate change in the response of sap flux with decreasing soil moisture	Pataki *et al.*, 2000

P. deltoides *P. fremontii* *P. trichocarpa* *P. balsamifera* *P. angustifolia*	Adult trees	British Columbia/ Missouri/ Montana/ Nevada/ California/ Utah, Canada/ USA	AV	$-1.7 < \psi_{leaf} < -0.8$ MPa	Investigation of xylem cavitation and precocious senescence and branch dieback in cottonwood	PLC/Cc	'Branch sacrifice' describes the cavitation-associated senescence and branch dieback that may provide a drought adaptation for the prairie and Fremont cottonwoods	Rood *et al.*, 2000a
S. psammophila	Adult trees	Mu Us Sandland, Inner Mongolia, China	D	Water spender versus water saver species	Use of the leaf water potential components to elucidate the major types of functional adaptations of the dominant shrubs to drought stress	BEM/π_O/ LAnat	Drought adaptations can to some extent be interpreted within the framework of the water-spending/ water-saving paradigm	Dong and Zhang, 2001
P. angustifolia *P. tremuloides*	Adult trees	Salt Lake City, Utah, USA	AV	$-4.0 < \psi_{xp} < 0$ MPa	Determination of the effect of a cavitation and refilling cycle on cavitation resistance	PLC	Considerable reduction in cavitation resistance was observed after cavitation-refilling cycle for *P. angustifolia* and *P. tremuloides*	Hacke *et al.*, 2001

Growth conditions: AV, alluvial valley; D, desert; F, forest; GC, growth chamber; Gh, greenhouse; P, plantation. Traits examined: AE, activation energy; BEM, bulk elastic modulus; Cc, stem circumference/diameter; D, population density; g, conductance; GE, gas exchange; g_h, hydraulic conductance; g_s, stomatal conductance; Ht, stem height; k, specific conductivity; LA, leaf area; LAI, leaf area index; LAnat, leaf anatomy; LW, leaf dry weight; PLC, percentage loss of conductivity; PNE, percentage native embolism; π_o, osmotic potential; R_{root}, root respiration; RVFD, root volume flow density; SCA, stem cross-sectional area; SD, stomatal density/dimension; SF, sap flux; SG, stem growth; SWC, soil water content; T, transpiration; WUE, water-use efficiency ψ, water potential.

Table 7.1d. Overview of recent available literature on drought response of *Populus* and *Salix* experiments in ISI (Thompson Institute) scientific journals (period 1998–2006) and of former keystone studies (1987–1997). (d) Leaf biochemistry – osmotic adjustment. Studies are ordered according to the age of the plant material and then chronologically (year of publication). Abbreviations used are listed at the end of the table.

Species	Plant material	Experiment location	Growth conditions	Substrate	Water deficit intensity	Objectives of the study	Traits examined	Main conclusions	References
P. deltoides	Young cuttings	Lincoln, Nebraska, USA	P	Vermiculite/ peat/silty clay loam	$-0.7 < \psi_{wp} < 0$ MPa	Comparison of seasonal and clonal variations in leaf water potential, leaf osmotic potential and dry weight fraction of three *P. deltoides* clones	π_O/RWC/II	All clones had drought hardened, indicating that they had some degree of drought tolerance	Gebre and Kuhns, 1991
P. deltoides *P. trichocarpa* *P. trichocarpa* × *P. deltoides*	Young cuttings	Oak Ridge, Tennessee, USA	Gh	Promix BX	$-2.0 < \psi_s < -0.03$ MPa	Characterization of the limits of water-stress tolerance	RGR/SV/ BA/π_O/ RWC/ GE/g_s/T	Three of the four hybrids displayed some degree of osmotic adjustment at saturation after bud set – screening clones of *Populus* should take into account the segregating tendency of hybrids to allocate carbon to lateral meristems under stress	Tschaplinski *et al.*, 1994
P. euphratica *P. alba* × *P.* × *tomentosa*	*In vitro* shoot	Tokyo, Japan	MpS	Murashige and Skoog (1962) medium	0, 200, 300, 400 mM of mannitol	Evaluation of osmotic and salt tolerance and the effects of salt and mannitol in the medium on proline and sugar accumulation	Biomass/ [Pro]/ [Sugar]	Accumulated proline and sugars promote osmotic and salt stress	Watanabe *et al.*, 2000

P. simonii	1-month-old cuttings	Maoxian Field Ecological Station, China	Gh		100, 50, 25% of field capacity	Understanding the mechanisms responsible for adaptation to drought and growth maintenance	BA/GE/OA/LAP/[ABA]	Large set of parallel changes in morphological, physiological and biochemical responses, possibly enhancing capability to survive and to maintain growth	Yin *et al.*, 2005a
P. tremuloides	3-month-old cuttings	Syracuse, New York, USA	Gh	Hydroponic conditions	$-3.0 < \psi_{leaf} < -1.0$ MPa	Description of an hydroponic droughting technique and demonstration that it causes responses in leaf water, solute and turgor potentials and growth	SG/[AA]/P_t	The growth inhibitions observed during drought could not have been the consequence of loss of turgor	Griffin *et al.*, 1991a
P. tremuloides	3-month-old cuttings	Syracuse, New York, USA	Gh	Hydroponic conditions	$-3.0 < \psi_{leaf} < -1.0$ MPa	Determination of the effects of diurnal drought stress on the free amino acid contents of leaves, stems and roots	[AA]	Drought stress caused both organ-specific and clone-specific changes in amino acid concentrations – the method of inducing drought stress affects the responses observed	Griffin *et al.*, 1991b
P. deltoides	3-month-old cuttings	Oak Ridge, Tennessee, USA	Gh	Vermiculite/peat/perlite/soil	$-1.02 < \psi_{leaf} < -0.03$ MPa	Comparison of organic solute accumulation in two poplar clones before and after re-watering	π_o/[Sugar]	There was a significant accumulation of glucose and fructose at predawn and either no accumulation or a significant reduction of these solutes at midday before and after re-watering	Gebre *et al.*, 1997

Continued

Table 7.1d. Continued.

Species	Plant material	Experiment location	Growth conditions	Substrate	Water deficit intensity	Objectives of the study	Traits examined	Main conclusions	References
P. deltoides	4-month-old cuttings	Oak Ridge, Tennessee, USA	Gh	Vermiculite/ peat/ perlite/soil	$-1.74 < \psi_{leaf} < -0.08$ MPa	Identification of solutes that contribute to osmotic adjustment in three *P. deltoides* clones	π_o/[Sugar]	The solutes that contribute at least –0.05 MPa each to leaf osmotic potential in any clone were sucrose, malic acid, glucose, fructose, myo-inositol and salicin	Gebre *et al.*, 1994
P. euphratica	1-year-old seedlings	Champenoux, France	Gh	Peat/sand	5% soil humidity during 10 days	Technical improvements allowing the MS analysis of carbohydrates and presentation of results obtained on drought-stressed poplar	[Salicin], [Gluc], [Suc], [Fruc], [Galac]	Different responses to water deficit and rehydration were obtained for several carbohydrates, suggesting different roles in osmoprotection processes	Guignard *et al.*, 2005
P. trichocarpa × *P. deltoides* *P. deltoides* × *P. nigra*	2- to 5-year-old trees	Wallula, Washington/ Boardman, Oregon, USA	P	Active dune land/ mixed sand	$-1.20 < \psi_{wp} < -0.65$ MPa	Characterization of the leaf osmotic potential at full turgor and the major solutes that contribute to osmotic potential	π_o/[Sugar]/ [Ion]/ [AA]	The extent of osmotic adjustment was small – other drought resistance mechanisms contributed to the clonal differences in field performance	Gebre *et al.*, 1998
P. nigra	Adult trees	Orléans, France	N		0 < [mannitol] < 0.5 M	Comparison of the ability of several solutes accumulated by stressed poplars to affect the GS activity	GS	Proline and putrescine were the most effective in alleviating the inhibitory effect of mannitol	Sibout and Guerrier, 1998

P. euphratica	Adult trees	Tarim River Basin, China	AV	Sandy loam	2.47 < groundwater level < 10.16 m	Study of the relationship between the proline accumulation in the bodies of *P. euphratica* and the change of groundwater level	[Pro]	The constraining groundwater depth for the normal growth and the critical one for the survival of *P. euphratica* are below 4.5 m and 10 m, respectively, at the lower reaches of the Tarim River	Chen *et al.*, 2003b
P. euphratica	Adult trees	Qira oasis, China	O		Seasonal groundwater variation in desert	Investigation of variation of leaf solutes in order to elucidate their adaptation to saline groundwater	[Cation]/ [Anion]/ [Acid]/ [Sugar]	All plants were well adapted to the moderate salinity of the NaCl-dominated groundwater and no signs of salt-related drought stress were observed	Arndt *et al.*, 2004

Growth conditions: AV, alluvial valley; Gh, greenhouse; MpS, micropropagation system; N, nursery; O, oasis; P, plantation. Traits examined: AA, amino acids; ABA, abscisic acid; BA, biomass allocation; Fruc, fructose; GE, gas exchange; Galac, galactose; Gluc, glucose; g_s, stomatal conductance; GS, glutamine synthase; II, injury index; LAP, leaf antioxidant properties; MS, mass spectrometry; OA, osmotic adjustment; π_o,osmotic potential; Pro, proline; P_t, turgor pressure; RGR, relative growth rate; RWC, relative water content; SG, stem growth; Suc, sucrose; SV, stem volume; T, transpiration; ψ, water potential; [ ], content/concentration; _, enzyme activity.

Table 7.1e. Overview of recent available literature on drought response of *Populus* and *Salix* experiments in ISI (Thompson Institute) scientific journals (period 1998–2006) and of former keystone studies (1987–1997). (e) Plant growth – biomass allocation. Studies are ordered according to the age of the plant material and then chronologically (year of publication). Abbreviations used are listed at the end of the table.

Species	Plant material	Experiment location	Growth conditions	Substrate	Water deficit intensity	Objectives of the study	Traits examined	Main conclusions	References
P. ×*canadensis* *P.* 'Tristis' × *P. balsamifera*	Young cuttings	East Lansing, Michigan, USA	Gh	Sandy-loam soil	$-0.5 < \psi_{soil\ matric} < 0$ MPa	Examination of growth and morphological responses induced by the onset and release of cycles of water stress under two levels of N availability	Biomass/BA/[Chl]/N/LN/LA/SLA/RSR	Soil moisture determined the amount of biomass that accumulated in roots – leaves became thinner as soil moisture decreased from flooding	Liu and Dickmann, 1992b
P. deltoides	Young seedlings	Fort Collins, Colorado, USA	OC	Coarse sand	20 < water level < 100 cm	Examination of germination and establishment of Russian-olive and cottonwood under a range of experimental moisture and light conditions	Biomass/BA	Russian-olive succeeds under conditions optimal for cottonwood establishment and under many conditions unfavourable for cottonwood	Shafroth *et al.*, 1995

P. balsamifera × *P. trichocarpa*	Young cuttings	Craigiebuckler, UK	Gh	Sand	Watering every 2 or 10 days	Study of the effects of nitrogen supply and drought stress	BA/Biomass/LA/LN/N/SLA/GE	Whole-plant respiration decreased in the water-stressed trees due to a reduction in total biomass and lower rates of respiration per unit tissue	Ibrahim, *et al.*, 1997
S. nigra	Young cuttings	Memphis, Tennessee, USA	Gh	Sand/Sharkey Clay Series soil	–200 < soil redox potential < 700 mV	Evaluation of the effects of static and dynamic flooding regime and drought on root and shoot development, pattern of root distribution, biomass production and its allocation	GE/g_s/LA/LW/Biomass/Chl/RSR/BA	Maximum photosynthesis and growth in willow cuttings required ample soil moisture and adequate drainage in the top 60 cm of soil	Pezeshki *et al.*, 1998
P. nigra	Young cuttings	Cambridgeshire, UK	AV	Various sediment types	$-1 < \psi_{soil} < 0$ MPa	Investigation of the differences in the responses of male and female black poplars to a variety of soil moisture conditions	LW/RGR/Ht/Biomass	Some limited spatial segregation of the sexes does occur in response to soil moisture availability	Hughes *et al.*, 2000

Continued

Table 7.1e. Continued.

Species	Plant material	Experiment location	Growth conditions	Substrate	Water deficit intensity	Objectives of the study	Traits examined	Main conclusions	References
P. angustifolia *P. balsamifera* *S. drummondiana* *S. exigua* *S. lutea*	Young saplings and seedlings	Lethbridge, Alberta, Canada	Rz/Gh	Coarse gravel/ medium sand/fine gravel	0, 1, 2, 3, 4 or 8 cm day^{-1} water table decline	Investigation of the relative tolerances of saplings and seedlings of different willow and cottonwood species to different rates of water table decline	Ht/RL/BA/S	Willow and cottonwood saplings similarly affected by abrupt water table decline, but willow seedlings slightly more vulnerable than cottonwood seedlings	Amlin and Rood, 2002
S. nigra	Young cuttings	Loosahatchie River, Tennessee, USA	Gh	Sand/field soil (Sharkey Clay)	6 < soil redox potential < 451 mV	Quantification of the overall effects of soaking across various soil moisture regimes	S/RSR/ Biomass	Soaking had significant effects on willow post success when evaluated across all soil moisture regimes	Schaff *et al.*, 2002

P. tremuloides	Young seedlings	Edmonton, Alberta, Canada	P	Slit/clay/loam	6 < volumetric soil moisture < 20%	Isolation and quantification of the relative competitive and facilitative effects present within mixtures of aspen seedlings, lucerne and marsh reedgrass	SG/LA/LN/BA/RGR/NPP	Evidence of facilitation was found when aspen was grown with lucerne, including increases of overall available soil N and transient increases in soil moisture with pulsed precipitation during drought	Powell and Bork, 2004
S. nigra	Young cuttings	Memphis, Tennessee, USA	Gh	Sand/soil	$-0.6 < \psi_{soil} < 0$ MPa	Quantification of the physiological responses of black willow to four soil moisture regimes	F/g_s/Ht/RSR/RG/SG	Photosynthesis and growth of black willow cuttings in response to both reduced soil conditions and moisture deficits were limited by stomatal closure	Li *et al.*, 2004b

Continued

Table 7.1e. Continued.

Species	Plant material	Experiment location	Growth conditions	Substrate	Water deficit intensity	Objectives of the study	Traits examined	Main conclusions	References
P. nigra *S. elaeagnos*	5- to 74-day-old cuttings		Rz/Gh	Sand or gravel	Water table decline from 0 to 3 cm day^{-1}	Investigation of the influence of different periods of inundation and rates of water table decline on the survival and growth rate	S/LN/LL/BA/RG	Restoration schemes utilizing *S. elaeagnos* cuttings may experience a greater success than those using *P. nigra* – *P. nigra* cuttings may establish more easily if situated where inundation does not occur often or for prolonged periods	Francis *et al.*, 2005
P. balsamifera × *P. deltoides* (= *P.* ×*acuminata*)	2-week-old seedlings	Lethbridge, Alberta, Canada	Rz/Gh	Sand/gravel	0, 1, 2, 4 and 8 cm day^{-1} water decline	Investigation of the influence of the rate of water table decline on poplar seedling growth and survival	S/LA/LN/RL/RW/Ht	Maximum root mass and length were observed in plants subjected to declines in water table level of 0 and 1 cm day^{-1}, respectively	Mahoney and Rood, 1991

P. 'Tristis' × *P. balsamifera* *P.* ×*canadensis*	1-month-old cuttings	East Lansing, Michigan, USA	GC	Sandy loam soil	$-1 < \psi_{soil} < -0.007$ MPa	Observation of early patterns of growth and water relations in response to changing conditions of water stress	g_s/BA/Ht/LA	The higher root: leaf weight ratio of the *P.* 'Tristis' × *P. balsamifera* clone enabled it to maintain a more favourable plant water status	Mazzoleni and Dickmann, 1988
P. 'Tristis' × *P. balsamifera* *P.* ×*canadensis*	1-month-old cuttings	Rhinelander, Wisconsin, USA	SRC		Seasonal variations of ψ_{soil}	Monitoring of growth and development of two *Populus* clones during their establishment year	LA/LN/SLA/Ht	The high shoot: root ratio developed by the *P.* ×*canadensis* clone limited its development during periods of low soil water potential	Michael *et al.*, 1988
S. alba *S. triandra* *S. viminalis* *P. nigra*	1-month-old seedlings	River Rhine, the Netherlands	Gh	Clay/sand	$0.10 < SWC < 0.30\ cm^3\ cm^{-3}$	Assessment of differences in drought-resistance properties among flood-plain species during their seedling stage	LA/SLA/RSR/BA/RL/T	*P. nigra* is most resistant to drought followed by *S. alba* and then *S. triandra* and *S. viminalis* – these differences are reflected in the distribution patterns of these species observed on the banks of the river Rhine	Van Splunder *et al.*, 1996

Continued

Table 7.1e. Continued.

Species	Plant material	Experiment location	Growth conditions	Substrate	Water deficit intensity	Objectives of the study	Traits examined	Main conclusions	References
P. deltoides *P. balsamifera* *P. angustifolia*	1-month-old cuttings	Lethbridge, Alberta, Canada	Rz/Gh	Commercial potting mixture	0, 4 and 10 cm day^{-1} water decline	Study of the influence of the rate of water table decline with three North American cottonwood species	Ht/RL/BA/T/LA/LN	The tolerance to water table decline varies across cottonwood genotypes and *P. balsamifera* saplings are the most vigorous	Kranjcec *et al.*, 1998
S. gooddingii	1-month-old seedlings	Flagstaff, Arizona, USA	Gh	Sand/river gravel	0.06 < volumetric soil water content < 0.30 m^3 m^{-3}	Investigation of the effects of differing rates of groundwater decline on growth and survival of *S. gooddingii*	S/Ht/RL/LA/RSR	Lateral root development may help to protect *Salix* seedlings from late season flood scour	Horton and Clark, 2001
S. dasyclados × *S. viminalis*	2-month-old cuttings	Uppsala, Sweden	Gh	Clay/peat	55% of the water given to the well-watered plants	Identification and localization of QTL for growth, nutrient and water economy – quantification of the effect of the QTL and evaluation of common QTL in two contrasting treatments: well watered and drought stressed	RGR/$\delta^{13}C$/N/Ht/Cc/RW/Biomass	QTL specific for each treatment were found, but QTL common across the treatments were also detected	Rönnberg-Wästljung *et al.*, 2005

S. myrsinifolia *S. myrsinites* × *S. myrsinifolia*	2-month-old plantlets	Punkaharju, Finland	Gh	Peat	20 < SWC < 50%	Study of the response of growth of willows to combinations of UVB radiation and drought stress	Ht/Biomass/BA/RSR	Drought stress produced family- and clone-specific reactions, whereas the response of studied willow genotypes to enhanced UVB and UVB × drought was mostly similar	Turtola *et al.*, 2006
P. balsamifera *P. trichocarpa* *P. nigra* × *P. nigra*	6- to 10-week-old cuttings	Columbia, Missouri, USA	Gh	Peat moss/sand/silt loam	$-3.5 < \psi_{leaf} < -0.5$ MPa	Identification of traits associated with superior growth potential on sites where water could be a limiting factor	GE/BA/LA/SLA/LAR	Rapid early leaf and root growth appear to be key attributes associated with productivity regardless of soil water availability	Rhodenbaugh and Pallardy, 1993
P. trichocarpa × *P. balsamifera*	1- to 4-month-old cuttings	Aberdeen, UK	Gh	Coarse, medium and fine sand	$-2 < \psi_{leaf} < -1$ MPa	Examination of the interaction of water and nitrogen availabilities on gas exchange and whole-plant carbon allocation	Biomass/BA/N/g_s/GE	Preferential allocation of biomass to roots under drought conditions and reduced rate of net photosynthesis associated with reductions in stomatal conductance and SLA	Ibrahim *et al.*, 1998

Continued

Table 7.1e. Continued.

Species	Plant material	Experiment location	Growth conditions	Substrate	Water deficit intensity	Objectives of the study	Traits examined	Main conclusions	References
P. deltoides	3-month-old cuttings	Mississippi State, USA	Gh	Pro-Mix/sand	$-0.5 < \psi_{wp} < 0$ MPa	Evaluation of the influence of various treatments on flowering in rooted cuttings from mature and juvenile trees	SG	The new set of modified treatments failed to induce precocious flowering in rooted cuttings from juvenile trees only	Yuceer *et al.*, 2003
P. nigra	2- to 6-month-old cuttings	Wooster, Ohio, USA	OC	Premier Pro-Mix BX	$-3 <$ PMMT < -25 kPa	Quantification of the effects of water and nutrient availability on carbon assimilation, growth and total foliar phenolic glycoside concentration and on constitutive and rapid-induced resistance to gypsy moth and whitemarked tussock moth	Biomass/BA/LA/SLA/RGR/GE/N/larval growth/[Phenols]	Drought decreased net assimilation rate and growth, while increasing total phenolic glycoside concentrations – drought decreased the growth of gypsy moth larvae but had no effect on whitemarked tussock moth	Hale *et al.*, 2005

P. fremontii *S. gooddingii*	4- to 5-month-old cuttings	Tucson, Arizona, USA	OC	Sand/potting mix	Withholding water	Determination if remotely sensed canopy temperatures could be used to estimate transpiration or water stress	Biomass/LA/LAI/LW/SW/SF/g_s	All species depleted soil moisture to the same extent and reached the wilt point at about the same time, salt cedar has adaptations that allow it to recover from drought better than the other species	Nagler *et al.*, 2003
P. trichocarpa *P. deltoides* *P. trichocarpa* × *P. deltoides*	1-year-old coppiced plants	Seattle, Washington, USA	P		Two consecutive contrasting years in terms of rainfall	Determination of the effects of irrigation on stem volume production and leaf expansion	SG/g_s/π_o	Unlike its effect on leaf area growth, irrigation increased stem volume growth of the hybrid and the parental species by a similar amount	Roden *et al.*, 1990
S. sericea *S. eriocephala* *S. sericea* × *S. eriocephala*	1-year-old seedlings	Williamstown, Massachusetts, USA	Gh	Soil/peat/vermiculite	$-2.5 < \psi_{stem} < -0.1$ MPa	Quantification of the relative performance of *S. sericea*, *S. eriocephala* and their F_1 hybrids to varying levels of water availability	LWC/SG/LN/LL/BA/LG	Temporal and spatial variability in water availability could determine the survivorship of hybrids and the frequency of introgression	Orians *et al.*, 1999

Continued

Table 7.1e. Continued.

Species	Plant material	Experiment location	Growth conditions	Substrate	Water deficit intensity	Objectives of the study	Traits examined	Main conclusions	References
P. nigra *P. euphratica* *S. nigra*	1- to 2-year-old trees	Khiva, Khorezm, Uzbekistan	P	Sandy site + loamy site	0 < monthly precipitations < 27 mm	Study over 24 months of the growth and development of ten trees differing in their tolerance to drought and salinity	Cc/RGR/SLA/LAI/LAR/RSR/LMR/GE	Ranking of all parameters concurrently showed the high potential of *Elaeagnus angustifolia* and *P. euphratica*, which matched previous rankings based on total biomass and financial added value	Lamers *et al.*, 2006
P. spp.	0 to 6-year-old trees	Arlington, Wisconsin, USA	P	Plano silt loam	Natural variations in rainfall between years	Examination of the survival, productivity, drought tolerance and pest status of 16 hybrid poplars	S/Ht/Cc/Biomass/PI	Differential responses among clones to drought	Robison and Raffa, 1998

P. trichocarpa × *P. deltoides* *P. deltoides* × *P. nigra*	2- to 3-year-old trees	Wallula, Washington, USA	D	Active dune land	$-1.2 < \psi_{wp} < -0.4$ MPa	Assessment of the role of the osmotic potential in determining drought tolerance	π_o/RGR/SG/RW	The higher drought resistance of clone 'DN' compared with clone 'TD' was the result of the maintenance of a more favourable water balance by stomatal regulation and greater carbon allocation to roots during the early stages of drought	Tschaplinski *et al.*, 1998
P. fremontii *S. gooddingii*	2- to 4-year-old seedlings and saplings	Central Arizona, USA	AV		$-3 <$ water table depth < 0 m	Quantification of the response of three riparian species to different water table dynamics	RL/SCA	Plant response is likely mediated by factors such as soil texture and stratigraphy, availability of precipitation-derived soil moisture, physiological and morphological adaptations to water stress and tree age	Shafroth *et al.*, 2000

Continued

Table 7.1e. Continued.

Species	Plant material	Experiment location	Growth conditions	Substrate	Water deficit intensity	Objectives of the study	Traits examined	Main conclusions	References
P. deltoides	Adult trees	Olentangy River, Ohio, USA	AV	Silt loam	Palmer Drought Severity Index	Investigation of the influence of climate and stream flow data on productivity	D/SG/SCA	Growth of eastern cottonwood was not related to stream flow, but was reduced by excess summer precipitation	Dudek *et al.*, 1998
P. deltoides	Adult trees	Denver, Colorado, USA	AV	Gravel or loam/gravel	$-0.8 < \psi_{wp} < -0.2$ MPa	Examination of the physiological and morphological response patterns of plains cottonwood to acute water stress imposed by groundwater pumping	LL/g_s	*P. deltoides* responded to abrupt alluvial water table decline with decreased shoot water potential followed by leaf mortality	Cooper *et al.*, 2003a
P. tremuloides	41- to 81-year-old trees	British Columbia to Manitoba, Canada	F	Sand/silt/clay	$-30 <$ climate moisture index $< +20$ cm year^{-1}	Examination of the cause and magnitude of inter-annual variation in aspen growth	Ht/Cc/SG/PI/Biomass	A major collapse in aspen productivity likely occurred during the severe drought that affected much of the region during 2001–2003	Hogg *et al.*, 2005

P. deltoides	80- to 200-year-old trees	Deerlodge Park, Colorado, USA	AV	Sands/silts/clays	0 < mean daily river discharge < 275 $m^3 s^{-1}$	Comparison of production and breakdown of leaf litter at matched flood plain sites on a regulated and an unregulated river	N/OM/MiC/SC	Factors reducing flood flow frequency and magnitude will reduce overall breakdown rates on the flood plain towards those found in drier upland environments	Andersen and Nelson, 2003

Growth conditions: AV, alluvial valley; D, desert; F, forest; GC, growth chamber; Gh, greenhouse; OC, outdoor containers; P, plantation; Rz, rhizopods; SRC, short-rotation culture. Traits examined: BA, biomass allocation; Cc, stem circumference/diameter; Chl, chlorophylls; $\delta^{13}C$, carbon isotope composition; F, fluorescence; GE, gas exchange; g_s, stomatal conductance; Ht, stem height; LA, leaf area; LAI, leaf area index; LAR, leaf area ratio; LG, leaf growth; LL, leaf loss; LMR, leaf mass ratio; LN, number of leaves; LW, leaf dry weight; LWC, leaf water content; MiC, macroinvertebrate collection; N, nitrogen content; NPP, net primary production; OM, organic matter; π_o, osmotic potential; PI, pathogen incidence; PMMT, potting medium moisture tension; QTL, quantitative trait loci; RG, root growth; RGR, relative growth rate; RL, root length; RSR, root to shoot ratio; RW, root dry weight; S, survival; SC, standing crop; SCA, stem cross-sectional area; SF, sap flux; SG, stem growth; SLA, specific leaf area; SW, stem dry weight; SWC, soil water content; T, transpiration; Ψ, water potential; [ ], content/concentration.

resulting first in the production of the superoxide radical (O_2^-) and then in the formation of various ROS such as the hydroxyl free radical (OH) and hydrogen peroxide (H_2O_2) (Edreva, 2005). ROS are highly toxic and can cause lipid peroxidation and, consequently, membrane injury, protein degradation, enzyme inactivation, pigment bleaching and disruption of DNA strands (Smirnoff, 1993). Allen (1995) reported that much of the injury to plants caused by exposure to various constraints was associated with oxidative damage at the cellular level. Plant cells are normally protected against the detrimental effects of reactive oxygen by a complex antioxidant system; active oxy-free radicals can be scavenged by both enzymatic and non-enzymatic detoxification mechanisms (Smirnoff, 1993; Edreva, 2005). Oxidative stress can occur when the scavenging of ROS is overwhelmed by the production. In poplar, it has been shown that protection against oxidative stress generated by elevated CO_2, paraquat and ozone mainly involved superoxide dismutase (SOD), catalase and peroxidase (Arisi *et al.*, 1998; Strohm *et al.*, 1999, 2002; Schwanz and Polle, 2001). During drought, an enhancement of the activity of most of the antioxidant enzymes has been observed in poplar, but no clear link with the level of drought tolerance of the tested clones has been established (Courtois *et al.*, 1999; Guerrier *et al.*, 2000; Marron *et al.*, 2006). On the other hand, the non-enzymatic pathway seems to have a limited influence under drought conditions in poplar (Marron *et al.*, 2002).

Relationships between water-use efficiency and drought resistance

Some authors consider the ability to present and to maintain a high WUE under drought conditions an important mechanism for plant resistance to water deficit, entirely independent from the widely described tolerance and avoidance strategies (Jones, 1992, 2004) (Table 7.1h). However, the links between WUE and drought resistance are still unclear and only a few studies have tried to clarify the links between these two properties along with the physiological determinism of WUE.

In *P.* ×*canadensis* hybrids, a negative correlation has been observed between WUE determined from gas exchange rates (intrinsic WUE) and carbon isotope discrimination (Δ), while a positive correlation has been demonstrated between Δ and stomatal conductance under moderate drought. This suggested that the diversity for Δ was driven mainly by stomatal conductance (Monclus *et al.*, 2006). On the other hand, no relationship between Δ and biomass production could be established among different kinds of poplar hybrids, providing opportunities for selecting poplar clones combining high productivity and high WUE (Rae *et al.*, 2004; Marron *et al.*, 2005; Monclus *et al.*, 2005). Likewise, no direct relationship was found between WUE and drought resistance in *P.* ×*canadensis* hybrids (Monclus *et al.*, 2006). But it is noteworthy that, in this latter study, none of the poplar genotypes for which Δ tended to increase in response to drought was a drought-resistant genotype. This result suggested that the genotypic ability to increase WUE was necessary to produce a high level of drought resistance. However, the whole range of drought resistance levels was observed among the genotypes for which Δ decreased or tended to decrease, suggesting that the ability to increase WUE was necessary, but not sufficient, to explain genotypic diversity of drought resistance among *P.* ×*canadensis* hybrids. It is also clear that drought resistance is not only governed by WUE but also probably includes a lot of other traits that also contribute to the overall drought-resistance capacity of a particular genotype (Liu and Dickmann, 1992b; Gebre *et al.*, 1994; Chen *et al.*, 1997; Ibrahim *et al.*, 1997; Tschaplinski *et al.*, 1998; Marron *et al.*, 2002, 2003).

Conclusions and perspectives

From many studies, it has become clear that the members of the *Salicaceae* family are useful and elegant model trees for drought studies. All aspects of the response of poplars and willows to drought have been studied extensively during the period 1998–2008, from the biochemical response to whole population and ecosystem behaviour (Table 7.1i). There is, however, an imbalance between drought studies conducted on poplars and drought studies on willows during the period 1998–2006 (about 85% of the total number of studies on poplars versus 15% on willows; Fig. 7.3). Within the *Populus* genus, the disequilibrium between species is also striking: 24%

Table 7.1f. Overview of recent available literature on drought response of *Populus* and *Salix* experiments in ISI (Thompson Institute) scientific journals (period 1998–2006) and of former keystone studies (1987–1997). (f) Leaf biochemistry – antioxidant ability. Studies are ordered according to the age of the plant material and then chronologically (year of publication). Abbreviations used are listed at the end of the table.

Species	Plant material	Experiment location	Growth conditions	Substrate	Water deficit intensity	Objectives of the study	Traits examined	Main conclusions	References
P. ×*canadensis*	1-month-old cuttings	Orléans, France	Gh		–0.336 < osmotic potential < 0 MPa	Assessment of the relationships between drought response and anatomical/ physiological properties	GE/[Rub]/ [Chl]/[Car]/ [Asc]/[Glu]/ AP/GR	The drought-tolerant poplar exhibited higher net photosynthetic rate and contents of antioxidants and antioxidative enzymes following an exposure to 150 mM mannitol than the drought-sensitive one	Courtois *et al.*, 1999
P. ×*canadensis*	1-month-old cuttings	Orléans, France	Gh		Leaf and root fresh weight loss of 85% of the original fresh weight	Study of the redox status and the activities of antioxidant enzymes in leaves and roots exposed to wilting or to osmotic stress	LAP	Enhanced capability to dismutase superoxide and of ascorbate peroxidase activity in wilted leaves	Morabito and Guerrier, 2000
P. ×*canadensis*	1-month-old cuttings	Orléans, France	IvC	Murashige and Skoog (1962) solution	150 mmol l^{-1} mannitol (–0.336 MPa)	Appreciation of the mechanisms by which drought-induced oxidative stress is tolerated in poplar	MDH/SOD/ Cat/AP/GR	Exposure to osmotic stress resulted in a decrease in Cat and GR activities, in an enhancement of SOD and AP activities, but did not affect the contents of spermine, spermidine and ascorbate	Guerrier *et al.*, 2000

Continued

Table 7.1f. Continued.

Species	Plant material	Experiment location	Growth conditions	Substrate	Water deficit intensity	Objectives of the study	Traits examined	Main conclusions	References
P. ×*canadensis*	1-month-old cuttings	Orléans, France	Gh	Hydroponic conditions	Addition of mannitol	Determination whether a drought-tolerant *P.* ×*canadensis* clone exhibits a more efficient reactive oxygen species (ROS) scavenging system than a drought-sensitive one	Pod/GR	The efficiency of the ascorbate–glutathione cycle operating in cytosol and chloroplast seemed to be sufficient for avoiding the deleterious effects of ROS on the plastidial GR activity of the tolerant clone	Edjolo *et al.*, 2001
P. ×*canadensis*	2-month-old cuttings	Orléans, France	Gh	Sand/clay/ brown peat/ composted cluster pine bark/ wood fibre/ blond peat/ moss	$-1.6 < \psi_{wp} < -0.8$ MPa	Association of variations in leaf polyphenols and antioxidant activity with variations in water relation parameters	LG/SLA/g_s/N/ [Chl]/ [Sugar]/ [Flav]/LAP/ [Phe]	Earlier stomatal closure for the drought-sensitive clone and better ψ_{wp} maintenance – antioxidant activity decrease in response to drought for the drought-sensitive clone only	Marron *et al.*, 2002
P. ×*canadensis*	2-month-old cuttings	Orléans, France	Gh	Blond peat/ brown peat/horse manure/ compost	$-2.52 < \psi_{wp} < -0.59$ MPa	Investigation of the impact of mild and severe constraints on leaf protein content and activities of SOD, catalase and peroxidase	[Prot]/SOD/ Cat/Pod	Activities of the three enzymes were stimulated in response to drought depending on constraint intensity, leaf age and clone	Marron *et al.*, 2006

S. viminalis × *S. dasyclados*	3-month-old cuttings	Uppsala, Sweden	Gh	Clay/peat	55% irrigation of the well-watered plants	Examination of the effect of willow genotype and irrigation regime on growth parameters, foliar N, phenolic content and on preference and performance of the blue leaf beetle	Biomass/N/SLA/[Phe]/FP	There was significant genotypic variation in each of the phenolic substances – there was no effect of irrigation treatment on larval performance	Glynn *et al.*, 2004
S. myrsinifolia *S. myrsinites* × *S. myrsinifolia*	2-month-old plantlets	Punkaharju, Finland	Gh	Peat	20 < SWC < 50%	Study of the effects of enhanced UVB radiation and drought stress on willow secondary phenolics	[Phe]	The response of salicylates, flavonoids and phenolic acids to enhanced UVB and drought stress was clone specific, which may indicate that climatic changes will alter the genetic composition of northern forests	Turtola *et al.*, 2005
P. euphratica	Adult trees	Tarim River Basin, China	AV		5 < groundwater level < 12 m	Determination of the optimal depth of groundwater for restoring the local ecosystem and provide a scientific basis for efficient use of limited water resources	[Pro]/[SOD]/[Pod]	The stress groundwater depths for the *Phragmites communis*, *Tamarix* spp. and *P. euphratica* are 3.5, 5 and 4.5 m, respectively	Chen *et al.*, 2004

Growth conditions: AV, alluvial valley; Gh, greenhouse; IvC, *in vitro* culture. Traits examined: AP, ascorbate peroxidase; Asc, ascorbate; Car, carotenoids; Cat, catalase; Chl, chlorophylls; Flav, flavanols; FP, feeding preference; GE, gas exchange; Glu, glutathione; GR, glutathione reductase; g_s, stomatal conductance; LAP, leaf antioxidant properties; LG, leaf growth; MDH, malate dehydrogenase; N, nitrogen content; Phe, phenolic compounds; Pod, peroxidase; Pro, proline; Prot, protein; ROS, reactive oxygen species; Rub, rubisco; SLA, specific leaf area; SOD, superoxide dismutase; SWC, soil water content; Ψ, water potential; [ ], content/concentration; _, enzyme activity.

Table 7.1g. Overview of recent available literature on drought response of *Populus* and *Salix* experiments in ISI (Thompson Institute) scientific journals (period 1998–2006) and of former keystone studies (1987–1997). (g) Leaf biochemistry – other biochemical aspects. Studies are ordered according to the age of the plant material and then chronologically (year of publication). Abbreviations used are listed at the end of the table.

Species	Plant material	Experiment location	Growth conditions	Substrate	Water deficit intensity	Objectives of the study	Traits examined	Main conclusions	References
P. 'Popularis' *P.* ×*tomentosa*	Softwood cuttings	Rehovot, Israel	Gh	Perlite/ vermiculite	$-3.2 < \psi_{leaf} < -0.8$ MPa	Investigation of the relationship between the presence of proteins and water stress tolerance	EL/[Prot]/ [Sugar]	Increased water stress tolerance is correlated positively with accumulation of BspA, dehydrin homologue and sucrose synthase	Pelah *et al.*, 1997
P. ×*canadensis*	Young cuttings	Orléans, France	GC	Peat	$-1.5 < \psi_{wp} < -0.5$ MPa	Analysis of dehydrin gene expression under dehydration conditions	LG/g_s/EG	A cDNA encoding a dehydrin was characterized – it was induced by withholding water, salt stress, cold and osmotic stress	Caruso *et al.*, 2002
P. tremula	4- to 5-week-old plantlets	Rehovot, Israel	IvC		20 and 30% plant water loss	Description of the cloning, sequence analysis, isolation and characterization of the SP1 protein and its stress responsiveness, boiling solubility and oligomeric structure	EG/[SP1]	SP1 proteins are hydrophilic and remain soluble on boiling – they represent a new class of protein involved in the plant's response to abiotic stress	Wang *et al.*, 2002

P. tremula	2-month-old shoot explants	Rehovot, Israel			Weight loss to 80 or 60% of the shoot original fresh weight	Report of the temporal expression, response to ABA and partial characterization of a boiling-stable protein found in gradually water-stressed shoot cultures	[Prot]	The BspA protein was the only major water stress-responsive boiling-stable protein detected in aspen	Pelah *et al.*, 1995
P. trichocarpa	2-month-old cuttings	Lethbridge, Alberta, Canada	Gh	Natural river valley sand and gravel	Reservoir water decline from 0 to 4 cm day^{-1}	Investigation of phytohormonal involvement in the growth reallocation that follows changes in water table depth	SG/[Gib]/ψ_{xp}	Little evidence that endogenous gibberellins play a primary role in the regulation of root elongation in response to water table decline	Rood *et al.*, 2000b
S. sericea	3-month-old cuttings	Waltham, Massachusetts, USA	N	Loam/peat moss/ vermiculite	Dry/field capacity/ flooded	Investigation of the effects of soil nutrient and water availability on the growth and chemistry of the silky willow and on the performance of the imported willow leaf beetle	RWC/N/ Biomass/ LG/ [Salicortin]	Nutrient–water interactions influence plant traits that are potentially important for insect performance	Lower and Orians, 2003

Continued

Table 7.1g. Continued.

Species	Plant material	Experiment location	Growth conditions	Substrate	Water deficit intensity	Objectives of the study	Traits examined	Main conclusions	References
P. ×*canadensis* *P.* 'Popularis'	1-year-old cuttings	Beijing, China	N	Loam	70 and 30% of field capacity	Investigation of the effects of endogenous and exogenous ABA on polyamines and ethylene synthesis in a drought-sensitive and a drought-tolerant poplar genotype	LL/[ABA]$_{xylem}$/ EER/ [Polya]	Inhibitory effect of ABA on polyamine synthesis more pronounced in the drought-sensitive genotype – more important production of ethylene in the drought-sensitive genotype	Chen *et al.*, 2002b
P. alba						Investigation of the temporal changes in the emission of ethylene and oxygen of water-stressed leaves of *P. alba* using photoacoustic techniques	EER	Information obtained could be used to analyse the stomatal closure dynamics and processes related to photosynthesis and cell death	Acosta-Avalos *et al.*, 2005

Growth conditions: GC, growth chamber; Gh, greenhouse; IvC, *in vitro* culture; N, nursery. Traits examined: ABA, abscisic acid; EG, expression of genes; EER, ethylene emission rate; EL, electrolyte leakage; Gib, gibberelines; g_s, stomatal conductance; LG, leaf growth; LL, leaf loss; N, nitrogen content; Polya, polyamines; Prot, protein; RWC, relative water content; SG, stem growth; Ψ, water potential; [ ], content/concentration.

Table 7.1h. Overview of recent available literature on drought response of *Populus* and *Salix* experiments in ISI (Thompson Institute) scientific journals (period 1998–2006) and of former keystone studies (1987–1997). (h) Water-use efficiency. Studies are ordered according to the age of the plant material and then chronologically (year of publication). Abbreviations used are listed at the end of the table.

Species	Plant material	Experiment location	Growth conditions	Substrate	Water deficit intensity	Objectives of the study	Traits examined	Main conclusions	References
P. ×*canadensis* *P.* 'Tristis' × *P. balsamifera*	Young cuttings	East Lansing, Michigan, USA	Gh	Sandy-loam soil	$-0.5 < \psi_{soil\ matric} < 0$ MPa	Investigation of how plants with low- or high-N status respond physiologically to drought and flooding	GE/g_s/WUE	Drought-stressed plants gained full and quick recovery of photosynthesis on relief from stress	Liu and Dickmann, 1993
P. ×*canadensis* *P.* 'Tristis' × *P. balsamifera*	Young cuttings	East Lansing, Michigan, USA	Gh	Sandy-loam soil	0, –0.02, –0.05, –0.1 and –0.5 MPa of soil matric potential	Examination of the interactions of water and nitrogen availability	GE/WUE	The highest water-use efficiency was found on the high-N/severe drought zone for clone 'Eugenei', whereas it was found on the high-N/mild drought zone for 'Tristis'	Liu and Dickmann, 1996
P. fremontii *S. gooddingii*	Young cuttings	Tucson, Arizona, USA	Gh	Washed river and potting medium	Treatment with NaCl solutions	Determination of the interaction of salt and water stress on survival, salt tolerance, WUE and growth	RGR/WUE/π_o	At the control salinity level, the two species were able to extract water from a drying soil equal to that of *Tamarix* species	Vandersande *et al.*, 2001

Continued

Table 7.1h. Continued.

Species	Plant material	Experiment location	Growth conditions	Substrate	Water deficit intensity	Objectives of the study	Traits examined	Main conclusions	References
S. viminalis *S. schwerinii* × *S. viminalis*	Young cuttings	Uppsala, Sweden	OC	Pure sand	Low and high irrigation	Test of the hypothesis that fast-growing breeds of willow are more sensitive to nutrient and water stress and less efficient in nutrient- and water-use than slower-growing clones	BA/N/$\delta^{13}C$/SLA/LA/LAR/RGR	Fast-growing hybrids are more sensitive to nutrient and water stress – natural clones have higher water-use efficiency than fast-growing hybrids	Weih, 2001
P. balsamifera *P. tremuloides*	Young cuttings or seedlings	Edmonton, Alberta, Canada	Gh	Coarse sand/peat	Water table at 15 or 30 cm below the soil surface	Monitoring of the water use under different water table conditions and soil temperatures	ODR/GE/WUE/LW/LA/LMR/BA	*P. balsamifera* is likely to be a good hydrological nurse crop to lower the water table when soils are warm	Landhäusser *et al.*, 2003
P. cathayana *P. simonii*	Young cuttings	Maoxian Field Ecological Station, China	Gh		100, 25% of field capacity	Determination of drought effects on two sympatric species	BA/WUE	Existence of interspecific genetic differences as affected by drought, useful as criteria for selection and tree improvement	Yin *et al.*, 2005b
P. simonii *P. cathayana*	Young cuttings	Maoxian Field Ecological Station, China	Gh		100, 25% of field capacity	Investigation of the physiological basis involved in drought tolerance and the interrelationships between ABA and WUE	BA/GE/[ABA]/WUE	Evidence for adaptive differentiation between the two contrasting species	Yin *et al.*, 2004

P. przewalski	Young cuttings	Maoxian Field Ecological Station, China	Gh		100, 25% of field capacity	Investigation of the photosynthetic responses of *Populus przewalski* subjected to drought stress	g_s/WUE/Y/ q_{np}/T/GE	Drought stress decreased plant assimilation and increased dissimilation through affected gas exchange, the diurnal pattern of gas exchange and the photosynthesis–PAR response curve, thereby reducing plant growth and productivity	Yin *et al.*, 2006
P. trichocarpa *P. deltoides* *P. trichocarpa* × *P. deltoides*	3-week-old cuttings	Pullman, Washington, USA	Gh	Sand	$-2.40 < \psi_{xp} < -0.75$ MPa	Evaluation of some physiological characteristics that may be related to productivity on dry continental sites	GE/LA/WUE	Introduction of clones of eastern cottonwood into breeding programmes is likely to yield lines with favourable growth characteristics combined with enhanced WUE and adaptation to soil water deficits	Bassman and Zwier, 1991

Continued

Table 7.1h. Continued.

Species	Plant material	Experiment location	Growth conditions	Substrate	Water deficit intensity	Objectives of the study	Traits examined	Main conclusions	References
P. tremula	1-month-old seedlings	Helsinki, Finland	Gh	Sand	100, 50% of field capacity	Investigation of the physiological basis of climatic adaptation involved in drought tolerance and the interrelationship between endogenous ABA concentration and water-use efficiency	Ht/LA/BA/GE/[ABA]/WUE	Compared with the wet climate population, the dry climate population was more responsive to exogenous ABA application	Li *et al.*, 2004a
P. davidiana	1-month-old seedlings	Helsinki, Finland	Gh	Peat	100, 80, 60, 40, 20% of field capacity	Underlining of differences between three contrasting ecotypes	BA/GE/[ABA]/WUE	Prodigal WUE and quick growth of the wet climate ecotype – conservative WUE and slow growth of the dry climate ecotype	Zhang *et al.*, 2004 Zhang *et al.*, 2005
S. dasyclados × *S. viminalis*	5- to 9-week-old cuttings	Uppsala, Sweden	Gh	Clay/peat	55% of the water given to the well-watered plants	Analysis of phenotypic correlations and QTL for important growth traits and a surrogate of WUE in a willow pedigree	RGR/LAR/SLA/N/Biomass/$\delta^{13}C$	The results demonstrate a genetic basis for phenotypic correlations among growth traits and provide evidence for the existence of 'master switches' regulating some of the traits	Weih *et al.*, 2006

P. trichocarpa × *P. deltoides* *Salix* sp.	2-month-old cuttings	Gainesville, Florida, USA	Gh	Sand/peat moss	$-1.2 < \psi_{wp} < -0.4$ MPa	Characterization of the responses to elevated CO_2 and water stress and determination of elevated CO_2 mitigated drought stress effects	GE/WUE/Ht/Cc	Elevated CO_2 did mitigate the effects of water stress in willow, but not in poplar	Johnson *et al.*, 2002
P. trichocarpa *P. deltoides* *P. trichocarpa* × *P. deltoides*	3-month-old cuttings	Victoria, British Columbia, Canada	Gh	Sand	$-1.91 < \psi_{xp} < -0.76$ MPa	Examination of N and K nutrition on drought and cavitation resistance	GE/WUE/PLC/LA/SD	Drought-resistant clones exhibited similar WUE to drought-susceptible clones, but had smaller, more numerous stomata and greater leaf retention under drought conditions	Harvey and van den Driessche, 1999
P. ×*canadensis*	3-month-old coppiced plants	Orléans, France	P	Loamy sand	$-0.6 < \psi_{xp} < 0$ MPa	Study of the relationships between productivity, WUE and drought tolerance and test whether some leaf traits could be used as predictors for productivity, WUE and drought tolerance	SLA/N/C/LA/$\delta^{13}C$/SG/Cc/Biomass/GE/g_s/SD	Most of the productive genotypes displayed a low level of drought tolerance, and conversely – the ability to increase WUE in response to water deficit was necessary but not sufficient to explain the genotypic diversity of drought tolerance	Monclus *et al.*, 2006

Continued

Table 7.1h. Continued.

Species	Plant material	Experiment location	Growth conditions	Substrate	Water deficit intensity	Objectives of the study	Traits examined	Main conclusions	References
P. trichocarpa × *P. deltoides* *P. deltoides* × *P. nigra* *P. trichocarpa* × *P. trichocarpa*	1- and 2-year-old trees	Bedfordshire, UK	Gh	Sandy loam/pit mix	Medium drought: ψ_{soil} ~ –0.1 MPa Severe drought: ψ_{soil} ~ –1MPa	Determination of clonal differences in drought response and quantification of relationships between WUE and biomass production	BA/LA/WUE	Highest reduction in biomass accumulation for clone 'Trichobel' – relationships between WUE and biomass production constant over both years	Souch and Stephens, 1998
P. deltoides	3-year-old coppiced plants	Oracle, Arizona, USA	P		25 < volumetric soil water content < 39%	Examination of the effects of atmospheric vapour pressure deficit and soil moisture stress on leaf- and stand-level CO_2 exchange	GE/WUE	The limitation of assimilation because of canopy light environment switched to a predominantly individual leaf limitation in response to water stress	Murthy *et al.*, 2005
P. fremontii	Adult trees	New Mexico to Northern California, USA	F	Sand/gravel	0.2 < precipitation < 299.5 mm	Examination of $\delta^{13}C$ variation among populations from central New Mexico to Northern California	$\delta^{13}C$	Water availability can account for some of the $\delta^{13}C$ variation among populations but, given the large residual variance, other factors are important	Leffler and Evans, 2001

P. fremontii *S. gooddingii*	Adult trees	Bill Williams River, Arizona, USA	AV		$-1.5 < \psi_{wp} < -0.3$ MPa	Investigation of the environmental water sources used in midsummer at sites that differed in water table depth	$\delta^{13}C$	This study highlights the difficulty in sampling all water sources in large-scale studies of riparian ecosystems with complex subsurface hydrogeology	Horton *et al.*, 2003
P. fremontii	15- to 50-year-old trees	San Pedro River, Arizona, USA	AV	Coarse, alluvial soil	40 < August precipitations < 150 mm	Assessment of intra- and inter-annual variation in integrated leaf gas exchange responses to monsoonal moisture inputs	$\delta^{13}C$	Intra- and inter-annual variation in integrated photosynthetic response exists at the population-scale – changes in monsoonal precipitation and stream flow may alter photosynthetic gas exchange differentially	Potts and Williams, 2004
P. deltoides	30- to 40-year-old trees	Rio Grande basin, New Mexico, USA	AV	Sand/clay/silt	0.5 < depth of the groundwater < 2 m	Quantification of variability among four study sites in both soil and groundwater conditions – examination of physiological and morphological characters on trees at these sites	GE/g_s/T/WUE/ψ_{xp}/LA/[Chl]/SLA/SD	Physiological and morphological trait variability among populations is ecologically important and may be of use in present reclamation and conservation efforts in these areas	Rowland *et al.*, 2001

Continued

Table 7.1h. Continued.

Species	Plant material	Experiment location	Growth condi-tions	Substrate	Water deficit intensity	Objectives of the study	Traits examined	Main conclusions	References
Populus spp. *Salix* spp.	Review				$-2.53 < \psi_{wp} < -1.15$ MPa	Review of the water relations of riparian plants from desert regions of North America	GE/N/$\delta^{13}C$/WUE/SF/SG/g_s	Management efforts aimed at maintaining native forests must include at least partial reintroduction of historical flow regimes which favour the recruitment of native riparian species and reverse long-term desiccation of desert flood-plain environments	Smith *et al.*, 1998

Growth conditions: AV, alluvial valley; F, forest; Gh, greenhouse; OC, outdoor containers; P, plantation. Traits examined: ABA, abscisic acid; BA, biomass allocation; C, carbon; Cc, stem circumference/diameter; Chl, chlorophylls; $\delta^{13}C$, carbon isotope composition; GE, gas exchange; g_s, stomatal conductance; Ht, stem height; LA, leaf area; LAR, leaf area ratio; LMR, leaf mass ratio; LW, leaf dry weight; N, nitrogen content; ODR, oxygen diffusion rate; PLC, percentage loss of conductivity; q_{np}, non-photochemical quenching; QTL, quantitative trait loci; RGR, relative growth rate; SD, stomatal density/dimension; SF, sap flux; SG, stem growth; SLA, specific leaf area; T, transpiration; WUE, water-use efficiency; Y, quantum yield; Ψ, water potential; π_o, osmotic potential; [ ], content/concentration.

Table 7.1i. Overview of recent available literature on drought response of *Populus* and *Salix* experiments in ISI (Thompson Institute) scientific journals (period 1998–2006) and of former keystone studies (1987–1997). (i) Population biology and management. Studies are ordered according to the age of the plant material and then chronologically (year of publication). Abbreviations used are listed at the end of the table.

Species	Plant material	Experiment location	Growth conditions	Substrate	Water deficit intensity	Objectives of the study	Traits examined	Main conclusions	References
P. deltoides	2- to 10-year-old trees	Western USA	F			Modelling to aid in planning prescribed floods by projecting how altered flow regimes might affect populations	S/D	Mature cottonwood forest should be most abundant near the observed natural flow regime – high flood frequencies result in stable population sizes, while stable flows result in highly variable population sizes	Lytle and Merritt, 2004
P. deltoides	Adult trees	South Platte River, Colorado, USA	AV		−1.5 < relative water table elevation < 1 m	Quantification of the changes in morphology, growth and mortality in response to measured declines in alluvial water table	S/D/Ht	Gradual water table declines had no measurable effect on mortality, stem growth or live crown volume and produced significant declines only in annual branch growth increment	Scott *et al.*, 1999

Continued

Table 7.1i. Continued.

Species	Plant material	Experiment location	Growth conditions	Substrate	Water deficit intensity	Objectives of the study	Traits examined	Main conclusions	References
Salix spp.	Adult trees	Stanley Creek, Idaho, USA	AV	Silty clay loam/sandy clay loam	Years of moderate drought	Examination of the response of a cold mountain meadow riparian system to three intensities of controlled late June cattle grazing	D/stream characteristics	A decrease in wet-site rhizomatous graminoids during drought suggests a depletion of meadow conditions	Clary, 1999
P. deltoides *S. exigua*	Adult trees	Green River, Utah/ Yampa River, Colorado, USA	AV		50 < instantaneous peak discharge < 1100 $m^3\ s^{-1}$	Evaluation of the effects of river damming on geomorphic processes and riparian vegetation	D	The probable long-term effects of channel and hydrologic changes include the eventual replacement of *Populus*-dominated riparian forest by drought-tolerant desert shrublands	Merritt and Cooper, 2000
P. fremontii	Adult tress	Rio Grande Basin, New Mexico, USA	AV		0 < distance from river < 450 m	Estimation of the variability among five riparian cottonwood populations in terms of size, density and spatial distribution	D/Ht	Tree size of the five populations appeared to be affected by environmental factors	Rowland *et al.*, 2000

P. fremontii *S. gooddingii*	Adult trees	Mojave River, California, USA	AV	Gravel/sand/silt	–2 < relative water table depth < 0 m	Test of a set of predictions based on the inference that cottonwood decline was an indirect result of lowered water table levels following flood-related channel incision	S/SCA	The integrity of riparian forests along arid region rivers with coarse alluvial flood-plain soils can be threatened by physical processes and human activities that result in sustained groundwater declines as small as 1 m	Scott *et al.*, 2000
P. euphratica	Adult trees	Qira Oasis, China	O		0 < soil water content < 100%	To provide an ecological basis for the regeneration and the sustainable use of the vegetation at the transition between oases and sandy desert	SCA/SW	Presentation and objectives of the joint European–Chinese project 'Ecological basis for a sustainable management of the indigenous vegetation in a Central Asian desert'	Thomas *et al.*, 2000
P. tremuloides	Adult trees	Jackson Hole, Wyoming, USA	F		–5 < Palmer drought severity index < 5	Analysis of the relationship between the rates of aspen regeneration, biophysical factors and human land use since 1830	Ht/D	No apparent relationship between periods of aspen regeneration and drought conditions was found	Hessl and Graumlich, 2002

Continued

Table 7.1i. Continued.

Species	Plant material	Experiment location	Growth conditions	Substrate	Water deficit intensity	Objectives of the study	Traits examined	Main conclusions	References
P. tremuloides	Adult trees	North-western Alberta, Canada	F		–20 < climate moisture index < 30 cm	Tree-ring analyses and forest health assessments in healthy and stressed aspen forests as a means of determining the onset, magnitude and causes of reduced stem growth, mortality and crown dieback	SG/PI	Under global change, the severity of the stressors may increase, which would pose a serious concern for the future health, productivity and carbon sequestration of aspen forests in the region	Hogg *et al.*, 2002
P. deltoides	Adult trees	North-western Colorado/ Eastern Utah, USA	AV		River flow variations	Study of the recruitment along broad, alluvial valley segments and canyon segments	Ht	The current regime mimics drought in a canyon setting, accelerating *Tamarix* invasion whereas in valleys the ongoing geomorphic adjustment of the channel, combined with reduced flow variability, has nearly eliminated *Populus* establishment	Cooper *et al.*, 2003b

P. fremontii *S. gooddingii*	Adult trees	San Pedro River, Arizona, USA	AV		Depth to groundwater: 3.8–4.2 m	Quantification of changes in population stand structure traits across gradients of groundwater depth and fluctuation and surface flow permanence	D/Ht/SCA	The altered vegetation structure along dewatered rivers may lower wildlife habitat quality	Lite and Stromberg, 2005
P. tremula	Adult trees	Lithuania	F			Investigation of growth-independent tree mortality in central Europe	S	More than 60% of dead trees had symptoms of wind damage – significant fluctuations in tree mortality rate over time were caused by extreme meteorological conditions (wind, drought)	Ozolincius *et al.*, 2005a
P. tremula	Adult trees	Lithuania	F		142 < annual precipitations < 393 mm	Presentation of the changes in the health of the dominant tree species in Lithuania between 1991 and 2001	LL	Average temperature, amount of precipitation and background air pollution during the active growing season can be regarded as a complex of factors influencing crown condition	Ozolincius *et al.*, 2005b

Continued

Table 7.1i. Continued.

Species	Plant material	Experiment location	Growth conditions	Substrate	Water deficit intensity	Objectives of the study	Traits examined	Main conclusions	References
P. tremuloides	Adult trees	Rocky Mountains, Colorado, USA	F		2002 drought	Investigation of the combined effects of past disturbances, current vegetation and topography on spatial variability of the severity of a fire during the extreme drought of 2002	Fire severity	Pre-burn stand conditions are important influences on burn severity, even for fires burning during extreme drought	Bigler *et al.*, 2005
P. tremuloides	Adult trees	South-western Yukon, Canada	F	Sand to clay loam and silty clay	Annual precipitations < 300 mm	Determination of the most important climatic factors governing growth and regeneration of trembling aspen and white spruce in Yukon	Ht/SG	Growth of both species was related most strongly to variation in precipitation – the regenerating aspen had a wide age-class distribution (1959–2000) and their growth was also related positively to precipitation	Hogg and Wein, 2005
S. caroliniana	Adult trees	St John River, Florida, USA	AV	Peat soils	–0.5 < water depth < 0 m	Evaluation of the use of roller chopping as a management technique to reduce willow cover and restore herbaceous marsh in an area where fire was no longer a viable option	D	Under dry conditions followed by flooding, roller chopping can be an effective method of willow control	Ponzio *et al.*, 2006

Populus spp. *Salix* spp.	Review	Examination of the potential for producing biomass on restored landfills using willow and poplar species in short-rotation energy forestry	The responses of poplars and willows to drought may be the key constraint to productivity since their natural distribution and productivity are closely related to the seasonal availability of soil water	Nixon *et al.*, 2001
Populus spp.	Review	Review of studies conducted after 1990 on cottonwood water relations, impacts of damming and life history and conservation	The conservation and restoration of cottonwoods will rely on the provision of river flow regimes that satisfy the ecophysiological requirements for survival, growth and reproduction	Rood *et al.*, 2003
P. tremuloides	Review	Identification of potential processes that contribute reduced vigour and dieback of aspen stands	Among inciting factors, severe episodes of drought have been implicated in the decline and death of aspen trees in western Canada	Frey *et al.*, 2004

Continued

Table 7.1i. Continued.

Species	Plant material	Experiment location	Growth conditions	Substrate	Water deficit intensity	Objectives of the study	Traits examined	Main conclusions	References
Populus spp. *Salix* spp.	Review					Comparison of the ecophysiology of *Tamarix ramosissima* and native trees in western US riparian zones		An effective management strategy for salt cedar must include the return of a more dynamic hydrological regime to regulated rivers, allowing salt cedar and native trees to coexist	Glenn and Nagler, 2005
P. deltoides	Review					Development of two conceptual models synthesizing processes affecting vegetation under fluctuating groundwater and identification of information gaps	Biomass	Such models provide a valuable tool for managing vegetation and groundwater use in areas where groundwater is important to both plants and humans, particularly in the context of climate change	Naumburg *et al.*, 2005

Growth conditions: AV, alluvial valley; F, forest; O, oasis. Traits examined: D, population density; Ht, stem height; LL, leaf loss; PI, pathogen incidence; S, survival; SCA, stem cross-sectional area; SG, stem growth; SW, stem dry weight.

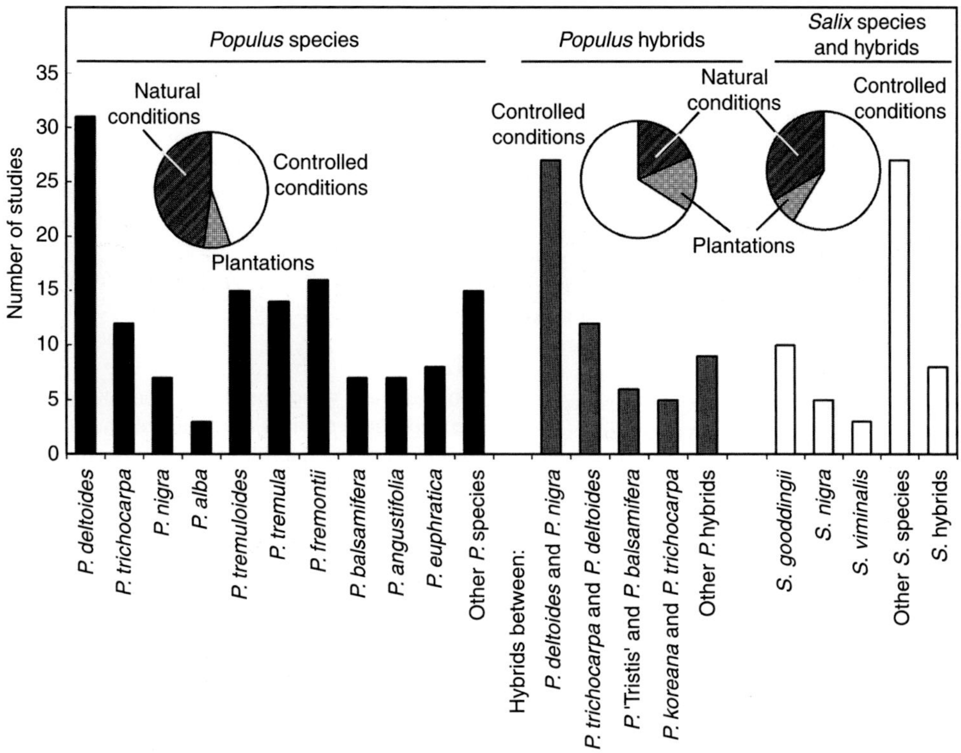

Fig. 7.3. Absolute number of studies performed on different *Populus* and *Salix* species and interspecific hybrids (histograms) and under different conditions (controlled conditions, plantations, natural conditions; sectors) during the period 1998–2006 to investigate responses to drought.

of the drought studies concern cottonwoods (*P. deltoides*), because of the many studies carried out on riverside forests in the USA. This phenomenon also explains the equal proportions of experiments with pure *Populus* species in controlled versus natural conditions, with plantation studies being minor in all cases. With regard to *Populus* hybrids, almost 43% of all studies concern *P.* ×*canadensis* hybrids, primarily under controlled conditions, versus the predominance of pure species in natural stands. Drought studies conducted with *P. nigra* and *P. alba* remain anecdotal in comparison with those of more exotic species, such as *P. euphratica*. For a more comprehensive insight into the response of poplar and willow to drought, efforts have to be made to relate controlled environment studies to the field, and to encompass a larger range of species within the *Populus* and *Salix* genera.

Poplar cultivation may lead to high biomass production under adequate conditions. The *Populus* genus remains an interesting model for the study of the mechanisms of response to drought and an invaluable pool of diversity in drought resistance levels and WUE. However, drought resistance appears to be a very complex and multigenic property, resulting from the combination of various mechanisms. From a practical viewpoint, the overall sensitivity of poplar to drought represents a severe limit to the future development of its cultivation. In the light of the present climatic and economic contexts, selection criteria for commercial poplar genotypes need to be reoriented in order to take into account resistance to abiotic constraints in general, and to drought in particular. In the near future, selection will have to focus on genotypes combining an optimal water use with a maximum biomass production.

7.2.2 Desertification
W.R. Schroeder

Desertification is a global environmental problem. It is defined in the United Nations Convention to Combat Desertification (UNCCD) as land degradation in dry lands resulting from various factors, including climate variations and human activities (UNCCD, 2004). According to the UNCCD, over 250 million people are directly affected by desertification. In addition, some 1 billion people in over 100 countries are at risk. There are a number of examples of how desertification is being addressed in various world situations. China has been recognized for implementation of tree planting programmes, based to a significant degree on the genus *Populus*, as a means of addressing desertification; therefore, a discussion of the role of poplar in China's desertification control programme is worthwhile.

Deserts cover an area of over 1.5 million km^2 in China. It is estimated that the mean annual rate of desertification in China is approximately 2100 km^2 (Zhu and Chen, 1994). Desertification has led to a significant decline in agricultural productivity in northern China, and as a result the national and local governments have implemented a number of practices to restore vegetation on affected lands and reduce the rapid expansion of deserts (Xang *et al.*, 2005). To address the issue of desertification, as well as other environmental issues, in 1978 the Chinese government implemented the Three North Shelterbelt Development Programme (FAO, 2002). The overall objective of the programme is to improve environmental conditions, control dust storms as well as soil and water erosion and fundamentally improve the local livelihood and economy through the formation of what is popularly referred to as the 'Great Green Wall' (Lu *et al.*, 2000).

The Three North Shelterbelt Programme in China is the largest multi-purpose protection forestry project in the world, with plans to plant more than 35 million ha of trees from 1978 to 2050 (FAO, 2002). The programme covers a total area of 4 million km^2, accounting for over 40% of the total land surface of China, including 551 counties of 13 provinces, autonomous regions and municipalities in the Three North Region (Anon., 2000). By the end of 2003, a cumulative area of 22 million ha had been added to the forested land in north China since 1978, leading to an increase in forest cover in the area from 5% to 10% (Yin and Lu, 2005). The *Populus* genus has a significant role in the Three North Shelterbelt Development Programme. From 1990 to 2000, the area covered by poplar (natural and planted) in the Three North Region increased from 4.8 million ha to 6.0 million ha (Lu *et al.*, 2000).

Planting poplars has been a common practice used in China to reclaim sandy lands and prevent desertification (Zhu and Chen, 1994). The rapid growth and protective role of *Populus* in reducing wind speed and controlling wind erosion have made it the genus of choice in the majority of tree-planting programmes to control desertification in northern China (Lu and Zhang, 2000). The poplars used in the Three North Region are mainly clones and hybrids of *P. simonii*, *P. szechuanica*, *P.* ×*tomentosa* and the exotic *P. alba* var. *pyramidalis* 'Bolleana' (Lu and Zhang, 2000). There are two common approaches in which poplars are used to combat desertification (Anon., 1977):

1. Arable land protection involves the protection of farmland from desert encroachment through the establishment of poplar forest belts on the periphery, together with the establishment of shelterbelt networks within the farming area. Poplar shelterbelts are first established on the fringe of the farmland and then on low-lying land between the major dunes encroaching the arable land. The goal is to surround the dunes with vegetation and reduce their encroachment on arable land by reducing wind velocity. Within the arable farmland, shelterbelt systems are established. The system is made up of main tree belts 10–12 m wide, consisting of five or six rows of poplar and other species, and secondary tree belts 6–8 m wide with three or four rows of poplar. The main belts follow main irrigation canals and roads, whereas the secondary belts are located on the periphery of individual fields.
2. To stabilize moving sands, poplars are planted on low-lying land among sand dunes. Once dunes have been fixed using stabilizing plant species (grasses and shrubs), poplars are planted in these areas. In some cases, irrigation may be used to assist with the establishment of the poplars.

The project 'Afforestation, Research, Planning and Development in the Three North Region of

China', jointly financed by the Belgian and Chinese governments and the FAO, is an example of a successful poplar-based project to address the problems of ever-increasing soil erosion, decreasing soil fertility and lack of wood products needed by local inhabitants, with emphasis on the genetic development and utilization of poplar to accomplish afforestation objectives (Anon., 2000). In this project, a major goal was to expand the genetic base of poplar clones used in desertification planting projects in order to increase productivity, resistance to frost and drought and overall resistance to insects and disease. The project followed three strategies to accomplish these goals: (i) the development of a long-term poplar breeding and selection programme; (ii) *ex situ* conservation of remaining natural populations (Wang *et al.*, 2000); and (iii) a short-term poplar evaluation and selection programme to identify adapted poplar clones with superior vigour, form and resistance to frost and pests (Lu *et al.*, 2001). The project also developed mechanized afforestation techniques suitable for large-scale poplar planting in desertification projects. The technique developed involves mechanical planting of 80-cm-long cuttings 70 cm into the sand. This technique proved to be more efficient and cost-effective than traditional methods of hand-planting rooted poplars.

Northern China has a number of native poplar species that are rarely found elsewhere in the world. For many years, desertification, drought, excessive firewood collection and water diversion projects have caused serious decline of the natural poplar resource in China. For example, in the Tarim Basin of Xinjiang Autonomous Region, *P. euphratica* forests that once covered 5200 km^2 are now reduced to an area of only 3500 km^2 (Lu *et al.*, 2000). The disappearance of these natural poplar resources is serious and proper measures are required to protect them. It is important that concerted efforts are made to establish poplar gene banks and poplar clone arboreta in order to preserve these valuable poplar germplasm resources.

Given the massive planting efforts of poplar in northern China since 1990 in what previously was a treeless landscape, it is not surprising that significant pest problems have developed. Large monocultures of any plant are inevitably linked to pest outbreaks, especially in environments less than optimal for tree growth. The most serious pest that has been encountered to date is the Asian longhorn beetle (*Anoplophora glabripennis*), which is seriously hindering the development of poplar resources in north China (Lu *et al.*, 2004). Since 1998, the damage to poplar tree plantations and shelterbelts caused by the Asian longhorn beetle has become critical. This insect pest is now distributed in 240 counties of 13 provinces in northern China (Pan, 2005). According to Yin and Lu (2005), Asian longhorn beetle infestation has reduced the lifespan of poplar shelterbelts significantly, from 20 years to 10 years or less. The overall impact of the infestation has been devastating, resulting in the demise of thousands of hectares of poplars. The reasons for the epidemic have been attributed to a number of factors: first, the limited number of poplar varieties used in China and the clones used are highly susceptible or have limited resistance to Asian longhorn beetle; second, most affected areas are not source areas of longhorned beetles, so in these areas the beetles have few natural enemies; and finally, poor tree growth due to environmental stresses, i.e. drought and extreme temperatures, predisposes poplars to pest infestation (Lu *et al.*, 2004).

The choice of poplar clones used in planting programmes and their spatial deployment are crucial considerations when combating pest problems such as Asian longhorn beetles. It makes sense to select the most resistant lines for planting; however, these resistant lines should not be planted in monocultures. The use of clonal mixes, where several clones varying in their resistance to Asian longhorn beetles are used in the construction of shelterbelts, is recommended. The use of three to four clones should be considered the minimum. Using clonal mixtures is desirable because of the increased heterogeneity within the shelterbelt system. The goal of clonal mixes is to prevent pests from responding uniformly to the resistant traits of the most resistant clones and in so doing developing countermeasures through mutation or other adaptation responses. In addition, by retaining some clones with limited or little resistance, there will likely be enough beetles around to sustain any populations of natural enemies. These more susceptible trees also provide a focal point for control measures. For example, the *P. szechuanica* clones 'Opera' and 'Popularis' have been widely used for tree planting in northern China (Lu *et al.*, 2004). These varieties were highly susceptible to the Asian longhorn beetle and were replaced by varieties of the *Leuce* section that had always proved resistant to the longhorn beetle. However, with the reduction in tree species diversity due to high mortality resulting from longhorn beetle infestations, clones such as 'Bolleana' poplar (*P. alba* var. *pyramidalis*) and 'Hebei' poplar (*P.* ×*hopeiensis*), which had originally been considered resistant to the longhorn beetle, are now being infested (Yin and Lu, 2005). Yin and Lu (2005) reported that in farmland shelterbelts in the Ningxia irrigation area, when *P. szechuanica* 'Opera' was planted as bait species along with 'Bolleana' poplar, which was the primary shelterbelt species in this region, longhorned beetle infestation of 'Bolleana' poplar was reduced from 60–70% to less than 15%.

Poplars have an important role in China's efforts to combat desertification. Massive tree plantings have been completed since 1950, but overall success has been impacted by pest infestations and poor growing conditions, characteristics of landscapes prone to desertification. Recently, poplar breeding strategies for shelterbelts have been implemented that focus on improving adaptation to regional environmental conditions with greater emphasis on native Chinese species. The main criteria for consideration in these programmes are cold and drought hardiness, pest resistance and adaptation to poor growing conditions. Genetic gains in these traits will increase the survival of trees and reduce the risk of future catastrophic pest infestation. In addition, research and development of new transgenic poplar varieties is being actively carried out in China.

7.2.3 Salinity
J. Gao and A. Polle

Background: changing environmental conditions, risks and countermeasures

Worldwide, almost 1 billion ha of land are affected by soil salinity (Szabolcs, 1994). Today, soil salinization is still increasing, mainly because of unsuitable irrigation practices causing secondary salinization (Williams, 1999; Munns, 2005). Salinity has a major impact on plant growth and productivity. It has been estimated that the world is losing at least 3 ha of arable land every minute because of soil salinity (Kundzewicz *et al.*, 2007). If recultivation measures are lacking, the land is prone to soil erosion and desertification. To cope with this enormous problem, efforts are undertaken to increase the salt tolerance of economically important plants, including tree species, with the goal of obtaining genotypes able to cope with excess salinity. Currently, different strategies are employed, e.g. exploring the natural diversity, traditional breeding and genetic engineering for higher salt tolerance.

The genus *Populus* is of interest in this respect because it contains a wide variety of species (~30) from different areas around the world, displaying a range of different growth characteristics and tolerance towards various stress conditions (Gielen and Ceulemans, 2001; Taylor, 2002). This includes significant variability in salt tolerance between poplar species and within species (Fung *et al.*, 1998; Singh *et al.*, 1999; Chen *et al.*, 2002a; Sixto *et al.*, 2005). Apparently, this genus is an interesting resource of traits relevant for salinity tolerance. Furthermore, *Populus* is amenable to transgenesis, and thus improving salt tolerance by biotechnological approaches is a feasible option.

Identification and occurrence of salt-tolerant poplar species

Sixto *et al.* (2005) tested the physiological responses and survival of 13 poplar genotypes under salinity stress. *P.* ×*canadensis* (= *Populus* × *euramericana*) was the most sensitive species; intermediate sensitivity was found in *P. alba* and in hybrids of *P. deltoides* and *P. alba*; highest tolerance was present in *P. euphratica* (Table 7.2). Among these species, *P.* ×*canadensis* is currently of the highest and *P. euphratica* of the lowest commercial interest. *P. alba*, which is widespread in the Mediterranean Basin, Central Europe and the Middle East, is gaining commercial importance. In optimal conditions, it can grow in dense plantations that give high biomass volumes in short rotations. *P. alba* is a typical riparian tree species; still, some provenances are adapted to drought, saline soil conditions and high temperatures, such as those occurring in south Mediterranean areas close to the sea

Table 7.2. Mean survival rate and confidence interval ($P = 0.95$) of poplar clones exposed for 2 months to 138 mM NaCl. Table adapted from Sixto *et al.* (2005).

Species	Survival (% of control)	Confidence interval	No. of genotypes tested
P. ×canadensis	2.5	0–13	4
P. deltoides × P. alba	45	22–67	2
P. alba	50	37–63	6
P. euphratica	100	70–100	1

(Sekawin, 1975; Gellini, 1980; Beritognolo *et al.*, 2003; Paolucci *et al.*, 2003; Piazzai *et al.*, 2003). In Italy, the native populations of *P. alba* are shrinking. Nevertheless, the contrasting ecological conditions to which this species is adapted have favoured the maintenance of intraspecific genetic variability, as assessed by phenological traits and DNA polymorphism (Paolucci *et al.*, 2003). Neutral molecular markers able to distinguish between a salt-tolerant ('14P11') and a salt-susceptible ('6K3') *P. alba* ecotype have been identified (Beritognolo *et al.*, 2003). The tolerant ecotype was from the south of Italy close to the sea ('14P11') and the sensitive one from the inland north of Italy ('6K3'). It is possible that exposure to sea spray resulted in adaption of *P. alba* '14P11' to elevated salinity. This shows that even within one poplar species, ecotypes for salt tolerance can be identified.

P. euphratica is another example of a highly stress-resistant species. Its native distribution ranges from the semi-arid areas of north-west China to Western Morocco and from north to south from Kazakhstan to Kenya (Browicz, 1977; Xu, 1988). It grows under unfavourable conditions such as in saline and alkaline soils (Kang *et al.*, 1996; Watanabe *et al.*, 2000; Chen *et al.*, 2001). The total salt content in the soil, where the forests of *P. euphratica* occur, is about 1%, but can reach 2–3% or more (Ma *et al.*, 1997). *P. euphratica* can also be found growing around the Mediterranean Sea on sites with salty water. In *in vitro* tests, *P. euphratica* can tolerate up to 450 mM NaCl (Gu *et al.*, 2004). Existing *P. euphratica* forests survive very cold winters (down to –40°C) and hot summers (up to 43°C) and exist in areas with low rainfall and high evaporation rates. At first glance, it appears paradoxical that *P. euphratica* is very drought sensitive, since its water transport system is damaged by cavitation even at mild water deficits (Hukin *et al.*, 2005). However, there is evidence that natural *P. euphratica* stands in desert sites, which can occur at great distances from the nearest river, still have access to the groundwater (Gries *et al.*, 2003). *P. euphratica* develops an extremely deep root system connecting the tree to the water table, thereby avoiding drought stress. In the Taklamakan desert (Xinjiang Uygur Autonomous Region, north-west China), natural populations of *P. euphratica* are endangered (Wang *et al.*, 1996) because a growing population with increased water consumption leads to a decreasing water table, and in consequence causes a decline in *P. euphratica* stands whose survival depends on access to groundwater (Fig. 7.4).

P. euphratica is not abundant in Europe, though some populations are found, for example in Bulgaria and Spain, and also in Morocco. However, the European populations are probably not native but have been introduced into Spain and multiplied by clonal propagation (Fay *et al.*, 1999). Due to early bud flush, *P. euphratica* is sensitive to late frost events in spring, which have prevented its cultivation in middle and northern Europe.

P. euphratica has been used successfully for large-scale afforestation projects on saline desert sites in China (Wei, 1993), where these trees play very important roles in stabilizing fragile environments such as the riparian ecosystems and protecting oases from invading wandering dunes. Maintenance and restoration of the ecological balance is, therefore, an important environmental issue. The capacity of *P. euphratica* to regenerate through root suckers makes it an excellent species for shelterbelts and for sand dune stabilization, particularly since *P. euphratica* also tolerates strong winds (Ball *et al.*, 1996). However, for large-scale multiplication, it is a disadvantage that cuttings are difficult to root and that seeds lose viability extremely rapidly.

Fig. 7.4. A declining stand of *Populus euphratica* at a distance of about 20 km from the Tarim River bed in Xinjiang Uygur Autonomous Region (north-west China). Photo courtesy of X. Jiang.

These problems can be circumvented by the establishment of *P. euphratica* tissue cultures and micropropagation systems, for which reliable protocols have been established (Huang *et al.*, 1992; Phan *et al.*, 2004). A drawback of these methods is that they require expert know-how and advanced laboratory facilities, which makes them expensive on the one hand and more difficult to implement in underdeveloped areas on the other hand.

In conclusion, *P. alba* and *P. euphratica* are promising species for breeding programmes aimed at increasing salt tolerance and as pioneering species for tree plantations on salty soils. An acute concern is that native populations of these two species are shrinking quickly because of strong competition from agriculture and other land uses (Ball *et al.*, 1996; Sabatti *et al.*, 1998). The conservation of these genetic resources is an important priority that requires fast and reproducible experimental protocols for vegetative propagation. Since *P. euphratica* grows relatively slowly, a further goal is to combine traits of fast growth and salt tolerance in hybrids for stabilization and reclamation of saline areas.

Mechanisms of salt tolerance in Populus

Salt imposes several kinds of stresses on plants. Increases in sodium chloride in the soil solution cause hypo-osmotic stress by decreasing the water potential. Colloquially speaking, this makes water uptake more difficult, causing symptoms similar to drought stress in salt-exposed plants (Polle *et al.*, 2006). At the whole-plant level, non-halophytes, to which the *Salicaceae* belong, try to exclude salt to avoid overaccumulation of sodium and disturbances in the cellular ion balance (Munns, 2005). If salt uptake cannot be avoided, excess salt leads to membrane disorganization and ion toxicity at the cellular level (Hasegawa *et al.*, 2000; Zhu, 2001). Both mechanisms, drought and ion imbalances, may cause the production of ROS as a secondary stress (Gueta-Dahan *et al.*, 1997; Savouré *et al.*, 1999; Borsani *et al.*, 2001). The consequences of oxidative stress are: damage to important macromolecules such as proteins, lipids and DNA, subsequently reduced or complete loss of growth and eventually cell death. This typical response pattern to excess salinity occurred, for example, in the salt-sensitive poplar hybrid *P.* ×*canescens* (Bolu and Polle, 2004).

To cope with salinity, defence responses are activated at different levels. Common metabolic answers to salt stress are the synthesis of stress-related enzymes like antioxidant systems, chaperones, for example salt-shock proteins, and compatible solutes (Hasegawa *et al.*, 2000; Wang *et al.*, 2003). While these reactions are important to protect metabolic functions under increasing salt concentrations, a primary defence is the avoidance of excessive Na^+ accumulation in the cytosol (Blumwald *et al.*, 2000). In this context, so-called sodium/proton antiporters play crucial roles. Once sodium reaches the plasma membrane, it will be taken up readily by unspecific ion transporters, leading to increases in cytosolic Na^+ (Fig. 7.5). This activates Na^+/H^+ antiporters, which are localized in the plasma membrane and in the tonoplast. These transporters remove Na^+ from the cytosol, either exporting it into the vacuole or into the apoplastic space of the cell wall (Fig. 7.5). The counter-ion for Na^+ export is a proton (H^+). Therefore, the function of Na^+/H^+ antiporters requires the maintenance of a pH gradient across the membrane (Fig. 7.5). The pH gradient is generated by adenosine triphosphate (ATP)ases under the consumption of energy provided by the hydrolysis of ATP (Fig. 7.5). The key role of Na^+/H^+ antiporters in salt protection was shown by constructing transgenic plants. Overexpression of *SOS1* (a so-called NHX-type Na^+/H^+ antiporter) resulted in increased salt tolerance in various species, underlining the importance of sodium export for cytosolic ion homeostasis (Apse *et al.*, 1999; Shi *et al.*, 2000, 2003; Hamada *et al.*, 2001). Such functional approaches are still lacking in trees, but in *P. euphratica* a new NhaD-type Na^+/H^+ antiporter (Pe*NhaD1*) has been detected (Ottow *et al.*, 2005a). The transcription of this antiporter was, in contrast to a salt-sensitive poplar, maintained during salt stress (Ottow *et al.*, 2005a) and it was possible to complement a salt-sensitive bacterial mutant (*Escherichia coli* mutants) with Pe*NhaD1* and to restore the salt tolerance of these bacteria (Ottow *et al.*, 2005a). The function of these Na^+/H^+ antiporters is keeping salt out of the cell. This example shows that genes regulated in *P. euphratica* in response to excess salinity can be used to find traits that are important for salt tolerance.

It is remarkable that at the whole-plant level, Na^+-uptake into roots of *P. euphratica* is similar to that observed in salt-sensitive poplar species (Chen *et al.*, 2001). However, xylem loading and translocation to leaves was more restricted than in salt-sensitive poplar species (Chen *et al.*, 2003a). Under field conditions, salt exudation was also found. These exudates form a crust of salt on bark cracks of *P. euphratica*. Salt exudation may also be an important factor contributing to the salt tolerance of this species.

A key issue in salt adaptation is osmotic adjustment. Salinity decreases soil water potential and thus acts like drought stress, leading to turgor loss in non-acclimated plants. To maintain water uptake, adjustment of the osmotic potential of the cells is required by compensatory changes in bulk solutes. Typically involved in these responses are sugars, sugar alcohols,

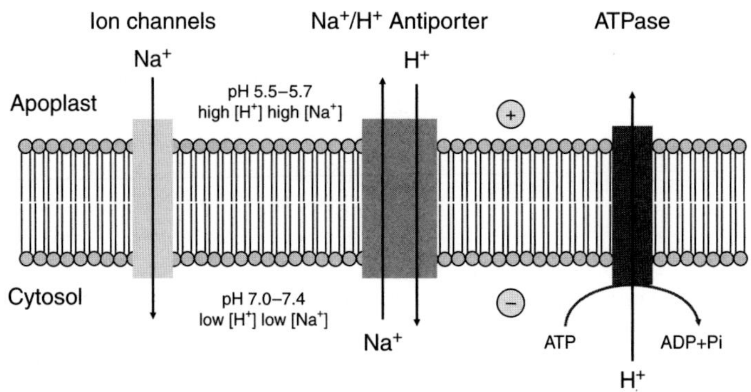

Fig. 7.5. Scheme of Na^+ transport across the plasma membrane (after Ottow, 2005). See text for details.

amino acids, organic acids or inorganic ions (Munns, 2005). Plants with increased concentrations of proline, mannitol or other products of sugar metabolism displayed increased salt tolerance (Kishor *et al.*, 1995; Karakas *et al.*, 1997; Garg *et al.*, 2002; Taji *et al.*, 2002). However, the significance of these osmolytes in conferring increased salt tolerance by osmotic adjustment has been questioned (Blum *et al.*, 1996). Sodium itself may act as an osmoticum (Munns, 2005) but it displaces other cations such as Ca^{2+} and K^+, which are important for membrane integrity and ion selectivity, and which also function as plant osmolytes (Epstein, 1998; Tester and Davenport, 2003).

In leaves of *P. euphratica*, high sodium concentrations are tolerated because the salt accumulates in the apoplastic space of the cell wall and not in the cytosol (Ottow *et al.*, 2005b). Osmotic adjustment was attained by the uptake of sodium, moderate increases in amino compounds and decreases in calcium, glucose and fructose (Ottow *et al.*, 2005b). Apparently, *P. euphratica* employs the strategy of using mainly sodium ions for osmotic adjustment. This is energetically favourable since the synthesis of compatible solutes is metabolically expensive, whereas Na is a cheap and abundant osmoticum. To unravel the regulatory steps which enable *P. euphratica* to cope with excess salt, molecular techniques such as differential display, EST (expressed sequence tag) sequencing and microarray analysis have been employed (Gu *et al.*, 2004; Brosché *et al.*, 2005). The identification of genes involved in acclimatory responses and their functional analysis is a promising way to increase the understanding of the complex molecular mechanisms involved in mediating salt tolerance. Furthermore, these strategies are likely to unravel candidate genes for improving salt tolerance by transgenesis.

Biotic interactions with mycorrhizal fungi affect salt resistance of trees

Mycorrhizae are mutualistic fungal symbionts. Associations of fungi with plant roots are found in up to 80% of plant species worldwide. Seven types of mycorrhizae exist, with differences in structure and function depending on the plant and fungus involved (Smith and Read, 1997). In woody species, only two groups are commonly found, the ectomycorrhizae and the arbuscular mycorrhizae. Ectomycorrhizal fungi form extramatrical hyphae and rhizomorphs, up to 200 μm wide, that extend well into the surrounding soil (Marschner, 1995). The main function of these symbiotic interactions is improved nutrient supply to the host, especially that of phosphorus and nitrogen (Smith and Read, 1997). But mycorrhizae can also protect plants from soilborne stresses, e.g. heavy metals, drought, etc. The protective effect of mycorrhizae against environmental constraints has been reviewed (Hall, 2002; Schützendübel and Polle, 2002; Polle and Schützendübel, 2003).

Populus is one of the genera known to form both arbuscular and ectomycorrhizal associations (Vozzo and Hackaylo, 1974; Lodge, 1989). In natural environments, arbuscular colonization is often variable and depends on soil moisture (Lodge, 1989; Al-Agely and Reeves, 1995), soil nutrient status (Silva *et al.*, 1993), season (Khan, 1974) and fungal species (Abbot and Robson, 1991). Khasa *et al.* (2002) reported arbuscular mycorrhizal colonization in different poplar clones to be between 20 and 50%.

Only a few studies on trees have addressed the role of mycorrhizae during salt stress. In crops such as maize, arbuscular mycorrhizae improved resistance against salt, probably through increased concentrations of osmotically active sugars in the roots (Feng *et al.*, 2002). Similar results have been reported for lettuce and tomato inoculated with arbuscular mycorrhizae (Al-Karki, 2000; Ruiz-Lozano and Azcon, 2000). Musin and Zwiazek (2002) showed that ectomycorrhizae of *Hebeloma crustuliniforme–Picea glauca* inhibited sodium uptake significantly and improved nutrition and transpiration compared with non-inoculated, salt-exposed seedlings. For poplar, very little information is available. In a survey of the mycorrhization of *P. euphratica* under field conditions (Ein Avdat Canyon, Israel), only vesicular arbuscular mycorrhizae were found (D.L. Godbold, 2008, unpublished results). The degree of colonization varied between 14% and 28% of the roots and was generally lower in roots exposed to higher salinity than in those exposed to lower salinity (D.L. Godbold, 2008, unpublished results). However, controlled experiments addressing the role of these mycorrhizae in salt tolerance of *P. euphratica* have not been conducted. Ectomycorrhizae of *Paxillus involutus* stimulated growth of the poplar hybrid *P.* ×*canescens* (Fig. 7.6) and partially protected its biomass formation against the negative influence of excess salinity and improved the nutrition of poplar (Langenfeld-Heyser *et al.*, 2007). Although the current data are very fragmentary, they suggest that inclusion of mycorrhizal management in reclamation strategies of salinity-affected land may increase the success of such measures. It is obvious that more information is needed on the interaction and possible ameliorative influence of mycorrhizae for poplar under salt stress.

7.2.4 Soil nutrition

J.D. Johnson

Poplars and willows evolved in soils affected by water, known as riparian areas. Fertility of riparian soils ranges from highest in alluvial flood plains to lowest in newly scoured gravel bars in stream channels. Alluvial flood plains derive their fertility from periodic flooding that deposits nutrient-laden sediments on these sites. Some of the most productive agriculture as well as poplar and willow plantations occur in these soils. At the other extreme in fertility are the gravel bars that consist of large soil particles with little or no organic material present (Fig. 7.7). Nearly all of the fertility in such situations is supplied by dissolved nutrients in the stream waters, which are usually quite low. As a result of evolving in the variable soil fertility found in riparian areas, poplars and willows are able to grow in a wide range of soil fertility. In high-fertility soils, both poplars and willows show some of the fastest growth of all tree species, whereas in the nearly nutrient-devoid gravel bars, the trees survive, but grow very slowly. Their survival in these gravel bars, usually at very high stem densities, has the effect over many years of slowing stream flow, which increases the deposition of nutrient-containing sediments. This is the first step in the process of developing rich alluvial soils (Plate 22A). As the site fertility slowly increases, poplars and willows possess the ability to respond to this by increasing their growth.

Within riparian zones, poplars and willows occupy distinct, but overlapping, niches. For example, willows can withstand growing season floods for up to 110 days compared to less than 50 days for poplars (FAO, 1980). Hence, willows are usually found in wetter areas closer to streams and lakes than poplars. Conversely, poplars are able to grow in drier, more upland soils

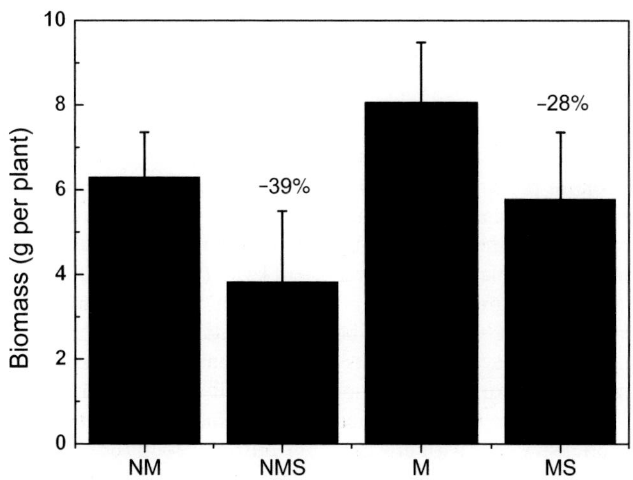

Fig. 7.6. Biomass of young poplar plantlets in the presence (M) and absence (NM) of mycorrhizae exposed to salt (MS and NMS). Poplar plantlets were multiplied by micropropagation and grown either with or without *Paxillus involutus*, strain 'MAJ' (as described by Gafur *et al.*, 2004). After establishment of mycorrhizae, the plants were exposed to 100 mM NaCl for 6 weeks. Subsequently, fresh biomass was recorded ($n = 5 \pm$ SD).

Fig. 7.7. Succession of black cottonwood (*Populus trichocarpa*) on a gravel bar in the Carbon River, Pierce County, Washington, USA. Photo courtesy of Jon D. Johnson.

than willows. Dickmann and Stuart (1983) reported that poplars – and the statement is undoubtedly true also for willows – can be planted on almost any site and, with proper care, will survive and can actually grow better than other tree species.

Characteristics of good poplar and willow sites

In the 1980 FAO book on poplars and willows, the discussion on site selection concentrated primarily on root access to groundwater, stating 'euramerican poplars can be planted wherever the roots have access to underground water, no matter the soil texture ...' and 'willows can be planted on bottom lands, with accessible underground water, unsuitable for euramerican poplars ...' (FAO, 1980). Soil texture was mentioned as reducing growth on heavy soils with groundwater deeper than 2 m and on light soils (sands and gravels) that were saline. Schreiner (1959) and Heilman *et al.* (1995) identified five important characteristics when selecting a site to support good tree growth: soil depth, soil fertility, soil pH and soil moisture and aeration. Along with deep soils (>1 m) and a water table 1–2 m deep (0.5–1.5 m for willow), soil fertility traits should include an undisturbed site or one with <5 years of cultivation, organic matter >3% in the A-horizon, i.e. >15 cm, a source of calcareous (basic) parent material in the rooting zone and pH ranging between 5.5 and 7.5. In contrast, any soil trait not meeting these criteria can cause low site fertility, leading to reduced tree growth.

Poor fertility and nutrient deficiencies

In most soils, it is not uncommon to have nutrient levels that limit tree growth potential, and in many of these situations, classic deficiency symptoms such as leaf chlorosis may not be evident. It has been shown that trees cope with limiting soil nutrients by reducing overall tree growth rate and thereby reducing the quantity of nutrients required by the trees (Fig. 7.8); the lower the nutrient availability, the slower the trees grow. As they grow more slowly, it has been found that trees also reallocate carbon from aboveground organs such as leaves, branches and stem to the root system, to enable greater soil exploration for the limiting nutrient.

Severe nutrient deficiencies can be encountered in highly altered or disturbed soils such as mine spoils, or overirrigated soils leached of soluble nutrients like Mg and K, but this topic is outside of the scope of this section. Erickson *et al.* (1992) provided a comprehensive review of the nutritional requirements of short-rotation forests. The nutrients that appear to be most limiting in natural soils are nitrogen and phosphorus, though sulfur has received some attention in the literature recently (van den Driessche, 2000; Brown and van den Driessche, 2002; Liang and Chang, 2004).

Fig. 7.8. Growth response to different levels of nitrogen nutrition. The hybrid poplar cutting on the right was grown at a nitrogen addition rate (NAR) of 8% day^{-1}, whereas the cutting on the left was grown at an NAR of 1.5%. The photo was taken after 42 days of growth. Note the differences in overall leaf size, leaf number, the lack of branches in the low N treatment and lack of any nitrogen deficiency symptoms. Photo courtesy of Jon D. Johnson.

Tree response to poor fertility

When growing in poor soils, poplars and willows reduce their growth, especially diameter growth, by decreasing their leaf area. This is accomplished in several ways. First, there is a general reduction in individual leaf area resulting from changes in cell wall properties during leaf development caused by the nutrient limitation (Taylor *et al.*, 1993). Two hybrid poplar clones grown for 42 days at two nitrogen addition rates, 1.5 and 8%, showed increases in mean individual leaf area of 50 and 82%, respectively, whereas the whole plant leaf area increased between 854 and 1098%, depending on the hybrid (Table 7.3, Fig. 7.8 and Plate 22C).

Under low fertility, tree canopy leaf area is reduced by the presence of fewer, smaller leaves, which reduce light interception and photosynthesis. In addition, trees grown under low fertility have fewer branches, and those branches are thinner, having less biomass (Table 7.4). Bowman and Conant (1994) reported similar responses in *S. glauca* growing along a soil fertility and moisture gradient. During early stand development, height growth has been found to be less affected by site fertility than diameter growth. Maximum height, however, is ultimately controlled by site fertility and is the basis for site index curves.

Table 7.3. Total leaf area and mean individual leaf area of two hybrid poplar clones in response to two nitrogen addition rates (NAR). Clone 1 was a *Populus deltoides* × *Populus trichocarpa* cross, whereas clone 2 was a reciprocal cross between a *P. trichocarpa* female and a *P. deltoides* male.

	1.5% NAR		8% NAR	
	Total leaf area (cm^2)	Mean leaf area (cm^2)	Total leaf area (cm^2)	Mean leaf area (cm^2)
Clone 1 (D × T)	628	34.4	7525	62.7
Clone 2 (T × D)	620	33.2	5915	49.6

Table 7.4. Per cent change in branch number, total and mean branch biomass after the first and second years of a field nitrogen study using six hybrid poplar clones. The per cent increases represent the increase from the lowest N to highest N treatment.

Response variable	Year 1 (%)	Year 2 (%)
Number of branches	+28	+16
Branch biomass	+82	+36
Mean branch biomass	+47	+27
Leaf biomass	+31	+32

Associated with the reduction in above-ground biomass, there is a reallocation of carbon to the root system that enables the tree to explore the soil for additional nutrients (Ibrahim *et al.*, 1998). In the same study in Table 7.4, the per cent biomass in roots increased from 25% in the high N treatment to over 30% in the low N treatment. Woolfolk and Friend (2003) found that the form of nitrogen – nitrate or ammonium – affected the development of *P. deltoides* roots. The ratio of 20:80 (NH_4:NO_3) resulted in the greatest root length, specific root length and root N concentration. In a study comparing the sprouting of *S. viminalis* grown under two nitrogen levels, von Fircks and Sennerby-Forsse (1998) reported higher starch levels in the roots of the low-N trees, which they found to support greater sprouting the following spring.

Other consequences of reduced growth from poor fertility

In addition to the direct effects of low fertility discussed above, there have been reports of other secondary responses. Because of the slower growth, the density of the wood produced under limiting nutrition is higher and has been shown to be a result of the production of vessels with smaller diameters (Harvey and van den Driessche, 1997, 1999). A physiological consequence of smaller vessel diameter was an increased resistance to cavitation during drought. The slowest-growing native willows were found to be less sensitive to both nutrient and water stress (Weih, 2001). Glynn *et al.* (2003) found altered insect resistance in poplar resulting from nutrient availability. Under low fertility, constitutive resistance to gypsy moth increased, as well as the rapid induced resistance to whitemarked tussock moth.

7.3 Atmospheric Stresses

Most of the observed increase in global average temperatures since the mid-20th century is very likely due to the increase in anthropogenic greenhouse gas concentrations. Global atmospheric concentrations of carbon dioxide, methane, ozone and nitrous oxide have indeed increased markedly as a result of human activities since 1750, and now far exceed pre-industrial values determined from ice cores spanning many thousands of years (IPCC, 2007). If future evolution is difficult to predict, this phenomenon is likely to continue to increase during the next decades and is not without consequences for plant growth and physiology.

7.3.1 Tropospheric ozone levels
D.F. Karnosky

Concurrently with increasing CO_2 concentration in the earth's atmosphere, tropospheric ozone (O_3) has risen sharply in the post-industrialization period and is expected to be at toxic levels for sensitive vegetation (where peak concentrations exceed 60 ppb) for some 50% of the world's forests by 2100 (Fowler *et al.*, 1999). It is well known that poplars are generally sensitive to O_3 (Karnosky *et al.*, 1996; Dickson *et al.*, 1998; Isebrands *et al.*, 2001). Although there is a large amount of genetic variation in poplar responses to O_3 (Karnosky, 1977; Berrang *et al.*, 1986; Karnosky *et al.*, 2005), overall the poplars are among the tree species most sensitive to elevated O_3. Adverse effects of O_3 on poplar foliage include visible foliar symptoms (Karnosky, 1976; Karnosky *et al.*, 1996), decreased chlorophyll content (Gagnon *et al.*, 1992), decreased leaf size (Oksanen *et al.*, 2001), increased peroxisome accumulation (Oksanen *et al.*, 2003) and premature leaf abscission (Karnosky *et al.*, 1996). Physiologically, O_3 decreases maximal photosynthetic levels in both sun and shade leaves (Coleman *et al.*, 1995), alters stomatal conductance and respiration (Noormets *et al.*, 2001), decreases height and diameter growth

(Karnosky *et al.*, 1996; Isebrands *et al.*, 2001), changes crown architecture (Dickson *et al.*, 2001), affects competitive ability (McDonald *et al.*, 2002) and decreases fitness (Karnosky *et al.*, 2003a).

In addition, and importantly for risk analysis for global change, elevated O_3 can modify poplar responses to important insect and disease pests. For example, with the important cyclic pest, the forest tent caterpillar, developing larval growth rates were increased and the fecundity of the adult female moths was increased under elevated O_3 (Holton *et al.*, 2003; Kopper and Lindroth, 2003). Also, aphid populations were significantly larger on aspen trees under elevated O_3 (Percy *et al.*, 2002; Awmack *et al.*, 2004; Mondor *et al.*, 2004).

For diseases, Karnosky *et al.* (2002) have reported a three- to fivefold increase in *Melampsora* leaf rust under elevated O_3 at the AspenFACE experiment over several growing seasons. Ozone-induced changes in the leaf surface waxes are likely predisposing aspen to the important leaf rust (Karnosky *et al.*, 2002; Percy *et al.*, 2002).

7.3.2 Effects of elevated CO_2 and combinations with ozone

B. Gielen, D.F. Karnosky, G. Scarascia-Mugnozza and R. Ceulemans

During the last centuries, we have increasingly been using natural non-renewable resources to satisfy our needs; fossil fuel is one of those resources. Among others, emission from fossil fuel burning is causing a steady increase of CO_2 concentration in the earth's atmosphere, which may reach 550 μmol mol^{-1} by the middle of the 21st century (Schimel *et al.*, 1996). This rising CO_2 is the main driver of ongoing climatic changes (IPCC, 1996). Table 7.5 gives an overview of the available literature on elevated CO_2 effects on poplar investigated using field experiments (Gielen and Ceulemans, 2001). In contrast to other greenhouse gases, CO_2 is a plant fertilizer rather than a pollutant. Elevated atmospheric CO_2 initially causes a photosynthetic stimulation, resulting in faster growth and higher production (Long and Drake, 1992). As a consequence, the capacity of ecosystems to store part of the excess carbon is of main interest as a possible feedback mechanism to global climate change.

Populus is characterized by a high growth rate, high plasticity and unusually high sink strength (Scarascia-Mugnozza *et al.*, 1997) related to an indeterminate growth pattern and the continuous production of sylleptic branches (not all species) during periods of active growth. Therefore, a sustained photosynthetic enhancement under elevated CO_2 may be expected for poplar, which could make short-rotation forestry interesting as a mitigation strategy for the increase of atmospheric CO_2 concentration. Indeed, light-saturated net photosynthesis at the leaf level was consistently enhanced by elevated CO_2 among different *Populus* species, clones and experiments (Gielen and Ceulemans, 2001). After long-term exposure to elevated CO_2, a reduced stimulation of photosynthesis may occur due to acclimation processes, considered to improve plant performance through increased resource use efficiency, or due to accumulation of excess carbohydrates (Sage, 1994). Acclimation of C_3 photosynthesis has been frequently observed (Ainsworth and Long, 2005), and it may cause a reduction of growth stimulation under elevated CO_2. Two free-air CO_2 enrichment (FACE) experiments have been set up to increase our understanding of the effects of elevated atmospheric CO_2 on poplar trees, POP-EUROFACE (http://www.unitus.it/euroface, Fig. 7.9) and AspenFACE (http://aspenface.mtu.edu, Plate 22B). In POP-EUROFACE, photosynthesis per unit land surface, i.e. gross primary production (GPP), was stimulated by elevated CO_2 in all 3 years of the first rotation cycle, with the largest stimulation during the first year (Wittig *et al.*, 2005). In the second and third years, there was a decline in stimulation due to canopy closure rather than to photosynthetic acclimation (Wittig *et al.*, 2005). Similarly, results from AspenFACE showed a sustained stimulation of photosynthesis in the upper, but not in the lower, canopy (Takeuchi *et al.*, 2001).

In view of the effects of rising atmospheric CO_2 on productivity and carbon balance, respiratory carbon losses need to be accounted for. At present, no clear consensus has emerged concerning respiration of poplar trees under elevated CO_2 concentrations, but in general, respiration is expected to increase because of the

Table 7.5. Overview of available literature on elevated CO_2 responses of *Populus* in field experiments (period 1990–2005). Table adapted from Gielen and Ceulemans (2001). Abbreviations used are listed at the end of the table.

Species	Clone	Location	Facility	Age at start	Duration of enrichment	Interaction	References
P. tremuloides		Siena, Italy	Natural CO_2 spring				Tognetti *et al.*, 1999a, b
P. grandidentata		Pellston, Michigan, USA	OTC (pots)	Seedlings (60 days)	10 weeks		Curtis and Teeri, 1992
P. grandidentata		Pellston, Michigan, USA	OTC (OBRB)	Rooted cuttings	1 growing season	N	Zak *et al.*, 1993; Curtis *et al.*, 1994
P. tremuloides		Pellston, Michigan, USA	OTC (OBRB)	Rooted cuttings	2 growing seasons	N	Kubiske *et al.*, 1997, 1998; Mikan *et al.*, 2000
P. tremuloides	Early/late leaf drop	Pellston, Michigan, USA	OTC (OBRB)	Rooted cuttings	2.5 growing seasons	N	Curtis *et al.*, 2000; Pregitzer *et al.*, 2000; Wang *et al.*, 2000; Zak *et al.*, 2000a, b; King *et al.*, 2001a
P. tremuloides		Pellston, Michigan, USA	OTC (OBRB)	4-year old	1 growing season		Rier *et al.*, 2002
P. tremuloides	O_3 tolerant O_3 sensitive	Alberta, Michigan, USA	OTC	Young trees (8 months)	3 growing seasons	O_3	Kull *et al.*, 1996; Karnosky *et al.*, 1998; Ambus and Robertson, 1999
P. ×*canadensis*	'Eugenei'	Pellston, Michigan, USA	OTC (OBRB)	Cuttings	1 growing season	N	Curtis *et al.*, 1995; Pregitzer *et al.*, 1995; Randlett *et al.*, 1996; Lussenhop *et al.*, 1998
P. ×*canadensis* *P. nigra* × *P. maximowiczii*	'DN-33', 'DN-44', 'DN-70', 'DN-74', 'NM-6'	Alberta, Michigan, USA	OTC (pots)	Cuttings	1 growing season	O_3	Dickson *et al.*, 1998
P. ×*canadensis* *P. trichocarpa* × *P. deltoides*	'Robusta' 'Beaupré'	Antwerp, Belgium	OTC	Cuttings	2 growing seasons + 1 growing season after coppice		Ceulemans *et al.*, 1995a, b, c, 1996, 1997; Kalina and Ceulemans, 1997; Will and Ceulemans, 1997
P. trichocarpa × *P. deltoides*	'Boelare'	Hampshire, UK	OTC (pots)	Cuttings	2 growing seasons (coppice after first)	O_3	Taylor *et al.*, 2001.
P. ×*canadensis* *P. deltoides*	'I-214' 'Lux'	Siena, Italy	FACE-system rings	Cuttings	1 growing season		Tognetti *et al.*, 1999a, b

P. alba *P. nigra* *P. ×canadensis*	'2AS-11' 'Jean_Pourtet' 'I-214'	Viterbo, Italy	FACE	Cuttings	3 growing seasons + 3 growing seasons after coppice	N (second rotation)	http://www.unitus.it/euroface Scarascia-Mugnozza *et al.*, 2000; Calfapietra *et al.*, 2001, 2003a, b, 2005; Ferris *et al.*, 2001; Gielen *et al.*, 2001, 2002, 2003a, b; Miglietta *et al.*, 2001; Taylor *et al.*, 2001, 2003, 2005; Bernacchi *et al.*, 2003; Lukac *et al.*, 2003; Hoosbeek *et al.*, 2004; King *et al.*, 2004; Liberloo *et al.*, 2004, 2005; Tricker *et al.*, 2004, 2005; Cotrufo *et al.*, 2005; Luo *et al.*, 2005; Moscatelli *et al.*, 2005a, b; Wittig *et al.*, 2005
P. tremuloides	6 different clones	Rhinelander, Wisconsin, USA	FACE	<1 year	Start: 1998	O_3	http://aspenface.mtu.edu Isebrands *et al.*, 2001; King *et al.*, 2001b; Noormets *et al.*, 2001; Oksanen *et al.*, 2001; Takeuchi *et al.*, 2001; Karnosky *et al.*, 2002, 2003b, 2005; Holmes *et al.*, 2003
P. deltoides	'Bartr.'	Biosphere 2 Laboratory, Oracle, Arizona, USA	Bays at 430, 800 and 1200 μmol mol^{-1} CO_2	Cuttings	Start: 1998 (coppiced yearly)	Drought	Pegoraro *et al.*, 2004, 2005a, b; Turnbull *et al.*, 2004; Ananyev *et al.*, 2005; Barron-Gafford *et al.*, 2005; Murthy *et al.*, 2005

FACE, Free-air CO_2 enrichment; N, nitrogen; OBRB, open-bottom root boxes; OTC, open-top chamber.

Fig. 7.9. Aerial view of one plot at the POPFACE site with three different *Populus* clones. Photo courtesy of S. Bunn, University of Southampton.

increase in biomass. Soil respiration will likely increase, as has been observed in large-scale field experiments (King *et al.*, 2004) and at the Biosphere 2 Laboratory in Arizona, USA (Barron-Gafford *et al.*, 2005).

The most striking and important effect of elevated CO_2 concentrations is the stimulating effect on above- and belowground growth (Gielen and Ceulemans, 2001; Calfapietra *et al.*, 2003a; Karnosky *et al.*, 2003b). Gielen and Ceulemans (2001) reported a mean biomass stimulation of 33% in elevated CO_2 conditions across poplar studies published between 1990 and 2000. Recent results from the two field FACE experiments showed a consistent increase in biomass production in elevated CO_2 (Calfapietra *et al.*, 2003b; Karnosky *et al.*, 2003b; Liberloo *et al.*, 2005).

Plant productivity is determined by photosynthetic efficiency in combination with light interception, which in turn depends on tree architecture and leaf area (Monteith, 1977). Particularly in high-density poplar plantations, leaf area development is considered to be extremely important for productivity (Stettler *et al.*, 1993); thus, responses of leaf area to elevated CO_2 will definitely influence wood productivity. Overall, an increase of total leaf area under elevated CO_2 concentration has been reported in poplar, though this is dependent on nitrogen availability and the developmental stage of the canopy (Curtis *et al.*, 2000; Gielen *et al.*, 2001, 2003a; Karnosky *et al.*, 2003b; Liberloo *et al.*, 2005). Not only total leaf area and leaf area distribution but also the length of the growing season determine productivity and may be influenced by rising CO_2. Previously, advanced budset of poplar has been observed in open-top-chamber studies, indicating a shorter growing season (Ceulemans *et al.*, 1995a; Sigurdsson, 2001). This was not confirmed in the two FACE studies, as budset was not affected significantly in POP-EUROFACE and was delayed in AspenFACE (Calfapietra *et al.*, 2003b; Karnosky *et al.*, 2003b). Alternatively, the leaf area duration can be compared between treatments; Curtis *et al.*

(1995) reported a stimulation of 39%, indicating a prolongation of the growing season under elevated CO_2. Research in this field was still ongoing in 2010 at both FACE sites.

Further, an increase in belowground biomass in response to elevated CO_2 was observed (Zak *et al.*, 1993; Dickson *et al.*, 1998; Mikan *et al.*, 2000; Pregitzer *et al.*, 2000; King *et al.*, 2001a; Calfapietra *et al.*, 2003b; Lukac *et al.*, 2003). Although the relative stimulation of fine root production is often larger than aboveground biomass stimulation, recent field studies provide evidence for unchanged partitioning between above- and belowground biomass (Calfapietra *et al.*, 2003b). Despite a stimulation of above- and belowground biomass production, the net carbon storage capacity of a short-rotation forest plantation in the first rotation cycle was not enhanced by elevated CO_2 due to increased losses of soil carbon (Gielen *et al.*, 2005).

Predicting terrestrial carbon sequestration and understanding ecosystem functioning under rising atmospheric CO_2 concentrations ultimately requires an integrated understanding of carbon and nitrogen cycling. Changes in root turnover, root nitrogen concentrations and leaf litter chemistry could alter microbial substrate availability and consequently microbial communities, decomposition and carbon–nitrogen cycles. Zak *et al.* (2000a) did not support the hypothesis of increased nitrogen cycling in future terrestrial ecosystems because neither the amount of biomass and community composition of soil microorganisms nor nitrogen mineralization were affected by elevated CO_2 in open-top chambers. Unchanged gross rates of nitrification and nitrate immobilization were also confirmed in the AspenFACE field experiment (Holmes *et al.*, 2003). At the Biosphere 2 Laboratory, elevated CO_2 accelerated depletion of soil nutrients but not of nitrogen (Barron-Gafford *et al.*, 2005). At the POP-EUROFACE site, a decrease of soil nitrogen was observed under elevated CO_2 at the end of the first rotation cycle, probably due to a decreased input from leaf litter and decreased decomposition rate (Calfapietra *et al.*, 2007). Further, FACE treatment significantly increased nitrogen-use efficiency (NUE), i.e. the woody biomass produced per unit of nitrogen, decreased nitrogen concentration in all plant tissues as a result and did not change the total nitrogen uptake of poplar trees (Calfapietra *et al.*, 2007). For a discussion of poplar litter decomposition under elevated CO_2 concentration, we refer to Cotrufo *et al.* (2005).

Typically, global change studies are performed in a controlled environment, without constraints on water and nutrient availability. It has been emphasized, however, that ecosystem responses to elevated CO_2 will be influenced by variable climate and environmental conditions (Luo *et al.*, 1999; Scholes *et al.*, 1999). Thus, local climate or site conditions complicate predictions about production and yield responses of natural and managed poplar stands to rising atmospheric CO_2. Therefore, efforts have been and are being made to study tree responses to CO_2 in interaction with ozone (O_3), nutrient availability, drought, temperature or radiation.

When investigating poplar trees under concurrent exposure to CO_2 and O_3, the gases tend to counteract one another. Damages caused by O_3 are generally offset under elevated CO_2 (Volin and Reich, 1996; Dickson *et al.*, 1998; Noormets *et al.*, 2001; Taylor *et al.*, 2001), but sometimes elevated CO_2 can exacerbate the O_3 effects (Kull *et al.*, 1996; Wustman *et al.*, 2001). For the longest running experiment (8 years of AspenFACE), examining the combined effects of elevated CO_2 and O_3 and the effects on trembling aspen growth (Isebrands *et al.*, 2001; Percy *et al.*, 2002; Karnosky *et al.*, 2003b, 2005) and productivity (King *et al.*, 2005) have largely cancelled one another, so that the combined treatments are not significantly different from the controls.

Besides various abiotic factors, a number of biotic factors also interact directly or indirectly with the effect of increased atmospheric CO_2 on poplar growth, productivity and physiology. A brief literature review on biotic interactions through insect and mycorrhizal activities on the CO_2 responses of poplar can be found in Gielen and Ceulemans (2001). In summary, soil conditions and especially sensitivity of certain poplar clones to insects, fungi and air pollutants, for example O_3, will influence the growth and functioning of natural and managed plantations, and interact with CO_2. None the less, only a minority of studies supported the strong interactive effects of biotic variables with atmospheric CO_2.

From a forestry viewpoint, possible alterations in wood quality under elevated CO_2 conditions have to be accounted for, but research is still ongoing. The responses of poplar to elevated CO_2 can also have practical implications. For example, Tupker *et al.* (2003) suggested the application of elevated CO_2 concentrations in the greenhouse in order to facilitate the early selection of fast-growing clones under future atmospheric conditions.

In brief, ample evidence has been found to suggest that poplar will profit considerably from CO_2 enrichment, at least in the short-term, but it is uncertain whether this will improve the long-term C storage capacity of poplar in natural stands or plantations.

7.3.3 Increased temperature
N. Marron

Context

Although trees have responded to global warming in the past, the rate of change currently observed and predicted for the next decades is likely to be unprecedented. Continued greenhouse gas emissions at or above current rates would cause further warming and induce many changes in the global climate system during the 21st century that would very likely be larger than those observed during the 20th century. For the next two decades, a warming of about 0.2°C per decade is projected for a range of IPCC scenarios (IPCC, 2007). The photosynthetic optimum of plants is generally between 20 and 30°C, but the range can be enlarged from 15 to 45°C according to the temperature experienced during growth or the biological type to which the plants belong. However, 50°C is generally considered as the limiting temperature for photosynthetic activity of vascular plants (Berry and Björkman, 1980; Saugier, 1983). Trees belonging to the *Populus* genus are known to be particularly sensitive to temperature rise.

Physiological basis

Tolerance of the photosynthetic apparatus to high temperature is thought to be related mainly to the thermostability of Photosystem II (PSII) photochemistry, which seems to be one of the most sensitive components of the electron-transport chain of the thylakoids to most of the major environmental constraints, such as ultraviolet radiation or heat (Trabaud and Méthy, 1992; Schumaker *et al.*, 1997). PSII deactivation induced by heat leads to damages such as protein denaturation and dissociation of pigments. These alterations result in changes in the interactions between lipids and proteins associated with an increase in membrane fluidity (Sundby *et al.*, 1986; Thompson *et al.*, 1989). The critical temperature (T_C) depends on both species and genotypes (Epron, 1997; Knight and Ackerly, 2002), but is also modulated to a large extent by climate. It increases after short periods of exposure to higher temperatures and in response to drought stress, ABA application or growth in high CO_2 concentration (Ivanov *et al.*, 1992; Havaux, 1993; Ladjal *et al.*, 2000; Taub *et al.*, 2000).

Experimental assessment

Three different kinds of experiments have been conducted in order to estimate the impact of high temperature on poplar plants: (i) plant growth under heated soil conditions; (ii) plant growth in heated growth chambers; and (iii) direct heating of the leaves.

1. A minimum soil temperature of 14°C during several days is needed to sustain rooting in poplar (Zalesny *et al.*, 2005). As compared with several conifer species, trembling aspen (*P. tremuloides*) has been shown to be particularly sensitive to soil temperature. Its optimal soil temperature for total biomass was 19.4°C, leaves showing a higher optimum temperature than roots (Tryon and Chapin, 1983; Peng and Dang, 2003). Specific leaf area showed a large decrease with the rise of soil temperature. Physiologically, photosynthesis was progressively less limited by stomatal conductance (relative to mesophyll limitation) as soil temperature increased (Dang and Cheng, 2004). Both the maximal rate of carboxylation (V_{cmax}) and the maximum rate of electron transport (J_{max}) increased with soil temperature, up to an optimum (generally around 25°C), and then decreased with further increases in soil temperature (Cai and Dang, 2002).

2. Acclimation is commonly observed for poplar plants grown under high temperature. Notably, respiration of roots and shoots has been observed to be largely inferior for plants grown under elevated temperature (Tjoelker *et al.*, 1999). This acclimation of respiration could play an important role in determining the relationships between respiration and photosynthesis (Loveys *et al.*, 2003).

3. With regard to the last type of procedure, the critical temperature (T_C) for the stability of PSII can be estimated: (i) as the point at which basal chlorophyll *a* fluorescence (F_0) begins to increase under gradually increasing temperatures (generally between 42 and 50°C); or (ii) as the temperature for which the PSII quantum yield, monitored in darkness, is decreased by 15% in response to a gradual increase in temperature (generally between 35 and 40°C) (Bilger *et al.*, 1987; Froux *et al.*, 2004). For quaking aspen (*P. tremuloides*), the critical temperature for the stability of PSII, estimated by monitoring initial fluorescence, has been shown to be around 49°C (Logan and Monson, 1999). On the other hand, for several clones of *P.* ×*canadensis*, it has been shown that T_C could vary from 43 to 47°C among leaves of a given plant differing in age or in stage of development (Marron, 2003). Structural changes in leaves during growth and ageing have been evidenced as a possible cause for this wide variation in terms of T_C.

As for many of the abiotic stresses described in this chapter, poplar and willow show, on the one hand, a very large range of diversity in their response to warming depending on growth conditions and plant material, and on the other hand, they are apparently able to acclimate to quite extreme conditions. With regard to the current and future changes in climate, the diversity in the heat responses observed within the *Salicaceae* family is encouraging for the selection of varieties adapted to upcoming climate conditions.

7.3.4 High irradiances (photoinhibition)

B. Gielen

The photosynthetic system of tree leaves can be damaged by very high irradiances. In fact, photoinhibition, which includes photodamage and protective downregulation of photosynthesis, is characteristic for the photosynthetic process (Long *et al.*, 1994; Osmond, 1994; Niinemets and Kull, 2001). Photoinhibition occurs at all light intensities, but increases with increasing irradiance (Tyystjärvi and Aro, 1996). In the field, the largest part of the decline in photosynthetic efficiency during the day, for example at midday when light intensity is high, results from downregulation without irreversible damage to the photosynthetic system (Demmig-Adams and Adams, 1996; Demmig-Adams *et al.*, 1996). Ögren and Sjöström (1990) estimated that upper canopy leaves of *Salix* lost about 6.3–12.6% of potential C gain because of photoinhibition, while Niinemets and Kull (2001) pointed out that this was an overestimation because stomatal closure was not taken into account. They found that photoinhibition was indeed common in upper canopies of *Populus tremula*, whereas the influence on daily photosynthesis was moderate (Niinemets and Kull, 2001). Differences in resistance to light stress among poplar or willow clones have not often been investigated, e.g. Zhang and Gao (2000). The sensitivity of trees to high irradiance depends primarily on the degree of shade adaptation, and on other stresses. Therefore, photoinhibition is important when poplar or willow trees are subjected to drought or extreme temperatures.

7.4 Conclusion

Worldwide, poplars and willows have colonized numerous kinds of environments and subsequently can be subjected to conditions far from their growth optima. These conditions are likely to be aggravated in response to the current global changes affecting atmospheric as well as soil compartments. However, a very wide range of diversity exists within the *Salix* and *Populus* genera, both in terms of level of resistance (defined as the ability to survive and/or to sustain production of biomass under constraining conditions) and in terms of mechanisms used to face these constraints occurring both at edaphic and atmospheric levels. Moreover, members of the *Salicaceae* family show an impressive acclimation ability in response to most stresses.

References

Aasamaa, K. and Sõber, A. (2001) Hydraulic conductance and stomatal sensitivity to changes of leaf water status in six deciduous tree species. *Biologia Plantarum* 44, 65–73.

Aasamaa, K., Sõber, A., Hartung, W. and Niinemets, Ü. (2002) Rate of stomatal opening, shoot hydraulic conductance and photosynthetic characteristics in relation to leaf abscisic acid concentration in six temperate deciduous trees. *Tree Physiology* 22, 267–276.

Aasamaa, K., Sõber, A., Hartung, W. and Niinemets, Ü. (2004) Drought acclimation of two deciduous tree species of different layers in a temperate forest canopy. *Trees – Structure and Function* 18, 93–101.

Abbot, L.K. and Robson, A.D. (1991) Factors influencing the occurrence of vesicular arbuscular mycorrhizae. *Agriculture, Ecosystems and Environment* 35, 121–150.

Abrams, M.D. and Mostoller, S.A. (1995) Gas exchange, leaf structure and nitrogen in contrasting successional tree species growing in open and understory sites during a drought. *Tree Physiology* 15, 361–370.

Acosta-Avalos, D., Alvarado-Gil, J. and Vargas, H. (2005) Monitoring ethylene and oxygen emission during water stress of *Populus alba* leaves. *Journal de Physique IV France* 125, 817–819.

Ainsworth, E.A. and Long, S.P. (2005) What have we learned from 15 years of free-air CO_2 enrichment (FACE)? A meta-analytic review of the responses of photosynthesis, canopy properties and plant production to rising CO_2. *New Phytologist* 165, 351–372.

Al-Agely, A.K. and Reeves, F.B. (1995) Inland sand dune mycorrhizae: effects of soil depth and pH on colonisation of *Oryzopsis hymenoides. Mycologia* 87, 54–60.

Al-Karki, G.N. (2000) Growth of mycorrhizal tomato and mineral acquisition under salt stress. *Mycorrhiza* 10, 51–54.

Allen, R.D. (1995) Dissection of oxidative stress tolerance using transgenic plants. *Plant Physiology* 107, 1049–1054.

Allen, S.J., Hall, R.L. and Rosier, P.T.W. (1999) Transpiration by two poplar varieties grown as coppice for biomass production. *Tree Physiology* 19, 493–501.

Ambus, P. and Robertson, G.P. (1999) Fluxes of CH_4 and N_2O in aspen stands grown under ambient and twice-ambient CO_2. *Plant and Soil* 209, 1–8.

Amlin, N.M. and Rood, S.B. (2002) Comparative tolerances of riparian willows and cottonwoods to water-table decline. *Wetlands* 22, 338–346.

Ananyev, G., Kolber, Z.S., Klimov, D., Falkowski, P.G., Berry, J.A., Rascher, U., *et al.* (2005) Remote sensing of heterogeneity in photosynthetic efficiency, electron transport and dissipation of excess light in *Populus deltoides* stands under ambient and elevated CO_2 concentrations, and in a tropical forest canopy, using a new laser-induced fluorescence transient device. *Global Change Biology* 11, 1195–1206.

Andersen, D.C. and Nelson, S.M. (2003) Effects of river flow regime on cottonwood leaf litter dynamics in semi-arid northwestern Colorado. *Southwestern Naturalist* 48, 188–201.

Anon. (1977) *Combating Desertification.* Institute of Glaciology, Cryopedology, and Desert, Chinese Academy of Forestry, Lanzhou, China.

Anon. (2000) *Tackling Desertification in the Korqin Sandy Lands through Integrated Afforestation.* Yachen Printing Ltd, Tongliao, Inner Mongolia, China.

Apse, M.A., Aharon, G.S., Snedden, W.A. and Blumwald, E. (1999) Salt tolerance conferred by overexpression of a vacuolar Na^+/H^+ antiport in *Arabidopsis. Science* 285, 1256–1258.

Arisi, A.-C., Cornic, G., Jouanin, L. and Foyer, C.H. (1998) Overexpression of FeSOD in transformed poplar modifies the regulation of photosynthesis at low CO_2 partial pressures or following exposure to the prooxidant herbicide methyl viologen. *Plant Physiology* 117, 565–574.

Arndt, S.K., Arampatsis, C., Foetzki, A., Li, X.Y., Zeng, F.J. and Zhang, X.M. (2004) Contrasting patterns of leaf solute accumulation and salt adaptation in four phreatophytic desert plants in a hyperarid desert with saline groundwater. *Journal of Arid Environments* 59, 259–270.

Awmack, C.S., Harrington, R. and Lindroth, R.L. (2004) Individual performance does not predict aphid population responses to elevated atmospheric CO_2 or O_3. *Global Change Biology* 10, 1414–1423.

Ball, J., Russo, L. and Thomson, L. (1996) Status of *Populus euphratica* and proposal for its conservation. Working Party on Breeding and selection, XX Session, International Poplar Commission, Budapest, Hungary, 1–12 October, 1996. FAO, Rome.

Barron-Gafford, G., Martens, D., Grieve, K., Biel, K., Kudeyarov, V., MClain, J.E.T., *et al.* (2005) Growth of eastern cottonwoods (*Populus deltoides*) in elevated [CO_2] stimulates stand-level respiration and

rhizodeposition of carbohydrates, accelerates soil nutrient depletion, yet stimulates above- and belowground biomass production. *Global Change Biology* 11, 1220–1233.
Bassman, J.H. and Zwier, J.C. (1991) Gas exchange characteristics of *Populus trichocarpa, Populus deltoides* and *Populus trichocarpa* × *P. deltoides* clones. *Tree Physiology* 8, 145–159.
Beritognolo, I., Piazzai, M., Muleo, R., Paolucci, I., Kuzminsky, E., Sabatti, M., *et al.* (2003) Molecular response to salt stress of three Italian *Populus* genotypes from different latitudes. Proceedings of the XLVII Italian Society of Agricultural Genetics – SIGA Annual Congress, Verona, Italy, 24–27 September 2003. Poster Abstract – 5.18.
Bernacchi, C.J., Calfapietra, C., Davey, P.A., Wittig, V.E., Scarascia-Mugnozza, G.E., Raines, C.A., *et al.* (2003) Photosynthesis and stomatal conductance responses of poplars to free-air CO_2 enrichment (PopFACE) during the first growth cycle and immediately following coppice. *New Phytologist* 159, 609–621.
Berrang, P.C., Karnosky, D.F., Mickler, R.A. and Bennett, J.P. (1986) Natural selection for ozone tolerance in *Populus tremuloides. Canadian Journal of Forest Research* 16, 1214–1216.
Berry, J. and Björkman, O. (1980) Photosynthetic response and adaptation to temperature in higher plants. *Annual Review of Plant Physiology* 31, 491–543.
Bigler, C., Kulakowski, D. and Veblen, T.T. (2005) Multiple disturbance interactions and drought influence fire severity in Rocky Mountain subalpine forests. *Ecology* 86, 3018–3029.
Bilger, W., Schreiber, U. and Lange, O.L. (1987) Chlorophyll fluorescence as an indicator of heat induced limitation of photosynthesis in *Arbutus unedo* L. *Plant Responses to Cellular Dehydration during Environmental Stress* 15, 391–399.
Blake, T.J., Sperry, J.S. and Tschaplinski, T.J. (1996) Water relations. In: Stettler, R.F., Bradshaw, H.D.J., Heilman, P.E. and Hinckley, T.M. (eds) *Biology of Populus and Its Implications for Management and Conservation.* NRC Research Press, National Research Council of Canada, Ottawa, pp. 401–422.
Blum, A., Munns, R., Passioura, J.B. and Turner, N.C. (1996) Genetically engineered plants resistant to soil drying and salt stress: how to interpret osmotic relations? *Plant Physiology* 110, 1051–1053.
Blumwald, E., Aharon, G.S. and Apse, M.P. (2000) Sodium transport in plant cells. *Biochimica et Biophysica Acta* 1465, 140–151.
Bolu, W.H. and Polle, A. (2004) Growth and stress reactions in roots and shoots of a salt-sensitive poplar species (*Populus* ×*canescens*). *Tropical Ecology* 45, 161–171.
Borsani, O., Valpuesta, V. and Botella, M. (2001) Evidence for a role of salicylic acid in the oxidative damage generated by NaCl and osmotic stress in *Arabidopsis* seedlings. *Plant Physiology* 126, 1024–1030.
Bovard, B.D., Curtis, P.S., Vogel, C.S., Su, H.B. and Schmid, H.P. (2005) Environmental controls on sap flow in a northern hardwood forest. *Tree Physiology* 25, 31–38.
Bowman, W.D. and Conant, R.T. (1994) Shoot growth dynamics and photosynthetic response to increased nitrogen availability in the alpine willow *Salix glauca. Oecologia* 97, 93–99.
Braatne, I.H., Hinckley, T.M. and Stettler, R.F. (1992) Influence of soil water on the physiological and morphological components of plant water balance in *Populus trichocarpa, Populus deltoides* and their F1 hybrids. *Tree Physiology* 11, 325–339.
Brignolas, F., Thierry, C., Guerrier, G. and Boudouresque, E. (2000) Compared water deficit response of two *Populus* ×*euramericana* clones, Luisa Avanzo and Dorskamp. *Annals of Forest Science* 57, 261–266.
Brosché, M., Vinocur, B., Alatalo, E.R., Lamminmäki, A., Teichmann, T., Ottow, E.A., *et al.* (2005) Gene expression and metabolite profiling of *Populus euphratica* growing in the Negev desert. *Genome Biology* 6, R101.
Browicz, K. (1977) Chorology of *Populus euphratica* Olivier. *Arboretum Kórnickie* 22, 5–27.
Brown, K.R. and van den Driessche, R. (2002) Growth and nutrition of hybrid poplars over 3 years after fertilization at planting. *Canadian Journal of Forest Research* 32, 226–232.
Cai, T. and Dang, Q.L. (2002) Effects of soil temperature on parameters of a coupled photosynthesis-stomatal conductance model. *Tree Physiology* 22, 819–827.
Calfapietra, C., Gielen, B., Sabatti, M., De Angelis, P., Scarascia-Mugnozza, G. and Ceulemans, R. (2001) Growth performance of *Populus* exposed to 'Free Air Carbon dioxide Enrichment' during the first growing season in the POPFACE experiment. *Annals of Forest Science* 58, 819–828.
Calfapietra, C., Gielen, B., Sabatti, M., De Angelis, P., Miglietta, F., Scarascia-Mugnozza, G., *et al.* (2003a) Do above-ground growth dynamics of poplar change with time under CO_2 enrichment? *New Phytologist* 160, 305–318.

Calfapietra, C., Gielen, B., Galema, A.N.J., Lukac, M., De Angelis, P., Moscatelli, M.C., *et al.* (2003b) Free-air CO_2 enrichment (FACE) enhances biomass production in a short-rotation poplar plantation (POPFACE). *Tree Physiology* 23, 805–814.

Calfapietra, C., Tulva, I., Eensalu, E., Perez, M., De Angelis, P., Scarascia-Mugnozza, G., *et al.* (2005) Canopy profiles of photosynthetic parameters under elevated CO_2 and N fertilization in a poplar plantation. *Environmental Pollution* 137, 525–535.

Calfapietra, C., De Angelis, P., Gielen, B., Lukac, M., Moscatelli, M.C., Avino, G., *et al.* (2007) Increased nitrogen-use efficiency of a short-rotation poplar plantation in elevated CO_2 concentration. *Tree Physiology* 27, 1153–1163.

Caruso, A., Morabito, D., Delmotte, F., Kahlem, G. and Carpin, S. (2002) Dehydrin induction during drought and osmotic stress in *Populus. Plant Physiology and Biochemistry* 40, 1033–1042.

Ceulemans, R., Jiang, X.N. and Shao, B.Y. (1995a) Effects of elevated atmospheric CO_2 on growth, biomass production and nitrogen allocation of two *Populus* clones. *Journal of Biogeography* 22, 261–268.

Ceulemans, R., Jiang, X.N. and Shao, B.Y. (1995b) Growth and physiology of one-year old poplar (*Populus*) under elevated atmospheric CO_2 levels. *Annals of Botany* 75, 609–617.

Ceulemans, R., Van Praet, L. and Jiang, X.N. (1995c) Effects of CO_2 enrichment, leaf position and clone on stomatal index and epidermal cell density in poplar (*Populus*). *New Phytologist* 131, 99–107.

Ceulemans, R., Shao, B.Y., Jiang, X.N. and Kalina, J. (1996) First- and second-year aboveground growth and productivity of two *Populus* hybrids grown at ambient and elevated CO_2. *Tree Physiology* 16, 61–68.

Ceulemans, R., Taylor, G., Bosac, C., Wilkins, D. and Besford, R.T. (1997) Photosynthetic acclimation to elevated CO_2 in poplar grown in glasshouse cabinets or in open top chambers depends on duration of exposure. *Journal of Experimental Botany* 48, 1681–1689.

Chaves, M.M. and Oliveira, M.M. (2004) Mechanisms underlying plant resilience to water deficits: prospects for water-saving agriculture. *Journal of Experimental Botany* 55, 2365–2384.

Chen, S., Wang, S., Altman, A. and Hüttermann, A. (1997) Genotypic variation in drought tolerance of poplar in relation to abscisic acid. *Tree Physiology* 17, 797–803.

Chen, S., Li, J., Wang, S., Hüttermann, A. and Altman, A. (2001) Salt, nutrient uptake and transport, and ABA of *Populus euphratica*; a hybrid in response to increasing soil NaCl. *Trees – Structure and Function* 15, 186–194.

Chen, S., Li, J., Fritz, E., Wang, S. and Hüttermann, A. (2002a) Sodium and chloride distribution in roots and transport in three poplar genotypes under increasing NaCl stress. *Forest Ecology and Management* 168, 217–230.

Chen, S.L., Wang, S.S., Hüttermann, A. and Altman, A. (2002b) Xylem abscisic acid accelerates leaf abscission by modulating polyamine and ethylene synthesis in water-stressed intact poplar. *Trees – Structure and Function* 16, 16–22.

Chen, S., Li, J., Wang, S., Fritz, E., Hüttermann, A. and Altman, A. (2003a) Effects of NaCl on shoot growth, transpiration, ion compartimentation, and transport in regenerated plants of *Populus euphratica* and *Populus tomentosa. Canadian Journal of Forest Research* 33, 967–975.

Chen, Y.N., Chen, Y.P., Li, W.H. and Zhang, H.F. (2003b) Response of the accumulation of proline in the bodies of *Populus euphratica* to the change of groundwater level at the lower reaches of Tarim River. *Chinese Science Bulletin* 48, 1995–1999.

Chen, Y.N., Li, W.H., Chen, Y.P., Zhang, H.F. and Zhuang, L. (2004) Physiological response of natural plants to the change of groundwater level in the lower reaches of Tarim River, Xinjiang. *Progress in Natural Science* 14, 975–983.

Clary, W.P. (1999) Stream channel and vegetation responses to late spring cattle grazing. *Journal of Range Management* 52, 218–227.

Cochard, H., Ridolfi, M. and Dreyer, E. (1996) Responses to water stress in an ABA-unresponsive hybrid poplar (*Populus koreana* × *trichocarpa* cv. Peace). II. Hydraulic properties and xylem embolism. *New Phytologist* 134, 455–461.

Coleman, M.D., Dickson, R.E., Isebrands, J.G. and Karnosky, D.F. (1995) Photosynthetic productivity of aspen clones varying in sensitivity to tropospheric ozone. *Tree Physiology* 15, 585–592.

Cooper, D.J., D'Amico, D.R. and Scott, M.L. (2003a) Physiological and morphological response patterns of *Populus deltoides* to alluvial groundwater pumping. *Environmental Management* 31, 215–226.

Cooper, D.J., Andersen, D.C. and Chimner, R.A. (2003b) Multiple pathways for woody plant establishment on floodplains at local to regional scales. *Journal of Ecology* 91, 182–196.

Cooper, D.T. and Van Haverbeke, D.F. (1990) *Populus deltoides* Bartr. ex Marsh.: eastern cottonwood. In: Burns, R.M. and Honkala, B.H. (eds) *Silvics of North America: Volume 2, Hardwoods.* Agriculture Handbook 654. USDA Forest Service, Washington, DC, pp. 530–543.
Cosgrove, D.J. (1993) How do plant cell walls extend? *Plant Physiology* 102, 1–6.
Costa, P., Bahrman, N., Frigerio, J.M., Kremer, A. and Plomion, C. (1998) Water-deficit-responsive proteins in maritime pine. *Plant Molecular Biology* 38, 587–596.
Cotrufo, M.F., De Angelis, P. and Polle, A. (2005) Leaf litter production and decomposition in a poplar short rotation coppice exposed to free air CO_2 enrichment (POPFACE). *Global Change Biology* 11, 971–982.
Courtois, M., Boudouresque, E. and Guerrier, G. (1999) Comparative anatomical, morphological, and physiological parameters controlling photosynthesis in two *Populus* ×*euramericana* clones during short-term osmotic treatment. *Photosynthetica* 37, 87–96.
Cox, G., Fischer, D., Hart, S.C. and Whitham, T.G. (2005) Non-response of native cottonwood trees to water additions during summer drought. *Western North American Naturalist* 65, 175–185.
Curtis, P.S. and Teeri, J.A. (1992) Seasonal responses of leaf gas exchange to elevated carbon dioxide in *Populus grandidentata. Canadian Journal of Forest Research* 22, 1320–1325.
Curtis, P.S., Zak, D.R., Pregitzer, K.S. and Teeri, J.A. (1994) Above- and belowground response of *Populus grandidentata* to elevated atmospheric CO_2 and soil N availability. *Plant and Soil* 165, 45–51.
Curtis, P.S., Vogel, C.S., Pregitzer, K.S., Zak, D.R. and Teeri, J.A. (1995) Interacting effects of soil fertility and atmospheric CO_2 on leaf area-growth and carbon gain physiology in *Populus* ×*euramericana* (Dode) Guinier. *New Phytologist* 129, 253–263.
Curtis, P.S., Vogel, C.S., Wang, X., Pregitzer, K.S., Zak, D.R., Lussenhop, J., *et al.* (2000) Gas exchange, leaf nitrogen, and growth efficiency of *Populus tremuloides* in a CO_2-enriched atmosphere. *Ecological Applications* 10, 3–17.
Dale, J.E. (1988) The control of leaf expansion. *Annual Review of Plant Physiology and Plant Molecular Biology* 39, 267–295.
Dang, Q.L. and Cheng, S. (2004) Effects of soil temperature on ecophysiological traits in seedlings of four boreal tree species. *Forest Ecology and Management* 194, 379–387.
Davies, W.J. and Zhang, J. (1991) Root signal and the regulation of growth and development of plant in drying soil. *Annual Review of Plant Physiology and Plant Molecular Biology* 42, 55–76.
DeBell, D.S. (1990) *Populus trichocarpa* Torr. and Gray: black cottonwood. In: Burns, R.M. and Honkala, B.H. (eds) *Silvics of North America: Volume 2, Hardwoods.* Agriculture Handbook 654. USDA Forest Service,Washington, DC, pp. 570–576.
Demmig-Adams, B. and Adams, W.W. III (1996) Xanthophyll cycle and light stress in nature: uniform response to excess direct sunlight among higher plant species. *Planta* 198, 460–470.
Demmig-Adams, B., Adams, W.W. III, Barker, D.H., Logan, B.A., Bowling, D.R. and Verhoeven, A.S. (1996) Using chlorophyll fluorescence to assess the fraction of absorbed light allocated to thermal dissipation of excess excitation. *Physiologia Plantarum* 98, 253–264.
Dickmann, D.I. and Stuart, K.W. (1983) *The Culture of Poplars in Eastern North America.* Michigan State University, East Lansing, Michigan.
Dickson, R.E., Coleman, M.D., Riemenschneider, D.E., Isebrands, J.G., Hogan, G.E. and Karnosky, D.F. (1998) Growth of five hybrid poplar genotypes exposed to interacting elevated CO_2 and O_3. *Canadian Journal of Forest Research* 28, 1706–1716.
Dickson, R.E., Coleman, M.D., Pechter, P. and Karnosky, D.F. (2001) Growth and crown architecture of two aspen genotypes exposed to interacting ozone and carbon dioxide. *Environmental Pollution* 115, 319–334.
Dong, X. and Zhang, X. (2001) Some observations of the adaptations of sandy shrubs to the arid environment in the Mu Us Sandland: leaf water relations and anatomic features. *Journal of Arid Environments* 48, 41–48.
Dreyer, E., Bogeat-Triboulot, M.B., Le Thiec, D., Guehl, J.M., Brignolas, F., Villar, M., *et al.* (2004) Tolérance des peupliers à la sécheresse: peut-on espérer l'améliorer? *Biofutur* 247, 54–58.
Dudek, D.M., McClenahen, J.R. and Mitsch, W.J. (1998) Tree growth responses of *Populus deltoides* and *Juglans nigra* to streamflow and climate in a bottomland hardwood forest in central Ohio. *American Midland Naturalist* 140, 233–244.
Dunlap, J.M. and Stettler, R.F. (2001) Variation in leaf epidermal and stomatal traits of *Populus trichocarpa* from two transects across the Washington Cascades. *Canadian Journal of Botany* 79, 528–536.
Edjolo, A., Laffray, D. and Guerrier, G. (2001) The ascorbate-glutathione cycle in the cytosolic and chloroplastic fractions of drought-tolerant and drought-sensitive poplars. *Journal of Plant Physiology* 158, 1511–1517.

Edreva, A. (2005) Generation and scavenging of reactive oxygen species in chloroplasts: a submolecular approach. *Agriculture, Ecosystems and Environment* 106, 119–133.
El-Khatib, R.T., Hamerlynck, E.P., Gallardo, F. and Kirby, E.G. (2004) Transgenic poplar characterized by ectopic expression of a pine cytosolic glutamine synthetase gene exhibits enhanced tolerance to water stress. *Tree Physiology* 24, 729–736.
Engel, V.C., Griffin, K.L., Murthy, R., Patterson, L., Klimas, C. and Potosnak, M. (2004) Growth CO_2 concentration modifies the transpiration response of *Populus deltoides* to drought and vapor pressure deficit. *Tree Physiology* 24, 1137–1145.
Epron, D. (1997) Effects of drought on photosynthesis and on the thermotolerance of photosystem II in seedlings of cedar (*Cedrus atlantica* and *C. libani*). *Journal of Experimental Botany* 48, 1835–1841.
Epstein, E. (1998) How calcium enhances plant salt tolerance. *Science* 280, 1906–1907.
Erickson, T., Rytter, L. and Linder, S. (1992) Nutritional dynamics and requirements of short rotation forests. In: Mitchell, C.P., Ford-Robertson, J.B., Hinckley, T. and Sennerby-Forsse, L. (eds) *Ecophysiology of Short Rotation Forest Crops.* Elsevier Applied Science, London, pp. 35–58.
FAO (1980) *Poplars and Willows in Wood Production and Land Use.* Food and Agriculture Organization of the United Nations, Rome.
FAO (2002) *Afforestation, Forestry Research, Planning and Development in the Three North Region of China.* Technical Project Review Document (1991–2002). Food and Agriculture Organization of the United Nations, Rome (http://www.fao.org/DOCREP/005/AC613E/AC613E00.HTM, accessed 20 December 2011).
Fay, M.F., Lledó, M.D., Kornblum, M.M. and Crespo, M.B. (1999) From the waters of Babylon? *Biodiversity and Conservation* 8, 769–778.
Feng, G., Zhang, F.S., Li, X.L., Tian, C.Y., Tang, C. and Rengel, Z. (2002) Improved tolerance of maize plants to salt stress by arbuscular mycorrhiza is related to higher accumulation of soluble sugars in roots. *Mycorrhiza* 12, 185–190.
Ferris, R., Sabatti, M., Miglietta, F., Mills, R. and Taylor, G. (2001) Leaf area is stimulated in *Populus* by free-air CO_2 enrichment (POPFACE), through increased cell expansion and production. *Plant, Cell and Environment* 24, 305–316.
Fowler, D., Cape, J.N., Coyle, M., Flechard, C., Kuylenstierna, J., Hicks, K., *et al.* (1999) The global exposure of forests to air pollutants. *Water Air and Soil Pollution* 116, 5–32.
Francis, R.A., Gurnell, A.M., Petts, G.E. and Edwards, P.J. (2005) Survival and growth responses of *Populus nigra*, *Salix elaeagnos* and *Alnus incana* cuttings to varying levels of hydric stress. *Forest Ecology and Management* 210, 291–301.
Frey, B.R., Lieffers, V.J., Hogg, E.H. and Landhausser, S.M. (2004) Predicting landscape patterns of aspen dieback: mechanisms and knowledge gaps. *Canadian Journal of Forest Research* 34, 1379–1390.
Froux, F., Ducrey, M., Epron, D. and Dreyer, E. (2004) Seasonal variations and acclimation potential of the thermostability in four Mediterranean conifers. *Annals of Forest Science* 61, 235–241.
Fung, L.E., Wang, S., Altman, A. and Hüttermann, A. (1998) Effect of NaCl on growth, photosynthesis, ion and water relations of four poplar genotypes. *Forest Ecology and Management* 107, 135–146.
Funk, J.L., Mak, J.E. and Lerdau, M.T. (2004) Stress-induced changes in carbon sources for isoprene production in *Populus deltoides. Plant Cell and Environment* 27, 747–755.
Gafur, A., Schützendübel, A., Langenfeld-Heyser, R., Fritz, E. and Polle, A. (2004) Compatible and incompatible *Paxillus involutus* isolates for ectomycorrhization *in vitro* with poplar (*Populus* ×*canescens*) differ in H_2O_2 production. *Plant Biology* 6, 91–99.
Gagnon, Z.E., Karnosky, D.F., Dickson, R.E. and Isebrands, J.G. (1992) Effect of ozone on chlorophyll content in *Populus tremuloides. American Journal of Botany* 79, 107–121.
Garg, A.K., Kim, J.K., Owens, T.G., Ranwala, A.P., Choi, Y.D., Kochian, L.V., *et al.* (2002) Trehalose accumulation in rice plants confers high tolerance levels to different abiotic stresses. *Proceedings of the National Academy of Sciences of the USA* 99, 15898–15903.
Gebre, G.M. and Kuhns, M.R. (1991) Seasonal and clonal variations in drought tolerance of *Populus deltoides. Canadian Journal of Forest Research* 21, 910–916.
Gebre, G.M., Kuhns, M.R. and Brandle, J.R. (1994) Organic solute accumulation and dehydration tolerance in three water-stressed *Populus deltoides* clones. *Tree Physiology* 14, 575–587.
Gebre, G.M., Brandle, J.R. and Kuhns, M.R. (1997) Influence of rewatering and time of sampling on solute accumulation of two *Populus deltoides* clones. *Tree Physiology* 17, 341–346.
Gebre, G.M., Tschaplinski, T.J., Tuskan, G.A. and Todd, D.E. (1998) Clonal and seasonal differences in leaf osmotic potential and organic solutes of five hybrid poplar clones grown under field conditions. *Tree Physiology* 18, 645–652.

Gellini, R. (1980) *Botanica Forestale.* CEDAM, Padova, Italy.
Gielen, B. and Ceulemans, R. (2001) The likely impact of rising atmospheric CO_2 on natural and managed *Populus*: a literature review. *Environmental Pollution* 115, 335–358.
Gielen, B., Calfapietra, C., Sabatti, M. and Ceulemans, R. (2001) Leaf area dynamics in a closed poplar plantation under free-air carbon dioxide enrichment. *Tree Physiology* 21, 1245–1255.
Gielen, B., Calfapietra, C., Claus, A., Sabatti, M. and Ceulemans, R. (2002) Crown architecture of *Populus* spp. is differentially modified by free-air CO_2 enrichment (POPFACE). *New Phytologist* 153, 91–100.
Gielen, B., Liberloo, M., Bogaert, J., Calfapietra, C., De Angelis, P., Miglietta, F., *et al.* (2003a) Three years of free-air CO_2 enrichment (POPFACE) only slightly affect profiles of light and leaf characteristics in closed canopies of *Populus. Global Change Biology* 9, 1022–1037.
Gielen, B., Scarascia-Mugnozza, G. and Ceulemans, R. (2003b) Stem respiration of *Populus* species in the third year of free-air CO_2 enrichment. *Physiologia Plantarum* 117, 500–507.
Gielen, B., Calfapietra, C., Lukac, M., Wittig, V.E., De Angelis, P., Janssens, I.A., *et al.* (2005) Net carbon storage in a poplar plantation (POPFACE) after three years of free-air CO_2 enrichment. *Tree Physiology* 25, 1399–1408.
Glenn, E.P. and Nagler, P.L. (2005) Comparative ecophysiology of *Tamarix ramosissima* and native trees in western US riparian zones. *Journal of Arid Environments* 61, 419–446.
Glynn, C., Herms D.A., Egawa, M., Hansen, R. and Mattson, W.J. (2003) Effects of nutrient availability on biomass allocation as well as constitutive and rapid induced herbivore resistance in poplar. *Oikos* 101, 385–397.
Glynn, C., Rönnberg-Wästljung, A.C., Julkunen-Tiitto, R. and Weih, M. (2004) Willow genotype, but not drought treatment, affects foliar phenolic concentrations and leaf-beetle resistance. *Entomologia Experimentalis et Applicata* 113, 1–14.
Gries, D., Zeng, F., Foetzki, A., Arndt, S.K., Bruelheide, H., Thomas, F.M., *et al.* (2003) Growth and water relations of *Tamarix ramosissima* and *Populus euphratica* on Taklamakan desert dunes in relation to depth to a permanent water table. *Plant, Cell and Environment* 26, 725–736.
Griffin, D.H., Schaedle, M., DeVit, M.J. and Manion, P.D. (1991a) Clonal variation of *Populus tremuloides* responses to diurnal drought stress. *Tree Physiology* 8, 297–304.
Griffin, D.H., Schaedle, M., Manion, P.D. and DeVit, M. (1991b) Clonal variation in amino acid contents of roots, stems, and leaves of aspen (*Populus tremuloides* Michx.) as influenced by diurnal drought stress. *Tree Physiology* 8, 337–350.
Gu, R., Fonseca, S., Puskas, L.G., Hackler, L. Jr, Zvara, A., Dudits, D., *et al.* (2004) Transcript identification and profiling during salt stress and recovery of *Populus euphratica. Tree Physiology* 24, 265–276.
Guerrier, G., Brignolas, F., Thierry, C., Courtois, M. and Kahlem, G. (2000) Organic solutes protect drought-tolerant *Populus* ×*euramericana* against reactive oxygen species. *Journal of Plant Physiology* 156, 93–99.
Gueta-Dahan, Y., Yaniv, Z., Zilinskas, B.A. and Ben-Hayyim, G. (1997) Salt and oxidative stress: similar and specific responses and their relation to salt tolerance in *Citrus. Planta* 203, 460–469.
Guignard, C., Jouve, L., Bogeat-Triboulot, M.B., Dreyer, E., Hausman, J.F. and Hoffmann, L. (2005) Analysis of carbohydrates in plants by high-performance anion-exchange chromatography coupled with electrospray mass spectrometry. *Journal of Chromatography A* 1085, 137–142.
Hacke, U.G., Stiller, V., Sperry, J.S., Pittermann, J. and McCulloh, K.A. (2001) Cavitation fatigue. Embolism and refilling cycles can weaken the cavitation resistance of xylem. *Plant Physiology* 125, 779–786.
Hale, B.K., Herms, D.A., Hansen, R.C., Clausen, T.P. and Arnold, D. (2005) Effects of drought stress and nutrient availability on dry matter allocation, phenolic glycosides, and rapid induced resistance of poplar to two Lymantriid defoliators. *Journal of Chemical Ecology* 31, 2601–2620.
Hall, J.L. (2002) Cellular mechanisms for heavy metal detoxification and tolerance. *Journal of Experimental Botany* 366, 1–11.
Hall, R.L., Allen, S.J., Rosier, P.T.W. and Hopkins, R. (1998) Transpiration from coppiced poplar and willow measured using sap-flow methods. *Agricultural and Forest Meteorology* 90, 275–290.
Hamada, A., Hibino, T., Nakamura, T. and Takabe, T. (2001) Na^+/H^+ antiporter from *Synechocystis* species PCC 6803, homologous to SOS1, contains an aspartic residue and long C-terminal tail important for the carrier activity. *Plant Physiology* 125, 437–446.
Harvey, H.P. and van den Driessche, R. (1997) Nutrition, xylem cavitation and drought resistance in hybrid poplar. *Tree Physiology* 17, 647–654.
Harvey, H.P. and van den Driessche, R. (1999) Nitrogen and potassium effects on xylem cavitation and water-use efficiency in poplars. *Tree Physiology* 19, 943–950.
Hasegawa, P.M., Bressan, R.A., Zhu, J.K. and Bohnert, H.J. (2000) Plant cellular and molecular responses to high salinity. *Annual Review of Plant Physiology and Plant Molecular Biology* 51, 463–499.

Havaux, M. (1993) Rapid photosynthetic adaptation to heat stress triggered in potato leaves by moderately elevated temperatures. *Plant, Cell and Environment* 16, 461–467.

Heilman, P.E., Stettler, R.F., Hanley, D.P. and Carkner, R.W. (1995) High yield hybrid poplar plantations in the Pacific Northwest. Cooperative Extension Bulletin PNW356. Washington State University, Pullman, Washington.

Hessl, A.E. and Graumlich, L.J. (2002) Interactive effects of human activities, herbivory and fire on quaking aspen (*Populus tremuloides*) age structures in western Wyoming. *Journal of Biogeography* 29, 889–902.

Hinckley, T.M. and Braatne, J.H. (1994) Stomata. In: Wilkinson, R.E. (ed.) *Plant–Environment Interactions.* Marcel Dekker, New York, pp. 323–355.

Hogg, E.H. and Wein, R.W. (2005) Impacts of drought on forest growth and regeneration following fire in southwestern Yukon, Canada. *Canadian Journal of Forest Research* 35, 2141–2150.

Hogg, E.H., Saugier, B., Pontailler, J.Y., Black, T.A., Chen, W., Hurdle, P.A., *et al.* (2000) Responses of trembling aspen and hazelnut to vapor pressure deficit in a boreal deciduous forest. *Tree Physiology* 20, 725–734.

Hogg, E.H., Brandt, J.P. and Kochtubajda, B. (2002) Growth and dieback of aspen forests in northwestern Alberta, Canada, in relation to climate and insects. *Canadian Journal of Forest Research* 32, 823–832.

Hogg, E.H., Brandt, J.P. and Kochtubajda, B. (2005) Factors affecting interannual variation in growth of western Canadian aspen forests during 1951–2000. *Canadian Journal of Forest Research* 35, 610–622.

Holmes, W.E., Zak, D.R., Pregitzer, K.S. and King, J.S. (2003) Soil nitrogen transformations under *Populus tremuloides, Betula papyrifera* and *Acer saccharum* following 3 years exposure to elevated CO_2 and O_3. *Global Change Biology* 9, 1743–1750.

Holton, M.K., Lindroth, R.L. and Nordheim, E.V. (2003) Foliar quality influences tree-herbivore-parasitoid interactions: effects of elevated CO_2, O_3, and genotype. *Oecologia* 137, 233–244.

Hoosbeek, M.R., Lukac, M., van Dam, D., Godbold, D.L., Velthorst, E.J., Biondi, F.A., *et al.* (2004) More new carbon in the mineral soil of a poplar plantation under Free Air Carbon Enrichment (POPFACE): cause of increased priming effect? *Global Biogeochemical Cycles* 18, GB1040.

Horton, J.L. and Clark, J.L. (2001) Water table decline alters growth and survival of *Salix gooddingii* and *Tamarix chinensis* seedlings. *Forest Ecology and Management* 140, 239–247.

Horton, J.L., Kolb, T.E. and Hart, S.C. (2001a) Physiological response to groundwater depth varies among species and with river flow regulation. *Ecological Applications* 11, 1046–1059.

Horton, J.L., Kolb, T. and Hart, S.C. (2001b) Leaf gas exchange characteristics differ among Sonoran Desert riparian tree species. *Tree Physiology* 21, 233–241.

Horton, J.L., Kolb, T.E. and Hart, S.C. (2001c) Responses of riparian trees to interannual variation in ground water depth in a semi-arid river basin. *Plant, Cell and Environment* 24, 293–304.

Horton, J.L., Hart, S.C. and Kolb, T.E. (2003) Physiological condition and water source use of Sonoran Desert riparian trees at the Bill Williams River, Arizona, USA. *Isotopes in Environmental and Health Studies* 39, 69–82.

Huang, Q., Yuan, Q. and Dong, M. (1992) Tissue culture of *Populus euphratica. Forest Research* 5, 125–130. (In Chinese)

Hughes, F.M.R., Barsoum, N., Richards, K.S., Winfield, M. and Hayes, A. (2000) The response of male and female black poplar (*Populus nigra* L. subspecies *betulifolia* (Pursh) W. Wettst.) cuttings to different water table depths and sediment types: implications for flow management and river corridor biodiversity. *Hydrological Processes* 14, 3075–3098.

Hukin, D., Cochard, H., Dreyer, E., Le Thiec, D. and Bogeat-Triboulot, M.B. (2005) Cavitation vulnerability in roots and shoots: does *Populus euphratica* Oliv., a poplar from arid areas of Central Asia, differ from other poplar species? *Journal of Experimental Botany* 56, 2003–2010.

Ibrahim, L., Proe, M.F. and Cameron, A.D. (1997) Main effects of nitrogen supply and drought stress upon whole-plant carbon allocation in poplar. *Canadian Journal of Forest Research* 27, 1412–1419.

Ibrahim, L., Proe, M.F. and Cameron, A.D. (1998) Interactive effects of nitrogen and water availabilities on gas exchange and whole-plant carbon allocation in poplar. *Tree Physiology* 18, 481–487.

IPCC (1996) Climate change 1995. In: Houghton, J.T., Meira Filho, L.G., Callader, B.A., Harris, N., Kattenberg, A. and Maskell, K. (eds) *The Science of Climate Change.* Cambridge University Press, New York, 572 pp.

IPCC (2007) Summary for policymakers. In: Solomon, S., Qin, D., Manning, M., Chen, Z., Marquis, M., Averyt, K.B., *et al.* (eds) *Climate Change 2007: the Physical Science Basis. Contribution of Working Group I to the Fourth Assessment Report of the Intergovernmental Panel on Climate Change.* Cambridge University Press, New York, 996 pp.

Isebrands, J.G., McDonald, E.P., Kruger, E., Hendrey, G., Percy, K., Pregitzer, K., *et al.* (2001) Growth responses of *Populus tremuloides* clones to interacting elevated carbon dioxide and tropospheric ozone. *Environmental Pollution* 115, 359–371.

Ivanov, A.G., Kitcheva, M.I., Christov, A.M. and Popova, L.P. (1992) Effects of abscisic acid treatment on the thermostability of the photosynthetic apparatus in barley chloroplasts. *Plant Physiology* 98, 1228–1232.

Johnson, J.D., Tognetti, R. and Paris, P. (2002) Water relations and gas exchange in poplar and willow under water stress and elevated atmospheric CO_2. *Physiologia Plantarum* 115, 93–100.

Jones, H.G. (1992) Drought and drought tolerance. In: Jones, H.G. (ed.) *Plants and Microclimate: a Quantitative Approach to Environmental Plant Physiology.* Cambridge University Press, Cambridge, UK, pp. 212–237.

Jones, H.G. (1993) Drought tolerance and water-use efficiency. In: Smith, J.A.C. and Griffiths, H. (eds) *Water Deficit: Plant Responses From Cell To Community.* Bios Scientific Publishers, Oxford, UK, pp. 193–201.

Jones, H.G. (2004) What is water use efficiency? In: Bacon, M.A. and Roberts, J.A. (eds) *Water Use Efficiency in Plant Biology.* Blackwell Publishing, CRC Press, Oxford, UK, pp. 27–41.

Kalina, J. and Ceulemans, R. (1997) Clonal differences in the response of dark and light reactions of photosynthesis to elevated atmospheric CO_2 in poplar. *Photosynthetica* 33, 51–61.

Kang, J.M., Kojima, K., Ide, Y. and Sasaki, S. (1996) Growth response to the stress of low osmotic potential, salinity and high pH in cultured shoot of Chinese poplars. *Journal of Forest Research* 1, 27–29.

Karakas, B., Ozias-Akins, P., Stushnoff, C., Suefferheld, M. and Rieger, M. (1997) Salinity and drought tolerance in mannitol-accumulating transgenic tobacco. *Plant, Cell and Environment* 20, 609–616.

Karnosky, D.F. (1976) Threshold levels for foliar injury to *Populus tremuloides* Michx. by sulfur dioxide and ozone. *Canadian Journal of Forest Research* 6, 166–169.

Karnosky, D.F. (1977) Evidence of genetic control of response to sulfur dioxide and ozone in *Populus tremuloides* Michx. *Canadian Journal of Forest Research* 7, 437–440.

Karnosky, D.F., Gagnon, Z.E., Dickson, R.E., Coleman, M.D., Lee, E.H. and Isebrands, J.G. (1996) Changes in growth, leaf abscission, and biomass associated with seasonal tropospheric ozone exposures of *Populus tremuloides* clones and seedlings. *Canadian Journal of Forest Research* 26, 23–37.

Karnosky, D.F., Podila, G.K., Gagnon, Z., Pechter, P., Akkapeddi, A., Sheng, Y., *et al.* (1998) Genetic control of responses to interacting tropospheric ozone and CO_2 in *Populus tremuloides. Chemosphere* 36, 807–812.

Karnosky, D.F., Percy, K.E., Xiang, B., Callan, B., Noormets, A., Mankovska, B., *et al.* (2002) Interacting elevated CO_2 and tropospheric O_3 predisposes aspen (*Populus tremuloides* Michx.) to infection by rust (*Melampsora medusae* f. sp. *tremuloidae*). *Global Change Biology* 8, 329–338.

Karnosky, D.F., Percy, K., Mankovska, B., Prichard, T., Noormets, A., Dickson, R.E., *et al.* (2003a) Ozone affects the fitness of trembling aspen. In: Karnosky, D.F., Percy, K.E., Chappelka, A.H., Simpson, C. and Pikkarainen, J.M. (eds) *Air Pollution, Global Change and Forests in the New Millennium.* Elsevier Press, Amsterdam, pp. 199–209.

Karnosky, D.F., Zak, D.R., Pregitzer, K.S., Awmack, C.S., Bockheim, J.G., Dickson, R.E., *et al.* (2003b) Tropospheric O_3 moderates responses of temperate hardwood forests to elevated CO_2: a synthesis of molecular to ecosystem results from the Aspen FACE project. *Functional Ecology* 17, 289–304.

Karnosky, D.F., Pregitzer, K.S., Zak, D.R., Kubiske, M.E., Hendrey, G.R., Weinstein, D., *et al.* (2005) Scaling ozone responses of forest trees to the ecosystem level in a changing climate. *Plant, Cell and Environment* 28, 965–981.

Khan, D. (1974) The occurrence of mycorrhizae in halophytes, hydrophytes and xerophytes of endogone spores in adjacent soils. *Journal of General Microbiology* 81, 7–14.

Khasa, P.D., Chackravarty, P., Robertson, A., Thomas, B.R. and Dancik, B.P. (2002) The mycorrhizal status of selected poplar clones introduced into Alberta. *Biomass and Bioenergy* 22, 99–104.

King, J.S., Pregitzer, K.S., Zak, D.R., Kubiske, M.E., Ashby, J.A. and Holmes, W.E. (2001a) Chemistry and decomposition of litter from *Populus tremuloides* Michaux grown at elevated atmospheric CO_2 and varying N availability. *Global Change Biology* 7, 65–74.

King, J.S., Pregitzer, K.S., Zak, D.R., Sober, J., Isebrands, J.G., Dickson, R.E., *et al.* (2001b) Fine-root biomass and fluxes of soil carbon in young stands of paper birch and trembling aspen as affected by elevated atmospheric CO_2 and tropospheric O_3. *Oecologia* 128, 237–250.

King, J.S., Hanson, P.J., Bernhardt, E., De Angelis, P., Norby, R.J. and Pregitzer, K.S. (2004) A multiyear synthesis of soil respiration responses to elevated atmospheric CO_2 from four forest FACE experiments. *Global Change Biology* 10, 1027–1042.

King, J.S., Kubiske, M.E., Pregitzer, K.S., Hendrey, G.R., McDonald, E.P., Giardina, C.P., *et al.* (2005) Tropospheric O_3 compromises net primary production in young stands of trembling aspen, paper birch and sugar maple in response to elevated atmospheric CO_2. *New Phytologist* 168, 623–636.

Kishor, P.B.K., Hong, Z., Miao, G.H., Hu, C.A.A. and Verma, D.P.S. (1995) Overexpression of D-pyrroline-5-carboxylate synthetase increases proline production and confers osmotolerance in transgenic plants. *Plant Physiology* 108, 1387–1394.

Knight, C.A. and Ackerly, D.D. (2002) An ecological and evolutionary analysis of photosynthetic thermotolerance using the temperature-dependent increase of fluorescence. *Oecologia* 130, 505–514.

Kopper, B.J. and Lindroth, R.L. (2003) Effects of elevated carbon dioxide and ozone on the phytochemistry of aspen and performance of an herbivore. *Oecologia* 134, 95–103.

Kranjcec, J., Mahoney, J.M. and Rood, S.B. (1998) The responses of three riparian cottonwood species to water table decline. *Forest Ecology and Management* 110, 77–87.

Kubiske, M.E., Pregitzer, K.S., Mikan, C.J., Zak, D.R., Maziasz, J.L. and Teeri, J.A. (1997) *Populus tremuloides* photosynthesis and crown architecture in response to elevated CO_2 and soil N availability. *Oecologia* 110, 328–336.

Kubiske, M.E., Pregitzer, K.S., Zak, D.R. and Mikan, C.J. (1998) Growth and C allocation of *Populus tremuloides* genotypes in response to atmospheric CO_2 and soil N availability. *New Phytologist* 140, 251–260.

Kull, O., Sober, A., Coleman, M.D., Dickson, R.E., Isebrands, J.G., Gagnon, Z., *et al.* (1996) Photosynthetic responses of aspen clones to simultaneous exposures of ozone and CO_2. *Canadian Journal of Forest Research* 26, 639–648.

Kundzewicz, Z.W., Mata, L.J., Arnell, N.W., Döll, P., Kabat, P., Jiménez, B., *et al.* (2007) Freshwater resources and their management. In: Parry, M.L., Canziani, O.F., Palutikof, J.P., van der Linden, P.J. and Hanson, C.E. (eds) *Climate Change 2007: Impacts, Adaptation and Vulnerability. Contribution of Working Group II to the Fourth Assessment Report of the Intergovernmental Panel on Climate Change.* Cambridge University Press, Cambridge, UK, pp. 173–210.

Ladjal, M., Epron, D. and Ducrey, M. (2000) Effects of drought preconditioning on thermotolerance of photosystem II and susceptibility of photosynthesis to heat stress in cedar seedlings. *Tree Physiology* 20, 1235–1241.

Lambs, L. and Muller, E. (2002) Sap flow and water transfer in the Garonne River riparian woodland, France: first results on poplar and willow. *Annals of Forest Science* 59, 301–315.

Lamers, J.P.A., Khamzina, A. and Worbes, M. (2006) The analyses of physiological and morphological attributes of 10 tree species for early determination of their suitability to afforest degraded landscapes in the Aral Sea Basin of Uzbekistan. *Forest Ecology and Management* 221, 249–259.

Landhäusser, S.M., Silins, U., Lieffers, V.J. and Liu, W. (2003) Response of *Populus tremuloides*, *Populus balsamifera*, *Betula papyrifera* and *Picea glauca* seedlings to low soil temperature and waterlogged soil conditions. *Scandinavian Journal of Forest Research* 18, 391–400.

Langenfeld-Heyser, R., Gao, J., Ducic, T., Tachd, P., Lu, C.F., Fritz, E., *et al.* (2007) *Paxillus involutus* mycorrhiza attenuate NaCl-stress responses in the salt-sensitive hybrid poplar *Populus ×canescens*. *Mycorrhiza* 17, 221–231.

Leffler, A.J. and Evans, A.S. (2001) Physiological variation among *Populus fremontii* populations: short- and long-term relationships between $\delta^{13}C$ and water availability. *Tree Physiology* 21, 1149–1155.

Leffler, A.J., England, L.E. and Naito, J. (2000) Vulnerability of fremont cottonwood (*Populus fremontii* Wats.) individuals to xylem cavitation. *Western North American Naturalist* 60, 204–210.

Li, C.Y., Yin, C.Y. and Liu, S.R. (2004a) Different responses of two contrasting *Populus davidiana* populations to exogenous abscisic acid application. *Environmental and Experimental Botany* 51, 237–246.

Li, S., Pezeshki, S.R., Goodwin, S. and Shields, P.D. Jr (2004b) Physiological responses of black willow (*Salix nigra*) to a range of soil moisture regimes. *Photosynthetica* 42, 585–590.

Li, Y.G., Li, L.H., Jiang, G.M., Niu, S.L., Liu, M.Z., Gao, L.M., *et al.* (2004c) Traits of chlorophyll fluorescence in 99 plant species from the sparse-elm grassland in Hunshandak Sandland. *Photosynthetica* 42, 243–249.

Liang, H. and Chang, S.X. (2004) Response of trembling and hybrid aspens to phosphorus and sulfur fertilization in a Gray Luvisol: growth and nutrient uptake. *Canadian Journal of Forest Research* 34, 1391–1399.

Liberloo, M., Gielen, B., Calfapietra, C., Veys, C., Pigliacelli, R., Scarascia-Mugnozza, G., *et al.* (2004) Growth of a poplar short rotation coppice under elevated atmospheric CO_2 concentrations (EUROFACE) depends on fertilization and species. *Annals of Forest Science* 61, 299–307.

Liberloo, M., Dillen, S., Calfapietra, C., Marinari, S., Luo, Z.B., De Angelis, P., *et al.* (2005) Elevated CO_2 concentration, fertilization and their interaction: growth stimulation in a short-rotation poplar coppice (EUROFACE). *Tree Physiology* 25, 179–189.

Lite, S.J. and Stromberg, J.C. (2005) Surface water and ground-water thresholds for maintaining *Populus-Salix* forests, San Pedro, Arizona. *Biological Conservation* 125, 153–167.

Liu, Z. and Dickmann, D.I. (1992a) Abscisic acid accumulation in leaves of two contrasting hybrid polar clones affected by nitrogen fertilization plus cyclic flooding and soil drying. *Tree Physiology* 11, 109–122.

Liu, Z. and Dickmann, D.I. (1992b) Responses of two hybrid *Populus* clones to flooding, drought, and nitrogen availability. I. Growth and morphology. *Canadian Journal of Botany* 70, 2265–2270.

Liu, Z. and Dickmann, D.I. (1993) Responses of two hybrid *Populus* clones to flooding, drought, and nitrogen availability. II. Gas exchange and water relations. *Canadian Journal of Botany* 71, 927–938.

Liu, Z. and Dickmann, D.I. (1996) Effects of water and nitrogen interaction on net photosynthesis, stomatal conductance, and water-use efficiency in two hybrid poplar clones. *Physiologia Plantarum* 97, 507–512.

Lockhart, J.A. (1965) Cell extension. In: Bonner, J. and Varner, J.E. (eds) *Plant Biochemistry*. Academic Press, New York, pp. 826–849.

Lodge, D.J. (1989) The influence of soil moisture and flooding on formation of VA, endo- and ectomycorrhizae in *Populus* and *Salix*. *Plant and Soil* 117, 243–253.

Loewenstein, N.J. and Pallardy, S.G. (1998) Drought tolerance, xylem sap abscisic acid and stomatal conductance during soil drying: a comparison of canopy trees of three temperate deciduous angiosperms. *Tree Physiology* 18, 431–439.

Logan, B.A. and Monson, R.K. (1999) Thermotolerance of leaf disks from four isoprene-emitting species is not enhanced by exposure to exogenous isoprene. *Plant Physiology* 120, 821–825.

Long, S.P. and Drake, B.G. (1992) Photosynthetic CO_2 assimilation and rising atmospheric CO_2 concentrations. In: Baker, N.R. and Thomas, H. (eds) *Topics in Photosynthesis*. Elsevier, Amsterdam, pp. 69–103.

Long, S.P., Humphries, S. and Falkowski, P.G. (1994) Photoinhibition of photosynthesis in nature. *Annual Review of Plant Physiology and Plant Molecular Biology* 45, 633–662.

Loveys, B.R., Atkinson, L.J., Sherlock, D.J., Roberts, R.L., Fitter, A.H. and Atkin, O.K. (2003) Thermal acclimation of leaf and root respiration: an investigation comparing inherently fast- and slow-growing plant species. *Global Change Biology* 9, 895–910.

Lower, S.S. and Orians, C.M. (2003) Soil nutrient and water availability interact to influence willow growth and chemistry but not leaf beetle performance. *Entomologia Expermentalis et Applicata* 107, 69–79.

Lu, W. and Zhang, W. (2000) The role and position of poplar development in the creation of protective forests in the Three North Region of China. *Protection Forest Science Technology* 2(43), 84–86. (In Chinese)

Lu, W., Zhang, W. and Bao, J. (2000) Poplar development and the Three North Shelterbelt project. *Protection Forest Science Technology* 5(360), 7–9. (In Chinese)

Lu, W., Song, B., Han, Y., Zhang, W. and Feng, Z. (2001) Preliminary study on evaluation of poplar clones for introduction to the Korgin sandy lands. *Scientia Silvae Sinica* 37(1), 72–82. (In Chinese)

Lu, W., Hu, M., Hu, J. and Yao, X. (2004) Discussion on severity and control of Asian longhorn beetle of poplar trees in the Three North Shelterbelt Program. *Protection Forest Science Technology* 1, 39–41. (In Chinese)

Lukac, M., Calfapietra, C. and Godbold, D.L. (2003) Production, turnover and mycorrhizal colonization of root systems of three *Populus* species grown under elevated CO_2 (POPFACE). *Global Change Biology* 9, 838–848.

Luo, Y., Canadell, J. and Mooney, H.A. (1999) Interactive effects of carbon dioxide and environmental stress on plants and ecosystems: a synthesis. In: Luo, Y. and Mooney, H.A. (eds) *Carbon Dioxide and Environmental Stress*. Academic Press, San Diego, California, pp. 393–408.

Luo, Z.B., Langenfeld-Heyser, R., Calfapietra, C. and Polle, A. (2005) Influence of free air CO_2 enrichment (EUROFACE) and nitrogen fertilization on the anatomy of juvenile wood of three poplar species after coppicing. *Trees – Structure and Function* 19, 109–118.

Lussenhop, J., Treonis, A., Curtis, P.S., Teeri, J.A. and Vogel, C.S. (1998) Response of soil biota to elevated atmospheric CO_2 in poplar model systems. *Oecologia* 113, 247–251.
Lytle, D.A. and Merritt, D.M. (2004) Hydrologic regimes and riparian forests: a structured population model for cottonwood. *Ecology* 85, 2493–2503.
Ma, H.C., Fung, L., Wang, S.S., Altman, A. and Hüttermann, A. (1997) Photosynthetic response of *Populus euphratica* to salt stress. *Forest Ecology and Management* 93, 55–61.
McDonald, E.P., Kruger, E.L., Riemenschneider, D.E. and Isebrands, J.G. (2002) Competitive status influences tree-growth responses to elevated CO_2 and O_3 in aggrading aspen stands. *Functional Ecology* 16, 792–801.
Mahoney, J.M. and Rood, S.B. (1991) A device for studying the influence of declining water table on poplar growth and survival. *Tree Physiology* 8, 305–314.
Marron, N. (2003) Ecophysiologie des peupliers euraméricains en réponse à la sécheresse. PhD thesis, University of Orléans, Orléans, France.
Marron, N., Delay, D., Petit, J.-M., Dreyer, E., Kahlem, G., Delmotte, F.M., *et al.* (2002) Physiological traits of two *Populus* ×*euramericana* clones, Luisa Avanzo and Dorskamp, during a water stress and re-watering cycle. *Tree Physiology* 22, 849–858.
Marron, N., Dreyer, E., Boudouresque, E., Delay, D., Petit, J.-M., Delmotte, F.M., *et al.* (2003) Impact of successive drought and re-watering cycles on growth and specific leaf area of two *Populus* ×*canadensis* (Moench) clones, 'Dorskamp' and 'Luisa Avanzo'. *Tree Physiology* 23, 1225–1235.
Marron, N., Villar, M., Dreyer, E., Delay, D., Boudouresque, E., Petit, J.-M., *et al.* (2005) Diversity of leaf traits related to productivity in 31 *Populus deltoides* × *Populus nigra* clones. *Tree Physiology* 25, 425–435.
Marron, N., Maury, S., Rinaldi, C. and Brignolas, F. (2006) Impact of drought and leaf development stage on enzymatic antioxidant system of two *Populus deltoides* × *nigra* clones. *Annals of Forest Science* 63, 323–327.
Marschner, H. (1995) *Mineral Nutrition of Higher Plants,* 2nd edn. Academic Press, London.
Mazzoleni, S. and Dickmann, D.I. (1988) Differential physiological and morphological responses of two hybrid *Populus* clones to water stress. *Tree Physiology* 4, 61–70.
Merritt, D.M. and Cooper, D.J. (2000) Riparian vegetation and channel change in response to river regulation: a comparative study of regulated and unregulated streams in the Green River Basin, USA. *Regulated Rivers: Research and Management* 16, 543–564.
Michael, D.A., Isebrands, J.G., Dickmann, D.I. and Nelson, N.D. (1988) Growth and development during the establishment year of two *Populus* clones with contrasting morphology and phenology. *Tree Physiology* 4, 139–152.
Miglietta, F., Peressotti, A., Vaccari, F.P., Zaldei, A., De Angelis, P. and Scarascia-Mugnozza, G. (2001) Free-air CO_2 enrichment (FACE) of a poplar plantation: the POPFACE fumigation system. *New Phytologist* 150, 465–476.
Mikan, C.J., Zak, D.R., Kubiske, M.E. and Pregitzer, K.S. (2000) Combined effects of atmospheric CO_2 and N availability on the belowground carbon and nitrogen dynamics of aspen mesocosms. *Oecologia* 124, 432–445.
Mittler, R. (2006) Abiotic stress, the field environment and stress combination. *Trends in Plant Science* 11, 15–19.
Monclus, R., Dreyer, E., Delmotte, F.M., Villar, M., Delay, D., Boudouresque, E., *et al.* (2005) Productivity, leaf traits and carbon isotope discrimination in 29 *Populus deltoides* × *P. nigra* clones. *New Phytologist* 167, 53–62.
Monclus, R., Dreyer, E., Villar, M., Delmotte, F.M., Delay, D., Petit, J.M., *et al.* (2006) Impact of drought on productivity and water-use efficiency in 29 genotypes of *Populus deltoides* Bartr. Ex Marsh. × *P. nigra* L. *New Phytologist* 169, 765–777.
Mondor, E.B., Tremblay, M.N., Awmack, C.S. and Lindroth, R.L. (2004) Divergent pheromone-mediated insect behaviour under global atmospheric change. *Global Change Biology* 10, 1820–1824.
Monteith, J.L. (1977) Climate and efficiency of crop production in Britain. *Philosophical Transactions of the Royal Society B: Biological Sciences* 281, 277–294.
Morabito, D. and Guerrier, G. (2000) The free oxygen radical scavenging enzymes and redox status in roots and leaves of *Populus* ×*euramericana* in response to osmotic stress, dessication and rehydration. *Journal of Plant Physiology* 157, 74–80.
Morgan, J.M. (1984) Osmoregulation and water stress in higher plants. *Annual Review of Plant Physiology* 35, 299–319.

Moscatelli, M.C., Lagomarsino, A., De Angelis, P. and Grego, S. (2005a) Seasonality of soil biological properties in a poplar plantation growing under elevated atmospheric CO_2. *Applied Soil Ecology* 30, 162–173.
Moscatelli, M.C., Lagomarsino, A., Marinari, S., De Angelis, P. and Grego, S. (2005b) Soil microbial indices as bioindicators of environmental changes in a poplar plantation. *Ecological Indicators* 5, 171–179.
Munns, R. (2005) Genes and salt tolerance: bringing them together. *New Phytologist* 167, 645–663.
Murashige, T. and Skoog, F. (1962) A revised medium for rapid growth and bioassays with tobacco tissue cultures. *Physiologia Plantarum* 15, 473–497.
Murthy, R., Barron-Gafford, G., Dougherty, P.M., Engel, V.C., Grieve, K., Handley, L., *et al.* (2005) Increased leaf area dominates carbon flux response to elevated CO_2 in stands of *Populus deltoides* (Bartr.). *Global Change Biology* 11, 716–731.
Musin, T.M. and Zwiazek, J.J. (2002) Colonization with *Hebeloma crustuliniforme* increases water conductance and limits shoot sodium uptake in white spruce (*Picea glauca*) seedlings. *Plant and Soil* 238, 217–225.
Nagler, P.L., Glenn, E.P. and Thompson, T.L. (2003) Comparison of transpiration rates among saltcedar, cottonwood and willow trees by sap flow and canopy temperature methods. *Agricultural and Forest Meteorology* 116, 73–89.
Naumburg, E., Mata-Gonzalez, R., Hunter, R.G., Mclendon, T. and Martin, D.W. (2005) Phreatophytic vegetation and groundwater fluctuations: a review of current research and application of ecosystem response modeling with an emphasis on Great Basin vegetation. *Environmental Management* 35, 726–740.
Nautiyal, P.C., Rachaputi, N.R. and Joshi, Y.C. (2002) Moisture-deficit-induced changes in leaf-water content, leaf carbon exchange rate and biomass production in groundnut cultivars differing in specific leaf area. *Field Crop Research* 74, 67–79.
Neuman, D.S., Wagner, M., Braatne, J.H. and Howe, J. (1996) Stress physiology – abiotic. In: Stettler, R.F., Bradshaw, H.D. Jr, Heilman, P.E. and Hinckley, T.M. (eds) *Biology of Populus and Its Implications for Management and Conservation.* National Research Council of Canada Research Press, Ottawa, pp. 423–458.
Niinemets, Ü. (1999) Components of leaf dry mass per area – thickness and density – alter leaf photosynthetic capacity in reverse directions in woody plants. *New Phytologist* 144, 35–57.
Niinemets, Ü. (2001) Global-scale climatic controls of leaf dry mass per area, density, and thickness in trees and shrubs. *Ecology* 82, 453–469.
Niinemets, Ü. and Kull, O. (2001) Sensitivity of photosynthetic electron transport to photoinhibition in a temperate deciduous forest canopy: Photosystem II center openness, non-radiative energy dissipation and excess irradiance under field conditions. *Tree Physiology* 21, 899–914.
Niinemets, Ü., Bilger, W., Kull, O. and Tenhunen, J.D. (1999a) Responses of foliar photosynthetic electron transport, pigment stoichiometry, and stomatal conductance to interacting environmental factors in a mixed species forest canopy. *Tree Physiology* 19, 839–852.
Niinemets, Ü., Kull, O. and Tenhunen, J.D. (1999b) Variability in leaf morphology and chemical composition as a function of canopy light environment in coexisting deciduous trees. *International Journal of Plant Sciences* 160, 837–848.
Niinemets, Ü., Sober, A., Kull, O., Hartung, W. and Tenhunen, J.D. (1999c) Apparent controls on leaf conductance by soil water availability and via light-acclimation of foliage structural and physiological properties in a mixed deciduous, temperate forest. *International Journal of Plant Sciences* 160, 707–721.
Niinemets, Ü., Sonninen, E. and Tobias, M. (2004) Canopy gradients in leaf intercellular CO_2 mole fractions revisited: interactions between leaf irradiance and water stress need consideration. *Plant, Cell and Environment* 27, 569–583.
Nixon, D.J., Stephens, W., Tyrrel, S.F. and Brierley, E.D.R. (2001) The potential for short rotation energy forestry on restored landfill caps. *Bioresource Technology* 77, 237–245.
Noormets, A., Sôber, A., Pell, E.J., Dickson, R.E., Podila, G.K., Sôber, J., *et al.* (2001) Stomatal and non-stomatal limitation to photosynthesis in two trembling aspen (*Populus tremuloides* Michx.) clones exposed to elevated CO_2 and/or O_3. *Plant, Cell and Environment* 24, 327–336.
Ögren, E. and Sjöström, M. (1990) Estimation of the effect of photoinhibition on the carbon gain in leaves of a willow canopy. *Planta* 181, 560–567.
Oksanen, E., Sober, J. and Karnosky, D.F. (2001) Impacts of elevated CO_2 and/or O_3 on leaf ultrastructure of aspen (*Populus tremuloides*) and birch (*Betula papyrifera*) in the Aspen FACE experiment. *Environmental Pollution* 115, 437–446.

Oksanen, E., Häikiö, E., Sober, J. and Karnosky, D.F. (2003) Ozone-induced H_2O_2 accumulation in field-grown aspen and birch is linked to foliar ultrastructure and peroxisomal activity. *New Phytologist* 161, 791–799.
Orians, C.M., Bolnick, D.I., Roche, B.M., Fritz, R.S. and Floyd, T. (1999) Water availability alters the relative performance of *Salix sericea*, *Salix eriocephala*, and their F_1 hybrids. *Canadian Journal of Botany* 77, 514–522.
Osmond, C.B. (1994) What is photoinhibition? Some insights from comparison of shade and sun plants. In: Baker, N.R. and Bowyer, J.R. (eds) *Photoinhibition of Photosynthesis: From Molecular Mechanisms to the Field.* Environmental Plant Biology, Bios Scientific Publishers, Oxford, UK, pp. 1–24.
Ottow, E.A. (2005) Molecular and ecophysiological responses of *Populus euphratica* (Oliv.) and *Arabidopsis thaliana* (L.) to salt stress. Thesis, Cuvillier Verlag, Göttingen, Germany.
Ottow, E.A., Polle, A., Brosche, M., Kangasjärvi, J., Dibrov, P., Zörb, C., *et al.* (2005a) Molecular characterisation of Pe*NhaD*: the first member of the *NhaD* type Na^+/H^+ antiporter of plant origin. *Plant Molecular Biology* 58, 73–86.
Ottow, E.A., Brinker, M., Teichmann, T., Fritz, E., Kaiser, W., Brosché, M., *et al.* (2005b) *Populus euphratica* displays apoplastic sodium accumulation, osmotic adjustment by decreases in calcium and soluble carbohydrates, and develops leaf succulence under salt stress. *Plant Physiology* 139, 1625–1634.
Ozolincius, R., Miksys, V. and Stakenas, V. (2005a) Growth-independent mortality of Lithuanian forest tree species. *Scandinavian Journal of Forest Research* 20, 153–160.
Ozolincius, R., Stakenas, R. and Serafinaviciute, B. (2005b) Meteorological factors and air pollution in Lithuanian forests: possible effects on tree condition. *Environmental Pollution* 137, 587–595.
Pallardy, S.G. and Kozlowski, T.T. (1981) Water relations in *Populus* clones. *Ecology* 62, 159–169.
Pan, H.Y. (2005) Review of the Asian longhorned beetle – research, biology, distribution and management in China. Forest Health and Biosecurity Working Paper FBS/6E. Food and Agriculture Organization of the United Nations, Rome.
Paolucci, I., Pagnotta, M.A., Sabatti, M., Dominaci, L., Tanzarella, O.A. and Scarascia-Mugnozza, G. (2003) Analisi genetica e della variabilità in Pioppo bianco (*Populus alba* L.) mediante marcatori RFLP e RAPD. *Atti Società Italiana di Selvicoltura ed Ecologia Forestale* 3, 29–36.
Passioura, J.B. (2002) Environmental biology and crop improvement. *Functional Plant Biology* 29, 537–546.
Passioura, J.B. and Fry, S.C. (1992) Turgor and cell expansion: beyond the Lockhart equation. *Australian Journal of Plant Physiology* 19, 565–576.
Pataki, D.E., Oren, R. and Smith, W.K. (2000) Sap flux of co-occurring species in a western subalpine forest during seasonal soil drought. *Ecology* 81, 2557–2566.
Pavlović, P., Mitrović, M. and Djurdjević, L. (2004) An ecophysiological study of plants growing on the fly ash deposits from the 'Nikola tesla-A' thermal power station in Serbia. *Environmental Management* 33, 654–663.
Pegoraro, E., Rey, A., Bobich, E.G., Barron-Gafford, G., Grieve, K.A., Malhi, Y., *et al.* (2004) Effect of elevated CO_2 concentration and vapour pressure deficit on isoprene emission from leaves of *Populus deltoides* during drought. *Functional Plant Biology* 31, 1137–1147.
Pegoraro, E., Rey, A., Barron-Gafford, G., Monson, R., Malhi, Y. and Murthy, R. (2005a) The interacting effects of elevated atmospheric CO_2 concentration, drought and leaf-to-air vapour pressure deficit on ecosystem isoprene fluxes. *Oecologia* 2, 1–11.
Pegoraro, E., Abrell, L., Van Haren, J., Barron-Gafford, G., Grieve, K.A., Malhi, Y., *et al.* (2005b) The effect of elevated atmospheric CO_2 and drought on sources and sinks of isoprene in a temperate and tropical rainforest mesocosm. *Global Change Biology* 11, 1234–1246.
Pelah, D., Shoseyov, O. and Altman, A. (1995) Characterization of BspA, a major boiling-stable, water-stress-responsive protein in aspen (*Populus tremula*). *Tree Physiology* 15, 673–678.
Pelah, D., Wang, W., Altman, A., Shoseyov, O. and Bartels, D. (1997) Differential accumulation of water stress-related proteins, sucrose synthase and soluble sugars in *Populus* species that differ in their water stress response. *Physiologia Plantarum* 99, 153–159.
Peng, Y.Y. and Dang, Q.L. (2003) Effects of soil temperature on biomass production and allocation in seedlings of four boreal tree species. *Forest Ecology and Management* 180, 1–9.
Percy, K.E., Awmack, C.S., Lindroth, R.L., Kubiske, M.E., Kopper, B.J., Isebrands, J.G., *et al.* (2002) Altered performance of forest pests under CO_2- and O_3-enriched atmospheres. *Nature* 420, 403–407.
Pezeshki, S.R., Anderson, P.H. and Shields, F.D. Jr (1998) Effects of soil moisture regimes on growth and survival of black willow (*Salix nigra*) posts (cuttings). *Wetlands* 18, 460–470.

Phan, C.T., Jörgensen, J., Jouve, L., Hausman, J.F., Polle, A. and Teichmann, T. (2004) Micropropagation of *Populus euphratica* Olivier. *Belgian Journal of Botany* 137, 175–180.

Piazzai, M., Beritognolo, I., Abruzzese, G., Kuzminsky, E., Sabatti, M., Muleo, R., *et al.* (2003) Strategie di risposta allo stress salino in differenti genotipi di *Populus alba*. IV congresso della Società Italiana di Selvicoltura ed Ecologia Forestale. Rifreddo, Italy, 7–10 October 2003 (Poster).

Pockman, W.T. and Sperry, J.S. (2000) Vulnerability to xylem cavitation and the distribution of Sonoran desert vegetation. *American Journal of Botany* 87, 1287–1299.

Polle, A. and Schützendübel, A. (2003) Heavy metal signalling in plants: linking cellular and organismic responses. In: Hirt, H. and Shinozaki, K. (eds) *Plant Responses to Abiotic Stresses, Volume 4, Topics in Current Genetics.* Springer Verlag, Berlin, pp.167–215.

Polle, A., Altman, A. and Jiang, X. (2006) Towards genetic engineering for drought tolerance in trees. In: Fladung, M. and Ewald, D. (eds) *Tree Transgenesis: Recent Developments. Part C.* Springer Verlag, Heidelberg, Germany, pp. 275–297.

Ponzio, K.J., Miller, S.J., Underwood, E., Rowe, S.P., Voltolina, D.J. and Miller, T.D. (2006) Responses of a willow (*Salix caroliniana* Michx.) community to roller-chopping. *Natural Areas Journal* 26, 53–60.

Potts, D.L. and Williams, D.G. (2004) Response of tree ring holocellulose delta $\delta^{13}C$ to moisture availability in *Populus fremontii* at perennial and intermittent stream reaches. *Western North American Naturalist* 64, 27–37.

Powell, G.W. and Bork, E.W. (2004) Competition and facilitation in mixtures of aspen seedlings, alfalfa, and marsh reedgrass. *Canadian Journal of Forest Research* 34, 1858–1869.

Pregitzer, K.S., Zak, D.R., Curtis, P.S., Kubiske, M.E., Teeri, J.A. and Vogel, C.S. (1995) Atmospheric CO_2, soil nitrogen and turnover of fine roots. *New Phytologist* 129, 579–585.

Pregitzer, K.S., Zak, D.R., Maziasz, J., DeForest, J., Curtis, P.S. and Lussenhop, J. (2000) Interactive effects of atmospheric CO_2 and soil-N availability on fine roots of *Populus tremuloides. Ecological Applications* 10, 18–33.

Rae, A.M., Robinson, K.M., Street, N.R. and Taylor, G. (2004) Morphological and physiological traits influencing biomass productivity in short-rotation coppice poplar. *Canadian Journal of Forest Research* 34, 1488–1498.

Randlett, D.L., Zak, D.R., Pregitzer, K.S. and Curtis, P.S. (1996) Elevated atmospheric carbon dioxide and leaf litter chemistry: influences on microbial respiration and net nitrogen mineralization. *Soil Science Society of America Journal* 60, 1571–1577.

Rhodenbaugh, E.J. and Pallardy, S.G. (1993) Water stress, photosynthesis and early growth patterns of cuttings of three *Populus* clones. *Tree Physiology* 13, 213–226.

Ridolfi, M. and Dreyer, E. (1997) Responses to water stress in an ABA-unresponsive hybrid poplar (*Populus koreana* × *trichocarpa* cv. Peace). III. Consequences for photosynthetic carbon assimilation. *New Phytologist* 135, 31–40.

Ridolfi, M., Fauveau, M.L., Label, P., Garrec, J.P. and Dreyer, E. (1996) Response to water stress in an ABA-unresponsive hybrid poplar (*Populus koreana* × *trichocarpa* cv. Peace). I. Stomatal function. *New Phytologist* 134, 445–454.

Rier, S.T., Tuchman, N.C., Wetzel, R.G. and Teeri, J.A. (2002) Elevated-CO_2-induced changes in the chemistry of quaking aspen (*Populus tremuloides* Michaux) leaf litter: subsequent mass loss and microbial response in a stream ecosystem. *Journal of the North American Benthological Society* 21, 16–27.

Robison, D.J. and Raffa, K.F. (1998) Productivity, drought tolerance and pest status of hybrid *Populus*: tree improvement and silvicultural implications. *Biomass and Bioenergy* 14, 1–20.

Roden, J., Van Volkenburgh, E. and Hinckley, T.M. (1990) Cellular basis for limitation of poplar leaf growth by water deficit. *Tree Physiology* 6, 211–219.

Rönnberg-Wästljung, A.C., Glynn, C. and Weih, M. (2005) QTL analyses of drought tolerance and growth for a *Salix dasyclados* × *Salix viminalis* hybrid in contrasting water regimes. *Theoretical and Applied Genetics* 110, 537–549.

Rood, S.B., Patino, S., Coombs, K. and Tyree, M.T. (2000a) Branch sacrifice: cavitation-associated drought adaptation of riparian cottonwoods. *Trees – Structure and Function* 14, 248–257.

Rood, S.B., Zanewich, K., Stefura, C. and Mahoney, J.M. (2000b) Influence of water table decline on growth allocation and endogenous gibberellins in black cottonwood. *Tree Physiology* 20, 831–836.

Rood, S.B., Braatne, J.H. and Hughes, F.M.R. (2003) Ecophysiology of riparian cottonwoods: stream flow dependency, water relations and restoration. *Tree Physiology* 23, 1113–1124.

Rowland, D.L., Biagini, B. and Evans, A.S. (2000) Variability among five riparian cottonwood (*Populus fremontii* Wats.) populations: an examination of size, density and spatial distribution. *Western North American Naturalist* 60, 384–393.

Rowland, D.L., Beals, L., Chaudhry, A.A., Evans, A.S. and Grodeska, L.S. (2001) Physiological, morphological, and environmental variation among geographically isolated cottonwood (*Populus deltoides*) populations in New Mexico. *Western North American Naturalist* 61, 452–462.

Ruiz-Lozano, J.M. and Azcon, R. (2000) Symbiotic efficiency and infectivity of an autochthonous arbuscular mycorrhizal *Glomus* sp. from saline soils and *Glomus deserticola* under salinity. *Mycorrhiza* 10, 137–143.

Sabatti, M., Scarascia-Mugnozza, G.E. and Anselmi, N. (1998) Conservazione e caratterizzazione del germoplasma di popolazioni naturali italiane di *Populus alba* L. In: SISEF (eds) *La Ricerca Italiana per le Foreste e la Selvicoltura, Atti I.* SISEF, Legnaro, Padova, Italy, 4–6 Giugno, 1997, pp. 165–170.

Sage, R.F. (1994) Acclimation of photosynthesis to increasing atmospheric CO_2 – the gas-exchange perspective. *Photosynthesis Research* 39, 351–368.

Saugier, B. (1983) Plant growth and its limitations in crop and natural communities. In: Mooney, H.A. and Godron, M. (eds) *Disturbance and Ecosystems, Components of Response. Ecological Studies 44.* Spinger-Verlag, Berlin, pp. 159–174.

Savouré, A., Thorin, D., Davey, M., Hua, X.J., Mauro, S., Van Montagu, M., *et al.* (1999) NaCl and $CuSO_4$ treatments trigger distinct oxidative defence mechanisms in *Nicotiana plumbaginifolia* L. *Plant, Cell and Environment* 22, 387–396.

Scarascia-Mugnozza, G.E., Ceulemans, R., Heilman, P.E., Isebrands, J.G., Stettler, R.F. and Hinckley, T.M. (1997) Production physiology and morphology of *Populus* species and their hybrids grown under short rotation. II. Biomass components and harvest index of hybrid and parental species clones. *Canadian Journal of Forest Research* 27, 285–294.

Scarascia-Mugnozza, G., De Angelis, P., Sabatti, M., Calfapietra, C., Ceulemans, R., Peressotti, A., *et al.* (2000) A FACE experiment on short rotation, intensive poplar plantation: objective and experimental set up of POPFACE. In: Sutton, M.A., Moreno, J.M., van der Putten, W.H. and Struwe, S. (eds) *Terrestrial Ecosystem Research in Europe: Successes, Challenges and Policy.* Office for Official Publications of the European Communities, Luxembourg, pp. 136–140.

Schaff, S.D., Pezeshki, S.R. and Shields, F.D. Jr (2002) Effects of pre-planting soaking on growth and survival of black willow cuttings. *Restoration Ecology* 10, 267–274.

Schimel, D., Alves, D., Enting, D., Heimann, M. and Joos, F. (1996) Radiative forcing of climate change. In: Houghton, J.T., Filho, L.G.M., Callander, B.A., Harris, N., Kattenberg, A. and Maskell, K. (eds) *Climate Change 1995: the Science of Climate Change.* Cambridge University Press, Cambridge, UK, pp. 65–132.

Scholes, R.J., Schulze, E.-D., Pitelka, L.F. and Hall, D.O. (1999) Biogeochemistry of terrestrial ecosystems. In: Walker, B., Steffen, W., Canadell, J. and Ingram, J. (eds) *The Terrestrial Biosphere and Global Change. Implications for Natural and Managed Ecosystems.* Cambridge University Press, Cambridge, UK, pp. 271–303.

Schreiner, E.J. (1959) *Production of Poplar Timber in Europe and Its Significance and Application in the U.S.* Agriculture Handbook 150. USDA Forest Service, Washington, DC.

Schulte, P.J. and Hinckley, T.M. (1987) Abscisic acid relations and the response of *Populus trichocarpa* stomata to leaf water potential. *Tree Physiology* 3, 103–113.

Schumaker, M.A., Bassman, J.H., Robberecht, R. and Radamaker, G.K. (1997) Growth, leaf anatomy, and physiology of *Populus* clones in response to solar ultraviolet-B radiation. *Tree Physiology* 17, 617–626.

Schützendübel, A. and Polle, A. (2002) Plant responses to abiotic stresses: heavy metal-induced oxidative stress and protection by mycorrhization. *Journal of Experimental Botany* 53, 1351–1365.

Schwanz, P. and Polle, A. (2001) Growth under elevated CO_2 ameliorates defences against photo-oxidative stress in poplar (*Populus alba* × *tremula*). *Environmental and Experimental Botany* 45, 43–53.

Scott, M.L., Shafroth, P.B. and Auble, G.T. (1999) Responses of riparian cottonwoods to alluvial water table declines. *Environmental Management* 23, 347–358.

Scott, M.L., Lines, G.C. and Auble, G.T. (2000) Channel incision and patterns of cottonwood stress and mortality along the Mojave River, California. *Journal of Arid Environments* 44, 399–414.

Sekawin, M. (1975) La génétique du *Populus alba* L. *Annales Forestales, Zagreb* 6, 157–189.

Shafroth, P.B., Auble, G.T. and Scott, M.L. (1995) Germination and establishment of the native plains cottonwood (*Populus deltoides* Marshall subsp. *monilifera*) and the exotic Russian-olive (*Elaeagnus angustifolia* L.). *Conservation Biology* 9, 1169–1175.

Shafroth, P.B., Stromberg, J.C. and Patten, D.T. (2000) Woody riparian vegetation response to different alluvial water table regimes. *Western North American Naturalist* 60, 66–76.

Shi, H., Ishitani, M., Kim, C. and Zhu, J.K. (2000) The *Arabidopsis thaliana* salt tolerance gene *SOS1* encodes a putative Na^+/H^+ antiporter. *Proceedings of the National Academy of Sciences of the USA* 97, 6896–6901.

Shi, H., Lee, B.H., Wu, S.J. and Zhu, J.K. (2003) Overexpression of a plasma membrane Na^+/H^+ antiporter gene improves salt tolerance in *Arabidopsis thaliana. Nature Biotechnology* 21, 81–85.

Sibout, R. and Guerrier, G. (1998) Solute incompatibility with glutamine synthetase in water-stressed *Populus nigra. Environmental and Experimental Botany* 40, 173–178.

Siemens, J.A. and Zwiazek, J.J. (2003) Effects of water deficit stress and recovery on the root water relations of trembling aspen (*Populus tremuloides*) seedlings. *Plant Science* 165, 113–120.

Siemens, J.A. and Zwiazek, J.J. (2004) Changes in root water flow properties of solution culture-grown trembling aspen (*Populus tremuloides*) seedlings under different intensities of water-deficit stress. *Physiologia Plantarum* 121, 44–49.

Sigurdsson, B.D. (2001) Elevated (CO_2) and nutrient status modified leaf phenology and growth rhythm of young *Populus trichocarpa* trees in a three-year field study. *Trees – Structure and Function* 15, 403–413.

Silva, D.M., Hammond, L.C., Bennet, J.M., Hass, J.H. and Linda, S.B. (1993) Field response of maize; *Zea mays* L. to a VAM fungus and water management. *Agronomy Journal* 85, 193–198.

Singh, M., Jain, M. and Pant, R.C. (1999) Clonal variability in photosynthetic and growth characteristics of *Populus deltoides* under saline irrigation. *Photosynthetica* 36, 605–609.

Sixto, H., Grau, J.M., Alba, N. and Alia, R. (2005) Response to sodium chloride in different species and clones of genus *Populus* L. *Forestry* 78, 93–1004.

Smirnoff, N. (1993) The role of active oxygen in the response of plants to water deficit and desiccation. *New Phytologist* 125, 27–58.

Smith, S.D., Devitt, D.A., Sala, A., Cleverly, J.R. and Busch, D.E. (1998) Water relations of riparian plants from warm desert regions. *Wetlands* 18, 687–696.

Smith, S.E. and Read, D.J. (1997) *Mycorrhizal Synthesis,* 2nd edn. Academic Press, San Diego, California.

Souch, C.A. and Stephens, W. (1998) Growth, productivity and water use in three hybrid poplar clones. *Tree Physiology* 18, 829–835.

Sparks, J.P. and Black, R.A. (1999) Regulation of water loss in populations of *Populus trichocarpa*: the role of stomatal control in preventing xylem cavitation. *Tree Physiology* 19, 453–459.

Sperry, J.S., Hacke, U.G., Oren, R. and Comstock, J.P. (2002) Water deficits and hydraulic limits to leaf water supply. *Plant, Cell and Environment* 25, 251–263.

Stettler, R.F., Braatne, J., Ceulemans, R., Clum, P., Dunlap, J., Newman, D., *et al.* (1993) Growth dynamics and canopy structure. In: Mitchell, C.P., Ford-Robertson, J.B., Hinkley, T.M. and Sennerby-Forsse, L. (eds) *Ecophysiology of Short-rotation Forest Crops.* Elsevier, London, pp. 1–34.

Strohm, M., Eiblmeier, M., Langebartels, C., Jouanin, L., Polle, A., Sandermann, H., *et al.* (1999) Responses of transgenic poplar (*Populus tremula* × *P. alba*) overexpressing glutathione synthetase or glutathione reductase to acute ozone stress: visible injury and leaf gas exchange. *Journal of Experimental Botany* 50, 365–374.

Strohm, M., Eiblmeier, M., Langebartels, C., Jouanin, L., Polle, A., Sandermann, H., *et al.* (2002) Responses of antioxidative systems to acute ozone stress in transgenic poplar (*Populus tremula* × *P. alba*) over-expressing glutathione synthetase or glutathione reductase. *Trees – Structure and Function* 16, 262–273.

Sundby, C.A., Melis, P., Mäenpää, P. and Andersson, B. (1986) Temperature-dependent changes in the antenna size of photosystem II. *Biochimica et Biophysica Acta* 851, 475–483.

Szabolcs, I. (1994) Soils and salinization. In: Pessarakli, M. (ed.) *Handbook of Plant and Crop Stress.* Marcel Dekker, New York, pp. 3–11.

Taji, T., Oshumi, C., Luchi, S., Seki, M., Kasuga, M., Kobayashi, M., *et al.* (2002) Important roles of drought- and cold-inducible genes for galactinol synthase in stress tolerance in *Arabidopsis thaliana. The Plant Journal* 29, 417–426.

Takeuchi, Y., Kubiske, M.E., Isebrands, J.G., Pregitzer, K.S., Hendrey, G. and Karnosky, D.F. (2001) Photosynthesis, light and nitrogen relationships in a young deciduous forest canopy under open-air CO_2 enrichment. *Plant, Cell and Environment* 24, 1257–1268.

Tang, Y.H. and Liang, N.S. (2000) Characterization of the photosynthetic induction response in a *Populus* species with stomata barely responding to light changes. *Tree Physiology* 20, 969–976.

Tardieu, F. and Simonneau, T. (1998) Variability among species of stomatal control under fluctuating soil water status and evaporative demand: modelling isohydric and anisohydric behaviours. *Journal of Experimental Botany* 49, 419–432.

Taub, D.R., Seemann, J.R. and Coleman, J.S. (2000) Growth in elevated CO_2 protects photosynthesis against high-temperature damage. *Plant, Cell and Environment* 23, 649–656.
Taylor, G. (2002) *Populus*: *Arabidopsis* for forestry. Do we need a model tree? *Annals of Botany* 90, 681–689.
Taylor, G., McDonald, J., Stadenberg, I. and Freer-Smith, P.H. (1993) Nitrate supply and the biophysics of leaf growth of *Salix viminalis*. *Journal of Experimental Botany* 44, 155–164.
Taylor, G., Ceulemans, R., Ferris, R., Gardner, S.D.L. and Shao, B.Y. (2001) Increased leaf area expansion of hybrid poplar in elevated CO_2. From controlled environments to open-top chambers and to FACE. *Environmental Pollution* 115, 463–472.
Taylor, G., Tricker, P.J., Zhang, F.Z., Alston, V.J., Miglietta, F. and Kuzminsky, E. (2003) Spatial and temporal effects of free-air CO_2 enrichment (POPFACE) on leaf growth, cell expansion, and cell production in a closed canopy of poplar. *Plant Physiology* 131, 177–185.
Taylor, G., Street, N.R., Tricker, P.J., Sjödin, A., Graham, L., Skogström, O., *et al.* (2005) The transcriptome of *Populus* in elevated CO_2. *New Phytologist* 167, 143–154.
Tester, M. and Davenport, R. (2003) Na^+ tolerance and Na^+ transport in higher plants. *Annals of Botany* 91, 503–527.
Thomas, F.M., Arndt, S.K., Bruelheide, H., Foetzki, A., Gries, D., Huang, J., *et al.* (2000) Ecological basis for a sustainable management of the indigenous vegetation in a Central Asian desert: presentation and first results. *Journal of Applied Botany* 74, 212–219.
Thompson, L.K., Blaylock, R., Sturtevant, J.M. and Brudvig, G.W. (1989) Molecular basis of heat denaturation of photosystem II. *Biochemistry* 28, 6686–6695.
Tjoelker, M.G., Oleskyn, J. and Reich, P.B. (1999) Acclimation of respiration to temperature and CO_2 in seedlings of boreal tree species in relation to plant size and relative growth rate. *Global Change Biology* 49, 679–691.
Tognetti, R., Longobucco, A. and Raschi, A. (1999a) Seasonal embolism and xylem vulnerability in deciduous and evergreen Mediterranean trees influenced by proximity to a carbon dioxide spring. *Tree Physiology* 19, 271–277.
Tognetti, R., Longobucco, A., Raschi, A., Miglietta, F. and Fumagalli, I. (1999b) Responses of two *Populus* clones to elevated atmospheric CO_2 concentration in the field. *Annals of Forest Science* 56, 493–500.
Trabaud, L. and Méthy, M. (1992) Effets de températures sub-latérales sur l'appareil photosynthétique du chêne vert (*Quercus ilex* L). *Annals of Forest Science* 49, 637–649.
Tricker, P.J., Calfapietra, C., Kuzminsky, E., Puleggi, R., Ferris, R., Nathoo, M., *et al.* (2004) Long-term acclimation of leaf production, development, longevity and quality following 3 years exposure to free-air CO_2 enrichment during canopy closure in *Populus*. *New Phytologist* 162, 413–426.
Tricker, P.J., Trewin, H., Kull, O., Clarkson, G.J.J., Eensalu, E., Tallis, M.E., *et al.* (2005) Stomatal conductance and not stomatal density determines the long-term reduction in leaf transpiration of poplar in elevated CO_2. *Oecologia* 143, 652–660.
Tryon, P.R. and Chapin, F.S. III (1983) Temperature control over root growth and root biomass in taiga forest trees. *Canadian Journal of Forest Research* 13, 827–833.
Tschaplinski, T.J., Tuskan, G.A. and Gunderson, C.A. (1994) Water-stress tolerance of black and eastern cottonwood clones and four hybrid progeny. I. Growth, water relations and gas exchange. *Canadian Journal of Forest Research* 24, 364–371.
Tschaplinski, T.J., Tuskan, G.A., Gebre, G.M. and Todd, D.E. (1998) Drought resistance of two hybrid *Populus* clones grown in a large-scale plantation. *Tree Physiology* 18, 653–658.
Tupker, K.A., Thomas, B.R. and Macdonald, S.E. (2003) Propagation of trembling aspen and hybrid poplar for agroforestry: potential benefits of elevated CO_2 in the greenhouse. *Agroforestry Systems* 59, 61–71.
Turnbull, M.H., Tissue, D.T., Murthy, R., Wang, X., Sparrow, A.D. and Griffin, K.L. (2004) Nocturnal warming increases photosynthesis at elevated CO_2 partial pressure in *Populus deltoides*. *New Phytologist* 161, 819–826.
Turtola, S., Rousi, M., Pusenius, J., Yamaji, K., Heiska, S., Tirkkonen, V., *et al.* (2005) Clone-specific responses in leaf phenolics of willows exposed to enhanced UVB radiation and drought stress. *Global Change Biology* 11, 1655–1663.
Turtola, S., Rousi, M., Pusenius, J., Yamaji, K., Heiska, S., Tirkkonen, V., *et al.* (2006) Genotypic variation in drought response of willows grown under ambient and enhanced UV-B radiation. *Environmental and Experimental Botany* 56, 80–86.

Tyree, M.T., Alexander, J. and Machado, J.L. (1992) Loss of conductivity due to water stress in intact juveniles of *Quercus rubra* and *Populus deltoides*. *Tree Physiology* 10, 411–415.
Tyree, M.T., Kolb, K.J., Rood, S.B. and Patiño, S. (1994) Vulnerability to drought-induced cavitation of riparian cottonwoods in Alberta: a possible factor in the decline of the ecosystem? *Tree Physiology* 14, 455–466.
Tyystjärvi, E. and Aro, E.M. (1996) The rate constant of photoinhibition, measured in lincomycin-treated leaves, is directly proportional to light intensity. *Proceedings of the National Academy of Sciences of the USA* 93, 2213–2218.
UNCCD (2004) *Preserving Our Common Ground. UNCCD 10 Years On.* United Nations Convention to Combat Desertification, Bonn, Germany.
van den Driessche, R. (2000) Phosphorus, copper and zinc supply levels influence growth and nutrition of a young *Populus trichocarpa* (Torr. & Gray) × *P. deltoides* (Bartr.ex Marsh) hybrid. *New Forests* 19, 143–157.
Vandersande, M.W., Glenn, E.P. and Walworth, J.L. (2001) Tolerance of five riparian plants from the lower Colorado River to salinity drought and inundation. *Journal of Arid Environments* 49, 147–159.
Van Splunder, I., Voesenek, L.A.C.J., Coops, H., De Vries, X.J.A. and Blom, C.W.P.M. (1996) Morphological responses of seedlings of four species of *Salicaceae* to drought. *Canadian Journal of Botany* 74, 1988–1995.
Volin, J.C. and Reich, P.B. (1996) Interaction of elevated CO_2 and O_3 on growth, photosynthesis and respiration of three perennial species grown in low and high nitrogen. *Physiologia Plantarum* 97, 674–684.
von Fircks, Y. and Sennerby-Forsse, L. (1998) Seasonal fluctuations of starch in root and stem tissues of coppiced *Salix viminalis* plants grown under two nitrogen regimes. *Tree Physiology* 50, 1797–1806.
Vozzo, J.A. and Hackaylo, E. (1974) Endo- and ectomycorrhizal association in five poplar species. *Bulletin of the Torrey Botanical Club* 101, 182–186.
Wang, S., Chen, B. and Li, H. (1996) *Euphrates Poplar Forest.* China Environmental Science Press, Beijing.
Wang, W.X., Pelah, D., Alergand, T., Shoseyov, O. and Altman, A. (2002) Characterization of SP1, a stress-responsive, boiling-soluble, homo-oligomeric protein from aspen. *Plant Physiology* 130, 865–875.
Wang, W., Vinocur, B. and Altman, A. (2003) Plant responses to drought, salinity and extreme temperatures: towards genetic engineering for stress tolerance. *Planta* 218, 1–14.
Wang, Y., Lu, W., Zhang, W., Bao, J., van Slycken, J. and Sigaud, P. (2000) Research advances with *Populus simonii* Carr. *Protection Forest Science and Technology* 3, 66–70. (In Chinese)
Watanabe, S., Kojima, K., Ide, Y. and Sasaki, S. (2000) Effects of saline and osmotic stress on proline and sugar accumulation in *Populus euphratica in vitro. Plant Cell, Tissue and Organ Culture* 63, 199–206.
Wei, Q.J. (1993) *Euphratica Poplar.* Chinese Forestry Press, Beijing, China. (In Chinese)
Weih, M. (2001) Evidence for increased sensitivity to nutrient and water stress in a fast-growing hybrid willow compared with a natural willow clone. *Tree Physiology* 21, 1141–1148.
Weih, M., Rönnberg-Wästljung, A.C. and Glynn, C. (2006) Genetic basis of phenotypic correlations among growth traits in hybrid willow (*Salix dasyclados* × *S. viminalis*) grown under two water regimes. *New Phytologist* 170, 467–477.
Wikberg, J. and Ögren, E. (2004) Interrelationships between water use and growth traits in biomass-producing willows. *Trees – Structure and Function* 18, 70–76.
Will, R.E. and Ceulemans, R. (1997) Effects of elevated CO_2 concentration on photosynthesis, respiration and carbohydrate status of coppice *Populus* hybrids. *Physiologia Plantarum* 100, 933–939.
Williams, C.A. and Cooper, D.J. (2005) Mechanisms of riparian cottonwood decline along regulated rivers. *Ecosystems* 8, 382–395.
Williams, W.D. (1999) Salinization: a major threat to water resources in the arid and semi-arid regions of the world. *Lake and Reservoir Management* 4, 85–91.
Wittig, V.E., Bernacchi, C.J., Zhu, X., Ceulemans, R., De Angelis, P., Gielen, B., *et al.* (2005) Gross primary production is stimulated for three *Populus* species grown under free-air CO_2 enrichment from planting through canopy closure. *Global Change Biology* 11, 644–656.
Woolfolk, W.T.M. and Friend, A.L. (2003) Growth response of cottonwood roots to varied NH_4:NO_3 ratios in enriched patches. *Tree Physiology* 23, 427–432.
Wustman, B.A., Oksanen, E., Karnosky, D.F., Sober, J., Isebrands, J.G., Hendrey, G.R., *et al.* (2001) Effects of elevated CO_2 and O_3 on aspen clones varying in O_3 sensitivity: can CO_2 ameliorate the harmful effects of O_3? *Environmental Pollution* 115, 473–481.

Xang, X., Zhang, K., Jia, B. and Ci, L. (2005) Desertification assessment in China: an overview. *Journal of Arid Environments* 63, 517–531.

Xiao, C.W., Zhou, G.S., Zhang, X.S., Zhao, J.Z. and Wu, G. (2005) Responses of dominant desert species *Artemisia ordosica* and *Salix psammophila* to water stress. *Photosynthetica* 43, 467–471.

Xu, W.H. (1988) *Poplar.* Heilongjiang People's Press, Harbin, China. (In Chinese)

Yang, J., Zhao, H. and Zhang, T. (2004) Heat and drought tolerance of two willow species, *Salix gordejevii* and *Salix babylonica*: a comparative study. *Israel Journal of Plant Sciences* 52, 301–306.

Yin, C.Y., Duan, B.L., Wang, X. and Li, C.Y. (2004) Morphological and physiological responses of two contrasting *Poplar* species to drought stress and exogenous abscisic acid application. *Plant Science* 167, 1091–1097.

Yin, C.Y., Peng, Y.H., Zang, R.G., Zhu, Y.P. and Li, C.Y. (2005a) Adaptive responses of *Populus kangdingensis* to drought stress. *Physiologia Plantarum* 123, 445–451.

Yin, C.Y., Wang, X., Duan, B.L., Luo, J.X. and Li, C.Y. (2005b) Early growth, dry matter allocation and water use efficiency of two sympatric *Populus* species as affected by water stress. *Environmental and Experimental Botany* 53, 315–322.

Yin, C.Y., Berninger, F. and Li, C.Y. (2006) Photosynthetic responses of *Populus przewalski* subjected to drought. *Photosynthetica* 44, 62–68.

Yin, W. and Lu, W. (2005) Review of tree selection and afforestation for control of Asian longhorned beetle in north China. Forest Health and Biosecurity Working Paper FBS/7E. Food and Agriculture Organization of the United Nations, Rome.

Yuceer, C., Kubiske, M.E., Harkess, R.L. and Land, S.B. (2003) Effects of induction treatments on flowering in *Populus deltoides. Tree Physiology* 23, 489–495.

Zak, D.R., Pregitzer, K.S., Curtis, P.S., Teeri, J.A., Fogel, R. and Randlett, D.L. (1993) Elevated atmospheric CO_2 and feedback between carbon and nitrogen cycles. *Plant and Soil* 151, 105–117.

Zak, D.R., Pregitzer, K.S., Curtis, P.S., Vogel, C.S., Holmes, W.E. and Lussenhop, J. (2000a) Atmospheric CO_2, soil-N availability, and allocation of biomass and nitrogen by *Populus tremuloides. Ecological Applications* 10, 34–46.

Zak, D.R., Pregitzer, K.S., Curtis, P.S. and Holmes, W.E. (2000b) Atmospheric CO_2 and the composition and function of soil microbial communities. *Ecological Applications* 10, 47–59.

Zalesny, R.S. Jr, Hall, R.B., Bauer, E.O. and Riemenschneider, D.E. (2005) Soil temperature and precipitation affect the rooting ability of dormant hardwood cuttings of *Populus. Silvae Genetica* 54, 47–58.

Zhang, H.P., Morison, J.I.L. and Simmonds, L.P. (1999) Transpiration and water relations of poplar trees growing close to the water table. *Tree Physiology* 19, 563–573.

Zhang, S.-R. and Gao, R.-F. (2000) Effects of light stress on oxygen evolution and photochemical energy storage of hybrid poplar clones determined by photoacoustic technique. *Acta Botanica Sinica* 42, 818–823.

Zhang, X., Zang, R. and Li, C. (2004) Population differences in physiological and morphological adaptations of *Populus davidiana* seedlings in response to progressive drought stress. *Plant Science* 166, 791–797.

Zhang, X., Wu, N. and Li, C. (2005) Physiological and growth responses of *Populus davidiana* ecotypes to different soil water contents. *Journal of Arid Environments* 60, 567–579.

Zhu, J.K. (2001) Plant salt tolerance. *Trends in Plant Science* 6, 66–71.

Zhu, Z. and Chen, G.T. (1994) *Sandy Desertification in China.* Science Press, Beijing. (In Chinese)

Zsuffa, L., Giordano, E., Pryor, L.D. and Stettler, R.F. (1996) Trends in poplar culture: some global and regional perspectives. In: Stettler, R.F., Bradshaw, H.D. Jr, Heilman, P.E. and Hinckley, T.M. (eds) *Biology of Populus and Its Implications for Management and Conservation.* National Research Council of Canada Research Press, Ottawa, pp. 515–539.

8 Diseases of Poplars and Willows

M. Ostry,[1]* M. Ramstedt,[2] G. Newcombe[3] and M. Steenackers[4]
[1]Northern Research Station, USDA Forest Service, St Paul, Minnesota, USA; [2]Department of Forest Mycology and Pathology, Swedish University of Agricultural Sciences, Uppsala, Sweden; [3]College of Natural Resources, University of Idaho, Moscow, Idaho, USA; [4]Research Institute for Nature and Forest (INBO), Geraardsbergen, Belgium

8.1 Introduction

Across their native and planted ranges, *Populus* and *Salix* species and hybrids are host to a wide array of microorganisms. Fortunately, only a relatively small number of these organisms are pathogens that cause economically damaging diseases (Ostry and McNabb, 1985). No single pathogen is equally important throughout the entire range of *Populus* and *Salix*, and many organisms are highly specific to only some species and hybrids (Newcombe, 1996; Newcombe *et al.*, 2001a). Since poplars and willows are similar ecologically and closely related through evolution (Argus, 2010; Eckenwalder, 2010), there are many parallels among their collective pathogens. Many of the genera of pathogens that affect one also affect the other (Newcombe *et al.*, 2010).

Poplars and willows are affected by pathogens that typically specialize on leaves, stems or roots. Diseases incited by these pathogens are often responsible for reduced growth and wood quality of affected trees; complete planting failures due to pathogens are also known (Steenackers *et al.*, 1996). The threat of disease is a major consideration in developing sustainable poplar and willow cropping systems for renewable sources of fibre and energy. Since intensive cultivation of poplars and willows is still new to many parts of the world, the disease challenges are often significant, ranging from selecting species and clones adequately adapted to the planting site, to developing and applying cultural practices to minimize disease. Genetic resistance to disease has been a major objective of poplar and willow selection and breeding programmes throughout the world. Unfortunately, pathogenic organisms continue to evolve rapidly and adapt to new hosts. Breeding for resistance is thus likely to continue to be a high priority in the future.

In many regions of the world where poplars and willows have been introduced, successful pathogen-free cultivation has initially been possible. Release from pathogens outside native ranges is an ecologically significant phenomenon (Mitchell and Power, 2003), but pathogens from native ranges can eventually be reunited with their hosts (Newcombe and Dugan, 2010). Then, the disease outbreaks that follow these reunions may be especially severe, since pathogens may also benefit from release from biotic or abiotic constraints that limit disease in native ranges. The planting of new varieties of poplars and willows that are poorly adapted to the site

*Corresponding author: mostry@fs.fed.us

may result in diseases caused by opportunistic pathogens (Trench and Churchill, 1987; Spiers, 1989, 1998).

Willow and poplar species used in industrial production are selected for specific traits for a variety of uses, including production of energy, matches, veneer or wastewater treatment and heavy metal uptake from contaminated soils. However, species and hybrids with these traits may not have optimal growth or adaptive capacity. They may also display poor resistance to disease and insect pests, especially when they are planted in areas where they have not been tested (Pinon and Valadon, 1997).

Intensifying the culture of either poplars or willows can, in general, lead to disease problems. For example, willows that had been grown for basket production for centuries were scaled up rapidly for large, monoclonal plantations when short-rotation forestry (SRF) for energy production was introduced in Europe in the 1970s. Promising clones with good yield potential and initially low disease susceptibility were suddenly much more common in the landscape. Within 10–15 years, these clones were severely affected by new species and races of *Melampsora* rust. A very similar scenario developed in the Pacific Northwest of the USA, where *Melampsora* responded to intensive culture of initially resistant *Populus* hybrids with significant increases in pathogenic variation (Newcombe *et al.*, 2001b).

Although most pathogens of *Populus* and *Salix* are fungi, there are exceptions. Bacterial diseases of *Populus* have been important in Europe, where *Xanthomonas populi* is a serious pathogen of susceptible varieties. Another important exception is the water mark disease of tree-forming willows (*Salix* subgenus *Salix*), caused by the bacterium, *Brenneria salicis* (*Erwinia salicis*). This is a serious disease, although tree willows are not generally used in SRF. Recent research, however, has shown that the bacterium could reside in symptomless tissue in many different tree species and thereby be transmitted unintentionally to new sites.

Other bacterial species in recent years have also had large impacts. For example, species of *Pseudomonas* have recently caused substantial dieback, resulting in willow crop losses, especially in combination with frost damage. Poplar plantations are planted at wider spacing, resulting in lower humidity than in willow plantings, which in turn reduces damage by these pathogens.

Poplar breeding has a long history of hybridization involving parental species from Europe, North America and Asia. Willow breeding has a shorter history, with traditional selection and breeding of superior individuals in the early years to using the new biotechnologies in the breeding programmes of today. In both poplar and willow breeding, productivity ideally is accompanied by durable disease resistance.

The increasing interest in breeding and growing poplars and willows for fibre and energy has resulted in the movement of planting materials around the world. Poplar hybrids of European and North American origin have been grown widely in plantations during the past 100 years (Eckenwalder, 2001), and selected clones have frequently been exchanged among countries, widely planted on a large range of sites and managed under various cultural systems.

It remains impossible to predict the outcome of any encounter between a novel pathogen and an evolutionarily naïve host (Newcombe and Dugan, 2010). Some exotic poplars have been severely damaged by indigenous pathogens in their new locations, while native poplars are relatively unaffected. But the reverse outcome is also commonly observed (Newcombe, 2005), underscoring the need for regional clonal testing as opposed to theory-based prediction. Given this unpredictability, vigilance is needed to prevent the additional, inadvertent movement of pathogens on tolerant or resistant plant material into new geographic areas, where these pathogens could become damaging to native poplars or result in the evolution of new pathogenic races or hybrid species (Spiers and Hopcroft, 1998; Newcombe *et al.*, 2000). The international exchange of planting material has increased disease occurrence worldwide, and only strict phytosanitary practices will prevent future introductions of potentially damaging pathogens into new habitats (Rossman, 2001, 2009).

Worldwide, potentially the most damaging disease is perhaps leaf rust, caused by species and races of *Melampsora* that can cause premature leaf fall and reduce growth; rust can also predispose highly susceptible trees to other biotic and abiotic damaging agents (Royle and Ostry, 1995). Premature leaf fall associated with rust and other foliar pathogens can lead to dieback or outright mortality (Newcombe *et al.*, 2001a).

Some pathogens, in addition to causing a leaf disease, are also able to invade twigs, branches and main stems, resulting in cankers and dieback. Branch and stem cankers caused by *Mycosphaerella populorum* (with the better known asexual state called *Septoria musiva*) have a far greater impact than the foliar disease caused by the fungus in many regions of North America. The bacterium *Xanthomonas populi* has been a pathogen of high priority for resistance breeding due to its severe effects on the main trunks of susceptible poplars in Europe. A recent problem is infection by ice nucleation active (INA) bacteria that initiate early freezing in the twigs and stems of both willows and poplars at temperatures normally harmless, predisposing the plant to further infection by pathogenic bacteria or fungi.

This chapter provides an overview of the range of fungal and bacterial diseases affecting poplars and willows, including their distribution and importance in different regions of the world. We have chosen to highlight briefly those major pathogens that overall have been the most damaging.

It is important to note that many pathogens of *Populus* and *Salix* still await discovery. For example, a recent study (Busby *et al.*, 2012) of the foliar pathogens of *P. angustifolia* (the narrowleaf cottonwood of higher elevations in the western USA) revealed many surprises: (i) the absence of many of the foliar pathogens found on *P. trichocarpa* in the western USA; (ii) the relative absence of *Melampsora* rust; (iii) the presence of *Phyllactinia populi* that is commonly found on poplars in Asia but which appears to be restricted to *P. angustifolia* and its hybrids in North America; (iv) the presence of *Mycosphaerella angustifoliorum*, which is also restricted to *P. angustifolia* and its hybrids in North America; (v) new species of *Mycosphaerella* that will need to be described (Busby *et al.*, 2012).

Over the years, the literature on *Populus* and *Salix* pathogens has become voluminous. We provide a few recent general references that include numerous citations to the pertinent literature. Our intent is to bring attention to select pathogens that have the potential to threaten the health of willows and poplars and severely limit their usefulness and productivity. Not mentioned in this section are many foliage and stem diseases that normally are of minor importance but that opportunistically can become quite damaging under some environmental and site conditions. There are many diagnostic guides to the pests of poplars or willows available to those interested in distinguishing major and minor problems, and some of these citations are provided as well (Ostry *et al.*, 1989; Callan and Ring, 1994; Callan, 1998; Sinclair and Lyon, 2005; Horst, 2008).

8.2 Disease Prevention

8.2.1 Disease resistance

Prevention of disease problems is always a better management strategy than control after disease outbreaks (Ostry and McNabb, 1990). Planting disease-free stock, applying suitable cultural techniques, sanitation, avoiding conditions leading to plant stress and monitoring for unfavourable conditions are among the commonly suggested preventative practices to avoid disease.

Planting species and hybrids resistant to the major diseases is the best long-term strategy economically and environmentally. Recognizing this, breeding programmes for producing improved poplar clones have been active for many years in many countries. The focus in Europe has long been on producing high-yielding clones with emphasis on selecting for resistance to *Xanthomonas* bacterial canker and *Melampsora* leaf rust. In North America, resistance to leaf rust and *Septoria* leaf spot and canker have been among the targeted traits. Although willows have been cultivated for centuries, especially for basket production, willow improvement work to select fast-growing and highly productive clones started on a large scale only in the late 20th century.

Planting clones that have been screened for resistance to the local populations of pathogens is the most effective tactic for minimizing risk of damage (Ostry and McNabb, 1986). Plantations are expensive to establish and few, if any, direct control measures are available or cost-effective. Continually evolving races of pathogens on poplar and willow crops require a continuing effort to select and breed resistance to these new threats.

Effective and efficient disease control strategies and tactics, including the production of improved planting stocks, will require a continued investment in research and development. The temptation to plant the latest 'super clone' exclusively over large areas must be avoided; locally untested clones should also not be widely planted. Maintaining a broad genetic base in planting stock will require ongoing selection and breeding programmes that include collaboration among plant pathologists, breeders and entomologists.

8.2.2 Clonal mixtures

Successful experiments have been conducted to develop a practical, cost-effective disease control strategy to suppress leaf rust on willows within plantations using clonal mixtures (McCracken and Dawson, 1997). A random mixture of clones with somewhat different levels of resistance but similar growth form and rate has been demonstrated to slow disease increase during the growing season, resulting in less rust and increasing yields (Dawson *et al.*, 2005). Clearly, for clonal mixtures to succeed, the clones must vary specifically in genetic resistance.

When willows were planted in intimate, random mixtures, the onset of *Melampsora* spp. rust was delayed. The build-up of disease was slowed and the final rust level was reduced significantly (McCracken and Dawson, 2003; Samils *et al.*, 2003). Studies have indicated that the use of genotype mixtures has not led to the development of rust 'super' races or pathotypes capable of infecting all mixture components (McCracken *et al.*, 2000).

Climate can have a significant effect on new pathotype development. In maritime conditions where winters are mild and damp, *Melampsora* rust can survive asexually, whereas in countries with severe winter temperatures the pathogen must pass through a sexual stage. The latter begins with meiosis in the telial state on overwintered poplar leaves. Haploid basidiospores then infect the alternate host, for example the needles of a species of *Larix*, where insects can cross-fertilize the haploid spermogonia. Dikaryotic aecia can then form on the alternate host and the aeciospores can infect the leaves of *Populus*.

An important advantage of diverse mixtures is that if a particular genotype is eliminated over time due to disease or other cause, the remaining genotypes can capture the growing space and compensate for the loss (McCracken and Dawson, 2003). The greater the variation in genetic resistance among the clones in the mixture, the better the mixture will perform. In a mixture made up of seven straight *S. viminalis* genotypes, disease reduction was only evident in the early years after plantation establishment (Begley *et al.*, 2009). Yield from diverse mixtures is also greater than the equivalent yield of their components grown in monocultures (McCracken *et al.*, 2001), possibly because plant diversity in general is associated with greater productivity (Tilman *et al.*, 2001).

In North America, poplars have generally been planted in mosaics of pure clonal blocks, mimicking the clonal patterns of native stands of trembling aspen (*P. tremuloides*). The sizes of blocks vary depending on the size and configuration of the plantation. This allows for the independent management of the individual blocks should a problem arise in one or more clones and avoids large gaps resulting from variable growth rates and tree mortality among clones.

8.2.3 Biological control

Environmental concerns with chemical control and the limited number of available clones proven resistant to many diseases make the biological control of pathogens a highly desirable goal. However, thus far, few biological control agents for major poplar pathogens have been identified, and a practical approach for their use has not been developed. A common problem in using biological control agents concerns the formulation and application of the agents in order to have a significant effect on disease development in the field. Although not in practical use yet, the few examples that follow illustrate the potential use of biological control agents for the control of important pathogens of poplars and willows.

Eudarluca caricis (with the asexual state known as *Sphaerellopsis filum*) is a hyperparasite of rust fungi that is associated with the suppression of rust pustule development in the UK and Germany. In the USA, *E. caricis* is commonly seen hyperparasitizing *Melampsora medusae* on *P. deltoides* in the south-eastern region of the country (Nischwitz *et al.*, 2004), but it is absent or uncommon elsewhere. Foliar endophytes show considerable promise in reducing rust severity (Newcombe *et al.*, 2010) and selections are at present being tested in poplar bioenergy plantations in the Pacific Northwest (G. Newcombe *et al.*, 2012, unpublished information).

Under laboratory and greenhouse conditions, a spore suspension and culture filtrate of *Phaeotheca dimorphospora* inhibited *Septoria musiva* (Yang *et al.*, 1994). In addition, *Streptomyces* species have been demonstrated to suppress the incidence and severity of *Septoria* leaf spot on poplars in the field. Applied weekly, bimonthly or monthly to susceptible hybrid poplars, strains of *Streptomyces* significantly reduced leaf spot caused by *Septoria musiva* compared to control trees (Gyenis *et al.*, 2003).

8.3 Selected Leaf, Stem and Root Diseases of Poplars and Willows

As we have mentioned already, a large number of diseases have been recorded on poplars and willows in natural stands and in plantations around the world. The diseases affecting poplars and willows are caused by a very wide range of microorganisms and abiotic factors. Often, multiple foliar and stem pathogens co-occur, resulting in multiple diseases that can damage trees. Many foliar pathogens can invade twigs, branches and stems, resulting in cankers and branch dieback and breakage. Their impact is largely dependent on tree age, local microclimate, timing of infection and conditions for pathogen spread and disease development.

Native pathogens not normally important on trees in natural stands can become damaging on both native species and exotic species and hybrids when these are grown as an agricultural crop under intensive cropping systems. Following are brief descriptions of the most damaging and important diseases affecting poplar and willow crops. These diseases were selected based on their potential risk for large economic impacts and the critical need to prevent further spread of the causal agents into new regions.

8.3.1 *Melampsora* leaf rust: *Melampsora* spp.

Distribution

Leaf rust is the most widespread and serious disease of poplars and willows throughout the world. Some species of *Melampsora* have been introduced outside their native ranges within the northern hemisphere (Newcombe and Chastagner, 1993a; Innes *et al.*, 2004), and these and other taxa have been introduced inadvertently into the southern hemisphere along with their deliberately introduced hosts (Spiers, 1998).

Hosts/host specificity

Species of *Melampsora* are specialized on either *Populus* or *Salix*, and this specialization has evolved not once but multiple times in different parts of the northern hemisphere (Bennett *et al.*, 2011). Identification of species on the basis of uredinial morphology alone is often difficult. This is particularly true of species of *Melampsora* on willows, because many more species need to be described in the first place (Bennett *et al.*, 2011). Movement of *Melampsora* species into new geographic regions and the presence of *Melampsora* races, *formae speciales*, mixed infections on single leaves and interspecific hybrid species contribute to the complexity (Shain, 1988; Pataky, 1990; Pei and McCracken, 2005; Chen and Harrington, 2006). The potential evolution of new rust pathotypes affecting previously resistant clones underscores the critical need of monitoring poplar rust populations. Specialization within species of *Melampsora* is common (Ramstedt *et al.*, 2002).

Willow rusts

Melampsora epitea, a species native to Europe, was described on the basis of specimens from the Eurasian tree, *S. alba* (Plate 23A). Even in Europe, *M. epitea* is an ambiguous taxon (Pei and McCracken, 2005), but in the past, the taxon unfortunately was applied to willow rusts elsewhere in the northern hemisphere. It is now clear that this taxon does not apply to the willow rusts of the western USA (Bennett *et al.*, 2011), where many species will have to be described once their life cycles are completely understood. In contrast, knowledge of the willow rusts of Europe (Helfer, 1992; Pei *et al.*, 1993, 2005), Australasia (Spiers and Hopcroft, 1994) and Asia (Tian *et al.*, 2004) is relatively advanced, and a recent summary (Pei and McCracken, 2005) need not be repeated in its entirety here. In brief, Pei and others have recognized 33 taxa apart from *M. epitea*. Willow rust pathotypes or races specialize even at the level of host clones (Pei *et al.*, 1997), and much of this information has also been summarized in Pei and McCracken (2005).

The population structure of willow rusts has been studied by amplified fragment length polymorphism (AFLP) screening of rust diversification within fields and between countries, including the UK, Northern Ireland, Sweden, France and Chile (Hurtado and Ramstedt, 2002, 2003). No significant genetic differentiation between countries and continents was found, although pathotypes varied. This result implied that the willow rusts of Chile were introduced from Western Europe, just as willow rusts were introduced elsewhere in the southern hemisphere (Spiers, 1998).

Poplar rusts

Melampsora laricis-populina and its races are the most common and serious of the Eurasian poplar rusts. This species is also now the most widespread poplar rust, since it has been introduced into Australia and South and North America (Newcombe and Chastagner, 1993a; Innes *et al.*, 2004; Grondin *et al.*, 2005; Barrès *et al.*, 2008). *Melampsora allii-populina* is a closely related species also found on a range of poplars in Europe, but it has not been introduced elsewhere. *Melampsora medusae* is widely distributed in its native range in eastern North America and has now been found in western North America (Newcombe and Chastagner, 1993b). *Melampsora occidentalis* is native to western North America (Hsiang and Chastagner, 1993) and has been introduced in the central USA (Stack and Ostry, 1989; Moltzan *et al.*, 1993). In the western USA, *M. medusae* and *M. occidentalis* have hybridized and the hybrid taxon is known as *M.* ×*columbiana* (Newcombe *et al.*, 2000). This hybrid is currently generating the pathogenic variation that is proving challenging to poplar breeders in the Pacific Northwest region of the USA (Newcombe *et al.*, 2001b). *Melampsora abietis-canadensis* is found on hemlock and poplars in North America.

Biology

Most *Melampsora* species on poplars require an alternate host in the *Pinaceae* to complete their life cycle. Spermogonia and aecia are formed on the alternate host. Uredinia, telia and basidia form on *Populus*. The Eurasian species, *M. allii-populina*, is unusual in that its spermogonia and aecia regularly form on *Allium* or on *Arum*, but *M. laricis-populina* also occasionally forms these spore states on *Allium* or even *Corydalis* (Farr *et al.*, no date).

Melampsora species on willows collectively exhibit a much broader aecial host range in that *Tsuga*, *Abies*, *Larix*, *Ribes*, *Saxifraga*, *Allium*, *Euonymus*, *Viola*, *Galanthus*, *Corydalis*, and even some members of the *Orchidaceae* can all be infected (Pei and McCracken 2005; Farr *et al.*, no date), although no single species of *Melampsora* is likely to have this broad a range. In spring, teliospores in infected, dead leaves of *Populus* or *Salix* germinate and produce basidiospores, which are actively discharged and then windblown to the aecial host, where infection takes place. In summer, aeciospores produced in the needles may be carried by the wind back to poplar or willow leaves, where uredinial pustules develop and release urediniospores that intensify the disease on the hosts during the growing season. With a cycle of 7–12 days, new uredinia appearing on the leaves theoretically could give rise to 15–20 generations during the growing season. Normally, however, the disease intensifies during late summer due to the cooler and more humid periods required for spore germination and infection.

In the UK, and especially Northern Ireland with its more humid climate, rust infection can occur in early summer, so epidemics have longer to develop. Some rust fungi are also believed to survive within perennating structures, buds for example, of poplars or willows, and again the effect would be to increase the period during which an epidemic might develop.

Symptoms and damage

Uredinia (Plate 23D) commonly develop on the lower surface of leaves, giving rise to yellow or necrotic spots and flecks on the upper surface

(Plate 23C). Severely affected trees are defoliated prematurely (Plate 23B), reducing growth potential and predisposing them to environmental stresses and invasion by secondary damaging agents. Early leaf fall also increases the risk for winter injury, dieback and infection by secondary pathogens.

Control

There are no practical direct control measures for leaf rust. Preventative strategies include planting species and clones resistant to local populations of *Melampsora* (Pinon, 1992). Plantation densities will affect humidity and microclimate and this can be problematic, especially in bioenergy plantations that use close spacing.

Planting large, monoclonal blocks should be avoided as a precaution in the event that a new species or pathotype of rust becomes damaging. For willow cultivation, avoidance of large monoclonal blocks has proven successful, as have planting clone mixtures in Northern Ireland. As already discussed, the clonal mixtures are believed to reduce the selection pressure of a single gene on rust fungi and are thus more sustainable than monocultures (Ramstedt, 2003).

Removal of alternate hosts, for example *Larix*, will probably not reduce disease severity. In a study of two plantations in Sweden, one with larch trees growing within the plantation and the other separated by 25 km from the nearest larch, the endpoint for disease severity was the same, even though there was a delay in the latter (M. Ramstedt, 2010, unpublished data).

Biological control, as already discussed, is currently limited to *E. caricis* (Pei *et al.*, 2010). Some foliar endophytes are, however, very promising.

8.3.2 *Marssonina* leaf spot and blight: *Marssonina brunnea* (telemorph *Drepanopeziza tremulae*), *M. castagnei* (*D. populi-albae*), *M. populi* (*D. populorum*) and *M. balsamiferae* (telemorph unknown)

Marssonina is a damaging foliar pathogen of poplar. Several *Marssonina* species can cause premature defoliation, reduce growth potential and predispose affected trees to secondary damaging agents and stress (Plate 23E). Some clones of willows, usually tree willows not grown for biomass, are seriously affected, but generally willows escape significant damage.

Distribution

Four, relatively well-studied species of *Marssonina* on poplar are found throughout Europe, North America, Asia, Australia and New Zealand: *Marssonina brunnea* (teleomorph *Drepanopeziza tremulae*), *Marssonina castagnei* (*Drepanopeziza populi-albae*), *Marssonina populi* (*Drepanopeziza populorum*) and *Marssonina balsamiferae* (teleomorph unknown). On willows, *Marssonina brunnea* and *Marssonina salicicola* are the major species of importance. A fifth taxon, *Marssonina populicola*, is known to affect poplars in China (Zhuang, 2005).

Symptoms and damage

Leaf spots are small (1 mm), circular to angular, brownish to black. The small spots can coalesce into irregularly shaped larger flecks (Plate 23F). Whitish masses of conidia can be seen in the centre of the spots. On highly susceptible clones, lens-shaped lesions develop on petioles and current-year shoots. Severe disease on susceptible clones results in premature defoliation, dieback and predisposition to secondary organisms and abiotic stress.

Host specificity

Species in section *Populus* are susceptible to *M. brunnea* f. sp. *trepidae*, while species in the *Aigeiros* are susceptible to *M. brunnea* f. sp. *brunnea* (Newcombe and Callan, 1997). *M. balsamiferae* was reported occurring on *P. balsamifera* in Ontario and Manitoba. *M. castagnei* infects *P. alba* and *M. populi* is common on *P. tremuloides* (Spiers, 1983, 1984, 1988; Spiers and Hopcroft, 1998). *M. populicola* has been reported in China from *P. davidiana*, *P. euphratica*, *P.* ×*euramericana*, *P. laurifolia*, *P. pseudosimonii*, *P. simonii* and *P. tomentosa* (Zhuang, 2005).

Biology

The fungus overwinters in fallen leaves and in infected shoots (Plate 23G). In spring, ascospores produced in the leaves and conidia from lesions on the shoots are released in wet weather. Leaves and new shoots are infected throughout the growing season by rain-splashed conidia, intensifying disease severity.

CONTROL. Use of resistant clones is the best control measure. The fungus can infect branches and twigs without causing symptoms, so avoiding the spread of the pathogen by movement of diseased cuttings is important. It is also known to be seedborne, so care needs to be taken to avoid inadvertent shipment of infested seed. Pruning diseased twigs and branches as well as raking and burning fallen leaves may also provide disease control. Fungicidal control has proven to be successful, but no fungicides are labelled specifically for use in plantations at this time.

8.3.3 *Venturia* leaf and shoot blight

Leaf and shoot blight of poplars and willows are caused by a total of 17 taxa of *Venturia*, 12 taxa on *Populus* and an additional five on *Salix* (Farr *et al.*, no date). *Venturia* species produce either *Fusicladium* or *Pollaccia* anamorphs, or asexual states (Beck *et al.*, 2005).

Distribution

Northern hemisphere. Although species of *Populus* native to both Eurasia and North America have been reported as hosts of *Venturia*, this does not appear to be the case for *Salix* (Farr *et al.*, no date). *Venturia* has been reported from *Salix* in North America, but for the four major subgenera (*Vetrix*, *Longifoliae*, *Salix* and *Protitea*) the reports are only from Eurasian species. In other words, it is not clear that there are North American species of *Venturia* that parasitize the North American species of these four subgenera.

Hosts/host specificity

Studies of the host specificity of *Venturia* on *Populus* and *Salix* have been limited.

Symptoms and damage

Wilting shoot tips and leaves turn black and can often resemble frost damage (Plate 24C). Irregular necrotic blotches develop, with a layer of olive-green spores covering the diseased area of the leaves. Severe disease can reduce potential growth and stunt and deform young trees, resulting in a shrubby appearance. Willow scab caused by *Venturia saliciperda* can be serious on some ornamental species of Eurasian origin, e.g. *S. alba* or *S. babylonica*. Young plants may be killed. *V. saliciperda* appears to be a minor pathogen on willow clones cultivated for SRF.

Biology

A high incidence of leaf and shoot blight often occurs during wet spring weather. In early spring, conidia produced in shoots killed the previous year and ascospores from infected overwintered leaves infect developing leaves and shoots. Conidia that develop on newly killed shoots (Plate 24A) and infected leaves can infect additional leaves and shoots in wet weather during late spring and early summer (Plate 24B).

Control

Trees growing in dense stands under high humidity are often more susceptible to damage than those growing at wider spacing. Disease severity varied among host species and hybrid clones (Newcombe and van Oosten, 1997), and resistance was inherited from the exotic parent in a field experiment with interspecific hybrid pedigrees on Vancouver Island (Newcombe, 2003). Breeding and selection for resistance is practised in the Pacific Northwest of the USA. Chemical control is possible, but is not practical for large plantations (Anderson and Anderson, 1980).

8.3.4 Bronze leaf disease: *Apioplagiostoma populi*

Distribution

This disease is known only in North America in Alberta, Saskatchewan, Quebec, Ontario and Manitoba, Canada, and in the north-eastern and mid-western USA (Northover and Desjardins, 2003; Kawchuk *et al.*, 2010).

Hosts/host specificity

Bronze leaf disease affects several *Populus* species and is particularly damaging to hybrids in section *Populus* (Cash and Waterman, 1957; Heimburger, 1966). Disease severity and damage varies among clones. Hybrids of *P. grandidentata* and *P. tremuloides* can be severely damaged. *P.* ×*canescens* planted as an ornamental or in windbreaks (Plate 24D) has also been severely damaged in many areas. Susceptibility of *P. alba* has been reported, but in both reports, authors noted that affected individuals might have been misidentified hybrids with aspen.

Symptoms and damage

Developing leaves on affected branches in spring are often small and chlorotic, later turning to the distinctive bronze and reddish-brown colour (Plate 24E). Disease initially is most severe in the lower crown, eventually developing throughout the crown. Dieback of affected branches leads to a general decline in vigour over time, and eventual tree death.

Biology

A. populi overwinters as immature perithecia in infected leaves on the ground or leaves remaining on branches (Dance, 1957). Ascospores are discharged in spring. The resulting scattered infected leaves turn orange-brown or reddish-brown from the margins inward in midsummer, and become dry and turn dark reddish-brown in autumn. Acervuli with unicellular spores develop in early autumn and are thought to function as a spermatial stage. The fungus also overwinters as mycelium in infected branches and systemically invades developing tissues the following year (Smith *et al.*, 2002). This results in the uniform infection of all leaves on affected branches, in contrast to the scattered infected leaves resulting from ascospore infection.

Control

Prevention of the disease by avoiding propagating and planting highly susceptible clones in areas where the disease is known to be present is the only practical measure. Care needs to be exercised not to move systemically infected plant material inadvertently.

8.3.5 *Septoria* leaf spot and canker

On *Populus*, taxa include the following: *Septoria musiva* (teleomorph *Mycosphaerella populorum*), *Septoria populicola* (teleomorph: *Mycosphaerella populicola*) and *Septoria populi* (teleomorph *Mycosphaerella populi*) (Farr *et al.*, no date). On *Salix*, a dozen taxa have been reported (Farr *et al.*, no date).

S. musiva is the most economically important species, causing both leaf spot and canker diseases of poplars. Defoliation and stem breakage can result in the complete loss of highly susceptible poplar clones.

Willows are affected by *Septoria didyma*, *Septoria salicina* and *Septoria salicicola*, among others, but these fungi are not regarded as economically important pathogens of willow. Recently *S. musiva* was reported from *Salix lucida* in Quebec, Canada (Feau and Bernier, 2004). This potentially could be a serious disease if there were further spread and establishment of the pathogen on other important willow species and clones.

Distribution

In North America, *S. musiva* is widely distributed in the mid-west and eastern regions. Recently, it was confirmed in commercial hybrid plantations and nurseries in British Columbia (Callan *et al.*, 2007). There has been to date no confirmation of its presence on native *P. trichocarpa* in the area around the affected plantations. *S. musiva* is not known to be present in Europe and is regarded as a high priority quarantine pathogen.

Hosts/host specificity

Species and hybrids in sections *Populus*, *Aigeiros* and *Tacamahaca* vary widely in susceptibility to leaf spot; however, many *Tacamahaca* hybrids are highly susceptible to both leaf spot and canker in eastern North America. *S. musiva* has been recovered from leaf spots on shining willow (*Salix lucida* spp. *lucida*) in Quebec. *Septoria populicola* (teleomorph *Mycosphaerella populicola*) causes leaf spots on *P. balsamifera* and *P. trichocarpa*, but generally is not thought to cause cankers in nature. *S. musiva* has been reported causing cankers in plantings of *P. balsamifer* (Leboldus *et al.*, 2008a, b). *Septoria populi* is of periodic importance in causing leaf spot disease in Europe. *Mycosphaerella angustifoliorum* causes leaf spots on *P. angustifolia* in the mountains of western North America (Ramaley, 1991). It is likely that with more study new species of *Mycosphaerella* will be discovered on both *Salix* and *Populus* (Busby *et al.*, 2012).

Symptoms and damage

On native *P. trichocarpa*, *S. musiva* causes a minor leaf disease resulting in necrotic spots (Newcombe *et al.*, 1995) (Plate 25B); however, on many hybrid poplars, severe disease can result in premature defoliation, reduced growth potential and may subject severely affected trees to invasion by secondary damaging fungi and insect pests. In mid-western and eastern North America, *S. musiva* is the common cause of branch and stem cankers (Plate 25A) on planted *P. trichocarpa* hybrids and many other hybrid poplars, which can result in tree breakage and windbreak and plantation failures (Plate 25C).

Biology

The fungus primarily overwinters in fallen leaves. Ascospores are windblown and rain-splashed to developing leaves, stems and branches in the spring. Conidia produced in these infected leaves and young branch cankers are rain-splashed to adjacent leaves, stems and branches throughout the growing season, intensifying the foliage and canker disease within individual trees and infecting additional trees. Infection courts include leaf scars and stipules, lenticels or various wounds.

Control

Although research on chemical and biological control treatments has been encouraging in nurseries, direct control in plantations is more difficult. Successful biological control of *Septoria* leaf spot in the field using *Streptomyces* spp. applications has been demonstrated (Gyenis *et al.*, 2003). Planting resistant poplar species and clones is the best preventative strategy (Newcombe and Ostry, 2001; Ward and Ostry, 2005). Care needs to be exercised to prevent movement of infected planting stock into areas where the pathogen is not known to be present.

8.3.6 Hypoxylon canker: *Entoleuca mammata (= Hypoxylon mammatum)*

Distribution

Entoleuca mammata is widely distributed throughout North America. Hypoxylon canker is particularly damaging in Minnesota, Wisconsin and Michigan, but is less common in the western USA and has not been reported from Alaska. The disease has been reported from Finland, Germany, Sweden and France. The pathogen has been collected in Italy and Switzerland (Kasanen *et al.*, 2004).

Hosts/host specificity

P. tremuloides is the most commonly and severely affected species in North America, and *P. grandidentata* is occasionally infected. *P. tremula* in Europe, and when planted in North America, is susceptible. In Europe, *P. alba*, *P. trichocarpa* and *P. tremula* × *P. tremuloides* have been reported as hosts as well. In North America, Hypoxylon cankers have been reported occurring on *P. balsamifera*, *P. nigra* var. *betulifolia* × *P. nigra* 'Volga', *P. nigra* var. *betulifolia* × *P. trichocarpa*, *P. maximowiczii* × (*P.* ×*berolinensis*), *P.* 'Candicans' × (*P.* ×*berolinensis*) and *P. deltoides* × *P. nigra* 'Incrassata'. Several other hardwoods, including *Salix* spp., have been reported as hosts, but conclusive evidence for confirming saprophytic or pathogenic relationships on these species is lacking. *E. mammata* isolates vary in virulence, and somatic incompatibility keeps them genetically isolated and unique.

Symptoms and damage

Cankers can girdle and kill branches (Plate 25E) and main-stem cankers can cause top dieback, stem breakage or kill trees, depending on the height of the canker on the stem (Plate 25D).

Biology

Ascospores, but not conidia, are infective (Plate 25F). The fungus requires a wound through the bark for infection. Several insects common on the branches of poplars cause oviposition wounds that are often infection sites. A number of interacting pre- and post-infection host, pathogen and environmental factors are critical for infection, disease progression and subsequent disease severity (Ostry and Anderson, 2009).

Control

No direct control is known. Maintaining high stem densities in stands promotes self-pruning of branches, reduces insect oviposition wounds and minimizes branch and lethal lower-stem infections. Aspen clones and species vary widely in susceptibility to the disease. It has been proposed in native aspen stands that highly susceptible clones not be regenerated but allowed to break up, so that the surrounding superior canker-resistant clones replace the susceptible clone. Improvement of *P. tremuloides* through selection and breeding has been hampered by the many non-specific wound responses of aspen clones that account for poor correlation of resistance screening results to natural disease incidence or prevalence.

8.3.7 Miscellaneous branch and stem cankers

Cryptodiaporthe (*Dothichiza*) canker

Cryptodiaporthe populea (*Discosporium populeum*). *Cryptodiaporthe* canker is a damaging canker pathogen of poplars in the northern hemisphere and in South America. Trees under stress are highly susceptible to infection. Susceptibility varies by poplar species and hybrid, with members of sections *Aigeiros* and *Tacamahaca* being especially vulnerable. Early infection in nurseries can destroy seedlings or reduce their growth potential. Cankers can impact the growth and form of older trees and can girdle stems and subject infected trees to wind breakage.

Leucostoma canker

Cytospora chrysosperma. *Cytospora* species are predominantly opportunistic secondary pathogens that infect poplars and willows through wounds, scars or cracks in the bark on the main stem, twigs and branches. Infected bark becomes discoloured, and the affected areas may coalesce and girdle the stem. Trees under stress, especially water stress, are most susceptible. Damaged stems are susceptible to wind breakage.

8.3.8 Bacterial diseases: *Pseudomonas* spp., *Xanthomonas* spp., *Sphingomonas* spp., *Erwinia* spp.

Stem diseases of willows have not been considered a serious problem until recent years. Bacterial canker and bark necrosis combined with freezing stress has largely been a neglected problem within SRF. Several bacterial species, for example *Pseudomonas syringae*, *Erwinia* spp. and *Xanthomonas* spp., can be involved and are usually ubiquitously present within plantings. *P. syringae* is an INA bacterium that, together with several other bacterial pathogens, is found on willows (Plate 26A). It generally only infects and damages trees under frost stress.

Xanthomonas populi (= *Aplanobacter populi*)

The bacterium *X. populi* has been a priority pathogen for resistance breeding in poplars for many years in Europe. *Xanthomonas* pv. *salicis* has been reported from the Netherlands, killing and damaging *Salix dasyclados* (Gremmen and de Kam, 1974).

Distribution. The poplar canker disease caused by this bacterium is common in Europe but is not known to be present in North America.

Hosts/host specificity. Several physiologic races of *X. populi* pv. *populi* infect species and clones in sections *Aigeiros* and *Tacamahaca* (Ridé and Ridé, 1978; de Kam, 1981; Nesme *et al.*, 1993) and *X. populi* pv. *salicis* infects willows (de Kam, 1977, 1978).

Symptoms and damage. Bacterial canker (Plate 26B) is one of the most damaging diseases of poplars throughout much of Europe, resulting in branch and stem cankers and dieback (Plate 26C).

Biology. The bacterium is spread by rain splash, insects, animals and equipment. Infection occurs in spring through bud scales, stipules, leaf scars and fresh wounds (Plate 26D).

Control. Sanitation to avoid the spread of the bacterium into new areas on planting stock or equipment and planting disease-resistant poplar clones are the best preventative measures (Lonsdale and Rose, 1998).

Pseudomonas syringae

Distribution. *P. syringae* is a ubiquitous bacterium with numerous varieties and pathovars. In SRF plantings, this bacterium was first reported on poplars in the Netherlands, but was not considered important economically. Since 2000, increasing dieback of poplars and willows caused by this bacterium has been reported in Sweden.

Hosts/host specificity. *P. syringae* strains isolated from poplars or willows can infect and cause disease on trees of both genera, although clone specificity has not been found among different isolates. Susceptibility differs among willow clones, and epiphytic populations of the bacterium differ significantly among clones.

Biology. As an opportunistic pathogen, it can grow epiphytically on stems and leaves in many environments and crops without causing damage until suitable conditions for disease develop.

Symptoms and damage. *P. syringae* is a pathogen that can seriously damage trees throughout a plantation. Dieback of shoots and large areas of affected bark can be invaded by damaging secondary pathogens. Strains with ice-nucleating properties will initiate freezing at an early stage and predispose the plants to further infection and disease development, as well as infection by other pathogens (de Kam, 1982; Ramstedt *et al.*, 1994; Nejad *et al.*, 2004, 2006a, b; Cambours *et al.*, 2005).

Control. Use of resistant clones and wider spacing in plantations are the best management practices. Using clonal mixtures and maintaining vigorous growth of plants are not effective, since the pathogen is not host-specific and is favoured by high humidity.

8.3.9 Root diseases

Brenneria salicis (syn. *Erwinia salicis*, *Pseudomonas saliciperda*)

Root disease research of poplars and willows has not yet been a high priority. One exception is the water mark disease, caused by the Gram-negative bacterium, *B. salicis*, that is especially damaging to *S. alba* (Day, 1924). The disease was first described in the early 1900s in England and the Netherlands, and it was recently reported from Belgium and Japan. The bacterium infects the roots and vessels where it resides, and can be asymptomatic in less susceptible hosts or non-hosts such as poplar, alder and other willows species. Disease symptoms include wilting leaves and discoloured wood with a high moisture content. An important agent for spreading the pathogen is infested propagative material (Gremmen and de Kam, 1975). The potential presence of latent infections in roots and vessels of a wide range of species underscores the importance of careful screening of propagation stocks to avoid further spread of this pathogen.

Phytophthora spp.

Although *Phytophthora* root rot has been reported only occasionally on poplars, with three of the five current records from the southern hemisphere (Farr *et al.*, no date), vigilance is warranted. *Phytophthora* diseases have been causing increasing problems for a number of tree species. *Phytophthora* is able to adapt rapidly to new hosts. A new, more aggressive population has appeared on alder in recent years (Brasier *et al.*, 2004). These emergent species hybrids are a new and serious threat to *Alnus* in Europe. A similar microevolutionary development in *Phytophthora* in dense willow or poplar plantings in wet areas or in plantations treated with wastewater could be devastating.

References

Anderson, N.A. and Anderson, R.L. (1980) Leaf and shoot blight of aspen caused by *Venturia macularis* in northern Minnesota. *Plant Disease* 64, 558–559.

Argus, G.W. (2010) *Salix*. In: Editorial Committee (ed.) *Flora of North America North of Mexico*. Oxford University Press, New York and Oxford, UK, pp. 23–51.

Barrès, B., Halkett, F., Dutech, C., Andrieux, A., Pinon, J. and Frey, P. (2008) Genetic structure of the poplar rust fungus *Melampsora larici-populina*: evidence for isolation by distance in Europe and recent founder effects overseas. *Infection, Genetics and Evolution* 8, 577–587.

Beck, A., Ritschel, A., Schubert, K., Braun, U. and Triebel, D. (2005) Phylogenetic relationships of the anamorphic genus *Fusicladium* s. lat. as inferred by ITS nrDNA data. *Mycological Progress* 4, 111–116

Begley, D., McCracken, A.R., Dawson, W.M. and Watson, S. (2009) Interaction in short rotation coppice willow, *Salix viminalis* genotype mixtures. *Biomass and Bioenergy* 33, 163–173.

Bennett, C.M., Aime, M.C. and Newcombe, G. (2011) Molecular and pathogenic variation within *Melampsora* on *Salix* in western North America reveals numerous cryptic species. *Mycologia* 103, 1004–1018.
Brasier, C.M., Kirk, S.A., Delcan, J., Cooke, D.E.L., Jung, T. and Man In't Veld, W.A. (2004) *Phytophthora alni* sp. nov. and its variants: designation of emerging heteroploid hybrid pathogens spreading on *Alnus* trees. *Mycological Research* 108, 1172–1184.
Busby, P.E., Aime, M.C. and Newcombe, G. (2012) Foliar pathogens of *Populus angustifolia* are consistent with a hypothesis of Beringian migration into North America. *Fungal Biology* 116, 792–801.
Callan, B.E. (1998) *Diseases of Populus in British Columbia: a Diagnostic Manual.* Natural Resources Canada, Canadian Forest Service, Pacific Forestry Centre, Victoria, British Columbia, Canada.
Callan, B.E. and Ring, F.M. (1994) An annotated host fungus index for *Populus* in British Columbia. FRDA Report 222. Canadian Forest Service, Pacific Forestry Centre, Victoria, British Columbia, Canada.
Callan, B.E., Leal, I., Foord, B., Dennis, J.J. and van Oosten, C. (2007) *Septoria musiva* isolated from cankered stems in hybrid poplar stool beds, Fraser Valley, British Columbia. *Pacific Northwest Fungi* 2, 1–9.
Cambours, M.A., Nejad, P., Granhall, U. and Ramstedt, M. (2005) Frost-related dieback of willows; comparison of epiphytic and endophytic bacteria of different *Salix* clones, with emphasis on ice nucleation activity, pathogenic properties and seasonal variation. *Biomass and Bioenergy* 28, 15–27.
Cash, E. and Waterman, A. (1957) A new species of *Plagiostoma* associated with a leaf disease of hybrid aspens. *Mycologia* 49, 756–760.
Chen, W. and Harrington, T.C. (2006) Genetic diversity of poplar leaf rust populations in the north-central United States. *Canadian Journal of Forest Res*earch 36, 2047–2057.
Dance, B.W. (1957) A fungus associated with blight and dieback of hybrid aspen. Canada Department of Agriculture, Science Service, Forest Biology Division. *Bi-monthly Progress Report* 13, 1–2.
Dawson, W.M., McCracken, A.R. and Carlisle, D. (2005) Short rotation coppice willow mixtures and yield. In: Pei, M.H. and McCracken, A.R. (eds) *Rust Diseases of Willow and Poplar.* CAB International, Wallingford, UK, pp. 195–208.
Day, W.R. (1924) The watermark disease of the cricket-bat willow (*Salix caerculea*). Clarendon Press, Oxford, UK. *Oxford Forestry Memoirs* 3, 1–30.
de Kam, M. (1977) A bacterial disease of *Salix dasyclada* caused by a *Xanthomonas* species and its relation to *Aplanobacter populi. European Journal of Forest Pathology* 7, 257–262.
de Kam, M. (1978) *Xanthomonas populi* subsp. *salicis*, cause of bacterial canker in *Salix dasyclados. European Journal of Forest Pathology* 8, 334–337.
de Kam, M. (1981) The identification of the two subspecies of *Xanthomonas populi in vitro. European Journal of Forest Pathology* 11, 25–29.
de Kam, M. (1982) Damage to poplar caused by *Pseudomonas syringae* in combination with frost and fluctuating temperatures. *European Journal of Forest Pathology* 12, 203–209.
Eckenwalder, J.E. (2001) Descriptions of clonal characteristics. In: Dickmann, D.I., Isebrands, J.G., Eckenwalder, J.E. and Richardson, J. (eds.) *Poplar Culture in North America.* National Research Council of Canada Research Press, Ottawa, pp. 331–382.
Eckenwalder, J.E. (2010) *Populus.* In: Editorial Committee (ed.) *Flora of North America North of Mexico.* Oxford University Press, New York and Oxford, UK, pp. 5–23.
Farr, D.F., Rossman, A.Y., Palm, M.E. and McCray, E.B. (no date) Fungal Databases. Systematic Mycology and Microbiology, US Department of Agriculture, Agriculture Research Service (http://nt.ars-grin.gov/fungaldatabases/, accessed 30 May 2012).
Feau, N. and Bernier, L. (2004) First report of shining willow as a host plant for *Septoria musiva. Plant Disease* 88, 770.
Gremmen, J. and de Kam, M. (1974) Research on poplar canker (*Aplanobacter populi*) in the Netherlands. II. *European Journal of Forest Pathology* 4, 175–181.
Gremmen, J. and de Kam, M. (1975) The necessity of using healthy propagation material of *Salix alba* in connection with the spread of watermark disease in the Netherlands. *European Journal of Forest Pathology* 5, 376–383.
Grondin, J., Bourassa, M. and Hamelin, R.C. (2005) First report of the aecial state of *Melampsora larici-populina* on *Larix* spp. in North America. *Plant Disease* 89, 1242.
Gyenis, L., Anderson, N.A. and Ostry, M.E. (2003) Biological control of *Septoria* leaf spot disease of hybrid poplar in the field. *Plant Disease* 87, 809–813.
Heimburger, C. (1966) Susceptibility to a serious fungus attack as a genetic barrier between aspen species. In: Gerhold, H.D., Schreiner, E.J., McDermott, R.E. and Winieske, J.A. (eds) *Breeding Pest-Resistant Trees.* Pergamon Press, Oxford, UK, pp. 391–393.

Helfer, S. (1992) The rust diseases of willow in Britain. *Proceedings of the Royal Society of Edinburgh* 98B, 119–134.
Horst, R.K. (2008) Plant diseases and their pathogens. In: Horst, R.K. (ed.) *Westcott's Plant Disease Handbook*, 7th edn. Springer, Dordrecht, Berlin, Heidelberg, New York, pp. 573–574.
Hsiang, T. and Chastagner, G.A. (1993) Variation in *Melampsora occidentalis* rust on poplars in the Pacific Northwest. *Canadian Journal of Plant Pathology* 15, 175–181.
Hurtado, S. and Ramstedt, M. (2002) AFLP comparison of distant *Melampsora epitea* (willow rust) populations. *Mycological Research* 106, 1400–1407.
Hurtado, S. and Ramstedt, M. (2003) Comparing Chilean and Swedish rust populations: possible impact of long-distance transport of *Melampsora* rust. *Forest Pathology* 33, 69–80.
Innes, L., Marchand, L., Frey, P., Bourassa, M. and Hamelin, R.C. (2004) First report of *Melampsora larici-populina* on *Populus* spp. in Eastern North America. *Plant Disease* 88, 85.
Kasanen, R., Hantula, J., Ostry, M., Pinon, J. and Kurkela, T. (2004) North American populations of *Entoleuca mammata* are genetically more variable than populations in Europe. *Mycological Research* 108, 766–774.
Kawchuk, L.M., Howard, R.J., Kalischuk, M.L., Northover, P.R., Desjardins, M. and Spencer, R.C.J. (2010) First report of bronze leaf disease on poplar in Alberta, Canada and sequence of *Apioplagiostoma populi. Plant Disease* 94, 377.
Leboldus, J.M., Blenis, P.V. and Thomas, B.R. (2008a) An epidemic of *Septoria* canker on *Populus balsamifera* in northern Alberta. *Phytopathology* 98 (Supplement), S87 (abstract).
Leboldus, J.M., Blenis, P.V. and Thomas, B.R. (2008b) Clone by isolate interaction in the hybrid poplar–*Septoria musiva* pathosystem. *Canadian Journal of Forest Research* 38, 1888–1896.
Lonsdale, D. and Rose, J. (1998) Resistance of new Belgian poplar clones to British isolates of the bacterial canker pathogen *Xanthomonas populi. European Journal of Forest Pathology* 28, 227–232.
McCracken, A.R. and Dawson, W.M. (1997) Using mixtures of willow clones as a means of controlling rust disease. In: Bullard, M.J., Ellis, R.G., Health, M.C., Knight, J.D., Lainsbury, M.A. and Parker, S.R. (eds) *Aspects of Biology 49 : Biomass and Energy Crops.* Association of Applied Biologists, Cirencester, UK, pp. 97–103.
McCracken, A.R. and Dawson, W.M. (2003) Rust disease (*Melampsora epitea*) of willow (*Salix* spp.) grown in short rotation coppice (SRC) in inter- and intra-species mixtures. *Annals of Applied Biology* 143, 381–393.
McCracken, A.R., Dawson, W.M., Watson, S. and Allen, C.Y. (2000) Pathotype composition in *Melampsora epitea* populations occurring on willow (*Salix*) grown in mixed and monoculture plantations. *European Journal of Plant Pathology* 106, 879–886.
McCracken, A.R., Dawson, W.M. and Bowden, G. (2001) Yield response of willow (*Salix*) grown in mixtures in short rotation coppice (SRC). *Biomass and Bioenergy* 21, 311–319.
Mitchell, C.E. and Power, A.G. (2003) Release of invasive plants from fungal and viral pathogens. *Nature* 421, 625–627.
Moltzan, B.D., Stack, R.W., Mason, P.A. and Ostry, M.E. (1993) First report of *Melampsora occidentalis* on *Populus trichocarpa* in the Central United States. *Plant Disease* 77, 953.
Nejad, P., Ramstedt, M. and Granhall, U. (2004) Pathogenic ice-nucleation active bacteria in willows for short rotation forestry. *Forest Pathology* 34, 369–381.
Nejad, P., Granhall, U. and Ramstedt, M. (2006a) Factors influencing pathogenic and Ice-Nucleation Active (INA) bacteria isolated from *Salix* plants, soil and litter. *International Journal of Agricultural Technology* 2, 207–222.
Nejad, P., Ramstedt, M., Granhall, U., Roos, S. and McIvor, I. (2006b) Biochemical characterization and identification of Ice-Nucleation-Active (INA) willow pathogens by means of BIOLOG® MicroPlate, INA gene primers and a PCR based 16S rRNA analysis. *Journal of Plant Diseases and Protection* 113, 97–106.
Nesme, X., Steenackers, M., Steenackers, V., Picard, Ch., Ménard, M. and Ridé, M. (1993) Differential host–pathogen interactions among clones of poplar and strains of *Xanthomonas populi* pv. *populi. Phytopathology* 84, 101–107.
Newcombe, G. (1996) The specificity of fungal pathogens of *Populus*. In: Stettler, R.F., Bradshaw, H.D. Jr, Heilman, P.E. and Hinckley, T.M. (eds) *Biology of Populus and Its Implications for Management and Conservation.* National Research Council of Canada Research Press, Ottawa, pp. 223–246.
Newcombe, G. (2003) Native *Venturia inopina* sp. nov., specific to *Populus trichocarpa* and its hybrids. *Mycological Research* 107, 108–116.

Newcombe, G. (2005) Genes for parasite-specific, nonhost resistance in *Populus. Phytopathology* 95, 779–783.
Newcombe, G. and Callan, B.E. (1997) First report of *Marssonina brunnea* f. sp. *brunnea* on hybrid poplar in the Pacific Northwest. *Plant Disease* 81, 231.
Newcombe, G. and Chastagner, G.A. (1993a) First report of the Eurasian poplar leaf rust fungus *Melampsora larici-populina*, in North America. *Plant Disease* 77, 532–535.
Newcombe, G. and Chastagner, G.A. (1993b) A leaf rust epidemic of hybrid poplar along the lower Columbia River caused by *Melampsora medusae. Plant Disease* 77, 528–531.
Newcombe, G. and Dugan, F. (2010) Fungal pathogens of plants in the homogocene. In: Gherbawy K.V.Y. (ed.) *Molecular Identification of Fungi.* Springer Verlag, Berlin, Heidelberg, New York, pp. 3–35.
Newcombe, G. and Ostry, M. (2001) Recessive resistance to *Septoria* stem canker of hybrid poplar. *Phytopathology* 91, 1081–1084.
Newcombe, G. and van Oosten, C. (1997) Variation in resistance to *Venturia populina*, the cause of poplar leaf and shoot blight in the Pacific Northwest. *Canadian Journal of Forest Research* 27, 883–889.
Newcombe, G., Chastagner, G.A., Callan, B.E. and Ostry, M.E. (1995) An epidemic of *Septoria* leaf spot on *Populus trichocarpa* in the Pacific Northwest in 1993. *Plant Disease* 79, 212.
Newcombe, G., Stirling, B., McDonald, S. and Bradshaw, H.D. Jr (2000) *Melampsora ×columbiana*, a natural hybrid of *M. medusae* and *M. occidentalis. Mycological Research* 104, 261–274.
Newcombe, G., Ostry, M., Hubbes, M., Périnet, P. and Mottet, M.-J. (2001a) Poplar diseases. In: Dickmann, D.I., Isebrands, J.G., Eckenwalder, J.E. and Richardson, J. (eds) *Poplar Culture in North America.* National Research Council of Canada Research Press, Ottawa, pp. 249–276.
Newcombe, G., Stirling, B. and Bradshaw, H.D. Jr (2001b) Abundant pathogenic variation in the new hybrid rust *Melampsora ×columbiana* on hybrid poplar. *Phytopathology* 91, 981–985.
Newcombe, G., Martin, F. and Kohler, A. (2010) Defense and nutrient mutualisms in *Populus.* In: Jansson, S. (ed.) *Genetics and Genomics of Populus.* Springer-Verlag, Berlin, Heidelberg, Germany, pp. 247–278.
Nischwitz, C., Newcombe, G. and Anderson, C.L. (2004) Host specialization of the mycoparasite *Eudarluca caricis* and its evolutionary relationship to *Ampelomyces. Mycological Research* 109, 421–428.
Northover, P.R. and Desjardins, M. (2003) First report of bronze leaf disease on hybrid poplar (*Populus ×canescens* 'Tower') caused by *Apioplagistoma populi* in Manitoba, Canada. *Plant Disease* 87, 1538.
Ostry, M.E. and Anderson, N.A. (2009) Genetics and ecology of the *Entoleuca mammata–Populus* pathosystem: implications for aspen improvement and management. *Forest Ecology and Management* 257, 390–400.
Ostry, M.E. and McNabb, H.S. Jr (1985) Susceptibility of *Populus* species and hybrids to disease in the North Central United States. *Plant Disease* 69, 755–757.
Ostry, M.E. and McNabb, H.S. Jr (1986) *Populus* species and clones resistant to *Melampsora, Marssonina,* and *Septoria.* United States Department of Agriculture, Forest Service, North Central Forest Experiment Station. Research Paper NC-272, 7p.
Ostry, M.E. and McNabb, H.S. Jr (1990) Minimizing disease injury to hybrid poplars. *Journal of Environmental Horticulture* 8, 96–98.
Ostry, M.E., Wilson, L.F., McNabb, H.S. and Moore, L.M. (1989) *A Guide to Insect, Disease, and Animal Pests of Poplars.* Agriculture Handbook 677. US Department of Agriculture, Washington, DC.
Pataky, N.R. (1990) Leaf rust on poplars and willows in the Midwest. *Report on Plant Disease*, RPD No 605, University of Illinois, Champaign, Illinois.
Pei, M.H. and McCracken, A.R. (2005) *Rust Diseases on Willow and Poplar.* CAB International, Wallingford, UK.
Pei, M.H., Royle, D.J. and Hunter, T. (1993) Identity and host alteration of some willow rusts (*Melampsora* spp.) in England. *Mycological Research* 97, 845–851.
Pei, M.H., Whelan, M.J., Halford, N.G. and Royle, D.J. (1997) Distinction between stem- and leaf-infecting forms of *Melampsora* rust on *Salix viminalis* using RAPD markers. *Mycological Research* 101, 7–10.
Pei, M.H., Bayon, C. and Ruiz, C. (2005) Phylogenetic relationships in some *Melampsora* rusts on *Salicaceae* assessed using rDNA sequence information. *Mycological Research* 109, 401–409.
Pei, M.H., Ruiz, C., Hunter, T. and Bayon, C. (2010) Interaction between *Melampsora larici-epitea* pathotypes and the mycoparasite *Sphaerellopsis filum* from willow rusts. *Forest Pathology* 40, 33–42.
Pinon, J. (1992) Variability in the genus *Populus* in sensitivity to *Melampsora* rusts. *Silvae Genetica* 41, 25–34.
Pinon, J. and Valadon, A. (1997) Comportement des cultivars de peupliers commercialisables dans l'Union européenne vis-à-vis de quelques parasites majeurs. *Annals of Forest Science* 54, 19–38.

Ramaley, A.W. (1991) *Clypeispora* and its *Mycosphaerella* teleomorph. *Mycotaxon* 40, 13–22.
Ramstedt, M. (2003) Willow rust pathotypes in clonal mixtures as compared to monoculture plantations. In: Xu, M.-Q., Walla, J.A. and Zhao, W.-X. (eds) *Proceedings of the Second IUFRO Rusts of Forest Trees Working Party Conference*, 19–23 August 2002, Yangling, China. *Forest Research* 16 (Suppl.), 91–94.
Ramstedt, M., Åström, B. and Von Fircks, H. (1994) Dieback of poplar and willow caused by *Pseudomonas syringae* in combination with freezing stress. *European Journal of Forest Pathology* 24, 305–315.
Ramstedt, M., Hurtado, S. and Åström, B. (2002) Pathotypes of *Melampsora* rust on *Salix* in short-rotation forestry plantations. *Plant Pathology* 51, 185–190.
Ridé, M. and Ridé, S. (1978) *Xanthomonas populi* (Ridé) comb. Nov. (syn. *Aplanobacter populi* Ridé), specificity, variability and absence of relationship with *Erwinia cancerogena* Ur. *European Journal of Plant Pathology* 8, 310–333.
Rossman, A.Y. (2001) A special issue on global movement of invasive plants and fungi. *BioScience* 51, 93–94.
Rossman, A.Y. (2009) The impact of invasive fungi on agricultural ecosystems in the United States. *Biological Invasions* 11, 97–107.
Royle, D.J. and Ostry, M.E. (1995) Disease and pest control in the bioenergy crops poplar and willow. *Biomass and Bioenergy* 9, 69–79.
Samils, B., McCracken, A.R., Dawson, W.M. and Gullberg, U. (2003) Host specific genetic composition of *Melampsora larici-epitea* populations of two *Salix viminalis* varieties in a mixture trial. *European Journal of Plant Pathology* 109, 183–190.
Shain, L. (1988) Evidence for *formae speciales* in the poplar leaf rust fungus, *Melampsora medusae*. *Mycologia* 80, 729–732.
Sinclair, W.A. and Lyon, H.H. (2005) *Diseases of Trees and Shrubs*. Cornell University Press, Ithaca, New York.
Smith, J.A., Blanchette, R.A., Ostry, M.E. and Anderson, N.A. (2002) Etiology of bronze leaf disease of *Populus*. *Plant Disease* 86, 462–469.
Spiers, A.G. (1983) Host range and pathogenicity studies of *Marsonina brunnea* to poplars. *European Journal of Forest Pathology* 13, 181–196.
Spiers, A.G. (1984) Comparative studies of host specificity and symptoms exhibited by poplars infected with *Marssonina brunnea*, *Marssonina castagnei*, and *Marssonina populi*. *European Journal of Forest Pathology* 14, 202–218.
Spiers, A.G. (1988) Comparative studies of type and herbarium specimens of *Marssonina* species pathogenic to poplars. *European Journal of Forest Pathology* 18, 140–156.
Spiers, A.G. (1989) Introduction of poplar and willow pathogens into New Zealand and their effects. *New Zealand Journal of Forestry Science* 19, 347–352.
Spiers, A.G. (1998) *Melampsora* and *Marssonina* pathogens of poplars and willows in New Zealand. *European Journal of Forest Pathology* 28, 233–240.
Spiers, A.G. and Hopcroft, D.H. (1994) Comparative studies of the poplar rusts *Melampsora medusae*, *M. larici-populina* and their interspecific hybrid *M. medusae-populina*. *Mycological Research* 98, 889–903.
Spiers, A.G. and Hopcroft, D.H. (1998) Morphology of *Drepanopeziza* species pathogenic to poplars. *Mycological Research* 102, 1025–1037.
Stack, R.W. and Ostry, M.E. (1989) *Melampsora* leaf rust on *Populus* in the North Central United States. In: Merrill, W. and Ostry, M.E. (eds) *Proceedings IUFRO Recent Research on Foliage Diseases*. General Technical Report GTR-WO-56. United States Department of Agriculture, Forest Service, Washington, DC, pp. 119–124.
Steenackers, J., Steenackers, M., Steenackers, V. and Stevens, M. (1996) Poplar diseases, consequences on growth and wood quality. *Biomass and Bioenergy* 10, 267–274.
Tian, C.-M., Shang, Y.-Z., Zhuang, J.-Y., Wange, Q. and Kakishima, M. (2004) Morphological and molecular phylogenetic analysis of *Melampsora* species on poplars in China. *Mycoscience* 45, 56–66.
Tilman, D., Reich, P.B., Knops, J., Wedin, D., Mielke, T. and Lehman, C. (2001) Diversity and productivity in a long-term grassland experiment. *Science* 294, 843–845.
Trench, T.N. and Churchill, H. (1987) Report of *Melampsora medusae* on *Populus deltoides* in Southern Africa. *Plant Disease* 71, 761.
Ward, K.T. and Ostry, M.E. (2005) Variation in *Septoria musiva* and the implications for disease resistance screening of poplars. *Plant Disease* 89, 1077–1082.
Yang, D., Bernier, L. and Dessureault, M. (1994) Biological control of *Septoria* leaf spot of poplar by *Phaeotheca dimorphospora*. *Plant Disease* 78, 821–825.
Zhuang, W.-Y. (2005) *Fungi of Northwestern China*. Mycotaxon Ltd, Ithaca, New York.

9 Insect and Other Pests of Poplars and Willows

J.G. Charles,[1]* L. Nef,[2] G. Allegro,[3] C.M. Collins,[4] A. Delplanque,[5] R. Gimenez,[6] S. Höglund,[7] H. Jiafu,[8] S. Larsson,[7] Y. Luo,[8] P. Parra,[9] A.P. Singh,[10] W.J.A. Volney[11] and S. Augustin[12]

[1]The New Zealand Institute for Plant and Food Research Limited, New Zealand; [2]Université catholique de Louvain, Belgium; [3]CRA – Unità di Ricerca per le Produzioni legnose fuori foresta, Italy; [4]Imperial College, UK; [5]Ligny le Ribault, France; [6]University of Buenos Aires, Argentina; [7]Swedish University of Agricultural Sciences, Sweden; [8]Forest University Beijing, People's Republic of China; [9]Instituto de Investigación Forestal de Chile – INFOR, Chile; [10]Rain Forest Research Institute, India; [11]Northern Forestry Centre, Canada; [12]INRA – UR633, Zoologe Forestière, France

9.1 Introduction

Poplars and willows in their native environments around the world provide habitats of rich biodiversity. They are eaten by animals, birds and insects. Large mammals browse on saplings and shoots, rodents destroy buds and bark and woodpeckers excavate holes for nests. However, by sheer numbers, diversity and impact, insects usually command the greatest ecological attention and attack all parts of the tree. They eat leaves and buds, suck sap, induce galls, bore into shoots and wood, attack roots and transmit plant diseases.

Many species of insects and mites attack both poplars and willows. *Populus* species are attacked by at least 300 species of insects and mites in North America (Mattson *et al*., 2001) and about 525 species in Europe (Schwenke, 1972–1986; Delplanque, 1998). Significant numbers of pests have also been identified in India (Pandey *et al*., 2007) and Iran (Akbarian *et al*., 2006; Babmorad *et al*., 2007) and elsewhere in Asia. Cultivated species of *Salicaceae* have also been introduced to many regions where they were previously absent, especially during the past 250 years of European colonization. The insects found on these trees are often restricted to a few species that arrived with the plants themselves, a few more recent immigrants and an equally small number of native insects that have adapted to the novel opportunities provided by these new trees.

Although about 1000 species of insects have been found on willows and poplars around the world, not all of them should be thought of as pests. An insect collected from a willow or poplar does not necessarily feed on it, or stay there long enough to cause any harm. Only a relatively small number of species regularly cause such severe physical damage to trees that they reduce their economic or environmental value and are justifiably called pests. These insects may be present in large numbers, but

*Corresponding author: john.charles@plantandfood.co.nz

even so, measuring the economic damage caused by insects is not easy and requires considerable skill. In addition, the genetic diversity of insects may be so great that different populations of the same species feed on different host plants. For example, the leaf beetle, *Chrysomela lapponica* Linnaeus (Coleoptera: Chrysomelidae), usually feeds on willows in northern Europe, but some populations in central Europe have become specialized feeders on birch (*Betula pubescens* Ehrhart) (Gross *et al.*, 2007).

In this chapter, it is not feasible to present details of every pest found on *Salicaceae*. Nor is it necessary, since many species, or groups of closely related species, have similar feeding habits and eat both willows and poplars. A list of nearly 70 species illustrates the range of insects and mites that attack willows and poplars, the damage they cause and the severity of their attack in different parts of the world (Table 9.1). The table shows that insect pests can be grouped largely according to their lifestyle, and so we provide a brief overview of the damage caused by insects as bud and leaf feeders, leaf miners, young shoot feeders, wood borers, root feeders and those that lead to the formation of often very distinctive galls. A few species are so widespread, or so consistently damaging, that they warrant more detailed description, and so a series of case studies of 16 representative insect species, or groups of similar species, is also included.

The economic and/or environmental impacts of these functionally similar pests, and the appropriate management responses to them, are also usually similar. An overview of integrated pest management in cultivated crops, in which emphasis is placed on the primary roles of plant breeding and silviculture (as opposed to pesticides) as key tools for managing insects pests, is also provided.

Finally, the importance of the global economy, which affects us all, is recognized with the inclusion of a section on invasive species, quarantine issues and pest ecology in exotic environments, e.g. countries in which endemic *Salicaceae* are either absent or are outnumbered by exotic species.

The structure adopted in this chapter, with the inclusion of willow pests, is a natural extension of 'Damaging poplar insects – internationally important species', the FAO's Internet guide to poplar insects (de Tillesse *et al.*, 2007). This guide was first conceived as an update to the original FAO books on poplar production.

9.2 Insect Pests of Poplars and Willows

When an insect species causes sufficient damage to affect production or other values that are important to humans it is called a 'pest', but it is important to recognize that the term 'pest' does not have any natural or ecological meaning. A 'pest' is an entirely human construct, and it does not describe an insect's abundance or even the relative abundance of one insect compared with others. A species known as a 'pest' may be quite uncommon for long periods of time at any one location, and might be a pest in one region or country but not in another (Table 9.1).

Of the many hundreds of insects that eat *Salicaceae*, perhaps fewer than 10% regularly cause sufficient damage to be called serious pests. Insects that eat or mine buds, leaves and shoots are among the most common pests, and most willow and poplar trees have many partially eaten leaves. Under the most severe attacks, trees may be completely defoliated, growth rates may decline and trees may even die. Some of these insects induce the tree to produce very distinctive galls in which they live. Wood-boring insects usually cause physical damage. They can weaken the trunk and make them vulnerable to windthrow. They may, in particular, make the trees unsuitable for many of the most valuable uses, such as furniture and veneering. On the other hand, trunk borers may have no measurable impact on paper or pulp production.

Insects may also greatly affect the functional or aesthetic attributes of willows and poplars planted for reasons other than wood production. A pest outbreak can reduce, if not destroy, the function of a row of trees planted as a windbreak or for sunshade. Conversely, windbreak trees may be a problem if they are hosts of pests of neighbouring food crops. Urban specimen trees lose their aesthetic value following attacks by leaf miners, which spoil the leaves and cause premature leaf drop. Finally, insects can transmit or facilitate the introduction of fungal or other plant pathogens

Table 9.1. Important insect pests of poplars (P) and willows (W) from around the world.

			Geographical distribution[b]									
Species	P or W	Damage[a]	Amer	Neotr	Atl	MEur	Medit	AraCp	Chin	IndM	AusPp	Afr
Acarina												
Aceria parapopuli	P	SG	++									
Hemiptera												
Mordwilkoja vagabunda	P	G	+									
Parthenolecanium corni	PW	S	++		+						+	
Phloeomyzus passerinii	P	S	+		++	+++	+++	+	+			
Tuberolachnus salignus	W	S	++	++	++	++	++	++	++	++		++
Coleoptera												
Altica populi	P	D	+									
Anoplophora glabripennis	P	B							+++			
Anoplophora malasiaca	P	B							+			
Anoplophora nobilis	P	B							+++			
Apriona cinerea	P	B						++		++		
Apriona germari	P	B							+++			
Apriona japonica	P	B							++			
Batocera horsfieldi	P	B						+++	+++	+++		
Batocera lineolata	P	B							++			
Byctiscus populi	PW	D			+	+++	++					
Capnodis miliaris	P	B					+++	+++				
Chrysomela* (= *Melasoma*) *tremulae	P	D			+++	++						
Chrysomela* (= *Melasoma*) *populi	P	D		+++	+++	+++		+	+			
Chrysomela vigintipunctata	P	D				++						
Cryptorhynchus lapathi	PW	B	++		+++	+++	+++	++	+++			
***Megaplatypus mutatus* (= *sulcatus*)**	P	B		+++			+					
Melanophila picta	P	B			++	++	+++	+++				
Phratora (= *Phyllodecta*) spp.	PW	D	++		+++	++	++					
Phyllobius spp.	PW	D			++	+	++	++				
Polydrusus (= *Polydrosus*) spp.	PW	D	++			+	+++					
Saperda* (= *Anaerea*) *carcharias	P	B		+++	+++	+++						
Saperda* (= *Compsidia*) *populnea	PW	BG	++		+++	+++	++		+++			
Xyleborus (= *Anisandrus*) *dispar*	PW	B	+++		+++		+++					

Continued

Table 9.1. Continued.

Species	P or W	Damage[a]	Geographical distribution[b]									
			Amer	Neotr	Atl	MEur	Medit	AraCp	Chin	IndM	AusPp	Afr
Lepidoptera												
Cerura (= *Dicranura*) *menciana*	PW	D							++			
Cerura (= *D.*) *vinula*	PW	D			++	++	++					
Choristoneura conflictana	P	D	+++									
Clostera (= *Pygaera*) *anachoreta*	PW	D					++	++	++			
Clostera (= *P.*) *anastomosis*	PW	D			+	++	++		++			
Clostera* (= *P.*) *cupreata	PW	D						++		++		
Clostera* (= *P.*) *fulguriata	PW	D								++		
Cossus cossus	PW	B			+++	++	+++		++			
Epinotia solandriana	P	D	++		+							
Gypsonoma aceriana	P	BLG			+++	+++	+++					
Gypsonoma haimbachiana	P	BLG	++									
Gypsonoma riparia	P	BLG						++				
Hyphantria cunea	PW	D				+++	+++		++			
Leucoma (= *Stilpnotia*) *salicis*	PW	D			+	+++	+++		++			
Leucoma (= *St.*) *candida*	PW	D							++			
Lithocolletis spp.	PW	L	++		++	++	++	++		++		
Malacosoma disstria	P	D	+++									
Operophtera brumata + *bruceata*	PW	D	++		++	++	+++					
Orgyia thyellina	PW	D							++			
Orgyia antiqua	PW	D	++		++	++			++			
Paranthrene tabaniformis	P	BG			+++	+++	+++		+++			
Phassus (= *Endoclita*) *excrescens*	PW	B							++			
Phyllocnistis unipunctella	PW	L			+++	++	++					
Phyllocnistis labyrinthella	P	L			++	++	++					
Phyllocnistis populiella	P	L	++									
Lymantria dispar	PW	D	+++		+	+++	++	++	+	+		
Lymantria obfuscate	PW	D								+++		
Sesia* (= *Aegeria*) *apiformis	P	B			+++	++	++					
Yponomeuta rorrella	PW	D				+++						
Zeuzera pyrina	PW	B			++		+++					

Hymenoptera												
Janus abbreviates	PW	D	++									
Janus luteipes	PW	D			++		++					
Nematus oligospilus	W	D	++	++	++	++	++	++	++	++	++	++
Trichiocampus populi	P	D							++			
Trichiocampus viminalis	PW	D			++							
Tremex fuscicornis	PW	B		++	+	+	+	+	+		+	
Diptera												
***Dasineura* spp.**	PW	G	+			++	+		+			
Phytobia spp.	PW	B			++	++	+					
Rhabdophaga salicis	W	G	++		+							

[a]D, Leaf feeder/defoliator; L, leaf miner; S, sucking insect/phloem feeder; G, gall-former; B, wood borer; more than one category often indicates damage is caused by different life stages.
[b]Geographical distribution received from the following regions: Amer, Canada, USA; Neotr, South America – especially Argentina, Chile; Atl, Eurosiberian region – Atlantic Domain: Belgium, Germany, Finland, France, UK, the Netherlands; MEur, Eurosiberian region – Mid-European Domain: Austria, Bulgaria, Hungary, Romania; Medit, Mediterranean region – Egypt, Italy, the Maghreb, Portugal, Spain, Syria, Turkey, ex-Yugoslavia; AraCp, Aralo-Caspian region – Iran, Iraq, Pakistan; Chin, Chinese region – China, Japan, Korea; IndM, Indian region – India, Pakistan; AusPp, Australo-Papua region – Australia, New Zealand; Afr, African region – South Africa.
Pest status: +, important species, but local or found occasionally; ++, important species, occurring quite frequently; +++, frequently occurring and particularly harmful species.
Species in bold italicized text: see 'Selected Examples' (Section 9.7) for additional information on these pests.

that also attack willows and poplars (see Chapter 8 on the diseases of poplars and willows).

9.2.1 Leaf feeders – the defoliators (Fig. 9.1)

The most obvious damage – and often the most spectacular – is caused by insects that eat leaves. Large numbers of some species, usually in the insect orders Coleoptera (beetles), Lepidoptera (moths and caterpillars), Hymenoptera (wasps) and Diptera (flies and midges) can completely defoliate wide areas of trees. Even if the insects themselves cannot be seen or have left the tree, the damage they cause can often be used to identify them.

Leaf rollers produce silken threads to 'web' leaves together to provide shelter and protection while they feed. The adult feeding patterns of many Coleoptera on leaves may be quite distinctive, and some larvae 'skeletonize' one side of the leaf only. Sawfly wasp larvae such as *Nematus oligospilus* Förster (Hymenoptera: Tenthredinidae) (see section 9.7.1) eat whole leaves. Adults lay eggs in a 'pouch' under the surface of willow leaves and young larvae are often found in small 'shot-holes' in the leaves next to the hatched egg. The empty eggshell and the hole in the leaf are typical signs of sawfly presence, even if the larva itself is no longer present.

The ability of trees to compensate for leaf damage, the size of trees attacked and the difficulties in assessing insect populations and environmental variables make it very difficult to measure the economic or environmental impact of defoliators. The timing of attack can determine the consequences for the tree. If defoliation occurs in spring, both willows and poplars can

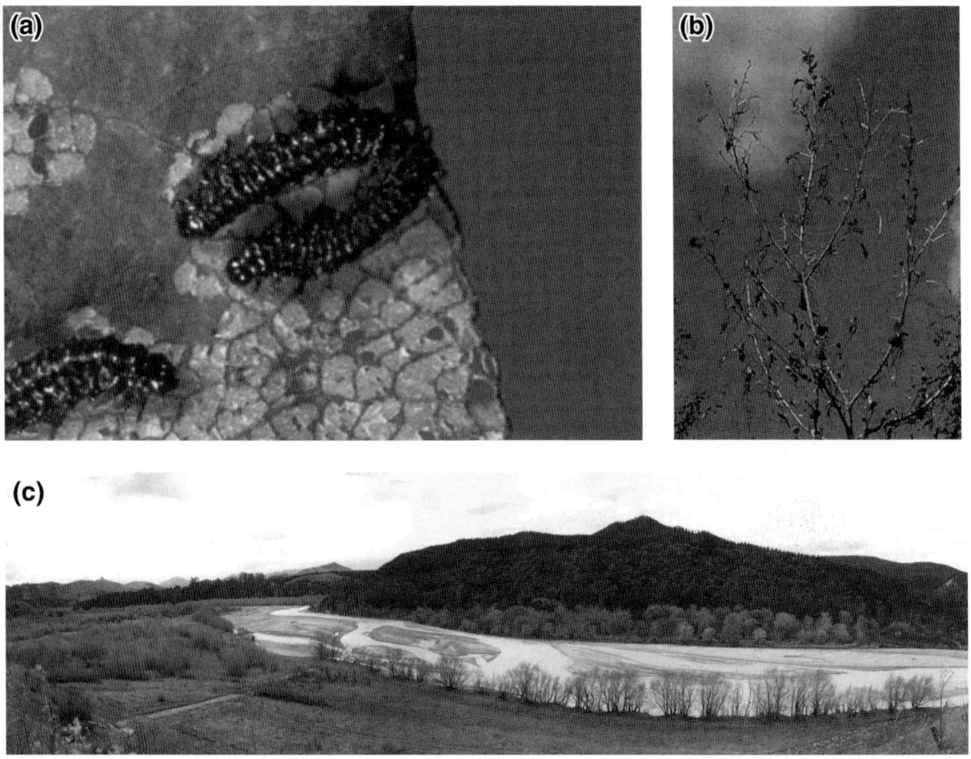

Fig. 9.1. Damage caused by leaf feeders – skeletonizers, rollers and defoliators. (a) *Altica populi* larvae. (b) Poplar defoliated by *Clostera anachoreta*. (c) Riverbank willows in New Zealand defoliated by *Nematus oligospilus* (© 2011, Plant and Food Research). Photos courtesy of L. Nef (a), FAO (b), J. Charles (c). See also Plates 27A, B, C and D, 34E, 35A5, 35B5.

compensate by growing new leaves in the summer, and tree productivity may hardly be affected. By contrast, summer defoliation, particularly if repeated several times within a season or over several consecutive years, can reduce tree growth dramatically and may even kill the tree. Despite these difficulties, many studies have shown that damage can be severe. In the USA, attack by *Chrysomela scripta* Fabricius (Coleoptera: Chrysomelidae) (see section 9.7.2) reduced the growth of seriously defoliated clones by 80% (Caldbeck *et al.*, 1978), and *Choristoneura conflictana* (Walker) (Lepidoptera: Tortricidae) (see section 9.7.3) reduced radial increment and killed twigs (Cerezke and Volney, 1995). In Europe, repeated defoliation by *Hyphantria cunea* Drury (Lepidoptera: Arctiidae) through the year can affect the growth of whole plantations severely (Allegro, 1987), and in China, a single artificial defoliation of 75%, mimicking that of *Clostera* (=*Pygaera*) *anachoreta* (Fabricius) (see section 9.7.4), reduced tree growth by approximately 20%; defoliation of 100% reduced it by about 50% (Gao *et al.*, 1985).

9.2.2 Leaf miners (Fig. 9.2)

Species of leaf miner occur within the orders Lepidoptera, Diptera, Hymenoptera and occasionally Coleoptera. Larvae eat a leaf between the upper and lower epidermis, forming a visible narrow 'mine', 'tunnel' or 'track' that charts its progress. The mine usually starts at the point at which the young larva emerged from the egg and becomes progressively wider as the larva grows. At the end of the mine is either a larva, pupa or, sometimes, a corpse and evidence of natural enemy attack. Leaf miners can be extremely numerous, and large populations may affect photosynthesis significantly. Seventy-five per cent of the leaves of some poplar clones in Belgium were mined by *Phyllocnistis unipunctella* Stephens (Lepidoptera: Phyllocnistidae) (see section 9.7.5) (Nef, 1988a). Photosynthesis of mined leaves was 25% lower than that of undamaged leaves. Both mineral elements and polyphenols in the leaves were modified. The latter, a defence response induced by the plant, reduced the number of insects and reduced the growth, weight and fecundity of the survivors (Nef, 1988a). In the USSR, attack by *Phyllonorycter populifoliella* (Treitschke) (Lepidoptera: Gracillariidae) hastened leaf fall, delayed growth and diminished the aesthetic value of poplars (Belova and Vorontsov, 1987). Urban specimen trees lost their aesthetic value following attacks by the leaf miner, *Zeugophora turneri* Power (Coleoptera: Megalopodidae), which caused premature leaf drop (Tomilova and Kusnetsova, 1975). *Paraleucoptera sinuella* Reutti (Lepidoptera: Lyonetiidae), a species with gregarious larvae in a single mine, may totally defoliate trees and require insecticide treatments (Arru, 1967a).

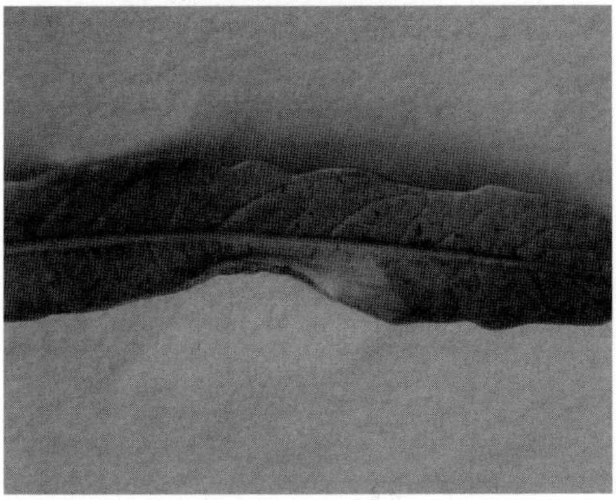

Fig. 9.2. Damage caused by *Phyllonorycter* sp. leaf miner on *Salix viminalis*. Photo courtesy of S. Augustin. See also Plate 27C.

9.2.3 Sucking insects/phloem feeders (Fig. 9.3)

Most sucking insects (such as aphids, scale insects and leafhoppers) feed on and extract phloem sap from the tree or feed on bark tissues. The damage they inflict on the tree can vary from inconsequential to lethal, especially if the insects vector plant pathogens such as viruses. If the trunk is targeted, the tree may react by modifying the structure of the bark. *Phloeomyzus passerinii* (Signoret) (Hemiptera: Aphididae) (see section 9.7.6) feeds on bark parenchymal tissues and modifies the tree structure, causing cracks in the bark and wood necrosis. Large populations of this aphid may kill trees (Allegro and Cagelli, 1996; Allegro, 1997a). Another aphid, *Tuberolachnus salignus* Gmelin (Hemiptera: Aphididae) (see section 9.7.7), reduced both the growth and survival of infested trees (Collins *et al.*, 2001). *T. salignus* infested 40–100% of willow plantations and killed many trees in Himachal Pradesh, India (Chakrabarti *et al.*, 2005). By contrast, *Aphis maculatae* Oestlund (Hemiptera: Aphididae) in the USA hardly affected the growth of various hybrids, despite major differences in degrees of attack (Wilson and Moore, 1986). Leaf and planthoppers can also be pests. In Iran, *Monosteira unicostata* Mulsant & Rey (Hemiptera: Tingidae), the poplar lace bug, seriously damages poplars in nurseries, plantations and wood lots (Babmorad *et al.*, 2007), and *Chaitophorus leucomelas* Koch (Hemiptera: Drepanosiphidae) is considered a pest of poplars (Yali *et al.*, 2006). In China, the armoured scale insect *Quadraspidiotus gigas* (Thiem & Gerneck) (Hemiptera: Diaspididae), living on the trunks and branches, reduced poplar growth by 17% (Hu *et al.*, 1985).

In similar locations, other insects, such as thrips (Thysanoptera), feed on soft plant tissues rather than phloem and do not usually cause significant damage.

9.2.4 Gall formers (Fig. 9.3)

Feeding or oviposition by some insects induces rapid, localized growth of plant cells to produce a gall on a branch, shoot, petiole or leaf lamina. The young stage of the insect usually lives in the gall, which provides both food and protection from natural enemies. Galls are formed by some species of mites (Acari), moths (Lepidoptera), wasps (Hymenoptera), aphids and sap-sucking plant bugs (Hemiptera) and flies (Diptera). The galls themselves can be simple or complex in structure and are often sufficiently distinctive to identify the insect that caused them. *Aceria parapopuli* (Keifer) (Acari: Eriophyoidea) induces galls in the axillary buds, and its population dynamics are affected by plant genotype (Gom and Rood, 1999; McIntyre and Whitham, 2003). Sawflies of the genus *Euura* (Hymenoptera: Tenthredinidae) form galls on the stem, petiole, buds and leaf midribs of *Salix* (Price *et al.*, 2004). Studies over 20 years have shown a general preference of *Euura* species for long vigorous shoots, and other Tenthredinidae, for example *Pontania* and *Nematus*, have showed similar preference (Price, 2003). Aphids, for example *Pemphigus* spp., that attack leaf petioles or stalks may induce galls with very complex structures. *Pemphigus spirothecae* Passerini induces a corkscrew-like gall. *Pemphigus bursarius* (Linnaeus), which may be host specific to *Populus nigra* Linnaeus, is even used as an index of the genetic purity of poplars of this species within its natural distribution (Tittle, 1972). Different species of *Dasineura* gall midges (Diptera: Cecidomyiidae) (see section 9.7.8) attack the terminal buds and

Fig. 9.3. Damage caused by phloem feeders and gall formers. (a) *Chaitophorus leucomelas* on *Populus nigra*. (b) Galls caused by *Rhabdophaga salicis* on *Salix cinerea*. Photos courtesy of K. Tuba (a), T. Nyman, (b). See also Plates 28A, B and C, 39D, 43B3.

young leaves of *Salix* (Gagné, 2004) and oviposit in willow buds, where the developing larvae induce the formation of galls.

9.2.5 Bud and young shoot feeders and borers (Fig. 9.4)

Bud feeders

'Leaf rollers' are the caterpillars of moths in the family Tortricidae. Different species of leaf rollers eat buds in the spring before moving on to the leaves. Adult Coleoptera (such as *Peritelus sphaeroides* Germar (Coleoptera: Curculionidae)) may eat whole buds, or the females may lay their eggs in the buds. Emerging larvae then feed within the bud, leading to desiccation and bud drop. The weevil, *Pselaphorynchites tomentosus* Gyll. (Coleoptera: Curculionidae), lays eggs in the buds of poplars and willows. Larvae fall to the ground when the buds drop, and they complete their development in the soil (Kippenberg, 1975). A few sawflies in the genus *Pontania* also oviposit in the terminal buds of willow species, leading to gall formation as a habitat for their larvae.

Shoot borers

Shoot-boring insects usually drill straight tunnels that are smaller than those of wood borers. *Paranthrene* spp. (Lepidoptera: Sesiidae) and some *Saperda* spp. (Coleoptera: Cerambycidae) are regularly observed feeding and boring into shoots. Shoot borers can be particularly harmful in nurseries, where their feeding on terminal shoots induces stem forking or ramification, or causes the shoots to break. The damage distorts the shape of the growing tree and can reduce its commercial value significantly. Shoot borers can also allow entry by stem rots. *Paranthrene tabaniformis* Rottemburg (see section 9.7.9) larvae feed within the stems, shoots and branches of many species of poplars and willows and cause distinctive swelling of plant tissue (Bertucci, 1986). Larvae of *Gypsonoma aceriana* Duponchel (Lepidoptera: Tortricidae) (see section 9.7.10) feed initially on leaves, but continue their development in shoots close to the terminal buds. Repeated insecticide treatments are required in nurseries (Attard, 1979). On the other hand, *Saperda inornata* Say (Coleoptera: Cerambycidae) (see section 9.7.11) in the USA may attack up to 60% of shoots, but few are broken and ultimately the attacks have little impact on growth (Moore and Wilson, 1986). Female leaf-rolling beetles, *Byctiscus* spp. (Coleoptera), puncture young shoots or leaf stems and then 'roll' the damaged tissues into which they oviposit. The damaged shoots or leaves then fall to the ground (Gruppe *et al.*, 1999).

Cicadas also cause physical damage that weakens young shoots and causes them to

Fig. 9.4. Damage caused by bud and young shoot feeder *Earias chlorana* on *Salix viminalis*. Photo courtesy of S. Augustin. See also Plates 28D, E and F, 40D.

break. However, this damage results from adult females drilling into soft shoots to lay often long lines of eggs.

9.2.6 Wood borers (Fig. 9.5)

Wood-boring insects are usually members of the Coleoptera and Lepidoptera, and occasionally Diptera and Hymenoptera. Larvae typically drill tunnels or galleries in the trunk or branches, which interrupt sap circulation, cause cracks in the bark and lead to wood necrosis. The damage may also allow the establishment of plant pathogenic diseases, which cause significantly more damage than the original insect. Tunnels of wood borers can be quite large and are often more sinuous than those of shoot borers. The physical appearance of the tunnel may often signal the identity of the insect that caused it. Large numbers of big borers, for example *Saperda calcarata* Say (Coleoptera: Cerambycidae) (see section 9.7.11), can cause severe structural damage to a tree, with large limbs or even whole trees subject to breaking under severe weather conditions (especially wind and snow) (Ostry *et al.*, 1989). The Asian longhorn beetle, *Anoplophora glabripennis* (Motschulsky) (Coleoptera: Cerambycidae) (see section 9.7.12), which has recently spread from China and Korea to the USA and Western Europe, is particularly destructive because it attacks healthy trees. Larvae bore into trunks and large branches, severely damaging the structural integrity of the wood and eventually killing the tree. Adult *Megaplatypus mutatus*

Fig. 9.5. Damage caused by wood borers. (a) *Saperda carcharias* tunnels in wood. (b) Windthrow resulting from *Cryptorhynchus lapathi* damage. (c) *Phytobia cambii* injury to wood – cross section. (d) *Cossus cossus* exit hole at the base of the trunk. Photos courtesy of A. Delplanque (a, b and d), M. Martinez (c). See also Plates 45D and E, 47F.

(Chapuis) (Coleoptera: Platypodidae) (see section 9.7.13) bore large gallery systems into living poplars, which significantly weaken tree stems and devalue the timber (Alfaro *et al.*, 2007). *Cryptorhynchus lapathi* (Linnaeus) (Coleoptera: Curculionidae) (see section 9.7.14) can cause severe economic losses, for example 25% in the Netherlands (Moraal, 1996a), whereas, despite their abundance, *Sesia apiformis* (Lepidoptera: Sesiidae) (see section 9.7.15) larvae cause hardly any damage to large poplars because their galleries are restricted to the lower parts of the trunk. *Tremex fuscicornis* Fabricius (Hymenoptera: Siricidae) (see section 9.7.16) is also known as a wood-boring wasp, with a broad host range. Boring larvae can damage the timber severely, and the fungus associated with *T. fuscicornis* causes wood decay (Parra *et al.*, 2007). However, changing crop values may change the pest status of an insect. For example, *Paranthrene robiniae* (Hy. Edwards), an endemic sesiid of the north-western USA, could be tolerated when poplar plantations were harvested for pulp, but became of much greater concern when crop rotation was increased for the production of saw timber (Brown *et al.*, 2006). Even small insects can cause economic damage that may appear quite disproportionate to the size of the tunnel. Hence, the tiny vertical galleries made by *Phytobia* spp. (Diptera: Agromyzidae) in cambium can subsequently cause visible black lines or cracks when veneering (Martinez *et al.*, 1985).

9.2.7 Root feeders

Some soil-living beetle larvae feed on roots and can be particularly harmful to young willow and poplar plantations. The larvae of the weevil, *Lepyrus palustris* (Scopoli) (Coleoptera: Curculionidae), eat fine, softer roots, while adults defoliate poplars and willows (Hoffman, 1954). *Melolontha melolontha* (Linnaeus) (Coleoptera: Scarabaeidae) larvae may also feed on the roots (Bougard, 1977). Lepidoptera larvae may also colonize roots. Galleries of *Sesia apiformis* (Clerck) and *Cossus cossus* Linnaeus formed in the lower parts of the trunk may extend downwards into the roots. After harvest, the larvae may continue to develop in the remaining roots and attack a replacement plantation, causing considerable economic damage.

9.2.8 Disease vectors (Fig. 9.6)

Willows and poplars are attacked by a wide range of fungal and bacterial diseases that bypass the trees' physical defences through their close association with wood-boring insects. Consequently, many insects are considered pests more because of the diseases they introduce than the physical damage they themselves cause. Many plant pathogens rely on insects for their transfer from tree to tree, and they are also often a food source for the young insect larvae themselves. *Megaplatypus mutatus* (see section 9.7.13) introduces *Ambrosia* to poplars in the same way that bark beetles in elm trees are associated with Dutch elm disease (Guerrero, 1966), and *Saperda calcarata* (see section 9.7.11) in Canada facilitates the establishment of disease (Drouin and Wong, 1975). A quantitative relationship between the weevil *Cryptorhynchus lapathi* (see section 9.7.14) and the canker *Mycosphaerella populorum* G.E. Thompson has been demonstrated in the USA (Abebe *et al.*, 1990), while in China, there is a relationship between the establishment of canker *Cytospora chrysosperma* (Pers.) Fr. and *Melanophila decastigma* (Fabricius) (Coleoptera: Buprestidae) (Liu and Jia, 1988). In Italy, initial attacks by *Paranthrene tabaniformis* (see section 9.7.9) are followed by infestations by various fungi, such as *Xanthomonas* (= *Aplanobacter*) *populi* (Ride), and borers such as *C. cossus* (Bertucci, 1986).

Other plant pathogens are transmitted by phloem-feeding insects. Some of these relationships are very specific. Many leaf hoppers, and particularly *Rhytidodus decimusquartus* (Schrank) and *Populicerus populi* (Linnaeus), are vectors of phytoplasma (Berges *et al.*, 1997; Cousin *et al.*, 1999). The bacterium *Xanthomonas populi*, which causes bacterial canker, is found in the tunnels made by *Agromyzidae* midges (Martinez and Gumez, 1998), and in Canada, many bacteria are introduced when trunks are attacked by *Lygus lineolaris* (Palisot de Beauvois) (Juzwik and Hubbes, 1986).

Ecological studies to understand the tri-trophic relationships between diseases, insect vectors and plant hosts should be regarded as

Fig. 9.6. Disease vectors. (a) *Xanthomonas populi.* (b) Beetle injury and *Ambrosia* fungus in a mature poplar, and vectors: (c) *Phytoplasma* on *Populus nigra*. Photos courtesy of M. Villar (a), A. Delplanque (b and c). See also Plate 29A, B and C.

some of the most useful research to improve the commercial value of *Salicaceae* in the future.

9.3 Other Animal Pests of Poplars and Willows

9.3.1 Mammals (Fig. 9.7)

Poplars and willows are damaged by both large and small mammalian browsers – from moose, elk and deer, to beavers, rabbits, mice and voles. Young trees are especially susceptible to bark-eating mammals during winter, when other food is scarce, and may be killed by ring-barking (girdling). Large mammals injure stems, and rabbits destroy young shoots. Secondary fungi often enter trees through wounds caused by animal feeding and can cause branch dieback. Feeding by reindeer decreases biomass production, reducing the height of willows and accelerating dieback of the shoots (Den Herden *et al.*, 2004). Feeding by mammals may also affect the shape of the plants because damage leads to apical breaking and resprouting increases the numbers of lateral branches. The appearance of such young growth may then attract insects, leading, for example, to increased densities of the galling sawflies, *Phyllocolpa* sp. (Roininen *et al.*, 1997).

Willows and poplars are a favourite food of beavers, which cut down smaller trees for their dams. Beavers also eat the bark and leaves of trees, and can girdle and kill larger trees that are left standing. Of course, beavers usually also significantly modify their environment and any economic damage should be analysed against the environmental benefits that they provide (e.g. Pollock *et al.*, 2003, and many other references).

Browsing mammals may modify or determine the structure of native stands of *Salicaceae* substantially by destroying young seedlings or

Fig. 9.7. Damage caused by mammals and birds. (a) Beaver damage to *Populus nigra*. (b) Elk feeding on bark of aspen poplar. (c) Woodpecker damage to poplar. Photos courtesy of M. Villar (a and c), FAO (b). See also Plate 29D.

saplings. Similarly, they have the potential to destroy young nurseries or commercial plantations. On the other hand, they are generally easier to control at a population level than insect pests, and their damage decreases as the trees grow older.

There are some options available for small mammal control, especially in nurseries and young plantations. Good grass control can minimize damage by mice and voles effectively. Physical barriers can be applied around young trees to protect them from small mammals, and repellents may discourage animal feeding effectively (Ostry *et al.*, 1989). Finally, marked differences in winter browsing damage among poplar clones have been observed (Dickmann, 1978), so favoured clones should not be planted in areas where animal damage is common or unavoidable (Ostry *et al.*, 1989).

9.3.2 Birds (Fig. 9.7)

Some birds can be a nuisance in poplar plantations, especially woodpeckers (which nest in trunks) and corvids (which nest in the canopy). The nests of crows are sometimes sufficiently heavy to break branches. Anyway, the damage caused by woodpeckers is usually negligible, as they prefer to nest in dead or rotten wood (Allegro, 1993).

9.4 Integrated Pest Management (IPM) of Insect Pests of Poplars and Willows

9.4.1 What is IPM?

IPM has become widely accepted as both an environmentally and economically sustainable strategy for pest control. Many books and papers have been written about IPM, which nowadays is often regarded as a holistic and sustainable approach to the management of all pests, including insects, other animals, plant diseases and weeds. IPM accepts that a diversity of tactics is more effective over time than a single tactic (especially if that tactic is use of broad-spectrum pesticides), and it generally leads to a reduction in pesticide use. It is worth remembering that IPM was conceived in the 1970s to address problems in the capital-intensive agricultural systems of the USA. Here, the new, cheap and reliable broad-spectrum insecticides (especially organochlorines, organophosphates and carbamates) had become the only real strategy for pest control and had led to the then 20-year-old 'pesticide treadmill'. Under the new IPM model, 'economic thresholds' (often measured by complicated and knowledge-intensive pest monitoring systems) were used to restrict or modify pesticide use, while other tactics such as biological control and plant resistance were encouraged. A somewhat different IPM model was developed elsewhere that relied on training farmers to observe and understand the interactions between pests and their natural enemies as a basis for reducing pesticides (Dent, 2000). In both cases, it has become clear over the past 40 years that it is often much easier to measure a change in pest numbers than it is to determine why the change has occurred. Nevertheless, IPM did introduce significant scientific discipline to pest control, as well as recognizing the requirements for knowledge of the biology and ecology of crops, pests and natural enemies.

In forestry, the size and longevity of plantation or natural production forests dictate that large-scale inputs such as pesticides are both uneconomic and impractical, although they have a place in commercial nurseries or occasionally in high-value plantations. Pest management aims to control populations, rather than individual insects. Poking a wire into a hole may kill an individual borer very effectively, but is a completely impractical method for borer control throughout a forest. Forest managers also want to control all pests, and a good technique for controlling wood borers may leave defoliators, for example, untouched. Adopting a holistic approach to pest management means addressing the plantation as a whole and understanding that soil, site aspect and climate, tree species and cultivars, pests and diseases, other plants and other insects (especially natural enemies) all have a potential effect on pest numbers. There are so many connections between these components that a full knowledge of the interactions between them can never be achieved. The evolution of very rapid growth rates in the *Salicaceae* may inevitably have been accompanied by an increased susceptibility to attack by phytophagous insects (Mattson *et al.*, 2001). On the other hand, the nutritional constraints of 'poor' host plants can have a very significant effect on the population dynamics of insects that feed on them, and may influence strongly whether populations remain at low or outbreak levels (Price *et al.*, 2005). Such dynamic adaptations have ensured that there is an ecological balance between the growth and death rates of trees and herbivores, even though the 'balance' may fluctuate wildly over ecological time frames. Extraordinary environmental conditions may also affect both natality and mortality dramatically, and together with population dynamics, can lead to the outbreaks of pests that are a feature of forest entomology. Time is then required for the pre-existing 'balance' to return. Hence, IPM for poplars and willows has developed largely as an attempt to build in as much ecological stability as possible by increasing environmental biodiversity, using pest-resistant trees and so on.

In principle, the numbers of a particular species of insect at any one time and place are determined by the balance between natality (the birth rate of the insect) and mortality (its death rate). Pest management activities aim to promote the factors that minimize natality and maximize mortality. Natality is perhaps managed most easily by manipulating the nutritional attributes of the host plant, while mortality is most manageable by encouraging natural enemies or (occasionally) by applying pesticides. Host plants are best manipulated in the long term by breeding for pest resistance, and in the short term by selecting the best planting environments and silvicultural techniques. The introduction or encouragement of natural enemies can be a very effective and ecologically sustainable control weapon, while pesticides should be recognized as a very powerful but short-term and temporary solution to a pest problem – to be used sparingly and with caution, and, in forestry, usually restricted to nurseries.

These human-led pest management activities are inevitably augmented by natural environmental factors that may be more or less predictable, with more or less predictable impacts. Low temperatures may provide a vital mortality factor in overwintering insects, and extended winter freezing is often more reliable than the occasional storms that can also reduce insect numbers dramatically in some years. While severely adverse weather usually impacts directly on mortality, benign weather usually acts indirectly on natality through improved larval nutrition that increases survival rates and adult longevity and fecundity. A simple 'armchair' exercise illustrates how apparently small changes may have dramatic impacts. Let us suppose that a population consists of one pair of insects in a tree and the female lays 100 eggs, half of which are males and half females. Then, in order to start the next generation with the same size population, i.e. one pair of insects, 98 of the mother's offspring must die before they reproduce. Such 98% mortality within a generation is quite normal for insects, which is one reason why insects have not taken over the earth's resources; but if only 96 offspring die (an apparently small reduction to 96% mortality), then the next generation doubles in size. Because many insects can breed so quickly, it takes only a relatively small reduction in mortality over one or two generations for insect populations to 'explode'. Conversely, of course, relatively small increases in mortality can reduce populations dramatically, which is what IPM tries to exploit. By applying a range of 'additive' tactics, IPM aims to increase mortality at different stages of insects' life cycle throughout the year to reduce pest populations to very low levels.

Successful IPM in *Salicaceae* uses combinations of any appropriate control methods to manage pest populations within acceptable levels. Practices may differ markedly between nurseries and production forests for timber, pulp or veneer. IPM systems are designed for local conditions and should respond to actual pest densities and to actual or expected damage. Control decisions take account of economic, ecological and social aspects. IPM is dynamic over time, tends not to rely on one or the same control method and is tailored most successfully to regional requirements (e.g. Wu *et al.*, 2006). Not surprisingly, IPM systems vary considerably among countries, and even among different regions within countries, but successful IPM inevitably requires significant knowledge of the biology and population dynamics of pests as well as silviculture. Traditional IPM approaches used for agriculture may not always be appropriate for forestry, as illustrated by a review of insect pest management in intensively managed hardwood forest systems in North America (Coyle *et al.*, 2005).

Pest control methods in willow and poplar plantations are dominated by those that can be described as either 'cultural control' or 'biological control' strategies. These also apply in nurseries, where the addition of insecticides is also more likely to be justifiable. Rationally, however, pesticides should be applied only when the cost is lower than the damage being caused. This requires considerable knowledge of the pest identity, biology and economics, as well as the ability to measure and monitor changes in pest populations over time.

Pheromones (Fig. 9.8)

Insect sex pheromones are an increasingly valuable tool in IPM programmes. Sex pheromones are species-specific chemicals produced by the virgin females of many (but not all) species of insects that attract males to them for mating. The chemical structures of many pheromones, especially for species of Lepidoptera, have now been identified and synthesized (El-Sayed, 2005) and are commercially available. Pheromones have a wide range of uses in IPM. They are commonly used in monitoring programmes, to signal the arrival of a pest in a new region or to measure its dispersal, distribution or relative population size. Pheromones have been used to monitor populations of the Lepidoptera, *Cossus cossus* and *Zeuzera pyrina* Linnaeus (Pasqualini and Natale, 1999). In small areas, they have been used successfully to control pests by disrupting mating (by artificially flooding an area with pheromone so that the males are unable to locate real females). Attempts to control the pest by 'mass trapping', i.e. removing so many males from the population that females are not mated, are not usually successful. Mass trapping of *Paranthrene tabaniformis* to restrict populations has only been achieved in isolated plantations (Wu *et al.*, 1987; Moraal *et al.*, 1993).

Pheromones can also be used in 'lure-and-infect' and 'lure-and-kill' strategies. In the former, traps attract males to sources of insect viruses or bacteria, so that females or their offspring become infected and die. In the latter, the males are attracted to insecticides and are killed. The many potentially valuable monitoring and control uses of pheromones in IPM programmes are covered in many books (e.g. Dent, 2000; Schowalter, 2000) and is beyond the scope of

Fig. 9.8. Integrated pest management: pheromone trap. Photo courtesy of S. Augustin.

this chapter. They certainly are poised to become important tools in willow and poplar IPM programmes.

9.4.2 Cultural control of insect pests

Cultural control is the most important strategy for managing most insect pests of poplars and willows. Cultural control techniques make use of the physical, chemical and structural attributes of trees that limit pest populations by either making themselves completely unpalatable to herbivores or by decreasing the number of offspring of the pests that do feed on them. Effective cultural control is often the essence of good silviculture, but there is no single or universal method for predicting or reducing insect damage. Certainly, the most appropriate willow and poplar cultivars should be selected for the local environmental conditions and for resistance or tolerance to the prevailing pests and disease species. Tree stand density and agroforestry may also have significant influence on the spatial distribution and impact of pests (and diseases) (Heidari *et al.*, 2006).

Interactions between the environment and tree species or clonal susceptibilities are well recognized. It is not unknown for the most damaged tree types in one plantation to be the healthiest in another, so the choice of the most appropriate (best-for-purpose) cultivars or species for a given site may depend crucially on local knowledge. In principle, foresters should identify the most harmful pests of willows or poplars in the area or site to be planted and select the most appropriate growing techniques and cultivars to minimize the threats from these pests.

Resistant species and tolerant clonal selection

There is a large body of data on the resistance of willow and poplar species and cultivars to insects. Many experiments carried out under varied planting regimes have shown clearly that the severity of insect attacks is linked to genetic factors that determine either physical or chemical defences in trees (Tomescu and Nef, 2007). One significant disadvantage of traditional breeding is that a research programme to create or define resistant clones and then to test them under various growing conditions can often take up to 15 years to complete. Recent advances in genomics research are providing new insights and perspectives into the chemical ecology and biochemistry of poplar defences against insects (Philippe and Bohlmann, 2007). Resistance may come either from natural species or clonal characteristics that may be improved by plant breeders (Figs 9.9 and 9.10), or those provided by biotechnology, for example by introducing genes that induce toxin formation for *Bacillus thuringiensis* (*Bt*) Berliner (see below). In both cases, resistance ideally should not be derived from the activity of a single gene but from the joint impact of a series of genes. Such polygenic resistance usually provides a broader defence against pests because evolution to overcome multiple resistance mechanisms usually takes far longer to occur.

Whatever the resistance mechanism, the objective is not to eliminate the pest entirely but to

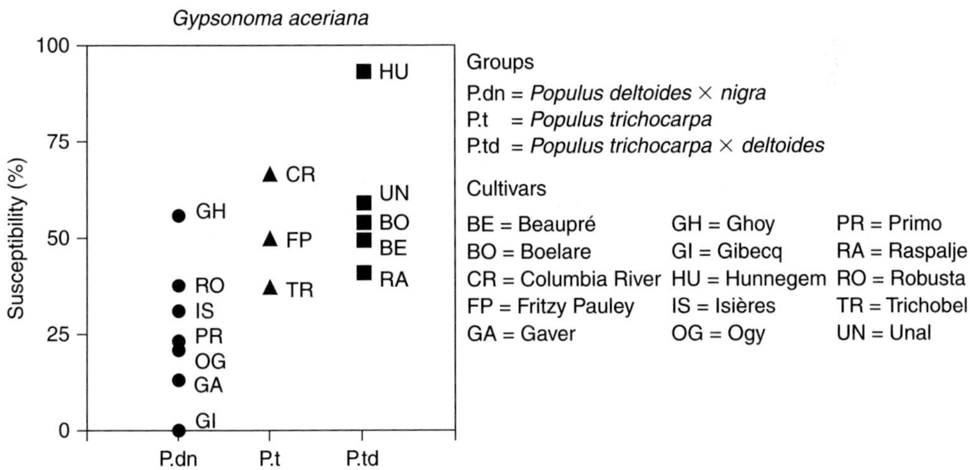

Fig. 9.9. Relative susceptibility to *Gypsonoma aceriana* of some Belgian poplar cultivars (from de Tillesse and Nef, 1998).

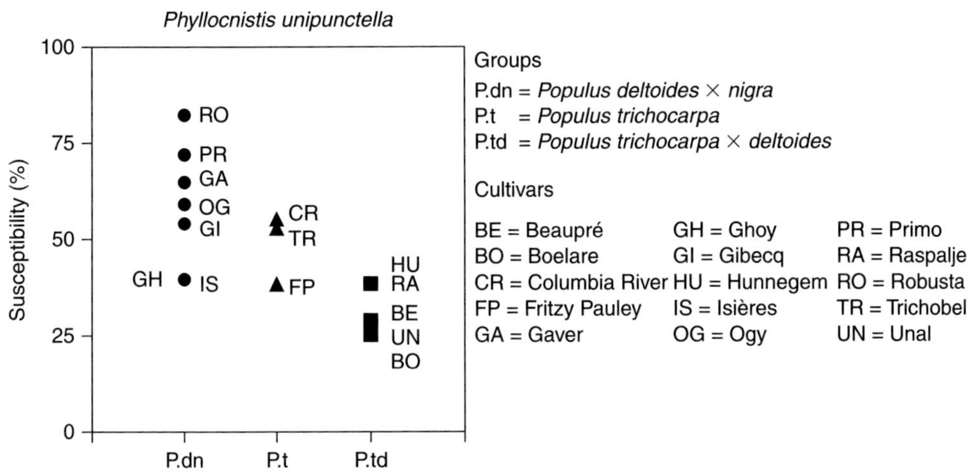

Fig. 9.10. Relative susceptibility to *Phyllocnistis unipunctella* of some Belgian poplar cultivars (from de Tillesse and Nef, 1998).

reduce numbers to acceptable levels. The advantages of selective cross-breeding are that it can lead to persistent resistance across large geographical areas. In addition, after the preliminary research effort, the resistance is 'free'. Research with poplars has shown that the cultivar 'Robusta' (*Populus deltoides* × *P. nigra*), obtained about 1910, is more susceptible to some insects than other *P. deltoides* × *P. nigra* crosses of more recent origin. Damage caused by defoliating caterpillars in Belgium is generally more abundant on *P. deltoides* × *P. nigra* hybrids than on *P. trichocarpa* (Torr. & Gray) clones (Tomescu and Nef, 2007). Hybrids originating from *P. trichocarpa* or from *P. deltoides* Bartram ex Marshall have different characteristics of resistance to *Phyllocnistis unipunctella*. The differences can be linked to the fact that the former is a forest species and the latter is a pioneer poplar (Nef, 2007). In Finland, *Saperda carcharias* (Linnaeus) (see section 9.7.11) is a serious pest of 'regular' and 'hybrid' aspens. Investigation showed that both genotypes were equally susceptible to the beetle and so planting decisions must be based on other criteria (Välimäki and Heliövaara, 2007). *P. alba* Linnaeus is effectively resistant to the borer *Cryptorhynchus lapathi*, and clones are available in Italy for cultivation for biomass production (Allegro *et al.*, 2007). Damage to *P. simonii* Carrière by the poplar leaf bug, *Monosteira unicostata* (Mulsant & Rey), was greater than that to *P. trichocarpa*, *P. ciliata* and *P. alba* '44/9'. Other clones of *P. alba* and *P. nigra* were less damaged (Babmorad *et al.*, 2007). Seven of 12 poplar species and clones in Iran were immune to woolly poplar aphid, *Phloeomyzus passerinii* (see section 9.7.6), but five clones of *P. nigra* were susceptible to the pest to varying degrees (Sadeghi *et al.*, 2007). Poplar breeding programmes in Italy have, for a long time, selected clones which are resistant to *Ph. passerinii* (Arru, 1974; Lapietra and Allegro, 1990a; Allegro and Cagelli, 1996; Allegro *et al.*, 1996a). Pest and disease resistance in a range of genotypes from different countries of origin imported into a new region can be evaluated and compared against local pests to help determine the 'best' genotypes for that region (Casa *et al.*, 2007).

Polycultures versus monocultures

Willow and poplar growers are well aware that pest outbreaks are often associated with large monocultures of species or clones. The use of monoclonal plantations is probably a major cause of the proliferation of insect damage in the Euro-Mediterranean area and in China. Evidence from a wide range of agricultural ecosystems indicates that increasing the diversity of plant species within a system often also increases the ecological stability of insect populations within it. More ecological niches are available for different plant-feeding insects and their natural enemies, and a combination of competition, interference and predation and parasitism may often be sufficient to prevent pest outbreaks altogether. Additionally, when an outbreak does occur, the increased ecological stability in the system may limit the intensity and duration of the outbreak. Hence, the use of mixed plantations and a species-diverse understorey should be promoted. Three or more cultivars or species of tree, varying in their susceptibility to the major pests, can prevent insects from responding uniformly to the resistance traits of the most resistant cultivars and hence from adapting to the defences of the host plants.

The impact of space on insect populations and persistence is also important. Mosaics of isolated small plantations can minimize pest outbreaks. Isolation may be provided by planting stands at some distance from each other or by separating them by different plant species that are unlikely to share insects with *Salicaceae*. One problem for plantation design is that the best spatial scale for pest control differs among pest species, according to their biology and ability to disperse. Hence, there will never be a single spatial design to suit all pests. In addition, interstand or understorey plants may be either attractive or repellent to pests. *Leucoma salicis* (Linnaeus) (Lepidoptera: Lymantriidae) outbreaks are very rare in poplar plantations with an abundant and varied understorey. Caterpillars migrate regularly from tree to tree, but with a varied understorey they more frequently find themselves on a non-host plant on which they cannot survive. Low plant diversity may be one reason why linear plantations, for example by the roadside, are more vulnerable to attacks by this insect (Nef, 1978). *Gypsonoma aceriana* (see section 9.7.10) attacks are rarer if the understorey cover is not very dense (Cavalcaselle, 1972a; Sekendiz, 1973; Heymans *et al.*, 1984).

'Tree traps' (non-economic plants which are more attractive to a pest than the crop in which they are planted) have also been used successfully. *Lygus lineolaris* in the USA is highly attracted to *Erigeron canadensis*, which can be planted as a tree trap among poplars, reducing damage to poplars by 90% (Sapio *et al.*, 1982); and in Canada, the introduction of *Melia azedarach* and *Acer negundo* as tree traps reduced damage by *Anoplophora glabripennis* by 60–70% (Sun *et al.*, 1990). In a Chilean *P. nigra* plantation, a comparison of the effectiveness of treated trees and logs as traps to detect and control *Tremex fuscicornis* showed that logs attracted 83% of the total number of ovipositing females caught, and all of the oviposition sites were found in the logs. Trees and logs used as traps were later destroyed by chipping or burning (Parra and González, 2011).

Soil nutrition and other environmental conditions

Soil nutrition and structure (hence, water retention and microfauna and flora) play a major role in determining the growth rates and health of willows and poplars. Tree health can be a key factor determining the severity of insect damage, so site selection for soil characteristics is an important aspect of pest management. In fact, appropriate site selection for a nursery or plantation may prove to be the most important decision for limiting pest damage. Well-aerated soils with good water retention and pH (KCl) from 5.5 to 7 should be sought to promote poplar growth. Comparative experiments showed that *Phyllocnistis unipunctella* (see section 9.7.5) was less abundant on sandy ground than on muddy and wet ground. These soils probably modified the chemical composition of the leaf (especially tannins) and hence increased insect mortality (Nef, 1988b, 2007). Trees in the centre of plantations, and those on rich soils, were more subject to attack by *Cryptorhynchus lapathi* (see section 9.7.14) (Attard, 1978).

Trees that are physiologically stressed through drought, poor site selection and plantation techniques are less able to tolerate insect attack than healthy trees. Low numbers of a pest that might cause no damage to a healthy tree may damage or even kill a stressed one. Preliminary research showed that the intensity of defoliation was correlated with various chemical elements in the soil, such as the C/N relationship and potassium, or iron and calcium (Hakizimana and Nef, 1998). In sunny and dry positions, broader tree spacing reduced attacks by *Phloeomyzus passerinii* (see section 9.7.6) (Arzone and Vidano, 1984).

Poplars may grow well in industrial areas with potentially contaminated soils, but there may still be impacts on pests. Poplar growth in the USSR was satisfactory near industrial complexes that produced iron oxides or fluoride, but pest damage, especially by tetranychid mites and aphids, was more abundant (Makhovskaya *et al.*, 1984). The use of wastewater did not increase attack from borers (*Cryptorhynchus lapathi*) (see section 9.7.14) or defoliators (such as *Chrysomela scripta*) (see section 9.7.2) unless it was excessive or salted (Szontagh, 1981; Augustin *et al.*, 1994). Effects of wastewater treatment were found on leaf damage in willow plantations, but no significant differences were found for chrysomelid beetles (Ahman and Wilson, 2008).

9.4.3 Biological control of insect pests (Plate 30)

Biocontrol by insect natural enemies

Perhaps the most pervasive form of insect pest control is through the activity of natural enemies. These are the many species of animals (especially other insects) and diseases that kill herbivorous insects. The term 'biological control' (often shortened to 'biocontrol') was first proposed and defined in 1919 by an American entomologist, Harry S. Smith, as the 'control or regulation of pest populations by natural enemies'. Natural enemies of insects include predators and parasitoids, pathogens such as viruses and fungi and bacteria such as *Bt* (although, strictly, only the spores and not its protein derivatives). Most natural enemies of herbivores are other insects that can be divided into two groups – predators and parasitoids. Predators can be thought of as the 'lions' of the insect world, and one individual predator (such as a ladybird or lacewing) usually catches and eats many insects. Parasitoids are unique to the insect world and are a real wonder of evolution. These insects are usually small (or tiny) wasps or flies. Many are almost invisible and they are quite harmless to humans, but are among the most important regulators of insect populations everywhere.

Adult parasitoids lay an egg (or eggs) on or in a host insect and the young parasitoid larvae then eat their prey alive, finally killing the host just as they complete their own development.

Biocontrol provided by natural enemies is perhaps the most important 'ecosystem service' for pest control available to poplar and willow growers. Many species of natural enemy often attack a single species of pest insect (e.g. Sadeghi *et al.*, 2007), but it is usually difficult to determine which ones provide the 'best' control. Nevertheless, effective biocontrol agents kill large numbers of pests and prevent their populations from expanding, and a successful biocontrol programme uses natural enemies to keep pests below the numbers that cause economic loss. However, there is no biological rule that states that predators and parasitoids will 'naturally' control all pests to the economic level required by humans. This is partly because the term 'pest' has no ecological meaning – it is a purely human construct, and as discussed earlier, an insect's pest status varies according to crop, time and location.

Three types of biological control programmes are generally used in insect pest management:

1. Classical biological control: the importation and establishment of a natural enemy into a new habitat. Many insects become pests because they arrive in a new country without their natural enemies. Classical biocontrol programmes identify the key natural enemies in a pest's native habitat and introduce them to the pest's new environment. Where *Salicaceae* are native, both the pest insects and their natural enemies are usually also native, so there is no reason to import new natural enemies – except, of course, when new invasive species arrive.
2. Augmentative biocontrol: managed releases of more natural enemies to supplement those already present. Here, natural enemies are added to supplement existing populations and improve their impact. Natural enemies (usually from commercial insectaries) may be released by 'inoculation' or 'inundation'. For inoculative biocontrol, timing is usually critical and relatively small numbers are released to ensure that subsequent generations of natural enemies achieve control. Thus, if natural enemies are known to have been killed by a cultural activity, for example spraying or pruning, or by an unavoidable action, for example bad weather, then steps can be taken immediately to rectify the situation. For inundative biocontrol, very large numbers are released at once to provide rapid control of an immediate pest problem. Apart from nurseries, most willow and poplar plantations are usually unsuitable for augmentative releases of natural enemies, because the spatial and temporal distribution of pests is too irregular to justify costly natural enemy rearing programmes. *Tremex fuscicornis* (see section 9.7.16), which is native to Europe and Asia, was found in the central area of Chile in 2000. In the same year, *Megarhyssa praecellens* Tosquinet (Hymenoptera: Ichneumonidae), a parasitoid of late-instar larvae of the pest, was introduced to Chile deliberately. Parasitism levels reached close to 30%, but its low natural dispersal rate led to the development of a programme to breed and release large numbers of the parasitoid where *T. fuscicornis* was active in the V, IV and Metropolitan regions of Chile (Parra and González, 2007).
3. Conservation biocontrol: the preservation and/or manipulation of the habitat or environment of natural enemies to improve or enhance their effectiveness over long periods. Two key features of conservation biocontrol are: (i) removing or mitigating external management hazards, especially pesticides; and (ii) providing and preserving a favourable environment for the survival of natural enemies. Conservation biocontrol recognizes that many pesticides kill natural enemies as well as the target pest, so if insecticides must be used, they should be chosen and applied carefully to minimize any impact on natural enemies. For many (or most) willow and poplar plantations, more effective biocontrol may be achieved by developing a favourable environment for natural enemies (see 'Polycultures versus monocultures' above).

In practice, the more artificial the willow or poplar plantation, the greater are the opportunities for managing natural enemies in some way, especially through conservation or classical biocontrol. Good pest control from natural enemies known to be present can often be achieved by using common-sense management activities, but knowledge of the ecology of natural enemies may be a key component for the success of any biological control programme. *Telenomus nitidulus* (Thomson) (Hymenoptera: Proctotrupidae) is an egg parasitoid of the satin moth, *Leucoma salicis*. Studies showed that it was rare at the beginning of a moth outbreak, but numbers increased rapidly as moth numbers increased. Features of its life history, such as the fact that adult females emerged with some mature eggs and a preference for lower parts of the tree, helped to explain its impact at different parts of the outbreak cycle and confirmed that chemical applications against the egg stage of the pest moth would be detrimental to biocontrol by the parasitoid (Nef, 1976).

Biological control of insect pests by birds and mammals

Large numbers of defoliators and bud- and leaf-feeding insects are eaten by many species of birds, but birds are rarely important biocontrol agents because they kill relatively few individuals of pest species. Birds and small mammals may, however, be effective against long-lived or slow-breeding pests, or those pests that are attacked during dormancy (especially as pupae in the soil), or those that can cause significant economic damage at very low populations. Hence, various species of woodpeckers (Picidae) eat significant numbers of xylophagous insects, especially in winter, and in Italy this may compensate largely for damage caused by their nesting (Allegro, 1993). Predation of pupae by northern orioles (*Icterus galbula* (Sibley and Monroe)) may be responsible for maintaining low densities of the forest tent caterpillar, *Malacosoma disstria* Hübner (Lepidoptera: Lasiocampidae), in North America (Parry *et al.*, 1997a). Small rodents eat large numbers of forest insects such as Lymantriidae, and they may make important contributions to limiting pest populations (Hanski, 1992).

Biological control of insect pests by entomopathogens and nematodes

Microbiological control uses microorganisms such as the fungus, *Beauveria bassiana*, viruses and insect-destroying nematodes, or products derived from microorganisms such as the proteins and spores from *Bt*. Many of these very small organisms occur naturally in willow and poplar environments, but are often not in sufficient quantities at the right place or time to control pests adequately. Hence, commercially supplied formulations, traditionally applied as sprays, are used when required. Although they are quite safe environmentally, they are effectively insecticides with the same problems as any chemical spray, for example cost and practical difficulties of application to tall trees across large areas. They may be particularly useful in nurseries and may be the only way to target wood-boring larvae effectively. Novel techniques for using microorganisms, such as those that combine pheromones and viruses or fungi, for example 'lure-and-infect', are starting to be developed. Their practical value is likely to increase in the future. In the meantime, there are many reports on their potential as biocontrol agents against several willow and poplar pests. Mortality from viruses that infect larvae of *Leucoma salicis* in Belgium was additive through each development stage, providing up to 80% mortality of the population over time (Nef, 1978). In the USA, the entomopathogenic nematode, *Neoaplectana carpocapsae* Weiser, associated with the bacterium, *Xenorhabdus nematophilus*, killed 90% of *Paranthrene robiniae* when applied at the first signs of attack (Kaya and Lindegren, 1983), and using the same organisms in Italy, Cavalcaselle and Deseo (1984) obtained very positive results against *Paranthrene tabaniformis* but less encouraging ones against *Cryptorhynchus lapathi*. A local strain of *B. bassiana* at concentrations of 10^7 and 10^8 conidia ml^{-1} killed 100% of *Saperda populnea* (Linnaeus) larvae in Turkey, after median lethal times of 4.6 and 4.4 days, respectively (Eken *et al.*, 2006).

Finally, it should be noted that biocontrol specifically excludes many ecologically rational methods of control, such as use of hormones, pheromones, ryania, neem and other organic pesticides. Also, all environments are regularly hit by changes and perturbations, such as wind, rain, hail and dust storms. Although these natural but non-biological (abiotic) disturbances may kill many insects, they are not called biological control.

9.4.4 Chemical control of insect pests

The modern use of manufactured broad-spectrum insecticides against agricultural pests began with the introduction of Paris Green in the 1870s. This chemical, which was originally used as a dye and rodenticide, contained 28% arsenic and was almost as lethal to humans as to insects. Since then, insecticides have become accepted as cheap and highly effective means of pest control, especially since the development of organochlorines, organophosphates and pyrethroids in the 1950s and 1960s. They do, however, have many drawbacks. In particular, those with broad-spectrum activity kill natural enemies and non-target insects as well as the pests they are applied against; and many pest populations have developed genetically inherited resistance to one or more groups of insecticides. Unfortunately, natural enemies have not developed resistance to nearly the same extent as have the pests. Repeated use of some insecticides can even induce pest outbreaks, as shown for *Paranthrene tabaniformis* (Giorcelli and Allegro, 1999). Since about 1960, our knowledge of the environmental pollution of these broad-spectrum insecticides and their implication in human health issues has led to an increasing social backlash against their use. These pressures have driven (and continue to drive) the development of new generations of insecticides that are relatively safe to humans and to the environment, and are targeted against specific groups of pests.

Pesticides continue to be extremely powerful weapons and undoubtedly should be retained as part of the arsenal of weapons for insect control. However, they should be used sparingly and only when a number of conditions are met. Growers should never apply any pesticide without knowing that the pest is present, and they should also quantify the extent of insect damage before an insecticide is used. The potential impact on natural enemies should also be considered before application. The damage (or predicted future damage) of the pest should exceed a reasonable economic threshold, so that the estimated economic loss from a pest attack is greater than the cost of insecticide treatment. Even substantial damage in plantations can be tolerated if it does not kill the trees, but insecticides will be justified more easily in nurseries if the growth or shape of plants for commercial use is compromised by pest attacks.

Insecticides should be selected carefully for optimum control of the target pest. It is important to read and understand the product label before spraying. Instructions must be followed, particularly those relating to the environment, the development stage of the target insect, density and condition of the vegetation, dose rates and the places to be treated. It goes without saying that national and/or state regulations for chemical control must be adhered to at all times.

9.4.5 Genetic engineering

Poplars were the first trees to be genetically modified by insecticidal genes. McCown *et al.* (1991) reported the transformation of poplar with cry1a delta-endotoxin from *Bt*, which provided protection against *Lymantria dispar* (Linnaeus) (Lepidoptera: Lymantriidae) and *Malacosoma disstria* (Robison *et al.*, 1994). Transgenic poplar (*P. nigra* cv. 'Jean Pourtet') coding for a trypsine proteinase inhibitor were also obtained; however, no insect pest resistance was obtained (Confalonieri *et al.*, 1998). Transgenic poplars targeted against Coleoptera, carrying either a plant protease inhibitor (Leplé *et al.*, 1995; Delledonne *et al.*, 2001) or a cry3a delta-endotoxin *Bt* gene (Strauss *et al.*, 2001; Génissel *et al.*, 2003) with insecticide activity, have also been produced. While transgenic poplars engineered with proteinase inhibitors showed a moderate activity against *Chrysomela tremulae* Fabricius (Coleoptera: Chrysomelidae) in young trees, toxicity of transgenic poplars producing *Bt* cry3a toxin against *C. tremulae* was high (Augustin *et al.*, 2004). These *Bt* poplars can provide an attractive alternative to conventional insecticide sprays to control some of the key pests (Carrière *et al.*, 2003). However, one of the main risks associated with the widespread use of the *Bt* plants is the rapid development of genetically inherited resistance in pest populations targeted by the toxins (Gould, 1998). Field resistance to *Bt* cotton has been reported for a few moth species feeding on this crop, but field resistance to *Bt* toxin sprays has been detected in several populations of *Plutella xylostella* (Linnaeus), and many targeted pests have been selected for *Bt* resistance in the laboratory (reviews in Ferré and Van Rie, 2002; Tabashnik *et al.*, 2003; Griffiths and Aroian, 2005). The speed at which the evolution of resistance to transgenic plants is likely to occur depends on key factors supported by the 'high dose refuge management' strategy originally proposed by Georghiou and Taylor (1977). To delay resistance, the initial frequency of alleles conferring resistance in insect field populations has to be rare, the resistance to *Bt* plants must be recessive so that only resistant homozygotes can complete their life cycle on *Bt* plants, and a biological 'cost' should be associated with resistance so that the fitness of resistant individuals on non-*Bt* plants will be lower than that of susceptible individuals. In the case of *C. tremulae*, it has been shown that the resistance allele frequency at the start of *Bt* plant cultivation was higher than originally presumed, suggesting that resistance might be selected rapidly in target pests in the absence of refuges. However, *Bt* resistance was counter-selected in *Bt*-free environments, making it possible to delay or to prevent the evolution of resistance if enough refuges were planted (Wenes *et al.*, 2006).

Other pest inhibitor genes, such as plant cysteine proteinase inhibitor gene (CPI) and spider neurotoxin gene (BGT) are also being investigated in poplars by molecular biologists (Lin *et al.*, 2006; Zhang *et al.*, 2007).

In China, any field transgenic testing (whether large or small) and commercial releases must be approved by the State Forest Administration. At the end of 2008, 400 ha were planted with four transgenic lines of genetically modified (GM)-poplars (three lines of *Bt P. nigra* and one line of *Bt*Cry IAc and arrowhead proteinase inhibitor (API) white hybrid poplar) (Hu *et al.*, 2010).

9.5 Invasive Species and International Exchanges of Plant Material

9.5.1 Immigration along man-made pathways

The international mobility of insects through trade and other human activities is of increasing concern. Many important pests in one part of the world have been transported across countries or continents to new environments. In their new 'homes', they may add to or displace existing pests on commercial crops, or rapidly alter the ecological balance of native insect fauna. Many insects that cause no damage in their country of origin have become serious economic or environmental pests elsewhere. In addition, global warming is enabling alien species to expand into regions that were previously too inhospitable for survival (Walther *et al.*, 2009).

Species can enter a new country along a number of different pathways, and hugely damaging pests have been transferred accidentally between continents by poorly considered (or unconsidered) shipments of plants or timber. Many incursions happened during the period of European colonialism, for example *Sirex* spp. from Europe to New Zealand and Australia in the late 19th century (Nuttall, 1989). Many have also occurred recently – for example *Anoplophora glabripennis* from China to the USA, Canada and Europe (see section 9.7.12); and, in the other direction, *Megaplatypus mutatus* from South America to Europe (see section 9.7.13) (Alfaro *et al.*, 2007); and *Tremex fuscicornis* from the northern hemisphere to Australia and Chile (see section 9.7.16). Invasions within continents also commonly occur. Alien invasive species are being viewed increasingly as a long-term threat to global sustainable forestry – in both *Salicaceae* and other forest species (see FAO, 2010) – and various international trade mechanisms and protocols exist to help block these pathways. Effective actions almost always result from combinations of well-defined and regulated import health standards, stringent quarantine measures and ongoing education.

An obvious pathway for new pests is on *Salicaceae* germplasm during the deliberate transfer of plants between countries. The threat of new pest incursions *via* international exchanges of plant material cannot be overstated. This pathway is usually travelled by scientists and foresters – those working with *Salicaceae* for breeding or other studies – who thus have a particular responsibility to ensure that the plants they import are free of pests and diseases. Techniques for ensuring that pollen, seeds, cuttings or rooted plants are pest and disease free should be employed at all times. Some countries, or groups of countries, have already produced rigorous regulations governing the exchange of plant material, e.g. various EU directives. Even if no such regulations exist, willow and poplar researchers should develop and adhere to a set of principles designed to minimize the risk of pest transfer between countries.

Monitoring and treatment techniques, the required expertise and responsibilities of programme participants and the structure and operational procedures of quarantine laboratories should all be defined and adhered to.

9.5.2 Seeds

Pests of willow or poplar seeds themselves are not known. Insects do occur in the fruit capsules, but there is really no need for whole capsules to be exchanged as there is little impact on seed viability if they are removed for transfer. Nevertheless, extreme caution should still be exercised. The Douglas-fir seed chalcid, *Megastigmus spermotrophus* Wachtl (Hymenoptera: Torymidae), is a rather innocuous insect in North America, where it damages Douglas-fir seeds. Yet it has become a major pest in Europe, capable of destroying 90% of seeds (Roques, 1981; but see Rappaport *et al.*, 1993). Similar examples for willow seeds are quite possible. So, before shipment, all seeds should be inspected very carefully visually and/or irradiated to ensure they are not carrying any insect. Any exchange should be accompanied, both before and after dispatch, by pesticide treatment.

9.5.3 Cuttings and seedlings

Exchanges of cuttings and seedlings pose a greater challenge as they may be hosts to many more types of insects, especially sap suckers and stem borers. For small quantities of cuttings, obligatory and thorough visual inspections should ensure that most adult insects or large larvae are detected and then destroyed. Eggs and small larvae or small adults, for example mites, may escape visual detection and so additional heat and chemical disinfestation treatments (such as fumigation or immersion for several hours in a suspension of insecticide and fungicide) should be carried out both before and after transport.

In vitro growing techniques may offer the best possible solutions to the problems of exchanging new plant material. These techniques enable new individual plants to be produced rapidly and in great number, but while they ensure freedom from insects, bacteria can sometimes avoid disinfection, and viral diseases remain a threat.

9.5.4 Bark or cut timber

Exchanges of, or trade in, bark or other chipped wood or cut timber can be accompanied by many insect species in a dormant, for example eggs or caterpillars hibernating in bark cavities, or an active stage, for example on shoots. Insects living internally in wood or shoot exports are practically undetectable. There have been disastrous precedents, such as nematodes and fungi associated with *Monochamus* spp. (Coleoptera: Cerambycidae) introduced accidentally on conifers into Japan, and *Sirex noctilio* Fabricius (Hymenoptera: Siricidae) into Australia and New Zealand. *Anoplophora chinensis* (Forster) (Coleoptera: Cerambycidae) has been introduced in different European countries through bonsais imported from Asian countries. Debarking before shipment reduces the pest risks, although increasing the risk of timber desiccation.

Associated insects exported from temperate to tropical climates, for example accompanying exports of industrial bark to regions that are too hot for poplar growing, may not be able to establish in such different environments. In such cases, it may be argued that precautionary measures are superfluous. However, insects are so adaptable that this remains a very high-risk strategy.

9.5.5 Response to invasion

Unfortunately, and perhaps inevitably, despite all precautions, new insects that can feed on *Salicaceae* do continue to arrive in new countries. The ecological and economic impacts of these new invaders are often unpredictable because of the numbers of vacant ecological niches, the absence of natural enemies and different climatic or environmental conditions. A minor (or even almost unknown) pest in one country may become a major pest in another. *Nematus oligospilus* (see section 9.7.1) is almost unknown in its native northern hemisphere, but is a significant pest in South America and New Zealand.

New pests may be particularly damaging when they arrive in a country where native *Salicaceae* are rare or absent. Conversely (but perhaps less commonly), native polyphagous species may move on to *Salicaceae* for the first time.

Experience has shown that the effectiveness of responses to new invaders depends on the state of preparedness of society, the research community and the industries that use *Salicaceae*. Some of the questions that need to be answered before (or soon after) the arrival of a new pest are:

- What are the most common invasive species from a world perspective (in an attempt to predict the most likely invaders and their pathways)? A central and regularly updated list of potentially damaging immigrants is useful.
- Can a new arrival be identified? Maintaining sufficient taxonomic expertise to identify new pests and to recognize them when they do arrive is crucial.
- Will the new species survive and become established? Ecological expertise to carry out appropriate biological studies of individual species may be an essential part of determining whether they will establish. Studies of the new invader or its close relatives in their region of origin may be required.
- Will the new species cause sufficient damage to be considered a pest? This is a complex question and the answer may have to consider analyses of economic, environmental and social costs. It will need to take into account the potential host range (especially of native plants) of the insect and a prediction, if they do survive, of the population development in new habitats. The 'new' ecology may be crucially different to that in its country of origin.
- Has the new pest arrived with any of its natural enemies, or do existing species attack the new pest?

Only then can a sensible decision be made on whether to attempt eradication or accept it as a new pest to be dealt with through normal pest management programmes. If eradication is considered to be an option, then early detection, identification and impact analysis is vital, as is rapid access to appropriate eradication technology (such as fixed-wing aeroplanes or helicopters and support crews for aerial spraying). Time is of the essence, as the more time the pest is given to establish, the larger the population, the more widespread it becomes and the more technically difficult it is to run even the most efficient eradication programme.

9.6 Conclusions

Poplars and willows are attacked by a huge array of diverse animals, especially insects. Over the course of evolutionary history, dynamic adaptations have ensured that there is an ecological balance between the growth and death rates of trees and herbivores. The balance that allows willows and poplars to flourish in the wild must be replicated, at least in part, in managed plantations in order to prevent outbreaks of pests that may cause economic or environmental damage. IPM is a key management tool for foresters, but the successful implementation of IPM requires significant knowledge of the ecologies of the pests and their host trees. Cultural control and biological control by natural enemies are two important pest management techniques that can regulate the birth and death rates, respectively, of herbivores, but the relative importance of either is often site and tree-use specific. The increase in international trade threatens the production of willows and poplars everywhere through the introduction of new and unexpected insect pests. Willow and poplar researchers and end-users have particular responsibilities to ensure that they themselves are aware of the risks and take informed and appropriate care to avoid or manage them.

9.7 Selected Examples of Insect Pests of Poplars and Willows

9.7.1 Willow sawfly, *Nematus oligospilus* (Hymenoptera: Tenthredinidae) (Plate 33)

J.G. Charles

Global distribution

Species of the subfamily *Nematinae* form the dominant group of sawflies in the boreal and arctic regions of the northern hemisphere. The Holarctic genus *Nematus* contains more than 120 species, with host plants dominated by *Salix* spp. and *Populus* spp. *N. oligospilus* Förster is native to north and central Europe and the Caucasus. It is also widely distributed through North America and Eurasia, where it may possibly represent a complex of closely related species (Koch and Smith, 2000). Only a few species of *Nematinae*, and no *Nematus*, are native to the southern hemisphere.

In the southern hemisphere, *N. oligospilus* was first recorded from Argentina in 1980 (as *N. desantisi* (Smith, 1983; Koch and Smith, 2000)). It then established in Chile in 1984 (Gonzalez *et al.*, 1986), Lesotho and South Africa in 1992/1993 (Urban and Eardley, 1995), New Zealand in 1997 (Berry, 1997) and Australia in 2003 (Bruzzese and McFadyen, 2006). *N. oligospilus* in the wild appears to live exclusively on *Salix* spp. Laboratory trials in New Zealand have shown that larvae can complete development on 20 species of *Salix* and two species of *Populus* (J. Charles, 2011, unpublished data).

Description

Adult *N. oligospilus* are 6–9 mm long, with a dark brown head and rather pale straw-coloured legs and thorax. The abdomen of a young female is green, which is a consequence of the colour of mature eggs in her ovaries. Hence, it becomes straw coloured as her eggs are laid. The wings are colourless, but with distinct venation. Larvae are green, with a brown head capsule in early instars and a pale head capsule with brown marking above the eye in later instars. This coloration, together with its close alignment along the leaf edge, provides extraordinarily effective camouflage. Cocoons are usually either pale green or dark brown.

Biology

Females lay kidney-shaped eggs under the upper or lower lamina of leaves, although the upper surface is often preferred. The neonate larva rapidly eats a 'shot-hole' in the leaf and then straddles the edge of the hole, curling its body around the hole's perimeter while continuing to feed. After one to two moults, the larva moves to the edge of the leaf, where the opposable grip of the tibial claws prevents dislodgement by adverse weather conditions. Biotic disturbances from either above or below are avoided by swinging its body rapidly to the other side of the leaf. In the laboratory, a larva eats about two leaves during its lifespan, about 1.5 leaves during the final two instars. The larva develops through five or six instars before spinning a tough cocoon. During the summer generations in the southern hemisphere, the pre-pupa is followed quickly by pupation and adult emergence. Shortening day length in late summer (with a critical photoperiod of just over 11 h), induces an obligatory diapause during which the sawfly remains as a pre-pupa through the winter. Pupation occurs in spring (or, rarely, the following spring) and adults emerge several weeks after *Salix* bud break, when there are mature leaves available for oviposition. In its Holarctic range, *N. oligospilus* is bisexual, with one or two generations per year. In the southern hemisphere, all populations are unisexual (females only) and develop through up to six or seven generations a year. Females are strong fliers, and dispersal in both New Zealand and South America has been measured at about 300 km year^{-1} (Ovruski and Fidalgo, 1991; Charles and Allan, 2000).

Impact

Although apparently insignificant in the northern hemisphere, *N. oligospilus* has become a chronic pest of willows in New Zealand. Sawfly-infested trees in the eastern regions of the North Island are often defoliated by early January. Re-growth is also defoliated once or twice more before autumn, resulting in poor growth, limb loss and even death of mature trees within one to two seasons. Manual defoliation of potted willows to mimic sawfly damage seriously compromises root growth. Such losses critically reduce the effectiveness of willows for erosion control and riverbank protection. Similar losses have restricted the use of willows in parts of Argentina. By contrast, *N. oligospilus* can be viewed as a useful biocontrol agent in Australia (Caron *et al.*, 2011), where willows are more likely to be regarded as exotic weeds; and in parts of South Africa it is considered by recreational fishermen to be a useful insect because it provides a novel food source for freshwater fish such as trout.

Control

Where willows are planted primarily to prevent soil erosion or for riverbank stabilization, sawfly control by insecticides is not practical, economically feasible or environmentally acceptable. In New Zealand, *N. oligospilus* shows some host preferences for oviposition (Charles *et al.*, 1998), and species or cultivars showing any resistance to *N. oligospilus* are sought for inclusion in a willow breeding programme. Biological control by natural enemies imported from Europe remains a possible option (Alderete *et al.*, 2002), while mixed plantings to reduce the reliance of river protection on willows alone are also being investigated.

Invasive risk

N. oligospilus probably spread through the southern hemisphere from Argentina, rather than by multiple introductions from the northern hemisphere. Some evidence for this route includes the simultaneous arrival in South Africa of *N. oligospilus* and its South American parasitoid, *Dibrachys cavus* (Urban and Eardley, 1995). In addition, populations in the northern hemisphere are bisexual, while all in the southern hemisphere are females, with thelytokous parthenogenic reproduction.

None of the widespread and common *Nematus* species in the northern hemisphere is considered to be anything other than an occasional minor pest. Yet in the southern hemisphere, *N. oligospilus* illustrates many potential issues associated with invasive insects. For example:

- There was no prior indication from its biology in the Holarctic that *N. oligospilus* was a potential invader.
- The biology of *N. oligospilus* is significantly different from that in its home range, especially the development of parthenogenesis and multi-voltinism permitted by warmer/longer seasons.
- Populations are often relatively huge, due either to lack of natural enemies (top-down effects) or improved host quality (bottom-up effects), or both.
- The presence of a new invader may have unpredicted ecological consequences that raise unexpected problems for communities.

9.7.2 Poplar and willow leaf beetles, *Chrysomela* spp. (Coleoptera: Chrysomelidae) (Plates 34 and 35)

S. Augustin

Chrysomela (= *Melasoma*) *populi* Linnaeus
Chrysomela (= *Melasoma*) *tremulae* Fabricius
Chrysomela (= *Melasoma*) *scripta* Fabricius
Chrysomela (= *Melasoma*) *vigintipunctata* Scopoli, willow leaf beetle
Chrysomela crotchi Brown, aspen leaf beetle
Chrysomela aeneicollis Schaeffer

Global distribution

All *Chrysomela* species associated with poplar and willow are from the northern hemisphere. The most important are: *C.* (= *M.*) *populi*, *C.* (= *M.*) *tremulae*, *C.* (= *M.*) *scripta*, *C.* (= *M.*) *vigintipunctata*, *C.* (= *M.*) *lapponica*, *C. crotchi* and *C. aeneicollis* Schaeffer. *C. populi* and *C. tremulae* have a Euro-Asiatic distribution. Their native range is from Western Europe (UK, France, Spain and Portugal) to China and Japan, and both species occur among poplars in the whole range of their occurrence, except in the extreme northern area. They were introduced into the eastern maritime region of North America (Novak *et al.*, 1976). *C. scripta* is distributed through the natural range of nearly all poplar species in North America, from Canada to northern Mexico. The aspen leaf beetle, *C. crotchi*, is widespread in North America, including Canada. Of the willow leaf beetles, *C. lapponica* is found through northern Eurasia and *C. vigintipunctata* through Europe and Palaearctic Asia. *C. aeneicollis* is found mainly at altitude throughout north-western North America.

Description

These leaf beetle species are all similar in appearance. Adults are sexually dimorphic, with males smaller than females. The largest, *C. populi* (10–12 mm long) and *C. tremulae* (6–9 mm), have a metallic-red sheen. *C. scripta* (5.4–9 mm) is greenish- to reddish-yellow in colour and its elytra or wing covers are marked with black, irregularly shaped spots. *C. vigintipunctata* (6.5–8.5 mm) are overall yellow to brownish-green in body colour, marked with black spots on the head and thorax and broad black stripes on the elytra. *C. crotchii* on aspens is black with brown elytra. *C. lapponica* (5–8 mm) differ in the colouring pattern of their red and black morphs, described by Gross *et al.* (2004a) as a thermic adaptation. *C. aeneicollis* is brown, with light green to bright red spots. Larvae of all these species are dark brown at emergence and generally paler when older, and are difficult to identify to species before the adult stage.

Biology

C. populi, *C. tremulae* and *C. scripta* feed and reproduce on many poplar species and hybrids of sections *Populus*, *Aigeros* and *Tacamahaca*, and on willow (Brown, 1956; Augustin *et al.*, 1993a, b; Mattson *et al.*, 2001). *C. crotchi* feeds primarily on trembling aspen, *P. tremuloides* Michx., and occasionally on largetooth aspen, *P. grandidentata* Michx. (Smereka, 1965), *P. balsamifera* Linnaeus and *Salix* spp. (Brown, 1956). *C. vigintipunctata* and *C. aeneicollis* feed on many willow species, and *C. lapponica* feeds on willow and beech (Gross *et al.*, 2004a).

All these species have similar life cycles. Adults overwinter under dead leaves or in the soil, both within and near plantations, and emerge from diapause in spring when poplar and willow leaves begin to unfold. Mating occurs after a feeding phase on the new foliage and couples can be found on shoots and leaves, the females continuing to crawl and feed during mating. The fertilized females generally lay eggs in groups on the underside of leaves. *C. scripta* females lay an average of 823 eggs (Head *et al.*, 1977), *C. tremulae* up to 1000 eggs and *C. crotchi* up to 326 eggs (Smereka, 1965). Incubation lasts from a few days to 2 weeks, depending on temperature. Young larvae just after emergence usually feed gregariously on the lower epidermis and skeletonize the leaf. As they grow, they disperse progressively into smaller groups and feed individually. Large larvae and adults eat whole leaves, but generally leave the midrib and veins. Both adults and larvae generally prefer younger leaves (Augustin and Lévieux, 1993; Fang and Hart, 2000). Large populations of larvae or adults feed also on growing shoots, buds and on the tender bark of the tips of twigs. The larvae develop through three instars and pupation occurs after about 1 month. At the end of the late third stage, larvae move to various parts of the tree or to the undergrowth, fasten themselves with a posterior adhesive pad and pupate.

Voltinism depends on climate. *C. tremulae* and *C. populi* develop through two to three generations a year (Jodal, 1973; Augustin and Lévieux, 1993). In northern latitudes, *C. scripta* has only one generation, but in the southern USA as many as seven generations have been recorded (Ostry *et al.*, 1989). *C. crotchi* and *C. aeneicollis* are monovoltine in the USA and Canada (Brown, 1956; Smereka, 1965; Rose and Lindquist, 1982).

When disturbed, larvae of all species of *Chrysomela* discharge a defensive secretion from nine pairs of dorsal glands. The secretion is mostly salicylaldehyde, and the larvae use phenolic glycosides (salicin) from the leaves of poplar and willow as the precursor to its production (Pasteels *et al.*, 1983, 1984; Smiley *et al.*, 1985; Soetens *et al.*, 1998). This secretion is an effective defence against some generalist predators and an antimicrobial effect has also been shown for *C. vigintipuntata* and *C. lapponica* (Gross *et al.*, 2002). However, the host plant-derived defensive secretions may also be used by a specialized fly parasitoid (*Megaselia opacicornis* Schmitz, Diptera: Phoridae) to locate its host (Gross *et al.*, 2004b; Zvereva and Rank, 2004). The volume of secretions by *C. vigintipunctata* larvae is reduced when they are on host plants with low phenolic glycoside content, such as *Salix purpurea*, *S. myrsinifolia* and *S. fragilis*.

Many natural enemies of *C. scripta*, *C. populi* and *C. tremulae* have been observed (Loi, 1970; Head *et al.*, 1977; Burkot and Benjamin, 1979; Augustin and Lévieux, 1993), but although they evidently do affect host populations, insufficient information is available to make recommendations for pest management. However, parasitoids and predators may have driven *C. lapponica* from willows to pioneer birches, where the risk of predation and parasitism is lower (Gross *et al.*, 2004a).

Impact

Chrysomelid beetles are major pests of poplars and willows. The impact of defoliation is particularly severe in short-rotation coppice (SRC) plantation in the first 1–3 years of growth, when trees have a high ratio of the young foliage preferred by all *Chrysomela* species. The beetles can reduce biomass or kill young nursery plants and destroy new growth on old trees (which themselves are not as prone to attack).

C. scripta is the most important defoliating poplar pest in the USA, causing economic damage in young plantations (Burkot and Benjamin, 1979; Harrell *et al.*, 1981; Reichenbacker *et al.*, 1996; Coyle *et al.*, 2008). Coyle *et al.* (2002) showed that *C. scripta* defoliation reduced tree volumes of certain *Populus* clones by more than 70% after three growing seasons. In France, natural defoliation by *C. tremulae* and *C. populi* induced a biomass reduction of more than 30% on *P. tremula* Linnaeus × *P. tremuloides* hybrids after two growing seasons (S. Augustin, 2011, unpublished data).

Clonal susceptibility to *C. populi*, *C. tremulae* and *C. scripta* varies, but hybrids or clones having a *P. alba* parent are generally not preferred for feeding or oviposition by adults (Caldbeck *et al.*, 1978; Harrell *et al.*, 1981; Augustin *et al.*, 1993a, b). Within the *Populus* section, the preferences for adult oviposition and larval development of *C. populi* and *C. tremulae* are: *P. tremula* × *P. tremuloides* > *P. tremula* × *P. alba* (*P.* ×*canescens*) > *P. alba* (Augustin *et al.*, 1993a, b). The main hosts of *C. scripta* are in *Aigeiros* and *Tacamahaca*, but clones with *P. deltoides* parentage are preferred for oviposition and development (Caldbeck *et al.*, 1978; Harrell *et al.*, 1981; Bingaman and Hart, 1992). *C. lapponica* individuals prefer *Salix* species with high salicyl glycoside (SG) content, such as *S.borealis* (Zvereva and Rank, 2003), but a relationship between host-plant chemistry and larval survival does not appear to exist for *C. aeneicollis* (Rank, 1994).

Control

Nursery infestation can be reduced by removing leaf litter around young trees, depriving the adults of their overwintering habitat, although some species hibernate in the soil. Some insecticides can be applied successfully when the eggs begin to hatch, although pest control by insecticides is not allowed in forests in many countries and is economically and environmentally not acceptable. Several *Bt* formulations have provided control of *C. scripta*. Transgenic poplars have been created both with proteinase inhibitors and *Bt* genes. *Bt* poplars have shown a strong insecticidal effect on *C. tremulae*, although resistance to *Bt* plants may be present before selection at frequencies higher than originally thought, suggesting that selection for resistance may be rapid in the absence of refuges (Wenes *et al.*, 2006).

Several alternative control methods can be considered. These include using resistant plant cultivars, avoiding monocultures, establishing mosaics of small plantations rather than a single large one and maintaining the ratio of natural enemies as high as possible (see Coyle *et al.*, 2005).

Invasive risk

These pests are not present in the 'alert lists' of phytosanitary organizations. Although they have the capacity for active movements and dispersal, international trade appears to be the most likely pathway for passive introduction.

9.7.3 Large aspen tortix, *Choristoneura conflictana* (Lepidoptera: Tortricidae) (Plate 36)

W.J.A. Volney

Global distribution

The large aspen tortrix (LAT), *C. conflictana* (Walker) (Tortricinae), is native to North America and feeds principally on trembling aspen (*P. tremuloides*). A common associate, also feeding on aspen during outbreaks and with which it may be confused, is the olethreutine moth, *Sciaphila duplex* Walsingham. The range of LAT is transcontinental in the boreal forests from Labrador to Alaska. Its southern limit in eastern and southwestern USA is probably aspen forests in the mountains of these regions (Beckwith, 1973), but it would not be surprising to find it in Mexico.

Description

A female's entire complement of between 60 and 700 (Prentice, 1955; Beckwith, 1973; Evenden *et al.*, 2006) pale green eggs may often be laid in a single layered mass, appressed to the upper surfaces of aspen leaves. Prentice (1955) indicated that, on hatching, the larvae are pale yellow-green, with light brown heads, prothoracic shield and anal plates. The coloration darkens with each moult, and by the fifth instar the body is dark green, with a black head capsule, setal bases and anal plate. The thoracic shield becomes reddish-brown to black, with a light brown anterior margin in the final (fifth) instar. Its length varies from 15 to 21 mm. Pupae are initially bright green but darken to reddish-brown or black. Male pupae have five abdominal segments, whereas females have only four. Female moths have a wingspan between 27 and 35 mm, and males are smaller. The base coloration of the forewings is light grey with dark grey basal, median and distal irregular bands that complete the forewing colour pattern. Hindwings are light grey, with a feathery fringe of scales on the posterior margins.

Biology

On hatching in mid-July, larvae disperse to feed by skeletonizing upper leaf surfaces. In late August, larvae migrate to the base of trees, where they spin hibernacula in bark crevices, dead bark and mosses, and where they overwinter in the second instar. Larvae emerge when aspen buds begin to swell in early to mid-May. They then mine these buds, feed and moult to the third instar. As the leaves expand, the larvae leave the mines, eventually webbing leaves together to form a shelter in which they feed. More than one feeding structure may be constructed in the larval stage. Feeding is completed by mid- to late June, with pupation lasting 1–2 weeks. Moth emergence begins in late June to early July and lasts from 5 to 8 days. Females release a sex pheromone apparently consisting only of (Z)-11-tetradecenal (Evenden and Gries, 2006). Mating typically occurs on the first evening after female emergence and she lays her eggs almost immediately. Delayed mating results in a reduction of a female's realized fecundity (Evenden *et al.*, 2006). Eggs take 7–10 days to hatch under field conditions. There is one generation per year.

Impact

The damage caused by larvae feeding on buds and foliage is often accompanied by silk webbing. When outbreaks occur and defoliation is severe, this webbing may be found in the understorey (Ives and Wong, 1988). Outbreaks seldom last for more than 2 years (Cerezke and Volney, 1995). Thus, a reduction in radial increment of defoliated trees is most often the result of these episodes. However, some top-kill and branch mortality has been observed. Tree mortality rarely occurs because of this feeding, but mortality would be restricted to suppressed trees if it does occur (Thomas, 1978). The feeding also elicits a short-term resistance to the insect with the production of salicortin and tremulacin in damaged leaves (Clausen *et al.*, 1989). Long-term induced resistance has not been demonstrated. There are clonal differences in host phenology (Parry *et al.*, 1997b) and these induced responses (Clausen *et al.*, 1991) play largely uninvestigated roles in the success of LAT populations and the damage sustained as a consequence of outbreaks.

Control

A wide range of natural enemies has been observed to attack LAT outbreak populations. These include birds, such as the downy woodpecker and red-eyed vireos and chickadees. Insect predators include an anthocorid bug (Beckwith, 1973) two species of ants (Prentice, 1955) and a carabid. All immature stages of this insect are attacked by parasitoids. Over 26 different genera of insect parasitoids have been recovered from LAT from studies in Canada and Alaska (Prentice, 1955; Torgersen and Beckwith, 1974; Wong, 1979). The array of microorganisms infecting LAT includes a granulosis virus, an entomopoxvirus, several microsporidian species and several species of fungi (Burke, 1982). An industrial preparation of *Bt* (Berliner) showed promise in reducing field populations (Holsten and Hard, 1985).

Management of these populations on an extensive scale has not been necessary. However, if an intensive, short-rotation aspen plantation culture is to be undertaken, direct control techniques for LAT populations may warrant development (Mattson *et al.*, 2001).

9.7.4 Poplar leaf defoliators, *Clostera* spp. (Lepidoptera: Notodontidae) (Plate 37)

A.P. Singh

Global distribution

The genus *Clostera* has a worldwide distribution (Costa, 2006). *Clostera fulgurita* Walker is widely distributed in the Oriental tropics including China, Pakistan,

India, Sri Lanka, Borneo, Sumatra and Java (Southdene Sadirian Berhad, 2011). *Clostera cupreata* Butler occurs in northern India and adjoining Pakistan (Singh and Singh, 1986).

Food plants of *C. fulgurita* and *C. cupreata* include: *Xylosma longifolium* (*Flacourtiaceae*); *Elaeodendron glaucum* (*Celastraceae*); *S. tetrasperma*, *S. babylonica*, *P. alba*, *P. ciliata*, *P. nigra* and *P. deltoides* (*Salicacaceae*); and *Gymnosporia falconeri* (*Celastraceae*) (Beeson, 1941; Mathur and Singh, 1960; Browne, 1968; Singh and Singh, 1986; Singh *et al.*, 2004). Both species are major pests of poplar and willows.

Description

C. fulgurita is a greyish-brown moth. The forewings have irregular white markings on the anterior half, a reddish-brown spot on the apical half and an indistinct series of submarginal black spots. Wing expanse is 36–46 mm. Eggs are spherical and bright yellow in colour. A full-grown larva is 30–35 mm long, with lateral tufts of sparse, whitish hairs on each segment. Two broad white lines run almost the length of the body dorsally, interrupted by two distinctive black 'spots', one posteriorly and the other on body segment four. Both of these spots have distinctive yellow setae, while the setae on the rest of the body are white. The spiracles are white with black rims. There are also two narrow, pale dorsolateral lines along the length of the body. The head capsule of the earlier instars (one to three) is black, but gradually turns orange-brown in the fifth instar, measuring 4 mm. The colour of the larva also turns from dark to pale brown as it grows from the second to the fifth instar. The pupa is reddish-brown in colour, elongated, cylindrical in shape, rounded at the cephalic end and pointed at the posterior end. In *C. cupreata*, the head, thorax and forewings of the moth are pale reddish-brown in colour, while the abdomen and hindwings are much paler as compared to *C. fulgurita*. The male moth has characteristic anal tufts arising on the dorsal side of the anal segment. Wingspan varies from 25 to 30 mm. Freshly laid eggs are light green to yellow in colour, but after some time, longitudinal, curved, reddish-brown lines originating from the centre appear on the egg. The larva is light brown with a greenish mid-dorsal portion, the head light brown speckled with dark brown spots and with two reddish tufts of hairs each on the mid-dorsal region of the first abdominal segment and on the eighth abdominal segment. The mature larva is 28 mm long. The pupa is reddish-brown in colour, elongated and cylindrical in shape, with the cephalic end rounded and the posterior end pointed (Singh and Singh, 1986).

Biology

C. fulgurita has 10–12 generations in the plains of northern India. The moth lays eggs in groups on the under surface of a poplar leaf, usually on the terminal part. One female may lay 500–700 eggs. Incubation period is 4 days. The first two instars feed gregariously on the upper and lower epidermis of the leaf, thus skeletonizing it, while the later stages consume the whole leaf, leaving only the larger veins. There are five larval instars and the larval period varies from 17 to 23 days. The mature larva spins a thin hairy pupal cocoon in between two leaves, fastening them together with silk, or in a leaf fold. In winter, the larva pupates in leaf debris on the ground or crevices in the rocks and tree trunk close to the ground. The pupal period varies from 7 to 10 days in May–June. The biology of *C. cupreata* is quite similar to that of *C. fulgurita*, except that the eggs are laid on both the lower and upper surface of leaves in groups of 200–300. The larval period varies from 11 to 13 days, the pupal period from 4 to 6 days and the entire life cycle is from 19 to 21 days, with eight or nine generations a year in the plains of northern India (Singh and Singh, 1986). The pupae of the two species can also be identified and differentiated from *Clostera anastomosis* (Arru, 1965) by the shape of the cremaster, the distribution of punctures on the abdominal segments and the presence or absence of folds around the anus (Singh and Sharma, 1979a). In addition, the shape of the uncus, harpé and anellus in the male genitalia and the size of the signum in the female genitalia are characters of taxonomic importance (Singh and Sharma, 1979b) that distinguish these two species from *C. anastomosis* genitalia as described by Arru (1965).

Impact

Clostera are common in poplar nurseries. The attacks start in March–April and continue until leaf fall in October. Symptoms of infestation are crown death, with leafless crowns leading to loss of apical dominance, resulting in forking and sometimes plant death in nurseries when the attack is severe. Mortality is due mainly to rapid and repeated infestation and deformities in the main stem, otherwise there is only loss in growth. Alternate/collateral hosts growing naturally in the vicinity of poplar plantations play a vital role in the survival and population growth of the insect. The loss in radial increment is appreciable when more than half the foliage is lost (Singh and Singh, 1986). However, even 25% defoliation by *Clostera* spp. is known to decrease the growth increment of poplar trees significantly (Gao *et al.*, 1985). Severe and repeated defoliation results only rarely in the death of plants. Complete loss of leaves and defoliation late in the growing season, for instance in August, are generally the most damaging because they leave the plant in a weakened condition and open to attack by other pests and diseases. Severe and repeated defoliation can kill young plants (Singh *et al.*, 2004).

Control

Serious outbreaks of both *C. cupreata* and *C. fulgurita* can occur on *P. deltoides* grown in agroforestry in northern India (Singh *et al.*, 2004; Costa, 2006). A wide range of natural enemies native to India have been recorded to check the pest populations effectively (Singh *et al.*, 2004). These include a nuclear polyhedrosis virus (NPV) and a large number of egg, larval and pupal parasitoids, including: *Telenomus colemani* (Hymenoptera: Sceleonidae); *Trichogramma chilonis* and *Trichogramma poliae* (Hymenoptera: Trichogrammatidae); *Aleoides percurrens* (Hymenoptera: Braconidae); *Sturmia* sp. (Diptera: Tachinidae); *Canthecona furcellata* (Hemiptera: Pentatomidae); and *Sycanus collaris* (Hemiptera: Reduviidae). These are potential biocontrol agents in India that can be used for the biocontrol of *Clostera* (Singh *et al.*, 2004). Clonal susceptibility is also very important. *P. deltoides* clones 'IC', 'G-3', 'G-48', 'D-100' and *Populus* ×*euramericana* that were planted extensively under agroforestry in the 1980s were heavily attacked during epidemics (Singh and Singh, 1986), while a number of promising tolerant *P. deltoides* clones have been evaluated (Singh and Pandey, 2002).

Clostera can be managed in both nurseries and plantations by foliar insecticide sprays, for example 0.1% carbaryl or fenitrothion in water (Singh *et al.*, 2004). Aerial spraying of 1 kg (a.i.) of carbaryl (Sevin 85S) at 1 kg ha^{-1} in water solution gave 99.1% kill of the target pest in the Terai region of northern India during an outbreak in 1981 covering over 1100 ha (Singh *et al.*, 1983).

Invasive risk

Both species of *Clostera* pose a threat to commercial plantations of poplar throughout northern India and Pakistan in the absence of natural enemies and with large-scale monoculture poplar plantations. The spread is caused mainly by commercial trade, with eggs, larvae and pupae carried on young plants to new sites from nurseries. Management of *Clostera* populations on an extensive scale is not always necessary, but is advised locally in nurseries. These two species are also limited by the colder climates in temperate regions of the northern hemisphere.

9.7.5 Poplar leaf miner, *Phyllocnistis unipunctella* (Lepidoptera: Phyllocnistidae) (Plate 38)

L. Nef

Global distribution

Most species of *Phyllocnistis* are associated with poplars and willows. Among the most important are *Ph. unipunctella* Stephens (= *Ph. suffusella* Zeller) on poplars and *Ph. saligna* Zeller on willows in Europe, *Ph. labyrinthella* (Bjerk.) on aspens in Asia and *Ph. populiella* on aspens in North America (de Tillesse *et al.*, 2007).

Description

Phyllocnistis spp. are small moths, with a 7–8 mm wingspan. The forewings are narrow and fringed. The hindwings are narrower still, with a long fringe. The eggs of *Ph. unipunctella* are greenish-grey, and the flat, tiny larvae are yellowish-white. The larvae are almost legless, but mobility is enhanced by lateral protuberances.

Biology

Ph. unipunctella can have as many as three generations a year, with the first adults appearing at the start of summer. The eggs are laid individually on the lower surface of new leaves. The young larva hatches after 6–9 days, penetrates the leaf epidermis and makes a tortuous mine, which often follows a main vein or the leaf edge. The larva feeds in the leaf for about 2 months. The shiny white leaf surface above the mine looks like a snail trail. Pupation occurs in a cocoon under a silken membrane in the folded edge of the leaf (Nef and Janssens, 1982).

Impact

Physical and soil conditions at a site may be the most important factors determining population densities, although they also vary from year to year with prevailing weather conditions. Clonal selection also has an impact on individual growth and population densities. The length of mines varies from 14 to 34 cm in different clones, averaging 17 cm in *P. deltoides* × *P. nigra*, 27 cm in *P. trichocarpa* and 24 cm in *P. trichocarpa* × *P. deltoides*. Mined young leaves will grow to only about 75% of the size of healthy leaves. According to extensive research (Nef, 1988a), 20% of the leaves are attacked, and the insect reduces the average leaf area (and thus probably the growth) by about 5%. However, up to 70% of the leaves of some poplars are attacked. The frequency of the insect on a clone is linked significantly to the quantities of each of the elements P, Mn, K or Zn, and the quantities of K, Mo, Cu and Zn are modified significantly – either positively or negatively – after attack (Bouyaiche and Nef, 1987).

The numbers and development of *Ph. unipunctella* larvae depend also on foliar tannins (polyphenols). Tannin-rich leaves, typical of forest poplars (*P. trichocarpa* and its hybrids), support longer but fewer mines, and the rate of larval development and pupal weights are reduced. In leaves of pioneer poplars (*P. deltoides* × *P. nigra*), these parameters are reversed, and moreover, the levels of secondary polyphenols produced after attacks are almost 100% higher than in *P. trichocarpa*. It can be concluded that resistance to *Ph. unipunctella* is markedly lower in clones derived from pioneer poplars (Nef, 1988b, 2007).

The reduction in photosynthesis caused by *Ph. populiella* is attributable primarily to the failure of stomata to open normally on bottom-mined leaves (Wagner *et al.*, 2008).

Control

Despite its significant impact, chemical control of *Ph. unipunctella* is not economically justified, even in forest nurseries, because the damage is not persistent. A cultural control strategy should be adopted, based on planting resistant clones in physically suitable sites that avoid humid or heavy soils. A good choice of sites and clones will reduce populations of this insect significantly. The *P. deltoides* × *P. nigra* hybrids show great variability, the most susceptible being 'Robusta' (to which the insect is expected to be best adapted because of its ancient co-evolutionary history). *P. trichocarpa* is passably resistant, and *P. trichocarpa* × *P. deltoides* hybrids are the most resistant. Common clones of decreasing susceptibility used in Belgium are: 'Robusta', 'Primo', 'Gaver', 'Ogy', 'Columbia River', 'Gibecq', 'Trichobel', 'Ghoy', 'Isières', 'Fritzi Pauley', 'Hunnegem', 'Raspalje', 'Beaupré', 'Unal' and 'Boelare' (Nef, 1985).

In North America, the susceptibility of willows to *Phyllocnistis* spp. larvae was also dependent on the willow species and their hybrids, and increased with fertilizer use. Physiological stress and susceptibility to the pest can thus be inversely related (Orians and Floyd, 1997).

9.7.6 Woolly poplar aphid, *Phloeomyzus passerinii* (Hemiptera: Aphididae) (Plate 39)

G. Allegro

Phloeomyzus passerinii (Signoret)
(= *Phloeomyzus redelei*, *Phloeomyzus dearborni* (Eastop and Hille Ris Lambers, 1976; Masutti, 1982))

Global distribution

Ph. passerinii has a Eurasian distribution. Its native range stretches from Europe (UK, Netherlands, Belgium, France, Germany, Spain, Italy, Greece, Hungary, Russia) and the Middle East (Israel, Egypt, Syria, Iran) to Pakistan and China, where a new subspecies has been recorded (Zhang *et al.*, 1982). It has been introduced to North America, probably in the recent past, and recorded (as *Ph. dearborni*) in the USA (Maine) and Canada (Smith, 1974). Further studies are needed to define the taxonomic status of the different populations.

Description (Theobald, 1929; Arzone and Vidano, 1984)

The wingless, virginoparous females are a pale yellowish-green to dull white, covered lightly with a mealy wax, with dusky legs and lateral tufts of white or bluish-white woolly wax on each side posteriorly. The head is dusky, with five-segmented brown or smoky coloured antennae. The rostrum tapers to a very long point, reaching the metacoxae. The eyes are small and black, and the legs short and thick. There are two wax plates each side in front of the cauda. The body is oval in shape and 1.6–1.7 mm long.

The first-instar larvae are yellowish-green, with dusky antennae, legs and tip of rostrum. They have an oval-elongated body shape, with parallel sides and are 0.5 mm long. The antennae have four segments, and the rostrum is thick and very long (exceeding the body length). The second-, third- and fourth-instar larvae have, in comparison with the first instar, more slender legs and antennae and a shorter rostrum, the tip of which hardly reaches the apex of the abdomen. Their body is darker and more oval in shape, and 0.7, 1.1 and 1.4 mm long, respectively. The third- and fourth-instar nymphs are yellowish-green, with wing pads that are very small in the third instar and well developed in the fourth instar. The antennae are five- and six-segmented in the third and fourth instars, respectively.

The winged, oviparous females are about 1.5 mm long, green, with a dark head and thorax and dusky antennae and legs. The six-segmented antennae are slender and without secondary sensoria. The eyes are large and the wing veins clouded on each side with brown. The alate males are smaller than the females (length about 1.0 mm) but also have six-segmented antennae. Each female generally carries two eggs at a time. The eggs are subcylindrical in shape, three times longer than wide and a shiny, reddish-orange in colour. They are 0.3–0.4 mm long.

Biology

As far as we know, *Ph. passerinii* feeds exclusively on *Populus* species. *P. nigra* and *P.* ×*canadensis* (*P. nigra* × *P. deltoides* hybrids) are its favourite hosts, but *P. deltoides* can also be affected, even though its susceptibility to the pest is generally lower (Lapietra and Allegro, 1990a). *P. ciliata* was recorded as a potential host in Pakistan (Habib and Ghani, 1970) and in Iran (Shojai and Lotfian, 1988). The pest was also recorded on *P. alba* (Theobald, 1929; Della Beffa, 1936). *P. tremuloides* can be affected in North America (Smith, 1974).

Very little is known of the physiology of feeding of *Ph. passerinii*, which is thought to occur on the parenchymal tissues of the bark that are rich in carbohydrates and nitrogen compounds (Allegro, 1997a). A morphological study was carried out on the digestive system of *Ph. passerinii* (Ponsen, 1982), without investigating the physiology of digestion.

Populations of *Ph. passerinii* usually develop anholocyclically on poplar through apterous virginoparous females. Colonies begin to grow in May–June, when temperatures become favourable, and disappear at the latest in August–September, even if overwintering might be anticipated by the presence of adverse environmental conditions such as early cold weather or violent storms. Infestations are revealed by the waxy flocks, woolly and whitish in colour, which are produced in abundance mainly by apterous females. Over 12 generations per year can occur in the field. Each generation lasts 9–31 days (on average 11 days), depending on air temperature and host. Each female can produce about 170 larvae over an average lifespan of 36 days. Overwintering occurs in the second or third larval instar (rarely in the fourth instar), generally in the crevices of the bark of superficial roots. The first-instar larva is the only dispersal stage, both actively (by crawling to a new site on the plant) and passively (being carried by the wind that catches the waxy tufts). Once the larva has reached its new location and has inserted the stylet into the bark tissues, it develops to the adult stage without moving. Adult females also generally spend their life without moving from the original position (Arzone and Vidano, 1984). Winged forms, both males and oviparous females, are frequently observed in laboratory colonies when the host plants or cuttings start to deteriorate. However, they have been observed in the field only occasionally, mainly during autumn. Egg laying has never been observed in wild populations, and a life cycle with a secondary host has never been demonstrated. Most biological observations (as well as those on host-resistance behaviour) have been carried out in the laboratory, using colonies established by artificial inoculations of poplar cuttings kept in jars with water at the bottom, at a constant temperature of 20°C and a relative humidity of 70% (Arru, 1971, 1974; Lapietra and Allegro, 1990a; Allegro *et al.*, 1996b). Development of *Ph. passerinii* is favoured by moderately high air temperatures (20–25°C), high air humidity (>70% RH), limited air movement and shady habitats. A number of natural enemies (predators) of *Ph. passerinii* have been quoted in the literature (Vidotti, 1960; Arzone, 1987; Raspi, 1996), even though the effectiveness of the complex is normally inadequate to control the pest beneath the damage threshold in poplar plantations.

Impact

Ph. passerinii is considered an economically important pest of cultivated poplars in Italy, Syria, Belgium and Hungary, but there is evidence of an increasing virulence in other countries such as France (Maugard and Chauvel, 1997; Bankhead-Dronnet *et al.*, 2008) and Spain (Aparisi, 1971). It is periodically (at intervals of 5–10 years) very harmful to poplar plantations in northern Italy, where most of the cultivated clones (about 80% of trees) are susceptible to its attack (Arru, 1975; Lapietra and Allegro, 1990b).

Ph. passerinii feeding activity, which involves the injection of a toxic saliva, causes bark suberization and, during the heaviest attacks, cracks and necroses. Death of the bark disrupts water and nutrient circulation, and hence increasingly stresses the roots. In some cases, the tree can collapse during the summer period or can be windthrown (Lapietra and Allegro, 1981). *Ph. passerinii* is generally present at low population levels on *Populus* species in the natural riparian woodlands of the Mediterranean area. By contrast, its virulence is enhanced greatly by cultivation, mainly due to the great reduction of genetic variability of trees (plantations are usually monoclonal) and to high plant density (more than 300 plants ha^{-1}) (Lapietra and Allegro, 1981; Arzone and Vidano, 1984; Allegro, 1997a). Therefore, *Ph. passerinii* can be considered as a typical 'cultural pest' (Allegro and Cagelli, 1996).

Control

The pest can be controlled by growing resistant or moderately resistant poplar clones. The resistance to *Ph. passerinii* among different *Populus* species and cultivated clones may vary consistently and is genetically based (Arru, 1967b). For this reason, the breeding of resistant poplars may lead to appreciable benefits in a reasonably short time (Arru and Lapietra, 1979). A laboratory test based on the artificial inoculation of poplar cuttings (Arru, 1971, 1974) proved to be a very useful method for early evaluation of resistance in progenies from *P. deltoides* and *P. nigra* crosses, which in Italy were bred in order to obtain hybrid clones of cultural interest. *P. nigra* genotypes showed a wide range of resistance levels from very low to very high, with a higher ratio of resistant genotypes among those from southern Italy (Allegro and Cagelli, 1996). *P. deltoides*, on the other hand, exhibited generally high resistance levels (Arru and Lapietra, 1979). The laboratory test was validated by comparing those results with field data from the same clones in natural infestations of the aphid (Lapietra and Allegro, 1986a). It was shown that laboratory tests carried out with cuttings collected during the vegetative period gave more reliable results (Allegro *et al.*, 1996b). The newly selected hybrid clones in Italy range from highly resistant to highly susceptible (Lapietra and Allegro, 1986a, 1990a; Allegro *et al.*, 1996b). Iranian poplar species and cultivars were also tested by field and laboratory methods (Shojai and Lotfian, 1988, 1990; Shojai *et al.*, 1998; Sadeghi *et al.*, 2002). A study of the biochemical markers of resistance, based on analysis of phenolic compounds in three groups of *P. nigra*, each characterized by a different resistance behaviour, showed that some phenolics were linked to specific resistance groups; so, they are likely to play a role in the tree resistance mechanism (Allegro *et al.*, 1996a).

Chemical control is needed generally when outbreaks occur in poplar plantations. Spraying trunks with white mineral oils alone or mixed with organophosphates always proved to be effective when carried out at the beginning of the outbreaks (Aparisi, 1971; Lapietra and Allegro, 1981; Allegro, 1989, 1997a; Giorcelli and Allegro, 1999).

Monitoring older (>5-year-old) poplar plantations during the risk period (May–August) is necessary to detect outbreaks early and to plan a rational control strategy. As a rule, the intervention threshold is reached when colonies start to spread from the bark crevices and cover large areas of the trunk. When colonies cover the whole trunk circumference, a delay of about 2 weeks in applying control methods can be enough to cause serious damage to the tree.

The probability of heavy infestations and of economic damage can be decreased significantly by planting trees more than 6 m from their nearest neighbour (Lapietra and Allegro, 1981).

Invasive risk

Ph. passerinii outbreaks are increasing in frequency in northern Italy (Allegro, 1997a), with heavy economic losses caused both by the death of thousands of trees and by the need for expensive chemical control. In France, there is evidence that pest populations are increasing (Bankhead-Dronnet *et al.*, 2008; S. Augustin, France, 2011, personal communication). The pest is not present in the 'alert lists' of phytosanitary organizations. *Ph. passerinii* has a very low capacity for active movement and dispersal, but attention must be paid to the possibility of its passive introduction into southern hemisphere countries by the transport of freshly felled poplar trunks, if not debarked.

9.7.7 Giant willow aphid, *Tuberolachnus salignus* (Hemiptera: Aphididae) (Plate 40)
C.M. Collins

Global distribution

The giant willow aphid, *T. salignus* Gmelin (Lachninae: Lachnini), is thought to originate from the Asian subcontinent, and its distribution reflects that of willows; it is virtually cosmopolitan, being absent only from Australasia. It is a specialist willow feeder, but very occasionally has been recorded on poplar.

Description

The giant willow aphid is one of the largest aphids recorded, with an adult body length of 5.0–5.8 mm. It occurs in both winged and wingless parthenogenetic forms, producing live offspring. All morphs are mid-grey-brown to dark grey-brown, with a lighter underside and darker spots on the abdomen. The most distinctive feature of the aphid is a large dark brown dorsal tubercule of unknown function. The species is strongly aggregative, forming vast brownish-black colonies on infested trees. Accidental brushing against colonies of *T. salignus* can stain clothing a rusty-reddish colour, and the abundant honeydew production of this species attracts many wasps.

Biology

T. salignus has an unusual temporal distribution among aphids. It first appears in early summer, when trees are considered to be nutritionally poor hosts. Populations peak in late autumn, and aphids can continue to feed and reproduce long after the host tree has shed its leaves. The species is rarely seen in spring, when trees are considered to be high-quality hosts to aphids, due to nutrient translocation associated with the rapid growth stage.

The aphid is asexual. No sexual morphs have ever been found and studies of the ribosomal DNA arrays associated with X-chromosomes confirm it to be parthenogenetic (Blackman and Spence, 1996). In the absence of sexual reproduction, and thus the lack of an egg stage, *T. salignus* is thought to spend late winter and spring viviparously. Whether there exists a specialist morph, such as a hiemalis (an overwintering asexual stage), remains unknown and a mystery surrounds the location and occurrence of the aphid during this period. Adults can survive several days at temperatures of –5°C and can live for up to 3 months at temperatures of 10°C. Both winged and wingless forms spend a significant proportion of the adult lifespan in the post-reproductive stage, which may benefit their clonal siblings and progeny by improving the nutritional quality of the host by increasing 'sink' characteristics.

Phylogenetic studies link this species to conifer-feeding aphids, which suggests that it may be linked ancestrally to gymnosperm host plants (Normark, 2000).

Impact

Infestation by the giant willow aphid reduces the shoot and root biomass of host trees substantially and increases water use and photosynthetic rate (Collins *et al.*, 2001). Infestation in the first year after planting can cause a 50% reduction in root and shoot productivity. The presence of *T. salignus* can exacerbate seriously the effects of summer drought stress of host trees. It also leads to an increased proportion of nitrogen in the leaves of infested plants, and this may potentially improve the host's suitability as food to other herbivores. There are also indirect effects of aphid feeding mediated by honeydew deposition. Trees flower more and are more branched, which has consequences for SRC harvesting, as branched trees do not cut and bundle compactly. There are few data available on the long-term duration of infestation, so it may be that although the impact is substantial within a year, this is mitigated by it being occasional between years.

Control

As this aphid is primarily a pest to SRC willow crops, there are huge problems with insecticide application. Conventional pest control such as pesticide sprays are hard to apply to a willow SRC crop, as the canopy can reach several metres high and the subcanopy is sufficiently dense to impede the passage of tractors and spraying booms. The use of helicopters and aeroplanes to apply pesticide to mature crops is unlikely to be economic and may well be inefficient, as canopy foliage will intercept sprays before they reach the stem-feeding aphids.

There are no significant arthropod natural enemies recorded, and as the species appears to sequester salicylic acid, its foul taste deters vertebrates.

Future research into the identity (whether viral or fungal) and ecology of the entomopathogen that reduces numbers dramatically in late autumn may hold promise.

Miscellanea

The giant willow aphid has made several appearances in the scientific literature. Thomas Mittler used this species in his classic studies of aphid feeding and nutrition. Each aphid's stylet (the hypodermic-like mouthparts) was severed during feeding and provided direct access to plant phloem for the analysis of its composition (Mittler, 1957). *T. salignus* also made an appearance in the chemical literature when, having been noted for its staining, it was investigated as a potential source of dyes for military uniforms. In 1881, William Curtis suggested that the abundant honeydew produced by *T. salignus* 'be gathered and, by purification, converted into the choicest sugar-candy'. He also mentioned its special attraction for wasps, but not bees.

9.7.8 Willow gall midges, *Dasineura* spp. (Diptera: Cecidomyiidae) (Plate 40)

S. Höglund and S. Larsson

Global distribution

Dasineura is the largest genus in Cecidomyiidae and includes species that make simple and complex galls, roll leaves and live freely in flowers or flower heads. The genus *Rhabdophaga* is considered synonymous with the genus *Dasineura* by some taxonomists (Gagné, 2004). Numerous species of *Dasineura* are known from *Salicaceae*, in particular from *Salix* spp. and a few from *Populus* spp. (Gagné, 1989). Gall-forming *Dasineura* spp. are known to attack willows and poplars in Europe, East Asia and North America.

Description

The adults are small (2–3 mm) and short-lived, typically living only long enough to mate and lay eggs. Larvae are legless and may be white, yellow, orange or red, depending on species and age. Larvae are the only feeding stage and they induce modifications of the plant tissue. The outcome of this interaction is typically an abnormal plant structure – the gall – the structure of which is often typical of the midge species that caused it. Identification of the gall midge species is often based on the form and position of the gall. There are several types of galls: (i) leaf galls appear on leaf blades or petioles; (ii) stem galls on stems and twigs, ranging from slight swelling to large knot-like growths; and (iii) bud/flower galls on bud or flower structures.

Biology

The female lays her eggs close to the feeding site where the larva, although rather immobile, selects the final place for gall initiation. The larva attacks plant parts that are actively growing, such as buds and undeveloped leaves, and develops through three instars. Certain species have several generations per year, depending on whether or not the host plant has an indeterminate growth. Gall midge biology is complex, in particular regarding diapause and when to leave the gall. Larvae of some species leave the host when fully grown, whereas others pupate in the gall and leave as adults (Yukawa and Rohfritsch, 2005). Some species have extended diapause, such that a fraction of the population remains in diapause over the next generation(s). At the population level, sex ratio is close to 1:1. Individual females, however, can produce progeny of a single sex.

Impact

Galls act as sinks for resources, and therefore compete with the plant's own sinks, i.e. buds, leaves and shoots. Galling insects can change plant characteristics such as photosynthesis, phenology, leaf longevity, architecture and biomass production. The effect of galling insects on plant growth depends on the density of galls and on the importance of the infested plant part for growth. For example, one *Dasineura ingeris* (Sylvén and Lövgren) larva can kill the apical bud of *S. viminalis* Linnaeus. This leads to loss of apical dominance, followed by rearrangement of growth by stimulating development of many new shoots and subsequent reduction of biomass production. By comparison, *D. marginemtorquens* (Bremi) lives in the leaf margin of *S. viminalis* and hundreds of galls are required to reduce biomass growth significantly. In addition, galling insects can affect the density of other insects, such as leaf beetles, indirectly by stimulating regrowth and thereby enhancing the food quality of the plant.

Control

The population density of gall midges is influenced strongly by parasitoids and predators. Parasitism by both endo- and ectoparasitoids can be very high. The ability of parasitoids to control gall midge populations is probably higher in perennial plants such as poplars and willows than in annual plants, because perennials are not harvested every year. In Sweden, parasitoids and predators commonly seem to control populations of *D. marginemtorquens* (Strong and Larsson, 1994); three species of parasitoids (*Aprostocetus abydenus* Walker and *Aprostocetus torquentis* Graham (Hymenoptera: Eulophidae) and *Synopeas myles* Wallner (Hymenoptera: Platygastridae)) attack *D. marginemtorquens* larvae. The larvae and pupae are also killed by the predatory bug, *Anthochoris nemoralis* Linnaeus.

Chemical control is difficult because the gall midge larva is well protected within the gall and the adult only lives for a very short period. The most effective control appears to be to use resistant plant varieties. In a semi-natural crop such as *S. viminalis*, great genotypic variation in resistance has been revealed against *D. marginemtorquens* (Strong *et al.*, 1993). The resistance is expressed as high neonate larval mortality without gall formation. Both hypersensitive and non-hypersensitive plant responses have been associated with resistance (Höglund *et al.*, 2005). Interestingly, females do not discriminate against resistant genotypes and therefore lay as many eggs on resistant as on susceptible genotypes (Larsson *et al.*, 1995).

Invasive risk

The risk for outbreaks of *Dasineura* spp. is linked strongly to the occurrence of parasitoids and predators. With several generations per year, some *Dasineura* species can reach high densities rapidly. When present, *D. marginemtorquens* can be spread easily during the establishment of new plantations for short-rotation forestry because the overwintering cocoons are on the cuttings that are planted in the field.

9.7.9 Dusky clearwing, *Paranthrene tabaniformis* (Lepidoptera: Sesiidae) (Plate 41)

A. Delplanque

= *Sciapteron tabaniformis*

Global distribution

P. tabaniformis Rottemburg (Paranthreninae) is widespread in central and southern Europe, North Africa and Asia, especially in China, the north of India and Pakistan. It can also be found

Plate 33. *Nematus oligospilus* on willow. **(A)** Adult female. **(B)** Eggs. **(C)** Mature larva feeding on edge of a leaf. **(D)** Pupal cocoon. Photos courtesy of J. Charles. All photos © 2011 Plant & Food Research.

Plate 34. Poplar leaf beetle. *Chrysomela populi*. **(A)** Adult. **(B)** Eggs. **(C)** Larva. **(D)** Pupa. **(E)** Damage. Photos courtesy of A. Delplanque (A, C, E) and S. Augustin (B, D).

Plate 35. Poplar and willow leaf beetles. **(A)** *Chrysomela tremulae*. **(B)** *Chrysomela scripta*. 1, Adults; 2, eggs; 3, larvae; 4, pupae;. 5, damage. Photos courtesy of A. Delplanque (A4, A5, B1, B2, B3), S. Augustin (A1, A2, A3), Gerald J. Lenhard, Louisiana State University, Bugwood.org (B4) and James Solomon, USDA Forest Service, Bugwood.org (B5).

Plate 36. *Choristoneura conflictana* on poplars. **(A)** Adult. **(B)** Hatched egg mass. **(C)** Leaves webbed together by larva. **(D)** Mature larva. **(E)** Overwintering larvae in hibernacula. **(F)** Leaves webbed together by feeding larva. **(G)** Pupal case protruding from webbed leaves. **(H)** Several webbed leaf clusters with feeding damage. **(I)** Severely defoliated trembling aspen stands. Note understorey plants festooned with silk from wandering larvae. Photos courtesy of W.J.A. Volney, Natural Resources Canada.

A B C D E

Plate 37. Poplar defoliator *Clostera fulgurita.* **(A)** Moth. **(B)** Eggs with 1st instar larvae. **(C)** Parasitized eggs (black) by *Trichogramma* sp. **(D)** 5th instar larvae. **(E)** Pupa. Photos courtesy of A.P. Singh.

Plate 38. *Phyllocnistis unipunctella* on poplar. **(A)** Adult. **(B)** Larva. **(C)** Rolled leaf containing pupa. **(D)** Mine. **(E)** *Phyllocnistis saligna* mine on *Salix* sp. Photos courtesy of Tom Tams, www.northumberlandmoths.org.uk (A), A. Delplanque (B, C, D) and Gyorgy Csoka, Hungary Forest Research Institute, Bugwood.org (E).

Plate 39. *Phloeomyzus passerinii* on poplar. **(A)** Adult. **(B)** Adult females and neanids. **(C)** First infestation symptoms on the trunk. **(D)** Heavily infested poplar trunk. Photos courtesy of G. Allegro.

Plate 40. *Tuberolachnus salignus* on willow. **(A)** Apterous adults on young shoot. **(B)** Adult. **(C)** Winged adults. **(D)** Gall on leaf margin. **(E)** Injury to bark. **(F)** *Dasineura marginemtorquens* on willow. *Dasineura ingeris* gall with pupae in leaf bud. Photos courtesy of T. Collins (A), S. Höglund (B, D, F) and S. Augustin (C, E).

A B C D E F

Plate 41. *Paranthrene tabaniformis.* **(A)** Adult female. **(B)** Adult male. **(C)** Larva in gallery. **(D)** Exuvia of chrysalis. **(E)** Gall on young shoot. **(F)** Gallery emptied by European green woodpecker *Picus viridis.* Photos courtesy of A. Delplanque.

Plate 42. *Gypsonoma aceriana* on poplar. **(A)** Adult. **(B)** Exuvia of chrysalis hanging from a branch. **(C)** Frass of young larva on leaf. **(D)** Branch reaction to attack. **(E)** Result of repeated attacks on branches. Photos courtesy of A. Delplanque (A, B, C, D) and L. Nef (E).

Plate 43. *Saperda* spp. on poplar. **(A)** *Saperda carcharias*. **(B)** *Saperda populnea*. **(C)** *Saperda calcarata*. 1, Adult; 2, chrysalis in gallery; 3, damage. Photos courtesy of A. Delplanque (A, B, C1), James Solomon, USDA Forest Service, Bugwood.org (C2) and Minnesota Department of Natural Resources Archive, Bugwood.org (C3).

Plate 44. *Anoplophora glabripennis.* **(A)** Adult. **(B)** Oviposition pits and exit holes. **(C)** Pupa in its pupal chamber. **(D)** Galleries with mature larvae. **(E)** Infested poplar in Ningxia, China. Photos courtesy of F. Hérard (A, B, C), A. Roques (D) and Michael Bohne, Bugwood.org (E).

A B C D E F

Plate 45. *Megaplatypus mutatus*. **(A)** Adults. **(B)** Hole with sawdust produced by adult. **(C)** Trap over a hole for adult emergence. **(D, E)** Damage to trunk. **(F)** Windthrow damage in poplar plantation. Photos courtesy of G. Allegro (A, D), R. Gimenez (B, C, E) and L. Nef (F).

Plate 46. *Cryptorhynchus lapathi*. **(A)** Adults. **(B)** Larva in gallery. **(C)** Pupa in gallery. **(D)** Damage to branch. **(E)** Damage to young branch. **(F)** Broken tree with *C. lapathi* gallery. Photos courtesy of G. Allegro (A, B, C, D) and A. Delplanque (E, F).

Plate 47. *Sesia apiformis* on poplar. **(A)** Adults mating. **(B)** Eggs on trunk. **(C)** Larva in gallery. **(D)** Chrysalis surrounded by frass. **(E)** Wood injury. **(F)** Exit holes at the base of the trunk. Photos courtesy of A. Delplanque.

Plate 48. *Tremex fuscicornis*. **(A)** Female. **(B)** Larva. **(C)** Fallen tree with damage. **(D)** Larval galleries. **(E)** Windthrow damage in poplar plantation. Photos courtesy of P. Parra.

across Canada and in Russia and Finland. It is a pest of *Populus* and *Salix* in many regions (Brizzi, 1962; Templado, 1964; Postner, 1978), but also feeds on *Alnus*. In North America, other clearwing moths of the genus *Paranthrene* attack the Salicaceae family, e.g. *P. dollii* (Neumoegen), *P. robiniae* (Hy. Edwards) and the closely related *P. tricincta* Harris. (See also section 9.7.15, *Sesia apiformis*.)

Description

The adult insect has a wasp-like appearance. The head is black and appears to be separated from the thorax by a yellow collar. The body is black, with three yellow rings on the abdominal segments and an expandable anal tuft on the last segment. The posterior yellow ring in males is divided into two. The front wings are brown and the back ones are transparent. The wingspan is 25–35 mm (Postner, 1978).

Eggs are brown-red in colour and have a fine reticulation. Caterpillars are creamy white, with slightly darker dorsal and lateral stripes. They have three pairs of functional legs that readily distinguish them from beetle wood-boring larvae.

Biology

Adults emerge from May to July and mate immediately. Egg laying occurs throughout the summer, but primarily in July and August in central Europe and Canada. Females lay numerous eggs on branches, close to wounds or cracks in the bark of young trees, or in old galleries of other insects. Caterpillars hatch after about 10 days and immediately bore a gallery under the bark. Trees react by cell multiplication, which forms a swelling on 1–2-year-old seedlings in nurseries (Postner, 1978; Delplanque, 1998; Allegro, 2008b). Larvae continue boring into the wood until they reach the centre of the stem, where they cease feeding during winter. In the following spring, the larvae bore a longitudinal gallery, always ascending, in the heart of the stem, which can become 120–150 mm long and 7–8 mm in diameter. At maturity, the caterpillar can be up to 40 mm long. Larvae may utilize old galleries of *Saperda populnea* Inglebert or *Cimbex lutea* Thomson. Galleries can also be found at the base of seedlings in nurseries, close to *Sesia apiformis* galleries, which also exploit the base of stems and the roots.

Larvae may hibernate once or twice, depending on the climate, before pupating within a cocoon close to the tunnel opening. The pupa is cylindrical, and the anterior third appears from the tunnel opening to allow the adult to emerge. Pupal exuviae can be found hanging from the exit hole after the emergence of the imago.

Impact

P. tabaniformis lives on all cultivated poplars, but willows are seldom attacked. The insects are common in nurseries, where they cause particularly heavy damage on 1- and 2-year-old plants. They are also common in young plantations, but are rarely found in natural forests. The presence of *P. tabaniformis* can be detected by a gall-shaped swelling of the trunk, sometimes pierced with an exit hole and frass. Tree growth is disturbed, and they also become much more fragile and can break when exposed to the wind.

Control

Some natural enemies may regulate the population at a relatively low level. The parasitoid complex of *P. tabaniformis* in Europe includes numerous species (Georgiev, 2000, 2001), and it can decrease the pest level up to 55–65% (Postner, 1978; Moraal, 1996b). In nurseries, *P. tabaniformis* can be controlled chemically if applications are made during the oviposition period when the caterpillars begin boring into the bark. Insecticide coverage must be achieved throughout the flight period. Adult flight periods can be determined, and insecticide applications timed against them, by using pheromone traps to monitor for the presence of adult males (Lapietra and Allegro, 1994).

No clonal preferences have been found, but the balsam poplars (section *Tacamahaca*) are particularly affected. Plants must be checked before planting, to verify the absence of galls. Another means of prevention is to avoid trunk wounding or pruning during the flight period. Old infested poplars, willows or alders in the vicinity of plantations must be removed before the adults hatch, in order to avoid further infestation.

Invasive risk

P. tabaniformis is a widespread species that has colonized almost all the northern hemisphere and threatens most poplar species. In the USA, it coexists with the well-adapted native *P. tricincta* that develops through two generations per year. Consequently, it seems that *P. tabaniformis* is a species that can adapt easily to new environments.

9.7.10 Poplar twig borers, *Gypsonoma* spp. (Lepidoptera: Tortricidae) (Plate 42)

L. Nef

Global distribution

The 15 species in the genus *Gypsonoma*, all from the northern hemisphere, are associated mostly with poplars and willows. The most important are: *G. aceriana* Duponchel (from Europe to Iraq, North Africa, and recently introduced in North America), *G. haimbachiana* Kearfott (on *P. deltoides* in North America) and *G.* (= *Eucosma*) *hapalosarca* Meyr (Iraq, Pakistan) (de Tillesse *et al.*, 2007).

Description

Gypsonoma spp. are small, pale coloured moths, with a 12–15 mm wingspan. The tips of the forewings of *G. aceriana* are spotted brown. The yellowish larvae, 8–12 mm long when mature, are almost hairless. The head and the cephalic capsule vary from black to light brown in different species.

Biology

G. aceriana has two generations a year in southern Europe and in North Africa, but only one in central Europe, where adults appear in June–July. The female lays groups of one to three eggs near the central vein of the lower leaf surface. Eight to 15 days later, the neonate larvae start to tunnel a tiny mine, feeding on the leaf parenchyma. Some time before the leaf falls in autumn, the larvae disperse to hibernate in small crevices in the branch or trunk. In the spring, they leave this shelter and bore into young twigs or buds. The damage is noticeable only 4–10 days later, when a characteristic conical tube of frass and sawdust, about 5 mm long, appears at the entrance of the mine. The most important impact, however, is that the larvae induce galls to form on the branches, which are weakened. The size of the galls varies markedly from clone to clone. Mature larvae move from the twigs to the trunk, or fall to the ground, to pupate. The adult emerges 12–17 days later (Heymans, 1984). The southern, bivoltine populations of *G. aceriana* also overwinter as larvae. *G. haimbachiana* has only one annual generation, whereas *G. hapalosarca* can be multivoltine.

Impact

G. aceriana causes negligible damage to adult poplars. In nurseries, or in young plantations, damage can become much more severe, mainly in the apical bud, which is the preferred site for the larvae. When apical shoots are weakened or destroyed, they are replaced by a deformed top, which makes the plant unmarketable. More than 80% of nursery plants attacked by *G. aceriana* may be deformed, of which 50% are ‘double tops’ (Heymans, 1984). *G. haimbachiana* can cause similar damage to young *P. deltoides*. The larvae of *G. hapalosarca* web together and skeletonize leaves of *P. euphratica*.

Control

Natural enemies can sometimes kill large numbers of these pests and may be useful in integrated control programmes. There has been considerable research on different ways to control *G. aceriana*. Attractants can be used to capture adults (Booij and Voerman, 1984; Allegro, 1992). Damaged branches can be removed. Chemical control is not nowadays justified ecologically or economically, except maybe in nurseries. Here, well-timed applications may be effective against young or hatching larvae, during the autumn dispersal or against ovipositing adults. Application of a birdlime strip around the lower part of the apical bud is very effective (Heymans, 1984) and is particularly useful in protecting more susceptible clones. *G. aceriana* damage is negligible in nurseries that support favourable poplar growth, with a not too acid pH, appropriate groundwater and a soil rich in mineral elements but without vegetation (may be linked to insect mortality). Dry and sandy soils should be avoided.

Clonal susceptibility is very important. The clones of *P. trichocarpa* × *P. deltoides* or of *P. trichocarpa* are more susceptible to *G. aceriana* than those of *P. deltoides* × *P. nigra*. Decreasing ranks of sensibility for most clones used in Belgium are: 'Hunnegem', 'Columbia River', 'Unal', 'Ghoy', 'Boelare', 'Fritzi Pauley' = 'Beaupré', 'Raspalje', 'Robusta', 'Trichobel', 'Isières', 'Primo', 'Ogy', 'Gaver' and 'Gibecq' (Nef, 1985).

Damage by *G. haimbachiana* does not appear to be significantly different among the non-hybrid clones of *P. deltoides*. (Payne *et al.*, 1972), but the insects prefer non-irrigated poplars (Woessner and Payne, 1971).

P. alba and *P. ciliata* are quite resistant to *G. hapalosarca* and *P. euramericana* and *P. nigra* are highly susceptible, but the differences are very dependent on environmental conditions (Chaudry and Ahmad, 1973).

9.7.11 Poplar borers, *Saperda* spp. (Coleoptera: Cerambycidae) (Plate 43)

S. Augustin

Saperda carcharias (Linnaeus) (= *Anaerea carcharias* Inglebert) (large poplar longhorned beetle)
Saperda populnea (Linnaeus) (= *Compsidia populnea* Inglebert) (small poplar longhorned beetle)
Saperda calcarata Say (poplar borer)
Saperda inornata Say (poplar gall borer)

Origin and global distribution

Several species of *Saperda* are known to attack *Populus* and *Salix* spp. The main species in Europe and Asia are *S. carcharias* and *S. populnea*, and in North America, *S. calcarata* and *S. inornata*. *S. carcharias* ranges from Europe (Sweden, Finland to the Mediterranean) to the Far East (eastern Siberia, northern China and North Korea). *S. populnea* is a common species in the Holarctic region, whereas *S. calcarata* and *S. inornata* are widely distributed in North America.

Description

S. carcharias and *S. calcarata* are *c.*30 mm long. *S. carcharias* is black and covered with a yellowish or greyish pubescence. *S. calcarata* is greyish blue and heavily striped, with fine brown dots that overlay a faint yellow pattern. *S. populnea* and *S. inornata* are smaller and more elongate than *S. calcarata*. *S. inornata* is greyish, while *S. populnea* is grey-black, with faint spots of yellowish pubescence. Larvae of these species are elongate, cylindrical, legless and yellowish white, with variation in chitinization of the thorax according to species.

Biology

In Europe and Asia, *S. carcharias* may attack all poplar and willow species, but prefers aspen (*P. tremula*), including hybrids *P. tremula* × *P. tremuloides* (Välimäki and Heliövaara, 2007). Aspen is also the main host of *S. populnea*, which also attacks other poplar species and willows. In North America, *S. inornata* favours quaking aspen (*P. tremuloides*), whereas *S. calcarata* prefers quaking aspen in the northern part of its range (Peterson, 1947) and eastern cottonwood (*P. deltoides*) in the southern part (Morris, 1963), while other poplar species and cultivars are also attacked (Peterson, 1947).

The life cycle of *S. carcharias* lasts for 2–4 years, depending on climate. Adults emerge from infested trees in summer and oviposition occurs from July to September. Females prefer to lay eggs under the bark of 5- to 10-year-old trees on any part of the bole. On young trees, eggs are usually laid in the lower part of the trunk. The larva does not hatch until late spring of the following year and bores a horizontal tunnel towards the centre of the tree. The larval galleries of this large wood borer are long, and some descend from the trunk to the roots then rise again to the centre of the sapwood. After extensive summer feeding, the larva hibernates until the spring of the following year. It pupates at the beginning of summer, and about 2 weeks later the young adult emerges from the larval gallery.

The life cycle of *S. populnea* lasts 1 year in temperate areas, but 2–3 years in colder regions. Adults emerge in spring and oviposition occurs from the beginning of May until July. Eggs are laid in horseshoe-shaped notches carved out by the females, preferably on small, lignified branches from the previous year. The larvae hatch after 10–15 days. The young larva bores a circular gallery under the bark, causing a characteristic gall at the base of the branch. It subsequently penetrates the wood and then the heartwood, which it follows upwards. The larva hibernates in autumn and either emerges the following year after pupation or spends another 1 or 2 years as a larva while continuing to tunnel.

The life cycle of *S. calcarata* varies from 2 years in Mississippi (Morris *et al.*, 1975) to 3–5 years in the northern part of its range (Peterson, 1947; Drouin and Wong, 1975). Adults emerge in May and June in the southern part of North America and as late as July and August in the north (Hofer, 1920; Morris *et al.*, 1975). Mating begins approximately 1 week after emergence, and a few days later, females start to lay single eggs (occasionally two or three at a time) in niches cut into the bark of the trunk and branches of trees that are at least 3 years old. The eggs hatch after 2–3 weeks and larvae begin boring into the bark tissue of stems and branches. Later, they move into the sapwood and the heartwood. Larvae hibernate in the galleries until spring, and then they feed actively and enlarge and extend their tunnels (Drouin and Wong, 1975). Pupation occurs from April to July, depending on climatic conditions, and 2–3 weeks later, the adults emerge from the gallery entrances.

The life cycle of *S. inornata* varies from 1 to 2 years (Nord *et al.*, 1972). In late spring and summer, the female lays eggs in incisions made in the bark of young stems and branches of 1- to 3-year-old trees. Eggs are often laid at intervals around the stem at the same height. Generally, one larva develops in each location and begins feeding and boring in irregular galleries under the bark. As a result of this tunnelling activity, a globulose gall appears around the injured area. Part of the population, apparently resulting from eggs laid early in the oviposition period, bore a central tunnel parallel to the axis of the stem, where they overwinter and pupate in spring. Adults emerge in the following spring. The remainder of the population resumes feeding in the spring and spends a second winter as larvae, which emerge in the following spring, 2 years after hatching from eggs.

Impact

The adults of the *Saperda* species treated here feed on leaves and the bark of new shoots, although damage is generally negligible. Larval damage is much more severe. Galls and sap runs mixed with frass on the trunk are the first signs of larval presence.

S. carcharias is considered to be the principal poplar pest in the Mediterranean region. Generally, it does not kill the trees, but it predisposes host trees to other mortality agents, e.g. fungi (Cherepanov, 1988). Poplars and willows of all ages are attacked. Larval tunnelling into the sapwood reduces wood quality, but economic damage from timber degradation and reduction in growth is regarded as moderate in Europe (Evans *et al.*, 2004). However, large numbers of boring larvae weaken trees, which may snap during heavy winds, and this may be an important factor in some regions.

S. populnea is a primary pest in many areas around the world. *P. tremula* is especially affected, but other species, for example *P. alba* and *P. nigra*, may be attacked severely. Trees weakened by drought stress and poor soils are particularly susceptible. Branches and shoots are deformed and break easily at the level of galling when exposed to wind. *S. populnea* is also a vector of fungal and bacterial diseases, such as *A. populi* and *Dotichiza populnea* (Grechkin and Vorontzov, 1962; Roques, 1998b). In young plantations and nurseries, the death of branches and shoots may lead to important economic losses. For older trees planted under good conditions, *S. calcarata* is a pest of secondary importance.

S. calcarata is a serious pest of poplars throughout most of North America (Morris *et al.*, 1975). Young trees are attacked at and below the ground level and larger trees via the bole (Solomon, 1995). Damage is more obvious in open than in dense stands and along the edge of stands, with higher light levels preferred by females for oviposition (Mattson *et al.*, 2001). Large trees are rarely killed, but larval tunnelling weakens the stem and attacks occurring high up on the stem increase susceptibility to wind breakage (Coyle *et al.*, 2005). Small trees are occasionally killed by *S. calcarata*. In addition to larval damage, excavations from woodpeckers searching for larvae, and Hypoxylon canker (caused by fungi (*Hypoxylon* spp.)) that infect the wounds, further weaken trees and increase the risks of wind breakage.

S. inornata damage is mainly a problem in young stands and is worst in nurseries and in 1- to 3-year-old trees in plantations. Stems and branches occasionally break off or die above the gall, although most trees overgrow the gall. Economic loss caused by *S. inornata* in natural stands of aspen is negligible, but some mortality can occur in test clones (Nord *et al.*, 1972). Woodpeckers and infection by Hypoxylon canker disease can amplify seriously the weakness of trees damaged by *S. inornata* (Nord and Knight, 1972; Ostry and Anderson, 1995).

Control

The natural mortality of *S. carcharias* is high, and the trees' chemical defences kill up to 80% of eggs and young larvae (Allegro, 1991). Among the natural enemies, fungi and hymenopteran parasitoids in the Ichneumonidae, Braconidae and Chalcidoidea have mainly been recorded (Roques, 1998a). Woodpeckers (especially the great spotted woodpecker, *Dendrocopos major* (*Picidae*) in Europe) may also destroy many larvae (Allegro, 1991). In young poplar plantations, most eggs are laid at the base of plants and a coating of birdlime (a sticky barrier) on the lower part of the trunk may prevent oviposition. Insecticides can also be sprayed effectively when larvae are present on the bark (just after hatching) or can be injected into the larval holes (Allegro, 1998). However, there is no completely effective control method and infested poplars should be removed to avoid contamination of healthy trees nearby.

Population densities of *S. populnea* are limited by natural enemies (Hellrigl, 1974; Pulkkinen and Yang, 1984; Georgiev *et al.*, 2004), but these generally do not control the pests adequately. Infestations can be reduced by avoiding siting new plantations in poor and drought-stress affected soil, by selecting clones adapted to the site conditions or by applying fertilizer during the first year of planting. In nurseries and young plantations, tree monitoring is necessary in winter, and shoot swellings must be destroyed before adult emergence.

Natural control of *S. calcarata* by parasitoids, predators, fungi and woodpeckers is important (Hofer, 1920; Peterson, 1947). Avoiding poor sites for plantations and maintaining healthy trees can reduce the incidence of *S. calcarata* markedly. Borer infestations vary directly with stem diameter and inversely with stocking rate (Solomon, 1995). In aspen stands, periodic removal of infested trees has not been effective because the reduction in stand density may result in higher infestations (Peterson, 1947). In managing aspen, therefore, the recommended practice is to maintain well-stocked stands and to clear cut at maturity (Ostry *et al.*, 1989). Insecticides may be necessary for protecting trees in parks and urban areas.

For *S. inornata* control, planting resistant clones on good poplar sites, followed by good arboricultural practices such as pruning of rooted nursery stock to remove galls at harvest and eliminating infested stems, are recommended. Spraying nursery stool beds or rooting beds when adults begin oviposition may also be necessary (Ostry *et al.*, 1989).

9.7.12 Asian longhorn beetle, *Anoplophora glabripennis* (Coleoptera: Cerambycidae) (Plate 44)
H. Jiafu and Y. Luo

Global distribution

A. glabripennis (Motschulsky) was originally restricted to China and Korea (Cavey *et al.*, 1998; Lingafelter and Hoebeke, 2002; Williams *et al.*, 2004) and was detected outside Asia for the first time in 1996 in New York City. It subsequently spread to other places in North America (Chicago 1998, New Jersey 2002 and Toronto, Ontario, Canada, 2003). Elsewhere, it was first discovered in 2001 at Braunau am Inn, Austria (Tomiczek *et al.*, 2002) and then in 2002 at Yokohama, Japan (Takahashi and Ito, 2005), 2003 at Gien in France (Hérard *et al.*, 2006), 2004 at Neukirchen am Inn in Germany (EPPO, 2008) and 2007 at Corbetta in Italy (Maspero *et al.*, 2007). Models have demonstrated that *A. glabripennis* could become established in many locations worldwide (MacLeod *et al.*, 2002; Townsend Peterson *et al.*, 2004; Keena, 2006).

Description

Eggs are off-white, 5–7 mm long, cylindrical, tapering towards each end. The larvae are off-white, up to 50 mm long, soft-bodied with a hard, brown head. The pupae are about 30–33 mm long and 10 mm wide. Pupae are initially off-white, but gradually turn to light then dark brown. Adults are black, shiny, 20–35 mm long and 7–12 mm wide. Each wing case has about 20 distinctive, irregular white spots. Females are generally larger than males. The antennae are very long: on females they are about the same length as the body; on males they are about twice the length of the body (Li and Wu, 1993).

Biology

In China, *A. glabripennis* requires 1–2 years to develop from egg to adult. It generally overwinters as a larva. The number of generations per year may vary as a function of local climatic conditions. In Inner Mongolia (northern China), a single generation takes 2 years to develop, while in Taiwan, one generation per year has been documented (Li and Wu, 1993). To complete metamorphosis, *A. glabripennis* needs 1264.2 ± 188.3 accumulated degree-days (DD), at an average lower development threshold of 13.4 ± 0.3°C (Yang *et al.*, 2000).

The eggs need between 8 and 12 days to complete their development. Neonate larvae begin to feed on the phloem layers around the oviposition site. The early larvae tunnel laterally into the phloem and cambium layers under the tree bark. Larger larvae tunnel deeper into the heartwood, where they are well protected. Each mature larva creates a chamber near the outer bark in which to pupate. After emergence and melanization, adults spend several days resting before chewing a 6–18 mm exit hole (Lingafelter and Hoebeke, 2002). In China, adults emerge from April/May to October (Zhao and Yoshida, 1999). In New York City and Chicago, adult *A. glabripennis* have been recorded from July to November (Haack *et al.*, 1996). Under laboratory conditions, the highest recorded fecundity was 66.8 ± 5.0 eggs per female at 25°C, but in nature fecundity may vary from 30 to 178 viable eggs per female (Keena, 2005, 2006). Fecundity is correlated positively with beetle body size and negatively with beetle age. Adult beetles can fly several hundred metres in a single flight to locate suitable host trees.

A. glabripennis larvae have been reared in laboratories on an artificial diet (Zhao *et al.*, 1999; Dubois *et al.*, 2002; Keena, 2005).

The major economic damage reported all over the world due to *A. glabripennis* is to poplars (*Populus*), maples (*Acer*), willows (*Salix*) and elms (*Ulmus*) (Sawyer, 2003; Haack *et al.*, 2006). In China, *A. glabripennis* has caused the greatest damage to poplar species. Although *A. glabripennis* populations predominantly increase on poplar plantings from the sections *Tacamahaca* and *Aigeiros*, it prefers poplar species and hybrids of the section *Aigeiros*. In the *Aigeiros* section, *P. nigra*, which is native to Europe, south-west and central Asia and north-western Africa, is considered the most vulnerable species. In particular, the popular *P. nigra* 'Italica' and 'Thevestina' are very susceptible to *A. glabripennis*. *P. deltoides* is less susceptible, followed by *P.* ×*canadensis*. Poplar species belonging to the sections *Tacamahaca* (balsam poplars) and *Leucoides* are also attacked by *A. glabripennis*, but at lower levels (Li and Wu, 1993). Poplar species belonging to the sections *Turanga*, for example *P. euphratica* Oliv., *P. pruinosa* Schrenk., and *Populus*, for example *P. alba* L., *P.* ×*hopeiensis* Hu et Chou, *P.* ×*tomentosa* and *P. tremula*, are considered less susceptible or slightly resistant (Bao *et al.*, 1999). No records have been found regarding the reactions of *A. glabripennis* to poplars from the section *Abaso*.

Despite the above-mentioned economic damage caused to poplars by *A. glabripennis*, it is very polyphagous. The host list includes the genera *Acer*, *Betula*, *Elaeagnus*, *Fraxinus*, *Hedysarum*, *Hippophae*, *Koelreuteria*, *Platanus*, *Populus*, *Robinia*, *Salix*, *Sophora*, *Tilia* and *Ulmus* (Li *et al.*, 1999). *Acer* is the most attractive genus, followed by several *Populus* species, such as *P. simonii* Carr. and *P. cathayana* Rehd.(Gao *et al.*, 1997). In North America, *A. glabripennis* has been reported to attack 18 deciduous tree species belonging to 12 genera (Haack *et al.*, 1997; Lingafelter and Hoebeke, 2002).

Impact

A. glabripennis has an enormous destructive potential because it attacks healthy trees and spends most of its life as a larva, boring inside tree trunks and large branches. This compromises the tree's vascular system, causes severe damage to the wood's structural properties and eventually leads to the death of the tree (Cavey *et al.*, 1998). It poses an enormous threat to urban, suburban and rural forests in areas where it is introduced (e.g. Haugen, 2000; Nowak *et al.*, 2001). In China alone, *A. glabripennis* causes an estimated annual loss of more than

10 billion Chinese yuan (Su *et al.*, 2004), and it has been listed as one of that country's most dangerous forest pests. In Chicago and New York City, the efforts to control or eradicate the exotic *A. glabripennis* have led to the removal and destruction of all trees having symptoms of *A. glabripennis* attack (Haack *et al.*, 1997). These eradication efforts already cost many millions of dollars annually. The estimated maximum potential national urban impact of *A. glabripennis* in the USA is a value loss of US$669 billion, with tree mortality of about 30% and a reduction of total canopy cover of about 35% (Nowak *et al.*, 2001).

Control

The life cycle of *A. glabripennis* combines concealed immature stages and a tendency to lay small numbers of eggs in several trees, so it is very difficult to prevent its spread. The USDA's Animal and Plant Health Inspection Service (APHIS) has established domestic quarantine regulations that prohibit the local transport of potentially infested wood or wood products from areas where *A. glabripennis* infestations have been found.

In areas where *A. glabripennis* has been detected as a new invasive species, eradication programmes have been adopted (Haack *et al.*, 1997). These programmes include felling, removal and chipping or incineration of infested trees (Smith *et al.*, 2001). In the USA, trees are usually surveyed for beetles within an 800 m radius of each infestation point, and trunk or soil injections with imidacloprid are applied to each potential host tree within this radius (USDA-APHIS, 2006).

The most widely adopted method for controlling high populations of *A. glabripennis* in China consists of spraying chemical pesticides into the canopies of host trees to kill adults (Liu *et al.*, 1999). Another commonly used chemical control strategy uses bamboo or wooden sticks containing aluminium phosphide that are inserted into larval frass holes, where the gas that is produced kills the *A. glabripennis* larvae (Zhao *et al.*, 1995a). In addition, injections of systemic insecticides into the trunks of infested trees and the application of trunk-coating insecticides are effective measures that are considered to have low environmental impact (Zhao *et al.*, 1995b). In particular, injecting tree trunks with the organophosphate insecticide, methamidophos, controls not only *A. glabripennis* but also piercing-sucking insect pests (Zhang *et al.*, 1994). Finally, local forestry organizations have promoted other physical control measures such as catching adults, killing eggs and young larvae and blocking frass holes (Gao and Li, 2001). These physical control measures can be effective in maintaining the *A. glabripennis* population below pest thresholds, especially in the case of young trees and in urban areas.

For biological control measures, entomopathogenic fungi have been developed for *A. glabripennis* control, and entomopathogenic nematodes, coleopteran and hymenopteran parasitoids and predatory woodpeckers have been investigated (Ogura *et al.*, 1999; Wang *et al.*, 1999; Zhang *et al.*, 1999; Li *et al.*, 2000; Solter *et al.*, 2001; Dubois *et al.*, 2004a, b, 2008; Fallon *et al.*, 2004; Hajek *et al.*, 2006).

Ecological control of *A. glabripennis* in China involves planting mixtures of preferred and non-preferred tree species, and this practice can prevent outbreaks successfully (Yan and Yan, 1999; Luo *et al.*, 2003).

Invasive risk

Based on information concerning the current distribution, biology and economic impact of *A. glabripennis* in Asia and North America, together with recent European interceptions, there is a significant risk that *A. glabripennis* could become established and cause damage to important forest and fruit tree species all over the world (MacLeod *et al.*, 2002). Particular attention should be paid to preventing the possible transportation of eggs and larvae of the pest via untreated solid wood packing material (SWPM) used in international cargo.

A. glabripennis is considered a dangerous quarantine pest in North America and the European Community (EPPO, 2004b).

9.7.13 Ambrosia beetle, *Megaplatypus mutatus* (Coleoptera: Platypodidae) (Plate 45)

R. Gimenez

= *Platypus sulcatus*, *Platypus mutatus*

Global distribution

M. mutatus (Chapuis) (Platypodinae: Platipodini) is native to South America and is found in Argentina, Bolivia, Brazil, French Guiana, Paraguay, Peru, Uruguay and Venezuela. It is especially prevalent in Argentina, where infestations have caused severe damage to species of *Populus*, *Quercus*, *Eucalyptus* and *Pinus* (Bascialli *et al.*, 1996; Alfaro, 2003; Giménez and Etiennot, 2003). Willows may be attacked, but are not severely damaged.

M. mutatus was discovered in 2000 in *Populus* plantations at Caserta, Italy (Tremblay *et al.*, 2000; Allegro and Della Beffa, 2001; EPPO, 2004a; Allegro, 2008a; Servizio fitosanitario regionale, 2011).

Description

The adult beetle is small (7–8 mm long) and a dark brown to black colour. The elytra are striated with four longitudinal ridges. The head is as long as the pronotum, and the antennae are short (Brethes, 1908; Santoro, 1957; Brugnoni, 1980). The female body has rounded terminal segments, while the male abdomen appears to be truncate. The eggs are oblong, smooth, 0.5–0.9 mm long and a brilliant white colour. The legless larvae are white initially and 1.5–4 mm long, growing to 9–11 mm in the final instar. The free pupa is 7.5–9.2 mm long (Santoro, 1963, 1965).

Biology

Most adults leave their galleries in a tree from November to January (spring and summer), although they do not fly until the beginning of autumn. The adults emerge from their parents' entrance hole. In the River Plate Delta in Argentina only 3–5% of the entry holes show adult emergence, compared with 7–15% in Neuquén Province (Thomas, 2005). This low rate of survival indicates that currently unknown mortality factors are involved. Traps catch a mean of 20 adults per trap, and a few catch over 300 adults.

Infestation is initiated by the male, which bores a short gallery in the bark and waits for a female with which to copulate. The couple (mostly the female) continues to bore radial galleries, where eggs are laid (Santoro, 1957, 1963). A pair of adults, eggs, larvae and pupae can be found in the galleries throughout the year, with inactive adults at the end of winter.

M. mutatus is an 'ambrosia beetle' that carries a symbiotic fungus into a mycangium (Alfaro, 2003). The fungal mycelium stains the wood and bark of infested trees, but is not pathogenic. As the beetle moves through the tunnels, the walls are smeared with the fungus (*Raffaellea santoroi* Von Arx), which is cultivated for larval feeding (Guerrero, 1966). First- and second-instar larvae are mycetophagous and xylophagous and those from the second instar are xylophagous (Santoro, 1963).

Impact

M. mutatus attacks many species of native and exotic broadleaf and coniferous trees (Allegro, 1990a; Giménez and Etiennot, 2003). Unlike most ambrosia beetles, it only attacks living, standing trees and is unable to live in standing dead timber. Of all the host trees, only the poplars have shown a propensity to be windthrown. *M. mutatus* tunnels weaken the tree trunks and seriously damaged poplars are frequently felled by strong winds. 'Catfish' clones seem to be more susceptible than others cultivated in Argentina. Windthrow damage begins 2 years after the initial infestation in the plantation, and only if the trees are older than 15 years. In poplars, 86% of attacks are in trees with diameters at breast height greater than 15 cm. In the Paraná River Delta, 15–40% of trees were attacked from 1997 to 2000 and 3–10% after 2000. This is a chronic pest with a very variable abundance. In 1999, 72% of harvested timber was damaged by *M. mutatus* (Gimenez *et al.*, 2004).

Control

Infested trees must be removed before the adult flight period. The trunks should be cut as low as possible and the stump quickly surface dried (Toscani, 1991). In addition, the entrance holes should be blocked with a stick (Santoro, 1962, 1967). This manual control is a very effective alternative, as the insects die because they cannot remove the sawdust and are deprived of oxygen in the tunnels.

No biological control agents have been identified.

For chemical control, bark spraying to a height of 8–10 m was effective with carbaryl 4.25 g a.i. l^{-1}, cypermethrin 0.05 g a.i. l^{-1}, lambdacihalothrine 0.0167 ml a.i. l^{-1}, chlorpyrifos 1.275 ml a.i. l^{-1} or calcium polysulphide 3% (Giménez and Etiennot, 2002; Giménez and Panzardi, 2003; Giménez *et al.*, 2003; Thomas, 2011). Azadarichtin (0.3 and 1.2%) prevented *M. mutatus* attack, but higher concentrations were not effective (Giménez and Kocsis, 2007).

The economic injury level (EIL) and economic threshold (ET) were calculated at 0.29 and 0.086 attacks per tree, respectively, for poplars of the Paraná River Delta, Argentina, where they are planted at a ratio of 70:10:20% for sawing, peeling and cellulose pulp production, respectively (Gimenez and Moya, 2011).

Pheromones detected and tested in laboratory and field conditions to attract *M. mutatus* were: (+)-6-methyl-5-hepten-2-ol ((+)-sulcatol, or retusol) and 6-methyl-5-hepten-2-one (sulcatone) (Gonzalez Audino, 2011).

Barbosa and Wagner (1989) discussed the importance of moisture for the associated fungi. *M. mutatus* are unlikely to survive if logs or wood products are dried sufficiently (Santoro, 1963; Davis *et al.*, 2005). Wood products may be shipped safely from infested areas following drying by heat treatments or fumigation with methyl bromide (SAGPyA, 2003; FAO, 2009) or sulfuryl fluoride (Mizobuchi *et al.*, 1996).

Invasive risk

Although the flight capacity of this pest is not well known, it is very improbable that *M. mutatus* can disperse naturally to new countries. It is most likely to be transported over long distances in woody plants, planting and wood products. Its presence in Italy (Allegro and Della Beffa, 2001) may be attributed to an introduction of infested wood from South America (Tremblay *et al.*, 2000; Allegro, 2008a).

Unspecified '*Platypus* spp.' have been intercepted at ports of entry in the USA at least 46 times from 1985 to 2004, and 2.3 interceptions of *Platypodidae* have been reported annually (Davis *et al.*, 2005). Most of the insect interception reports are taxonomically incomplete, so specimens identified as '*Platypus* sp.' or 'Platypodidae' are not necessarily *M. mutatus*. However, even if all of the '*Platypus* sp.' or 'Platypodidae' reported were *M. mutatus*, the arrival rate would still be low compared with other insect pests (Davis *et al.*, 2005).

Even so, the presence of a new invader may have unpredicted ecological consequences, and concerns have been raised at the prospect of this insect becoming a global pest.

9.7.14 Poplar and willow borer, *Cryptorhynchus lapathi* (Coleoptera: Curculionidae) (Plate 46)

G. Allegro

Global distribution

C. lapathi (Linnaeus) is native to the entire Palaearctic region from Europe to Japan, but can now be considered Holarctic as, in the 18th century, it was introduced to eastern North America, where it spread rapidly north into Canada and then west to British Columbia.

C. lapathi is host specific to *Populus* spp. and *Salix* spp. (although *Alnus* and *Betula* are occasional hosts). *C. lapathi* shows a marked preference for humid riparian habitats, where its main host plants, poplars and willows, are commonly found. It is also widespread in commercial poplar and willow plantations, particularly in those growing on humid soils, where it can cause severe economic losses. *C. lapathi* can feed and reproduce on a large number of *P.* and *Salix* species. *P. nigra*, *P. deltoides*, *P.* ×*berolinensis* and their hybrids, as well as *P. deltoides* × *P. trichocarpa* hybrids, may be severely attacked (Johnson and Johnson, 2003), while *P. alba* and *P. simonii* are generally less affected (Cadahia, 1965; Dafauce, 1976; Jodal, 1987; Broberg *et al.*, 2005). Among willows, *S. viminalis*, *S. caprea* Linnaeus, *S. triandra* Linnaeus, *S. purpurea* Linnaeus and *S. fragilis* Linnaeus are potential hosts (Roques, 1998a), as well as *S. alba* Linnaeus (G. Allegro, Italy, 2011, personal observation).

Description

The ovoid eggs are about 1 mm long, soft, white and laid in the bark. The larvae are white with a brown head, subcylindrical, legless and more or less curved in shape. Body length ranges from 1 mm for first-instar larvae to 13 mm just prior to pupation. The pupae average 10 mm in length and have a visible developing snout. The legs and wings are pressed closely to the body. At first, pupae are white but gradually turn to brown, pink and grey (Garbutt and Harris, 1994). The adults are hard-bodied, rough-appearing, winged weevils about 8 mm long. The head is largely hidden by the thorax; the forward end tapers into a narrow snout which, when folded, tucks into a ventral thoracic groove. Adults are readily distinguished by their colour pattern. Tiny black and either grey or pink scales on the back contrast with the rest of the body, which is predominantly black.

Biology

The pest is univoltine in Asia and in the southern areas of Europe and the USA. It is biennial in countries with a colder climate (Canada and the northern parts of Europe and the USA). Where *C. lapathi* is univoltine, the winter is spent mainly in the larval stage, although a few eggs and adults may overwinter together with larvae (Zocchi, 1951; Szalay-Marzsò, 1962; Dafauce *et al.*, 1963; Cavalcaselle, 1966). There is general agreement that larval feeding starts in spring after budburst and lasts until May or June. Adults appear in the field from June and oviposition occurs until autumn, with eggs hatching normally before the onset of cold winter weather. In countries with a biennial cycle, the first winter is spent at the egg or larval stage and the second one at the adult stage, sometimes in the pupal chamber and sometimes in the soil and litter (Francke-Grosmann, 1960; Smith and Stott, 1964). Larval feeding occurs in the spring period. In very cold environments, a 3-year cycle is possible (Garbutt and Harris, 1994).

The adults possess functional wings but movement by flight has never been observed in the field (Smith and Stott, 1964), so active dispersal by walking on the ground is limited to, at most, 100 m year^{-1} (G. Allegro, Italy, 2011, personal observation). The adults are most active in the evening or early morning, but they readily drop to the ground feigning death when disturbed. Sometimes, they can be heard producing distinctive squeaking sounds. They feed by puncturing the succulent bark of new shoots. After mating, females lay single eggs in small holes excavated in the bark of young stems, usually at lenticels or at branch bases. Trees are chosen independently of their health, but a relationship between attack incidence and soil P and K content has been observed (Abebe *et al.*, 1990). The eggs hatch in about 2–3 weeks, but larvae soon enter diapause, overwintering under the bark, and start feeding activity only in the spring of the following year. They initially tunnel around the circumference of the stem but, at about the fourth instar, they turn and bore radially into the wood. The frass is at first pushed out by the larvae, but later it is allowed to accumulate in the tunnels. Six or seven larval instars have been observed (Cavalcaselle, 1966; Garbutt and Harris, 1994). The physiological aspects of larval feeding were investigated by Chararas (1969). At the end of larval development, which can last from 2 to many months, depending on climatic conditions, pupation occurs at the end of the tunnel.

Larvae have been reared in the laboratory on artificial media (Cavalcaselle, 1972b; Hou *et al.*, 1992b).

Many natural enemies of *C. lapathi* have been quoted in the literature (Ratzeburg, 1839; Schmiedeknecht, 1914; Muesebeck, 1931; Sweetman, 1936; Strojny, 1954; Szalay-Marzsò, 1962; Mrkva, 1963; Cavalcaselle, 1966; Yang, 1984), but the effectiveness of the complex is normally limited and inadequate to control the pest below the damage threshold in poplar and willow commercial plantations.

Impact

Young stems (ranging in diameter from 2 to 8 cm) can be severely damaged by larvae. Young trees can be killed or drastically weakened by galleries, and wind often breaks damaged branches.

C. lapathi is considered an economically important pest in Turkey, Italy, Belgium, France, Spain, Yugoslavia, Hungary, China, Korea, Japan, USA and Canada. It is most injurious in countries where poplars (and secondarily willows) are grown to produce high-quality wood, e.g. for plywood and furniture industries (Italy, Spain, France, Hungary). In Italy, it has been estimated that 180 t of insecticides are sprayed annually in poplar stands against the pest, at a cost of about €1,000,000 (Allegro, 1997b). *C. lapathi* may spread throughout poplar and willow SRF plantations, seriously reducing their productivity and also presenting a hazard for the neighbouring traditional poplar plantations (Allegro *et al.*, 2007).

C. lapathi is thought to have little or no impact on natural poplar and willow populations, as damaged (or even broken) branches of *Salicaceae* are very capable of sprouting and regenerating. Moreover, natural poplars or willows rarely form pure communities, thus preventing an epidemic spread of the pest.

C. lapathi has been indicated as a possible vector of bacterial willow diseases (Lindeijer, 1932; Callan, 1939).

Control

Elimination of sources of infestation is considered to be almost impossible, as the pest is widely distributed on wild poplars and willows and on the crowns of older trees under culture.

Poplars and willows in nurseries must be protected in order to prevent passive transfer of *C. lapathi* to commercial plantations. Young plants can be chemically protected and checked to eliminate the infested ones. Chemical protection is generally also needed in young poplar stands when high-quality wood has to be obtained. Protection in older stands is not warranted, as trunks over 15–20 cm in diameter are unlikely to be attacked and damaged. Organophosphate or pyrethroid insecticides are commonly used. Control of adults (Dafauce, 1965; Lapietra and Arru, 1973) or young larvae (Schvester and Bianchi, 1961; Lapietra, 1972; Cavalcaselle and De Bellis, 1983; Allegro, 1997b) can be achieved by spraying trunks, ensuring that they are thoroughly wetted. Very high larval mortality is achieved while they still feed on bark tissues and before they penetrate into the wood. In countries where the pest overwinters as larvae, nursery plants can be treated effectively in the winter period by using pyrethroid insecticides (Lapietra and Allegro, 1986b). This technique is particularly useful to prevent the passive distribution of the pest to stands by infested nursery stock.

Biological control of the pest is possible by injecting entomopathogenic nematodes (Cavalcaselle and Deseo, 1984; Hou *et al.*, 1992a) or fungi (Cavalcaselle, 1975) into the larval galleries. Experiments to control *C. lapathi* larvae by X-rays have been carried out (Cavalcaselle and De Bellis, 1968, 1970). Physical barriers proved to be effective in preventing *C. lapathi* adults from climbing trunks (Allegro, 1990b).

Invasive risk

C. lapathi has been designated a quarantine pest by El Comité de Sanidad Vegetal del Cono Sur (COSAVE) of South America (Argentina, Brazil, Chile, Paraguay and Uruguay). Particular attention should be paid to preventing the possible movement of eggs and larvae of the pest by commercial trade of nursery plantlets.

9.7.15 Poplar hornet clearwing, *Sesia* (= *Aegeria*) *apiformis* (Lepidoptera: Sesiidae) (Plate 47)

A. Delplanque

Global distribution

S. apiformis (Clerck) is widespread in the northern hemisphere. It is found across Europe, from Finland to the Mediterranean basin, in the Middle East, Asia Minor and China. It was introduced into North America around 1880 in Massachusetts and is present now in Connecticut, New York, New Jersey, Pennsylvania and California (Morris, 1986). It attacks poplars and willow and has been recorded from lime, birch and ash. *S. bembeciformis* (Hübner) and *S. tibialis* (Harris) are closely related species.

Description

Members of this family are known as clearwing moths because most of one or both pairs of wings lack any scales, leaving them clear or transparent. *S. apiformis* bears a striking resemblance to the giant hornet, *Vespa crabro* (Hymenoptera: Vespidae), with predominantly yellow abdomen and legs and yellow and brownish-black head and thorax. The wingspan is 34–44 mm. The forewings are long and narrow, with an apparent brown border. The hindwings are somewhat broader than the forewings.

Eggs are spherical, orange to dark purple. They are laid apparently randomly by the female, either individually or in small clusters in the bark at the base of the trunk or among the roots. This oviposition behaviour leads to significant mortality that is compensated for by high fecundity.

Larvae are yellowish-white, with brown pronotal plate and anal shield. Mature larvae can grow to 65 mm long (Postner, 1978). They are covered by short hairs and have three pairs of functional legs that readily distinguish them from beetle wood-boring larvae.

Biology

The colourful adults fly during the day from May to August. They mate immediately after emergence and females oviposit without delay, each laying up to 1000 eggs. Eggs are laid at the base of poplar trees and hatch after approximately 2–4 weeks. Larvae are very mobile. They eat through the bark and penetrate into the base of the trunk or the roots, forming galleries from 20 to 50 cm long. Mature larvae are 15–25 mm long. The galleries usually descend towards the roots and increase in diameter as the larva grows. In large trees, the galleries never quite reach the heartwood (Postner, 1978; Heath and Emmet, 1985).

The life cycle lasts 2 or occasionally 3 years. In the first winter, the larvae hibernate in their galleries under the bark from October and continue to develop in the following spring, boring galleries farther into the wood. When mature, the larva migrates from the roots to the base of the tree trunk and weaves a cocoon of wood debris, where it spends the second winter. It pupates in the following spring and the young adult emerges about 3 weeks later. The pupal chamber is made very close to the bark and the chrysalis is orientated with its head to the outside of the tree. The head of the chrysalis is very heavily sclerotized into cutting plates to bore through the final layer of bark and allow the adult to emerge. Empty pupal cases are quite noticeable near the base of the trunk or around the root collar, and up to 20 of these exuviae have been seen hanging from one tree after the adults have emerged. The circular exit holes are about 8 mm in diameter (Postner, 1978; Delplanque, 1998).

Damage

In contrast to other borers, larvae of *S. apiformis* do not eject frass from their galleries and so they are difficult to detect. However, the exit holes are easily seen after adult emergence, especially when the pupal exuvia is hanging from them.

Impact

S. apiformis has been regarded as a major pest of poplar in central and eastern Europe (Postner, 1978). Galleries in shoots can reduce the growth of young trees, and may also kill them. Severe infestations can occur in both old and young plantations (Chrystal, 1937; Speight, 1986). *S. apiformis* occurs frequently in nurseries and especially in commercial plantations that provide the seedlings for nurseries. It also occurs in new plantations, especially those replanted after harvesting poplars, with roots hosting large insect populations. Poplars of all ages are attacked, even very large ones. Larval galleries result in timber loss in large trees. Root damage can encourage penetration of fungi and other xylophagous insects. Damage to older trees affects the trunk base and thus is of lesser importance than on young trees. Attacks seem more frequent where the understorey vegetation is dense, particularly where it consists of shrub, tall weed or rough grass (Coleman and Boyle, 2000; Arundell and Straw, 2001; Straw *et al.*, 2007).

In certain areas, there is a preference for young trees. Only poplars grown on stony ground are spared as the eggs need relatively high humidity to hatch. The percentage of sapwood moisture also strongly influences caterpillar growth. There is very strong caterpillar mortality when wood becomes too dry.

Control

No resistant poplar clones have yet been found and so preventative control is advisable. Old infested roots should be removed before planting, as they may be a source of infestation for 2 years after a tree has been removed. Damaged poplar, willow and alder trunks situated in the proximity of the plantation should be removed. IPM strategies developed for other species of clearwing moths may also be applicable to *S. apiformis*. For example, *Paranthrene robiniae* is an endemic sesiid of the north-western USA, where it has recently become a significant pest in extensive poplar plantations. Brown *et al.* (2006) recommends that contact insecticides are not used against adult moths and are replaced by mating disruption and, eventually, planting clones that are not as susceptible to sesiid attack.

Weed control in young plantations and coating the base of poles with Stockholm tar up to about 20 cm above the ground level before planting are also very effective methods of protection by preventing oviposition. If necessary, tree necks can be sprinkled with systemic insecticides or the trunk can be treated with contact insecticides to a height of 1 m.

Invasive risk

S. apiformis is a widespread species and has been able to colonize new environments already inhabited by well-adapted, closely related species (such as *S. tibialis*). It seems that *S. apiformis* is a species that can adapt easily to new environments and so its eradication or control in countries where it has been recently introduced should be undertaken.

9.7.16 Tremex wasp, *Tremex fuscicornis* (Hymenoptera: Siricidae) (Plate 48)
P. Parra

Global distribution

T. fuscicornis (Fabricius) is native to Europe and Asia. In Europe, it is found in Austria, the Czech Republic, Denmark, Finland, France, Germany, Hungary, Italy, Latvia, Norway, Poland, European Russia, Armenia and the Ukraine. In Asia, it is widely distributed across China, Japan, Korea, Asian Russia (including Kamchatka, Kurile Islands, Sakhalin, Siberia) and Taiwan (Smith, 1978).

It has established in the southern hemisphere (possibly through transportation in dunnage during international trade), with first records in Australia in 1996 and Chile in 2000.

It attacks weakened or moribund trees in the genera *Betula*, *Pterocarya*, *Fagus*, *Pyrus*, *Robinia*, *Juglans*, *Acer*, *Ulmus*, *Alnus*, *Quercus*, *Prunus*, *Zelkova*, *Celtis*, *Carpinus*, *Salix* and *Populus*. Successive attacks kill the trees (Viitasaari, 1984).

In Chile, it has been detected in poplars (*P. nigra*, *P. alba* and *P. deltoides*) and occasionally in willows (*S. babylonica* Linnaeus and *S. humboldtiana* Willd.), false acacia (*Robinia pseudoacacia* Linnaeus), boxelder maple (*Acer negundo* Linnaeus) and walnut (*Juglans regia* Linnaeus). In Australia, it attacks poplars and willows.

Description

Adult females have a cylindrical body from 14 to 40 mm long. The abdomen is yellow-orange, with black transverse bands. The black, 13-segmented antennae vary between 4 and 10 mm long. Wings (up to 26 mm long) are a transparent honey colour, with the first pair of superior marginal and the interior veins darkest. Legs are brown to yellow. The abdomen terminates with a short thorn-like tergite, above the very robust and protruding ovipositor, from 9 to 21 mm in length.

Males are totally black (including legs and antennae), with a bright metallic sheen, and are from 11 to 29 mm long. Wings are similar to females but a darker amber colour and from 8 to 22 mm long. The antennae are 12-segmented and 3–8 mm long. The abdomen also terminates in a short thorn-like tergite.

Eggs are whitish, from 1 to 1.2 mm long and 0.2–0.25 mm in diameter. The female inserts her ovipositor into the wood fibre perpendicularly and obliquely, so that eggs are laid separately in the wood, but in groups. Larvae have a cylindrical, cream-coloured body, a semi-spherical head with strong jaws and one-segmented antennae, and three pairs of prethoracic legs. The largest larvae can be 3.7 cm long and 0.7 cm in diameter. Pupae are initially cream-coloured but develop pigmentation progressively from the head towards the rest of the body and their appendages until taking the adult's colour (Parra *et al.*, 2005).

Biology

Relatively little is known of *T. fuscicornis* in its native habitat. In Chile, mating occurs in the crown of trees, where the highest number of males is concentrated. Females are strong fliers and can disperse several kilometres in search of a suitable host. They select weakened, stressed or recently cut trees in which to oviposit, taking advantage of the natural fissures of the trunk to insert the ovipositor more easily into less lignified tissues.

During oviposition, the female injects a phytotoxin and also spores of a symbiotic fungus (*Cerrena unicolor* (Bull.) Murrill) that cause a white rot in the timber (Palma *et al.*, 2005). The ovaries of ten field-collected females held between 398 and 901 eggs (Palma and San

Martin, 2004), but females generally lay 300–400 eggs, usually dying in the oviposition posture (González, 2000; Servicio Agricola y Ganadero, 2000). Larvae appear after 3–4 weeks and initially feed exclusively on the hyphae of *C. unicolor*. Later, they begin to construct cylindrical galleries toward the xylem. Frequently, the damage causes sap to leak, resulting in patches of dark, oily-looking stains on the bark. In its final instar, the larva tunnels close to the surface of the tree and pupates at the end of the gallery. The adults emerge, leaving a circular exit hole of 5–6 mm diameter.

T. fuscicornis is univoltine, but under advantageous environmental conditions it may develop through a generation in 5 months, especially in dead trees. The life-cycle duration depends on the climatic conditions and the moisture content of the wood. In the Metropolitan Region of Chile, 95% of the population emerges between October and February. Emergence declines from March and adult activity ceases in July, when only immature stages exist. The sex ratio is usually 1:1 (Parra *et al.*, 2007).

Impact

In its natural range, populations of *T. fuscicornis* are typically low and difficult to measure. In Chile, the insect infests trees that present evident symptoms of weakness caused by various stress factors. Water deficiency is the main factor, besides fire damage to the base of trees and nutritional deficiencies. *T. fuscicornis* also lives in dead or fallen trees and crop waste.

The physical results of *T. fuscicornis* damage are tree mortality and the deterioration of wood quality, with loss of volume resulting from larval galleries, emergence holes and the white wood rot caused by *C. unicolor*.

Control

In Chile, an IPM strategy against *T. fuscicornis* is used, combining monitoring and detection methods and phytosanitary measures including cultural and biological control (Parra *et al.*, 2007). *T. fusciformis* is detected in the forest through surveillance and a system of trap trees (weakened by herbicide application) that aims to attract and to concentrate populations of the insect to facilitate its control. Phytosanitary measures (through chipping, incineration or burial of timber with immature stages of the insect) are obligatory during an infestation in order to develop an efficient cultural control strategy by public or private entities affected by the pest. Silvicultural management aims to maintain or improve the vigour of the trees to prevent damage or to contain it within acceptable levels. Application of biological control is possible using the parasitoids *Megarhyssa praecellens* and *Ibalia jakowlewi* Jacobson (Hymenoptera: Ibaliidae). *M. praecellens* attacks final-instar larvae and *Ibalia* attacks eggs and young larvae. Levels of parasitism of 30% by *M. praecellens*, plus natural death and the action of predators, has led to high expectations for the success of the biological control programme.

References

Abebe, G., Hart, J.H. and Adams, R.D. (1990) The relationship of site factors to the incidence of *Cytospora* and *Septoria* cankers and poplar and willow borer in hybrid poplar plantations. In: Adams, R.D. (ed.) *Aspen Symposium '89 Proceedings. General Technical Report NC-140.* USDA Forest Service, North Central Forest Experiment Station, St Paul, Minnesota, pp. 163–171.

Ahman, I. and Wilson, F. (2008) Symptoms of pest, rust and other disorders on leaves of willow fertilised with wastewater, urine or sewage sludge. *Biomass and Bioenergy* 32, 1001–1008.

Akbarian, J., Pourmirza, A.A., Khajeiy, E. and Valizadegan, O. (2006) Efficiency of ethological based methods in reduction of poplar stem borer, *Melanophila picta* Pall. (Col. Buprestidae) population. *Agricultural Sciences and Technology, Ferdowsi University of Mashhad, Iran* 20, 3–12.

Alderete, M.C., Fidalgo, P. and Ovruski, S. (2002) Perspectivas en el control biologico de *Nematus oligospilus* Foerster (= *N. desantisi* Smith) (Hymenoptera: Tenthredinidae), plaga de sauces en la Argentina y Chile. *Acta Entomologica Chilena* 26, 7–16.

Alfaro, R.I. (2003) El 'taladrillo grande de los forestales' *Platypus mutatus* (= *sulcatus*): importante plaga de la populicultura Argentina, un plan de acción. *SAGPyA Forestal* 28, 11–18.

Alfaro, R.I., Humble, L.M., Gonzalez, P., Villaverde, R. and Allegro, G. (2007) The threat of the ambrosia beetle *Megaplatypus mutatus* (Chapuis) (= *Platypus mutatus* Chapuis) to world poplar resources. *Forestry* 80, 471–479.
Allegro, G. (1987) Un nemico Nuevo per la pioppicoltura italiana. L' Ifantria Americana (*Hyphantria cunea* Drury). *Economia Montana* 19, 50–53.
Allegro, G. (1989) La difesa contro gli insetti parassiti del pioppo: un aggiornamento tecnico. *L' Informatore Agrario* 45, 93–96.
Allegro, G. (1990a) I parassiti animali del pioppo e del salice in Argentina. *Cellulosa e Carta* 41, 18–22.
Allegro, G. (1990b) Lotta meccanica contro i principali insetti xilofagi del pioppo mediante impiego di sbarramenti sui tronchi. *L' Informatore Agrario* 46, 91–95.
Allegro, G. (1991) Il Picchio rosso maggiore (*Picoides major* L.) nella limitazione naturale delle popolazioni della Saperda maggiore del pioppo (*Saperda carcharias* L.). *Avocetta* 15, 33–41.
Allegro, G. (1992) Trappole a feromoni per il monitoraggio delle popolazioni della Gemmaiola e del Tarlo vespa del pioppo. *Cellulosa e Carta* 43, 54–58.
Allegro, G. (1993) Un prezioso alleato della pioppicoltura: il picchio rosso maggiore (*Picoides major*). *L' Informatore Agrario* 49, 73–76.
Allegro, G. (1997a) Nuove infestazioni di Afide lanigero del pioppo (*Phloeomyzus passerinii* Sign.) in Pianura Padana. *Sherwood – Foreste ed alberi oggi* 27, 41–45.
Allegro, G. (1997b) Conoscere e combattere il Punteruolo del pioppo (*Cryptorhynchus lapathi* L.). *Sherwood – Foreste ed alberi oggi* 29, 33–38.
Allegro, G. (1998) Biologia e controllo della Saperda Maggiore del pioppo (*Saperda carcharias* L.). *Sherwood – Foreste ed alberi oggi* 34, 35–40.
Allegro, G. (2008a) PLATIPO. *Platypus mutatus* Chapuis (Coleoptero, Platypodidae). Instituto di Sperimentazione per le Pioppicoltura (www.populus.it/public/platipo.pdf, accessed 25 March 2013).
Allegro, G. (2008b) Tarlo-vespa del pioppo *Paranthrene tabaniformis* Rott. (Lepidoptera, Sesiidae) (http://www.populus.it/public/tarlo_vespa.pdf, accessed 1 February 2012).
Allegro, G. and Cagelli, L. (1996) Susceptibility of *Populus nigra* L. to the woolly poplar aphid (*Phloemyzus passerinii* Sign). *Forest Genetics* 3, 23–26.
Allegro, G. and Della Beffa, G. (2001) Un nuovo problema entomologico per la pioppicoltura Italiana: *Platypus mutatus* Chapuis (Coleoptera, Platypodidae). *Sherwood – Foreste ed alberi oggi* 66, 31–34.
Allegro, G., Picco, F. and Bianco, B. (1996a) Resistance behaviour to *Phloeomyzus passerinii* of some recently-selected Italian poplar clones. In: Bach, I. (ed.) *Proceedings 20th Session of the International Poplar Commission*, Budapest, 1–4 October 1996. International Poplar Commission, Budapest, Hungary, pp. 199–208.
Allegro, G., Augustin, S., Delplanque, A., Giorcelli, A., Steenackers, M. and Pinon, J. (1996b) Interdisciplinary research for poplar improvement. Task 2 'Genetics of resistance and virulence: molecular, biochemical and conventional approaches'. Sub-Task 2.3 'Biochemical markers'. AIR1-CT92-0349. Final Technical Report, pp. 79–110.
Allegro, G., Picco, F. and Bazzani, R. (2007) Il punteruolo del pioppo negli implanti da biomassa: un rischio da non sottovalutare. *Sherwood – Foreste ed alberi oggi* 129, 13–16.
Aparisi, C. (1971) Noticia sobre el Afido lanigero del chopo, *Phloeomyzus passerinii* Signoret, y ensaios para su tratamiento. *Boletin del Servicio de Plagas Forestales* 14, 3–11.
Arru, G.M. (1965) *Pygaera anastomosis* (L.) (Lepidoptera: Notodontidae). Studio morfologico ed etologico. *Bolletino di Zoologia Agraria e di Bachicoltura Serie II* 6, 207–271.
Arru, G.M. (1967a) I piu importanti insetti minatori delle foglie di pioppo. *Bolletino di Zoologia Agraria e di Bachicoltura Serie II* 8, 41–74.
Arru, G.M. (1967b) Resistance to insects in poplars grown in Italy. In: *Proceedings, 14th Congress, International Union of Forestry Research Organizations*, Munich, Germany, 1967, 3, pp. 861–866.
Arru, G.M. (1971) A method for the evaluation of the resistance of poplars to *Phloeomyzus passerinii* Sign. Publication FAO/CIP/71/24. 14th Session of International Poplar Commission, Bucarest, Romania, 1967. International Poplar Commission, FAO, Rome, Italy.
Arru, G.M. (1974) Metodo per valutare la resistenza dei pioppi all'Afide lanigero (*Phloeomyzus passerinii* Sign.). *Cellulosa e Carta* 25, 45–49.
Arru, G.M. (1975) Elenco commentato dei piu importanti insetti dannosi alla coltura del pioppo in Italia. *Cellulosa e Carta* 26, 47–50.
Arru, G.M. and Lapietra, G. (1979) Breeding poplars for resistance to insect pests. In: *Proceedings of the IUFRO Working Parties S2.02.10 and S2.03.07*, Orléans-Geraardsbergen, September 17–22, 1979.

Arundell, J.C. and Straw, N.A. (2001) Hornet clearwing moth (*Sesia apiformis* [Clerck]) and dieback of poplars in eastern England. *Arboricultural Journal* 25, 235–253.
Arzone, A. (1987) Contributo alla conoscenza dei limitatori di *Phloeomyzus passerinii* (Sign.). *Annali della Accademia d'Agricoltura di Torino* 129, 139–158.
Arzone, A. and Vidano, C. (1984) Indagini su *Phloemyzus passerinii* (Sign.) in Piemonte. *Annali della Facolta di Scienze Agrarie della Universita degli Studi di Torino* 13, 337–356.
Attard, G. (1978) Extension des dégâts du charançon du peuplier sur les peupleraies d'Aquitaine. *Phytoma* 298, 26.
Attard, G. (1979) La tordeuse des pousses du peuplier et son cycle biologique dans le Sud-Ouest. *Phytoma* 305, 23–28.
Augustin, S. and Lévieux, J. (1993) Life history of the poplar beetle *Chrysomela tremulae* in the Central Region of France. *Canadian Entomologist* 125, 399–401.
Augustin, S., Courtin, C. and Delplanque, A. (1993a) Poplar clones effect on development, mortality, fecundity and longevity of *Chrysomela* (= *Melasoma*) *populi* L. and *C. tremulae* F. (Col., Chrysomelidae). *Journal of Applied Entomology* 116, 39–49.
Augustin, S., Courtin, C. and Delplanque, A. (1993b) Preferences of *Chrysomela* (= *Melasoma*) *populi* L. and *C. tremulae* F. (Col., Chrysomelidae) for Leuce section poplar clones. *Journal of Applied Entomology* 115, 370–378.
Augustin, S., Wagner, M.R. and Clancy, K.M. (1994) *Chrysomela scripta*. Performance on five poplar clones. *Norwegian Journal of Agricultural Sciences* 18, 111–117.
Augustin, S., Courtin, C., Réjasse, A., Lorme, P., Génissel, A. and Bourguet D. (2004) Genetics of resistance to transgenic *Bacillus thuringiensis* poplars in the poplar leaf beetle, *Chrysomela tremulae* F. (Coleoptera: Chrysomelidae). *Journal of Economic Entomology* 97, 1058–1064.
Babmorad, M., Azizkhani, E., Omid, R. and Farashiani, E. (2007) Poplar lace bug (*Monosteira unicostata*) damage on different poplar species and clones in Karaj. *Iranian Journal of Forest and Range Protection Research* 4, 103–112.
Bankhead-Dronnet, S., Allegro, G. and Lieutier, F. (2008) The poplar woolly aphid *Phloeomyzus passerinii* Signoret in the Mediterranean area: relations with host trees and spreading risks. In: *International Symposium of Entomological Research in Mediterranean Forest Ecosystems*, Estoril, Portugal, 5–9 May 2008, Programs and Abstracts, pp. 25.
Bao, S., Li, F., Li, Z. and Wu, T. (1999) The resistance of fourteen poplar species to *Anoplophora glabripennis* (Motsch.). *Journal of Beijing Forestry University* 21, 97–100. (In Chinese)
Barbosa, P. and Wagner, M.R. (1989) *Introduction to Forest and Shade Tree Insects*. Academic Press, San Diego, California, USA.
Bascialli, M.E., Giménez, R.A., Etiennot, A.E. and Toscani, H. (1996) Manejo de la población de *Platypus sulcatus* Chapuis, durante tres años en la región del Delta del Río Paraná mediante control químico. *Investigación Agraria: Sistemas y Recursos Forestales* 5, 129–140.
Beckwith, R.C. (1973) The large aspen tortrix. Forest Pest Leaflet 139. USDA Forest Service, Washington, DC.
Beeson, C.F.C. (1941) *The Ecology and Control of Forest Insects of India and the Neighbouring Countries*. Vasant Press, Dehra Dun, India.
Belova, N.K. and Vorontsov, A.I. (1987) The poplar moth. *Zashchita Rastenii Moskva* 7, 32–35.
Berges, R., Cousin, M.-T., Roux, J., Mäurer, R. and Seemüller, E. (1997) Detection of phytoplasma infections in declining *Populus nigra* 'Italica' trees and molecular differentiation of the aster yellows phytoplasmas identified in various *Populus* species. *European Journal of Forest Pathology* 27, 33–43.
Berry, J.A. (1997) *Nematus oligospilus* (Hymenoptera: Tenthredinidae), a recently introduced sawfly defoliating willows in New Zealand. *New Zealand Entomologist* 20, 51–54.
Bertucci, B.M. (1986) Il tarlo vespa del pioppo. *Informatore Fitopatologico* 36, 29–34.
Bingaman, B.R. and Hart, E.R. (1992) Feeding and oviposition preferences of adult cottonwood leaf beetles (Coleoptera: Chrysomelidae) among *Populus* clones and leaf age classes. *Environmental Entomology* 21, 508–517.
Blackman, R.L. and Spence, J.M. (1996) Ribosomal DNA is frequently concentrated on only one X chromosome in permanently apomictic aphids, but this does not inhibit male determination. *Chromosome Research* 4, 314–320.
Booij, C. and Voerman, S. (1984) Sex attractant for the poplar shoot-borer, *Gypsonoma aceriana* (Duponchel) (Lepidoptera, Tortricidae). *Zeitschrift fur Angewandte Entomologie* 97, 176–179.
Bougard, M. (1977) Les insects nuisibles et nematodes des peupliers de la section Aegeiros et leur divers hybrides. *Bulletin Trimestriel du Centre de Populiculture du Hainault* 2, 7–30.

Bouyaiche, M. and Nef, L. (1987) Relation entre les attaques de *Phyllocnistis suffusella* Zell., la croissance et la composition minerale des feuilles de peuplier. *Mededelingen van de Faculteit Landbouwwetenschappen, Universiteit-Gent.* (*International Symposium on Crop Protection, Proceedings*) 52(2a), 259–266.

Brethes, J. (1908) Dos nuevos *Platypus* (Col.) argentinos. *Anales del Museo Nacional de Historia Natural* 17, 225–227.

Brizzi, G. (1962) *Paranthrene tabaniformis* dannoso alle giovani piante del pioppo. *Informatore Fitopatologico* 12, 81–85.

Broberg, C.L., Borden, J.H. and Gries, R. (2005) Olfactory and feeding preferences of *Cryptorhynchus lapathi* L. (Coleoptera Curculionidae) among hybrid clones and natural poplars. *Environmental Entomology* 34, 1606–1613.

Brown, J.J., Kittelson, N.T., Hannon, E.R. and Walsh, D.B. (2006) An endemic population of Western Poplar clearwing moths (Lepidoptera: Sesiidae) invades a monoculture of hybrid poplar. *Journal of Economic Entomology* 99, 771–779.

Brown, W.J. (1956) The new world species of *Chrysomela* L. (Coleoptera: Chrysomelidae). *Canadian Entomologist* 88, 5–54.

Browne, F.G. (1968) *Pests and Diseases of Plantation Trees.* Clarendon Press, Oxford, UK.

Brugnoni, H.C. (1980) *Plagas forestales.* Ed. Hemisferio Sur S.A., Buenos Aires, Argentina.

Bruzzese, E. and McFadyen, R. (2006) Arrival of leaf-feeding willow sawfly *Nematus oligospilus* Förster in Australia – pest or beneficial? *Plant Protection Quarterly* 21, 43–44.

Burke, J.M. (1982) Micro-organisms infecting aspen tortrix, *Choristoneura conflictana* Wlk. (Lepidoptera: Tortricidae). Information Report FPM-X-61. Environment Canada, Canadian Forest Service, Forest Pest Management Institute, Sault Ste Marie, Ontario, Canada.

Burkot, T.R. and Benjamin, D.M. (1979) The biology and ecology of the cottonwood leaf beetle, *Chrysomela scripta* (Coleoptera: Chrysomelidae), on tissue cultured hybrid Aigeiros (*Populus* ×*euramericana*) subclones in Wisconsin. *Canadian Entomologist* 111, 551–556.

Cadahia, D. (1965) Preferencias clonales del Gorgojo perforador del chopo (*Cryptorrhynchus lapathi* L. (Col. Curculionidae). *Boletin del Servicio de Plagas Forestales* 8, 115–125.

Caldbeck, E.S., McNabb, H.S. and Hart, E.R. (1978) Poplar clonal preference of the cottonwood leaf beetle. *Journal of Economic Entomology* 71, 518–520.

Callan, E.M. (1939) *Cryptorrhynchus lapathi* L. in relation to the watermark disease of the cricket-bat willow. *Annals of Applied Biology* 26, 135–137.

Caron, V., Moslih, F., Ede, F.J. and O'Dowd, D.J. (2011) An accidental biological control agent? Host specificity of the willow sawfly *Nematus oligospilus* (Hymenoptera: Tenthredinidae) in Australia. *Australian Journal of Entomology* 50, 290–295.

Carrière, Y., Ellers-Kirk, C., Sisterson, M., Antilla, L., Whitlow, M., Dennehy, T.J., *et al.* (2003) Long-term regional suppression of pink bollworm by *Bacillus thuringiensis* cotton. *Proceedings of the National Academy of Sciences of the USA* 100, 1519–1523.

Casa, J., Boff, M., Rech, T.D. and Boff, P. (2007) Resistencia do vimeiro, *Salix* spp. (*Salicaceae*), a pragas e doencas. *Ciência Florestal* 17, 1–8.

Cavalcaselle, B. (1966) Contributo alla conoscenza del *Cryptorrhynchus lapathi* L. (Coleoptera Curculionidae). *Pubblicazioni del Centro di Sperimentazione Agricola e Forestale Roma* 8, 135–173.

Cavalcaselle, B. (1972a) Ecologia ed etologia di alcuni Buprestidi nocivi al pioppo nell'Italia centro-meridionale. *Ente nazionale per la Cellulosa e per la Carta. Redia* 53, 67–122.

Cavalcaselle, B. (1972b) Esperienze preliminari sull'allevamento del *Cryptorrhynchus lapathi* L. con diete artificiali. *Pubblicazioni del Centro di Sperimentazione Agricola e Forestale Roma* 11, 85–93.

Cavalcaselle, B. (1975) Possibilité d'emploi de produits à base de *Beauveria bassiana* (Bals.) Vuill. contre les larves de quelques insectes xylophages. *Mededelingen van de Faculteit Landbouwwetenschappen, Rijksuniversiteit-Gent* 40, 437–442.

Cavalcaselle, B. and De Bellis, E. (1968) Esperienze preliminari per la lotta contro il *Cryptorhynchus lapathi* L. con radiazioni ionizzanti. *Pubblicazioni del Centro di Sperimentazione Agricola e Forestale Roma* 10, 89–101.

Cavalcaselle, B. and De Bellis, E. (1970) Ulteriori esperienze per la lotta contro il Crittorrinco con raggi X e Afolato. *Pubblicazioni del Centro di Sperimentazione Agricola e Forestale Roma* 11, 47–53.

Cavalcaselle, B. and De Bellis, E. (1983) Sperimentazione di nuovi insetticidi a bassa tossicità contro le larve subcorticali di Crittorrinco e Saperda. *Cellulosa e Carta* 5, 29–34.

Cavalcaselle, B. and Deseo, K.V. (1984) Prove di lotta contro le larve di due insetti xilofagi del pioppo con nematodi entomopatogeni. *Atti Giornate Fitopatologiche* (*Sorrento*) 2, 393–402.

Cavey, J.F., Hoebeke, E.R., Passoa, S. and Lingafelter, S.W. (1998) A new exotic threat to North American hardwood forests: an Asian longhorned beetle, *Anoplophora glabripennis* (Motschulsky) (Coleoptera: Cerambycidae). I. Larval description and diagnosis. *Proceedings of the Entomological Society of Washington* 100, 373–381.
Cerezke, H.F. and Volney, W.J.A. (1995) Forest insect pests in the northwest region. In: Armstrong, J.A. and Ives, W.G.H. (eds) *Forest Insect Pests in Canada*. Natural Resources Canada, Canadian Forest Service, Ottawa, Canada, pp. 59–72.
Chakrabarti, S., Kapoor, K.S., Kumar, S. and Kumar, S. (2005) Outbreak of *Tuberolachnus salignus* Gmelin (Homoptera: Lachnidae) infesting willows in the Lahaul valley of Himachal Pradesh: a case study. In: Verma, K.S., Khurana, D.K. and Christersson, L. (eds) *Short Rotation Forestry for Industrial and Rural Development*. Proceedings of the IUFRO-ISTS-UHF International Conference on World Perspective on Short Rotation Forestry for Industrial and Rural Development. ISTS, Nauni, Solan, India, pp. 394–398.
Chararas, C. (1969) Etude du preferendum, de la nutrition et de l'activité enzymatique de *Cryptorhynchus lapathi* L. (Col. Curculionidae xylophage). *Bulletin du Service de Cultre et d'Etudes du Peuplier et du Saule* 1969, 155–170.
Charles, J.G. and Allan, D.J. (2000) Development of the willow sawfly, *Nematus oligospilus*, at different temperatures, and an estimation of voltinism throughout New Zealand. *New Zealand Journal of Zoology* 27, 197–200.
Charles, J.G., Allan, D.J. and Fung, L. (1998) Susceptibility of willows to oviposition by the willow sawfly, *Nematus oligospilus*. In: Proceedings, 51st New Zealand Plant Protection Conference 1998. NZ Plant Protection Society (Inc), Christchurch, New Zealand, pp. 230–234.
Chaudhry, M. and Ahmad, M. (1973) Population dynamics of two poplar defoliators. Final technical report. Pakistan Forest Institute, Peshawar, Pakistan.
Cherepanov, A.I. (1988) *Cerambycidae of Northern Asia. Volume 3. Lamiinae Part III*. [Usachi Severnoi Azii (Lamiinae)]. Ed. S. Otdelenie. Translated from Russian. Amerind Publishing Co Pvt Ltd, New Delhi, India.
Chrystal, R.N. (1937) *Insects of the British Woodlands*. Frederick Warne and Co, London, UK.
Clausen, T.P., Reichardt, P.B., Bryant, J.P., Werner, R.A., Post, K. and Frisby, K. (1989) Chemical model for short-term induction in quaking aspen (*Populus tremuloides*) foliage against herbivores. *Journal of Chemical Ecology* 15, 2335–2346.
Clausen, T.P., Reichardt, P.B., Bryant, J.P. and Werner, R.A. (1991) Long-term and short-term induction in quacking aspen: related phenomena? In: Tallamy, D.W. and Raupp, M.J. (eds) *Phytochemical Inductions by Herbivores*. Wiley, New York, USA, pp. 71–83.
Coleman, D.A. and Boyle, M.K. (2000) The status and ecology of the hornet moth, *Sesia apiformis* (Clerck) (Lepidoptera: Sesiidae), in suburban south London. *British Journal of Entomology and Natural History* 13, 99–106.
Collins, C.M., Rosado, R.G. and Leather, S.R. (2001) The impact of the aphids *Tuberolachnus salignus* and *Pterocomma salicis* on willow trees. *Annals of Applied Biology* 138, 133–140.
Confalonieri, M., Allegro, G., Balestrazzi, A., Fogher, C. and Delledonne, M. (1998) Regeneration of *Populus nigra* transgenic plants expressing a Kunitz proteinase inhibitor (KTi3) gene. *Molecular Breeding* 4, 137–145.
Costa, J.T. (2006) *The Other Insect Societies*. Belknap Press of Harvard University Press, Cambridge, Massachusetts, USA.
Cousin, M.-T., Berges, R., Roux, J., Moreau, J.-P., Hiruki, C. and Seemüller, E. (1999) *Populus nigra* L. *italica* decline in France, variability of the Phytoplasma responsible for the disease in Europe, results and perspectives. *Acta Horticulturae* 496, 77–86.
Coyle, D.R., McMillin, J.D., Hall, R.B. and Hart, E.R. (2002) Cottonwood leaf beetle (Coleoptera: Chrysomelidae) defoliation impact on *Populus* growth and aboveground volume in a short-rotation woody crop plantation. *Agricultural and Forest Entomology* 4, 293–300.
Coyle, D.R., Nebeker, T.E., Hart, E.R. and Mattson, W.J. (2005) Biology and management of insect pests in North American intensively managed hardwood forest systems. *Annual Review of Entomology* 50, 1–29.
Coyle, D.R., Hart, E.R., McMillin, J.D., Rule, L.C. and Hall, R.B. (2008) Effects of repeated cottonwood leaf beetle defoliation on *Populus* growth and economic value over an 8-year harvest rotation. *Forest Ecology and Management* 255, 3365–3373.
Dafauce, C. (1965) Combate del *Cryptorrhynchus lapathi* L. (Coleoptera Curculionidae) insecto perforador del chopo. *Boletin del Servicio de Plagas Forestales* 8, 77–96.

Dafauce, C. (1976) Susceptibilidad de clones de chopo al ataque de *Cryptorhynchus lapathi* L. (Col. Curculionidae). *Boletin de la Estacion Central de Ecologia* 5, 39–66.
Dafauce, C., Astiaso, F. and Bachiller, P. (1963) Aspectos biològicos del gorgojo perforador del chopo (*Cryptorhynchus lapathi* L., Curculionidae). *Boletin del Servicio de Plagas Forestales* 6, 85–97.
Davis, E.E., French, S. and Venette, R.C. (2005) Mini Risk Assessment Ambrosia beetle: *Platypus quercivorus* Murayama [Coleoptera: Platypodidae]. US Department of Agriculture (http://www.aphis.usda.gov/plant_health/plant_pest_info/pest_detection/downloads/pra/pquercivoruspra.pdf, accessed 1 February 2012).
Della Beffa, G. (1936) Contributo alla conoscenza degli insetti parassiti dei pioppi. Il *Phloeomyzus passerinii* Sign. (Afide lanigero dei pioppi). *Bolletino del Laboratorio Sperimentale R. Oss. Fitopologica Torino* 18, 17–23.
Delledonne, M., Allegro, G., Belenghi, B., Balestrazzi, A., Picco, F., Levine, A., *et al.* (2001) Transformation of white poplar (*Populus alba* L.) with a novel *Arabidopsis thaliana* cysteine proteinase inhibitor and analysis of insect pest resistance. *Molecular Breeding* 7, 35–42.
Delplanque, A. (1998) *Les insectes associés aux peupliers.* Editions Memor, Bruxelles, Belgium.
Den Herden, M., Virtanen, R. and Roininen, H. (2004) Effects of reindeer browsing on tundra willow and its associated herbivores. *Journal of Applied Ecology* 41, 870-879.
Dent, D. (2000) *Insect Pest Management,* 2nd edn. CAB International, Wallingford, UK.
de Tillesse, V. and Nef, L. (1998) Les insectes dommageables au peuplier – espèces d'importance internationale. International Poplar Commission/FAO (www.fao.org/docrep/004/ac489f/ac489f00.htm, accessed 5 August 2013).
de Tillesse, V., Nef, L., Charles, J.G., Hopkin, A. and Augustin, S. (2007) Damaging poplar insects – internationally important species. International Poplar Commission, FAO, Rome, Italy (http://www.fao.org/forestry/ipc/69946@158688/en/, accessed 5 August 2013).
Dickmann, D.I. (1978) Marked differences among poplar clones in winter browsing damage by cottontail rabbits. *Canadian Journal of Forest Research* 8, 351–354.
Drouin, J.A. and Wong, H.R. (1975) Biology, damage and chemical control of the poplar borer (*Saperda calcarata*) in the junction of the root and stem of Balsam poplar in western Canada. *Canadian Journal of Forest Research* 5, 433–439.
Dubois, T., Hajek, A.E. and Susan, S. (2002) Methods for rearing the Asian longhorned beetle (Coleoptera: Cerambycidae) on artificial diet. *Annals of the Entomological Society of America* 95, 223–230.
Dubois, T., Hajek, A.E., Hu, J.F. and Li, Z.Z. (2004a) Evaluating the efficiency of entomopathogenic fungi against the Asian longhorned beetle, *Anoplophora glabripennis* (Coleoptera: Cerambycidae), by using cages in the field. *Environmental Entomology* 33, 62–74.
Dubois, T., Hajek, A.E., Hu, J.F. and Li, Z.Z. (2004b) Efficacy of fiber bands impregnated with *Beauveria brongniartii* cultures against the Asian longhorned beetle, *Anoplophora glabripennis* (Coleoptera: Cerambycidae). *Biological Control* 31, 320–328.
Dubois, T., Lund, J., Bauer, L.S. and Hajek, A.E. (2008) Virulence of entomopathogenic hypocrealean fungi infecting *Anoplophora glabripennis. BioControl* 53, 517–528.
Eastop, W.F. and Hille Ris Lambers, D. (1976) *Survey of the World's Aphids.* W. Junk b.v. Publishers, The Hague, Netherlands.
Eken, C., Tozlu, G., Dane, E., Coruh, S. and Demirci, E. (2006) Pathogenicity of *Beauveria bassiana* (Deuteromycotina: Hypomycetes) to larvae of the small poplar longhorn beetle, *Saperda populnea* (Coleoptera: Cerambycidae). *Mycopathologia* 162, 69–71.
El-Sayed, A.M. (2005) The Pherobase: database of pheromones and semiochemicals (http://www.pherobase.com, accessed 2 February 2012).
EPPO (European and Mediterranean Plant Protection Organization) (2004a) Data sheets on quarantine pests – *Anoplophora glabripennis* (http://www.eppo.org/QUARANTINE/insects/Anoplophora_glabripennis/ANOLGL_ds.pdf, accessed 2 February 2012).
EPPO (2004b) First report of *Platypus mutatus* in Italy: addition to the EPPO Alert List. Eppo reporting service 2004, No 4. 2004/061 (http://www.invasive.org/library/eppo/Rse-0404.pdf, accessed 2 February 2012).
EPPO (2008) EPPO Reporting Service. No 5 (http://archives.eppo.org/EPPOReporting/2008/Rse-0805.pdf, accessed 2 February 2012).
Evans, H., Moraal, L. and Pajares, J. (2004) Biology, ecology and economic importance of Buprestidae and Cerambycidae. In: Lieutier, F., Day, K., Battisti, A., Grégoire, J. and Evans, H. (eds) *Bark and Wood Boring Insects in Living Trees in Europe, A Synthesis.* Kluwer Academic Publishers, Dordrecht, Netherlands.

Evenden, M.L. and Gries, R. (2006) Sex pheromone of the large aspen tortrix, *Choristoneura conflictana* (Lepidoptera: Tortricidae). *Chemoecology* 16, 115–122.

Evenden, M.L., Lopez, M.S. and Keddie, B.A. (2006) Body size, age, and disease influence female reproductive performance in *Choristoneura conflictana* (Lepidoptera: Tortricidae). *Annals of the Entomological Society of America* 99, 837–844.

Fallon, D.J., Solter, L.F., Keena, M., McManus, M., Cate, J.R. and Hanks, L.M. (2004) Susceptibility of Asian longhorned beetle, *Anoplophora glabripennis* (Motschulsky) (Coleoptera: Cerambycidae) to entomopathogenic nematodes. *Biological Control* 30, 430–438.

Fang, Y. and Hart, E.R. (2000) Effect of cottonwood leaf beetle (Coleoptera: Chrysomelidae) larval population levels on *Populus* terminal damage. *Environmental Entomology* 29, 43–48.

FAO (2009) Regulation of wood packaging material in international trade. International Standards for Phytosanitary Measures, Revision of ISPM No 15 (https://www.ispm15.com/ISPM15_Revised_2009.pdf, accessed 5 August 2013).

FAO (2010) Invasive species: impacts on forests and forestry (www.fao.org/forestry/site/27082/en, accessed 2 February 2012).

Ferré, J. and Van Rie, J. (2002) Biochemistry and genetics of insect resistance to *Bacillus thuringiensis*. *Annual Review of Entomology* 47, 501–533.

Francke-Grosmann, H. (1960) Zur lebenweise und Bekämpfung des Erlenrüsslers. *Die Hollzucht* 14, 32–33.

Gagné, R.J. (1989) *The Plant-feeding Gall Midges of North America*. Cornell University Press, Ithaca, New York, USA.

Gagné, R.J. (2004) A catalog of the Cecidomyiidae (Diptera) of the world. *Memoirs of the Entomological Society of Washington* 25.

Gao, H.Z., Yang, X.Y, Wei, J.N. and Lang, X.R. (1997) An investigation on the resistance of major forestation species to *Anoplophora glabripennis* and *A. nobilis*. *Journal of Northwest Forestry College* 12(Supplement), 42–46. (In Chinese)

Gao, R.T. and Li, G.H. (2001) Review and prospect of research on *Anoplophora glabripennis* in China. *Entomological Knowledge* 38, 252–257. (In Chinese)

Gao, R.T., Qin, X.X., Li, J.Z., Hao, W.Q. and Wang, X.Z. (1985) A preliminary study on the relationship between artificial defoliation of poplar trees and their growth. *Scientia Silvae Sinicae* 21, 199–205.

Garbutt, R. and Harris, J.W.E. (1994) Poplar and willow borer. Forest Pest Leaflet 7. Natural Resources Canada, Canadian Forest Service, Pacific Forestry Centre, Victoria, British Columbia, Canada.

Génissel, A., Leplé, J.-C., Millet, N., Augustin, S., Jouanin, L. and Pilate, G. (2003) High tolerance against *Chrysomela tremulae* of transgenic poplar plants expressing a synthetic cry3aA gene from *Bacillus thuringiensis* ssp *tenebrionis*. *Molecular Breeding* 11, 103–110.

Georghiou, G.P. and Taylor, C.E. (1977) Operational influences in the evolution of insecticide resistance. *Journal of Economic Entomology* 70, 653–658.

Georgiev, G. (2000) Studies on larval parasitoids of *Paranthrene tabaniformis* (Rott.) (Lepidoptera: Sesiidae) on urban poplars (*Populus* spp.) in Sofia, Bulgaria. *Annals of Forest Science* 57, 181–186.

Georgiev, G. (2001) Notes on the biology and ecology of the parasitoids of the poplar clearwing moth, *Paranthrene tabaniformis* (Rott.) (Lep., Sesiidae) in Bulgaria. II. *Eriborus terebrans* (Gravenhorst, 1826) (Hym., Ichneumonidae). *Journal of Applied Entomology* 125, 289–292.

Georgiev, G., Raikova, M., Ljubomirov, T. and Ivanov, K. (2004) New parasitoids of *Saperda populnea* (L.) (Col. Cerambycidae) in Bulgaria. *Journal of Pest Science* 77, 179–182.

Giménez, R.A. and Etiennot, A.E. (2002) Control químico de *Platypus sulcatus* (Coleoptera: Platypodidae) en chopos. *Investigación Agraria: Sistemas y Recursos Forestales* 11, 227–232.

Giménez, R.A. and Etiennot, A.E. (2003) Host range of *Platypus mutatus* (Chapuis, 1865) (Coleoptera; Platypodidae). *Entomotropica* 18, 89–94.

Giménez, R. and Kocsis, G. (2007) Eficacia del extracto de Neem para el control de *Megaplatypus mutatus*. XXII Jornadas Forestaleas de Entre Ríos, Concordia, Argentina, 25–26 October 2007 (http://anterior.inta.gov.ar/concordia/info/Forestales/contenido/pdf/2007/posters07/370.51.GIMENEZ.pdf, accessed 2 February 2012).

Giménez, R.A. and Moya, A. (2011) Nivel y Umbral de daño económico de *Megaplatypus mutatus*. 3° Congreso Internacional de Salicáceas. Neuquén, Argentina (http://64.76.123.202/new/0-0/forestacion/salicaceas/jornadas%20salicaceas%202011/Actas/Trabajos_completos_Formato%20pdf/Gimenez_Rosana_1_TC.pdf, accessed 2 February 2012).

Giménez, R.A. and Panzardi, S.R. (2003) Ambrosia beetle, *Platypus mutatus*: review of their management in Argentina. *IDESIA (Chile)* 21, 53–56.

Giménez, R., Moya, M. and Michetti, M. (2003) Control de *Megaplatypus mutatus* (Coleptera, Platypodidae) en álamos: pulverización de carbaril sobre la corteza de los árboles de filas perimetrales. *IDESIA (Chile)* 21(2), 97–102.

Giménez, R.A., Etiennot, A.E., Frank, R. and Alvarez, G. (2004) Incidence of the damage of *Platypus sulcatus* Chapuis in the quality of *Populus deltoides* woods. Poster presentation. Abstract book. FOREST 2004. Biosfera (Brazilian Inst. for the Environment), Brasilia, Brazil pp. 84.

Giorcelli, A. and Allegro, G. (1999) I trattamenti per una corretta difesa fitosanitaria del pioppeto. *Sherwood – Foreste ed alberi oggi* 45, 39–44.

Gom, L.A. and Rood, S.T. (1999) The discrimination of cottonwood clones in a mature grove along the Oldman River in southern Alberta. *Canadian Journal of Botany* 77, 1084–1094.

González, P.E. (2000) Para detectar y controlar la avispa taladradora en álamo (http://www.inia.cl/medios/biblioteca/ta/NR27955.pdf, accessed 2 February 2012).

Gonzalez, R.H., Barria, G. and Guerrero, M.A. (1986) *Nematus desantisi* Smith, nueva especie de importancia forestal en Chile (Hymenoptera: Tenthredinidae). *Revista Chilena de Entomologia* 14, 13–15.

Gonzalez Audino, P. (2011) Ecología química de *Megaplatypus sulcatus* y su aplicación al control de la plaga. 3° Congreso Internacional de Salicáceas. Disertación. Neuquén, Argentina (http://www.minagri.gob.ar/new/0-0/forestacion/salicaceas/jornadas%20salicaceas%202011/Actas/Trabajos_completos_Formato%20pdf/Gonzalez%20Audino_D.pdf, accessed 2 February 2012).

Gould, F. (1998) Sustainability of transgenic insecticidal cultivars: integrating pest genetics and ecology. *Annual Review of Entomology* 43, 701–726.

Grechkin, V.P. and Vorontzov, A.I. (1962) *Insect Pests and Diseases of the Poplars and Their Control.* Goslesbumizdat, Moscow, Russia. (In Russian)

Griffiths, J.S. and Aroian, R.V. (2005) Many roads to resistance: how invertebrates adapt to Bt toxins. *BioEssays* 27, 614–624.

Gross, J., Podsiadlowski, L. and Hilker, M. (2002) Antimicrobial activity of.exocrine glandular secretion of *Chrysomela* larvae. *Journal of Chemical Ecology* 28, 317–331.

Gross, J., Schmolz, E. and Hilker, M. (2004a) Thermal adaptations of the leaf beetle *Chrysomela lapponica* (Coleoptera: Chrysomelidae) to different climes of Central and Northern Europe. *Environmental Entomology* 33, 799–806.

Gross, J., Fatouros, N.E., Neuvonen, S. and Hilker, M. (2004b) The importance of specialist natural enemies for *Chrysomela lapponica* in pioneering a new host plant. *Ecological Entomology* 29, 584–593.

Gross, J., Fatouros, N.E., Neuvonen, S. and Hilker, M. (2007) The role of competitors for *Chrysomela lapponicus*, a north Eurasian willow pest, in pioneering a new host plant. *Journal of Pest Science* 80, 139–143.

Gruppe, A., Fußeder, M. and Schop, R. (1999) Short rotation plantations of aspen and balsam poplar on former arable land in Germany: defoliating insects and leaf constituents. *Forest Ecology and Management* 121, 113–122.

Guerrero, R.T. (1966) Una nueva especie de hongo imperfecto asociado con el *Platypus sulcatus* Chapuis. *Revista de Investigaciones Agropecuarias INTA,* Serie 5, Vol III, No 8, 97–103.

Haack, R.A., Cavey, J.F., Hoebeke, E.R. and Law, K.L. (1996) *Anoplophora glabripennis*: a new tree-infesting exotic cerambycid invades New York. *Newsletter of the Michigan Entomology Society* 41(2/3), 1–3.

Haack, R.A., Law, K.R., Mastro, V.C., Ossenbruggen, H.S. and Raimo, B.J. (1997) New York's battle with the Asian long-horned beetle. *Journal of Forestry* 95, 11–15.

Haack, R.A., Bauer, L.S., Gao, R.-T., McCarthy, J.J., Miller, D.L., Petrice, T.R. and Poland, T.M. (2006) *Anoplophora glabripennis* within-tree distribution, seasonal development, and host suitability in China and Chicago. *Great Lakes Entomologist* 39, 169–183.

Habib, R. and Ghani, M.A. (1970) Eriosomatinae on poplars and their natural enemies in West Pakistan. *Technical Bulletin of the Commonwealth Institute of Biological Control* 13, 43–58.

Hajek, A.E., Huang, B., Dubois, T., Smith, M.T. and Li, Z.Z. (2006) Field studies of control of *Anoplophora glabripennis* (Coleoptera: Cerambycidae) using fiber bands containing the entomopathogenic fungi *Metarhizium anisopliae* and *Beauveria brongniartii. Biocontrol Science and Technology* 16, 329–343.

Hakizimana, E. and Nef, L. (1998) Attaques d'insectes sur peupliers, et caractéristiques stationnelles et foliaires. Rapport. Commission Régionale Wallonne de Peuplier, Liège, Belgium.

Hanski, I. (1992) Insectivorous mammals. In: Crawley, M.J. (ed.) *Natural Enemies. The Population Biology of Predators, Parasites and Diseases.* Blackwell Scientific Publications, London, UK, pp. 163–187.

Harrell, M.D., Benjamin, D.M., Berbee, J.G. and Burkot, T.R. (1981) Evaluation of adult cottonwood leaf beetle, *Chrysomela scripta* (Coleoptera: Chrysomelidae), feeding preference for hybrid poplars. *Great Lakes Entomologist* 14, 181–184.

Haugen, D.A. (2000) Update on Asian longhorned beetle infestations in the US. *Newsletter of the Michigan Entomology Society* 45, 2–3.

Head, R.B., Neel, W.W. and Morris, R.C. (1977) Seasonal occurrence of the cottonwood leaf beetle *Chrysomela scripta* (Fab.) and its principal insect predators in Mississippi and notes on parasites. *Journal of the Georgia Entomological Society* 12, 157–163.

Heath, J. and Emmet, A.M. (eds) (1985) *The Moths and Butterfies of Great Britain and Ireland. Vol. 2 Cossidae – Heliodinidae.* Harley Books, Colchester, UK.

Heidari, S., Fathipour, Y. and Sadeghi, S.E. (2006) The effect of poplar stand density on the spatial distribution pattern of pests and their natural enemies in Karaj. *Journal of the Entomological Society of Iran* 26, 13–31.

Hellrigl, K. (1974) Cerambycidae. In: Schwenke, W. (ed.) *Die Forstschädlinge Europas. Band 2, Käfer.* Parey, Hamburg, Germany, pp. 130–202.

Hérard, F., Ciampitti, M., Maspero, M., Krehan, H., Benker, U., Boegel, C., *et al.* (2006) *Anoplophora* species in Europe: infestations and management processes. *EPPO Bulletin* 36, 470–474.

Heymans, P. (1984) Cycle biologique et nuisance de *Gypsomona aceriana* Dup. vis-à-vis de divers clones de peupliers. Thesis, Université libre de Bruxelles, Brussels, Belgium.

Heymans, P., Deligne, J. and Nef, L. (1984) Détermination des lieux d'hivernage et de nymphose et perspectives de control de la tordeuse du peuplier, *Gypsonoma aceriana* Dup. (Lepidoptera: Tortricidae). *Mededelingen van de Faculteit Landbowwetenschappen* 49, 709–717.

Hofer, G. (1920) The aspen borer and how to control it. USDA Farmer's Bulletin 1154. US Department of Agriculture, Washington, DC, USA.

Hoffman, A. (1954) *Faune de France: Coléoptères Curculionidae. 2ème partie.* Lechevalier Ed., Paris, France, 59, pp. 487–1208.

Höglund, S., Larsson, S. and Wingsle, G. (2005) Both hypersensitive and non-hypersensitive responses are associated with resistance in *Salix viminalis* against the gall midge *Dasineura marginemtorquens. Journal of Experimental Botany* 56, 3215–3222.

Holsten, E.H. and Hard, J. (1985) Efficacy of *Bacillus thuringiensis* Berliner for suppressing populations of large aspen tortrix in Alaska. *Canadian Entomologist* 117, 587–591.

Hou, A., Liu, J. and Man, S. (1992a) Biological control of *Cryptorrhynchus lapathi* L. In: Padró, A. (ed.) *Proceedings 19th Session of the International Poplar Commission, Zaragoza, Spain, 22–25 September 2002. Volume 1.* International Poplar Commission, pp. 361–365.

Hou, A., Liu, J. and Xue, S. (1992b) Artificial diet, rearing and behavior of Osier weevil (*Cryptorrhynchus lapathi* L.). In: Padró, A. (ed.) *Proceedings 19th Session of the International Poplar Commission, Zaragoza, Spain, 22–25 September 2002. Volume 1.* International Poplar Commission, pp. 368–369.

Hu, J., Yang, M. and Lu, M. (2010) Advances in biosafety studies on transgenic insect-resistant poplars in China. *Biodiversity Science* 18, 336–345. (In Chinese, with English abstract)

Hu, Y.Y., Liu, K.Y., Wang, L.C., Zhang, J.W., Jiang, Y.H., Huang, Z.F., *et al.* (1985) A discussion on the control threshold of poplar scale insect *Quadraspidiotus gigas. Journal of Northeast Forestry University, China* 13, 43–49.

Ives, W.G.H. and Wong, H.R. (1988) Tree and shrub insects of the prairie provinces. Information Report NOR-X-292. Canadian Forest Service, Northern Forestry Centre, Edmonton, Alberta, Canada.

Jodal, I. (1973) On the biology of *Melasoma tremulae. Topola* 17, 3–7.

Jodal, I. (1987) A study of the susceptibility of poplar clones to the attack of poplar and willow borer (*Cryptorhynchus lapathi* L.). In: Guzina, V. (ed.) Proceedings, Symposium on improvement of the production and processing of poplar and willow wood in Croatia, Novi Sad, 5 November 1987. *Radovi – Institut za topolarstvo* 18, 188–196.

Johnson, J.D. and Johnson, K.R. (2003) Hybrid poplar genotype affects attack incidence by the poplar-and-willow borer (*Cryptorhynchus lapathi*). *Western Journal of Applied Forestry* 18, 276–280.

Juzwik, J. and Hubbes, M. (1986) Bacteria associated with tarnished plant bug stem lesions on hybrid poplars in Ontario. *European Journal of Forest Pathology* 16, 390–400.

Kaya, H.K. and Lindegren, J.E. (1983) Parasitic nematode controls western poplar clearwing moth. *California Agriculture* 37, 31–32.

Keena, M.A. (2005) Pourable artificial diet for rearing *Anoplophora glabripennis* (Coleoptera: Cerambycidae) and methods to optimize larval survival and synchronize development. *Annals of the Entomological Society of America* 98, 536–547.

Keena, M.A. (2006) Effects of temperature on *Anoplophora glabripennis* (Coleoptera: Cerambycidae) adult survival, reproduction, and egg hatch. *Environmental Entomology* 35, 912–921.

Kippenberg, H. (1975) *Die Käfer Mitteleuropas. Volume 11*. Groecke and Evers, Krefeld, Germany.

Koch, F. and Smith, D.R. (2000) *Nematus oligospilus* Förster (Hymenoptera: Tenthredinidae), an introduced willow sawfly in the southern hemisphere. *Proceedings of the Entomological Society of Washington* 102, 292–300.

Lapietra, G. (1972) Insetticidi a moderata tossicità verso gli animali a sangue caldo nella lotta contro le larve di *Cryptorhynchus lapathi* L. (Col. Curculionidae). *Bolletino di Zoologia Agraria e di Bachicoltura Serie II* 11, 11–18.

Lapietra, G. and Allegro, G. (1981) La difesa del pioppeto: cenni sulla biologia e sui metodi di controllo degli insetti più dannosi. *L'Informatore Agrario* 37, 16995–16998.

Lapietra, G. and Allegro, G. (1986a) Susceptibility to *Phloeomyzus passerinii* Sign. of poplar clones selected in Italy. In: *Proceedings, XIth Session, International Poplar Commission Working Party on Insects and Other Animal Pests*, Louvain-la-Neuve, Belgium. FAO, Rome, Italy, FO:CIP:I/86/6, pp. 165–177.

Lapietra, G. and Allegro, G. (1986b) Nuove possibilità di intervento durante il periodo autunno-invernale contro il Punteruolo del pioppo (*Cryptorhynchus lapathi*). *Atti Giornate Fitopatologiche* (*Riva del Garda*) 1, 111–120.

Lapietra, G. and Allegro, G. (1990a) Suscettibilità a *Phloeomyzus passerinii* Sign. dei cloni di pioppo coltivati in Italia. *Informatore Fitopatologico* 40, 41–44.

Lapietra, G. and Allegro, G. (1990b) Insects damaging poplars in Italy during 1987–89, control strategies and future perspectives. In: *Proceedings, International Poplar Commission Working Party on Insects and Other Animal Pests*, Buenos Aires, Argentina. FAO, Rome, Italy, FO:CIP:I/90/4, pp. 31–36.

Lapietra, G. and Allegro, G. (1994) Monitoraggio delle popolazioni del Tarlo-vespa del pioppo contrappole a feromoni. *Cellulosa e Carta* 45, 12–17.

Lapietra, G. and Arru, G.M. (1973) Lotta chimica contro gli adulti di *Cryptorhynchus lapathi* L. (Col. Curculionidae). *Atti Giornate Fitopatologiche* 1973, 261–270.

Larsson, S., Glynn, C. and Höglund, S. (1995) High oviposition rate of *Dasineura marginemtorquens* on *Salix viminalis* genotypes unsuitable for offspring survival. *Entomologia Experimentalis et Applicata* 77, 263–270.

Leplé, J.C., Bonadé-Bottino, M., Augustin, S., Pilate, G., Le Tan Dumanois, V., Delplanque, A., *et al.* (1995) Toxicity to *Chrysomela tremulae* (Coleoptera: Chrysomelidae) of transgenic poplars expressing a cysteine proteinase inhibitor. *Molecular Breeding* 1, 319–328.

Li, F., Liu, R.G., Bao, S. and Wu, T. (1999) Selection of trap trees for controlling *Anoplophora glabripennis* and *A. nobilis*. *Journal of Beijing Forestry University* 21, 85–89. (In Chinese)

Li, G., Re, X. and Zhang, J.Y. (2000) Primary study on controlling *Anoplophora glabripennis* (Motsch.) with woodpecker. *Journal of Inner Mongolia Forest Investigation and Design* 4, 34–36. (In Chinese)

Li, W. and Wu, C. (1993) *Integrated Management of Longhorn Beetles Damaging Poplar Trees*. Forest Press, Beijing, China. (In Chinese)

Lin, T., Wang, Z.Y., Liu, K.Y., Jing, T.Z. and Zhang, C.X. (2006) Transformation of spider neurotoxin gene with prospective insecticidal properties into hybrid poplar *Populus simonii* × *P. nigra*. *Acta Entomologica Sinica* 49, 593–598.

Lindeijer, E.J. (1932) Een bacterie-ziekte van der wilg (II.). *Tijdschrift over Plantenziekten* 38, 9–11.

Lingafelter, S.W. and Hoebeke, E.R. (2002) *Revision of the Genus* Anoplophora (*Coleoptera: Cerambycidae*). Entomological Society of Washington, Washington, DC, USA.

Liu, X.D. and Jia, X.Z. (1988) A grey relatedness analysis of *Cytosporia chrysosperma* with *Melanophila decastigma* of poplar. *Forest Pest and Disease* 4, 26–27.

Liu, Z.F., Fan, J. and Zhang, K. (1999) The experiments of prevention and control of *Anoplophora glabripennis* in *Salix matsudana* with insecticide. *Journal of Hubei Agricultural University* 19, 313–314. (In Chinese)

Loi, G. (1970) Note biologiche sulla *Melasoma populi* L. (Col., Chrys.) in Toscana. *Frustula Entomologica* 9, 1–57.

Luo, Y.Q., Wen, J.B. and Xu, Z.C. (2003) Poplar longhorned beetle. In: Zhang, X.Y. and Luo, Y.Q. (eds) *Major Forest Disease and Insect Pests in China*. Forest Press, Beijing, P.R. China pp. 30–55. (In Chinese)

McCown, B.H., McCabe, D.E., Russel, D.R., Robison, D.J., Bareton, K.A. and Raffa, K.F. (1991) Stable transformation of *Populus* and incorporation of pest resistance by electric discharge particle acceleration. *Plant Cell Reports* 9, 590–594.

McIntyre, P.J. and Whitham, T.G. (2003) Plant genotype affects long-term herbivore population dynamics and extinction: Conservation implications. *Ecology* 84, 311–322.

MacLeod, A., Evans, H.F. and Baker, R.H.A. (2002) An analysis of pest risk from an Asian longhorn beetle (*Anoplophora glabripennis*) to hardwood trees in the European community. *Crop Protection* 21, 635–645.
Makhovskaya, M.A., Zemkova, R.I. and Kruglikov, S.A. (1984) Effects of industrial pollutants on the ash composition of leaves from woody plants and their infestation by arthropods. *Soviet Journal of Ecology* 15, 115–119.
Martinez, M. and Gumez, J.L. (1998) Les Agromyzidae. In: Delplanque, A. (ed.) *Les insectes associés aux peupliers*. Editions Memor, Bruxelles, Belgium, pp.117–127.
Martinez, M., Gumez, J.L. and Munnier, P. (1985) Un ravageur mal connu: la mouche mineuse du cambium des peupliers. *Phytoma* 372, 51–53.
Maspero, M., Jucker, C. and Colombo, M. (2007) First record of *Anoplophora glabripennis* (Motschulsky) Coleoptera: Cerambycidae Lamiinae Lamiini) in Italy. *Bolletino di Zoologia Agraria e di Bachicoltura* 39, 161–164. (In Italian)
Masutti, L. (1982) Gli insetti dei pioppi euramericani e le pianti ospiti originarie. Aspetti ecologici dei problemi di prevenzione e controllo. *Arboricoltura da Legno* 25, 2–8.
Mathur, R.N. and Singh, B. (1960) A list of insect pests of forest plants in India and adjacent countries. *Indian Forest Bulletin (Entomology)* 171(7), 1–130.
Mattson, W.J., Hart, E.A. and Volney, W.J.A. (2001) Insect pests of *Populus*: coping with the inevitable. In: Dickmann, D.I., Isebrands, J.G., Eckenwalder, J.E. and Richardson, J. (eds) *Poplar Culture in North America*. National Research Council of Canada Research Press, Ottawa, Canada, pp. 219–248.
Maugard, F. and Chauvel, G. (1997) Enseignements et consequences d'une premiere forte attaque de *Phloeomyzus passerinii* dans les peupleraies de Tarn et Garonne. In: *Annales, ANPP International Conference on Pests in Agriculture*, Montpellier, France, 6–8 January 1997. Volume 3. ANPP, Paris, France, pp. 919–926.
Mittler, T.E. (1957) Studies on the feeding and nutrition of *Tuberolachnus salignus*. I. – The uptake of phloem sap. *Journal of Experimental Biology* 34, 334–341.
Mizobuchii, M., Matsuoka, I., Soma, Y., Kishino, H., Yabuta, S., Imamura, M., *et al.* (1996) Susceptibility of forest insect pests to sulfuryl fluoride: 2. ambrosia beetles. *Research Bulletin of the Plant Protection Service, Japan* 32, 77–82.
Moore, L.M. and Wilson, L.F. (1986) Impact of the poplar-gall saperda, *Saperda inornata*, on a hybrid *Populus* plantation in Michigan. *Great Lakes Entomologist* 19, 163–167.
Moraal, L.G. (1996a) Aantasting door insecten en mijten in 1995 in bossen, natuurgebieden en wegbeplantingen. *Nederlands Bosboustijdschrift* 68, 111–120.
Moraal, L.G. (1996b) Evaluation of infestations by the poplar clearwing moth, *Paranthrene tabaniformis* Rott. In: Bach, I. (ed.) *Proceedings of the 20th Session of the International Poplar Commission, Budapest, 1–4 October 1996. Volume 1*. International Poplar Commission, pp. 250–259.
Moraal, L.G., Van der Kraan, C. and Van der Voet, H. (1993) Studies on the efficacy of the sex attractant of *Paranthrene tabaniformis* Rott. *Journal of Applied Entomology* 116, 364–370.
Morris, R.C. (1963) Trunk borers in cottonwood. *Mississippi Farm Research* 26(10), 8.
Morris, R.C., Filer, T.H., Solomon, J.D., McCracken, F.I., Overgaard, N.A. and Weiss, M.J. (1975) Insects and diseases of cottonwood. General Technical Report SO-8. USDA Forest Service, Southern Forest Experiment Station, New Orleans, Louisiana, USA.
Morris, R.F. (1986) Notes on the occurrence of the hornet moth, *Sesia apiformis* (Lepidoptera: Sesiidae). *Canadian Field Naturalist* 100, 253–254.
Mrkva, R. (1963) The biology and control of *Cryptorrhynchus lapathi* L. *Lesnictvi cas Praha* 9, 551–568.
Muesebeck, C.F.W. (1931) Description of a new genus and eight new species of Ichneumon-flies, with taxonomic notes. *Proceedings of the United States National Museum* 79, 1–16.
Nef, L. (1976) Etude écologique de *Telonomus nitidulus*, parasite des oeufs de *Stilpnotia* (*Leucoma*) *salicis*. *Zeitschrift für Pflanzenkrankheiten und Pflanzenschutz* 83,109–119.
Nef, L. (1978) *Leucoma salicis*. In: Schenke, W. (ed.).*Die Forstschädlinge Europas. Vol 3*. Parey, Hamburg, Germany, pp. 375–380.
Nef, L. (1985) Relation entre certaines caractéristiques des peupliers et l'abondance d'insectes phytophages. *Zeitschrift fur Angewandte Entomologie* 99, 160–170.
Nef, L. (1988a) Etude quantitative de populations de *Phyllocnistis suffusella* Z. (Lep., Lithocolletidae) en Belgique. *Annales des Sciences Forestières* 45, 255–263.
Nef, L. (1988b) Interactions between the leaf miner *Phyllocnistis suffusella* and poplars. In: Mattson, W., Levieux, J. and Bernard-Dagan, C. (eds) *Mechanisms of Woody Plant Defenses Against Insects*. Springer Verlag, New York, USA pp. 239–251.

Nef, L. (2007) Stratégies de défenses et de compétition de peupliers forestiers et pionniers. *Revue Forestière Française* 54, 119–128.
Nef, L. and Janssens, F. (1982) *Les principaux insectes nuisibles au peuplier en Belgique et en Europe occidentale.* C.L.I.I.F., Université Catholique de Louvain, Louvain-la-Neuve, Belgium.
Nord, J.C. and Knight, F.B. (1972) The importance of *Saperda inornata* and *Oberea schaumii* (Coleoptera: Cerambycidae) galleries as infection courts of *Hypoxylon pruinatum* in trembling aspen, *Populus tremuloides. Great Lakes Entomologist* 5, 87–92.
Nord, J.C., Grimble, D.G. and Knight, F.B. (1972) Biology of *Saperda inornata [S. concolor]* (Coleoptera: Cerambycidae) in trembling aspen, *Populus tremuloides. Annals of the Entomological Society of America* 65, 127–135.
Normark, B.B. (2000) Molecular systematics and evolution of the aphid family Lachnidae. *Molecular Phylogenetics and Evolution* 14, 131–140.
Novak, V., Hrozinka, F. and Stary, B. (1976) *Atlas of Insects Harmful to Forest Trees, Volume I.* Elsevier Scientific Publishing Company, Amsterdam, the Netherlands.
Nowak, D.J., Pasek, J.E., Sequeira, R.A., Crane, D.E. and Mastro, V.C. (2001) Potential effect of *Anoplophora glabripennis* (Coleoptera: Cerambycidae) on urban trees in the United States. *Journal of Economic Entomology* 94, 116–122.
Nuttall, M.J. (1989) *Sirex noctilio* F., sirex wood wasp (Hymenoptera: Siricidae). In: Cameron, P.J., Hill, R.L., Bain, J. and Thomas, W.P. (eds) *A Review of Biological Control of Insect Pests and Weeds in New Zealand 1874 to 1987.* Technical communication 10. CAB International Institute of Biological Control 10. CABI, Wallingford, UK, pp. 299–306.
Ogura, N., Tabata, K. and Wang, W. (1999) Rearing of the colydiid beetle predator, *Dastarcus helophoroides,* on artificial diet. *Biological Control* 44, 291–299.
Orians, C. and Floyd, T. (1997) The susceptibility of parental and hybrid willows to plant enemies under contrasting soil nutrient conditions. *Oecologia* 109, 407–413.
Ostry, M.E. and Anderson, N.A. (1995) Infection of *Populus tremuloides* by *Hypoxylon mammatum* ascospores through *Saperda inornata* galls. *Canadian Journal of Forest Research* 25, 813–816.
Ostry, M.E., Wilson, L.F., McNabb, H.S. and Moore, L.M. (1989) *A Guide to Insect, Disease, and Animal Pests of Poplars.* Agricultural Handbook 677. US Department of Agriculture, Washington, DC, USA.
Ovruski, S. and Fidalgo, P. (1991) Distribucion geografica de *Nematus desantisi* Smith (Hymenoptera: Tenthredinidae), plaga de salicaceas. *Ciência e Cultura* (Supplement) 43, 36–37.
Palma, M.A. and San Martín, J. (2004) Producción de carpóforos de Cerrena sp. II Fase Informe No 2. Laboratorio de Fitopatología del Servicio Agrícola y Ganadero, V Región, Valparaiso, Chile.
Palma, M.A., Valenzuela, E., Parra, P., Gutierrez, M. and Torelli, L. (2005) *Cerrena unicolor* (Bull.) Murr. (Basidiomycota) aislado de micangio de *Tremex fuscicornis* Fabr. (Hymenoptera Siricidae) asociado a decaimiento y pudrición del álamo (*Populus* sp) en Chile. *Boletín micológico* 20, 57–61.
Pandey, A.K., Namgyal, D., Mir, M.S. and Ahmed, S.B. (2007) Major insect pest associated with forest plantations in cold arid region, Ladakh of Jammu and Kashmir. *Journal of Entomological Research* 31, 155–162.
Parra, P., González, M., Soto, D. and Salinas, A. (2005) La avispa taladradora de la madera *Tremex fuscicornis* (Fabr.) [monografías]. Informativo Sanitario Forestal No 4. INFOR, Subgerencia de Tecnologías Silvícolas, Santiago, Chile.
Parra, S.P. and González, G.M. (2007) *Megarhyssa praecellens,* parasitoide de la avispa de la madera *Tremex fuscicornis* en Chile. *Ciencia e Investigación Forestale* 13, 473–490.
Parra, S.P. and González, G.M. (2011) Eficacia de tratamientos con árboles y trozas cebo para la detección y control de *Tremex fuscicornis* Fabr., en la zona central de Chile. Jornadas de Salicáceas 2011. Tercer Congreso Internacional de Salicaceas en Argentina (http://64.76.123.202/new/0-0/forestacion/salicaceas/jornadas%20salicaceas%202011/Actas/Trabajos_completos_Formato%20pdf/Parra_Patricio_1_TC.pdf, accessed 2 February 2012).
Parra, S.P., González, G.M., Soto, A.D. and Salinas, R.A. (2007) Estrategia de Manejo integrado de *Tremex fuscicornis* (Fabr.) en Chile. Informativo Sanitario Forestal INFOR No 5. Instituto Forestal, Santiago, Chile.
Parry, D., Spence, J.R. and Volney, W.J.A. (1997a) Responses of natural enemies to experimentally increased populations of the forest tent caterpillar, *Malacosoma disstria. Ecological Entomology* 22, 97–108.
Parry, D., Volney, W.J.A. and Currie, C.R. (1997b) The relationship between trembling aspen phenology and larval development of the large aspen tortrix. Information Report NOR-X-350. Natural Resources Canada, Canadian Forest Service, Northern Forestry Centre, Edmonton, Alberta, Canada.

Pasqualini, E. and Natale, D. (1999) *Zeuzera pyrina* and *Cossus cossus* (Lepidoptera: Cossidae) control by pheromones: four years advances in Italy. *International Organisation for Biological Control of Noxious Animals and Plants, West Palaearctic Regional Section Bulletin* 22, 115–124.
Pasteels, J.M., Rowell-Rahier, M., Braekman, J.C. and Dupont, A. (1983) Salicin from host plant as precursor of salicylaldehyde in defensive secretion of Chysomeline larvae. *Physiological Entomology* 8, 307–314.
Pasteels, J.M., Rowell-Rahier, M., Braekman, J.C. and Daloze, D. (1984) Chemical defences in leaf beetles and their larvae: the ecological, evolutionary and taxonomic significance. *Biochemical Systematics and Ecology* 12, 395–406.
Payne, T., Woessner, R. and Mastro, V. (1972) Evaluation of cottonwood clonal selections for resistance to cottonwood twig borer attack. *Journal of Economic Entomology* 65, 1178–1179.
Peterson, L.O.T. (1947) Some aspects of poplar borer, *Saperda calcarata* Say, (Cerambycidae) infestations under parkbelt conditions. *Annual Report of the Entomological Society of Ontario* 78, 56–61.
Philippe, R.N. and Bohlmann, J. (2007) Poplar defense against insect herbivores. *Canadian Journal of Botany* 85, 1111–1126.
Pollock, M.M., Heim, M. and Werner, D. (2003) Hydrologic and geomorphic effects of beaver dams and their influence on fishes. *American Fisheries Society Symposium* 37, 1–17
Ponsen, M.B. (1982) The digestive system of *Phloeomyzus passerinii* (Signoret) (Homoptera: Aphidoidae). *Mededelingen van de Landbouw-hogeschool Wageningen* 10, 1–6.
Postner Von, M. (1978) Familienreiche Aegerioidea. In: Schwenke, W. (ed.) 1972–1986. *Die Forstschädlinge Europas. 3 Band, Schmetterlinge.* Paul Parey, Hamburg, Germany, pp. 188–205.
Prentice, R.M. (1955) The life history and some aspects of the ecology of the large aspen tortrix, *Choristoneura conflictana* (Wlkr.) (N.Comb.) (Lepidoptera: Tortricidae). *Canadian Entomologist* 87, 461–473.
Price, P.W. (2003) *Macroevolutionary Theory on Macroecological Patterns.* Cambridge University Press, Cambridge, UK.
Price, P.W., Ohgushi, T., Roininen, H., Ishihara, M., Craig, T.P., Tahvanainen, J., *et al.* (2004) Release of phylogenetic constraints through low resource heterogeneity: the case of gall-inducing sawflies. *Ecological Entomology* 29, 467–481.
Price, P.W., Roininen, H. and Ohgushi, T. (2005) Adaptive radiation into ecological niches with eruptive dynamics: a comparison of tenthredinid and diprionid sawflies. *Journal of Animal Ecology* 74, 369–377.
Pulkkinen, M. and Yang, Z. (1984) The parasitoids and predators of *Saperda populnea* (Linnaeus) (Coleoptera, Cerambycidae) in Finland. *Annales Entomologici Fennici* 50, 7–12.
Rank, N.E. (1994) Host plant effects on larval survival in a salicin-using leaf beetle *Chrysomela aeneicollis* Schaeffer (Coleoptera: Chrysomelidae). *Oecologia* 97, 342–353.
Rappaport, N., Mori, S. and Roques, A. (1993) Estimating effect of *Megastigmus spermotrophus* (Hymenoptera Torymidae) on Douglas-fir seed production: the new paradigm. *Journal of Economic Entomology* 86, 845–849.
Raspi, A. (1996) *Thaumatomyia elongatula* (Becker) (Chloropidae) and *Leucopis annulipes* Zetterstedt (Chamaemyiidae): two Diptera preying on *Phloeomyzus passerinii* (Signoret) (Homoptera: Phloeomyzidae) in Italy. *Proceedings of the Entomological Society of Washington* 98, 509–516.
Ratzeburg, T.C. (1839) *Die Forst-Insecten.* Nicolaiaschen, Berlin, Germany.
Reichenbacker, R.R., Schultz, R.C. and Hart, E.R. (1996) Impact of artificial defoliation on *Populus* growth, biomass production, and total non-structural carbohydrate concentration. *Environmental Entomology* 25, 632–642.
Robison, D.J., McCown, B.H. and Raffa, K.F. (1994) Responses of gypsy moth (Lepidoptera: Lymantridae) and forest tent caterpillar (Lepidoptera: Lasiocampidae) to transgenic poplar *Populus* spp., containing a *Bacillus thuringiensis* delta-endotoxin gene. *Environmental Entomology* 23, 1030–1041.
Roininen, H., Price, P.W. and Bryant, J.P. (1997) Response of galling insects to natural browsing by mammals in Alaska. *Oikos* 80, 481–486.
Roques, A. (1981) Biologie et répartition de *Megastigmus spermotrophus* Wachtl (Hymenoptera: Chalcidoidea Torymidae) et des autres insectes liés aux cônes dans les peuplements forestières et vergers à graines français de sapin de Douglas *Pseudotsuga menziesii* [Mirb]. Franco. *Acta Oecolica-Oecologia Applicata* 2, 161–180.
Roques, A. (1998a) Les Curculionidae. In: Delplanque A. (ed.) *Les insectes associés aux peupliers.* Éditions Memor, Bruxelles, Belgium, pp. 47–64.
Roques, A. (1998b) Les Cerambicidae ou longicornes. In: Delplanque A. (ed.) *Les insectes associés aux peupliers.* Éditions Memor, Bruxelles, Belgium, pp. 27–37.

Rose, A.H. and Lindquist, O.H. (1982) *Insectes des feuillus de l'est du Canada.* Environnemment Canada, Service canadien des forêts, Ottawa, Canada.
Sadeghi, S.E., Tarasi, J. and Askary, H. (2002) Resistance and susceptibility of eleven poplar clones to poplar woolly aphid; *Phloeomyzus passerinii* Signoret in Zanjan province of Iran. *Applied Entomology and Phytopathology* 69, 57–67.
Sadeghi, S.E., Rajabi-Mazhar, N.A. and Moharramipour, S. (2007) A study on the incidence of woolly poplar aphid, *Phloemyzus passerinii* (Hom: Aphididae) on poplar species and clones in Hamedan province, Iran. *Journal of Entomological Society of Iran* 26, 47–59.
SAGPyA (Secretaría de Agricultura, Pesca y Alimentación) (2003) Implementación de control sanitario de los embalajes de madera. Resolución 626/03. SAGPyA, Buenos Aires, Argentina.
Santoro, F.H. (1957) Contribución al conocimiento de la biología de *Platypus sulcatus* Chapuis. *Revista de Investigaciones Forestales* 1(3), 7–19.
Santoro, F.H. (1962) Fundamentos para el control manual de *Platypus sulcatus. Revista de Investigaciones Forestales* 3(1), 17–23.
Santoro, F.H. (1963) Bioecología de *P. sulcatus* Chapuis (Col. Platypodidae). *Revista de Investigaciones Forestales* 4(1), 47–79.
Santoro, F.H. (1965) Descripción de 5 estadios larvales y de la pupa de *Platypus sulcatus. IDIA, Suplemento Forestal* 16, 49–58.
Santoro, F.H. (1967) Nuevo antecedente sobre la lucha manual contra *Platypus sulcatus* Chapuis. *IDIA, Suplemento Forestal* 4, 70–74.
Sapio, F.J., Wilson, L.F. and Ostry, M.E. (1982) A split-stem lesion on young *Populus* trees caused by the tarnished plant bug, *Lygus lineolaris. Great Lakes Entomologist* 15, 237–246.
Sawyer, A.J. (2003) Annotated categorization of ALB host trees. (Revised February 2010) USDA-APHIS-PPQ, Otis Plant Protection Laboratory (http://www.uvm.edu/albeetle/hosts.htm, accessed 2 February 2012).
Schmiedeknecht, O. (1914) *Opuscula Ichneumonologica.* Blankenburg, Thüringen, Germany.
Schowalter, T.D. (2000) *Insect Ecology. An Ecosystem Approach.* Academic Press, Waltham, Massachusetts.
Schvester, D. and Bianchi, H. (1961) Sur une méthode de lutte contre le charançon de la patience (*Cryptorrhynchus lapathi* L., Col. Curculionidae) dans les oseraies. *Procès-verbal de l'Académie d'Agriculture de France*, pp. 494–497.
Schwenke, W. (ed.) (1972–1986) *Die Forstschädlinge Europas. Ein Handbuch in 5 Bänden* [Forest Insects of Europe; a handbook in 5 volumes; in German]. Volume 1: *Würmer, Schnecken, Spinnentiere, Tausendfüßler und hemimetabole Insekten.* 1972. Volume 2: *Käfer.* 1974. Volume 3: *Schmetterlinge.* 1978. Volume 4: *Hautflügler und Zweiflügler.* 1982. Volume 5: *Wirbeltiere.* 1986. Paul Parey, Hamburg, Germany.
Sekendiz, O. (1973) The damage caused by *Agrilus ater* in the Marmara region and central Anatolia. *Istanbul Universitesi Orman Fakultesi Dergisi* 23, 128–137.
Servicio Agrícola y Ganadero (2000) Plan de contingencia para la detección y control de la avispa de la madera de las latifoliadas *Tremex fuscicornis* Fabricius (Hymenoptera: Siricidae). Servicio Agrícola y Ganadero, Santiago, Chile.
Servizio fitosanitario regionale (2011) *Megaplatypus mutatus* (http://www.agricoltura.regione.campania.it/difesa/platypus.htm, accessed 14 February 2012).
Shojai, M. and Lotfian, H. (1988) Resistance of Iranian poplar clones to woolly Aphid *Phloeomyzus passerinii* Sign., Homoptera, Aphididae. FO:CIP:I/88/12. Paper presented to 18th Session, International Poplar Commission, Beijing. Available from IPC Secretariat, FAO, Rome, Italy.
Shojai, M. and Lotfian, H. (1990) Studies about the resistance of poplar species, cultivars and clones to *Phloeomyzus passerinii* Sign. (Hom. Aphididae). FO:CIP:I/90/6. Working paper presented at 35th Session of the Executive Committee, International Poplar Commission, Buenos Aires. Available from IPC Secretariat, FAO, Rome, Italy.
Shojai, M., Lotfian, H., Nasrollahi, A., Yassini, A., Azma, M. and Gasmi, R. (1998) Bioecology of poplar wooly aphid *Phloeomyzus passerini* Sign (Hom.: Aphididae) and it's host resistance conditions in poplar plantations of Iran. *Journal of Agricultural Sciences, Islamic Azad University* 4, 13–14, 21–49, 83–85.
Singh, A.P. and Pandey, R. (2002) Natural resistance in *Populus deltoides* clones selection to defoliator *Clostera cupreata* in northern India: relative pupal weight as an easier criteria for accurate evaluation. *Journal of Applied Entomology* 129, 475–480.
Singh, A.P., Bhandari, R.S. and Verma, T.D. (2004) Important insect pests of poplars in agroforestry and strategies for their management in northwestern India. *Agroforestry Systems* 63, 15–26.
Singh, P. and Sharma, K.K. (1979a) External morphology of the pupae of poplar defoliators *Pygaera fulgurita* Walker and *P. cupreata* Butler (Lepidoptera: Notodontidae). *Entomon* 4, 81–88.

Singh, P. and Sharma, K.K. (1979b) Genitalia of *Pygaera fulgurita* Walker and *Pygaera cupreata* Butler (Lepidoptera: Notodontidae). *Entomon* 4, 255–261.
Singh, P. and Singh, S. (1986) *Insect Pests and Diseases of Poplars*. Forest Research Institute Press, Dehradun, India.
Singh, P., Rawat, D.S., Mishra, R.M., Fasih, M., Prasad, G. and Tyagi, B.D.S. (1983) Epidemic defoliation of polar and its control in Tarai Central Forest Division, Uttar Pradesh. *Indian Forester* 109, 675–693.
Smereka, E.P. (1965) The life history and habits of *Chrysomela crotchi* Brown (Coleoptera: Chrysomelidae) in Northwestern Ontario. *Canadian Entomologist* 97, 541–549.
Smiley, J.T., Rank, N.E. and Horn, J.M. (1985) Ecological effects of salicin at three trophic levels: new problems from old adaptations. *Science* 229, 649–651.
Smith, B.D. and Stott, K.G. (1964) The life history and behaviour of the willow weevil *Cryptorrhynchus lapathi* L. *Annals of Applied Biology* 54, 141–151.
Smith, C.F. (1974) The genus *Phloeomyzus* with the description of *P. dearborni* n. sp. from *Populus tremuloides* Michx. (Homoptera; Aphididae). *Proceedings of the Entomological Society of Washington* 76, 66–72.
Smith, D.R. (1978) Pars 14, Suborder Symphyta. In: *Hymenopterum Catalogus* (*Nova editio*). Dr. W. Junk, B.V., The Hague, Netherlands.
Smith, D.R. (1983) The first record of *Nematus* Panzer from South America: a new species from Argentina (Hymenoptera: Tenthredinidae). *Proceedings of the Entomological Society of Washington* 85, 260–262.
Smith, M.T., Bancroft, J. and Li, G.H. (2001) Dispersal of *Anoplophora glabripennis* (Cerambycidae). *Environmental Entomology* 30, 1036–1040.
Soetens, P., Pasteels, J.M., Daloze, D. and Kaisin, M. (1998) Host plant influence on the composition of the defensive secretion of *Chrysomela vigintipunctata* larvae (Coleoptera: Chrysomelidae). *Biochemical Systematics and Ecology* 26, 703–712.
Solomon, J.D. (1995) *Guide to Insect Borers of North American Broadleaf Trees and Shrubs*. Agricultural Handbook 706. US Department of Agriculture, Forest Service, Washington, DC, USA.
Solter, L.F., Keena, M., Cate, J.R., McManus, M.L. and Hanks, L.M. (2001) Infectivity of four species of nematodes (Rhabditoidea: Steinernematidae, Heterorhabditidae) to the Asian longhorned beetle, *Anoplophora glabripennis* (Motschulsky) (Coleoptera: Cerambycidae). *Biocontrol Science and Technology* 11, 547–552.
Southdene Sadirian Berhad (2011) The moths of Borneo (http://www.mothsofborneo.com/part-4/clostera/notodontidae_63_1.php, accessed 17 February 2012).
Speight, M.R. (1986) Tree pests – 13. Goat and clear wing moths. (Lepidoptera: Cossidae and Sesiidae). *Arboricultural Journal* 10, 113–116.
Strauss, S.H., DiFazio, S.P. and Meilan, R. (2001) Genetically modified poplars in context. *The Forestry Chronicle* 77, 271–279.
Straw, N.A., Green, G. and Williams, D.T. (2007) Dieback and recovery in poplar and attack by the hornet clearwing moth, *Sesia apiformis* (Clerck) (Lepidoptera: Sesiidae). *Bulletin of Entomological Research* 97, 555–567.
Strojny, W. (1954) Szkodniki drewna drzew szybko przyrastajacych-Czesc II. Krytoryjek olszowiec. *Cryptorrhynchus lapathi* L. (Coleoptera Curculionidae). *Polski Pismo Entomologiczne* 24, 71–131.
Strong, D.R. and Larsson, S. (1994) Is the evolution of herbivore resistance influenced by parasitoids? In: Hawkins, B.A. and Sheehan, W. (eds) *Parasitoid Community Ecology*. Oxford University Press, Oxford, UK, pp. 261–276.
Strong, D.R, Larsson, S. and Gullberg, U. (1993) Heritability of host plant resistance to herbivory changes with gall midge density during an outbreak on willow. *Evolution* 47, 291–300.
Su, H.J., Zhao, J., You, D., Chang, G., Chai, S. and Qu, T. (2004) Evaluation of economic losses caused by forest pests disasters in China. *Forest Pest and Disease* 23(5), 1–6. (In Chinese)
Sun, J.Z., Zhao, Z.Y., Ru, T.Q., Qiang, Z.G. and Song, X.J. (1990) Control of *Anoplophora glabripennis* by using cultural methods. *Forest Pest and Disease* 2, 10–12.
Sweetman, H. (1936) *The Biological Control of Insects*. Comstock Publishers, Ithaca, New York, USA.
Szalay-Marzsò, L. (1962) Zur Morphologie, Biologie und Bekämpfung des Erlenwürgers *Cryptorrhynchus lapathi* L. (Col. Curculionidae) in Ungarn. *Zeitschrift für Angewandte Entomologie* 49, 163–194.
Szontagh, P. (1981) Phytopathological evaluation of poplar stands irrigated with waste water. *Erdeszeti Kutatasok* 74, 369–377.
Tabashnik, B.E., Carrière, Y., Dennehy, T.J., Morin, S., Sisterson, M., Roush, R.T., *et al.* (2003) Insect resistance to transgenic Bt crops: lessons from the laboratory and the field. *Journal of Economic Entomology* 96, 1031–1038.

Takahashi, N. and Ito, M. (2005) Detection and eradication of the Asian longhorned beetle in Yokohama, Japan. *Research Bulletin of the Plant Protection Service* 41, 83–85. (In Japanese)

Templado, J. (1964) *Paranthrene tabaniformis* Rott., importante plaga del chopo, *Publicaciones del Instituto de Biologia Aplicada, Barcelona* 37, 33–52.

Theobald, F.V. (1929) Genus *Phloeomyzus* Horváth. *The Plant Lice or Aphididae of Great Britain.* Volume 3. Headley Brothers, Ashford, UK, pp. 268–272.

Thomas, E. (2005) Monitoreo de *Megaplatypus mutatus* 'Taladrillo de los Forestales' en los Departamentos Confluencia y Añelo de la Provincia del Neuquén. Report 2004/5. Provincia del Neuquen, Ministerio de Producción y Turismo, Subsecretaría de Producción, Delegación Forestal Neuquén, Argentina.

Thomas, E. (2011) Biología y manejo integrado de *Megaplatypus mutatus* en los valles del norte de la Patagonia. 3° Congreso Internacional de Salicáceas, Disertaciones. Neuquén, Argentina (http://64.76.123.202/new/0-0/forestacion/salicaceas/jornadas%20salicaceas%202011/Actas/Trabajos_completos_Formato%20pdf/Thomas_E_D.pdf, accessed 2 February 2012).

Thomas, J.B. (1978) A review of the economic impact of insects on the genus *Populus* in Ontario. Information Report O-X-271. Environment Canada, Canadian Forest Service, Great Lakes Forestry Centre, Sault Ste Marie, Ontario, Canada.

Tittle, S.A.W. (1972) Suborder Aphidina. In: Schwenke, W. (ed.) *Die Forstschädlinge Europas. Ein Handbuch in 5 Bänden. Bänd 1: Würmer, Schnecken, Spinnentiere, Tausendfüßler und hemimetabole Insekten.* Paul Parey, Hamburg, pp. 162–369. (In German)

Tomescu, R. and Nef, L. (2007) Leaf eating insect damage to poplar clones. *Annals of Forest Science* 64, 99–108.

Tomiczek, C., Krehan, H. and Menschhorn, P. (2002) Dangerous Asiatic longicorn beetle found in Austria: new danger for our trees? *Allgemeine Forst Zeitschrift fur Waldwirtschaft und Umweltvorsorge* 57, 52–54.

Tomilova, V.N. and Kusnetsova, N.P. (1975) The mining chrysomelid of poplar. *Zashchita Rastenii* 9, 45.

Torgersen, T.R. and Beckwith, R.C. (1974) Parasitoids associated with the aspen tortrix, *Choristoneura conflictana* (Lepidoptera: Tortricidae), in interior Alaska. *Canadian Entomologist* 106, 1247–1265.

Toscani, H. (1991) Manual para la protección de los cultivos forestales en la región del Delta del Paraná. FO:CIP:I/90/1. Working paper presented at 35th Session of the Executive Committee, International Poplar Commission, Buenos Aires. Available from IPC Secretariat, FAO, Rome, Italy.

Townsend Peterson, A., Scachetti Pereira, R. and Hargrove, W.W. (2004) Potential geographic distribution of *Anoplophora glabripennis* (Coleoptera: Cerambycidae) in North America. *American Midland Naturalist* 151, 170–178.

Tremblay, E., Espinosa, B., Manzini, D. and Caprio, G. (2000) Un coleottero proveniente dal Sudamérica minaccia i pioppi. *L'Informatore Agrario* 48, 89–90.

Urban, A.J. and Eardley, C.D. (1995) A recently introduced sawfly, *Nematus oligospilus* Förster (Hymenoptera: Tenthredinidae), that defoliates willows in southern Africa. *African Entomology* 3, 23–27.

USDA-APHIS (USDA Animal Plant Health Inspection Service) (2006) The Asian longhorned beetle cooperative eradication program. ALB Newsletter 1 (http://www.aphis.usda.gov/plant_health/plant_pest_info/asian_lhb/alb_pdfs/newsletters/alb_newsletter_9_06.pdf, accessed 2 February 2012).

Välimäki, S. and Heliövaara, K. (2007) Hybrid aspen is not preferred by the large poplar borer (*Saperda carcharias*). *Arthropod–Plant Interactions* 1, 205–211.

Vidotti, G. (1960) Osservazioni sull'etologia ed ecologia del *Phloeomyzus passerinii* Sign. nella zona di Este (Hemiptera Aphididae Aphidinae). *Atti del Istituto Veneto di Scienze, Lettere ed Arti* 118, 203–218.

Viitasaari, M. (1984) Siricoidea, Orussoidea ja Cephoidea. Reports 6, Sahapistiäiset 3. University of Helsinki. Department of Agricultural and Forest Zoology (http://www.funet.fi/pub/sci/bio/life/insecta/hymenoptera/symphyta/siricoidea/siricidae/tremicinae/tremex/index.html, accessed 2 February 2012).

Wagner, D., Defoliart, L., Doak, P. and Schneiderheinze, J. (2008) Impact of epidermal leaf-mining by the aspen leaf miner (*Phyllocnistis populiella*) on the growth, physiology, and leaf longevity of quaking aspen. *Oecologia.* 157, 259–267.

Walther, G.-R., Roques, A., Hulme, P., Sykes, M.T., Pyšek, P., Kuhn, I., *et al.* (2009) Alien species in a warmer world: risks and opportunities. *Tree* 24, 686–693.

Wang, W.D., Liu, Y.N., Bao, S., Ogura, N. and Maruda, H. (1999) Research of the enemy of *Anoplophora glabripennis* and *A. nobilis* in Ningxia. *Journal of Beijing Forestry University* 21, 90–93. (In Chinese)

Wenes, A.-L., Bourguet, D., Andow, D.A., Courtin, C., Carré, G., Lorme, P., *et al.* (2006) Frequency and fitness cost of resistance to *Bacillus thuringiensis* in *Chrysomela tremulae* (Coleoptera: Chrysomelidae). *Heredity* 97, 127–134.

Williams, D.W., Lee, H. and Kim, I. (2004) Distribution and abundance of *Anoplophora glabripennis* (Coleoptera: Cerambycidae) in natural *Acer* stands in South Korea. *Environmental Entomology* 33, 540–545.

Wilson, L.F. and Moore, L.M. (1986) Preference for some nursery-grown hybrid *Populus* trees by the spotted poplar aphid and its suppression by insecticidal soaps. *Great Lakes Entomologist* 19, 21–26.

Woessner, R. and Payne, T. (1971) An assessment of cottonwood twig-borer attacks. In: *Proceedings, 11th Southern Forest Tree Improvement Conference*, June 1971, Atlanta, GA, USA. Southern Forest Tree Improvement Committee Sponsored Publication 33. Eastern Tree Seed Laboratory, Macon, Georgia, USA, pp. 98–107.

Wong, H.R. (1979) Biological observations on overwintering larvae of the large aspen tortrix in Alberta. Environment Canada, Forest Service. *Bi-monthly Research Notes* 35, 21.

Wu, P.H., Li, Z.Y. and Wei, K.N. (1987) Studies on the biological characteristics and sex pheromones utilized for the control of the poplar twig clearwig moth. *Scientia Sylvae Sinicae* 23, 491–497.

Wu, X.Q., Zheng, M.C., Wang, Y., Ye, J.R. and Qian, F.J. (2006) Integrated pest management strategy to ecological forest in Shanghai. *Journal of Nanjing Forestry University* (*Natural Sciences Edition*) 30, 109–112.

Yali, M.P., Sadeghi, S.E., Moharramipour, S., Zeinali, S. and Ali, B. (2006) Comparing pre-mature and post mature periods, longevity and fecundity of poplar leaf aphid, *Chaitophorus leucomelas* (Hom: Aphididae) on poplar clones in laboratory conditions. *Iranian Journal of Forest and Range Protection Research* 4, 1–6.

Yan, J.J. and Yan, Y.H. (1999) The study on the model of ecological control of longhorned beetle. *Journal of Agricultural University of Hebei* 22(4), 83–87. (In Chinese)

Yang, Z.M., Wang, X.N., Yao, W.S., Chu, X.M. and Li, P. (2000) Generation differentiation and effective accumulated temperature of *Anoplophora glabripennis* (Motsch.). *Forest Pest and Disease* 19(6), 12–14. (In Chinese)

Yang, Z.Q. (1984) Notes on the larva and puparium of *Odinia xanthocera* Collin (Diptera, Odiniidae). *Annales Entomologici Fennici* 50, 93–94.

Yukawa, J. and Rohfritsch, O. (2005) Biology and ecology of gall-inducing Cecidomyiidae (Diptera). In: Raman, A., Schaefer, C.W. and Withers, T.M. (eds) *Biology, Ecology, and Evolution of Gall-inducing Arthropods*. Science Publishers Inc, Enfield, New Hampshire, USA, pp. 273–304.

Zhang, B., Liu, Y., Bai, Y. and Shimazu, M. (1999) Pathogenic fungi of *Anoplophora* spp. (Coleoptera: Cerambycidae) in Ningxia Hui Autonomous Region and their virulence. *Journal of Beijing Forestry University* 21(4), 67–72. (In Chinese)

Zhang, G.X., Zhong, T.S. and Chang, G.S. (1982) New species and subspecies of Chinese Aphidoidea. *Sinozoologica* 2, 19–28.

Zhang, S.H., Xia, X. and Shu, H. (1994) An experimental study on the control of *Anoplophora glabripennis* (Motschulsky) with stem injection of insecticides. *Journal of Inner Mongolia Institute of Agriculture and Animal Husbandry* 15(4), 15–20. (In Chinese)

Zhang, X.Y., Qin, Q.F., He, W., Liang, J. and Liu, H.X. (2007) Cloning and sequence analysis on cDNA of cysteine proteinase inhibitor gene from *Populus tomentosa* Carr. *Journal of Beijing Forestry University* 29(6), 23–28.

Zhao, J., Ogura, N. and Isono, M. (1999) Artificial rearing of *Anoplophora glabripennis* (Motsch.) (II). *Journal of Beijing Forestry University* 21(4), 62–66. (In Chinese)

Zhao, X. and Yoshida, N. (1999) The relationship of emergence, oviposition duration of *Anoplophora* and climate factors. *Journal of Beijing Forestry University* 21(4), 52–57. (In Chinese)

Zhao, X.R., Li, G., Zhang, J., Li, Z.Z., Zhao, J.Z. and Guan, X.Q. (1995a) Study on harm behavior of *Anoplophora glabripennis* Motsch. and control effect by new type of insecticide – Mothcide II, IV. *Journal of Ningxia Agricultural College* 16(3), 24–26. (In Chinese)

Zhao, X.R., Li, G. and Li, Z.Z. (1995b) Study on selection new type of insecticide by the new technique of injecting trunk strongly for controlling adults of capricorn beetle. *Journal of Ningxia Agricultural College* 16(1), 71–73. (In Chinese)

Zocchi, R. (1951) Contributi alla conoscenza degli insetti delle foreste. I. *Cryptorhynchus lapathi* L. (Coleoptera Curculionidae). *Bolletino dell'Istituto di Entomologia della Università di Bologna* 18, 245–258.

Zvereva, E.L. and Rank, N.E. (2003) Host plant effects on parasitoid attack on the leaf beetle *Chrysomela lapponica*. *Oecologia* 135, 258–267.

Zvereva, E.L. and Rank, N.E. (2004) Fly parasitoid *Megaselia opacicornis* uses defensive secretions of the leaf beetle *Chrysomela lapponica* to locate its host. *Oecologia* 140, 516–522.

10 Properties, Processing and Utilization*

John Balatinecz,[1] Patrick Mertens,[2] Lieven De Boever,[3] Hua Yukun,[4] Juwan Jin[5] and Joris van Acker[6]

[1]North York, Ontario, Canada; [2]Département de l'étude du milieu naturel et agricole – Direction du milieu forestier, Gembloux, Belgium; [3]Belgian Institute of Wood Technology (CTIB/TCHN), Bruxelles, Belgium; [4]Wood-based Composites Institute, Nanjing Forestry University, Jiangsu, PR of China; [5]College of Wood Science and Technology, Nanjing Forestry University, Jiangsu, PR of China; [6]Ghent University, Laboratory of Wood Technology, Gent, Belgium

10.1 Introduction

The processing and utilization of both poplars and willows for different products are strongly influenced by their anatomical, physical and chemical wood properties. That is why utilization cannot be discussed without consideration of specific wood properties. In addition, the overall quality of the wood, such as the presence, size and frequency of growth-related defects, for example knots, spiral grain, tension wood, etc., the presence or absence of discoloured heartwood and decay, log size and shape also have an impact on various utilization options. These are factors which are inherent to the raw material. Finally, there are important parameters external to wood which also play a role in the utilization process.

The utilization of wood species for a variety of products is also affected by market forces, such as supply and demand, competition from other species and other materials for specific end uses. In addition, the quantity and price of the raw material also play an important role in different utilization scenarios. Increasingly, the cost and availability of energy and skilled labour, the distance from markets, technology trends and the cost and availability of capital also have an influence on utilization. These factors need to be considered when investment decisions are made about both the production and the utilization of poplars and willows.

The main objective of this chapter is to focus on the general discussion of the wood property–processing–utilization relationships for poplars and willows. But from time to time, reference is made to other factors which also have a significant bearing on utilization: for example, availability of appropriate process technology, supply and cost of adhesives, competition from other species and the 'carbon footprint' of wood and competing materials. In the past, the primary emphasis on the utilization of poplars and willows was on traditional wood and paper products (FAO, 1980). In the present chapter, new and emerging utilization scenarios, as well as challenges and impediments to expanded utilization, are addressed.

*Correspondence for this chapter should be directed to: jrichardson@on.aibn.com

(eds J.G. Isebrands and J. Richardson)

Globally, about 91% of the current poplar resources grow in natural forests and woodlands, 6% is found in plantations and 3% in agroforestry production systems. The largest proportion (97%) of natural poplar stands occurs in Canada, the Russian Federation and the USA. China accounts for about 73% of the world's total area of poplar plantations, followed by India, France, Turkey, Italy and Argentina. The largest area of natural willow forests is in the Russian Federation, followed by France, China, Italy and Croatia (Ball *et al.*, 2005).

The knowledge and experience concerning the utilization of poplars and willows are tilted heavily in favour of poplars, because the global poplar resources are far greater than the willow resources (as indicated above). Thus, most of the information presented in the chapter is based on the 'poplar experience'. But, in general, that experience can be applied to willows also, because the wood of the two genera is quite similar. However, willows are rarely available as large-size trees in sufficient volume in natural forests or plantations. So, the industrial utilization of willows remains relatively minor, except in South America. Thus, the utilization section of this chapter covers principally poplars; however, some alternatives which are unique to willows, are also presented. One final point: because of the broad nature of each subject in this chapter, only an overview of the utilization potential of poplar and willow wood is presented.

10.2 Macroscopic and Microscopic Wood Features

The anatomical features of the wood of poplars and willows are very similar, making their positive identification and separation quite challenging. As far as microscopic features are concerned, only the ray structure is slightly different between the two groups. Poplars have homocellular rays, whereas the rays are heterocellular in willows (Panshin and de Zeeuw, 1980). In general, the wood of willows has not received nearly as much scientific study as the wood of poplars. For this reason, and because of the greater economic importance of poplars, the presentation focuses on poplars, but some information on willows is also presented.

10.2.1 Macroscopic anatomy and general characteristics

The anatomy and general wood characteristics of all poplar species (including hybrids and cultivars) are essentially the same. But, there are differences between species in the texture, density, strength, wood quality traits and sometimes the colour of the wood. For example, aspens usually have a finer texture than cottonwoods, which is due to the somewhat larger pores in the latter group. The wood of fast-growing hybrid poplars often has a lower density and coarser texture than wood from trees grown in natural forests. Therefore, information about wood density and strength properties is essential for decisions concerning wood utilization. The grain is generally straight. The rays are narrow (uniseriate) and not visible to the naked eye (Panshin and de Zeeuw, 1980).

The colour of the sapwood of poplars is usually white to light yellow, merging gradually into the heartwood, which is whitish-grey to grey-brown (Fig. 10.1a). The growth rings are visible to the naked eye but not always conspicuous because of the fine uniform texture of the wood. The wood is diffuse to semi-diffuse porous (Fig. 10.1b, c and d). In some new cultivars, growth rings may be quite broad, often exceeding 2 cm and sometimes can reach 3 cm or even more. Growth rate and ring width are a reflection both of genetic improvement of individual hybrids as well as favourable environmental conditions and silvicultural treatment. When fresh, poplar wood has a characteristic disagreeable odour, but it disappears completely when the wood is dry. The absence of odour and taste make poplar wood very suitable for containers for various food products, such as fruit, vegetables and cheese.

The sapwood of willows is white to whitish grey, whereas the heartwood is most often reddish-brown, sometimes with darker streaks along the grain. The wood is diffuse to semi-diffuse porous, usually with a straight grain and fine texture (Fig. 10.2) and without characteristic odour or taste.

10.2.2 Microscopic anatomy

The volumetric composition of both poplars and willows is dominated by fibres or the supportive

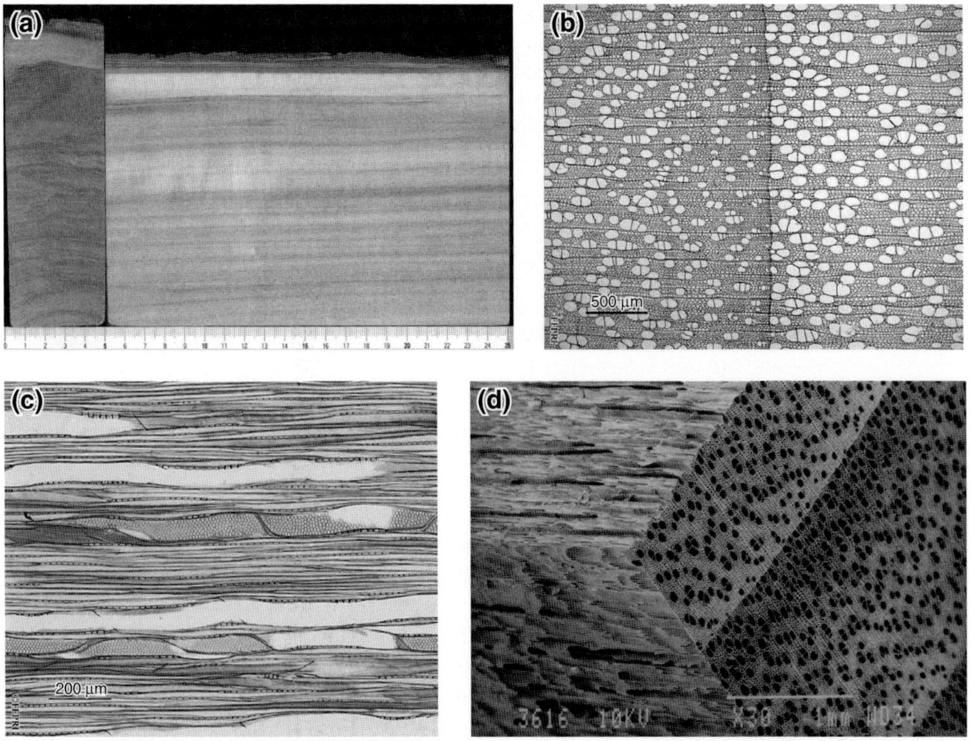

Fig. 10.1. Photograph (a) and photomicrographs (b–d) of poplar wood. Face and end grain of *Populus maximowiczii* (a). Cross (b) and tangential (c) sections of *P. maximowiczii*. Scanning electron micrograph of *P. maximowiczii* (d). Photos courtesy of the Forestry and Forest Products Research Institute of Japan.

tissue (in the range of 50–65%), followed by the vessel elements or conductive tissue (28–38%) and the parenchyma cells or storage tissue (7–15%). The horizontal parenchyma cells make up the rays, and only a small fraction (about 0.1%) is present as longitudinal or axial parenchyma (Table 10.1).

The *support tissue* of the wood in both poplars and willows is made up of fibres, giving rigidity and flexibility to the stem. The fibres are the libriform type, with relatively thin walls in normal wood, whereas in tension wood, they have a thick gelatinous layer adjacent to the cell lumen. The average fibre length (FL) in the mature wood of poplars is around 1.3 mm. Fibre diameter is in the range of 20–30 µm. On the other hand, willows usually have shorter fibres than poplars, often less than 1 mm (Table 10.1). In juvenile wood, FL is considerably shorter in both poplars and willows. It can range from 0.4 to 0.6 mm (depending on cultivars and position in the stem) during the first few years of growth and will increase gradually to the 'mature length' as the tree and the cambium age (Panshin and de Zeeuw, 1980; Hernandez *et al.*, 1998; Schoch *et al.*, 2004). The juvenile period in poplar ranges from 8 to 12 years.

The *conductive tissue* is composed of the vessel elements in both poplars and willows, which are responsible for the movement of sap in the sapwood. The vessel elements of the heartwood also contribute to tree support. The vessels, called pores when viewed in the cross section, are abundant (30–100 mm^{-2} in poplars). Their diameters are in the range of 50–150 µm, depending on their location in the growth ring. They are invisible to the naked eye. The diameter of vessels is generally larger in the cottonwood-type poplars as opposed to aspens, giving the wood somewhat coarser texture. The pore distribution is diffuse to semi-diffuse. In mature wood, the vessel cells have an average length of about 0.6 mm in poplars and 0.4 mm in willows (Table 10.1). As is the case with fibres, vessel elements are shorter

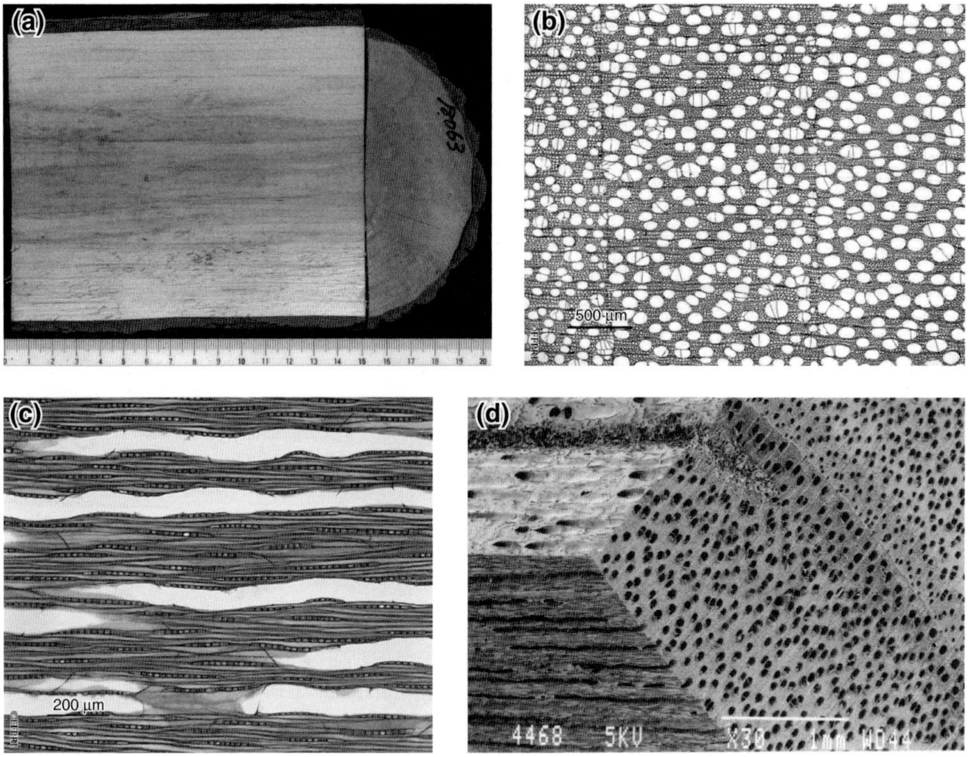

Fig. 10.2. Photograph (a) and photomicrographs (b–d) of willow wood. Face and end grain of *Salix sachalinensis* (a). Cross (b) and tangential (c) section of *Salix pierotii*. Scanning electron micrograph of *S. sachalinensis* (d). Photos courtesy of the Forestry and Forest Products Research Institute of Japan.

Table 10.1. A few important anatomical characteristics of mature wood of three North American poplar and one willow species from natural forests. (Source: Panshin and de Zeeuw, 1980.)

	Percentage of total volume				Average length (mm)	
Species	Vessels	Fibres	Rays	Axial parenchyma	Fibres	Vessel elements
Populus deltoides	33.0	53.1	13.7	0.2	1.38	0.58
Populus grandidentata	33.8	55.0	11.1	0.1	1.33	0.64
Populus tremuloides	33.2	55.3	11.4	0.1	1.35	0.63
Salix nigra	38.1	54.4	7.4	0.1	0.85	0.42

in juvenile wood than in mature wood. Their length increases gradually with the maturation of the cambium.

In cross section, the vessels may appear individually or in short radial files of two to four elements side by side or in small clusters (Figs 10.1 and 10.2, b and d). The vessel elements possess simple perforation plates. Their walls have bordered pits, crowded and arranged in staggered rows. In the cross-fields between the vessels and ray cells, the pits are simple. The lumens of some vessels may sometimes be obstructed by tyloses, which may affect chemical treatability.

The *storage tissue* is made up of ray parenchyma cells and limited axial parenchyma, usually at the end of growth rings. They are used for the storage of the various metabolic substances in the sapwood while the tree is alive. The rays are uniseriate and very thin, abundant (10–13 rays per tangential mm) and not visible to the naked eye. The rays are homocellular in poplars and heterocellular in willows. This specific feature is the main difference between the two genera in the microstructure of their wood (Panshin and de Zeuuw, 1980). The height of rays in poplars consists of between 10 and 30 stacked cells and measures between 200 and 400 μm. Ray height in willows rarely exceeds 20 cells. Sometimes, crystals of calcium oxalate, produced by parenchyma cells, may be present in poplars.

10.2.3 Growth-related defects

The term 'defect' refers to those tissue imperfections or irregularities that develop during tree growth (Panshin and de Zeeuw, 1980). Defects diminish the overall quality and often the strength of the wood, and thus also reduce the value and product suitability of logs. The main growth-related defects include knots, grain abnormalities (e.g. spiral grain, microfibril angle), tissue separations caused by growth stresses (e.g. ring shake, heart cracks) and reaction wood (e.g. tension wood in hardwoods).

A *knot* is a branch base that is embedded in the tree trunk. While the branch is living, there is a tissue connection between the stem wood and the branch wood. This type of knot is referred to as a 'tight knot'. On the radial cut surface, it appears as a slightly conical spike (hence the term 'spike knot'), whereas on the tangential surface, knots appear as round or oval. Once the branch dies, the stem continues to grow around it, until the branch falls off, which results in an 'encased' or 'loose' knot. The better alternative is when the branch is pruned off, resulting in a knot-free, high-quality trunk and logs. Knots reduce the strength of wood by causing grain deviation in the main stem. They can also cause machining defects during wood processing. When breeding poplar clones, it is important to select those cultivars which have good branching characteristics and are well disposed to early self-pruning in plantations.

Spiral grain refers to the helical orientation of longitudinal tissues in a tree stem. A slight degree of spiral grain occurs fairly widely among different species and trees, therefore some wood anatomists consider this the 'normal' grain pattern (Panshin and de Zeeuw, 1980). The problem with logs having such a grain pattern is that during sawing the fibres are cut across, resulting in cross or diagonal grain, which reduces the strength properties of wood. For example, an angle of 10° will decrease strength properties by 30%. In subsequent drying, diagonal grain induces uneven deformation in lumber, resulting in warping defects. In poplars, spiral grain seems to be inherited and cultivar dependent.

Less is known about other grain defects in poplar and willow wood, particularly in relation to microfibril angle. This will become an issue in machine stress grading. Knots are responsible for local grain deviations.

The formation of *tension wood* in poplars is induced by gravitational stimuli (Isebrands and Parham, 1974; Holt and Murphy, 1978; Clair *et al.*, 2006). This was proven experimentally by Jourez *et al.* (2003) for *Populus* ×*euramericana* cv. 'Ghoy'. Badia *et al.* (2005, 2006) reported different patterns of tension wood distribution between poplar clones. They also stated that the extent of tension wood was highest at the tree base. Tension wood appears generally on the upper face of leaning stems and branches, although sometimes it can also form in upright stems due to internal stresses caused by sustained winds or uneven crown (Isebrands and Bensend, 1972). Anatomically, it is characterized by the presence of a thick gelatinous layer (also called G layer) next to the lumen in fibres. This wall layer is often partially separated from the adjacent secondary wall. The G layer is oriented nearly parallel to the fibre axis and it is composed of concentric lamellae of cellulose microfibrils without lignin (Cote *et al.*, 1969). Because of this fact, tension wood contains more cellulose and less lignin than normal wood. The lack of connection between the G layer and the S-2 layer of the secondary wall, together with increased microfibrillar orientation in the S-2, causes increased longitudinal shrinkage in tension wood (Arganbright *et al.*, 1970). This leads to warping in lumber. On lumber and veneer, tension wood causes a woolly surface, which presents problems in drying, sanding and finishing.

The amount of tension wood fibres in a cross section can vary by as much as 22–63% (Kroll *et al.*, 1992; Leclercq, 1997; De Boever *et al.*, 2007). De Boever *et al.* (2007) also proved that clones with spatially more diffuse tension wood showed lower amounts of waviness after veneer drying compared to other clones. In the selection process, attention should be given to the amount of tension wood that is formed, but also more specifically to the spatial pattern (aggregated or diffuse).

Growth stresses are unique internal stresses in the growing tree, caused mainly by the deposition of new xylem and phloem tissues over the existing stem. If large enough to exceed the strength of the wood, growth stresses may cause tissue separations, for example heart splits, cracks and ring shakes, when the tree is cut or converted into lumber. Apart from tissue separations, growth stresses may also cause form defects, such as bowing, crooking and twisting, at the time of sawing. Growth stresses tend to be released and equilibrate somewhat during log storage following the harvesting of trees, especially when the logs are kept moist by water spray. Often, a period of storage for a few months is sufficient to prevent the development of defects which may be caused by growth stresses.

10.3 Physical, Mechanical and Chemical Properties and Natural Durability in Relation to Wood Quality

The physical, mechanical and chemical properties are the main wood characteristics which strongly influence the suitability of the material for various products and applications, because they, in turn, determine the properties of the finished product. Thus, wood property information is important to wood processors and end-users. In addition, tree growers also have an interest, since some wood characteristics, for example density and chemical composition, may be influenced during active tree growth, both through genetic selection as well as silvicultural treatments. The possibility of genetic improvement is especially important for poplars and willows, which are generally fast-growing species and are regarded as vital potential fibre sources for the future. Average wood properties are typical for a species, and they can be quantified. However, attention should be given to the variability of these wood characteristics, and this variation will determine the suitability of specific willow or poplar clones for particular end uses.

10.3.1 Physical properties

In a broad sense, physical properties include those characteristics of wood which define its physical nature as a material. The most important parameters in relation to products and processing are density, moisture content (MC) and dimensional stability. To a lesser extent, appearance (such as grain, texture and colour) and thermal and electrical properties are also important (US Department of Agriculture, 1999). The discussion here is limited to density, MC and dimensional changes.

Density

Poplars and willows are ranked among the species with low specific gravity (Table 10.2). The terms 'specific gravity' and 'density' are often used interchangeably. Density is the ratio of the wood's mass to its volume, expressed as kg m^{-3} (or g cm^{-3}). The MC basis of density needs to be stated because both the mass (weight) and the volume of wood vary with MC. To standardize species comparisons, wood science and industry often use specific gravity (which is essentially a density index). The traditional definition of specific gravity is the ratio of the density of wood to the density of water, an index without units. Commonly used bases for expressing specific gravity of wood are oven-dry (OD) weight and volume at: (i) green; (ii) OD; and (iii) 12% (or air-dry) MC. Specific-gravity values at OD weight and green volume are most often used in databases (Alden, 1995; US Department of Agriculture, 1999; Monteoliva *et al.*, 2005). The density of the wood substance, i.e. cell wall material, is about 1.5 g cm^{-3}, which is similar for all species. Thus, density and/or specific gravity of wood give an estimation of the ratio of cell walls, i.e. wood substance, to cell lumens for a particular species. Density is important because of its strong positive correlation with strength, elasticity, hardness and pulp yield. Dimensional changes, for example shrinkage and swelling,

Table 10.2. Selected physical properties of mature wood of North American poplar and willow species from natural forests in comparison with two common softwood species. (Source: US Department of Agriculture, 1999.)

Species	Specific gravity (OD weight per green volume)	Shrinkage (green to OD) (%)			Side hardness (N) Air dry (12% MC)
		Radial	Tangential	Volumetric	
Populus deltoides	0.37	3.9	9.2	13.9	1900
Populus grandidentata	0.36	3.3	7.9	11.8	1870
Populus tremuloides	0.35	3.5	6.7	11.5	1600
Populus trichocarpa	0.31	3.6	8.6	12.4	1600
Salix nigra	0.36	3.3	8.7	13.9	–
Picea glauca	0.37	3.8	7.8	11.8	2100
Pinus banksiana	0.40	3.7	6.6	10.3	2500

Note: Hardness is a mechanical property. OD, Oven dry; MC, moisture content.

are also influenced by density. For tree improvement, it is important to know that density is under partial genetic control.

Density of poplar wood has a wide range between taxonomic sections and between clones. The Italian clone 'I-214' shows low density values (300 kg m^{-3}), while some 'Euramerican' clones (*Populus deltoides* × *Populus nigra*) such as 'Robusta' show high average density values of over 550 kg m^{-3}. The latter have the same density range as aspens. In addition, the variability within the poplar stem is significant. Ranges of 200 kg m^{-3} within one log are not exceptional. The large variability in density also implies significant differences in mechanical properties. Structural applications need a well-defined grading system.

It can be generally stated that the wood of poplar and willow is similar to that of softwoods, as far as density is concerned (Table 10.2). This comparison is based mainly on their potential for structural applications. Poplar and willow wood show high strength values in relation to their limited density. The higher-density poplars also show similar potential to, for example, birch or beech.

Moisture content (MC)

Wood is a hygroscopic material with a strong affinity for water. Moisture exists in wood in two forms and in two principal locations: (i) free water in cell cavities; and (ii) bound or adsorbed water in cell walls (held by hydrogen bonding). MC is expressed as a percentage based on the OD mass, i.e. weight, of wood. Two special MCs have practical significance. First, there is the fibre saturation point (FSP). It is the MC at which the cell walls are fully saturated and there is no free water in cell cavities. FSP varies among species, usually in the range of 25–35%, depending on such factors as chemical composition and types of extractives in wood. For poplars and willows, FSP is between 28% and 35%. It is important in freshly sawn wood and in initial drying. Second, and more important, there is the equilibrium moisture content (EMC). It is an MC below the FSP, which is reached as an equilibrium condition with the relative humidity (RH) of the surrounding atmosphere at a given temperature. Under outdoor conditions, in temperate climates, EMC is in the range of 12–15% (RH between 60% and 90%), whereas indoor EMC varies from 6% to 10% (RH between 30% and 60%) (Panshin and de Zeeuw, 1980; US Department of Agriculture, 1999; De Boever *et al.*, 2007).

MC affects the physical and mechanical properties, as well as the processing characteristics, for example machining, gluing, finishing, of wood. It is at the FSP and below where the influence of MC is primarily manifested. Thus, properties are normally published for air-dry, i.e. 12% MC, and green wood, i.e. MC at FSP or greater. The MC of standing trees is of interest. For poplars and willows, it is usually quite high. For example, average values of 95% (heartwood) and 113% (sapwood) have been reported for aspen, and 162% (heartwood) and 146% (sapwood) for cottonwood (Kellogg and Swan, 1985; US Department of Agriculture, 1999). For hybrid poplars, even higher green moisture levels are not uncommon. The consequence of higher initial MC in poplars and willows is that more water is transported from the forest to the primary processor and needs to be removed during drying, requiring more energy. One tangible benefit of high green MC is that freshly cut poplar logs can be processed by peeling and stranding without special preparation, e.g. soaking and steaming.

Dimensional stability (shrinkage and swelling)

Shrinkage and swelling are induced by changes in the quantity of water in the cell wall between the FSP and the OD state. Shrinkage is expressed as a percentage dimensional change based on the original (or green) dimension. With respect to dimensional changes, wood is an anisotropic material. Shrinkage is greatest in the tangential direction, i.e. along the annual rings, followed by radial shrinkage. Axial (or longitudinal) shrinkage is usually very low (0.1–0.2%), but can reach 1% when tension wood is present (Sassus *et al.*, 1995; Jourez *et al.*, 2001). Volumetric shrinkage is the combined product of the three. Average shrinkage values for several North American poplar and one willow species are presented in Table 10.2 (Alden, 1995; US Department of Agriculture, 1999). Shrinkage values are in the same range for fast-growing hybrid poplars (Koubaa *et al.*, 1998a). Considering their low density, poplars have relatively high shrinkage values, which is due mainly to their chemical composition, e.g. relatively high polysaccharide content. Generally, the occurrence of tension wood can increase total volumetric shrinkage by up to 20%.

It should also be noted that the ratio of tangential to radial shrinkage (shape factor) in poplars and willows is relatively high, for example 2.0 or higher, for the whole MC range from fresh to OD. One important consequence of the high shape factor is the creation of internal stress during wood drying. Rapid drying often creates large stresses, with accompanying drying defects, for example checks and splits, as well as form distortions, which is why empirically established drying schedules should be followed.

10.3.2 Mechanical properties

Prior to discussing the mechanical properties of poplars and willows, a brief overview of terminology is appropriate. Mechanical properties are those characteristics which define the behaviour of wood under applied forces or loads. They are grouped into strength and elastic properties (US Department of Agriculture, 1999). Common strength characteristics include bending strength (flexural strength or modulus of rupture, MOR), compression strength, shear strength, impact bending, toughness and tensile strength. All of these can be tested parallel or perpendicular to the grain. Sometimes, hardness is also defined as a mechanical property.

Elasticity implies that deformations caused by low stress levels are completely recoverable after the loads are removed. Besides being elastic, wood is also an orthotropic material, which means that it has independent and different mechanical properties along its three structural axes, for example longitudinal or parallel to grain, radial and tangential. That is why strength properties are determined and reported to reflect the orthotropic nature of wood. Of the elastic properties, it is the modulus of elasticity in bending (or MOE) which is generally reported.

Finally, it should be noted that standard test methods specify the use of small, clear, i.e. straight-grained and defect-free, specimens for the determination of mechanical properties. To allow proper comparison of published data in various reports and databases, standard conventions have been set concerning MC, duration of load, test speed, etc. For structural applications, new standards are being developed to incorporate full-size timber also (including defects).

The mechanical properties of the wood of poplars and willows are relatively low when compared to most other hardwoods and softwoods. However, when their low density is taken into account, then the ranking of poplars and willows improves significantly. For example, the strength to density ratio (or specific strength) is similar to that of most commercial softwoods (De Boever *et al.*, 2007). Average mechanical properties for mature wood of several North American poplars and one willow species are presented in Table 10.3, in comparison with two softwood species (Alden, 1995; US Department of Agriculture, 1999). The average bending strength (MOR) of the four poplar species is in a relatively narrow range, between 58 and 63 MPa, which is about 14% lower than the MOR of jack pine and white

Table 10.3. Selected mechanical properties of mature wood from natural forests of North American poplar and willow species in comparison with two common softwoods. These average values are for wood at 12% moisture content. (Source: US Department of Agriculture, 1999.)

	Flexural		Compression strength (MPa)		Shear strength (MPa)
Species	Strength (MOR) (MPa)	Modulus (MOE) (GPa)	‖ to grain	⊥ to grain	‖ to grain
Populus deltoides	59.0	9.40	33.9	2.60	6.40
Populus grandidentata	63.0	9.90	36.5	3.10	7.40
Populus tremuloides	58.0	8.10	29.3	2.60	5.90
Populus trichocarpa	59.0	8.80	31.0	2.10	7.20
Salix nigra	54.0	7.00	28.3	3.00	8.60
Picea glauca	68.0	9.20	37.7	3.20	7.40
Pinus banksiana	68.0	9.30	39.0	4.00	8.10

MOR, Modulus of rupture; MOE, modulus of elasticity.

spruce. Notably, the MOE values are closer between the four poplars and the two softwoods.

The mechanical properties of poplar hybrids grown in plantations are of special interest from the perspective of fast-growing fibre crops of the future. Various studies reported considerable variation in properties. For example, MOR values for 'Interamerican' poplar clones, grown in Belgium, range from 50.8 MPa to 55.7 MPa, whereas flexural MOE values range from 5.30 GPa to 6.25 GPa for the same clones. De Boever *et al.* (2007) indicated MOR values for 'Interamerican' poplar clones 'Beaupre', 'Hoogvorst' and 'Hazendans' were in the range of 62–74 MPa, whereas flexural MOE values were in the range 6.5–7.8 GPa for the same clones. Williams (1998) reported MOR of 57.2 MPa and MOE of 7.55 GPa for hybrid poplar grown in central Canada, whereas S.Y. Zhang, Quebec, Canada (2006, personal communication) indicated MOR values ranging from 43.7 MPa to 60.1 MPa and MOE ranging from 3.41 GPa to 4.70 GPa for cottonwood hybrids grown in British Columbia. Part of the reason for the lower flexural strength and modulus in hybrids when compared to wood grown in natural forests (Table 10.3) is that hybrids contain more juvenile wood, which is generally weaker. There is also general agreement that strength is correlated positively with wood density. Since density is under partial genetic control, i.e. heritable, it is possible to select clones for high density, and thereby high strength. However, De Boever *et al.* (2007) found relations between density and strength were mainly valid between clones. If clones are mixed, this relation loses significance. This problem is the main bottleneck in grading poplar wood from a mixture of clones.

10.3.3 Chemical properties

The chemical composition of poplars and willows is characterized by high polysaccharide and low lignin content, for example approximately 80% holocellulose and 20% lignin. For aspen (*Populus tremuloides*), Mullins and McKnight (1981) reported the following chemical profile: cellulose 53%, hemicelluloses 31% and lignin 16%, on an extractive-free wood basis. The amount of extractives was reported at 2.1% (hot water as solvent). For 14-year old hybrid willow (*Salix* spp.), Deka *et al.* (1992) reported average holocellulose, i.e. cellulose plus hemicelluloses, content of 79%, whereas lignin content was 21%. Szczukowski *et al.* (2002) studied the chemical composition of short-rotation hybrid willows in Poland and found that cellulose content increased with the length of the cutting cycle from 1 to 3 years; for example, from 46% at 1 year, to 48% at 2 years, to 56% at 3 years. These findings have important implications for the prospective utilization of short-rotation willow fibre crops for energy or pulp and paper. There are also significant differences in chemical properties between poplar clones (Dickson *et al.*, 1974; Anderson and Zsuffa, 1975; Blankenhorn *et al.*, 1985).

The acidity (pH) of poplar wood, determined on its aqueous extract, is in the range of 5.8–6.4. However, this level of acidity does not cause any corrosion phenomena with metals or any reactions with glues, preservatives or coating products.

10.3.4 Natural durability

The wood of poplars and willows is susceptible to decay, i.e. has low natural durability. In published references, for example the US Department of Agriculture (1999), poplars and willows are ranked as non-durable. This is also manifested in older natural poplar stands in Canada, where decay, for example heart rot, is common in overmature trees. For example, Thomas (1968) reported a 25% incidence of decay in 70-year-old aspen stands on good sites in Alberta. Such a long rotation age would not be contemplated for plantations. Decay resistance is important for poplar-based products where they may be exposed to adverse conditions that favour decay, for example prolonged exposure to high humidity or water. For such conditions, preservative treatment is the best option. However, outdoor exposure of poplar plywood or I-joists sometimes shows less decay than expected, even without any preservative.

In the 1960s, some poplar logs were stored for up to 1–2 years in flowing water, which removed the starch and gave the wood a higher resistance, especially against insect attack. The wood remained susceptible to fungal attack under higher moisture conditions. Boards of *Populus alba* × *tremula*, used mainly as cladding for farm sheds in central Europe, were given this treatment.

10.3.5 Wood quality considerations

The concept of wood quality has evolved since 1960 and has become very important to both wood users and tree growers. It is especially important for plantation forestry and the production and management of fast-growing species, such as poplars and willows. That importance results from the fact that both product quality and value are strongly influenced by wood quality.

Wood quality can be defined in terms of external characteristics for logs, for example size and form, and internal attributes of wood, for example density and its uniformity, FL, fibre coarseness, microfibril angle, slope of grain and knots. There are some differences in the ranking of important quality traits between different products (Gartner, 2005). For example, for poles, dimension lumber, laminated veneer lumber and glulam, high strength, i.e. high density, and good dimensional stability are important, whereas oriented strandboard (OSB) and Parallam require good compressibility, i.e. low to moderate density. Good-quality mouldings and millwork need good machinability and dimensional stability in their raw material. Finally, for pulp and paper products, several wood quality traits need to be considered. They include density (influencing pulp yield), FL, fibre coarseness and microfibril angle (affecting paper strength and smoothness), cellulose content and type and amount of extractives (for yield and bleaching).

Much research has been done on the wood quality of poplars and willows (Matyas and Peszlen, 1997; Koubaa *et al.*, 1998b; Fujiwara and Yang, 2000; Jourez *et al.*, 2001; DeBell *et al.*, 2002; Monteoliva *et al.*, 2005; Fang *et al.*, 2006; De Boever *et al.*, 2007; and others). It is not the intent to present a comprehensive review of wood quality in this chapter. Suffice to point out that researchers generally agree that density, FL and microfibril angle are under partial genetic control, with the heritability of density appearing to be most strongly controlled. Another factor to consider in wood quality assessment is the presence of juvenile wood (or core wood), which forms under the direct influence of the crown. The period of juvenility is shorter in hardwoods, for example poplars and willows, than in softwoods. In general, FL and diameter increase with age, whereas microfibril angle decreases. Juvenile wood is weaker and dimensionally less stable than mature wood.

10.4 Processing

Processing represents those steps which are involved with the conversion of the raw material to finished products. For wood products, they include storing, machining, drying, treating, bonding, sanding, finishing and mechanical fastening. Some of these are common to the manufacturing of all or most products, for example storing, machining and drying. Therefore, a short introduction to these topics is warranted prior to the presentation of specific products and utilization scenarios.

10.4.1 Storing

A certain volume of poplar logs needs to be stored in a log yard for the manufacture of all primary products, for example lumber, veneer, composites. Since poplar is susceptible to the development of stain and decay, the period of storage should be as short as possible, 6–8 weeks for example. Alternatively, methods may be employed to prevent deterioration, for example periodic water spray or coating the ends of high-value logs, such as veneer logs, with a sealant.

10.4.2 Machining

In general, the wood of poplars and willows machines well, with low energy requirements. Machining problems, for example fuzzy surface or wooliness, are usually related to the presence of tension wood. In a comprehensive study, Forintek Canada Corporation (Williams, 1998) investigated the machining properties of hybrid poplars in accordance with ASTM Standards. Planing, boring, shaping and sanding operations were involved. The study concluded that hybrid poplar performed moderately well in machining tests when compared to other commercial wood species. The only problem encountered was due to the occasional presence of fuzzy grain caused by tension wood.

10.4.3 Drying

The sapwood of poplars is permeable and can be dried relatively easily and rapidly. The heartwood, on the other hand, has poor permeability and often has wet pockets; therefore, its drying is difficult. De Boever *et al.* (2005) investigated clones from a multiclonal stand of *P. deltoides* × *nigra* dried with a conventional kiln-drying process at low temperature, followed by a 2-week homogenization period, and defined the variability in the end MC. The homogenization period also lowered the average end MC significantly, by 2–3%. Poplar clones show a different end MC after drying, meaning that the drying schedules need to be adapted to the clone or clonal group (Kretschmann *et al.*, 1999). This may directly affect the logistics for drying companies. In practice, the distinction between clones can rarely be made on the level of delivered logs.

The occurrence of wet pockets is a major issue of concern in drying poplar lumber. Because the literature does not give an objective and overall applicable definition of wet pockets, several definitions have been evaluated. The definition based on absolute differences between the locally measured MC and the MC corresponding to the 25th percentile value increased by 5% allows for discrimination of beams containing wet pockets with high certainty. Again, clonal differences have been found.

The presence of tension wood and the high tangential to radial shrinkage ratio may also cause drying defects. The best prescription for drying poplars seems to be to air-dry the lumber prior to kiln-drying for 6–8 weeks, and then to use a mild drying schedule in the kiln. However, for low-priced material, a long drying period is not always economically viable. Vansteenkiste *et al.* (1997) showed that high temperature drying with cyclic temperature variations could dry poplar wood with dimensional stability and a minimum of wet pockets. In veneer drying, the distribution of the tension wood zones will strongly determine the uniformity of developing drying stresses. Clonal differences have been observed in drying schedules for poplar veneers.

10.4.4 Treating

Although poplar and willow wood are considered to be permeable, differences in density, the possibility of fibre collapse during drying, the occurrence of tension wood and wood anatomical structure influence their permeability in regard to preservatives (Bauch *et al.*, 1976; Singh *et al.*, 1999). Members of the *Salicaceae* family, both poplar and willow, have considerable importance as fast-growing plantation species in a number of countries or regions throughout the world. Much research has been done to characterize the treatability of poplar wood, but few experiments are reported on willow wood (Cooper, 1976; Murphy *et al.*, 1991; Troya *et al.*, 1995; Van Acker and Stevens, 1995; Van Acker *et al.*, 1995). Poplar has been reported as not easily treatable. However, it is inaccurate to call poplar a refractory species. Practical experiences show an irregular impregnability due to an irregular penetration by preservatives. Van Acker *et al.* (1995) reported significant differences in treatability depending on the poplar clone or hybrid. The presence of a transition zone at the heartwood/sapwood boundary which showed refractory properties was shown earlier (Murphy *et al.*, 1991; Van Acker *et al.*, 1995). The rate of flow of liquids through the wood structure increases when pit membranes are destroyed or broken. Poplars are well known to be highly susceptible to bacterial wetwood or water heartwood. The bacteria causing wetwood may increase the permeability of the timber dramatically by degrading the pits (Clausen and Kaufert, 1952; Knuth, 1964). De Boever *et al.* (2008) showed similar behaviour in willow wood with a transition zone that was less permeable and treatable.

10.4.5 Bonding and finishing

Poplars and willows bond easily with all commercial adhesives. Some adjustments to the viscosity of adhesives may need to be made because of high wood porosity, in order to prevent the surface penetration of the glue causing a weak bond. Poplars and willows have good finishing properties, if prepared properly, and take stains, lacquers and paints well, but a primer-sealer is advisable.

10.4.6 Mechanical fastening

Poplars and willows can be fastened easily with all types of mechanical fasteners, for example staples, nails and screws. The wood does not split when stapled or nailed.

10.5 Utilization

Product options for the conversion and utilization of poplars and willows are numerous. They range from lumber to veneer, plywood and composites as wood-based products, as well as pulp and paper as fibre-based products. In addition, chemicals and energy may also be produced from poplars and willows. The primary wood products, in turn, can be used in many construction applications, as well as for containers and furniture. In addition, several new technologies and new alternative uses for poplar wood have emerged globally, especially in the engineered wood composites sector (Castro and Fragnelli, 2006). One of the major advantages of growing poplars and willows for various products is their rapid growth rate, enabling their production in relatively short rotations. Due to the broad and complex nature of the utilization topics for poplars and willows, and the limitations of time and space, it is not possible to give a comprehensive coverage to each product area, but only an overview.

10.5.1 Lumber

Poplar is well suited for the production of sawn wood (FAO, 1980; Hall *et al.*, 1982). For example, sawmills in Canada and the USA have been manufacturing poplar lumber since at least the 1960s. But production volumes have remained low because of economic factors. Due to small log diameters and the high incidence of decay, the average cost of sawing aspen is generally higher than for other hardwoods, and is much higher than for softwoods (Balatinecz and Kretschman, 2001).

In France, lumber recovery from poplar logs has been found to be 53% of the total log volume, i.e. bark on. The remaining material was made up of bark 11%, sawdust 14% and slabs and edgings 22%. In a Canadian report, lumber recovery was in the range of 45–50%, with 40–50% chips and 5% residual sawdust accounting for the rest (bark-free log basis). Besides yield, it is lumber grade and value which determine the competitive position of poplar. Taking grades into account, unfortunately poplar does not compare well to other species (Hall *et al.*, 1982). For example, only 15% of the lumber was in the high-value top grades (Select and No 1 Common), 25% was medium grades (No 2 and 3 Common) and 60% was in the low-value utility grades, e.g. pallet stock. In the USA,

Kretschmann *et al.* (1999) investigated the suitability of hybrid poplar for lumber. They found that hybrids had similar properties to native cottonwood and that hybrids could produce visually graded material, 65% of which would rank as either 'Standard' and better, or No 2. They suggested that in order to avoid excessive drying defects, the hybrid material should be dried in flitch or cant form and then re-sawn into lumber.

One way of improving the yield of defect-free lumber from poplar is through finger jointing. This process allows the removal of defects. Combining finger jointing with edge gluing allows the production of clear planks for shelving, furniture components and many other uses.

Besides its use as structural timber, poplar sawn wood is used in several other applications. Pallets for transport are an important worldwide market. Packaging materials account for a significant amount of poplar wood usage.

10.5.2 Wood-based composites and panel products

Introduction

There are several possible ways of categorizing wood-based composites and panel products, as well as glued structural products. One option is to use primary processing as a means of differentiating products. Another is to use end-use applications. In this chapter, we take the latter approach, as outlined in the US Department of Agriculture *Wood Handbook*, Chapter 10 (US Department of Agriculture, 1999). That publication identifies 'wood-based composites and panel products' as veneer and plywood, fibreboard, particleboard, OSB, wood–cement products and wood–plastic products.

Composites offer by far the largest and broadest scope for the utilization of poplars and willows. This is because many composites do not require large-diameter trees, and stem form is not critical. However, not all material is suitable as such. Specific requirements such as bark content, minimum diameter, FL or discoloration remain for some composites. Some composites can also use residues, and recovered wood is being used increasingly in Europe. Plus, the conversion efficiency from raw material to finished product is much higher with composites than with traditional wood products, for instance lumber. The conversion efficiency of logs to lumber rarely exceeds 45%, whereas the efficiency factor for composites can range from 50% up to 95%, depending on the type of product.

The quality and strength of composites are generally uniform and they do not exhibit common wood defects such as knots and splits found in lumber. The manufacture of composites also allows a degree of flexibility in processing whereby wood elements, such as veneers, strands, flakes and fibres, of the highest strength and quality can be incorporated into the outer surface of the product for both aesthetics and enhancement of strength.

In a composite, two or more component elements are combined to form a new material. Composites take advantage of the beneficial characteristics of each component material and often have more useful properties than any of the constituent materials on their own. In a broad sense, wood composites include a wide range of products, from composite panels, for example particleboard, hardboard, insulation board, medium-density fibreboard (MDF), waferboard and OSB, to composite lumber, for example laminated strand lumber (LSL), parallel strand lumber (PSL) and composite I-beams. In addition, mineral bonded wood composites, for example excelsior cement board, wood–cement particleboard, cement-bonded fibreboard, as well as wood–plastic composites, offer product opportunities for the utilization of poplars and willows. Those products which are bonded with a structural adhesive are referred to as structural or engineering composites. They may be produced in the form of panels, for example OSB, or lumber type profiles, for example PSL or I-beams.

One advantage of composite manufacturing plants is that they are highly automated and the technology for their manufacture is available 'off the shelf' from several major machinery suppliers located on all continents. On the other hand, a major handicap for most composite products is that their plants are designed for large capacities and their manufacture requires major capital investment, for example over US$200 million. Exceptions to this are wood–plastic and wood–cement composites, as well as some composite lumber products, such as laminated veneer lumber (LVL), or I-beams.

Adhesives are a vital ingredient in composites. Their manufacture presupposes the availability of adhesives. Conventional wood composites are made with a thermosetting or heat-curing synthetic adhesive (US Department of Agriculture, 1999). The most commonly used adhesives are: phenol formaldehyde (PF), urea formaldehyde (UF), melamine formaldehyde (MF) and isocyanate (MDI, or methylene-diphenyl-di-isocyanate). Of these, PF and MDI are used in the manufacture of structural products, for example OSB, LVL, LSL I-joists, construction plywood, etc., whereas UF is used in particleboard and MDF, for example products for interior applications. Resorcinol formaldehyde (RF) is used in glulam manufacture. There are two major uncertainties about the future of these synthetic adhesives. One is their rapidly escalating costs, since they are derived from petrochemicals. The second is more stringent regulations about formaldehyde-containing products, especially UF resins. Consequently, there is intensive R&D focusing on the development of new adhesive systems from renewable sources, for example lignin, tannins, starches, terpenes, vegetable oils, animal proteins, etc. When successful, these efforts will also have important potential benefits for developing countries.

Veneer and plywood

Poplar wood is well suited for the manufacture of veneer and plywood, but veneer and plywood production requires logs of the highest quality. Poplar peeler logs need the least preconditioning of any species, because of the wood's low density, good machining characteristics and its high green MC. The optimal peeling temperature for poplar is about 16°C, but acceptable results can be achieved in a temperature range of 7–30°C (Baldwin, 1981). Thus, during the winter months, the temperature of logs needs to be at least 7°C. A schematic flow diagram for the manufacture of veneer and plywood is presented in Fig. 10.3. Veneer is usually produced in the thickness range of 2.1–5 mm, depending on the type of plywood to be manufactured. As the veneer comes off the lathe, it is clipped and subsequently sorted, graded and dried. This is followed by edge gluing for the desired sheet size, glue application, lay-up into the plywood sandwich, hot pressing and finishing. The suitability of hybrid poplars for veneer production (to be used in plywood or LVL) has been investigated by Forintek Canada Corporation (S.Y. Zhang, Quebec, Canada, 2006, personal communication). The results were quite positive, the hybrids peeled and bonded well, although longer veneer drying times were required because of the higher initial MC. Similar results were found by Alvarez *et al.* (2004) for poplar hybrids grown in Spain. Belgian poplar hybrids (*Populus trichocarpa* × *P. deltoides*) were peeled successfully and processed into a technical plywood in accordance with the new CE-marked European guidelines (De Boever *et al.*, 2007).

Italy, on the other hand, has a long history of producing plywood for the furniture industry. These plywoods are lighter because the clone 'I-214' is used, which has a low intrinsic density. Also, very special boards with coloured veneers are produced, for example, for design furniture. Poplar veneer can also be used as core to make exotic plywood with imported decorative face veneers.

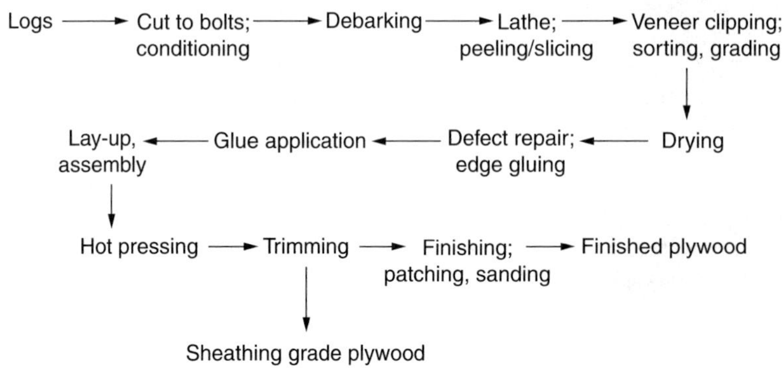

Fig. 10.3. Schematic flow diagram for the manufacture of veneer and plywood.

In addition, poplar lumber-core plywood can be produced as stiff shelving products. The match industry as well as a segment of the packaging industry, for example for fruit and vegetable baskets and cheese boxes, also use poplar veneer. Finally, manufacturers of several small specialty products for the food and health care sectors also prefer poplar veneer, for example popsicle sticks, chopsticks, tongue depressors (Fig. 10.4).

Between 1990 and 2010, new technical developments in the plywood industry helped to increase output and quality. For example, the positioning of the logs using a laser guided X-Y loader, the development of peeling technology without a spindle, and steam-heated knives are examples of technical innovation. The spindleless lathe allows the veneer bolt to be peeled to a core of about 50 mm, thereby increasing veneer yield significantly. The heated knife reduces friction between the knife edge and the wood, thus improving output and veneer quality.

The production of poplar plywood in North America has declined and been displaced by OSB. However, in Europe, and especially in China, the poplar plywood industry is thriving, largely because of the availability of good-quality peeler logs from plantations and good market demand (Fig. 10.4).

Fibreboard

The term 'fibreboard' includes a group of products in which the component material is fibres and fibre bundles, as opposed to particles, flakes and strands. Three major fibreboard products are recognized: insulation board, hardboard and MDF. Fibre material of willow and poplar wood is very suitable for the production of all different types of fibreboard (Scheithauer, 1999). The feedstock material usually comes from residues, such as planer shavings, sawdust and wood chips. The raw material is converted to fibres by thermomechanical pulping.

Fig. 10.4. Photographs showing poplar veneer production and products in Canada (a) and China (b–d). (d: 'blockboard' stacks). Photos courtesy of FPInnovations, Forintek Division (a), J. Richardson (b), Nanjing Forestry University (c) and J.G. Isebrands (d).

Insulation board has the lowest density of the various fibreboards, for example 190–380 kg m^{-3}, and its manufacture does not involve a hot press. The fibre mat for this product is wet-formed from aqueous pulp slurry on to a wire screen. Water is drained from the mat by gravity and a belt press operating at room temperature. The board is dried and baked in an oven, to an MC of 4%. The board then is trimmed, painted or asphalt coated, depending on its end use.

Hardboard products have the highest density, for example 625–1100 kg m^{-3}. They can be manufactured by wet, wet/dry and dry processes, which refer to the condition of the fibre mat as it enters the hot press. In the wet and wet/dry process, a continuous fibre mat is formed from pulp slurry on to a wire screen. Some water drainage takes place by gravity and belt pressing, but the final mat consolidation is achieved in the hot press, which is typically a multi-opening press. The bonding of fibres comes from added synthetic resin, usually PF, plus the lignin on the surface of the fibres. Hardboard had many applications, for example in construction, such as exterior siding and facing for garage and interior doors, as well as furniture and store fixtures, toys, automotive interiors, trunk liners, etc. However, due to the problems of water pollution and the cost of the clean-up of process water, wet process hardboard production has declined dramatically during the past 20 years.

Medium-density fibreboard (MDF) is a dry-formed composite product manufactured from lignocellulosic fibres combined with synthetic resin and other additives and consolidated into a rigid panel under heat and pressure. The density of MDF is in the range of 500–800 kg m^{-3}. The process steps involved in the manufacture of MDF are: thermomechanical pulping of wood furnish, drying, blending of fibres with resin and wax, mat forming and hot pressing. MDF can be machined, painted and printed on. It is the choice material for the manufacture of furniture, mouldings and cabinets. There is a special resin-fortified (by waterproof resin, e.g. melamine), high-density fibreboard, or HDF, which is used for decorative flooring.

S.Y. Zhang (Quebec, Canada, 2006, personal communication) reported on the suitability of several poplar hybrids for MDF in Canada. The conclusions were that in all properties the experimental panels met or exceeded the property requirements of existing standards, except in thickness swelling. Appropriate process modifications can help overcome this problem.

In Italy, Belgium and France, much poplar wood is used in MDF production (Giordano, 1974; van der Zwan, 1987; Otto, 1989; Rittershofer and Hofmann, 1999). It is easier to defibrillate than other comparable hardwoods at the same density, and even better than some softwoods. The greatest disadvantage of the willow and poplar wood is that the fibre material contains a substantial amount of fine dust. This dust slows down the drainage of water during the production process. Poplar-based, wet-formed MDF panels vary from 2 to 8 mm in thickness. The conversion efficiency is around 40–42%. These MDF panels have very smooth surfaces. Boards made homogeneously out of willow wood have a light yellow to brown colour. Densities range from 600 to 800 kg m^{-3}. The mechanical properties of the boards are strongly related to their thickness. The bending strength of willow-based panels is lower than those of poplar- and aspen-based panels. The tensile strength perpendicular to the grain varies below a density of 700 kg m^{-3}, especially for willow. On a vertical section, a zone of higher compression is found in the poplar-based MDF boards

Particleboard

Particleboard is a composite panel, manufactured from particles or flakes of wood and other lignocellulosic materials, where the particles are bonded together with a synthetic adhesive, for example UF resin. Particleboard is produced in densities ranging from 550 to 800 kg m^{-3}. The manufacturing process consists of particle preparation (flaking), particle classification by size for face and core, drying, blending with resin and wax, mat forming, hot pressing, cutting to size and finishing (sanding, overlaying). The mat is formed so that the fine particles are deposited to the two faces and the coarse material to the core. Particleboard is rarely used in this 'raw' form. It is usually overlaid with veneer, or paper for printing, or synthetic overlays (like melamine) for countertops. Press systems for particleboard manufacture may be single-opening, continuous-belt presses or multi-opening presses.

Poplars and willows are well suited for the manufacture of particleboard, especially because of their good bonding characteristics and compressibility (Geimer and Crist, 1980). Due to economic considerations, particleboard manufacture should be based on available residues. However, residues from poplar processing plants, for example sawmills and veneer mills, may not be available in sufficient volume to feed a particleboard plant. Under such circumstances, poplar or willow residues are blended with wood waste from other species. In Europe, the particleboard industry is using more and more recovered or recycled wood, sometimes referred to as 'urban wood'.

The volume shrinkage of poplar-based particleboard exceeds that of board produced from softwoods (Giordano, 1974; van der Zwan, 1987). Willow wood (*Salix alba*) has been evaluated for board production, together with grey poplar (*Populus* ×*canescens*) and aspen. All boards complied with standards for bending strength, tension strength and elasticity. None of the boards could comply with the standards for thickness swelling. For the poplar-based boards, a 1% addition of paraffin was needed to reach acceptable values, whereas willow needed only half of that amount (Scheithauer, 1999).

Oriented strandboard (OSB)

The Structural Board Association defines OSB as an engineered, mat-formed structural panel product made of strands, large flakes or wafers, usually sliced from small-diameter logs. The panels often have a layered construction much like plywood where, in the surface layer, the strands are aligned in the long panel direction, whereas the inner layer consists of cross-aligned or random strands. The strand dimensions used by most manufacturers are in the range of 150 × 25 mm. The average strand thickness in OSB is about 0.07 mm.

The OSB industry evolved from the waferboard industry in Canada in about 1980. The first plant was built in 1982 near Edson, Alberta. Since then, OSB manufacturing has become a global industry. More than any other product, OSB revolutionized aspen utilization in Canada and the USA, because it provided a major new outlet for the vast and under-utilized aspen resources in the two countries. The development of OSB coincided with the slow demise of the softwood plywood industry due to the diminishing availability of quality peeler logs. Currently, OSB is manufactured throughout the world, for example Europe, Asia and North and South America.

A simplified flow diagram for the manufacture of OSB is presented in Fig. 10.5. Two main press technologies evolved for OSB manufacture: (i) the multi-opening press system; and (ii) the single-opening, continuous-belt press system. The suitability of fast-growing hybrid poplars for OSB manufacture has been studied extensively in China (Hua and Zhou, 1994) and in Canada (S.Y. Zhang, Quebec, Canada, 2006, personal communication). The results in both countries indicated that good-quality OSB could be produced from hybrid poplars. Some adjustments may need to be made to strand drying because of the higher initial MC. OSB is used in a broad range of applications, including construction, such as sheathing, web in I-joists, stress skin panels, concrete forms, etc., as well as packaging and crating, pallets, furniture frames, shelving, hardwood floor core, etc. (Fig. 10.6).

Tröger and Wegener (1999) evaluated the possible use of 5- to 10-year-old willow and poplar stems in the production of OSB. They were evaluated against European standards for the OSB-4 type (heavy load bearing in wet conditions). Bending strength of both poplar and willow OSB was adequate. The elasticity of willow boards, however, was too low. Internal bonds were adequate in all boards. However, no poplar-based OSB could comply with the standards for thickness swelling for use in humid conditions. By contrast, the willow-based OSB showed very low shrinkage values.

Wood–cement composites

Cement-bonded wood composites are strands, particles, excelsior or fibres of wood mixed together with cement (usually Portland cement) and manufactured into panels, tiles, slabs, blocks, bricks and other products used in the construction industry. Commercially produced composites may contain between 5% and 70% by weight of particles, excelsior or fibres. The raw material ratios, the geometry of the wood material and the density of the composite are the main parameters that influence product properties, especially strength. Poplars are suitable for wood–cement composites (Fig. 10.7).

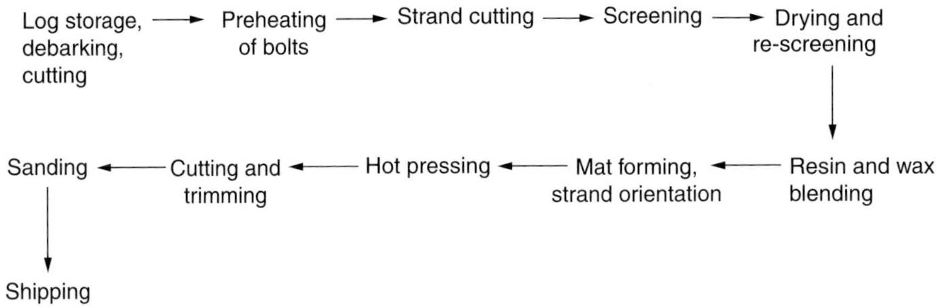

Fig. 10.5. Schematic process flow diagram for the manufacture of OSB.

Fig. 10.6. Photographs of different composite products and their applications for poplar wood. (a) LVL header beam and I-joists in flooring. (b) OSB wall and roof sheathing for homes. (c) I-joists in residential floors. (d) Large fruit boxes on pallets. Photos courtesy of APA – The Engineered Wood Association.

Wood–cement composites benefit from the positive attributes of each component material, resulting in a more desirable end product. Cement contributes high compressive strength, excellent fire resistance, enhanced durability and dimensional stability. Wood fibres and particles, on the other hand, add improved flexural strength, fracture toughness, lower density and superior thermal- and sound-insulation properties. Wood–cement composites have a strong attraction for use in warm, humid climates where termites and decay are a major concern for traditional wood products. Their good thermal- and sound-insulating properties and their resistance to fire are added benefits in any environment. The cement binder provides a durable surface that can be easily embossed and painted for an attractive low-maintenance finished product. One drawback is that the production of cement is very energy-intensive and expensive (Eltomation, 2007; Armtec, 2012).

Portland cement is a strongly alkaline material. Consequently, over time, the hemicelluloses and lignin components of the wood particles and

Fig. 10.7. Photographs showing a few applications of cement-bonded wood composites. These products can and do use poplar wood. (a) Examples of wood–wool–cement panels; (b) when used in wall sections; (c) in ceiling of a parking garage. (d) Woodparticle–cement composites as neighbourhood sound barriers. Photos courtesy of Eltomation BV, Holland (a–c) and Durisol Inc., Canada (d).

fibres in the composite may become embrittled and deteriorate. This can cause loss of strength and possibly product failure. An option is to incorporate ingredients into the matrix that reduce the alkalinity of the cement, by adding, for example, magnesia cement. This approach is employed in the manufacture of excelsior–cement products.

The basic process steps in the manufacture of wood–cement composites are: raw material preparation, mixing/blending of cement slurry with fibres/particles, forming, compacting/pressing, curing/autoclaving, finishing (Moslemi, 1988; Berkenkamp, 1997).

The properties and performance of wood–cement composites are superior to those of conventional wood composites in resistance to fire, fungi, insects and the weather. They have moderately good strength properties, but they are not suitable for heavy structural loads. Wood–cement composites can be cut, drilled and fastened like conventional wood products; however, carbide-tipped blades are recommended for cutting.

Wood wool (or excelsior)–cement panels typically use excelsior made from aspen and are produced in the density and thickness range of 250–500 kg m^{-3} and 25–100 mm, respectively. Both the wood particle and excelsior cement composites have gained wide acceptance in a broad range of applications, for example acoustic ceiling panels in institutional and commercial structures, cladding for industrial and warehouse buildings, fire-resistant, non-load-bearing partitions, low-cost housing elements, sound barriers, privacy fencing, etc. (Fig. 10.7). In addition, excelsior–cement panels have an attractive texture and they can be embossed and finished by painting or plastering. They are relatively light and have good machining and fastening characteristics.

Wood–plastic composites

Wood–plastic composites (WPCs) are intimate mixtures of wood fibres or particles dispersed and encased in a polymer matrix. Both thermoplastic and thermoset polymers are used in industry. Thermoplastics are easy to recycle and are therefore the preferred polymers for products in non-structural or semi-structural applications. The term 'fibre' in a broad sense encompasses fibres, fibre bundles and small particles of refined woody tissues. A variety of agricultural crops are also suitable. Since the fibres originate from renewable resources and can be derived from consumer and manufacturing waste streams, WPCs fit well into the framework of 'green' and sustainable materials. The fact that these composites are durable and suitable for many long-term applications, are easily reused and recycled, contain no harmful substances and can be manufactured into a wide range of consumer and industrial products with relatively low energy consumption gives these materials a strong competitive position in the marketplace when material choices are made. They can also use poplars and willows as feedstock materials (Balatinecz and Sain, 2007).

WPCs are different from the more traditional wood composite panels, such as particleboard, MDF and OSB, both in formulation and processing. In wood-based panels, the 'polymer' or 'plastic', in the form of an adhesive, is a relatively minor constituent of the finished product, for example 2–12%. On the other hand, in WPCs, the polymer to fibre ratio is much higher, often 50:50, depending on the product and end use, and consequently they are more expensive than traditional wood composites. But their uses are also more versatile. The technology to manufacture wood fibre polymer composites has been adopted primarily from the plastics industry.

The automotive industry has been using wood and other natural fibres, in the form of composite materials, for interior car panels in Europe, Japan and also in North America since about 1980. However, today, in the age of sustainability, 'green' materials and global warming, there are both marketing and economic advantages to the use of natural fibre polymer composites not only in cars but also in many other applications. Even Mercedes Benz and BMW automobiles have these composite materials in their interior parts. The broad range of current applications of WPCs in cars and trucks include door panels, seat backs, headliners, package trays, dashboards, instrument panels, arm rests, trunk liners, spare tyre covers and sun visors. Potential applications extend to body panels, fenders, cargo box and grill opening reinforcement. The benefits of the use of natural fibres in auto parts include weight reduction, with potential savings of up to 30%, and a potential cost advantage of 20% over the use of glass fibres. A lighter car with the same structural reliability also means significant savings in fuel costs.

In the period 1995–2005, WPCs also made significant inroads into the huge building materials markets in North America, and more recently also in Europe and Asia. This market penetration was driven not so much by price (because WPCs are more expensive than wood) but by the ease of maintenance, long service life and the 'green' image of WPCs. In about 13 years (from 1990 to 2003), the value of annual shipments of WPCs for building products in North America grew to nearly US\$1 billion year^{-1}, with projected growth rates of more than 15% year^{-1} for the foreseeable future. In North America, the main construction applications are exterior decking and railing systems (currently the largest uses), boardwalks, door and window profiles, flooring, decorative trim, window blinds, louvered doors and siding and roofing tiles. The manufacture of furniture components, both office and household, are the latest areas of application for WPCs (Fig. 10.8).

High-density polyethylene (HDPE), both virgin and recycled, is the most commonly used matrix polymer in industry. This is partly because of its general availability, including from recycled plastic waste streams. Other polymers with low melt temperatures, i.e. less than 200°C, are also used, principally polypropylene (PP), polyvinyl chloride (PVC) and polystyrene (PS). Matrix polymers are available in pellet, flake or powder form. The sensitivity of cellulosic fibres to thermal decomposition above 200°C sets that temperature as the practical upper limit to define the suitability of thermoplastic polymers for WPCs. Fibre loadings may range from 20% to 80%, but most commercial products contain about 50% filler.

Fig. 10.8. Photographs of wood–plastic composite profiles (a) and applications in decking and railing (b). Photos courtesy of Nexwood Industries Ltd.

Another important material asset of WPCs is their suitability for a broad range of products. This characteristic naturally means that these materials can be substituted for products that are made from non-renewable resources. Their applications may also expand into many types of farm construction and equipment. Their high durability, moisture resistance and non-splintering surface make WPCs eminently suitable for feeders, manure spreaders and horizontal silos in agriculture. To date, the greatest substitution was to replace pressure-treated lumber for environmental considerations. But, with the advent and imminent development of high-performance WPCs, considerably more substitution opportunities exist in construction, infrastructure and also in industrial applications (Clemons, 2002; Balatinecz and Sain, 2007).

10.5.3 Glued structural products

Glued structural members include LSL, PSL, LVL, glulam and glued members with lumber and panels, for example I-joists, as categorized in the USDA *Wood Handbook*, Chapter 11 (US Department of Agriculture, 1999).

Laminated strand lumber (LSL) and parallel strand lumber (PSL)

Both LSL and PSL are composite lumber products, both of which can be manufactured from poplar wood. The difference is that in LSL, strands (the same type as used in OSB, but longer, e.g. 300 mm) are the constituent wood elements used in the manufacture, whereas in PSL, strips ('whiskers') of veneer are the basic wood components. The original PSL product, 'Parallam', was developed by a former Canadian company (McMillan-Bloedel) during the 1970s. PSL makes it possible to use veneer scraps.

In LSL manufacture, after the strands are dried, screened, coated with PF or MDI resin plus wax, they are formed into a thick mat, with all strands having parallel orientation to the long axis of the mat. Subsequently, the mats are cut to length, conveyed into a single-opening, steam-injection press and then compressed into a rigid billet. The steam-injection press facilitates rapid heat transfer, which allows 6-min press cycles. Billet dimensions may be 2.4 m wide, 14 cm thick and 15 m long. Subsequent to hot pressing, the billets are cut into the desired lumber dimensions, sanded and packaged.

As mentioned earlier, in PSL manufacture, slender and long strips of veneer are used for the composite. With PSL, the thick composite profile is created in a continuous microwave press system. Billet cross-sectional dimensions of 280 × 480 mm are possible, with lengths of up to 20 m. Following pressing, the billets are cut into smaller sizes according to customer specifications, sanded and packaged. Chen *et al.* (1994) from China reported that hybrid poplar was suitable for the manufacture of PSL.

A product, by the name of 'Scrimber', which is somewhat similar to PSL, was developed in Australia during the 1970s, but the first commercial plant went into receivership. The technology was 'resurrected' in the USA under the brand name 'TimTek' in the south in the late 2000s, with Mississippi State University's involvement (Mississippi State University, 2009). A similar product has also been developed in

Japan by the name of SST ('Super Strength Timber') for the utilization of small-diameter plantation materials (Suzuki, 2005). Both SST and TimTek claim very high yields (over 85%) of finished product. If successful, these technologies may offer new opportunities for composite lumber production from small-diameter, short-rotation poplars and willows.

Laminated veneer lumber (LVL)

LVL is a layered, engineered material made up of laminated veneers, with the grain of all veneers running parallel with one another along the length of the finished product. The veneers are bonded together with a waterproof structural adhesive. A simple process flow diagram for the manufacture of LVL is presented in Fig. 10.9 (US Environmental Protection Agency, 1995; Canadian Wood Council, 2007).

The first step in the manufacturing process is log debarking, followed by cutting to peeler bolts, steaming or soaking and veneer cutting on a lathe. Veneer thickness may be in the range of 2.5–4.8 mm, and the length of veneer sheets is about 2640 mm. The veneers are clipped to 660 mm or 1320 mm width, dried and graded.

Veneers may be graded visually, but in modern mills grading is done automatically for stiffness and strength, using ultrasound. The lower-grade veneers are used for the core of the LVL, whereas the higher grades are incorporated into the face. This ensures the optimization of both aesthetics and strength in the finished product. The veneer dryers used in LVL plants are the same type as those in plywood mills.

The dry veneers are coated with a waterproof structural adhesive and are subsequently laid up into a long, thick sandwich, with parallel orientation between all layers of veneers. End joints between individual veneers are staggered along the length of the sandwich to disperse any veneer defects. The end joints between veneers may also be scarf joints or simply overlapped slightly to provide better stress transfer.

The veneer stack is fed to a hot press, where the sandwich is consolidated into a solid billet under heat and pressure. The LVL is manufactured to either a fixed length using a batch press or to an indefinite length using a continuous press. Most modern plants employ continuous press systems with radio frequency (RF) energy for resin cure, which reduces press times from 20 min in traditional presses to about 5 min in RF presses.

In addition to its availability in a range of sizes, including relatively large dimensions, the main competitive advantages of LVL are its uniformity, dimensional stability and highly predictable strength properties. Naturally, it commands a significantly greater price than lumber.

From the perspective of end use, there are two types of LVL, structural and non-structural LVL. Structural LVL must be manufactured with a waterproof adhesive. Its end uses cover both residential and non-residential construction in such applications as support beams, trusses, rafters and purlins (Fig. 10.6). Non-structural LVL may be used in windows, door frames, stairs, furniture and fixtures and kitchen cabinets. For aesthetic reasons, this type of LVL may be made

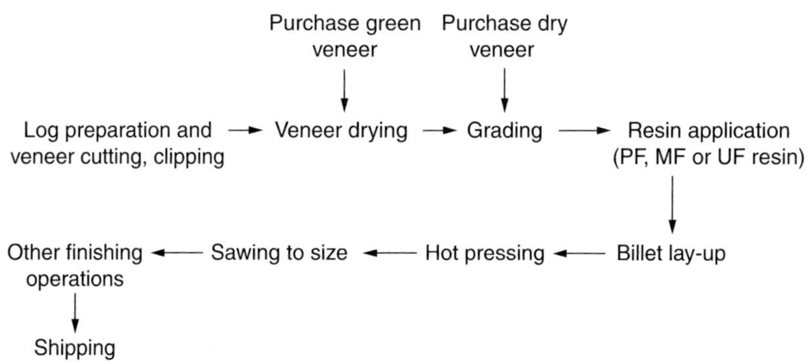

Fig. 10.9. Schematic process flow diagram for LVL manufacturing.

with UF or MF resins, which produce an invisible glue line.

Structural LVL consumption is concentrated in North America because of the strong tradition and preference for wood frame technology for residential construction. However, significant volumes are also used in Europe, primarily in architectural applications. On the other hand, non-structural LVL consumption is concentrated in Asia (Schuler, 2002). Most of the LVL in North America is produced from softwood species. This is because the LVL industry is integrated with the softwood plywood industry. The better-quality veneers are 'diverted' to LVL production, where they offer a better financial return than being converted to sheathing grade plywood. It is worthy to note that two LVL mills in Canada pioneered the manufacture of aspen LVL. The success of these mills demonstrate that poplars are very suitable for LVL production.

The suitability of hybrid poplar for LVL has been evaluated in Canada (Knudson and Brunette, 2002; S.Y. Zhang, Quebec, Canada, 2006, personal communication) (Fig. 10.10). Pilot-scale trials with 23-year-old hybrid poplar trees ('Walker') showed that good-quality LVL could be produced from young hybrid trees, especially if the veneers were stress graded. In this way, it was possible to extract approximately 25% of the hybrid poplar veneer, which was suitable for the face of LVL. Hua *et al.* (1994) also found that fast-growing poplar hybrids were suitable for the production of LVL.

Clearly, LVL offers excellent opportunities for the utilization of poplars for high-value products and uses both in structural and non-structural applications. But careful assessment of domestic and international market opportunities need to be made prior to investing in this option. One alternative is to consider the establishment of new LVL production facilities in conjunction with existing poplar veneer and plywood mills. This was the approach taken by the softwood LVL industry in the USA. This type of integration might reduce the size of the initial investment and maximize the recovery of product and value from the resource.

I-joists

Wood I-joists are a group of engineered wood products consisting of a web made from a structural panel, for example OSB or plywood, which is bonded to two flanges made of solid lumber or LVL (Fig. 10.11). There is a potential opportunity for the utilization of poplars in I-joists, where both the web and the flange may be made of poplar, for example poplar OSB and poplar LVL. Since hybrid poplars are suitable for the production of both OSB and LVL (Hua *et al.*, 1994; S.Y. Zhang, Quebec, Canada, 2006, personal communication), they could also be used in the manufacture of I-joists via OSB and LVL. One issue which needs to be evaluated is the stress rating of hybrid poplar I-joists, based on actual products, because of the lower density and structural strength of these materials.

The manufacture of I-joists can be highly automated, involving the rip-sawing and grooving of the flanges (LVL or lumber), the application of PRF (phenol-resorcinol formaldehyde) or MDI resin to the flange grooves and fitting the

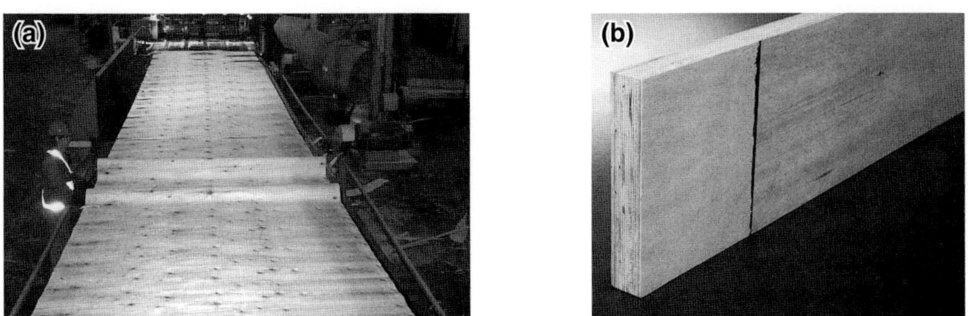

Fig. 10.10. Photographs of veneer (a) and LVL (b) produced from hybrid poplar. Note the numerous knots on the veneer surface (a), typical of small-diameter logs. Photos courtesy of FP Innovations, Forintek Division.

pre-cut web (OSB) and the two flanges (top and bottom) together in the I-joist assembler. I-joists are used in residential and commercial construction as floor joists, roof joists, headers and other structural applications (Fig. 10.6).

10.5.4 Pulp and paper

Poplar and willow wood can be pulped by all commercial pulping methods. Thus, mechanical, semi-chemical and chemical (e.g. sulfate or soda or kraft process and sulfite process) methods are now used (Fig. 10.12). The kraft process produces the highest-quality pulps, but its major disadvantage is that the consumption of pulping chemicals during the digestion is high; therefore, the chemicals have to be recovered. This recovery operation requires that kraft mills have to be built to a large scale with high capacity, requiring huge investment. Pulp mills designed for pulping hardwood can use up to 100% poplar (Stanton *et al.*, 2002). Poplar kraft pulps are particularly well suited to the manufacture of fine papers because of their high opacity, good bulk, sheet formation and good printability.

Fig. 10.11. Wood I-joists made of poplar. Photo courtesy of FPInnovations, Forintek Division.

Prior to conversion to paper products, poplar pulps are often blended with long-fibred softwood pulps to facilitate the development of wet strength on fast-running paper machines. The major uses of poplar pulps fall into three categories. First, they are used widely in specialty paper products, such as napkins, towels, absorbent tissues and fine paper grades. Second, with softwood pulp blends, poplar pulps are used for newsprint and other printing papers. Third, they are well suited and utilized for the manufacture of paperboard for packaging, as well as for building boards, for example insulation board and ceiling tiles, and for roofing felt.

The suitability of different hybrid poplar clones for papermaking has been investigated by various researchers. For example, Parham *et al.* (1977) studied the effects of tension wood (TW) on kraft paper. They found that during paper formation, the TW fibres resisted collapse in the sheet (even on extended refining) and prevented proper inter-fibre bonding, causing inferior strength. Thus, the high incidence of TW in the raw material will have a significant negative impact on paper quality. Labosky *et al.* (1983) and Law and Rioux (1997) investigated chemithermomechanical pulps (CTMP) produced from poplar hybrids. The pulps had a high proportion of very short fibres, for example <0.2 mm, but overall the CTMP pulps were of acceptable quality. Sierra-Alvarez and Tjeerdsma (1995) demonstrated that good-quality pulps, with

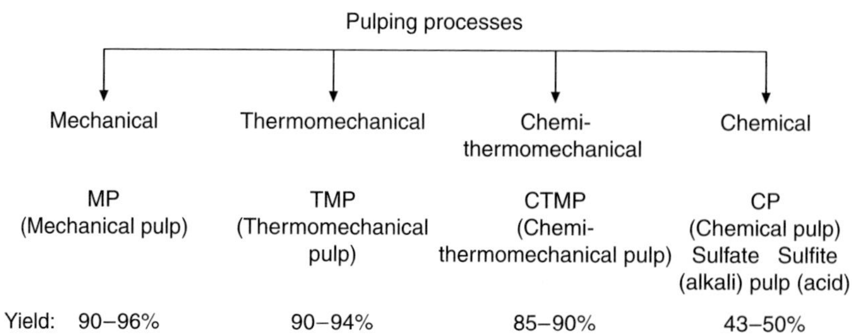

Fig. 10.12. Schematic outline for the major pulping processes (indicating pulp yield).

about 55% yield, could be produced from short-rotation hybrid poplars.

Yong (2005) reported on future market prospects for bleached hardwood kraft pulps (BHKP). He indicated that the robust global demand for this commodity was expected to continue. The growth is driven mainly by China's demand for writing and printing papers (which are the main uses for BHKP). Fast-growing plantations (mostly eucalypts) from Brazil and Indonesia are the main sources of BHKP, but there may also be a role for fast-growing poplar and willow plantations from other regions in supplying future BHKP demand.

In paper production, pulp based on hardwood is often referred to as the 'short fibre resource'. By blending different types of pulp, high-quality paper can be made. For this reason, poplar and willow play an important role in paper production.

10.5.5 Integrated poplar utilization

Integrated poplar utilization involves using all the wood from a harvest (Isebrands *et al.*, 1979). For example, small-diameter logs and wood residues can be used for the production of particleboard and MDF. These panels, in turn, are utilized by the furniture industry in value-added products. HDF may also be produced to be converted into laminated finish flooring (with a resin-impregnated paper overlay as the decorative surface). Ultra-thin and ultra-thick fibreboards are also produced, the former for decorative panels and the latter as core stock. In China, another innovation is the development of 'Sci-Tech' wood, which is a poplar wood veneer specially dyed and embossed to simulate more expensive exotic woods. The light-coloured (nearly white) poplar wood is well suited for colouring. 'Sci-Tech' wood finds applications in furniture, cabinets and decorative panelling. New adhesive systems have been developed to minimize and eliminate formaldehyde emission from finished products.

Another success story is the rapid development of the veneer and plywood industry, which is the largest user of high-quality poplar logs. Currently, there are over 600 small- and medium-size plywood manufacturers, which make China the largest producer of poplar plywood in the world. Besides standard grades, several specialty plywood types are also manufactured, for example fire-resistant plywood, blockboard or lumber-core plywood, decorative plywood with exotic face veneers, corrugated veneer core plywood, honeycomb core plywood, etc. Edge gluing and finger jointing permit the upgrading of low-grade material. These intermediate processing steps are an integral part of blockboard manufacture (Fig. 10.4d). The furniture industry is a major user of blockboard, whereas the low-density honeycomb plywood is used by the building industry for interior doors and non-load-bearing partitions. Significant volumes of plywood (particularly decorative specialty grades) are exported.

10.5.6 Special aspects of willow utilization

The structure and properties of willow wood are very similar to those of poplars. Therefore, they are equally suitable for most of the products for which poplars are utilized. However, willows are not always available in the size and volume of trees that poplars are, either in natural forests or in plantations. Therefore, their utilization has lagged behind that of poplars, except in two specific areas: basket willow and energy crops. Energy crops will be discussed briefly in the next section. Here, a brief overview is presented of basket willow and furniture, largely based on Abalos-Romero (2005).

The Chilean basket willow programme can serve as an example of a successful national strategy. During the 1990s, it was recognized in Chile that the global trade in consumer products, such as furniture, made from natural materials such as rattan had expanded rapidly. At the same time, some Chilean producers of basket willow were exporting willow shoots rather than finished products. Therefore, the Chilean Forest Research Institute, in collaboration with universities, initiated a project from 1997 to 2003 to upgrade and develop the basket willow sector. The project was comprehensive, including growing, improving product quality, producing furniture and developing domestic and export markets. Part of the Chilean project also involved the organization of training sessions. The production process following harvest involves stripping the bark from the switches (i.e. shoots), sorting, splitting the switches lengthwise into three or four sections, removing the pith, soaking the strips in water, weaving the product by hand into, for example, furniture, baskets, etc., over a frame, sanding and finishing (clear coating or paint) furniture items. The Chilean project was successful, and it clearly demonstrated the economic, social and environmental benefits of basket willow production.

The production and utilization of willow for baskets, furniture and other consumer products is practised in many other countries, including the UK, France, China, Argentina, Brazil, Canada, etc. (Plate 32A). Fig. 10.13 illustrates some of the baskets and furniture items made from willow. This type of willow production and utilization is especially suitable for improving the rural economy of developing countries. The industry does not require large capital investment, and the skills needed can be taught by expert artisans. Market opportunities exist both locally and globally for these products, fuelled by the growing desire of consumers to favour products made from natural materials using renewable resources.

A very specific application of willow in the UK is the production of cricket bats, for

Fig. 10.13. The production (a) and utilization of willow for various types of furniture and baskets (b and c). Photos (a and b) courtesy of English Willow Baskets, UK; (c) FAO/Jim Carle (Chile).

which very high-quality, specially bred willow is used. These bats are very expensive.

10.5.7 Biomass energy

The first energy crises of the 1970s prompted policy makers and researchers to consider and develop alternatives to fossil fuels based on renewable sources. From the perspective of forest biomass energy crops in countries with a temperate climate, poplars and willows were a logical choice for these endeavours because of their rapid growth rate and relative ease of cultivation (Anderson *et al.*, 1983). In addition to bioenergy initiatives by individual countries, in 1978 the International Energy Agency (IEA) set up the IEA Bioenergy organization with the aim of improving cooperation between countries that have national programmes in bioenergy research and development. Currently, 20 countries plus the European Commission participate in IEA Bioenergy (International Energy Agency, Bioenergy Agreement, 2008).

This section presents a brief overview of the role that poplars and willows might play in the emerging bioenergy picture. Since about 1980, a great deal of R&D and considerable industrial effort have been devoted to forest biomass production systems for energy, especially in short rotations (Isebrands *et al.*, 1979; Anderson *et al.*, 1983; Siren *et al.*, 1987; Zsuffa, 1988; Kenney *et al.*, 1996; Larsson *et al.*, 1998; Armstrong *et al.*, 1999; Volk *et al.*, 2000; Hall, 2002; Richardson *et al.*, 2002; González-García *et al.*, 2010). In a similar vein, research and industrial development to advance conversion technology options have also been initiated and discussed (McMillan, 2004). The biomass energy system embraces the process components of production, harvesting, feedstock preparation, conversion, distribution and consumption, i.e. consumer. The economic and technical viability of the system depends on the harmonization of all components, especially having a reliable feedstock supply, efficient conversion technologies and competitive cost/price structure (Hoffmann and Weih, 2005).

Sweden was one of the first countries to demonstrate the suitability of short-rotation biomass crops for energy generation during the late 1970s and early 1980s using willows (Siren *et al.*, 1987). Willows were selected because: (i) they offered high production potential; (ii) they were easily established with unrooted cuttings; (iii) they re-sprouted vigorously after each harvest; and (iv) they had limited pest problems, high genetic diversity and a short breeding cycle. Subsequent to the early trials, 25,000 ha of commercial coppice willow plantations were established and harvested repeatedly for energy generation by the late 1990s. Projections indicate that over 100,000 ha of energy plantations will be established in Sweden. Willow cultivation is fully mechanized, from planting to harvest, to feedstock preparation, for example chips and pellets. Willow biomass yields are between 6 and 12 t ha^{-1} $year^{-1}$, depending on site conditions. The biomass crop is used in district heating plants or for combined heat and power (CHP) production. The most likely reason for the success of biomass energy in Sweden is the enactment of long-term policies for fossil fuel substitution, coupled with investment subsidies and tax incentives.

In 1995, in the north-eastern and north-central USA, 20 public and private sector partners, including universities, major utilities, associations, environmental organizations, conservation groups and regional and national government agencies, formed the Salix Consortium as a major biomass energy demonstration project (Volk *et al.*, 2000). The objective was to demonstrate the commercial viability of growing willow biomass crops for energy and possible value-added bioproducts (such as pharmaceuticals and biodegradable plastics). Willows were selected for similar reasons as in Sweden, plus the extensive prior R&D work and experience of scientists at the State University of New York, one of the lead partners in the project. The near-term energy market strategy of the consortium is to use the willow biomass crop in co-firing coal power plants. Careful analysis of all production parameters indicates that willow biomass crops can be produced on a sustainable basis for energy generation, and they are CO_2 neutral (Volk *et al.*, 2004). A schematic outline for the major components of the project is presented in Fig. 10.14.

The production of poplar and willow biomass for energy is under development or consideration in several other countries, including Canada, the UK, Ireland, Belgium, Denmark, Germany, Poland, Finland and the Baltic States (Szczukowski *et al.*, 2002; Dibbelt, 2005; Ericsson and Nilsson, 2006).

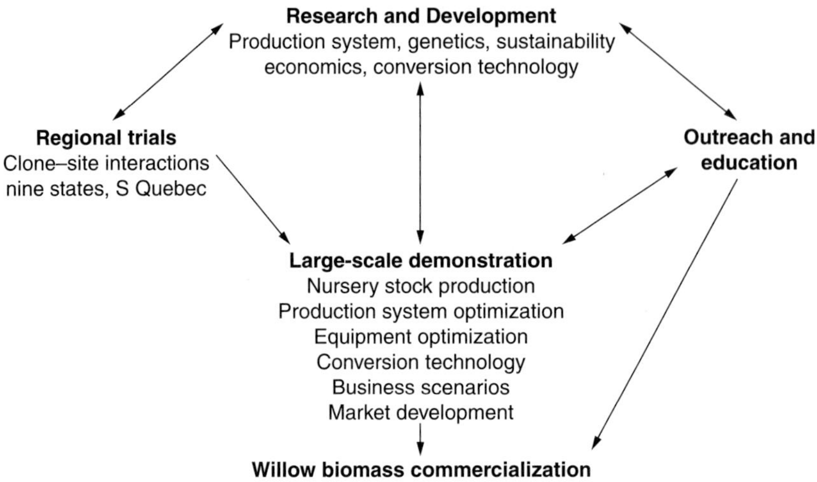

Fig. 10.14. Components of the Salix Consortium's biomass energy programme in the north-eastern and mid-western USA. Source: Volk *et al.* (2006).

For the conversion of biomass to energy, there are three basic pathways: (i) thermal; (ii) biological or biochemical; and (iii) physical (McMillan, 2004). The thermal or thermochemical processes may follow direct combustion (heat as the end product), gasification and pyrolysis. The biological pathways use enzymes and bacteria to decompose and convert the biomass into liquids, for example ethanol, and gases via anaerobic fermentation to CH_4. The physical or physicochemical pathway uses hydrolysis, with heat and pressure, to break down the biomass into sugars and lignin, for example aromatics, which can be further processed into desirable end products like ethanol. A promising bioconversion pathway to produce ethanol and co-products from hybrid poplar wood is the organosolv fractionation process (Pan *et al.*, 2006).

For direct combustion, woody biomass is not a very efficient fuel, due to its bulkiness, i.e. low energy density, and high MC. Densification to fuel pellets solves both problems. Fuel pellets have gained widespread use for space heating of homes and institutional buildings during the past two decades, especially in Europe, for example Sweden, Italy, Germany, Austria and Denmark. The US-based Pellet Fuels Institute estimates that there are about 800,000 homes using wood pellet stoves or furnaces for heating. Various models of pellet stoves and central heating furnaces, equipped with automatic feeding and control systems, are available. Biomass fuel pellets are cost-competitive with heating oil and natural gas, and much cheaper than electricity and LP gas.

At the industrial scale, in Europe poplar biomass crops are being used increasingly for power production. There is a trend towards co-firing poplar biomass with coal to decrease costs and improve air quality (Tillman, 2000; De and Assadi, 2009).

There is considerable controversy about diverting maize from agriculture to produce ethanol. Eaton (2008) argued that it was more profitable to produce cellulosic ethanol based on fast-growing poplar clones than it was from maize. For example, with poplar biomass, over 2300 l of cellulosic ethanol can be produced per hectare, compared to about 600 l of maize ethanol. Maize production also requires more herbicide and nitrogen fertilizer than any other biofuel crop. Patterson (2005) also supports the cost advantages of cellulosic ethanol, which is cheaper to produce than ethanol from grain. However, wood-based ethanol production requires large industrial plants.

To obtain maximum value yield from fast-growing hybrid poplars, Eaton (2008) recommended the use of an integrated multiple-product management concept, i.e. select rotation times to produce logs for veneer and wood products, chips for pulp and paper and the residual biomass for energy. Shifts in the allocation of the crop could be made, depending on economics and market conditions at a given time for each product. Eaton foresaw a new type of agroforestry practice, i.e. producing short-rotation forest biomass crops, like poplars and willows, on agricultural land. A major advantage of forest biomass over agricultural biomass is that the former is available for harvest year-round, and it can be stored on the stump at no added cost.

The ongoing escalation of energy prices in the 21st century places a new urgency on commercializing viable biomass energy options. Another key driver for biomass energy is the fact that it is CO_2 neutral and can contribute to mitigation of climate change. The international experience and achievements mentioned above demonstrate what can be accomplished in a relatively short time in fossil fuel substitution. What is needed is the right policy framework and the will to act.

10.6 Utilization Trends, Conclusions and Recommendations

10.6.1 Utilization trends – general discussion

Utilization trends for poplars and willows will be influenced not only by market forces and technology but also by new public policies which will recognize the benefits of using renewable materials and energy wherever possible. In general, the wood products industries have not always promoted the many environmental benefits of their products. Moreover, the general public, i.e. the consumer, is not aware of the fact that wise wood utilization and sustainable forest management practices can actually benefit the environment in tangible ways. Unfortunately, most often, forests and wood only get into the media when it is 'bad news', for example deforestation, destruction of wildlife habitat, home and forest fires, etc. Sustainable forest management can and should be practised on forest lands for the benefit of humanity.

> It has been estimated that an annual 4% increase to 2010 in Europe's wood consumption would sequester an additional 150 million tonnes CO_2 per year and that the market value of this environmental service would be about 1.8 billion Euros per year.
>
> (CEI-Bois, 2012)

This quote from the *Roadmap 2010* for the European woodworking industries foretold emerging policy changes. It was expected to have a profound influence on the growing and utilization of wood in general, and of poplars and willows in particular, not only in Europe but also throughout the world in the years to come. Appropriately, the title of the roadmap is: 'Tackle climate change – use wood!'

The roadmap emphasizes the significant advantages of wood over other building materials, such as steel and concrete. Wood has the lowest 'embodied energy', i.e. the energy used to create a material or product, and the best thermal efficiency of any building material. Thus, when wood is substituted for other materials in buildings, it contributes to the savings of CO_2 emissions throughout a building's life (CEI-Bois, 2012).

With a strong policy framework to reduce greenhouse gas emissions already in place, Europe will most likely lead the way to enact new policy measures to stimulate the substitution of wood to replace energy-intensive materials wherever possible. Similar measures will likely follow in other parts of the world. Fast-growing poplar and willow plantations can and will likely supply some of the needed raw material. Carbon taxes and carbon credits may become just two of the policy tools governments can use to bring about change in the way materials are used and specified. Wood products with long service life, which enable the sequestering of significant quantities of CO_2, should qualify for carbon credits, which might work as a 'tax incentive'.

The potential expansion in the use of poplar- and willow-based wood products in buildings and elsewhere will also bring new responsibilities for improved product quality and reliable, long-term performance, especially durability and fire performance. Since poplars and willows have low natural durability, this characteristic might be alleviated either through genetic engineering with growing trees to make the wood pest resistant or through post-manufacturing preservative treatment, or a combination of the two. Manufacturers of all products, especially composites, must follow stringent quality control procedures in their industrial plants and, for construction applications, comply with the specifications of building and fire codes. In order to compete in construction markets, manufacturers of engineered composites made of poplar have to provide not only a high-quality product but also a comprehensive design and specifications package to the users of their products, e.g. architects and engineers.

The application of science and technology will influence the utilization of poplars and willows during the coming years. This influence will probably extend from the application of genetic engineering to improve wood quality traits and pest resistance, to the acceleration of biomass production (so-called 'tailor-made' wood supply). On the resource conversion side, new technologies will be developed and put in place to derive energy, chemicals, for example polymers and adhesives, and speciality chemicals, for example pharmaceuticals, cosmetics, from poplar and willow biomass, for example 'bio-refinery'. In the area of adhesive chemistry, which is vital for the manufacture of all composites, novel isocyanate resins are under development. These can cure at room temperature and tolerate higher wood MC than is currently the practice, which will save energy in wood drying and pressing, as well as minimize the emissions of volatile organic compounds (VOCs). During the coming years, new types of high-performance composites will likely emerge which may incorporate cellulosic nanoparticles or whiskers in cross-linked polymer matrices. These composites may have foam-like structure to improve weight efficiencies and insulation performance. New and more efficient process technologies, such as radiation-assisted curing, will likely emerge for existing structural composites. Finally, composites based on new material formulations, fo example using wood components, polymers and inorganic constituents, will be developed to improve structural and fire performance in buildings.

10.6.2 Conclusions and recommendations

Poplars and willows are suitable for a broad range of products, including traditional wood products, for example lumber and plywood, composites, pulp and paper, energy and chemicals. These products in turn are converted and utilized in numerous end-use applications, such as building components, furniture, containers, transportation goods and equipment, tissue and printing papers, chemicals and energy to replace fossil fuels. One major advantage of poplars and willows is that they can be produced in relatively short rotation times to produce maximum yields of fibre on a sustainable basis.

Future poplar and willow utilization should be based on an integrated and total use concept, i.e. produce the highest-value products or product combinations from a given raw material input, so that nothing is wasted. Future growth in poplar and willow utilization will most likely come in the area of composite products for existing and new applications, and in bioenergy and chemicals. The wise application of biotechnology will likely make it possible to produce 'tailor-made' trees to suit different applications. Stable investment in appropriate research, development and innovation and further international cooperation will advance poplar and willow utilization for the benefit of society.

References

Abalos-Romero, M.I. (2005) Towards the development of the Chilean basket willow sector. *Unasylva* 221, 40–46.

Alden, H.A. (1995) Hardwoods of North America. General Technical Report FPL-GTR-83. US Department of Agriculture, Forest Service, Forest Products Laboratory, Madison, Wisconsin.

Alvarez, C., Romero, P. and Padro, A. (2004) Study of the quality of poplar wood for veneer peeling in an experimental plantation. Presented to International Poplar Commission 22nd Session, 28 November–2 December, 2004, Santiago. Available from IPC Secretariat, FAO, Rome.

Anderson, H.W. and Zsuffa, L. (1975) Yield and wood quality of hybrid cottonwood grown in two year rotation. Forestry Research Report 101. Ontario Ministry of Natural Resources, Forest Research Branch, Maple, Ontario, Canada.

Anderson, H.W., Papadopol, C.S. and Zsuffa, L. (1983) Wood energy plantations in temperate climates. *Forest Ecology and Management* 6, 281–306.

Arganbright, D.G., Bensend, D.W. and Manwiller, F.G. (1970) Influence of gelatinous fibers on the shrinkage of silver maple. *Wood Science* 3, 83–89.
Armstrong, A., Johns, C. and Tubby, I. (1999) Effects of spacing and cutting cycle on the yield of poplar grown as an energy crop. *Biomass and Bioenergy* 17, 305–314.
Armtec (2012) Armtec Limited Partnership, Guelph, Ontario, Canada (http://www.armtec.com, accessed 19 January 2012).
Badia, M.A., Mothe, F., Constant, T. and Nepveu, G. (2005) Assessment of tension wood detection based on shiny appearance for three poplar clones. *Annals of Forest Science* 62, 43–49.
Badia, M.A., Constant, T., Mothe, F. and Nepveu, G. (2006) Tension wood occurrence in three cultivars of *Populus* × *euramericana*. Part I: Inter-clonal and intra-tree variability of tension wood. *Annals of Forest Science* 63, 23–30.
Balatinecz, J. and Kretschmann, D. (2001) Properties and utilization of poplar wood. In: Dickmann, D.I., Isebrands, J.G., Eckenwalder, J.E. and Richardson, J. (eds) *Poplar Culture in North America*. National Research Council of Canada Research Press, Ottawa, pp. 277–291.
Balatinecz, J. and Sain, M. (2007) Cars made of wood and hemp fibres? Why not? *Forestry Chronicle* 83, 482–484.
Baldwin, R.F. (1981) *Plywood Manufacturing Practices,* revised 2nd edn. Miller Freeman, San Francisco, California.
Ball, J., Carle, J. and Del Lungo, A. (2005) Contribution of poplars and willows to sustainable forestry and rural development. *Unasylva* 221, 3–9.
Bauch, J., Liese, W. and Berndt, H. (1976) Biological investigations for the improvement of the permeability of softwoods. *Holzforschung* 24, 199–205.
Berkenkamp, R. (1997) Wood fibre–cement products, process and properties. In: Moslemi, A.A. (ed.) *Inorganic Bonded Wood and Fibre Composite Materials.* Proceedings 5th International Conference, Sun Valley, Idaho, USA, 2006. Forest Products Society, Madison, Wisconsin, pp. 8–13.
Blankenhorn, P.R., Bowersox, T.W., Kuklewski, K.M. and Stimely, G.L. (1985) Effects of rotation, site, and clone on the chemical composition of *Populus* hybrids. *Wood Science* 17, 351–360.
Canadian Wood Council (2007) Laminated veneer lumber (LVL). Canadian Wood Council (www.cwc.ca/index.php?option=com_content&view=article&id=203&Itemid=459, accessed 19 January 2012).
Castro, G. and Fragnelli, G. (2006) New technologies and alternative uses for poplar wood. *Boletin del Centro de Investigación y Documentación del Eucalipto* 2, 27–36.
CEI-Bois (2012) *Roadmap 2010.* European Confederation of Woodworking Industries. PowerPoint presentation (http://www.cei-bois.org/en/roadmap-2010/wood-in-sustainable-development, accessed 19 January 2012).
Chen, Z., Zhang, Q. and Hong, Z. (1994) Research on poplar parallel strand lumber. In: Chao, C., Sasaki, H., Hua, Y., Zhou, D., Li, R., Huang, Y., *et al.* (eds) *Properties and Utilization of Fast-growing Trees.* Proceedings, International Symposium on the Utilization of Fast-Growing Trees, Nanjing, China, 15–17 October, 1994. China Forestry Publishers, Beijing, pp. 350–356.
Clair, B., Almeras, T. and Sugiyama, J. (2006) Compression stress in opposite wood of angiosperms: observations in chestnut, mani and poplar. *Annals of Forest Science* 63, 507–510.
Clausen, V.H. and Kaufert, F.A. (1952) Occurrence and probable cause of heartwood degradation in commercial species of *Populus. Journal of Forest Products Research Society* 2, 62–67.
Clemons, C. (2002) Wood–plastic composites in the United States: the interfacing of two industries. *Forest Products Journal* 52, 10–18.
Cooper, P.A. (1976) Pressure-preservative treatment of poplar lumber. *Forest Products Journal* 26, 28–31.
Cote, W.A. Jr, Day, A.C. and Timell, T.E. (1969) A contribution to the ultra-structure of tension-wood fibers. *Wood Science and Technology* 3, 257–271.
De, S. and Assadi, M. (2009) Impact of cofiring biomass with coal in power plants – a techno-economic assessment. *Biomass and Bioenergy* 33, 283–293.
DeBell, D.S., Singleton, R., Harrington, C.A. and Gartner, B.L. (2002) Wood density and fiber length in young *Populus* stems: relation to clone, age, growth rate and pruning. *Wood and Fiber Science* 34, 529–539.
De Boever, L., Vansteenkiste, D. and Van Acker, J. (2005) Using poplar timbers in light constructions: the problem of non-uniform moisture distributions after drying. In: Teischinger, A. and Van Acker, J. (eds) *Proceedings of the COST Action E44 Wood Processing Strategy Conference on Broad Spectrum Utilization of Wood at BOKU, Vienna, Austria, June 14–15, 2005.* Lignovisionen, issue 9 (Special Edition). Book series of the Institute of Wood Science and Technology, Vienna, pp. 111–120.

De Boever, L., Vansteenkiste, D., Van Acker, J. and Stevens, M. (2007) End-use related physical and mechanical properties of selected fast-growing poplar hybrids (*Populus trichocarpa* × *P. deltoides*). *Annals of Forest Science* 64, 621–630.

De Boever, L., Van Acker, J., Vansteenkiste, D. and Stevens, M. (2008) Preservative treatment of willow wood (*Salix alba*): product retention and spatial distribution. *Wood Research* 53, 29–42.

Deka, G.C., Wong, B.M. and Roy, D.N. (1992) Suitability of hybrid willow as a source of pulp. *Journal of Wood Chemistry and Technology* 12, 197–211.

Dibbelt, R. (2005) Willow – a new crop for Alberta's landowners. *Western Farm and Forest*, December, 2005, pp. 32.

Dickson, R.E., Larson, P.R. and Isebrands, J.G. (1974) Differences in cell-wall chemical composition among eighteen three-year-old *Populus* hybrid clones. In: Bey, C.F. (ed.) *Proceedings 9th Central States Forest Improvement Conference, October 10–11, 1974, Ames, Iowa, USA.* USDA Forest Service, Upper Darby, Pennsylvania, pp. 21–34.

Eaton, J. (2008) Agroforestry and cellulosic ethanol from sustainable tree farms. Presentation at Harvesting Clean Energy VIII Conference, 27–29 January, 2008, Portland, Oregon (http://harvestcleanenergy.org/2010-conference-resources/2008-harvesting-clean-energy-conference/2008-conference-presentations/eaton-pdf.pdf, accessed 19 January 2012).

Eltomation (2007) Eltomation BV, Barneveld, Holland (http://www.eltomation.com, accessed 19 January 2012).

Ericsson, K. and Nilsson, L.J. (2006) Assessment of the potential biomass supply in Europe using a resource-focused approach. *Biomass and Bioenergy* 30, 1–15.

Fang, S., Yang, W. and Tian, Y. (2006) Clonal and within tree variation of microfibril angle in poplar clones. *New Forests* 31, 373–383.

FAO (1980) *Poplars and Willows in Wood Production and Land Use.* Forestry Series No 10. Food and Agriculture Organization of the United Nations, Rome.

Fujiwara, S. and Yang, K.C. (2000) The relationship between cell length and ring width and circumferential growth rate in five Canadian species. *International Association of Wood Anatomists Journal* 21, 335–345.

Gartner, B.L. (2005) Assessing wood characteristics and wood quality in intensively managed plantations. *Journal of Forestry* 103, 75–77.

Geimer, R.L. and Crist, J.B. (1980) Structural flakeboard from short-rotation intensively cultured hybrid *Populus* clones. *Forest Products Journal* 30, 42–48.

Giordano, G. (1974) Spaander-en vezelplaten vervaardigd uit populier. *Populier* 11, 73–75.

González-García, S., Gasol, C.M., Gabarrell, X., Rieradevall, J., Moreira, M.T. and Feijoo, G. (2010) Environmental profile of ethanol from poplar biomass as transport fuel in Southern Europe. *Renewable Energy* 35, 1014–1023.

Hall, J.P. (2002) Sustainable production of forest biomass for energy. *Forestry Chronicle* 78, 391–396.

Hall, R.B., Hilton, G.D. and Maynard, C.A. (1982) Construction lumber from hybrid aspen plantations in the central states. *Journal of Forest*ry 80, 291–294.

Hernandez, R.E., Koubaa, A., Beaudoin, M. and Fortin, Y. (1998) Selected mechanical properties of fast-growing poplar hybrid clones. *Wood Fiber Science* 30, 138–147.

Hoffmann, D. and Weih, M. (2005) Limitations and improvements of the potential utilization of woody biomass for energy derived from short rotation woody crops in Sweden and Germany. *Biomass and Bioenergy* 28, 267–279.

Holt, D.H. and Murphy, W.K. (1978) Properties of hybrid poplar juvenile wood affected by silvicultural treatments. *Wood Science* 10, 198–203.

Hua, Y. and Zhou, X. (1994) The development of the hybrid poplar processing industry in China. Report of Nanjing Forestry University, Nanjing, China.

Hua, Y., Wang, S. and Du, G. (1994) Research on the technology of laminated veneer lumber of fast-growing poplar. In: Chao, C., Sasaki, H., Hua, Y., Zhou, D., Li, R., Huang, Y., *et al.* (eds) *Properties and Utilization of Fast-growing Trees.* Proceedings, International Symposium on the Utilization of Fast-Growing Trees, Nanjing, China, 15–17 October, 1994. China Forestry Publishers, Beijing, pp. 379–385.

International Energy Agency, Bioenergy Agreement (2008) IEA Bioenergy (http://www.ieabioenergy.com, accessed 19 January 2012).

Isebrands, J.G. and Bensend, D.W. (1972) Incidence and structure of gelatinous fibers within rapid growing cottonwood. *Wood Science* 4, 61–71.

Isebrands, J.G. and Parham, R.A. (1974) Tension wood anatomy of short-rotation *Populus* spp. before and after kraft pulping. *Wood Science* 6, 256–265.

Isebrands, J.G., Sturos, J.A. and Crist, J.B. (1979) Integrated utilization of biomass: a case study of short rotation intensively cultured *Populus* raw material. *TAPPI Journal* 62, 67–70.
Jourez, B., Riboux, A. and Leclercq, A. (2001) Anatomical characteristics of tension wood and opposite wood in young inclined stems of poplar (*Populus euramericana* cv 'Ghoy'). *International Association of Wood Anatomists Journal* 22, 133–157.
Jourez, B., Vaianopoulos, C. and Hebert, J. (2003) Effect of intensity of gravitational stimulus artificially induced on growth and tension wood formation in young shoots of poplar (*P. euramericana* cv 'Ghoy'). *Annals of Forest Science* 60, 789–802.
Kellogg, R.M. and Swan, E.P. (1985) Physical properties of black cottonwood and balsam poplar. *Canadian Journal of Forest Research* 16, 491–496.
Kenney, W.A., Gambles, R.L. and Vanstone, B.J. (1996) Willow biomass prototype farms of the University of Toronto: status and future plans. In: Karau, J. (ed.). *Proceedings of the Canadian Energy Plantation Workshop*, Gananoque, Ontario, Canada, 2–4 May, 1995. Natural Resources Canada, Canadian Forest Service, Ottawa, pp. 19–22.
Knudson, B. and Brunette, G. (2002) Hybrid poplar for plywood and laminated veneer lumber (LVL). Technote 02-15W. Forintek Canada Corporation, Quebec, Canada.
Knuth, D.T. (1964) Bacteria associated with wet woods and their effect on certain properties of wood. *Dissertation Abstracts* 25, 175–182.
Koubaa, A., Hernandez, R.E. and Beaudoin, M. (1998a) Shrinkage of fast-growing hybrid poplar clones. *Forest Products Journal* 48, 82–87.
Koubaa, A., Hernandez, R.E., Beaudoin, M. and Poliquin, J. (1998b) Interclonal, intraclonal and within-tree variation in fiber length of poplar hybrid clones. *Wood Fiber Science* 30, 40–47.
Kretschmann, D.E., Isebrands, J.G., Stanosz, G., Dramm, J.R., Olstad, A., Cole, D. and Samsel, J. (1999) Structural lumber properties of hybrid poplar. Research Paper FPL-RP-573. US Department of Agriculture Forest Service, Forest Products Laboratory, Madison, Wisconsin.
Kroll, R.E., Ritter, D.C., Gertejansen, R.O. and Au, K.C. (1992) Anatomical and physical properties of balsam poplar (*Populus balsamifera* L.) in Minnesota. *Wood Fiber Science* 24, 13–24.
Labosky, R., Bowersox, T.W. and Blankenhorn, P.R. (1983) Kraft pulp yields and paper properties from first and second rotations of three hybrid poplar clones. *Wood Fiber Science* 15, 81–89.
Larsson, S., Melin, G. and Rosenqvist, H. (1998) Commercial harvest of willow wood chips in Sweden. In: Kopetz, H., Weber, T., Palz, W., Chartier, P. and Ferrero, G.L. (eds) *Proceedings of the 10th European Conference and Technology Exhibition, Biomass for Energy and Industry*, Würzburg, Germany, 8–11 June, 1998. CARMEN, Rimpar, Germany, pp. 200–203.
Law, K.N. and Rioux, S. (1997) Five short-rotation poplar clones grown in Quebec: wood and paper-making properties. In: Zhang, S.Y., Grosselin, R. and Chauret, G. (eds) *Timber Management Toward Wood Quality and End Product Value.* Proceedings of the CTIA–IUFRO International Wood Quality Workshop, Quebec, Canada, 18–22 August, 1997, pp. VII.19–VII.28.
Leclercq, A. (1997) Wood quality of white willow. *Biotechnologie, Agronomie, Société et Environnement* 1, 59–64.
McMillan, J.D. (2004) Biotechnological routes to biomass conversion. Presentation to DOE/NASULGC Biomass and Solar Energy Workshops, 3–4 August, 2004. US Department of Energy, National Renewable Energy Laboratory, Golden, Colorado (http://www.nrel.gov/docs/gen/fy04/36831g.pdf, accessed 20 January 2012).
Matyas, C. and Peszlen, I. (1997) Effect of age on selected wood quality traits of poplar clones. *Silvae Genetica* 46, 64–72.
Mississippi State University (2009) TimTek at Mississippi State University (http://www.cfr.msstate.edu/timtek/, accessed 20 January 2012).
Monteoliva, S., Senisterra, G. and Marlatas, R. (2005) Variation of wood density and fiber length in six willow clones. *International Association of Wood Anatomists Journal* 26, 197–202.
Moslemi, A.A. (1988) Wood–cement panel products: coming of age. In: Proceedings 1st International Conference on Fiber and Particleboards Bonded with Inorganic Binders, Moscow, Idaho. University of Idaho, Moscow, Idaho, pp. 2–18.
Mullins, E.J. and McKnight, T.S. (eds) (1981) *Canadian Woods: Their Properties and Uses,* 3rd edn. University of Toronto Press, Toronto, Ontario, Canada.
Murphy, R.J., Din, S.U. and Stone, M.J. (1991) Observations on preservative penetration in poplar. IRG/WP 3662. International Research Group on Wood Preservation, Stockholm.
Otto, H. (1989) Ergebnisse orientierendes untersuchungen zur verwertbarkheit von Baumwerde in der industrie. *Holztechnologie* 30, 85–89.

Pan, X., Gilkes, N., Kadia, J., Pye, K., Saka, S., Gregg, D., *et al.* (2006) Bioconversion of hybrid poplar to ethanol and co-products using an organosolv fractionation process: optimization of process yields. *Biotechnology and Bioengineering* 94, 851–861.

Panshin, A.J. and de Zeeuw, C. (1980) *Textbook of Wood Technology,* 4th edn. McGraw-Hill, New York.

Parham, R.A., Robinson, K.W. and Isebrands, J.G. (1977) Effect of tension wood on kraft paper from short rotation hardwoods. *Wood Science and Technology* 11, 291–303.

Patterson, D.J. (2005) Producing alternative fuels from woody biomass. Forest Engineering Research Institute of Canada (www.feric.ca/en/index.cfm?objectid=764736CD-C09F-3A58-EAEF4A1B-629F1F8B, accessed 20 January 2012).

Richardson, J., Björheden, R., Hakkila, P., Lowe, A.T. and Smith, C.T. (eds) (2002) *Bioenergy from Sustainable Forestry. Guiding Principles and Practice.* Kluwer Academic Publishers, Dordrecht, Netherlands.

Rittershofer, B. and Hofmann, G. (1999) Silberweide: Baum es Jahres 1999. *Allgemeine Forstzeitschrift für Waldwirtschaft und Umweltvorsorge* 23, 1256–1263.

Sassus, F., Fournier, M. and Thibaut, B. (1995) Longitudinal growth strains and drying shrinkage in tension wood of poplars (*Populus euramericana* cv. I-214*). International Association of Wood Anatomists Journal* 16, 16–17.

Scheithauer, M. (1999) Investigations on the application of poplar and white willow wood at IHD. In: Wegener, G., Berti, S., Macchioni, N., Negri, M. and Rachello, E. (eds) *Industrial End-uses of Fast-grown Species.* Proceedings of Eurowood Technical Workshop, Florence, Italy, 31 May–1 June, 1999. National Research Council, Wood Research Institute, Florence, Italy, pp. 103–123.

Schoch, W., Heller, I., Schweingruber, F.H. and Keinast, F. (2004) Wood anatomy of central European species (www.woodanatomy.ch, accessed 20 January 2012).

Schuler, A. (2002) Marketability of hardwood LVL. US Department of Agriculture, Forest Service, Princeton, West Virginia.

Sierra-Alvarez, R. and Tjeerdsma, B.F. (1995) Organosolv pulping of poplar wood from short-rotation intensive culture plantations. *Wood Fiber Science* 27, 395–401.

Singh, A., Dawson, B., Franich, R., Cowan, F. and Warnes, J. (1999) The relationship between pit membrane ultrastructure and chemical impregnability of wood. *Holzforschung* 53, 341–346.

Siren, G., Sennerby-Forsse, L. and Ledin, S. (1987) Energy plantations – short rotation forestry in Sweden. In: Hall, D.O. and Overend, R.P. (eds) *Biomass: Regenerable Energy.* Wiley, Chichester, UK.

Stanton, B., Eaton, J., Johnson, J., Rice, D., Schuette, B. and Moser, B. (2002) Hybrid poplar in the Pacific Northwest: the effects of market-driven management. *Journal of Forestry* 100, 28–33.

Suzuki, S. (2005) Using cedar plantation materials for wood-based-composites in Japan. In: Winandy, J.E., Wellwood, R.W. and Hiziroglu, S. (eds) *Using Wood Composites as a Tool for Sustainable Forestry.* General Technical Report FPL-GTR-163. Proceedings of Scientific Session 90, XXII IUFRO World Forestry Congress, 8–13 August, 2005, Brisbane, Australia. US Department of Agriculture, Forest Service, Forest Products Laboratory, Madison, Wisconsin, pp. 9–25.

Szczukowski, S., Tworkowski, J., Klasa, A. and Stolarski, M. (2002) Productivity and chemical composition of wood tissues of short rotation willow coppice cultivated on arable land. *Rostlinna Vyroba* 48, 413–417.

Thomas, G.P. (1968) Decay as a limiting factor on poplar utilization. In: Maini, J.S. and Cayford, J.H. (eds) *Growth and Utilization of Poplars in Canada.* Publication No 1205. Canada Department of Forestry and Rural Development, Ottawa, pp. 145–148.

Tillman, D.A. (2000) Biomass cofiring*:* the technology, the experience, the combustion consequences. *Biomass and Bioenergy* 19, 365–384.

Tröger, F. and Wegener, G. (1999) Oriented structural boards (OSB) from fast-growing poplar and willow. In: Wegener, G., Berti, S., Macchioni, N., Negri, M. and Rachello, E. (eds) *Industrial End-uses of Fast-grown Species.* Proceedings of Eurowood Technical Workshop, Florence, Italy, 31 May–1 June, 1999. National Research Council, Wood Research Institute, Florence, Italy, pp.135–145.

Troya, M.T., Navarette, A., Sanchez, E., Garcia de los Rios, J.E. and Rodriguez Trobajo, E. (1995) Evaluation of the permeability of fast grown wood species. Paper prepared for the 26th Annual Meeting, International Research Group on Wood Preservation, Helsingor, Denmark, 11–16 June, 1995. International Research Group on Wood Preservation, Stockholm.

US Department of Agriculture (1999) *Wood Handbook – Wood as an Engineering Material.* General Technical Report FPL-GTR113. US Department of Agriculture, Forest Service, Forest Products Laboratory, Madison, Wisconsin.

US Environmental Protection Agency (1995) EPA Office of Compliance Sector Notebook Project. Profile of the lumber and wood products industry. Publication EPA/310-R-95-006 (http://www.epa.gov/compliance/resources/publications/assistance/sectors/notebooks/lmbrwdsn.pdf, accessed 20 January 2012).

Van Acker, J. and Stevens, M. (1995) Investigation into the heterogeneous nature of the impregnability of some poplar hybrids. Paper prepared for 26th Annual Meeting, International Research Group on Wood Preservation, Helsingor, Denmark, 11–15 June, 1995. International Research Group on Wood Preservation, Stockholm, Sweden.

Van Acker, J., Stevens, M. and de Haas, C. (1995) Influence of clonal variability on the impregnability of some poplar hybrids. Paper prepared for 26th Annual Meeting, International Research Group on Wood Preservation, Helsingor, Denmark, 11–15 June, 1995. International Research Group on Wood Preservation, Stockholm.

van der Zwan, R.P. (1987) Populierehout een grondstof. *Populier* 24, 53–57.

Vansteenkiste, D., Stevens, M. and Van Acker, J. (1997) High temperature drying of fresh sawn poplar wood in an experimental convective dryer. *Holz als Roh- und Werkstoff* 55, 307–314.

Volk, T.A., Abrahamson, L.P., White, E.H., Neuhauser, E., Gray, E., Demeter, C., *et al.* (2000) Developing a willow biomass crop enterprise for bioenergy and bioproducts in the United States. Presented to 'Bioenergy 2000, Moving Technology Into the Marketplace', Buffalo, New York, 15–19 October, 2000. Northeast Regional Biomass Program, Albany, New York (https://bioenergy.ornl.gov/papers/bioen00/volk.html, accessed 18 March 2013).

Volk, T.A., Verwijst, T., Tharakan, P.J., Abrahamson, L.P. and White, E.H. (2004) Growing fuel: a sustainability assessment of willow biomass crops. *Frontiers in Ecology and the Environment* 2, 411–418.

Volk, T.A., Abrahamson, L.P., Nowak, C.A., Smart, L.B., Tharakan, P.J. and White, E.H. (2006) The development of short rotation willow in the northeastern United States for bioenergy and bioproducts, agroforestry and phytoremediation. *Biomass and Bioenergy* 30, 715–727.

Williams, D. (1998) *Machining, Laminating, Fastener Withdrawal and Finishing Properties of Hybrid Poplar.* Forintek Canada Corporation, Quebec, Canada.

Yong, P. (2005) Wood pulp sector: insatiable Chinese appetite for fiber and pulp. Sector Report. DBS Vickers Securities, Singapore.

Zsuffa, L. (1988) A review of progress in selecting and breeding North American *Salix* species for energy plantations at the Faculty of Forestry, University of Toronto, Canada. In: Eriksson, G. (ed.) *Proceedings, IEA Willow Breeding Symposium*, Uppsala, Sweden, 31 August–1 September, 1987. Research Notes 41. Swedish University of Agricultural Sciences, Department of Forest Genetics, Uppsala, Sweden, pp. 41–51.

11 Markets, Trends and Outlook

Qiang Ma* and Arvydas Lebedys
Forestry Department, FAO, Rome, Italy

11.1 Introduction

Over the past decade, the global wood-based products market has changed significantly, driven by overall economic development and more specifically by the globalization and the shifts of competitiveness from developed countries to developing countries in processing wood-based products. At the same time, environmental protection efforts, such as biodiversity conservation and mitigation of climate change, have been emphasized and strengthened in many countries. This chapter will explore the impacts of these overall changes to the markets, trends and outlook of poplars and willows.

11.2 Overall Global Market Trends for Wood Products

11.2.1 Transformation of market structures – towards emerging economies

Globalization has been a major driving force in recent years and has been supported by policies that have reduced barriers to the movement of goods, capital and technology across national boundaries. For the forest sector, the main direct effect of globalization has been access to lower labour and regulatory costs, coupled with a reduction in transport costs, that have led to increased trade in forest products. The main indirect effect of globalization has been changes in the competitiveness of wood products, which have resulted in the evolution of market structures.

The European Forest Sector Outlook Study (UNECE/FAO, 2005) revealed that globalization of the competitiveness of the European forest sector would include increased international flows of capital, cross-border mergers and relocation of companies across national boundaries. Globalization and innovation are perceived as having a more stimulating effect in Eastern Europe and the Commonwealth of Independent States (CIS) subregion[1] than in Western Europe. In fact, wood processing units have shifted from Western European countries to Eastern European countries, from where semi-finished or end products are exported to the Western European market. High and increasing wages and increasing transport costs have resulted in weaker competitiveness in Western European countries (R. Novakov and P. Mertens, Belgium, 2006, unpublished poster presentation).

China has taken advantage of globalization and gained competitiveness in many industrial products. It is said that China has become the

*Corresponding author: qiang.ma@fao.org

 Poplars and Willows: Trees for Society and the Environment (eds J.G. Isebrands and J. Richardson)

'world processing centre', producing and exporting end products by consuming imported raw material. In the forestry sector, both the production and trade of forest products have increased significantly over the past decade, which has changed the global market pattern of some wood-based products. Meanwhile, the domestic wood industry has shifted from resource-oriented development to market-oriented development. The wood products consuming area has become the wood products producing area. For example, three provinces (Hebei, Jiangsu and Shandong) produced more than half of the total wood-based panels while producing only slightly more than 3% of the total roundwood. On the other hand, five provinces (Fujiang, Jiangxi, Guangxi, Helongjiang and Hunan) produced nearly half of the total roundwood while producing only about 18% of the total wood-based panels in 2005 (SFA, 2006).

Many countries in Asia, Latin America and Eastern Europe are emerging as important exporters of wood products with an increasing share of high-value products in the world market. Privatization of wood industries, low wages, easy access to technology, availability of skilled and semi-skilled workers and improved efficiency of marketing chains have revolutionized the production of value-added products, creating new opportunities for a number of developing countries.

11.2.2 Global trends in production of wood products

Over the period 1990–2010 (FAO, 2011), the wood-based panels sector has experienced very dynamic growth; production of wood-based panels increased on average by 3.7% annually. The increase in paper and paperboard production over the same period is slightly lower, i.e. 2.5% $year^{-1}$. On the other hand, the production of sawn wood has shown a decreasing trend, declining slightly at an average rate of 0.9% $year^{-1}$. One key driving force for the increase of production of wood-based panels, paper and paperboard has been the overall macroeconomic conditions; specifically, world total GDP has increased at an average rate of about 3% $year^{-1}$ over the period 1990–2010. However, the worldwide collapse of the housing market and the economy starting in early 2008 has led to decreased production, cutbacks and unemployment in the wood products industry. In the USA, production cutbacks and consumption declines were significant in nearly every segment of the wood products industry through mid-2009. In Canada, the forest, paper and packaging industries are seeing extremely poor financial results from the recession (Taylor, 2009). Nevertheless, China's rapid and continuous economic expansion over the past two decades, which has become a major engine of global growth, has been generating tremendous increases in the production and trade of forest products. In addition, growth in other countries like India and some Eastern European countries has further resulted in a change to the global wood products market. For example, in paper and paperboard production, among the major producer regions North America declined by 17.2% between 2004 and 2009 and Europe by 6.7%. Strong production growth in Asia (mainly China) was unable to compensate fully for the falling output in 'old' paper continents. China passed the USA as the world's biggest paper and board producer in 2008. Asia held a 43% share of the global output in paper and board in 2009, ahead of Europe (27%) and North America (23%). International paper and paperboard companies are increasingly moving their production capacities into the fast-growth markets like China and India to tap local opportunities and the rising demand for paper of the emerging economies. Paper production is moving close to the largest consumer markets, unlike the pulp industry, which is expanding close to abundant wood resources (Tissari, 2011). The slight decrease in sawn wood consumption is attributed mainly to the change in consumption patterns, the development of new technology for wood-based panel production and the change in roundwood supply, i.e. the size of roundwood has become much smaller.

11.3 Poplar and Willow Resources and Products: Production and Trade

11.3.1 Poplar and willow resources for wood production

According to the syntheses of country progress reports of the 22nd and 23rd Sessions of the International Poplar Commission (IPC) (FAO, 2004,

2008), an estimated 70 countries grow poplars and willows in mixture with other natural forest species in planted forests and as individual trees in the landscape (including agroforestry systems). The total area of poplar and willow in the world is estimated at 79.1 million ha, of which the area of natural poplar and willow forests is 71.1 million ha and the area of poplar and willow plantations is 5.4 million ha. The area of poplar and willow in agroforestry systems is about 2.6 million ha. The largest areas of naturally occurring poplar are reported in Canada (40%), the Russian Federation (31%) and the USA (18%). China dominates the area of planted poplar forests and poplar trees in agroforestry in the world, accounting for 81% and 95%, respectively.

Of the total of about 5 million ha of productive poplar plantations in the world, China accounts for 83% and Europe accounts for 16% (Fig. 11.1). Significant expansion of poplar plantations has been achieved in China. Poplar plantations are the second largest forest plantations in China, accounting for 19% of the total (SFA, 2009b). In terms of forest area and timber production, poplars are one of the most important tree genera in China. The significant expansion of the poplar resource in China has been driven mainly by the fast development of forest plantations for both environmental and production purposes, and by the rapid development of the wood-based industry over the past decade. France, Turkey, Hungary, Germany, Spain, Italy, Serbia and Belgium are the main countries of productive poplar and willows in Europe (Fig. 11.2).

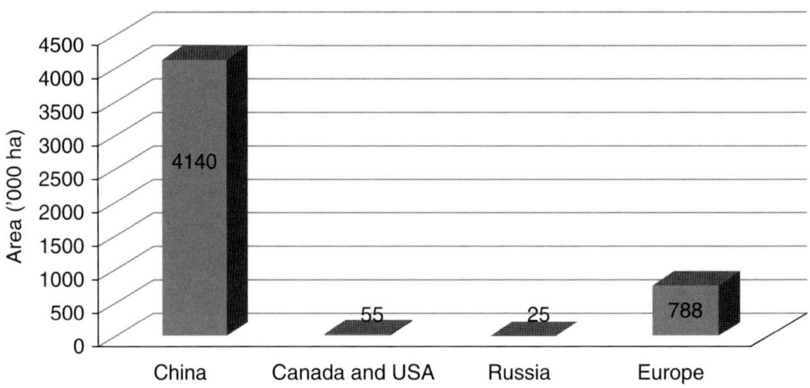

Fig. 11.1. Area of productive poplar plantations by region. Source: FAO (2008).

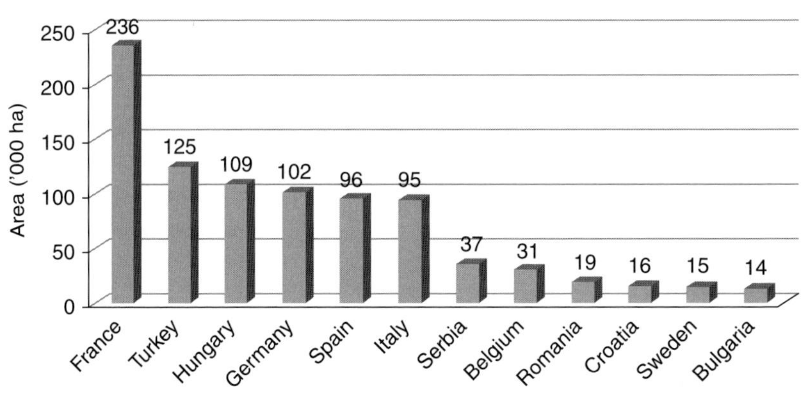

Fig. 11.2. Area of productive poplar and willow plantations in Europe. Source: FAO (2008).

11.3.2 Production and trade of poplar and willow products

According to available information (FAO, 2008), plantations and agroforestry systems are the main sources for the production of wood-based products from poplars and willows, accounting for 97% and 2%, respectively. The total annual production of various products was about 86.55 million m^3, of which plywood and veneer, reconstituted wood panels, pulp and paper and sawn wood were the four main products, with shares of 60%, 21%, 11% and 5%, respectively (Fig. 11.3). China dominates all types of poplar and willow wood-based products, accounting for 91% of the world total (FAO, 2008).

Poplar and willow roundwood and wood chips are traded mainly among European countries. In addition, the Republic of Korea imports roundwood from the Russian Federation, China, USA, Finland and Croatia. In world trade of poplar roundwood, Italy is the biggest importer, followed by France, Belgium, Bulgaria and Croatia. Belgium is the biggest exporter (FAO, 2008).

11.4 Evolution of the Poplar Products Market in China, Europe and North America

11.4.1 China has become the main engine in the production and trade of wood and poplar wood products in the world

China has become the most dynamic and hottest timber market in the world, with significant impacts on the global market. Over the period 2000–2010, China systematically has reduced imports of primary processed wood products such as panels and veneer and established domestic production based on imported logs and chips. Despite raw material imports, China remains competitive due to low labour and manufacturing costs. Domestic consumption of value-added wood products, such as furniture and flooring, is increasing, together with a rise in the standard of living. Much of the production is exported to North America and Europe. China exports wooden furniture to North America (46%), Europe (23%), Asia (23%), the Pacific (5%), Africa (2%) and Latin America (2%) (SFA, 2009a). The development of domestic production in roundwood, sawn wood and wood-based panels over the period 1990–2010 in China has changed dramatically. Roundwood and sawn wood production gradually decreased over the 1990s. This trend was driven mainly by the degradation of the natural forest, specifically the implementation of the 'logging ban' in 1998. Since 2003, there has been a slight increase in roundwood production. This trend has resulted from increased production from agroforestry and forest plantations (SFA, 2006). Wood-based panel development has been remarkable. Production reached nearly 94 million m^3 in 2008, which was more than ten times the level in 1998 (SFA, 2009b). China has become the largest producer of wood-based panels in the world (FAO, 2011). Wooden furniture, wood-based panels and paper and paperboard have achieved the fastest growth rates in the Chinese forestry industry over the period 2000–2010.

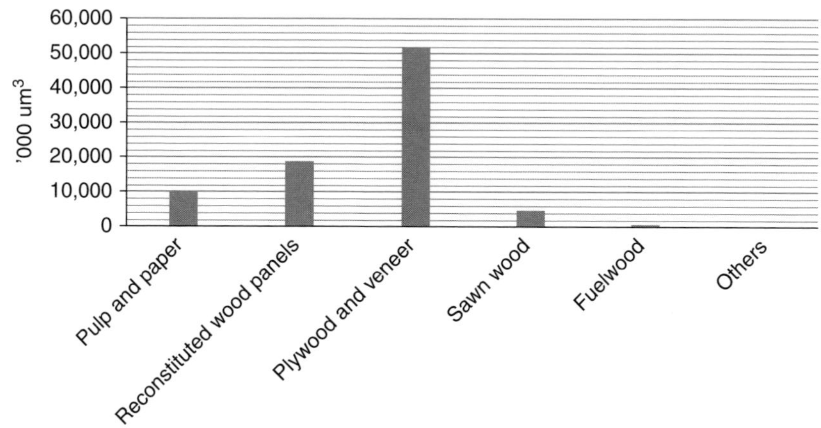

Fig. 11.3. Production of poplar and willow wood-based products. Source: FAO (2008).

These developments have been driven by domestic demand, as well as the significant increase in exports to other countries, especially the USA and the European Union.

Trade by the Chinese forest products industry shows two remarkable features: (i) high imports of primary wood products, i.e. roundwood and sawn wood; and (ii) high exports of processed/value-added wood products, i.e. furniture and plywood. Plate 31A provides the net trade trends of the main products traded between 1993 and 2008 (SFA, 2006, 2009a). The net import value of roundwood and sawn wood has increased more than ten times from 1993 to 2008 (SFA, 2006, 2009a). China has become one of the largest importers of roundwood and sawn wood in the world (FAO, 2011). Over the same period, furniture and plywood have experienced significant increases in net exports. Until 2005, wooden furniture exports from China exceeded those from Poland and approached those of Italy, which is the largest exporter of wooden furniture (European Commission, 2011). The amount of furniture the USA imports from China is eight times more than the amount of furniture it imports from Italy. Asian penetration (mainly China, Malaysia and Vietnam) into the European markets has continued, mainly at the cost of intra-European trade (UNECE/FAO, 2006).

According to country reports to the IPC in 2008, China also dominates the world production of poplar-based products, accounting for more than 90% of the total world production. Among various poplar products, plywood and veneer take the biggest share, about 60%. Particleboard and fibreboard, and wood pulp follow, accounting for 22% and 12% (Plate 31B).

Poplar wood is used mainly for the production of veneer, plywood, particleboard, fibreboard, sawn wood and wood pulp. It is estimated that the proportion of poplar raw material in veneer and plywood is 60% in China and 5–100% in Europe. The proportion of poplar raw material in particleboard and fibreboard, and wood pulp is 30% and 40%, respectively, in China (FAO, 2008, 2011). Over the period 1980–2010, China has been leading the global increase in the production of plywood, particleboard and fibreboard (Fig. 11.4). The production of plywood, fibreboard and particleboard in China accounted for 46%, 41% and 11%, respectively, of the world total in 2008 (SFA, 2009a; FAO, 2011).

The Chinese wood-based industry achieved tremendous development over the period 2000–2010, including significant increases in the production and export of furniture, wood-based panels, paper and paperboard. Obviously, both domestic and imported poplar wood as one of the main raw materials has made an essential contribution to this development. Specifically, plywood development has been supported mainly by poplar wood (accounts for an estimated 60%). Plywood production has increased four times, from 7.2 million m^3 to 35.4 million m^3 in 10 years (1999–2008) (SFA, 2009a). The trade in Chinese plywood has experienced a transition over the period 2000–2010 (Fig. 11.5). China used to depend heavily on imports of plywood. The turning point was in 2002, when the value of exports exceeded that of imports of

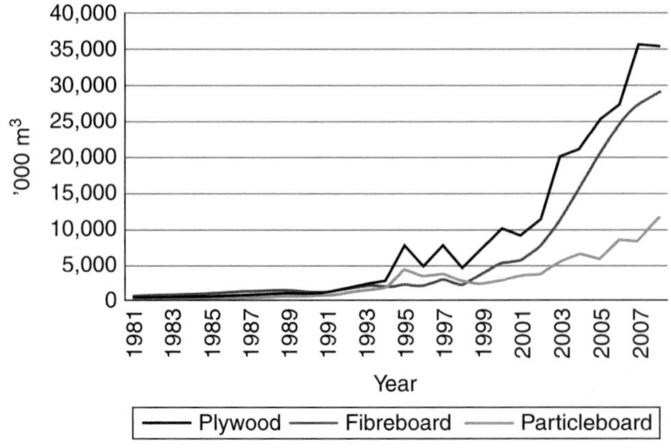

Fig. 11.4. Production trend of plywood, fibreboard and particleboard in China. Source: SFA (2009a).

plywood, and since then China has become a net exporter of plywood. This trend has grown quickly and is expected to continue (Fig. 11.5).

In China, Hebei, Jiangsu and Shandong provinces have become the top three provinces in producing wood-based panels. Plywood production in these three provinces accounted for 56% of the total Chinese output in 2005. Ironically, the three provinces are located in non-forest areas. Box 11.1 provides more detailed information on poplar-based panel industry development in Wenan county of Hebei province.

Driven by the fast development of wood-based panel industries in this area, fast-growing and high-yield forest plantations and agro-forestry have expanded quickly. Poplars are considered to be the most significant species providing raw material to the wood-based panel industry, especially plywood (SFA, 2006). It is reported that the expanding Chinese wood-based panel industry has benefited mainly from fast-growing forests in the northern plains. Regarded as 'plain areas with no forestry', Jiangsu, Shandong and Hebei provinces have made great strides in establishing fast-growing poplar plantation bases in recent years. These provinces have undergone a transition from traditional agriculture to modern forestry, and from traditional grain production to timber production and utilization. This has broadened the farmers' income base from agriculture and livestock husbandry to forestry and timber processing activities (ITTO, 2005).

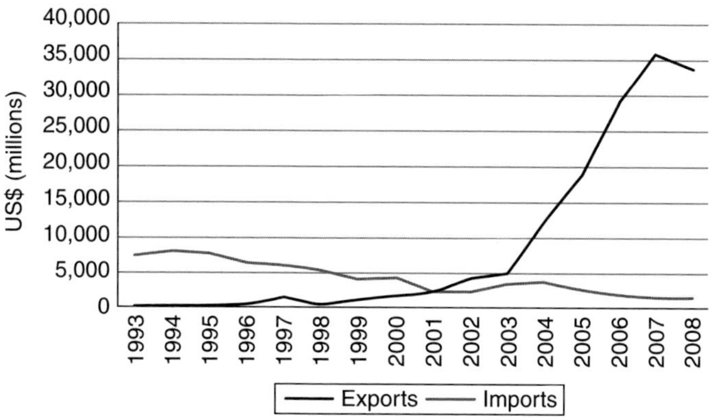

Fig. 11.5. Trends of plywood imports and exports in China. Source: SFA (2006, 2009a).

Box 11.1. A case of poplar-based panel industry development in Wenan county, Hebei province, China

Wenan county, located near Beijing, provides an overview of the pace of development of wood industries in China. The number of wood-based panel factories has increased to more than 2000 in 2005, from only two in 1986. Annual production is more than 5 million m^3, which results in income of about 6 billion yuan. The wood-based panel industry has about 120,000 employees, and is thus one of the three main sources of the county's finance. On the other hand, Wenan county has very poor forests. All the wood material supporting the wood-based panel industry comes from other areas or imports. The plywood produced consumes mainly poplar wood as core and birch as surface and back veneer. Driven by the recent development of the wood industry, poplar plantations have expanded rapidly during recent years. Forest cover has increased from 10% in 2000 to 28% in 2006. The main drivers of this significant development of the wood industry are the implementation of a non-public ownership policy in wood industry development, strong market demands (Wenan county is close to Beijing and Tianjin) and low labour costs. Wood-based panel industry development in Wenan county is a typical example of the development of the wood industry in China in the past decade.

Source: Q. Ma (2006), unpublished field trip report. FAO, Rome.

11.4.2 Evolution of the European poplar-based industrial market

Over the past decades, among industrial products using poplar, only the production of reconstituted panels (particleboard and fibreboard) has increased significantly in European countries. Production of other products has remained flat or declined over the period 1980–2010. Figure 11.6 shows European production trends in industrial product groups using poplars (among other species) as the raw material.

However, poplar accounts for only a small fraction of the raw material used in Europe. Poplar is important in only a few European countries: France, Italy, Spain, Hungary, Belgium, Bulgaria and Croatia. In these countries, poplar accounts for 30% and more of the raw material used in the production of veneer and plywood (FAO, 2008).

The total production of wood products from cultivated poplar in eight European countries is about 5 million m³. Veneer and plywood share the largest part, about 2 million m³, sawn wood is in second place, about 1.4 million m³, particleboard and fibreboard are about 0.7 million m³ and wood pulp production is about 0.6 million m³. France is the largest producer of poplar products, followed by Italy, Spain, Belgium, Romania and Germany (Plate 31C).

Italy is the largest importer of poplar industrial roundwood and sawn wood in Europe. Italy imported 300,000 m³ of roundwood and 300,000 m³ of sawn wood in 2010. France (roundwood) and Hungary (roundwood and sawn wood) have been the largest exporters of poplar primary products and the main suppliers to the Italian market. Austria and Germany are the other important suppliers to the Italian poplar roundwood and sawn wood market (Plate 31D). Obviously, Italy is the major factor determining prices on the European poplar wood market. Italy has been the major consumer of poplar primary products and the major processor of poplar secondary products. It is estimated that Italy consumes about 1.5 million m³ of poplar per year, including 24–30% imported poplar.

The European market has been experiencing transition in recent decades. The Western European market faces three main challenges and constraints. First, high labour costs and land prices in western countries reduce the competitiveness

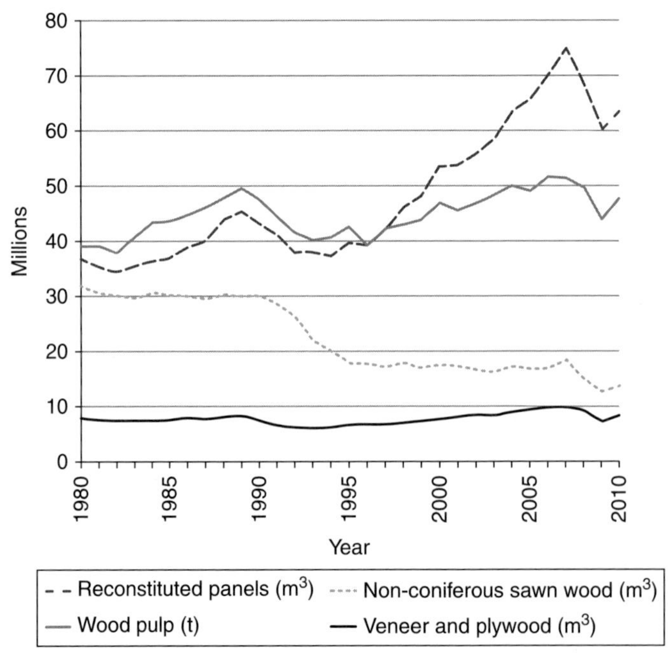

Fig. 11.6. Trends in production of wood processing products in Europe, 1980–2010.

of their products. Second, transportation costs have increased by 40%, resulting from the continuous increase in oil prices in recent years. Finally, the strong ecologic constraints in Belgium, Italy and France are not in favour of poplar resource development. For example, in Italy, many poplar forests are not replanted with poplar after being harvested because of the pressure from environmentalists. Therefore, the area of poplar plantations and the poplar wood supply have declined substantially over the last decades. For example, in France, the production of poplar logs decreased from 3 million m^3 in the late 1980s to 1.5 million m^3 in the late 2000s.

Driven by these challenges and constraints to Western European markets, more and more poplar processing units have moved to eastern countries, where semi-finished or final products are exported to the western market (R. Novakov and P. Mertens, Belgium, 2006, unpublished poster presentation). On the other hand, the Eastern European market is emerging and has great potential. The opportunities and advantages of the eastern countries are relatively low labour and land prices, more favourable climate for poplar growth and, more importantly, a favourable economic development environment. In addition, Eastern European countries have great potential in poplar plantation development, as it is now considered that poplar growing activity is profitable and can stimulate economic development in rural areas. More specifically, large areas of poplar forests were cleared during the transition period from a centralized to a market economy in eastern countries, which provided available land for planting poplar trees.

Presumably, the pattern of the importance of the poplar growing and processing regions will change according to the above analysis of strengths, weaknesses, opportunities and threats, and the overall economic, social and environmental development in European countries. It seems evident that there will be a shift from north to south and south-east. Poplar plantations will decrease in Belgium, the Netherlands, Germany and northern France, while increasing in Spain, Hungary and other Eastern European countries. Changes to the poplar processing base will take a similar approach. The shift is happening already, with Italian sawmills in poplar processing having moved to eastern countries and exporting semi-finished products to western countries for final processing.

11.4.3 The importance of poplar for wood pulp and paper and the marketing challenges of poplar products in North America

Poplar has long been used as a fibre source in the pulp and paper industry in North America. Native *Populus trichocarpa* (black cottonwood) is an important fibre source for operations in British Columbia, Canada and the Pacific Northwest part of the USA. Similarly, native *Populus deltoides* (eastern cottonwood) has been used as pulp and paper fibre in the southeast USA, as has *Populus tremuloides* (aspen) in the Lake States of the USA. As papermakers grew to favour the fibre characteristics of native cottonwood and aspen, hybridization programmes were launched to produce genotypes for plantation development (J. Eaton, GreenWood Resources, USA, 2008, unpublished information) (Box 11.2).

According to a USDA study on pulpwood production in the northern region of the USA, pulpwood is a large component of the industrial timber products harvested annually in the northern region, accounting for more than half of the total harvest in some states each year. Aspen is one of the main tree species for pulpwood production. In particular, aspen/balsam poplar has been the dominant species group harvested for pulpwood in the Lake States, accounting for 44% of the total (10.9 million m^3) in 2006. However, as the pulp products markets continue to decline, pulpwood production has shown a general downward trend since 2002. A total of 11 paper mills were idled or closed in the northern region during 2005 and 2006. This has resulted in a decreased trend in pulpwood production from aspen/balsam poplar. The production of aspen/balsam poplar pulpwood in the Lake States decreased from 14.5 million m^3 in 2002 to 10.9 million m^3 in 2006 (Piva, 2010).

In Canada, the vast majority of the *Populus* inventory consists of natural stands. The estimated poplar and aspen inventory is 4 billion m^3 (aspen accounts for 2.2 billion m^3) on 161 million ha. All plantations of poplar are located on forestland

Box 11.2. Poplar development oriented to wood production in North America

In Canada, poplars (other than aspens) are planted on both forest and farmland. Poplars on forestland are planted in reforestation projects. Poplars (other than aspens) planted on active farmland are managed as 'short-rotation, intensive-culture', or 'SRIC', crops and are considered 'prime agricultural production', which is not covered by provincial forest legislation or regulations.

As of 2011, the majority of poplar forest plantations and SRIC crops are controlled by vertically integrated forest companies producing pulp (kraft and mechanical), paper (tissue and printing papers), liner- and containerboard (packaging) or OSB (oriented strandboard). The main commercial poplar-related activities are located in the provinces of Quebec, Alberta and British Columbia. Small private landowners own and manage the balance in the above-mentioned provinces and also in the provinces of Saskatchewan, Manitoba and Ontario.

A total of 13,500 ha (83% industrial) of poplar forest plantations had been established by 2011, with annual planting averaging 600 ha, all of it industrial. The largest poplar forest plantation efforts are in the province of Quebec, where a breeding and selection programme by the provincial government has underwritten the Quebec efforts since 1969.

A total of 11,650 ha (84% industrial) of poplar SRIC crops had been established by 2011, with annual planting averaging 1300 ha, all of it industrial. The largest industrial SRIC crop programme is in Alberta; this programme is supported by recent industrial breeding and selection. To date, most poplar selections for SRIC in the Prairie Region of Canada (Alberta, Saskatchewan, north-east British Columbia and parts of Manitoba) have been based on poplars bred for shelterbelt use.

In the USA, three areas of development are prominent. In the south-east, the former Mead Westvaco holdings are located along the lower Mississippi River flood plain. Here, elite varieties of eastern cottonwood are used for pulp and paper production. In recent years, this development has become scattered and the long-term fate of this asset is unknown. In northern Minnesota, Verso Paper has established nearly 9000 ha of hybrid poplar plantations to supply wood to its paper mill at Sartell, Minnesota. This programme continues to be active, with breeding and nursery activities supporting the development that is scheduled to expand to 10,000 ha.

The third and largest area of development is located in the Pacific Northwest and boasts a multiple market strategy as compared to the other developments in North America. Along the Columbia River in Oregon and Washington, GreenWood Resources Inc (GWR) of Portland, Oregon, manages over 14,000 ha of hybrid poplar plantations on behalf of an investment fund. The GreenWood Tree Farm Fund LP (GTFF) is a US$210 million fund that has consolidated the former James River, Boise, and Potlatch poplar assets. In addition, the fund has capitalized a sawmill manufacturing complex that came on line in late 2008. The tree farms are being managed for multiple products, including saw logs, pulp chips and residuals for bioenergy. GWR's intensive management strategy includes state-of-the-art irrigation distribution systems, elite germplasm and applied silvicultural management systems. The plantations are certified by the Forest Stewardship Council (FSC) and will produce a sustainable supply of poplar logs for the sawmill.

Mill development will include a sawmill located in the centre of the tree farm, dry kilns and a finishing mill. Collins Companies of Portland, Oregon, will manage the sawmill and they will market poplar wood products under the 'Pacific Albus' trademark name. The mill will produce high-grade moulding, furniture stock, veneer and pallet stock. Annual wood use is projected to be 300,000 m^3, which will make the mill one of the largest hardwood mills in North America.

Current business development in the USA is centred on purpose-grown biomass plantations to provide feedstock to the combined heat and power industry and for conversion to liquid fuels. One example of this strategy is the poplar residuals that will be used by ZeaChem Corp at their demonstration biorefinery located adjacent to the Boardman Tree Farm in Oregon. This facility is the stepping stone to the first commercial-sized biorefinery that will produce ethanol and other bio-based chemicals from sustainable poplar feedstock.

Sources: J. Eaton, GreenWood Resources Inc, Portland, Oegon, USA (2011), personal communication; C. van Oosten, SilviConsult, Nanaimo, British Columbia, Canada (2011), personal communication.

and total approximately 11,420 ha. These crops produce mostly pulpwood for the pulp and paper industry; however, several recent crops were established to produce wood for the panelboard industry (OSB). Canada is a net importer of poplar and aspen logs, with a net import of 113,000 m^3 in 2007, and a net exporter of poplar and aspen lumber, with a net export of 92,000 m^3 in 2007 (Poplar Council of Canada, 2008).

According to a strategic analysis of aspen solid wood products in the province of Saskatchewan, Canada, the aspen is used as the major source of raw material in pulp and OSB production. The current aspen solid wood products industry in Canada can be described as small and fragmented. They have less value added and supply chains are weaker than in the USA. Furthermore, North America as a whole is facing strong competition from China.

In the USA, aspen solid wood production appears to be similarly small in scale and fragmented. The US producers' advantage of being located closer to the large markets of major urban centres provides them with the ability to focus on a higher percentage of finished products. In most cases, these products are manufactured and distributed through the traditional hardwood supply chain made up of many small-scale regional sawmills and a large network of secondary manufacturing and wholesale distribution that supports the manufacture of high-value interior finished products. Companies producing aspen lumber products usually offer aspen along with a full range of other complementary hardwood species (MW Friesen Consulting, 2011).

China has continued to play an increasingly stronger role in global furniture and finished wood product markets. Much of the hardwood furniture and finished interior wood products manufacturing that was once done in the USA has now been moved offshore to China. China, with its comparative advantages, has proven that it is able to buy raw material from North America in both logs and lumber, manufacture a wide range of finished products and export these products back to North America profitably. Where the USA was once considered one of the most developed markets for higher grades of hardwood lumber, China may now have replaced a significant portion of this demand, particularly in the middle range of the high grades. The number of US furniture and other finished wood product factories has decreased significantly over the past 10 years (MW Friesen Consulting, 2011).

11.5 Economic Importance (Comparative Advantages) of Poplar and Willow Plantations

11.5.1 Financial advantages of poplar plantations in China

Poplars make afforestation projects more economically attractive by their fast growth in comparison to other tree species. The financial advantages of poplar plantations have been assessed by an FAO project. As a contribution to China's Three-North Shelterbelt Programme (SFA, 2010), which is the largest afforestation programme in the world and aims to establish 35 million ha of forests between 1978 and 2050, the FAO assisted the government of China with the development and implementation of improved afforestation techniques in the Korqin Sandy Lands from 1991 to 2002. One of the outputs of the FAO project is that 16 afforestation models have been developed and assessed. These afforestation models mainly involve shelterbelts aimed at: (i) increasing vegetation cover and producing other environmental benefits; (ii) increasing and protecting agriculture production; and (iii) increasing wood and fodder production from the established forest plantations and shelterbelts (Ma, 2004) (see also Chapter 7, this volume).

Poplars, pines, willow and other native shrubs were the main species selected in the afforestation models. Four productive plantation models were described in order to compare the economic advantages of poplars in forest plantations. These four afforestation models are categorized as productive plantation models. The focus is on wood production as a primary objective and a secondary objective is the protection of neighbouring agricultural crops. In these models, the trees would be planted in large blocks, so only the costs and benefits of wood and fodder production were counted in the appraisal. However, these models also resulted in significant off-site benefits in terms of shelter for neighbouring agricultural crops. All four models were developed based on the same site class: slopes of hills with a water table 2.5–4.0 m from the surface. Model I (M1) is composed of block poplar plantations with rows of shrubs. Model II (M2) is composed of pure poplar shelterbelts. Model III (M3) is composed of block pine plantations with rows of shrubs. Model IV (M4) is composed of alternating belts of different species of poplar, pine and willow.

Appraisal of the different afforestation models was based on standard cost–benefit techniques, including calculating the net present value (NPV) of each model. According to the results of the cost–benefit analysis, M1, M2 and M4 are financially viable, while M3 is not because it has a negative NPV (–133 yuan[2] ha^{-1} $year^{-1}$). The reason why M3 is uneconomic is because the benefits are so low. This is due to the choice of pine in model M3, which requires a much longer time to mature (40 years). M2 has the highest NPV (378 yuan ha^{-1} $year^{-1}$). This is due to the choice of poplar in model M2, which requires a much shorter time to mature (15 years) and has a high production yield. M1 and M4 have medium NPVs (311 and 186 yuan ha^{-1} $year^{-1}$). This is due to the mixture of poplar with other species, which reduces the benefits to some extent. Poplar makes the afforestation projects more economically attractive by its fast growth in comparison to other tree species.

11.5.2 Cost-efficiency of willows for energy in Sweden

Cultivation of short-rotation willow coppice was introduced in Sweden after the oil crisis in the 1970s, with the intention of replacing fossil fuels by new energy sources. Extensive research to identify fast-growing species that could be grown intensively for use in energy production suggested that willows grown in coppice systems were the most suitable. Nutrient utilization and stand management were seen to be more cost-efficient for willow than for other woody species, and short-rotation willow coppice proved to be a sustainable way of producing fuels that were carbon dioxide neutral, since burning of the biomass would release into the atmosphere the carbon dioxide that the plants had taken from the air. About 16,000 ha of willows in short-rotation coppice systems are grown commercially, mainly on agricultural land, in Sweden. The willows are harvested every 3–5 years. The estimated economic lifespan of a short-rotation willow coppice stand is 20–25 years. Currently, biomass production of willow grown commercially in Sweden is about 6–12 t ha^{-1} $year^{-1}$, supporting heating plants for combined heat and power production (Dimitriou and Aronsson, 2005). These include:

11.5.3 Potential economic benefits of expanded basket willow cultivation in Chile

Chile has an optimal climate and soils for growing basket willow (*Salix viminalis*), which is well known for its qualities in the production of baskets, packaging and furniture. Although there are many challenges in revitalizing a declining sector of basket willow and its products – furniture and handicrafts – through improved quality and market development, it is considered that the expansion of the basket willow sector offers potential economic benefits, according to researchers in Chile (Abalos Romero, 2005). These include:

- higher rural production and income, linked to the production of raw material, the manufacture of products and the supply of services for industrialization;
- the possibility of increasing basket willow cultivation and production in rural areas, with the possible use of plantation yields also for industrial and domestic bioenergy uses;
- diversification of farm production;
- increase in property values; and
- development of a trade system for goods and services linked to basket willow cultivation and processing.

11.6 Policies, Outlook and Development Potential

11.6.1 Poplar and willow development and biodiversity conservation

From 2004 to 2007, increasing or stable trends of poplar and willow resources have been achieved in many countries and regions, particularly a positive trend for poplar plantations in China, Bulgaria, Germany and Sweden, and a positive trend for poplars and willows used in agroforestry systems or as trees outside forests in China, India, Spain, the USA and other countries. In contrast, a negative trend in poplar plantations is reported by Argentina, Belgium, Croatia, the Republic of Korea and Romania (FAO, 2008). Overall, the poplar and willow area has been increasing at the global level. The fast increases of poplar plantations and agroforestry in China have contributed greatly to this global increase. Driven by the increasing production of wood-based panels and pulp and paper in China, the rapid expansion of poplar plantations is expected to continue. Specifically, about half of the Chinese high-yield, short-rotation plantations (3.7 million ha, established during 2001 and 2005) are of poplar (CAF and ITTO, 2007).

Many countries have implemented a variety of biodiversity and nature conservation policies over the period 2000–2010. Governments have made specific commitments to enhance biodiversity and nature conservation through several international policy processes such as the Convention on Biological Diversity. Such commitments will result in diversification of species composition and structure of ecological communities in forests, as well as planting of endemic and indigenous species, increasing mixtures of coniferous and broadleaved species in forests and increasing rotation lengths. This emphasis may result in limiting the expansion of poplar and willow forests, and as in the case of northern European countries, reducing the existing area of poplar and willow forests (R. Novakov and P. Mertens, Belgium, 2006, unpublished poster presentation). This future trend is also shaped by the loss of competitiveness in poplar production and processing in comparison to Eastern European countries. China has the largest area of planted forests in the world, estimated at 62 million ha; however, its forest plantations are composed of only a few tree species, the three major ones being *Cunninghamia lanceolata*, *Populus* spp. and *Pinus massoniana*, which account for nearly 60% of the total planted forest area (SFA, 2009b). The lack of diversity of species composition has resulted in increased insect and disease problems, land degradation and loss of biodiversity, and has had negative impacts on the health and vitality of forest plantations. It is expected that diversification of species composition will be taken into consideration in the future development of forest plantations.

11.6.2 Changes in market patterns for poplar and willow products

Globalization has reduced the dependence of the wood processing industry on local supplies of raw materials. Companies can now utilize materials from regional sources and locate manufacturing facilities in international locations all along the production chain from the forest to the consumer. Thus, the location and development of the wood processing industry is influenced less by the local availability of forest resources, but more by the prevailing investment climate and general economic conditions in a country. In Eastern Europe and the CIS, the rapid and dramatic changes that have occurred over the last decade have placed these countries in a very competitive position in terms of wood supply and production costs. Production of forest products in these two subregions is expected to increase dramatically in the future. In contrast, the forest sector in Western Europe is likely to continue to expand to meet a broader set of objectives (UNECE/FAO, 2005). Similarly, the dramatic growth in production and exports of Chinese wooden furniture, wood-based panels and paper and paperboard are expected to continue in the future. It is expected that China will dominate continuously the production and market of poplar wood-based products.

Regarding the types of poplar- and willow-based products, the pattern of product types is expected to shift as well. Based on market demand, economic efficiencies and technology development in wood processing, wood-based panels, pulp, paper and paperboard are expected to be the most competitive and enjoy continued expansion in the future.

11.6.3 Potential contribution to climate change policies such as bioenergy policy

Increasing fossil fuel prices, combined with growing concerns that burning fossil fuels contributes to global warming, have led to increased interest in finding alternative, preferably renewable, energy sources in both developed and developing countries. In many developing countries, wood is already the primary source of energy for heating and cooking. In developed countries, it is likely that the use of wood for energy will continue to increase as fossil fuel prices continue to rise. In countries such as Austria, Finland, Germany and Sweden, biofuels are used increasingly for the production of electricity, attracting huge investments in wood-energy industries. In the USA, poplars are currently being used on a limited basis as an environmentally acceptable source of biomass for wood and energy. In addition, the amount of energy that can be produced from short-rotation, intensively cultured poplar trees is substantial (Box 11.3).

In the future, more wood-based fuels will be derived directly from forests and tree plantations, although currently most are derived from by-products (residues and wastes). Rapid expansion of bioenergy plantations is supported by many national governments, including Belgium, Italy, Sweden and the USA (FAO, 2008). In 2007, the Chinese authorities issued a strong directive to develop renewable energy from non-food crop sources. Wood-based fuel production and export could become key ingredients for the development and expansion of forest activities (FAO, 2007). However, the potential contribution of wood-based fuels, including poplar and willow biomass, to bioenergy policy is dependent on the following main issues:

- Competition for land between forest, food and energy sectors.
- Integrating bioenergy into national and regional energy policies.
- Policies, including incentives and taxes for the promotion of wood as fuel and for investment in wood energy. Demand for woody feedstock is increasing, motivated by public policies that have set ambitious targets for renewable energy: for instance, the EU's aim of meeting 20% of its overall energy requirement from renewable sources by 2020 and, to a certain extent, the renewable portfolio standards in the USA. Financial support to improve the cost competitiveness of wood energy with fossil fuels will be instrumental in sustaining growth in the sector. The spike in oil prices in 2011 has provided a further rationale for choosing wood rather than fossil fuels (UNECE/FAO, 2011).
- The competitiveness of wood-based fuels with other energy sources, especially fossil fuels. Although fast-growing shrub willows can be bred and selected for use as dedicated energy crops to provide a long-term, sustainable replacement for fossil fuels in temperate regions, they have not yet been widely adopted because of their high cost of production relative to currently used fossil fuels. The willow breeding programme at the College of Environmental Science and Forestry of the State University of New York has produced some crosses with growth improved

Box 11.3. Poplar biomass for energy in the USA

Poplars are currently being used as an environmentally acceptable source of biomass for wood and energy. Wood chips can be mixed with traditional fuels such as coal to produce electricity. This approach is cleaner, cheaper and more suitable than coal alone. Poplar growers like GreenWood Resources Inc, in the western USA, are creating multiple markets for the fibre. After producing saw logs and chipping the non-merchantable aboveground portion of whole trees, the residual biomass is used for hog fuel to generate a renewable, low-polluting source of electricity. GreenWood Resources is also partnering with several technology leaders to develop cellulosic ethanol from poplar feedstock. The US Department of Energy is supporting this effort by awarding grants to the most promising technologies.

The amount of energy that can be produced from short-rotation, intensively cultured poplar trees is substantial. Caloric values for poplar biomass components have been reported to be between 4.3 and 4.8 kcal g^{-1}, which is equivalent to approximately 27 barrels of oil ha^{-1} $year^{-1}$.

Source: Dickmann *et al.* (2001).

by 20–40% over that of a standard cultivar (Smart *et al.*, 2005).

- Competition with other wood products. For example, in European markets, roundwood that does not meet the standard for pulpwood traditionally ends up as fuelwood. This distinction has begun to change since 2006. Fuelwood prices have increased since 2000. The combination of rising fuelwood prices and falling pulpwood prices has resulted in broadly similar price levels. This implies the potential competition of raw material between wood pulp and wood-based fuels; actually, this competition is happening currently in some of the European countries (UNECE/FAO, 2006, 2011).

In addition, it is worth noting that poplar and willow have the potential to be carbon sinks, which is germane to the establishment and implementation of national climate change policies. For example, policies promoting the use of poplars and willows as carbon sinks have been implemented in some countries, such as Belgium and Romania (FAO, 2008).

11.7 Conclusions

Driven by overall economic development, and more specifically by globalization and shifts in competitiveness from developed countries to emerging economies, the production and market pattern of poplar and willow products have changed significantly. China dominates both the production and the export of poplar products in the world, mainly by its plywood, particleboard and fibreboard, along with the fast development in the production and market of the wood-based panels sector. China is shaping the trend, market and future development of poplar products.

In other aspects, driven by the environmental protection policies implemented in many countries, poplar and willow have great potential to contribute to bioenergy development and carbon sequestration efforts.

Finally, the characteristics of the fast growth and short rotations of poplar create its economic feasibility for industrial products and its low cost for environmental purposes. These are important factors determining the recent fast development of poplar products and its future potential, both for economic and environmental purposes.

Notes

[1] CIS subregion includes the countries of Belarus, the Republic of Moldova, the Russian Federation and Ukraine.

[2] US$1 equals 6.34 yuan.

References

Abalos Romero, M.I. (2005) Towards the development of the Chilean basket willow sector. *Unasylva* 221, 40–46.

Chinese Academy of Forestry (CAF) and ITTO (2007) China forest products market information. Bulletin No 3. ITTO, Beijing.

Dickmann, D.I., Isebrands, J.G., Eckenwalder J.E. and Richardson, J. (eds) (2001) *Poplar Culture in North America.* National Research Council of Canada Research Press, Ottawa.

Dimitriou, I. and Aronsson, P. (2005) Willows for energy and phytoremediation in Sweden. *Unasylva* 56, 47–50.

European Commission (2011) EUROSTAT statistical database (http://epp.eurostat.ec.europa.eu/portal/page/portal/statistics/search_database, accessed 21 November 2011).

FAO (2004) *The Contribution of Poplars and Willows to Sustainable Forestry and Rural Development. Synthesis of Country Progress Reports*, prepared for 22nd Session of the International Poplar Commission, Santiago, Chile, 28 November–1 December 2004. Working Paper IPC/3E. FAO, Forest Management Division, Rome.

FAO (2007) *State of the World's Forests 2007.* Food and Agriculture Organization (FAO) of the United Nations, Rome.

FAO (2008) *Synthesis of Country Progress Reports*, prepared for 23rd Session of the International Poplar Commission, Beijing, China, 27–30 October 2008. Working Paper IPC/6. FAO, Forest Management Division, Rome.

FAO (2011) FAOSTAT, ForesSTAT, forestry trade flows. Statistical database (http://faostat.fao.org/site/630/default.aspx, accessed 21 November 2011).

ITTO (2005) Report from China. International Tropical Timber Organization (ITTO), Yokohama, Japan. *Tropical Timber Market Report* 10(15), 13–15.

Ma, Q. (2004) Appraisal of tree planting options to control desertification: experiences from Three-North Shelterbelt Programme. *International Forestry Review* 6(3–4), 327–334.

MW Friesen Consulting (2011) A strategic analysis for aspen solid wood products in Saskatchewan. FPInnovations, Vancouver, Canada (http://www.er.gov.sk.ca/SaskAspenReport, accessed 5 January 2012).

Piva, R.J. (2010) *Pulpwood Production in the Northern Region, 2006*. Research Bulletin NRS-39, USDA Forest Service, Northern Research Station, Newtown Square, Pennsylvania.

Poplar Council of Canada (2008) Activities related to poplar and willow cultivation and utilization in Canada. Canadian report to the 23rd session of the International Poplar Commission (http://www.poplar.ca/upload/documents/IPCCan2008.pdf, accessed 29 July 2013).

SFA (2006) *China Forestry Development Report 2006*. State Forestry Administration, China Forestry Publishing House, Beijing.

SFA (2009a) *China Forestry Development Report 2009*. State Forestry Administration, China Forestry Publishing House, Beijing.

SFA (2009b) *China Forest Resources Report – The Seventh National Forest Inventory*. China Forestry Publishing House. Beijing.

SFA (2010) Three-north Shelterbelt Forest Program (http://english.forestry.gov.cn/web/article.do?action=readnew&id=201001141142307633, accessed 22 November 2011).

Smart, L.B., Volk, T.A., Lin, J., Kopp, R.F., Phillips, I.S., Cameron, K.D., *et al.* (2005) Genetic improvement of shrub willow (*Salix* spp.) crops for bioenergy and environmental applications in the United States. *Unasylva* 56, 51–55.

Taylor, R. (2009) Crisis in the wood products industry and markets: perspectives from North America. *Unasylva* 60, 13–22.

Tissari, J. (2011) Highlights on paper and paperboard: 1999–2009 (unpublished report). FAO, Forestry Department, Rome (http://faostat.fao.org/Portals/_Faostat/documents/pdf/Paper%20and%20paperboard.pdf, accessed 6 November 2011).

UNECE/FAO (2005) European Forest Sector Outlook Study 1960–2000–2020, Main Report. United Nations Economic Commission for Europe and Food and Agriculture Organization of the United Nations (http://www.fao.org/docrep/008/ae428e/ae428e00.htm, accessed 6 November 2011).

UNECE/FAO (2006) Forest Products Annual Market Review 2005–2006. Geneva Timber and Forest Study Paper 21. United Nations Economic Commission for Europe and Food and Agriculture Organization of the United Nations (http://www.unece.org/fileadmin/DAM/timber/docs/fpama/2006/fpamr2006.pdf, accessed 6 November 2011).

UNECE/FAO (2011) Forest Products Annual Market Review 2010–2011. Geneva Timber and Forest Study Paper 27. United Nations Economic Commission for Europe and Food and Agriculture Organization of the United Nations (http://www.unece.org/fileadmin/DAM/publications/timber/FPAMR_2010–2011_HQ.pdf, accessed 6 November 2011).

12 Poplars and Willows for Rural Livelihoods and Sustainable Development

Walter Kollert,[1]* Jim Carle[2] and Linda Rosengren[3]
[1]*Secretary of the International Poplar Commission (IPC), Forestry Department, FAO, Rome, Italy;* [2]*Forestry Department, FAO, Rome, Italy (retired) – Present address: Mount Maunganui, Bay of Plenty, New Zealand;* [3]*Forestry Department, FAO, Rome, Italy. Present address: WWF, Helsinki, Finland*

12.1 Introduction

Poplars and willows are multi-purpose species. They provide timber, fibre, fuelwood and other wood and non-wood forest products and services such as the rehabilitation of degraded lands, forest landscape restoration and climate change mitigation. These many attributes make poplars and willows ideally suited for supporting rural livelihoods and contributing to sustainable development, particularly in developing countries and countries with economies in transition. Case studies from around the world illustrate in this chapter the economic, social, cultural and environmental impacts that poplars and willows have on food security, poverty alleviation and sustainable land use. As such, poplar and willow cultivation is an outstanding tool to contribute to the achievement of Millennium Development Goal No 1 (eradicate extreme poverty and hunger), No 3 (promote gender equality and empower women) and No 7 (ensure environmental stability).

The term 'livelihood' as adopted by this chapter describes a multi-dimensional concept covering not only a person's economic means of supporting himself or herself, but also the cultural capabilities, social assets and ecological resources serving to making a living. These include, for instance, the provision of wood fuel and non-wood forest products for local use as well as for sale, contributions to poverty alleviation and to food security, formal and informal employment, as well as the planting of poplars and willows to protect the natural and social environment on which livelihoods depend (Chambers and Conway, 1991).

An estimated 70 countries grow poplars and willows in indigenous or planted forests, agroforestry production systems and as distinctive landscape components for protective and productive purposes. Country reports to the International Poplar Commission (IPC) (see Chapter 1, this volume) in 2008 indicated that the area of poplar forests and woodlands amounted to 78.5 million ha globally, an area equivalent to the size of Mozambique or Pakistan. The area of willows was substantially smaller. Of the total reported area of 0.58 million ha of willows, 76% were natural (0.44 million ha) and 24% (0.14 million ha) were classified as planted forests.

Perhaps more than any other family of woody plants, excluding fruit trees, the *Salicaceae* with the genera *Populus* (*c.*100 species) and *Salix* (*c.*450 species) have been 'domesticated' and brought into continuing, purposeful interaction with human cultures (Gordon, 2001). The Latin

*Corresponding author: walter. kollert@fao.org

word 'populus' (people or populace) and the name of the tree genus '*Populus*' (poplar) reveal the popular identity of the poplar tree. Poplars and willows have been used and modified by humans for thousands of years. The close association of poplars and willows with people is best expressed in China, where 'Yang' (杨, poplar tree) and 'Liu' (柳, willow tree) are common surnames. An early use of poplar in art was the utilization of wooden panels as the medium for paintings, until canvas became the more popular support in the 16th century. Artists would often use wood native to their region; in Italy, for example, these panels were usually made of seasoned poplar, willow or linden (*Tilia* sp.). Leonardo da Vinci painted the *Mona Lisa* (1503–1505, also known as *La Gioconda*) on a poplar wood panel; Albrecht Dürer used to paint on poplar when he was in Venice from 1505 to 1506 (Wikipedia 'Mona Lisa' and 'Panel Painting').

Today, the interest in poplars and willows is ever increasing. Poplars and willows form an important component of forestry and agricultural production systems worldwide, often for small-scale farmers. Many of them are adapted to a wide range of climatic and soil conditions, from the heat of the Chinese desert to the cold, windy conditions of the South American Andes. Some poplars and willows are undemanding and can grow on degraded, polluted and wet lands, an attribute which makes them superior on lands unsuitable for other species. Poplars and willows are among the fastest-growing trees in temperate, arid and subtropical regions, and are very effective for carbon sequestration. In planted forest stands, the reported average growth performance is up to 21 m^3 ha^{-1} $year^{-1}$ (FAO, 2006). Tree improvement and hybridization are straightforward in poplars and willows, and they are easy to cultivate and to propagate vegetatively, for example by scions or cuttings. Poplars produce soft, odourless, tasteless, white wood with a relatively low density and diffuse porous structure. The strength properties are relatively low but the bending strength high (Dickmann *et al.*, 2002; see also Chapter 10, this volume). These qualities make poplars and willows adaptable to a wide range of different uses. They provide a nearly endless list of wood and fibre products (sawn lumber, veneer, plywood, pulp and paper, packing crates, pallets, poles, furniture and small handicraft), non-wood products (animal fodder), environmental services (shelter, shade and protection of soil, water, agricultural crops, livestock and dwellings), and are grown increasingly in bioenergy plantations for wood fuel. The major forest products from poplar and willow wood as reported by 19 IPC member countries, in order of economic importance, were: (i) plywood and veneer; (ii) reconstituted wood panels; (iii) pulp, paper and cardboard; (iv) sawn wood; (v) wood fuel and biomass for bioenergy; and (vi) other uses. With respect to the three first-ranked products, it is noteworthy that China holds a share of more than 90% of the global production (FAO, 2008a). For a broader and more in-depth and technical treatment of the properties, processing and utilization of poplar and willow wood, please refer to Chapter 10, this volume.

A number of studies around the world have presented strong evidence of the significance of forests and trees outside forests as supporters of sustainable livelihoods, food security and poverty alleviation (Box 12.1). In many developing countries, smallholders and farmers own an increasing area of poplar and willow plantations and depend on them for their livelihoods, in particular if they are established on land that has either been abandoned or has previously been of marginal use. A large (but unknown) number of people, particularly women, may gain employment and income from poplar and willow cultivation, particularly in remote, undeveloped areas where forestry is the only economically viable land-use option. The multiplier effect on employment at secondary (wood processing) and tertiary (service sector) levels can be considerable, especially when a significant portion of the wealth of such value-added activity remains in the local economy and among the forest-dependent workforce.

12.2 Poplar and Willow Production Systems

12.2.1 Poplars and willows in native forests

Poplars and willows are native to northern and temperate regions of the northern hemisphere down to a latitude of about 30°N. However, they have been widely planted beyond their natural habitat in both the southern and the northern hemispheres (Heilman, 1999). The great majority

Box 12.1. The contribution of forests[1] to sustainable livelihoods, food security and poverty reduction

Household income from forests

- Smallholders living in forest margins in diverse parts of the world earn between 10% and 25% of their household income from non-timber forest products (Ndoye *et al.*, 1999; Wunder, 2000).
- In Ghana, 10% of the population generated some cash income from forest product activities in the early 1990s. Only a minority reported that it was a major source of income, but more than 70% stated that it was important in helping them meet particular needs (Townson, 1995).
- In Burkina Faso, farmers planted 'live fences' generating supplies of fuelwood and fodder that increased household income by 11–16% (Scherr *et al.*, 2004).
- In Sahelian countries, such as Burkina Faso, planting of 'live fences' has generated supplies of fuelwood and fodder while increasing farm incomes by US$40 year^{-1} – a sizeable bonus for farming families whose annual incomes range between US$250 and US$350 (World Bank, 2004).

Rural employment

- In Chile, half a million rural people depend on forestry activities stemming largely from plantations; job creation in forest plantations is higher on a per hectare basis than in traditional farming activities (Contreras-Hermosilla and Gregersen, 2001).
- In China, World Bank-funded plantation projects have provided incomes and temporary employment for 2 million poor people, while a total of 12 million people have been provided temporary employment through the National Afforestation Projects (Rozelle *et al.*, 2002).
- World wage employment in forestry is approximately 3 million people, about 1 million in industrialized countries. Unpaid subsistence work, primarily for fuelwood harvesting, probably occupies about 14 million people full-time, 90% of them in developing countries. While wage employment is largely a male domain, with women rarely exceeding 10% of the workforce, subsistence employment is dominated by women in many developing countries (International Labour Organization, 2004).
- Small-scale forest product enterprises are among the top three non-farm rural commercial activities in most countries (FAO and DFID, 2001).
- Small-scale forest product processing is one of the largest sources of rural non-farm employment and, unlike formal sector employment, appears to be increasing (Scherr *et al.*, 2004).

Fuelwood, fodder and other uses

- Wood energy is the dominant source of energy for over 2 billion people, or one-third of the world's population, particularly in households in developing countries (FAO website: http://www.fao.org/forestry/energy/en/).
- Low-income forest and farm producers supply the vast majority of fuelwood, which constitutes some 50% of total global wood product demand (Scherr *et al.*, 2004).
- Fuelwood is used for 58% of all the energy used in Africa, 15% in Latin America and 11% in Asia. In some 40 developing countries (many of them among the least developed), fuelwood accounts for more than 70% of all energy use. It is estimated that by 2050 demand for fuelwood will increase from 3.0 to 3.5 billion m^3 year^{-1} (WCFSD, 1999).

[1]In this context, the term 'forests' includes woodlands and trees outside forests, and all species, not only poplars and willows.

of poplars and willows grow in native forests and woodlands under public ownership. The total area of natural poplars reported to the IPC in 2008 was over 70 million ha, 96% of which occurred in Canada, the Russian Federation and the USA (Table 12.1). Canada had the world's largest area of native poplar stands at 28.3 million ha (40%), followed by the Russian Federation with 21.5 million ha (31%). The third largest area was found in the USA, at 17.7 million ha (25%). All three countries reported that the main purpose of the native poplar forests was wood production. The People's Republic of China, which had the fourth largest natural poplar area at 3 million ha, reported that the main management objective of these forests was environmental protection, as did the next three countries in terms of naturally occurring poplars: France (39,800 ha), Spain (25,000 ha) and Romania (24,300 ha) (FAO, 2008a).

Table 12.1. Area of natural poplar stands in main countries reporting to IPC (FAO, 2008a).

Country	Area (1000 ha)
Canada	28,300
Russian Federation	21,500
USA	17,700
China	3,000
France	40
Spain	25
Romania	24
Croatia	9
India	9
Republic of Korea	6

The major native poplar species in Canada were *Populus tremuloides* and *Populus balsamifera*. In the USA, *P. tremuloides* was the most widely distributed native poplar species and was by far the most important commercially used poplar species. Other important species were *Populus deltoides and Populus trichocarpa*. Native poplar stands in the USA were owned predominantly by private smallholders (57%). An important share was owned by public institutions (37%); only 6% was owned by corporate bodies (Isebrands, 2004). The history and potential uses of native poplar and willow stands in the Russian Federation, the USA and Canada are illustrated in Boxes 12.2 and 12.3.

Box 12.2. The use of native willow stands, Russian Federation

The area of native willow stands in the Russian Federation is about 2.85 million ha (all forms, including shrubs). Tree-like willow stands cover around 1.1 million ha, with a stock of 86.5 million m^3. The native willow stands grow along riverbanks and reservoirs on rich, humid alluvial soils. Particularly in regions where forests are scarce, the white willow (*Salix alba*) has prominent economic value due to its fast growth, its high content of chemical substances and its significance as a source of nutritious honey. The bark contains salicylate (4–5%) and tannin (up to 5%). An extract made of bark is used for dyeing silk, wool and goatskin in reddish-brown colours. Cords and ropes are produced from fibres of the bark. The wood is used for construction purposes and is especially valuable in manufacturing arches, hoops, washtubs and other products of daily use.

Source: Tsarev (2005a, b).

Box 12.3. Utilization of native poplar stands for various wood products, USA and Canada

Pulp and paper, lumber, hardboard and insulation board are the major products from poplar in the USA, which in most cases is naturally grown quaking aspen (*Populus tremuloides*). Aspen can be pulped by all commercial pulping methods, including mechanical, semi-chemical, kraft and sulfite processes. Many of the woodworking mills use exclusively poplar wood – aspen, eastern (*Populus deltoides*) and western (*Populus fremontii*) cottonwood and, more recently, hybrid poplars – to produce lumber and reconstituted wood panels. Hybrid poplars from plantations are gradually becoming a more important source of material as native poplar resources decline (Stanturf, 2008).

The history and potential uses of poplars are much more varied than the current utilization suggests. There is an interesting series of publications from the 1940s that presents a thorough analysis of the properties and potential uses of aspen in Canada and the northern Lake States of the USA (Dickmann *et al.*, 2002). This series describes uses such as house logs, veneer, lumber, chemicals, furniture core stock and others. There are numerous historical accounts of the importance of other *Populus* spp. for logs and lumber for home construction and other purposes. In the plains region of western North America, poplars were often the only trees of a sufficient size to use for building materials. The value of poplars even went far beyond the manufacturing of forest products. In all parts of their range they are important, in some cases critical, elements in the landscape for their role in reducing erosion on upland and flood-plain sites, providing habitat for wildlife and fish, and aesthetics. Historically, these species have been important to Native Americans as a source of food, medicine and building material. In recent times, parts of some trees, for example the bark of older *Populus trichocarpa* and *Populus balsamifera*, are used by carvers and artists for the creation of art objects.

Source: Dickmann *et al.* (2002).

Indigenous *Salix* forests occur primarily in riparian zones, and although they are not utilized commercially to any great extent, they represent important wildlife habitats. The Russian Federation has the world's largest area of natural willow stands (*Salix* spp.) at 242,000 ha, followed by France (66,600 ha), China (60,000 ha), Spain (25,000 ha), Romania (15,200 ha), India (14,000 ha) and Croatia (10,000 ha). Mixed stands of indigenous poplars and willows were mostly reported in Croatia (14,000 ha) and Spain (12,000 ha) (IPC, 2008).

12.2.2 Poplars and willows in planted forests

The total area of planted poplar reported to the IPC in 2008 was 5.3 million ha, of which 3.9 million ha (73%) were planted primarily for wood production, 1.2 million ha (23%) for environmental protection and 0.2 million ha (4%) for other purposes (FAO, 2008a). The People's Republic of China accounted for 82% of the world's planted poplar area (4.3 million ha). The ITTO (2007), from statistics of China's poplar industry, reported that the country had as much as 7 million ha of poplar plantations, the most in the world. Around 3.1 million ha of this total was for timber production, accounting for 40% of the national poplar plantation area. The plantations were reported to comprise a number of different poplar species that were distributed in most provinces throughout the country. Other countries that reported significant areas of planted poplar included France (236,000 ha), Turkey (125,000 ha), Italy (118,500 ha), Germany (100,000 ha), Spain (98,500 ha) and Romania (55,000 ha) (IPC, 2008) (Table 12.2).

Table 12.2. Area of planted poplar stands in main countries reporting to IPC (FAO, 2008a).

Country	Area (1000 ha)
China	4300
France	236
Turkey	125
Italy	119
Germany	100
Spain	99
Romania	55

The global area of planted willows was reported at 133,000 ha, of which 106,500 ha were for wood production (80%) and the balance for environmental protection (IPC, 2008). China had the largest area of willow planted for wood production (43,200 ha). In New Zealand, willows were planted mainly for riverbank stabilization (Ball *et al.*, 2005).

Nearly half (47%) of the poplar and willow plantations reported to the IPC were owned by private corporations (2.5 million ha). Around one-third was publicly owned (1.8 million ha). Yet private smallholders owned a significant portion of one-fifth (19%) of the total (1 million ha) (FAO, 2008a). For a more in-depth and technical treatment of industrial plantations, please also refer to Chapter 5, this volume.

Large-scale industrial plantations

Large-scale industrial poplar plantations form a major land use in Argentina, China, France, Iran, Italy, Pakistan, Spain and Turkey. The land where these plantations have been established is owned by either the government or a private corporation. In many cases, the government as the landowner allocates concessionary, management and/or harvest rights to a private company. Large-scale plantations are usually established in even-aged blocks of one single species and few clones that are managed in short rotations, with a focus on maximum productivity and on minimizing production costs. The level of technology applied is normally high, which increases production capacity as well as the quality and homogeneity of the product (Fig. 12.1). An additional benefit of large-scale poplar plantations is usually the establishment of new or improvement of the existing infrastructure in the plantation area. Plantation companies are usually obliged to invest in developing the local infrastructure by building roads, electricity and communication networks. In some cases, the companies also build educational and medical facilities (Cossalter and Pye-Smith, 2003).

The financial and technological thresholds to establish large-scale plantations are high for several reasons (Macqueen, 2005). There is a high cost of entry in the establishment of nurseries and plantations. Harvesting, transportation and processing facilities are highly mechanized and

Fig. 12.1. A well-managed, large-scale poplar plantation for wood production in Chile. Photo courtesy of FAO/P.J. Alvarez.

they require large capital investments that are beyond the capacity of small companies or farmers. Large-scale plantations of poplar and willow may contribute positively to social and cultural sustainability by offering employment and education possibilities for members of local communities. However, in general, forestry tends to be less labour-intensive than agriculture (Angelsen and Wunder, 2003). Whether large-scale poplar plantations increase or decrease employment and benefits to the livelihoods of the local population depends to a large extent on the activities they replace and on the way the wood is processed. If plantations are established on fertile agricultural land, the chances are high that the number of jobs is reduced, especially in developing countries. If plantations are established on land that has either been abandoned or has previously been of little use to farmers or others, then plantation establishment and management activities may create new job opportunities locally. High employment benefits can be generated where plantations replace degraded or unused land, where alternative agricultural employment is low or where rotation cycles require continuous replanting, maintenance and harvesting.

In some cases, the social and cultural dimensions of large-scale plantations are not addressed properly and combinations of conflicting interests, unfulfilled demands and inadequate responses have led to encroachment, illegal logging, vandalism and conflicts between local communities and plantation managers. To avoid such conflicts, it is crucial to recognize social and cultural values in planning, managing and using the poplar and willow plantations and to apply consultative and participatory decision-making processes, as well as social impact assessments (FAO, 2006). Preventing displacement or resettlement of communities without free, prior and informed consent, providing a safe and healthy working environment in compliance with national or international standards, respecting community ancestral rights and hunting grounds, and protecting sites and landscapes of archaeological, cultural, traditional, spiritual, scientific, aesthetic or other sociocultural significance helps to avoid the emergence of such conflicts.

Some of the large-scale plantations have undergone voluntary forest management certification schemes to demonstrate compliance with environmental, social and cultural regulations concerning local communities and biodiversity conservation. Large corporations also apply corporate social responsibility (CSR) frameworks which encourage organizations to consider the interests of society by taking responsibility for the impact of the organization's products and services on customers, employees, shareholders, communities and the environment in all aspects of its operations. This obligation extends beyond regulatory obligations to improve the quality of life of the employees and their families, as well as of the local community and society at large.

Boxes 12.4–12.7 illustrate the economic significance of large-scale poplar plantations and their outstanding contribution to sustainable rural development in Argentina, Chile, Iraq and the Islamic Republic of Iran.

Box 12.4. Production of pencils from poplar wood, Argentina

In the province of Mendoza in western Argentina, poplar is the major planted forest species. Its timber sustains a diversified wood industry that produces a number of products for different purposes. The major use of the poplar wood is for chipwood, which serves for the production of reconstituted wood panels. The growing resource base has also created opportunities for small, specialized niche markets, such as the production of small wooden slabs for the manufacturing of pencils. Every year, 1200 m^3 of these wooden slabs are exported to Brazil for pencil production. Easy processing, high homogeneity and, foremost, the lack of knots after pruning, are important properties of poplar wood, which make it particularly suitable for pencil production.

Source: Calderón *et al.* (2004).

Box 12.5. Large-scale poplar plantations supply pulpwood, Islamic Republic of Iran

The Islamic Republic of Iran is relying increasingly on the cultivation of fast-growing tree species to meet the country's demand for industrial wood. Iran's forest policy aims at achieving more wood production by targeting private planting programmes of 10,000 ha $year^{-1}$ through the government granting land, long-term low-interest loans, financial incentives, distribution of seedlings and technical assistance programmes. The forest inventory of 1992 found that 3 million m^3 of pulpwood and industrial roundwood were produced by planted forests, mainly from fast-growing poplar species. The total area of poplar plantations was 150,000 ha, of which 35% were young stands. The average standing volume of these stands was estimated at 155 m^3 ha^{-1}.

Source: Rouchiche and Haji Mirsadeghi (2003).

Box 12.6. The contribution of the poplar industry to sustainable rural development in the Maule Region of Linares Province, Chile

The El Alamo Agricultural and Forest Company Ltd (El Alamo, www.cafelalamo.cl/company.htm) has established poplar plantations on 2915 ha, the largest area in Chile planted with poplar (Fig. 12.2). Poplar was chosen mainly because of its good growth characteristics, short rotation period and the qualities of its timber, such as white colour, even texture and absence of resin. The plantations are established from genetically improved planting material that is produced in the company's own nurseries (40 ha). Plantation management comprises mainly weed control, pruning and irrigation carried out between ages 1 and 13 years. The rotation age is 12–14 years, the annual allowable cut on the total area is 38,000 m^3, which equates to 222 ha $year^{-1}$, with an average volume of 171 m^3 ha^{-1}. The company has adopted an integrated production system to maximize land use and rate of return through a combined production of wood, agricultural crops and livestock. This involves:

Continued

Box 12.6. Continued

Fig. 12.2. Poplar plantations with even-age class distribution from 1 to 14 years. Photo courtesy of FAO/El Alamo.

- the production of high-quality poplar roundwood in forest plantations (3.2 m logs with diameters of 16–60 or 70 cm), which are used for the manufacturing of safety matches, chopsticks, ice-cream sticks and paint stirrers;
- the planting of agricultural crops, for example maize, bilberries, between the tree rows during the first 2 years after planting; and
- the introduction of cattle starting in the third year after planting, to control weeds and undergrowth, to reduce the risk of fire and to improve soil fertility (Fig. 12.3).

El Alamo is a major social and economic player in its municipality, where the majority of the population (83.5%) is rural and some 37% live in poverty. The company continually offers employment opportunities, issues a community newsletter, offers training in agricultural practices and environmental protection, and contributes to the community's welfare. The identification of the community with El Alamo extends to the point that the municipality has adopted the slogan 'the Alamo municipality'. In July 2002, El Alamo obtained certification from the Forest Stewardship Council (FSC) in recognition of the company's efforts to support forest conservation and its commitment to contribute to sustainable rural development.

Source: Ulloa and Villacura (2005).

Box 12.6. Continued

Fig. 12.3. In the applied agro-silvo-pastoral production system, cattle help control weeds and undergrowth, fertilize the soil and haul harvested logs. Photo courtesy of FAO/E. Beuker.

Box 12.7. Poplars for the production of utility poles, Iraq

Iraq produces a very small portion of its wood supply in country and is largely dependent on timber imports. To alleviate this dependency, fast-growing poplar plantations of *Populus nigra* have been established on a large scale in the Duhok and Zakho Governorate in northern Iraq. Growth rates of up to 45 m^3 ha^{-1} $year^{-1}$ have been reported in a 5-year rotation cycle. The poplar timber is used for a wide range of products such as utility poles, construction timber, pulp, veneer, matches, etc.

The area under poplar is expanding rapidly, as private investors have found that the production of utility poles from poplar plantations offers good financial returns that surpass the returns from fruit orchards or agricultural crops. The internal rates of return (IRR) achieved under different management regimes range between 12 and 34%.

Source: Hassan and Salim (2004).

Small-scale planted forests

Poplar and willow forests owned by small landowners create a more diverse landscape, often in a mosaic of different land uses, than large-scale plantations, due to differing production goals, varying management techniques and lower technology input. Productivity and the quality of the product are usually lower than in intensely managed large-scale plantations. The smallholder's objectives and expectations concerning land use are commonly more diverse, as their livelihood depends partly, or in some cases entirely, on the land. Thus, the land has to produce a diverse portfolio of products to lower the investment risk. Besides producing wood as a main product, the land may also produce sequentially or simultaneously food crops or fodder, or fulfil protective purposes.

Smallholders who own planted forests with poplars or willows operate either as independent entrepreneurs or are business partners contracted by a private company within the framework of an outgrower scheme. Independent entrepreneurs operate in the free market and are not always in a fixed partnership with a company. In most cases, they produce a blend of forest and agricultural products from their land as cash crops and for subsistence purposes. This set-up offers both risks and opportunities. Independent smallholder entrepreneurs seldom have a formal education in management techniques or financial planning and control; meeting the demands of the markets poses large challenges for them. Even if the macroeconomic conditions are favourable, the smallholders still need to meet demands for quality, design, quantity and continuity of supply. The success of their operations is often dependent on the availability of an efficient forest and agricultural extension service, which may also be provided by a private sector processing industry.

Forest industries in many parts of the world, which draw a large part of their wood and fibre raw material from smallholders or farmers, have been contracted in an outgrower scheme (FAO, 2001). In such an arrangement, the contracting company outsources the production of raw material to smallholders and farmers in the vicinity of the processing plant. The potential advantages of such arrangements include the benefit to the industry of limiting the need to invest in land, labour and forest operations, and to the contracted smallholders of a secure market and access to technical services.

Outgrower schemes offer a wide range of formal partnership arrangements between smallholders/farmers and contracting companies. The FAO (2001) distinguishes between:

- partnerships in which smallholders/farmers are largely responsible for wood production, while the company guarantees the purchase of the product at harvest, sometimes at a prearranged price;
- partnerships in which the company is largely responsible for wood production, paying smallholders market prices for their wood allocation;
- land lease agreements in which smallholders grant user rights to the partner company and are not involved greatly in woodlot management; and
- partnerships in which the smallholder/farmer and the company share the risks and the responsibility for production and product marketing, dividing the returns in proportion to the level of inputs.

Outgrower schemes may have both positive and negative impacts on sustainable livelihoods and land use. They diversify farm production and offer an additional production opportunity on underutilized land, as well as additional income and employment for local communities. Environmental risks involved in planted forests are diminished when the plantations are spread to many different, smaller entities. Further, there is generally increased community support for the wood processing company when the partnership arrangement provides a broader social and economic development framework for the individuals and communities (FAO, 2001). However, conflicts may arise in a partnership arrangement if the prices offered are too low or the contract period too long, so that the smallholder makes insufficient profit to reinvest in the establishment of new plantations.

Some successful examples of poplar and willow farms and the creation of rural handicraft businesses are reported from Chile and India (Boxes 12.8–12.10), though these examples demonstrate that formalized long-term partnerships are not always the best model of interaction for both small landowners and wood processing companies.

12.2.3 Poplars and willows in agroforestry and trees outside forests

Agroforestry is a collective name for a land-use system where woody perennials (trees, shrubs, palms, bamboo, etc.) are produced and managed on the same land unit as agriculture (field crops or animal husbandry), horticulture and/or aquaculture in some form of spatial or temporal sequence (Kishwan *et al.*, 2005; World Agroforestry Centre, 2007). There is a wide range of different agroforestry production systems, including home gardens, plantation-based intercropping systems, shelterbelts, windbreaks, scattered trees on farmland and woodlots for supplementary fodder. Some examples are given in Plate 32E, Fig. 12.4 and Boxes 12.11–12.14.

The concept of agroforestry originated from tropical regions where population growth caused severe land shortage, prompting the need for efficient production systems to address the demand for food and wood resources (FAO, 1989; Wu and Zhu, 1997). Agroforestry production systems are better suited to conserve and maintain socio-cultural values than most other land-use systems,

Box 12.8. Employment generation through poplar farming and utilization, northern India

In the remote rural areas of Punjab, Haryana, Western Uttar Pradesh and Uttranchal, smallholders grow poplars (*Populus deltoides*) in agroforestry systems on fertile farmland. In the initial years of a plantation, the growing of agricultural crops together with the trees provides an immediate income. When the trees grow in size and the canopy closes, the crop yields decline. Poplars are grown over a rotation period of 8 years and produce peeler logs (75%), pulpwood (18%), roots and firewood (7%). The poplar wood provides an industrial raw material base for many wood-based industries producing plywood, matches, packaging material, pulp, sports items, ice-cream spoons, sticks and artificial limbs.

Nursery operations, land preparation, plantation establishment and maintenance, and the utilization of the poplar wood play a significant role in rural development by providing employment and generating income for many skilled, semi-skilled and unskilled workers, especially for women, throughout the poplar growing region. The annual manpower requirements to run a poplar nursery are 990 man-days ha^{-1} nursery area. For the establishment and maintenance of the poplar plantations over a rotation period of 8 years, an annual manpower input of 55 man-days ha^{-1} plantation area is estimated on average. The direct employment generation in the wood-based industry is estimated at 137,500 employees in the poplar growing region.

Source: Dhiman (2006).

Box 12.9. Basket willow production creates jobs for small farmers, Chile

The suitability of flexible shoots of basket willow (*Salix viminalis*) for making handicraft items was discovered in the small town of Chimbarongo, 200 km south of Santiago de Chile, in the early 20th century. People began to cultivate the species and artisans were trained to produce furniture that was sold in the capital and elsewhere in the country. However, wood, leather and plastic were slowly beginning to replace willow in the manufacture of furniture, packaging and other household articles. The quality of willow products was insufficient to compete in better markets, and designs were old-fashioned. The Chilean Forest Research Institute (Instituto de Investigación Forestal de Chile, INFOR), in collaboration with universities and other national institutions, carried out a project from 1997 to 2003 to promote the resurgence of the Chilean basket willow sector.

The focus was on: (i) technical and scientific expertise to improve product quality; (ii) developing the domestic market; and (iii) increasing exports of raw material and products, especially furniture. The project sponsored training courses in the rural areas where basket willow trials had been set up, and about 120 people from those areas were trained in basket willow manufacturing techniques. Most of the beneficiaries of these courses were women, who wanted to contribute to their family income and improve their role in the community. The results showed that women had considerable capacity for working in basket willow, an activity which allowed them to generate income while still caring for their home and children. In due course, female artisans formed organized groups, for whom workshops on the production and marketing of their output were held with the support of the local government. With a view to improving furniture design, the project established collaboration with the design schools of the country's main universities, to promote quality and design of the end products and create new applications for the material (see Fig. 10.13c). The designs were later shown at the first basket willow furniture fair in Chile. (For a more technical perspective on basket willow production, please refer to Chapter 10, this volume.)

Source: Abalos Romero (2005).

Box 12.10. Poplar matches are produced in an outgrower arrangement in India

In India, private companies are the main buyers and processors of wood fibre. Government policy, however, limits the area of land that can be owned by private entities and rules out any direct private sector roles on state forestland. Under this policy, framework wood processing companies are obliged to purchase wood fibre from smallholders and farmers. In the early 1980s, the Western India Match Company (Wimco) Ltd, the leading safety match manufacturing company in India, was experiencing a shortage in raw material and decided to experiment with outgrower schemes for poplar (*Populus deltoides*). Partnership arrangements started with the distribution of seedlings free of charge; however, the survival rates achieved after planting remained low. In the next step, Wimco and other companies increased inputs through bank loans, technical extension and buy-back guarantees. This was also largely unsuccessful, because smallholders either defaulted on their loans or found higher prices for their product on the open market.

In spite of these initial failures, the experience with outgrower arrangements in India has been largely positive, though formalized schemes have mostly been abandoned. Wood processing companies have moved on to focus on the business of developing and supplying high-quality clonal seedlings through local nurseries. They buy wood fibre on the open market at prevailing market prices, while smallholders benefit from market competition among fibre-buying companies. Farm forestry is now a viable land-use option for smallholders in many parts of the country that does not displace agriculture, as larger-scale farmers choose to spread their risk between agricultural and timber crops and small-scale farmers are unwilling to forego food security and mainly plant trees along field boundaries.

Source: Mayers and Vermeulen (2002).

Fig. 12.4. A riverine poplar plantation provides the ideal biotope for geese farming, China. Photo courtesy of FAO/A. Del Lungo.

Box 12.11. Pollarding poplars for supplementary fodder, New Zealand

Poplars and willows are exotic to New Zealand and have been introduced and cultivated over the past 160 years for soil erosion control on pastoral hill country, riverbank protection, provision of shelter and shade for farm livestock, windbreaks and woodlot forestry (Wilkinson, 1999). In the drought-prone areas of the country from Otago in the south to Gisborne in the north, poplars also serve as a source of supplementary fodder during dry summer months (Fig. 12.5). Poplars can tolerate regular defoliation and regenerate quickly to supply future supplementary fodder. The trees are pollarded at animal height within 4 or 5 years and continue to grow vigorously. Some 20-year-old trees have been pollarded at least six times, while farmers have observed that a mature tree with about 5 year's regrowth can feed up to 30 cows daily. Regrowth from a poplar tree pollarded (complete canopy removal) 5 years earlier was 29.3 kg dry matter, of which about 30% comprised edible foliage (leaves and stem less than 5 mm diameter).

Source: National Poplar and Willow Users Group (2007).

Fig. 12.5. Sheep feeding on poplars in New Zealand. Photo courtesy of FAO/D. Charlton.

Box 12.12. Poplar-based agroforestry for rural development, eastern China

Heze Prefecture in Shandong Province, with a population of 800,000 people, is one of China's major agricultural production bases for grain, cotton, groundnuts, fruit, vegetables, livestock and poplar wood. Heze Prefecture is located in the lower reaches of the Yellow River's old alluvial plain, which has serious problems from sedimentation and salinization. Since the 1950s, a well-organized network of forest shelterbelts, greenbelts and woodlots has been established along with the planting of trees along rivers and canals (Plate 32B). Poplars are the predominant species used in afforestation, along with *Paulownia* sp. The most common and economically most successful agroforestry production system observed was intercropping poplar with winter wheat, cotton, groundnuts, maize and shade-tolerant grasses such as lucerne, which can be used for livestock fodder. Wheat can only be intercropped with poplar during the first 3 years, after which the yield begins to decline due to the shade of the trees. Some farmers reported that their income had doubled, if not tripled, from intercropping poplars with lucerne. Trees planted as shelterbelts have further contributed to a reduction in the speed of hot, dry

Continued

Box 12.12. Continued

winds, and thus lower evaporation, and have resulted in an increase in humidity and soil moisture, and eventually in an increase in wheat yields. In general, farm life has improved due to the economic benefits and ecological services provided by the agroforestry production systems. The increasing availability of wood resources and the liberalization of the timber market have contributed to the emergence of a diversified wood industry. Products include plywood, veneer, fibreboard and laminated wood products. Heze Prefecture boasts more than 28,000 manufacturers of wood products, the majority being on a small-scale household basis. Twenty-seven woodworking mills are joint ventures with international investors. The industry has supplied 500,000 people with employment, in particular rural women, and continues to offer more jobs as the industry grows. In some villages, more than 90% of the population is involved in some way with the manufacturing of wood products.

Source: Wood (2005).

Box 12.13. Poplar for fodder in animal agroforestry systems, Islamic Republic of Iran

In the arid, semi-arid and subtropical zones of the Islamic Republic of Iran, trees and shrubs of poplar, *Acacia* and *Robinia* are important sources of fodder in traditional animal agroforestry systems. Poplars are the species preferred by farmers, due to their high growth performance, good adaptability to extreme site conditions and high nutrient content. They have proven to be an effective insurance against seasonal fodder shortages or the risk of drought. During dry seasons, tree fodder may contribute up to 30–45% of total livestock feed intake. The productivity of poplar clones of 1-year-old plants (*Populus xcanadensis, Populus euphratica, Populus deltoides*) was found to be between 16 and 18 t ha^{-1} biomass (leaves and edible branches).

Source: Calagari *et al.* (2003).

Box 12.14. Agroforestry with poplar in northern India

India's National Forestry Policy in 1988 declared wood from poplar plantations an agricultural product free of tax liabilities, and thus prompted a boom in the establishment of tree plantations in agroforestry systems. Poplar is the most popular tree species in these agroforestry production systems, due to quick and high financial returns. In the Yamunanagar district in Haryana, 79% of all planted trees are poplars, followed by eucalypts (19%). Wood produced from these plantations is now sustaining the raw material supply of hundreds of small and medium-scale wood-based industries in northern India, manufacturing around three dozen products such as plywood, paper pulp, laminated board, packing cases, pencils, ice-cream spoons, toothpicks, artificial limbs, sport articles, furniture and window frames. Lop-and-top and roots are used as firewood. Farmers have planted poplar in different designs (block, boundary and row planting), depending on their financial and technological capacity and management objectives. Usually, poplar trees are intercropped with agricultural crops like wheat, sugarcane, paddy rice and shade-tolerant fodder crops like barseem (*Trifolium* sp.) and chari (*Sorghum vulgare*). A financial cost–benefit analysis showed very favourable rates of return for the agroforestry production systems. The cost–benefit ratio of pure poplar plantations was 1:1.92 and 1:2.13 for poplar plantations with intercropping of agricultural crops. The higher returns of poplar with intercropping were due mainly to the higher productivity of the poplar trees in combination with agricultural crops. Farmers, however, favoured poplars without intercropping, due to the better timber quality (and thus the price they received) and the longer intervals between harvesting. The agroforestry production systems have brought about some significant socio-economic impacts. Agroforestry plantations contribute between 26% and 35% of local income, while agricultural crops contribute 34–60%. Landless people get 15 days of work a month on average in the agroforestry sector. Employment opportunities have further increased by training the rural population in skilled jobs like the manufacture of pencils and packing cases, wood carving and carpentry. New prosperity has increased awareness of the importance of education and has prompted villagers to send their children to school and college. Further, agroforestry has improved the availability of fodder, which has led to a significant increase of milk production by 84%.

Source: Dhillon *et al.* (2001); Kumar *et al.* (2003); Kumar (2004).

and are proven to contribute positively to food security, poverty alleviation and sustainable livelihoods (Kishwan *et al.*, 2005).

Poplars and willows are common trees in agroforestry in many countries. The IPC reported that globally 2.6 million ha of poplars and willows were established in agroforestry production systems, of which 0.8 million ha (31%) were planted primarily for wood production, 1.7 million ha (65%) for environmental protection and 0.1 million ha (4%) for other purposes. The People's Republic of China accounted for 96% of the world's poplar area established in agroforestry production systems (2.5 million ha). Other countries which reported significant areas of poplar in agroforestry included India (60,000 ha), Argentina (20,500 ha), Canada (14,000 ha) and New Zealand (11,000 ha) (IPC, 2008).

The specific properties of poplar species, such as a straight, clean bole, leaflessness during winter, multiple uses of wood and bark, soil enrichment qualities, compatibility with agricultural crops and high economic returns, make poplar a very suitable species for agroforestry. The use of poplars in combination with agricultural crops has been shown to improve the productivity of agriculture by providing shade and protection against wind for both the crops and the grazing livestock. The Poplar Council of Canada has reported quantitative evidence that poplar shelterbelts established in Saskatchewan have a significant positive impact on the yield of wheat and lucerne crops (Silversides, 1991). If combined with animal husbandry, agroforestry also has a positive impact on soil fertility, and large ruminants may provide draft power for land preparation, soil conservation practices and haulage. In India, poplar is the most preferred species in agroforestry, where plantations of over 30,000 ha have for many years been supplying poplar wood for matches and plywood, particularly from the 'G-48' clone of *P. deltoides*.

Agroforestry plays an important role in China, where both demand for wood and the pressure on land are high. A vast amount of the North China Plain has been converted into agroforestry systems such as farmland shelterbelts and forest networks, intercropping agricultural crops with poplar trees and so-called four-side agroforestry (planting trees around houses, roadsides and riverbanks, see Box 12.12). Due to increased wood availability and the liberalization of the timber market, the wood products manufacturing industry has been able to develop in northern China. Products include plywood, fibreboard, laminated wood products and veneer. The government is actively encouraging agroforestry: (i) by allocating smallholder land-use and crop ownership rights to farming families; (ii) by reducing or totally removing taxes paid by farmers; and (iii) by the provision of low-interest loans, reduced prices for electricity, improved infrastructure and minimal regulations for manufacturing (Carle and Ma, 2005). In the USA, poplars and willows are receiving increasing attention for tree planting along waterways to prevent soil erosion and runoff of agricultural chemicals, to enhance wildlife habitat and to provide bioenergy and wood products for the rural economy (Ball *et al.*, 2005) (see also Chapter 6, this volume).

12.2.4 Poplars and willows for bioenergy

Wood fuels comprising firewood and charcoal are the most important energy source and the most important forest product for many developing countries, particularly in Africa. There, wood fuels contribute from 50% to 90% of all energy consumed, and at the same time represent about 60–80% of all wood consumed. Globally, about half of the annual wood removals (about 1.7 billion m^3) are used for fuel (FAO, 2010) and 3 billion people rely on unsustainable biomass-based energy resources (UNDP/WHO, 2009).

The use of renewable energy resources and, foremost, the use of wood and woody biomass is promoted increasingly by industrialized countries. The global consumption of biomass for energy increased by 51% between 2006 and 2009. The global pulp industry in 2009 used an estimated 75 million t of biomass for energy generation to reduce dependence on fossil fuels (Wood Resources International, 2010). The significance of wood-based fuels is likely to increase even further due to high oil prices, which make wood fuels increasingly price-competitive with fossil-fuel alternatives, and the persistence of global poverty, which will increase dependency of the poor rural population on wood fuels as their primary source of energy (De Miranda *et al.*, 2010).

Fast-growing planted forests can be an efficient way to produce wood fuel. Currently, they supply an estimated 15–20% of the world's wood fuels, whether in the form of by-products or residues from industrial forest plantations or as whole trees from dedicated bioenergy plantations (Mead, 2005). Globally, around 8.6 million ha of planted forests, of which 6.7 million ha (78%) are in Asia, are being grown specifically for wood fuel. They consist mostly of fast-growing species such as *Populus*, *Salix*, *Eucalyptus* and *Acacia*. In Brazil, for example, 25% of the *Eucalyptus* plantations, which produce 4 million t of biomass annually, are grown specifically to provide charcoal for the pig-iron, steel and cement industries (Andersson *et al.*, 2002; Ceccon and Miramontes, 2008).

Wood fuel production in rural areas is reported to have a high economic significance and a high impact on employment generation (Gan and Smith, 2007). Wood fuels, though perceived as a cheap and primitive source of energy, can create commercial value chains that often grow to significant proportions, involve considerable amounts of money and provide a secure source of income for the rural population. Firewood, for example, constrained by its high transportation costs, has to be processed for the most part close to its production site, thus creating jobs at the local level. Research from Poland suggests that the cultivation of short-rotation willow species for bioenergy plantations offers a financially competitive land-use option on arable land, due to the favourable wood properties and the high-yielding potential of willows. *Salix viminalis*, for example, is reported to yield 26.4 t ha^{-1} $year^{-1}$ of biomass in a 3-year cutting cycle and to produce wood with a high content of cellulose and lignin coupled with low water content, resulting in a high calorific value of the willow wood (Szczukowski *et al.*, 2002).

Case studies from Sweden and the USA on the use of willows for bioenergy are presented in Boxes 12.15 and 12.16. For an extended scientific-technical description of the use of poplars and willows for bioenergy and wood fuel, please refer to Chapter 10, this volume.

Box 12.15. Willows for bioenergy and phytoremediation, Sweden

The cultivation of short-rotation bioenergy plantations was introduced in Sweden after the oil crisis in the 1970s, with the intention of replacing fossil fuels by renewable, carbon dioxide neutral sources of energy. Willows of different clones and hybrids of *Salix viminalis, Salix dasyclados and Salix schwerinii* grown in coppice systems were found to be the most efficient species in terms of biomass production, stand management and nutrient utilization. The produced biomass (6–12 t ha^{-1} $year^{-1}$) was used in district heating plants for combined heat and power production.

In recent years, a process known as phytoremediation has been added to the production of bioenergy from willow coppice. In this process, nutrient-rich waste products, such as urban wastewater, landfill leachate, industrial wastewaters (e.g. log-yard runoff), sewage sludge and wood ash, have been applied successfully to the willow coppice to reduce, through plant uptake, the content of pollutants and/or excess nutrients in water and soils, and to facilitate microbial degradation of organic pollutants. This method of treating waste products is more cost-effective than conventional treatments, and the nutrients contained in them serve as low-cost fertilizers to increase biomass production.

Enköping municipality in central Sweden, with a population of 20,000 people, has established a unique biocycle process that produces renewable energy via an advanced bioenergy system linking the municipal heating and sewage treatment networks with local farms. Almost all of the power plant's fuel comes from renewable biomass from short-rotation willow coppice and forest industry waste (branches, treetops, bark and sawmill dust). The willows grow in a closed loop-type system that pumps sewage water from the town into dams for purification. The water is then used to irrigate the willow plantations, which absorb the heavy metal cadmium and large quantities of nitrogen and phosphorus effluent that otherwise would enter a nearby freshwater lake. At the heart of the Enköping model is cooperation between the municipality and the local farmers, who grow the *Salix* crops on more than 1000 ha and use the ash left behind after biomass incineration as agricultural fertilizer. All the biofuel used for bioenergy production comes from the local neighbourhood and has an average transport distance of about 70 km. The Enköping plant produces around 220,000 MWh of heat and 100,000 MWh of electricity each year, enough to heat all the town's houses and its tap water.

Source: Dimitriou and Aronsson (2005); SymbioCity (2011).

Box 12.16. The *Salix* Consortium – willow biomass production for bioenergy, USA

Years of research on best management practices for short-rotation woody crops in New York State, combined with growing concern about environmental issues, prompted the formation of the Salix Consortium in 1994 of over 20 organizations from the science, energy and rural development sector. The goal of the Salix Consortium was to establish a rural-based enterprise to facilitate the development of willow biomass crops as a locally grown renewable feedstock for bioenergy and cellulose for the north-eastern and north-central regions of the USA. The project was one of three selected to demonstrate the development of biomass energy under the Biomass Power for Rural Development programme supported by the United States Department of Energy and Department of Agriculture. Willows were selected over other woody species because of their rapid juvenile growth rates, vigorous coppicing ability, ease of establishment from unrooted cuttings and high potential for rapid genetic improvement. Seedlings were planted mechanically in double rows at 15,300 plants ha^{-1} and harvested mechanically after 3- to 4-year cycles. Yields of fertilized and irrigated willow grown for 3 years have exceeded 27 t ha^{-1} $year^{-1}$ (oven-dry). First-rotation, non-irrigated trials in central New York have produced yields of 8.4–11.6 t ha^{-1} $year^{-1}$ (oven-dry). Initial assessments of the rural development benefits associated with a willow biomass enterprise indicate that about 76 direct and indirect jobs will be created for every 4040 ha of willow established. An extended description of the Salix Consortium is also given in the section on biomass energy in Chapter 10, this volume.

Source: Volk *et al.* (2000).

12.3 Poplars and Willows for Livelihood Protection

12.3.1 Protection and rehabilitation of the environment

Increased public awareness of air and water pollution, global climate change and soil erosion have prompted several countries to develop new techniques for the cultivation of poplars and willows focusing on the protection and conservation of the sources of livelihoods rather than on the production of wood and fibre. Springtime sandstorms, for instance, are a common event in northern China, as Siberian winds blow dust and sand off the Gobi desert across East Asia, sometimes as far as North America. The frequency of sandstorms has been exacerbated by desertification due to agricultural expansion, overgrazing and population growth, starting in the 1950s. In China, poplars are thus used extensively for shelterbelts, sandstorm mitigation and sand dune stabilization, and the country has invested heavily in planting trees and small shrubs, predominantly of poplar and willow species, on former croplands to prevent the spread of arid land (Ramzy, 2010). In Latin America, poplar and willow species serve as shelterbelts from sun and wind at high altitudes (Plate 32C). In Uzbekistan, widespread land degradation is overcome by the establishment of large-scale poplar plantations to restore the land's productive capacity and enable agricultural development. In Bulgaria and Chile, willows are planted along riverbanks to stabilize them with their strong root system and reduce surface water flow, and thus sedimentation. In the UK, they provide shelter and ground cover for free-range chickens, whose produce is sold as 'woodland eggs'.

In several countries, poplar and willow plantations are used in wastewater treatment and in phytoremediation (see Box 12.15). In New Zealand, for example, poplar clones that accumulate high concentrations of boron, a common contaminant in timber industry sites, are being used commercially for the remediation of a large 5 ha wood-waste dump. In Serbia and Montenegro, field studies have been carried out on the use of poplar for cadmium phytoextraction. Sweden is exploring the use of willows for the absorption of nitrogen arising from intensive livestock farming. Research results suggest that 1 ha of willow plantation can potentially absorb 150–200 kg N $year^{-1}$. In the Republic of Korea, research is being carried out on planting poplars and willows on landfill sites and irrigating them with livestock wastewater. A project in New Zealand investigated the efficacy of using poplars and willows in a coppicing system to reduce the amount of nitrate leaching from dairy-shed effluent that was normally applied to pasture. For a more technically oriented treatment of this topic, please refer to Chapter 6, this volume.

Boxes 12.17–12.20 present some outstanding examples of the significance poplars and willows have for environmental protection and the mitigation of climate change in India, Uzbekistan and China.

Box 12.17. Poplar–wheat-based agroforestry farms as carbon sinks, India

Removing atmospheric carbon and storing it in the terrestrial biosphere, for example through the planting of trees, is one of the methods accepted under the Kyoto Protocol for countries to meet their national carbon reduction targets. The emerging carbon market may provide a new agroforestry option for landowners, provided that carbon prices are high enough to make growing trees a more profitable investment than traditional land uses. A study conducted in northern India explored the carbon sequestration potential of poplar–wheat-based agroforestry systems (Fig. 12.6). In a 7-year rotation of *Populus deltoides* intercropped with wheat, the total CO_2 assimilation by the biomass (above- and belowground) was estimated at 200 t ha^{-1} or 28.6 t ha^{-1} $year^{-1}$. Monocropping of either poplar or wheat yielded considerably lower carbon stocks in the biomass. The high carbon sequestration potential of agroforestry results from the increased growth and assimilation rates of the intercropped components as compared to monocropping systems. In terms of carbon sequestration, agroforestry production systems provide the best land-use option and can be valued higher than traditional agricultural systems. For a broader treatment of poplars and willows for carbon sequestration, please refer to Chapter 6, this volume.

Source: Chauhan and Chauhan (2009).

Fig. 12.6. Intercropping poplar and wheat was found to have a high potential for carbon sequestration. Photo courtesy of FAO/S. Chauhan.

Box 12.18. Restoring degraded landscapes in the Aral Sea Basin, Uzbekistan

Land degradation is a serious hindrance to agricultural development in the Central Asian Republic of Uzbekistan, a country striving to rebuild its agricultural sector for national production. Land degradation in the irrigated areas is caused by land salinization, raising groundwater due to unsustainable agricultural practices, irrigation water scarcity and seasonal and long-term drought. Poplar is planted on a large scale on degraded land to restore its productive capacity, for which poplar has proved to be an exceptionally suitable species due to its extensive root system. Poplar additionally displays fast above-ground growth performance, which contributes to the production of scarce firewood and fodder. An estimated 95% of the country's annual wood removals are used as firewood, and its demand is expected to grow as the costs of other energy sources (gas, electricity) rise and the population growth continues at an annual rate of 2%.

Source: Khamzina *et al.* (2006).

Box 12.19. Combating desertification of the Korqin Sandy Lands in the Three-North Region, China

Desertification of the Korqin Sandy Lands in the eastern part of the Three-North Region was caused mainly by overgrazing, wood cutting, shifting agriculture and tillage agriculture. Very strong winds in spring and winter are the chief agents of desertification and are associated with the generation of dust storms. About 70% of rural income is based on agriculture (maize, legumes and rice) and 30% on animal husbandry (goats, sheep, cattle, pigs, geese and poultry). The Chinese Government recognized that to maintain crop and livestock productivity and the livelihoods of expanding communities, it was essential to integrate forests and trees for shelter and shade with agriculture and livestock management. From 1991 to 2002, over 10,000 ha of the sparsely wooded shrub grasslands were replanted with drought-tolerant and cold-resistant clones of *Populus simonii* × *Populus nigra* and *Pinus sylvestris* to establish a living wall to curb sandstorms and, at the same time, improve the socio-economic well-being of the population living in the area (Plate 32D). The project demonstrated land-use systems integrating tree, shrub, pasture and cash crops in agroforestry systems combined with horticulture and viticulture. It developed shelterbelt designs including the use of tree and shrub species to provide fodder for grazing animals. Shrubs were used as fences and hedgerows around and in pastureland, and shelterbelts were established to reduce wind erosion and to implement controlled rotational grazing. Local herdsmen and farmers were consulted and involved in the design, planning and implementation of the agroforestry systems. Wind erosion in the demonstration areas was reduced by 75% in comparison with adjoining lands, and the annual grain production increased considerably. For a broader treatment of this topic, please also refer to Chapter 7, this volume.

Source: Three-North Shelterbelt Bureau (1989); Carle and Ma (2005); Jacquot (2007).

Box 12.20. Environmental protection in Siyang County – a role model in poplar cultivation, China

Siyang County, with a population of 1 million inhabitants, is located on the northern plain of Jiangsu province, eastern China. It is known as the 'land of rivers and lakes', a name that alludes to the network of rivers, canals, lakes and reservoirs in the fertile, alluvial plain of the Yangtze and Huai rivers and the vast plains to the south of the Huang He (Yellow River). In the 1970s, Siyang imported 32 Italian poplar clones of seven species and embraced cultivation of poplars for smallholder plantings, agroforestry and watershed management, skilfully using the network of waterways as a framework for tree planting. Poplars have become the principal tree species planted on more than 100,000 ha of flood plains and marginal lands. The growing stock of these plantations is estimated at 6.5 million m^3. This massive afforestation effort has increased the forest cover in Siyang from the initial 7% to the current 48%, and presents an excellent example of how the cultivation of the versatile and multi-purpose poplar tree contributes to protecting the environment and water resources. Poplars stabilize riverbanks, mitigate soil erosion of over 20 middle- and small-sized rivers and help to lessen the impact of sandstorms and windstorms.

Continued

Box 12.20. Continued

They have increased the numbers and diversity of wildlife and present a space for open-air recreation of the local population.The sustainable management of smallholder woodlots and the integration of poplar into agroforestry production systems have further played a crucial role in developing a local wood industry and in contributing to rural livelihoods and people's well-being. The resources of the poplar plantations account for 50% of the GDP in Siyang and feed 720 large and medium-sized wood processing plants with an annual output capacity of over 2 million m^3. The farmers consider poplars as the green bank that supports a number of emerging activities, such as raising chickens and ducks and cultivating edible fungi. The large-scale planting of poplars has made possible a remarkable economic transformation of Siyang, showing how forestry and agroforestry can be a successful, alternative way of addressing poverty, hunger, malnutrition and deterioration of the environment. Poplars have made such a positive impact on the landscape, livelihoods and economy of Siyang County that local authorities have built the only poplar museum in the world. They hold a Poplar Festival biannually, which attracts more than 100,000 people and is broadcast by the major Chinese TV channels. The significance of poplar cultivation for the sustainable development of Siyang and for the benefit of about 1 million people was acknowledged by the provincial government of Jiangsu through an award for outstanding environmental achievements and the granting of the name 'Hometown of Poplar Culture'.

Source: FAO (2008b); Siyang County People's Government (2010).

12.3.2 Rural landscape and urban amenity

Fast-growing, tall and elegantly shaped poplars and willows are widely used in urban and rural areas as aesthetically pleasing landmarks beside crossroads, bridges and wells. The light and tender crowns of the willow make a decorative choice of tree on roadsides, riverbanks or by the shore. Poplars and willows show high tolerance of mechanical damage and manage to grow new roots and branches if they break or are damaged. Willows in particular tolerate a rise in the ground level in connection with building projects, as they have the ability to form new roots close to the new surface. Both poplars and willows can survive for some time in flooded areas and can contribute to the stabilization of riverbanks.

Since the beginning of the 20th century, farmers in the agricultural regions of western Canada have appreciated the use of poplars for protection and amenity purposes. Early settlers used the native *P. tremuloides* for their farmstead shelterbelts, because the species was readily available, adapted to the soil and climatic conditions and provided reasonably quick protection from strong winds and drifting snow (Schroeder, 1988). In China, poplars and willows are widely used for greening urban and rural areas (Puyang Forestry Bureau, 2008) (Fig. 12.7) (Box 12.21). In Canada, aesthetically appealing living walls of willows protect residential areas from traffic noise, dust, exhaust fumes and snow drifts along highways (Box 12.22).

Forests and woodlands make important contributions to rural livelihoods and sustainable development, and while it is not always possible to identify the particular share that the *Salicaceae* family makes to these benefits, it does appear that poplars and willows are true miracle species and extremely versatile and useful trees. They contribute significantly to people, their lives and livelihoods, through the provision of a large number of wood and non-wood products. Their outstanding performance in the phytoremediation of severely degraded sites, the rehabilitation of fragile ecosystems, the quick restoration of degraded landscapes and the combating of desertification is unmatched by any other species. As fast growers, they are effective at sequestering carbon and thus contribute to both adaptation and mitigation of the effects of climate change. In addition, poplars and willows can be used for energy purposes as a substitute for fossil fuels, contributing to greenhouse gas emission reductions.

Poplars and willows have become a significant natural resource in many countries, and their cultivation and uses are well established in their national economies. They can be grown as pure or mixed plantations, as well as in association

Fig. 12.7. A poplar alley provides shelter and shade in Puyang City, Henan Province, PR China. Photo courtesy of FAO/Puyang Forestry Bureau.

Box 12.21. Poplars and willows in Beijing's urban landscape, China

The urban forests of Beijing municipality and the planting of trees along roadsides and canals have been increasing significantly since the 1950s, as a result of large-scale tree-planting campaigns. They play a critical role in the ecology, aesthetics and socio-economic development of the municipality, which had a population of 16.3 million by the end of 2007, when its forest area had reached almost 1.1 million ha, covering 50.5% of the area. The economic and ecological services that Beijing's forests provide for wood supply, non-wood forest products, generation of employment and income, water conservation, soil protection, agricultural protection, air purification and temperature regulation, carbon sequestration and oxygen supply, forest ecotourism and biodiversity conservation have been valued at US\$6.3 billion year^{-1}, of which 86% are attributed to forest ecosystem services. Poplars and willows play a major role in the provision of these services and are a crucial element in the cityscape, decorating roadsides and parks (Fig. 12.8). Peking willow (*Salix matsudana*), Simon poplar (*Populus simonii*) and white poplar (*Populus alba*) have been used in street plantings, plantations and semi-natural stands since the 3rd century AD. Some major introduced urban species are Peking poplar (*Populus* ×*tomentosa*), Canadian poplar (*Populus* ×*canadensis*), eastern cottonwood (*Populus deltoides*), black poplar (*Populus nigra*) and Lombardy poplar (*P. nigra* var. *italica*), which was introduced in the 1940s and 1950s. All large roads and boulevards are planted with at least two tree species, often in two or more rows. Wider planting strips are often used along large boulevards. Forest bands 30 m wide have been planted along major thoroughfares throughout the city, as well as along roads leading out of the city. In suburban areas, there are concentric, though unconnected, bands of forests, while in the outer counties more than 10,000 ha of protection forests have been planted to reduce wind and dust in the city. In northern Beijing, an extensive green belt of different poplar and willow species has been planted, which, besides its ornamental function, separates industrial from residential areas, contributes to cleaner air in the city and also provides shelter from wind and sandstorms that sweep across the Mongolian Plateau, bringing sand and silt across the mountains from the Gobi desert.

Source: Profous (1992); Wu *et al.* (2010).

Continued

Box 12.21. Continued

Fig. 12.8. A greenbelt of poplars and willows at the outskirts of northern Beijing protects against wind, dust and sand, and provides recreational space for the city dwellers. Photo courtesy of FAO/J. Carle.

Box 12.22. Living willow walls increase amenity values in urban areas, Canada

Living walls from fast-growing willow species have been constructed and used as noise barriers along urban highways in Quebec and Ontario. The porous organic structure of plants and soil reportedly reduce traffic noise levels effectively and even absorb sound waves rather than reflecting them. Tests showed that the noise reduction properties of living walls were comparable to any existing type of metal, cement or concrete wall. Willows used in living wall barriers offer a number of environmental, aesthetic and economical advantages. They can be established quickly from cuttings and only require easily accessible material (plants, soil, wood, water). Being aesthetically appealing and green, they improve the urban landscape, limit dust dispersion in the atmosphere, decrease the presence of greenhouse gases by sequestering carbon dioxide and improve air quality. In addition, these structures can serve as visual screens to conceal unappealing urban structures, diversify fauna habitats, particularly for birds, and in comparison to conventional solid noise barriers, they are graffiti-proof and show a high level of social acceptance. Unramified straight cuttings of basket willow (*Salix viminalis*) with a minimal length of 2.5 m and a diameter at the base of between 4 and 6 cm constitute the ideal material for the construction of most living walls. To assure the best conditions for root development and growth, they require fertilization in the first year and irrigation throughout the growing season. About 8 weeks after the establishment of the wall, the developing foliage greens the wall structure completely, and at the end of the first year a green wall of 4.5 m height has grown (Fig. 12.9). The vigorous growth of the willow is estimated to absorb about 5–6 kg carbon dioxide m^{-1} living wall. So far, living walls established

Box 12.22. Continued

along highways are reported to be resistant to normal winter conditions, in particular strong winds and saline spray from the application of road de-icing salt. They are not, however, immune to diseases and pests, and show some winter mortality from severe cold.

Source: Labrecque (2010); Labrecque and Teodorescu (2010).

Fig. 12.9. Stem cuttings of *Salix viminalis* develop foliage after about 8 weeks of establishment and form a closed green wall after 1 year. Photo courtesy of FAO/Werther.

with agricultural crops. They form a valuable part of the rural landscape, support employment in rural areas and contribute substantially to integrated rural development, often in combination with agriculture, livestock production, cash crop production, horticulture and the manufacture of handicrafts. The planting of poplars and willows in small woodlots and in agroforestry production systems is common in Asia, especially China and India, and in South America, and is popular with rural farmers, as it provides regular and relatively secure financial returns.

References

Abalos Romero, M.I. (2005) Towards development of the Chilean basket willow sector. *Unasylva* 56, 40–46.

Andersson, G., Asikainen, A., Björheden, R., Hall, P.W., Hudson, J.B., Jirjis, R., *et al.* (2002) Production of forest energy. In: Richardson, J., Björheden, R., Hakkila, P., Lowe, A.T. and Smith, C.T. (eds) *Bioenergy from Sustainable Forestry: Guiding Principles and Practice.* Kluwer Academic Publishers, Dordrecht, Netherlands, pp. 49–123.

Angelsen, A. and Wunder, S. (2003) Exploring the forestpoverty link: key concepts, issues and research implications. Occasional Paper No 40. Center for International Forestry Research, Bogor, Indonesia.

Ball, J., Carle, J. and del Lungo, A. (2005) Contribution of poplars and willows to sustainable forestry and rural development. *Unasylva* 56, 3–9.

Calagari, M., Modirrahmati, A., Asadi, F. and Ghasemi, R. (2003) Utilization of poplar biomass as a source of fodder. *Iranian Journal of Forest and Poplar Research* 11, 365–384.

Calderón, A., Mastrantonio, L. and Perez, S. (2004) Rendimiento de madera de álamo podado para la fabricación de lápices. Poster presented in XIX Jornadas Forestales de Entre Ríos, Argentina. ISSN 1667-9253.

Carle, J. and Ma, Q. (2005) Challenges of translating science into practice: poplars and other species in the Three North Region of China. *Unasylva* 56, 31–37.

Ceccon, E. and Miramontes, O. (2008) Reversing deforestation? Bioenergy and society in two Brazilian models. *Ecological Economics* 67, 311–317.

Chambers, R. and Conway, R. (1991) Sustainable rural livelihoods: practical concepts for the 21st century. Discussion Paper 296. Institute of Development Studies, Brighton, UK.

Chauhan, S.K. and Chauhan, R. (2009) Exploring carbon sequestration in poplar–wheat-based integrated cropping system. *APA News, Asia-Pacific Agroforestry Newsletter* 35, 9–10.

Contreras-Hermosilla, A. and Gregersen, H. (2001) Trade and transnationals. In: Chipeta, M. and Joshi, M. (eds) *Investing in Sustainable Forest Management: the Private Sector Speaks.* Center for International Forestry Research, Bogor, Indonesia.

Cossalter, C. and Pye-Smith, C. (2003) *Fast-wood Forestry. Myths and Realities.* Center for International Forestry Research, Bogor, Indonesia.

De Miranda, R.C., Sepp, S., Ceccon, E., Mann, S. and Singh, B. (2010) Sustainable production of commercial woodfuel: lessons and guidance from two strategies. World Bank Working Paper 206. World Bank Publications, Washington, DC.

Dhillon, A., Sangwan, V., Malik, D.P. and Luhach, M.S. (2001) An economic analysis of poplar cultivation. *Indian Forester* 127, 86–90.

Dhiman, R.C. (2006) Employment generation through poplar farming and utilisation. Working Paper. WIMCO Seedlings Limited, Uttarakand, India. Available from IPC Secretariat, FAO, Rome.

Dickmann, D.I., Isebrands, J.G., Eckenwalder, J.E. and Richardson, J. (2002) *Poplar Culture in North America.* National Research Council of Canada Research Press, Ottawa.

Dimitriou, I. and Aronsson, P. (2005) Willows for energy and phytoremediation in Sweden. *Unasylva* 56, 47–55.

FAO (1989) *Forestry and Nutrition.* A reference manual. Regional Office for Asia and the Pacific (RAPA), FAO, Bangkok.

FAO (2001) Forestry out-grower schemes: a global view. Report based on the work of D. Race and H. Desmond. Forest Plantations Thematic Papers, Working Paper 11. Forest Resources Development Service, Forest Resources Division. FAO, Rome (ftp://ftp.fao.org/docrep/fao/006/ac131e/ac131e00.pdf, accessed 23 January 2012).

FAO (2006) Global planted forests thematic study. Results and analysis. FAO Working Paper FP/38E. FAO, Rome (www.fao.org/forestry/12139-03441d093f070ea7d7c4e3ec3f306507.pdf, accessed 12 January 2012).

FAO (2008a) Poplars, willows and people's wellbeing. Synthesis of Country Progress Reports, prepared for 23rd Session of the International Poplar Commission, Beijing, China, 27–30 October 2008. Working Paper IPC/6. FAO, Forest Management Division, Rome.

FAO (2008b) Report of the 23rd Session of the International Poplar Commission, 26–30 October 2008. FAO, Rome (ftp://ftp.fao.org/docrep/fao/011/k4541e/k4541e00.pdf, accessed 13 January 2012).

FAO (2010) *Global Forest Resources Assessment 2010. Key Findings.* FAO, Rome.

FAO and DFID (2001) How forests can reduce poverty. FAO, Rome and UK Department for International Development, London.

Gan, J. and Smith, C.T. (2007) Co-benefits of utilizing logging residues for bioenergy production: the case for East Texas, USA. *Biomass and Bioenergy* 31, 623–630.

Gordon, J.C. (2001) Poplars: trees of the people, trees of the future. *Forestry Chronicle* 77, 217–219.

Hassan, M.K. and Haji Salim, H. (2004) The financial analysis of poplar plantations. Working Paper. 23rd Session of the International Poplar Commission, Beijing. FAO, Rome.

Heilman, P.E. (1999) Planted forests: poplars. *New Forests* 17, 89–93.

International Labour Organization (2004) Protecting people, promoting jobs (http://www.ilo.org/wcmsp5/groups/public/@dgreports/@dcomm/documents/meetingdocument/wcms_113996.pdf, accessed 20 March 2013).

International Poplar Commission (IPC) (2008) Poplars, willows and people's wellbeing. Synthesis of country progress reports, 23rd Session of the IPC, 27–30 October, 2008, Beijing (ftp://ftp.fao.org/docrep/fao/011/k3380e/k3380e.pdf, accessed 21 March 2013)

Isebrands, J.G. (2004) Activities related to poplars and willow cultivation and utilization. National Poplar Commission of the USA. Available from IPC Secretariat, FAO, Rome.

ITTO (2007) Poplar is the dominant plantation species in China. International Tropical Timber Association, Yokohama, Japan. *Tropical Timber Market Report* 12(6), 10 (http://www.itto.int/mis_detail/id=32320000, accessed 20 March 2013).

Jacquot, J.E. (2007) Living, green wall fends off encroaching desert. Treehugger, a discovery company (http://www.treehugger.com/files/2007/06/living_green_wall.php, accessed 13 January 2012).

Khamzina, A., Lamers, J.P.A., Worbes, M., Botman, E. and Vlek, P.L.G. (2006) Assessing the potential of trees for afforestation of degraded landscapes in the Aral Sea Basin of Uzbekistan. *Agroforestry Systems* 66, 129–141.

Kishwan, J., Sharma, K.K. and Ratho, S.K. (eds) (2005) *Agroforestry: Manual for Asia Pacific Region.* Ministry of Environment and Forests, New Delhi.

Kumar, R. (2004) Poplar in agro-forestry: potential resource for socio-economic and environmental development. Unpublished working report. Available from IPC Secretariat, FAO, Rome.

Kumar, R., Gupta, P.K. and Gulati, A. (2003) Agroforestry extension and its impact on socio-economic scenario – a case study of Yamunanagar district of Haryana State (India). *Indian Forester* 129, 435–445.

Labrecque, M. (2010) Learning from our mistakes: achievements and misfortunes in phytotechnology. Paper presented at the Environmental Applications of Poplar and Willow Working Party Meeting, International Poplar Commission, 17–18 September 2010. Available from IPC Secretariat, FAO, Rome.

Labrecque, M. and Teodorescu, T.I. (2010) Willows: wonderful tools to remedy environmental problems in urban areas. Working report to the International Poplar Commission. Available from IPC Secretariat, FAO, Rome.

Macqueen, D.J. (2005) Time and temperance: how perceptions about time shape forest ethics and practice. *International Forestry Review* 7, 250–257.

Mayers, J. and Vermeulen, S. (2002) *Company–Community Forestry Partnerships. From Raw Deals to Mutual Gains? An International Review with Proposals for Improving Forests, Enterprise and Livelihoods.* International Institute for Environment and Development (IIED), London.

Mead, D.J. (2005) Forests for energy and the role of planted trees. *Critical Reviews in Plant Sciences* 2, 407–421.

National Poplar and Willow Users Group (2007) Growing poplar and willow trees on farms. Report of New Zealand Sustainable Farming Fund, Poplar and Willow Project (http://www.fao.org/forestry/21644-03ae5c141473930a1cf4b566f5963255f.pdf, accessed 23 January 2012).

Ndoye, O., Ruiz Perez, M., Mamoin, A.D. and Lema Ngono, L. (1999) Les effets de la crise économique et de la dévaluation sur l'utilization des plantes medicinales au Cameroun. Implications pour la gestion durable des forêts. In: Nasi, R., Amsallem, I. and Drouineau, S. (eds) *La gestion des forêts denses africaines aujourd'hui.* Seminaire Forafri, Libreville, Gabon. CIRAD-Forêt, Montpellier, France.

Profous, G.V. (1992) Trees and urban forestry in Beijing, China. *Journal of Arboriculture* 18, 145–154.

Puyang Forestry Bureau (2008) Puyang City and Forestry. Brochure. Available from IPC Secretariat, FAO, Rome.

Ramzy, A. (2010) Beijing: onslaught of the Mongolian cyclone. *Time* on-line edition (http://www.time.com/time/world/article/0,8599,1974013,00.html?xid=rss-topstories#ixzz0kQ6CYXjW, accessed 13 January 2012).

Rouchiche, S. and Haji Mirsadeghi, M.A. (2003) Role of planted forests and trees outside forests in sustainable forest management: Islamic Republic of Iran. Country Case Study. In: FAO (ed.) *Planted Forests and Trees Working Paper FP/32E.* Available from IPC Secretariat, FAO, Rome.

Rozelle, S., Huang, J. and Lele, U. (2002) China: the world's largest experiment in conservation and development. In: Lele, U. (ed.) *Managing a Global Resource: Challenges of Forest Conservation and Development.* World Bank Series on Evaluation and Development Vol 5. Transaction Publishers, New Brunswick, New Jersey, pp. 73–98.

Scherr, S.A., White, A. and Kaimowitz, D. (2004) *A New Agenda for Forest Conservation and Poverty Reduction: Making Markets Work for Low-income Producers.* Forest Trends. CIFOR and IUCN, Washington, DC.

Schroeder, W.R. (1988) Utilization of poplars in the prairies. Unpublished working report of the Poplar Council of Canada, Edmonton. Available from IPC Secretariat, FAO, Rome.

Silversides, R. (1991) Poplar and willow growing in combination with agriculture. Canadian Theme Report to the 19th Session of the International Poplar Commission. Poplar Council of Canada, Edmonton, Canada. Available from IPC Secretariat, FAO, Rome.

Siyang County People's Government (2010) Information notes for the third China poplar festival 28–30 May, 2010. International conference on sustainable management of multi-purpose poplar plantations. Available from IPC Secretariat, FAO, Rome.

Stanturf, J.A. (2008) Activities related to poplar and willow cultivation and utilization. Report of the National Poplar Commission of the USA prepared for the 23rd Session of the International Poplar Commission, Beijing. Available from IPC Secretariat, FAO, Rome.

SymbioCity (2011) Enköping municipality. Close cooperation fuels bioenergy production (http://symbiocity.se/Templates/pages/page.aspx?id=128&epslanguage=en#subpageanchor/, accessed 29 July 2013).

Szczukowski, S., Tworkowski, J., Klasa, A. and Stolarski, M. (2002) Productivity and chemical composition of wood tissues of short rotation willow coppice cultivated on arable land. *Rostlinna Výroba* 48, 413–417.

Three-North Shelterbelt Bureau (1989) *A Rising Green Great Wall.* Dadi Publishing House, Beijing.

Townson, I.M. (1995) *Forest Products and Household Incomes: a Review and Annotated Bibliography.* Tropical Forestry Papers 31. Centre for International Forestry Research and Oxford Forestry Institute. Oxford University Press, Oxford, UK.

Tsarev, A.P. (2005a) Industrial uses of willow in Russia. Unpublished working paper. Available from IPC Secretariat, FAO, Rome.

Tsarev, A.P. (2005b) Natural poplar and willow ecosystems on a grand scale: the Russian Federation. *Unasylva* 56, 10–11.

Ulloa, J. and Villacura, L. (2005) Contribution of a private poplar industry in Chile to sustainable rural development. *Unasylva* 56, 12–17.

UNDP/WHO (2009) The energy access situation in developing countries. A review focusing on the least developed countries and sub-Saharan Africa (http://www.who.int/indoorair/publications/energyaccesssituation/en/, accessed 23 January 2012).

Volk, T.A., Abrahamson, L.P., White, E.H., Neuhauser, E., Gray, E., Demeter, C., *et al.* (2000) Developing a willow biomass crop enterprise for bioenergy and bioproducts in the United States. Presented to 'Bioenergy 2000, Moving Technology Into the Marketplace', Buffalo, New York, 15–19 October, 2000. Northeast Regional Biomass Program, Albany, New York (https://bioenergy.ornl.gov/papers/bioen00/volk.html, accessed 18 March 2013).

WCFSD (World Commission on Forests and Sustainable Development) (1999) *Our Forests, Our Future.* Report of the World Commission on Forests and Sustainable Development. Cambridge University Press, Cambridge, UK.

Wilkinson, A.G. (1999) Poplars and willows for soil erosion control in New Zealand. *Biomass and Bioenergy* 16, 263–274.

Wood, Y. (2005) The role of poplar-based agroforestry in rural development: a case study of the Heze Prefecture, Shandong Province, China. PREMIUM Research Experience for Undergraduates, National Science Foundation and Michigan State University (http://www.pptuu.com/show_509791_1.html, accessed 20 March 2013).

Wood Resources International LLC (2010) Forest products market update. News Brief, February 2010. Wood Resources International LLC, Bothell, Washington.

World Agroforestry Centre (2007) *China-Agroforestry Programme: Strategic Framework and Medium-term Plan 2008–2012.* World Agroforestry Centre – ICRAF, Beijing.

World Bank (2004) *Sustaining Forests. A Development Strategy.* Washington, DC.

World Forestry Congress (1978) *Forestry and Nutrition. A Reference Manual.* FAO, Rome.

Wu, S., Hou, Y. and Yuan, G. (2010) Valuation of forest ecosystem goods and services and forest natural capital of the Beijing municipality, China. *Unasylva* 61, 28–36.

Wu, Y. and Zhu, Z. (1997) Temperate agroforestry in China. In: Gordon, A.M. (ed.) *Temperate Agroforestry Systems.* CAB International, Wallingford, UK, pp. 149–179.

Wunder, S. (2000) Poverty alleviation and tropical forests – what scope for synergies? *World Development* 29, 1817–1833.

13 Epilogue

J. Richardson[1]* and J.G. Isebrands[2]
[1]*Ottawa, Ontario, Canada;* [2]*Environmental Forestry Consultants LLC, New London, Wisconsin, USA*

Ecologists have warned about the impending effects of human domination of our earth's ecosystems for some time (Vitousek *et al.*, 1997). Soon our planet will have over 8 billion humans and the consequences of their activities on the planet's environment will be daunting. Human activity and the globalization of industrialization will have a major effect on water quality, air quality, biodiversity and even the global climate. We think the ecosystem services provided by poplar and willow culture can play an important role in mitigating the impacts of these activities. They can help prevent soil erosion from wind and water and thereby improve water quality in our streams and oceans. They help clean the air we breathe in urban and agricultural landscapes. And they can be used to restore land degraded by human activities. Poplars and willows are key in the use of green technologies such as phytoremediation to clean up contaminated soils and water. Bio-based wood products derived from them can help decrease our world dependence on fossil fuels (Chapter 10) and provide renewable energy for power generation while sequestering carbon to help mitigate climate change.

The emerging role of poplar and willow as a 'model biological system' and the recent sequencing of the poplar genome will offer new technologies for combating pests and diseases, improving wood quality and bioenergy systems and helping us adapt to the ever-changing global climate environment. We agree with Gordon (2001) that poplars and willows can play an important role in the future with an active and organized multi-disciplinary effort toward solving the aforementioned future environmental challenges facing our world. The opportunities for expanding the role in improving rural livelihoods are unlimited. These are truly exciting times for poplar and willow professionals and enthusiasts.

An ancient Chinese proverb holds that one picture is worth more than ten thousand words. The image shown in Plate 32E, also used in part on the cover of this volume, conveys in a very succinct manner many of the key topics of this book. A person, perhaps a field scientist or technologist and probably in a developing country, stands with notepad in hand observing a field of maize growing between rows of young planted poplar trees in an agroforestry setting, with an older block planting in the background. The poplars – they could equally well be willows – are clearly of a single, identified taxon (Chapter 2), selected originally from naturally occurring genetic resources (Chapter 3), but having undergone a process of domestication (Chapter 4) to enhance productivity and perhaps resistance to

*Corresponding author: jrichardson@on.aibn.com

diseases (Chapter 8) and damaging insects (Chapter 9). The procedures for operationally producing poplar planting material, and for ensuring successful establishment and growth once planted, have been developed, honed and adapted to different regions of the world (Chapter 5). The trees provide shelter – an environmental benefit – to the field crop (Chapter 6). The scientist needs to be aware of the stresses placed on the agroforestry ecosystem by abiotic factors such as drought, salinity and the changing global climate (Chapter 7). The trees in the older plantation in the photo will soon be ready to harvest for a variety of products (Chapter 10) and the person managing this agroforestry system will need to consider the market trends and future outlook for different poplar products (Chapter 11), as well as for the field crops. By its very nature, the scene is one of support for rural livelihoods and sustainable development (Chapter 12).

The scientist in the field is young, yet building on the solid groundwork laid by those who have gone before, like Dr Victor Steenackers, to whose memory this book is dedicated. A vast amount of information about poplars and willows is contained within the covers of this volume, but much remains to be learned. The future, as always, holds challenges. Poplars and willows, as trees for society and for the environment, can help us move forward sustainably and on a sound scientific base.

References

Gordon, J.C. (2001) Poplars: trees of the people, trees of the future. *The Forestry Chronicle* 77, 217–219.

Vitousek, P.M., Mooney, H.A., Lubchenco, J. and Melilo, J.M. (1997) Human domination of earth's ecosystems. *Science* 277, 494–499.

Glossary

Abiotic: Non-living.

Abiotic stress: Stress caused in living organisms by non-living environmental factors, such as drought, extreme temperatures, edaphic conditions and high winds that substantially limit plant growth and survival.

Abscisic acid (ABA): Plant growth regulator found in plants that acts as a signalling molecule which helps to control overall growth and physiology. It regulates responses to environmental stresses such as drought, cold, salinity and heat, but also leaf abscission, seed dormancy, stomatal closure and shoot growth.

Acclimation: Phenotypic adaptation to environmental fluctuations; the gradual and reversible adjustment of physiology or morphology as a result of changing environmental conditions.

Additive gene effects: Pertaining to genes that act in an additive fashion; the effect on trait variation is equal to the sum of the contribution of individual genes at different loci.

Adhesive (glue): A substance capable of bonding materials together by surface attachment.

Adventitious: Refers to an organ growing where it is not normally expected, e.g. roots growing from a stem or stems growing from a root (see root sucker).

Adventitious rooting: In *Populus*, most commonly used to describe roots that develop from stem sections or shoots buried in soil.

AFLP: Amplified fragment length polymorphism – a genetic marker that reveals the presence or absence of a specific fragment of DNA at a given genome location.

A-horizon: Top layer of the soil, commonly referred to as topsoil.

Air-dry moisture content: The average moisture content (MC) of wood in equilibrium with outdoor air (by convention it is taken as 12% MC).

Allelic variation: Variation among different gene forms (alleles) in a population.

Allozyme: Variable form of individual enzymes due to intra-locus allelic variation.

Alluvial: Sediment deposited by flowing water, as in a riverbed.

Ament: See catkin.

Amphistomatous: Stomata distributed on both upper (adaxial) and lower (abaxial) leaf surfaces, although in the *Salicaceae* they are typically more dense on the lower surface.

Anal plate: In caterpillars, the shield-like covering on the dorsum of the posterior segment.

Anamorph: The asexual or imperfect state of a fungus.

Anholocyclic: In aphids. Host plants are all of the summer type. No winged sexual forms are produced, and reproduction is always by parthenogenesis (asexual).

Annual ring: The layer of wood deposited during a given year (same as growth ring).
Antiporter: A transporter that moves two substrates in opposite directions across the membrane.
Apoplast: Region of the plant body outside of the living cell contents; typically limited to the cell wall and intercellular spaces.
Appressed: Closely pressed against.
Apterous: Without wings, wingless.
Ascospore: A sexually produced spore of an ascomycete fungus.
Asexual: Imperfect state of a fungus.
Asian longhorn beetle (*Anoplophora glabripennis*): A beetle from China that has been found in the USA and is a threat to hardwood trees; lives inside the tree; no natural predators in the USA.
Association genetics: The study of the relationship between single nucleotide polymorphisms and variation in phenotypes.
ATPase: An enzyme that hydrolyses ATP into ADP and phosphate.
Backcrossing: Crossing of an F_1 hybrid with an unrelated parent genotype belonging to either one of the F_1 parental species.
Backcross hybrid: A spontaneous or purposefully bred hybrid resulting from crossing a first-generation (F_1) hybrid back to one of the original parents, called the recurrent parent, e.g. the progeny of (*Salix purpurea* × *Salix viminalis*) × *S. viminalis*, the recurrent parent is usually a different genotype from the same species or population.
Bark: The tissues of trees outside the cambium (consists of living inner bark and dead outer bark).
Basal fluorescence: Fluorescence emission in the darkness.
Batture: Land built up by deposition of sediments between a levee and a river.
Biochemical marker: A genetic marker based on segregation of allelic forms of secondary metabolites (e.g. terpenes) or enzymes (allozymes).
Biosphere 2 Laboratory: A dome-shaped laboratory in Arizona which contains 3.15 acres (13,000 m^2) of a closed man-made ecosystem.
Biotic: Living.
Blight: Rapid death or dieback of plants.
Blotch: Irregular, necrotic area.
Bound water: Water held in the walls of wood cells by hydrogen bonding (i.e. adsorbed water).
Bract: A modified, usually reduced, leaf in the inflorescence.
Broad sense heritability: The proportion of total phenotypic variance in a quantitative trait that is controlled by additive and non-additive genetic factors.
C_3 plants: C_3 carbon fixation is a pathway for carbon fixation in photosynthesis. This process converts carbon dioxide and ribulose bisphosphate (RuBP, a six-carbon sugar) into phosphoglycerate (a C_3 carbohydrate). This reaction occurs in all plants as the first step of the Calvin cycle. In C_4 plants, carbon dioxide is drawn out of malate and into this reaction rather than directly from the air.
Cambium: A thin layer of growth cells between wood (xylem) and bark (phloem), responsible for the formation new cells for each tissue.
Candidate gene: A gene that has been implicated in causing or contributing to a particular function.
Candidate gene selection: Marker-assisted selection in which SNP markers associated with individual genes of known function are used in targeting quantitative trait loci.
Canker: A localized dead area surrounded by living tissues on branches or stems.
Capsule: The dry, dehiscent fruit of poplars and willows composed of two to four carpels (valves) that split open at maturity to release seed.
Carbohydrates: Biochemical name for sugar containing molecules including single sugar (monosaccharides) as glucose and galactose, but also polysaccharides (complex carbohydrates) as starch (polyglucose), cellulose (plant fibre material), chitin (hard shells of insects) and more complex carbohydrate component parts of lipids and proteins.
Carbon allocation: Process of distribution of carbon within the plant to different plant parts.

Carbon sequestration: Uptake and storage of carbon. Trees and plants absorb carbon dioxide, release the oxygen and store the carbon. Fossil fuels were at one time biomass and continue to store carbon until burned.

Carboxylation: Introduction of a carboxyl group (-COOH) or carbon dioxide into a compound.

Carpel: The floral organ that bears ovules; in the *Salicaceae* the compound pistil comprises two, three or four carpels (see valve).

Catalase: Enzyme catalysing the destruction of oxygen peroxide.

Catkin: A compact cluster of stalkless, unisexual flowers without a conspicuous perianth, in poplars pendulous and in willows mostly erect (syn. ament).

Cauda: The 'tail' of an insect; especially any extension of the anal segment of the abdomen resulting from, for example, wax extrusion or other modification of the terminal segments.

Cavitation: The rupture of a water column in xylem when tension overcomes the cohesive nature of water; an embolism forms.

Cellulose: The major polymeric constituent of the cell walls of higher plants (composed of a long chain of beta-glucose units).

Cell wall: The rigid outermost cell layer observed in plants and certain algae, bacteria and fungi, but characteristically absent from animal cells.

Centimorgan: A unit of measure quantifying the relative distance between genes within a linkage group based on their recombination frequency.

Chalcid: A wasp belonging to the family Chalcididae (Hymenoptera), usually a solitary internal parasitoid of a Diptera or Lepidoptera insect.

Chaperone (chaperon protein): Proteins that assist the non-covalent folding/unfolding and the assembly/disassembly of other macromolecular structures, but do not occur in these structures when the latter are performing their normal biological functions.

Chilling requirement: The amount of time a plant requires exposure to cold temperatures before vegetative and floral buds will develop.

Chitinization: The process of depositing or filling with chitin, the main substance that gives insects their hard body shell.

Chloroplast: The organelle that carries out photosynthesis and starch grain formation. A chlorophyll-containing organelle in plants that is the site of photosynthesis.

Chlorosis: Abnormal yellowing or bleaching of the leaves due to lack of chlorophyll.

Chromosome haplotyping: Determination of the combination of alleles on a chromosome following gametic recombination during reduction division.

Cladoptosis: A minor asexual (vegetative) reproduction process in riparian habitats whereby leafy twigs abscise, then become partially buried in sand or silt and take root, forming a new plant.

Clone: (i) To propagate a plant asexually (vegetatively) by grafting, rooting cuttings, microculture or apomictic seed; except for an extremely low level of mutation, all plants from a clone are genetically identical; (ii) a group of plants produced asexually (see ramet) from cuttings, stump or root sprouts, layering, cladoptosis, fragmentation, microculture or some other method that produces offspring genetically identical to the original plant (see ortet).

Cocoon: A protective cover for the pupa of an insect as it develops into an adult, usually spun of silk by the larval stage.

Coma: A tuft of soft hairs, usually terminal on a seed.

Composite: When elements of two or more materials are combined to make a new material.

Conidia: An asexually produced fungal spore.

Conservation: In *Populus*, the maintenance of biological resources that represent the totality of genetic variation composing the genus (species, varieties, populations, individuals).

Coppice: (i) To cut a plant or stand to the ground so as to produce a vigorous new plant or stand from stump, stool or root sprouts; (ii) a stand of trees of sprout origin; a thicket or copse of small trees.

Corvid: A bird in the family Corvidae (crows).

Cultivar: A named variety of plant produced by hybridization or selected from a wild population and maintained by mass vegetative propagation; usually associated with the commercial trade or production plantations.

Cyclophysis: Age effects in vegetative reproduction; for example, when cuttings taken from branches on old trees root less well than cuttings taken from branches on young trees or coppice sprouts.

Cytosol: The fluid portion of the cytoplasm, outside the organelles.

Decay: The decomposition of wood by fungi.

Decorticated: Having had the bark removed.

Decumbent: Reclining or prostrate, with shoot tips ascending.

Dehiscence (adj. dehiscent): Breaking open of a capsule at maturity to discharge seeds by way of valves.

Density: Mass per unit volume (for wood, the volume is based on the actual external dimensions, i.e. wood substance plus internal voids).

Determinate: A growth pattern whereby a dormant period intervenes between the initiation and elongation of stem units and leaf primordia, resulting in a fixed number of early, preformed leaves, cessation of stem elongation in early to mid summer and relatively slow growth rate.

Diapause: A genetically driven condition of suspended development in an insect – usually so as to survive long periods when food is not available and/or when physical conditions are harsh (e.g. winter).

Dieback: Progressive death of shoots, branches or roots from the tip.

Diffuse-porous wood: Hardwoods in which pore size varies little across annual rings (also semi-diffuse porous woods, like poplars and willows).

Dimorphic: The same species occurring in two distinct forms (sexually dimorphic = distinctive males and females).

Dioecious: Unisexual, with male (staminate) and female (pistillate) inflorescences borne on separate plants.

Distal: Near or towards the free end of an appendage or segment, farthest from the body.

DNA polymorphism: Difference in DNA sequence among individuals or lines that permits genetic linkage analysis or DNA marker-aided selection.

Domestication: Process of artificial selection in which the genetic make-up of a population of plants is changed to produce more utilitarian phenotypes.

Dormancy: The resting or inactive phase of plants or seeds. Dormancy of shoots is usually in response to unfavourable environmental conditions.

Durability: Resistance to deterioration (when against fungi: decay resistance).

Early leaves: The earliest leaves to appear on an elongating heterophyllous shoot and the only leaves that appear on determinate shoots; primordia are initiated the previous season and spend the dormant period in a bud.

Earlywood: Part of the annual ring which is formed during the first phase of the growing season.

Ecology: Study of relationships among organisms and the relationship between them and the physical environment.

Ecophysiology: Study of organisms and their functions and how they exist in the environment.

Ecotype: The smallest taxonomic subdivision of an ecospecies, consisting of populations adapted to a particular set of environmental conditions. The populations are infertile with other ecotypes of the same ecospecies.

Edaphic stress: Stress due to soil conditions.

Electron-transport chain: (also called the electron-transfer chain, ETC, e-train, or simply electron transport), is any series of protein complexes and lipid-soluble messengers that convert the reductive potential of energized electrons into a cross-membrane proton gradient.

Embolism: Blocking of vessels by a foreign material, air when it results from xylem cavitation.

Embryo rescue: The *in-vitro* propagation of interspecific hybrid embryos that would otherwise abort during ovular development.

Entomopathogen: A pathogen of insects; in pest control usually a fungus, virus or bacterium that kills pests.

Epicormic: New shoots borne on old tree stems derived from the release of dormant buds in the bark (also called water sprouts).

Epigenetics: The study of the mechanism by which gene expression determines phenotypic expression.

Epipedron: A soil layer with high organic matter near the surface that is saturated with water for more than 30 days.

Equilibrium moisture content (EMC): The condition of wood (at a given relative humidity and temperature) when moisture gain and loss are equal.

***Escherichia coli*:** Common bacterium that has been studied intensively by geneticists because of its small genome size, normal lack of pathogenicity and ease of growth in the laboratory.

Eutrophication: Excessive nutrients in a body of water due to runoff from the land causing a proliferation of plant life.

Evapotranspiration: The sum of evaporation and plant transpiration from the Earth's land surface to the atmosphere.

Excitation energy: The minimum energy required to change a system from its ground state to a particular excited state.

***Ex situ* conservation:** Conservation of genetic resources in artificial plantings or the storage of reproductive or vegetative materials for later controlled reproduction or vegetative regeneration.

Expressed sequence tag (EST): Small part of the active part of a gene, made from cDNA, which can be used to fish the rest of the gene out of the chromosome, by matching base pairs with part of the gene. The EST can be radioactively labelled in order to locate it in a larger segment of DNA.

Extractives: Chemical substances in wood, which are not part of the cell wall structure, and which are removable by neutral solvents (e.g. water, alcohol, benzene, etc.).

Extramatrical: Outside the body substance in which tissue cells are embedded.

Exuviae: The cast or discarded 'skins' of the different stages of an insect's growth.

F_1 hybrid: The first filial generation of a hybrid cross; the progeny of a spontaneous or controlled crossing of two pure species, e.g. the progeny of *Populus nigra* × *Populus trichocarpa*, or two geographically isolated genotypes of a single species.

Fastigiate: Erect, dense branching that forms a columnar-shaped crown, e.g. Lombardy poplar.

Fecundity: A measure of the reproductive rate of an insect, measured by the number of eggs it produces.

Fibre (US: 'fiber'): The long, tapered strengthening cells in hardwood xylem (exist in two forms: libriform fibres and fibre-tracheids).

Fibreboard: A composite panel made up of interfelted fibres and fibre bundles.

Fibre saturation point (FSP): Moisture content (MC) at which the cell walls are saturated but no liquid free water is present in cell cavities (usually about 30% MC). It is below the FSP when wood begins to shrink.

Fibril angle (microfibril angle): The angle between the long axis of wood cells and the strands of cellulosic microfibrils making up the thickest layer (S2 layer) of the cell wall.

Flood plain: Low area of land surrounding water bodies, which holds the overflow of water during a flood.

Fluorescence: Property of certain molecules that will emit light at another frequency than the frequency they absorb. This property is especially useful in biology due to indicator dyes whose amount of fluorescence changes under certain conditions, typically calcium concentration, acidity (pH) or voltage.

Forest tent caterpillar (*Malacosoma disstria*): Native North American insect and pest of hardwoods. The insect is widely distributed from coast to coast and defoliates trembling aspen (*Populus tremuloides*). During outbreaks, it will attack other hardwood species.

Fragmentation: An asexual (vegetative) reproduction process in riparian habitats whereby branches that have broken off a tree become partially buried in sand or silt and take root, forming a new plant or plants (see also cladoptosis).

Frass: The solid excrement from insects, often appearing as sawdust from wood-boring caterpillars.

Free-air CO_2 enrichment (FACE): Technique that is used to deliver CO_2-enriched air to entire ecosystems located under natural field settings. The delivery system is usually constructed out of pipes and tubes that are arranged in a circle. Because these units do not employ walls or enclosures, FACE experiments are considered to be on the cutting edge of CO_2 enrichment technology.

Fructose: A very sweet sugar occurring in many fruits and honey and used as a preservative for foodstuffs and as an intravenous nutrient. Also called *fruit sugar, levulose*.

Fruitbody: A reproductive structure of a fungus containing or bearing spores.

Gametic recombination: Exchange of genetic material between homologous chromosomes during reduction division of meiosis.

Gas exchange: The diffusion of gases from an area of higher concentration to an area of lower concentration, especially the exchange of oxygen and carbon dioxide between an organism and its environment. In plants, gas exchange takes place during photosynthesis and respiration.

Gelatinous fibre: A fibre having an unlignified inner cell wall layer of its secondary wall (found in tension wood).

Gene complexes: Tightly linked group of genes often inherited together as a unit.

Gene flow: The movement of genes between separate populations within a species or between sexually compatible species by means of pollen, seed or vegetative propagules.

Gene locus: The specific location or position of a gene on a chromosome.

Gene sequencing: The determination of the order of nucleotide bases within nuclear DNA that transmits heritable information.

Genet: The totality of all ramets of a clone, including the ortet.

Genetic distance: A measure of the dissimilarity of individuals of the same species or between different species within a genus based on molecular variation in nucleic acid composition.

Genetic maps: A map of the genome that positions genes within linkage groups, based on the relative distance between molecular markers estimated from the frequency of gametic recombination between the markers.

Genetic resources: In *Populus*, the totality of plant genotypes encompassed within the natural and cultivated stocks of species within the genus that have domestication or conservation value.

Genetic systems: The way individual species organize and transmit heritable information between generations through their manner of reproduction, population structure, genomic organization, etc.

Genetic transformation: Process of introducing DNA through asexual means into a plant that is subsequently incorporated into its genome. Genes transferred are those inaccessible to the host plant through normal hybridization methods.

Genome: All of the genes in a chromosome set contained in pollen and egg cells.

Genomics: The study of organisms in terms of their genomes (their full DNA sequences).

Genomic selection: Marker-assisted selection based on SNP markers covering the whole genome and the assumption of linkage disequilibrium between quantitative trait loci and the ubiquitous SNP markers.

Genotype: Assembly of all genetic constituents of an individual, whether expressed or not. A group of genetically identical organisms. The specific genetic make-up of an individual or a clone.

Girdle: Damage that encircles a stem, often killing the portion above.

Glucose: A type of sugar; the main source of energy for living organisms.

Grain (of wood): The direction, size and alignment of wood cells with respect to the long axis of the tree (or lumber). For example: straight grain, spiral grain, diagonal grain.

Green (wood, in reference to moisture content or MC): MC in standing trees, or MC above the fibre saturation point (FSP) when cell walls are fully saturated.

Gross primary production (GPP): Total amount of solar or geothermal energy converted into chemical energy over a certain period by organisms in a given region.

Growth stress: Internal stresses in living trees, caused by the deposition of new tissues.

Guard cell: Epidermal cells that open and close to let water, oxygen and carbon dioxide pass through the stomata.

Gypsy moth (*Lymantria dispar*): European race of defoliating insect introduced into North America in 1869. This alien species has become a serious pest of hardwoods and its range now includes the Maritimes, Quebec, Ontario and British Columbia.
Halophyte: Plant capable of living under salty conditions.
Hardwood: The wood of broadleaved trees, or Angiosperms.
Heartwood: The inner core of the tree stem where, in living trees, all the cells are dead. In many species, the heartwood has darker colour than the sapwood.
Hemicellulose(s): A group of polysaccharides (often having branched structure) making up part of the walls of wood cells in both hardwoods and softwoods.
Heteroblastic: With juvenile foliage distinctly different from adult foliage in size or shape.
Heterophyllic (adj. heterophyllous): With leaves of distinctly different sizes and shapes on a current shoot (see early leaves and late leaves).
Heterosis: Increased vigour or performance of hybrid offspring compared to both of their parents or the mean of the parents, a common occurrence in the *Salicaceae* (syn. hybrid vigour).
Heterozygosity: The condition in which different alleles are found at the same locus of homologous chromosomes.
Hibernacula: A physical shelter, often made by an insect larva incorporating leaves or other objects, in which it hibernates.
Holarctic: A faunal region encompassing all of Europe and North America (and more).
Homeostasis: The inherent tendency in an organism towards maintenance of physiological and psychological stability.
Hybrid breeding value: An assessment of a parent's ability to produce superior interspecific offspring based on the test performance of its hybrid progeny.
Hybrid swarm: A population of related taxa resulting from hybridization of two or more species followed by backcrossing in subsequent generations.
Hybrid vigour: See heterosis.
Hypersensitive response: In the *Populus–Melampsora* pathosystem, a host defence mechanism conditioned by major gene effects that control the death of cells in advance of leaf infection.
Hyphae: Filaments that make up the body of a fungus.
Hypostomatous: Stomata uniformly distributed only on the lower (abaxial) leaf surface.
Imago: The adult, sexually developed stage of an insect.
Ideotype: Listing of ideal characteristics – morphological, wood quality, yield, physiological, etc. – that define a *Populus* cultivar for a specified use.
Indeterminate: A growth pattern whereby the initiation and elongation of stem units proceed sequentially, resulting in continuous growth throughout a growing season; early leaves that are preformed in the dormant bud flush first followed by additional late (neoformed) leaves that are initiated by the growing apical meristem.
Index selection: A statistical procedure for combining phenotypic and genetic information from a multiplicity of traits into a single index to rank an array of varieties.
Indumentum: See pubescence.
Infructescence: In *Populus*, an inflorescence that has developed to the fruiting stage.
Inoculum: Fungal spores, mycelium, bacterial cells, nematodes, virus particles, etc.
***In situ* conservation:** Conservation of genetic resources as they occur in naturally established habitats.
Instar: The stage between moults in an insect larva, numbered to designate the different stages in a growing larva.
Intersectional hybrid: A cross between species in two different sections of a genus, e.g. the progeny of *Salix eriocephala* (section *Cordatae*) × *Salix petiolaris* (section *Geyerianae*).
Interspecific hybridization: In *Populus*, the combination under natural or controlled conditions of the genomes of two or more species into a distinct taxon.
Intrasectional hybrid: A cross between species within a given section of a genus, e.g. the progeny of *Populus deltoides* (section *Aigeiros*) × *Populus nigra* (section *Aigeiros*).

Introgression: Process of sustained gene flow between species by continuous backcrossing of interspecific hybrids.

Invasive: A species, often exotic, that spreads aggressively from the original site(s) of introduction, becoming a pest or weed.

***In vitro* culture:** Propagation method of growing cells, tissues or embryos under controlled conditions in a sterile nutrient medium.

Juvenile wood: Wood formed during the early years in a growing tree (characterized by short cells and relatively large fibril angle; e.g. >10°), generally weaker than mature wood.

Knot: A branch base which is embedded in the main wood tissue of the tree trunk.

Lamina: Of a leaf, the flat part of a leaf or leaflet; the blade.

Laminated strand lumber (LSL): A composite lumber-like product made up of long and thin wood strands, bonded together with near parallel orientation between strands.

Laminated veneer lumber (LVL): A composite lumber product, where sheets of veneer are aligned with their grain parallel to one another and bonded together to form a lumber-like profile.

Late leaves: The last leaves to appear on an indeterminate, heterophyllous shoot; they are initiated by an actively growing apical meristem and expand without a period of dormancy (syn. neoformed leaves).

Latewood: The portion of annual ring which is formed during the latter part of the growing season.

Layering: A type of asexual (vegetative) propagation in which part of a low branch becomes buried in soil or plant debris and forms adventitious roots, then turns upward to form a new clonal stem.

Light harvesting antenna: System of proteins that can trap photons over a wide spectral range and transfer them efficiently to the reaction centre.

Lignin: An amorphous phenolic polymer in wood, with irregular structure. Found in the secondary wall, but concentrated in the middle lamellae of wood cells.

Linkage disequilibrium: Condition in which a preferred allelic combination is maintained between loci on the same chromosome by virtue of a close proximity of their position and selection forces.

Linkage group: All of the known gene loci grouped together on a single chromosome and inherited as a unit because of relatively infrequent gametic recombination.

Lipid peroxidation (= lipoperoxidation): The process whereby free radicals 'steal' electrons from the lipids in cell membranes, resulting in cell damage and increased production of free radicals.

Loess: Loosely packed, windblown deposit of silt or clay.

Lumen: The cavity of a cell.

Major gene: A gene that has a pronounced effect on a phenotype, with variation occurring in discrete, easily recognizable classes.

Mannitol: A sugar alcohol widely distributed in plants.

Mature wood: Wood which is formed after the tree matures (e.g. 20+ years in softwoods and 10+ years in hardwoods), characterized by relatively constant cell dimensions and small fibril angle.

Medium density fibreboard (MDF): A dry-formed composite panel made of fibres and fibre bundles, bonded together with a synthetic adhesive under heat and pressure.

Meristem: Undifferentiated tissue capable of cell division into specialized tissue, i.e. cambium.

Mesic: A temperate, moderately moist, well-drained habitat that represents ideal conditions for growth of most plants.

Mesophyll: Plant tissue formed by the inner cells of a leaf, i.e. the tissue layer which lies between the upper and lower epidermis. The mesophyll is comprised of palisade cells and spongy cells.

Metacoxa (pl. metacoxae): The posterior coxa, which is the body segment of an insect that attaches the leg to the body.

Microarray: A method for profiling gene and protein expression in cells and tissues.

Microculture: Cultivation of plant cells, tissues or organs in a sterile, synthetic medium in a controlled environment; includes tissues excised from a plant, pollen grains or seeds (also called tissue culture).

Micropropagation: *In vitro* clonal propagation of plants from shoot tips or nodal explants, usually with an accelerated proliferation of shoots during subcultures. In *Populus*, propagation method usually used to multiply a selected cultivar rapidly, beginning with axillary meristems to proliferate shoots, followed by rooting under *in vitro* conditions.

Minor gene: Gene that exerts a relatively small effect on a phenotype that exhibits variation on a quantitative scale.

Modulus of elasticity (bending; MOE): A stiffness index, indicating resistance to deflection.

Modulus of rupture (MOR): The maximum bending strength of wood (i.e. stress at failure).

Molecular marker: Gene or DNA sequence that can be used to identify an organism, species or strain, or phenotypic trait(s) associated with it.

Monoecious: Bisexual, with male (staminate) and female (pistillate) inflorescences borne on the same plant.

Monotypic: Including a single representative.

Morph: One of two or more clearly different phenotypes in a species.

Mutant: A cell microorganism that manifests new characteristics due to a change in its genetic material.

Mutualistic symbiosis: Symbiosis in which all partners obtain an advantage.

Mycangium: Specialized structure on the body of an insect adapted for the transport of symbiotic fungi (usually in spore form).

Mycelium: Mass of hyphae forming the fungus vegetative body.

Mycetophagous: Feeding on fungi.

Mycorrhizae: A symbiotic relationship between a fungus and the roots of a plant.

Myo-inositol: A form of inositol that is a component of the vitamin B complex and occurs widely in microorganisms, higher plants and animals.

Natality: Birth rate (in an insect population). Counter to mortality or death rate.

Necrosis: Death of cells and animal or plant tissues leading to decay.

Nectary: An organ where nectar is secreted, as in the inflorescences of *Salix*.

Neoformed leaves: See late leaves.

Neonate: Of insect larvae, newly emerged from an egg or newly born.

Nitrification: The process whereby ammonia in wastewater is oxidized to nitrite (NO_2^-) and then to nitrate (NO_3^-) by bacterial or chemical reactions.

Nitrogen-use efficiency (NUE): The ratio between nitrogen consumption and produced biomass.

Non-additive gene effects: Condition in which the effects of individual genes do not control phenotypic expression in an independent or additive fashion but do so interactively between alleles at the same or at different loci.

Nothospecies: A named but unsubstantiated species, one that does not conform to accepted definitions of a species.

Ontogenetic development: Growth and development of an individual from embryonic stage to maturity.

Open-top chamber (OTC): Structures within which plants are grown in CO_2 enrichment experiments, generally consisting of transparent side walls and open tops, through which either ambient or CO_2-enriched air is continually pumped.

Organogenesis: Initiation and growth of roots and shoots from cells or tissue under *in vitro* conditions beginning directly from excised plant parts or indirectly after the proliferation of callus.

Oriented strand board (OSB): A mat-formed composite panel of wood strands (flakes), bonded together with a synthetic adhesive under heat and pressure. Strands in surface layers are oriented.

Ortet: The original plant from which a clone is propagated vegetatively through root suckering, rooted cuttings, grafting, microculture or other means.

Osmolyte: A neutral solute that reacts minimally with the contents of a cell while protecting it from drying out, or in a cell's response to salinity changes.

Osmotic adjustment: Physiological process of accumulation of solute molecules inside the cells in response to a decline in external water potential. This adjustment may postpone and contribute to lessen tissue death after desiccation by maintaining cell turgor pressure.

Osmotic potential: Potential brought about by dissolving a substance, especially in water.

Osmotic stress: Osmotic stress occurs when the concentration of molecules in solution outside of the cell is different from that inside the cell. When this happens, water flows either into or out of the cell by osmosis, thereby altering the intracellular environment. Hyper-osmotic stress causes water to diffuse out of the cell (while hypo-osmotic stress causes water to diffuse into the cell) resulting in cell shrinkage, which can lead to DNA and protein damage, cell cycle arrest and ultimately cell death.

Osmoticum: A molecule, such as mannitol, glucose or sucrose, employed to maintain the osmotic potential.

Oviparous: Females that lay eggs (as opposed to live young).

Oviposition: The act of laying eggs.

Oxy-free radical (OFR): A highly reactive chemical that contains oxygen (O*) and is produced when molecules are split to give products that have unpaired electrons (a process called oxidation). Free radicals can damage important cellular molecules such as DNA or lipids or other parts of the cell such as thylakoid membranes.

Palaearctic: The largest of the eight ecozones dividing the Earth's surface, including Europe, Asia north of the Himalaya foothills, northern Africa and the northern and central parts of the Arabian Peninsula.

Parallel strand lumber (PSL): A composite lumber product which consists of parallel strands of veneer compressed and bonded into a lumber-like profile.

Paraquat: A standard herbicide used to kill various types of crops, including marijuana. Causes lung damage if smoke from the crop is inhaled.

Parasitoid: An insect that develops through its immature stages within or attached to another single insect, which it ultimately kills. Parasitoids are unique to the insect world.

Parenchyma: Storage tissue in wood and bark, consisting of relatively short, thin-walled cells. In wood, they may be ray parenchyma and axial (i.e. longitudinal) parenchyma.

Parthogenesis: A form of asexual reproduction, where growth and development of embryos within a female occurs without fertilization by a male.

Particleboard: A composite panel made of particles and flakes, bonded together with an adhesive under heat and pressure.

Paternity analysis: The use of DNA markers – usually microsatellites – to determine the male parentage of a progeny.

Pathogen: An organism capable of causing disease.

Peroxidase: Any of a group of enzymes that occur especially in plant cells and catalyse the oxidation of a substance by a peroxide.

Peroxisome: Small, membrane-bounded organelle that uses molecular oxygen to oxidize organic molecules. Contains some enzymes that produce hydrogen peroxide and others that degrade hydrogen peroxide (H_2O_2).

Phenological adaptation: Adaptation of the timing of spring growth and flower initiation and the onset of autumnal dormancy to seasonal changes in temperature and photoperiod.

Phenology: Periodicity of biological phenomena, for example flowering or cessation of shoot growth, correlated with climatic conditions or time of year.

Phenotype evaluation: The testing and measurement of the physical manifestation of individual genotypes in specific environments.

Phenotypic assortative mating: Non-random mating of individuals in which those of similar phenotypes are paired together or, alternatively, those of contrasting phenotypes are brought together.

Photochemistry: Study of the chemical and physical changes occurring when a molecule or atom absorbs light.

Photoinhibition: State of physiological stress that occurs in all oxygen-evolving photosynthetic organisms exposed to light.

Photosynthesis: Synthesis by green plants of carbohydrates from carbon dioxide and water to provide food for plant processes.

Photosystem II (PSII): In the process of photosynthesis, light is absorbed by a photosystem (ancient Greek: *phos* = light and *systema* = assembly) to begin an energy-producing reaction. The photosytems are contained within the chloroplasts in the leaves of plants. Two types of photosystems exist: Photosystem I (P700) and Photosystem II (P680). Each photosystem is differentiated by the wavelength of light to which it is most reactive (700 and 680 nm, respectively), and the type of terminal electron acceptor.

Phreatophyte species: A plant that obtains its water supply either directly from the zone of saturation or through the capillary fringe.

Physical maps: A map of the genome that locates genes on chromosomes based on the number of base pairs between their loci positions.

Physiological seed dormancy: Dormancy imposed by internal compounds that inhibit germination as opposed to external factors such as temperature and moisture.

Physiology: In the context of trees, the study of a plant's life processes and their function in relation to the environment.

Phytohormone: Plant hormone, plant growth regulator.

Phytophagous: Feeding on plants.

Phytoplasma: Specialized bacteria and obligate parasites of plant phloem tissue. Transmitted by insects, usually leafhoppers.

Pistillate inflorescense: A floral structure composed of a multiplicity of female flowers.

Plasma membrane (cell membrane): The cell's outer membrane made up of a double layer of phospholipids with embedded proteins.

Ploidy: The number of copies of the basic haploid chromosome set (n) found in normal somatic cells of a species.

Plywood: A panel made up of three or more layers of veneer, where the plies are bonded together with perpendicular grain orientation between adjacent layers.

Pollarding: The process of severely cutting back the tops of trees each year to the same spots on branches, forcing the growth of large knobby stubs from which long shoots grow to form a bushy crown.

Polygenic: Relating to two or more genes.

Polygenic systems: Gene systems that control the inheritance of quantitative characteristics that exhibit continuous variation.

Polymix breeding: Controlled breeding procedure in which a mix of pollen from several males is used in pollinating females.

Polyploidy: Cellular condition in which there are more than two homologous chromosome sets, e.g. triploid ($3n$), tetraploid ($4n$), hexaploid ($6n$).

Pore: A vessel or vessels appearing in cross or transverse section of wood.

Porous wood: Wood containing pores or vessels (same as hardwood).

Positional cloning: Process in which a gene for a specific phenotype is identified and cloned.

Pre-pupa: A quiescent larva prior to moulting to the pupa.

Prolepsis (adj. proleptic): When lateral vegetative and reproductive buds leaf out or flower only after a period of dormancy or overwintering.

Proline: One of the 20 amino acids directly coded for in proteins.

Prothorax: The first segment of an insect's thorax.

Protoplast fusion: A technique by which cellular contents of two cells are fused to form a somatic hybrid cell.

Pubescence: (i) Covering of short, soft, spreading hairs, often giving a dull whitish, velvety appearance; (ii) fine, hair-like setae growing from an insect's cuticle.

Pupation: The act of becoming a pupa.

Putative: Natural spontaneous hybrid between two unlike individuals occurring in nature.

Quantum yield: The ratio of the amount of light emitted from a sample to the amount of light absorbed by the sample.

Ramet: A clonal copy of a plant produced by vegetative propagation (see clone).

RAPD: Random amplified polymorphic DNA – a molecular marker based on PCR technology used in the identification of individual genotypes.
Ray: A ribbon-like tissue strand extending to the cambium in the radial direction. In hardwoods, it is composed entirely of parenchyma cells. Rays may be homocellular (consisting of only one type of cell, either horizontal or upright) or heterocellular (consisting of both horizontal and upright cells). Rays may be one cell wide or uniseriate (e.g. poplars and willows) or two or more cells wide (multiseriate, e.g. maples).
Reactive oxygen species (ROS): Damaging molecules, including oxygen radicals (i.e. superoxide radical) and other highly reactive forms of oxygen (i.e. singlet oxygen) that can harm biomolecules and contribute to disease states. Other oxygen-containing radicals, such as the hydroxyl and peroxyl radicals, are often classified as ROS too.
Reciprocal hybrid cross: In *Populus*, the hybridization of two species in which the cross is made in two ways by reversing the role of female and male parents between the species.
Reciprocal recurrent selection: A breeding method used to improve the hybridizing value of parental species by recurrent, intraspecific breeding based on the performance of their interspecific hybrid progeny.
Recurrent breeding: A breeding method designed to increase the frequency of favourable alleles within a population by successive intermating of selected individuals across generations while balancing the need to maintain genetic diversity in the breeding population.
Reduction division: A process of nuclear division in which the number of chromosomes per cell is reduced by one half, leading to the production of haploid gametes. Also called first-stage meiosis.
Reproductive crossability: In *Populus*, the ability of distinct species to produce viable progeny when crossed under natural or artificial conditions.
Respiration: Oxidation of food in plant cells bringing about release of energy used in maintenance and growth.
Rhizomatous: Arising from underground stems that send out roots and shoots.
Rhizomorph: A root-like aggregation of hyphae, with a well-defined apical meristem and often differentiated into a rind of small, dark cells surrounding a central core of elongated hyaline cells.
Rhizosphere: The zone around plant roots with higher microbial numbers and activity than in bulk soil.
Riparian: Relating to or living on the bank of a natural watercourse.
Root sucker: A stem sprout that forms from adventitious buds on shallow, horizontal roots of a tree, usually after the tree has been cut or killed; a common occurrence in section *Populus* in the genus *Populus* and section *Longifoliae* in the genus *Salix*.
Rostrum: In insects; hard, beak-like structure that encloses the mouthparts.
Salinization: The condition in which the salt content of soil accumulates over time to above normal levels; occurs in some parts of the world where water containing high salt concentration evaporates from fields irrigated with standing water.
Sapwood: The outer portion of woody stem (between heartwood and the cambium) where in living trees the parenchyma cells are alive. Usually very light (nearly white) coloured.
Seed stratification: The exposure of seeds to low temperatures, sometimes in a moist environment to overcome the effects of physiological dormancy.
Selection intensity: The superiority of the mean value of a group of genotypes chosen for breeding or clonal propagation relative to the population of variable individuals from which they were selected, expressed in terms of standard deviations above the mean of the population.
Silvics: Knowledge of forest trees and forests, how they grow, reproduce and respond to changes in the environment.
Sink: Place where carbon is stored, mostly used for forests and underground/deep sea reservoirs of CO_2.
Sink strength: Demand for carbohydrates by a specific plant organ. Strength or magnitude of the carbon sink.
Site index: A measure of forest site quality (i.e. the actual or potential productivity of a site) based on the height of dominant trees at a specified age.

SNP: Single nucleotide polymorphism – a molecular marker based on discrepancies in the sequence of base pairs within a gene used in association genetic studies.
Soil texture: The relative proportions of clay, silt and sand (less than 2 mm in diameter) within the soil.
Somaclonal variation: Variation induced within a genotype by the process of organogenesis from callus culture.
Somatic embryogenesis: A form of clonal propagation in which somatic embryos are produced from vegetative tissue under sterile, *in vitro* conditions.
Specific gravity (of wood): The ratio between the density of wood and the density of water. By convention, the oven-dry weight of wood is used, and volume at a specific moisture content (MC) (e.g. green, oven-dry, air-dry).
Specific leaf area (SLA): The ratio between leaf area and leaf dry weight.
Spontaneous hybrid: A cross that occurs naturally between two species with a sympatric natural range or between an indigenous species and an exotic species planted within the former's natural range.
Spore: Reproductive structure of fungi and bacteria.
Staminate inflorescence: A floral structure composed of a multiplicity of male flowers.
Stipule: A pair of leaf-like appendages at the base of a petiole, one on each side, usually not persistent, except in certain species of *Salix*.
Stoma (pl. stomata): One of the minute pores in the epidermis of a leaf or stem through which gases and water vapour pass. Also called *stomate*.
Stomatal conductance: Measure of how easily water molecules can escape from inside of plant leaves through the tiny pores (stomata) on the leaf surface to the free air. It is highly positively correlated with transpiration rate.
Stool: A callused stump that when cut back produces numerous vegetative sprouts from adventitious buds; 1-year-old stool shoots are commonly used to make hardwood cuttings.
Suberization: Deposition of suberin on the walls of plant cells and their subsequent conversion into cork tissue.
Sucrose: Sugar composed of two molecules, one glucose molecule and one fructose molecule.
Superoxide dismutase (SOD): This enzyme catalyses the dismutation of superoxide into oxygen and hydrogen peroxide. As such, it is an important antioxidant defence in nearly all cells exposed to oxygen.
Syllepsis: Development of a lateral branch without a period of dormancy, i.e. simultaneous with its parent axis. Hence, a sylleptic branch is a branch developed by syllepsis.
Sympodial: A type of shoot growth whereby the terminal bud dies and a subtending segment of the shoot apex aborts during dormancy induction, as in *Salix*; the proximal lateral bud then produces the following season's shoot leader.
Taxon: A taxonomic group of individuals (population) recognized as a formal unit at any level of a hierarchical classification, e.g. an entire genus or a species.
Teleomorph: The sexual or perfect state of a fungus.
Tension wood: Reaction wood in hardwoods, usually formed on the upper side of leaning stems and branches. Anatomically, it is characterized by the presence of gelatinous fibres.
Tergite: A dorsal segment of an insect.
Thelytoky: Parthenogenic reproduction in which the progeny are all female (cf. arrhenotoky, which is parthenogenic reproduction in which the progeny are all male).
Thorax: The middle part of an insect's body, carrying the legs and wings.
Thylakoid: A small, closed membranous sac within a chloroplast; thylakoids are often stacked into piles called grana; the molecules that perform the light reactions of photosynthesis are all embedded within the membranes of the thylakoids.
Tibia: The fourth segment of an insect's leg, joined proximally to the femur and distally to the tarsi.
Tolerance to water deficit: The ability of the plant to survive and yield under periods of limited moisture.
Tonoplast: The limiting membrane surrounding a vacuole.

Topophysis: Location effects in vegetative propagation. For example, when plants grown from rooted cuttings (ramets) maintain for some time certain characteristics of the position on the ortet where the cuttings were taken.

Transgene: A gene from a donor organism that is introduced into a recipient plant via non-sexual means.

Transgenesis: The introduction of foreign genes into an organism to confer new properties on that organism. This method is generally used to study the function of genes in the context of a living organism.

Transgenic: Relating to or being an organism whose genome has been altered by the transfer of a gene or genes from another species.

Transgressive segregants: The offspring of two parents with a phenotypic value that exceeds or falls below the limits of its parents.

Tri-tropic: In ecology, relating to three different levels within a food chain, e.g. plant, herbivore, predator.

Turgor: The pressure that can build up in a space that is enclosed by a membrane which is permeable to a solvent of a solution, such as water, but not to the solutes of the solution.

Urediniospore: Asexual spores of rust fungi.

Vacuole: A cavity in a plant cell, bounded by a membrane, in which various plant products and by-products are stored.

Valve: One of the parts into which a dehiscing capsule splits (see carpel).

Vegetative cap (or cover): A long-term cap of plants growing in and/or over materials that pose environmental risk, e.g. landfills.

Veneer: A thin sheet of wood, sliced or rotary-peeled from a log.

Vessel element: A single cellular component of the conductive tissue, i.e. vessels, in hardwoods.

Virginoparous: Of aphids, giving birth to live young (as opposed to laying eggs) through parthenogenesis.

Voltinism: Indicator of the number of generations of an organism in a year. Hence uni-, bi- and multi-voltine, for one, two or many generations a year.

Water potential: Difference between the activity of water molecules in pure distilled water at atmospheric pressure and 30°C (standard conditions) and the activity of water molecules in any other system. The activity of these water molecules may be greater (positive) or less (negative) than the activity of the water molecules under standard conditions.

Water-use efficiency (WUE): A measure of the amount of water used by plants per unit of plant material produced. The term can be applied at the leaf, whole-plant and ecosystem levels. At the leaf level, it is referred to more precisely as the instantaneous transpiration efficiency, the CO_2 assimilation rate (photosynthesis), divided by the transpiration rate (the moles of CO_2 taken up divided by the moles of water lost through transpiration in a unit of time per unit leaf area).

Wetwood: A tissue zone in the wood of some living trees, with abnormally high moisture content. Usually a result of bacterial activity in the tree stem. May cause collapse in drying.

Whitemarked tussock moth (*Orgyia leucostigma*): Widely distributed moth throughout eastern North America, as far west as Texas and Colorado in the USA. The larvae are known to feed on foliage of a wide variety of trees, both conifers and hardwoods. Hardwood hosts include basswood, maple, sycamore, apple, oak, poplar, willow and elm.

Xeric: An extremely dry habitat, usually associated with sandy, excessively drained soils or areas of low rainfall and high evaporation.

Xylem: A term synonymous with wood. The supporting and water-conducting tissue of vascular plants, consisting primarily of tracheids and vessels.

Xylophagous: Feeding on woody tissue.

Index